PINSTRIPE®

United States CITY TO CITY ATLAS

For the traveling professional

Cover Photo: Mark E. Gibson.

Contents

State, City Center, & Vicinity Maps

(*denotes detailed city center map)

Mileage Chart

	Albany, NY	Albuquerque, NM	Amarillo, TX	Atlanta, GA	Austin, TX	Baltimore, MD	Billings, MT	Birmingham, AL	Boise, ID	Boston, MA	Brownsville, TX	Buffalo, NY	Charleston, SC	Charleston, WV	Charlotte, NC	Chicago, IL	Cincinnati, OH	Cleveland, OH	Columbia, SC	Columbus, OH	Dallas, TX	Daytona Beach, FL	Denver, CO	Des Moines, IA	Detroit, MI	El Paso, TX	Fargo, ND	Fort Lauderdale, FL	Fort Wayne, IN	Fort Worth, TX	Grand Rapids, MI	Greensboro, NC	Hartford, CT	Houston, TX	Indianapolis, IN	Jackson, MS	Jacksonville, FL	Kansas City, MO	Knoxville, TN	Las Vegas, NV	Lincoln, NE	Little Rock, AK
Albany, NY	0	2125	1825	1007	1882	332	2073	1112	2601	170	2007	292	932	639	795	795	729	496	823	657	1679	1209	1853	1193	690	2327	1463	1403	705	1682	710	661	106	1825	836	1320	1111	1279	836	2609	1336	1370
Albuquerque, NM	2125	0	300	1387	716	1881	1022	1260	970	2214	988	1801	1695	1583	1628	1346	1394	1606	1598	1468	673	1716	446	1013	1537	267	1314	1953	1410	632	1491	1677	2084	870	1289	1087	1678	811	1407	576	837	883
Amarillo, TX	1825	300	0	1087	485	1581	1037	965	1235	1914	784	1501	1517	1304	1338	1046	1094	1306	1298	1168	363	1467	454	806	1289	508	999	1670	1109	344	1191	1377	1822	608	989	787	1378	552	1107	876	596	607
Atlanta, GA	1007	1387	1087	0	884	669	1804	160	2252	1068	1175	912	300	495	251	695	438	692	211	543	805	446	1401	924	726	1453	1364	681	612	837	749	348	969	816	543	397	329	810	204	1947	1013	540
Austin, TX	1882	716	485	884	0	1550	1449	793	1716	1930	331	1566	1247	1251	1237	1100	1127	1371	1095	1233	203	1158	1009	897	1330	583	1333	1326	1200	192	1288	1281	1867	162	1111	519	1057	680	1051	1297	851	520
Baltimore, MD	332	1881	1581	669	1550	0	1875	804	2416	409	1825	352	567	339	430	697	510	355	513	405	1347	876	1692	997	511	1997	1339	1036	550	1379	624	346	308	1409	584	1006	770	1078	503	2408	1192	1037
Billings, MT	2073	1022	1037	1804	1449	1875	0	1759	586	2232	1771	1857	2222	1762	2027	1214	1479	1662	2075	1654	1395	2173	579	959	1579	1284	625	2466	1405	1406	1396	1958	2169	1739	1400	1743	2227	1078	1723	1060	836	1439
Birmingham, AL	1112	1260	965	160	793	804	1759	0	2101	1267	1065	941	460	539	411	669	468	722	359	584	637	505	1370	838	721	1304	1311	764	610	702	739	493	1058	676	497	251	472	724	255	1822	953	394
Boise, ID	2601	970	1235	2252	1716	2416	586	2101	0	2794	1921	2271	2503	2246	2408	1777	1983	2058	2289	2069	1610	2576	870	1402	2020	1241	1245	2820	1871	1598	1917	2408	2652	1854	1890	2091	2579	1476	2022	662	1205	1781
Boston, MA	170	2214	1914	1068	1930	409	2232	1267	2794	0	2255	454	989	728	828	965	875	632	928	738	1727	1257	1953	1305	795	2376	1623	1492	847	1761	908	739	105	1878	933	1395	1167	1435	871	2765	1500	1472
Brownsville, TX	2007	988	784	1175	331	1825	1909	1065	1921	2255	0	1865	1500	1479	1426	1430	1426	1670	1360	1533	526	1353	1251	1184	1694	806	1601	1542	1455	518	1585	1480	2094	357	1427	791	1264	1008	1320	1573	1216	819
Buffalo, NY	292	1801	1501	912	1566	352	1857	941	2271	454	1865	0	947	430	666	543	438	195	822	333	1363	1069	1602	867	366	2011	1185	1400	381	1395	419	641	397	1492	512	1119	1068	1007	671	2254	1057	1046
Charleston, SC	932	1695	1517	300	1247	567	2222	460	2503	989	1500	947	0	479	203	912	628	750	113	670	1164	351	1743	1185	874	1729	1557	586	740	1116	959	271	867	1027	743	702	248	1135	368	2247	1287	814
Charleston, WV	639	1583	1304	495	1251	339	1762	539	2246	728	1479	430	479	0	276	469	178	284	376	156	1134	726	1377	761	371	1672	1126	1004	284	1056	457	227	662	1246	328	790	676	777	284	2119	960	707
Charlotte, NC	795	1628	1338	251	1237	430	2027	411	2408	828	1426	666	203	276	0	738	446	543	100	453	1054	469	1548	1029	607	1710	1414	721	602	1061	791	89	763	1053	551	640	413	940	219	2173	1151	743
Chicago, IL	795	1346	1046	695	1100	697	1214	669	1777	965	1430	543	912	469	738	0	291	348	794	340	932	1096	1037	357	284	1435	649	1348	159	945	178	729	875	1160	186	762	999	543	537	1749	527	675
Cincinnati, OH	729	1394	1094	438	1127	510	1479	468	1983	875	1426	438	628	178	446	291	0	249	502	105	924	861	1199	583	260	1472	940	1086	184	956	357	458	746	1053	105	680	786	591	246	1921	715	608
Cleveland, OH	496	1606	1306	692	1371	355	1662	722	2058	632	1670	195	750	284	543	348	249	0	627	138	1168	952	1407	672	171	1716	997	1232	211	1200	284	486	539	1297	317	924	908	803	489	2059	867	851
Columbia, SC	823	1598	1298	211	1095	513	2075	359	2289	928	1360	822	113	376	100	794	502	627	0	513	1032	381	1616	1126	745	1668	1446	622	671	1057	858	188	836	1066	625	610	290	1025	267	2162	1199	759
Columbus, OH	657	1469	1168	543	1233	405	1654	584	2069	738	1533	333	670	156	453	340	105	138	513	0	1030	901	1270	657	195	1540	989	1126	150	1062	311	381	641	1159	179	786	850	665	351	1995	776	713
Dallas, TX	1679	673	363	805	203	1347	1395	637	1610	1727	526	1363	1164	1134	1054	932	924	1168	1032	1030	0	1123	806	714	1203	648	1131	1097	1030	33	1110	1122	1664	243	908	422	1005	511	820	1249	648	317
Daytona Beach, FL	1209	1716	1467	446	1158	876	2173	505	2576	1257	1353	1069	351	726	469	1096	883	952	381	901	1123	0	1823	1329	1103	1728	1714	227	1006	1126	1143	551	1138	952	880	688	97	1209	603	2316	1401	904
Denver, CO	1853	446	454	1401	1009	1692	579	1370	870	1953	1251	1602	1743	1377	1548	1037	1199	1407	1616	1270	806	1823	0	695	1321	705	915	2067	1186	773	1201	1621	1988	1060	1091	1246	1779	608	1341	743	507	992
Des Moines, IA	1193	1013	806	924	897	997	959	838	1402	1305	1184	867	1185	761	1029	357	583	672	1126	657	714	1329	695	0	600	1114	475	1581	516	747	502	1028	1283	930	478	846	1270	203	821	1399	203	562
Detroit, MI	690	1537	1289	726	1330	511	1579	721	2020	795	1694	366	874	371	607	284	260	171	745	195	1203	1103	1321	600	0	1701	922	1346	170	1240	162	567	701	1304	293	923	1046	791	506	2011	819	850
El Paso, TX	2327	267	508	1453	583	1997	1274	1304	1241	2376	806	2011	1729	1672	1710	1435	1372	1716	1668	1540	648	1728	705	1114	1701	0	1460	1869	1573	609	1554	1783	2263	743	1460	1070	1626	915	1488	722	946	960
Fargo, ND	1463	1314	999	1364	1333	1339	625	1311	1245	1625	1601	1185	1557	1126	1414	649	940	997	1446	989	1131	1714	915	475	922	1460	0	2007	808	1072	827	1412	1531	1334	835	1335	1704	609	1195	1535	451	1091
Fort Lauderdale, FL	1435	1953	1670	681	1326	1036	2466	764	2820	1492	1542	1400	586	1004	721	1348	1086	1232	622	1126	1097	227	2067	1581	1346	1869	2007	0	1271	1129	1342	786	1403	1191	1232	883	332	1459	860	2530	1670	1184
Fort Wayne, IN	705	1410	1109	612	1200	550	1405	610	1871	847	1455	381	740	284	602	159	184	219	671	150	1030	1006	1186	516	170	1573	808	1271	0	1053	172	551	768	1176	122	784	929	648	430	1878	686	711
Fort Worth, TX	1682	632	344	837	192	1379	1406	702	1598	1761	518	1395	1116	1056	1061	945	956	1200	1057	1062	33	1126	773	747	1236	609	1072	1129	1053	0	1121	1154	1696	264	912	446	1037	513	853	1203	648	349
Grand Rapids, MI	710	1491	1191	749	1288	624	1396	739	1917	908	1585	419	959	457	791	178	357	284	858	311	1110	1143	1201	502	162	1554	827	1342	172	1121	0	707	794	1196	263	957	1071	638	573	1889	699	799
Greensboro, NC	661	1677	1377	348	1281	346	1958	493	2408	739	1480	641	271	227	89	729	458	486	188	381	1122	551	1621	1028	567	1783	1412	786	551	1154	707	0	650	1167	563	770	483	1013	283	2237	1202	778
Hartford	106	2084	1822	969	1867	308	2169	1058	2652	105	2094	397	867	662	763	875	746	539	836	641	1664	1138	1988	1283	701	2263	1531	1403	768	1696	794	650	0	1773	805	1306	1071	1297	841	2675	1378	1344
Houston, TX	1825	870	608	816	162	1409	1639	676	1854	1878	357	1492	1027	1246	1053	1160	1053	1297	1066	1159	243	952	1060	930	1304	743	1334	1191	1176	264	1196	1167	1773	0	1041	406	891	754	922	1468	892	446
Indianapolis, IN	836	1289	989	543	1111	584	1400	497	1890	933	1427	512	743	328	551	186	105	317	625	179	908	980	1091	478	293	1460	835	1232	122	912	263	563	805	1041	0	681	867	486	351	1816	673	608
Jackson, MS	1320	1087	787	397	519	1006	1743	251	2091	1395	991	1119	702	790	640	762	680	924	610	786	422	688	1246	846	923	1070	1335	883	784	446	957	770	1306	406	681	0	591	716	506	1650	874	251
Jacksonville, FL	1111	1678	1378	329	1057	770	2227	472	2579	1167	1264	1068	248	676	413	999	786	908	290	850	1005	97	1779	1270	1046	1626	1704	332	929	1037	1071	483	1071	891	867	591	0	1110	555	2238	1321	843
Kansas City, MO	1279	811	552	810	680	1078	1078	724	1476	1435	1008	1007	1135	777	940	543	591	803	1025	665	511	1209	608	203	791	916	609	1459	648	513	638	1013	1297	754	486	716	1110	0	752	1345	211	409
Knoxville, TN	836	1407	1107	204	1051	503	1723	255	2022	871	1320	671	368	284	219	537	246	489	267	351	820	603	1341	821	506	1488	1195	860	430	872	573	283	841	922	351	506	555	752	0	1983	944	523
Las Vegas, NV	2609	576	876	1947	1297	2408	1060	1822	662	2765	1573	2254	2247	2119	2173	1749	1921	2059	2162	1995	1249	2316	743	1399	2011	722	1535	2530	1878	1203	1889	2237	2675	1468	1816	1650	2238	1345	1983	0	1224	1483
Lincoln, NE	1386	837	596	1013	851	1192	836	953	1205	1500	1216	1057	1287	960	1151	527	802	867	1199	776	648	1401	507	203	819	946	451	1670	686	648	699	1202	1378	892	673	874	1321	211	944	1224	0	616
Little Rock, AK	1370	883	607	540	520	1037	1439	394	1781	1472	819	1046	814	707	743	675	608	851	759	713	317	904	992	562	850	960	1091	1184	711	349	799	778	1344	446	608	251	843	409	523	1483	616	0
Los Angeles, CA	2911	823	1095	2197	1410	2676	1254	2067	837	2993	1678	2587	2521	2394	2617	1989	2164	2392	2426	2254	1401	2407	1009	1654	2270	818	1844	2704	2137	1361	2148	2478	2829	1581	2075	1880	2402	1589	2201	275	1476	1678
Louisville, KY	868	1332	1041	421	1022	608	1550	373	1908	976	1321	543	608	258	438	300	105	349	494	211	819	801	1127	591	365	1467	949	1078	222	851	373	462	867	948	129	575	729	519	246	1861	730	502
Memphis, TN	1232	1021	721	397	658	900	1557	239	1833	1379	957	908	689	653	604	551	469	713	612	575	455	749	1151	599	712	1103	1224	989	592	487	690	640	1209	584	470	211	697	470	385	1581	647	138
Miami, FL	1439	1994	1694	665	1338	1095	2580	788	2860	1516	1580	1524	630	1046	745	1338	1086	1264	658	1210	1321	259	2131	1582	1386	1958	1986	24	1326	1353	1356	810	1427	1207	1208	907	356	1475	859	2570	1673	1208
Milwaukee, WI	933	1443	1143	784	1203	794	1143	766	1777	1078	1530	640	1032	566	835	92	388	445	891	448	1013	1180	1070	365	389	1528	576	1443	256	1059	275	826	948	1155	283	884	1067	568	643	1752	560	772
Minneapolis, MN	1215	1256	1062	1105	1120	1105	812	1088	1488	1362	1456	948	1316	874	1143	405	696	753	1276	753	1013	1458	956	251	698	1520	244	1723	564	1001	583	1135	1257	1266	591	1123	1376	459	932	1630	409	881
Mobile, AL	1322	1265	965	340	656	990	1854	269	2343	1379	851	1184	607	825	575	908	745	989	555	834	592	502	1372	954	988	1236	1413	705	839	624	1006	681	1290	478	749	178	413	819	449	1841	1039	430
Montgomery, AL	1178	1345	1042	164	804	833	1836	93	2346	1232	1041	1076	464	632	405	762	561	815	379	707	677	458	1412	1311	814	1325	1521	671	686	701	819	512	1133	709	590	255	379	867	348	2015	975	470
Nashville, TN	993	1232	932	243	869	688	1640	195	2059	1062	1168	722	576	458	399	474	283	527	458	389	666	639	1167	712	536	1314	1136	900	385	698	534	429	973	795	302	422	592	590	174	1792	770	349
New Orleans, LA	1453	1187	875	493	535	1136	1820	352	2191	1526	730	1273	727	891	721	925	820	1078	701	940	530	632	1323	978	1077	1127	1494	843	916	519	1071	810	1436	367	857	193	551	857	607	1800	1114	430
New York City, NY	146	1995	1695	855	1728	201	1926	1019	2571	203	2002	390	786	524	625	794	628	446	715	551	1525	1054	1775	1070	620	2173	1450	1289	691	1557	706	561	101	1679	730	1192	957	1192	771	2520	1274	1235
Norfolk, VA	505	1905	1632	551	1403	237	2098	711	2551	660	1735	569	454	369	341	851	681	493	412	559	1359	702	1800	1202	711	1998	1581	964	709	1382	802	227	471	1362	669	948	632	1179	412	2534	1354	1025
Oakland, CA	2982	1134	1430	2488	1786	2864	1218	2321	671	3124	2034	2745	2788	2600	2755	2098	2317	2498	2703	2391	1803	2831	1223	1742	2350	1194	1870	3041	2257	1723	2308	2809	2909	1957	2212	2270	2771	1799	2509	582	1604	1984
Oklahoma City, OK	1523	559	267	863	414	1322	1188	701	1451	1659	680	1242	1176	1031	1069	804	835	1047	1091	909	211	1257	660	576	1030	708	989	1481	852	210	932	1118	1546	454	730	575	1181	373	847	1119	433	324
Omaha, NE	1308	905	754	989	847	1143	904	904	1274	1443	1249	1005	1303	899	1155	464	721	824	1283	795	693	1402	559	146	726	1256	464	1604	634	634	640	1208	1321	949	616	914	1305	195	930	1249	57	690
Orlando, FL	1249	1751	1451	446	1142	917	2277	545	2695	1297	2034	1306	401	814	559	1127	892	1046	440	997	1078	81	1896	1363	1143	1735	1826	208	1059	1110	1188	648	1208	964	989	697	138	1266	665	2311	1452	965
Philadelphia, PA	251	1922	1622	766	1630	97	2051	880	2488	327	1954	397	688	517	522	757	559	430	627	460	1427	914	1762	998	585	2075	1370	1127	604	1459	672	438	220	1581	624	1094	859	1134	646	2449	1260	1110
Phoenix, AZ	2512	446	746	1810	1030	2311	1220	1700	1022	2644	1289	2269	2222	2045	2061	1776	1808	2045	2025	1907	1013	2101	802	1497	2019	438	1791	2244	1831	983	1906	2099	2523	1110	1719	1500	2100	1277	1845	284	1236	1379
Pittsburgh, PA	471	1654	1354	712	1412	245	1681	778	2203	584	1713	219	778	211	504	470	291	128	572	186	1209	859	1475	770	304	1833	1112	1216	336	1241	397	438	486	1345	349	965	882	851	515	2181	965	899
Portland, Or	2869	1378	1636	2763	2059	2765	867	2571	439	3149	2468	2677	2952	2615	2757	2140	2474	2416	2972	2478	2051	3018	1283	1816	2368	1661	1484	3204	2299	2003	2251	2823	2877	2369	2335	2518	3042	1809	2550	991	1641	2270
Providence, RI	178	2156	1856	1027	1898	356	2238	1189	2701	41	2222	454	956	699	785	924	790	600	888	713	1695	1201	1961	1256	746	2335	1566	1459	799	1727	859	706	73	1849	892	1362	1127	1378	935	2683	1451	1395
Raleigh, NC	656	1759	1459	397	1355	324	2273	557	2560	713	1506	721	300	297	162	802	540	559	215	453	1152	559	1694	1123	643	1800	1485	794	603	1184	754	73	624	1233	635	815	486	1086	356	2319	1263	851
Reno, NV	2763	1056	1345	2411	1775	2562	1021	2363	404	2871	2068	2433	2765	2407	2570	1897	2221	2238	2626	2295	1731	2758	1030	1606	2190	1183	1660	3008	2056	1608	2067	2591	2749	1932	2116	2143	2716	1606	2363	478	1411	1986
Richmond, VA	482	1833	1533	527	1463	155	1655	699	2594	544	1646	552	462	251	280	788	356	583	390	498	1313	713	1904	1293	609	2023	1481	946	635	1333	722	191	455	2291	907	946	693	1205	440	2406	1249	963
Rochester, NY	219	1857	1557	1015	1623	300	1922	965	2352	381	1891	81	871	495	689	608	502	268	789	397	1420	1176	1637	932	424	2036	1249	1410	464	1452	499	600	324	1555	551	1183	1102	1062	710	2371	1135	1111
Saint Louis, MO	1028	1054	754	588	806	827	1381	539	1727	1184	1216	747	884	544	626	300	340	552	737	414	641	956	859	377	585	1238	872	1208	369	698	437	762	1030	835	251	495	859	251	497	1581	462	422
Saint Paul, MN	1215	1362	1062	1105	1120	1095	812	1118	1398	892	1565	948	1316	870	1143	405	696	753	1320	745	1013	1458	956	251	698	1541	244	1753	564	987	583	1138	1257	1266	591	1160	1376	459	951	1630	408	881
Salt Lake City, UT	2290	621	1324	1900	1341	2051	579	1825	349	2417	1775	1922	2254	1896	2059	1386	1710	1727	2115	1711	1287	2283	519	1085	1679	892	1172	2578	1527	1184	1556	2059	2238	1460	1605	1742	2286	1095	1766	413	900	1462
San Antonio, TX	1986	684	530	965	81	1646	1600	895	1709	2052	300	1638	1371	1419	1272	1208	1208	1443	1180	1305	284	1175	975	1022	1500	576	1402	1378	1269	283	1353	1321	1965	203	1200	649	1086	795	1150	1273	916	592
San Diego, CA	2855	787	1078	2174	1313	2714	1309	2034	1010	2992	1574	2613	2505	2402	2423	2334	2193	2334	2389	2277	1369	2418	1054	1760	2419	721	1934	2621	2189	1332	2269	2457	2944	1484	2067	1783	2329	1627	2269	349	1573	1743
San Francisco, CA	2966	1135	1396	2511	1776	2765	1239	2371	595	3133	2044	2403	2923	2616	2756	2108	2329	2408	2738	2461	1865	2827	1233	1832	2360	1184	1886	3073	2304	1735	2318	2740	3019	1947	2224	2183	2781	1869	2549	592	1614	1994
Seattle, WA	2855	1500	1805	2656	2157	2685	815	2475	524	2961	2521	2531	2960	2748	2765	2043	2334	2336	2971	2408	2203	3070	1371	1889	2299	1775	1440	3882	2202	2071	2221	2773	2918	2498	2229	2585	3090	1884	2553	1209	1636	2320
Shreveport, LA	1599	868	568	624	340	1229	1691	474	1912	1618	644	1265	945	899	885	916	826	1070	833	592	196	903	1112	827	1069	844	1229	1144	909	228	1018	937	1519	271	827	223	814	624	729	1468	727	219
Spokane, WA	2652	1346	1563	2367	1981	2417	541	2469	369	2693	2359	2263	2700	2503	2505	1775	2066	2068	2572	2115	1978	2811	1095	1556	2020	1686	1166	3014	1921	1978	1941	2503	2650	2222	1961	2205	2822	1727	2298	1119	1460	2092
Tallahassee, FL	1249	1508	1208	268	899	932	2306	302	2512	1312	1094	1155	364	868	559	957	706	949	408	828	835	259	1727	1216	957	1492	1598	462	871	867	1022	608	1223	721	818	421	170	1045	584	2068	1227	679
Tampa, FL	1281	1759	1459	476	1150	949	2143	553	2763	1329	1345	1346	479	903	584	1143	908	1091	497	1013	1086	141	1858	1460	1200	1743	1849	268	1092	1118	1219	673	1240	972	1069	672	195	1296	730	2319	1477	958
Toledo, OH	633	1526	1220	641	1315	454	1557	673	2020	742	1622	309	973	284	533	243	203	114	683	138	1112	1063	1272	567	65	1699	885	1289	105	1144	170	516	624	1249	243	876	989	717	449	1954	762	803
Tuscon, AZ	2442	486	656	1785	908	2246	1342	1621	1144	2571	1176	2166	2100	2039	1913	1711	1750	1971	2080	1833	964	1999	835	1484	1938	316	1897	2271	1783	932	1856	2059	2433	1079	1668	1378	2010	1281	1755	389	1246	1257
Tulsa, OK	1409	674	336	803	462	1208	1293	636	2020	1532	835	1128	1103	941	989	673	721	933	1018	795	259	1156	714	471	916	788	972	1409	738	305	818	1037	1411	504	616	510	1167	283	782	1224	416	259
Washington, DC	378	1864	1564	630	1509	41	2006	781	2441	430	1787	429	559	299	334	697	478	341	498	418	1306	802	1654	1054	506	1954	1322	1062	531	1338	608	309	341	1220	551	965	730	1046	554	2376	1184	832
West Palm Beach, FL	1396	1938	1638	632	1329	1046	2736	702	2942	1426	1524	1443	568	982	673	1289	1063	1192	586	1146	1265	195	2157	1646	1305	1922	2028	41	1157	1297	1476	737	1339	1151	1176	851	284	1052	806	2498	1598	1101
Youngstown, OH	462	1632	1346	719	1368	298	1632	738	2090	568	1695	190	709	251	502	413	275	74	561	170	1193	986	1421	737	239	1804	1038	1219	275	1225	340	482	470	1322	341	949	958	843	521	2124	923	876

City	Los Angeles, CA	Louisville, KY	Memphis, TN	Miami, FL	Milwaukee, WI	Minneapolis, MN	Mobile, AL	Montgomery, AL	Nashville, TN	New Orleans, LA	New York City NY	Norfolk, VA	Oakland,CA	Oklahoma City, OK	Omaha, NE	Orlando, FL	Philadelphia, PA	Phoenix, AZ	Pittsburgh, PA	Portland, OR	Providence, RI	Raleigh, NC	Reno, NV	Richmond, VA	Rochester, NY	Saint Louis, MO	Saint Paul, MN	Salt Lake City, UT	San Antonio, TX	San Diego, CA	San Francisco, CA	Seattle, WA	Shreveport, LA	Spokane, WA	Tallahassee, FL	Tampa, FL	Toledo, OH	Tucson, AZ	Tulsa, OK	Washington, DC	West Palm Beach, FL	Youngstown, OH
Albany, NY	2911	868	1232	1439	933	1215	1322	1178	993	1453	146	505	2982	1523	1308	1249	251	2512	471	2869	178	656	2763	482	219	1028	1215	2290	1986	2855	2966	2855	1599	2652	1249	1281	633	2442	1409	378	1396	462
Albuquerque, NM	823	1332	1021	1994	1443	1256	1265	1345	1232	1187	1995	1905	1134	559	905	1751	1922	446	1654	1378	2156	1759	1056	1833	1857	1054	1362	621	684	787	1135	1500	868	1346	1508	1759	1526	486	674	1864	1938	1646
Amarillo, TX	1095	1041	721	1694	1143	1062	965	1045	932	875	1695	1632	1430	267	754	1451	1622	746	1354	1636	1856	1458	1345	1533	1557	754	1062	917	530	1078	1396	1805	568	1563	1208	1459	1220	656	336	1564	1638	1346
Atlanta, GA	2197	421	397	665	784	1105	340	164	243	493	855	551	2488	863	989	446	766	1810	712	2763	1027	397	2411	527	1015	588	1105	1900	965	2174	2511	2656	624	2367	268	476	641	1785	803	630	632	719
Austin, TX	1410	1022	658	1338	1203	1120	656	804	869	535	1728	1403	1786	414	847	1142	1630	1030	1412	2059	1898	1355	1775	1463	1623	806	1120	1341	81	1313	1776	2157	340	1981	899	1150	1315	908	462	1509	1329	1368
Baltimore, MD	2676	608	900	1095	794	1105	990	833	688	1136	201	237	2864	1322	1143	917	97	2311	245	2765	356	324	2562	155	300	827	1095	2051	1646	2714	2765	2686	1229	2417	932	949	454	2246	1208	41	1046	298
Billings, MT	1254	1550	1557	2580	1143	812	1854	1836	1640	1820	1926	2098	1218	1168	904	2277	2051	1220	1681	867	2238	2273	1021	1655	1922	1381	812	579	1600	1309	1239	815	1691	541	2306	2143	1557	1342	1293	2006	2736	1632
Birmingham, AL	2067	373	239	788	766	1088	269	93	195	352	1019	711	2321	701	904	545	880	1700	778	2571	1189	557	2363	699	965	539	1118	1825	895	2034	2371	2475	474	2469	302	553	673	1621	636	781	702	738
Boise, ID	837	1908	1833	2860	1777	1488	2143	2346	2059	2191	2571	2551	671	1451	1274	2695	2498	1022	2203	439	2701	2560	404	2594	2352	1727	1398	349	1709	1010	595	524	1912	369	2512	2763	2020	1144	1582	2441	2492	2090
Boston, MA	2993	976	1379	1516	1078	1362	1379	1232	1062	1525	203	560	3124	1659	1443	1297	327	2644	584	3149	41	713	2871	544	381	1184	892	2417	2052	2992	3133	2961	1618	2693	1312	1329	742	2571	1532	430	1426	568
Brownsville, TX	1678	1321	957	1580	1530	1456	851	1041	1168	730	2002	1735	2034	680	1249	2034	1954	1289	1713	2468	2222	1506	2068	1646	1891	1216	1565	1609	300	1574	2044	2521	644	2359	1094	1345	1622	1176	835	1787	1524	1695
Buffalo, NY	2587	543	908	1424	640	948	1184	1076	722	1273	390	569	2745	1242	1005	1306	397	2269	219	2677	454	721	2433	552	81	747	948	1922	1638	2613	2667	2531	1265	2263	1155	1346	309	2166	1128	429	1443	190
Charleston, SC	2521	608	689	630	1032	1316	607	464	576	727	787	454	2788	1176	1303	401	688	2222	778	2952	956	300	2765	462	871	884	1316	2254	1371	2505	2923	2960	945	2700	364	479	973	2100	1103	559	568	709
Charleston, WV	2394	258	653	1046	566	874	825	632	458	891	524	369	2600	1031	899	814	517	2045	211	2615	699	297	2407	251	495	544	870	1896	1419	2402	2616	2748	899	2503	868	903	284	2039	941	299	982	251
Charlotte, NC	2417	438	592	604	835	1143	575	415	399	721	625	341	2755	1069	1135	559	522	2061	504	2757	785	162	2570	280	689	689	1143	2059	1272	2423	2756	2765	885	2505	559	584	583	1913	989	334	673	502
Chicago, IL	1989	300	551	1338	92	405	908	762	474	925	794	851	2098	804	454	1127	757	1776	470	2140	924	802	1897	788	608	292	405	1386	1208	2306	2108	2043	916	1775	957	1143	243	1711	673	697	1289	413
Cincinnati, OH	2164	105	469	1086	388	696	712	561	283	810	628	601	2317	835	721	892	559	1808	291	2369	790	540	2201	503	502	340	696	1671	1208	2193	2329	2334	813	2066	706	908	203	1756	721	478	1051	275
Cleveland, OH	2392	349	713	1264	445	753	989	815	527	1078	446	493	2498	1047	824	1046	430	2045	128	2416	600	559	2238	583	268	552	753	1727	1443	2384	2408	2336	1070	2068	949	1091	114	1971	933	341	1192	74
Columbia, SC	2426	494	612	658	891	1276	555	379	458	701	715	412	2703	1091	1283	440	627	2025	572	2972	888	215	2626	390	789	737	1320	2115	1180	2389	2738	2971	833	2572	408	497	683	2080	1018	498	586	561
Columbus, OH	2254	211	575	1210	448	753	834	707	389	940	551	559	2391	909	795	997	460	1907	186	2478	713	453	2295	498	397	414	745	1711	1305	2277	2461	2408	592	2115	828	1513	138	1833	795	418	1146	170
Dallas, TX	1401	819	455	1321	1013	1013	592	677	666	530	1525	1359	1803	211	693	1078	1427	1013	1209	2059	1695	1152	1731	1313	1420	641	1013	1287	284	1369	1865	2203	196	1978	835	1086	1112	964	259	1306	1265	1193
Daytona Beach, FL	2407	801	749	259	1180	1458	502	458	639	632	1054	702	2831	1257	1402	81	914	2102	859	3018	1201	559	2758	713	1176	956	1458	2283	1175	2418	2827	3070	903	2811	259	141	1063	1999	1156	802	195	986
Denver, CO	1009	1127	1151	2131	1070	956	1372	1412	1167	1323	1775	1800	1223	660	559	1896	1762	802	1475	1283	1961	1694	1030	1904	1637	859	956	519	975	1054	1233	1371	1112	1095	1727	1858	1272	835	714	1654	2157	1421
Des Moines, IA	1654	591	599	1582	365	251	954	1131	712	978	1070	1202	1742	576	146	1363	1037	1479	770	1816	1256	1123	1606	1293	932	377	251	1085	1022	1760	1832	1889	827	1556	1216	1460	567	1484	471	1054	1646	737
Detroit, MI	2270	365	712	1386	389	698	988	814	536	1077	620	711	2350	1030	726	1143	585	2019	304	2368	746	643	2190	609	424	535	698	1679	1500	2419	2360	2299	1069	2020	957	1200	65	1938	916	506	1305	239
El Paso, TX	818	1467	1103	1958	1528	1520	1236	1325	1314	1127	2173	1998	1194	708	1256	1735	2073	438	1833	1661	2335	1800	1185	2023	2036	1238	1541	894	576	721	1184	1775	844	1686	1492	1743	1699	316	788	1954	1922	1804
Fargo, ND	1844	949	1224	1987	576	244	1413	1525	1138	1521	1450	1581	1870	989	464	1826	1370	1791	1112	1484	1566	1485	1660	1481	1249	872	244	1172	1402	1934	1886	1440	1229	1166	1598	1849	885	1897	972	1322	2028	1038
Fort Lauderdale, FL	2704	1078	989	24	1443	1723	705	671	900	843	1289	964	3041	1481	1604	208	1127	2244	1216	3204	1459	794	3008	946	1410	1208	1753	2578	1378	2621	3073	3382	1144	3014	462	268	1289	2271	1409	1062	41	1219
Fort Wayne, IN	2137	222	592	1326	256	564	839	686	385	916	691	709	2257	852	634	1059	604	1831	336	2299	799	603	2056	635	464	369	564	1527	1269	2189	2304	2202	909	1921	871	1092	105	1783	738	531	1157	275
Fort Worth, TX	1361	851	487	1353	1059	1001	624	701	698	519	1557	1382	1723	210	634	1110	1459	983	1241	2003	1727	1184	1608	1333	1452	698	987	1184	283	1332	1735	2071	228	1978	867	1118	1144	932	305	1338	1297	1225
Grand Rapids, MI	2148	373	690	1356	275	583	1006	819	534	1071	706	802	2308	932	640	1188	672	1906	397	2251	859	754	2067	722	499	437	583	1556	1353	2269	2318	2221	1018	1941	1022	1219	170	1856	818	608	1476	340
Greensboro, NC	2478	462	640	810	826	1135	681	512	429	810	561	227	2809	1118	1208	648	438	2099	438	2823	706	73	2591	191	600	762	1138	2059	1321	2457	2740	2773	937	2503	608	673	516	2059	1037	309	737	482
Hartford, CT	2829	867	1209	1427	948	1257	1290	1133	973	1436	101	471	2909	1546	1321	1208	220	2523	486	2877	73	624	2749	455	324	1030	1257	2238	1965	2944	3019	2918	1529	2656	1223	1240	624	2433	1411	341	1339	470
Houston, TX	1581	948	584	1207	1155	1266	478	709	795	367	1679	1362	1957	454	949	964	1581	1110	1345	2369	1849	1233	1932	2291	1555	835	1266	1460	203	1484	1947	2498	271	2222	721	972	1249	1079	504	1420	1151	1322
Indianapolis, IN	2075	129	470	1208	283	591	749	590	302	857	730	669	2212	730	616	989	624	1719	349	2335	892	635	2116	907	551	251	591	1605	1200	2067	2224	2229	827	1961	818	1069	243	668	616	551	1176	341
Jackson, MS	1880	575	211	907	884	1123	178	255	422	193	1192	948	2270	575	914	697	1094	1500	965	2518	1362	815	2143	946	1183	495	1160	1742	649	1783	2183	2585	223	2205	421	672	876	1378	510	965	851	949
Jacksonville, FL	2402	729	697	356	1067	1374	413	379	592	551	957	632	2771	1181	1305	138	859	2100	882	3030	1127	584	2716	693	1102	867	1374	2286	1086	2329	2781	3042	814	2822	170	195	989	2010	1167	730	284	958
Kansas City, MO	1589	519	470	1475	568	459	819	867	590	857	1192	1179	1799	373	195	1266	1134	1277	851	1809	1378	1086	1606	1205	1062	251	459	1095	795	1627	1869	1834	624	1727	1045	1296	717	1281	283	1052	144	843
Knoxville, TN	2201	246	385	859	643	932	449	348	174	607	753	412	2509	847	930	665	646	1845	515	2550	951	356	2363	440	710	497	935	1766	1150	2269	2549	2553	729	2298	584	730	449	1755	782	554	806	521
Las Vegas, NV	275	1861	1581	2570	1752	1630	1841	2015	1792	1800	2520	2534	582	1119	1249	2311	2449	284	2181	991	2683	2319	478	2406	2371	1581	1630	413	1273	349	592	1209	1468	1119	2068	2319	1954	389	1224	2376	2498	2124
Lincoln, NE	1476	730	647	1673	560	409	1039	975	770	1014	1274	1354	1604	433	57	1452	1260	1236	965	1641	1451	1263	1411	1249	1135	462	408	900	916	1573	1614	1636	727	1460	1227	1477	762	1246	416	1184	1598	923
Little Rock	1678	502	138	1208	772	881	430	470	349	430	1235	1025	1984	324	690	965	1110	1379	899	2270	1395	851	1986	963	1111	422	881	1462	592	1743	1994	2368	219	2092	679	958	803	1257	259	832	1101	876
Los Angeles, CA	0	2136	1816	2828	2238	1905	2013	2035	2027	1883	2790	2809	372	1354	1508	2585	2717	389	2449	989	2902	2554	511	2641	2400	1849	1905	672	1378	121	382	1159	1687	1406	2342	2578	2213	502	1459	2659	2772	2424
Louisville, KY	2136	0	364	1102	397	705	626	466	178	729	707	607	2333	738	671	907	664	1782	381	2298	421	525	2132	686	608	290	705	1638	1103	2119	2388	2320	721	2075	664	915	300	1697	633	533	1094	373
Memphis, TN	1816	364	0	1013	673	949	389	332	211	397	1095	876	2122	462	705	770	989	1444	795	2408	1257	712	2124	825	973	300	949	1613	730	1881	2132	2506	357	2230	527	778	665	1370	397	875	957	738
Miami, FL	2828	1102	1013	0	1435	1743	729	695	908	867	1313	988	3087	1524	1670	227	1215	2448	1248	3366	1483	818	3032	988	1435	1231	1743	2602	1402	2645	3097	3406	1130	3138	486	292	1293	2326	1483	1086	65	1264
Milwaukee, WI	2238	397	673	1435	0	332	981	859	571	1045	851	948	2171	884	503	1224	834	1873	567	2002	1021	899	1930	832	697	389	332	1419	1305	2109	2175	1970	1013	1702	1053	1240	332	1808	770	770	1386	510
Minneapolis, MN	1905	705	949	1743	332	0	1227	1281	892	1346	1160	1337	2065	803	381	1673	1126	1679	868	1670	1322	1241	1776	1237	1005	628	1	1475	1265	1975	2075	1638	985	1370	1354	1589	641	1702	705	1078	1737	794
Mobile, AL	2013	626	389	729	981	1227	0	176	462	146	1176	884	2351	786	1011	486	1078	1593	1050	2611	1346	738	2278	867	1273	673	1158	1903	697	1971	2361	2710	401	2342	243	494	940	1542	713	949	673	1005
Montgomery, AL	2035	466	332	695	859	1281	176	0	288	322	1010	715	2295	794	1037	452	930	1655	843	2632	1191	561	2456	792	1179	632	1281	1918	904	2127	2464	2838	478	2562	209	460	766	1633	765	794	639	831
Nashville, TN	2027	178	211	908	571	892	462	288	0	551	859	665	2374	673	735	689	761	1671	575	2376	1029	502	2189	614	770	323	892	1678	941	2092	2375	2384	669	2124	486	737	478	1581	608	632	875	543
New Orleans, LA	1883	729	397	867	1045	1346	146	322	551	0	1322	1054	2317	681	1021	624	1224	1540	1078	2505	1492	876	2278	1114	1362	681	1346	1842	551	1824	2327	2574	309	2409	381	632	1005	1419	657	1095	811	1095
New York, NY	2790	707	1095	1313	851	1160	1176	1010	859	1322	0	389	2876	1436	1208	1094	101	2425	381	2880	162	510	2635	366	320	941	1160	2124	1832	2773	2886	2837	1415	2569	1109	1126	556	2360	1331	252	1223	403
Norfolk, VA	2809	607	876	988	948	1337	884	715	665	1054	389	0	2957	1349	1362	770	268	2393	365	2968	527	197	2789	98	535	927	1337	2262	1606	2669	2967	2975	1203	2700	802	827	592	2271	1234	190	916	438
Oakland, CA	372	2333	2122	3087	2171	2065	2351	2295	2374	2317	2876	2957	0	1660	1596	2844	2913	744	2528	614	2974	2865	201	2988	2681	2003	2065	545	1734	493	9	777	2052	979	2601	2836	2293	878	1766	2772	3055	2463
Oklahoma City, OK	1354	738	462	1524	884	803	786	794	673	681	1436	1349	1660	0	495	1281	1363	998	1095	1946	1597	1200	1662	1287	1298	495	803	1151	478	1345	1670	2044	389	1768	1038	1289	961	908	105	1305	1468	1087
Omaha, NE	1508	671	705	1670	503	381	1110	1037	735	1021	1208	1362	1596	495	0	1421	1200	1427	908	1605	1394	1281	1395	1400	1078	446	381	947	949	1641	1606	1679	729	1403	1241	1492	705	1400	401	1151	1671	875
Orlando, FL	2585	907	770	227	1224	1673	486	452	689	624	1094	770	2844	1281	1421	0	1005	2205	1025	3123	1273	608	2789	776	1200	1030	1551	2359	1159	2402	2854	3163	920	2895	243	89	1086	2083	1240	876	194	1052
Philadelphia, PA	2717	664	989	1215	834	1126	1078	930	761	1224	101	268	2913	1363	1200	1005	0	2374	292	2830	259	412	2627	252	324	868	1126	2209	1734	2911	2923	2780	1270	2512	1020	1037	502	2287	1249	133	1134	357
Phoenix, AZ	389	1782	1444	2448	1873	1679	1593	1655	1671	1540	2425	2393	744	998	1427	2205	2374	0	2084	1322	2586	2172	754	2285	2281	1484	1808	673	989	357	754	1492	1282	1386	1962	2213	1950	122	1103	2367	2392	2061
Pittsburgh, PA	2449	381	795	1248	567	868	1050	843	575	1078	381	365	2528	1095	908	1025	292	2084	0	2538	559	519	2335	333	284	600	868	1824	1468	2441	2538	2513	1087	2190	932	998	227	2021	981	219	1111	65
Portland, OR	985	2298	2408	3366	2002	1670	2611	2632	2376	2505	2880	2968	614	1946	1605	3123	2830	1322	2538	0	3000	2903	616	3012	2676	2068	1670	764	2168	1078	624	170	2255	365	2880	3131	2311	1483	1987	2846	3310	2481
Providence, RI	2902	421	1257	1483	1021	1322	1346	1191	1029	1492	162	527	2974	1597	1394	1273	259	2586	559	3000	0	673	2814	511	381	1102	1322	2303	2002	2978	2984	2967	1585	2644	1288	1322	689	252	1483	397	1402	538
Raleigh, NC	2554	525	713	818	899	1241	738	561	502	876	510	197	2865	1200	1281	608	412	2172	519	2903	673	0	2716	175	600	835	1241	2205	1427	2529	2797	2911	538	2651	624	657	584	2124	1103	283	754	535
Reno, NV	511	2132	2124	3032	1930	1776	2278	2456	2189	2278	2635	2789	201	1662	1395	2789	2627	754	2335	616	2814	2716	0	2803	2489	1881	1776	511	1768	535	211	810	1927	770	2546	2797	2133	867	1702	2579	2976	2303
Richmond, VA	2641	686	825	988	832	1237	867	792	614	1114	366	98	2988	1287	1400	776	252	2285	333	3012	511	175	2803	0	455	976	1237	2324	1555	2643	2989	3029	1173	2761	794	827	630	2425	1222	114	924	662
Rochester, NY	2400	608	973	1435	697	1005	1273	1179	770	1362	320	535	2681	1298	1078	1200	324	2281	284	2676	381	600	2489	455	0	803	1005	1978	1686	2674	2691	2660	1361	2384	1208	1257	381	2222	1398	341	1346	265
Saint Louis, MO	1849	290	300	1231	389	628	673	632	323	681	941	927	2003	495	446	1030	868	1484	600	2068	1102	835	1881	976	803	0	628	1370	916	1840	2075	2076	624	1881	795	1030	466	1419	381	795	1241	592
Saint Paul, MN	1905	705	949	1743	332	1	1158	1281	892	1346	1160	1337	2065	803	381	1551	1126	1808	868	1670	1322	1241	1776	1237	1005	628	0	1475	1265	1975	2075	1638	985	1370	1354	1589	641	1702	705	1078	1737	794
Salt Lake City, UT	672	1638	1613	2602	1419	1475	1903	1918	1678	1842	2124	2262	545	1151	947	2359	2209	673	1824	764	2303	2205	511	2324	1978	1370	1475	0	1447	762	714	851	1563	706	2116	2318	1715	795	1223	2141	2490	1805
San Antonio, TX	1378	1103	730	1402	1305	1265	697	904	941	551	1832	1606	1734	478	949	1159	1734	989	1468	2168	2002	1427	1768	1555	1686	916	1265	1447	0	1274	1774	2255	474	2110	916	1151	1387	876	535	1605	1370	1477
San Diego, CA	121	2119	1881	2645	2109	1975	1971	2127	2092	1824	2773	2669	493	1345	1641	2402	2911	357	2441	1078	2978	2529	535	2643	2674	1840	1975	762	1274	0	527	1276	1565	1403	2213	2464	2604	405	1403	2733	2072	2456
San Francisco, CA	382	2388	2132	3097	2175	2075	2361	2464	2375	2327	2886	2967	9	1670	1606	2854	2923	754	2538	624	2984	2797	211	2989	2691	2075	2075	714	1774	527	0	787	2061	852	2611	2846	2415	868	1776	2949	3065	2569
Seattle, WA	1159	2343	2506	3406	1970	1638	2710	2838	2384	2574	2837	2975	777	2044	1679	3163	2780	1492	2513	170	2967	2911	810	3029	2660	2076	1638	851	2255	1276	787	0	2335	274	2920	3155	2231	1573	2034	2684	3388	2456
Shreveport, LA	1687	721	357	1130	1013	985	401	478	669	309	1415	1203	2052	389	729	920	1270	1282	1087	2255	1585	538	1927	1173	1361	624	985	1563	474	1565	2061	2335	0	2038	644	895	1099	1160	356	953	507	1081
Spokane, WA	1406	2075	2230	3138	1702	1370	2342	2562	2124	2409	2569	2700	979	1768	1403	2895	2512	1386	2190	365	2644	2651	770	2761	2384	1816	1370	706	2110	1403	852	274	2038	0	2652	2887	2100	1496	1727	2417	3196	2125
Tallahassee, FL	2342	664	527	486	1053	1354	243	209	486	381	1109	802	2601	1038	1241	243	1020	1962	932	2880	1288	624	2546	794	1208	795	1354	2116	916	2213	2611	2920	644	2652	0	251	900	1840	955	891	430	1005
Tampa, FL	2578	915	778	292	1240	1589	494	460	737	632	1126	827	2836	1289	1492	89	1037	2213	998	3131	1322	657	2797	827	1251	1030	1589	2318	1151	2464	2846	3155	895	2887	251	0	1143	2076	1182	908	219	1088
Toledo, OH	2213	300	665	1293	332	641	940	766	478	1005	556	592	2293	961	705	1086	502	1950	227	2311	689	584	2133	630	381	466	641	1715	1387	2604	2415	2231	1099	2100	900	1143	0	2051	867	438	1248	170
Tuscon, AZ	502	1697	1370	2326	1808	1702	1542	1633	1581	1419	2360	2271	878	908	1400	2083	2287	122	2021	1483	252	2124	867	2425	2222	1419	1702	795	876	405	868	1573	1160	1496	1840	2076	2051	0	1038	2254	2270	2003
Tulsa, OK	1459	633	397	1483	770	705	713	765	608	657	1331	1234	1766	105	401	1240	1249	1103	981	1987	1483	1103	1702	1222	1398	381	705	1223	535	1403	1776	2034	356	1727	955	1182	867	1038	0	1284	1167	964
Washington, DC	2659	533	875	1086	770	1078	949	796	632	1095	252	190	2772	1305	1151	876	133	2367	219	2846	397	283	2579	114	341	795	1078	2141	1605	2733	2949	2684	953	2417	891	908	438	2254	1284	0	1005	283
West Palm Beach, FL	2772	1094	957	65	1386	1737	673	639	875	811	1223	916	3055	1468	1671	194	1134	2392	1111	3310	1402	754	2976	924	1346	1241	1737	2490	1370	2072	3065	3388	507	3196	430	219	1248	2270	1167	1005	0	1167
Youngstown, OH	2424	373	738	1238	510	794	1005	831	543	1095	403	438	2463	1087	875	1052	357	2076	65	2481	527	535	2303	662	265	592	794	1877	1477	2432	2569	2456	1181	2125	1005	1088	170	2003	964	283	1167	0

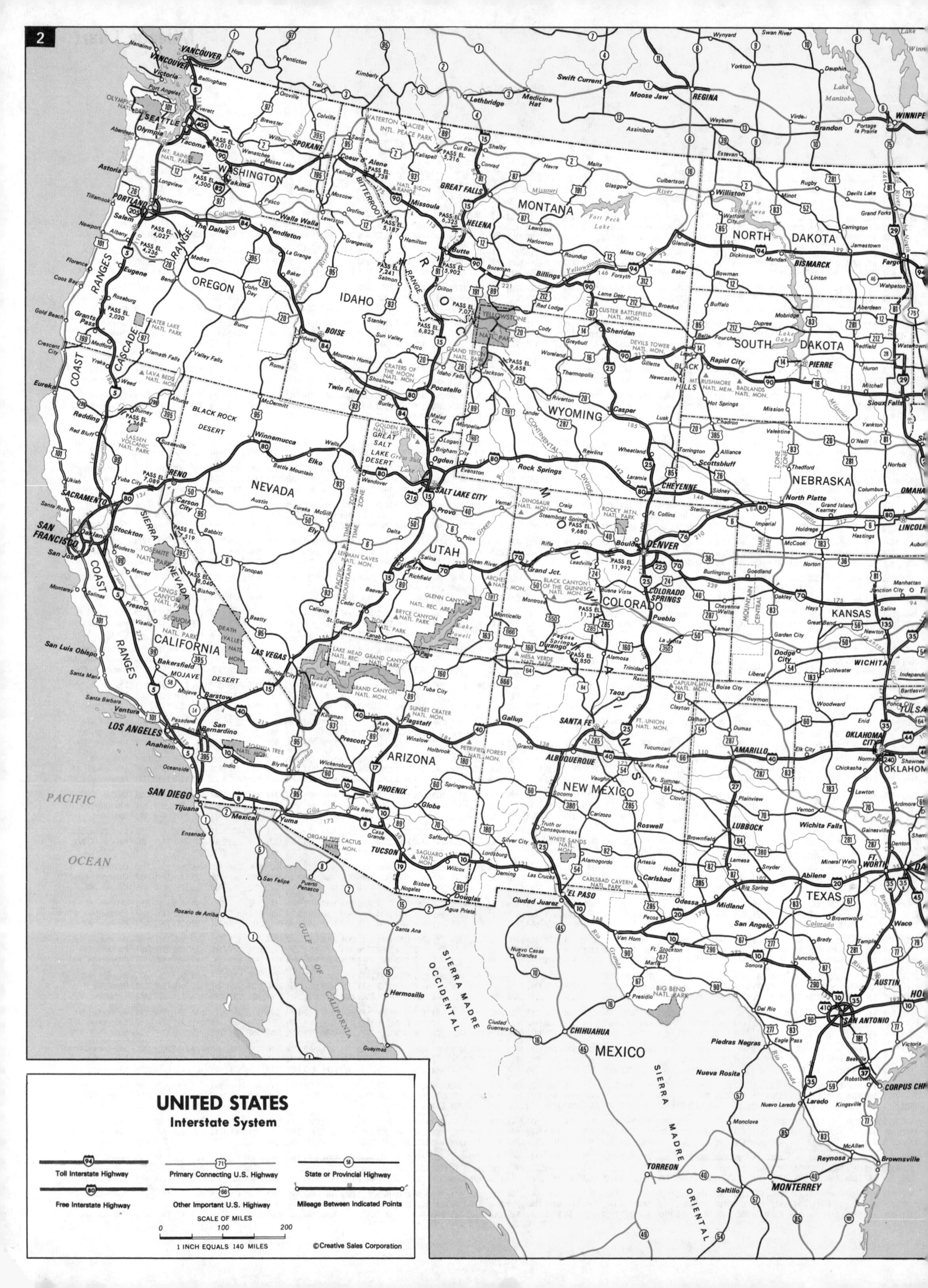
UNITED STATES
Interstate System
Toll Interstate Highway
Primary Connecting U.S. Highway
State or Provincial Highway
Free Interstate Highway
Other Important U.S. Highway
Mileage Between Indicated Points
SCALE OF MILES
0 100 200
1 INCH EQUALS 140 MILES
©Creative Sales Corporation
WASHINGTON
OREGON
IDAHO
MONTANA
NORTH DAKOTA
SOUTH DAKOTA
WYOMING
NEBRASKA
NEVADA
UTAH
COLORADO
KANSAS
CALIFORNIA
ARIZONA
NEW MEXICO
OKLAHOMA
TEXAS
MEXICO
PACIFIC
OCEAN
GULF OF CALIFORNIA
SIERRA MADRE OCCIDENTAL
SIERRA MADRE ORIENTAL
COAST RANGES
CASCADE RANGE
SIERRA NEVADA
BITTERROOT RANGE
ROCKY MOUNTAINS
BLACK ROCK DESERT
GREAT SALT LAKE DESERT
MOJAVE DESERT
BLACK HILLS
VANCOUVER
SEATTLE
SPOKANE
PORTLAND
BOISE
HELENA
GREAT FALLS
BISMARCK
PIERRE
SALT LAKE CITY
RENO
SACRAMENTO
SAN FRANCISCO
LOS ANGELES
SAN DIEGO
LAS VEGAS
PHOENIX
TUCSON
ALBUQUERQUE
SANTA FE
DENVER
CHEYENNE
EL PASO
AMARILLO
LUBBOCK
OKLAHOMA CITY
WICHITA
SAN ANTONIO
AUSTIN
MONTERREY
CHIHUAHUA
TORREON
REGINA
WINNIPEG

AREA CODES

ALABAMA

Birmingham 205
Montgomery 334

ALASKA

all locations 907

ARIZONA

Flagstaff 520
Phoenix 602
Tucson 520

ARKANSAS

all locations 501

CALIFORNIA

Bakersfield 805
Barstow 619
Encino 818
Eureka 707
Fresno 209
Los Angeles 213
Oakland 510
Orange 714
Redding 916
Sacramento 916
San Bernardino 909
San Diego 619
San Francisco 415
San Jose 408
San Luis Obispo 805
Stockton 209

COLORADO

Colorado Sprs 719
Denver 303
Durango 303
Pueblo 719

CONNECTICUT

all locations 203

DELAWARE

all locations 302

DISTRICT OF COLUMBIA

Washington 202

FLORIDA

Avon Park 813
Daytona Beach 904
Ft. Lauderdale 305
Ft. Myers 813
Jacksonville 904
Key West 305
Miami 305
Orlando 407
St. Petersburg 813
Tallahassee 904
Tampa 813
West Palm Beach ... 407
Winter Haven 813

GEORGIA

Atlanta 404
Augusta 706
Columbus 706
Rome 706
Savannah 912
Waycross 912

HAWAII

all locations 808

IDAHO

all locations 208

ILLINOIS

Aurora * 708, 630
Bloomington 309
Carbondale 618
Champaign 217
Chicago * 312, 630
Decatur 217
Elgin * 708, 630
Evanston * 708, 630
Joliet 815
La Grange * ... 708, 630
La Salle 815
Lincoln 217
Mt. Vernon 618
Peoria 309
Rockford 815
Schaumburg * .. 708, 630
Springfield 217
Waukegan * ... 708, 630

INDIANA

Angola 219
Bloomington 812
Connersville 317
Elkhart 219
Evansville 812
Fort Wayne 219
Gary 219
Green Castle 317
Hammond 219
Indianapolis 317
Kokomo 317
Lafayette 317
Michigan City 219
Muncie 317
Richmond 317
South Bend 219
Terre Haute 812
Warsaw 219

IOWA

Cedar Rapids 319
Council Bluffs 712
Davenport 319
Des Moines 515
Dubuque 319
Mason City 515
Sioux City 712

KANSAS

Dodge City 316
Ottawa 913
Topeka 913
Wichita 316

KENTUCKY

Ashland 606
Lexington 606
Louisville 502
Shelbyville 502
Winchester 606

LOUISIANA

Baton Rouge 504
Lake Charles 318
New Orleans 504
Shreveport 318

MAINE

all locations 207

MARYLAND

Annapolis 410
Baltimore 410
Cumberland 301
Rockville 301
Salisbury 410

MASSACHUSETTS

Boston 617
Framingham 508
Plymouth 508
Springfield 413

MICHIGAN

Adrian 517
Ann Arbor 313
Battle Creek 616
Cadillac 616
Detroit 313
Escanaba 906
Flint 313
Grand Rapids 616
Jackson 517
Lansing 517
Kalamazoo 616
Monroe 313
Port Huron 810
Saginaw 517

MINNESOTA

Albert Lea 507
Duluth 218
Ely 218
Minneapolis 612
Rochester 507
St. Paul 612

MISSISSIPPI

all locations 601

MISSOURI

Jefferson City 314
Joplin 417
Kansas City 816
St. Joseph 816
St. Louis 314
Springfield 417

MONTANA

all locations 406

NEBRASKA

Lincoln 402
North Platte 308
Omaha 402
Scottsbluff 308

NEVADA

all locations 702

NEW HAMPSHIRE

all locations 603

NEW JERSEY

Atlantic City 609
Hackensack 201
New Brunswick 908
Newark 201
Patterson 201
Trenton 609

NEW MEXICO

all locations 505

NEW YORK

Albany 518
Bronx * 718, 917
Brooklyn * 718, 917
Buffalo 716
East Hampton 516
Elmira 607
Greenwich 518
Manhattan * ... 212, 917
Mt. Vernon 914
Niagara Falls 716
Poughkeepsie 914
Queens * 718, 917
Rochester 716
Schenectady 518
Staten Island * . 718, 917
Syracuse 315

NORTH CAROLINA

Charlotte 704
Greensboro 919
Greenville 910
Raleigh 910
Salisbury 704
Williamston 910
Winston-Salem 919

NORTH DAKOTA

all locations 701

OHIO

Akron 216
Canton 216
Cincinnati 513
Cleveland 216
Columbus 614
Dayton 513
Massillon 216
Niles 216
Salem 216
Toledo 419
Youngstown 216

OKLAHOMA

Muskogee 918
Oklahoma City 405
Tulsa 918

OREGON

all locations 503

PENNSYLVANIA

Allentown
(Lehigh Co.) 610
Erie 814
Harrisburg 717
Hershey 717
Philadelphia 215
Pittsburgh 412
Reading 610
Scranton 717

RHODE ISLAND

all locations 401

SOUTH CAROLINA

all locations 803

SOUTH DAKOTA

all locations 605

TENNESSEE

Chattanooga 615
Knoxville 615
Memphis 901
Nashville 615

TEXAS

Amarillo 806
Austin 512
Brownsville 210
Corpus Christi 512
Dallas 214
El Paso 915
Ft. Worth 817
Galveston 409
Houston 713
Lubbock 806
Odessa 915
San Angelo 915
San Antonio 210
Temple 817
Texarkana 903
Tyler 903
Waco 817
Wichita Falls 817

UTAH

all locations 801

VERMONT

all locations 802

VIRGIN ISLANDS

all locations 809

VIRGINIA

Charlottesville 804
Fredericksburg 703
Norfolk 804
Richmond 804
Roanoke 703

WASHINGTON

Olympia 360
Seattle 206
Spokane 509
Vancouver 360
Walla Walla 509
Yakima 509

WEST VIRGINIA

all locations 304

WISCONSIN

Beloit 608
Eau Claire 715
Green Bay 414
Madison 608
Milwaukee 414
Racine 414
Wausau 715

WYOMING

all locations 307

800 SERVICE

all locations 800

CANADA

Chicoutimi, Quebec . 418
Calgary, Alb 403
Edmonton, Alb 403
Edmunston, N.B. ... 506
London, Ont 519
Montreal, Que 514
North Bay, Ont 705
Ottawa, Ont 613
Prince Albert, Sask . 306
Quebec, Que 418
Regina, Sask 306
Sault Ste. Marie, Ont ... 705
St. John's, Nfd 709
Sudbury, Ont 705
Thunder Bay, Ont ... 807
Toronto, Ont 905
Vancouver, B.C. 604
Winnepeg, Man 204

MEXICO

all locations ....
52 + city code + digits

PUERTO RICO

all locations 809

*** Note:** Cities with separate cellular area code overlay. The second number listed represents the cellular overlay.

1 • 2 • 3 • 4 • 5 • 6 • 7

A • B • C • D • E • F • G • H • J • K

BRITISH COLUMBIA
ALBERTA
SASKATCHEWAN
MANITOBA
CANADA
UNITED STATES
Vancouver
Calgary
Saskatoon
Moose Jaw
Regina
Winnipeg
Pacific Rim Nat'l. Park
San Juan Island Nat'l Hist Park
Olympic Nat'l. Park
Olympic Nat'l Forest
Mt. Baker Nat'l. Forest
N. Cascades Nat'l Park
Okanogan Nat'l Forest
Colville Nat'l Forest
Kaniksu Nat'l Forest
Kootenai Nat'l Forest
Waterton-Glacier Int'l Peace Park
Glacier Nat'l Park
Rocky Mtns. Forest Preserve
Riding Mountain Nat'l Park
Lake Manitoba
PACIFIC TIME ZONE
MOUNTAIN TIME ZONE
CENTRAL TIME ZONE
CONTINENTAL DIVIDE
Seattle
Olympia
Tacoma
Spokane
Coeur d'Alene
Mt. Rainier Nat'l. Park
Wenatchee Nat'l Forest
Ft. Vancouver Nat'l Hist. Site
Astoria
WASHINGTON
Yakima
Walla Walla
Gifford Pinchot Nat'l Forest
Mt. Saint Helens Nat'l Volcanic Monument
Portland
Salem
Mt. Hood Nat'l Forest
Pendleton
Umatilla Nat'l Forest
Wallowa Whitman Nat'l Forest
John Day Fossil Beds Nat'l Mon.
Ochoco Nat'l Forest
Malheur Nat'l Forest
Willamette Nat'l Forest
Siuslaw Nat'l Forest
Eugene
Umpqua Nat'l Forest
Crater Lake Nat'l Park
Deschutes Nat'l Forest
Winema Nat'l Forest
Fremont Nat'l Forest
Rogue River Nat'l Forest
Siskiyou Nat'l Forest
Grants Pass
Oregon Caves Nat'l Mon.
OREGON
CASCADE RANGE
COAST RANGES
St. Joe Nat'l Forest
Clearwater Nat'l Forest
Nezperce Nat'l Forest
Payette Nat'l Forest
Salmon Nat'l Forest
Boise Nat'l Forest
Sawtooth Nat'l Forest
Challis Nat'l Forest
Boise
IDAHO
Craters of the Moon Nat'l Mon.
Twin Falls
Pocatello
Caribou Nat'l Forest
Targhee Nat'l Forest
BITTERROOT RANGE
Flathead Nat'l Forest
Lolo Nat'l Forest
Bitterroot Nat'l Forest
Deerlodge Nat'l Forest
Beaverhead Nat'l Forest
Helena
Great Falls
Lewis and Clark Nat'l Forest
Butte
Gallatin Nat'l Forest
Custer Nat'l Forest
Billings
Custer Battlefield Nat'l Mon.
MONTANA
Missouri River
Ft. Peck Lake
Ft. Union Nat'l Hist. Site
Williston
Lake Sakakawea
Theodore Roosevelt Nat'l Park North
Theodore Roosevelt Nat'l Park South
Knife River Indian Village Nat'l Hist. Site
NORTH DAKOTA
Bismarck
Fargo
Lake Oahe
SOUTH DAKOTA
Pierre
Rapid City
Black Hills Nat'l Forest
Mt. Rushmore Nat'l Mem.
Badlands Nat'l Park
Wind Cave Nat'l Park
Hot Springs
Sioux Falls
Devil's Tower Nat'l Mon.
Thunder Basin Nat'l Grassland
Bighorn Nat'l Forest
Shoshone Nat'l Forest
Yellowstone Nat'l Park
Grand Teton Nat'l Park
Jackson
Bridger-Teton Nat'l Forest
WYOMING
Casper
Rock Springs
Ashley Nat'l Forest
Medicine Bow Nat'l Forest
Cheyenne
Laramie
Ft. Laramie Nat'l Hist. Site
Scottsbluff
Nebraska Nat'l Forest
Samuel R. McKelvie Nat'l Forest
North Platte
NEBRASKA
Lincoln
Redwood Nat'l Park
Klamath Nat'l Forest
Lava Beds Nat'l. Mon.
Modoc Nat'l Forest
Six Rivers Nat'l Forest
Shasta Nat'l Forest
Trinity Nat'l Forest
Eureka
Redding
Lassen Volcanic Nat'l Park
Lassen Nat'l Forest
Mendocino Nat'l Forest
Plumas Nat'l Forest
Tahoe Nat'l Forest
BLACK ROCK DESERT
Humboldt Nat'l Forest
Winnemucca
Reno
Carson City
Sacramento
Eldorado Nat'l Forest
Pt. Reyes Nat'l Seashore
Muir Woods
San Francisco
Oakland
San Jose
John Muir Nat'l Hist. Site
Stanislaus Nat'l Forest
Yosemite Nat'l Park
Toiyabe Nat'l Forest
Sierra Nat'l Forest
Inyo Nat'l Forest
Devils Postpile Nat'l Mon.
Kings Canyon Nat'l Park
Sequoia Nat'l Park
Sequoia Nat'l Forest
Fresno
Monterrey
Pinnacles Nat'l Mon.
Los Padres Nat'l Forest
San Luis Obispo
Death Valley Nat'l Mon.
SIERRA NEVADA
NEVADA
Ely
Lehman Caves Nat'l Mon.
Great Basin Nat'l Park
GREAT SALT LAKE
GREAT SALT LAKE DESERT
Golden Spike Nat'l Hist. Site
Wasatch Nat'l Forest
Ogden
Cache Nat'l Forest
Montpelier
Fossil Butte Nat'l Mon.
Salt Lake City
Provo
Uinta Nat'l Forest
UTAH
Dinosaur Nat'l Mon.
Manti-La Sal Nat'l Forest
Fishlake Nat'l Forest
Capitol Reef Nat'l Park
Arches Nat'l Park
Canyonlands Nat'l Park
Natural Bridges Nat'l Mon.
Hovenweep Nat'l Mon.
Cedar Breaks Nat'l Mon.
Dixie Nat'l Forest
Zion Nat'l Park
Bryce Canyon Nat'l Park
Lake Powell
Rainbow Bridge Nat'l Mon.
Las Vegas
Lake Mead
Bakersfield
MOJAVE DESERT
CALIFORNIA
Barstow
Ventura
Channel Islands Nat'l Park
Los Angeles
Angeles Nat'l Forest
San Bernardino Nat'l Forest
San Bernardino
Anaheim
Joshua Tree Nat'l Mon.
Cabrillo Nat'l Mon
Cleveland Nat'l Forest
San Diego
Tijuana
Mexicali
Ensenada
BAJA CALIFORNIA
PACIFIC OCEAN
Grand Canyon Nat'l Park
Kaibab Nat'l Forest
Canyon De Chelly Nat'l Mon
Wupatki Nat'l Mon
Sunset Crater Nat'l Mon
Flagstaff
Walnut Canyon Nat'l Mon
Coconino Nat'l Forest
Prescott Nat'l Forest
Prescott
Petrified Forest Nat'l Park
Tonto Nat'l Forest
Tonto Nat'l Mon.
Apache Nat'l Forest
Phoenix
Globe
Casa Grande Ruins Nat'l Mon.
ARIZONA
Yuma
Gila River
Organ Pipe Cactus Nat'l Mon
Saguaro Nat'l Mon
Tucson
Coronado Nat'l Forest
Chiricahua Nat'l Mon.
Tumacacori Nat'l Mon.
Douglas
Nogales
Coronado Nat'l Mon
UNITED STATES
MEXICO
SONORA
Santa Ana
SIERRA MADRE OCCIDENTAL
Ruins Nat'l Mon
Colorado Nat'l Mon
Grand Mesa Nat'l Forest
White River Nat'l Forest
Black Canyon of the Gunnison Nat'l Mon
Gunnison
Rio Grande Nat'l Forest
San Juan Nat'l Forest
Mesa Verde Nat'l Park
Durango
Aztec Ruins Nat'l Mon
Chaco Canyon Nat'l Mon
Carson Nat'l Forest
Santa Fe Nat'l Forest
Bandelier Nat'l Mon
Santa Fe
Pecos Nat'l Mon
Capulin Mtn. Nat'l Mon
Kiowa Nat'l Grasslands
Ft. Union Nat'l Mon
El Morro Nat'l Mon
Cibola Nat'l Forest
Albuquerque
Gran Quivira Nat'l Mon
Gila Nat'l Forest
Gila Cliff Dwellings Nat'l Mon.
White Sands Nat'l Mon
Lincoln Nat'l Forest
Carlsbad
Carlsbad Caverns Nat'l Park
Guadalupe Mtns. Nat'l Park
NEW MEXICO
El Paso
Ciudad Juarez
Nuevo Casas Grandes
CHIHUAHUA
ROCKY MOUNTAINS
Roosevelt Nat'l Forest
Rocky Mountain Nat'l Park
Arapaho Nat'l Forest
Pike Nat'l Forest
Pawnee Nat'l Grassland
Boulder
Denver
COLORADO
Colorado Springs
Pueblo
Comanche Nat'l. Grassland
KANSAS
Dodge City
Wichita
OKLAHOMA
Oklahoma City
Platt Nat'l Park
Amarillo
Lubbock
Wichita Falls
Ft. Worth
Dallas
Abilene
Odessa
Midland
San Angelo
TEXAS
Waco
Austin
Old Ft. Davis Nat'l Hist. Site
Big Bend Nat'l Park
San Antonio
COAHUILA
Piedras Negras
Nueva Rosita
Laredo
NUEVO LEON
Monterrey
Corpus Christi
Padre Island Nat'l Seashore
Brownsville
TAMAULIPAS
U.S.S.R.
U.S.S.R. TIME ZONE
INT'L DATE LINE
Wainwright
Barrow
Arctic Ocean
Point Hope
Prudhoe Bay
Noatak
Noatak Nat'l. Preserve
CONTINENTAL DIVIDE
Gates of the Arctic Nat'l Park & Preserve
Noorvik
Bering Land Bridge Nat'l. Preserve
Kobuk Valley Nat'l Park
Nome
Fort Yukon
Norton Sound
Bering Sea
Unalakleet
ALASKA
Fairbanks
College
Yukon-Charley Nat'l. Preserve
Delta Junction
Tok
Hooper Bay
Denali Nat'l. Park and Preserve
BERING TIME ZONE
ALASKA-HAWAII TIME ZONE
Glennallen
Palmer
Anchorage
Wrangell-St. Elias Nat'l Park
Valdez
Cordova
Lake Clark Nat'l. Park & Preserve
Togiak
Seward
Kenai Fjords Nat'l. Park
Chugach Nat'l. Forest
Katmai Nat'l. Park & Preserve
Bristol Bay
Aniakchak Nat'l. Park & Preserve
Kodiak
Kodiak Island
Gulf of Alaska
Aleutian Islands
Pacific Ocean
U.S.
CANADA
YUKON
PACIFIC TIME ZONE
YUKON TIME ZONE
Kluane Nat'l Park
Tongass Nat'l Forest
Glacier Bay Nat'l. Park
Juneau
Sitka
B.C.
Ketchikan
Lihue
Wahiawa
Kaneohe
Pearl City
Honolulu
Lahaina
Wailuku
Haleakala Nat'l. Park
HAWAII
Pacific Ocean
Hilo
Hawaii Volcanoes Nat'l Park
City of Refuge Nat'l Hist. Park

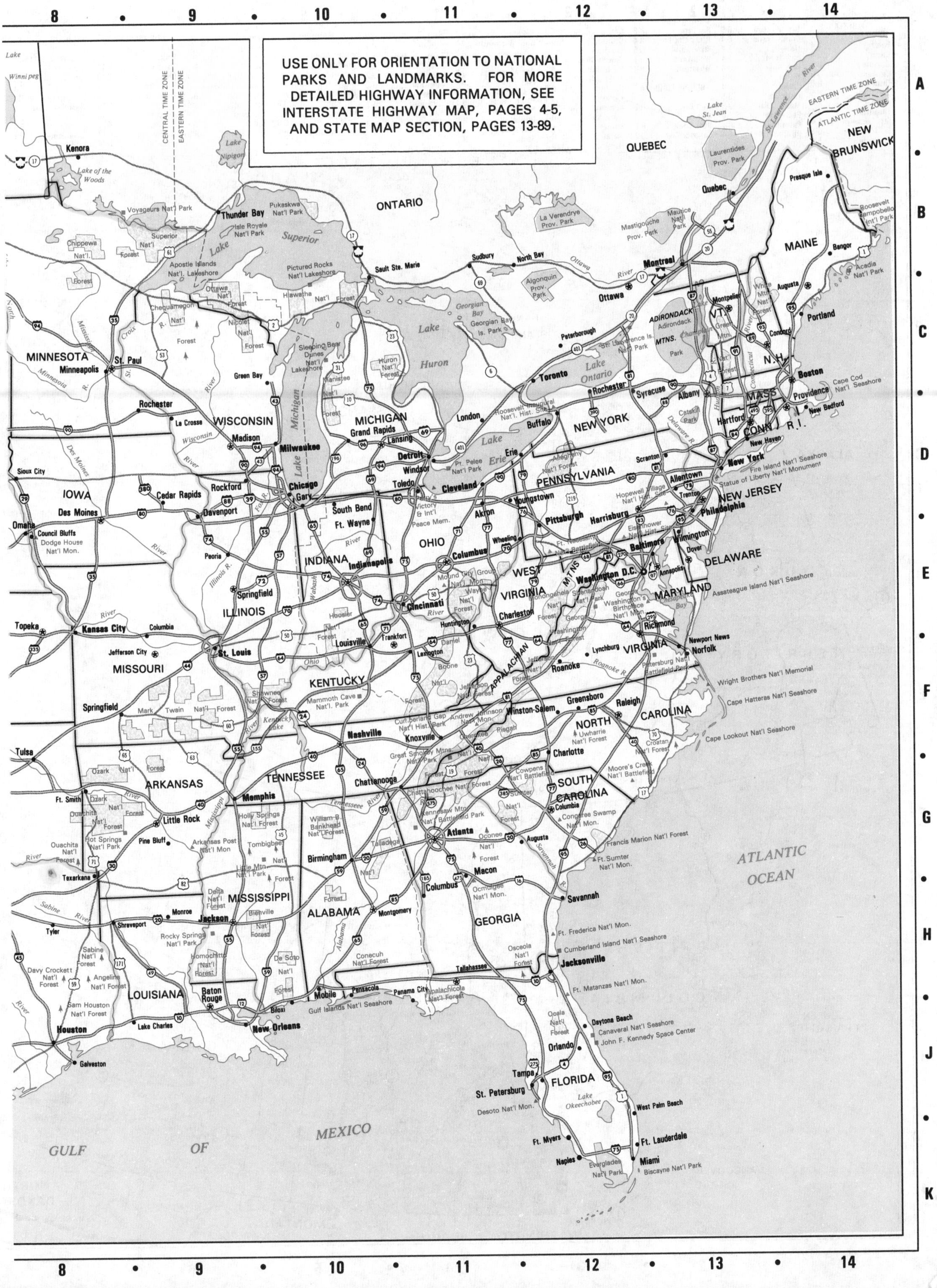

USE ONLY FOR ORIENTATION TO NATIONAL PARKS AND LANDMARKS. FOR MORE DETAILED HIGHWAY INFORMATION, SEE INTERSTATE HIGHWAY MAP, PAGES 4-5, AND STATE MAP SECTION, PAGES 13-89.

CENTRAL TIME ZONE
EASTERN TIME ZONE
ATLANTIC TIME ZONE

ONTARIO
QUEBEC
NEW BRUNSWICK
MAINE
MINNESOTA
WISCONSIN
MICHIGAN
IOWA
ILLINOIS
INDIANA
OHIO
PENNSYLVANIA
NEW YORK
VT.
N.H.
MASS.
CONN.
R.I.
NEW JERSEY
DELAWARE
MARYLAND
WEST VIRGINIA
VIRGINIA
KENTUCKY
MISSOURI
ARKANSAS
TENNESSEE
NORTH CAROLINA
SOUTH CAROLINA
GEORGIA
ALABAMA
MISSISSIPPI
LOUISIANA
FLORIDA

Lake Superior
Lake Michigan
Lake Huron
Lake Erie
Lake Ontario
Lake of the Woods
Lake Nipigon
Lake St. Jean
Lake Okeechobee
ATLANTIC OCEAN
GULF OF MEXICO

Voyageurs Nat'l Park
Isle Royale Nat'l Park
Pukaskwa Nat'l Park
Apostle Islands Nat'l. Lakeshore
Pictured Rocks Nat'l Lakeshore
Sleeping Bear Dunes Nat'l Lakeshore
Pt. Pelee Nat'l Park
Georgian Bay Is. Park
Algonquin Prov. Park
La Verendrye Prov. Park
Laurentides Prov. Park
Mastigouche Prov. Park
Mauricie Nat'l Park
Acadia Nat'l Park
Roosevelt Campobello Int'l Park
St. Lawrence Is. Nat'l Park
Roosevelt Inaugural Nat'l. Hist. Site
Cape Cod Nat'l Seashore
Fire Island Nat'l Seashore
Statue of Liberty Nat'l Monument
Hopewell Village Nat'l Hist. Site
Dodge House Nat'l Mon.
Perry's Victory & Int'l Peace Mem.
Mound City Group Nat'l Mon.
Mammoth Cave Nat'l. Park
Cumberland Gap Nat'l Hist. Park
Andrew Johnson Nat'l Mon.
Great Smokey Mtns Nat'l Park
Assateague Island Nat'l Seashore
Wright Brothers Nat'l Memorial
Cape Hatteras Nat'l Seashore
Cape Lookout Nat'l Seashore
Cowpens Nat'l Battlefield
Moore's Creek Nat'l Battlefield
Congaree Swamp Nat'l Mon.
Francis Marion Nat'l Forest
Ft. Sumter Nat'l Mon.
Kennesaw Mtn. Nat'l Battlefield Park
Ocmulgee Nat'l Mon.
Ft. Frederica Nat'l Mon.
Cumberland Island Nat'l Seashore
Ft. Matanzas Nat'l Mon.
Canaveral Nat'l Seashore
John F. Kennedy Space Center
Desoto Nat'l Mon.
Everglades Nat'l Park
Biscayne Nat'l Park
Gulf Islands Nat'l Seashore
Hot Springs Nat'l Park
Arkansas Post Nat'l Mon
Rocky Springs Nat'l Park
Sam Houston Nat'l Forest

ALBERTA

- Banff J-4
- Calgary J-4
- Camrose H-5
- Cardston J-5
- Drumheller H-5
- Edmonton H-5
- Edson H-4
- Fitzgerald F-5
- Fort Chipewyn F-4
- Ft. McMurray G-5
- Ft. Vermillion F-4
- Grande Centre G-5
- Grande Prairie G-4
- Habay F-4
- High Level F-4
- High River J-4
- Hinton H-4
- Jasper H-4
- Lethbridge J-5
- Medicine Hat J-5
- Peace Point F-4
- Peace River G-4
- Rainbow Lake F-4
- Red Deer H-5
- Rocky Mtn. House H-4
- Waterways G-5
- Wetaskiwin H-4
- Whitecourt H-4

BRITISH COLUMBIA

- Alert Bay J-2
- Atlin F-2
- Bella Coola H-2
- Campbell River J-2
- Cassiar F-2
- Courtnay J-2
- Cranbrook J-4
- Dawson Creek G-3
- Ft. Nelson F-3
- Ft. St. John G-3
- Hazelton G-2
- Kamloops J-3
- Kelowna J-3
- Kimberley J-4
- Kitimat H-2
- Masset G-1
- Nanaimo J-2
- Nelson J-4
- New Westminster J-3
- Ocean Falls H-2
- Penticton J-4
- Port Alberni J-2
- Port Alice J-2
- Powell River J-3
- Prince George H-3
- Prince Rupert G-2
- Quesnel H-3
- Revelstoke J-4
- Sandpit H-1
- Smithars G-2
- Stewart G-2
- Telegraph Creek F-2
- Trail J-4
- Ucluelet J-2
- Vancouver J-3
- Vernon J-4
- Victoria J-3
- Ware G-3
- Williams Lake H-3

MANITOBA

- Berens River G-7
- Brandon H-7
- Brochet F-6
- Churchill F-7
- Dauphin H-7
- Flin Flon G-6
- Gillam F-7
- Gods Lake G-8
- Grand Rapids G-7
- Island Lake G-7
- Lynn Lake F-6
- Norway House G-7
- Portage la Prairie H-7
- Selkirk H-8
- Snow Lake G-7
- S. Indian Lake F-7
- Swan River H-7
- The Pas G-6
- Thompson G-7
- Virden J-7
- Winnipeg H-8

NEW BRUNSWICK

- Bathurst E-13
- Campbellton E-12
- Chatam E-13
- Edmundston F-12
- Fredricton F-13
- Moncton E-13
- Newcastle E-13
- St. John F-13
- St. Stephen F-13
- Tracadie E-13
- Woodstock F-13

NEWFOUNDLAND

- Battle Harbour C-13
- Bonavista C-14
- Cartwright C-12
- Channel-Port aux Basques D-14
- Churchill Falls D-12
- Corner Brook D-13
- Gander C-14

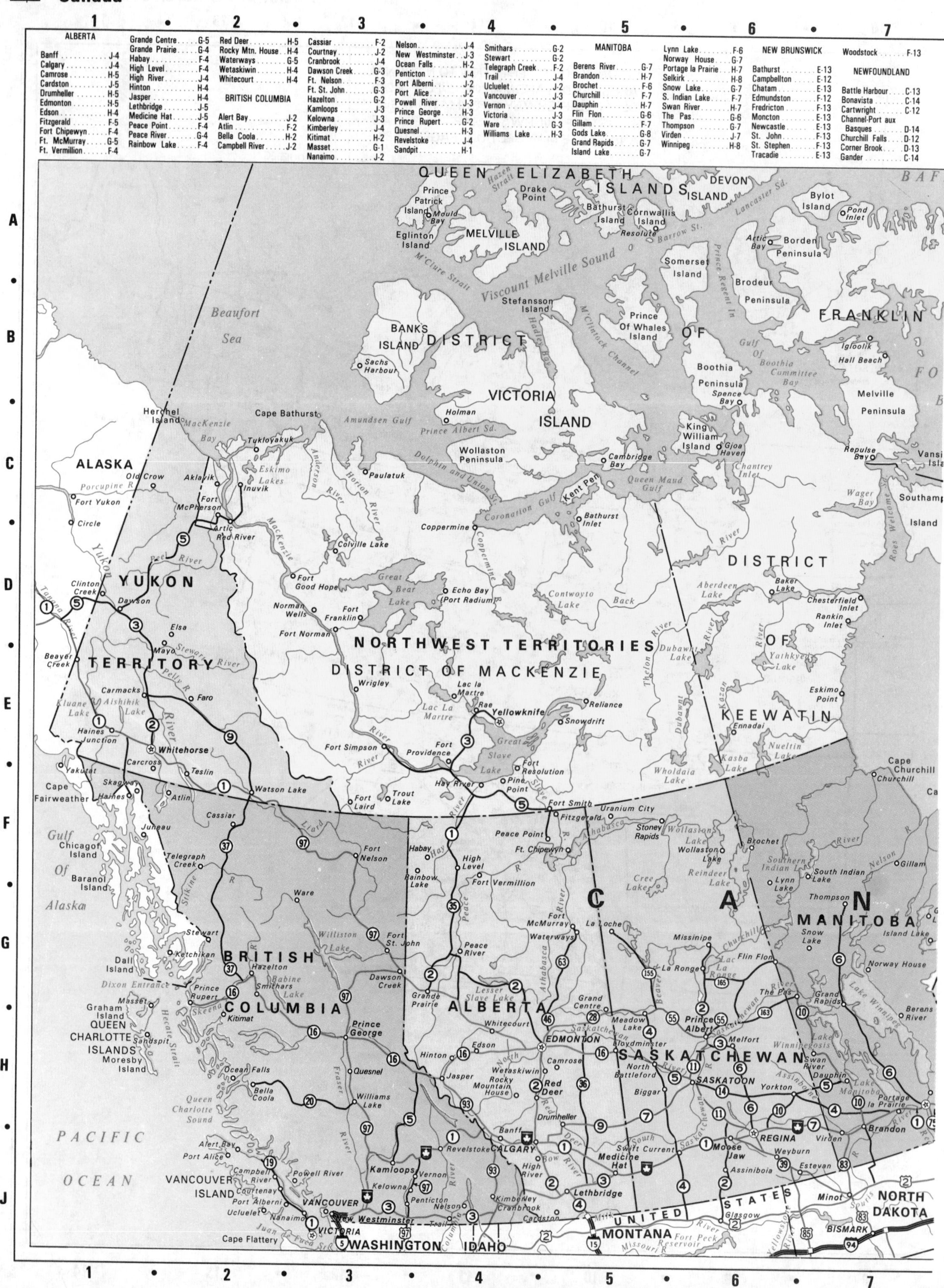

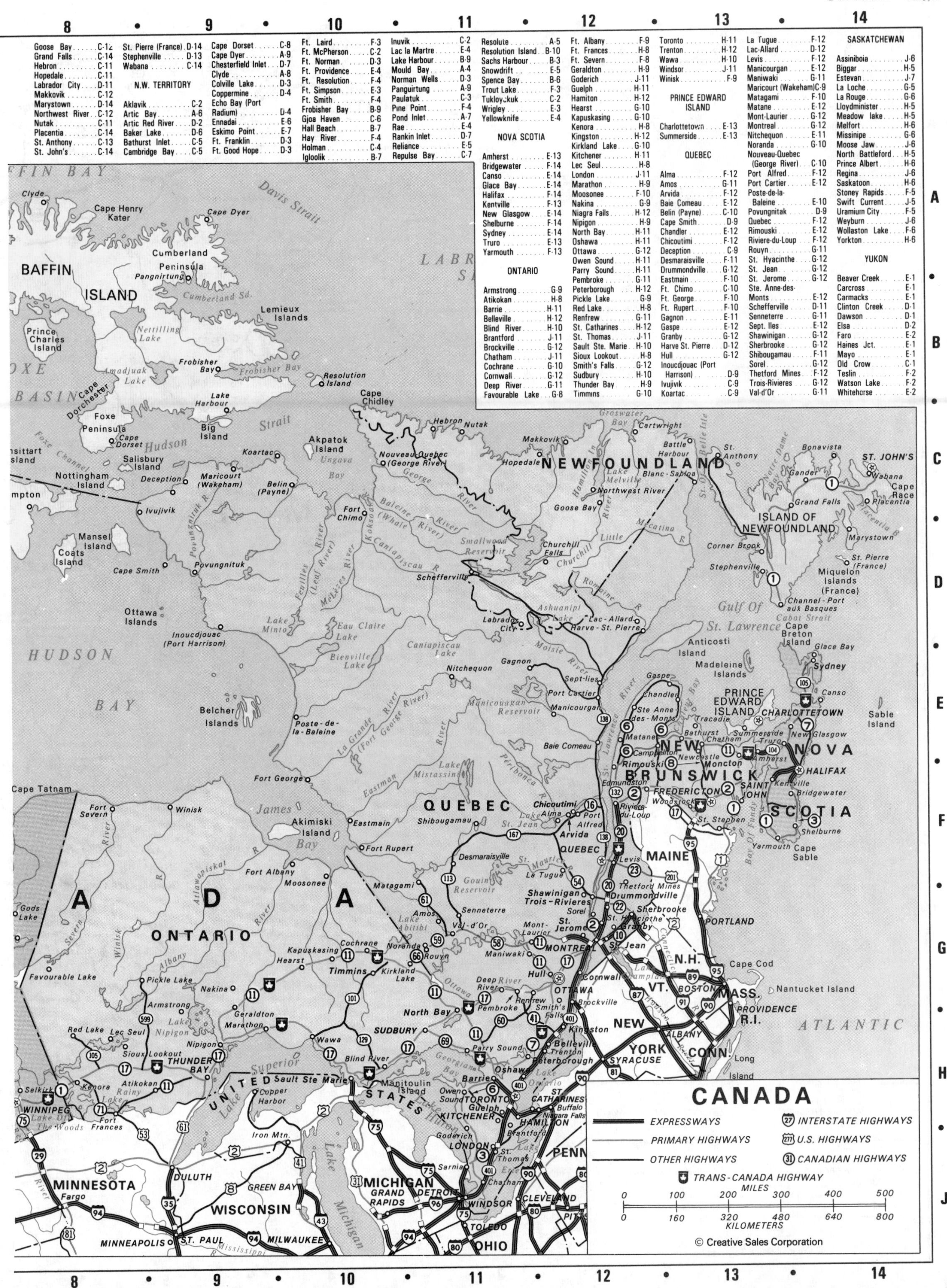

Goose Bay C-12
Grand Falls C-14
Hebron C-11
Hopedale C-11
Labrador City D-11
Makkovik C-12
Marystown D-14
Northwest River C-12
Nutak C-11
Placentia C-14
St. Anthony C-13
St. John's C-14
St. Pierre (France) D-14
Stephenville D-13
Wabana C-14
N.W. TERRITORY
Aklavik C-2
Artic Bay A-6
Artic Red River D-2
Baker Lake D-6
Bathurst Inlet C-5
Cambridge Bay C-5
Cape Dorset C-8
Cape Dyer A-9
Chesterfield Inlet D-7
Clyde A-8
Colville Lake D-3
Coppermine D-4
Echo Bay (Port Radium) D-4
Ennadai E-6
Eskimo Point E-7
Ft. Franklin D-3
Ft. Good Hope D-3
Ft. Laird F-3
Ft. McPherson C-2
Ft. Norman D-3
Ft. Providence E-4
Ft. Resolution F-4
Ft. Simpson E-3
Ft. Smith F-4
Frobisher Bay B-9
Gjoa Haven C-6
Hall Beach B-7
Hay River F-4
Holman C-4
Igloolik B-7
Inuvik C-2
Lac la Martre E-4
Lake Harbour B-9
Mould Bay A-4
Norman Wells D-3
Panguirtung A-9
Paulatuk C-3
Pine Point F-4
Pond Inlet A-7
Rae E-4
Rankin Inlet D-7
Reliance E-5
Repulse Bay C-7
Resolute A-5
Resolution Island B-10
Sachs Harbour B-3
Snowdrift E-5
Spence Bay B-6
Trout Lake F-3
Tukloyakuk C-2
Wrigley E-3
Yellowknife E-4
NOVA SCOTIA
Amherst E-13
Bridgewater F-14
Canso E-14
Glace Bay E-14
Halifax F-14
Kentville F-13
New Glasgow E-14
Shelburne F-14
Sydney E-14
Truro E-13
Yarmouth F-13
ONTARIO
Armstrong G-9
Atikokan H-8
Barrie H-11
Belleville H-12
Blind River H-10
Brantford J-11
Brockville G-12
Chatham J-11
Cochrane G-10
Cornwall G-12
Deep River G-11
Favourable Lake G-8
Ft. Albany F-9
Ft. Frances H-8
Ft. Severn F-8
Geraldton H-9
Goderich J-11
Guelph H-11
Hamilton H-12
Hearst G-10
Kapuskasing G-10
Kenora H-8
Kingston H-12
Kirkland Lake G-10
Kitchener H-11
Lec Seul H-8
London J-11
Marathon H-9
Moosonee F-10
Nakina G-9
Niagra Falls H-12
Nipigon H-9
North Bay H-11
Oshawa H-11
Ottawa G-12
Owen Sound H-11
Parry Sound H-11
Pembroke G-11
Peterborough H-12
Pickle Lake G-9
Red Lake H-8
Renfrew G-11
St. Catharines H-12
St. Thomas J-11
Sault Ste. Marie H-10
Sioux Lookout H-8
Smith's Falls G-12
Sudbury H-10
Thunder Bay H-9
Timmins G-10
Toronto H-11
Trenton H-12
Wawa H-10
Windsor J-11
Winisk F-9
PRINCE EDWARD ISLAND
Charlottetown E-13
Summerside E-13
QUEBEC
Alma F-12
Amos G-11
Arvida F-12
Baie Comeau E-12
Belin (Payne) C-10
Cape Smith D-9
Chandler E-12
Chicoutimi F-12
Deception C-9
Desmaraisville F-11
Drummondville G-12
Eastmain F-10
Ft. Chimo C-10
Ft. George F-10
Ft. Rupert F-10
Gagnon E-11
Gaspe E-12
Granby G-12
Harve St. Pierre D-12
Hull G-12
Inoucdjouac (Port Harrison) D-9
Ivujivik C-9
Koartac C-9
La Tugue F-12
Lac-Allard D-12
Levis F-12
Manicougan E-12
Maniwaki G-11
Maricourt (Wakeham) C-9
Matagami F-10
Matane E-12
Mont-Laurier G-12
Montreal G-12
Nitchequon E-11
Noranda G-10
Nouveau-Quebec (George River) C-10
Port Alfred F-12
Port Cartier E-12
Poste-de-la-Baleine E-10
Povungnitak D-9
Quebec F-12
Rimouski E-12
Riviere-du-Loup F-12
Rouyn G-11
St. Hyacinthe G-12
St. Jean G-12
St. Jerome G-12
Ste. Anne-des-Monts E-12
Schefferville D-11
Senneterre G-11
Sept. Iles E-12
Shawinigan G-12
Sherbrooke G-12
Shibougamau F-11
Sorel G-12
Thetford Mines F-12
Trois-Rivieres G-12
Val-d'Or G-11
SASKATCHEWAN
Assiniboia J-6
Biggar H-5
Estevan J-7
La Loche G-5
La Rouge G-6
Lloydminister H-5
Meadow lake H-5
Melfort H-6
Missinipe G-6
Moose Jaw J-6
North Battleford H-5
Prince Albert H-6
Regina J-6
Saskatoon H-6
Stoney Rapids F-5
Swift Current J-5
Uramium City F-5
Weyburn J-6
Wollaston Lake F-6
Yorkton H-6
YUKON
Beaver Creek E-1
Carcross E-1
Carmacks E-1
Clinton Creek D-1
Dawson D-1
Elsa D-2
Faro E-2
Haines Jct. E-1
Mayo E-1
Old Crow C-1
Teslin F-2
Watson Lake F-2
Whitehorse E-2
CANADA
EXPRESSWAYS
PRIMARY HIGHWAYS
OTHER HIGHWAYS
INTERSTATE HIGHWAYS
U.S. HIGHWAYS
CANADIAN HIGHWAYS
TRANS-CANADA HIGHWAY
MILES
0 100 200 300 400 500
0 160 320 480 640 800
KILOMETERS
© Creative Sales Corporation

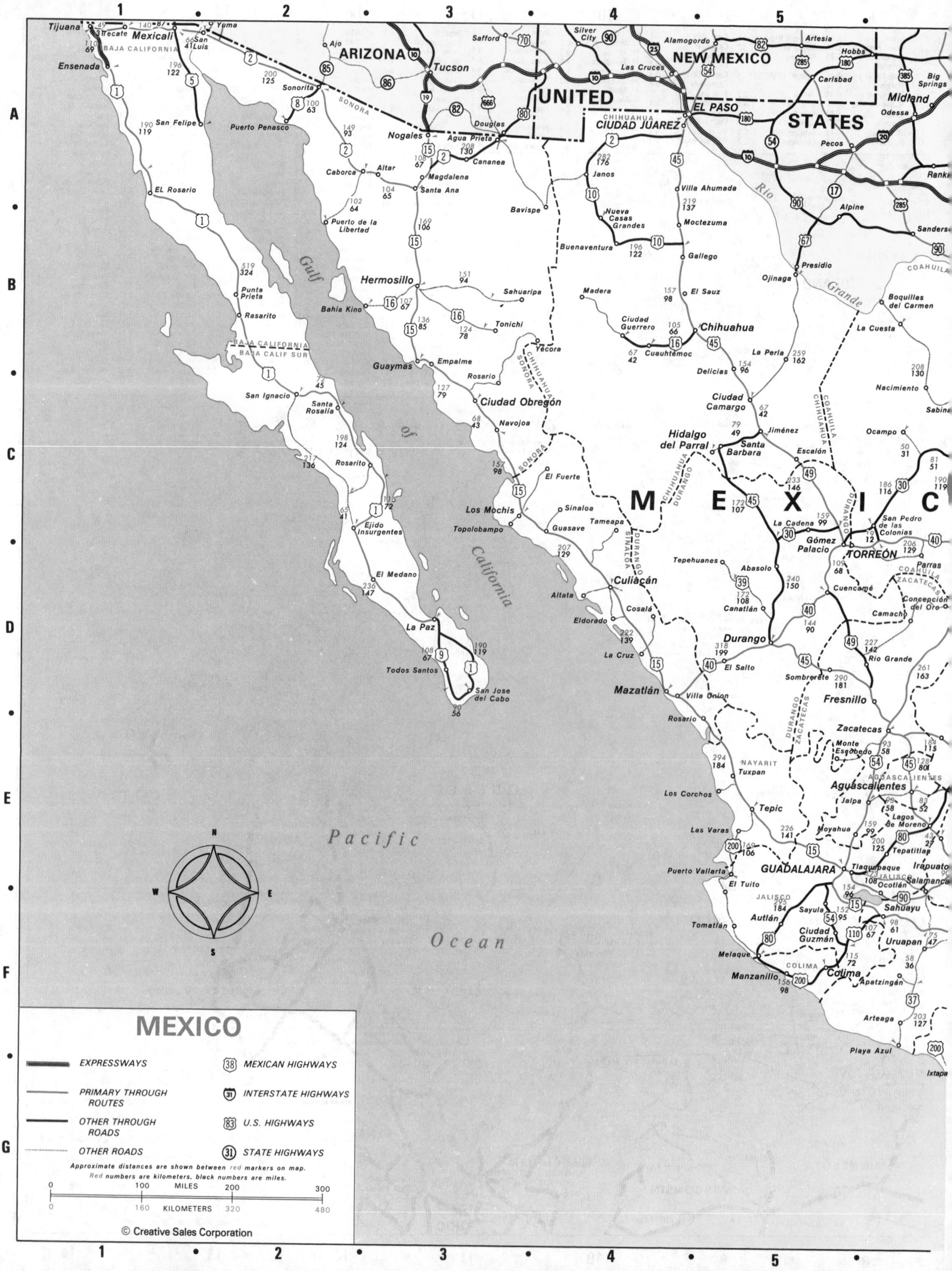

MEXICO

EXPRESSWAYS

PRIMARY THROUGH ROUTES

OTHER THROUGH ROADS

OTHER ROADS

38 MEXICAN HIGHWAYS

31 INTERSTATE HIGHWAYS

83 U.S. HIGHWAYS

31 STATE HIGHWAYS

Approximate distances are shown between red markers on map.
Red numbers are kilometers, black numbers are miles.

0 100 MILES 200 300

0 160 KILOMETERS 320 480

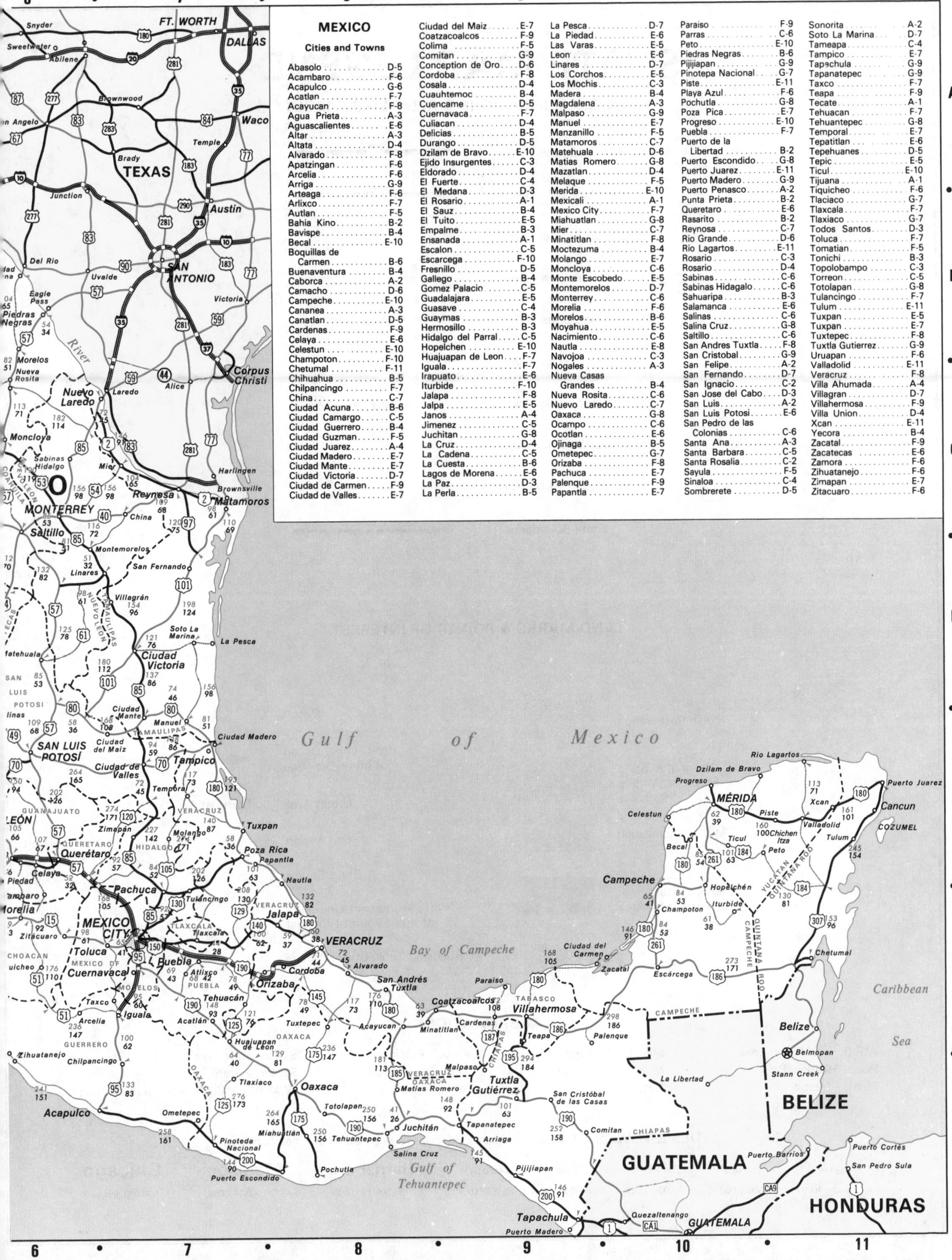

MEXICO

Cities and Towns

STATE MAP LEGEND

ROAD CLASSIFICATIONS & RELATED SYMBOLS

Free Interstate Hwy. — 90
Toll Interstate Hwy. — 76
Divided Federal Hwy. — 14
Federal Hwy. — 20
Divided State Hwy. — 31
State Hwy. — 147
Other Connecting Road — 258
Trans - Canada Hwy.
Point to Point Milage — 17
State Boundaries

LAND MARKS & POINTS OF INTEREST

Indian Reservation
National & State Forest or Wildlife Preserve
Military Installation
National & State Park or Recreation Area
Grassland
Desert
River, Lake, Ocean or other Drainage
Urban Area — Denver
Airport
State Capital
Park, Monument, University or other Point of Interest
Roadside Table or Rest Areas

ABBREVIATIONS

A.F.B. - Air Force Base
Hist. - Historical
Mem. - Memorial
Mgmt. - Management
Mon. - Monument
Nat. - Natural
Prov. - Province
Rec. - Recreation
Ref. - Refuge
S. F. - State Forest
St. Pk. - State Park
W.M.A. - Wildlife Management Area

CITIES & TOWNS - Type size indicates the relative population of cities and towns

Mapleton	Kenhorst	Somerset	Butler	Auburn	Harrisburg	Madison	Chicago
under 1000	1000-5,000	5,000-10,000	10,000-25,000	25,000-50,000	50,000-100,000	100,000-500,000	500,000 and over

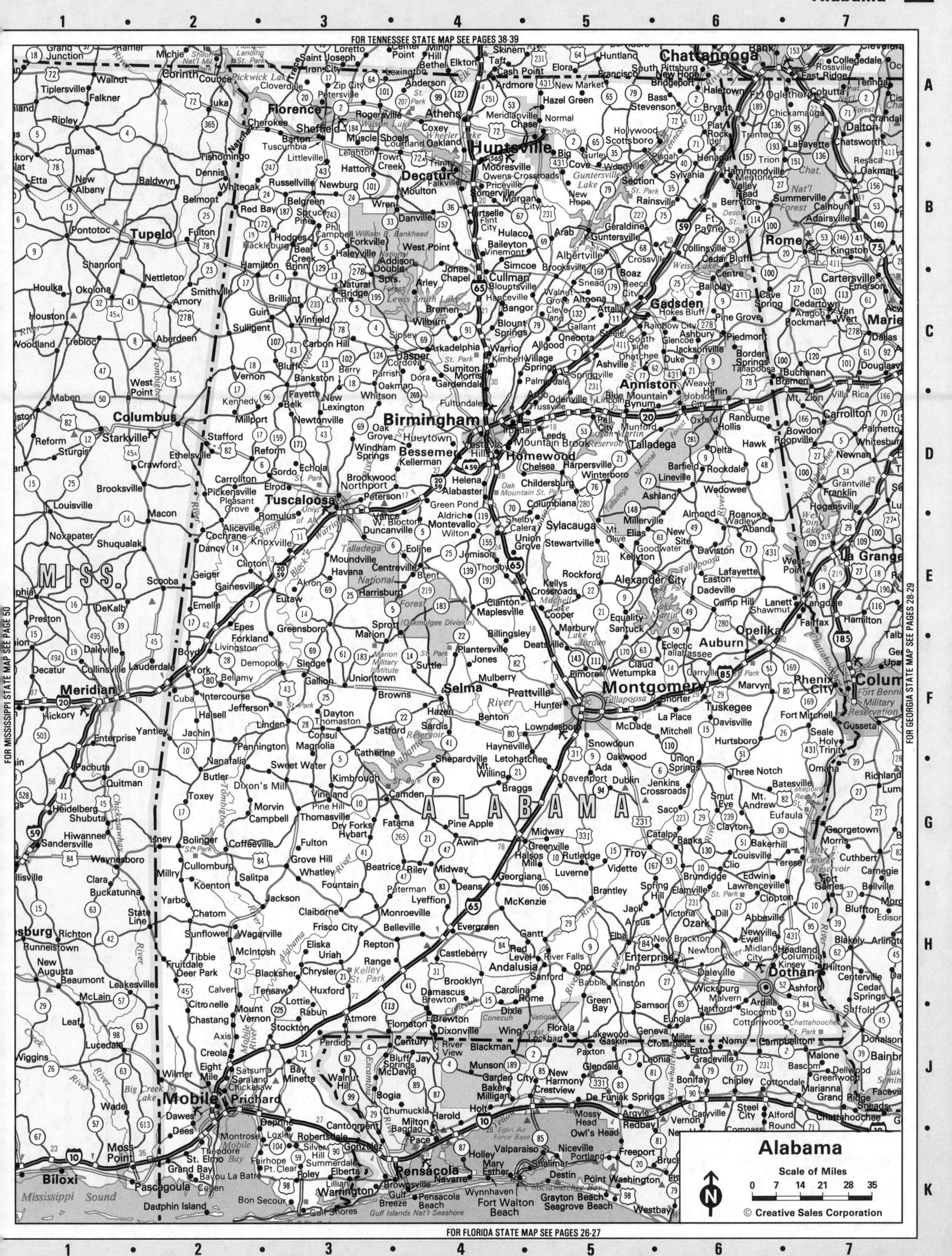
1 • 2 • 3 • 4 • 5 • 6 • 7
A B C D E F G H J K
FOR TENNESSEE STATE MAP SEE PAGES 38-39
FOR MISSISSIPPI STATE MAP SEE PAGE 50
FOR GEORGIA STATE MAP SEE PAGES 28-29
FOR FLORIDA STATE MAP SEE PAGES 26-27
MISS.
ALABAMA
Chattanooga
Florence
Sheffield
Athens
Huntsville
Decatur
Cullman
Gadsden
Anniston
Birmingham
Bessemer
Homewood
Tuscaloosa
Talladega
Sylacauga
Selma
Montgomery
Auburn
Opelika
Phenix City
Tuscaloosa
Tupelo
Columbus
Meridian
Rome
Dalton
La Grange
Troy
Enterprise
Dothan
Andalusia
Mobile
Prichard
Pensacola
Biloxi
Pascagoula
Fort Walton Beach
Mississippi Sound
Dauphin Island
Gulf Shores
Alabama
Scale of Miles
0 7 14 21 28 35
N
© Creative Sales Corporation

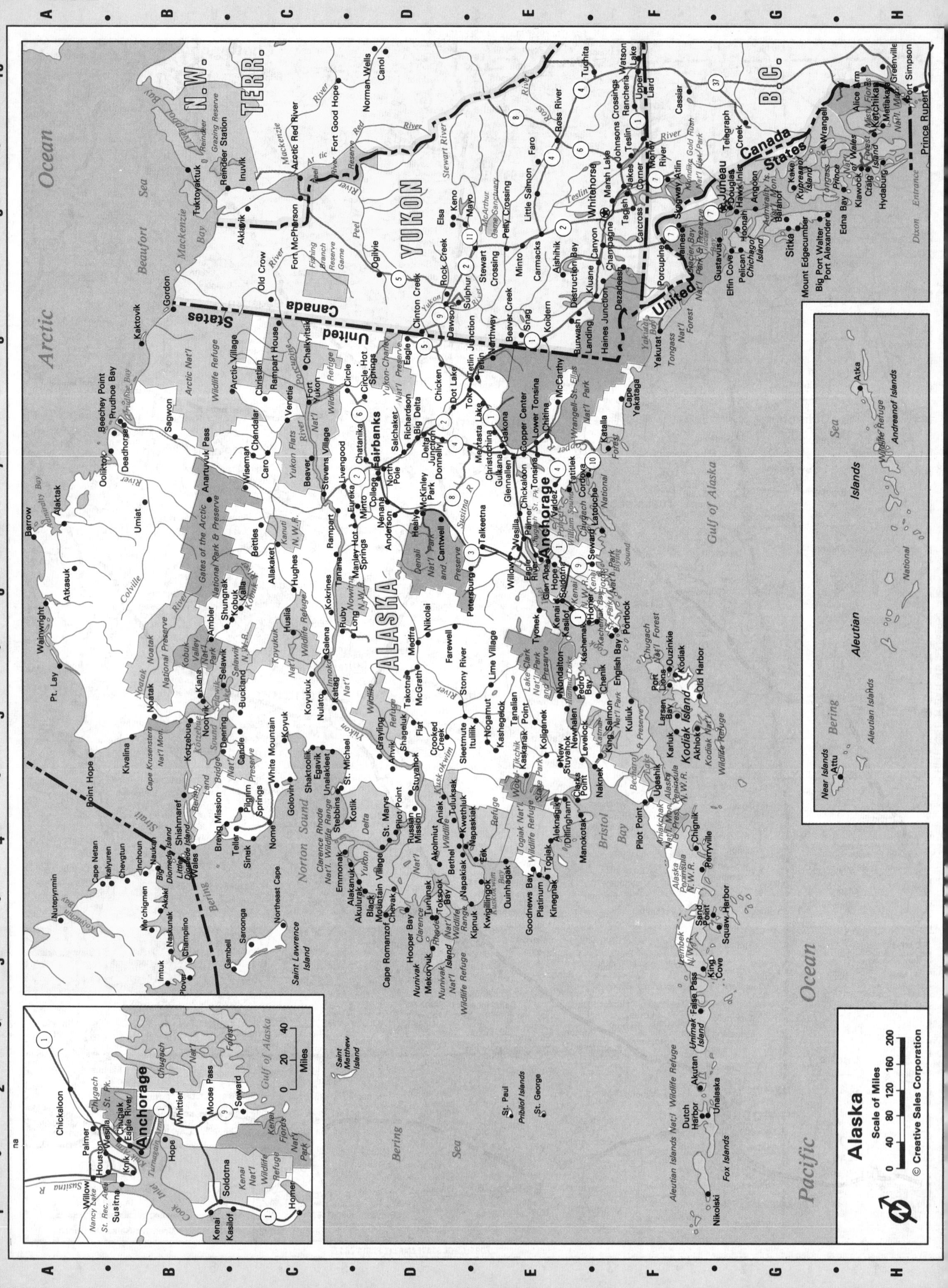
Alaska
Scale of Miles
0 40 80 120 160 200
© Creative Sales Corporation
Arctic Ocean
Beaufort Sea
Bering Sea
Gulf of Alaska
Pacific Ocean
Norton Sound
Bristol Bay
ALASKA
YUKON
N.W. TERR.
B.C.
United States
Canada
Anchorage
Fairbanks
Juneau
Barrow
Nome
Kodiak
Kodiak Island
Aleutian Islands
Pribilof Islands
Saint Lawrence Island
Saint Matthew Island
Near Islands
Fox Islands
Andreanof Islands
Whitehorse
Dawson
Prince Rupert
Gulf of Alaska
Kenai Nat'l Wildlife Refuge
Chugach St. Pk.
Miles 0 20 40

FOR MISSOURI STATE MAP SEE PAGES 48-49
FOR TENNESSEE STATE MAP SEE PAGES 38-39
FOR MISSISSIPPI STATE MAP SEE PAGE 50
FOR LOUISIANA STATE MAP SEE PAGE 40
FOR OKLAHOMA STATE MAP SEE PAGE 68-69
FOR TEXAS STATE MAP SEE PAGES 75-79

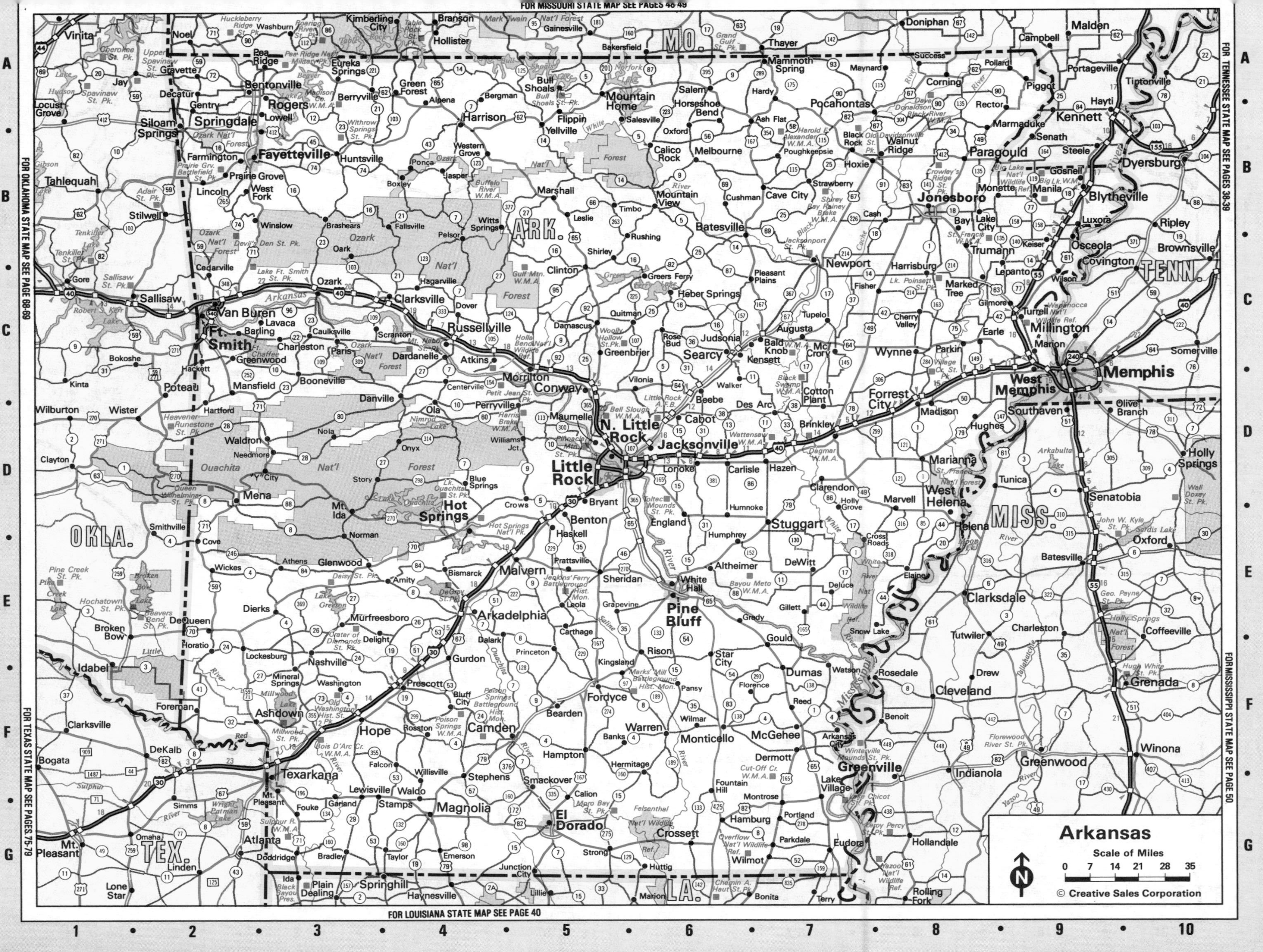

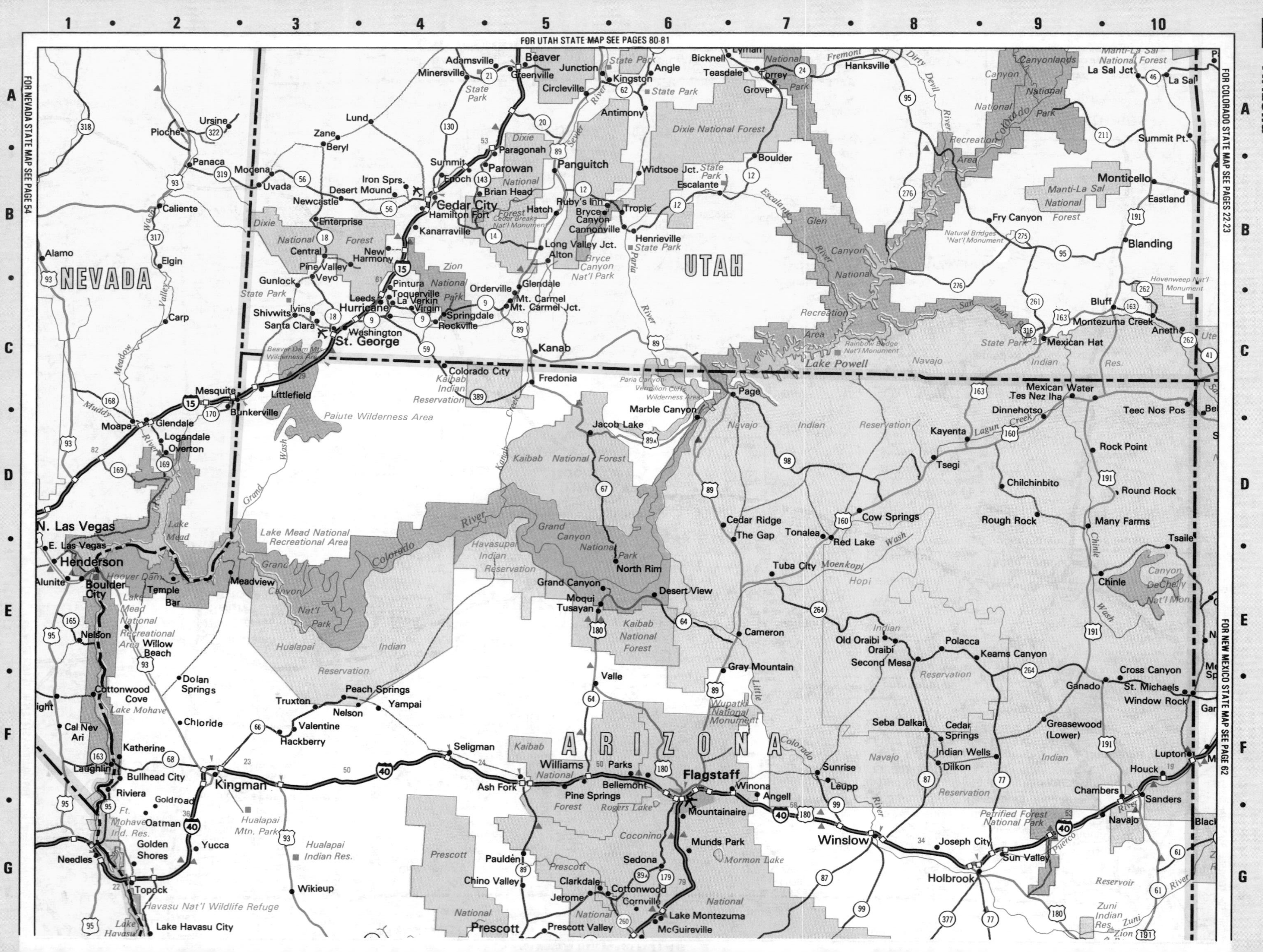
FOR UTAH STATE MAP SEE PAGES 80-81
FOR NEVADA STATE MAP SEE PAGE 54
FOR COLORADO STATE MAP SEE PAGES 22-23
FOR NEW MEXICO STATE MAP SEE PAGE 62
UTAH
ARIZONA
NEVADA
Beaver
Greenville
Adamsville
Minersville
Junction
Circleville
Kingston
Antimony
Angle
Bicknell
Teasdale
Torrey
Grover
Hanksville
La Sal Jct.
La Sal
Summit Pt.
Monticello
Eastland
Blanding
Bluff
Montezuma Creek
Aneth
Mexican Hat
Hovenweep Nat'l Monument
Canyonlands National Park
Manti-La Sal National Forest
Fry Canyon
Natural Bridges Nat'l Monument
Glen Canyon National Recreation Area
Rainbow Bridge Nat'l Monument
Lake Powell
Boulder
Widtsoe Jct.
Escalante
Tropic
Henrieville
Cannonville
Bryce Canyon
Bryce Canyon Nat'l Park
Ruby's Inn
Panguitch
Dixie National Forest
Paragonah
Parowan
Brian Head
Summit
Enoch
Cedar City
Hamilton Fort
Cedar Breaks Nat'l Monument
Hatch
Long Valley Jct.
Alton
Glendale
Mt. Carmel
Mt. Carmel Jct.
Orderville
Kanab
Zion National Park
Springdale
Rockville
Virgin
La Verkin
Toquerville
Pintura
Leeds
Hurricane
Washington
St. George
Santa Clara
Ivins
Shivwits
Gunlock
Veyo
Pine Valley
Central
New Harmony
Kanarraville
Enterprise
Newcastle
Desert Mound
Iron Sprs.
Lund
Zane
Beryl
Modena
Uvada
Pioche
Ursine
Panaca
Caliente
Alamo
Elgin
Carp
Mesquite
Bunkerville
Littlefield
Beaver Dam Mts. Wilderness Area
Moapa
Glendale
Logandale
Overton
Colorado City
Kaibab Indian Reservation
Fredonia
Paiute Wilderness Area
Paria Canyon-Vermilion Cliffs Wilderness Area
Page
Marble Canyon
Jacob Lake
Kaibab National Forest
Lake Mead National Recreational Area
Havasupai Indian Reservation
Grand Canyon National Park
North Rim
Grand Canyon
Moqui
Tusayan
Desert View
N. Las Vegas
E. Las Vegas
Henderson
Alunite
Boulder City
Hoover Dam
Temple Bar
Meadview
Nelson
Willow Beach
Dolan Springs
Cottonwood Cove
Lake Mohave
Cal Nev Ari
Chloride
Katherine
Laughlin
Bullhead City
Riviera
Goldroad
Ft. Mohave Ind. Res.
Oatman
Golden Shores
Needles
Topock
Havasu Nat'l Wildlife Refuge
Lake Havasu City
Yucca
Kingman
Hualapai Mtn. Park
Hualapai Indian Res.
Wikieup
Hackberry
Valentine
Truxton
Nelson
Peach Springs
Yampai
Seligman
Ash Fork
Williams
Parks
Bellemont
Pine Springs
Rogers Lake
Valle
Cameron
Gray Mountain
Wupatki National Monument
Cedar Ridge
The Gap
Tonalea
Red Lake
Cow Springs
Tuba City
Kayenta
Tsegi
Dinnehotso
Tes Nez Iha
Mexican Water
Teec Nos Pos
Rock Point
Round Rock
Chilchinbito
Rough Rock
Many Farms
Tsaile
Chinle
Canyon DeChelly Nat'l Mon.
Old Oraibi
Oraibi
Second Mesa
Polacca
Keams Canyon
Ganado
Cross Canyon
St. Michaels
Window Rock
Seba Dalkai
Cedar Springs
Indian Wells
Dilkon
Greasewood (Lower)
Lupton
Houck
Sanders
Chambers
Navajo
Petrified Forest National Park
Sunrise
Leupp
Flagstaff
Winona
Angell
Mountainaire
Munds Park
Mormon Lake
Coconino
Winslow
Joseph City
Holbrook
Sun Valley
Sedona
Cottonwood
Cornville
Clarkdale
Jerome
Lake Montezuma
McGuireville
Paulden
Chino Valley
Prescott
Prescott Valley
Zuni Indian Res.

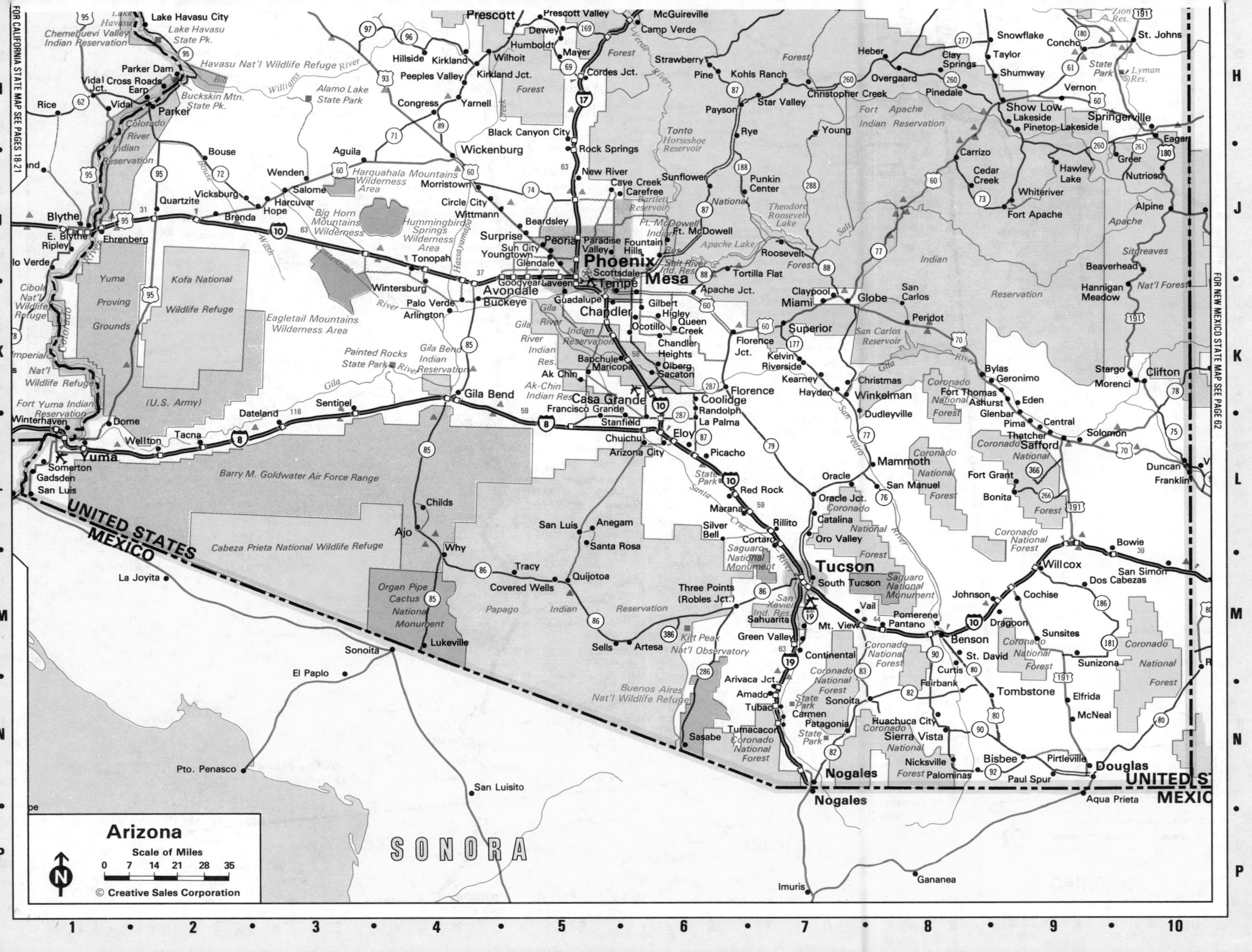
FOR NEW MEXICO STATE MAP SEE PAGE 62
FOR CALIFORNIA STATE MAP SEE PAGES 18-21
Arizona
Scale of Miles
0 7 14 21 28 35
© Creative Sales Corporation
UNITED STATES
MEXICO
SONORA
Lake Havasu City
Lake Havasu State Pk.
Chemehuevi Valley Indian Reservation
Havasu Nat'l Wildlife Refuge
Parker Dam
Vidal Cross Roads Jct.
Earp
Vidal
Parker
Rice
Buckskin Mtn. State Pk.
Colorado River Indian Reservation
Blythe
E. Blythe
Ripley
Ehrenberg
Quartzsite
Bouse
Vicksburg
Brenda
Hope
Harcuvar
Salome
Wenden
Aguila
Alamo Lake State Park
Hillside
Kirkland
Peeples Valley
Kirkland Jct.
Congress
Yarnell
Wilhoit
Humboldt
Dewey
Prescott
Prescott Valley
Mayer
Cordes Jct.
Camp Verde
McGuireville
Black Canyon City
Rock Springs
New River
Cave Creek
Carefree
Wickenburg
Morristown
Circle City
Wittmann
Beardsley
Surprise
Sun City
Youngtown
Peoria
Glendale
Paradise Valley
Fountain Hills
Ft. McDowell
Phoenix
Scottsdale
Tempe
Mesa
Goodyear
Laveen
Avondale
Buckeye
Tonopah
Wintersburg
Palo Verde
Arlington
Guadalupe
Chandler
Gilbert
Higley
Ocotillo
Queen Creek
Chandler Heights
Olberg
Sacaton
Bapchule
Maricopa
Ak Chin
Casa Grande
Francisco Grande
Stanfield
Chuichu
Arizona City
Eloy
Picacho
Coolidge
Randolph
La Palma
Florence
Florence Jct.
Apache Jct.
Tortilla Flat
Roosevelt
Superior
Miami
Claypool
Globe
San Carlos
Peridot
Kelvin
Riverside
Kearney
Hayden
Christmas
Winkelman
Dudleyville
Mammoth
San Manuel
Oracle
Oracle Jct.
Catalina
Oro Valley
Red Rock
Marana
Rillito
Cortaro
Silver Bell
Tucson
South Tucson
Vail
Mt. View
Pantano
Pomerene
Benson
St. David
Three Points (Robles Jct.)
Sahuarita
Green Valley
Continental
Arivaca Jct.
Amado
Tubac
Carmen
Tumacacori
Sonoita
Patagonia
Nogales
Sasabe
Huachuca City
Sierra Vista
Fairbank
Curtis
Tombstone
Bisbee
Nicksville
Palominas
Paul Spur
Pirtleville
Douglas
Aqua Prieta
Elfrida
McNeal
Sunizona
Sunsites
Dragoon
Cochise
Johnson
Willcox
Dos Cabezas
San Simon
Bowie
Bonita
Fort Grant
Safford
Thatcher
Pima
Central
Solomon
Glenbar
Ashurst
Eden
Fort Thomas
Geronimo
Bylas
Duncan
Franklin
Clifton
Morenci
Stargo
Hannigan Meadow
Beaverhead
Alpine
Nutrioso
Greer
Eagar
Springerville
St. Johns
Concho
Vernon
Snowflake
Taylor
Shumway
Show Low
Lakeside
Pinetop-Lakeside
Hawley Lake
Whiteriver
Fort Apache
Carrizo
Cedar Creek
Heber
Clay Springs
Pinedale
Overgaard
Christopher Creek
Kohls Ranch
Star Valley
Payson
Pine
Strawberry
Young
Rye
Punkin Center
Sunflower
Gila Bend
Sentinel
Dateland
Tacna
Wellton
Dome
Yuma
Winterhaven
Somerton
Gadsden
San Luis
Ajo
Childs
Why
Lukeville
Tracy
Covered Wells
Quijotoa
San Luis
Anegam
Santa Rosa
Sells
Artesa
Kofa National Wildlife Refuge
Yuma Proving Grounds
(U.S. Army)
Barry M. Goldwater Air Force Range
Cabeza Prieta National Wildlife Refuge
Organ Pipe Cactus National Monument
Papago Indian Reservation
Gila River Indian Reservation
Ak-Chin Indian Res.
Gila Bend Indian Reservation
Fort Apache Indian Reservation
San Carlos Reservoir
Theodore Roosevelt Lake
Tonto Horseshoe Reservoir
Coronado National Forest
Saguaro National Monument
Kitt Peak Nat'l Observatory
Buenos Aires Nat'l Wildlife Refuge
Eagletail Mountains Wilderness Area
Painted Rocks State Park
Harquahala Mountains Wilderness Area
Big Horn Mountains Wilderness
Hummingbird Springs Wilderness Area
La Joyita
Sonoita
El Paplo
Pto. Penasco
San Luisito
Imuris
Gananea

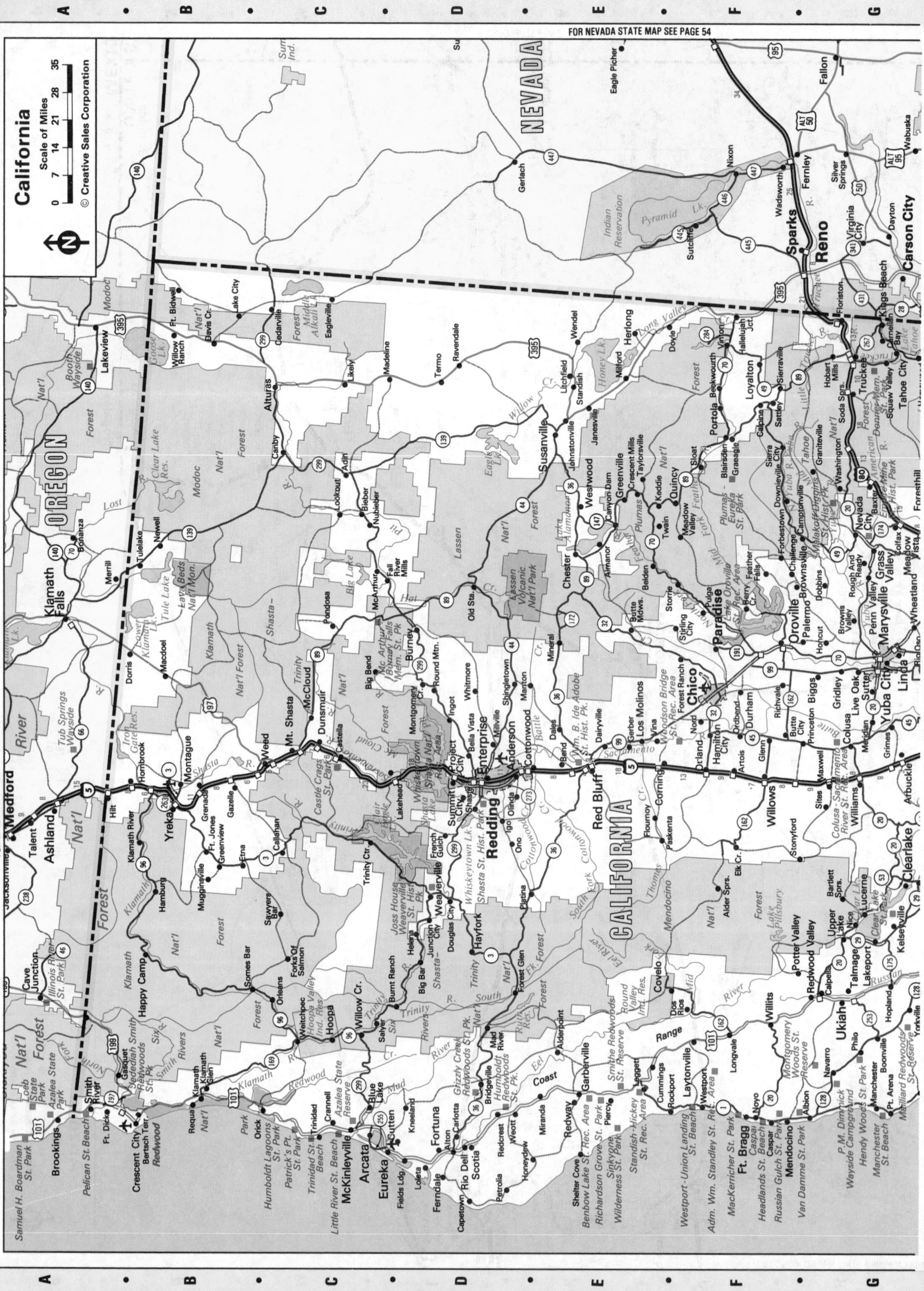

FOR NEVADA STATE MAP SEE PAGE 54

FOR OREGON STATE MAP SEE PAGES 70-71

California

Scale of Miles

0 7 14 21 28 35

OREGON

NEVADA

CALIFORNIA

A B C D E F G

1 2 3 4 5 6 7 8 9 10

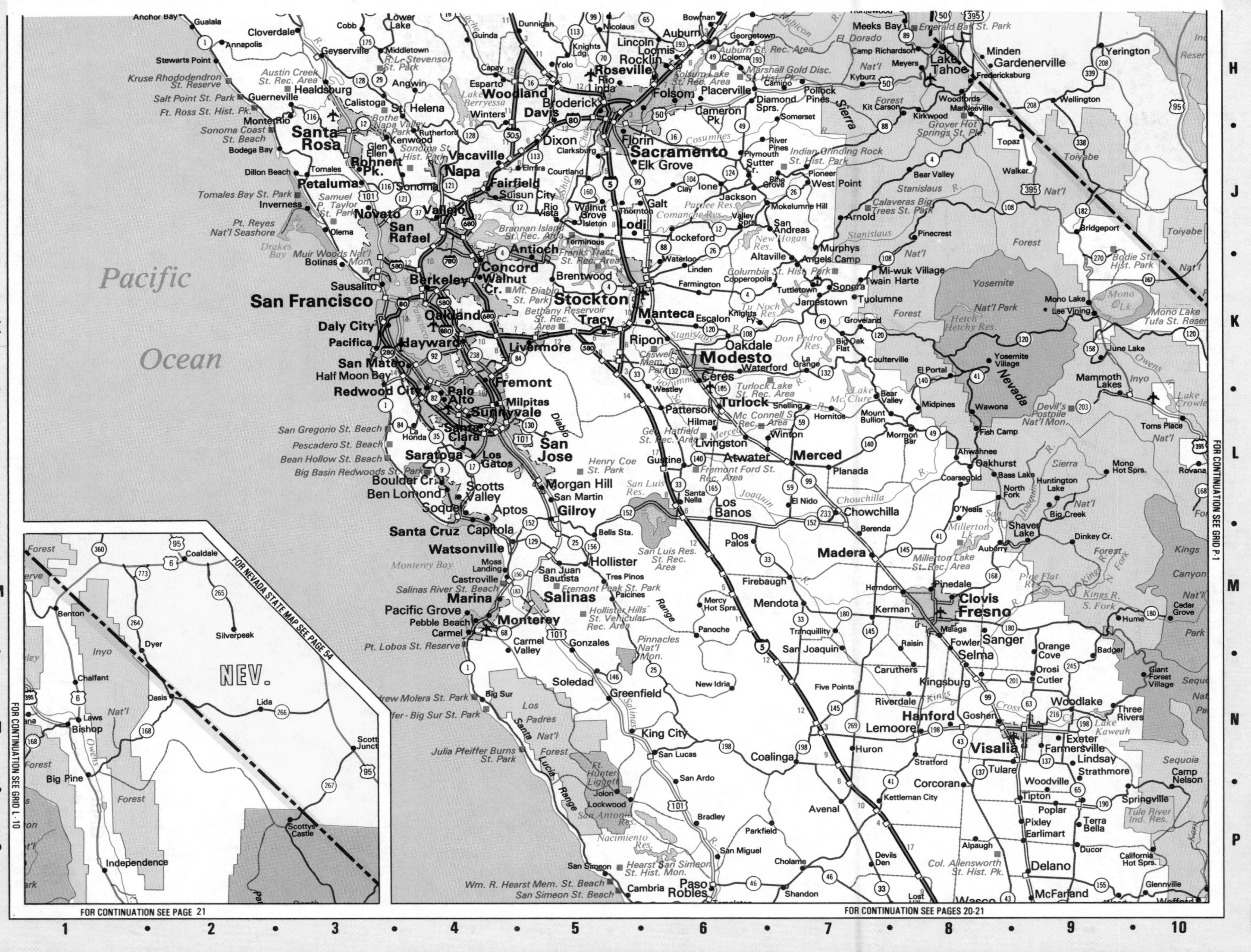

FOR CONTINUATION SEE GRID P-1
FOR CONTINUATION SEE PAGES 20-21
FOR CONTINUATION SEE PAGE 21
FOR CONTINUATION SEE GRID L-10
FOR NEVADA STATE MAP SEE PAGE 54
NEV.
Pacific
Ocean
San Francisco
Oakland
Berkeley
Daly City
Hayward
Fremont
San Jose
Sacramento
Stockton
Modesto
Merced
Madera
Fresno
Clovis
Visalia
Hanford
Salinas
Monterey
Santa Cruz
Watsonville
Santa Rosa
Napa
Vallejo
Lake Tahoe
Yosemite
Nevada
Sierra
Bishop
Independence
Paso Robles

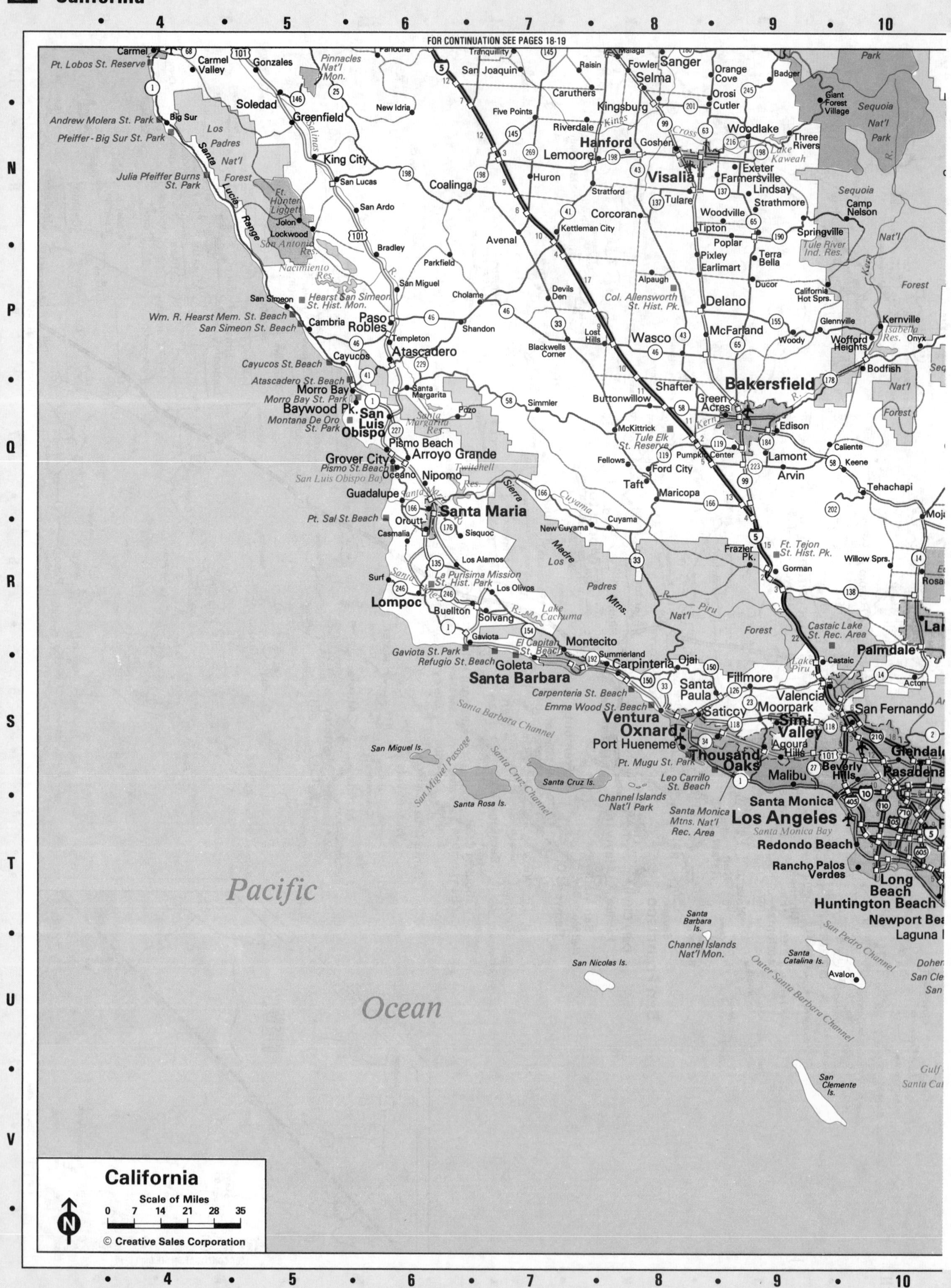

4 5 6 7 8 9 10
N P Q R S T U V
FOR CONTINUATION SEE PAGES 18-19
Carmel
Pt. Lobos St. Reserve
Carmel Valley
Gonzales
Pinnacles Nat'l Mon.
Panoche
Tranquillity
San Joaquin
Raisin
Malaga
Fowler
Sanger
Selma
Orange Cove
Badger
Park
Soledad
Greenfield
New Idria
Five Points
Caruthers
Kingsburg
Orosi
Cutler
Giant Forest Village
Sequoia Nat'l Park
Andrew Molera St. Park
Big Sur
Pfeiffer - Big Sur St. Park
Los Padres Nat'l Forest
Santa Lucia Range
Salinas
King City
San Lucas
Riverdale
Kings
Hanford
Lemoore
Goshen
Cross
Woodlake
Three Rivers
Lake Kaweah
Julia Pfeiffer Burns St. Park
Coalinga
Huron
Visalia
Exeter
Farmersville
Lindsay
Strathmore
Ft. Hunter Liggett
Jolon
Lockwood
San Antonio Res.
San Ardo
Stratford
Corcoran
Tulare
Woodville
Camp Nelson
Sequoia
Bradley
Avenal
Kettleman City
Tipton
Poplar
Pixley
Earlimart
Terra Bella
Springville
Tule River Ind. Res.
Nat'l
Nacimiento Res.
Parkfield
San Miguel
Alpaugh
Ducor
California Hot Sprs.
Kern
Forest
San Simeon
Hearst San Simeon St. Hist. Mon.
Cholame
Devils Den
Col. Allensworth St. Hist. Pk.
Delano
Wm. R. Hearst Mem. St. Beach
San Simeon St. Beach
Cambria
Paso Robles
Templeton
Shandon
Lost Hills
Wasco
McFarland
Glennville
Woody
Kernville
Isabella Res.
Wofford Heights
Onyx
Atascadero
Blackwells Corner
Cayucos
Cayucos St. Beach
Bodfish
Atascadero St. Beach
Morro Bay
Morro Bay St. Park
Santa Margarita
Simmler
Buttonwillow
Shafter
Green Acres
Bakersfield
Baywood Pk.
San Luis Obispo
Montana De Oro St. Park
Santa Margarita Res.
Pozo
McKittrick
Tule Elk St. Reserve
Edison
Pismo Beach
Arroyo Grande
Fellows
Pumpkin Center
Lamont
Caliente
Keene
Grover City
Pismo St. Beach
Oceano
San Luis Obispo Bay
Twitchell Res.
Nipomo
Ford City
Taft
Arvin
Tehachapi
Guadalupe
Sierra
Cuyama
Maricopa
Santa Maria
Pt. Sal St. Beach
Orcutt
Casmalia
Sisquoc
New Cuyama
Madre
Los
Frazier Pk.
Ft. Tejon St. Hist. Pk.
Gorman
Willow Sprs.
Los Alamos
Surf
La Purisima Mission St. Hist. Park
Los Olivos
Padres
Mtns.
Piru
Lompoc
Buellton
Solvang
Lake Cachuma
Nat'l
Forest
Castaic Lake St. Rec. Area
Palmdale
Gaviota
Gaviota St. Park
Refugio St. Beach
El Capitan St. Beach
Montecito
Summerland
Carpinteria
Ojai
Lake Piru
Castaic
Goleta
Santa Barbara
Santa Paula
Fillmore
Acton
Carpenteria St. Beach
Emma Wood St. Beach
Valencia
Moorpark
San Fernando
Santa Barbara Channel
Ventura
Saticoy
Simi Valley
Oxnard
Port Hueneme
Agoura Hills
Glendale
San Miguel Is.
San Miguel Passage
Santa Cruz Channel
Santa Cruz Is.
Pt. Mugu St. Park
Thousand Oaks
Leo Carrillo St. Beach
Beverly Hills
Pasadena
Malibu
Santa Rosa Is.
Channel Islands Nat'l Park
Santa Monica
Santa Monica Mtns. Nat'l Rec. Area
Los Angeles
Santa Monica Bay
Redondo Beach
Rancho Palos Verdes
Long Beach
Pacific
Huntington Beach
Newport Bea
Laguna
Santa Barbara Is.
Channel Islands Nat'l Mon.
San Pedro Channel
San Nicolas Is.
Santa Catalina Is.
Avalon
Outer Santa Barbara Channel
Ocean
San Clemente Is.
Gulf
California
Scale of Miles
0 7 14 21 28 35
N
© Creative Sales Corporation

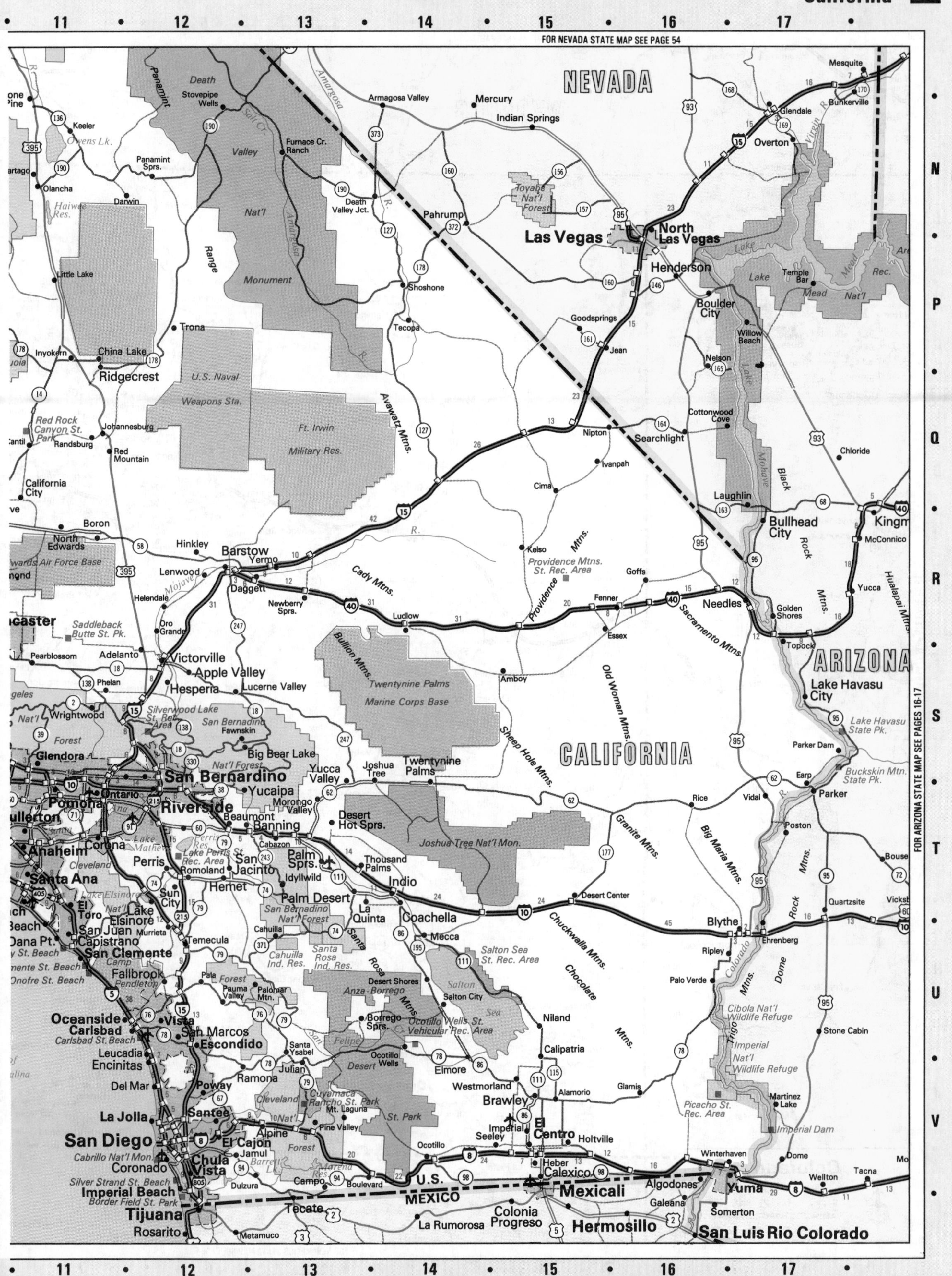
11 12 13 14 15 16 17
FOR NEVADA STATE MAP SEE PAGE 54
NEVADA
Death Valley Nat'l Monument
Panamint Range
Stovepipe Wells
Furnace Cr. Ranch
Death Valley Jct.
Armagosa Valley
Amargosa R.
Mercury
Indian Springs
Toyabe Nat'l Forest
Pahrump
Las Vegas
North Las Vegas
Henderson
Boulder City
Lake Mead Nat'l Rec.
Temple Bar
Mesquite
Bunkerville
Glendale
Overton
Virgin R.
Lone Pine
Keeler
Owens Lk.
Olancha
Cartago
Panamint Sprs.
Darwin
Haiwee Res.
Little Lake
Trona
Inyokern
China Lake
Ridgecrest
U.S. Naval Weapons Sta.
Shoshone
Tecopa
Goodsprings
Jean
Willow Beach
Nelson
Lake Mohave
Cottonwood Cove
Searchlight
Nipton
Ivanpah
Cima
Laughlin
Bullhead City
Kingman
McConnico
Black Mtns.
Chloride
Red Rock Canyon St. Park
Johannesburg
Randsburg
Cantil
Red Mountain
Ft. Irwin Military Res.
Avawatz Mtns.
California City
Boron
North Edwards
Edwards Air Force Base
Hinkley
Barstow
Yermo
Lenwood
Daggett
Helendale
Mojave R.
Cady Mtns.
Newberry Sprs.
Ludlow
Kelso
Providence Mtns. St. Rec. Area
Providence Mtns.
Goffs
Fenner
Essex
Needles
Golden Shores
Topock
Rock Mtns.
Yucca
Hualapai Mtns.
Lancaster
Saddleback Butte St. Pk.
Oro Grande
Pearblossom
Adelanto
Victorville
Apple Valley
Hesperia
Lucerne Valley
Phelan
Wrightwood
Angeles Nat'l Forest
Silverwood Lake St. Rec. Area
San Bernardino Nat'l Forest
Fawnskin
Big Bear Lake
Bullion Mtns.
Twentynine Palms Marine Corps Base
Amboy
Sheep Hole Mtns.
Old Woman Mtns.
Sacramento Mtns.
ARIZONA
Lake Havasu City
Lake Havasu State Pk.
Parker Dam
Buckskin Mtn. State Pk.
Parker
Earp
Vidal
Rice
CALIFORNIA
Glendora
San Bernardino
Ontario
Pomona
Yucaipa
Riverside
Fullerton
Anaheim
Santa Ana
Corona
Lake Mathews
Perris
Lake Perris St. Rec. Area
Cleveland Nat'l Forest
Beaumont
Banning
Cabazon
Morongo Valley
Yucca Valley
Joshua Tree
Twentynine Palms
Desert Hot Sprs.
Joshua Tree Nat'l Mon.
Palm Sprs.
Thousand Palms
Granite Mtns.
Big Maria Mtns.
Poston
Bouse
Romoland
San Jacinto
Hemet
Idyllwild
Indio
Palm Desert
La Quinta
Coachella
Desert Center
Chuckwalla Mtns.
Blythe
Ehrenberg
Quartzsite
Dome Rock Mtns.
Lake Elsinore
Sun City
El Toro
Laguna Beach
Dana Pt.
San Juan Capistrano
San Clemente
Murrieta
Temecula
Cahuilla
Mecca
Santa Rosa Mtns.
Santa Rosa Ind. Res.
Cahuilla Ind. Res.
Salton Sea St. Rec. Area
Salton Sea
Chocolate Mtns.
Ripley
Palo Verde
Colorado R.
Trigo Mtns.
Cibola Nat'l Wildlife Refuge
Imperial Nat'l Wildlife Refuge
Stone Cabin
Camp Pendleton
Fallbrook
Pala
Pauma Valley
Palomar Mtn.
Desert Shores
Anza-Borrego Desert St. Park
Salton City
Oceanside
Carlsbad
Carlsbad St. Beach
Vista
San Marcos
Escondido
Borrego Sprs.
Ocotillo Wells St. Vehicular Rec. Area
Niland
Calipatria
Leucadia
Encinitas
Santa Ysabel
Julian
Ocotillo Wells
Elmore
Del Mar
Poway
Ramona
Cuyamaca Rancho St. Park
Mt. Laguna
Westmorland
Brawley
Alamorio
Glamis
Picacho St. Rec. Area
Martinez Lake
Imperial Dam
La Jolla
Santee
Alpine
Pine Valley
Imperial
Seeley
El Centro
Holtville
San Diego
El Cajon
Cabrillo Nat'l Mon.
Coronado
Chula Vista
Jamul
Barrett Lk.
Morena Res.
Ocotillo
Heber
Calexico
Winterhaven
Dome
Wellton
Tacna
Silver Strand St. Beach
Imperial Beach
Border Field St. Park
Dulzura
Campo
Boulevard
U.S.
MEXICO
Mexicali
Algodones
Yuma
Tijuana
Tecate
Colonia Progreso
Galeana
Somerton
Rosarito
La Rumorosa
Hermosillo
San Luis Rio Colorado
Metamuco
FOR ARIZONA STATE MAP SEE PAGES 16-17
N P Q R S T U V

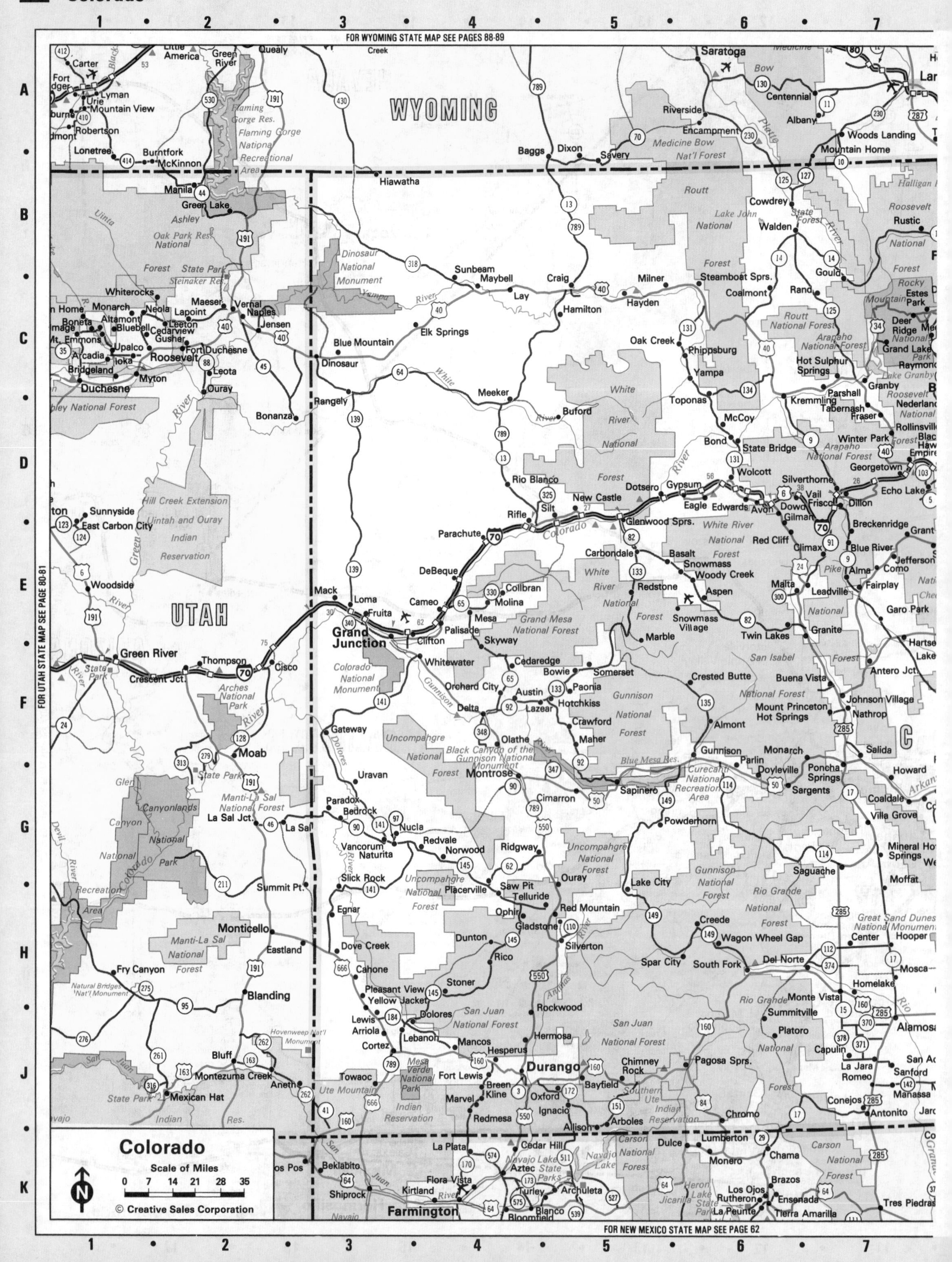
FOR WYOMING STATE MAP SEE PAGES 88-89
FOR UTAH STATE MAP SEE PAGE 80-81
FOR NEW MEXICO STATE MAP SEE PAGE 62
WYOMING
UTAH
Colorado
Scale of Miles
0 7 14 21 28 35
N
© Creative Sales Corporation
1 2 3 4 5 6 7
A B C D E F G H J K
Saratoga
Riverside
Encampment
Centennial
Albany
Woods Landing
Mountain Home
Baggs
Dixon
Savery
Medicine Bow
Nat'l Forest
Carter
Lyman
Urie
Mountain View
Robertson
Lonetree
Burntfork
McKinnon
Little America
Green River
Flaming Gorge Res.
Flaming Gorge
National
Recreational
Area
Hiawatha
Manila
Green Lake
Ashley
Oak Park Res.
National
Forest
State Park
Steinaker Res.
Dinosaur
National
Monument
Sunbeam
Maybell
Craig
Lay
Hamilton
Milner
Hayden
Steamboat Sprs.
Coalmont
Walden
Cowdrey
Routt
Lake John
National
Forest
State Forest
Gould
Rand
Rocky
Mountain
Estes Park
Whiterocks
Monarch
Neola
Lapoint
Maeser
Vernal
Naples
Jensen
Altamont
Bluebell
Leeton
Cedarview
Gusher
Upalco
Arcadia
Ioka
Fort Duchesne
Roosevelt
Bridgeland
Duchesne
Myton
Leota
Ouray
Bonanza
Elk Springs
Blue Mountain
Dinosaur
Rangely
Meeker
White River
Buford
White
River
National
Forest
Oak Creek
Phippsburg
Yampa
Toponas
McCoy
Bond
State Bridge
Wolcott
Kremmling
Parshall
Tabernash
Fraser
Granby
Hot Sulphur Springs
Grand Lake
Arapaho
National Forest
Winter Park
Georgetown
Silverthorne
Dillon
Frisco
Vail
Echo Lake
Rio Blanco
New Castle
Rifle
Silt
Dotsero
Gypsum
Eagle
Edwards
Avon
Glenwood Sprs.
Parachute
Colorado
DeBeque
Carbondale
Basalt
Snowmass
Woody Creek
Redstone
Aspen
Snowmass Village
Marble
Red Cliff
Climax
Breckenridge
Blue River
Alma
Fairplay
Leadville
Malta
Twin Lakes
Granite
Collbran
Molina
Mesa
Cameo
Palisade
Skyway
Grand Mesa
National Forest
Mack
Loma
Fruita
Grand Junction
Clifton
Whitewater
Colorado
National
Monument
Cedaredge
Bowie
Somerset
Orchard City
Austin
Paonia
Hotchkiss
Lazear
Delta
Crawford
Maher
Olathe
Gunnison
National
Forest
Crested Butte
Almont
Gunnison
Parlin
Buena Vista
Johnson Village
Nathrop
Mount Princeton Hot Springs
Salida
Monarch
Doyleville
Poncha Springs
Sargents
Howard
Coaldale
Sunnyside
East Carbon City
Hill Creek Extension
Uintah and Ouray
Indian
Reservation
Woodside
Green River
Thompson
Crescent Jct.
Cisco
Arches
National
Park
Moab
Gateway
Uncompahgre
National
Forest
Black Canyon of the Gunnison National Monument
Montrose
Blue Mesa Res.
Sapinero
Curecanti
National
Recreation
Area
Cimarron
Uravan
Paradox
Bedrock
Nucla
Redvale
Norwood
Vancorum
Naturita
Ridgway
Powderhorn
Villa Grove
Canyonlands
National
Park
La Sal Jct.
La Sal
Manti-La Sal
National Forest
Slick Rock
Placerville
Saw Pit
Telluride
Ouray
Lake City
Saguache
Moffat
Summit Pt.
Egnar
Ophir
Red Mountain
Gladstone
Silverton
Creede
Wagon Wheel Gap
Rio Grande
National
Forest
Center
Hooper
Monticello
Eastland
Dove Creek
Dunton
Rico
Spar City
South Fork
Del Norte
Mosca
Fry Canyon
Cahone
Stoner
Natural Bridges Nat'l Monument
Blanding
Pleasant View
Yellow Jacket
Dolores
San Juan
National Forest
Rockwood
Monte Vista
Homelake
Summitville
Platoro
Alamosa
Lewis
Arriola
Cortez
Lebanon
Mancos
Hesperus
Hermosa
Capulin
Hovenweep Nat'l Monument
Bluff
Montezuma Creek
Aneth
Towaoc
Mesa Verde National Park
Fort Lewis
Durango
Bayfield
Chimney Rock
Pagosa Sprs.
La Jara
Romeo
Sanford
Manassa
Mexican Hat
Ute Mountain
Indian
Reservation
Marvel
Breen
Kline
Oxford
Ignacio
Southern Ute
Indian Reservation
Chromo
Conejos
Antonito
Redmesa
Allison
Arboles
La Plata
Cedar Hill
Navajo Lake State Parks
Aztec
Dulce
Lumberton
Monero
Chama
Carson
National
Forest
Beklabito
Shiprock
Kirtland
Flora Vista
Farmington
Turley
Archuleta
Blanco
Bloomfield
Los Ojos
Rutheron
La Puente
Brazos
Ensenada
Tierra Amarilla
Tres Piedras

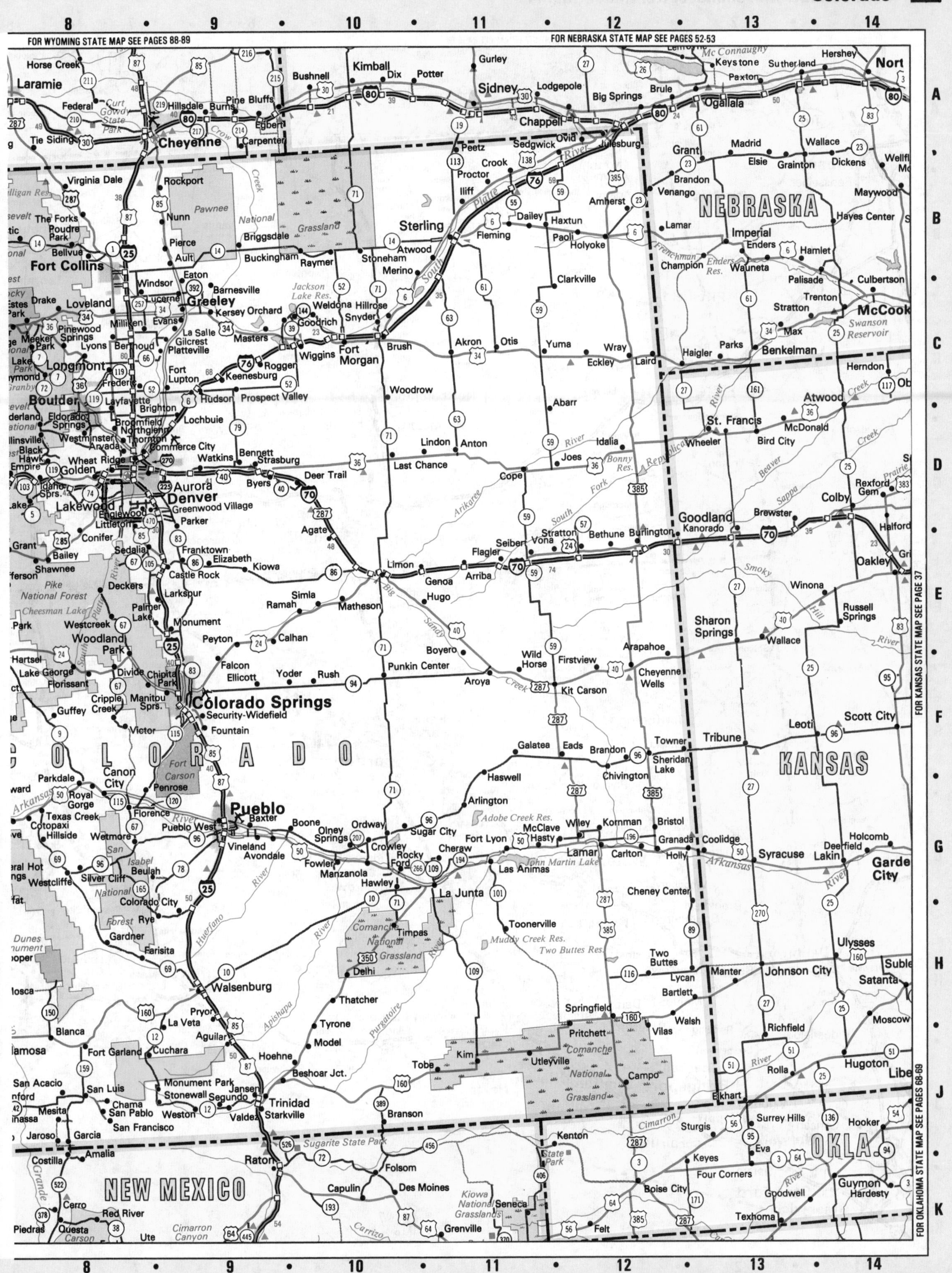

8
9
10
11
12
13
14
FOR WYOMING STATE MAP SEE PAGES 88-89
FOR NEBRASKA STATE MAP SEE PAGES 52-53
FOR KANSAS STATE MAP SEE PAGE 37
FOR OKLAHOMA STATE MAP SEE PAGES 68-69
A
B
C
D
E
F
G
H
J
K
NEBRASKA
KANSAS
OKLA.
NEW MEXICO
C O L O R A D O
Horse Creek
Laramie
Federal
Curt Gowdy State Park
Tie Siding
Cheyenne
Hillsdale
Burns
Pine Bluffs
Egbert
Carpenter
Bushnell
Kimball
Dix
Potter
Gurley
Sidney
Lodgepole
Chappell
Big Springs
Brule
Mc Connaughy
Keystone
Sutherland
Hershey
Paxton
Ogallala
Nort
Virginia Dale
Rockport
Pawnee
National
Grassland
Peetz
Crook
Proctor
Iliff
Sedgwick
Ovid
Julesburg
Grant
Madrid
Wallace
Elsie
Grainton
Dickens
Wellfl
Brandon
Venango
Maywood
Amherst
Dailey
Haxtun
Paoli
Holyoke
Hayes Center
Lamar
Imperial
Enders
Hamlet
Champion
Enders Res.
Wauneta
Palisade
Culbertson
Trenton
Stratton
McCook
Swanson Reservoir
Max
Benkelman
Haigler
Parks
Laird
Wray
Eckley
Yuma
Clarkville
Otis
Akron
The Forks
Poudre Park
Bellvue
Fort Collins
Nunn
Pierce
Ault
Briggsdale
Buckingham
Raymer
Stoneham
Atwood
Sterling
Fleming
Merino
Windsor
Eaton
Barnesville
Lucerne
Greeley
Kersey
Orchard
Jackson Lake Res.
Weldona
Hillrose
Snyder
Goodrich
Drake
Loveland
Estes Park
Milliken
Evans
Pinewood Springs
Meeker Park
Lyons
Berthoud
La Salle
Gilcrest
Platteville
Masters
Wiggins
Fort Morgan
Brush
Longmont
Frederick
Fort Lupton
Keenesburg
Roggen
Prospect Valley
Woodrow
Boulder
Eldorado Springs
Layfayette
Brighton
Hudson
Lochbuie
Broomfield
Northglenn
Westminster
Thornton
Arvada
Commerce City
Wheat Ridge
Golden
Empire
Watkins
Bennett
Strasburg
Lindon
Anton
Last Chance
Abarr
Idalia
Joes
Bonny Res.
Cope
Herndon
Atwood
St. Francis
McDonald
Wheeler
Bird City
Rexford
Gem
Colby
Brewster
Goodland
Kanorado
Halford
Oakley
Aurora
Denver
Byers
Deer Trail
Lakewood
Englewood
Littleton
Greenwood Village
Parker
Agate
Conifer
Bailey
Sedalia
Franktown
Elizabeth
Kiowa
Castle Rock
Limon
Genoa
Arriba
Flagler
Seibert
Vona
Stratton
Bethune
Burlington
Shawnee
Pike National Forest
Cheesman Lake
Deckers
Larkspur
Palmer Lake
Monument
Simla
Ramah
Matheson
Hugo
Winona
Russell Springs
Sharon Springs
Wallace
Westcreek
Woodland Park
Peyton
Calhan
Boyero
Wild Horse
Firstview
Arapahoe
Cheyenne Wells
Hartsel
Lake George
Florissant
Divide
Chipita Park
Falcon
Ellicott
Yoder
Rush
Punkin Center
Aroya
Kit Carson
Cripple Creek
Manitou Sprs.
Colorado Springs
Security-Widefield
Fountain
Guffey
Victor
Leoti
Scott City
Tribune
Galatea
Eads
Brandon
Towner
Sheridan Lake
Parkdale
Canon City
Fort Carson
Penrose
Haswell
Chivington
Royal Gorge
Texas Creek
Cotopaxi
Hillside
Florence
Pueblo
Baxter
Pueblo West
Arlington
Adobe Creek Res.
Boone
Olney Springs
Ordway
Sugar City
McClave
Wiley
Hasty
Kornman
Bristol
Granada
Coolidge
Holcomb
Deerfield
Lakin
Garden City
Wetmore
Vineland
Avondale
Crowley
Rocky Ford
Cheraw
Fort Lyon
John Martin Lake
Lamar
Carlton
Holly
Syracuse
Westcliffe
Silver Cliff
Beulah
Fowler
Manzanola
Las Animas
Hawley
La Junta
Colorado City
Rye
Gardner
Cheney Center
Toonerville
Comanche National Grassland
Timpas
Delhi
Muddy Creek Res.
Two Buttes Res.
Farisita
Walsenburg
Two Buttes
Lycan
Manter
Johnson City
Ulysses
Sublette
Satanta
Bartlett
Thatcher
Tyrone
Model
Springfield
Walsh
Pritchett
Vilas
Richfield
Moscow
Blanca
La Veta
Pryor
Aguilar
Cuchara
Fort Garland
Hoehne
Beshoar Jct.
Kim
Utleyville
Tobe
Campo
Hugoton
Rolla
Elkhart
Alamosa
San Luis
Monument Park
Stonewall
Jansen
Segundo
Trinidad
Starkville
Weston
Valdez
Branson
San Acacio
Chama
San Pablo
San Francisco
Mesita
Jaroso
Garcia
Costilla
Amalia
Sugarite State Park
Raton
Kenton
Sturgis
Surrey Hills
Hooker
Eva
Keyes
Four Corners
Folsom
Capulin
Des Moines
Boise City
Goodwell
Guymon
Hardesty
Cerro
Red River
Questa
Carson
Piedras
Ute
Cimarron Canyon
Kiowa National Grasslands
Seneca
Grenville
Felt
Texhoma
Platte River
South Platte
Arkansas River
Republican River
Arikaree
Big Sandy Creek
Huerfano River
Apishapa
Purgatoire River
Cimarron
Carrizo
Smoky Hill River
Sappa Creek
Beaver
Prairie
Frenchman

FOR VERMONT STATE MAP SEE PAGE 55

FOR NEW YORK STATE MAP SEE PAGES 58-61

N.Y.
VT.
N.H.
MASS.
CONN.
N.Y.

Albany
Rensselaer
Troy
Pittsfield
North Adams
Adams
Williamstown
Greenfield
Amherst
Northampton
Holyoke
Chicopee
Springfield
West Springfield
Agawam
Worcester
Fitchburg
Athol
Orange
Gardner
Southbridge
Webster
Enfield
Windsor Locks
Windsor
East Hartford
West Hartford
Hartford
Manchester
Vernon
Wethersfield
Newington
New Britain
Bristol
Torrington
Waterbury
Southington
Meriden
Middletown
Naugatuck
Wallingford
Hamden
New Haven
West Haven
East Haven
Milford
Stratford
Bridgeport
Fairfield
Westport
Norwalk
Stamford
Greenwich
Danbury
Norwich
New London
Groton
Long Island Sound

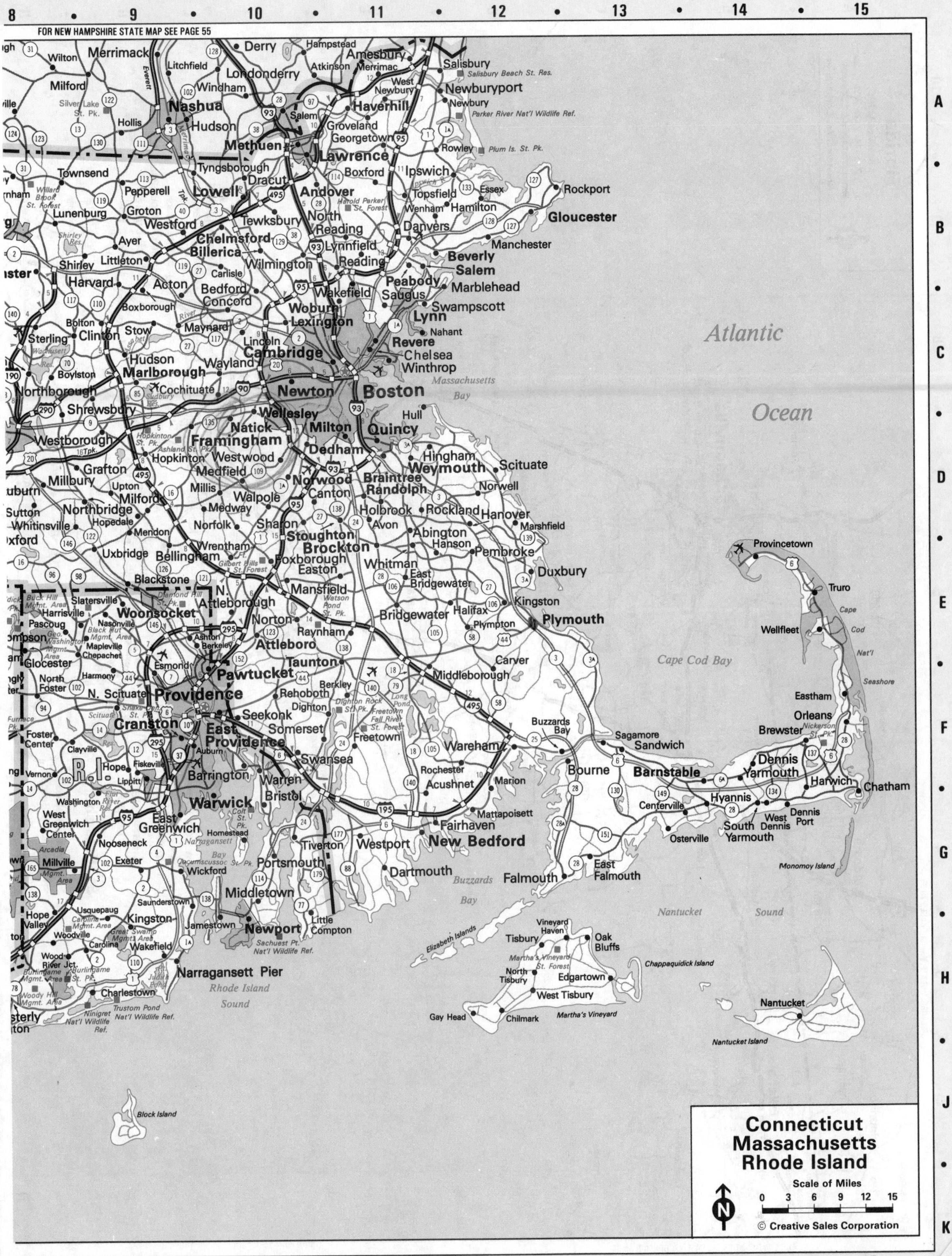
FOR NEW HAMPSHIRE STATE MAP SEE PAGE 55
Connecticut
Massachusetts
Rhode Island
Scale of Miles
0 3 6 9 12 15
© Creative Sales Corporation
Atlantic
Ocean
Massachusetts Bay
Cape Cod Bay
Buzzards Bay
Nantucket Sound
Rhode Island Sound
Cape Cod Nat'l Seashore
Boston
Cambridge
Newton
Quincy
Lowell
Lawrence
Haverhill
Nashua
Methuen
Brockton
Providence
Pawtucket
Cranston
East Providence
Warwick
Woonsocket
Taunton
Attleboro
New Bedford
Fall River
Plymouth
Barnstable
Newport
Gloucester
Rockport
Salem
Beverly
Lynn
Peabody
Framingham
Marlborough
Falmouth
Hyannis
Chatham
Orleans
Provincetown
Truro
Wellfleet
Eastham
Brewster
Dennis
Yarmouth
Harwich
Sandwich
Bourne
Wareham
Tisbury
Oak Bluffs
Edgartown
West Tisbury
Chilmark
Gay Head
Martha's Vineyard
Nantucket
Nantucket Island
Chappaquidick Island
Monomoy Island
Elizabeth Islands
Block Island
Narragansett Pier
Kingston
Westerly
R.I.

4/18-1149

804-8366

Florida

Scale of Miles

0 7 14 21 28 35

FOR GEORGIA STATE MAP SEE PAGES 28-29

FOR CONTINUATION SEE PAGE 27, GRID N-6

Atlantic Ocean

Gulf of Mexico

Apalachee Bay

GEORGIA

Okefenokee Nat'l Wildlife Refuge

Osceola Nat'l Forest

Ocala Nat'l Forest

Apalachicola Nat'l Forest

Jacksonville, Atlantic Beach, Neptune Beach, Jacksonville Beach, Fernandina Beach, St. Augustine, St. Augustine Beach, Palatka, Flagler Beach, Ormond by the Sea, Ormand Beach, Daytona Beach, Port Orange, New Smyrna Beach, Edgewater, DeLand, Deltona, Sanford, Winter Springs, Casselberry, Winter Pk., Orlando, Titusville, Mims, Cocoa, Cocoa Beach, Merritt Island, Cape Canaveral, Rockledge, Satellite Beach, Indian Harbour Beach, Melbourne, W. Melbourne, Palm Bay, Sebastian, Gifford, Vero Beach, Fellsmere, Kissimmee, St. Cloud, Haines City, Winter Haven, Lakeland, Bartow, Avon Park, Frostproof, Lake Wales, Tampa, Temple Terrace, Brandon, Plant City, Clearwater, Largo, St. Petersburg, Treasure Island, St. Petersburg Beach, Tarpon Springs, New Port Richey, Hudson, Brooksville, Inverness, Crystal River, Ocala, Gainesville, Lake City, Live Oak, Madison, Perry, Tallahassee, Quincy, Chattahoochee, Crawfordville, Carrabelle, Valdosta, Thomasville, Bainbridge, Cairo, Quitman, Lakeland, Homerville, Folkston, Kingsland, St. Marys, Cedar Key, Chiefland, Williston, Apopka, Winter Garden, Lake Buena Vista, Leesburg, Eustis, Mt. Dora, Clermont, Wildwood, Bushnell, Dade City, Zephyrhills, Baldwin, Orange Park, Green Cove Springs, Starke, Macclenny, Callahan, Hilliard

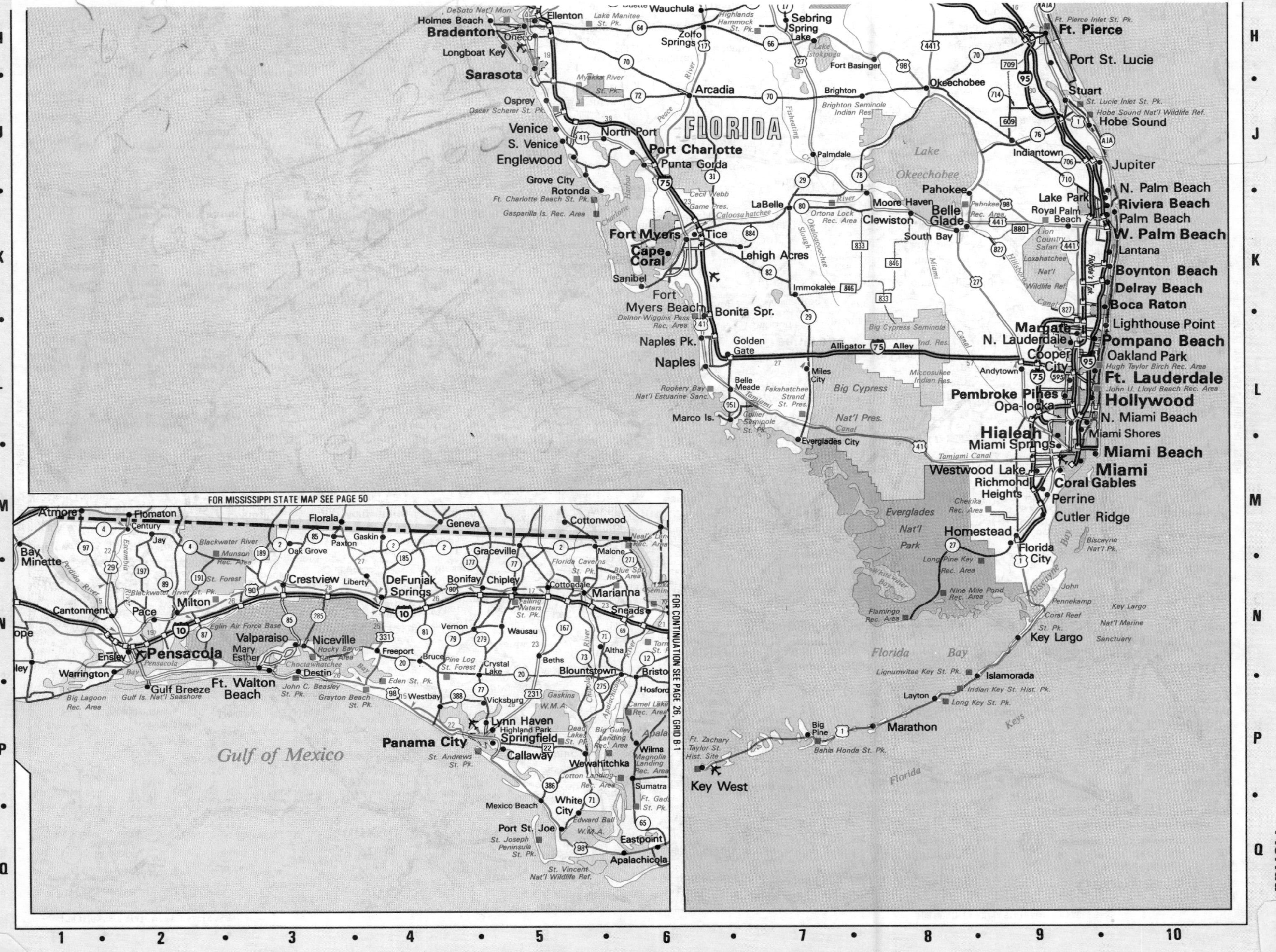
FLORIDA
Ft. Pierce
Port St. Lucie
Stuart
Hobe Sound
Jupiter
N. Palm Beach
Riviera Beach
Palm Beach
W. Palm Beach
Lantana
Boynton Beach
Delray Beach
Boca Raton
Lighthouse Point
Pompano Beach
Oakland Park
Ft. Lauderdale
Hollywood
N. Miami Beach
Miami Shores
Miami Beach
Miami
Coral Gables
Perrine
Cutler Ridge
Homestead
Florida City
Key Largo
Islamorada
Layton
Marathon
Big Pine
Key West
Bradenton
Holmes Beach
Longboat Key
Sarasota
Osprey
Venice
S. Venice
Englewood
Grove City
Rotonda
North Port
Port Charlotte
Punta Gorda
Arcadia
Wauchula
Zolfo Springs
Sebring
Spring Lake
Okeechobee
Lake Okeechobee
Pahokee
Belle Glade
Clewiston
Moore Haven
South Bay
LaBelle
Fort Myers
Cape Coral
Tice
Lehigh Acres
Sanibel
Fort Myers Beach
Bonita Spr.
Naples Pk.
Naples
Golden Gate
Marco Is.
Immokalee
Everglades City
Alligator Alley
Big Cypress Nat'l Pres.
Everglades Nat'l Park
Florida Bay
Margate
N. Lauderdale
Cooper City
Pembroke Pines
Opa-locka
Hialeah
Miami Springs
Westwood Lake
Richmond Heights
Gulf of Mexico
FOR MISSISSIPPI STATE MAP SEE PAGE 50
FOR CONTINUATION SEE PAGE 26, GRID B-1
Atmore
Flomaton
Century
Bay Minette
Cantonment
Pace
Milton
Pensacola
Ensley
Warrington
Gulf Breeze
Ft. Walton Beach
Valparaiso
Niceville
Mary Esther
Destin
Crestview
Florala
Geneva
DeFuniak Springs
Bonifay
Chipley
Graceville
Marianna
Cottonwood
Malone
Sneads
Freeport
Vernon
Wausau
Blountstown
Bristol
Lynn Haven
Springfield
Panama City
Callaway
Wewahitchka
White City
Port St. Joe
Mexico Beach
Eastpoint
Apalachicola
1
2
3
4
5
6
7
8
9
10
H
J
K
L
M
N
P
Q

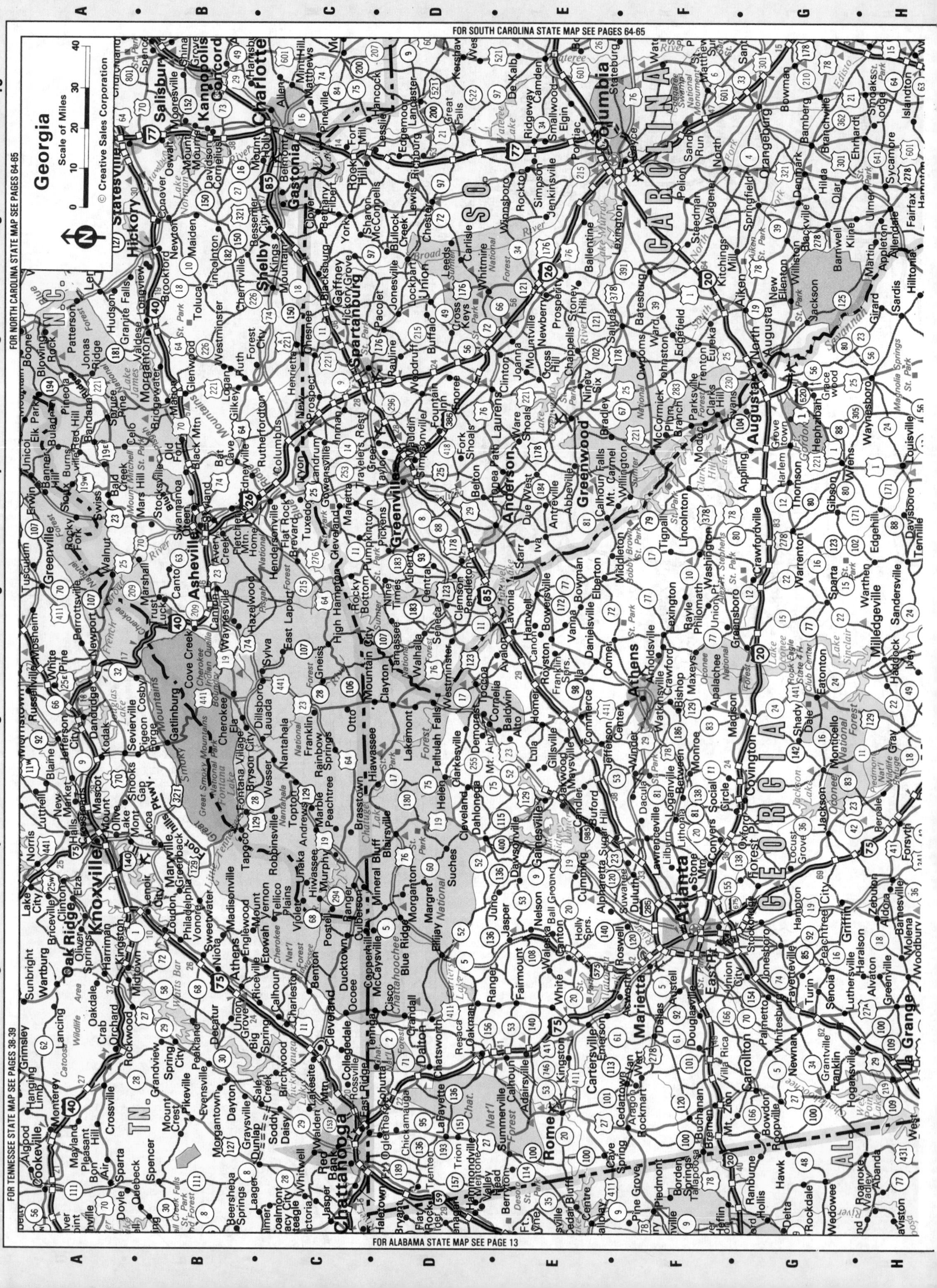

FOR SOUTH CAROLINA STATE MAP SEE PAGES 64-65

FOR NORTH CAROLINA STATE MAP SEE PAGES 64-65

Georgia

Scale of Miles

0 10 20 30 40

FOR TENNESSEE STATE MAP SEE PAGES 38-39

FOR ALABAMA STATE MAP SEE PAGE 13

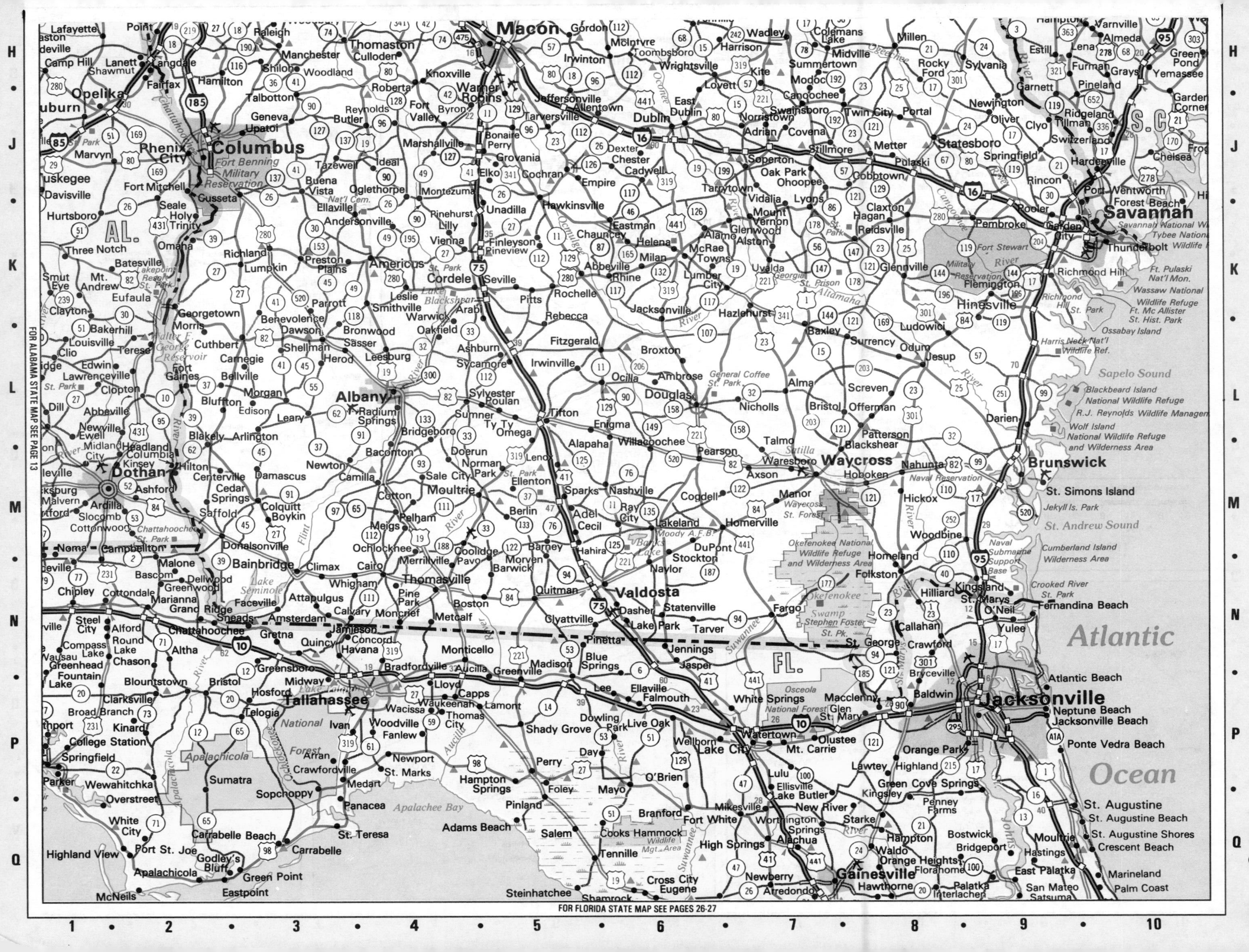

FOR ALABAMA STATE MAP SEE PAGE 13

FOR FLORIDA STATE MAP SEE PAGES 26-27

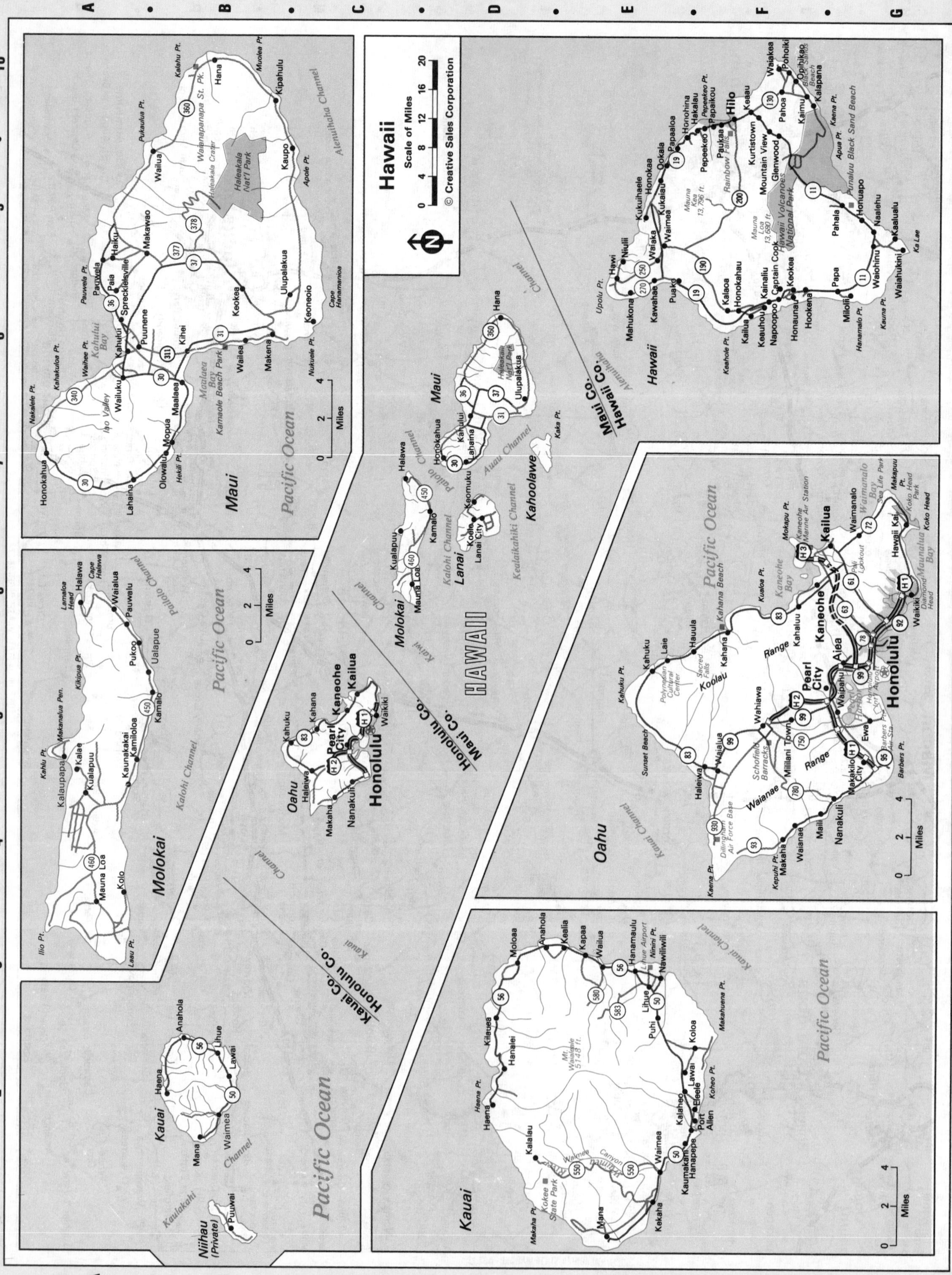

Hawaii

Scale of Miles

0 4 8 12 16 20

© Creative Sales Corporation

HAWAII

Niihau (Private)

Kauai

Oahu

Molokai

Lanai

Maui

Kahoolawe

Hawaii

Kauai Co.
Honolulu Co.

Honolulu Co.
Maui Co.

Maui Co.
Hawaii Co.

Pacific Ocean

Honolulu

Hilo

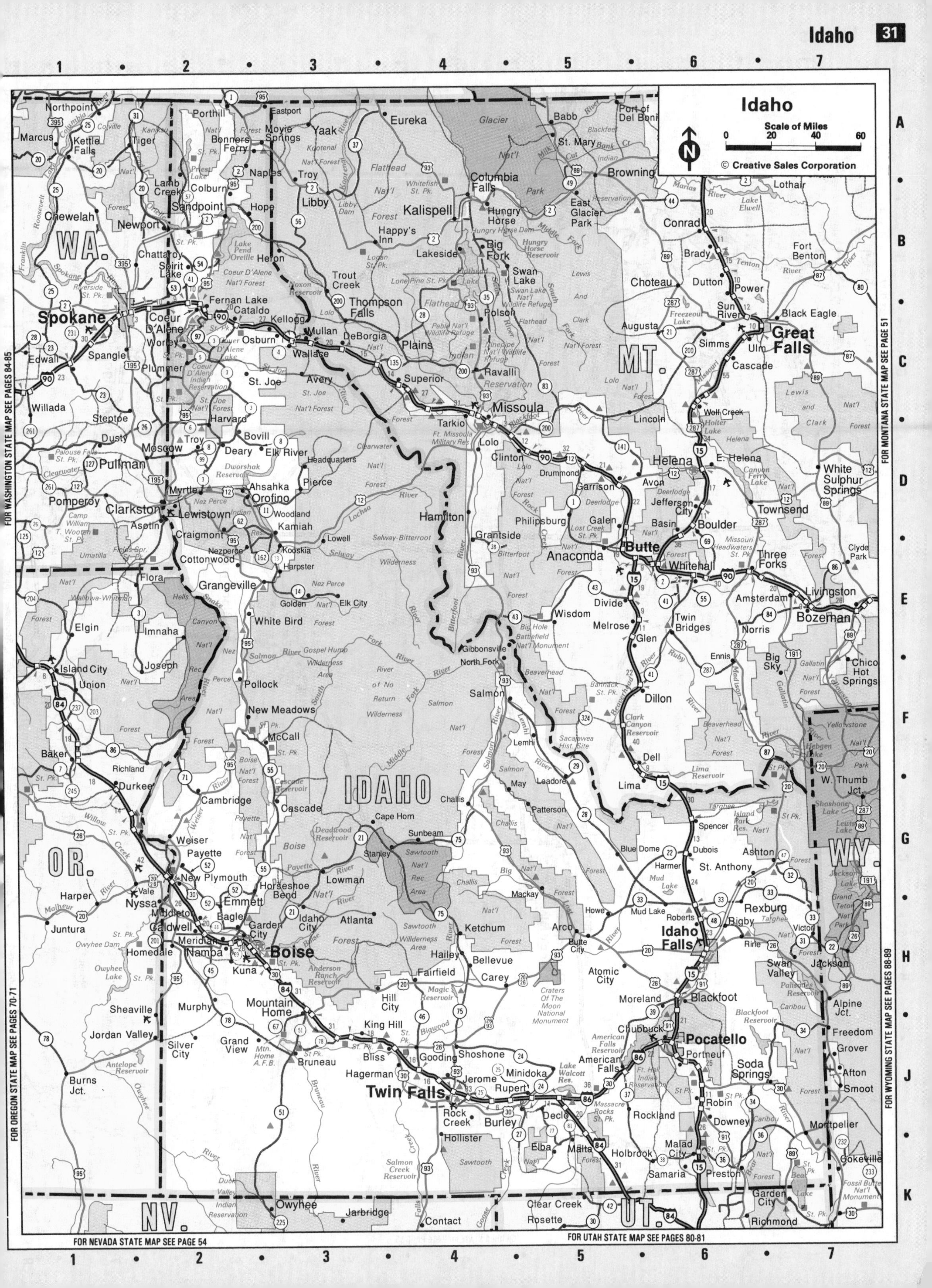
Idaho
Scale of Miles
0 20 40 60
N
© Creative Sales Corporation
FOR WASHINGTON STATE MAP SEE PAGES 84-85
FOR MONTANA STATE MAP SEE PAGE 51
FOR OREGON STATE MAP SEE PAGES 70-71
FOR WYOMING STATE MAP SEE PAGES 88-89
FOR NEVADA STATE MAP SEE PAGE 54
FOR UTAH STATE MAP SEE PAGES 80-81
IDAHO
WA.
MT.
OR.
WY.
NV.
UT.
Spokane
Coeur D'Alene
Sandpoint
Bonners Ferry
Moscow
Pullman
Lewiston
Clarkston
Orofino
Grangeville
McCall
Cascade
Weiser
Payette
Emmett
Caldwell
Nampa
Boise
Meridian
Kuna
Mountain Home
Twin Falls
Jerome
Burley
Rupert
Pocatello
Blackfoot
Idaho Falls
Rexburg
Ashton
St. Anthony
Salmon
Challis
Ketchum
Hailey
Arco
Soda Springs
Preston
Montpelier
Missoula
Kalispell
Helena
Butte
Great Falls
Bozeman
Dillon
Anaconda
Hamilton
Jackson
Baker
Ontario
Nyssa
Burns Jct.
Owyhee
Jarbridge
Contact
Malad City
American Falls
Shoshone
Gooding
Bliss
Hagerman
Craters Of The Moon National Monument
Sawtooth Nat'l Rec. Area
Yellowstone Nat'l Park
Glacier Nat'l Park
Lake Pend Oreille
Coeur D'Alene Lake
Bitterroot
Salmon River
Snake River
Hells Canyon
Kootenai
Clearwater
Palisades Reservoir
Bear Lake

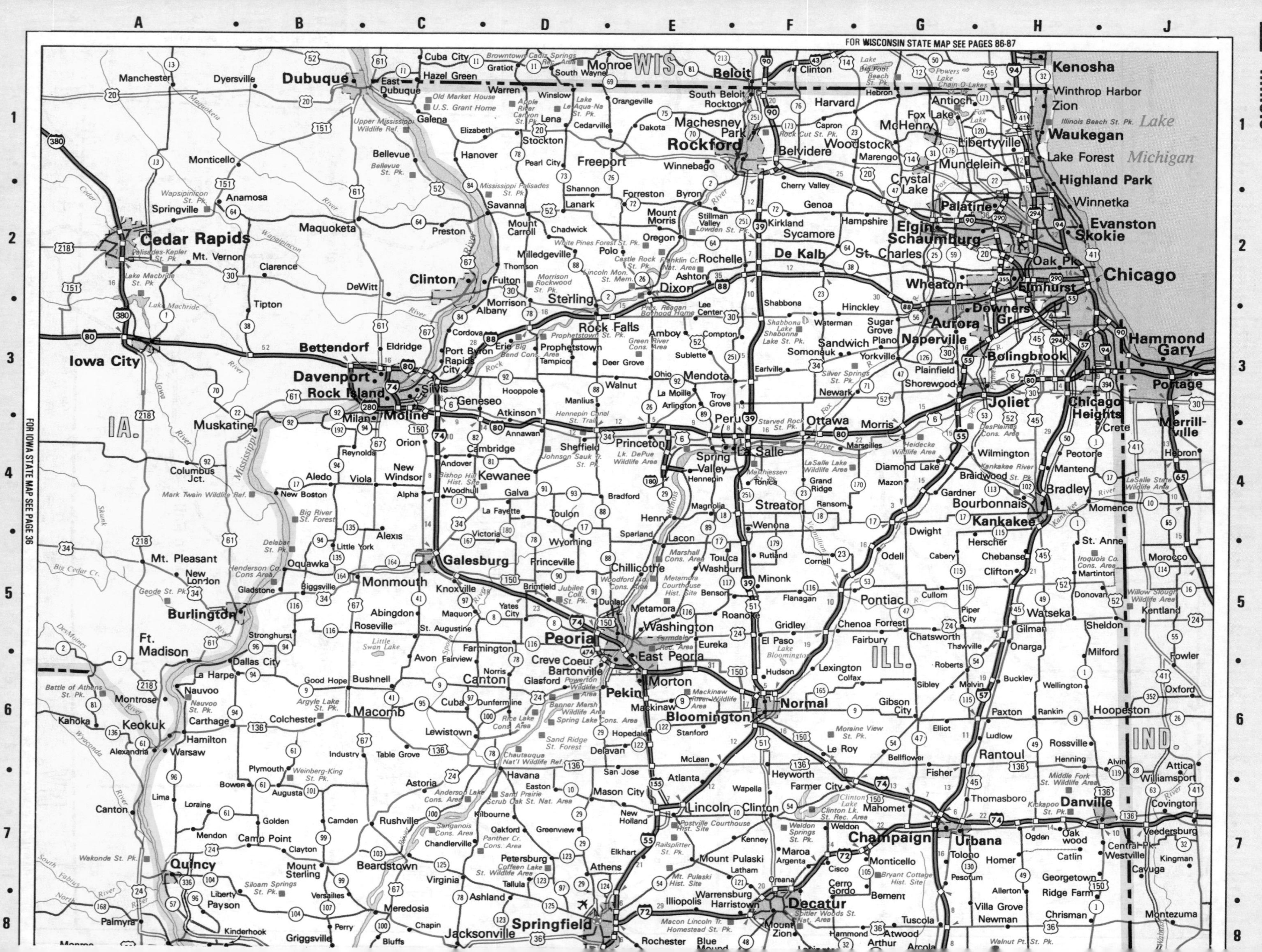
FOR WISCONSIN STATE MAP SEE PAGES 86-87
FOR IOWA STATE MAP SEE PAGE 36
Lake Michigan
WIS.
ILL.
IND.
IA.
Kenosha
Winthrop Harbor
Zion
Waukegan
Lake Forest
Highland Park
Winnetka
Evanston
Skokie
Chicago
Oak Pk.
Elmhurst
Hammond
Gary
Portage
Merrillville
Chicago Heights
Joliet
Bolingbrook
Downers Gr.
Aurora
Naperville
Wheaton
Elgin
Schaumburg
Palatine
Mundelein
Libertyville
Crystal Lake
Woodstock
McHenry
Rockford
Belvidere
Beloit
Freeport
Dixon
Sterling
Rock Falls
Rochelle
De Kalb
Sycamore
Mendota
Ottawa
Peru
La Salle
Kankakee
Bradley
Bourbonnais
Pontiac
Streator
Peoria
East Peoria
Washington
Pekin
Morton
Bloomington
Normal
Lincoln
Clinton
Champaign
Urbana
Rantoul
Danville
Decatur
Springfield
Jacksonville
Quincy
Galesburg
Kewanee
Macomb
Canton
Monmouth
Moline
Rock Island
Davenport
Bettendorf
Clinton
Dubuque
Cedar Rapids
Iowa City
Muscatine
Burlington
Keokuk
Ft. Madison
Mississippi
Illinois

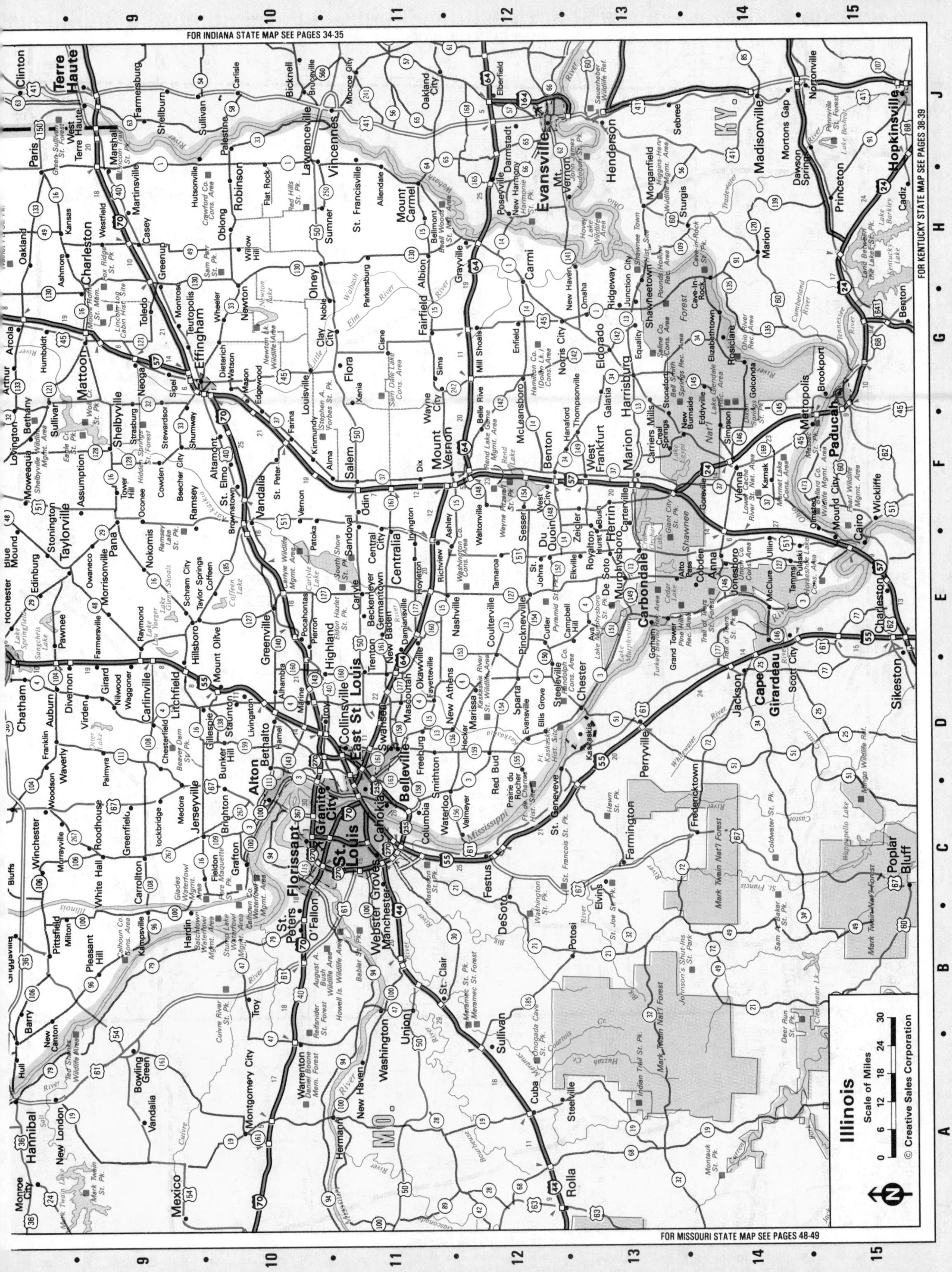

FOR INDIANA STATE MAP SEE PAGES 34-35

FOR KENTUCKY STATE MAP SEE PAGES 38-39

FOR MISSOURI STATE MAP SEE PAGES 48-49

FOR KENTUCKY STATE MAP SEE PAGES 38-39

Illinois

Scale of Miles

0 6 12 18 24 30

FOR OHIO STATE MAP SEE PAGES 66-67

FOR MICHIGAN STATE MAP SEE PAGES 44-45

FOR ILLINOIS STATE MAP SEE PAGES 32-33

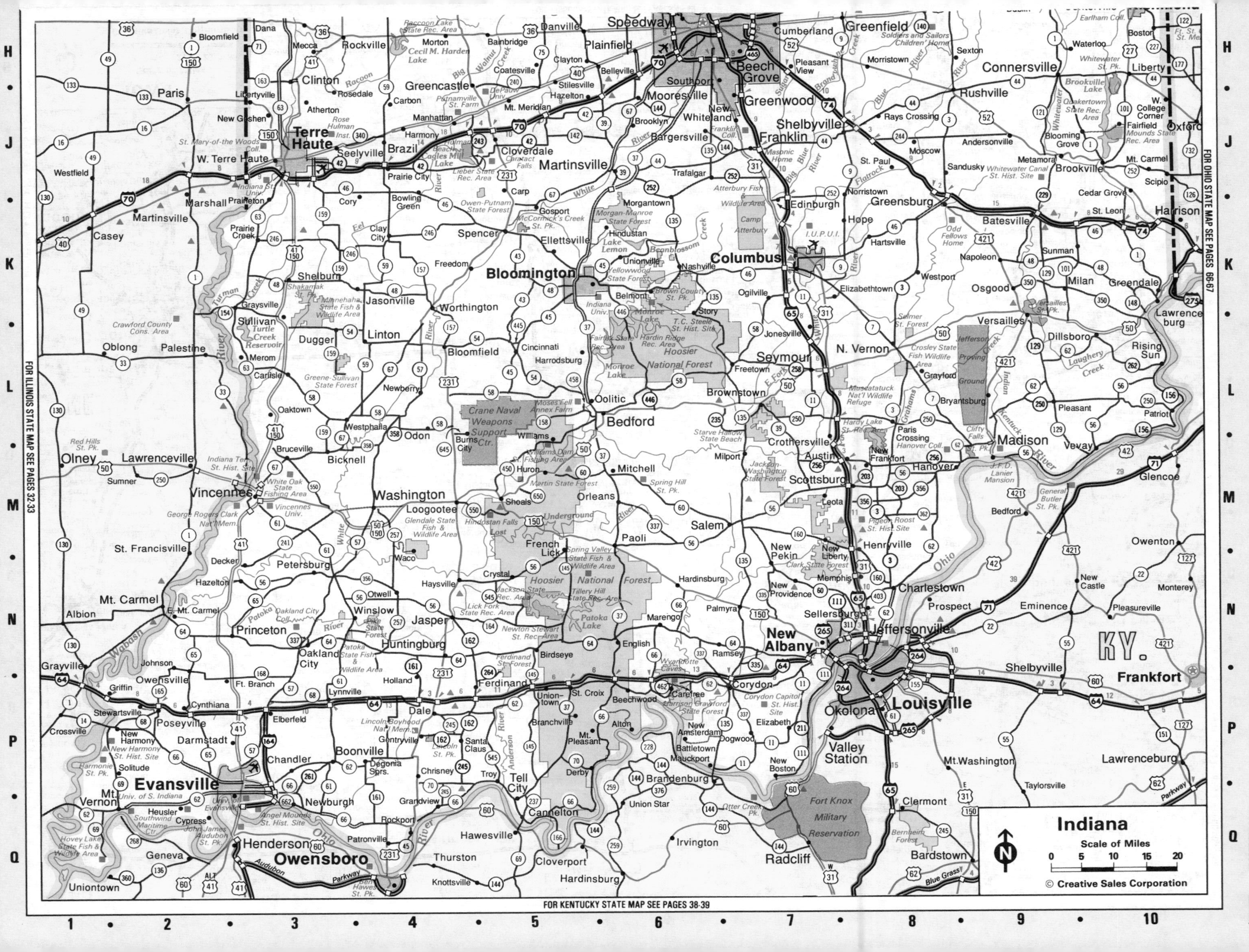
FOR OHIO STATE MAP SEE PAGES 66-67
FOR ILLINOIS STATE MAP SEE PAGES 32-33
FOR KENTUCKY STATE MAP SEE PAGES 38-39
Indiana
Scale of Miles
0 5 10 15 20
© Creative Sales Corporation
KY.
Frankfort
Louisville
Jeffersonville
New Albany
Evansville
Owensboro
Henderson
Terre Haute
Bloomington
Columbus
Bedford
Vincennes
Washington
Madison
Seymour
Greensburg
Shelbyville
Franklin
Martinsville
Greenwood
Beech Grove
Speedway
Greenfield
Connersville
Rushville
Brookville
Lawrenceburg
Jasper
Paoli
Salem
Corydon
Princeton
Sullivan
Linton
Brazil
Greencastle
Crane Naval Weapons Support Ctr.
Hoosier National Forest
Fort Knox Military Reservation
Camp Atterbury
Jefferson Proving Ground
Ohio
Wabash
Olney
Lawrenceville
Mt. Carmel
Paris
Marshall
Casey
Radcliff

FOR MINNESOTA STATE MAP SEE PAGES 46-47

FOR WISCONSIN STATE MAP SEE PAGES 86-87

FOR ILLINOIS STATE MAP SEE PAGES 32-33

FOR MISSOURI STATE MAP SEE PAGES 48-49

FOR NEBRASKA STATE MAP SEE PAGES 52-53

FOR SOUTH DAKOTA STATE MAP SEE PAGE 74

MINN. WIS. ILL. MO. NE. IOWA

Lismore Luverne Magnolia Adrian Worthington Brewster Heron Lake Jackson Welcome Sherburn Fairmont Truman Northrop Easton Winnebago Blue Earth Frost Elmore Wells Alden Albert Lea Manchester Clarks Grove Hayward Austin Maple view Brownsdale Waltham Rose Creek Adams Taopi Le Roy Grand Meadow Stewartville Spring Valley Preston Fountain Chatfield Rushford Harmony Mabel Lanesboro Houston Spring Grove Caledonia La Crescent Hokah Brownsville La Crosse Coon Valley Westby Stoddard Genoa Viroqua Melvina Cashton Ontario Hillsboro Kendall Elroy Mauston Union Center Wonewoc Reedsburg Cazenovia Loganville

Rock Rapids George Sibley Ashton Ocheyedan Allendorf Round Lake Lake Park Spirit Lake Arnolds Park Milford May City Melvin Superior Estherville Armstrong Ledyard Swea City Bancroft Buffalo Center Lake Mills Thompson Joice Northwood Lyle Glenville Stacyville McIntire Riceville St. Ansgar Lime Springs Saratoga Cresco Kendallville Ridgeway Decorah Burr Oak Hesper New Albin Dorchester Lansing Waukon Harpers Ferry De Soto Ferryville Readstown Soldiers Grove Mount Sterling Gays Mills Richland Center Lone Rock Muscoda Avoca Boscobel Steuben Eastman Prairie Du Chien Highland Barneveld Dodgeville Fennimore Montfort Livingston Lancaster Linden Mineral Point Platteville Bloomington Patch Grove Potosi Tennyson Dickeyville Belmont Darlington Benton Cuba City Hazel Green Shullsburg Gratiot

Sheldon Hull Boyden Perkins Sanborn Hartley Spencer Fostoria Terril Ringsted Graettinger Wallingford Ruthven Emmetsburg Fenton Whittemore Algona Wesley Titonka Garner Forest City Hanlontown Britt Ventura Klemme Clear Lake Mason City Manly Osage Charles City Nora Springs Rockford Floyd New Hampton Nashua Jerico Jackson Jct. Calmar Postville Monona McGregor Marquette Elkader Garnavillo Guttenberg Strawberry Pt. Colesburg Luxemburg Dyersville Dubuque Epworth Farley Asbury East Dubuque Galena Zwingle Bellevue Elizabeth Hanover Savanna Stockton Apple River

Hawarden Orange City Alton Paullina Primghar Sutherland Larrabee Cherokee Marcus Remsen Le Mars Akron Chatsworth Kingsley Pierson Washta Quimby Holstein Schaller Sac City Storm Lake Alta Aurelia Laurens Pocahontas Rolfe Humboldt Dakota City Fort Dodge Webster City Eagle Grove Goldfield Clarion Dows Belmond Iowa Falls Ackley Aplington Parkersburg Cedar Falls Waterloo Evansdale Independence Manchester Delhi Hopkinton Cascade Monticello Anamosa Maquoketa Preston Sabula Clinton

Sioux City Sergeant Bluff Correctionville Moville Anthon Ida Grove Battle Creek Odebolt Lake View Auburn Lohrville Rockwell City Manson Gowrie Harcourt Dayton Lehigh Webster City Jewell Williams Eldora Hubbard Steamboat Rock Grundy Cen. Dike Hudson La Porte City Vinton Shellsburg Urbana Center Point Cedar Rapids Marion Hiawatha Springville Mt. Vernon Lisbon Mechanicsville Tipton Lowden Wheatland De Witt Camanche

Onawa Mapleton Danbury Schleswig Charter Oak Ute Denison Vail Arcadia Carroll Glidden Scranton Jefferson Grand Jct. Boone Ogden Ames Nevada State Center Marshalltown Melbourne Tama Toledo Belle Plaine Marengo Amana Tiffin Iowa City Coralville West Branch West Liberty Durant Wilton Walcott Davenport Bettendorf Rock Island Moline Le Claire Port Byron

Missouri Valley Logan Woodbine Dunlap Harlan Manning Audubon Exira Coon Rapids Guthrie Ctr. Panora Perry Dallas Center Adel Waukee Grimes Urbandale Polk City Ankeny Altoona Des Moines West Des Moines Pleasant Hill Colfax Newton Grinnell Brooklyn Victor Montezuma Malcom Williamsburg

Council Bluffs Omaha Bellevue Papillion Glenwood Red Oak Atlantic Anita Adair Stuart Casey Earlham Winterset Indianola Norwalk Carlisle Knoxville Pella Oskaloosa Sigourney Washington Wellman Kalona Lone Tree Muscatine Columbus Jct. Wapello Aledo Galesburg Monmouth Kewanee

Shenandoah Clarinda Villisca Corning Creston Afton Osceola Chariton Albia Ottumwa Fairfield Mt. Pleasant Burlington Fort Madison Keokuk Lamoni Leon Corydon Centerville Bloomfield Keosauqua Donnellson Hamburg Sidney Tabor Nebraska City Maryville Bethany Lancaster Kahoka Memphis Macomb Carthage Nauvoo

Iowa

Scale of Miles

0 7 14 21 28 35

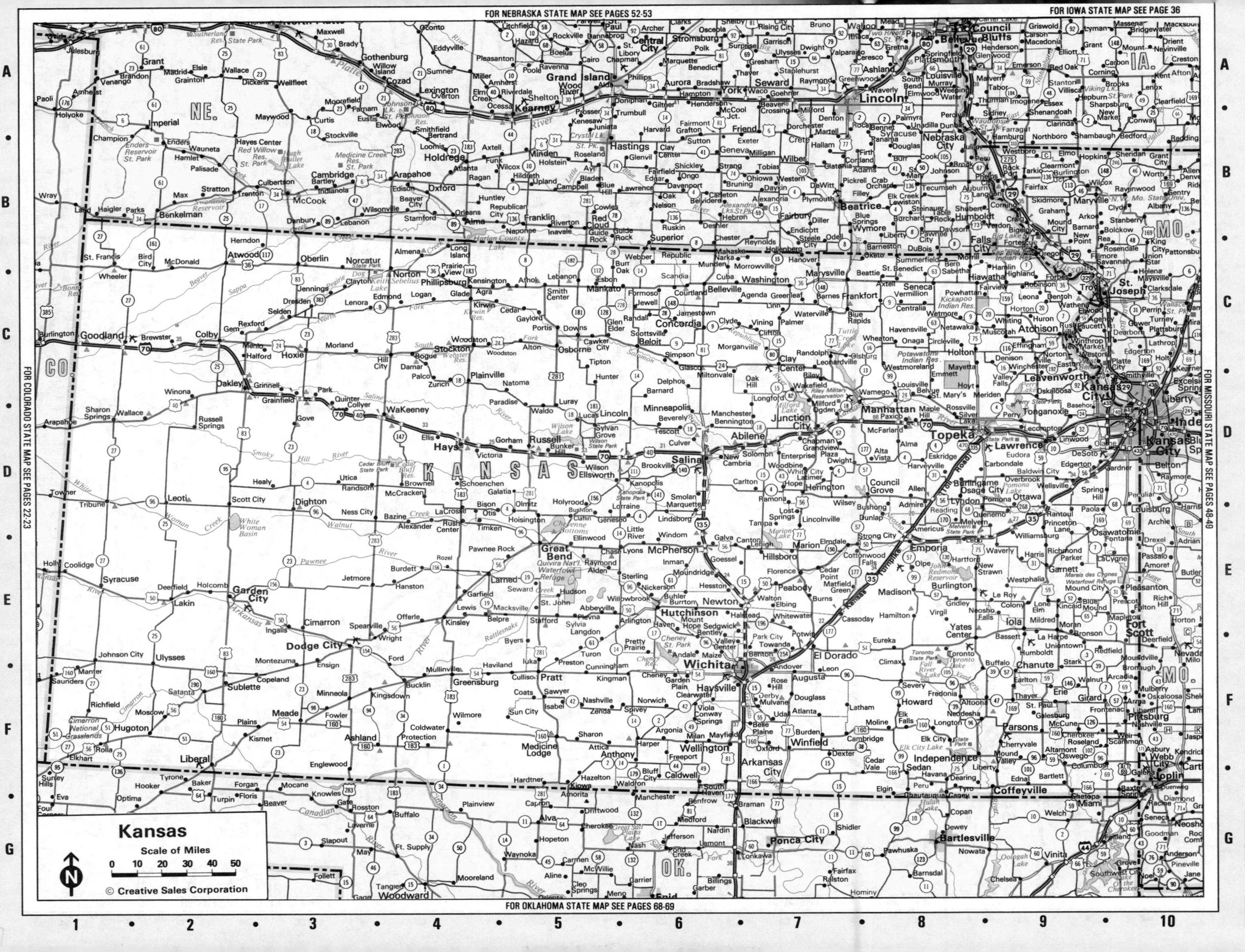
FOR NEBRASKA STATE MAP SEE PAGES 52-53
FOR IOWA STATE MAP SEE PAGE 36
FOR MISSOURI STATE MAP SEE PAGES 48-49
FOR COLORADO STATE MAP SEE PAGES 22-23
FOR OKLAHOMA STATE MAP SEE PAGES 68-69
KANSAS
Kansas
Scale of Miles
0 10 20 30 40 50
© Creative Sales Corporation
Topeka
Wichita
Kansas City
Salina
Dodge City
Garden City
Hutchinson
Lawrence
Manhattan
Emporia
Hays
Goodland
Liberal
Lincoln
St. Joseph
Joplin

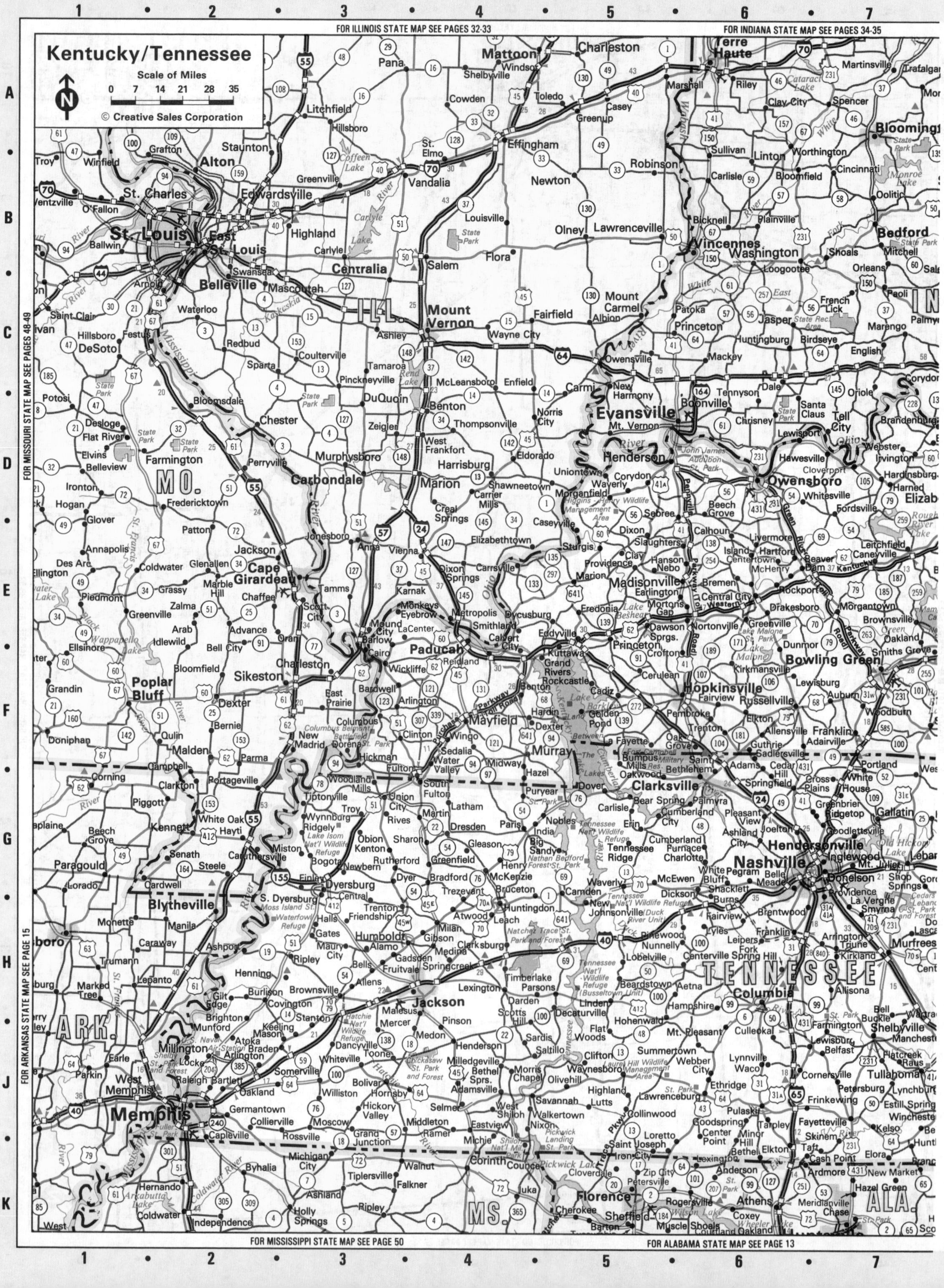

Kentucky/Tennessee
Scale of Miles
0 7 14 21 28 35
N
© Creative Sales Corporation
FOR ILLINOIS STATE MAP SEE PAGES 32-33
FOR INDIANA STATE MAP SEE PAGES 34-35
FOR MISSOURI STATE MAP SEE PAGES 48-49
FOR ARKANSAS STATE MAP SEE PAGE 15
FOR MISSISSIPPI STATE MAP SEE PAGE 50
FOR ALABAMA STATE MAP SEE PAGE 13
1 2 3 4 5 6 7
A B C D E F G H J K
ILL.
MO.
IN
ARK.
MS.
ALA.
TENNESSEE
St. Louis
East St. Louis
St. Charles
Alton
Belleville
Mattoon
Charleston
Terre Haute
Vincennes
Mount Vernon
Centralia
Carbondale
Cape Girardeau
Evansville
Henderson
Owensboro
Paducah
Hopkinsville
Bowling Green
Madisonville
Clarksville
Nashville
Hendersonville
Jackson
Memphis
Dyersburg
Blytheville
Poplar Bluff
Sikeston
Murray
Mayfield
Columbia
Florence
Corinth
Bloomington
Bedford

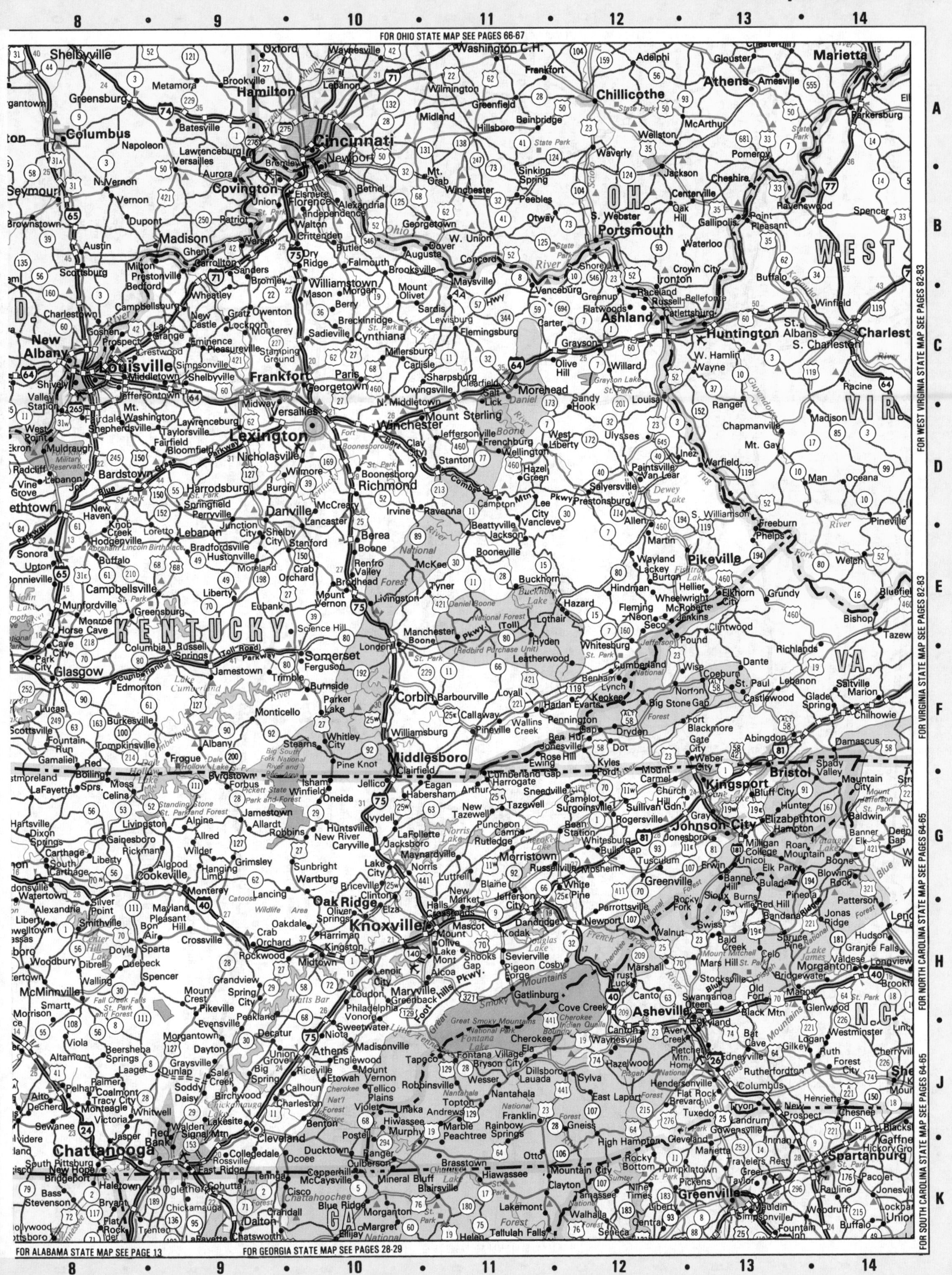
8
9
10
11
12
13
14
A
B
C
D
E
F
G
H
J
K
FOR OHIO STATE MAP SEE PAGES 66-67
FOR WEST VIRGINIA STATE MAP SEE PAGES 82-83
FOR VIRGINIA STATE MAP SEE PAGES 82-83
FOR NORTH CAROLINA STATE MAP SEE PAGES 64-65
FOR SOUTH CAROLINA STATE MAP SEE PAGES 64-65
FOR ALABAMA STATE MAP SEE PAGE 13
FOR GEORGIA STATE MAP SEE PAGES 28-29
KENTUCKY
OH.
WEST
VIR.
VA.
N.C.
GA
Cincinnati
Covington
Louisville
Frankfort
Lexington
Ashland
Huntington
Charleston
Chillicothe
Portsmouth
Marietta
Athens
Columbus
Hamilton
Madison
New Albany
Bardstown
Richmond
Danville
Somerset
Glasgow
Pikeville
Hazard
Middlesboro
Corbin
London
Bristol
Kingsport
Johnson City
Morristown
Knoxville
Oak Ridge
Cookeville
Chattanooga
Cleveland
Asheville
Greenville
Spartanburg
Gatlinburg
Daniel Boone National Forest
Great Smoky Mountains
Cumberland Gap

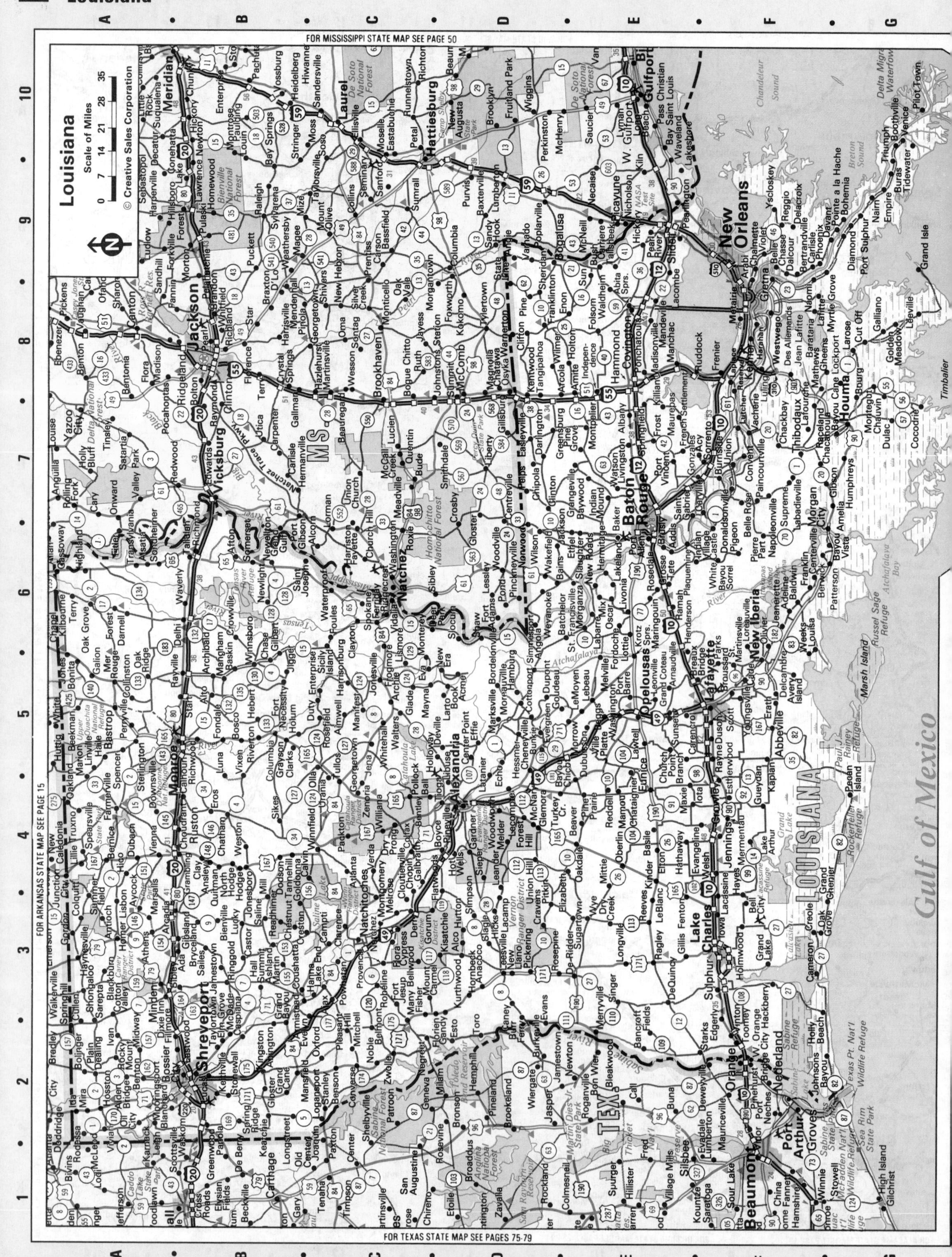
FOR MISSISSIPPI STATE MAP SEE PAGE 50
FOR ARKANSAS STATE MAP SEE PAGE 15
FOR TEXAS STATE MAP SEE PAGES 75-79
Louisiana
Scale of Miles
0 7 14 21 28 35
© Creative Sales Corporation
Shreveport
Monroe
Alexandria
Jackson
Vicksburg
Natchez
Hattiesburg
Baton Rouge
Lafayette
Lake Charles
New Orleans
Beaumont
Gulf of Mexico
LOUISIANA

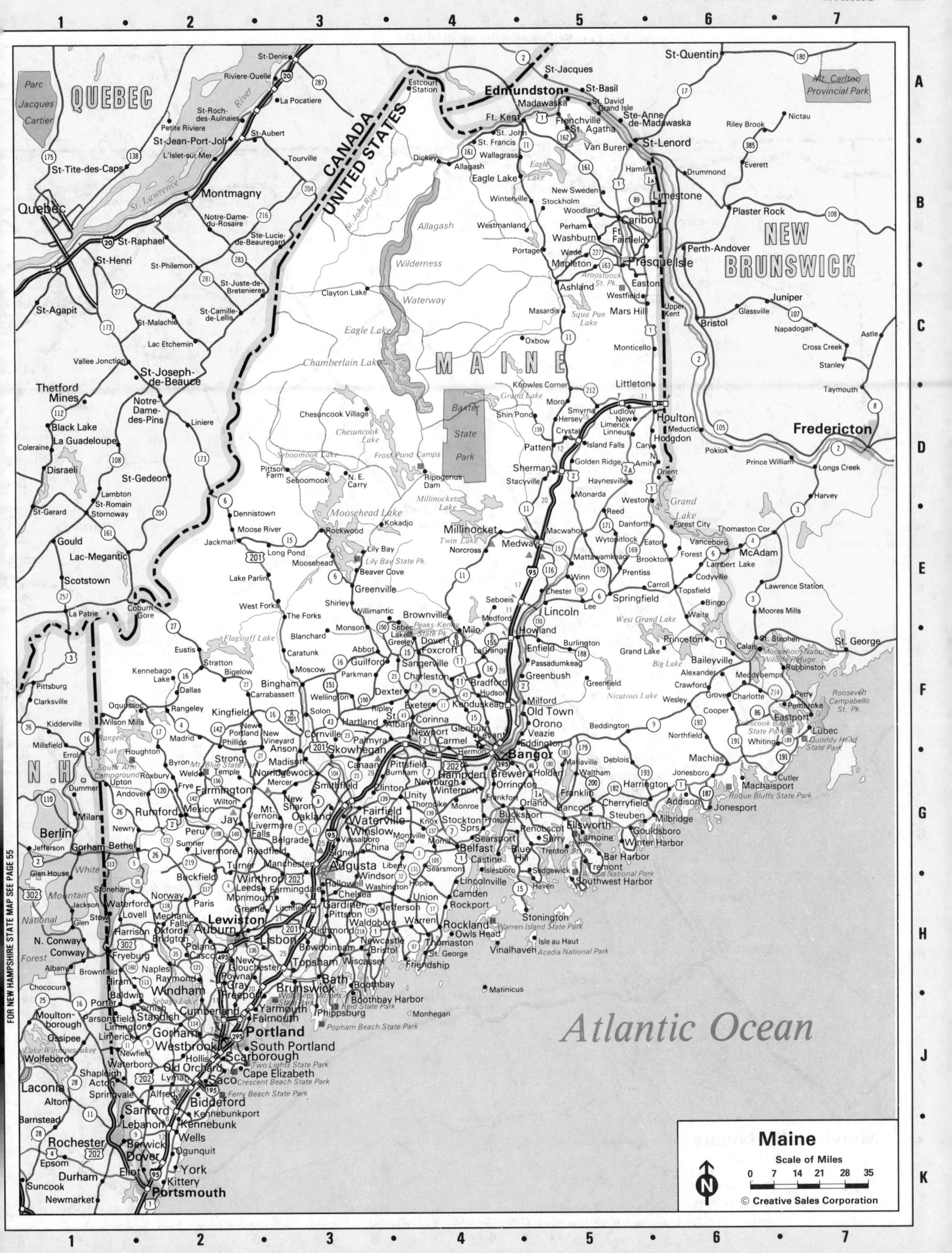
QUEBEC
CANADA
UNITED STATES
MAINE
NEW BRUNSWICK
N.H.
Atlantic Ocean
Parc Jacques Cartier
Mt. Carlton Provincial Park
Quebec
Montmagny
St-Jean-Port-Joli
St-Tite-des-Caps
La Pocatiere
Riviere-Ouelle
St-Denis
Estcourt Station
Edmundston
St-Jacques
St-Basil
St-Quentin
Madawaska
Ft. Kent
Frenchville
St. Agatha
Ste-Anne-de-Madawaska
St-Lenord
Van Buren
Allagash
Eagle Lake
Limestone
Caribou
Presque Isle
Perth-Andover
Plaster Rock
Ashland
Mars Hill
Bristol
Juniper
Allagash Wilderness Waterway
Eagle Lake
Chamberlain Lake
Chesuncook Lake
Baxter State Park
Houlton
Fredericton
Littleton
Patten
Sherman
Millinocket
Moosehead Lake
Rockwood
Jackman
Greenville
Lincoln
Howland
Springfield
Lee
Milo
Dover Foxcroft
Old Town
Orono
Bangor
Brewer
Calais
St. Stephen
St. George
Princeton
Eastport
Lubec
Machias
Jonesport
Ellsworth
Bar Harbor
Southwest Harbor
Bucksport
Belfast
Camden
Rockland
Thomaston
Vinalhaven
Stonington
Acadia National Park
Waterville
Skowhegan
Augusta
Gardiner
Bath
Brunswick
Boothbay Harbor
Lewiston
Auburn
Rumford
Farmington
Bingham
Kingfield
Rangeley
Berlin
Gorham
Bethel
Portland
South Portland
Westbrook
Scarborough
Old Orchard
Saco
Biddeford
Kennebunkport
Kennebunk
Sanford
Wells
Ogunquit
York
Kittery
Portsmouth
Dover
Rochester
Laconia
Wolfeboro
Conway
Monhegan
Matinicus
FOR NEW HAMPSHIRE STATE MAP SEE PAGE 55
Maine
Scale of Miles
0 7 14 21 28 35
N
© Creative Sales Corporation

FOR PENNSYLVANIA STATE MAP SEE PAGES 72-73

FOR WEST VIRGINIA STATE MAP SEE PAGES 82-83

FOR VIRGINIA STATE MAP SEE PAGES 82-83

Maryland/Delaware

Scale of Miles

0 3 6 9 12 15

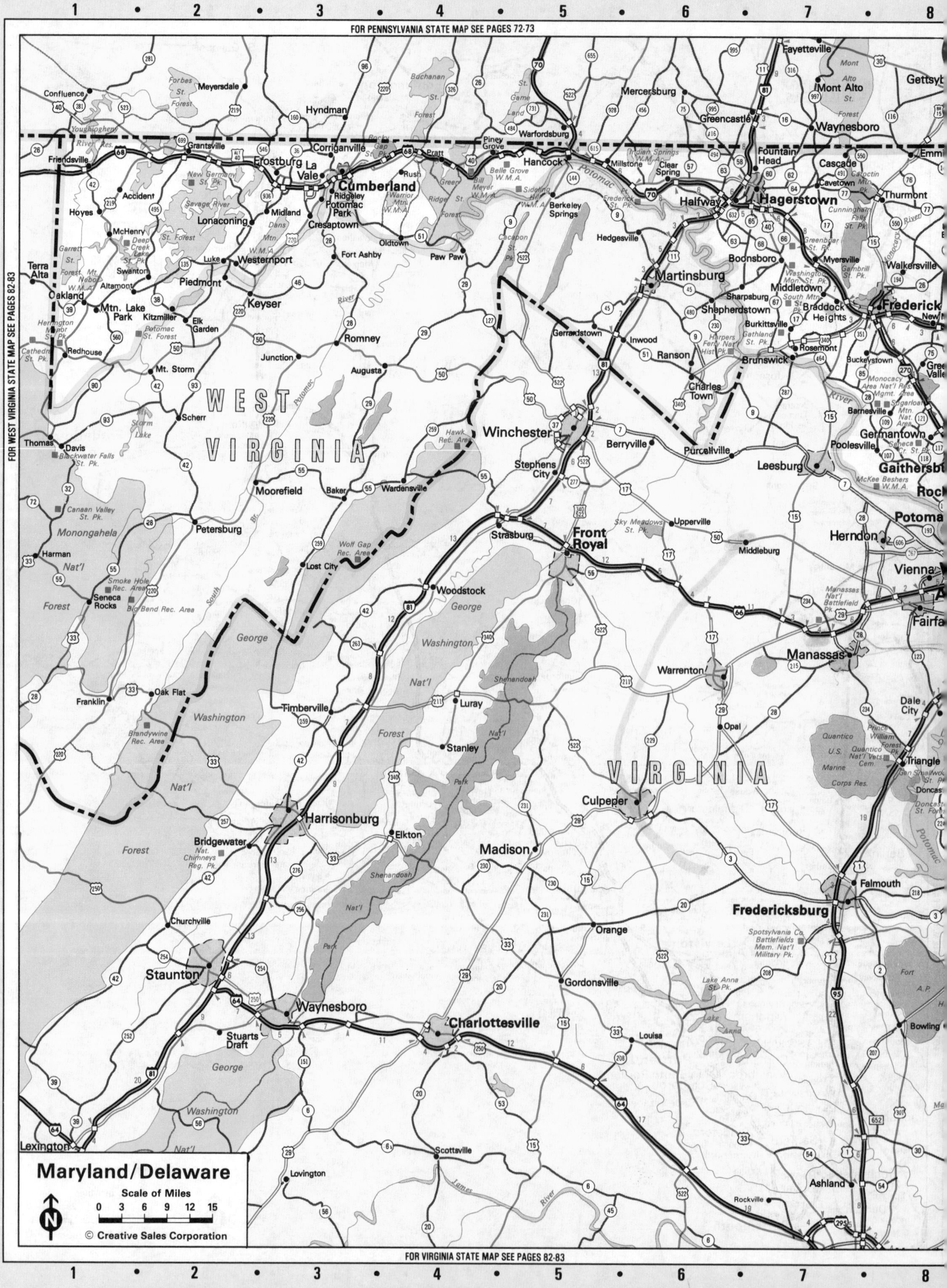

8 • 9 • 10 • 11 • 12 • 13 • 14 • 15

FOR PENNSYLVANIA STATE MAP SEE PAGE 72-73

FOR NEW JERSEY STATE MAP SEE PAGES 56-57

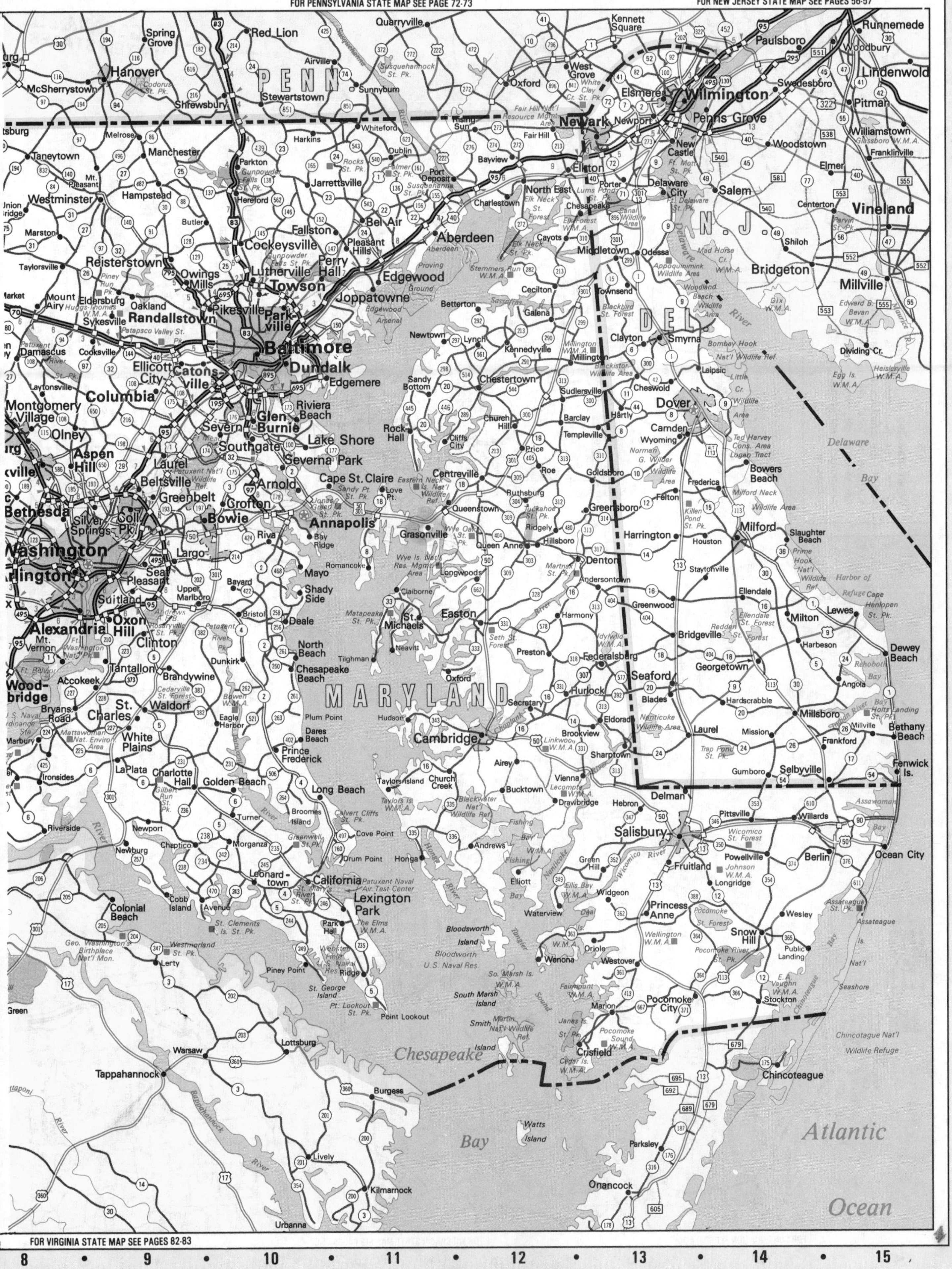

PENN
N. J.
DEL.
MARYLAND

Spring Grove
Red Lion
Quarryville
Kennett Square
Paulsboro
Runnemede
Woodbury
Hanover
McSherrystown
Codorus St. Pk.
Shrewsbury
Stewartstown
Airville
Susquehannock St. Pk.
Sunnyburn
Oxford
West Grove
White Clay Cr. St. Pk.
Elsmere
Wilmington
Swedesboro
Lindenwold
Pitman
Fair Hill Nat'l Resource Mgmt. Area
Rising Sun
Fair Hill
Newark
Newport
Penns Grove
New Castle
Williamstown
Glassboro W.M.A.
Franklinville
Woodstown
Taneytown
Melrose
Manchester
Whiteford
Harkins
Dublin
Parkton
Gunpowder Falls St. Pk.
Rocks St. Pk.
Palmer St. Pk.
Port Deposit
Bayview
Elkton
Ft. Mott St. Pk.
Elmer
Mt. Pleasant
Westminster
Hampstead
Hereford
Jarrettsville
Susquehanna St. Pk.
Charlestown
North East
Elk Neck St. Forest
Lums Pond St. Pk.
Porter
Delaware City
Ft. Delaware St. Pk.
Salem
Centerton
Vineland
Union Bridge
Marston
Butler
Bel Air
Fallston
Aberdeen
Chesapeake City
Elk Forest W.M.A.
Canal Wildlife Area
Parvin St. Pk.
Shiloh
Taylorsville
Reisterstown
Cockeysville
Pleasant Hills
Perry Hall
Gunpowder Falls St. Pk.
Aberdeen Proving Ground
Elk Neck St. Pk.
Cayots
Middletown
Odessa
Appoquinimink Wildlife Area
Mad Horse Cr. W.M.A.
Bridgeton
Millville
Owings Mills
Lutherville
Towson
Edgewood
Stemmers Run W.M.A.
Cecilton
Townsend
Woodland Beach Wildlife Area
Dix W.M.A.
Edward B. Bevan W.M.A.
Market
Mount Airy
Eldersburg
Hugg-Thomas W.M.A.
Piney Run Pk.
Sykesville
Oakland
Randallstown
Pikesville
Parkville
Joppatowne
Edgewood Arsenal
Betterton
Sassafras
Galena
Blackbird St. Forest
Delaware River
Patapsco Valley St. Pk.
Patuxent River
Damascus
Cooksville
Ellicott City
Catonsville
Baltimore
Dundalk
Edgemere
Newtown
Lynch
Kennedyville
Millington W.M.A.
Millington
Clayton
Smyrna
Bombay Hook Nat'l Wildlife Ref.
Dividing Cr.
Heislerville W.M.A.
Egg Is. W.M.A.
Laytonsville
Columbia
Sandy Bottom
Chestertown
Blackiston Wildlife Area
Cheswold
Leipsic
Little Cr. Wildlife Area
Dover
Montgomery Village
Olney
Aspen Hill
Laurel
Severn
Glen Burnie
Riviera Beach
Lake Shore
Southgate
Severna Park
Church Hill
Sudlersville
Barclay
Templeville
Hartly
Camden
Wyoming
Norman G. Wilder Wildlife Area
Ted Harvey Cons. Area Logan Tract
Delaware Bay
Bowers Beach
Rock Hall
Cliffs City
Price
Roe
Goldsboro
Patuxent Nat'l Wildlife Ref.
Beltsville
Greenbelt
Arnold
Crofton
Cape St. Claire
Sandy Pt. St. Pk.
Eastern Neck Is. Nat'l Wildlife Ref.
Love Pt.
Centreville
Ruthsburg
Tuckahoe St. Pk.
Greensboro
Felton
Frederica
Killens Pond St. Pk.
Milford Neck Wildlife Area
Bethesda
Silver Springs
Coll. Pk.
Bowie
Annapolis
Bay Ridge
Riva
Queenstown
Grasonville
Wye Oak St. Pk.
Queen Anne
Ridgely
Hillsboro
Harrington
Houston
Milford
Slaughter Beach
Washington
Arlington
Largo
Seat Pleasant
Upper Marlboro
Bayard
Mayo
Romancoke
Wye Is. Nat'l Res. Mgmt. Area
Longwoods
Denton
Martinak St. Pk.
Andersontown
Staytonville
Prime Hook Nat'l Wildlife Ref.
Harbor of Refuge
Cape Henlopen St. Pk.
Ellendale
Suitland
Shady Side
Claiborne
Alexandria
Oxon Hill
Andrews A.F.B.
Rosaryville St. Pk.
Bristol
Deale
Matapeake St. Pk.
St. Michaels
Easton
Harmony
Greenwood
Ellendale St. Forest
Milton
Lewes
Mt. Vernon
Ft. Washington Nat'l Pk.
Clinton
Patuxent River Pk.
Neavitt
Seth St. Forest
Preston
Idylwild W.M.A.
Federalsburg
Bridgeville
Redden St. Forest
Harbeson
Dewey Beach
Ft. Belvoir
Tantallon
Brandywine
Dunkirk
North Beach
Tilghman
Oxford
Georgetown
Rehoboth Bay
Woodbridge
Accokeek
Cedarville St. Forest
Chesapeake Beach
Hurlock
Seaford
Blades
Angola
U.S. Naval Ordinance Sta.
Bryans Road
St. Charles
Waldorf
Bowen W.M.A.
Eagle Harbor
Plum Point
Hudson
Secretary
Choptank River
Eldorado
Nanticoke Wildlife Area
Hardscrabble
Laurel
Millsboro
Indian River Bay
Holts Landing St. Pk.
Millville
Bethany Beach
Marbury
Mattawoman Nat. Enviro. Area
White Plains
Dares Beach
Cambridge
Brookview
Linkwood W.M.A.
Mission
Frankford
Prince Frederick
Sharptown
Trap Pond St. Pk.
Gumboro
Selbyville
Fenwick Is.
Ironsides
LaPlata
Charlotte Hall
Golden Beach
Taylors Island
Church Creek
Airey
Vienna
Lecompte W.M.A.
Delmar
Gilbert Run St. Pk.
Long Beach
Taylors Is. W.M.A.
Bucktown
Drawbridge
Hebron
Pittsville
Assawoman Bay
Willards
Riverside
Newport
Turner
Broomes Island
Calvert Cliffs St. Pk.
Blackwater Nat'l Wildlife Ref.
Fishing Bay W.M.A.
Salisbury
Wicomico St. Forest
Chaptico
Morganza
Greenwell St. Pk.
Cove Point
Drum Point
Honga
Honga River
Andrews
Green Hill
Wicomico River
Powellville
Berlin
Ocean City
Newburg
Leonardtown
California
Patuxent Naval Air Test Center
St. Mary's River St. Pk.
Lexington Park
Fishing Bay
Elliott
Nanticoke River
Fruitland
Johnson W.M.A.
Longridge
Colonial Beach
Cobb Island
Avenue
Ellis Bay W.M.A.
Widgeon
Princess Anne
Pocomoke St. Forest
Wesley
Assateague St. Pk.
Assateague Is.
St. Clements Is. St. Pk.
Park Hall
The Elms W.M.A.
Bloodsworth Island
Bloodworth U.S. Naval Res.
Tangier Sound
Waterview
Deal Is. W.M.A.
Snow Hill
Public Landing
Geo. Washington's Birthplace Nat'l Mon.
Westmorland St. Pk.
Lerty
Piney Point
Webster Field U.S. Naval Res.
Ridge
Wenona
Oriole
Wellington W.M.A.
Pocomoke River St. Pk.
Westover
So. Marsh Is. W.M.A.
E.A. Vaughn W.M.A.
Assateague Bay
Nat'l Seashore
Green
St. George Island
Pt. Lookout St. Pk.
Point Lookout
South Marsh Island
Fairmount W.M.A.
Marion
Pocomoke City
Stockton
Chincoteague Bay
Smith Island
Martin Nat'l Wildlife Ref.
Janes Is. St. Pk.
Pocomoke Sound W.M.A.
Crisfield
Cedar Is. W.M.A.
Chincoteague Nat'l Wildlife Refuge
Warsaw
Lottsburg
Chesapeake Bay
Tappahannock
Chincoteague
Mattaponi River
Rappahannock River
Burgess
Watts Island
Atlantic Ocean
Parksley
Lively
Kilmarnock
Onancock
Urbanna

A • B • C • D • E • F • G • H • J • K

FOR VIRGINIA STATE MAP SEE PAGES 82-83

8 • 9 • 10 • 11 • 12 • 13 • 14 • 15

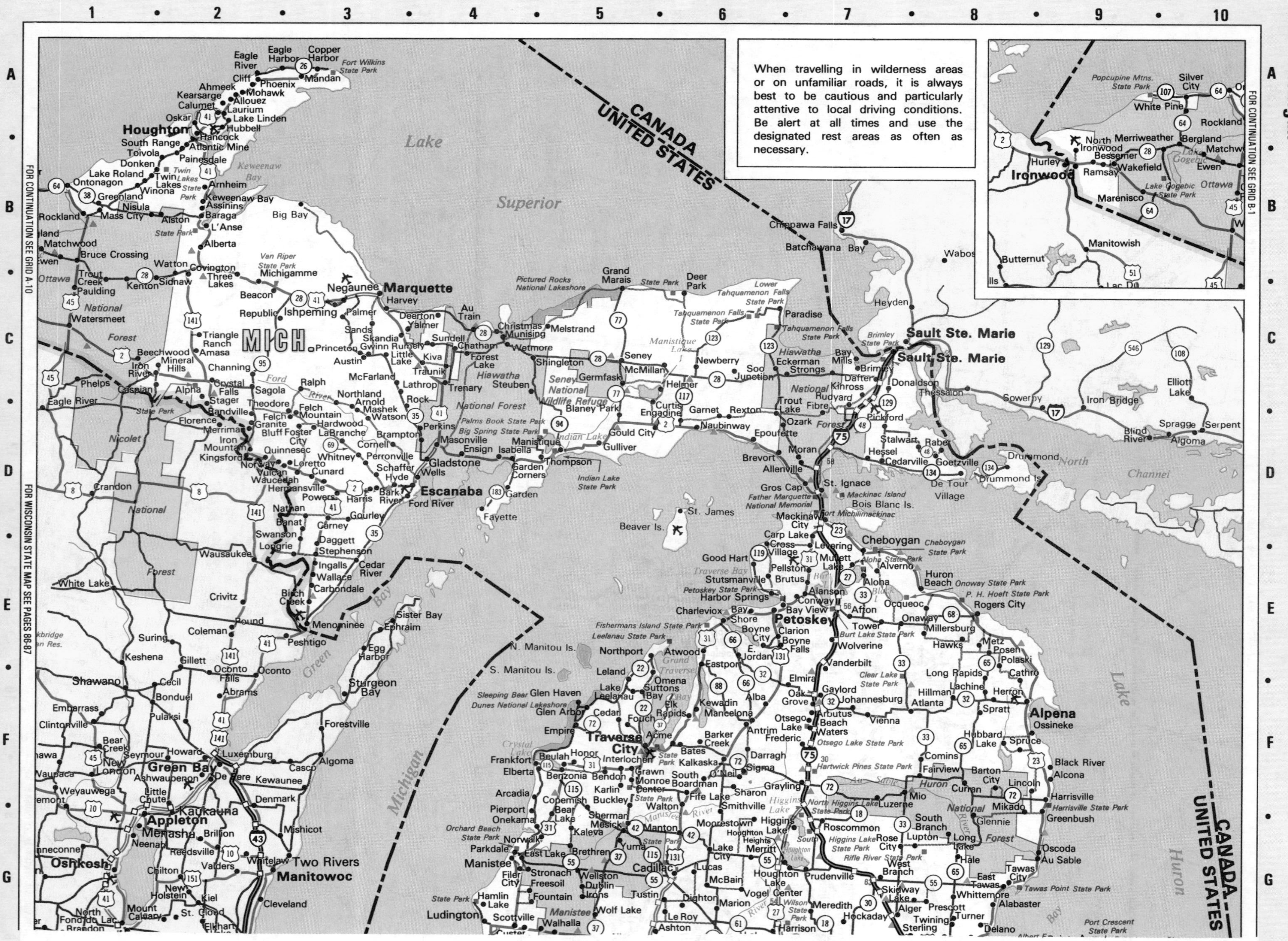

FOR CONTINUATION SEE GRID B-1
FOR CONTINUATION SEE GRID A-10
FOR WISCONSIN STATE MAP SEE PAGES 86-87
When travelling in wilderness areas or on unfamiliar roads, it is always best to be cautious and particularly attentive to local driving conditions. Be alert at all times and use the designated rest areas as often as necessary.
CANADA
UNITED STATES
Lake Superior
Lake Huron
Lake Michigan
Green Bay
North Channel
MICH
Sault Ste. Marie
Marquette
Escanaba
Houghton
Ironwood
Traverse City
Petoskey
Cheboygan
Alpena
Manistee
Ludington
Cadillac
Green Bay
Appleton
Manitowoc
Two Rivers
Oshkosh
St. Ignace
Mackinaw City
Menominee
Ishpeming
Negaunee
Munising
Newberry
Pictured Rocks National Lakeshore
Sleeping Bear Dunes National Lakeshore
Hiawatha National Forest
Ottawa National Forest
Nicolet National Forest
Huron National Forest
Manistee National Forest
Seney National Wildlife Refuge
Beaver Is.
Drummond Is.

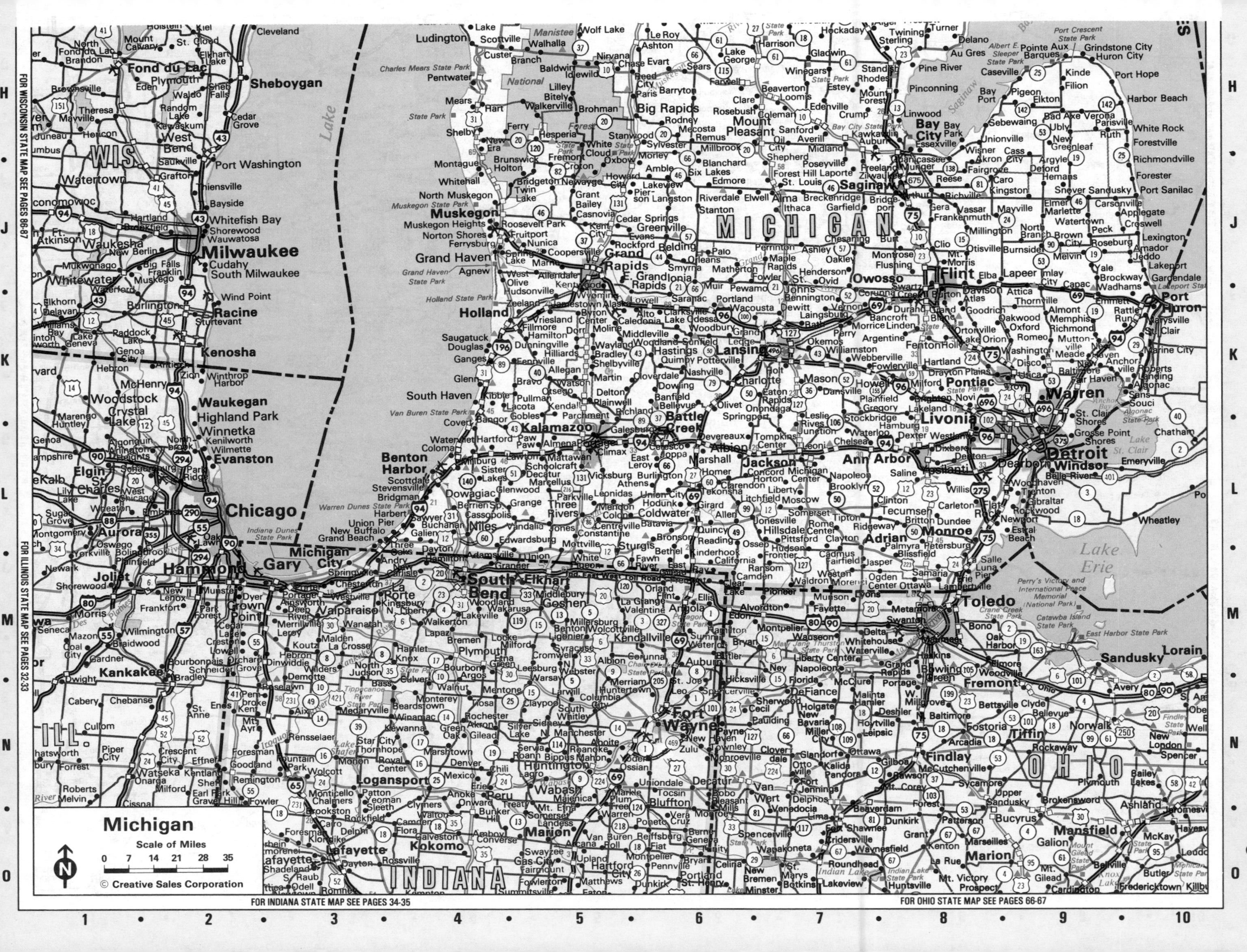
Michigan
Scale of Miles
0 7 14 21 28 35
© Creative Sales Corporation
FOR WISCONSIN STATE MAP SEE PAGES 86-87
FOR ILLINOIS STATE MAP SEE PAGES 32-33
FOR INDIANA STATE MAP SEE PAGES 34-35
FOR OHIO STATE MAP SEE PAGES 66-67
MICHIGAN
INDIANA
OHIO
WIS.
ILL.
Lake Erie
Chicago
Milwaukee
Detroit
Windsor
Toledo
Grand Rapids
Lansing
Kalamazoo
Flint
Saginaw
Ann Arbor
Fort Wayne
South Bend
Gary
Sheboygan
Kenosha
Racine
Waukegan
Evanston
Sandusky
Lorain
Port Huron
Muskegon
Holland
Jackson
Battle Creek
Bay City
Pontiac
Warren
Livonia
Owosso
Lapeer
Monroe
Adrian
Elkhart
Goshen
Kokomo
Lafayette
Logansport
Marion
Findlay
Mansfield
Kankakee
Joliet
Aurora
Elgin
Hammond

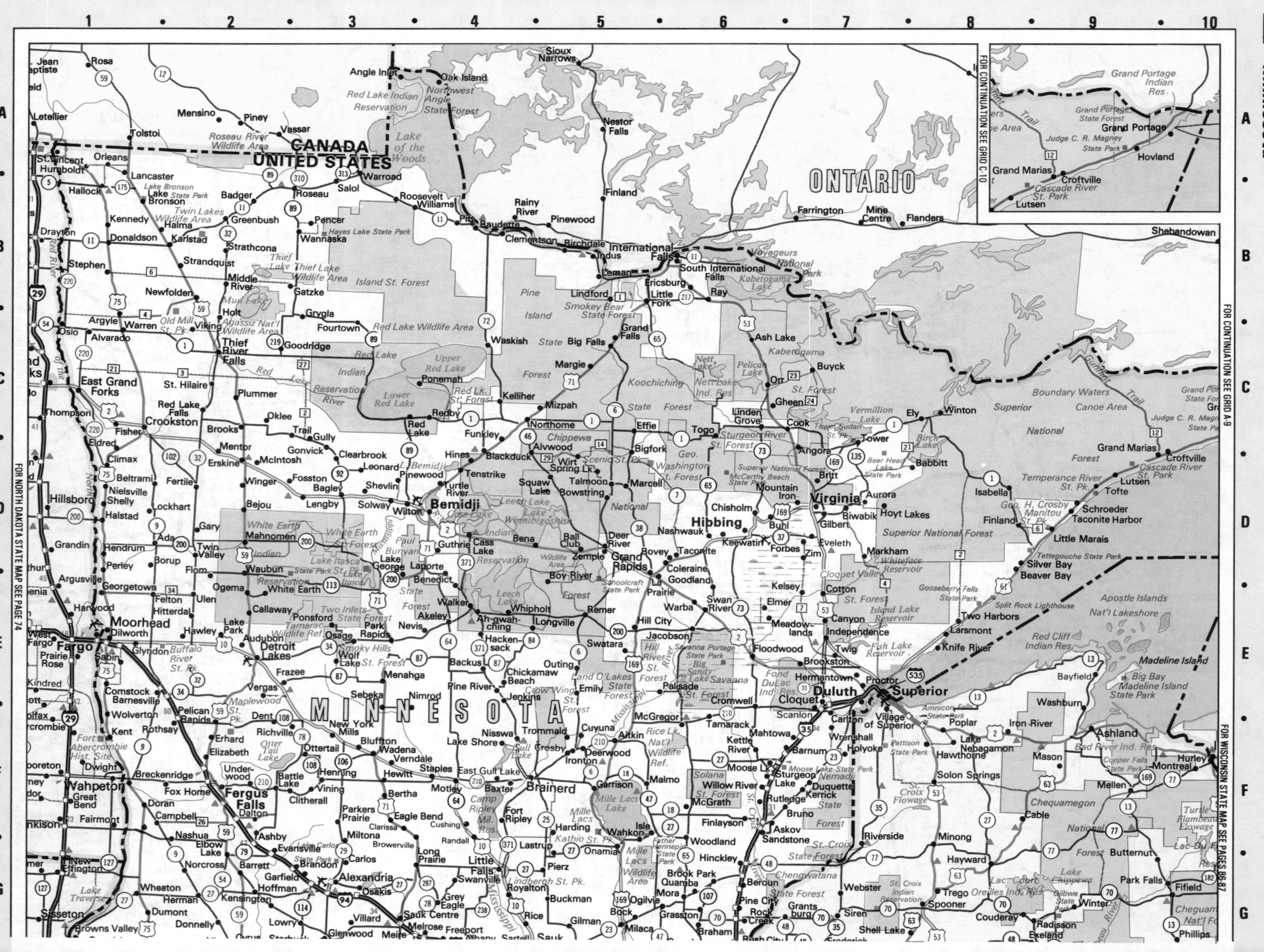
FOR CONTINUATION SEE GRID A-9
FOR WISCONSIN STATE MAP SEE PAGES 86-87
FOR CONTINUATION SEE GRID C-10
FOR NORTH DAKOTA STATE MAP SEE PAGE 74
ONTARIO
CANADA
UNITED STATES
MINNESOTA
Superior National Forest
Boundary Waters Canoe Area
Duluth
Superior
Bemidji
Moorhead
Fargo
Brainerd
Hibbing
Virginia
Grand Rapids
International Falls

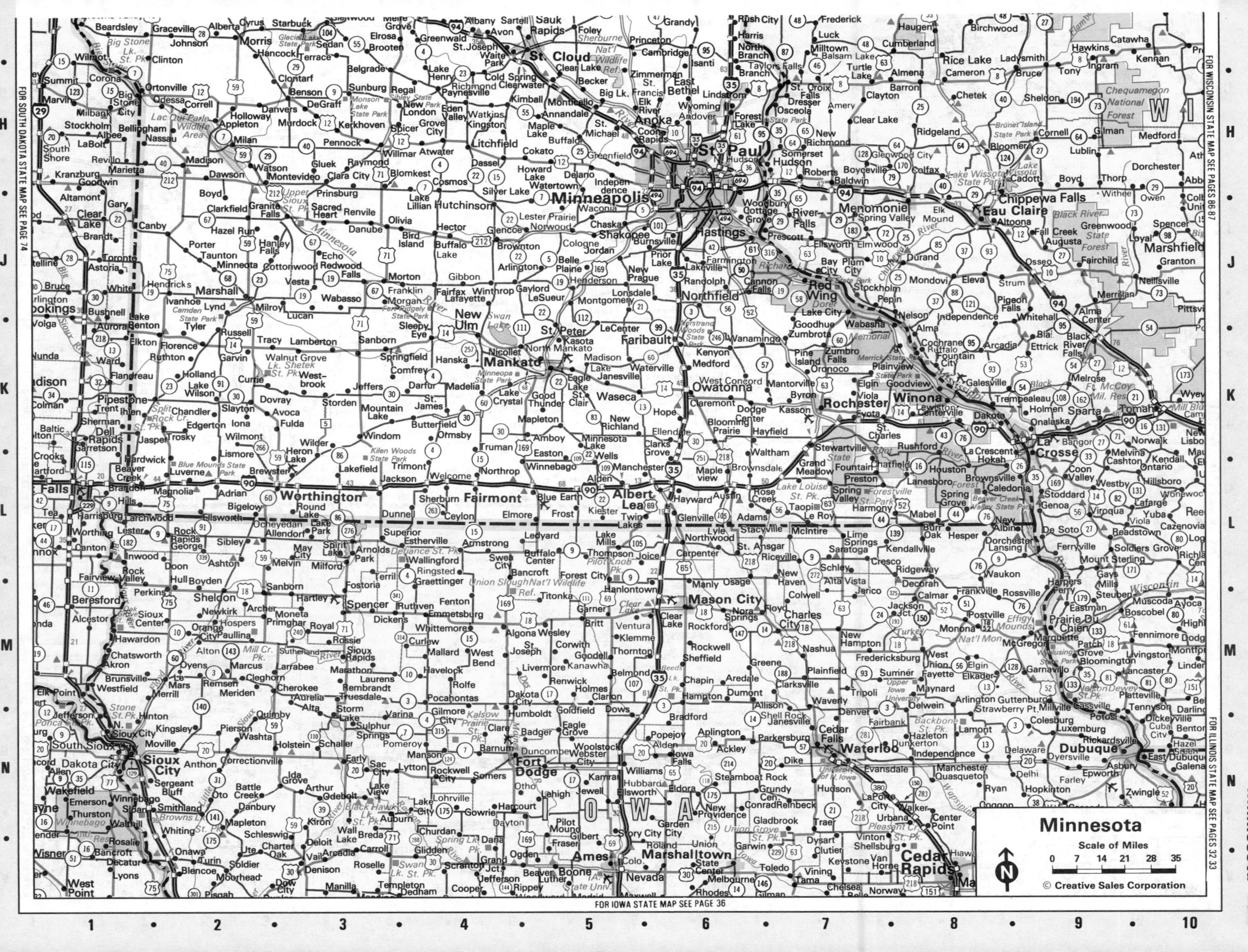
FOR WISCONSIN STATE MAP SEE PAGES 86-87
FOR ILLINOIS STATE MAP SEE PAGES 32 33
FOR IOWA STATE MAP SEE PAGE 36
FOR SOUTH DAKOTA STATE MAP SEE PAGE 74
Minnesota
Scale of Miles
0 7 14 21 28 35
© Creative Sales Corporation
Minneapolis
St. Paul
St. Cloud
Rochester
Mankato
Winona
Red Wing
Hastings
Faribault
Owatonna
Northfield
Albert Lea
Austin
Fairmont
Worthington
New Ulm
Marshall
Hutchinson
Eau Claire
Chippewa Falls
Menomonie
La Crosse
Marshfield
Dubuque
Waterloo
Cedar Rapids
Mason City
Fort Dodge
Sioux City
Marshalltown
Ames
Spencer
Decorah
Prairie Du Chien

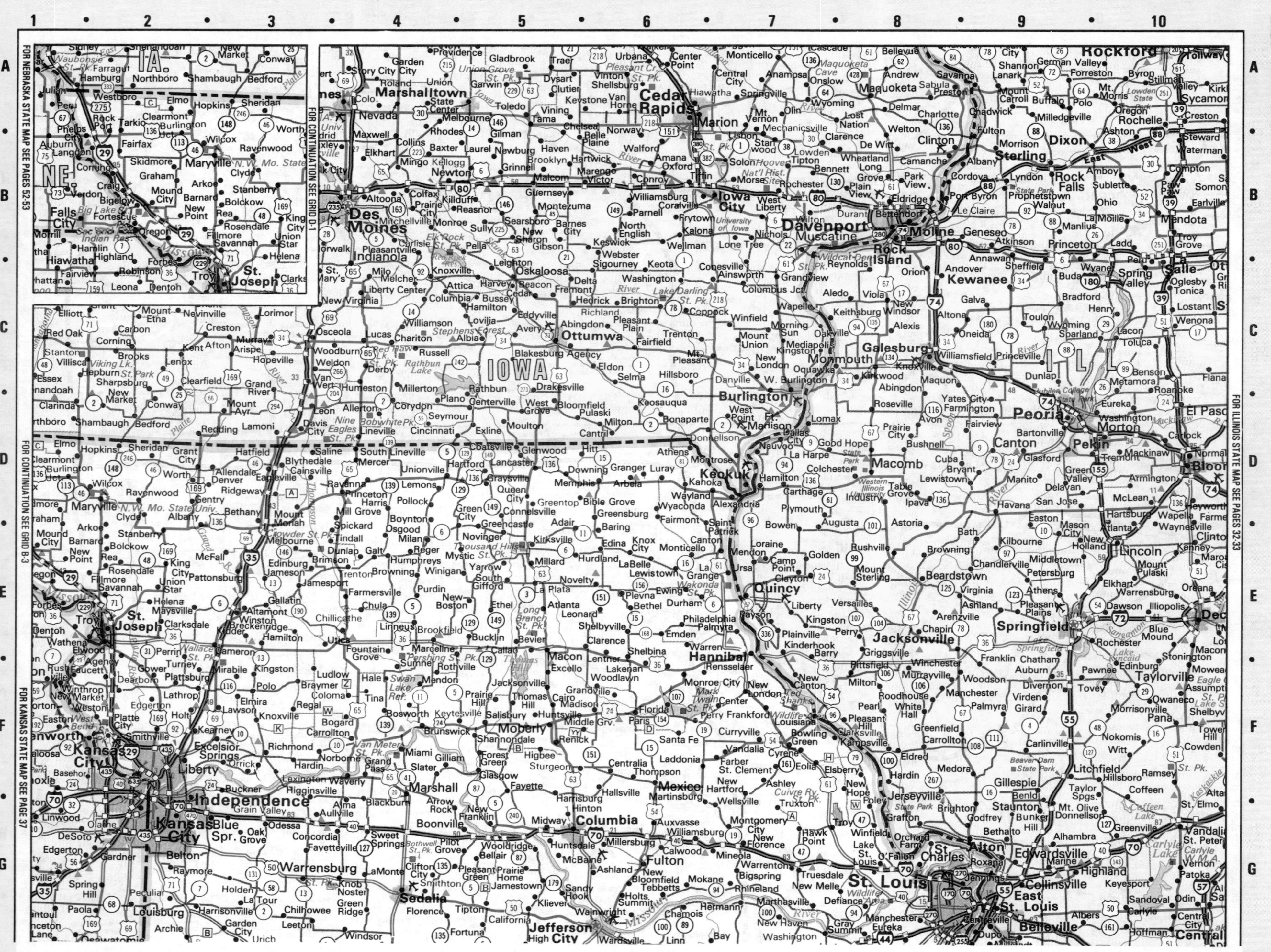
FOR ILLINOIS STATE MAP SEE PAGES 32-33
FOR CONTINUATION SEE GRID D-1
FOR NEBRASKA STATE MAP SEE PAGES 52-53
FOR CONTINUATION SEE GRID B-3
FOR KANSAS STATE MAP SEE PAGE 37
IOWA
Des Moines
Cedar Rapids
Davenport
Rockford
Peoria
Springfield
Jacksonville
Quincy
Hannibal
Kirksville
Columbia
Jefferson City
St. Louis
Kansas City
Independence
St. Joseph

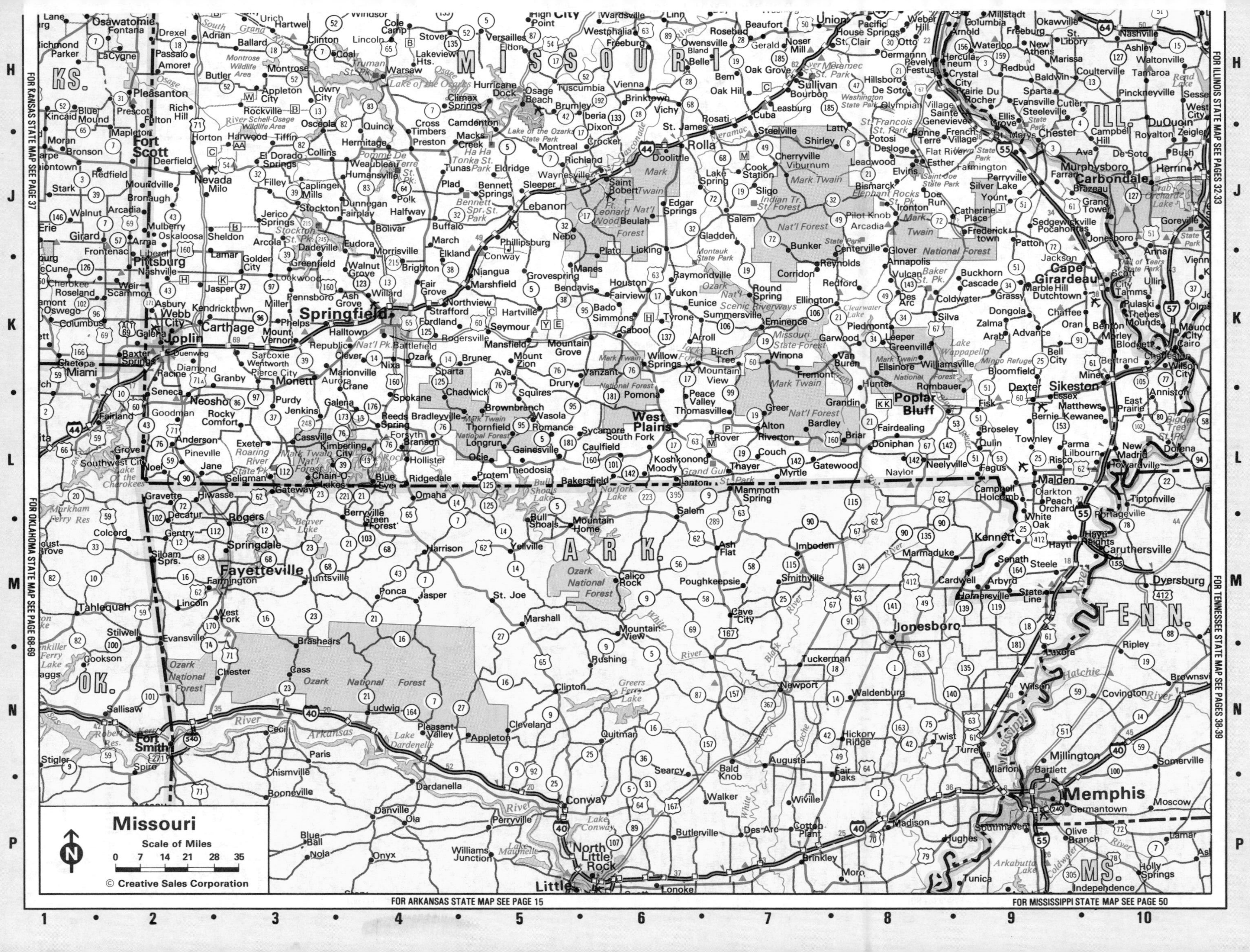
FOR ILLINOIS STATE MAP SEE PAGES 32-33
FOR TENNESSEE STATE MAP SEE PAGES 38-39
FOR MISSISSIPPI STATE MAP SEE PAGE 50
FOR ARKANSAS STATE MAP SEE PAGE 15
FOR OKLAHOMA STATE MAP SEE PAGE 68-69
FOR KANSAS STATE MAP SEE PAGE 37
Missouri
Scale of Miles
0 7 14 21 28 35
© Creative Sales Corporation
MISSOURI
ARK.
TENN.
ILL.
KS.
OK.
MS.
Springfield
Joplin
Rolla
Cape Girardeau
Poplar Bluff
Sikeston
West Plains
Jonesboro
Memphis
Little Rock
North Little Rock
Fayetteville
Fort Smith
Fort Scott
Carbondale
Pittsburg
Mark Twain National Forest
Ozark National Forest

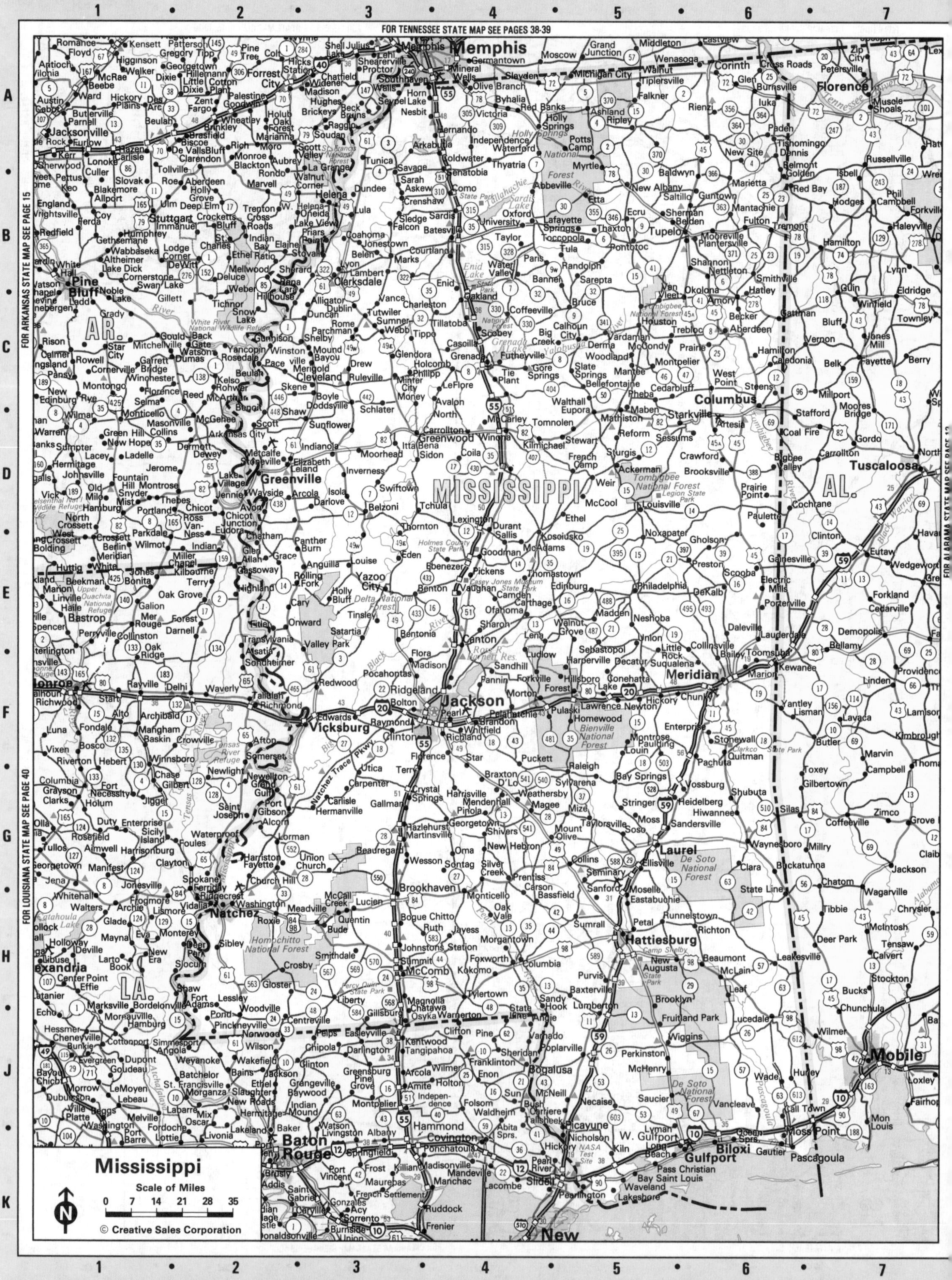

FOR TENNESSEE STATE MAP SEE PAGES 38-39

FOR ARKANSAS STATE MAP SEE PAGE 15

FOR LOUISIANA STATE MAP SEE PAGE 40

FOR ALABAMA STATE MAP SEE PAGE 12

1 2 3 4 5 6 7

A B C D E F G H J K

MISSISSIPPI

AR.

LA.

AL.

Mississippi

Scale of Miles

0 7 14 21 28 35

N

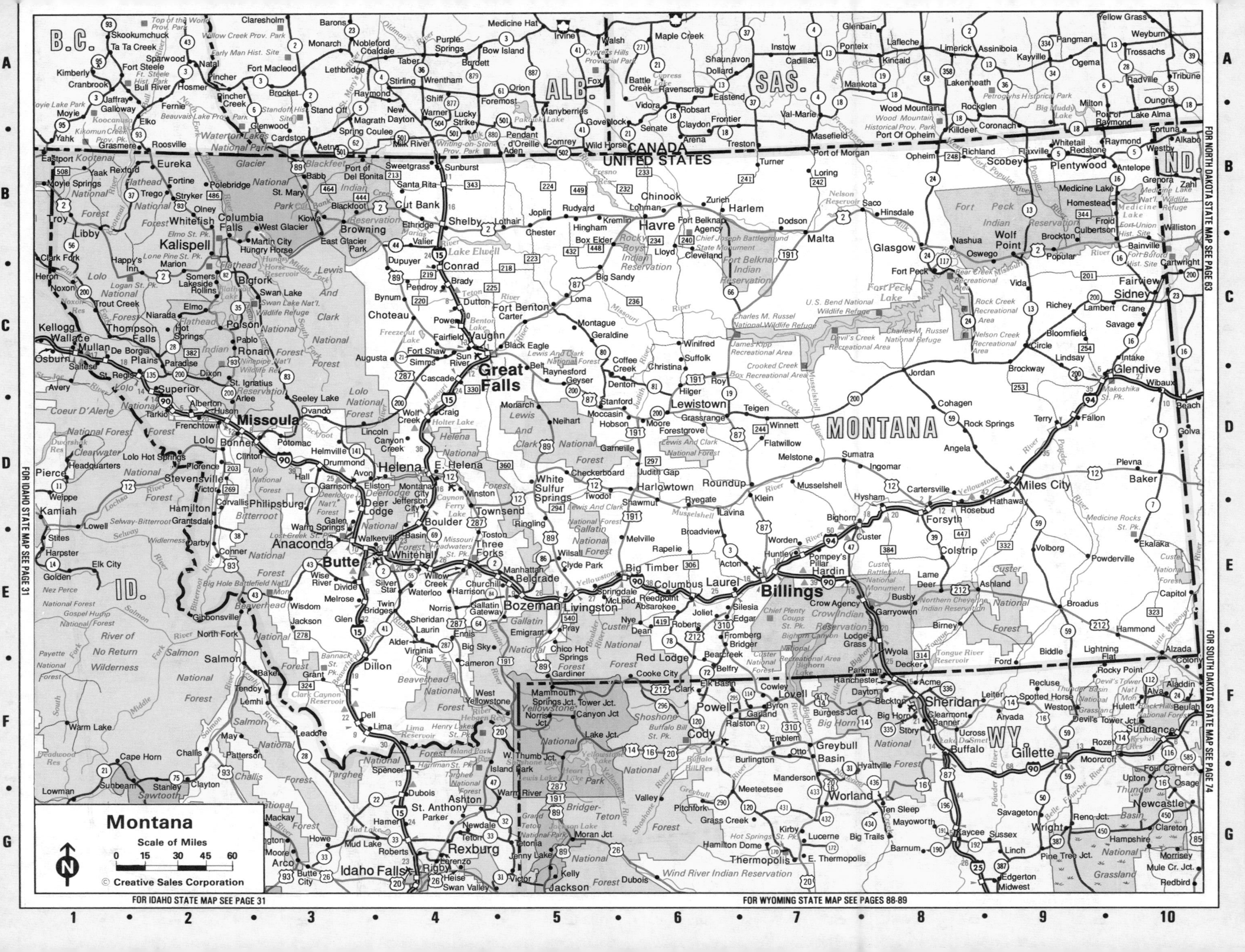
FOR NORTH DAKOTA STATE MAP SEE PAGE 63
FOR SOUTH DAKOTA STATE MAP SEE PAGE 74
FOR WYOMING STATE MAP SEE PAGES 88-89
FOR IDAHO STATE MAP SEE PAGE 31
CANADA
UNITED STATES
MONTANA
B.C.
ALB.
SAS.
ND.
ID.
WY.
Montana
Scale of Miles
0 15 30 45 60
© Creative Sales Corporation
Great Falls
Billings
Helena
Missoula
Butte
Bozeman
Kalispell
Havre
Miles City
Glendive
Lewistown
Glacier National Park
Yellowstone
Sheridan
Idaho Falls

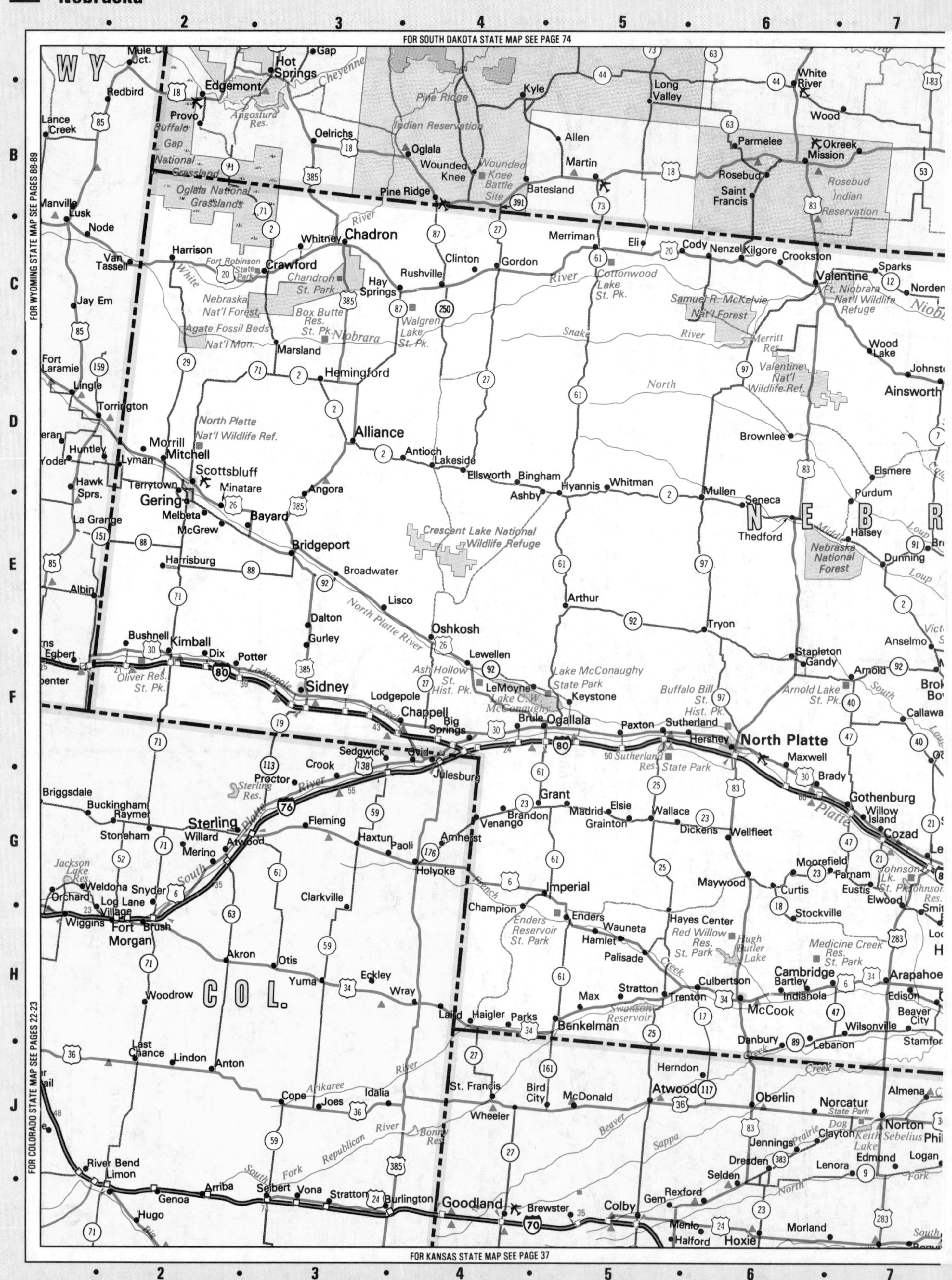
FOR SOUTH DAKOTA STATE MAP SEE PAGE 74
FOR WYOMING STATE MAP SEE PAGES 88-89
FOR COLORADO STATE MAP SEE PAGES 22-23
FOR KANSAS STATE MAP SEE PAGE 37
WY
NEBR
COL.
Mule Cr. Jct.
Hot Springs
Gap
Cheyenne
Edgemont
Redbird
Provo
Angostura Res.
Lance Creek
Buffalo Gap National Grassland
Oglala National Grasslands
Pine Ridge Indian Reservation
Kyle
Long Valley
White River
Wood
Oelrichs
Oglala
Wounded Knee
Wounded Knee Battle Site
Allen
Martin
Parmelee
Okreek
Mission
Rosebud
Saint Francis
Rosebud Indian Reservation
Manville
Lusk
Node
Pine Ridge
Batesland
Chadron
Whitney
Merriman
Eli
Cody
Nenzel
Kilgore
Crookston
Harrison
Van Tassell
Fort Robinson State Park
Crawford
Chadron St. Park
Hay Springs
Rushville
Clinton
Gordon
Cottonwood Lake St. Pk.
Valentine
Sparks
Norden
Ft. Niobrara Nat'l Wildlife Refuge
Jay Em
Nebraska Nat'l Forest
Box Butte Res. St. Pk.
Walgren Lake St. Pk.
Samuel R. McKelvie Nat'l Forest
Niobrara
Agate Fossil Beds Nat'l Mon.
Marsland
Snake
River
Merritt Res.
Wood Lake
Fort Laramie
Lingle
Hemingford
Valentine Nat'l Wildlife Ref.
Johnstown
Ainsworth
North
Torrington
North Platte Nat'l Wildlife Ref.
Alliance
Brownlee
Huntley
Yoder
Morrill
Mitchell
Lyman
Antioch
Lakeside
Scottsbluff
Terrytown
Gering
Minatare
Ellsworth
Bingham
Hyannis
Whitman
Ashby
Mullen
Seneca
Elsmere
Purdum
Hawk Sprs.
Angora
Melbeta
Bayard
McGrew
La Grange
Crescent Lake National Wildlife Refuge
Thedford
Halsey
Nebraska National Forest
Dunning
Bridgeport
Harrisburg
Broadwater
Albin
Lisco
North Platte River
Arthur
Tryon
Dalton
Oshkosh
Gurley
Bushnell
Kimball
Egbert
Dix
Potter
Oliver Res. St. Pk.
Lodgepole
Lewellen
Ash Hollow St. Hist. Pk.
Lake McConaughy State Park
Stapleton
Gandy
Anselmo
Arnold
Sidney
Lodgepole
LeMoyne
Lake C.W. McConaughy
Keystone
Buffalo Bill St. Hist. Pk.
Arnold Lake St. Pk.
Callaway
Chappell
Big Springs
Brule
Ogallala
Paxton
Sutherland
Hershey
North Platte
Maxwell
Sedgwick
Ovid
Sutherland Res. State Park
Brady
Crook
Proctor
Julesburg
Sterling Res.
Grant
Gothenburg
Briggsdale
Buckingham
Raymer
Sterling
Willard
Fleming
Brandon
Venango
Madrid
Elsie
Grainton
Wallace
Dickens
Wellfleet
Willow Island
Cozad
Stoneham
Merino
Atwood
Haxtun
Paoli
Amherst
Jackson Lake Res.
Holyoke
Moorefield
Farnam
Johnson Lk. St. Pk.
Weldona
Snyder
Orchard
Log Lane Village
Clarkville
Imperial
Maywood
Curtis
Eustis
Elwood
Champion
Enders
Enders Reservoir St. Park
Wauneta
Stockville
Wiggins
Fort Morgan
Brush
Hamlet
Hayes Center
Red Willow Res. St. Park
Hugh Butler Lake
Medicine Creek Res. St. Park
Palisade
Akron
Otis
Cambridge
Bartley
Arapahoe
Yuma
Eckley
Culbertson
Indianola
Edison
Woodrow
Wray
Stratton
Trenton
McCook
Beaver City
Max
Swanson Reservoir
Laird
Haigler
Parks
Benkelman
Wilsonville
Danbury
Lebanon
Stamford
Last Chance
Lindon
Anton
Herndon
St. Francis
Bird City
McDonald
Atwood
Oberlin
Norcatur State Park
Almena
Arikaree
Cope
Joes
Idalia
Wheeler
Beaver
Sappa
Norton
Keith Sebelius Lake
Clayton
Jennings
Prairie Dog
Republican
Bonny Res.
South Fork
River Bend
Limon
Dresden
Lenora
Edmond
Logan
Selden
Genoa
Arriba
Seibert
Vona
Stratton
Burlington
Goodland
Brewster
Colby
Gem
Rexford
Hugo
Menlo
Halford
Hoxie
Morland
Beaver
Elsie

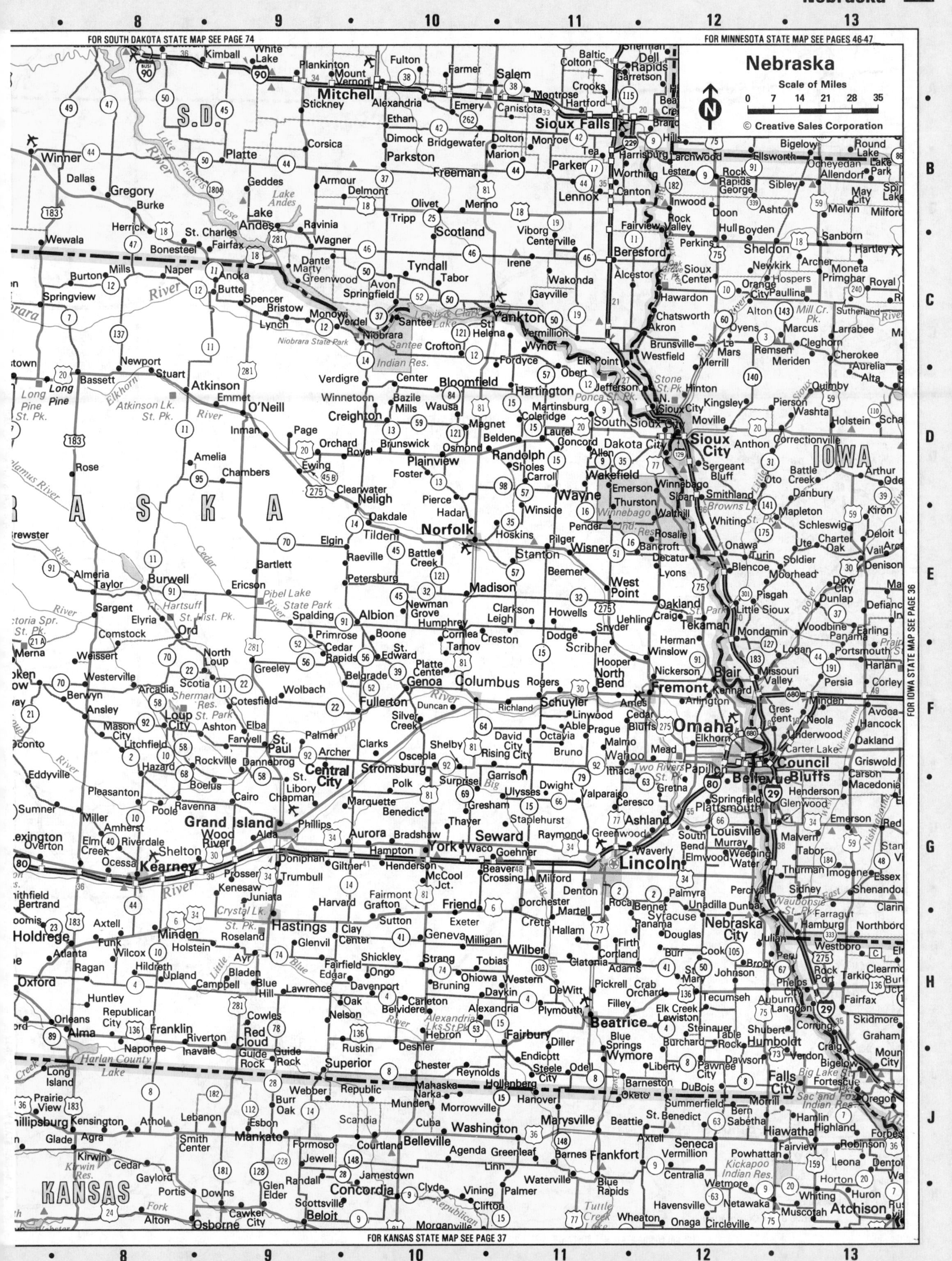
FOR SOUTH DAKOTA STATE MAP SEE PAGE 74
FOR MINNESOTA STATE MAP SEE PAGES 46-47
Nebraska
Scale of Miles
0 7 14 21 28 35
© Creative Sales Corporation
FOR IOWA STATE MAP SEE PAGE 36
FOR KANSAS STATE MAP SEE PAGE 37
S.D.
IOWA
KANSAS
Sioux Falls
Mitchell
Yankton
Norfolk
Sioux City
Columbus
Fremont
Omaha
Council Bluffs
Lincoln
Grand Island
Kearney
Hastings
Beatrice
Nebraska City
Falls City
Atchison
Concordia
Washington
Marysville
Belleville
Seward
York
Wayne
O'Neill
Holdrege
Alma
Superior
Fairbury
Wahoo
Blair
Tekamah
West Point
Madison
Albion
Central City
Aurora
Red Cloud
Minden

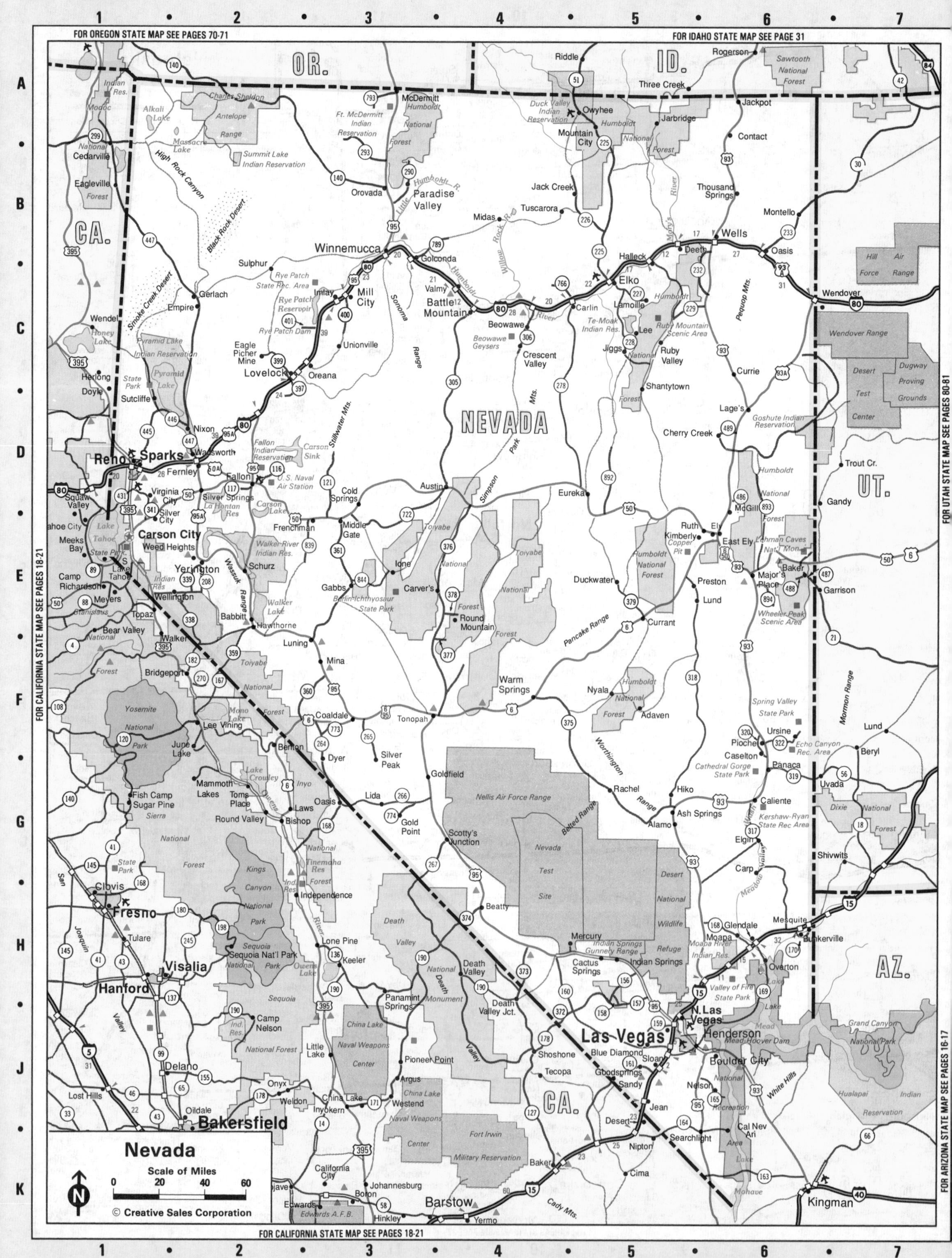

FOR OREGON STATE MAP SEE PAGES 70-71
FOR IDAHO STATE MAP SEE PAGE 31
FOR UTAH STATE MAP SEE PAGES 80-81
FOR ARIZONA STATE MAP SEE PAGES 16-17
FOR CALIFORNIA STATE MAP SEE PAGES 18-21
FOR CALIFORNIA STATE MAP SEE PAGES 18-21
OR.
ID.
CA.
UT.
AZ.
CA.
NEVADA
Riddle
Rogerson
Sawtooth National Forest
Three Creek
Jackpot
McDermitt
Humboldt National Forest
Ft. McDermitt Indian Reservation
Duck Valley Indian Reservation
Owyhee
Mountain City
Jarbridge
Contact
Charles Sheldon Antelope Range
Alkali Lake
Massacre Lake
Summit Lake Indian Reservation
Cedarville
Modoc National Forest
Eagleville
High Rock Canyon
Black Rock Desert
Orovada
Paradise Valley
Little Humboldt R.
Jack Creek
Tuscarora
Thousand Springs
Montello
Midas
Wells
Oasis
Winnemucca
Golconda
Sulphur
Rye Patch State Rec. Area
Rye Patch Reservoir
Rye Patch Dam
Imlay
Mill City
Valmy
Battle Mountain
Halleck
Deeth
Elko
Lamoille
Carlin
Beowawe
Beowawe Geysers
Crescent Valley
Wendover
Hill Air Force Range
Gerlach
Empire
Smoke Creek Desert
Wendel
Honey Lake
Pyramid Lake Indian Reservation
Pyramid Lake
Herlong
Doyle
Sutcliffe
State Park
Eagle Picher Mine
Unionville
Lovelock
Oreana
Sonoma Range
Te-Moak Indian Res.
Lee
Ruby Mountain Scenic Area
Ruby Valley
Jiggs
Currie
Wendover Range
Desert Test Center
Dugway Proving Grounds
Shantytown
Lage's
Goshute Indian Reservation
Cherry Creek
Pequop Mts.
Stillwater Mts.
Nixon
Wadsworth
Reno
Sparks
Fernley
Fallon
Fallon Indian Reservation
Carson Sink
U.S. Naval Air Station
Cold Springs
Austin
Simpson Park Mts.
Eureka
Trout Cr.
Gandy
Ruth
Kimberly
Copper Pit
Ely
McGill
East Ely
Lehman Caves Nat'l Mon.
Baker
Major's Place
Garrison
Wheeler Peak Scenic Area
Squaw Valley
Virginia City
Silver Springs
Silver City
Lahontan Res.
Carson Lake
Frenchman
Middle Gate
Toiyabe National Forest
Carson City
Lake Tahoe
Weed Heights
Walker River Indian Res.
Schurz
Yerington
Wassuk Range
Meeks Bay
S. Lake Tahoe
State Park
Camp Richardson
Meyers
Stanislaus
Wellington
Ione
Gabbs
Carver's
Berlin-Ichthyosaur State Park
Round Mountain
Duckwater
Preston
Lund
Humboldt National Forest
Walker Lake
Babbitt
Hawthorne
Currant
Pancake Range
Topaz
Bear Valley
Walker
Luning
Mina
Bridgeport
Warm Springs
Nyala
Adaven
Spring Valley State Park
Mormon Range
Yosemite National Park
Mono Lake
Coaldale
Tonopah
Lee Vining
June Lake
Benton
Dyer
Silver Peak
Worthington Range
Ursine
Pioche
Echo Canyon Rec. Area
Caselton
Lund
Beryl
Cathedral Gorge State Park
Panaca
Uvada
Goldfield
Lake Crowley
Inyo
Mammoth Lakes
Toms Place
Fish Camp
Sugar Pine
Sierra
Round Valley
Laws
Oasis
Lida
Gold Point
Nellis Air Force Range
Rachel
Hiko
Ash Springs
Alamo
Caliente
Kershaw-Ryan State Rec Area
Dixie National Forest
Bishop
Scotty's Junction
Belted Range
Nevada Test Site
Elgin
Shivwits
Carp
Meadow Valley
Clovis
Fresno
Independence
Beatty
Desert National Wildlife Refuge
Death Valley
Tulare
Sequoia Nat'l Park
Kings Canyon National Park
Visalia
Hanford
Lone Pine
Keeler
Owens Lake
Death Valley National Monument
Mercury
Indian Springs Gunnery Range
Indian Springs
Cactus Springs
Glendale
Moapa
Moapa River Indian Res.
Mesquite
Bunkerville
Overton
Valley of Fire State Park
Panamint Springs
Death Valley Jct.
Camp Nelson
Sequoia National Forest
China Lake Naval Weapons Center
N. Las Vegas
Las Vegas
Henderson
Mead-Hoover Dam
Lake Mead
Grand Canyon National Park
Boulder City
Blue Diamond
Little Lake
Pioneer Point
Shoshone
Tecopa
Goodsprings
Sloan
Sandy
Nelson
Hualapai Indian Reservation
White Hills
Delano
Lost Hills
Argus
Onyx
Weldon
Westend
Inyokern
Jean
Desert
Nipton
Searchlight
Cal Nev Ari
Lake Mohave
Oildale
Bakersfield
Fort Irwin Military Reservation
Baker
Cima
Kingman
California City
Johannesburg
Boron
Edwards
Edwards A.F.B.
Hinkley
Barstow
Yermo
Cady Mts.
Nevada
Scale of Miles
0 20 40 60
© Creative Sales Corporation

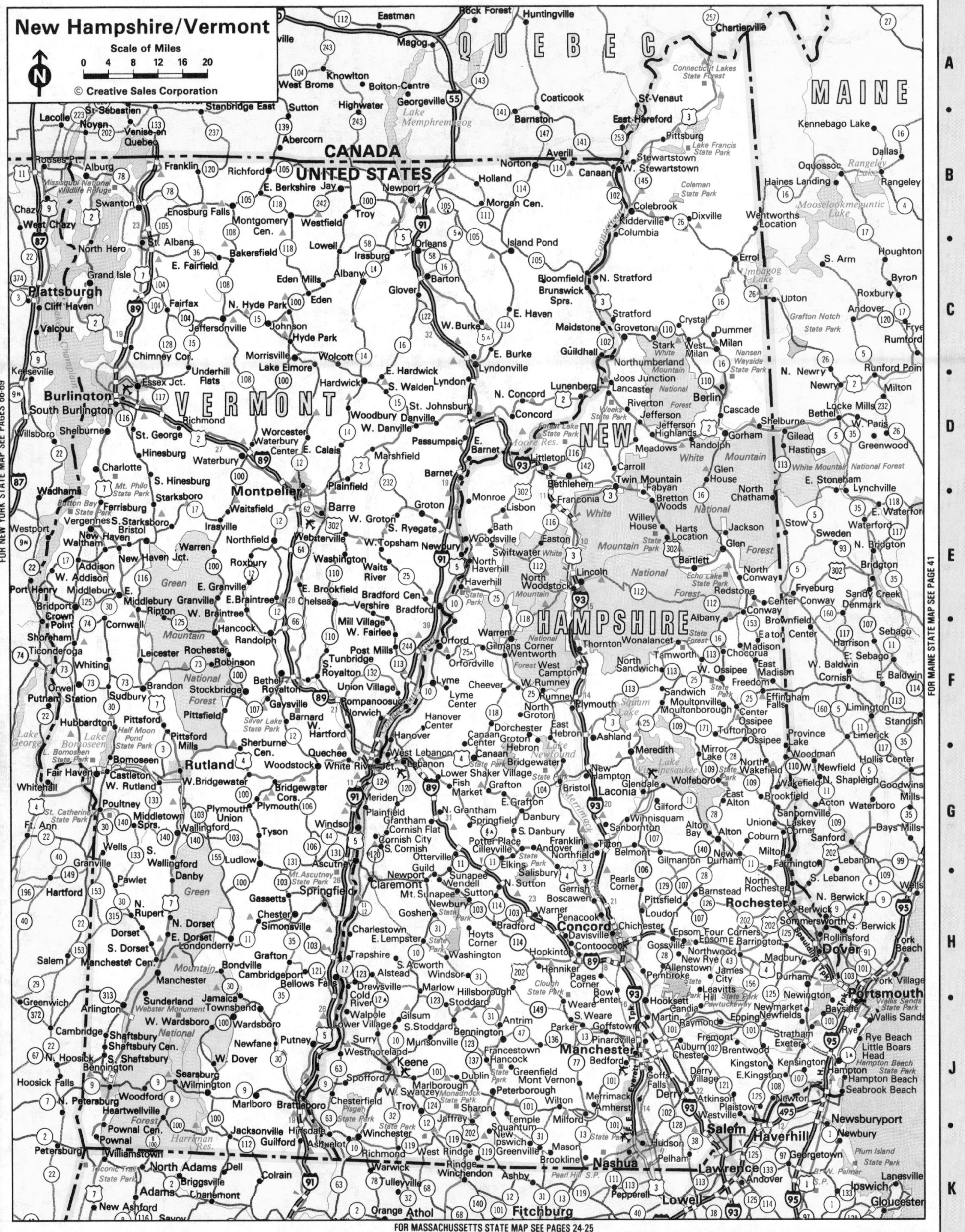

FOR NEW YORK STATE MAP SEE PAGES 66-69

FOR MAINE STATE MAP SEE PAGE 41

FOR MASSACHUSSETTS STATE MAP SEE PAGES 24-25

FOR NEW YORK STATE MAP SEE PAGES 58-61

FOR PENNSYLVANIA STATE MAP SEE PAGES 72-73

FOR PENNSYLVANIA STATE MAP SEE PAGES 72-73

NEW YORK

PENNSYLVANIA

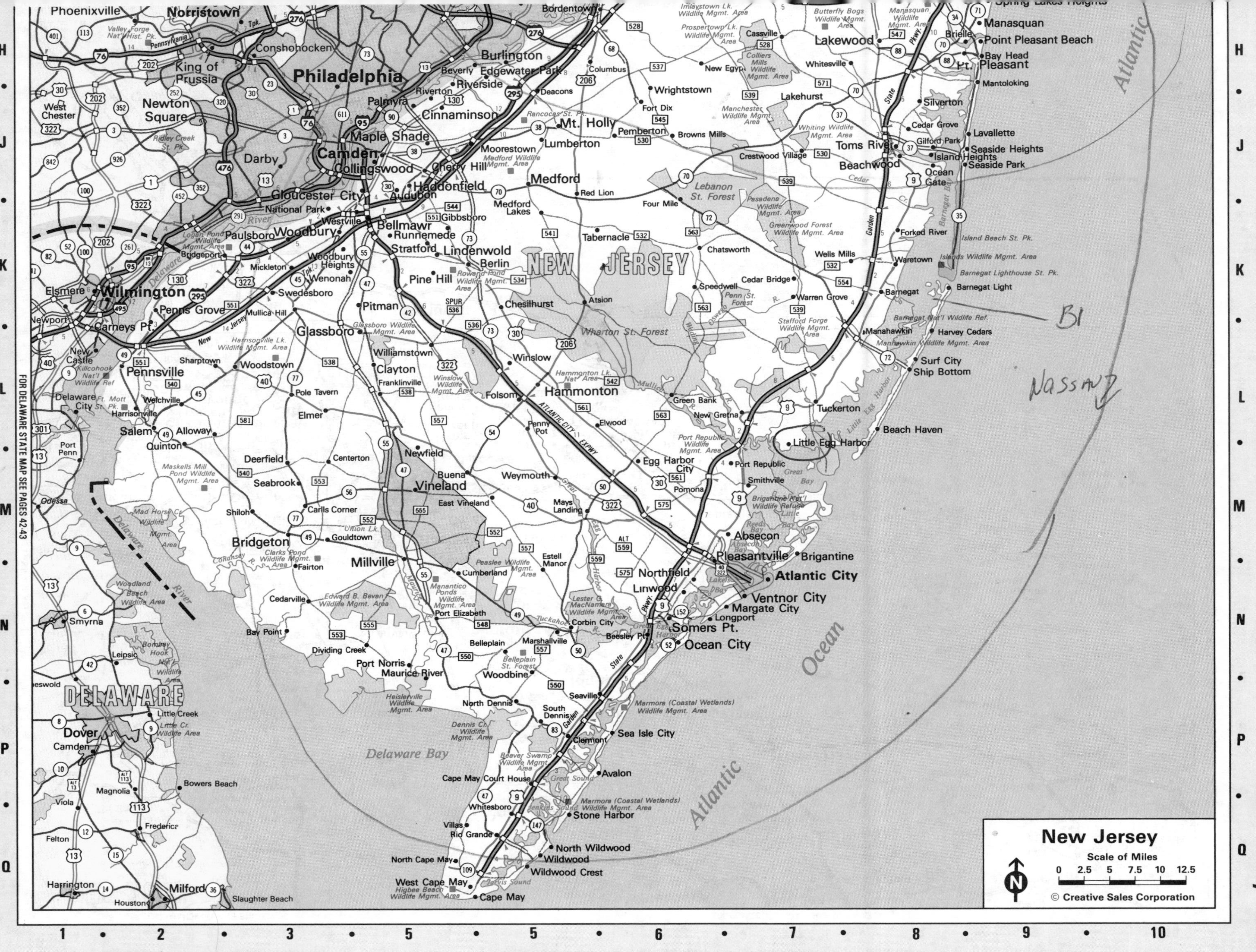

FOR DELAWARE STATE MAP SEE PAGES 42-43

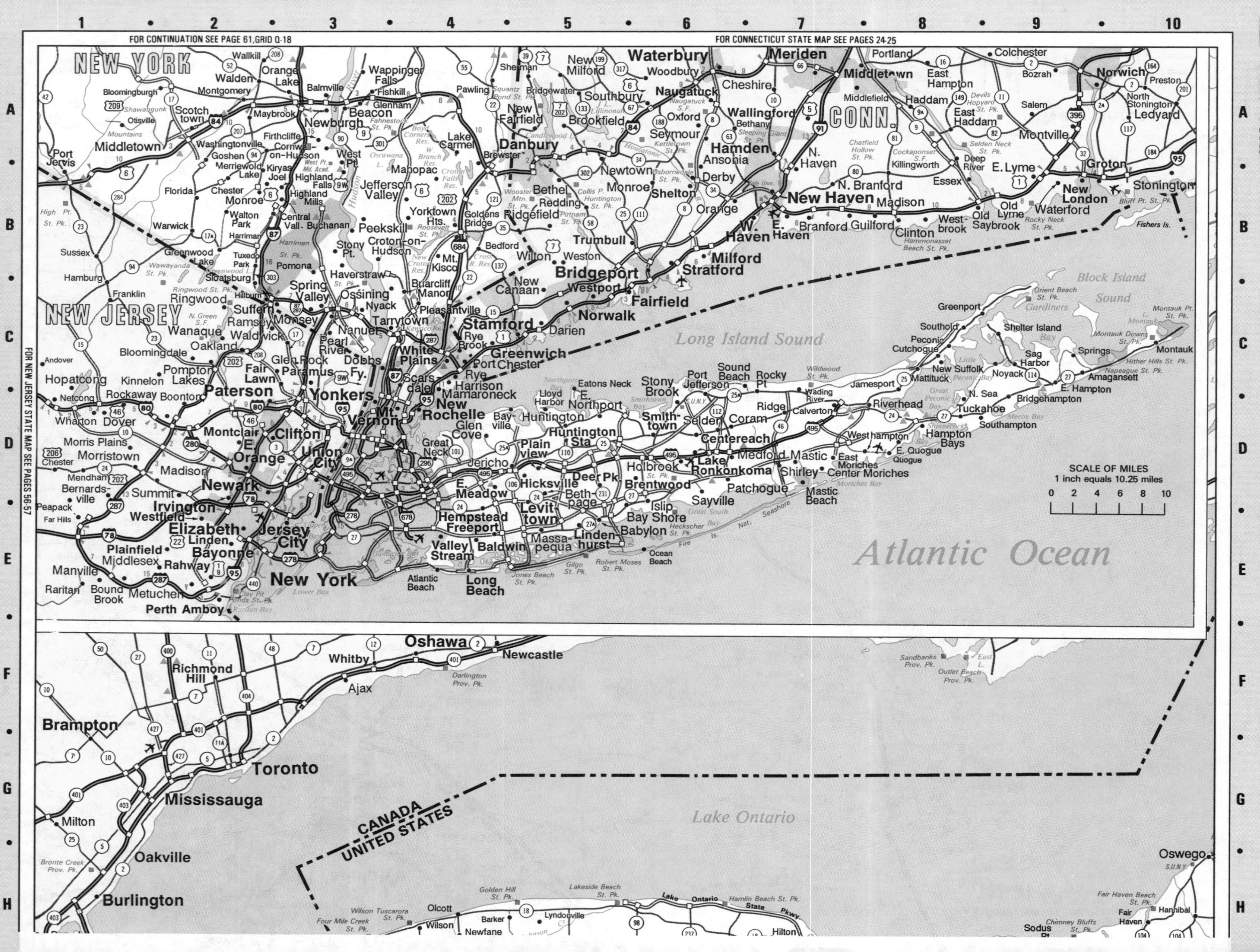

FOR CONTINUATION SEE PAGE 61, GRID 0-18
FOR CONNECTICUT STATE MAP SEE PAGES 24-25
FOR NEW JERSEY STATE MAP SEE PAGES 56-57
NEW YORK
NEW JERSEY
CONN.
Long Island Sound
Block Island Sound
Atlantic Ocean
Lake Ontario
CANADA
UNITED STATES
SCALE OF MILES
1 inch equals 10.25 miles
New York
Jersey City
Newark
Elizabeth
Paterson
Yonkers
New Rochelle
White Plains
Stamford
Norwalk
Bridgeport
New Haven
Waterbury
Meriden
Middletown
New London
Groton
Norwich
Danbury
Newburgh
Middletown
Port Jervis
Peekskill
Ossining
Greenwich
Stratford
Milford
Fairfield
Hempstead
Freeport
Babylon
Huntington
Riverhead
Southampton
Montauk
Toronto
Mississauga
Brampton
Oakville
Burlington
Oshawa
Whitby
Ajax
Oswego

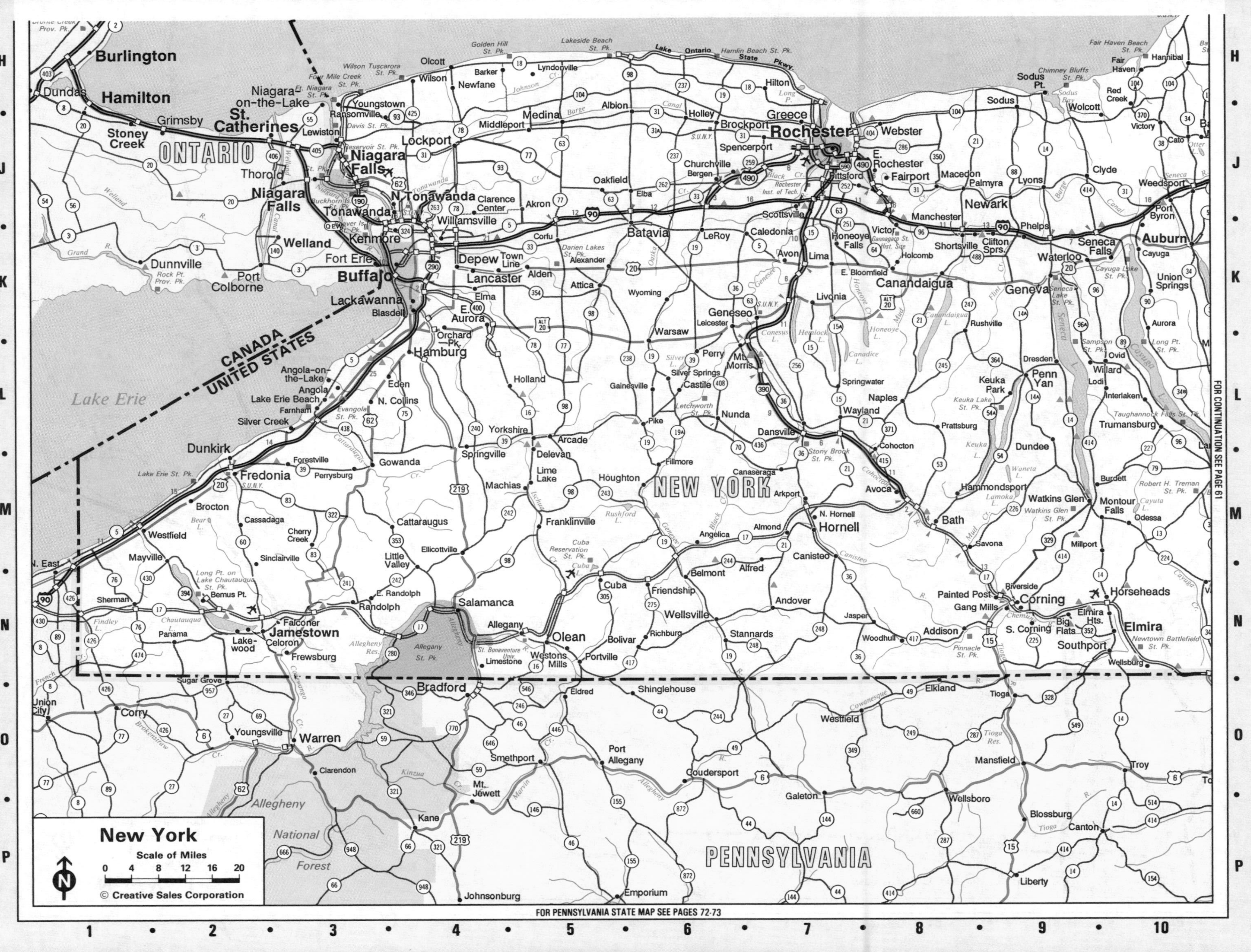
FOR CONTINUATION SEE PAGE 61
FOR PENNSYLVANIA STATE MAP SEE PAGES 72-73
New York
Scale of Miles
0 4 8 12 16 20
© Creative Sales Corporation
NEW YORK
PENNSYLVANIA
ONTARIO
CANADA
UNITED STATES
Lake Erie
Lake Ontario
Rochester
Buffalo
Niagara Falls
Lockport
Batavia
Geneseo
Canandaigua
Geneva
Auburn
Seneca Falls
Penn Yan
Corning
Elmira
Horseheads
Hornell
Bath
Olean
Salamanca
Jamestown
Dunkirk
Fredonia
Westfield
St. Catherines
Hamilton
Burlington
Welland
Bradford
Warren
Allegheny
National
Forest
Wellsboro
Coudersport
Emporium
Mansfield

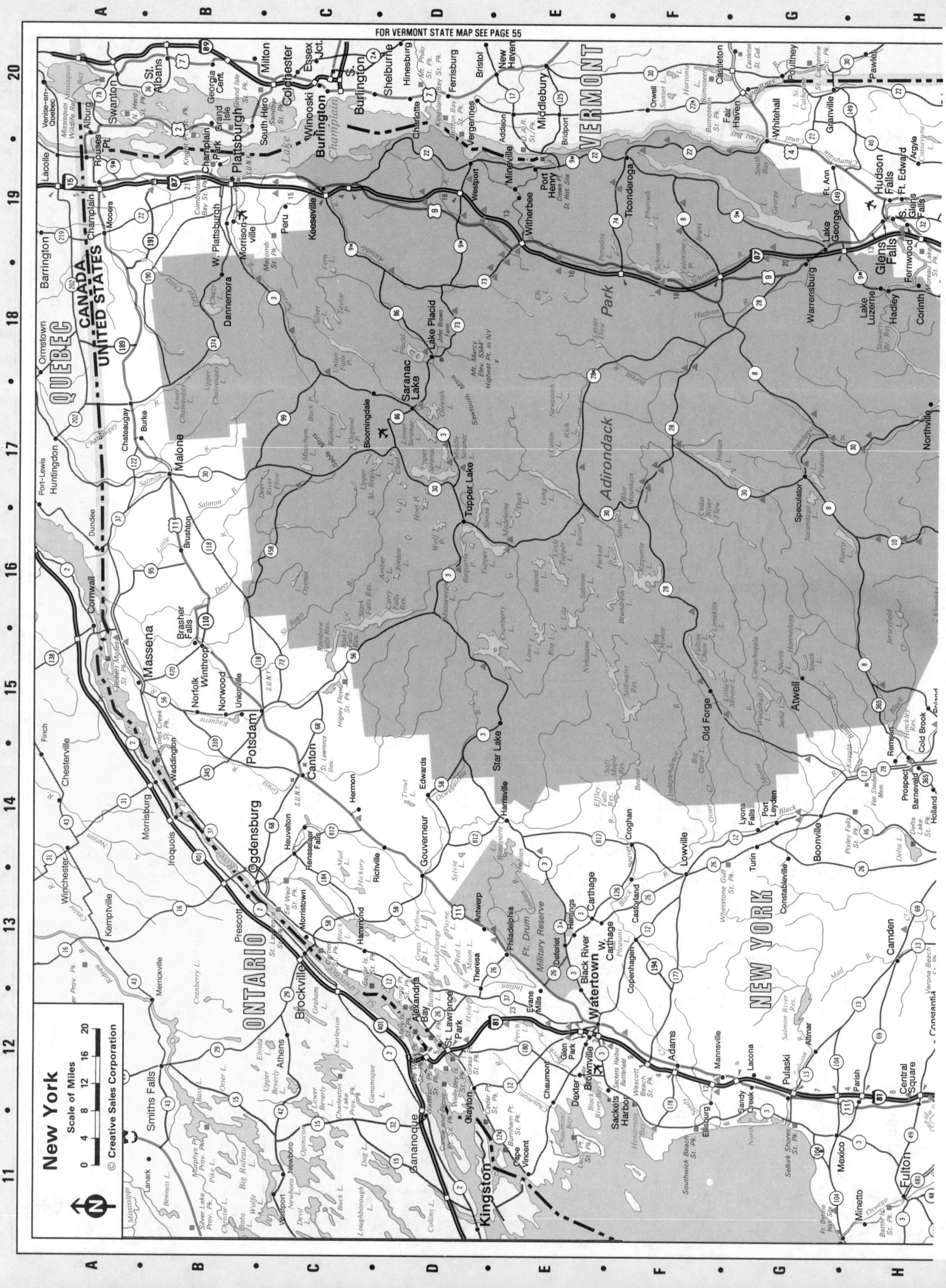

FOR VERMONT STATE MAP SEE PAGE 55

New York

Scale of Miles

0 4 8 12 16 20

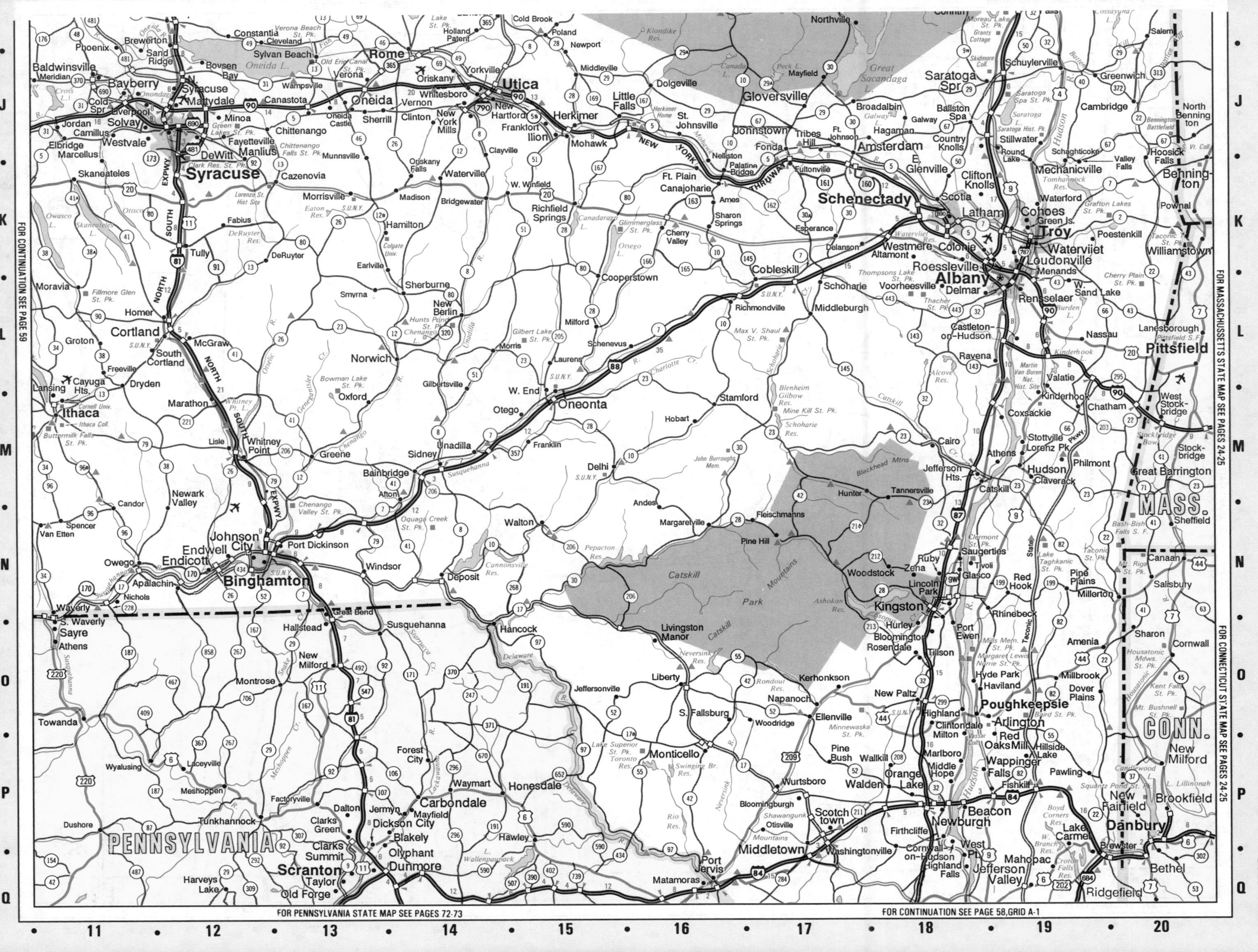
FOR MASSACHUSSETTS STATE MAP SEE PAGES 24-25
FOR CONNECTICUT STATE MAP SEE PAGES 24-25
FOR CONTINUATION SEE PAGE 58,GRID A-1
FOR PENNSYLVANIA STATE MAP SEE PAGES 72-73
FOR CONTINUATION SEE PAGE 59
MASS.
CONN.
PENNSYLVANIA
NEW YORK
Syracuse
Utica
Rome
Schenectady
Albany
Troy
Binghamton
Pittsfield
Poughkeepsie
Kingston
Scranton
Danbury
Middletown
Newburgh
Oneonta
Ithaca
Cortland
Amsterdam
Saratoga Spr
Gloversville
Catskill Park

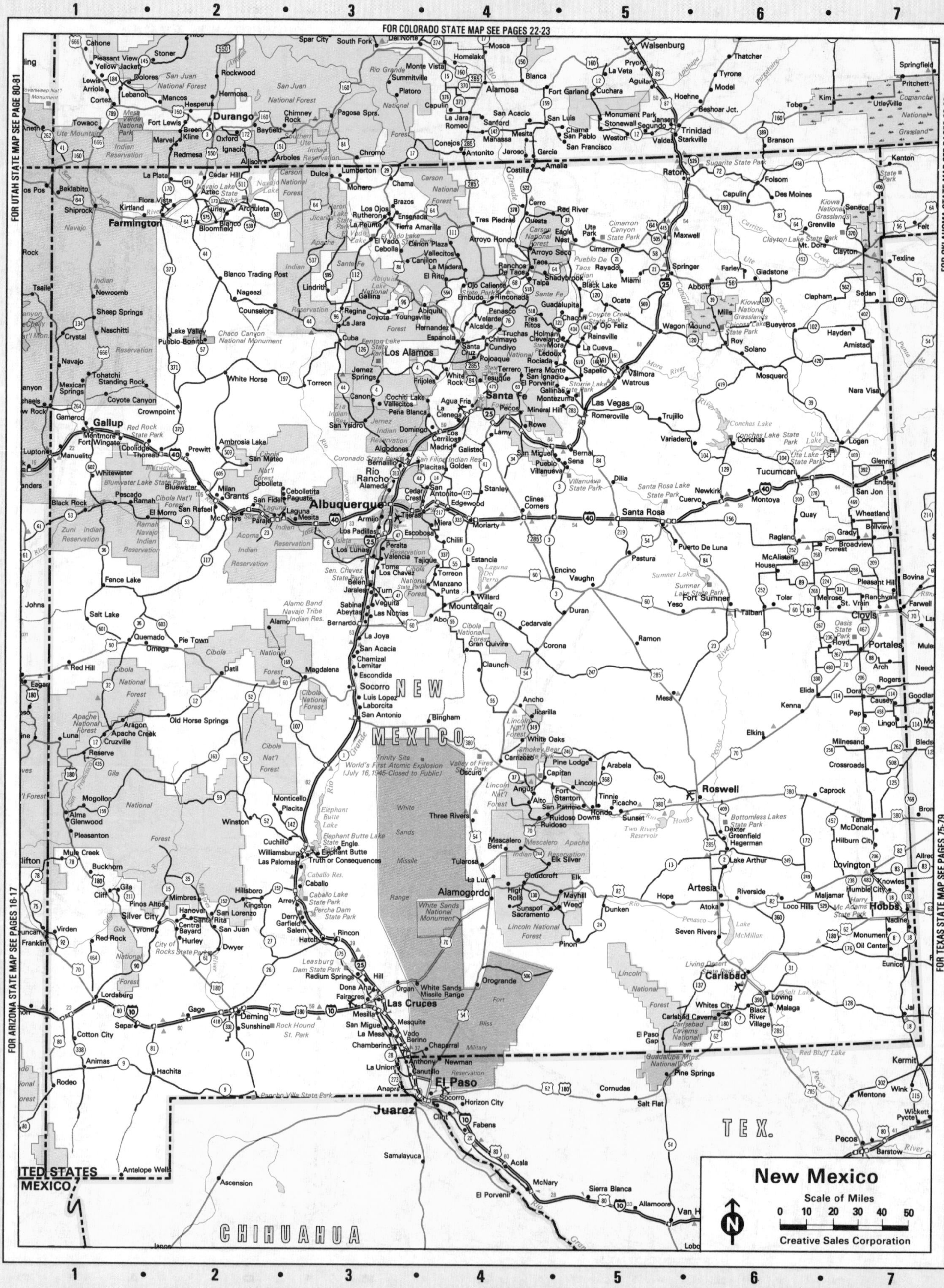
FOR COLORADO STATE MAP SEE PAGES 22-23
FOR UTAH STATE MAP SEE PAGE 80-81
FOR OKLAHOMA STATE MAP SEE PAGES 68-69
FOR TEXAS STATE MAP SEE PAGES 75-79
FOR ARIZONA STATE MAP SEE PAGES 16-17
NEW MEXICO
TEX.
CHIHUAHUA
UNITED STATES
MEXICO
New Mexico
Scale of Miles
0 10 20 30 40 50
Creative Sales Corporation
Santa Fe
Albuquerque
Los Alamos
Farmington
Gallup
Grants
Taos
Raton
Las Vegas
Santa Rosa
Tucumcari
Clovis
Portales
Roswell
Artesia
Carlsbad
Hobbs
Alamogordo
Las Cruces
El Paso
Juarez
Deming
Lordsburg
Silver City
Socorro
Truth or Consequences
Ruidoso
Durango
Trinidad
Alamosa
Trinity Site
World's First Atomic Explosion
(July 16, 1945-Closed to Public)
White Sands Missile Range
White Sands National Monument
Carlsbad Caverns National Park
Guadalupe Mtns. National Park
Chaco Canyon National Monument
Elephant Butte Lake State Park

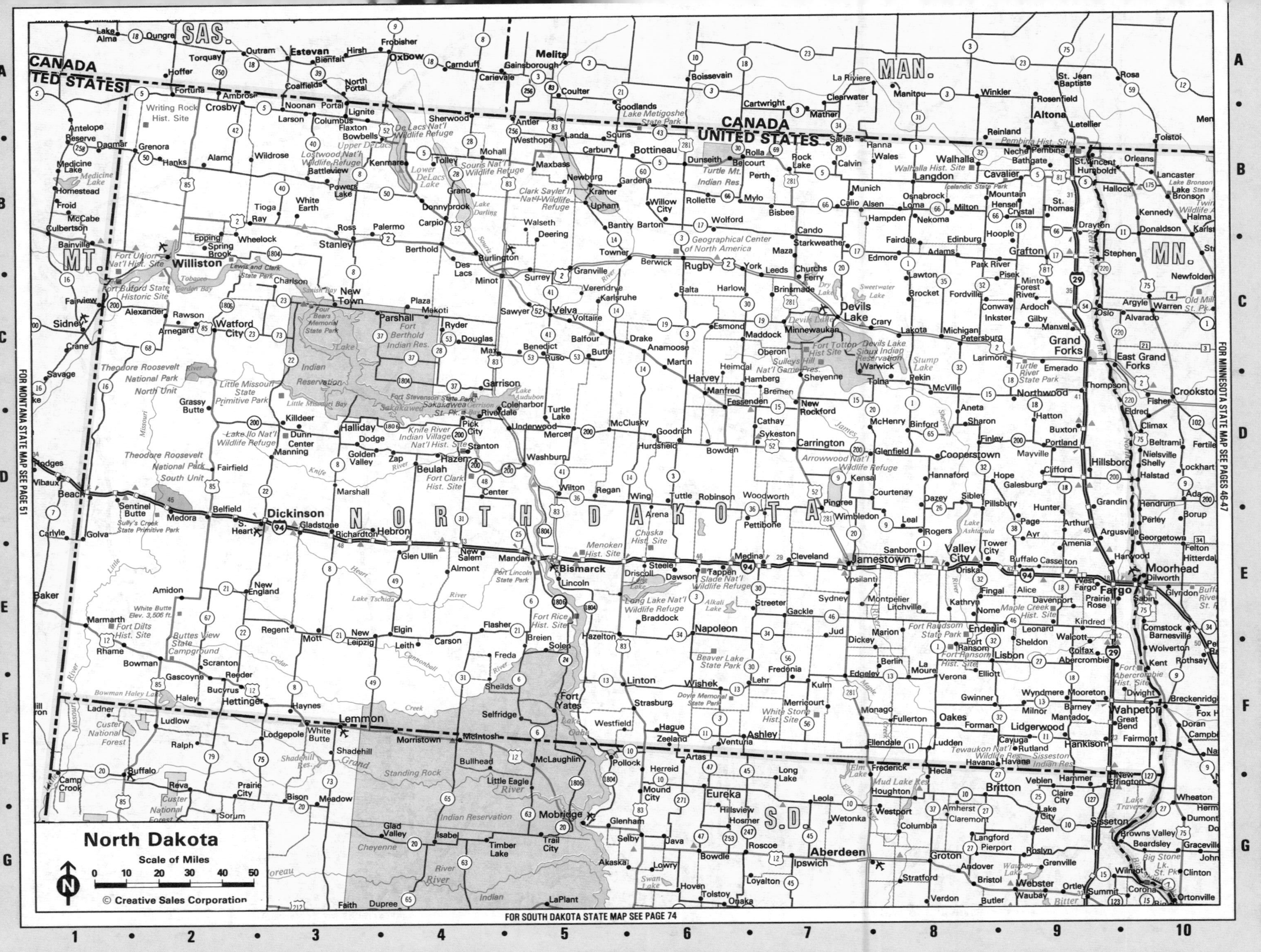
FOR MINNESOTA STATE MAP SEE PAGES 46-47
FOR SOUTH DAKOTA STATE MAP SEE PAGE 74
FOR MONTANA STATE MAP SEE PAGE 51
North Dakota
Scale of Miles
0 10 20 30 40 50
© Creative Sales Corporation
CANADA
UNITED STATES
SAS.
MAN.
MN.
MT.
S.D.
NORTH DAKOTA
Bismarck
Fargo
Grand Forks
Minot
Williston
Dickinson
Jamestown
Devils Lake
Valley City
Wahpeton
Mandan
Moorhead
Aberdeen

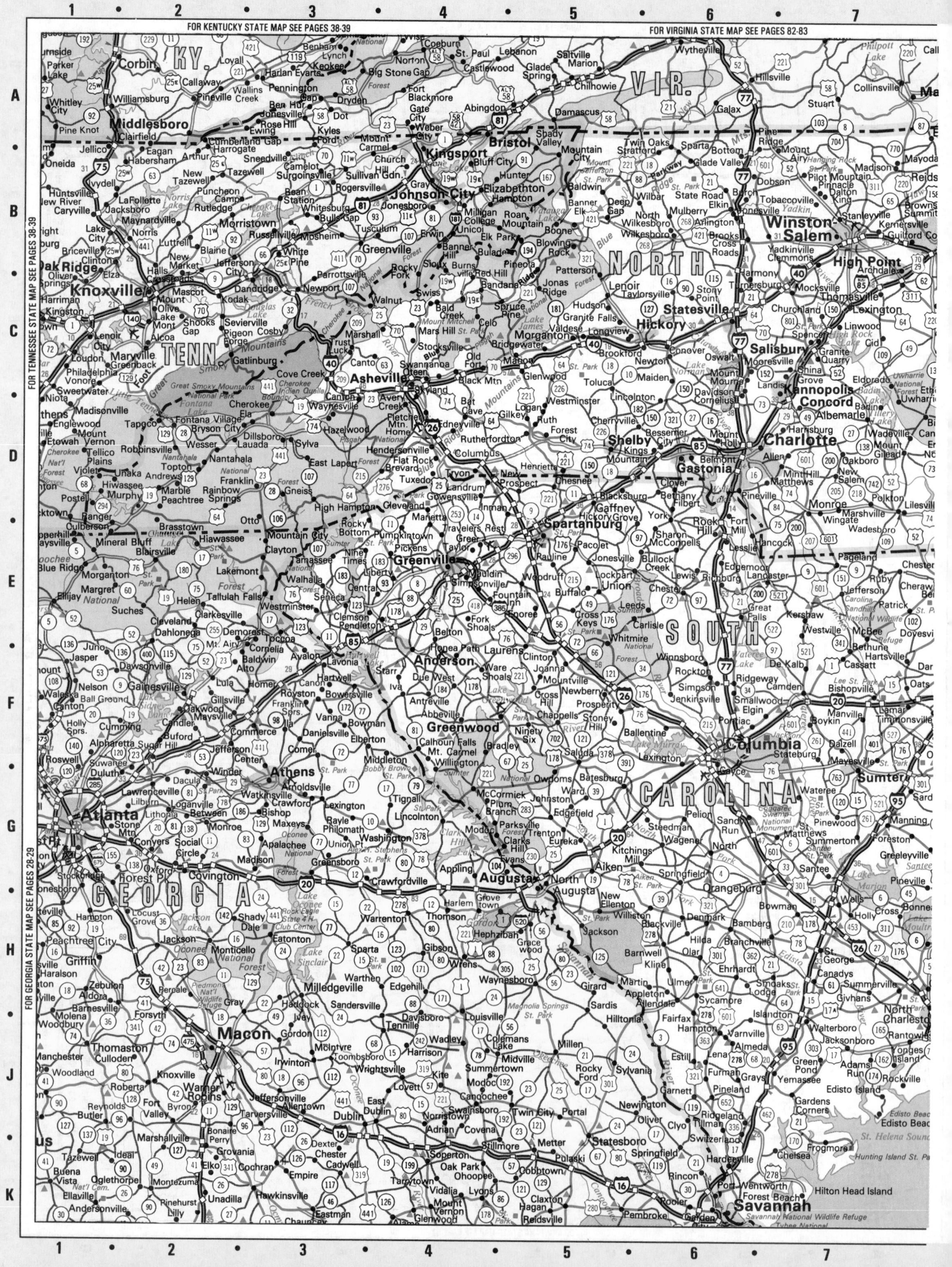

FOR KENTUCKY STATE MAP SEE PAGES 38-39

FOR VIRGINIA STATE MAP SEE PAGES 82-83

FOR TENNESSEE STATE MAP SEE PAGES 38-39

FOR GEORGIA STATE MAP SEE PAGES 28-29

1 2 3 4 5 6 7

A B C D E F G H J K

KY.

VIR.

TENN.

NORTH

SOUTH

CAROLINA

GEORGIA

Middlesboro

Kingsport

Bristol

Johnson City

Knoxville

Winston-Salem

High Point

Statesville

Hickory

Asheville

Salisbury

Kannapolis

Concord

Charlotte

Gastonia

Shelby

Spartanburg

Greenville

Anderson

Greenwood

Columbia

Sumter

Athens

Atlanta

Augusta

Macon

Savannah

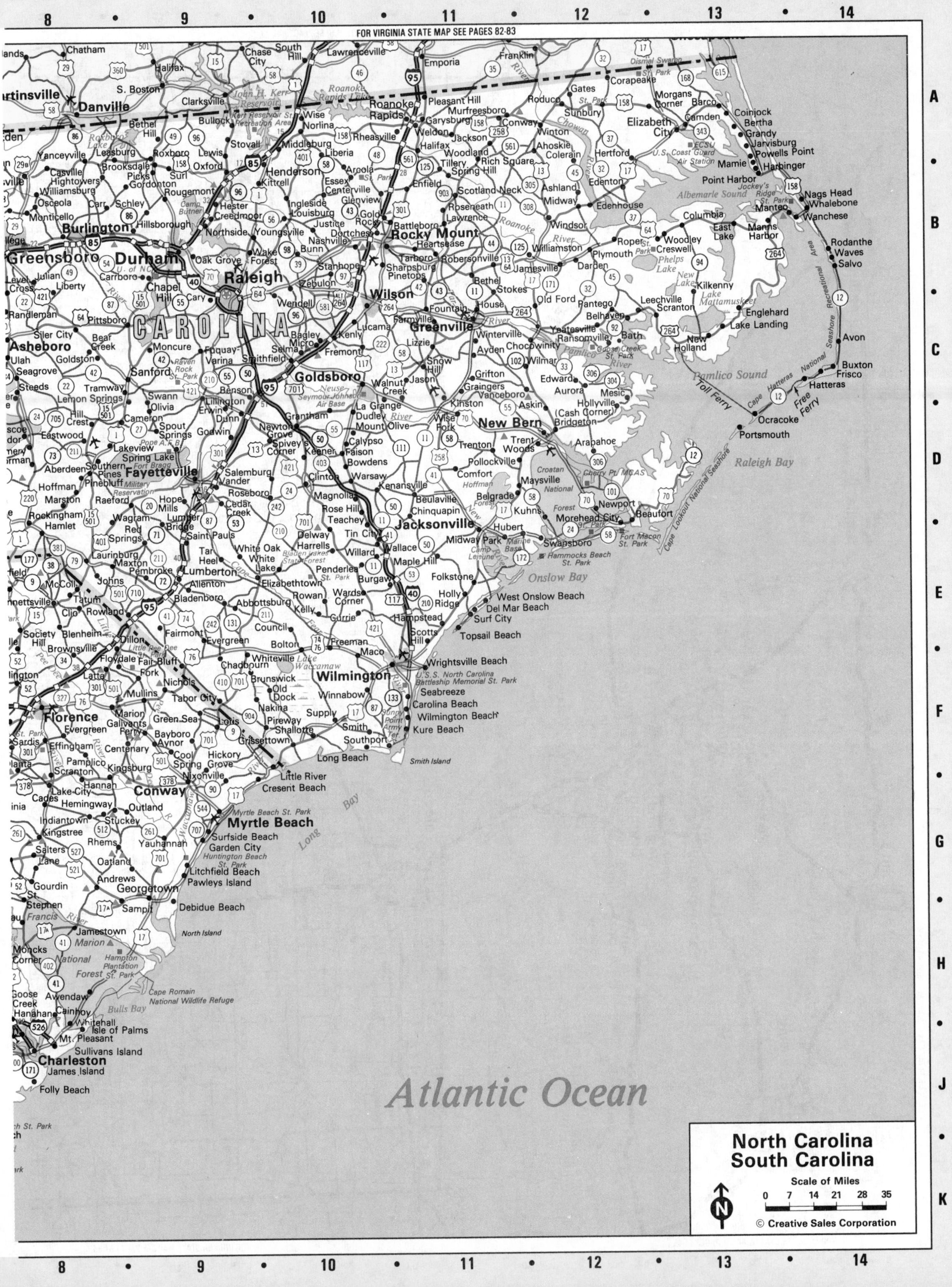

FOR VIRGINIA STATE MAP SEE PAGES 82-83
Chatham
Danville
S. Boston
Halifax
Clarksville
Chase City
South Hill
Lawrenceville
Emporia
Franklin
Dismal Swamp St. Park
Gates
Corapeake
Roanoke Rapids Lake
John H. Kerr Reservoir
Kerr Reservoir St. Recreation Area
Wise
Norlina
Roanoke Rapids
Pleasant Hill
Murfreesboro
Garysburg
Weldon
Jackson
Conway
Roduco
Sunbury
Chowan River
Morgans Corner
Barco
Camden
Elizabeth City
Coinjock
Bertha
Grandy
Jarvisburg
Powells Point
Harbinger
Mamie
Point Harbor
ECSU
U.S. Coast Guard Air Station
Bethel Hill
Roxboro Lake
Yanceyville
Leasburg
Roxboro
Stovall
Middleburg
Rheasville
Halifax
Woodland
Rich Square
Winton
Ahoskie
Colerain
Hertford
Caswell
Brooksdale
Hightowers
Picks
Gordonton
Surl
Oxford
Henderson
Kittrell
Liberia
Arcola
Tillery
Spring Hill
Enfield
Scotland Neck
Ashland
Midway
Edenton
Albemarle Sound
Jockey's Ridge St. Park
Nags Head
Whalebone
Wanchese
Manteo
Williamsburg
Osceola
Monticello
Carr
Schley
Rougemont
Camp Butner
Hester
Creedmoor
Essex
Centerville
Glenview
Ingleside
Louisburg
Gold Rock
Battleboro
Roseneath
Lawrence
Roanoke River
Windsor
Edenhouse
Columbia
East Lake
Manns Harbor
Burlington
Hillsborough
Northside
Youngsville
Justice
Dortches
Nashville
Rocky Mount
Heartsease
Williamston
Roper
Woodley
Creswell
Plymouth
Phelps Lake
Rodanthe
Waves
Salvo
Greensboro
Durham
Oak Grove
Wake Forest
Bunn
Stanhope
Tarboro
Sharpsburg
Pinetops
Robersonville
Jamesville
Darden
Julian
Carrboro
U. of N.C.
Raleigh
Zebulon
Wilson
Bethel
Stokes
Old Ford
Pantego
New Lake
Kilkenny
Lake Mattamuskeet
Level Cross
Liberty
Chapel Hill
Cary
Wendell
Fountain
House
Leechville
Scranton
Belhaven
Englehard
Lake Landing
Randleman
Pittsboro
CAROLINA
Farmville
Greenville
Tar River
Yeatesville
Ransomville
Bath
New Holland
Siler City
Bear Creek
Asheboro
Ulah
Goldston
Moncure
Fuquay-Varina
Raven Rock St. Park
Smithfield
Selma
Micro
Bagley
Kenly
Lucama
Fremont
Lizzie
Winterville
Ayden
Chocowinity
Wilmar
Goose Creek St. Park
Pamlico River
Avon
Buxton
Frisco
Hatteras
Seagrove
Steeds
Tramway
Sanford
Benson
Goldsboro
Walnut Creek
Snow Hill
Jason
Grifton
Graingers
Vanceboro
Edward
Aurora
Mesic
Hollyville
Cash Corner
Pamlico Sound
Toll Ferry
Cape Hatteras National Seashore
Recreational Area
Free Ferry
Ocracoke
Portsmouth
Lemon Springs
Swann
Olivia
Lillington
Erwin
Dunn
Neuse River
Seymour Johnson Air Base
La Grange
Dudley
Mount Olive
Kinston
Askin
Bridgeton
Arapahoe
Hill Crest
Eastwood
Cameron
Spout Springs
Godwin
Newton Grove
Spivey's Corner
Keener
Calypso
Faison
Bowdens
Wise Fork
Trenton
New Bern
Trent Woods
Pollockville
Raleigh Bay
Pope A.F.B.
Lakeview
Spring Lake
Aberdeen
Southern Pines
Pinebluff
Fort Bragg Military Reservation
Fayetteville
Salemburg
Vander
Clinton
Warsaw
Kenansville
Comfort
Hoffman Forest
Maysville
Croatan National Forest
Cherry Pt. MCAS
Hoffman
Marston
Raeford
Roseboro
Cedar Creek
Magnolia
Rose Hill
Teachey
Beulaville
Chinquapin
Belgrade
Kuhns
Newport
Morehead City
Beaufort
Cape Lookout National Seashore
Rockingham
Hamlet
Wagram
Hope Mills
Lumber Bridge
Red Springs
Saint Pauls
Tar Heel
Delway
Harrells
Tin City
Jacksonville
Hubert
Midway Park
Camp Lejeune Marine Base
Swansboro
Fort Macon St. Park
Hammocks Beach St. Park
Onslow Bay
Laurinburg
Maxton
Pembroke
Lumberton
White Oak
White Lake
Bladen Lakes State Forest
Willard
Wallace
Maple Hill
Folkstone
McColl
Johns
Allenton
Penderlea
Burgaw
Elizabethtown
Rowan
Wards Corner
Holly Ridge
West Onslow Beach
Del Mar Beach
Surf City
Topsail Beach
Tatum
Rowland
Bladenboro
Abbottsburg
Kelly
Currie
Hampstead
Clio
Blenheim
Society Hill
Brownsville
Dillon
Fairmont
Evergreen
Council
Bolton
Freeman
Maco
Scotts Hill
Floydale
Fair Bluff
Little Pee Dee St. Park
Whiteville
Chadbourn
Lake Waccamaw
Wrightsville Beach
Latta
Fork
Nichols
Brunswick
Old Dock
Wilmington
U.S.S. North Carolina Battleship Memorial St. Park
Seabreeze
Mullins
Tabor City
Winnabow
Carolina Beach
Florence
Marion
Gallivants Ferry
Green Sea
Loris
Nakina
Pireway
Shallotte
Supply
Smith
Sunny Point Army Ter.
Wilmington Beach
Kure Beach
Evergreen
Sardis
Effingham
Centenary
Bayboro
Aynor
Grissettown
Southport
Long Beach
Smith Island
Pamplico
Scranton
Kingsburg
Cool Spring
Hickory Grove
Nixonville
Little River
Cresent Beach
Lake City
Cades
Hannah
Conway
Hemingway
Outland
Myrtle Beach St. Park
Myrtle Beach
Long Bay
Indiantown
Stuckey
Kingstree
Rhems
Yauhannah
Surfside Beach
Garden City
Huntington Beach St. Park
Salters
Lane
Oatland
Litchfield Beach
Pawleys Island
Gourdin
Andrews
Georgetown
St. Stephen
Sampit
Debidue Beach
Francis Marion National Forest
Santee River
Jamestown
North Island
Moncks Corner
Hampton Plantation St. Park
Awendaw
Goose Creek
Hanahan
Cainhoy
Cape Romain National Wildlife Refuge
Bulls Bay
Whitehall
Isle of Palms
Mt. Pleasant
Sullivans Island
Charleston
James Island
Folly Beach
Atlantic Ocean
North Carolina
South Carolina
Scale of Miles
0 7 14 21 28 35
N
© Creative Sales Corporation

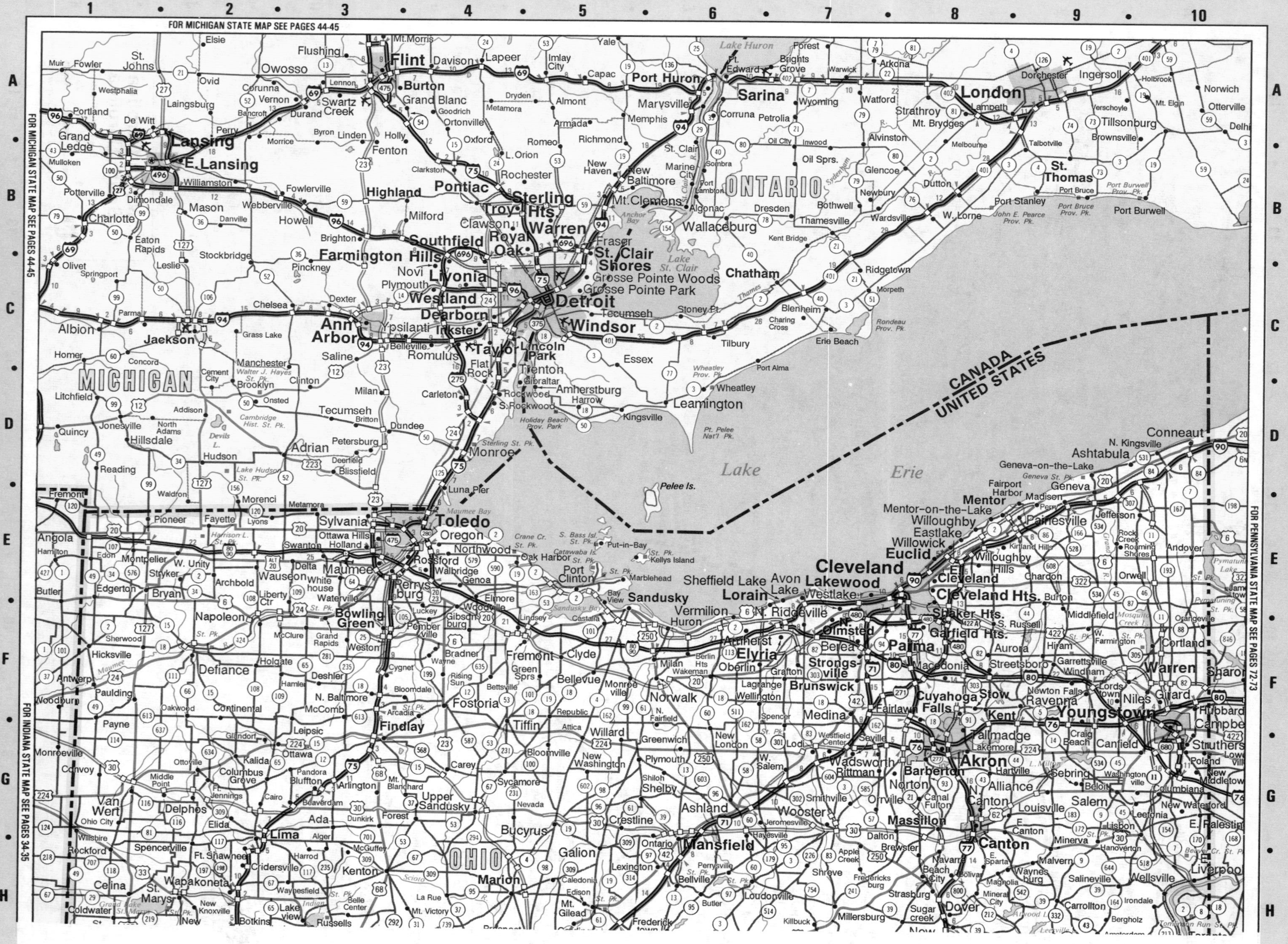
FOR MICHIGAN STATE MAP SEE PAGES 44-45
FOR PENNSYLVANIA STATE MAP SEE PAGES 72-73
FOR INDIANA STATE MAP SEE PAGES 34-35
CANADA
UNITED STATES
ONTARIO
MICHIGAN
OHIO
Lake Erie
Lake St. Clair
Lake Huron
Flint
Port Huron
Sarina
London
St. Thomas
Lansing
E. Lansing
Pontiac
Sterling Hts.
Warren
Southfield
Farmington Hills
Livonia
Westland
Dearborn
Detroit
Windsor
Chatham
Leamington
Ann Arbor
Jackson
Toledo
Sandusky
Lorain
Elyria
Cleveland
Lakewood
Euclid
Cleveland Hts.
Parma
Akron
Canton
Youngstown
Warren
Mansfield
Lima
Findlay
Marion
Ashtabula
Conneaut
Pelee Is.

FOR PENNSYLVANIA STATE MAP SEE PAGES 72-73

FOR WEST VIRGINIA STATE MAP SEE PAGES 82-83

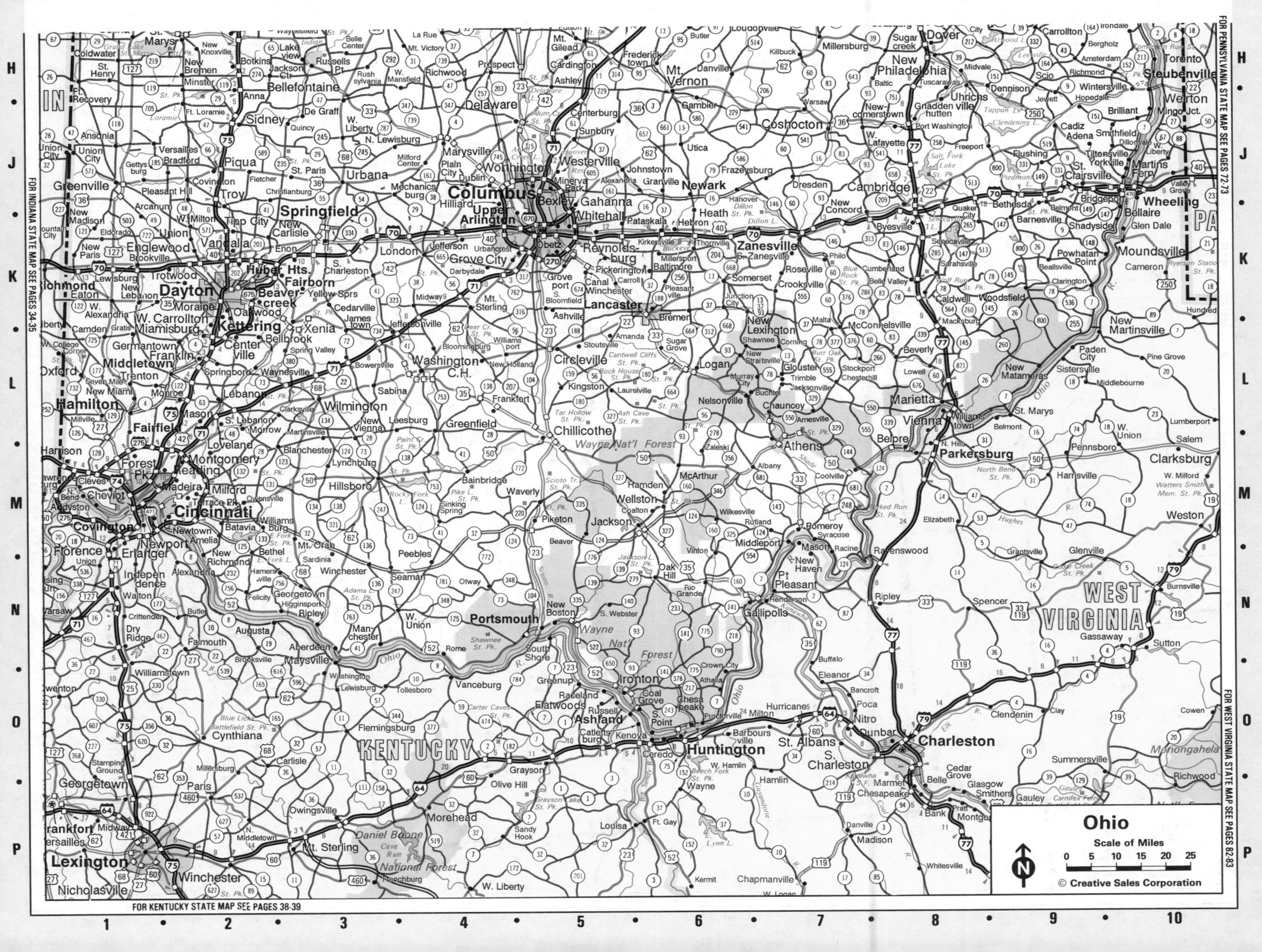

FOR KENTUCKY STATE MAP SEE PAGES 38-39

FOR INDIANA STATE MAP SEE PAGES 34-35

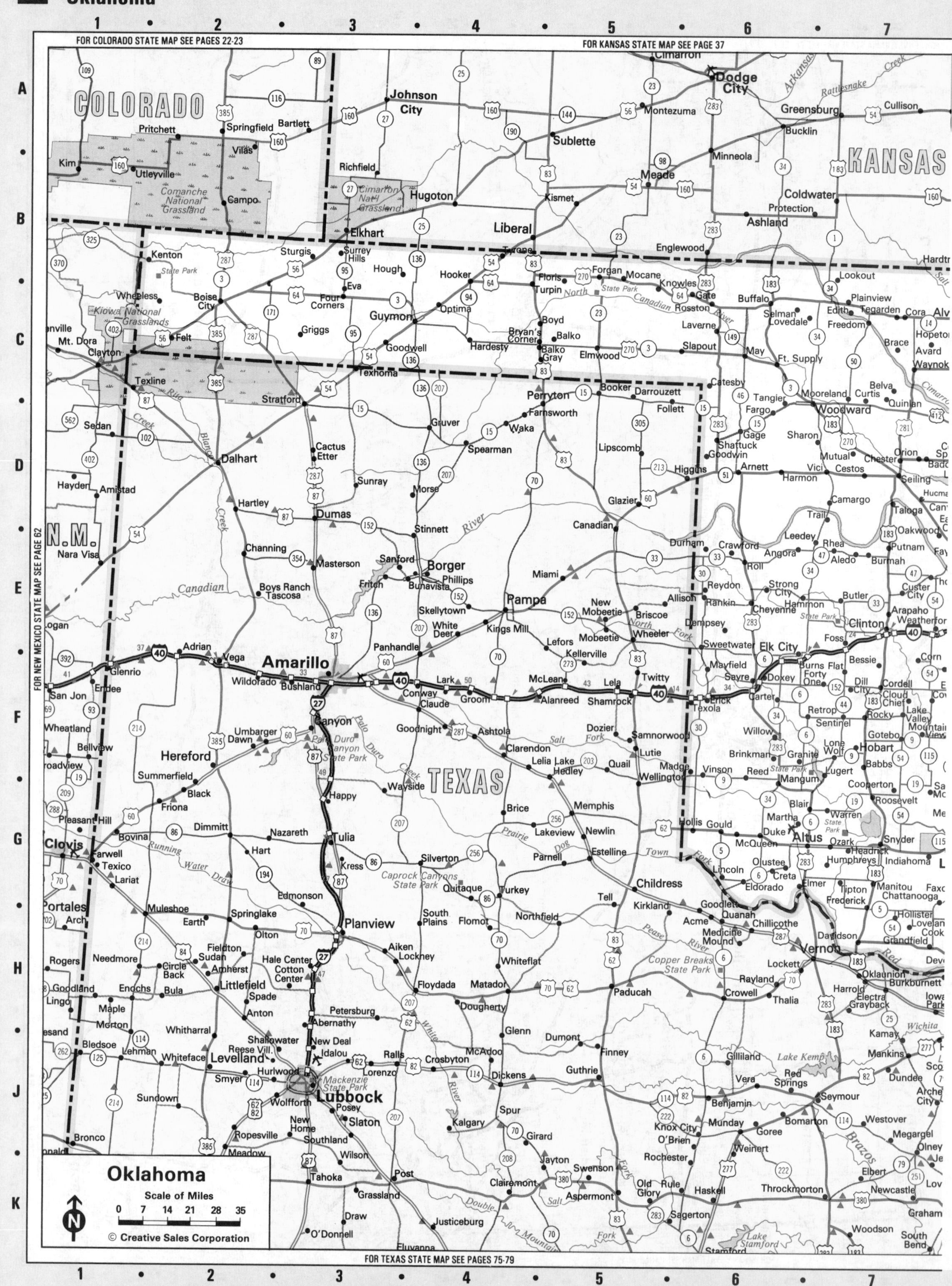

FOR COLORADO STATE MAP SEE PAGES 22-23
FOR KANSAS STATE MAP SEE PAGE 37
FOR NEW MEXICO STATE MAP SEE PAGE 62
FOR TEXAS STATE MAP SEE PAGES 75-79
COLORADO
KANSAS
N.M.
TEXAS
Comanche National Grassland
Cimarron Nat'l Grassland
Kiowa National Grasslands
Dodge City
Liberal
Guymon
Dalhart
Dumas
Borger
Pampa
Amarillo
Canyon
Hereford
Clovis
Plainview
Lubbock
Levelland
Littlefield
Childress
Vernon
Altus
Woodward
Clinton
Palo Duro Canyon State Park
Caprock Canyons State Park
Copper Breaks State Park
Mackenzie State Park
Lake Kemp
Lake Stamford
Oklahoma
Scale of Miles
0 7 14 21 28 35
© Creative Sales Corporation

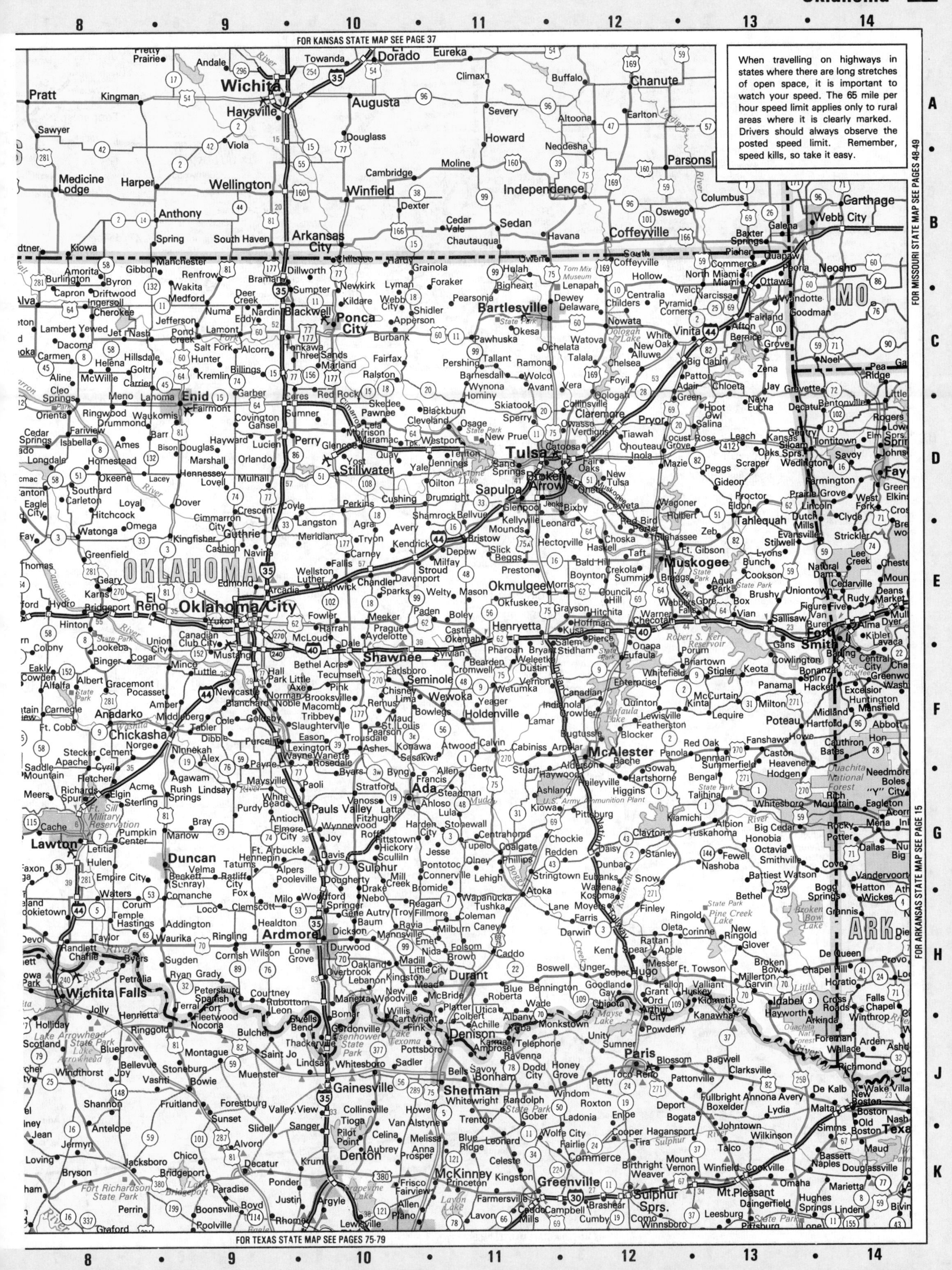

FOR KANSAS STATE MAP SEE PAGE 37
FOR MISSOURI STATE MAP SEE PAGES 48-49
FOR ARKANSAS STATE MAP SEE PAGE 15
FOR TEXAS STATE MAP SEE PAGES 75-79
When travelling on highways in states where there are long stretches of open space, it is important to watch your speed. The 65 mile per hour speed limit applies only to rural areas where it is clearly marked. Drivers should always observe the posted speed limit. Remember, speed kills, so take it easy.
OKLAHOMA
MO
ARK
Wichita
Oklahoma City
Tulsa
Enid
Ponca City
Bartlesville
Stillwater
Muskogee
Lawton
Shawnee
McAlester
Ardmore
Wichita Falls
Denison
Sherman
Paris
Durant
Fort Smith
Coffeyville
Independence
Arkansas City
Winfield
Wellington
Chickasha
Duncan
Ada
Okmulgee
Sapulpa
Broken Arrow
Claremore
Norman
Guthrie
Denton
Gainesville
Greenville
McKinney
Sulphur Sprs.
Mt. Pleasant
Tahlequah
Vinita
Hugo
Idabel
Poteau
Seminole
Henryetta
Holdenville
Pauls Valley
Sulphur
Anadarko
El Reno
Perry
Pryor

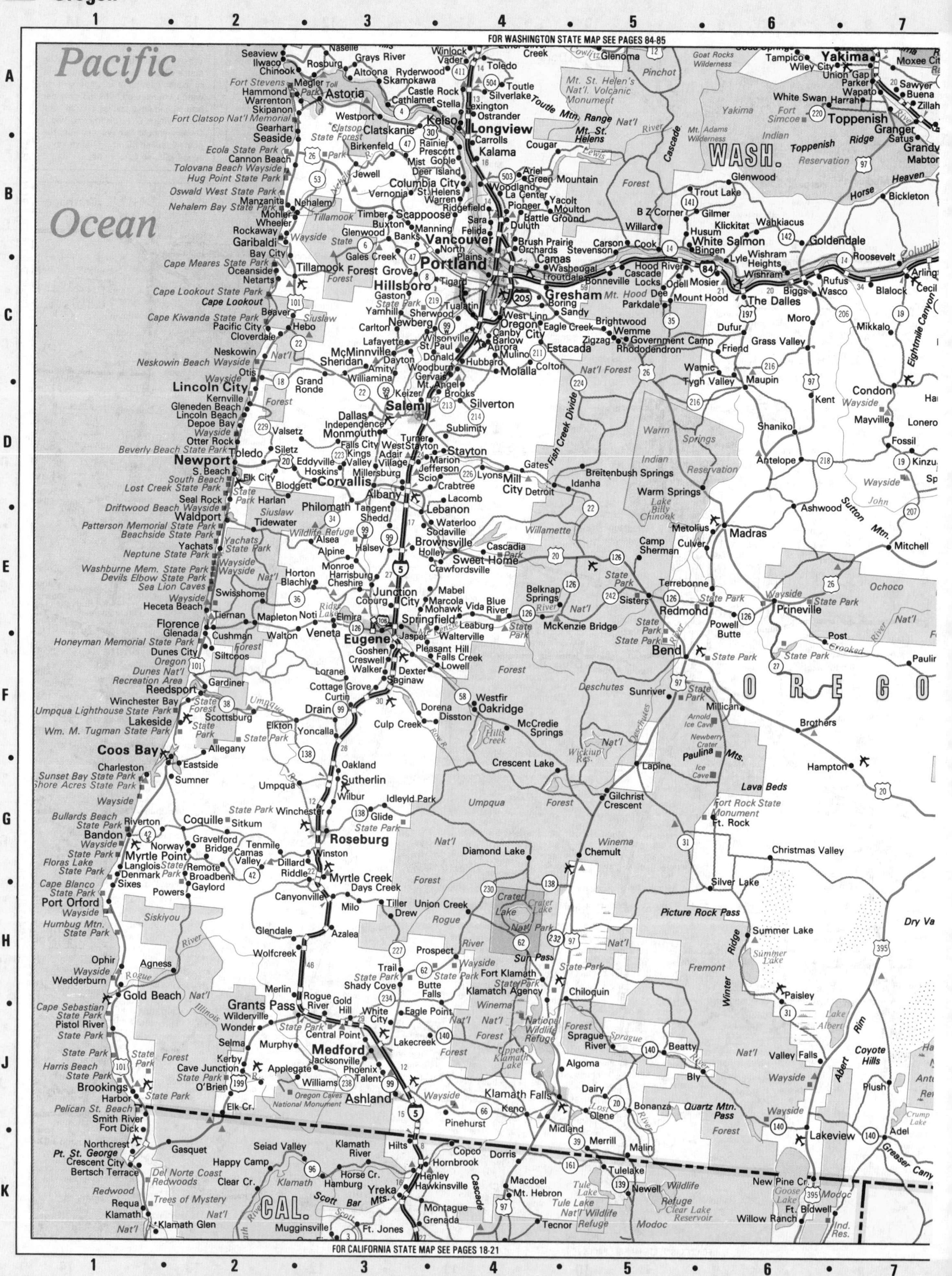
1 2 3 4 5 6 7
A B C D E F G H J K
FOR WASHINGTON STATE MAP SEE PAGES 84-85
Pacific
Ocean
WASH.
OREGO
CAL.
Seaview
Ilwaco
Chinook
Rosburg
Grays River
Altoona
Ryderwood
Skamokawa
Winlock
Vader
Toledo
Toutle
Silverlake
Glenoma
Pinchot
Goat Rocks Wilderness
Tampico
Wiley City
Yakima
Union Gap
Parker
Wapato
Moxee City
Sawyer
Buena
Zillah
Fort Stevens
Megler
Hammond
Warrenton
Skipanon
Astoria
Castle Rock
Cathlamet
Stella
Lexington
Ostrander
Toutle Mtn. Range
Mt. St. Helen's Nat'l. Volcanic Monument
Mt. St. Helens
Nat'l
Yakima
Fort Simcoe
White Swan
Harrah
Toppenish
Granger
Satus
Grandview
Mabton
Fort Clatsop Nat'l Memorial
Gearhart
Seaside
Westport
Clatsop State Forest
Clatskanie
Kelso
Longview
Carrolls
Kalama
Cougar
Mt. Adams Wilderness
Cascade
Indian
Toppenish Ridge
Reservation
Ecola State Park
Cannon Beach
Tolovana Beach Wayside
Hug Point State Park
Oswald West State Park
Nehalem Bay State Park
Manzanita
Nehalem
Mohler
Wheeler
Rockaway
Garibaldi
Bay City
Birkenfeld
Rainier
Prescott
Mist
Goble
Deer Island
Jewell
Columbia City
Vernonia
St. Helens
Warren
Ridgefield
Woodland
La Center
Ariel
Green Mountain
Yacolt
Pioneer
Moulton
Battle Ground
Forest
Lewis River
Trout Lake
Glenwood
Bickleton
Horse Heaven
B Z Corner
Gilmer
Willard
Husum
Klickitat
Wahkiacus
White Salmon
Bingen
Goldendale
Tillamook
Timber
Scappoose
Buxton
Manning
Banks
Glenwood
Sara
Felida
Duluth
Brush Prairie
Orchards
Vancouver
Gales Creek
Forest Grove
Hillsboro
Portland
Camas
Washougal
Carson
Stevenson
Cook
Hood River
Cascade Locks
Bonneville
Troutdale
Gresham
Lyle
Wishram
Wishram Heights
Rufus
Wasco
Roosevelt
Biggs
Blalock
Arlington
Cecil
Cape Meares State Park
Oceanside
Netarts
Cape Lookout State Park
Cape Lookout
Cape Kiwanda State Park
Beaver
Hebo
Pacific City
Cloverdale
Siuslaw
Gaston
Tigard
State Park
Yamhill
Sherwood
Tualatin
Boring
Sandy
Mt. Hood
Parkdale
Dee
Odell
Mosier
Mount Hood
The Dalles
Carlton
Newberg
West Linn
Oregon City
Canby
Barlow
Aurora
Eagle Creek
Brightwood
Wemme
Zigzag
Government Camp
Rhododendron
Dufur
Moro
Mikkalo
Grass Valley
Eightmile Canyon
Lafayette
Wilsonville
St. Paul
McMinnville
Dayton
Donald
Hubbard
Mulino
Estacada
Friend
Neskowin
Neskowin Beach Wayside
Sheridan
Amity
Woodburn
Gervais
Molalla
Colton
Wamic
Tygh Valley
Maupin
Williamina
Mt. Angel
Otis
Grand Ronde
Brooks
Nat'l Forest
Lincoln City
Kernville
Gleneden Beach
Lincoln Beach
Depoe Bay
Otter Rock
Wayside
Keizer
Salem
Silverton
Condon
Kent
Mayville
Lonerock
Valsetz
Dallas
Independence
Monmouth
Sublimity
Shaniko
Fossil
Beverly Beach State Park
Siletz
Falls City
West Stayton
Turner
Aumsville
Stayton
Fish Creek Divide
Warm Springs
Indian Reservation
Newport
Toledo
Eddyville
Kings Valley
Adair Village
Marion
Jefferson
Gates
Antelope
Kinzua
S. Beach
South Beach
Elk City
Hoskins
Millersburg
Scio
Lyons
Mill City
Detroit
Breitenbush Springs
Idanha
Lost Creek State Park
Blodgett
Corvallis
Albany
Crabtree
Seal Rock
Harlan
Lacomb
Driftwood Beach Wayside
Philomath
Tangent
Lebanon
Waldport
Tidewater
Shedd
Waterloo
Sodaville
Lake Billy Chinook
Ashwood
Sutton Mtn.
Patterson Memorial State Park
Beachside State Park
Yachats
Alsea
Wildlife Refuge
Brownsville
Willamette
Metolius
Madras
Mitchell
Neptune State Park
Alpine
Halsey
Cascadia
Holley
Sweet Home
Camp Sherman
Culver
Washburne Mem. State Park
Devils Elbow State Park
Sea Lion Caves
Monroe
Harrisburg
Cheshire
Crawfordsville
Horton
Blachly
Swisshome
Junction City
Mabel
Marcola
Mohawk
Vida
Blue River
Belknap Springs
Terrebonne
Sisters
Redmond
Prineville
Ochoco
Heceta Beach
Mapleton
Noti
Coburg
Ridge Lake
Elmira
Springfield
Leaburg
McKenzie Bridge
Powell Butte
Florence
Glenada
Tiernan
Cushman
Walton
Veneta
Eugene
Jasper
Walterville
Bend
Post
Crooked
Honeyman Memorial State Park
Dunes City
Oregon Dunes Nat'l Recreation Area
Siltcoos
Goshen
Creswell
Pleasant Hill
Falls Creek
Lowell
Paulina
Reedsport
Gardiner
Lorane
Walker
Dexter
Cottage Grove
Saginaw
Curtin
Westfir
Oakridge
Deschutes
Sunriver
Millican
Winchester Bay
Umpqua Lighthouse State Park
Lakeside
Scottsburg
Drain
Dorena
Disston
Culp Creek
McCredie Springs
Brothers
Wm. M. Tugman State Park
Elkton
Yoncalla
Hills Creek
Wickiup Res.
Arnold Ice Cave
Newberry Crater
Paulina Mts.
Coos Bay
Allegany
Eastside
Charleston
Sumner
Oakland
Sutherlin
Crescent Lake
Lapine
Hampton
Sunset Bay State Park
Shore Acres State Park
Umpqua
Wilbur
Idleyld Park
Ice Cave
Lava Beds
Winchester
Glide
Gilchrist
Crescent
Fort Rock State Monument
Ft. Rock
Bullards Beach State Park
Riverton
Coquille
Sitkum
Roseburg
Bandon
Norway
Gravelford
Bridge
Tenmile
Camas Valley
Dillard
Winston
Riddle
Diamond Lake
Winema
Chemult
Christmas Valley
Myrtle Point
Langlois
Floras Lake State Park
Remote
Broadbent
Denmark
Sixes
Gaylord
Powers
Myrtle Creek
Days Creek
Cape Blanco State Park
Port Orford
Canyonville
Milo
Tiller
Drew
Union Creek
Crater Lake
Crater Lake Nat'l Park
Silver Lake
Picture Rock Pass
Dry Valley
Humbug Mtn. State Park
Siskiyou
Glendale
Azalea
Rogue
Summer Lake
Winter Ridge
Ophir
Agness
Wayside
Wedderburn
Wolfcreek
Prospect
Trail
Sun Pass
State Park
Fort Klamath
Klamath Agency
Chiloquin
Fremont
Paisley
Gold Beach
Merlin
Shady Cove
Butte Falls
Cape Sebastian State Park
Pistol River
Grants Pass
Wilderville
Wonder
Rogue River
Gold Hill
White City
Eagle Point
Illinois
Lake Albert
Abert Rim
Coyote Hills
Selma
Murphy
Central Point
Lakecreek
National Wildlife Refuge
Sprague River
Beatty
Medford
Jacksonville
Phoenix
Talent
Kerby
Cave Junction
O'Brien
Applegate
Williams
Oregon Caves National Monument
Ashland
Upper Klamath Lake
Algoma
Bly
Valley Falls
Plush
Harris Beach State Park
Brookings
Harbor
Pelican St. Beach
Smith River
Fort Dick
Elk Cr.
Pinehurst
Keno
Klamath Falls
Dairy
Olene
Bonanza
Quartz Mtn. Pass
Lakeview
Adel
Crump Lake
Northcrest
Pt. St. George
Crescent City
Bertsch Terrace
Gasquet
Seiad Valley
Happy Camp
Klamath River
Hilts
Copco
Hornbrook
Midland
Merrill
Malin
Dorris
Tulelake
Greaser Canyon
Del Norte Coast Redwoods
Clear Cr.
Klamath
Horse Cr.
Hamburg
Henley
Hawkinsville
Yreka
Macdoel
Mt. Hebron
Tule Lake
Tule Lake Nat'l Wildlife Refuge
Newell
Wildlife Refuge
Clear Lake Reservoir
New Pine Cr.
Goose Lake
Modoc
Ft. Bidwell
Willow Ranch
Ind. Res.
Redwood
Requa
Klamath
Klamath Glen
Trees of Mystery
Scott Bar Mts.
Montague
Grenada
Cascade
Tecnor
Mugginsville
Ft. Jones
FOR CALIFORNIA STATE MAP SEE PAGES 18-21

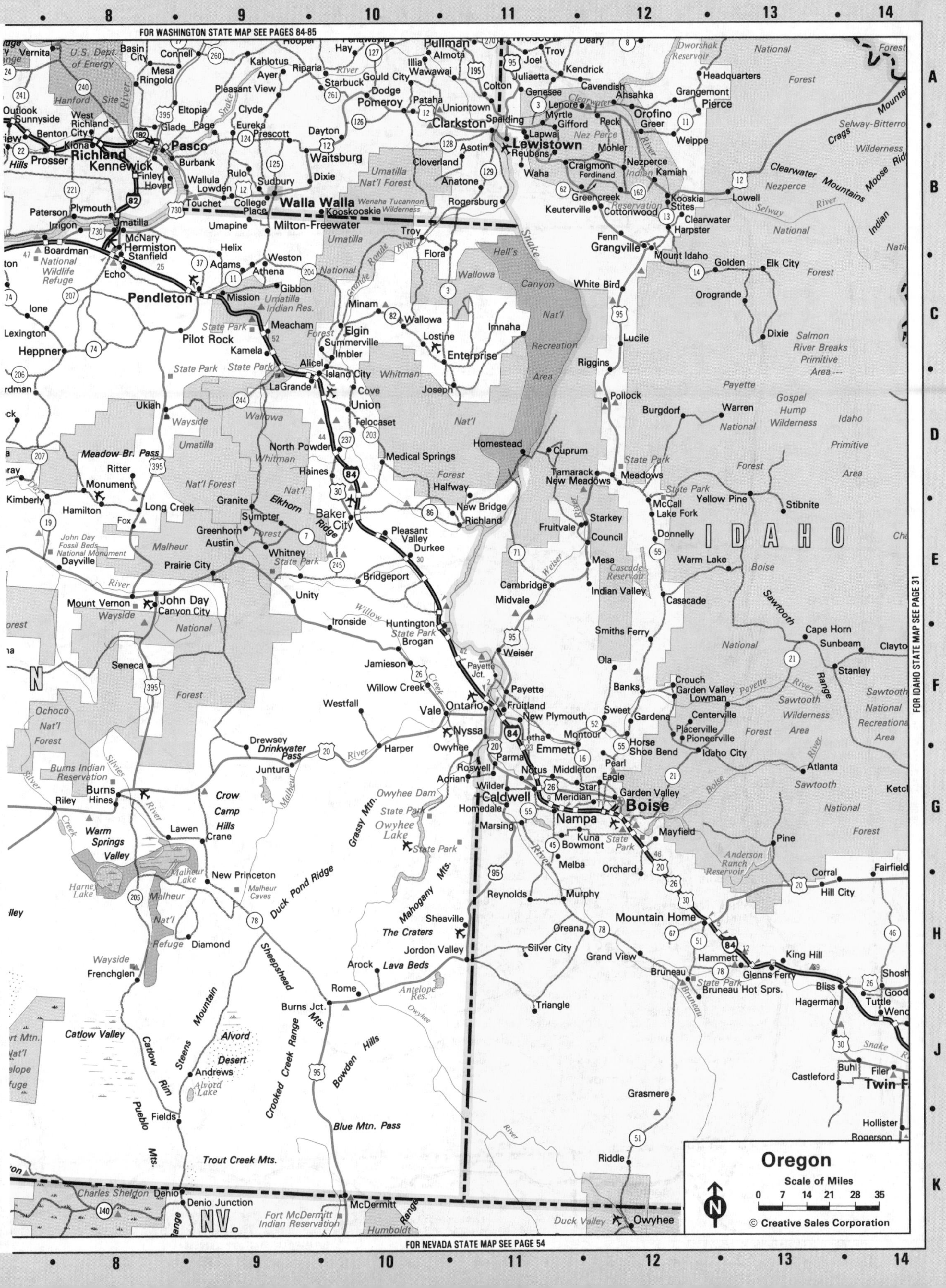

FOR WASHINGTON STATE MAP SEE PAGES 84-85

FOR IDAHO STATE MAP SEE PAGE 31

FOR NEVADA STATE MAP SEE PAGE 54

Oregon

Scale of Miles

0 7 14 21 28 35

FOR NEW YORK STATE MAP SEE PAGES 58-61

FOR OHIO STATE MAP SEE PAGES 66-67

FOR WEST VIRGINIA STATE MAP SEE PAGES 82-83

FOR MARYLAND STATE MAP SEE PAGES 42-43

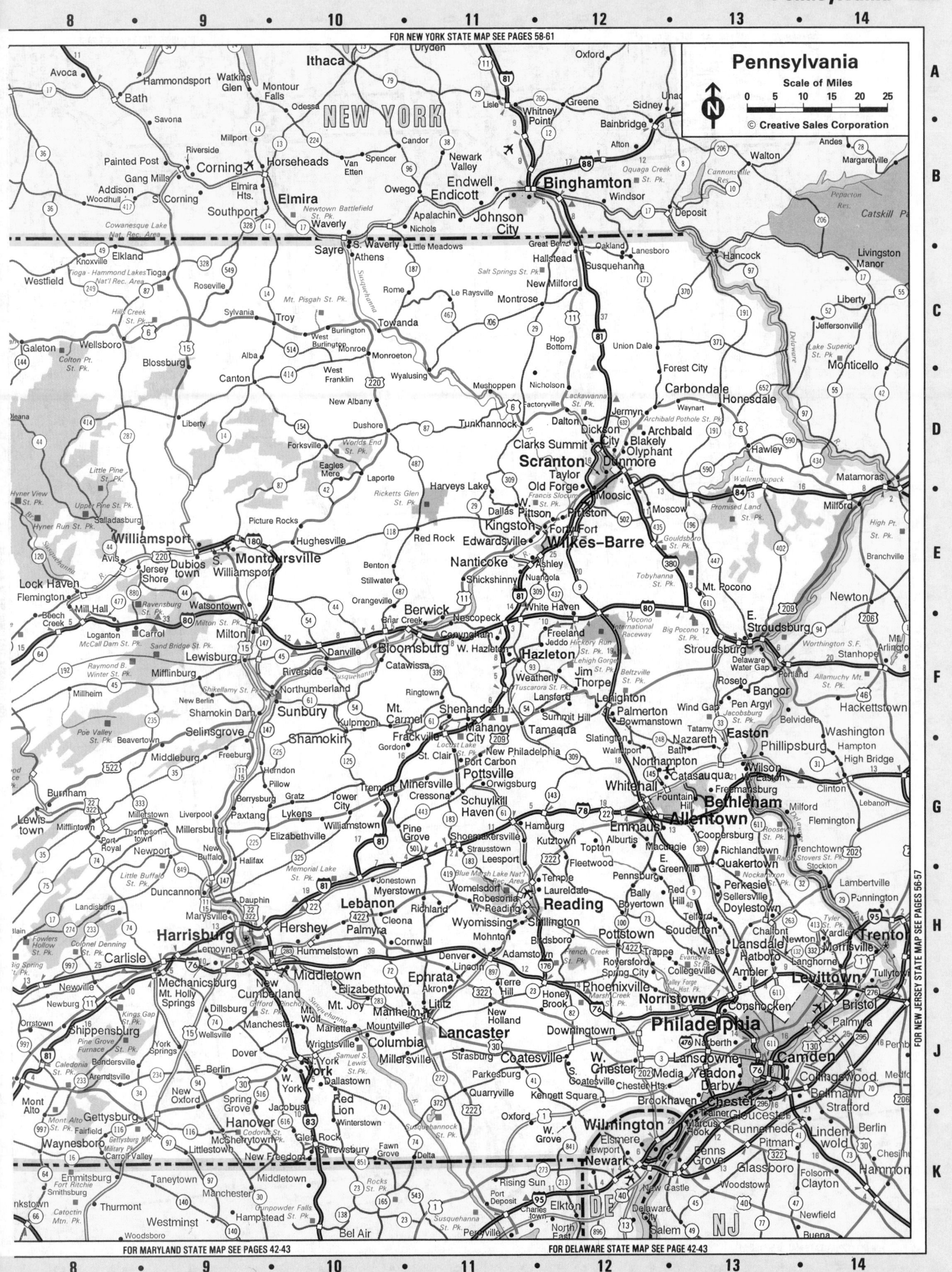
8 • 9 • 10 • 11 • 12 • 13 • 14
A B C D E F G H J K
FOR NEW YORK STATE MAP SEE PAGES 58-61
Pennsylvania
Scale of Miles
0 5 10 15 20 25
© Creative Sales Corporation
NEW YORK
Ithaca
Dryden
Oxford
Avoca
Bath
Hammondsport
Watkins Glen
Montour Falls
Odessa
Lisle
Whitney Point
Greene
Sidney
Bainbridge
Savona
Millport
Candor
Afton
Riverside
Painted Post
Corning
Horseheads
Van Etten
Spencer
Newark Valley
Walton
Margaretville
Andes
Gang Mills
Addison
Woodhull
S. Corning
Elmira Hts.
Elmira
Southport
Owego
Endwell
Endicott
Binghamton
Windsor
Deposit
Cannonsville Res.
Pepacton Res.
Catskill Park
Oquaga Creek St. Pk.
Newtown Battlefield St. Pk.
Apalachin
Johnson City
Nichols
Waverly
Cowanesque Lake Nat. Rec. Area
S. Waverly
Sayre
Athens
Little Meadows
Great Bend
Hallstead
Oakland
Lanesboro
Susquehanna
Hancock
Livingston Manor
Knoxville
Elkland
Westfield
Tioga - Hammond Lakes Nat'l Rec. Area
Tioga
Roseville
Salt Springs St. Pk.
New Milford
Rome
Le Raysville
Montrose
Liberty
Hills Creek St. Pk.
Mt. Pisgah St. Pk.
Sylvania
Troy
Towanda
Burlington
West Burlington
Monroe
Monroeton
Jeffersonville
Galeton
Wellsboro
Colton Pt. St. Pk.
Blossburg
Alba
Hop Bottom
Union Dale
Lake Superior St. Pk
Monticello
Canton
West Franklin
Wyalusing
Forest City
Meshoppen
Nicholson
Carbondale
Honesdale
New Albany
Factoryville
Lackawanna St. Pk.
Waynart
Jermyn
Archibald Pothole St. Pk.
Archbald
Blakely
Olyphant
Dushore
Tunkhannock
Dalton
Dickson City
Dunmore
Hawley
Liberty
Forksville
Worlds End St. Pk.
Clarks Summit
Scranton
Taylor
Old Forge
Moosic
Eagles Mere
Laporte
Little Pine St. Pk.
Wallenpaupack
Matamoras
Hyner View St. Pk.
Ricketts Glen St. Pk.
Harveys Lake
Dallas
Francis Slocum St. Pk.
Pittson
Kingston
Moscow
Promised Land St. Pk.
Milford
Upper Pine St. Pk.
Salladasburg
Hyner Run St. Pk.
Picture Rocks
Hughesville
Red Rock
Edwardsville
Forty Fort
Wilkes-Barre
Gouldsboro St. Pk.
High Pt. St. Pk.
Williamsport
Montoursville
Dubios town
S. Williamsport
Jersey Shore
Avis
Benton
Nanticoke
Ashley
Nuangola
Branchville
Lock Haven
Flemington
Stillwater
Shickshinny
Tobyhanna St. Pk.
Mt. Pocono
Mill Hall
Ravensburg St. Pk.
Watsontown
Orangeville
Berwick
White Haven
Newton
Beech Creek
Milton St. Pk.
Nescopeck
Briar Creek
Pocono International Raceway
Big Pocono St. Pk.
E. Stroudsburg
Loganton
Carroll
Milton
Conyngham
Freeland
Jeddo
Hickory Run St. Pk.
Stroudsburg
Worthington S.F.
Mt. Arlington
McCall Dam St. Pk.
Sand Bridge St. Pk.
Lewisburg
Danville
Bloomsburg
W. Hazleton
Hazleton
Lehigh Gorge St. Pk.
Delaware Water Gap
Stanhope
Raymond B. Winter St. Pk.
Mifflinburg
Riverside
Catawissa
Weatherly
Jim Thorpe
Beltzville St. Pk.
Portland
Allamuchy Mt. St. Pk.
Millheim
Shikellamy St. Pk.
Northumberland
Ringtown
Tuscarora St. Pk.
Lansford
Roseto
Bangor
New Berlin
Shamokin Dam
Sunbury
Kulpmont
Mt. Carmel
Shenandoah
Summit Hill
Lehighton
Palmerton
Bowmanstown
Wind Gap
Pen Argyl
Jacobsburg St. Pk.
Hackettstown
Poe Valley St. Pk.
Beavertown
Selinsgrove
Shamokin
Frackville
Mahanoy City
Tamaqua
Tatamy
Easton
Belvidere
Gordon
St. Clair
Locust Lake St. Pk.
New Philadelphia
Slatington
Nazareth
Phillipsburg
Washington
Middleburg
Freeburg
Port Carbon
Pottsville
Walnutport
Bath
Northampton
Hampton
High Bridge
Herndon
Pillow
Catasauqua
Wilson
W. Easton
Clinton
Burnham
Tremont
Minersville
Orwigsburg
Whitehall
Freemansburg
Lebanon
Millerstown
Berrysburg
Gratz
Tower City
Cressona
Schuylkill Haven
Fountain Hill
Bethlehem
Milford
Lewis town
Mifflintown
Liverpool
Millersburg
Paxtang
Lykens
Williamstown
Hamburg
Emmaus
Allentown
Flemington
Thompson town
Pine Grove
Kutztown
Alburtis
Coopersburg
Port Royal
Newport
New Buffalo
Elizabethville
Shoemakersville
Strausstown
Topton
Macungie
Roosevelt St. Pk.
Frenchtown
Richlandtown
Halifax
Leesport
Fleetwood
E. Greenville
Quakertown
Ralph Stovers St. Pk.
Stockton
Little Buffalo St. Pk.
Memorial Lake St. Pk.
Blue Marsh Lake Nat'l Rec. Area
Temple
Pennsburg
Nockamixon St. Pk.
Perkasie
Lambertville
Duncannon
Jonestown
Myerstown
Womelsdorf
Robesonia
W. Reading
Laureldale
Reading
Bally
Boyertown
Red Hill
Sellersville
Doylestown
Pennington
Landisburg
Dauphin
Marysville
Lebanon
Richland
Cleona
Wyomissing
Shillington
Telford
Chalfont
Tyler St. Pk.
Yardley
Trenton
Harrisburg
Hershey
Palmyra
Pottstown
Souderton
Lansdale
Newton
Morrisville
Fowlers Hollow St. Pk.
Colonel Denning St. Pk.
Carlisle
Lemoyne
Hummelstown
Cornwall
Mohnton
Birdsboro
French Creek St. Pk.
Trappe
N. Wales
Hatboro
Langhorne
Big Spring St. Pk.
Denver
Lincoln
Adamstown
Royersford
Evansville
Collegeville
Ambler
Tullytown
Newville
Mechanicsburg
Mt. Holly Springs
New Cumberland
Middletown
Ephrata
Terre Hill
Spring City
Valley Forge Nat. Hist. Pk.
Levittown
Newburg
Elizabethtown
Akron
Phoenixville
Norristown
Bristol
Orrstown
Kings Gap St. Pk.
Dillsburg
Gifford Pinchot St. Pk.
Mt. Joy
Manheim
Lititz
Honey Brook
Marsh Creek St. Pk.
Conshohocken
Shippensburg
Manchester
Mt. Wolf
Marietta
Mountville
New Holland
Philadelphia
Palmyra
Pine Grove Furnace St. Pk.
Wellsville
York Springs
Wrightsville
Columbia
Lancaster
Downingtown
Narberth
Pembe
Dover
Samuel S. Lewis St. Pk.
Millersville
Strasburg
Coatesville
W. Chester
Lansdowne
Camden
Caledonia St. Pk.
Bendersville
E. Berlin
York
Parkesburg
S. Coatesville
Media
Yeadon
Collingswood
Medford
Arendtsville
W. York
Dallastown
Quarryville
Chester Hts.
Darby
Bellmawr
New Oxford
Spring Grove
Red Lion
Kennett Square
Brookhaven
Chester
Stratford
Mont Alto
Jacobus
Winterstown
Trainer
Marcus Hook
Gloucester
Mont Alto St. Pk.
Gettysburg
Hanover
Codorus St. Pk.
Oxford
Wilmington
Runnemede
Berlin
Fairfield
Gettysburg Nat. Military Pk.
McSherrytown
Susquehannock St. Pk.
W. Grove
Elsmere
Linden wold
Waynesboro
Carroll Valley
Littlestown
Glen Rock
Fawn Grove
Delta
Pitman
Shrewsbury
New Freedom
Newport
Penns Grove
Chesilhurst
Newark
Glassboro
Hammonton
Emmitsburg
Fort Ritchie
Smithsburg
Taneytown
Manchester
Middletown
Rising Sun
Folsom
Clayton
Woodstown
New Castle
Port Deposit
Elkton
Hampstead
Gunpowder Falls St. Pk.
Rocks St. Pk.
Charlestown
Thurmont
Catoctin Mtn. Pk.
Westminster
Woodsboro
Bel Air
Susquehanna St. Pk.
Perryville
North East
Delaware City
Salem
Newfield
Buena
DE
NJ
FOR MARYLAND STATE MAP SEE PAGES 42-43
FOR DELAWARE STATE MAP SEE PAGE 42-43
FOR NEW JERSEY STATE MAP SEE PAGES 56-57

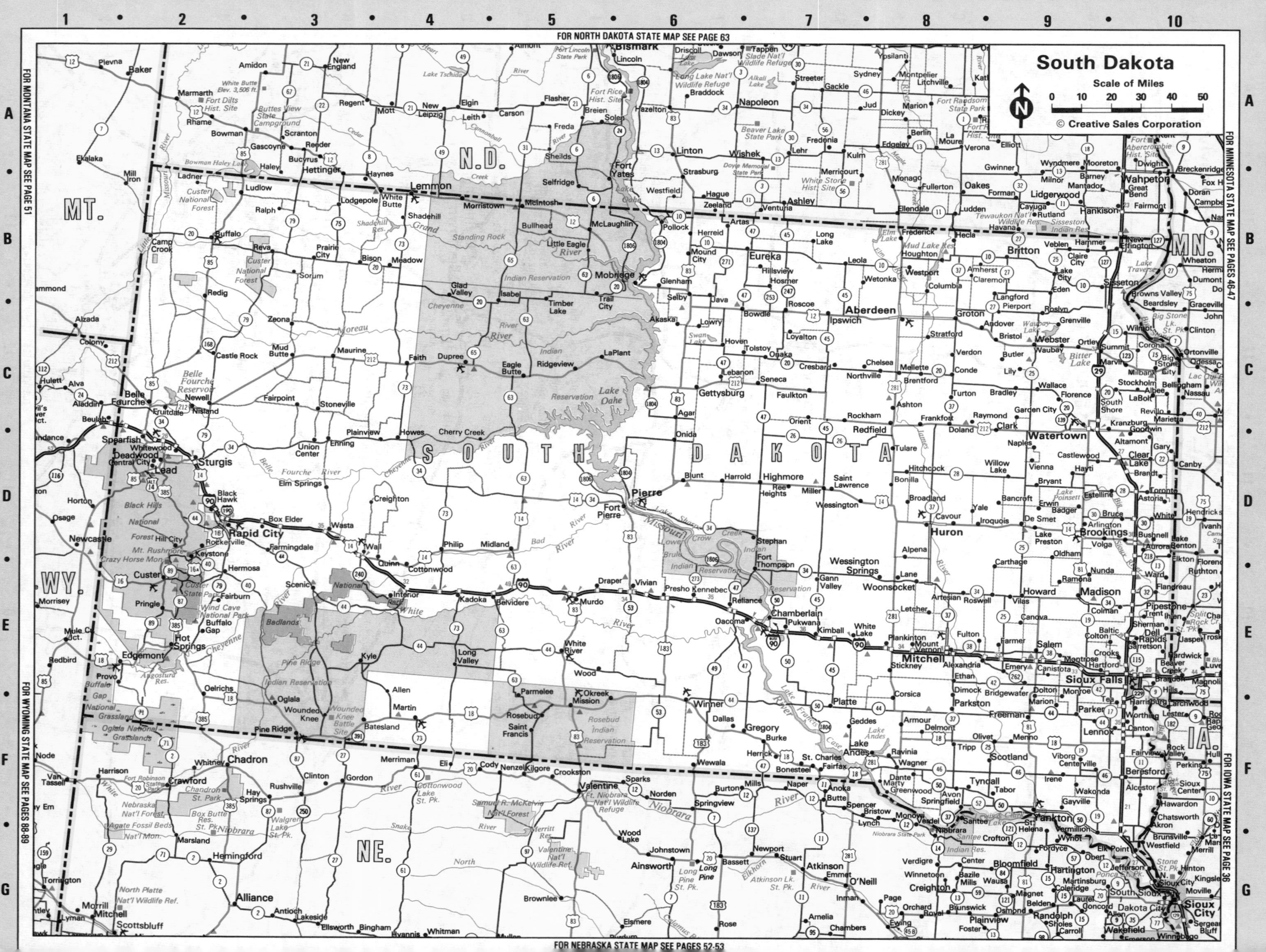

South Dakota
Scale of Miles
0 10 20 30 40 50
© Creative Sales Corporation
FOR NORTH DAKOTA STATE MAP SEE PAGE 63
FOR MINNESOTA STATE MAP SEE PAGES 46-47
FOR IOWA STATE MAP SEE PAGE 36
FOR NEBRASKA STATE MAP SEE PAGES 52-53
FOR MONTANA STATE MAP SEE PAGE 51
FOR WYOMING STATE MAP SEE PAGES 88-89
1 2 3 4 5 6 7 8 9 10
A B C D E F G
SOUTH DAKOTA
N.D.
MN.
IA.
NE.
WY.
MT.
Pierre
Rapid City
Sioux Falls
Aberdeen
Watertown
Huron
Mitchell
Brookings
Yankton

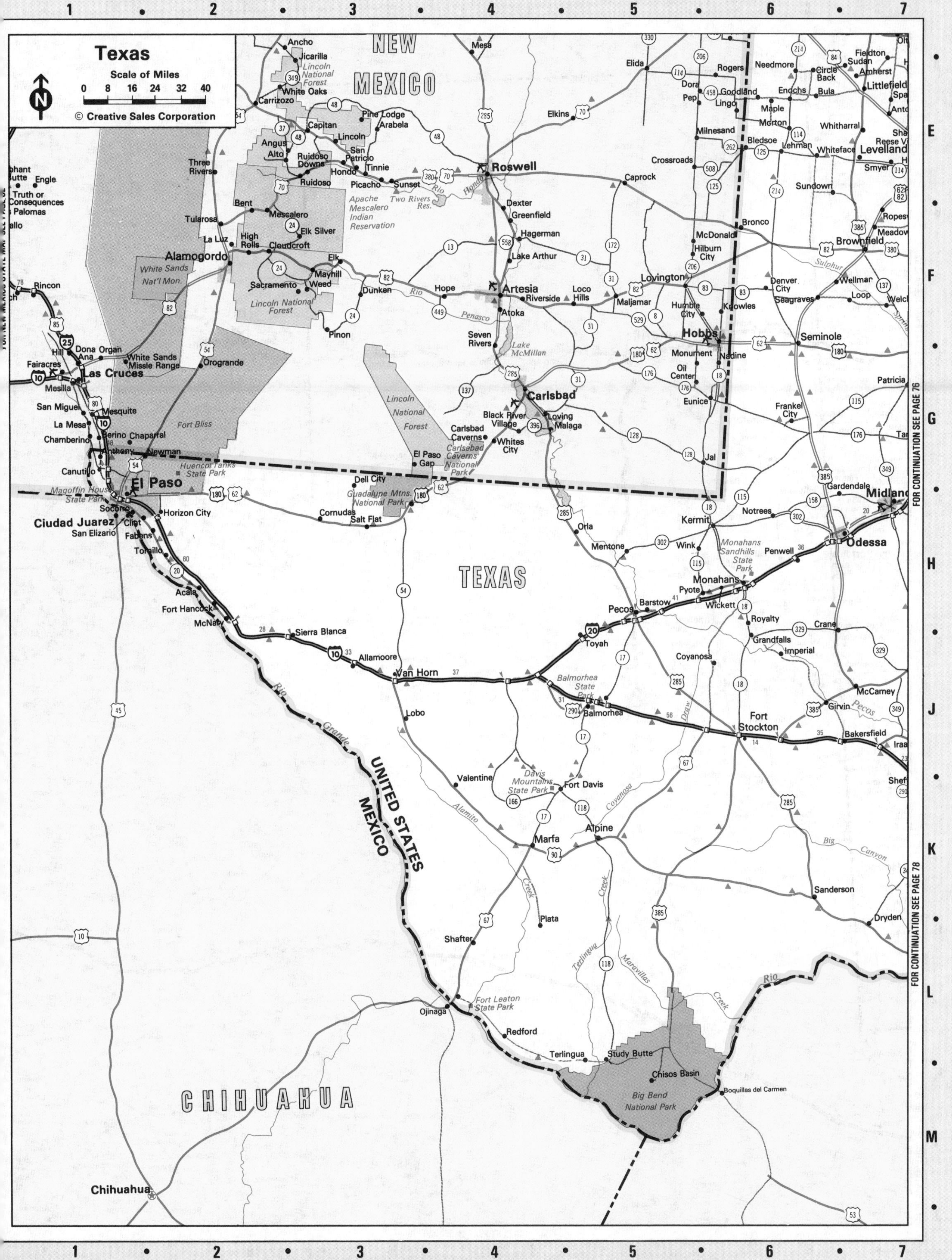
Texas
Scale of Miles
0 8 16 24 32 40
© Creative Sales Corporation
NEW MEXICO
TEXAS
CHIHUAHUA
UNITED STATES
MEXICO
Roswell
Artesia
Carlsbad
Hobbs
Alamogordo
Las Cruces
El Paso
Ciudad Juarez
Odessa
Midland
Pecos
Monahans
Kermit
Van Horn
Sierra Blanca
Fort Stockton
Alpine
Marfa
Fort Davis
Balmorhea
Seminole
Brownfield
Levelland
Lovington
Big Bend National Park
Chisos Basin
Terlingua
Study Butte
Boquillas del Carmen
Redford
Ojinaga
Presidio
Shafter
Fort Leaton State Park
Davis Mountains State Park
Balmorhea State Park
Monahans Sandhills State Park
Guadalupe Mtns. National Park
Carlsbad Caverns National Park
Lincoln National Forest
White Sands Nat'l Mon.
Fort Bliss
Apache Mescalero Indian Reservation
Chihuahua
FOR CONTINUATION SEE PAGE 76
FOR CONTINUATION SEE PAGE 78

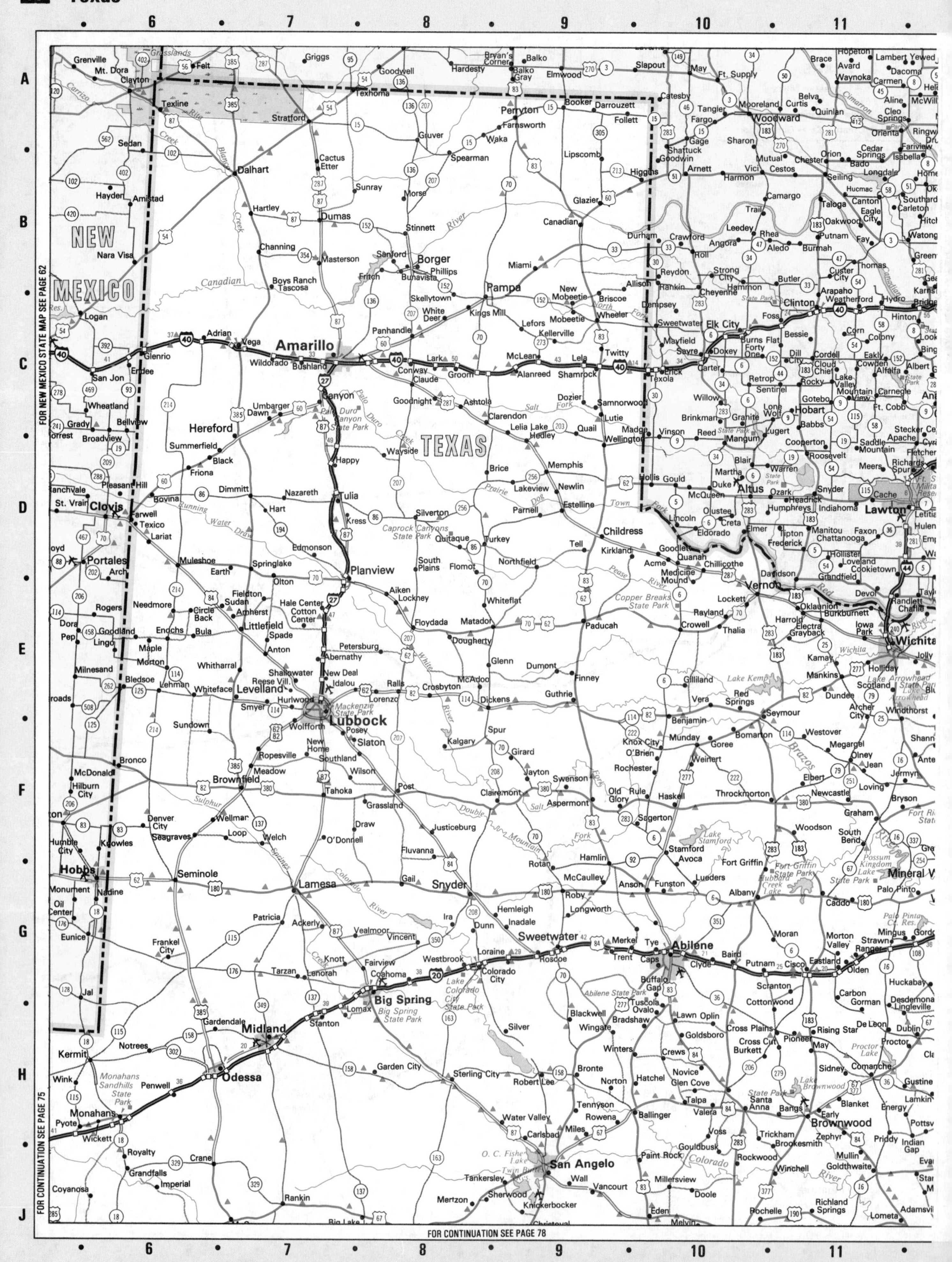

FOR NEW MEXICO STATE MAP SEE PAGE 62
FOR CONTINUATION SEE PAGE 75
FOR CONTINUATION SEE PAGE 78

FOR OKLAHOMA STATE MAP SEE PAGES 68-69

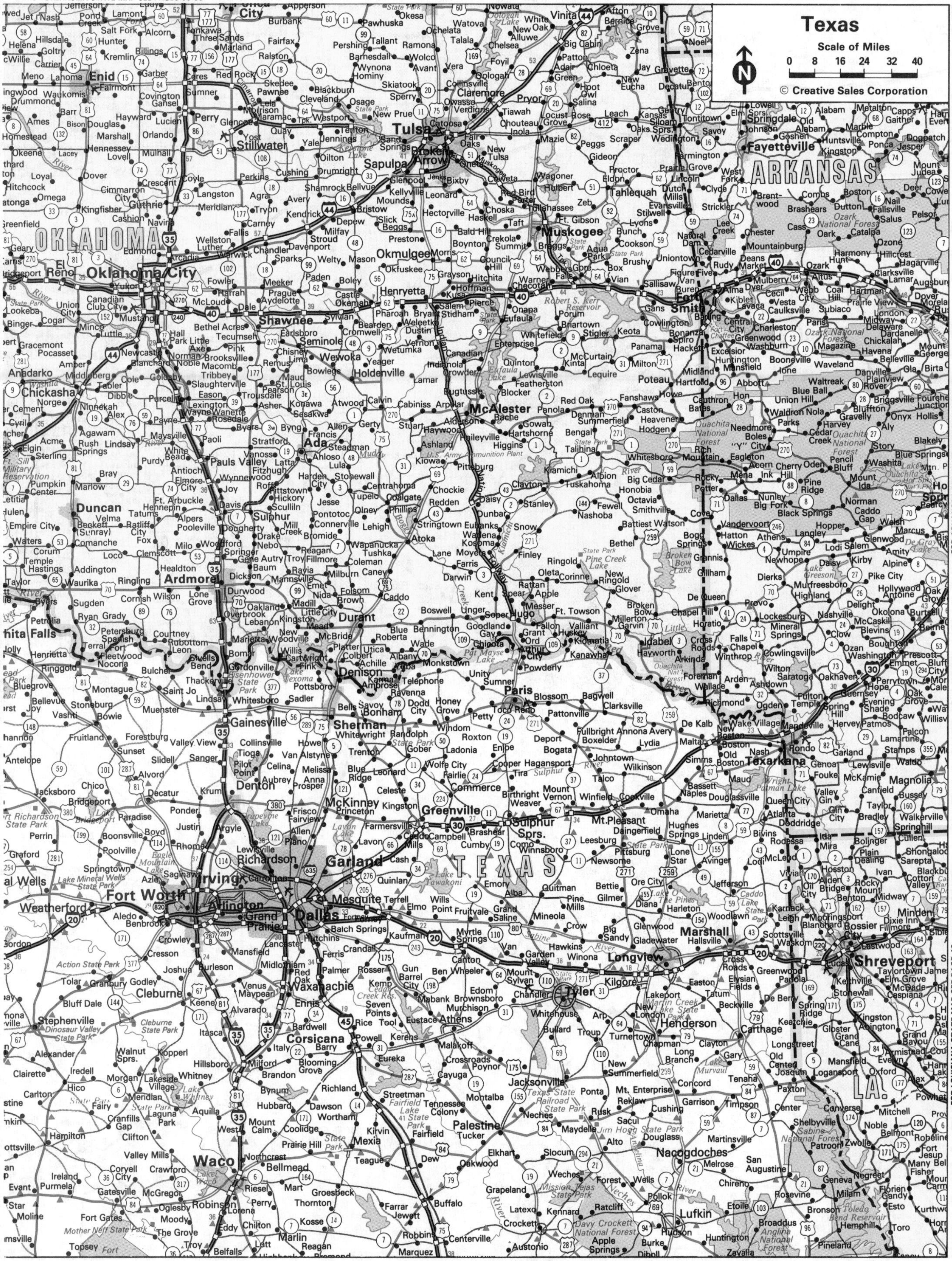

Texas

Scale of Miles

0 8 16 24 32 40

FOR ARKANSAS STATE MAP SEE PAGE 15

FOR LOUISIANA STATE MAP SEE PAGE 40

FOR CONTINUATION SEE PAGE 79

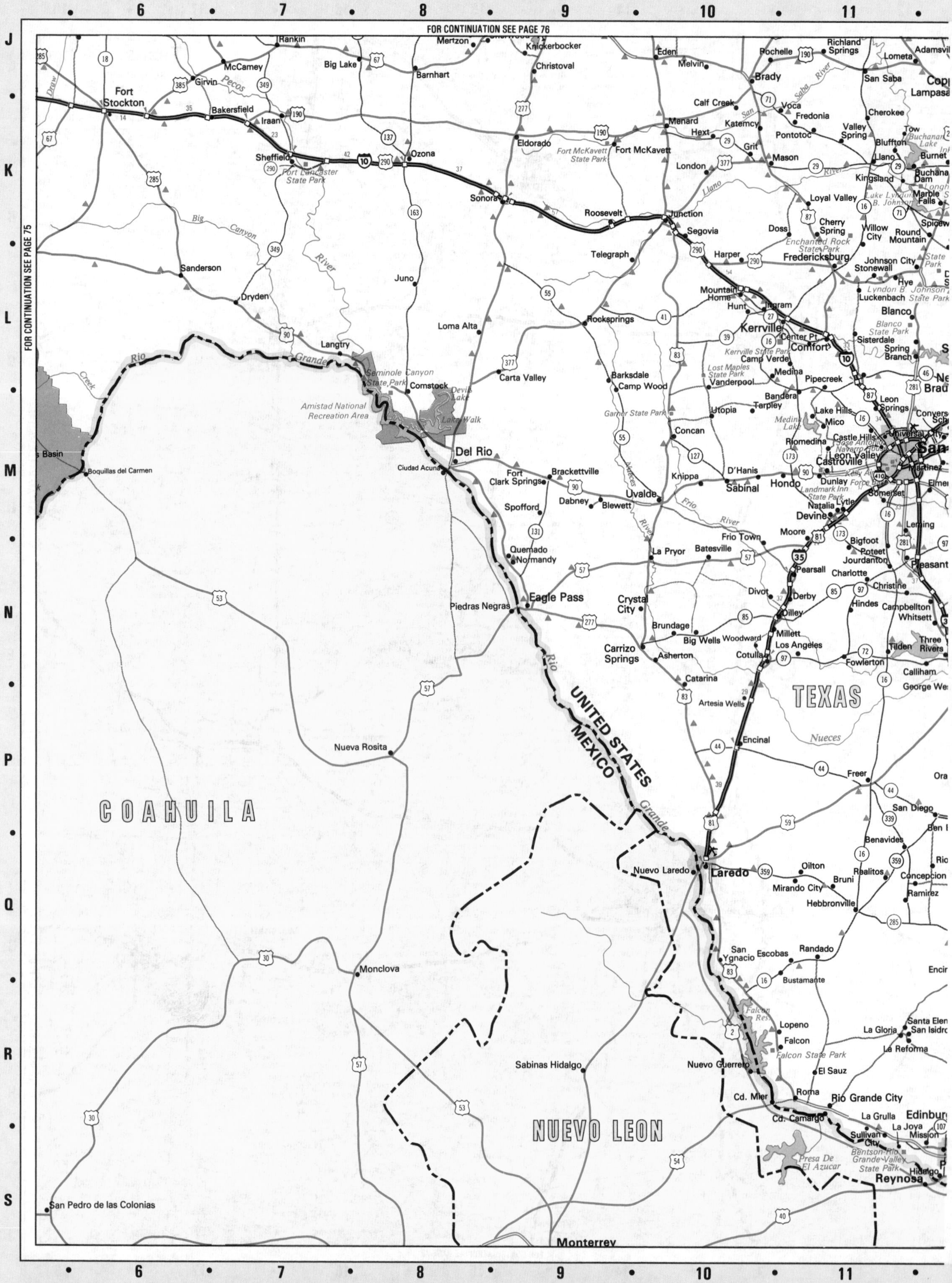

6 7 8 9 10 11

FOR CONTINUATION SEE PAGE 76

FOR CONTINUATION SEE PAGE 75

J K L M N P Q R S

Rankin
Mertzon
Knickerbocker
Christoval
Eden
Melvin
Rochelle
Richland Springs
Adamsville
Lometa
McCamey
Big Lake
Barnhart
Brady
San Saba
Lampasas
Fort Stockton
Girvin
Pecos
Bakersfield
Iraan
Calf Creek
Voca
Fredonia
Cherokee
Menard
Hext
Katemcy
Pontotoc
Valley Spring
Tow
Buchanan Lake
Bluffton
Eldorado
Fort McKavett State Park
Fort McKavett
Grit
Mason
Llano
Burnet
Sheffield
Ozona
Fort Lancaster State Park
London
Kingsland
Buchanan Dam
Marble Falls
Sonora
Roosevelt
Junction
Loyal Valley
Lake Lyndon B. Johnson
Big Canyon
Segovia
Cherry Spring
Doss
Willow City
Round Mountain
Spicewood
Enchanted Rock State Park
Harper
Fredericksburg
Johnson City
Stonewall
State Park
Sanderson
Telegraph
River
Juno
Dryden
Mountain Home
Hunt
Ingram
Hye
Lyndon B. Johnson State Park
Luckenbach
Blanco
Blanco State Park
Rocksprings
Loma Alta
Langtry
Rio Grande
Kerrville
Center Pt.
Comfort
Sisterdale
Spring Branch
Kerrville State Park
Camp Verde
Carta Valley
Seminole Canyon State Park
Comstock
Devils Lake
Lake Walk
Barksdale
Camp Wood
Lost Maples State Park
Vanderpool
Medina
Pipecreek
New Braunfels
Amistad National Recreation Area
Bandera
Tarpley
Utopia
Leon Springs
Converse
Garner State Park
Lake Hills
Mico
Medina Lake
Universal City
Big Bend Basin
Boquillas del Carmen
Concan
Rio Medina
Castle Hills
Jose Antonio Navarro
San Antonio
Del Rio
Ciudad Acuna
Leon Valley
Castroville
Fort Clark Springs
Brackettville
Knippa
D'Hanis
Sabinal
Hondo
Dunlay
Kelly AFB
Landmark Inn State Park
Martinez
Elmendorf
Spofford
Dabney
Blewett
Uvalde
Frio River
Nueces River
Natalia
Lytle
Somerset
Devine
Leming
Frio Town
Moore
Bigfoot
Poteet
Jourdanton
Pleasanton
Quemado
Normandy
La Pryor
Batesville
Pearsall
Charlotte
Christine
Eagle Pass
Piedras Negras
Crystal City
Divot
Derby
Hindes
Campbellton
Whitsett
Dilley
Brundage
Big Wells
Woodward
Millett
Los Angeles
Three Rivers
Tilden
Carrizo Springs
Asherton
Cotulla
Fowlerton
Calliham
George West
Catarina
TEXAS
Artesia Wells
UNITED STATES
MEXICO
Nueces
Encinal
Nueva Rosita
Freer
Orange Grove
COAHUILA
San Diego
Ben Bolt
Benavides
Laredo
Nuevo Laredo
Oilton
Realitos
Rio Grande
Concepcion
Mirando City
Bruni
Ramirez
Hebbronville
Monclova
San Ygnacio
Escobas
Randado
Bustamante
Encino
Falcon Res.
Lopeno
Falcon
Santa Elena
San Isidro
La Gloria
La Reforma
Falcon State Park
Sabinas Hidalgo
Nuevo Guerrero
El Sauz
Cd. Mier
Roma
Rio Grande City
Cd. Camargo
La Grulla
La Joya
Mission
Edinburg
Sullivan City
Bentsen-Rio Grande Valley State Park
NUEVO LEON
Presa De El Azucar
Reynosa
Hidalgo
San Pedro de las Colonias
Monterrey

6 7 8 9 10 11

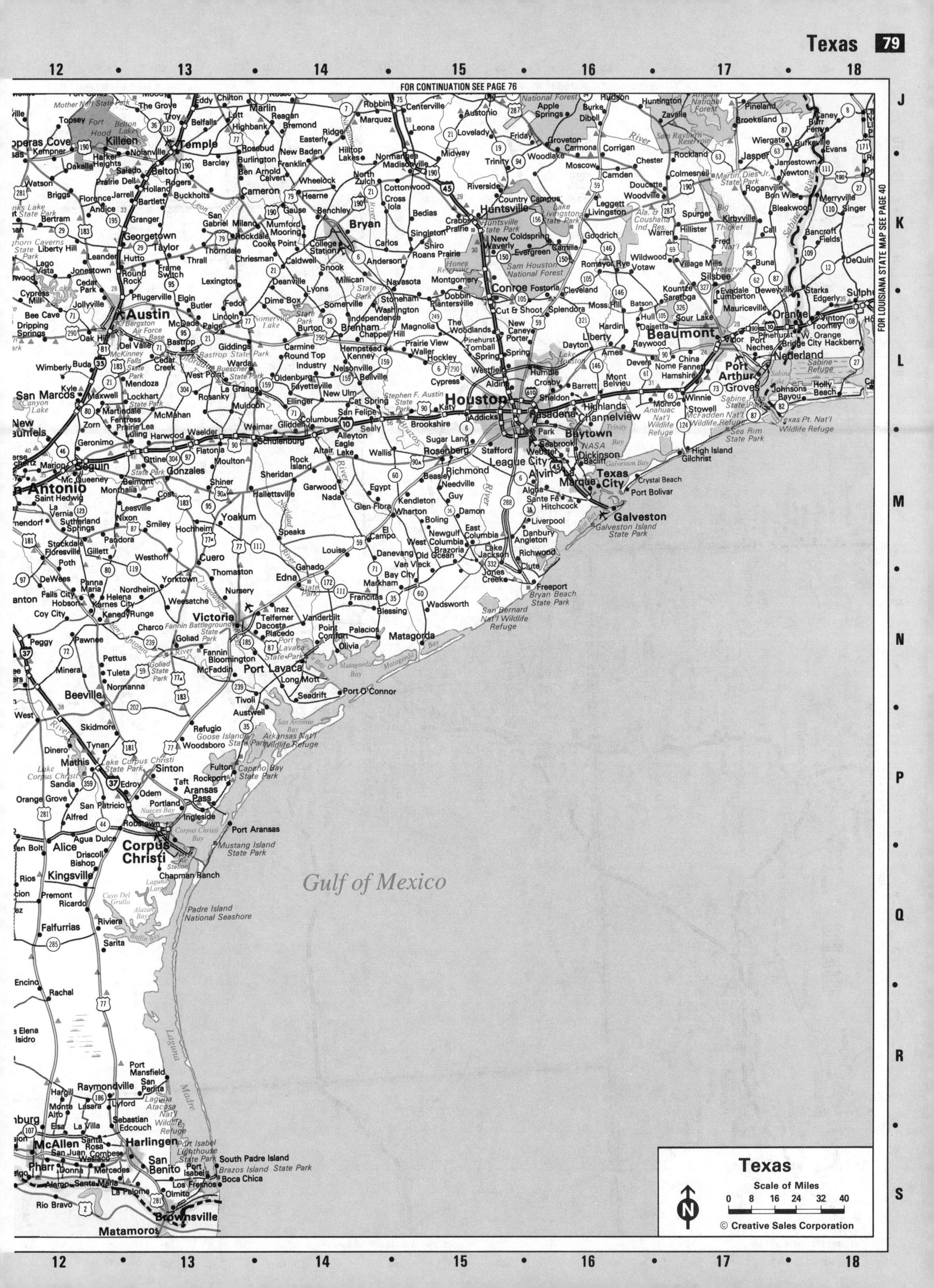

12 13 14 15 16 17 18
J K L M N P Q R S
FOR CONTINUATION SEE PAGE 76
FOR LOUISIANA STATE MAP SEE PAGE 40
Gulf of Mexico
Houston
Austin
San Antonio
Beaumont
Port Arthur
Galveston
Texas City
Pasadena
Baytown
Bryan
Huntsville
Conroe
Killeen
Temple
Victoria
Port Lavaca
Corpus Christi
Kingsville
Harlingen
McAllen
Brownsville
Matamoros
Padre Island National Seashore
Laguna Madre
South Padre Island
Sam Houston National Forest
Texas
Scale of Miles
0 8 16 24 32 40
N
© Creative Sales Corporation

FOR WYOMING STATE MAP SEE PAGES 88-89

FOR COLORADO STATE MAP SEE PAGES 22-23

FOR IDAHO STATE MAP SEE PAGE 31

FOR IDAHO STATE MAP SEE PAGE 31

FOR NEVADA STATE MAP SEE PAGE 54

Utah

Scale of Miles

0 7 14 21 28 35

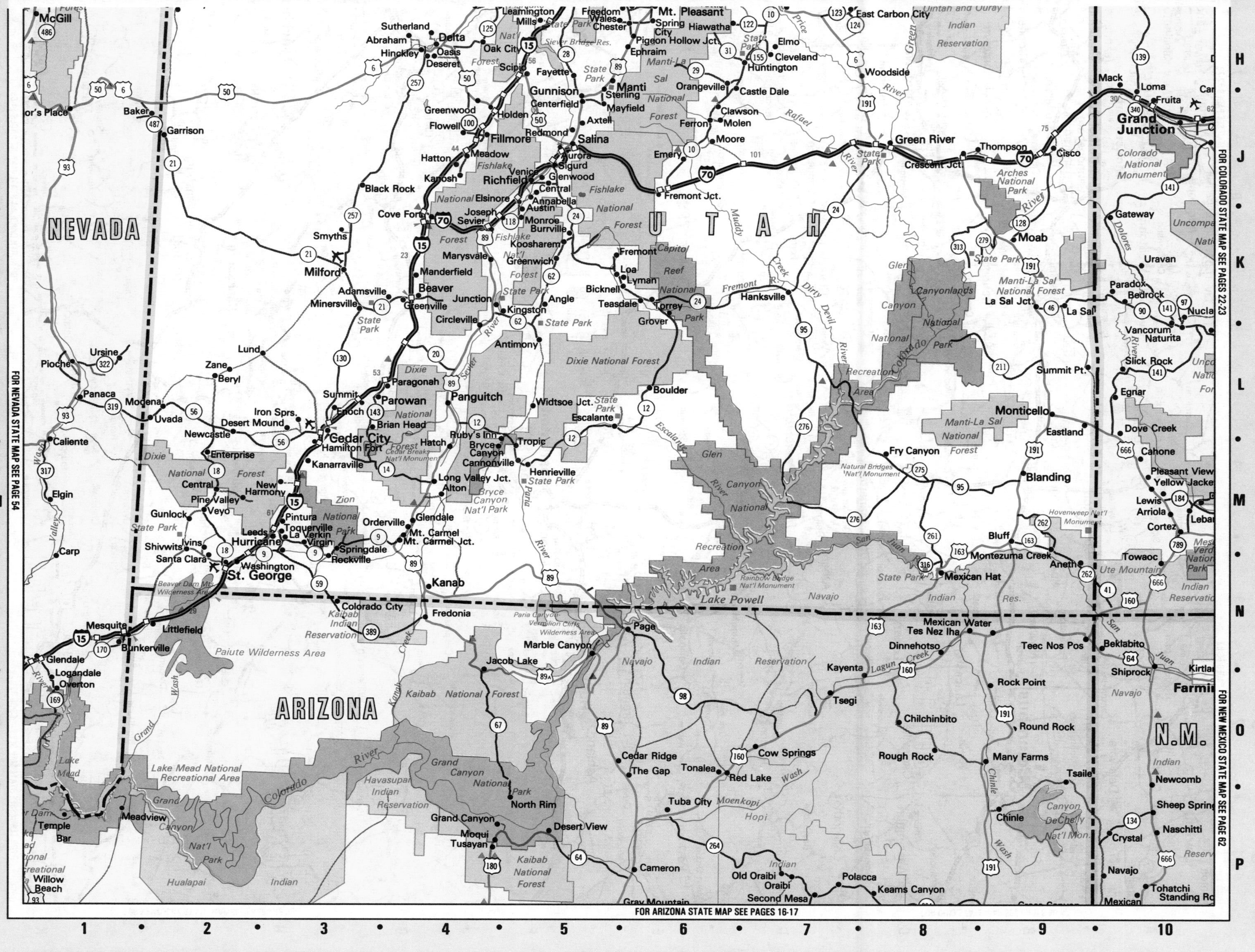
FOR COLORADO STATE MAP SEE PAGES 22-23
FOR NEW MEXICO STATE MAP SEE PAGE 62
FOR ARIZONA STATE MAP SEE PAGES 16-17
FOR NEVADA STATE MAP SEE PAGE 54
U T A H
NEVADA
ARIZONA
N.M.
Green River
Moab
Monticello
Blanding
Bluff
Mexican Hat
Hanksville
Richfield
Salina
Manti
Fillmore
Delta
Beaver
Milford
Panguitch
Parowan
Cedar City
St. George
Kanab
Page
Lake Powell
Grand Junction
Cortez
Shiprock
Kayenta
Tuba City
Mesquite
Littlefield
Fredonia
Escalante
Boulder
Torrey
Loa
Capitol Reef National Park
Canyonlands National Park
Arches National Park
Glen Canyon National Recreation Area
Bryce Canyon Nat'l Park
Zion National Park
Grand Canyon National Park
Dixie National Forest
Fishlake National Forest
Manti-La Sal National Forest
Kaibab National Forest
Lake Mead National Recreational Area

FOR OHIO STATE MAP SEE PAGES 66-67

FOR PENNSYLVANIA STATE MAP SEE PAGES 72-73

FOR OHIO STATE MAP SEE PAGES 66-67

FOR KENTUCKY STATE MAP SEE PAGES 38-39

FOR TENNESSEE STATE MAP SEE PAGES 38-39

FOR NORTH CAROLINA STATE MAP SEE PAGES 64-65

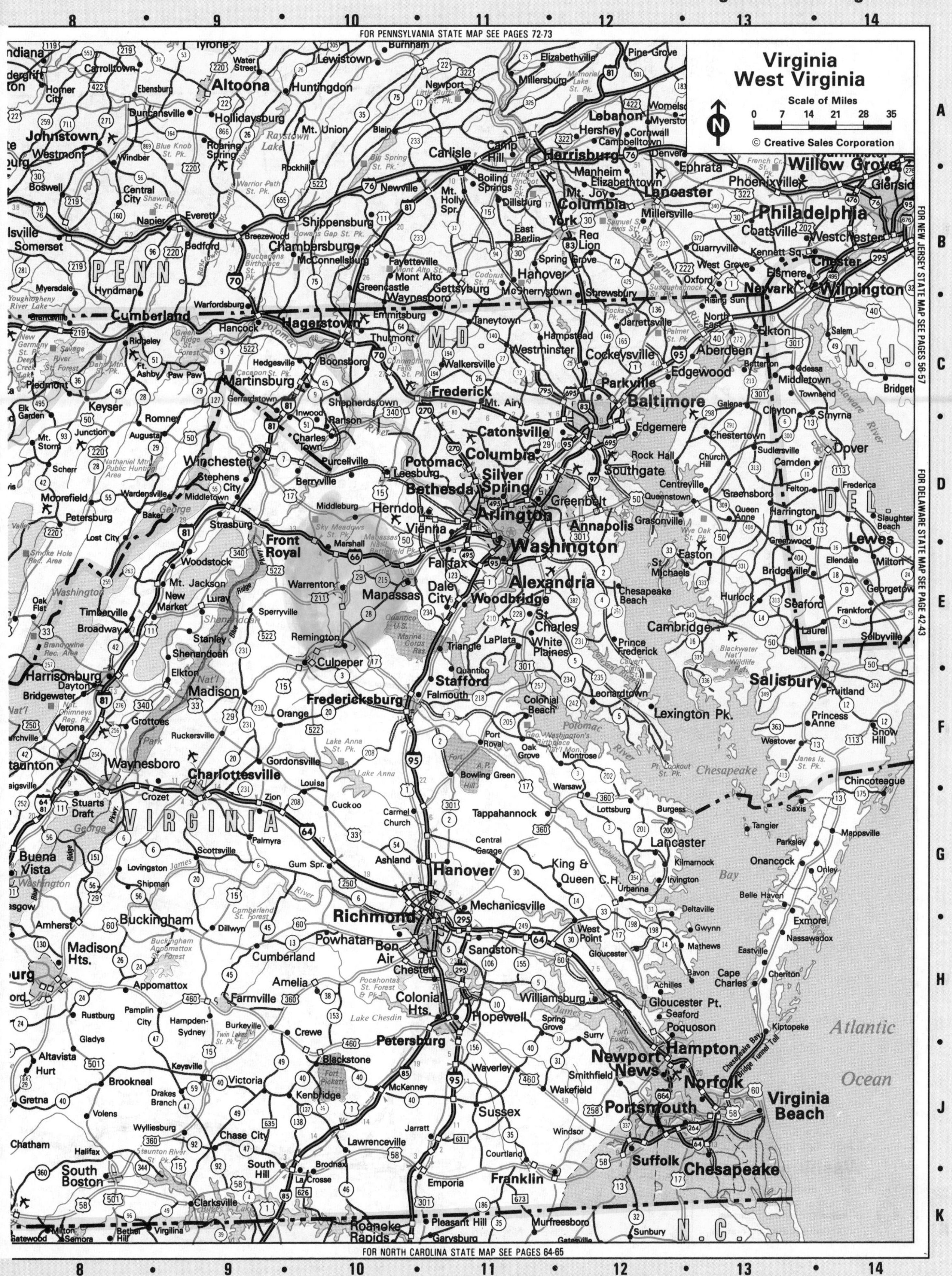
FOR PENNSYLVANIA STATE MAP SEE PAGES 72-73
FOR NEW JERSEY STATE MAP SEE PAGES 56-57
FOR DELAWARE STATE MAP SEE PAGE 42-43
FOR NORTH CAROLINA STATE MAP SEE PAGES 64-65
Virginia
West Virginia
Scale of Miles
0 7 14 21 28 35
© Creative Sales Corporation
8
9
10
11
12
13
14
A
B
C
D
E
F
G
H
J
K
PENN
MD
N. J.
DEL
VIRGINIA
N. C.
Altoona
Johnstown
Harrisburg
Lancaster
Philadelphia
York
Chambersburg
Hagerstown
Cumberland
Wilmington
Baltimore
Frederick
Martinsburg
Winchester
Washington
Arlington
Alexandria
Annapolis
Dover
Salisbury
Front Royal
Harrisonburg
Charlottesville
Fredericksburg
Richmond
Petersburg
Williamsburg
Newport News
Hampton
Norfolk
Portsmouth
Virginia Beach
Chesapeake
Suffolk
Franklin
Emporia
South Boston
Chesapeake Bay
Atlantic
Ocean

Washington

Scale of Miles

0 6 12 18 24 30

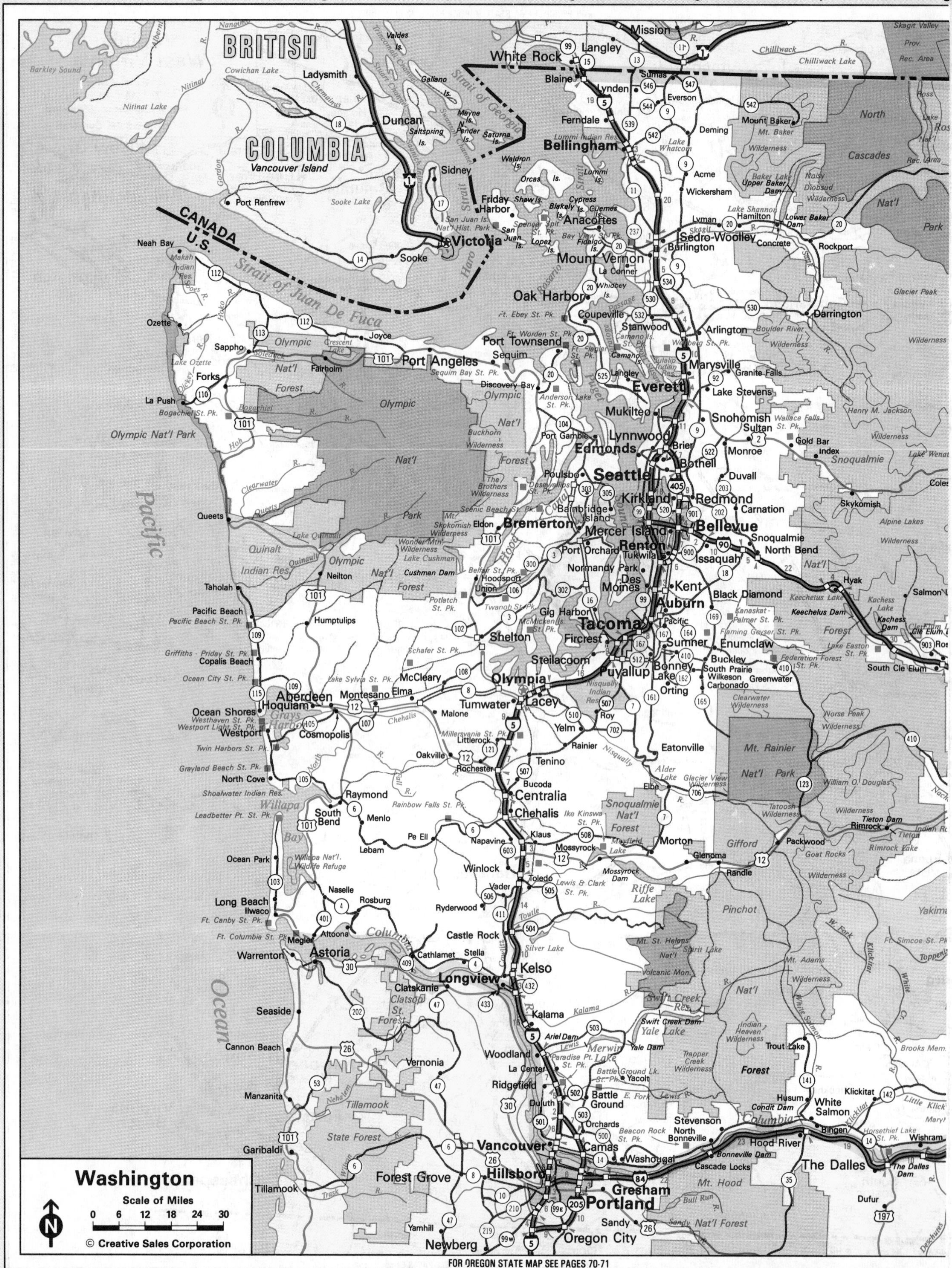

FOR OREGON STATE MAP SEE PAGES 70-71

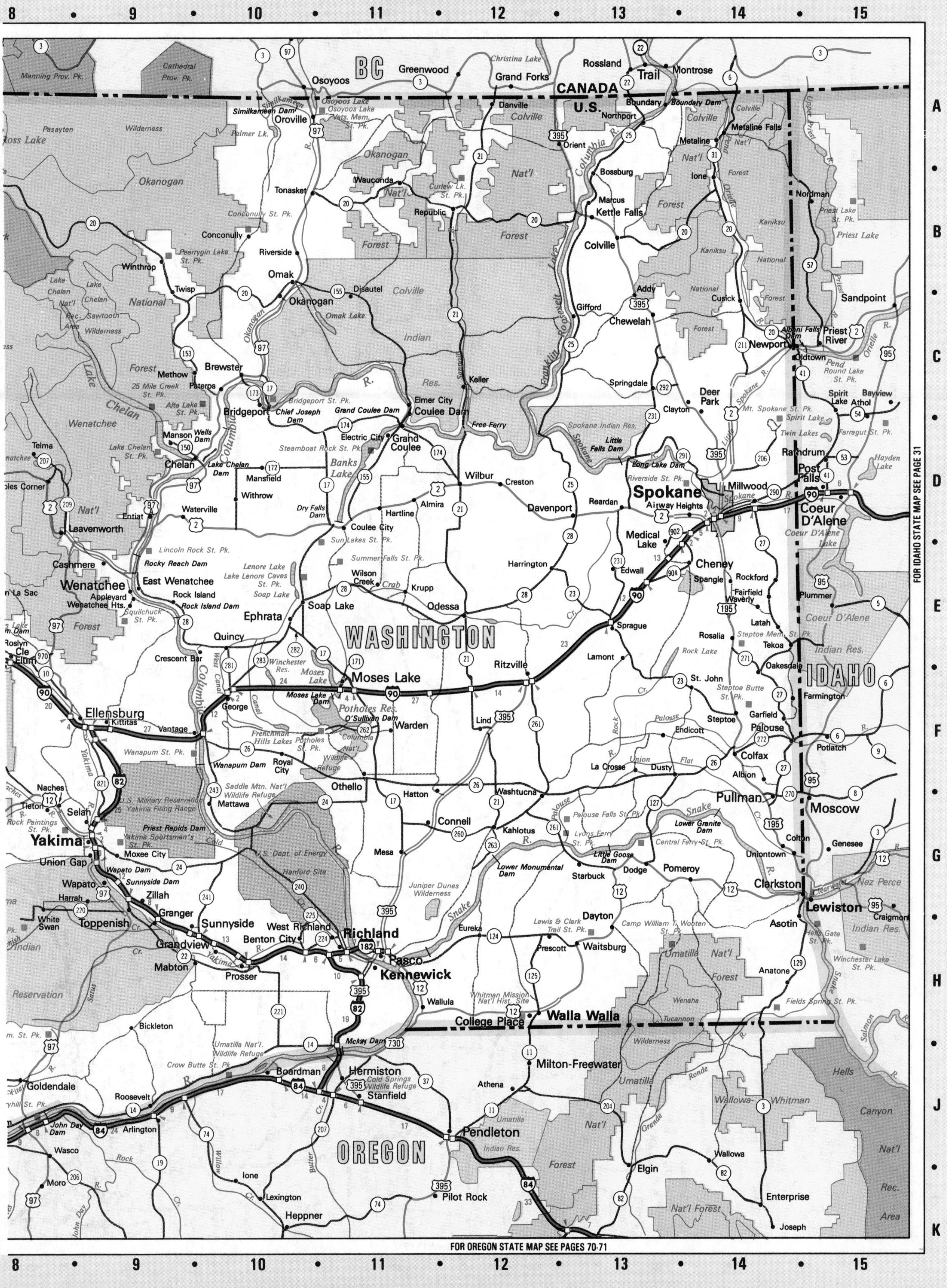

BC
CANADA
U.S.
WASHINGTON
IDAHO
OREGON

Osoyoos
Greenwood
Grand Forks
Rossland
Trail
Montrose
Manning Prov. Pk.
Cathedral Prov. Pk.
Christina Lake
Oroville
Similkameen Dam
Osoyoos Lake Vets. Mem. St. Pk.
Danville
Boundary
Boundary Dam
Northport
Orient
Metaline Falls
Metaline
Ione
Colville Nat'l Forest
Okanogan Nat'l Forest
Pasayten Wilderness
Ross Lake
Palmer Lk.
Tonasket
Wauconda
Curlew Lk. St. Pk.
Republic
Bossburg
Marcus
Kettle Falls
Colville
Conconully St. Pk.
Conconully
Riverside
Omak
Okanogan
Disautel
Omak Lake
Colville Indian Res.
Pearrygin Lake St. Pk.
Winthrop
Twisp
Lake Chelan Sawtooth Wilderness
Gifford
Chewelah
Addy
Cusick
Kaniksu National Forest
Nordman
Priest Lake St. Pk.
Priest Lake
Sandpoint
Newport
Albeni Falls Dam
Priest River
Oldtown
Round Lake St. Pk.
Methow
Brewster
Pateros
25 Mile Creek St. Pk.
Alta Lake St. Pk.
Bridgeport
Bridgeport St. Pk.
Chief Joseph Dam
Elmer City
Coulee Dam
Grand Coulee Dam
Keller
Springdale
Deer Park
Clayton
Spirit Lake
Athol
Bayview
Mt. Spokane St. Pk.
Farragut St. Pk.
Wenatchee
Manson
Wells Dam
Lake Chelan St. Pk.
Chelan
Lake Chelan Dam
Electric City
Grand Coulee
Steamboat Rock St. Pk.
Banks Lake
Free Ferry
Spokane Indian Res.
Little Falls Dam
Long Lake Dam
Rathdrum
Hayden Lake
Post Falls
Telma
Mansfield
Withrow
Wilbur
Creston
Riverside St. Pk.
Spokane
Millwood
Coeur D'Alene
Coeur D'Alene Lake
Leavenworth
Entiat
Waterville
Dry Falls Dam
Hartline
Almira
Davenport
Reardan
Airway Heights
Coulee City
Sun Lakes St. Pk.
Lincoln Rock St. Pk.
Rocky Reach Dam
Summer Falls St. Pk.
Medical Lake
Cheney
Cashmere
Wenatchee
East Wenatchee
Lenore Lake
Lake Lenore Caves St. Pk.
Soap Lake
Wilson Creek
Krupp
Harrington
Edwall
Spangle
Rockford
Fairfield
Waverly
Appleyard
Wenatchee Hts.
Rock Island
Rock Island Dam
Squilchuck St. Pk.
Ephrata
Odessa
Sprague
Latah
Plummer
Coeur D'Alene Indian Res.
Quincy
Crescent Bar
Winchester Res.
Moses Lake
Ritzville
Lamont
Rosalia
Steptoe Mem. St. Pk.
Tekoa
Rock Lake
Oakesdale
Cle Elum
Roslyn
Ellensburg
Kittitas
Vantage
George
Moses Lake Dam
Potholes Res.
O'Sullivan Dam
Warden
Lind
St. John
Steptoe Butte St. Pk.
Steptoe
Garfield
Farmington
Frenchman Hills Lakes
Potholes St. Pk.
Columbia Nat'l Wildlife Refuge
Endicott
Palouse
Potlatch
Wanapum St. Pk.
Wanapum Dam
Royal City
La Crosse
Dusty
Colfax
Albion
Naches
Tieton
Selah
Rock Paintings St. Pk.
U.S. Military Reservation Yakima Firing Range
Saddle Mtn. Nat'l Wildlife Refuge
Mattawa
Othello
Hatton
Washtucna
Pullman
Moscow
Yakima
Priest Rapids Dam
Yakima Sportsman's St. Pk.
Union Gap
Moxee City
Connell
Kahlotus
Palouse Falls St. Pk.
Lyons Ferry St. Pk.
Lower Granite Dam
Central Ferry St. Pk.
Colton
Genesee
Uniontown
U.S. Dept. of Energy Hanford Site
Mesa
Lower Monumental Dam
Little Goose Dam
Starbuck
Dodge
Pomeroy
Clarkston
Lewiston
Nez Perce Indian Res.
Craigmont
Wapato Dam
Sunnyside Dam
Wapato
Harrah
Zillah
Granger
White Swan
Toppenish
Sunnyside
Juniper Dunes Wilderness
West Richland
Richland
Benton City
Grandview
Mabton
Prosser
Eureka
Lewis & Clark Trail St. Pk.
Dayton
Camp William T. Wooten St. Pk.
Asotin
Hells Gate St. Pk.
Pasco
Kennewick
Prescott
Waitsburg
Umatilla Nat'l Forest
Anatone
Winchester Lake St. Pk.
Wallula
Whitman Mission Nat'l Hist. Site
Wenaha
Fields Spring St. Pk.
Tucannon
Bickleton
College Place
Walla Walla
Umatilla Nat'l. Wildlife Refuge
McNary Dam
Wilderness
Hells Canyon Nat'l Rec. Area
Crow Butte St. Pk.
Boardman
Hermiston
Cold Springs Wildlife Refuge
Milton-Freewater
Goldendale
Roosevelt
Stanfield
Athena
Umatilla
Wallowa-Whitman
John Day Dam
Arlington
Pendleton
Umatilla Indian Res.
Wasco
Moro
Ione
Lexington
Heppner
Pilot Rock
Elgin
Wallowa
Enterprise
Joseph

FOR IDAHO STATE MAP SEE PAGE 31

FOR OREGON STATE MAP SEE PAGES 70-71

United States Citizens Visiting Canada

All persons entering Canada must report to the Canadian Immigration and Customs Office at the Port of Entry and secure required permits for admission for their person and posessions. The transportation of plants and produce is rigidly controlled. Check with customs officials for complete regulations and requirements.

Canadian Citizens Visiting the United States

Passports or visas are not required of Canadian citizens or British subjects residing in Canada entering the United States for a period of six months or less, however, evidence of citizenship is required. Check with customs officials for complete regulations and requirements.

The Interstate Highway System in and around the Chicago area is confusing to many people. It is helpful to remember that, in most cases, Interstate Highways running north and south have odd numbers, and Interstate Highways running east and west have even numbers

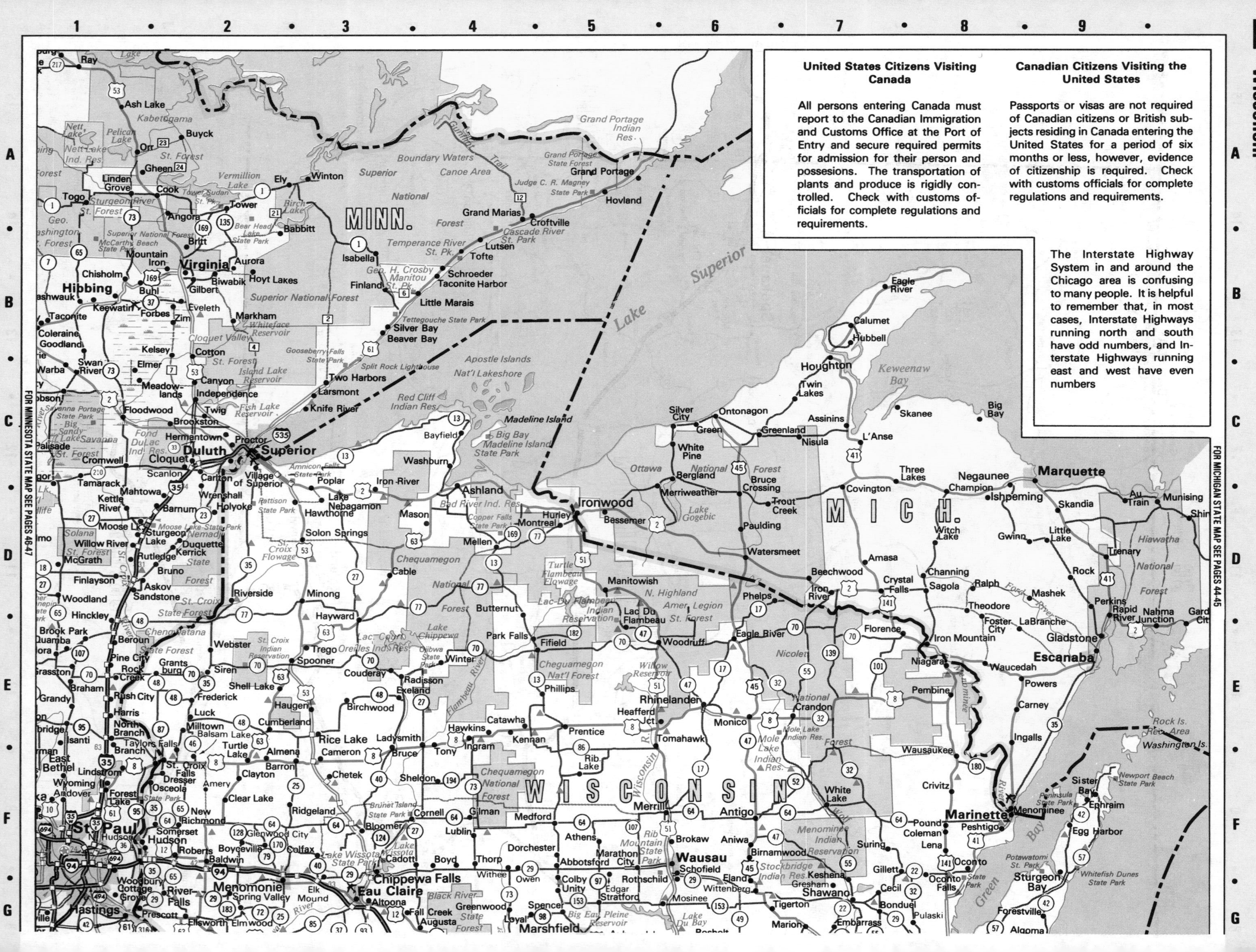

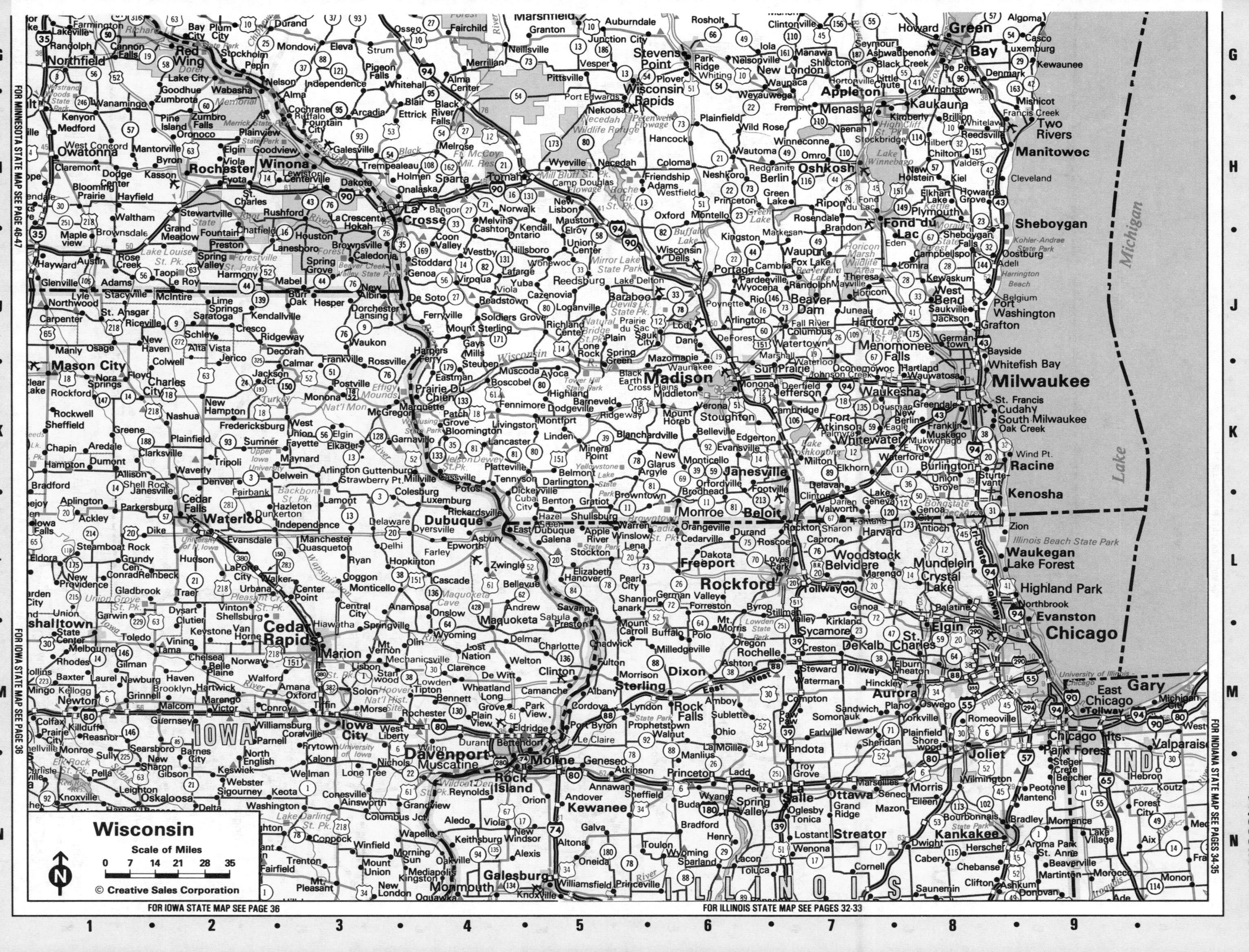
FOR INDIANA STATE MAP SEE PAGES 34-35
FOR ILLINOIS STATE MAP SEE PAGES 32-33
FOR IOWA STATE MAP SEE PAGE 36
FOR MINNESOTA STATE MAP SEE PAGE 46-47
Wisconsin
Scale of Miles
0 7 14 21 28 35
© Creative Sales Corporation
Lake Michigan
IOWA
ILLINOIS
Green Bay
Appleton
Oshkosh
Fond du Lac
Manitowoc
Sheboygan
Milwaukee
Waukesha
Madison
Racine
Kenosha
Janesville
Beloit
Rockford
Waukegan
Chicago
Gary
Joliet
Aurora
Dubuque
Davenport
Rock Island
Moline
Cedar Rapids
Waterloo
Iowa City
Rochester
Winona
La Crosse
Wisconsin Rapids
Stevens Point
Mason City

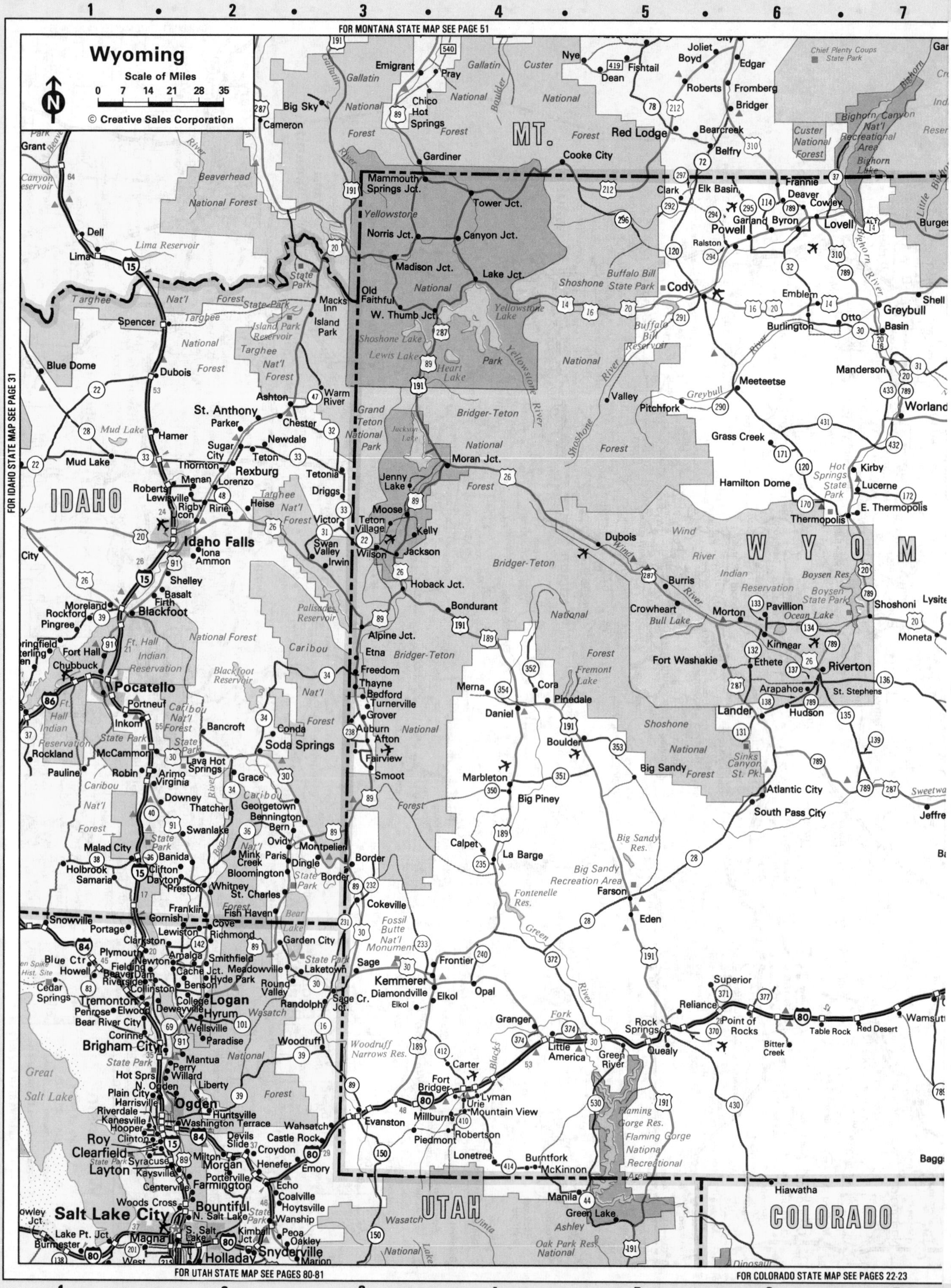

FOR MONTANA STATE MAP SEE PAGE 51

Wyoming

Scale of Miles

0 7 14 21 28 35

N

FOR IDAHO STATE MAP SEE PAGE 31

MT.

IDAHO

WYOM

UTAH

COLORADO

Emigrant
Pray
Gallatin National Forest
Custer National Forest
Nye
Fishtail
Dean
Joliet
Boyd
Edgar
Roberts
Fromberg
Bridger
Chief Plenty Coups State Park
Big Sky
Cameron
Chico Hot Springs
Red Lodge
Bearcreek
Belfry
Custer National Forest
Bighorn Canyon Nat'l Recreational Area
Bighorn Lake
Gardiner
Cooke City
Beaverhead National Forest
Grant
Canyon Reservoir
Mammoth Springs Jct.
Tower Jct.
Clark
Elk Basin
Frannie
Deaver
Cowley
Lovell
Garland
Byron
Powell
Ralston
Yellowstone
Norris Jct.
Canyon Jct.
Dell
Lima
Lima Reservoir
Madison Jct.
Lake Jct.
Buffalo Bill State Park
Cody
Shoshone
State Park
Macks Inn
Old Faithful
National
Emblem
Greybull
Shell
Basin
Targhee
Nat'l
Forest
State Park
W. Thumb Jct.
Yellowstone Lake
Spencer
Island Park Reservoir
Island Park
Shoshone Lake
Lewis Lake
Heart Lake
Park
Buffalo Bill Reservoir
Burlington
Otto
Blue Dome
Dubois
Targhee Nat'l Forest
National
River
Meeteetse
Manderson
Worland
Ashton
Warm River
Grand Teton National Park
Valley
Greybull
Pitchfork
St. Anthony
Parker
Chester
Mud Lake
Hamer
Jackson Lake
Bridger-Teton
Grass Creek
Newdale
Sugar City
Teton
Mud Lake
Thornton
Rexburg
Tetonia
Moran Jct.
National Forest
Kirby
Hot Springs State Park
Hamilton Dome
Lucerne
Menan
Lorenzo
Roberts
Lewisville
Rigby
Ririe
Heise
Driggs
Jenny Lake
Moose
E. Thermopolis
Thermopolis
Victor
Teton Village
Kelly
Idaho Falls
Iona
Ammon
Swan Valley
Irwin
Wilson
Jackson
Dubois
Wind River
City
Shelley
Hoback Jct.
Bridger-Teton
Indian Reservation
Boysen Res.
Boysen State Park
Basalt
Firth
Moreland
Rockford
Pingree
Blackfoot
Bondurant
Burris
Crowheart
Bull Lake
Morton
Pavillion
Ocean Lake
Shoshoni
Lysite
Moneta
Palisades Reservoir
Alpine Jct.
National Forest
Caribou
Etna
Bridger-Teton
Kinnear
Springfield
Sterling
Fort Hall
Ft. Hall Indian Reservation
Chubbuck
Blackfoot Reservoir
Freedom
Fort Washakie
Ethete
Riverton
Pocatello
Thayne
Bedford
Turnerville
Merna
Cora
Fremont Lake
Pinedale
Arapahoe
St. Stephens
Portneuf
Caribou Nat'l Forest
Grover
Auburn
Afton
National
Daniel
Lander
Hudson
Inkom
Bancroft
Conda
Soda Springs
Forest
Shoshone
Ft. Hall Indian Reservation
State Park
Rockland
McCammon
Lava Hot Springs
Fairview
Boulder
National
Sinks Canyon St. Pk.
Pauline
Robin
Arimo
Virginia
Grace
Smoot
Big Sandy
Marbleton
Forest
Atlantic City
Sweetwater
Caribou Nat'l Forest
Downey
Thatcher
Georgetown
Bennington
Big Piney
South Pass City
Jeffrey
Swanlake
Bern
Forest
Forest
Calpet
Big Sandy Res.
Malad City
Banida
Ovid
Paris
Montpelier
Mink Creek
Dingle
Bear
Border
La Barge
Holbrook
Samaria
Clifton
Dayton
Bloomington
State Park
Preston
Whitney
St. Charles
Big Sandy Recreation Area
Fontenelle Res.
Farson
Cokeville
Franklin
Fish Haven
Cornish
Snowville
Portage
Cove
Lewiston
Richmond
Eden
Fossil Butte Nat'l Monument
Garden City
Bear Lake
Clarkston
Plymouth
Amalga
Smithfield
Green
Blue Ctr.
Howell
Golden Spike Hist. Site
Fielding
Beaver Dam
Riverside
Newton
Cache Jct.
Meadowville
Laketown
State Park
Sage
Frontier
Kemmerer
Diamondville
Elkol
Opal
Superior
Cedar Springs
Collinston
Benson
Hyde Park
Round Valley
Tremonton
Penrose
Elwood
Logan
Dewyville
College
Hyrum
Randolph
Sage Cr. Jct.
Reliance
Bear River City
Corinne
Wellsville
Wasatch
Granger
Fork
Rock Springs
Point of Rocks
Table Rock
Red Desert
Wamsutter
Brigham City
Paradise
Woodruff
Woodruff Narrows Res.
Little America
Quealy
Bitter Creek
State Park
Mantua
National
Green River
Great Salt Lake
Hot Sprs.
Perry
Willard
Carter
Blacks
N. Ogden
Liberty
Fort Bridger
Plain City
Harrisville
Ogden
Forest
Lyman
Urie
Mountain View
Flaming Gorge Res.
Riverdale
Kanesville
Hooper
Huntsville
Washington Terrace
Wahsatch
Evanston
Millburne
Robertson
Flaming Gorge National Recreational Area
Roy
Clinton
Devils Slide
Castle Rock
Croydon
Piedmont
Clearfield
Syracuse
State Park
Milton
Morgan
Henefer
Emory
Lonetree
Burntfork
McKinnon
Baggs
Layton
Kaysville
Porterville
Centerville
Farmington
Echo
Coalville
Hiawatha
Woods Cross
Bountiful
Hoytsville
Wasatch
Manila
Green Lake
Ashley
Uinta
Salt Lake City
Cowley Jct.
N. Salt Lake
State Park
Wanship
Lake Pt. Jct.
Burmester
S. Salt Lake
Kimball Jct.
Peoa
Oakley
Magna
West
Holladay
Snyderville
Marion
National
Oak Park Res.
National
Dinosaur

FOR UTAH STATE MAP SEE PAGES 80-81

FOR COLORADO STATE MAP SEE PAGES 22-23

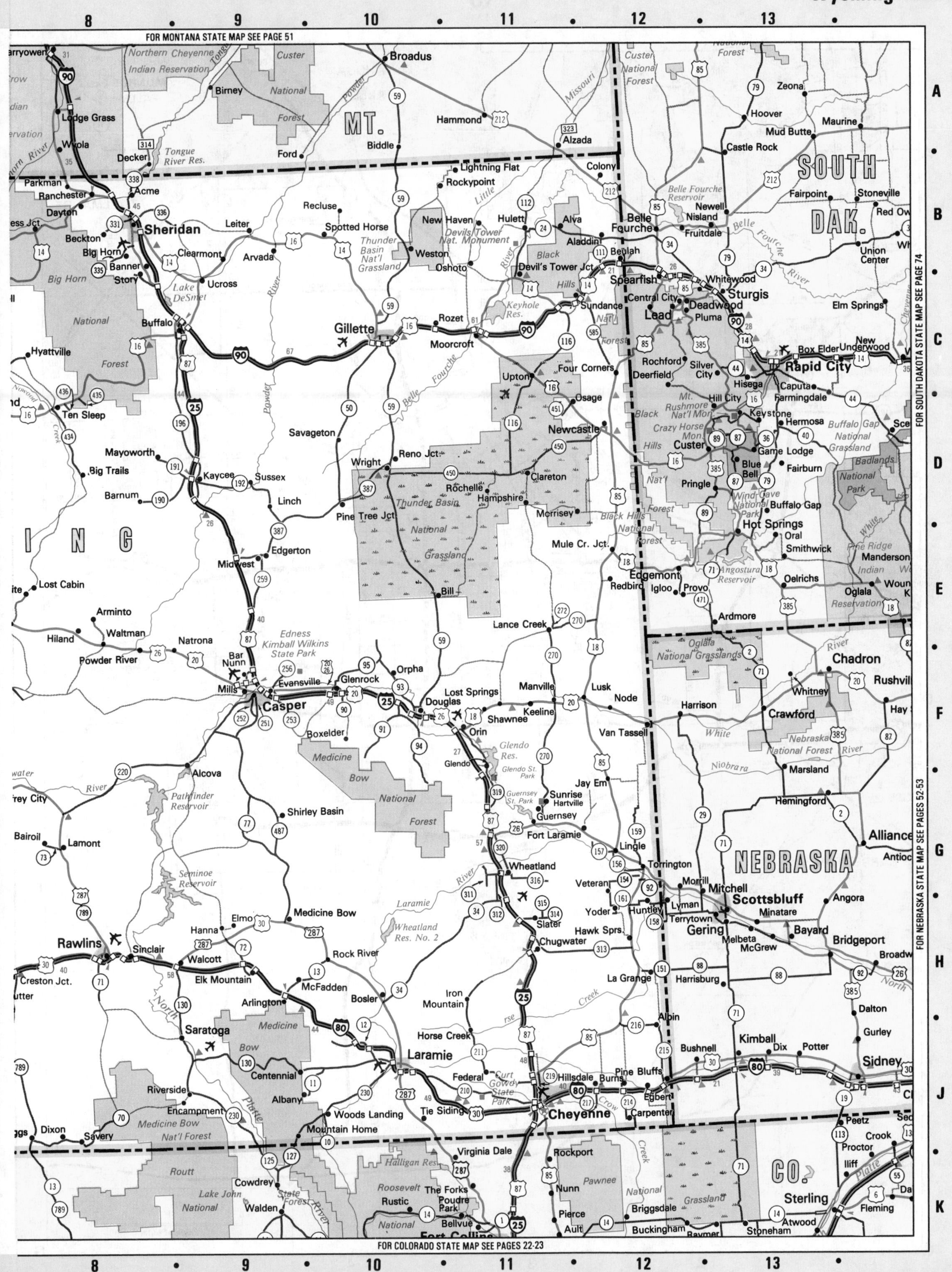

8
9
10
11
12
13
A
B
C
D
E
F
G
H
J
K
FOR MONTANA STATE MAP SEE PAGE 51
FOR SOUTH DAKOTA STATE MAP SEE PAGE 74
FOR NEBRASKA STATE MAP SEE PAGES 52-53
FOR COLORADO STATE MAP SEE PAGES 22-23
MT.
SOUTH DAK.
NEBRASKA
CO.
I N G
Northern Cheyenne Indian Reservation
Custer National Forest
Broadus
Birney
Decker
Tongue River Res.
Ford
Biddle
Hammond
Alzada
Lodge Grass
Wyola
Parkman
Ranchester
Dayton
Acme
Sheridan
Beckton
Big Horn
Banner
Story
Leiter
Clearmont
Arvada
Ucross
Lake DeSmet
Buffalo
Recluse
Spotted Horse
Thunder Basin Nat'l Grassland
Lightning Flat
Rockypoint
New Haven
Weston
Oshoto
Hulett
Alva
Devils Tower Nat. Monument
Aladdin
Colony
Belle Fourche
Beulah
Devil's Tower Jct.
Spearfish
Sundance
Keyhole Res.
Gillette
Rozet
Moorcroft
Big Horn National Forest
Hyattville
Ten Sleep
Mayoworth
Big Trails
Barnum
Kaycee
Sussex
Linch
Savageton
Wright
Reno Jct.
Pine Tree Jct.
Rochelle
Hampshire
Clareton
Morrisey
Upton
Four Corners
Osage
Newcastle
Mule Cr. Jct.
Black Hills National Forest
Midwest
Edgerton
Lost Cabin
Arminto
Waltman
Hiland
Powder River
Natrona
Bill
Lance Creek
Redbird
Edgemont
Igloo
Provo
Ardmore
Edness Kimball Wilkins State Park
Bar Nunn
Mills
Evansville
Casper
Glenrock
Orpha
Douglas
Lost Springs
Manville
Lusk
Node
Shawnee
Keeline
Van Tassell
Orin
Boxelder
Medicine Bow National Forest
Glendo
Glendo Res.
Glendo St. Park
Alcova
Pathfinder Reservoir
Shirley Basin
Sunrise
Hartville
Guernsey
Guernsey St. Park
Jay Em
Fort Laramie
Lingle
Bairoil
Lamont
Seminoe Reservoir
Wheatland
Torrington
Veteran
Yoder
Huntley
Elmo
Hanna
Medicine Bow
Rawlins
Sinclair
Walcott
Rock River
Wheatland Res. No. 2
Chugwater
Slater
Hawk Sprs.
Elk Mountain
McFadden
Creston Jct.
Arlington
Bosler
Iron Mountain
La Grange
Saratoga
Horse Creek
Albin
Laramie
Centennial
Riverside
Encampment
Albany
Federal
Curt Gowdy State Park
Hillsdale
Burns
Pine Bluffs
Egbert
Carpenter
Tie Siding
Cheyenne
Woods Landing
Mountain Home
Medicine Bow Nat'l Forest
Dixon
Savery
Virginia Dale
Rockport
Routt National Forest
Cowdrey
Walden
Lake John
Halligan Res.
Roosevelt National
Rustic
The Forks
Poudre Park
Bellvue
Fort Collins
Nunn
Pierce
Ault
Pawnee National Grassland
Briggsdale
Buckingham
Raymer
Stoneham
Atwood
Sterling
Fleming
Iliff
Proctor
Crook
Peetz
Sidney
Potter
Dix
Kimball
Bushnell
Gurley
Dalton
Bridgeport
Bayard
Minatare
Scottsbluff
Gering
Terrytown
Mitchell
Morrill
Lyman
Melbeta
McGrew
Harrisburg
Angora
Alliance
Hemingford
Marsland
Crawford
Nebraska National Forest
Harrison
Whitney
Chadron
Rushville
Oglala National Grasslands
Oelrichs
Oglala Indian Reservation
Angostura Reservoir
Hot Springs
Oral
Smithwick
Buffalo Gap
Wind Cave National Park
Pringle
Custer
Blue Bell
Game Lodge
Fairburn
Hermosa
Keystone
Hill City
Mt. Rushmore Nat'l Mon.
Crazy Horse Mon.
Rochford
Deerfield
Silver City
Hisega
Rapid City
Box Elder
New Underwood
Caputa
Farmingdale
Buffalo Gap National Grassland
Badlands National Park
Lead
Deadwood
Central City
Pluma
Sturgis
Whitewood
Elm Springs
Fruitdale
Nisland
Newell
Belle Fourche Reservoir
Fairpoint
Stoneville
Union Center
Castle Rock
Mud Butte
Maurine
Hoover
Zeona

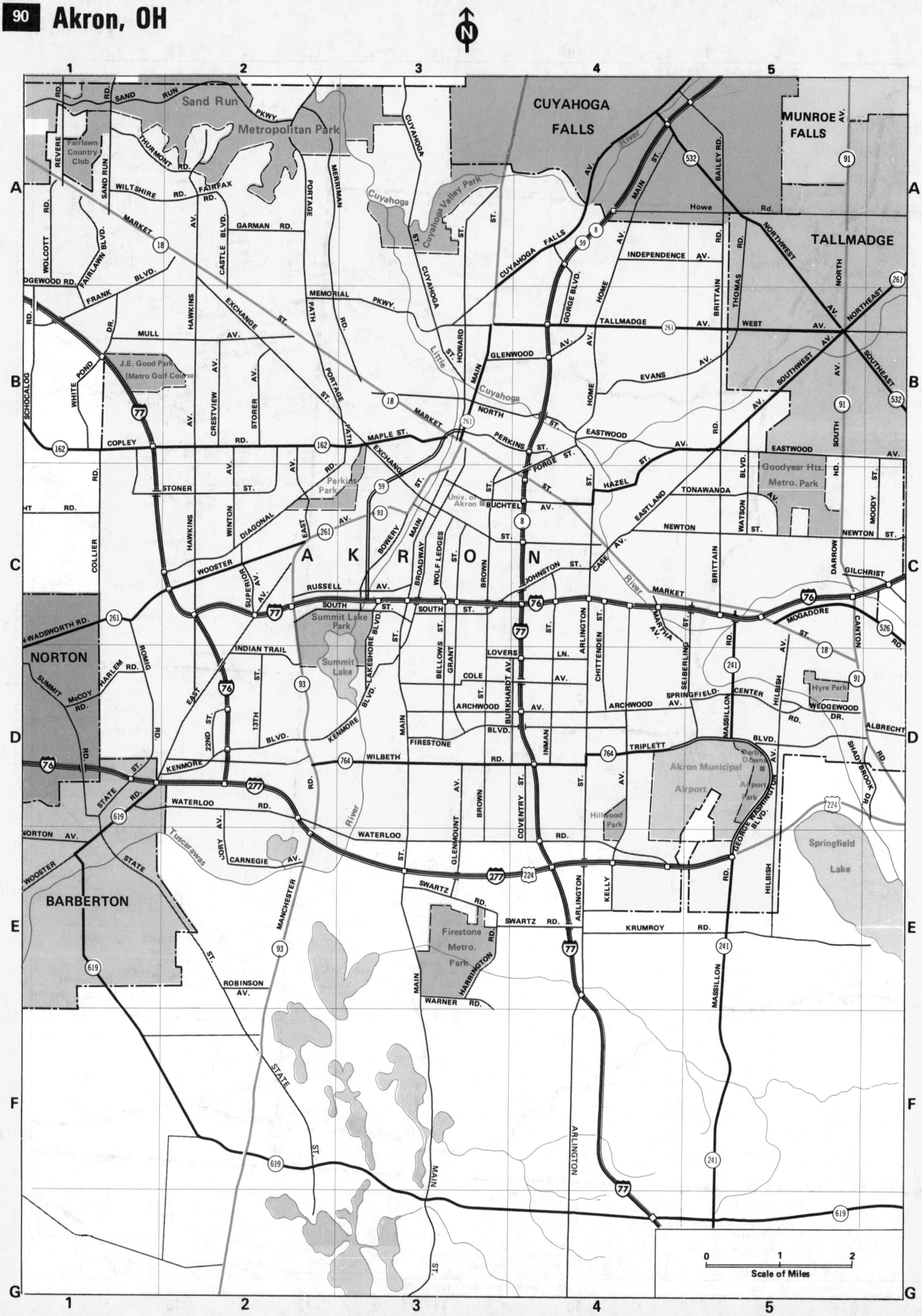

Akron, OH

CUYAHOGA FALLS

MUNROE FALLS

TALLMADGE

Sand Run Metropolitan Park

Fairlawn Country Club

Cuyahoga Valley Park

J.E. Good Park (Metro Golf Course)

Perkins Park

Univ. of Akron

Goodyear Hts. Metro. Park

AKRON

NORTON

Summit Lake Park

Summit Lake

Hyre Park

Derby Downs

Akron Municipal Airport

Airport Park

Hillwood Park

Springfield Lake

BARBERTON

Firestone Metro. Park

Scale of Miles

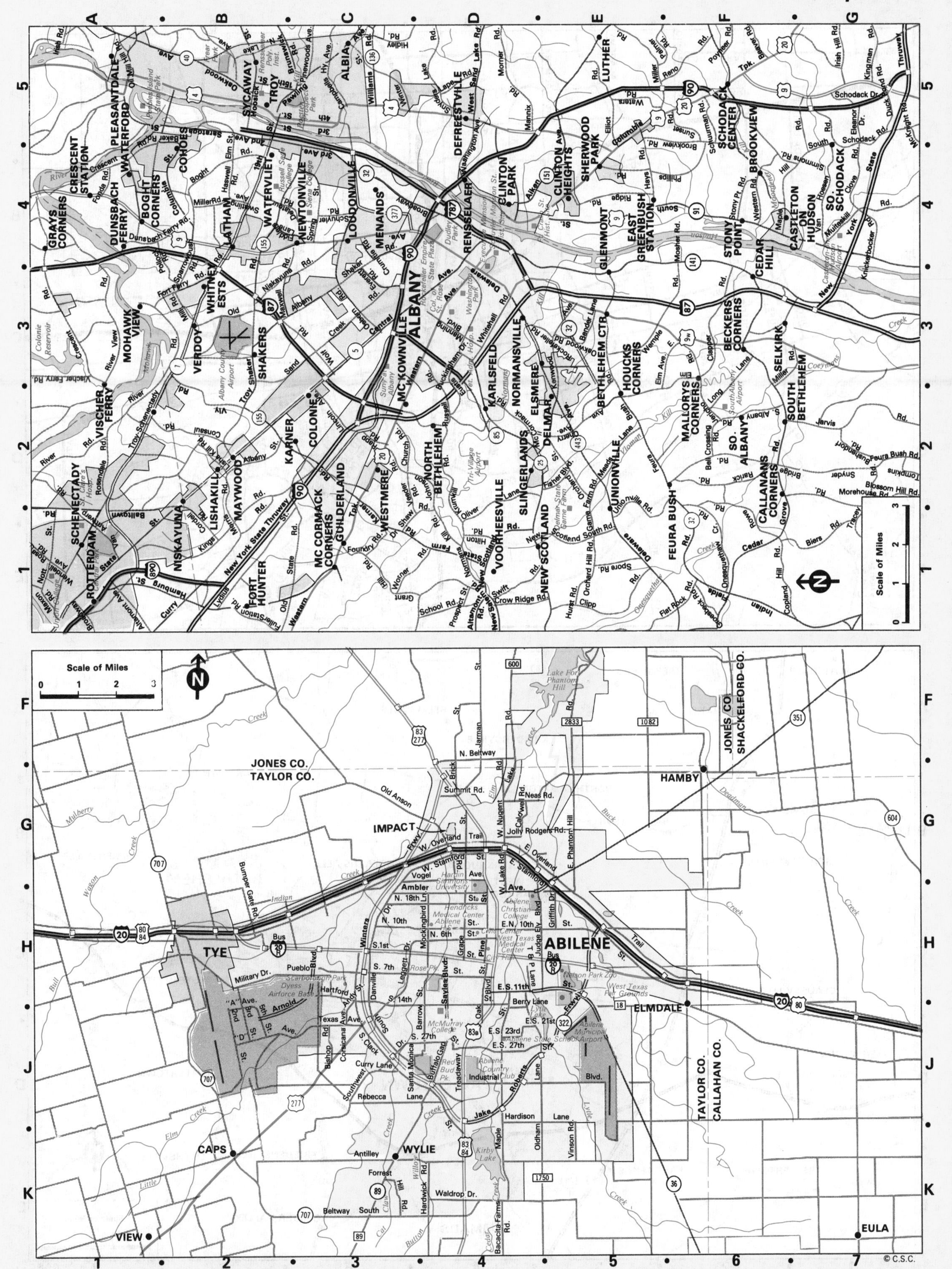
A
B
C
D
E
F
G
1
2
3
4
5
GRAYS CORNERS
CRESCENT STATION
DUNSBACH FERRY
BOGHT CORNERS
WATERFORD
PLEASANTDALE
COHOES
LATHAM
WATERVLIET
TROY
SYCAWAY
ALBIA
NEWTONVILLE
LOUDONVILLE
MENANDS
DEFREESTVILLE
RENSSELAER
CLINTON PARK
CLINTON HEIGHTS
SHERWOOD PARK
LUTHER
GLENMONT
EAST GREENBUSH STATION
SCHODACK CENTER
BROOKVIEW
STONY POINT
CEDAR HILL
CASTLETON ON HUDSON
SO. SCHODACK
SCHODACK
MOHAWK VIEW
VERDOY
WHITNEY ESTS
SHAKERS
ALBANY
MC KOWNVILLE
KARLSFELD
NORMANSVILLE
ELSMERE
DELMAR
BETHLEHEM CTR.
HOUCKS CORNERS
MALLORYS CORNERS
BECKERS CORNERS
SELKIRK
SOUTH BETHLEHEM
SO. ALBANY
VISCHER FERRY
SCHENECTADY
LISHAKILL
MAYWOOD
KARNER
COLONIE
NORTH BETHLEHEM
SLINGERLANDS
UNIONVILLE
FEURA BUSH
CALLANANS CORNERS
ROTTERDAM
NISKAYUNA
FORT HUNTER
MC CORMACK CORNERS
GUILDERLAND
WESTMERE
VOORHEESVILLE
NEW SCOTLAND
New York State Thruway
Albany County Airport
Colonie Reservoir
Scale of Miles
0
1
2
3
Scale of Miles
0
1
2
3
JONES CO.
TAYLOR CO.
JONES CO.
SHACKELFORD CO.
HAMBY
IMPACT
TYE
ABILENE
ELMDALE
CAPS
WYLIE
VIEW
EULA
TAYLOR CO.
CALLAHAN CO.
Dyess Airforce Base
Hardin Simmons University
Abilene Christian College
McMurray College
Abilene Country Club
Abilene Municipal Airport
Abilene State School Airport
West Texas Fair Grounds
Nelson Park Zoo
Lake Fort Phantom Hill
Kirby Lake
F
G
H
J
K
1
2
3
4
5
6
7
© C.S.C.

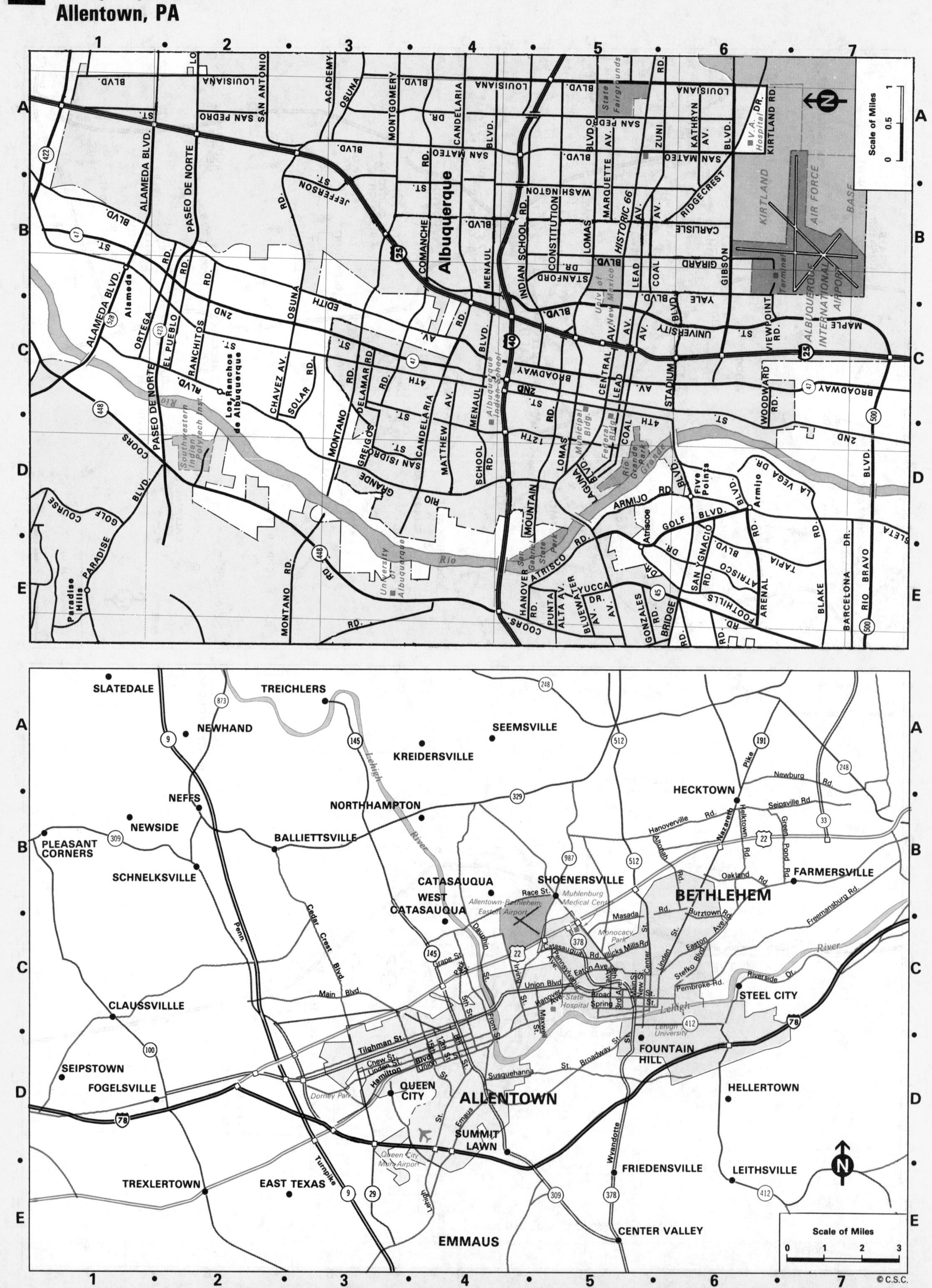
Albuquerque
Kirtland Air Force Base
Albuquerque International Airport
Univ. of New Mexico
University of Albuquerque
Rio Grande
Historic 66
ALLENTOWN
BETHLEHEM
CATASAUQUA
WEST CATASAUQUA
FOUNTAIN HILL
EMMAUS
STEEL CITY
HELLERTOWN
Lehigh River
Allentown-Bethlehem-Easton Airport
Queen City Muni Airport
Lehigh University
Scale of Miles
©C.S.C.

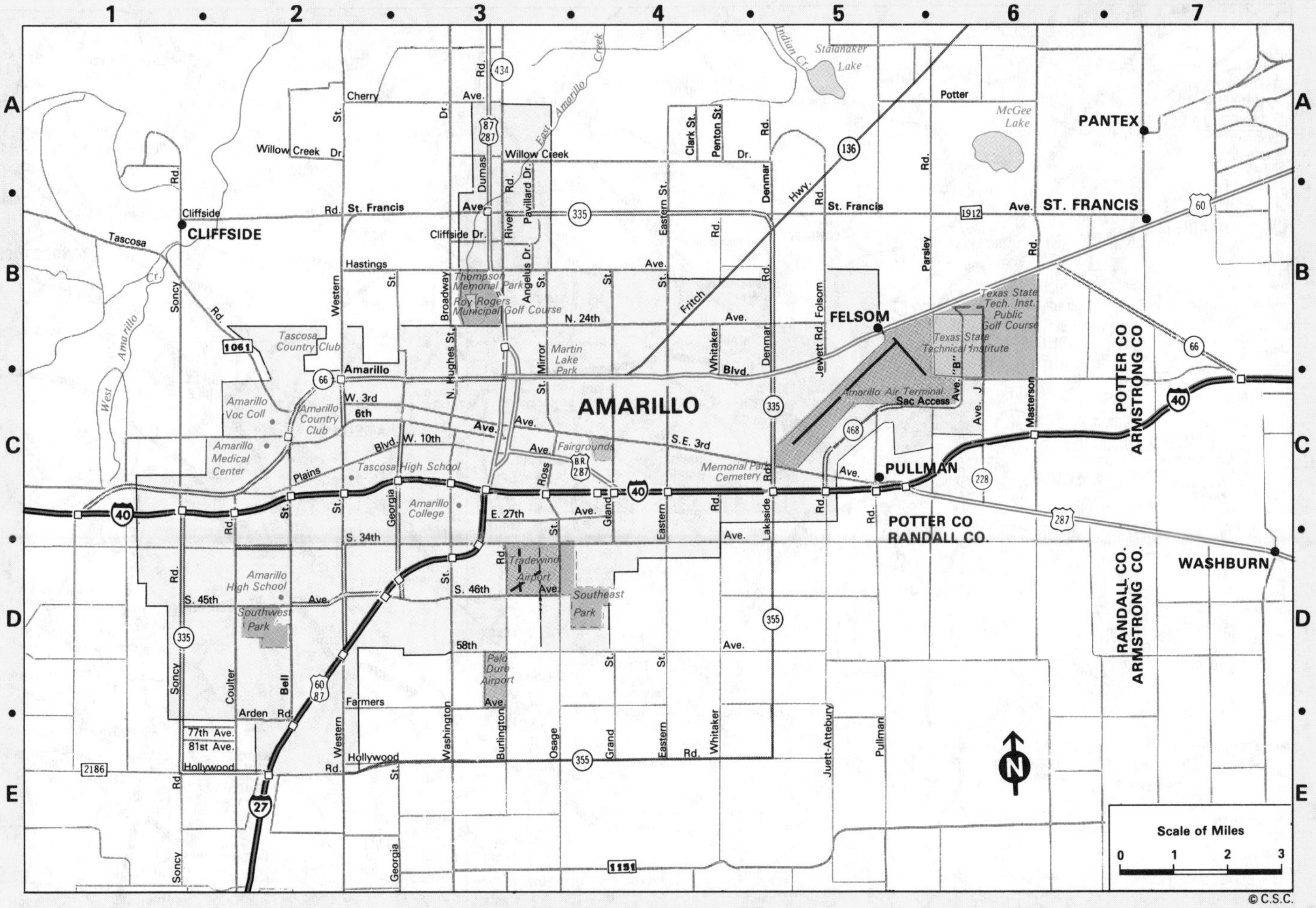
AMARILLO
CLIFFSIDE
FELSOM
PULLMAN
PANTEX
ST. FRANCIS
WASHBURN
POTTER CO
RANDALL CO.
POTTER CO
ARMSTRONG CO
RANDALL CO.
ARMSTRONG CO.
Stalanaker Lake
McGee Lake
Tascosa Country Club
Amarillo Voc Coll
Amarillo Country Club
Amarillo Medical Center
Tascosa High School
Amarillo College
Amarillo High School
Southwest Park
Thompson Memorial Park
Roy Rogers Municipal Golf Course
Martin Lake Park
Fairgrounds
Memorial Park Cemetery
Tradewind Airport
Southeast Park
Palo Duro Airport
Texas State Tech. Inst. Public Golf Course
Texas State Technical Institute
Amarillo Air Terminal
Sac Access
Willow Creek Dr.
Willow Creek
St. Francis
Cliffside Dr.
Hastings
N. 24th
Amarillo Blvd.
W. 3rd
6th
W. 10th Ave.
S.E. 3rd
Plains Blvd.
E. 27th Ave.
S. 34th
S. 45th Ave.
S. 46th Ave.
58th Ave.
Farmers Ave.
Arden Rd.
77th Ave.
81st Ave.
Hollywood Rd.
Soncy Rd.
Western St.
Georgia St.
Washington St.
Burlington St.
Osage St.
Grand St.
Eastern St.
Whitaker Rd.
Lakeside Rd.
Pullman Rd.
Masterson Rd.
Parsley Rd.
Jewett Rd.
Folsom
Denmar
Fritch Hwy.
Coulter
Bell
Juett-Attebury
Scale of Miles
0 1 2 3
© C.S.C.

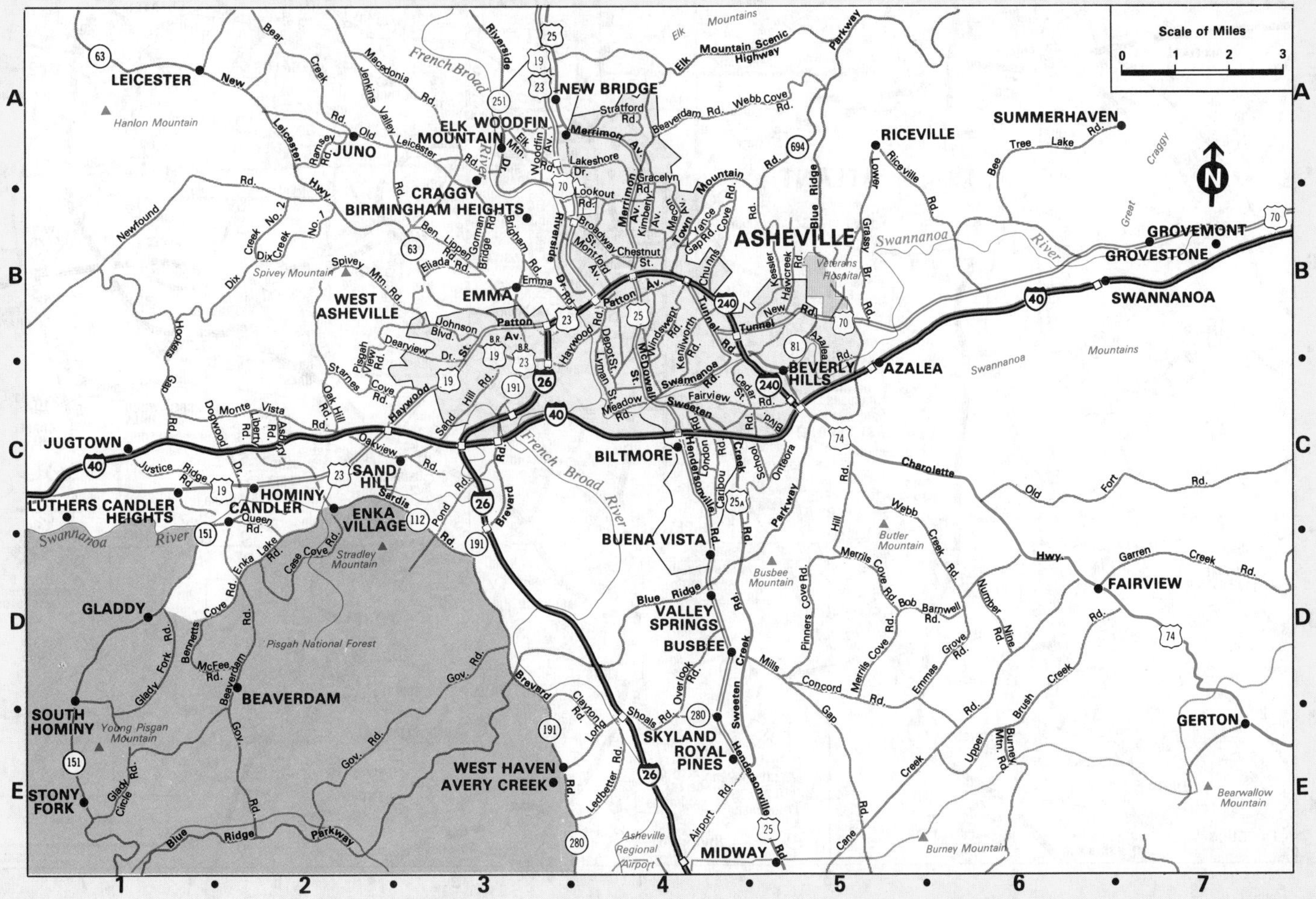
ASHEVILLE
LEICESTER
JUNO
NEW BRIDGE
ELK MOUNTAIN
WOODFIN
CRAGGY
BIRMINGHAM HEIGHTS
WEST ASHEVILLE
EMMA
RICEVILLE
SUMMERHAVEN
GROVEMONT
GROVESTONE
SWANNANOA
BEVERLY HILLS
AZALEA
JUGTOWN
SAND HILL
HOMINY
CANDLER
LUTHERS CANDLER HEIGHTS
ENKA VILLAGE
BILTMORE
BUENA VISTA
FAIRVIEW
GLADDY
VALLEY SPRINGS
BUSBEE
BEAVERDAM
SOUTH HOMINY
STONY FORK
WEST HAVEN
AVERY CREEK
SKYLAND
ROYAL PINES
GERTON
MIDWAY
Hanlon Mountain
Spivey Mountain
Stradley Mountain
Butler Mountain
Busbee Mountain
Young Pisgan Mountain
Pisgah National Forest
Bearwallow Mountain
Burney Mountain
Asheville Regional Airport
Veterans Hospital
Elk Mountains
Swannanoa Mountains
French Broad River
Swannanoa River
Elk Mountain Scenic Highway
Blue Ridge Parkway
Charlotte Hwy.
Hendersonville Rd.
Sweeten Creek Rd.
Tunnel Rd.
Patton Av.
Haywood Rd.
Merrimon Av.
Riverside Dr.
Brevard Rd.
Scale of Miles
0 1 2 3

Keenesaw Mountain National Battlefield Park
MARIETTA
DOBBINS AIR FORCE BASE
SMYRNA
Chattahoochee River National Recreation Area
SANDY SPRINGS
DORAVILLE
CHAMBLEE
NORCROSS
BERKELEY LAKE
Dekalb Peachtree Airport
NORTH BUCKHEAD
BROOKHAVEN
LENOX
BUCKHEAD
CLARKSTON
PINE LAKE
AVONDALE ESTATES
DECATUR
SCOTTDALE
EMORY
MIDTOWN
Fulton County Airport Brown Field
SIX FLAGS
ATLANTA
EAST POINT
COLLEGE PARK
HAPEVILLE
William B. Hartsfield Atlanta Intl Airport
FOREST PARK
FORT GILLEM
LAKE CITY
MORROW
UNION CITY
FAIRBURN
RIVERDALE

Scale of Miles
0 1/2 1 2 3

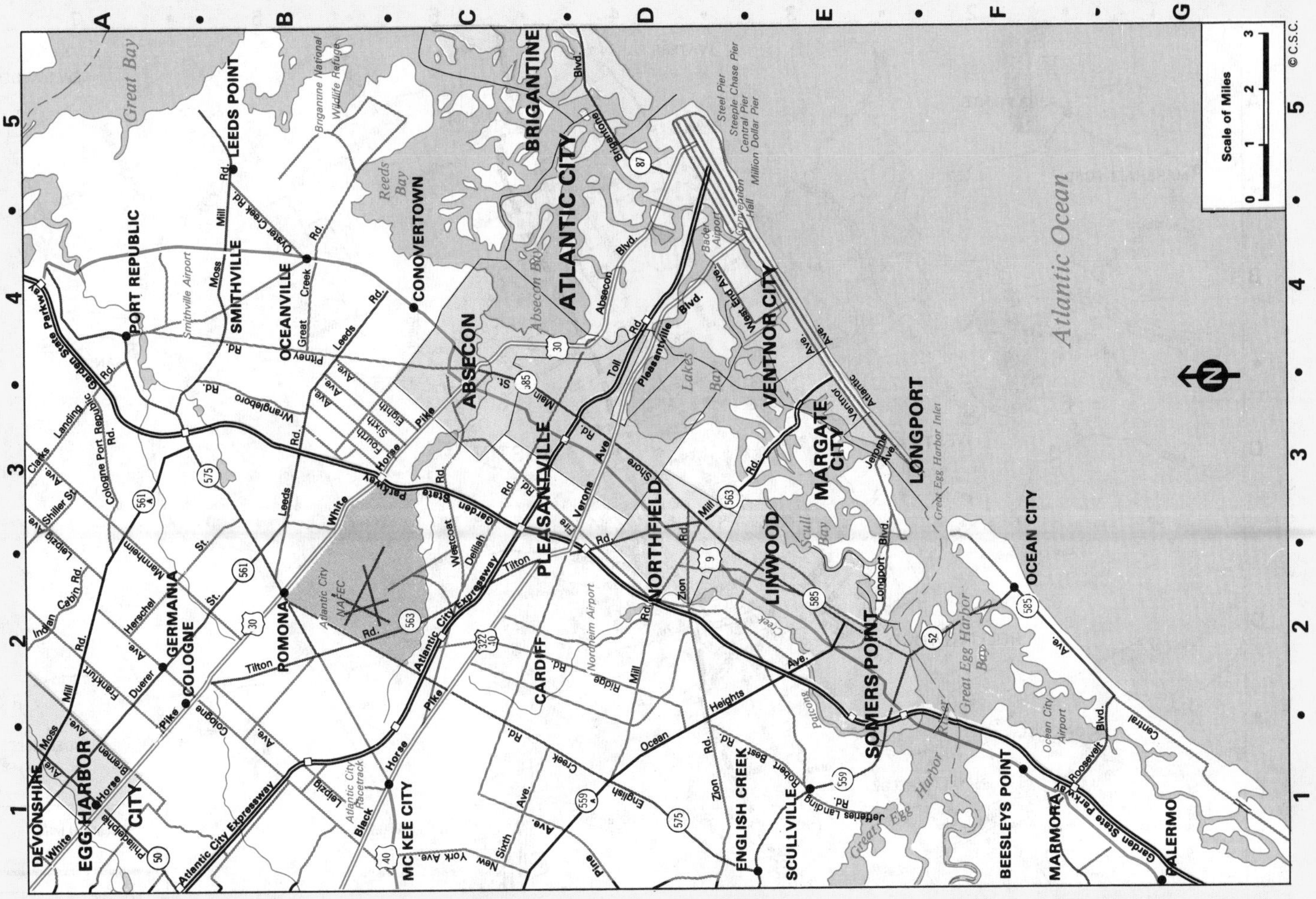

Scale of Miles
0 1 2 3

Atlantic Ocean

Great Bay

Reeds Bay

Absecon Bay

BRIGANTINE
ATLANTIC CITY
CONOVERTOWN
LEEDS POINT
SMITHVILLE
OCEANVILLE
PORT REPUBLIC
ABSECON
VENTNOR CITY
MARGATE CITY
LONGPORT
PLEASANTVILLE
NORTHFIELD
LINWOOD
OCEAN CITY
SOMERS POINT
CARDIFF
POMONA
GERMANIA
COLOGNE
EGG HARBOR CITY
DEVONSHIRE
MC KEE CITY
ENGLISH CREEK
SCULLVILLE
BEESLEYS POINT
MARMORA
PALERMO

Great Egg Harbor Inlet

Great Egg Harbor Bay

Garden State Parkway

Atlantic City Expressway

Steel Pier
Steeple Chase Pier
Central Pier
Million Dollar Pier

Brigantine National Wildlife Refuge

Smithville Airport

Atlantic City NAFEC

Nordheim Airport

Atlantic City Racetrack

Ocean City Airport

© C.S.C.

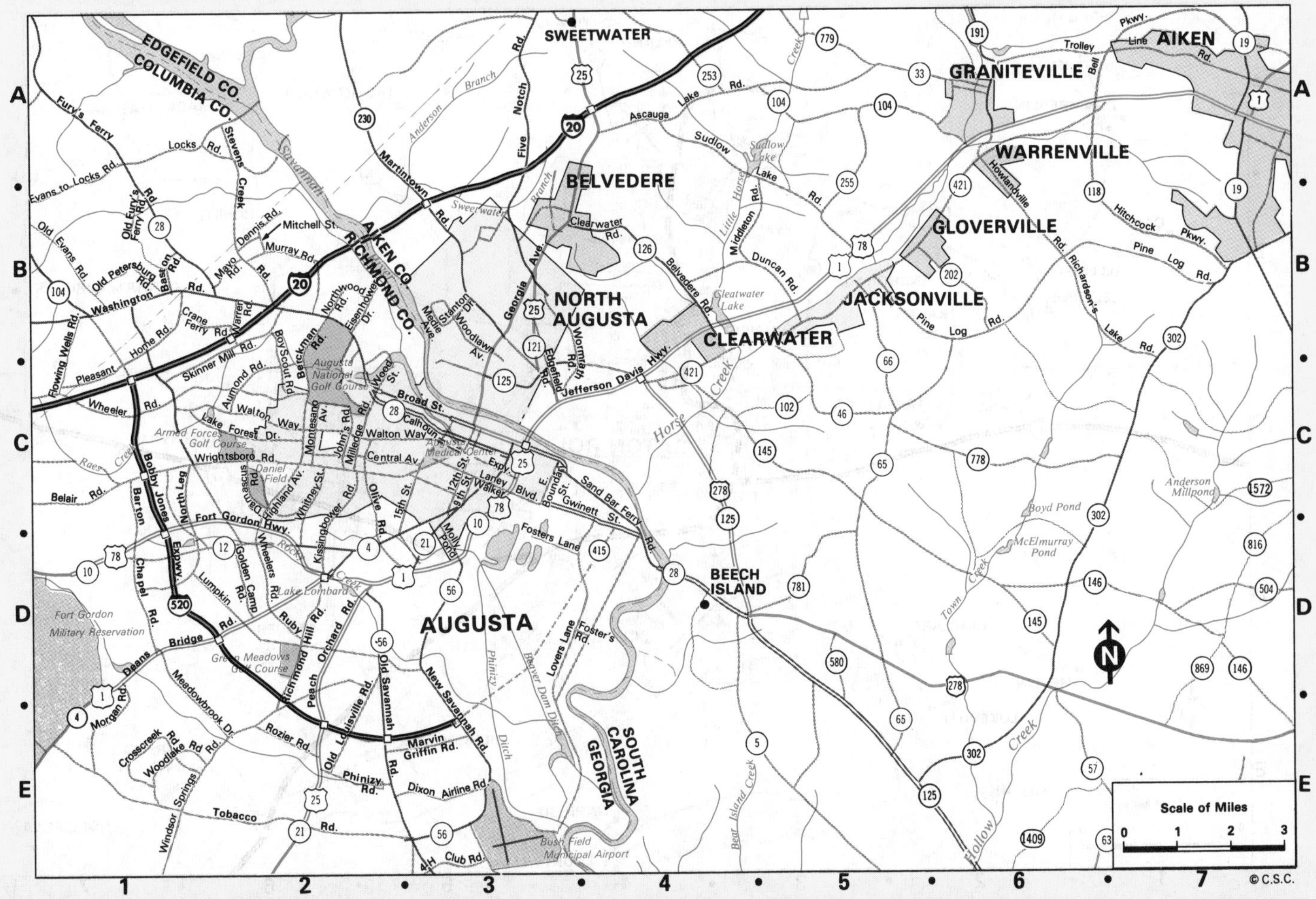

EDGEFIELD CO.
COLUMBIA CO.
AIKEN CO.
RICHMOND CO.

SWEETWATER
AIKEN
GRANITEVILLE
WARRENVILLE
GLOVERVILLE
BELVEDERE
NORTH AUGUSTA
JACKSONVILLE
CLEARWATER
AUGUSTA
BEECH ISLAND
SOUTH CAROLINA
GEORGIA

Savannah River

Augusta National Golf Course

Armed Forces Golf Course

Daniel Field

Fort Gordon Military Reservation

Green Meadows Golf Course

Bush Field Municipal Airport

Anderson Millpond

Boyd Pond

McElmurray Pond

Scale of Miles
0 1 2 3

© C.S.C.

Austin, TX
Baton Rouge, LA

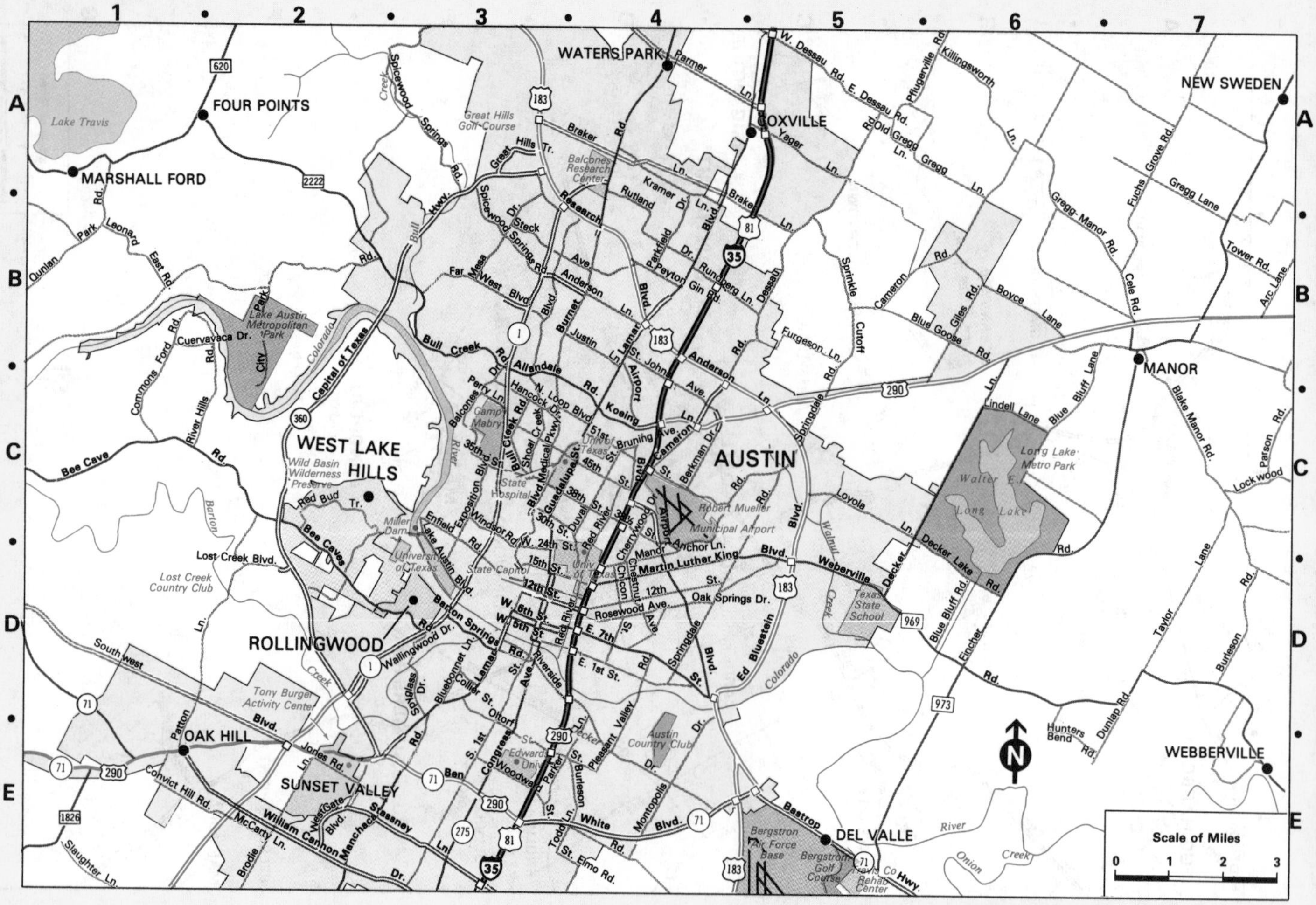

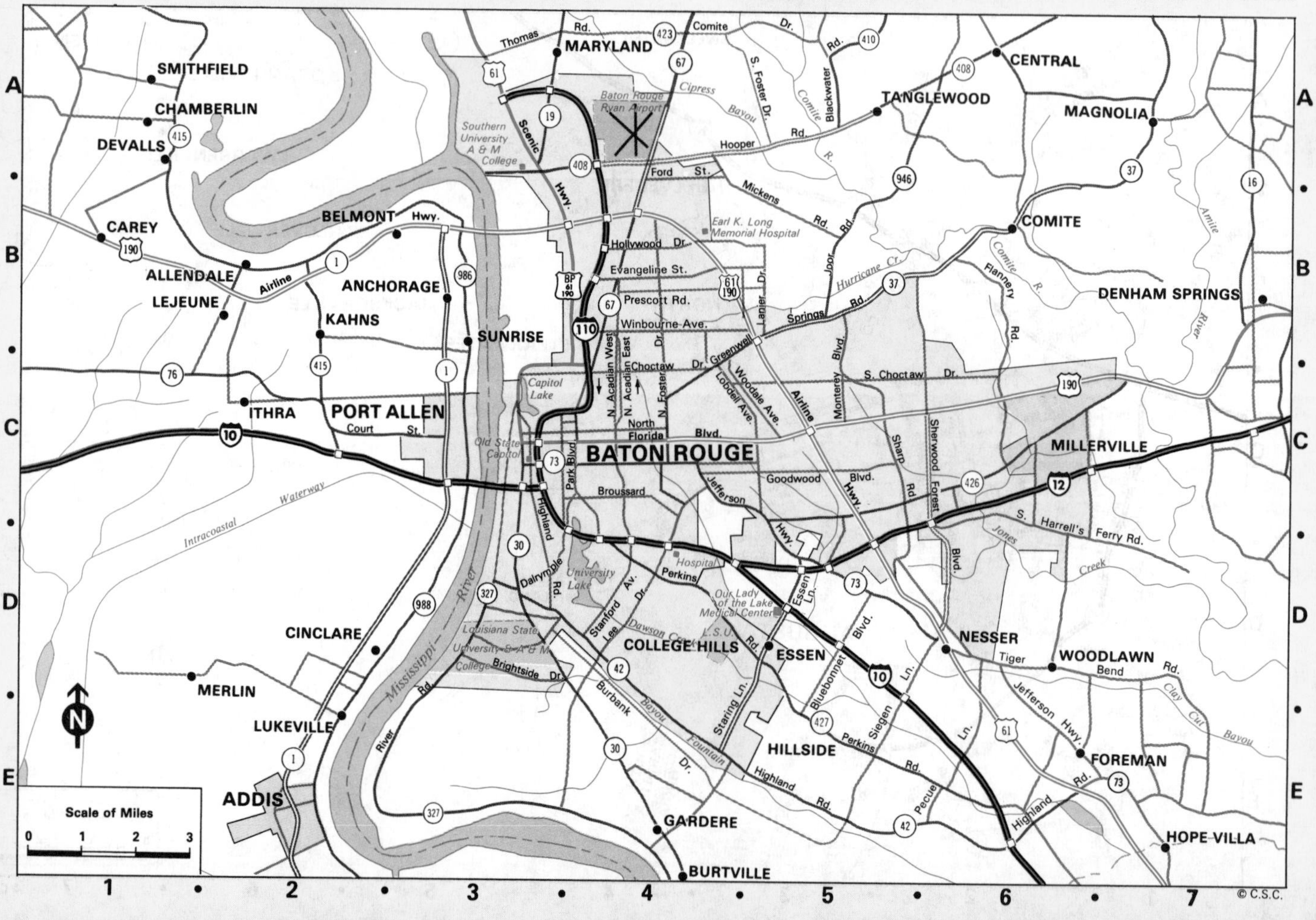

Austin, TX

WATERS PARK
FOUR POINTS
NEW SWEDEN
COXVILLE
MARSHALL FORD
Lake Travis
Great Hills Golf Course
Balcones Research Center
Lake Austin Metropolitan Park
MANOR
WEST LAKE HILLS
Wild Basin Wilderness Preserve
AUSTIN
Camp Mabry
State Hospital
Robert Mueller Municipal Airport
Long Lake Metro Park
Walter E. Long Lake
Miller Dam
University of Texas
State Capitol
Lost Creek Country Club
Texas State School
ROLLINGWOOD
Tony Burger Activity Center
OAK HILL
Austin Country Club
Hunters Bend
WEBBERVILLE
SUNSET VALLEY
DEL VALLE
Bergstrom Air Force Base
Bergstrom Golf Course
Rehab. Center

Scale of Miles
0 1 2 3

Baton Rouge, LA

MARYLAND
CENTRAL
SMITHFIELD
CHAMBERLIN
TANGLEWOOD
MAGNOLIA
DEVALLS
Baton Rouge Ryan Airport
Southern University A & M College
CAREY
BELMONT
COMITE
Earl K. Long Memorial Hospital
ALLENDALE
LEJEUNE
ANCHORAGE
DENHAM SPRINGS
KAHNS
SUNRISE
ITHRA
PORT ALLEN
Capitol Lake
Old State Capitol
BATON ROUGE
MILLERVILLE
Intracoastal Waterway
University Lake
Hospital
Our Lady of the Lake Medical Center
L.S.U.
Louisiana State University & A & M College
CINCLARE
COLLEGE HILLS
ESSEN
NESSER
WOODLAWN
MERLIN
LUKEVILLE
HILLSIDE
FOREMAN
ADDIS
GARDERE
HOPE VILLA
BURTVILLE
Mississippi River

Scale of Miles
0 1 2 3

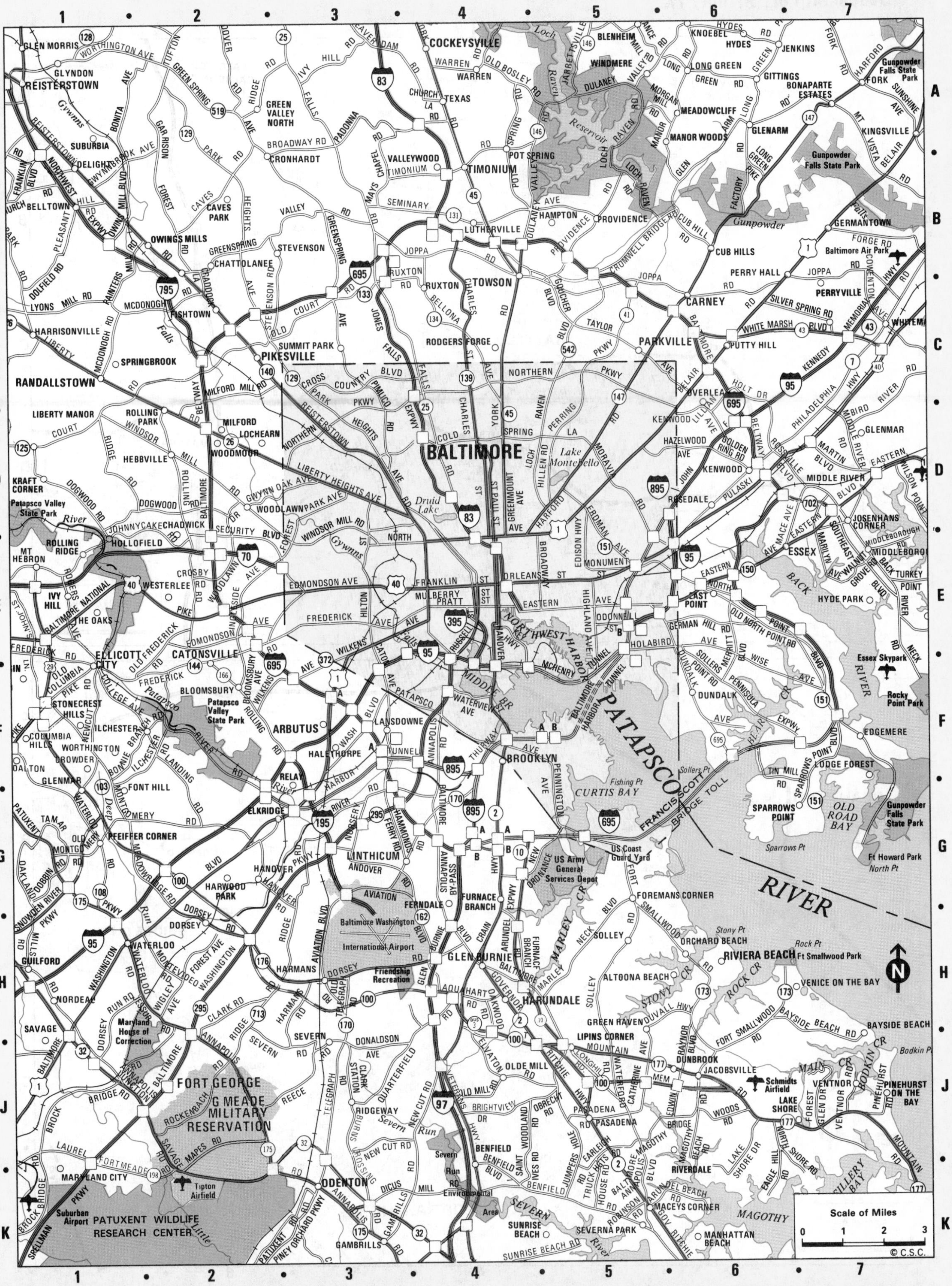
BALTIMORE
GLEN MORRIS
GLYNDON
REISTERSTOWN
COCKEYSVILLE
BLENHEIM
WINDMERE
KNOEBEL
HYDES
JENKINS
LONG GREEN
GITTINGS
BONAPARTE ESTATES
Gunpowder Falls State Park
FORK
TEXAS
WARREN
GREEN VALLEY NORTH
MEADOWCLIFF
GLENARM
KINGSVILLE
SUBURBIA
DELIGHT
CRONHARDT
VALLEYWOOD
TIMONIUM
POT SPRING
Loch Raven Reservoir
MANOR WOODS
BELLTOWN
CAVES PARK
LUTHERVILLE
HAMPTON
PROVIDENCE
Gunpowder
GERMANTOWN
OWINGS MILLS
STEVENSON
CHATTOLANEE
CUB HILLS
Baltimore Air Park
RUXTON
TOWSON
PERRY HALL
PERRYVILLE
FISHTOWN
CARNEY
HARRISONVILLE
SUMMIT PARK
RODGERS FORGE
PARKVILLE
PUTTY HILL
WHITE MARSH
SPRINGBROOK
PIKESVILLE
RANDALLSTOWN
OVERLEA
LIBERTY MANOR
ROLLING PARK
MILFORD
LOCHEARN
KENWOOD
GLENMAR
HEBBVILLE
WOODMOOR
Lake Montebello
HAZELWOOD
MIDDLE RIVER
KRAFT CORNER
Druid Lake
ROSEDALE
Patapsco Valley State Park
WOODLAWN
JOSENHANS CORNER
CHADWICK
SECURITY
ROLLING RIDGE
HOLLOFIELD
MT HEBRON
ESSEX
MIDDLEBOROUGH
WESTERLEE
MONUMENT
IVY HILL
THE OAKS
EAST POINT
HYDE PARK
TURKEY POINT
HOLABIRD
NORTHWEST HARBOR
MIDDLE BR
ELLICOTT CITY
CATONSVILLE
Essex Skypark
DUNDALK
BLOOMSBURY
PATAPSCO
Rocky Point Park
STONECREST HILLS
ARBUTUS
LANSDOWNE
ILCHESTER
COLUMBIA HILLS
WORTHINGTON
CROWDER
HALETHORPE
BROOKLYN
EDGEMERE
LODGE FOREST
DALTON
GLENMAR
FONT HILL
Fishing Pt
CURTIS BAY
Sollers Pt
RELAY
ELKRIDGE
SPARROWS POINT
OLD ROAD BAY
PFEIFFER CORNER
US Coast Guard Yard
LINTHICUM
US Army General Services Depot
Sparrows Pt
Ft Howard Park
North Pt
HANOVER
HARWOOD PARK
FERNDALE
FURNACE BRANCH
FOREMANS CORNER
RIVER
Baltimore Washington International Airport
Stony Pt
ORCHARD BEACH
Rock Pt
Ft Smallwood Park
RIVIERA BEACH
GUILFORD
WATERLOO
DORSEY
HARMANS
Friendship Recreation
GLEN BURNIE
ALTOONA BEACH
VENICE ON THE BAY
NORDEAL
HARUNDALE
Maryland House of Correction
SAVAGE
SEVERN
GREEN HAVEN
LIPINS CORNER
BAYSIDE BEACH
OLDE MILL
DUNBROOK
JACOBSVILLE
Bodkin Pt
FORT GEORGE
G MEADE MILITARY RESERVATION
Schmidts Airfield
VENTNOR
LAKE SHORE
PINEHURST ON THE BAY
PASADENA
RIDGEWAY
BENFIELD
RIVERDALE
MARYLAND CITY
Tipton Airfield
ODENTON
Severn Run Environmental Area
MACEYS CORNER
Suburban Airport
PATUXENT WILDLIFE RESEARCH CENTER
SUNRISE BEACH
SEVERNA PARK
MAGOTHY
MANHATTAN BEACH
GAMBRILLS
Scale of Miles
0 1 2 3
© C.S.C.

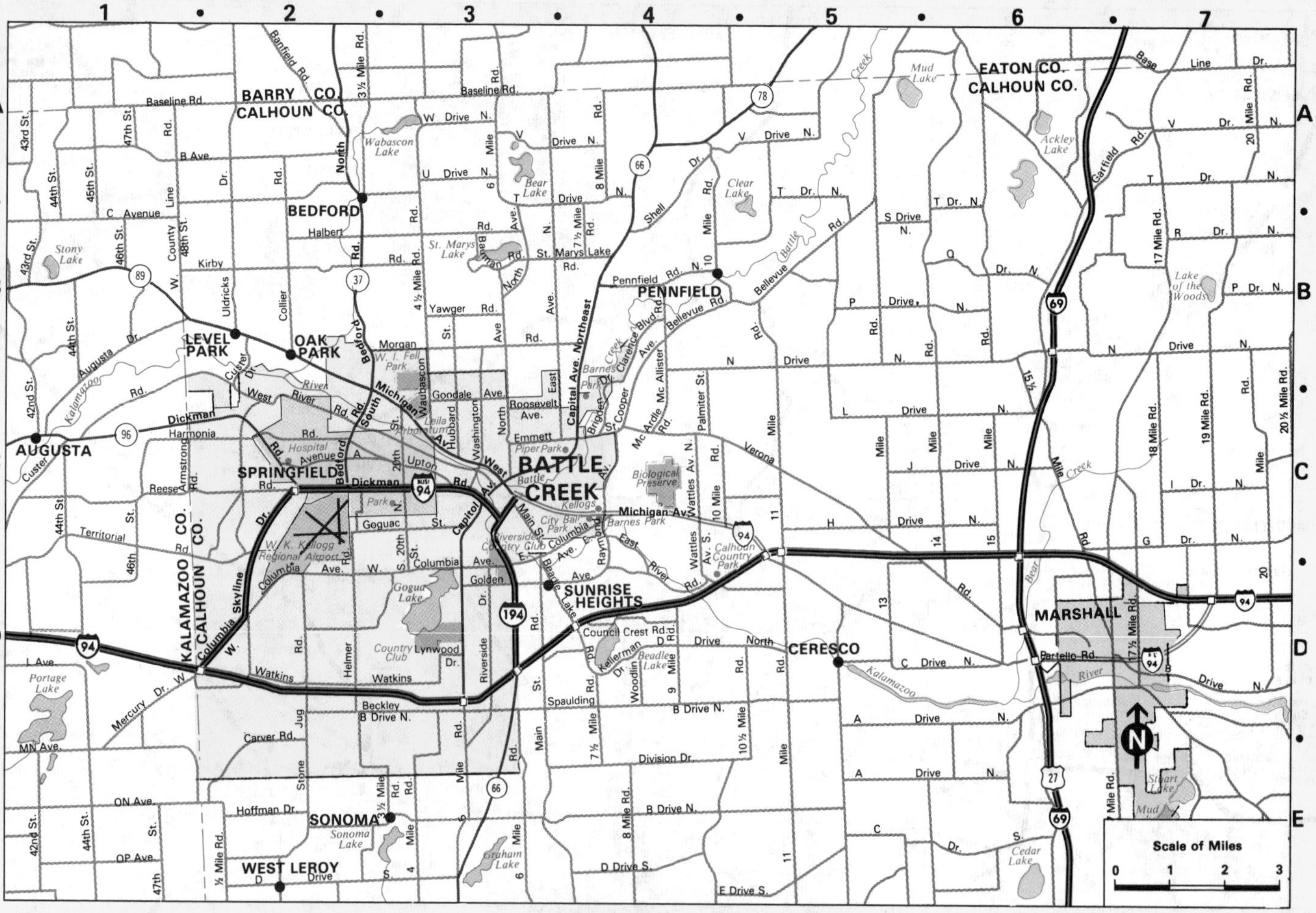
BARRY CO.
CALHOUN CO.
EATON CO.
CALHOUN CO.
KALAMAZOO CO.
CALHOUN CO.
BEDFORD
LEVEL PARK
OAK PARK
AUGUSTA
SPRINGFIELD
BATTLE CREEK
PENNFIELD
SUNRISE HEIGHTS
CERESCO
MARSHALL
SONOMA
WEST LEROY
Wabascon Lake
Bear Lake
St. Marys Lake
Clear Lake
Mud Lake
Ackley Lake
Stony Lake
Lake of the Woods
Portage Lake
Gogua Lake
Beadles Lake
Sonoma Lake
Graham Lake
Cedar Lake
Stuart Lake
W. K. Kellogg Regional Airport
Biological Preserve
Country Club
Kalamazoo River
Battle Creek
Baseline Rd.
Michigan Av.
Dickman Rd.
Columbia Ave.
Capital Ave. Northeast
Scale of Miles
0
1
2
3

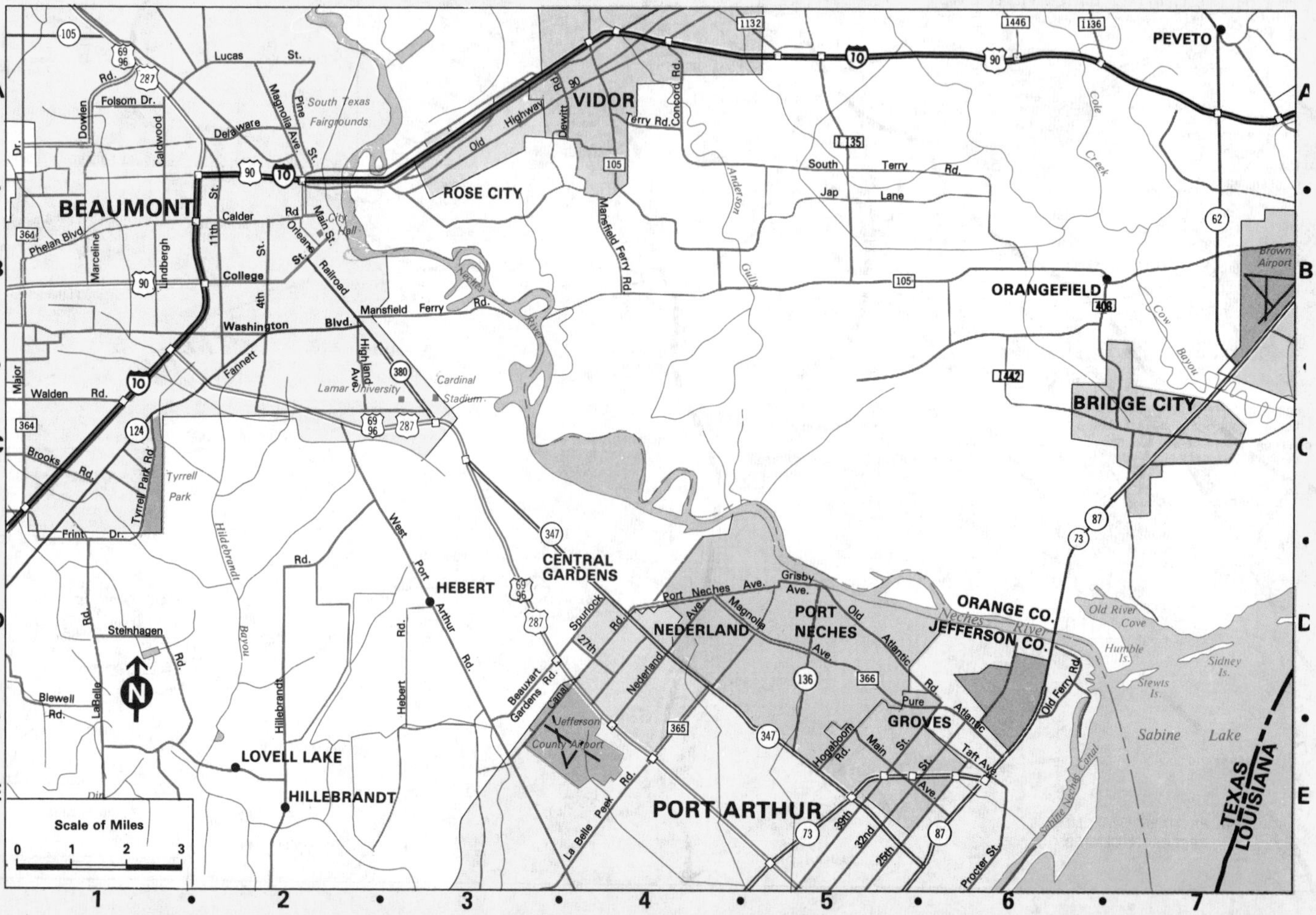
BEAUMONT
VIDOR
ROSE CITY
PEVETO
ORANGEFIELD
BRIDGE CITY
CENTRAL GARDENS
HEBERT
NEDERLAND
PORT NECHES
GROVES
PORT ARTHUR
LOVELL LAKE
HILLEBRANDT
ORANGE CO.
JEFFERSON CO.
TEXAS
LOUISIANA
South Texas Fairgrounds
Lamar University
Cardinal Stadium
Tyrrell Park
Jefferson County Airport
Brown Airport
Neches River
Sabine Lake
Old River Cove
Humble Is.
Sidney Is.
Stewts Is.
Sabine Neches Canal
Cow Bayou
Cole Creek
Washington Blvd.
College
Calder
Phelan Blvd
Mansfield Ferry Rd.
Old Highway 90
Port Neches Ave.
Scale of Miles
0
1
2
3

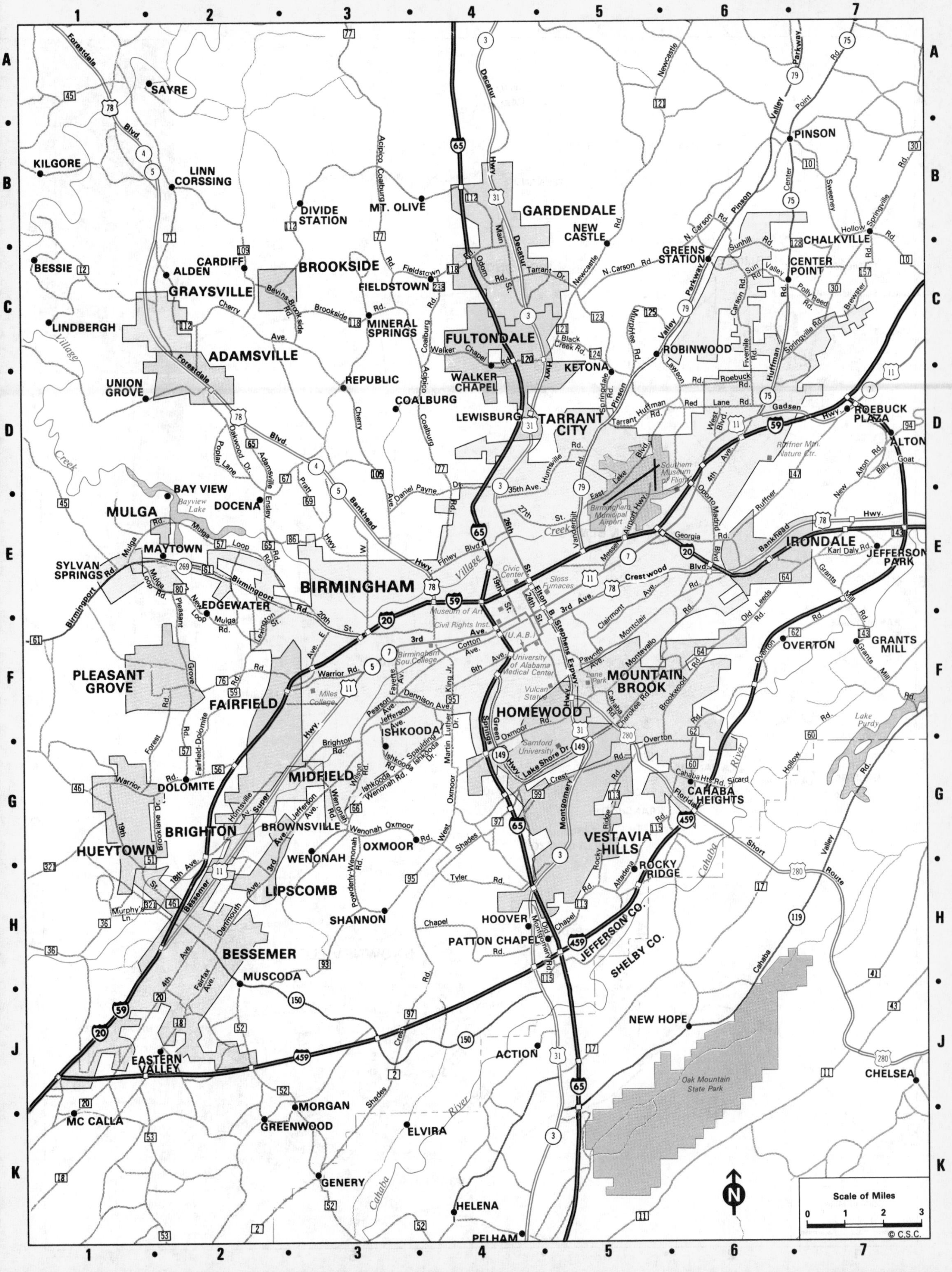

1 2 3 4 5 6 7

A B C D E F G H J K

SAYRE
KILGORE
LINN CORSSING
DIVIDE STATION
MT. OLIVE
GARDENDALE
NEW CASTLE
PINSON
CHALKVILLE
CENTER POINT
GREENS STATION
BESSIE
ALDEN
CARDIFF
BROOKSIDE
FIELDSTOWN
MINERAL SPRINGS
GRAYSVILLE
LINDBERGH
ADAMSVILLE
FULTONDALE
WALKER CHAPEL
ROBINWOOD
KETONA
UNION GROVE
REPUBLIC
COALBURG
LEWISBURG
TARRANT CITY
ROEBUCK PLAZA
ALTON
BAY VIEW
Bayview Lake
DOCENA
MULGA
MAYTOWN
SYLVAN SPRINGS
IRONDALE
JEFFERSON PARK
BIRMINGHAM
EDGEWATER
Birmingham Municipal Airport
Southern Museum of Flight
Ruffner Mtn. Nature Ctr.
Civic Center
Sloss Furnaces
Museum of Art
Civil Rights Inst.
U.A.B.
University of Alabama Medical Center
Birmingham Sou. College
Miles College
Vulcan Statue
Lane Park
Samford University
OVERTON
GRANTS MILL
PLEASANT GROVE
FAIRFIELD
MOUNTAIN BROOK
HOMEWOOD
ISHKOODA
MIDFIELD
DOLOMITE
CAHABA HEIGHTS
Lake Purdy
BRIGHTON
BROWNSVILLE
HUEYTOWN
OXMOOR
WENONAH
VESTAVIA HILLS
ROCKY RIDGE
LIPSCOMB
SHANNON
HOOVER
PATTON CHAPEL
BESSEMER
MUSCODA
JEFFERSON CO.
SHELBY CO.
NEW HOPE
EASTERN VALLEY
ACTION
CHELSEA
Oak Mountain State Park
MC CALLA
MORGAN
GREENWOOD
ELVIRA
GENERY
HELENA
PELHAM
Village Creek
Cahaba River

Scale of Miles
0 1 2 3

© C.S.C.

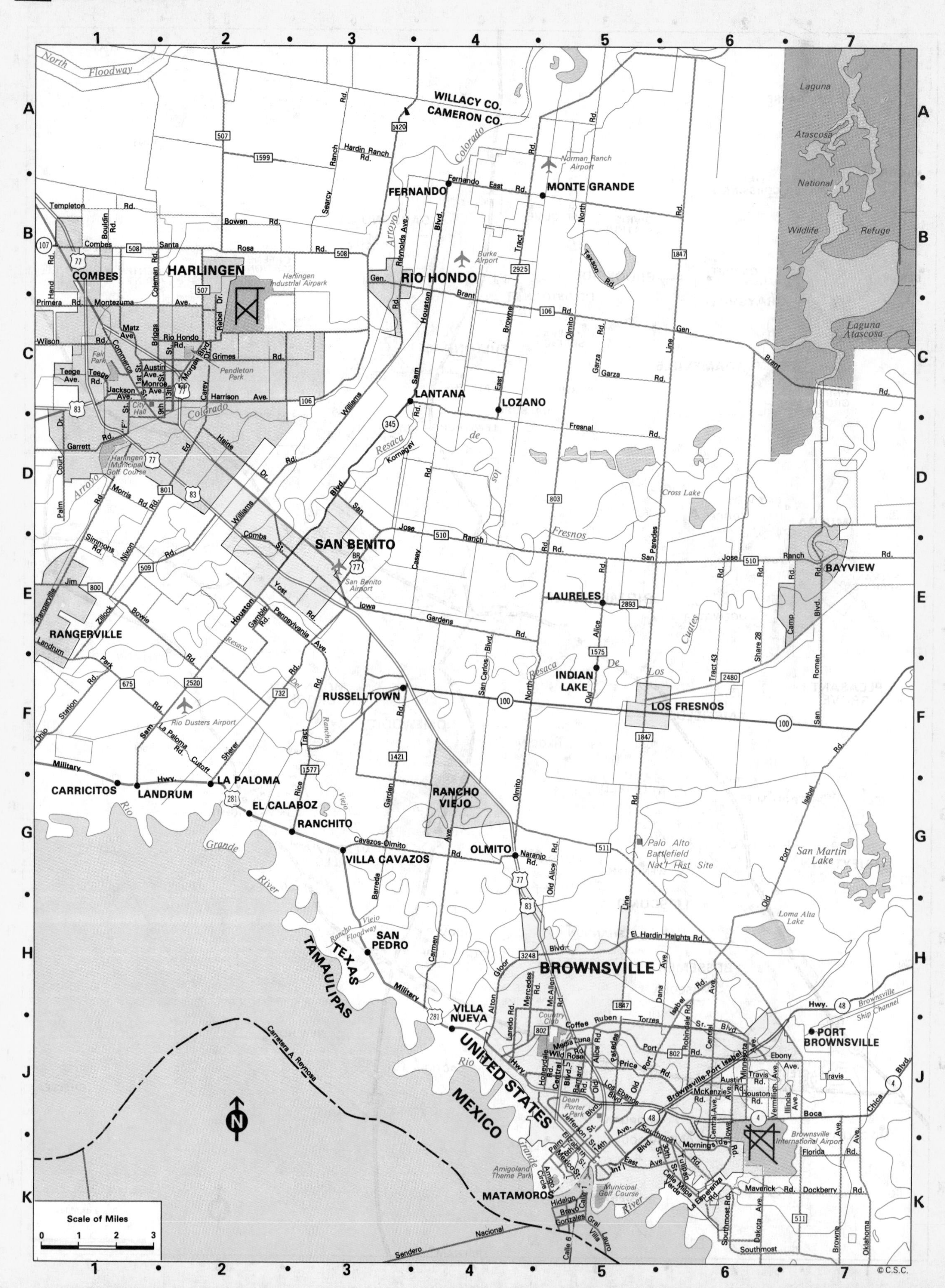
WILLACY CO.
CAMERON CO.
FERNANDO
MONTE GRANDE
Norman Ranch Airport
HARLINGEN
COMBES
Harlingen Industrial Airpark
RIO HONDO
Burke Airport
LANTANA
LOZANO
SAN BENITO
San Benito Airport
LAURELES
RANGERVILLE
INDIAN LAKE
LOS FRESNOS
RUSSELLTOWN
Rio Dusters Airport
CARRICITOS
LANDRUM
LA PALOMA
EL CALABOZ
RANCHITO
RANCHO VIEJO
VILLA CAVAZOS
OLMITO
Palo Alto Battlefield Nat'l Hist Site
SAN PEDRO
BROWNSVILLE
VILLA NUEVA
PORT BROWNSVILLE
Brownsville International Airport
MATAMOROS
Amigoland Theme Park
Municipal Golf Course
Country Club
Dean Porter Park
Laguna Atascosa National Wildlife Refuge
Laguna Atascosa
BAYVIEW
Cross Lake
San Martin Lake
Loma Alta Lake
Brownsville Ship Channel
TEXAS
TAMAULIPAS
UNITED STATES
MEXICO
Rio Grande River
Carretera A Reynosa
North Floodway
Arroyo Colorado
Resaca de los Fresnos
Resaca De Los Cuates
Harlingen Municipal Golf Course
Fair Park
Pendleton Park
City Hall
Scale of Miles
0 1 2 3
© C.S.C.

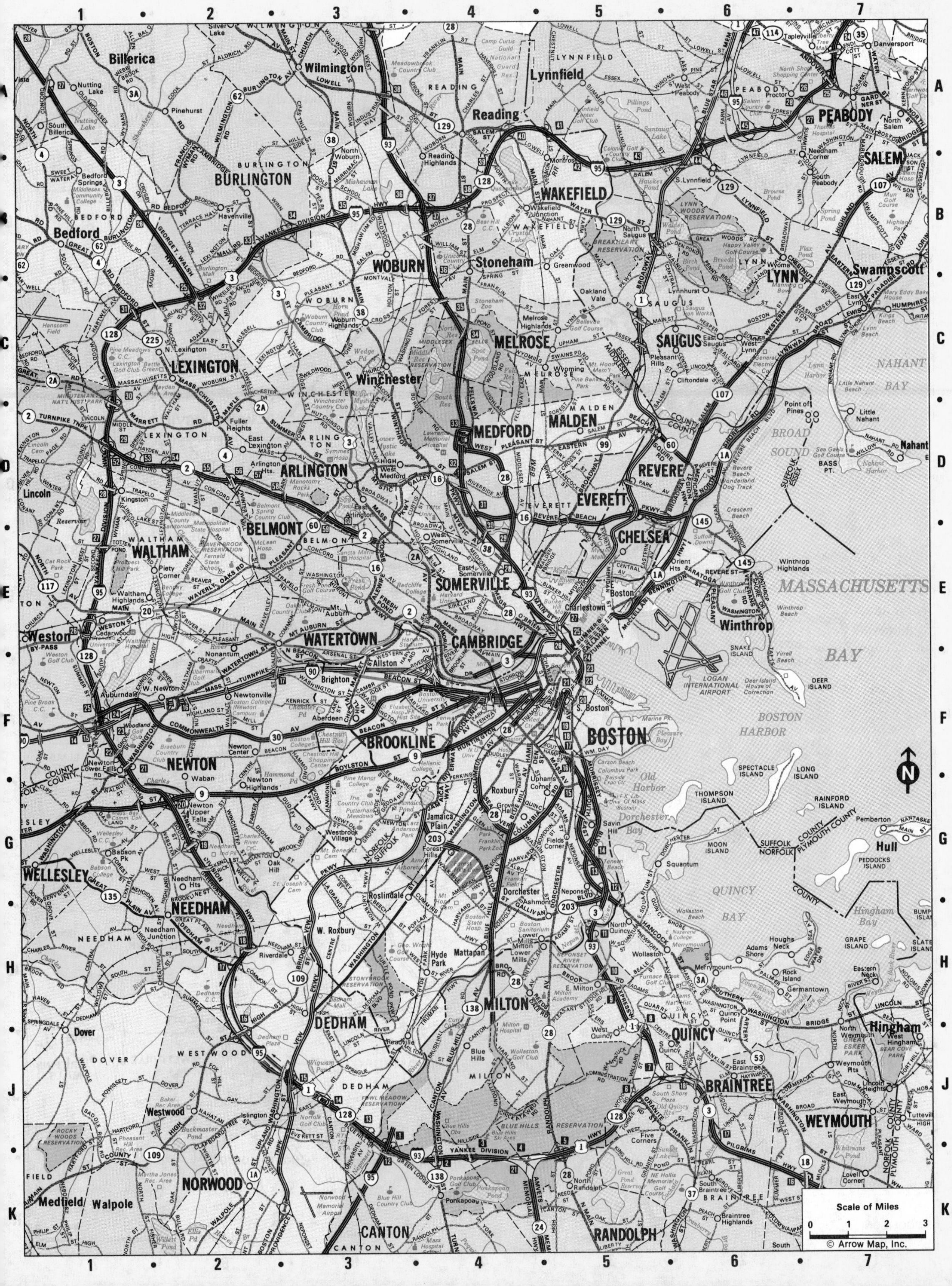

1 2 3 4 5 6 7

A B C D E F G H J K

Billerica
Wilmington
Lynnfield
Reading
Peabody
PEABODY
SALEM
BURLINGTON
WAKEFIELD
Bedford
WOBURN
Stoneham
LYNN
Swampscott
LEXINGTON
MELROSE
SAUGUS
Winchester
NAHANT BAY
MEDFORD
MALDEN
Nahant
ARLINGTON
REVERE
BROAD SOUND
Lincoln
EVERETT
BELMONT
CHELSEA
WALTHAM
SOMERVILLE
MASSACHUSETTS BAY
Weston
WATERTOWN
CAMBRIDGE
Winthrop
LOGAN INTERNATIONAL AIRPORT
BROOKLINE
BOSTON
BOSTON HARBOR
NEWTON
Roxbury
Jamaica Plain
Dorchester Bay
Hull
WELLESLEY
NEEDHAM
Dorchester
QUINCY BAY
Hingham Bay
Mattapan
Hyde Park
DEDHAM
MILTON
QUINCY
Hingham
Dover
BRAINTREE
WEYMOUTH
Westwood
BLUE HILLS RESERVATION
NORWOOD
Medfield
Walpole
CANTON
RANDOLPH

Scale of Miles
0 1 2 3

NIAGARA FALLS
ECHOTA
NIAGARA FALLS
LA SALLE
BERKHOLTZ
Niagara Falls International Airport
Oppenheim Zoo
ST. JOHNSBURG
NASHVILLE
BEACH RIDGE
Loveland
HOFFMAN
WENDELVILLE
PENDLETON
RAPIDS
MILLERSPORT
Steffen Airport
Amherst Airport
ELSERS CORNERS
Lundy's Lane Battleground
Convention Center
Niagara
CHIPPAWA
Chippawa Battle Monument
Navy Is.
Buckhorn State Park
PEACH HAVEN
NIAGARA CO.
ERIE CO.
WURLITZER PARK
SAWYER
MARTINSVILLE
SANDY BEACH
EDGEWATER
Big Six Mile Creek Park
Grand Island
NORTH TONAWANDA
Ellicott Ck Park
GETZVILLE
SWORMVILLE
EAST AMHERST
CLARENCE CENTER
TONAWANDA
S.U.N.Y. Buffalo
KENMORE
Sheridan Park
GRANDYLE VILLAGE
NORTH BAILEY
AMHERST
Park C.C.
Buffalo C.C.
FERRY VILLAGE
Beaver Island State Park
EGGERTSVILLE
SNYDER
S.U.N.Y. Buffalo
WILLIAMSVILLE
HARRIS HILL
SNYDER
STEVENSVILLE
Buffalo International Airport
Fort Erie Airport
FORT ERIE NORTH
Buffalo Zoo
SUNY Coll Buffalo
POWMANSVILLE
Fort Erie Race Track
DEPEW
LANCASTER
Buffalo Museum of Science
War Mem. Stadium
FORT ERIE
CHEEKTOWAGA
SLOAN
BELLEVUE
Como Park
POINT ABINO
RIDGEWAY
CRESCENT PARK
ERIE BEACH
Erie Beach
Crescent Beach
Waverly Beach
Airport
THUNDER BAY
CRYSTAL BEACH
Crystal Beach
Buffalo Air Park
BLOSSOM
GARDENVILLE
BUFFALO
ELMA
Buffalo Harbor
WEST SENECA
EAST SENECA
ELMA CENTER
Botanical Gardens
EBENEZER
SPRINGBROOK
LACKAWANNA
BLASDELL
Orchard Park Airport
Proner Airport
WELLAND CO.
ERIE CO.
ONTARIO
NEW YORK
CANADA
UNITED STATES
Lake Erie
WOODLAWN
WINDOM
BAY VIEW
EAST HAMBURG
WEBSTER CORNERS
ATHOL SPRINGS
BIG TREE
ORCHARD PARK
LOCKSLEY PARK
South Shore C.C.
Orchard Park C.C.
MT. VERNON
WANAKAH
CARNEGIE
DUELLS CORNERS
CLIFTON HEIGHTS
PINEHURST
SCRANTON
ARMOR
Erie Co. Fairgrounds
ELLICOTT
ELLICOTT HEIGHTS
HAMBURG
Chestnut Ridge Park
JEWETTVILLE
GRIFFINS MILLS
HIGHLAND-ON-THE-LAKE
LAKE VIEW
WATER VALLEY
Lakeview Airport
Eighteen Mile Creek
WEST FALLS
NORTH EVANS
NORTH BOSTON
JERUSALEM CORNERS
DERBY
EDEN VALLEY
ANGOLA-ON-THE-LAKE
EVANS
EAST EDEN
PATCHIN

Scale of Miles
0 1 2 3

© C.S.C.

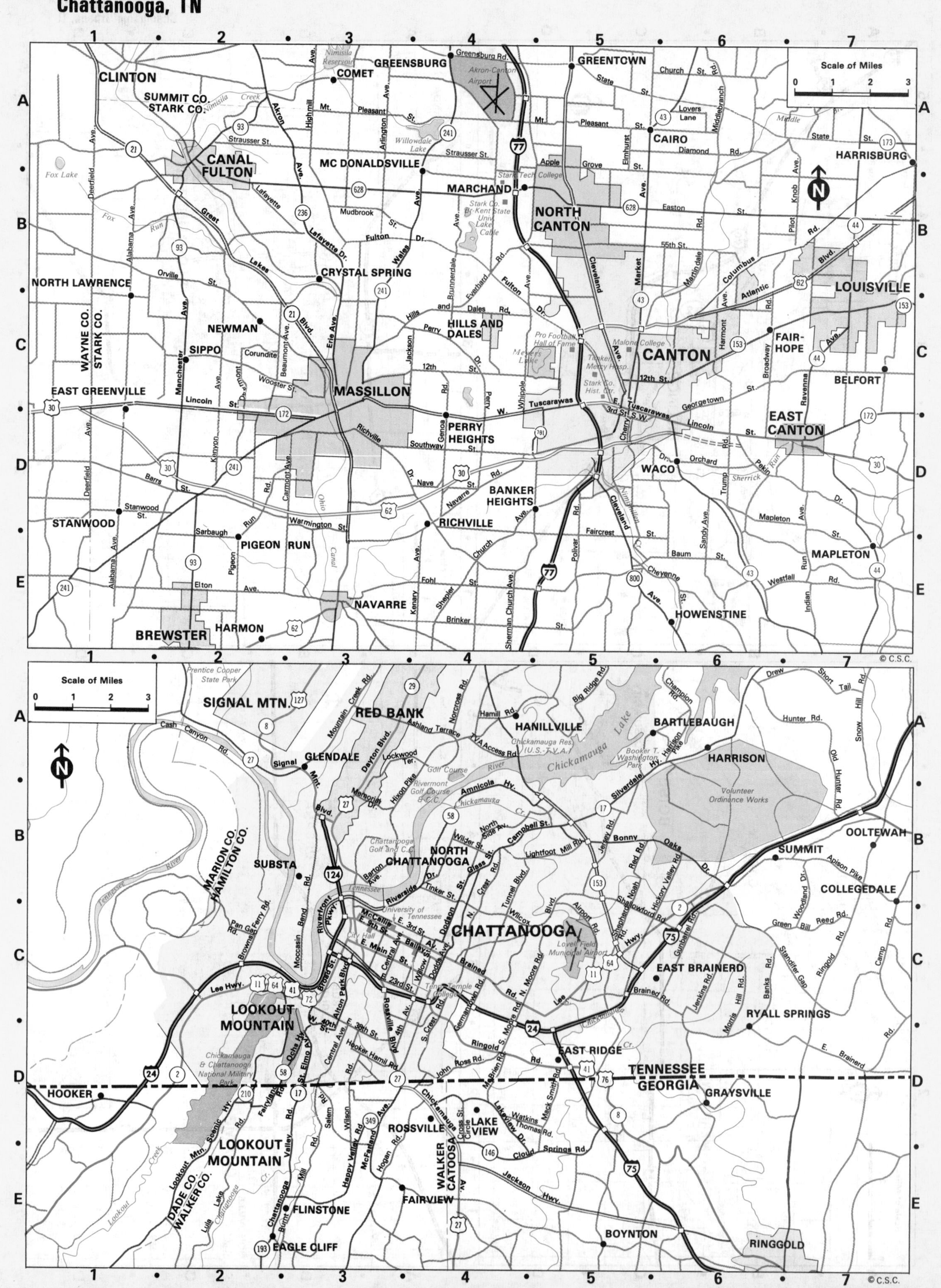
CANTON
MASSILLON
NORTH CANTON
CANAL FULTON
LOUISVILLE
EAST CANTON
CHATTANOOGA
NORTH CHATTANOOGA
RED BANK
SIGNAL MTN.
EAST RIDGE
LOOKOUT MOUNTAIN
ROSSVILLE
TENNESSEE
GEORGIA
Scale of Miles
© C.S.C.

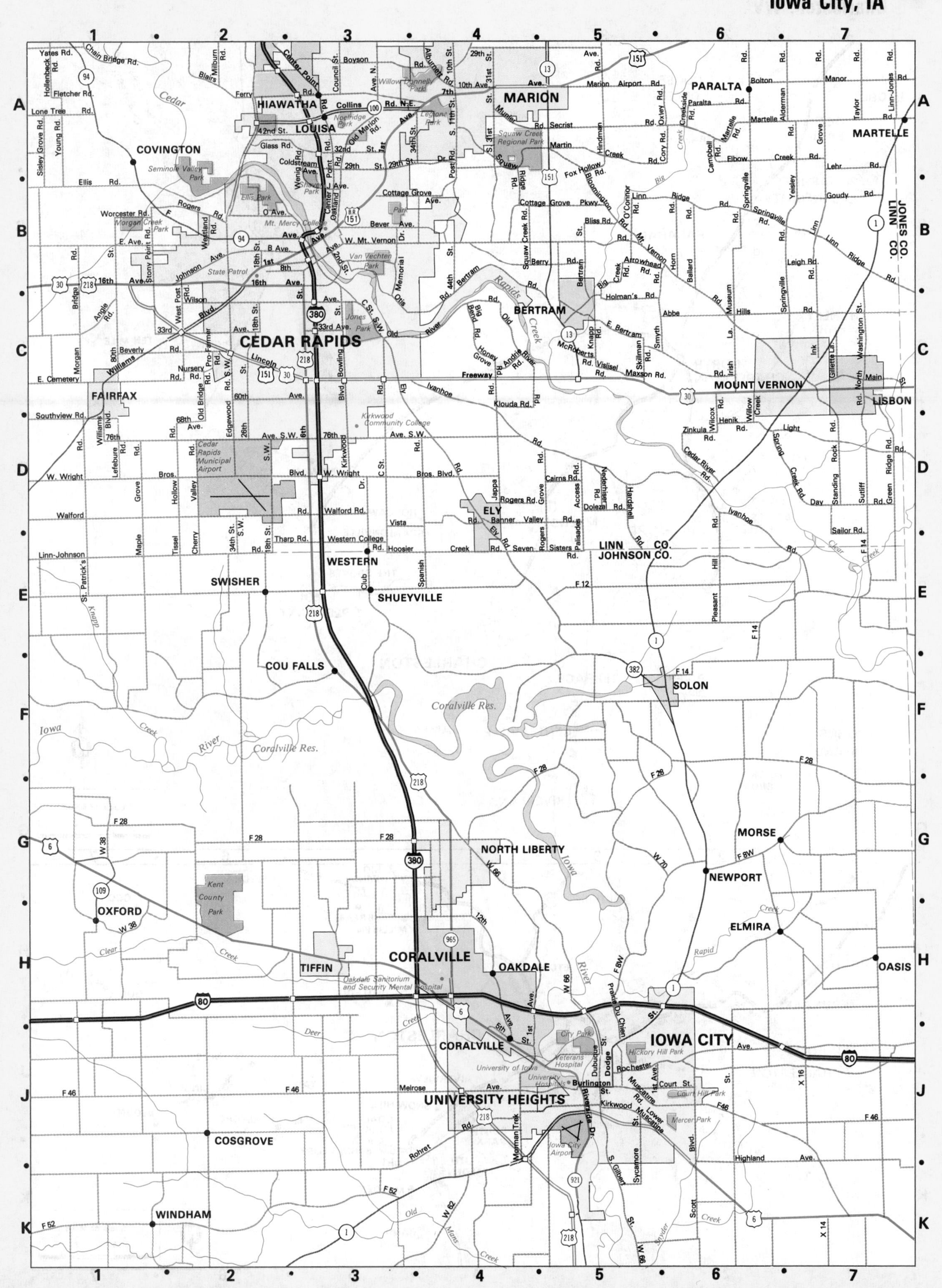
HIAWATHA
MARION
PARALTA
LOUISA
COVINGTON
MARTELLE
CEDAR RAPIDS
BERTRAM
FAIRFAX
MOUNT VERNON
LISBON
ELY
WESTERN
SWISHER
SHUEYVILLE
LINN CO.
JOHNSON CO.
JONES CO.
LINN CO.
COU FALLS
SOLON
Coralville Res.
Iowa River
NORTH LIBERTY
MORSE
NEWPORT
OXFORD
Kent County Park
ELMIRA
TIFFIN
CORALVILLE
OAKDALE
OASIS
Oakdale Sanitorium and Security Mental Hospital
IOWA CITY
City Park
Hickory Hill Park
Veterans Hospital
University of Iowa
University Hospitals
UNIVERSITY HEIGHTS
Court Hill Park
Mercer Park
Iowa City Airport
COSGROVE
WINDHAM
Seminole Valley Park
Ellis Park
Shaver Park
Morgan Creek Park
Mt. Mercy College
Van Vechten Park
Jones Park
Noelridge Park
Willow Donnelly Park
Legion Park
Squaw Creek Regional Park
State Patrol
Kirkwood Community College
Cedar Rapids Municipal Airport
Cedar River
Big Creek
Clear Creek
Deer Creek
Old Mans Creek
Snyder Creek
Rapid Creek
Indian Creek
Knapp Creek

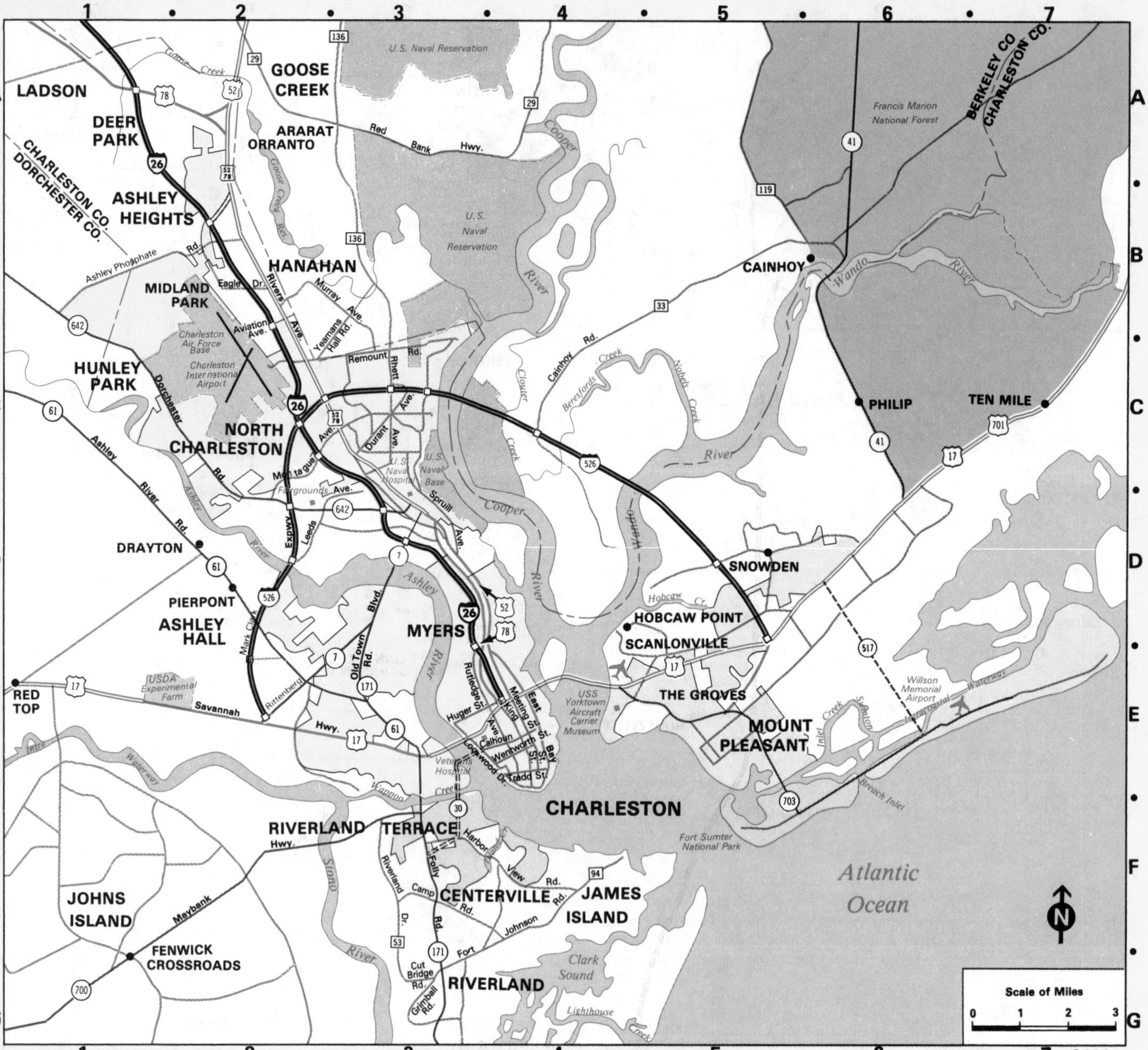

Charleston, SC

LADSON
DEER PARK
GOOSE CREEK
ARARAT
ORRANTO
CHARLESTON CO.
DORCHESTER CO.
ASHLEY HEIGHTS
HANAHAN
MIDLAND PARK
HUNLEY PARK
NORTH CHARLESTON
DRAYTON
PIERPONT
ASHLEY HALL
MYERS
RED TOP
CHARLESTON
RIVERLAND TERRACE
CENTERVILLE
JAMES ISLAND
JOHNS ISLAND
FENWICK CROSSROADS
RIVERLAND
CAINHOY
PHILIP
TEN MILE
SNOWDEN
HOBCAW POINT
SCANLONVILLE
THE GROVES
MOUNT PLEASANT
BERKELEY CO.
CHARLESTON CO.
U.S. Naval Reservation
Francis Marion National Forest
Charleston Air Force Base
Charleston International Airport
U.S. Naval Hospital
Naval Base
USDA Experimental Farm
USS Yorktown Aircraft Carrier Museum
Willson Memorial Airport
Veterans Hospital
Fort Sumter National Park
Atlantic Ocean
Clark Sound
Cooper River
Ashley River
Wando River
Stono River
Wappoo Creek
Intracoastal Waterway
Breach Inlet
Red Bank Hwy.
Ashley Phosphate Rd.
Rivers Ave.
Dorchester Rd.
Ashley River Rd.
Savannah Hwy.
Riverland Dr.
Folly Rd.
Maybank Hwy.
Cainhoy Rd.
Scale of Miles 0 1 2 3

Charleston, WV

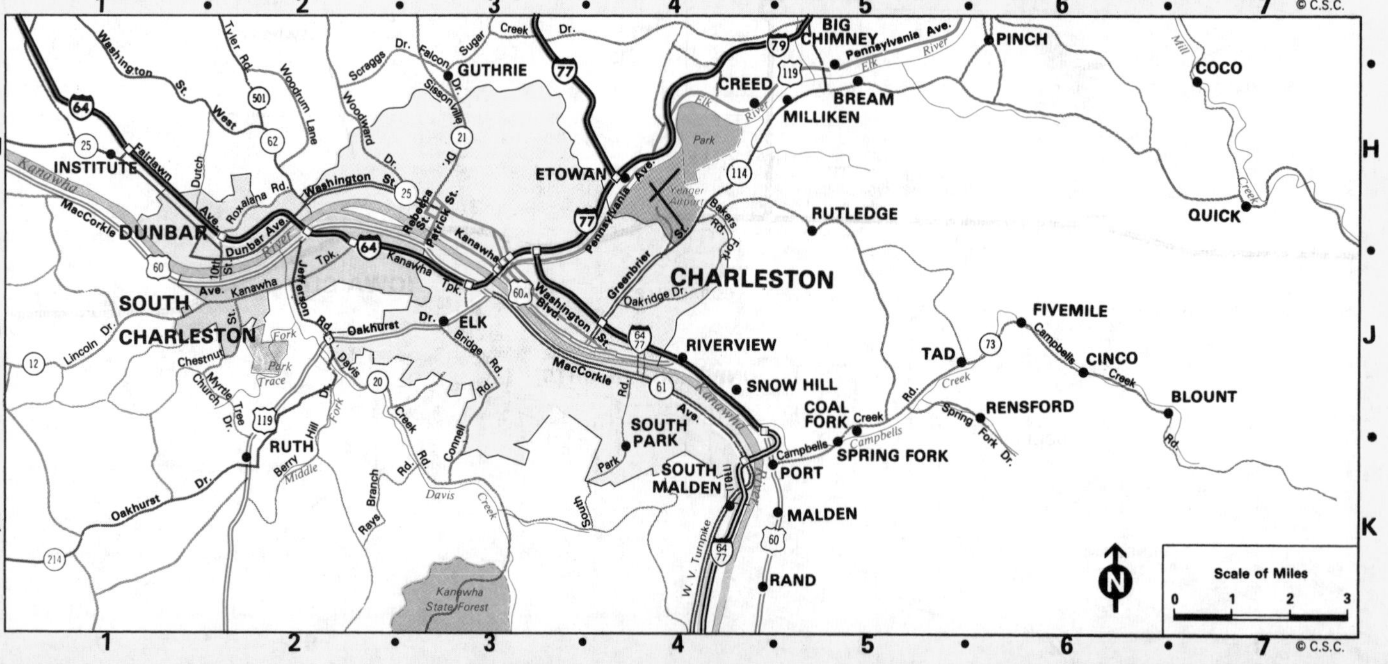

BIG CHIMNEY
PINCH
GUTHRIE
CREED
BREAM
MILLIKEN
COCO
INSTITUTE
ETOWAN
DUNBAR
RUTLEDGE
QUICK
SOUTH CHARLESTON
CHARLESTON
ELK
RIVERVIEW
FIVEMILE
TAD
CINCO
SNOW HILL
BLOUNT
COAL FORK
RENSFORD
SOUTH PARK
SPRING FORK
RUTH
SOUTH MALDEN
PORT
MALDEN
RAND
Kanawha River
Elk River
Yeager Airport
Kanawha State Forest
Washington St.
MacCorkle Ave.
Kanawha Tpk.
Pennsylvania Ave.
Oakhurst Dr.
W V Turnpike
Campbells Creek
Davis Creek
Scale of Miles 0 1 2 3

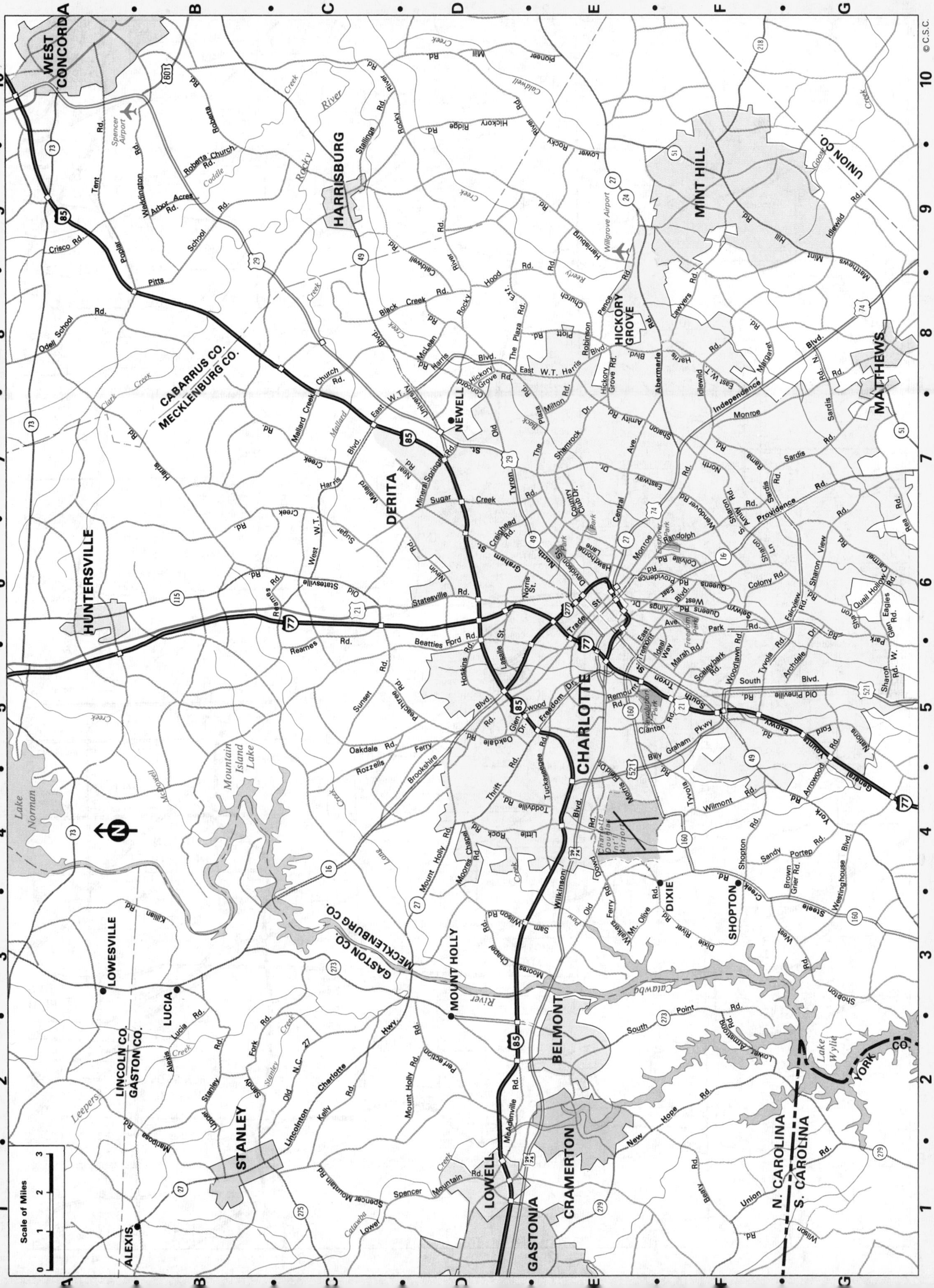

CHARLOTTE
WEST CONCORDA
HARRISBURG
MINT HILL
HICKORY GROVE
NEWELL
DERITA
MATTHEWS
HUNTERSVILLE
LOWESVILLE
LUCIA
STANLEY
ALEXIS
MOUNT HOLLY
BELMONT
LOWELL
CRAMERTON
GASTONIA
DIXIE
SHOPTON
CABARRUS CO.
MECKLENBURG CO.
UNION CO.
GASTON CO.
LINCOLN CO.
YORK CO.
N. CAROLINA
S. CAROLINA
Lake Norman
Mountain Island Lake
Lake Wylie
Catawba River
Spencer Airport
Willgrove Airport
Charlotte Douglas Int'l Airport
Scale of Miles
0 1 2 3
© C.S.C.

Wonder Lake
Ringwood
Johnsburg
Sunrise Ridge
MC CULLOM LAKE
Fox Lake
FOX LAKE
Fox Lake Hills
LAKE VILLA
Venetian Village
West Miltmore
ROUND LAKE HEIGHTS
ROUND LAKE BEACH
ROUND LAKE PARK
THIRD LAKE
Long Lake
ROUND LAKE
GRAYSLAKE
Wildwood
MC HENRY
Lilymoor
Volo
LAKEMOOR
HAINESVILLE
HOLIDAY HILLS
Fremont Center
Ivanhoe
PRAIRIE GROVE
Burtons Bridge
ISLAND LAKE
WAUCONDA
MUNDELEIN
LIBERTYVILLE
GREEN OAKS
Ridgefield
OAKWOOD HILLS
CRYSTAL LAKE
Gilmer
Diamond Lake
VERNON HILLS
METTAWA
FOX RIVER VALLEY GARDENS
TOWER LAKES
HAWTHORN WOODS
INDIAN CREEK
LAKEWOOD
FOX RIVER GROVE
CARY
LAKE BARRINGTON
NORTH BARRINGTON
LAKE IN THE HILLS
LAKE ZURICH
Prairie View
Half Day
LONG GROVE
Aptakisic
KILDEER
BUFFALO GROVE
ALGONQUIN
DEER PARK
MC HENRY CO.
KANE CO.
LAKE COUNTY
COOK COUNTY
BARRINGTON
WHEELING
BARRINGTON HILLS
CARPENTERSVILLE
INVERNESS
PALATINE
GILBERTS
EAST DUNDEE
ARLINGTON HEIGHTS
PROSPECT HEIGHTS
SLEEPY HOLLOW
WEST DUNDEE
SOUTH BARRINGTON
ROLLING MEADOWS
HOFFMAN ESTATES
MOUNT PROSPECT
ELGIN
SCHAUMBURG
STREAMWOOD
HANOVER PARK
ELK GROVE VILLAGE
SOUTH ELGIN
BARTLETT
Keeneyville
Medinah
ITASCA
ROSELLE
BENSENVILLE
WOOD DALE
VALLEY VIEW
BLOOMINGDALE
WAYNE
CHICAGO O'HARE

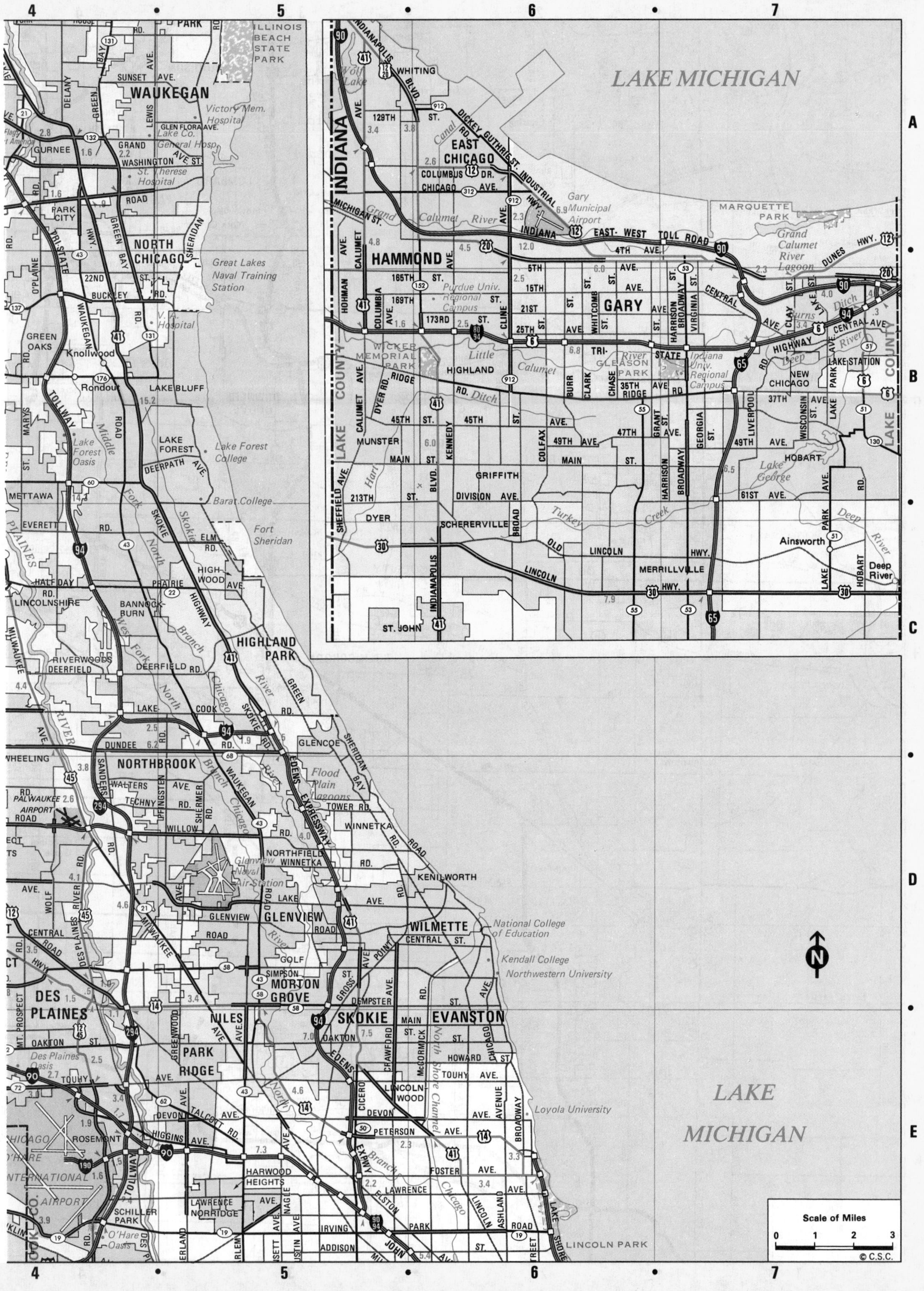
LAKE MICHIGAN
INDIANA
LAKE COUNTY
EAST CHICAGO
HAMMOND
GARY
WHITING
HIGHLAND
MUNSTER
GRIFFITH
DYER
SCHERERVILLE
MERRILLVILLE
ST. JOHN
NEW CHICAGO
HOBART
LAKE STATION
Ainsworth
Deep River
Gary Municipal Airport
Purdue Univ. Regional Campus
Indiana Univ. Regional Campus
Wicker Memorial Park
Gleason Park
Marquette Park
Grand Calumet River Lagoon
Lake George
EAST-WEST TOLL ROAD
ILLINOIS BEACH STATE PARK
WAUKEGAN
Victory Mem. Hospital
Lake Co. General Hosp.
St. Therese Hospital
GURNEE
PARK CITY
NORTH CHICAGO
Great Lakes Naval Training Station
V. A. Hospital
GREEN OAKS
Knollwood
Rondout
LAKE BLUFF
Lake Forest Oasis
LAKE FOREST
Lake Forest College
Barat College
METTAWA
Fort Sheridan
HIGHWOOD
LINCOLNSHIRE
BANNOCKBURN
HIGHLAND PARK
RIVERWOODS
DEERFIELD
GLENCOE
NORTHBROOK
WHEELING
Flood Plain Lagoons
PALWAUKEE AIRPORT
Glenview Naval Air Station
WINNETKA
NORTHFIELD
KENILWORTH
GLENVIEW
WILMETTE
National College of Education
Kendall College
Northwestern University
DES PLAINES
MORTON GROVE
NILES
SKOKIE
EVANSTON
PARK RIDGE
Des Plaines Oasis
LINCOLNWOOD
ROSEMONT
CHICAGO O'HARE INTERNATIONAL AIRPORT
HARWOOD HEIGHTS
NORRIDGE
SCHILLER PARK
O'Hare Oasis
Loyola University
LINCOLN PARK
LAKE MICHIGAN
Scale of Miles
0 1 2 3
© C.S.C.

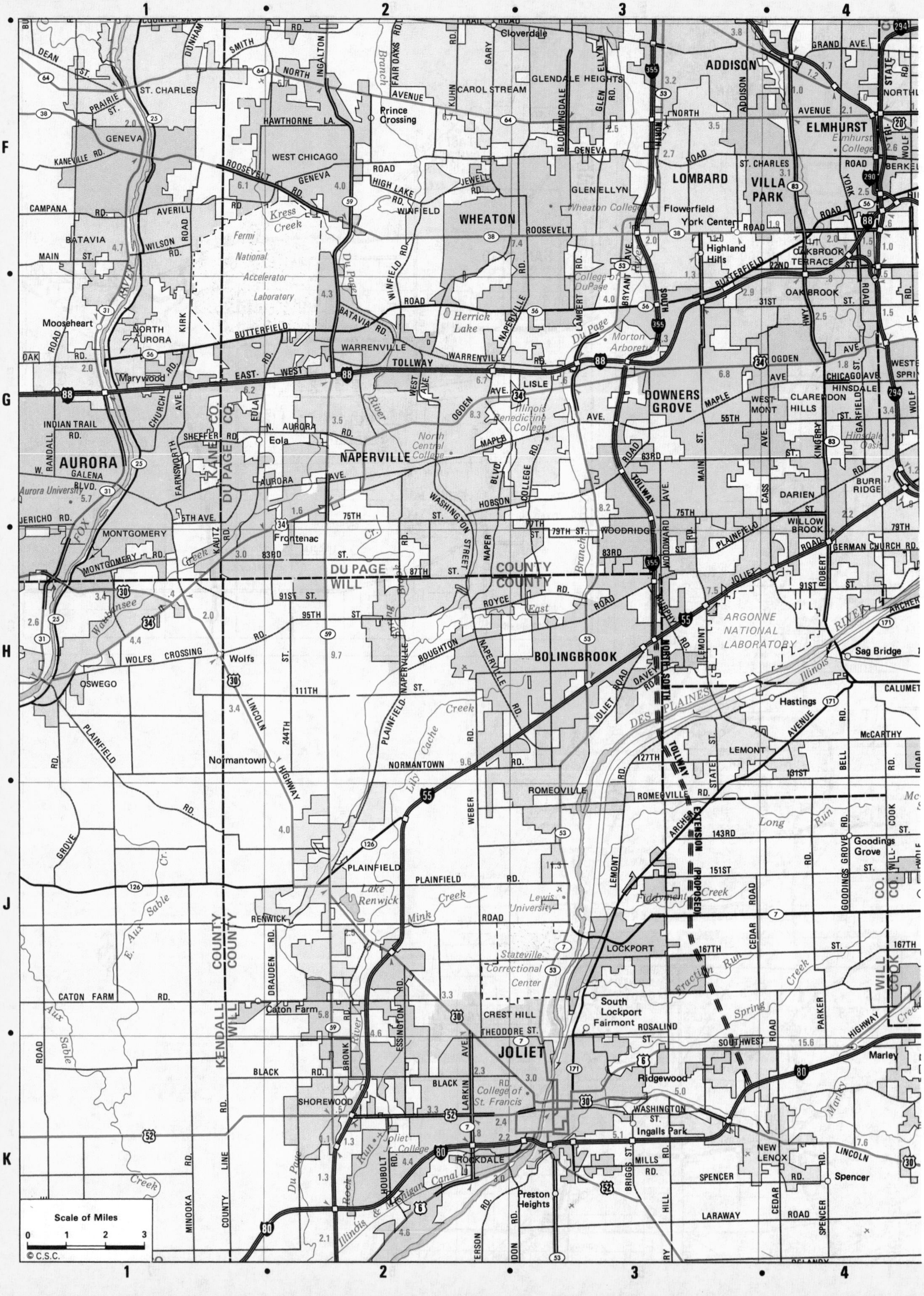

ADDISON
ELMHURST
Elmhurst College
LOMBARD
VILLA PARK
GLEN ELLYN
Wheaton College
Flowerfield
York Center
Highland Hills
OAKBROOK TERRACE
OAK BROOK
WHEATON
CAROL STREAM
GLENDALE HEIGHTS
Cloverdale
Prince Crossing
WEST CHICAGO
ST. CHARLES
GENEVA
BATAVIA
Mooseheart
NORTH AURORA
Marywood
Fermi National Accelerator Laboratory
Kress Creek
College of DuPage
Herrick Lake
Morton Arboretum
WARRENVILLE
LISLE
Illinois Benedictine College
North Central College
NAPERVILLE
Eola
AURORA
Aurora University
MONTGOMERY
Frontenac
DOWNERS GROVE
WESTMONT
CLARENDON HILLS
HINSDALE
Hinsdale Oasis
DARIEN
BURR RIDGE
WILLOWBROOK
WOODRIDGE
DU PAGE COUNTY
WILL COUNTY
KANE CO.
DU PAGE CO.
ARGONNE NATIONAL LABORATORY
Sag Bridge
BOLINGBROOK
Wolfs
OSWEGO
Hastings
LEMONT
Normantown
ROMEOVILLE
Goodings Grove
PLAINFIELD
Lake Renwick
Lewis University
Stateville Correctional Center
LOCKPORT
CREST HILL
South Lockport
Fairmont
KENDALL COUNTY
WILL COUNTY
Caton Farm
JOLIET
College of St. Francis
Joliet Jr. College
SHOREWOOD
ROCKDALE
Preston Heights
Ingalls Park
Ridgewood
NEW LENOX
Marley
Spencer
FOX RIVER
DU PAGE RIVER
DES PLAINES RIVER
Illinois & Michigan Canal
WILL CO.
COOK CO.
Scale of Miles
0 1 2 3
© C.S.C.

Navy Pier
GRANT PARK
Field Mus. of Natural History
Meigs Field
BURNHAM PARK
WASHINGTON PARK
JACKSON PARK
HUMBOLDT PARK
GARFIELD PARK
DOUGLAS PARK
COLUMBUS PARK
MARQUETTE PARK
Chicago Midway Airport
FRANKLIN PARK
NORTHLAKE
STONE PARK
MELROSE PARK
RIVER GROVE
ELMWOOD PARK
RIVER FOREST
OAK PARK
BERKELEY
BELLWOOD
MAYWOOD
FOREST PARK
WESTCHESTER
BROADVIEW
BERWYN
CICERO
NORTH RIVERSIDE
RIVERSIDE
LA GRANGE PARK
BROOKFIELD
STICKNEY
FOREST VIEW
WESTERN SPRINGS
LA GRANGE
LYONS
MC COOK
INDIAN HEAD PARK
HODGKINS
SUMMIT
BURR RIDGE
BEDFORD PARK
JUSTICE
BRIDGEVIEW
WILLOW SPRINGS
BURBANK
HOMETOWN
HICKORY HILLS
EVERGREEN PARK
OAK LAWN
CHICAGO RIDGE
PALOS HILLS
Sag Bridge
WORTH
MERRIONETTE PARK
ALSIP
PALOS PARK
PALOS HEIGHTS
BLUE ISLAND
CALUMET PARK
McGinnis Slough
CRESTWOOD
ROBBINS
DIXMOOR
RIVERDALE
Goodings Grove
ORLAND PARK
MIDLOTHIAN
POSEN
DOLTON
CALUMET CITY
OAK FOREST
HARVEY
PHOENIX
HAMMOND
MARKHAM
SOUTH HOLLAND
Alpine
ORLAND HILLS
TINLEY PARK
HAZEL CREST
EAST HAZEL CREST
THORNTON
Abraham Lincoln Oasis
LANSING
COUNTRY CLUB HILLS
HOMEWOOD
Tinley Park Mental Health Center
WICKER MEMORIAL PARK
MUNSTER
HIGHLAND
Marley
Aubury Hills
World Music Theater
FLOSSMOOR
GLENWOOD
LYNWOOD
MOKENA
ST. FRANCIS
OLYMPIA FIELDS
Prairie State College
FORD HEIGHTS
Lincoln Estates
MATTESON
CHICAGO HEIGHTS
DYER
SCHERERVILLE
FRANKFORT
PARK FOREST
SOUTH CHICAGO HEIGHTS
SAUK VILLAGE
RICHTON PARK
STEGER
Lake Calumet
WILLIAM W. POWERS CONSERVATION AREA
Wolf Lake
WHITING
EAST CHICAGO
Purdue Univ. Reg. Campus
ILLINOIS
INDIANA
WILL CO.
COOK CO.
LAKE CO.
KENNEDY EXPWY.
EISENHOWER EXPRESSWAY
STEVENSON EXPWY.
ADLAI E. STEVENSON
DAN RYAN EXPWY.
CHICAGO SKYWAY
TRI-STATE TOLLWAY
KINGERY EXPWY.
LAKE SHORE DRIVE
Sanitary and Ship Canal
Calumet Sag Channel
F
G
H
J
K
4
5
6
7

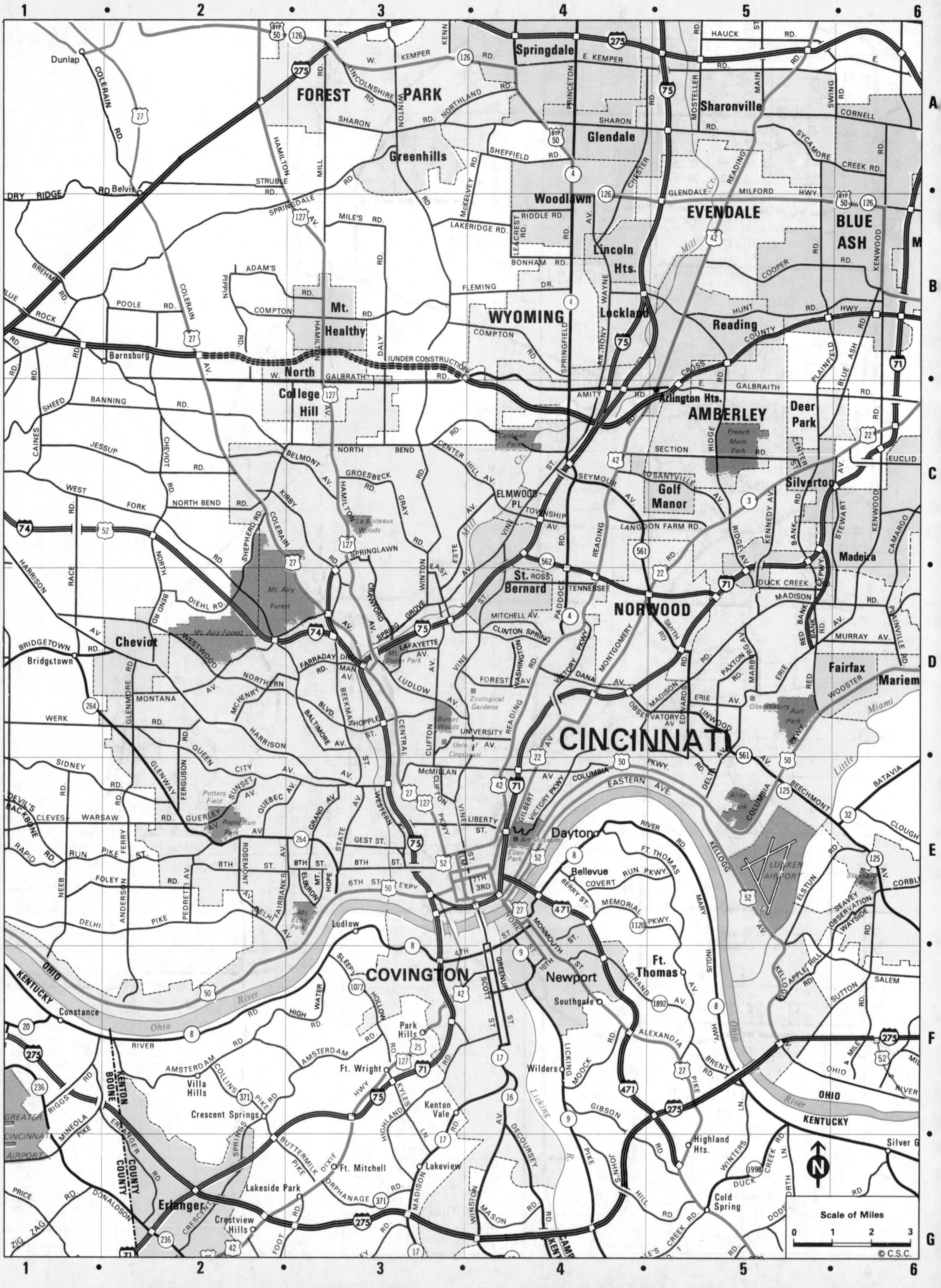

Dunlap
Springdale
FOREST PARK
Sharonville
Glendale
Greenhills
Woodlawn
EVENDALE
BLUE ASH
Belvis
Lincoln Hts.
Mt. Healthy
WYOMING
Lockland
Reading
Barnsburg
North College Hill
Arlington Hts.
AMBERLEY
Deer Park
Golf Manor
Silverton
Elmwood Pl
Madeira
St. Bernard
NORWOOD
Cheviot
Bridgetown
Fairfax
CINCINNATI
Dayton
Bellevue
Ludlow
COVINGTON
Newport
Ft. Thomas
Southgate
Constance
Park Hills
Ft. Wright
Villa Hills
Crescent Springs
Kenton Vale
Wilders
Highland Hts.
Ft. Mitchell
Lakeview
Lakeside Park
Cold Spring
Erlanger
Crestview Hills
Silver Grove
KENTON COUNTY
BOONE COUNTY
KENTUCKY
OHIO
Ohio River
Licking R.
Little Miami
Mt. Airy Forest
Zoological Gardens
Univ. of Cincinnati
Art Museum
Eden Park
LUNKEN AIRPORT
GREATER CINCINNATI AIRPORT
(UNDER CONSTRUCTION)

Scale of Miles
0 1 2 3

© C.S.C.

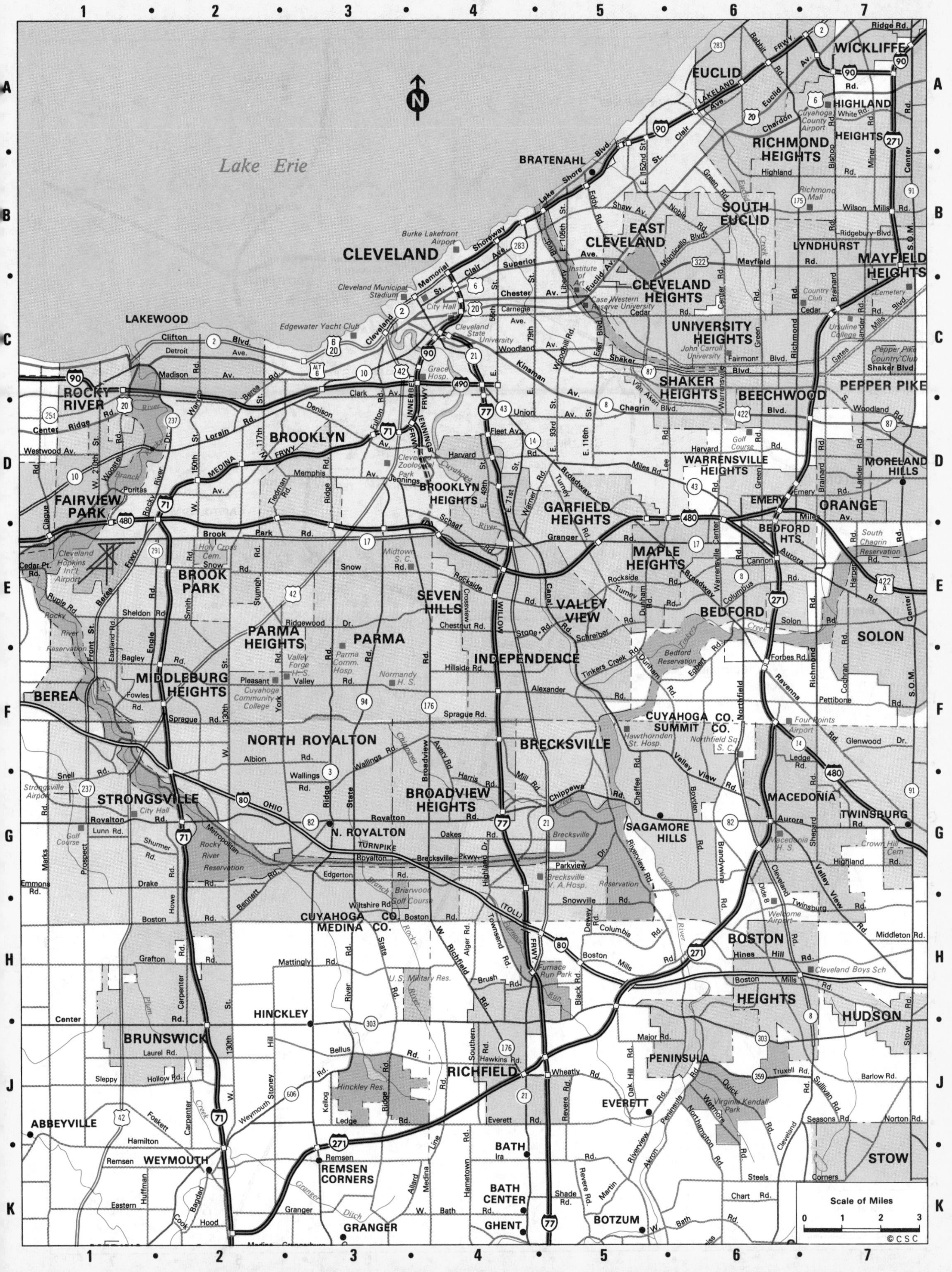

Lake Erie
CLEVELAND
LAKEWOOD
ROCKY RIVER
FAIRVIEW PARK
BROOKLYN
BROOKLYN HEIGHTS
BROOK PARK
PARMA HEIGHTS
PARMA
SEVEN HILLS
MIDDLEBURG HEIGHTS
BEREA
NORTH ROYALTON
STRONGSVILLE
BROADVIEW HEIGHTS
N. ROYALTON
INDEPENDENCE
BRECKSVILLE
VALLEY VIEW
GARFIELD HEIGHTS
MAPLE HEIGHTS
BEDFORD
BEDFORD HTS.
SOLON
EUCLID
WICKLIFFE
HIGHLAND HEIGHTS
RICHMOND HEIGHTS
BRATENAHL
SOUTH EUCLID
EAST CLEVELAND
LYNDHURST
MAYFIELD HEIGHTS
CLEVELAND HEIGHTS
UNIVERSITY HEIGHTS
SHAKER HEIGHTS
BEECHWOOD
PEPPER PIKE
WARRENSVILLE HEIGHTS
MORELAND HILLS
ORANGE
CUYAHOGA CO.
SUMMIT CO.
MEDINA CO.
SAGAMORE HILLS
MACEDONIA
TWINSBURG
BOSTON
BOSTON HEIGHTS
HUDSON
PENINSULA
HINCKLEY
BRUNSWICK
RICHFIELD
EVERETT
ABBEYVILLE
WEYMOUTH
REMSEN CORNERS
BATH
BATH CENTER
GHENT
GRANGER
BOTZUM
STOW
Burke Lakefront Airport
Cleveland Municipal Stadium
Edgewater Yacht Club
City Hall
Cleveland State University
Institute of Art
Case Western Reserve University
John Carroll University
Cleveland Hopkins Int'l Airport
Holy Cross Cem.
Cleveland Zoological Park
Midtown S.C.
Parma Comm. Hosp.
Normandy H.S.
Valley Forge H.S.
Cuyahoga Community College
Rocky River Reservation
Bedford Reservation
Hawthornden St. Hosp.
Northfield Sq. S.C.
Four Points Airport
Strongsville Airport
Brecksville V.A. Hosp.
Brecksville Reservation
Briarwood Golf Course
U.S. Military Res.
Furnace Run Park
Hinckley Res.
Virginia Kendall Park
Cleveland Boys Sch
Welcome Airport
Macedonia H.S.
Crown Hill Cem
South Chagrin Reservation
Pepper Pike Country Club
Ursuline College
Richmond Mall
Cuyahoga County Airport
Grace Hosp.
Scale of Miles
0 1 2 3
©CSC

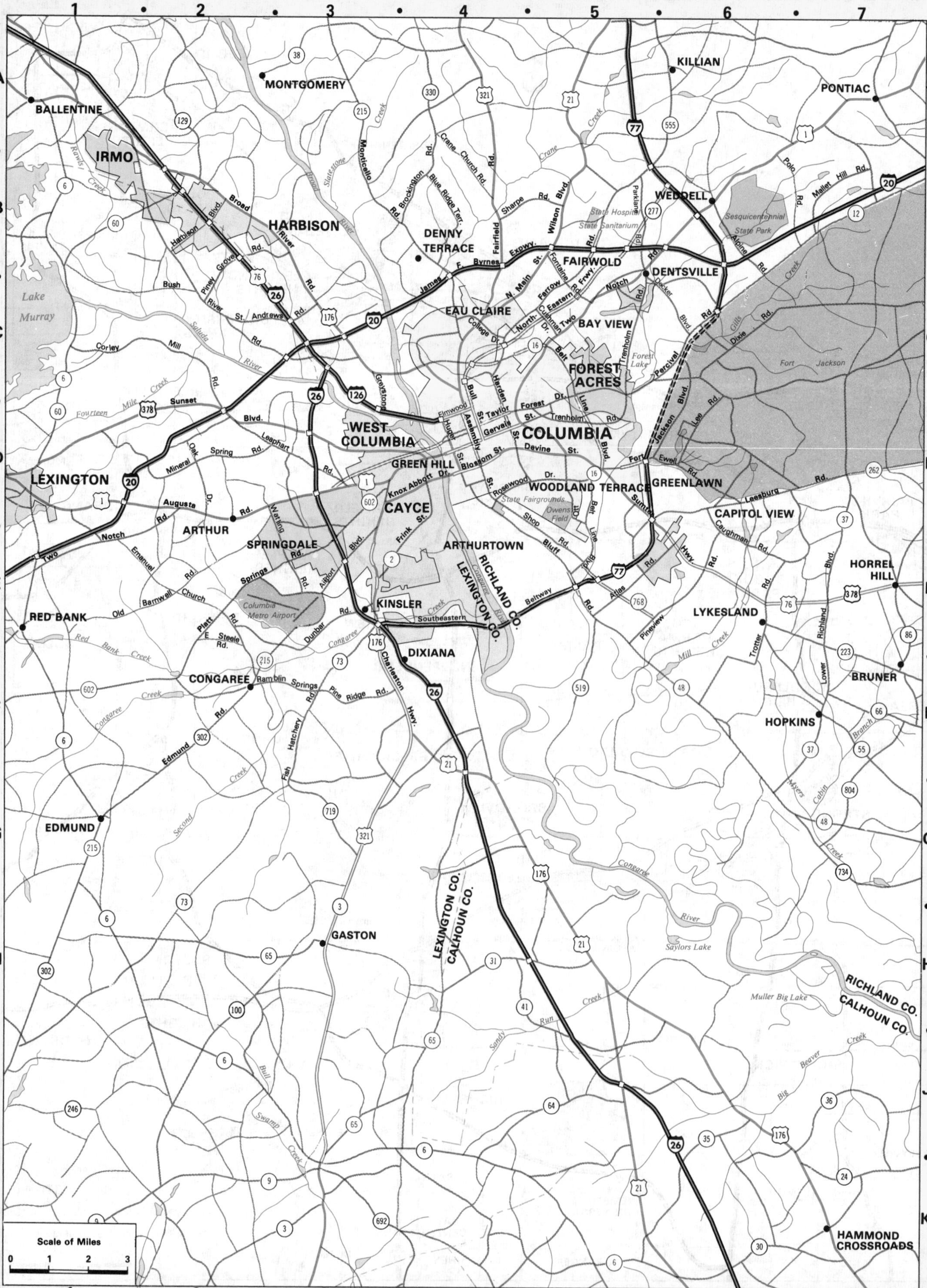

N

1 2 3 4 5 6 7

A B C D E F G H J K

MONTGOMERY
KILLIAN
PONTIAC
BALLENTINE
IRMO
HARBISON
DENNY TERRACE
WEDDELL
FAIRWOLD
DENTSVILLE
Sesquicentennial State Park
State Hospital
State Sanitarium
Lake Murray
EAU CLAIRE
BAY VIEW
FOREST ACRES
Forest Lake
Fort Jackson
WEST COLUMBIA
COLUMBIA
GREEN HILL
LEXINGTON
CAYCE
WOODLAND TERRACE
GREENLAWN
State Fairgrounds
Owens Field
CAPITOL VIEW
ARTHUR
SPRINGDALE
ARTHURTOWN
RICHLAND CO.
LEXINGTON CO.
Columbia Metro Airport
HORREL HILL
RED BANK
KINSLER
LYKESLAND
DIXIANA
CONGAREE
BRUNER
HOPKINS
EDMUND
GASTON
LEXINGTON CO.
CALHOUN CO.
Saylors Lake
Congaree River
RICHLAND CO.
CALHOUN CO.
Muller Big Lake
HAMMOND CROSSROADS

Scale of Miles
0 1 2 3

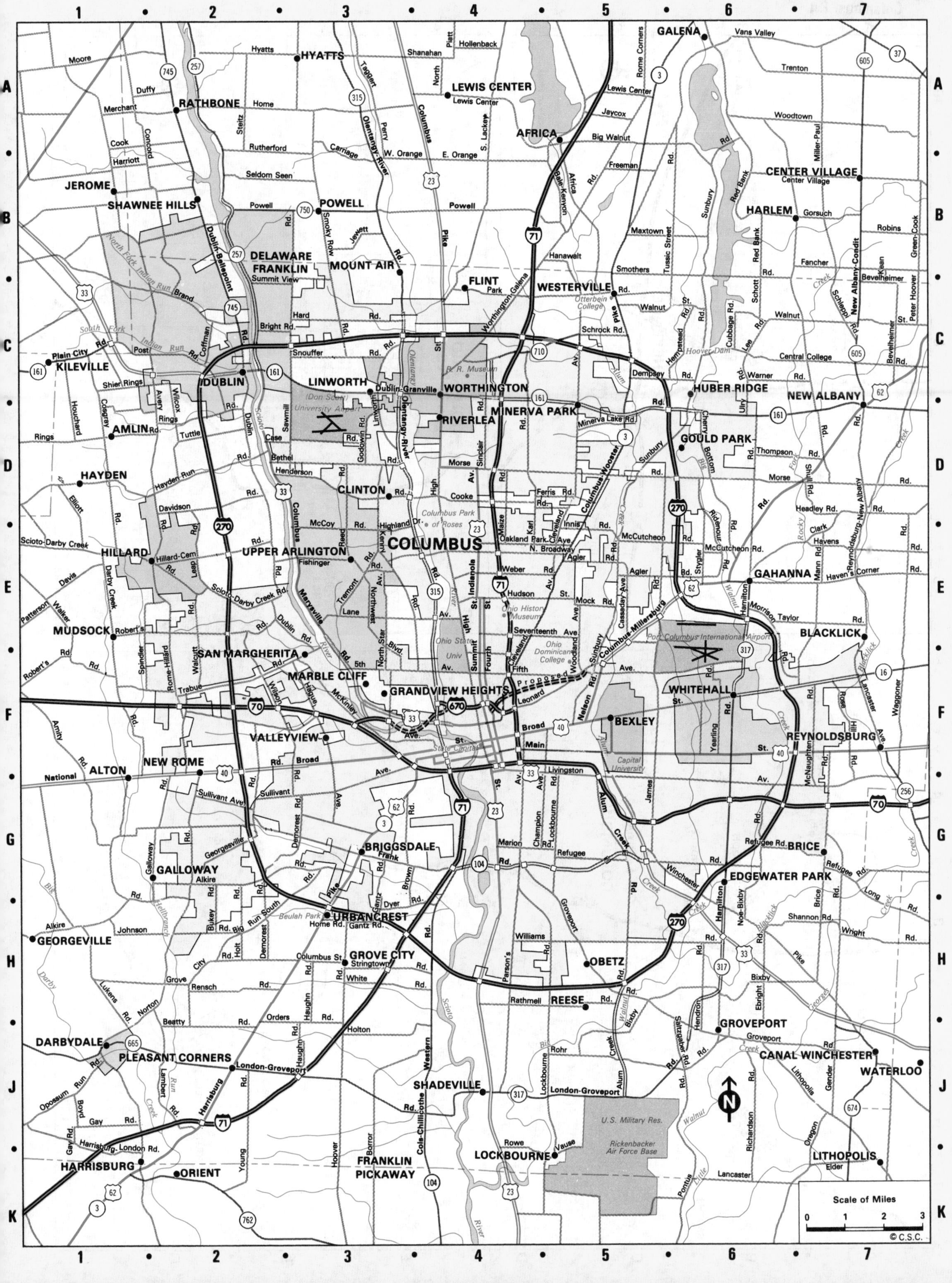

GALENA
HYATTS
LEWIS CENTER
RATHBONE
AFRICA
JEROME
SHAWNEE HILLS
POWELL
CENTER VILLAGE
HARLEM
DELAWARE
FRANKLIN
MOUNT AIR
FLINT
WESTERVILLE
KILEVILLE
DUBLIN
LINWORTH
WORTHINGTON
HUBER RIDGE
NEW ALBANY
RIVERLEA
MINERVA PARK
AMLIN
GOULD PARK
HAYDEN
CLINTON
COLUMBUS
HILLARD
UPPER ARLINGTON
GAHANNA
MUDSOCK
BLACKLICK
SAN MARGHERITA
MARBLE CLIFF
GRANDVIEW HEIGHTS
WHITEHALL
BEXLEY
VALLEYVIEW
REYNOLDSBURG
NEW ROME
ALTON
BRIGGSDALE
BRICE
GALLOWAY
EDGEWATER PARK
URBANCREST
GEORGEVILLE
GROVE CITY
OBETZ
REESE
GROVEPORT
DARBYDALE
PLEASANT CORNERS
CANAL WINCHESTER
WATERLOO
SHADEVILLE
LOCKBOURNE
HARRISBURG
ORIENT
FRANKLIN
PICKAWAY
LITHOPOLIS

U.S. Military Res.
Rickenbacker Air Force Base
Port Columbus International Airport
Ohio State Univ
Ohio Dominican College
Capital University
Otterbein College
Columbus Park of Roses
Ohio Historical Museum
State Capitol
R. R. Museum
(Don Scott) University Airport

Scale of Miles
0 1 2 3

© C.S.C.

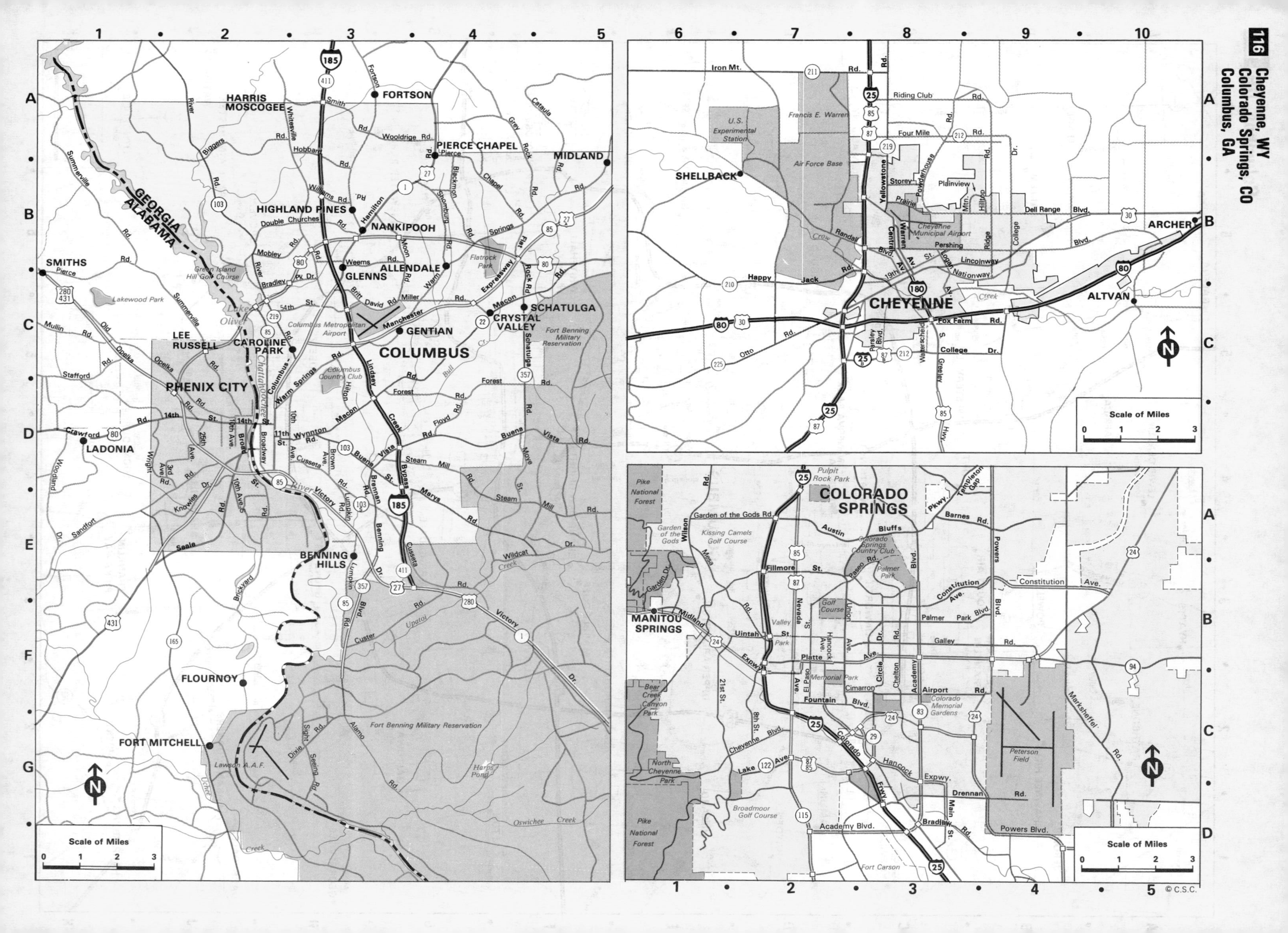
CHEYENNE
ARCHER
ALTVAN
SHELLBACK
Francis E. Warren Air Force Base
U.S. Experimental Station
Cheyenne Municipal Airport
Scale of Miles
COLORADO SPRINGS
MANITOU SPRINGS
Pike National Forest
Garden of the Gods
Peterson Field
Fort Carson
Broadmoor Golf Course
North Cheyenne Canyon Park
Bear Creek Canyon Park
Pulpit Rock Park
Colorado Springs Country Club
Memorial Park
Colorado Memorial Gardens
COLUMBUS
PHENIX CITY
FORTSON
MIDLAND
PIERCE CHAPEL
NANKIPOOH
HIGHLAND PINES
ALLENDALE
GLENNS
GENTIAN
CRYSTAL VALLEY
SCHATULGA
CAROLINE PARK
LEE RUSSELL
BENNING HILLS
FLOURNOY
FORT MITCHELL
LADONIA
SMITHS
HARRIS MOSCOGEE
GEORGIA
ALABAMA
Fort Benning Military Reservation
Columbus Metropolitan Airport
Columbus Country Club
Lakewood Park
Flatrock Park
Green Island Hill Golf Course
Lawson A.A.F.
Chattahoochee
Lake Oliver
© C.S.C.

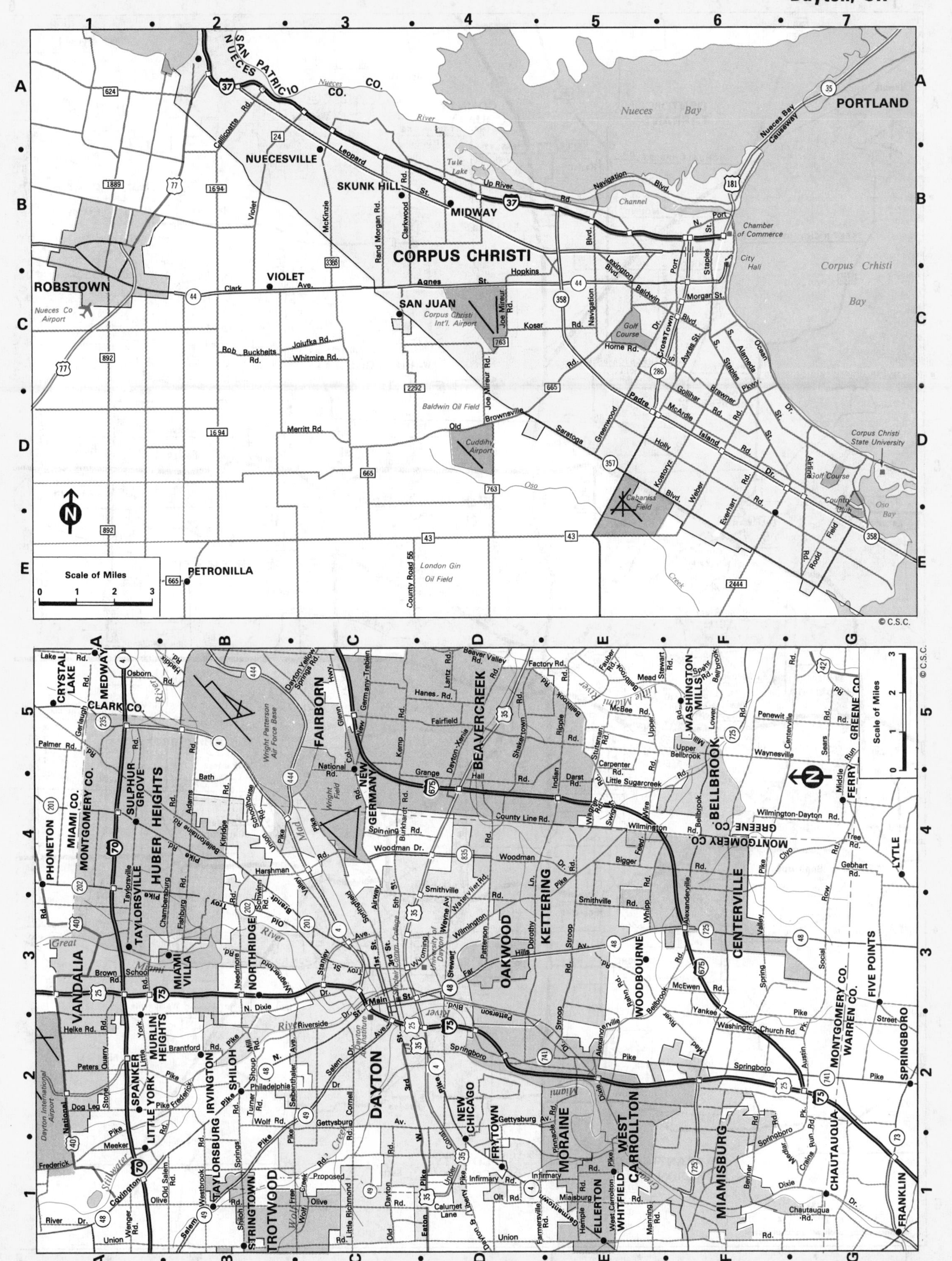
CORPUS CHRISTI
ROBSTOWN
NUECESVILLE
SKUNK HILL
MIDWAY
VIOLET
SAN JUAN
PORTLAND
PETRONILLA
Nueces Bay
Corpus Christi Bay
Corpus Christi Int'l. Airport
Nueces Co Airport
Cuddihy Airport
Cabaniss Field
Baldwin Oil Field
London Gin Oil Field
Corpus Christi State University
Oso Bay
Scale of Miles
© C.S.C.
DAYTON
FAIRBORN
BEAVERCREEK
KETTERING
OAKWOOD
CENTERVILLE
VANDALIA
HUBER HEIGHTS
MIAMISBURG
WEST CARROLLTON
MORAINE
TROTWOOD
NEW GERMANY
BELLBROOK
WASHINGTON MILLS
NORTHRIDGE
MIAMI VILLA
SULPHUR GROVE
TAYLORSVILLE
PHONETON
CRYSTAL LAKE
MEDWAY
SPANKER
LITTLE YORK
MURLIN HEIGHTS
IRVINGTON
SHILOH
TAYLORSBURG
STRINGTOWN
NEW CHICAGO
FRYTOWN
ELLERTON
WHITFIELD
WOODBOURNE
CHAUTAUQUA
FRANKLIN
SPRINGBORO
FIVE POINTS
FERRY
LYTLE
Wright Patterson Air Force Base
Dayton International Airport
MIAMI CO.
MONTGOMERY CO.
GREENE CO.
WARREN CO.
CLARK CO.
Scale of Miles

FT. WORTH
DENTON COUNTY
TARRANT COUNTY
TARRANT COUNTY
JOHNSON COUNTY
Newark
Roanoke
Westlake
Flower Mound
Grapevine Lake
Grapevine
Southlake
Keller
Colleyville
Haslet
EAGLE MOUNTAIN NATIONAL GUARD BASE
Eagle Mountain Lake
Watauga
North Richland Hills
Saginaw
Saginaw Airfield
Blue Mound
Fort Worth Nature Center and Refuge
Bedford
Euless
Haltom City
Richland Hills
Hurst
Lakeside
Lake Worth
Sansom Park
Meacham Field
River Oaks
CARSWELL A. F. B.
White Settlement
Westworth
Westover Hills
Arlington
Arlington Stadium
Pantego
Lake Arlington
Univ. of Texas at Arlington
Dalworthington Gardens
Texas Christian Univ.
Benbrook
Benbrook Lake
Edgecliff
Forest Hill
Kennedale
Everman
Arlington Municipal Airfield
Watsonville
Mansfield
Crowley
Rendon
Burleson
Retta
AVONDALE-HASLET RD.
KELLER-HASLET RD.
HASLET-ROANOKE RD.
BLUE MOUND RD.
KELLER-HICKS RD.
SKEET RICHARDSON RD.
HICKS RD.
WAGLEY-ROBERTSON RD.
HALSTON-BAILEY-BOSWELL RD.
PARK DR.
BOWMAN-ROBERTS RD.
OLD DECATUR RD.
SAGINAW RD.
ALTA VISTA RD.
DENTON HWY.
PRECINCT LINE RD.
WATAUGA SMITHFIELD RD.
DAVIS BLVD.
NORTHWEST PKWY.
COLLEYVILLE BLVD.
HARWOOD
INDUSTRIAL
JACKSON
AIRPORT FREEWAY
GLENVIEW
PIPELINE
EULESS BLVD.
JACKSBORO HWY.
AZLE AVE.
28TH ST.
MIDWAY
GRAPEVINE HWY.
NORTH FRWY.
BEACH ST.
BELKNAP ST.
RIVERSIDE FRWY.
RANDOL MILL
LN.
SANDY LN.
COOKS LN.
LAMAR
COLLINS
E. 1ST ST.
RANDOL MILL
NORTHSIDE DR.
HOUSTON
N. MAIN
COMMERCE
7TH ST.
LANCASTER
OAKLAND
DIVISION
E. ABRAM ST.
BOWIE BLVD.
WEST FRWY.
CAMP BOWIE BLVD.
ROSEDALE AVE.
PIONEER PKWY.
VILLAGE CREEK
FIELDER RD.
COOPER ST.
COLLINS AVE.
WATSON PKWY.
VICKERY BLVD.
UNIVERSITY DR.
FOREST PARK
8TH AVE.
S. MAIN
SOUTH FRWY.
RIVERSIDE DR.
MITCHELL
VAUGHN
BERRY ST.
WICHITA
POLY FRWY.
ARKANSAS LN.
LITTLE RD.
MANSFIELD HWY.
SEMINARY DR.
JAMES ST.
OLD GRANBURY RD.
MC CART
OAK GROVE
FOREST HILL
ANGLIN
MATLOCK RD.
MC KNIGHT RD.
DIRKS RD.
SCHOOL RD.
SYCAMORE
CLEBURNE RD.
EDEN RD.
HARRIS RD.
WEBB
Britton Rd.
MANSFIELD-
RENDON RD.
CROWLEY RD.
MAIN ST.
WILSHIRE
West Fork of the Trinity River
Mary's Creek
Clear Fork
Big Bear Creek
Little Bear Creek
Fish Creek
Scale of Miles 0 1 2 3 4 5
©C.S.C.

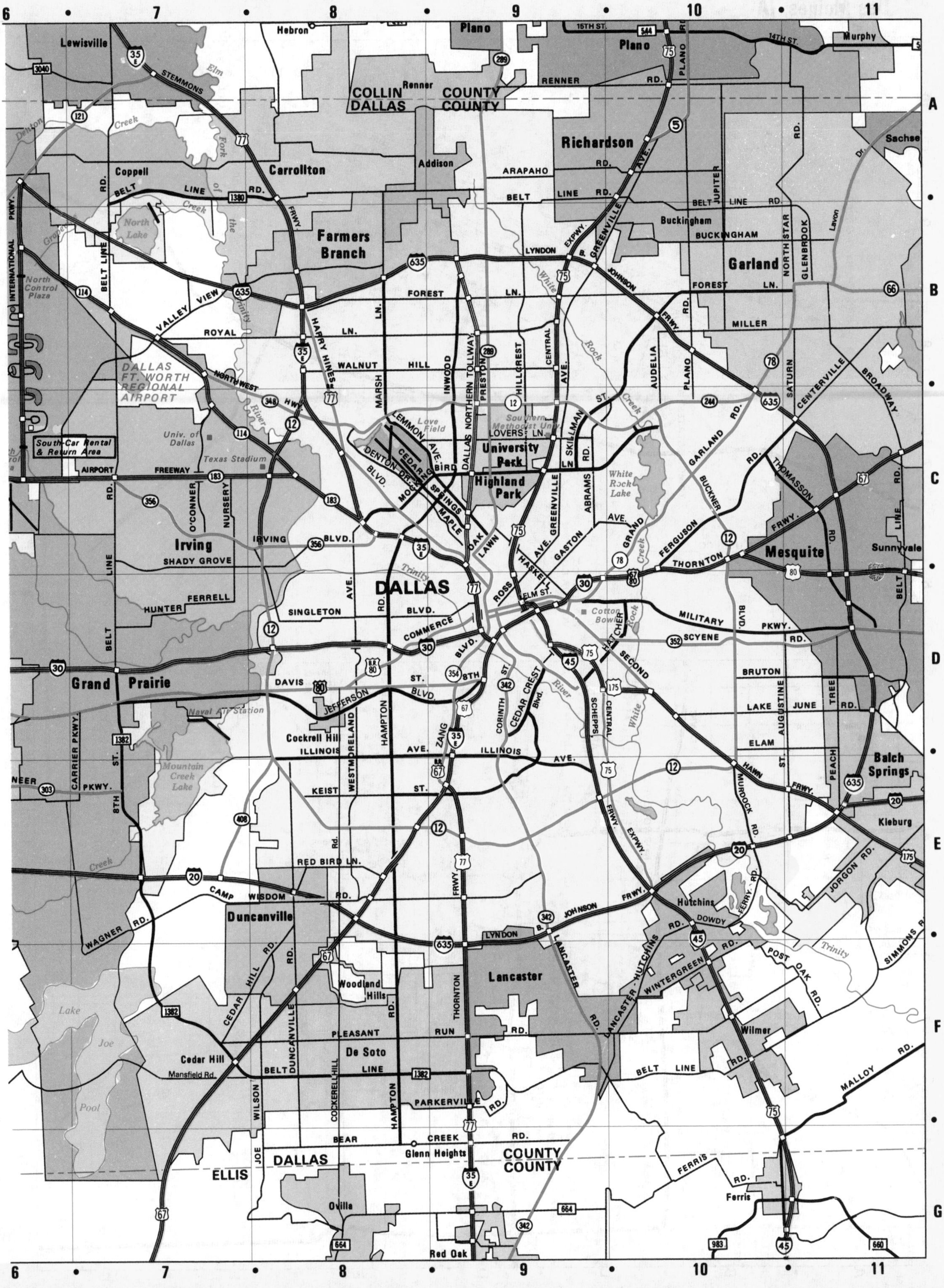

6 7 8 9 10 11

A B C D E F G

Lewisville
Hebron
Plano
Murphy
COLLIN COUNTY
DALLAS COUNTY
Renner
Richardson
Sachse
Coppell
Carrollton
Addison
Buckingham
Farmers Branch
Garland
North Control Plaza
DALLAS FT. WORTH REGIONAL AIRPORT
South-Car Rental & Return Area
Univ. of Dallas
Texas Stadium
Love Field
Southern Methodist Univ.
University Park
Highland Park
White Rock Lake
Irving
Mesquite
Sunnyvale
DALLAS
Cotton Bowl
Grand Prairie
Naval Air Station
Cockrell Hill
Mountain Creek Lake
Balch Springs
Kleburg
Duncanville
Hutchins
Lancaster
Woodland Hills
Wilmer
Cedar Hill
De Soto
Lake Joe Pool
Glenn Heights
ELLIS COUNTY
DALLAS COUNTY
Ovilla
Red Oak
Ferris

Elm Fork of the Trinity River
Denton Creek
Grapevine Creek
White Rock Creek
Trinity River
Mountain Creek

STEMMONS FRWY.
BELT LINE RD.
ARAPAHO RD.
LYNDON B. JOHNSON FRWY.
FOREST LN.
ROYAL LN.
WALNUT HILL
NORTHWEST HWY.
LOVERS LN.
VALLEY VIEW
AIRPORT FREEWAY
IRVING BLVD.
SHADY GROVE
HUNTER FERRELL
SINGLETON BLVD.
COMMERCE
DAVIS ST.
JEFFERSON BLVD.
ILLINOIS AVE.
KEIST ST.
RED BIRD LN.
CAMP WISDOM RD.
WAGNER RD.
PLEASANT RUN RD.
BELT LINE
PARKERVILLE RD.
BEAR CREEK RD.
MILITARY PKWY.
SCYENE RD.
BRUTON
LAKE JUNE RD.
ELAM
HAWN FRWY.
DOWDY FERRY RD.
POST OAK RD.
SIMMONS RD.
MALLOY RD.
FERRIS RD.
JORGON RD.
CARRIER PKWY.
INTERNATIONAL PKWY.
HARRY HINES
MARSH LN.
INWOOD
DALLAS NORTHERN TOLLWAY
PRESTON
HILLCREST
CENTRAL EXPWY.
GREENVILLE AVE.
SKILLMAN
ABRAMS RD.
AUDELIA
PLANO RD.
JUPITER RD.
NORTH STAR
GLENBROOK
SATURN
CENTERVILLE
BROADWAY
GARLAND RD.
BUCKNER
FERGUSON
THORNTON
THOMASSON
GASTON
GRAND AVE.
LEMMON AVE.
CEDAR SPRINGS
MOCKINGBIRD
DENTON DR.
MAPLE
OAK LAWN
ROSS
HASKELL
ELM ST.
HATCHER
SECOND
8TH
ZANG
CORINTH ST.
CEDAR CREST Blvd.
SCHEPPS
WESTMORELAND
HAMPTON RD.
COCKERELL HILL
DUNCANVILLE RD.
CEDAR HILL RD.
WILSON
JOE
LANCASTER RD.
LANCASTER-HUTCHINS RD.
WINTERGREEN RD.
MURDOCK RD.
PEACH TREE RD.
ST. AUGUSTINE
15TH ST.
14TH ST.
Mansfield Rd.
Lavon Dr.

Daytona Beach, FL
Des Moines, IA

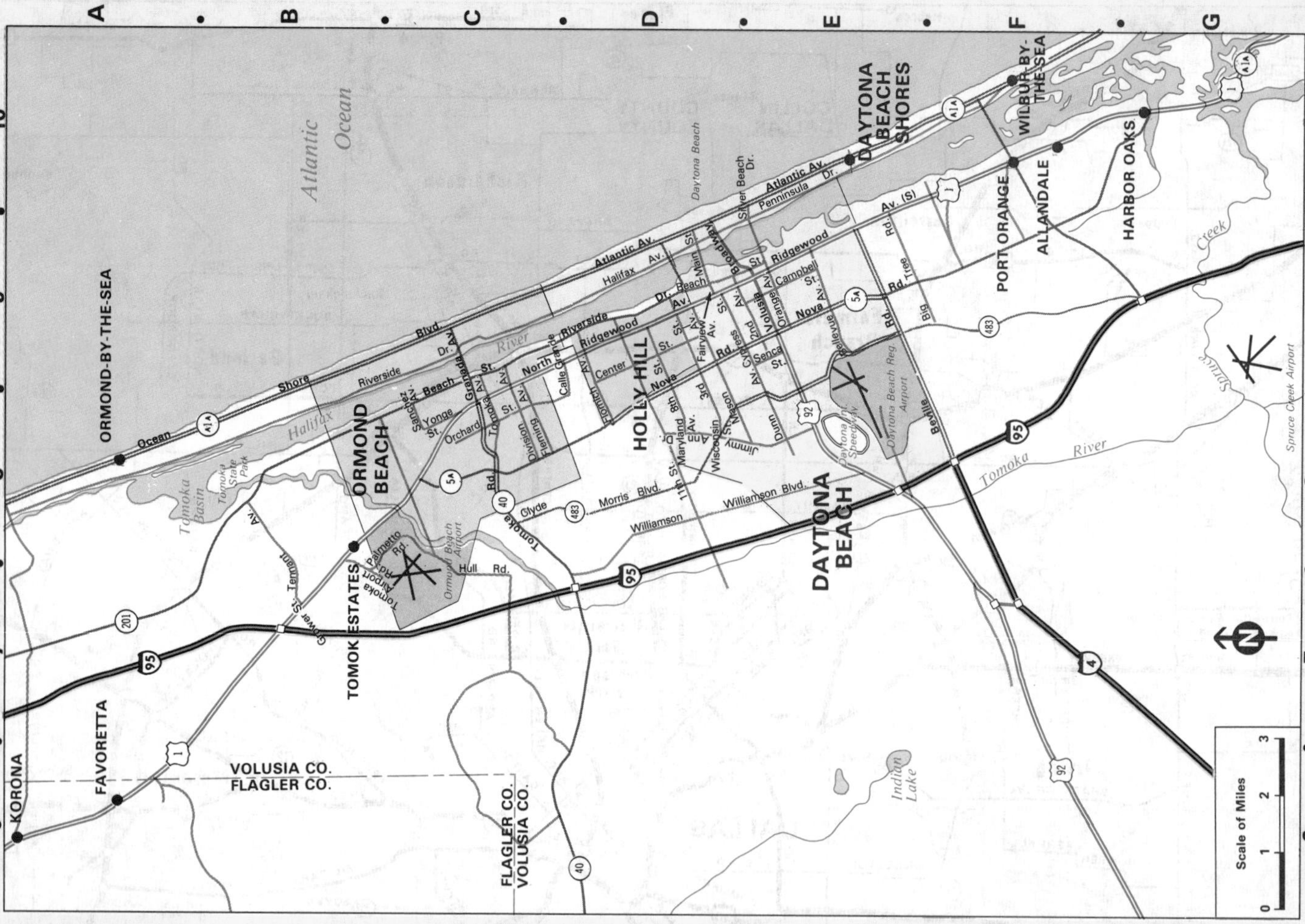

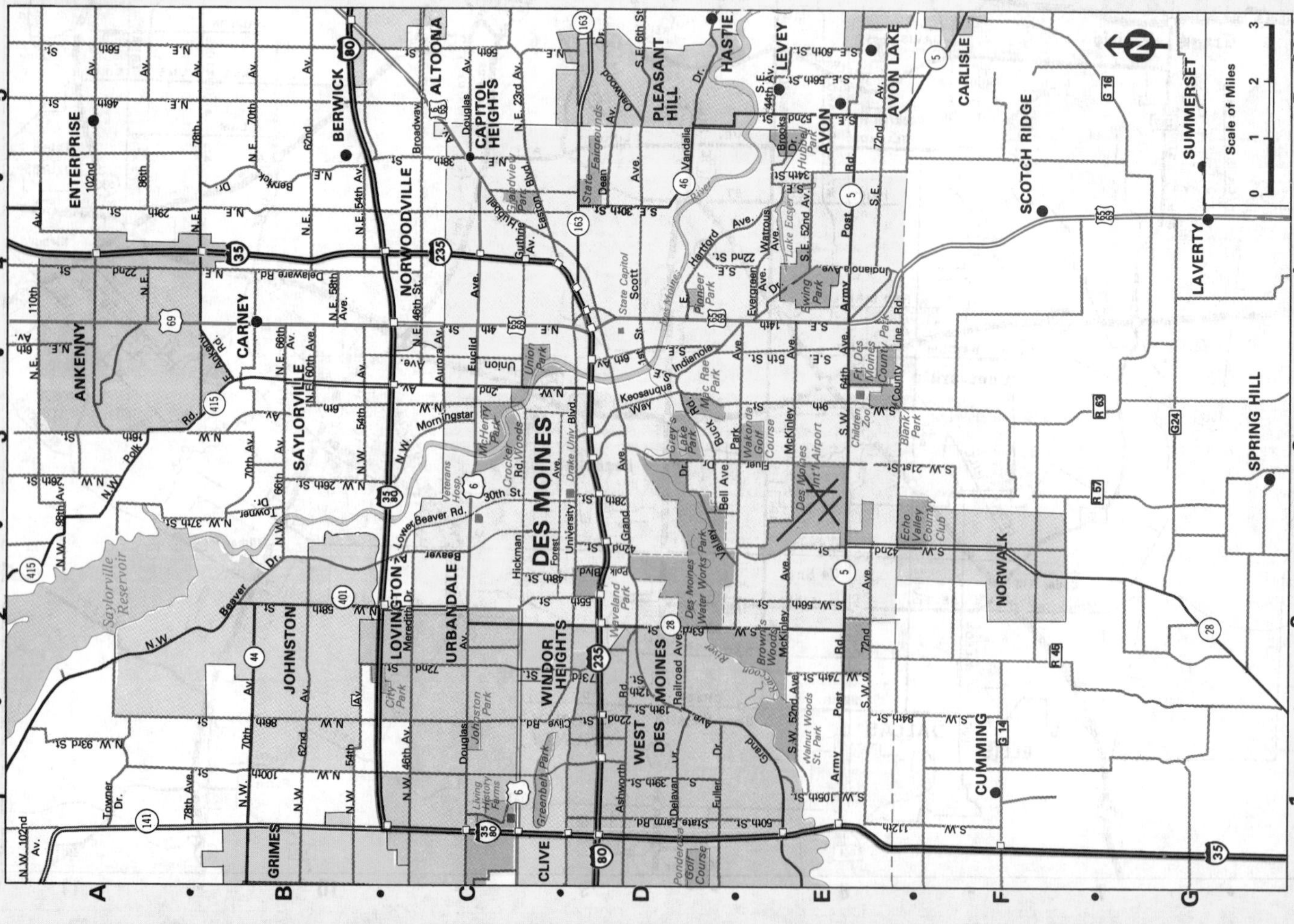

Scale of Miles 0 1 2 3

Scale of Miles 0 1 2 3

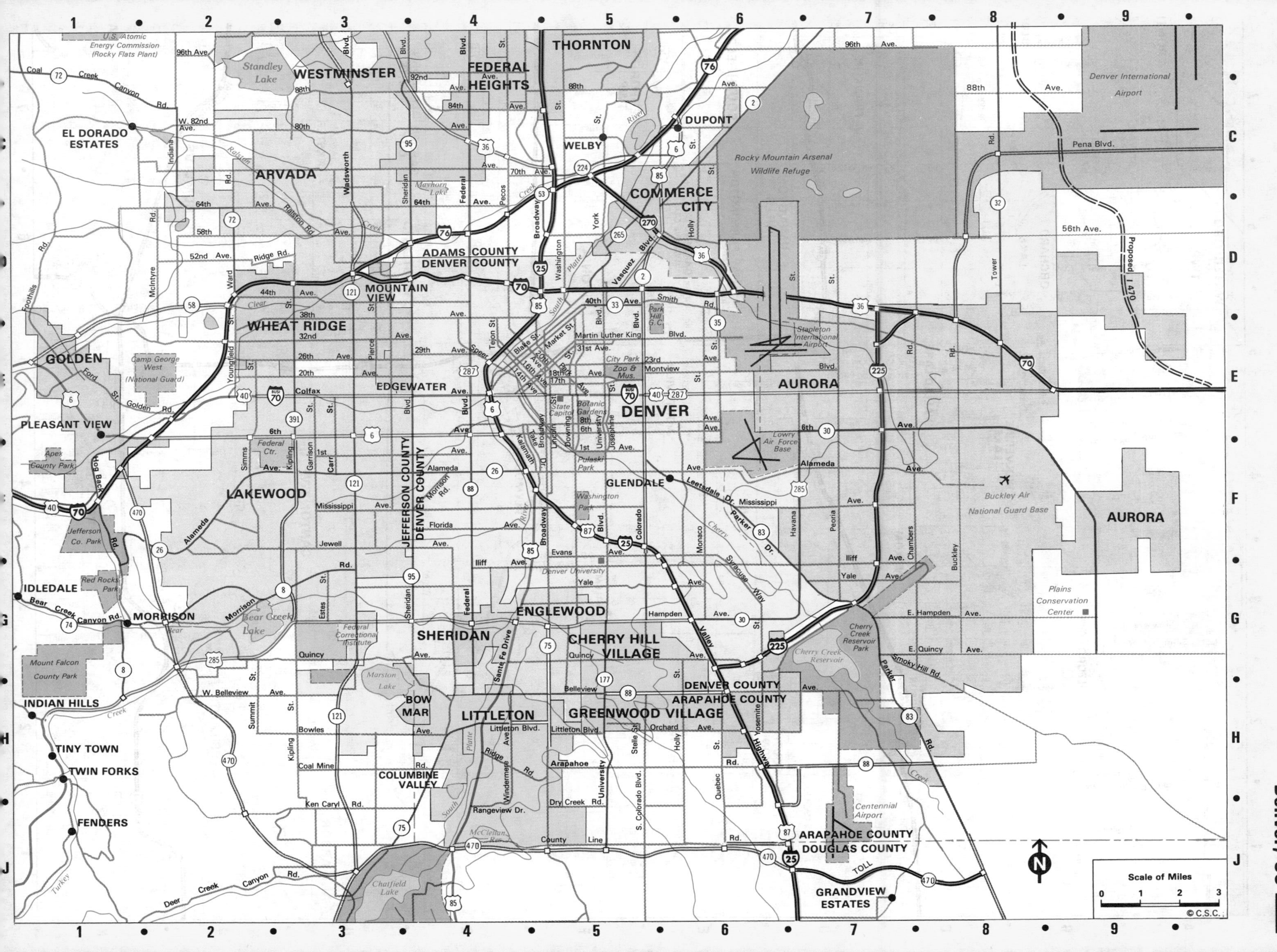
U.S. Atomic Energy Commission (Rocky Flats Plant)
Standley Lake
WESTMINSTER
FEDERAL HEIGHTS
THORNTON
DUPONT
WELBY
EL DORADO ESTATES
ARVADA
COMMERCE CITY
Rocky Mountain Arsenal Wildlife Refuge
Denver International Airport
ADAMS COUNTY
DENVER COUNTY
MOUNTAIN VIEW
WHEAT RIDGE
GOLDEN
Camp George West (National Guard)
EDGEWATER
Stapleton International Airport
AURORA
DENVER
PLEASANT VIEW
Apex County Park
Lowry Air Force Base
LAKEWOOD
JEFFERSON COUNTY
DENVER COUNTY
GLENDALE
Buckley Air National Guard Base
Jefferson Co. Park
Red Rocks Park
IDLEDALE
MORRISON
Bear Creek Lake
Denver University
ENGLEWOOD
Plains Conservation Center
SHERIDAN
CHERRY HILL VILLAGE
Federal Correctional Institute
Cherry Creek Reservoir Park
Cherry Creek Reservoir
Mount Falcon County Park
Marston Lake
BOW MAR
DENVER COUNTY
ARAPAHOE COUNTY
INDIAN HILLS
LITTLETON
GREENWOOD VILLAGE
TINY TOWN
TWIN FORKS
COLUMBINE VALLEY
FENDERS
Centennial Airport
ARAPAHOE COUNTY
DOUGLAS COUNTY
Chatfield Lake
GRANDVIEW ESTATES
Proposed 470
Scale of Miles
0 1 2 3
©C.S.C.

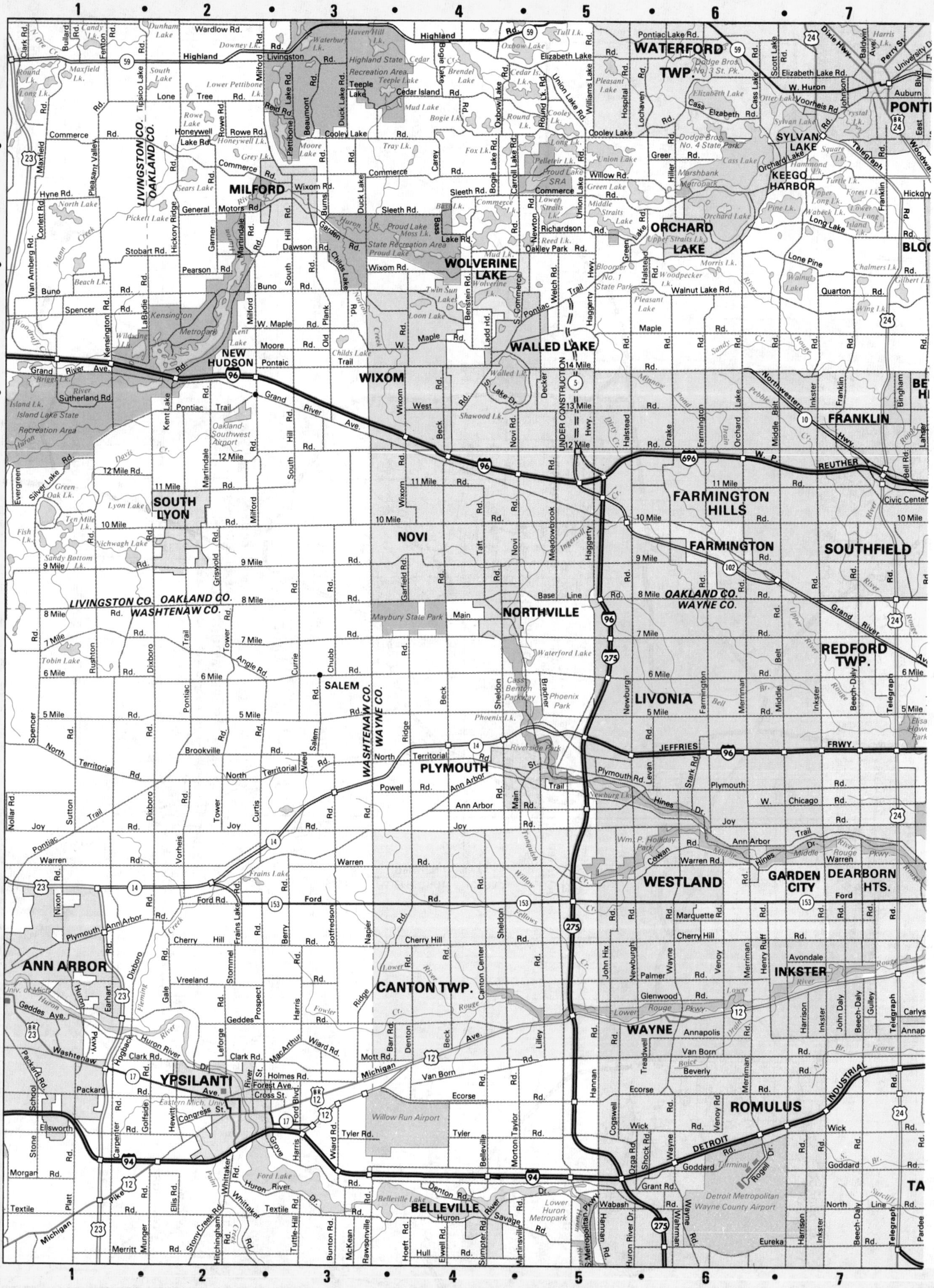
1
2
3
4
5
6
7
A
B
C
D
E
F
G
H
J
K
WATERFORD TWP.
PONTI
SYLVAN LAKE
KEEGO HARBOR
ORCHARD LAKE
MILFORD
WOLVERINE LAKE
WALLED LAKE
NEW HUDSON
WIXOM
FRANKLIN
BLOO
SOUTH LYON
NOVI
FARMINGTON HILLS
FARMINGTON
SOUTHFIELD
NORTHVILLE
REDFORD TWP.
LIVONIA
SALEM
PLYMOUTH
WESTLAND
GARDEN CITY
DEARBORN HTS.
ANN ARBOR
CANTON TWP.
INKSTER
WAYNE
YPSILANTI
ROMULUS
BELLEVILLE
LIVINGSTON CO.
OAKLAND CO.
WASHTENAW CO.
WAYNE CO.
Highland State Recreation Area
Proud Lake State Recreation Area
Kensington Metropark
Island Lake State Recreation Area
Oakland-Southwest Airport
Maybury State Park
Willow Run Airport
Detroit Metropolitan Wayne County Airport
Lower Huron Metropark
Eastern Mich. Univ.
Univ. of Mich.
Grand River Ave.
W. Maple
Pontiac Trail
12 Mile
11 Mile
10 Mile
9 Mile
8 Mile
7 Mile
6 Mile
5 Mile
JEFFRIES FRWY.
REUTHER
North Territorial Rd.
Plymouth Rd.
Ann Arbor Rd.
Joy Rd.
Warren Rd.
Ford Rd.
Cherry Hill
Michigan Ave.
Ecorse
Van Born
DETROIT INDUSTRIAL
Huron River Dr.
Telegraph

AUBURN HILLS
ROCHESTER HILLS
UTICA
WALDENBURG
Berz-Macomb Airport
Rochester-Utica State Rec. Area
Anchor Bay
TROY
STERLING HEIGHTS
CLINTON TWP.
MOUNT CLEMENS
BIRMINGHAM
BLOOMFIELD HILLS
BEVERLY HILLS
CLAWSON
OAKLAND CO.
MACOMB CO.
FRASER
Metro Beach Metropark
Oakland Troy Airport
ROYAL OAK
MADISON HEIGHTS
WARREN
ROSEVILLE
ST. CLAIR SHORES
LATHRUP VILLAGE
BERKLEY
HUNTINGTON WOODS
CENTER LINE
HAZEL PARK
EAST DETROIT
OAK PARK
FERNDALE
MACOMB CO.
WAYNE CO.
HARPER WOODS
GROSSE POINTE WOODS
GROSSE POINTE SHORES
State Fair Grounds
Derby Hill Park
Palmer Park
Univ. of Detroit
Lake St. Claire
HIGHLAND PARK
DETROIT
HAMTRAMCK
Detroit City Airport
Chandler Park
GROSSE POINT FARMS
GROSSE POINTE
GROSSE POINTE PARK
River Rouge Park
Memorial Park
Windmill Point
U.S.A.
CANADA
Wayne St. Univ.
Gabriel Richard Park
Belle Isle Park
Peche Is.
WAYNE CO.
ESSEX CO.
Tiger Stadium
DEARBORN
Ford Field
Patton Park
Greenfield Village
WINDSOR
TECUMSAH
MICH.
ONT.
Detroit River
Zug Is.
MELVINDALE
RIVER ROUGE
Windsor Airport
ALLEN PARK
ECORSE
LA SALLE
LINCOLN PARK
Grassy Is.
TAYLOR
SOUTH GATE
Grosse Is.
Fighting Is.
WYANDOTE

Scale of Miles
0 1 2 3

© C.S.C.

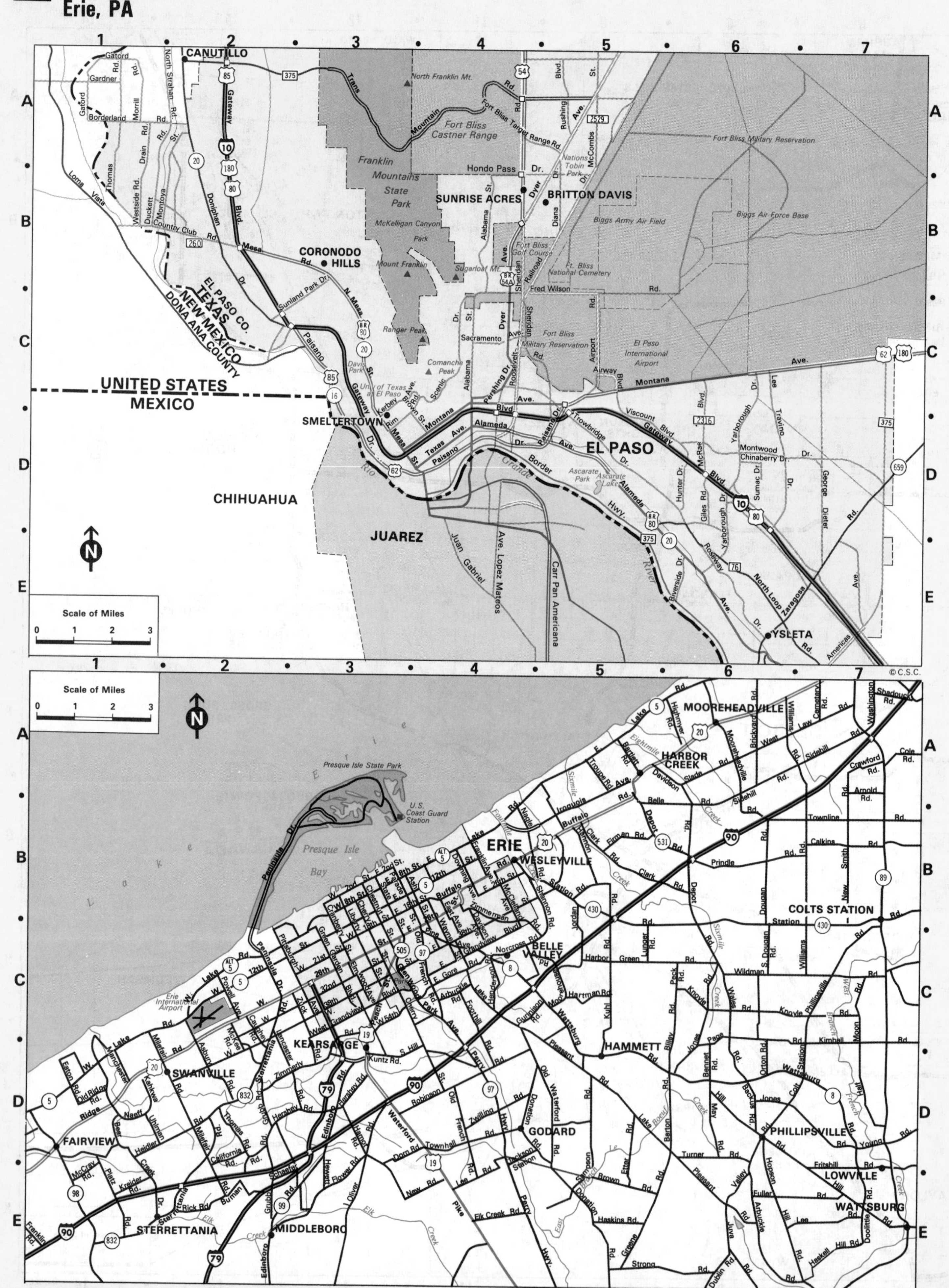
CANUTILLO
Gatord
Gardner
Borderland
Franklin Mountains State Park
North Franklin Mt.
Fort Bliss Castner Range
Fort Bliss Military Reservation
Trans Mountain Rd.
Fort Bliss Target Range Rd.
Hondo Pass Dr.
Nations Tobin Park
SUNRISE ACRES
BRITTON DAVIS
Biggs Army Air Field
Biggs Air Force Base
McKelligan Canyon Park
Mount Franklin
Sugarloaf Mt.
CORONODO HILLS
Fort Bliss Golf Course
Ft. Bliss National Cemetery
Fred Wilson Rd.
Ranger Peak
Comanche Peak
Sacramento
El Paso International Airport
Montana Ave.
EL PASO CO. TEXAS
NEW MEXICO DONA ANA COUNTY
UNITED STATES
MEXICO
SMELTERTOWN
Univ. of Texas at El Paso
Rio Grande
EL PASO
Ascarate Park
Ascarate Lake
CHIHUAHUA
JUAREZ
Juan Gabriel
Ave. Lopez Mateos
Carr Pan Americana
Border Hwy.
North Loop Rd.
Zaragosa Rd.
YSLETA
Americas Ave.
Scale of Miles
0 1 2 3
© C.S.C.
Scale of Miles
0 1 2 3
Lake Erie
Presque Isle State Park
U.S. Coast Guard Station
Presque Isle Bay
Peninsula Dr.
MOOREHEADVILLE
HARBOR CREEK
Eightmile Creek
Sixmile Creek
Fourmile Creek
ERIE
WESLEYVILLE
COLTS STATION
BELLE VALLEY
Erie International Airport
KEARSARGE
SWANVILLE
HAMMETT
FAIRVIEW
GODARD
PHILLIPSVILLE
LOWVILLE
WATTSBURG
STERRETTANIA
MIDDLEBORO
Elk Creek
Lake Pleasant
West Branch French Creek
© C.S.C.

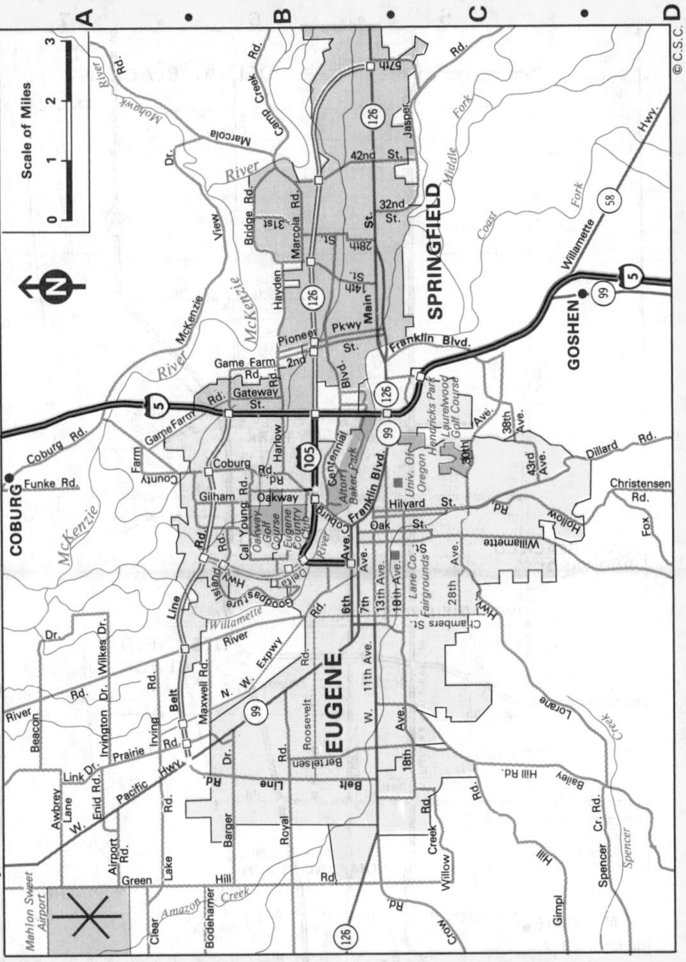
COBURG
EUGENE
SPRINGFIELD
GOSHEN
Scale of Miles
©C.S.C.

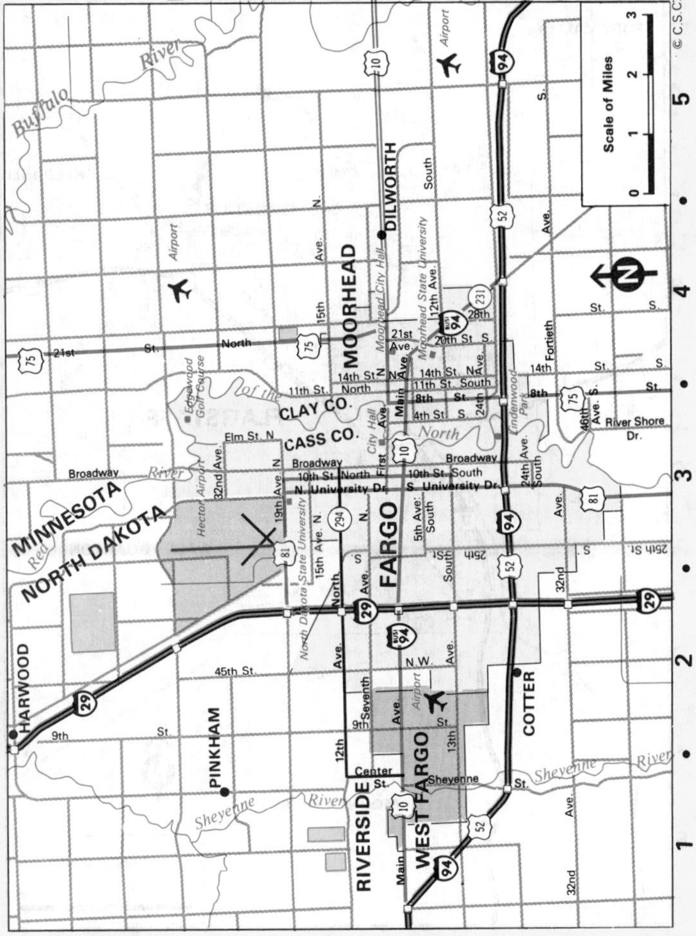
MOORHEAD
DILWORTH
FARGO
WEST FARGO
RIVERSIDE
HARWOOD
PINKHAM
COTTER
MINNESOTA
NORTH DAKOTA
CLAY CO.
CASS CO.
Scale of Miles
©C.S.C.

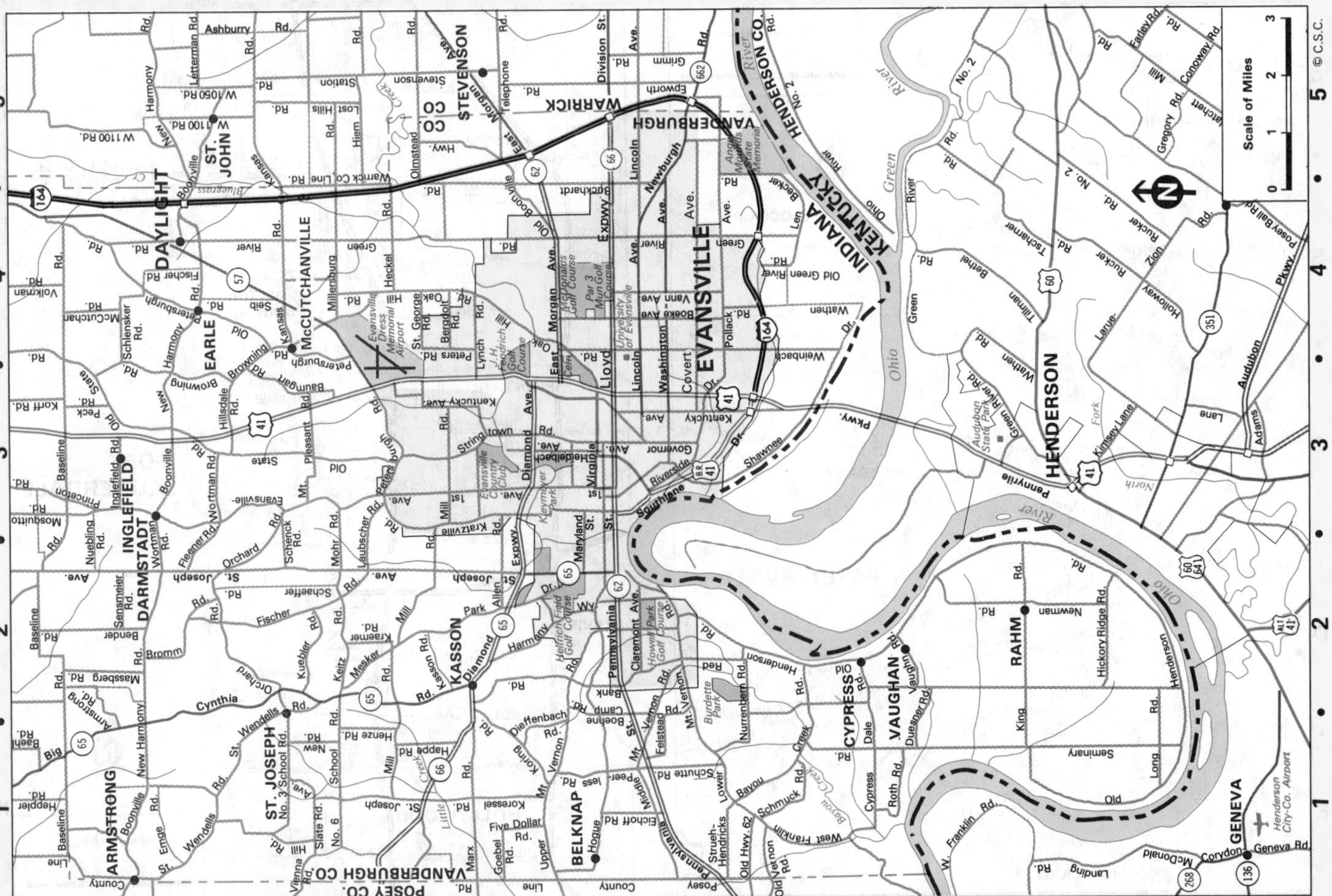
EVANSVILLE
HENDERSON
DAYLIGHT
ST. JOHN
McCUTCHANVILLE
EARLE
INGLEFIELD
DARMSTADT
ARMSTRONG
ST. JOSEPH
KASSON
STEVENSON
BELKNAP
CYPRESS
VAUGHAN
RAHM
GENEVA
VANDERBURGH CO.
WARRICK CO.
POSEY CO.
HENDERSON CO.
INDIANA
KENTUCKY
Scale of Miles
©C.S.C.

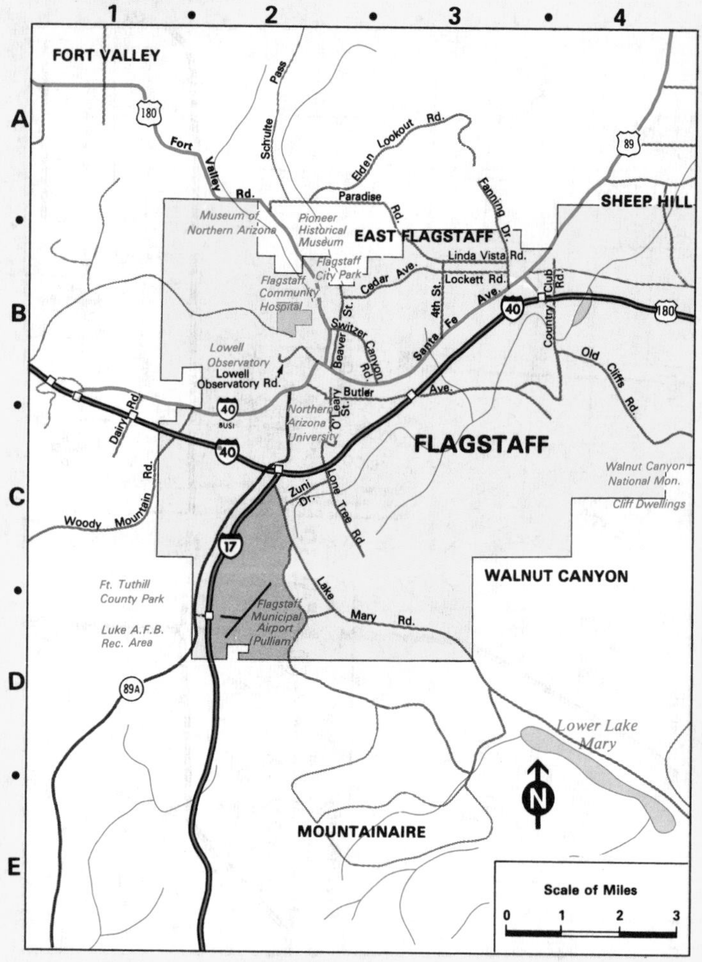

FORT VALLEY
Fort Valley Rd.
Schultz Pass
Elden Lookout Rd.
Paradise Rd.
Fanning Dr.
SHEEP HILL
Museum of Northern Arizona
Pioneer Historical Museum
EAST FLAGSTAFF
Linda Vista Rd.
Flagstaff City Park
Cedar Ave.
Lockett Rd.
Flagstaff Community Hospital
Beaver St.
4th St.
Country Club
Santa Fe Ave.
Switzer Canyon
Lowell Observatory
Lowell Observatory Rd.
Butler Ave.
O'Leary St.
Old Cliffs Rd.
Dairy Rd.
Northern Arizona University
FLAGSTAFF
Woody Mountain Rd.
Zuni Dr.
Lone Tree Rd.
Walnut Canyon National Mon.
Cliff Dwellings
WALNUT CANYON
Ft. Tuthill County Park
Luke A.F.B. Rec. Area
Lake Mary Rd.
Flagstaff Municipal Airport Pulliam
Lower Lake Mary
MOUNTAINAIRE
Scale of Miles
0 1 2 3

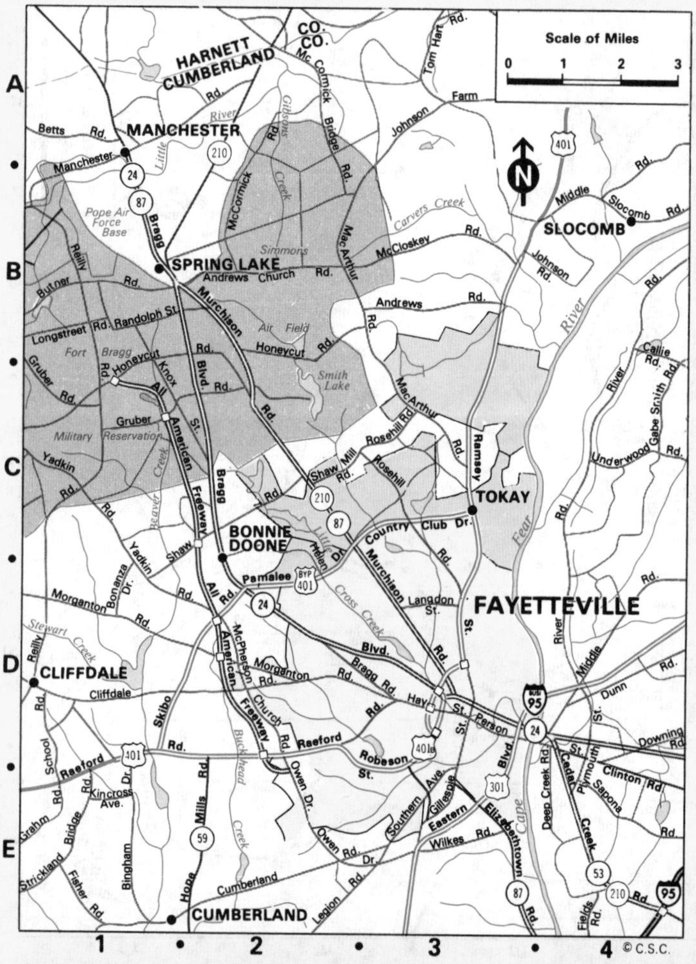

HARNETT CO.
CUMBERLAND CO.
Mc Cormick Bridge Rd.
Tom Hart Rd.
Farm
Scale of Miles
0 1 2 3
Betts Rd.
MANCHESTER
Little River
Gibsons Creek
Johnson
Manchester
McCormick
Pope Air Force Base
Bragg
Simmons
Mac Arthur
McCloskey Rd.
Middle Rd.
Slocomb Rd.
SLOCOMB
Carvers Creek
Reilly
SPRING LAKE
Andrews Church Rd.
Johnson Rd.
Butner Rd.
Murchison
Andrews Rd.
Longstreet Rd.
Randolph St.
Air Field
Fort Bragg
Honeycut Rd.
Honeycut
Knox
Smith Lake
Gruber Rd.
All American Fwy.
Blvd.
Callie
Gabe Smith Rd.
Gruber
Military Reservation
Yadkin Rd.
Beaver Creek
Rosehill Rd.
Shaw Mill
Ramsey St.
Underwood Rd.
Cape Fear River
TOKAY
Country Club Dr.
BONNIE DOONE
Shaw
Helen Dr.
Pamalee
Murchison
Bonanza Dr.
Morganton Rd.
Cross Creek
Langdon St.
FAYETTEVILLE
Stewart Creek
McPherson Church Rd.
Morganton
Bragg Rd.
Hay St.
Person
Dunn
CLIFFDALE
Cliffdale
Skibo Rd.
Raeford Rd.
Robeson St.
Downing Rd.
School Rd.
Owen Dr.
Southern Ave.
Gillespie
Clinton Rd.
Kincross Ave.
Mills Rd.
Buckhead Creek
Eastern Blvd.
Deep Creek Rd.
Plymouth
Sapona
Graham Rd.
Bridge
Bingham
Hope Mills Rd.
Wilkes Rd.
Elizabethtown Rd.
Strickland
Fisher Rd.
Cumberland
CUMBERLAND
Legion Rd.
Fields Rd.
© C.S.C.

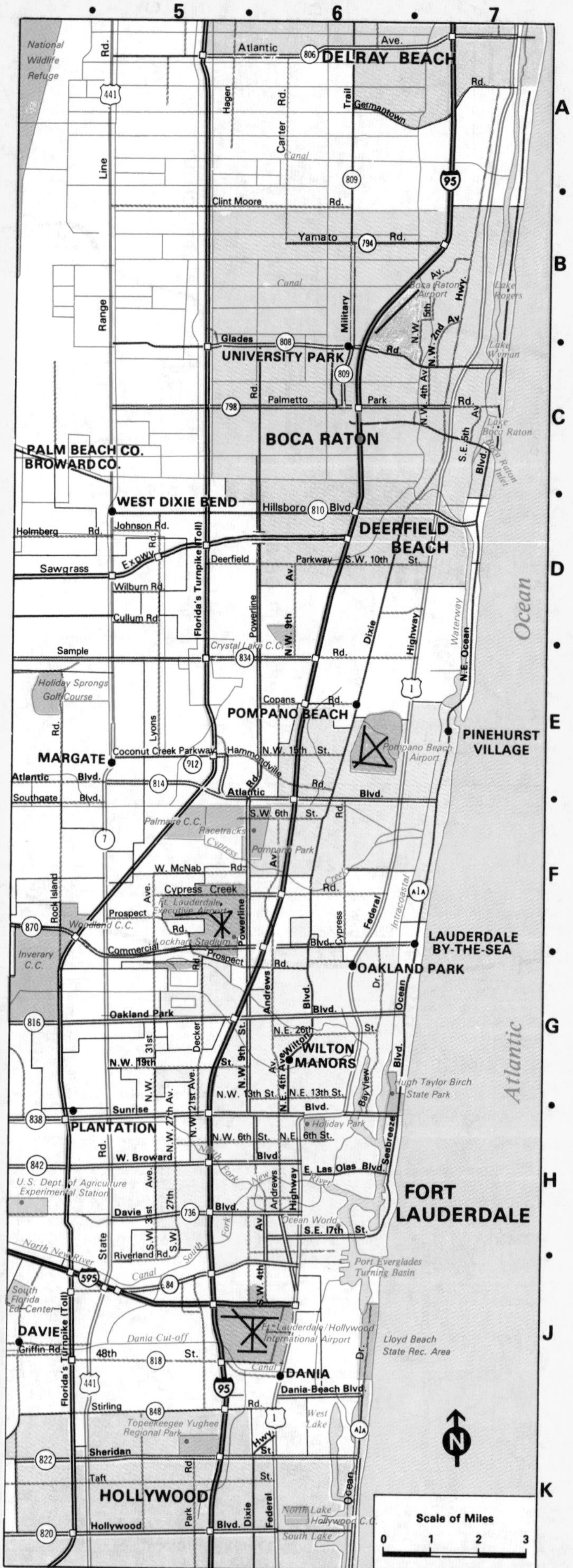

National Wildlife Refuge
Atlantic Ave.
DELRAY BEACH
Germantown Rd.
Hagen Rd.
Carter Rd.
Trail
Line Rd.
Canal
Clint Moore Rd.
Yamato Rd.
Range Rd.
Boca Raton Airport
Lake Rogers
Glades Rd.
Military Trail
UNIVERSITY PARK
N.W. 5th Av.
N.W. 2nd Av.
Hwy.
Lake Wyman
Palmetto Park Rd.
N.W. 4th Av.
BOCA RATON
Lake Boca Raton
S.E. 5th Av.
Boca Raton Inlet
PALM BEACH CO.
BROWARD CO.
WEST DIXIE BEND
Hillsboro Blvd.
DEERFIELD BEACH
Holmberg Rd.
Johnson Rd.
Sawgrass Expwy.
Deerfield Parkway
S.W. 10th St.
Wilburn Rd.
Florida's Turnpike (Toll)
Powerline
N.W. 9th Av.
Dixie Highway
Waterway
Ocean
Cullum Rd.
Crystal Lake C.C.
Sample Rd.
N.E. Ocean
Holiday Springs Golf Course
Copans Rd.
POMPANO BEACH
PINEHURST VILLAGE
Lyons
MARGATE
Coconut Creek Parkway
Hammondville
N.W. 15th St.
Pompano Beach Airport
Atlantic Blvd.
Southgate Blvd.
Palmaire C.C.
Racetracks
Cypress
S.W. 6th St.
Pompano Park
W. McNab Rd.
Rock Island
Cypress Creek
Ft. Lauderdale Executive Airport
Federal
Intracoastal
Prospect Rd.
Woodland C.C.
Commercial Blvd.
Lockhart Stadium
LAUDERDALE BY-THE-SEA
Inverary C.C.
OAKLAND PARK
Oakland Park Blvd.
N.E. 26th St.
N.W. 31st Av.
Decker
WILTON MANORS
N.W. 19th St.
N.W. 13th St.
N.E. 13th St.
Hugh Taylor Birch State Park
Atlantic Ocean
Sunrise Blvd.
PLANTATION
N.W. 6th St.
N.E. 6th St.
Holiday Park
W. Broward Blvd.
North Fork New River
E. Las Olas Blvd.
Seabreeze
U.S. Dept. of Agriculture Experimental Station
FORT LAUDERDALE
Davie Blvd.
Ocean World
S.E. 17th St.
North New River
Riverland Rd.
Port Everglades Turning Basin
South Florida Ed. Center
South Fork
DAVIE
Griffin Rd.
Dania Cut-off
48th St.
Ft. Lauderdale/Hollywood International Airport
Lloyd Beach State Rec. Area
DANIA
Dania-Beach Blvd.
Stirling Rd.
West Lake
Topeekeegee Yughee Regional Park
Sheridan St.
Taft St.
HOLLYWOOD
North Lake
Hollywood C.C.
South Lake
Hollywood Blvd.
Park Rd.
Scale of Miles
0 1 2 3
© C.S.C.

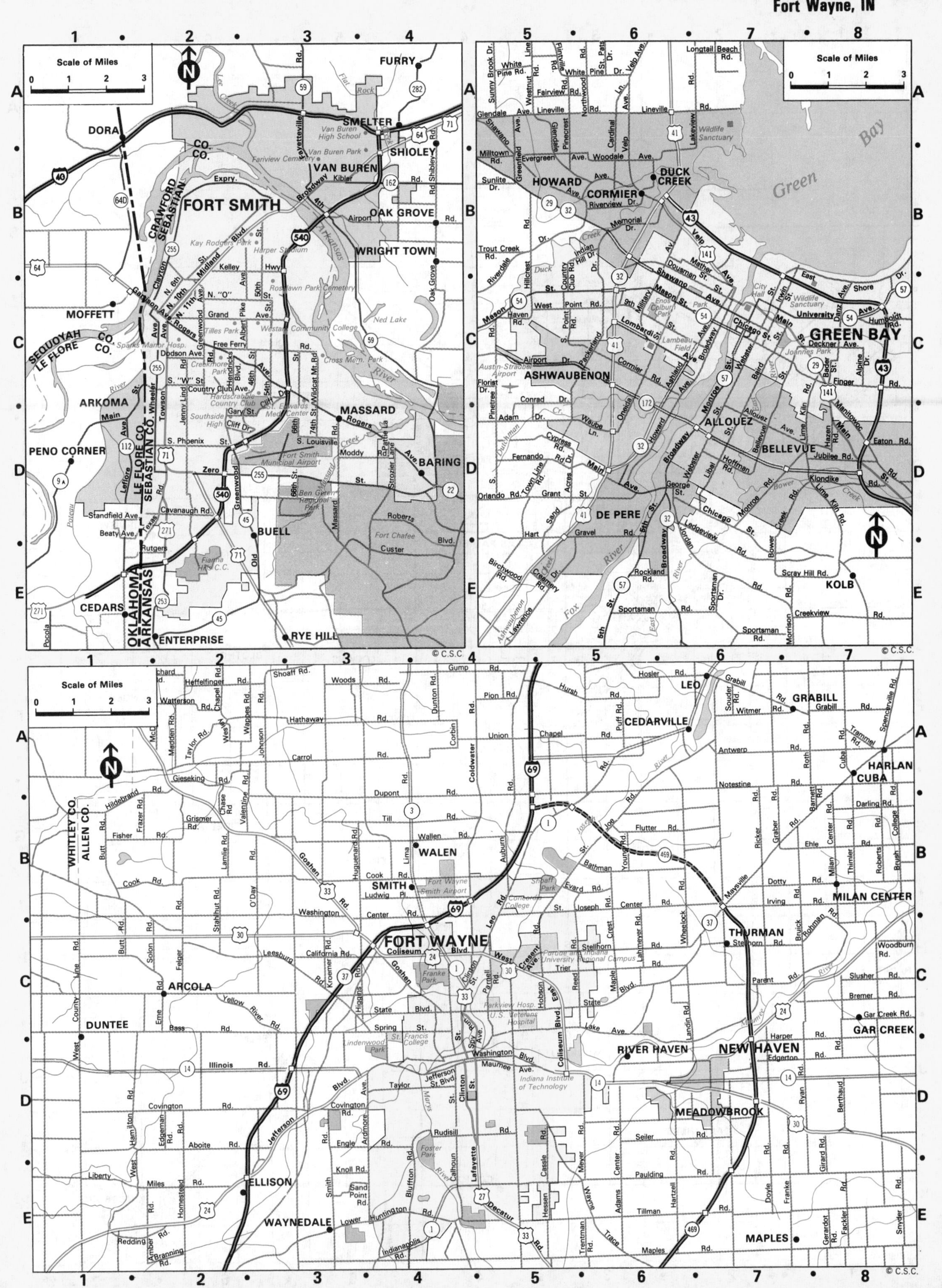
Scale of Miles
FURRY
DORA
SMELTER
SHIOLEY
VAN BUREN
CRAWFORD CO.
SEBASTIAN CO.
FORT SMITH
OAK GROVE
WRIGHT TOWN
MOFFETT
SEQUOYAH CO.
LE FLORE CO.
ARKOMA
PENO CORNER
MASSARD
BARING
BUELL
CEDARS
OKLAHOMA
ARKANSAS
ENTERPRISE
RYE HILL
Van Buren High School
Van Buren Park
Fariview Cemetery
Kay Rodgers Park
Harper Stadium
Rose Lawn Park Cemetery
Westark Community College
Ned Lake
Cross Mem. Park
Sparks Manor Hosp.
Creekmore Park
Hardscrabble Country Club
St. Edwards Med. Center
Southside High
Fort Smith Municipal Airport
Ben Geren Regional Park
Fort Chafee
Fiama Hills C.C.
© C.S.C.
HOWARD
DUCK CREEK
CORMIER
Green Bay
Wildlife Sanctuary
GREEN BAY
ASHWAUBENON
Austin-Straubel Airport
ALLOUEZ
BELLEVUE
DE PERE
KOLB
Fox River
City Hall
Lambeau Field
Enos Colburn Park
Joannes Park
© C.S.C.
LEO
GRABILL
CEDARVILLE
HARLAN
CUBA
WHITLEY CO.
ALLEN CO.
WALEN
SMITH
Fort Wayne Smith Airport
Shoaff Park
Concordia College
MILAN CENTER
THURMAN
FORT WAYNE
Franke Park
Purdue and Indiana University Regional Campus
ARCOLA
DUNTEE
Parkview Hosp.
U.S. Veterans Hospital
Lindenwood Park
St. Francis College
RIVER HAVEN
NEW HAVEN
GAR CREEK
Indiana Institute of Technology
MEADOWBROOK
Foster Park
ELLISON
WAYNEDALE
MAPLES
© C.S.C.

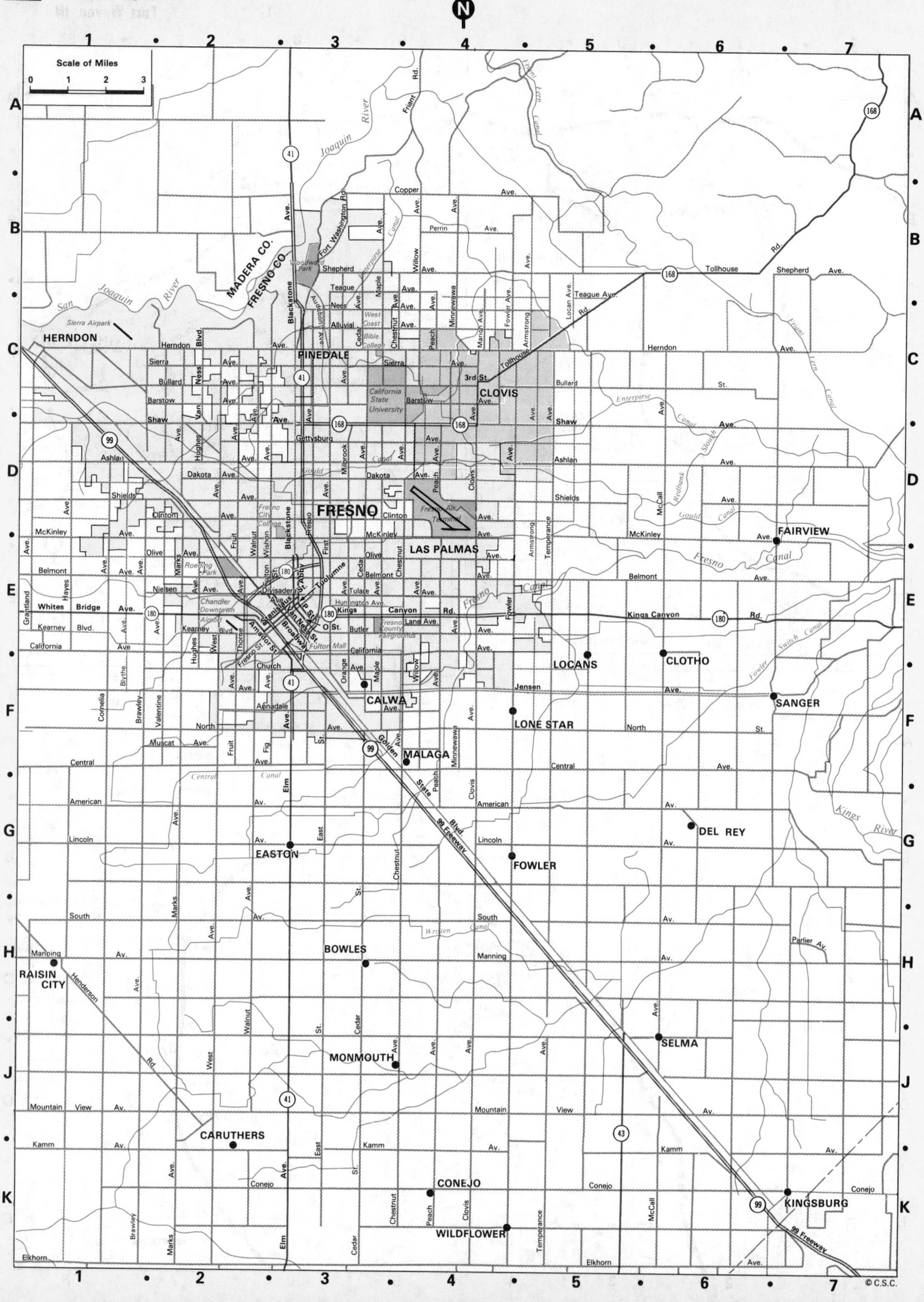

Fresno, CA

Scale of Miles
0 1 2 3

N

HERNDON
PINEDALE
CLOVIS
FRESNO
LAS PALMAS
FAIRVIEW
LOCANS
CLOTHO
CALWA
SANGER
LONE STAR
MALAGA
DEL REY
EASTON
FOWLER
BOWLES
RAISIN CITY
SELMA
MONMOUTH
CARUTHERS
CONEJO
KINGSBURG
WILDFLOWER

MADERA CO.
FRESNO CO.
Sierra Airpark
Woodward Park
West Coast Bible College
California State University
Fresno Air Terminal
Fresno City College
Roeding Park
Chandler Downtown Airport
Fresno County Fairgrounds
Fulton Mall

San Joaquin River
Kings River
Friant Kern Canal
Enterprise Canal
Gould Canal
Fresno Canal
Central Canal
Fowler Switch Canal
Wristen Canal
Redbank Slough

Golden State Blvd.
99 Freeway
Whites Bridge Ave.
Kings Canyon Rd.
Tollhouse Rd.
Fort Washington Rd.

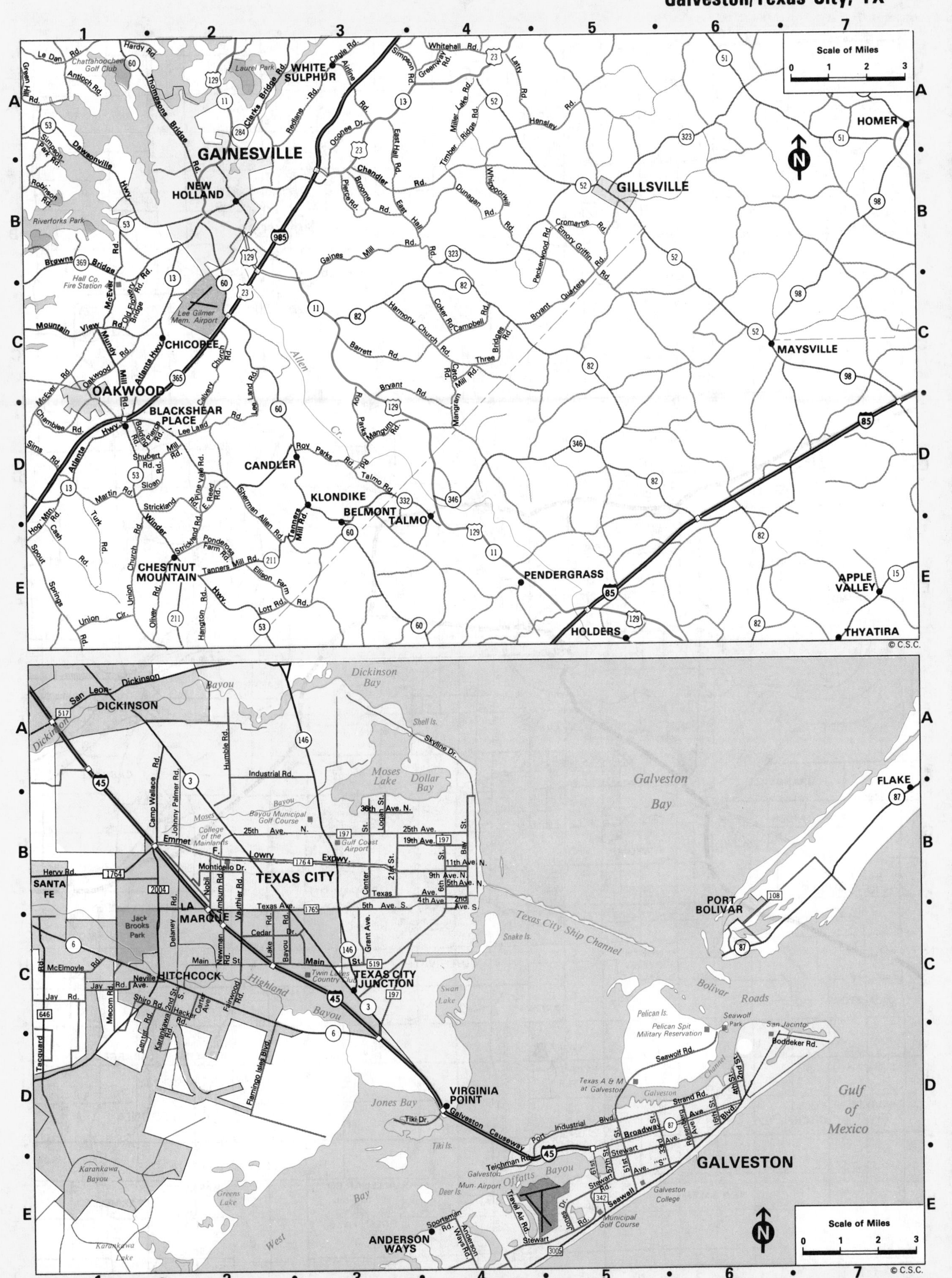
GAINESVILLE
NEW HOLLAND
WHITE SULPHUR
GILLSVILLE
HOMER
MAYSVILLE
CHICOPEE
OAKWOOD
BLACKSHEAR PLACE
CANDLER
KLONDIKE
BELMONT
TALMO
CHESTNUT MOUNTAIN
PENDERGRASS
HOLDERS
APPLE VALLEY
THYATIRA
Lee Gilmer Mem. Airport
Riverforks Park
Laurel Park
Chattahoochee Golf Club
Hall Co. Fire Station
Scale of Miles
© C.S.C.
DICKINSON
TEXAS CITY
SANTA FE
LA MARQUE
HITCHCOCK
TEXAS CITY JUNCTION
VIRGINIA POINT
GALVESTON
PORT BOLIVAR
FLAKE
ANDERSON WAYS
Galveston Bay
Dickinson Bay
Moses Lake
Dollar Bay
Jones Bay
West Bay
Gulf of Mexico
Texas City Ship Channel
Bolivar Roads
Pelican Is.
Pelican Spit Military Reservation
Texas A & M at Galveston
Galveston Mun. Airport
Galveston College
Jack Brooks Park
Twin Lakes Country Club
Bayou Municipal Golf Course
College of the Mainland
Gulf Coast Airport
Galveston Causeway
Scale of Miles
© C.S.C.

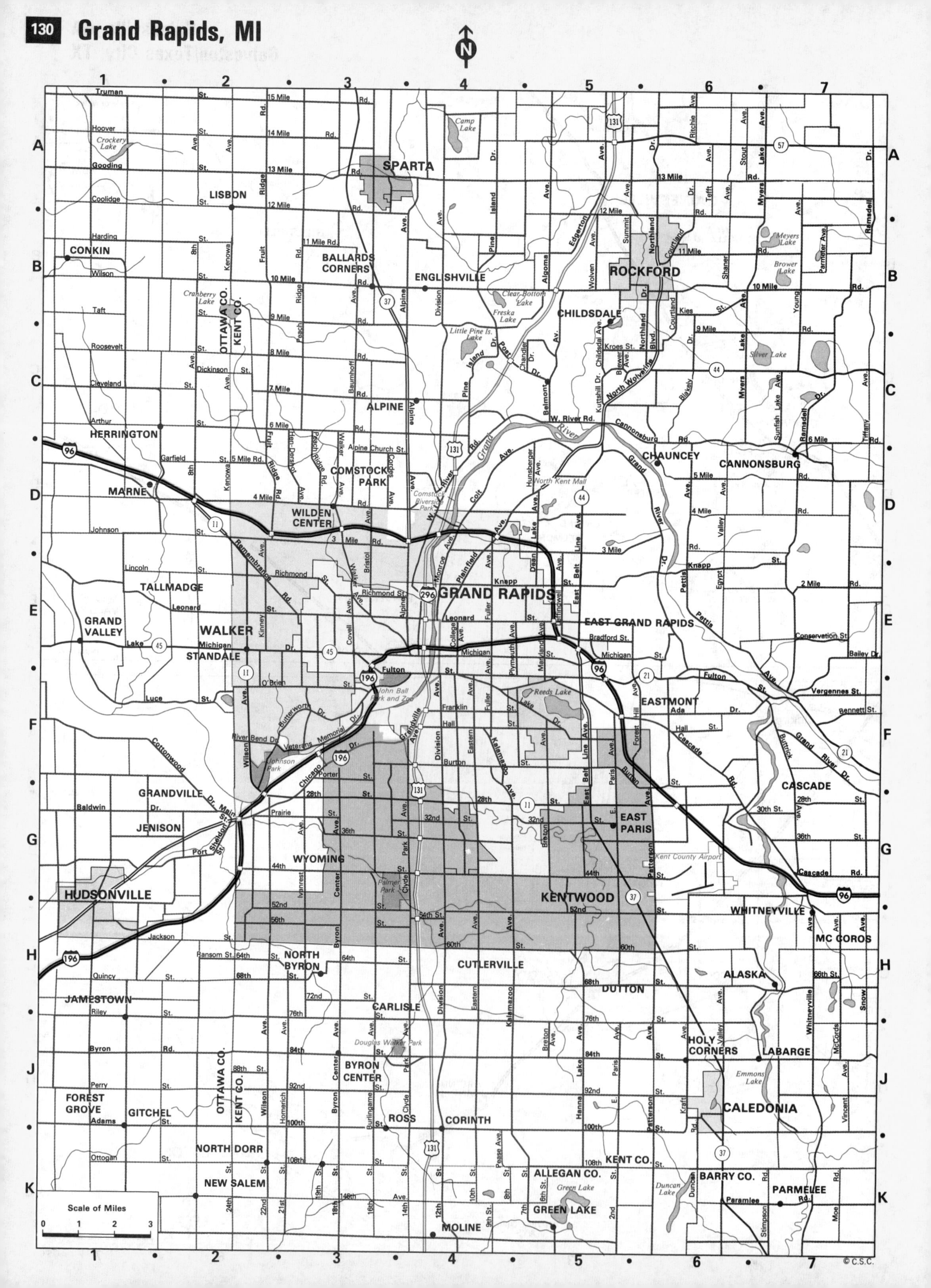
N
SPARTA
LISBON
CONKIN
BALLARDS CORNERS
ENGLISHVILLE
ROCKFORD
CHILDSDALE
ALPINE
HERRINGTON
COMSTOCK PARK
CHAUNCEY
CANNONSBURG
MARNE
WILDEN CENTER
TALLMADGE
GRAND RAPIDS
GRAND VALLEY
WALKER
EAST GRAND RAPIDS
STANDALE
EASTMONT
CASCADE
GRANDVILLE
JENISON
EAST PARIS
WYOMING
KENTWOOD
HUDSONVILLE
WHITNEYVILLE
MC COROS
NORTH BYRON
CUTLERVILLE
ALASKA
JAMESTOWN
DUTTON
CARLISLE
HOLY CORNERS
LABARGE
BYRON CENTER
FOREST GROVE
GITCHEL
ROSS
CORINTH
CALEDONIA
NORTH DORR
NEW SALEM
KENT CO.
ALLEGAN CO.
BARRY CO.
OTTAWA CO.
PARMELEE
GREEN LAKE
MOLINE
Crockery Lake
Camp Lake
Cranberry Lake
Clear Bottom Lake
Freska Lake
Little Pine Is. Lake
Meyers Lake
Brower Lake
Silver Lake
Reeds Lake
Johnson Park
John Ball Park and Zoo
Palmer Park
Douglas Walker Park
Comstock Riverside Park
North Kent Mall
Kent County Airport
Emmons Lake
Duncan Lake
Green Lake
Grand River
Scale of Miles
0 1 2 3
© C.S.C.

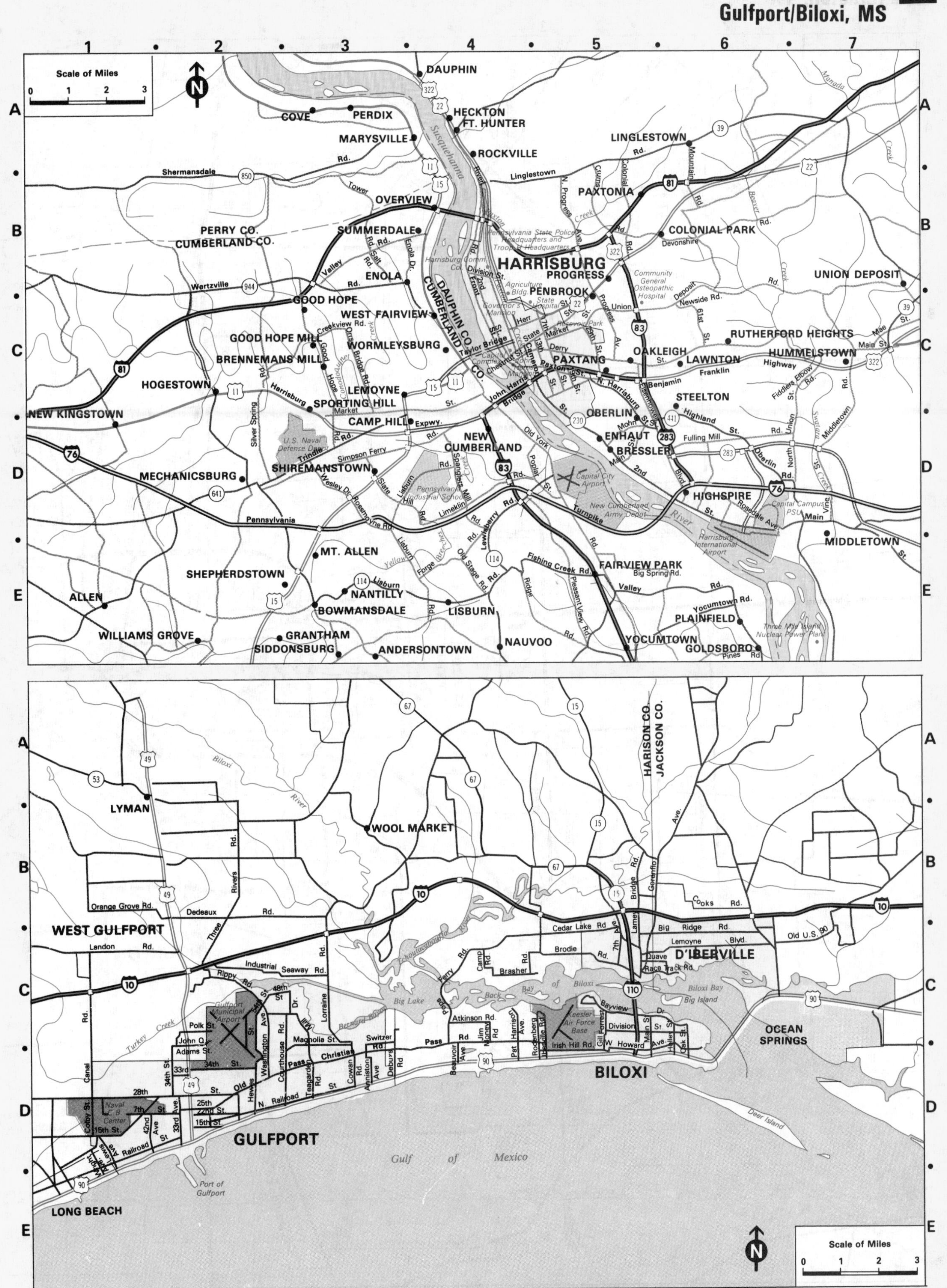
Scale of Miles
0 1 2 3
DAUPHIN
HECKTON
FT. HUNTER
COVE
PERDIX
MARYSVILLE
ROCKVILLE
Susquehanna
River
Shermansdale
Linglestown
LINGLESTOWN
PAXTONIA
Tower
OVERVIEW
PERRY CO.
CUMBERLAND CO.
SUMMERDALE
COLONIAL PARK
Devonshire
Pennsylvania State Police Headquarters and Troop H Headquarters
HARRISBURG
PROGRESS
Community General Osteopathic Hospital
UNION DEPOSIT
Wertzville
Valley
ENOLA
Enola Dr.
Harrisburg Comm. Col.
Division St.
Agriculture Bldg.
PENBROOK
State Hospital
Governor's Mansion
Newside Rd.
GOOD HOPE
WEST FAIRVIEW
DAUPHIN CO.
CUMBERLAND CO.
Reservoir Park
GOOD HOPE MILL
Creekview Rd.
Taylor Bridge
Capitol Complex
RUTHERFORD HEIGHTS
BRENNEMANS MILL
WORMLEYSBURG
OAKLEIGH
LAWNTON
HUMMELSTOWN
Main St.
PAXTANG
Franklin
Highway
HOGESTOWN
Harrisburg
LEMOYNE
John Harris Bridge
N. Harrisburg St.
Benjamin
STEELTON
NEW KINGSTOWN
SPORTING HILL
Market
CAMP HILL
Expwy.
OBERLIN
Highland
Silver Spring
U.S. Naval Defense Depot
Trindle
NEW CUMBERLAND
ENHAUT
BRESSLER
Fulling Mill
Simpson Ferry
SHIREMANSTOWN
Capital City Airport
MECHANICSBURG
Pennsylvania Industrial School
HIGHSPIRE
Capital Campus PSU
Wesley Dr.
Rossmoyne Rd.
Turnpike
New Cumberland Army Depot
Pennsylvania
Harrisburg International Airport
MIDDLETOWN
MT. ALLEN
Fishing Creek Rd.
FAIRVIEW PARK
Big Spring Rd.
SHEPHERDSTOWN
Yellow Breeches
Lewisberry
Old Stage Rd.
NANTILLY
Lisburn
ALLEN
BOWMANSDALE
LISBURN
Yocumtown Rd.
PLAINFIELD
Three Mile Island Nuclear Power Plant
WILLIAMS GROVE
GRANTHAM
SIDDONSBURG
ANDERSONTOWN
NAUVOO
YOCUMTOWN
GOLDSBORO
Pines
Biloxi
River
LYMAN
WOOL MARKET
HARISON CO.
JACKSON CO.
Orange Grove Rd.
Dedeaux
Three
Rivers
WEST GULFPORT
Cedar Lake Rd.
Big Ridge Rd.
Cooks Rd.
Old U.S. 90
Landon Rd.
Lemoyne Blvd.
Brodie
Quave
D'IBERVILLE
Industrial Seaway Rd.
Race Track Rd.
Rippy Rd.
Tchoutacabouffa River
Brasher
Back Bay of Biloxi
Biloxi Bay
Big Island
Gulfport Municipal Airport
Big Lake
Bayview
Turkey Creek
Bernard Bayou
Atkinson Rd.
Keesler Air Force Base
Division St.
OCEAN SPRINGS
Polk St.
John Q. Adams St.
Magnolia St.
Switzer Rd.
Pass Rd.
Irish Hill Rd.
W. Howard Ave.
34th St.
Washington Ave.
Courthouse Rd.
Christian
Canal Rd.
28th
Old Pass
BILOXI
25th St.
Naval C.B. Center
7th St.
22nd St.
15th St.
Railroad
GULFPORT
Deer Island
Gulf of Mexico
Port of Gulfport
LONG BEACH
Scale of Miles
0 1 2 3
© C.S.C.

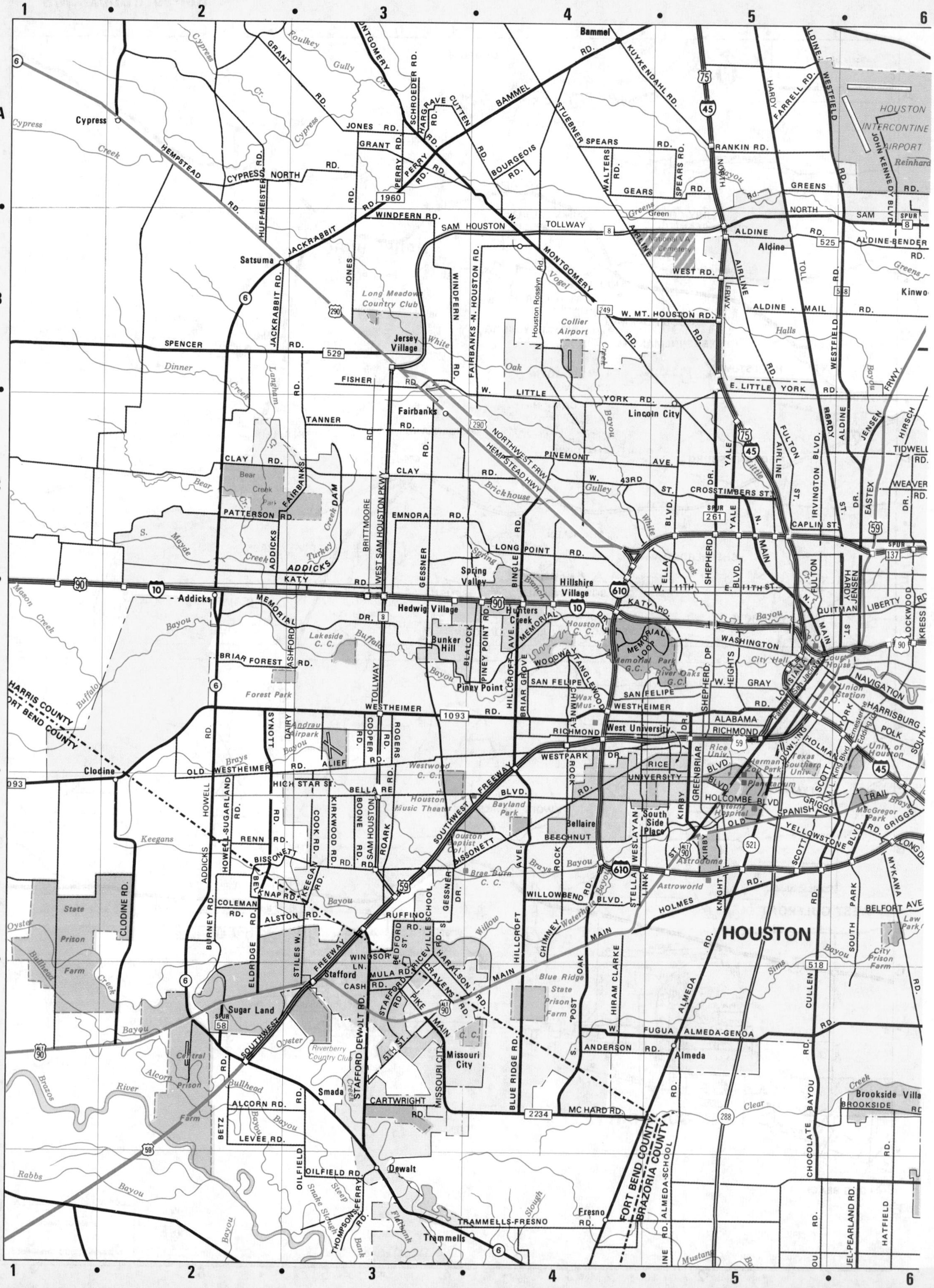

1 2 3 4 5 6

A B C D E F G

Bammel
Cypress
Satsuma
Jersey Village
Fairbanks
Lincoln City
Aldine
Addicks
Spring Valley
Hillshire Village
Hedwig Village
Hunters Creek
Bunker Hill
Piney Point
West University
Bellaire
South Side Place
Clodine
Stafford
Sugar Land
Missouri City
Smada
Dewalt
Trammells
Fresno
Almeda
HOUSTON
Kinwood
Brookside Village

HOUSTON INTERCONTINENTAL AIRPORT
Collier Airport
Andrau Airpark
Long Meadow Country Club
Lakeside C. C.
Houston C. C.
Memorial Park G.C.
River Oaks G.C.
Westwood C. C.
Bayland Park
Brae Burn C. C.
Riverberry Country Club
Houston Music Theater
Houston Baptist Col.
Rice Univ.
Texas Southern Univ.
Univ. of Houston
Hermann Park
Astrodome
Astroworld
City Hall
Court House
Union Station
MacGregor Park
Central Prison Farm
State Prison Farm
Blue Ridge State Prison Farm
City Prison Farm
Law Park
National VA Cemetery
Addicks Dam
Bear Creek Park
Forest Park
Wax Mus.

HARRIS COUNTY
FORT BEND COUNTY
BRAZORIA COUNTY

Cypress Creek
Brazos River
Buffalo Bayou
Brays Bayou
Sims Bayou
Greens Bayou
Halls Bayou
White Oak Bayou
Keegans Bayou
Oyster Creek
Bullhead Bayou
Alcorn Bayou
Clear Creek
Mustang Bayou
Rabbs Bayou
Mason Creek
Dinner Creek
Langham Cr.
Turkey Creek
Brickhouse Gulley
Vogel Creek
Spring Branch
Willow Waterhole Bayou
Snake Slough
Steep Bank
Flat Bank
Foulkey Gully
Mayde Creek

HEMPSTEAD RD.
JACKRABBIT RD.
SPENCER RD.
KATY RD.
MEMORIAL DR.
BRIAR FOREST DR.
WESTHEIMER RD.
OLD WESTHEIMER RD.
BELLAIRE
RICHMOND
SOUTHWEST FREEWAY
NORTHWEST FRWY.
HEMPSTEAD HWY.
SAM HOUSTON TOLLWAY
WEST SAM HOUSTON PKWY
NORTH FRWY.
EASTEX FRWY.
GULF FRWY.
KATY FRWY.
MAIN
ALMEDA
ALMEDA-GENOA
MC HARD RD.
TRAMMELLS-FRESNO
CARTWRIGHT RD.
OILFIELD RD.
ALCORN RD.
LEVEE RD.
BISSONETT
BEECHNUT
HOLMES
BELFORT AVE.
YELLOWSTONE BLVD.
HOLCOMBE BLVD.
WASHINGTON
SAN FELIPE
WOODWAY
LONG POINT RD.
CLAY RD.
PATTERSON RD.
EMNORA RD.
TANNER RD.
FISHER RD.
W. LITTLE YORK RD.
E. LITTLE YORK RD.
PINEMONT AVE.
W. 43RD ST.
CROSSTIMBERS ST.
TIDWELL RD.
W. MT. HOUSTON RD.
ALDINE MAIL RD.
ALDINE BENDER RD.
GREENS RD.
RANKIN RD.
SPEARS RD.
GEARS RD.
WEST RD.
WINDFERN RD.
BAMMEL RD.
CYPRESS NORTH HOUSTON RD.
JONES RD.
GRANT RD.
KUYKENDAHL RD.
STUEBNER AIRLINE
MONTGOMERY RD.
GESSNER
BLALOCK
BINGLE RD.
HILLCROFT AVE.
BLUE RIDGE RD.
KIRBY
SHEPHERD DR.
YALE
AIRLINE
IRVINGTON BLVD.
HARDY
JENSEN DR.
LOCKWOOD
SCOTT
CULLEN
MYKAWA
CHOCOLATE BAYOU RD.
HATFIELD
BURNEY RD.
HOWELL-SUGARLAND RD.
ELDRIDGE
DAIRY ASHFORD
KIRKWOOD RD.
BOONE RD.
COOK RD.
WILCREST
HIRAM CLARKE
POST OAK
FONDREN
STAFFORD DEWALT RD.
THOMPSONS FERRY RD.
ALMEDA-SCHOOL RD.
FUQUA
ANDERSON RD.

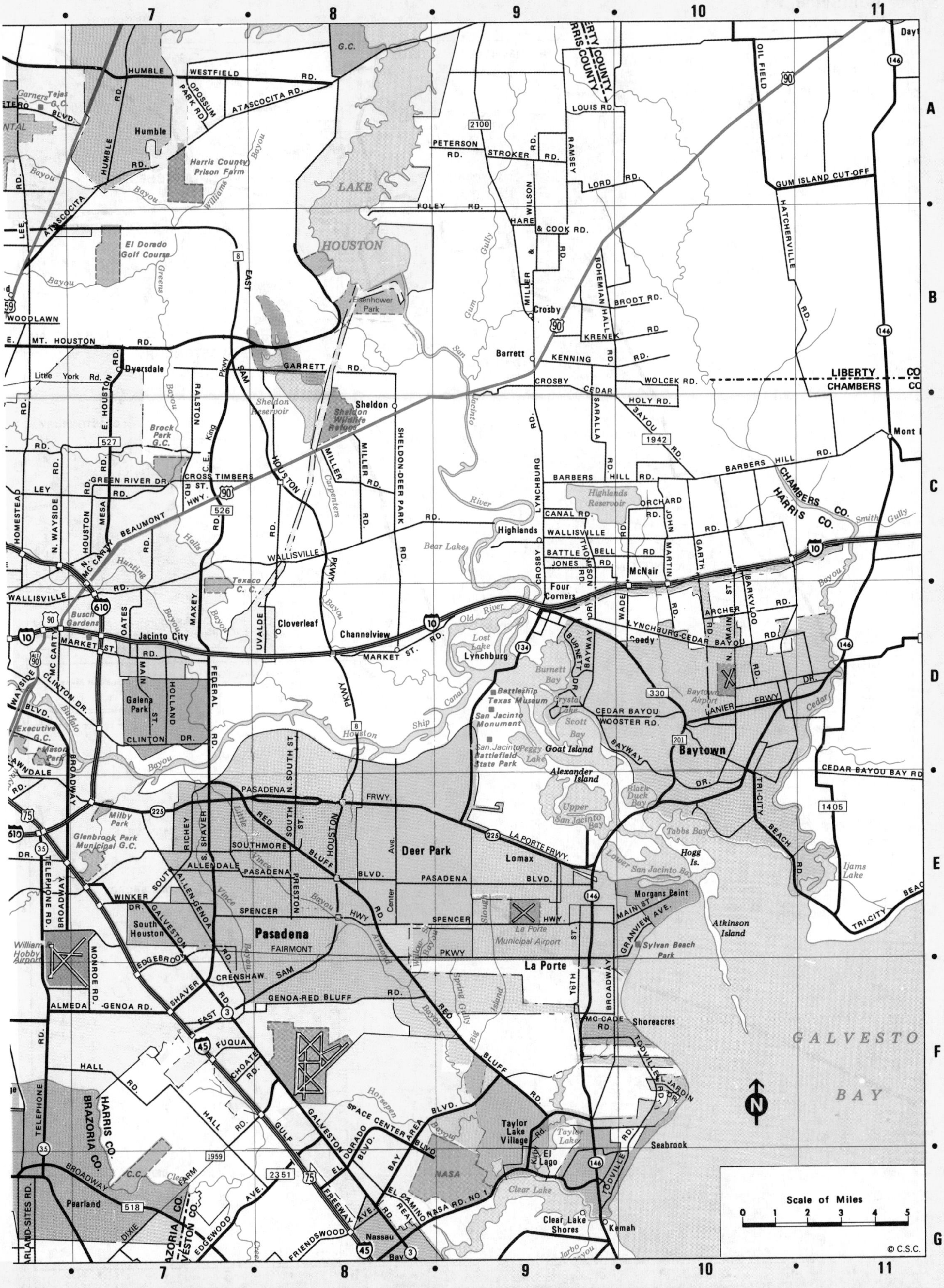
7
8
9
10
11
A
B
C
D
E
F
G
Humble
Harris County Prison Farm
LAKE HOUSTON
El Dorado Golf Course
Eisenhower Park
Crosby
Barrett
Dyersdale
Sheldon
Sheldon Reservoir
Sheldon Wildlife Refuge
Brock Park G.C.
Highlands
Highlands Reservoir
Bear Lake
McNair
Four Corners
Cloverleaf
Channelview
Jacinto City
Busch Gardens
Texaco C.C.
Lynchburg
Lost Lake
Cedy
Baytown
Baytown Airport
Galena Park
Executive G.C.
Mason Park
Milby Park
Glenbrook Park Municipal G.C.
Battleship Texas Museum
San Jacinto Monument
San Jacinto Battlefield State Park
Goat Island
Alexander Island
Black Duck Bay
Upper San Jacinto Bay
Lower San Jacinto Bay
Tabbs Bay
Hogg Is.
Ijams Lake
Deer Park
Lomax
Pasadena
South Houston
William P. Hobby Airport
Morgans Point
Atkinson Island
Sylvan Beach Park
La Porte
La Porte Municipal Airport
Shoreacres
GALVESTON BAY
Taylor Lake Village
Taylor Lake
El Lago
NASA
Seabrook
Clear Lake
Clear Lake Shores
Kemah
Nassau Bay
Pearland
HARRIS CO.
BRAZORIA CO.
LIBERTY CO.
CHAMBERS CO.
HARRIS CO.
LIBERTY COUNTY
HARRIS COUNTY
Mont Belvieu
HUMBLE RD.
WESTFIELD RD.
ATASCOCITA RD.
OPOSSUM PARK RD.
LOUIS RD.
PETERSON RD.
STROKER RD.
LORD RD.
FOLEY RD.
HARE & COOK RD.
MILLER & WILSON RD.
BOHEMIAN HALL RD.
BRODT RD.
KRENEK RD.
KENNING RD.
WOLCEK RD.
CROSBY CEDAR BAYOU RD.
HOLY RD.
BARBERS HILL RD.
MT. HOUSTON RD.
GARRETT RD.
WOODLAWN
HATCHERVILLE RD.
GUM ISLAND CUT-OFF
OIL FIELD
SHELDON-DEER PARK RD.
WALLISVILLE RD.
BEAUMONT HWY.
CROSS TIMBERS ST.
GREEN RIVER DR.
CANAL RD.
BATTLE BELL RD.
JONES RD.
LYNCHBURG-CEDAR BAYOU RD.
MARKET ST.
CLINTON DR.
LANIER FRWY
CEDAR BAYOU WOOSTER RD.
BAYWAY DR.
CEDAR BAYOU BAY RD.
TRI-CITY BEACH RD.
PASADENA FRWY.
LA PORTE FRWY.
PASADENA BLVD.
SOUTHMORE
ALLENDALE
SPENCER HWY.
FAIRMONT PKWY
RED BLUFF RD.
GENOA-RED BLUFF RD.
ALMEDA-GENOA RD.
EDGEBROOK
CRENSHAW
FUQUA
CHOATE RD.
SPACE CENTER BLVD.
NASA RD. NO 1
EL CAMINO REAL
BAY AREA BLVD.
GALVESTON RD.
GULF FREEWAY
TELEPHONE RD.
BROADWAY
FRIENDSWOOD
EDGEWOOD AVE.
DIXIE FARM RD.
ORCHARD RD.
TODVILLE RD.
EL JARDIN DR.
MC CABE RD.
GRANVIEW AVE.
MAIN ST.
Scale of Miles
0 1 2 3 4 5
© C.S.C.

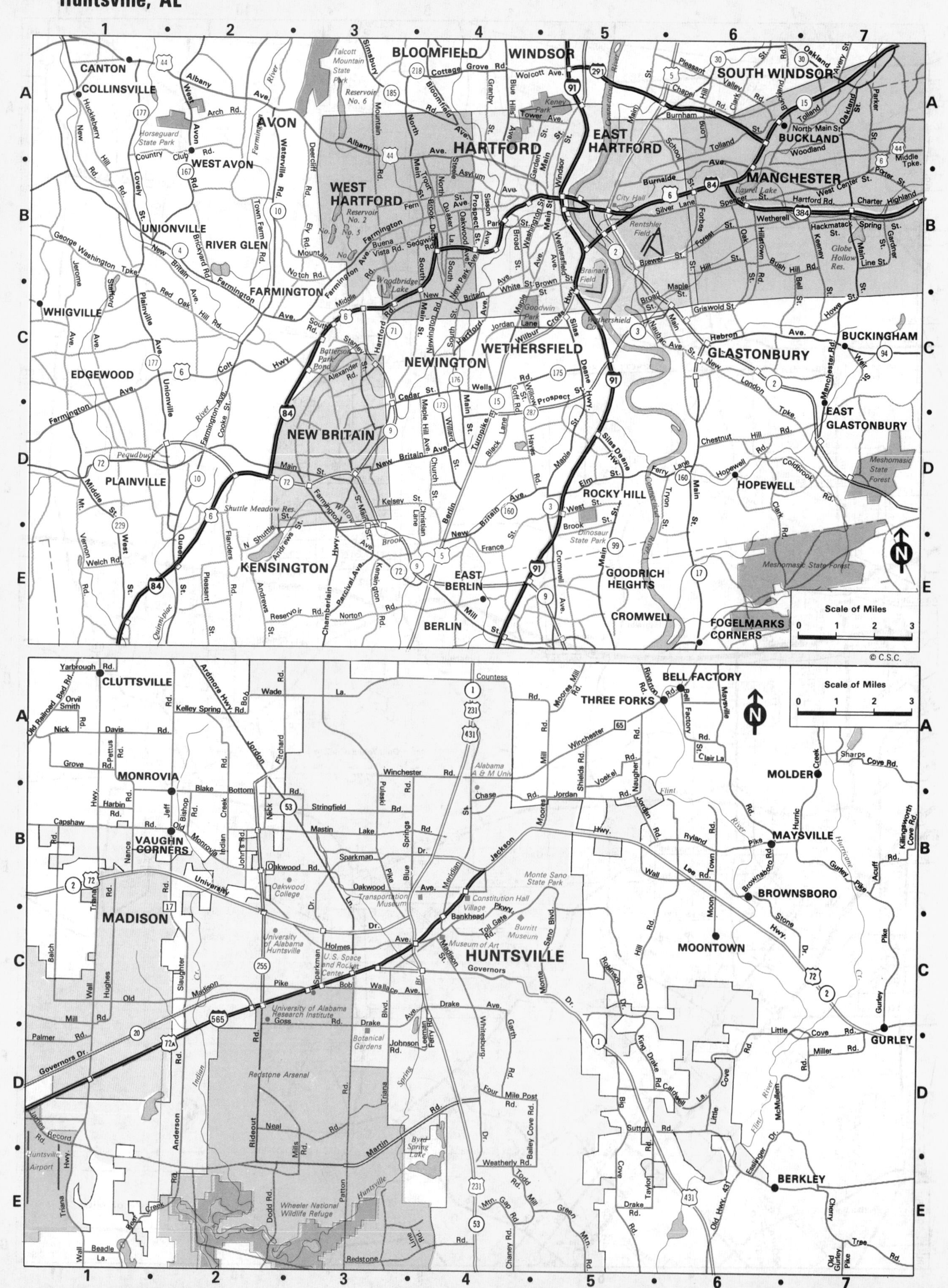
CANTON
COLLINSVILLE
AVON
WEST AVON
BLOOMFIELD
WINDSOR
SOUTH WINDSOR
HARTFORD
EAST HARTFORD
BUCKLAND
MANCHESTER
WEST HARTFORD
UNIONVILLE
RIVER GLEN
FARMINGTON
WHIGVILLE
EDGEWOOD
NEWINGTON
WETHERSFIELD
GLASTONBURY
BUCKINGHAM
EAST GLASTONBURY
NEW BRITAIN
PLAINVILLE
ROCKY HILL
HOPEWELL
KENSINGTON
EAST BERLIN
BERLIN
GOODRICH HEIGHTS
CROMWELL
FOGELMARKS CORNERS
Talcott Mountain State Park
Horseguard State Park
Keney Park
Meshomasic State Forest
Scale of Miles
© C.S.C.
CLUTTSVILLE
MONROVIA
VAUGHN CORNERS
MADISON
HUNTSVILLE
THREE FORKS
BELL FACTORY
MOLDER
MAYSVILLE
BROWNSBORO
MOONTOWN
GURLEY
BERKLEY
Alabama A & M Univ.
Oakwood College
University of Alabama Huntsville
U.S. Space and Rocket Center
Monte Sano State Park
Burritt Museum
Botanical Gardens
Redstone Arsenal
Wheeler National Wildlife Refuge
Huntsville Airport
Scale of Miles

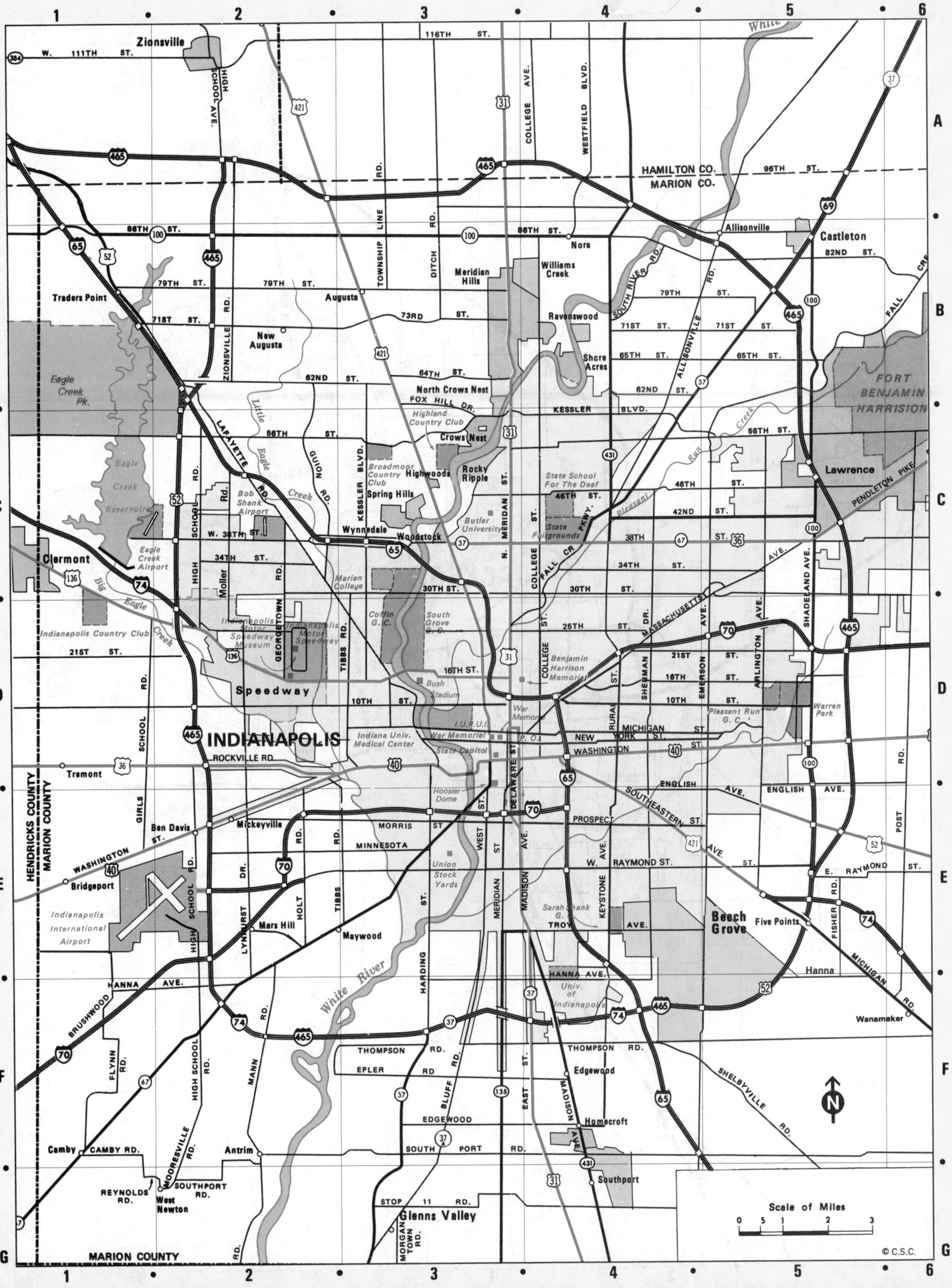

Zionsville
W. 111TH ST.
116TH ST.
HIGH SCHOOL AVE.
COLLEGE AVE.
WESTFIELD BLVD.
HAMILTON CO.
MARION CO.
96TH ST.
Allisonville
Castleton
86TH ST.
82ND ST.
Nora
Williams Creek
Meridian Hills
TOWNSHIP LINE RD.
DITCH RD.
79TH ST.
Traders Point
Augusta
73RD ST.
71ST ST.
New Augusta
Ravenswood
Shore Acres
65TH ST.
64TH ST.
62ND ST.
North Crows Nest
FOX HILL DR.
KESSLER BLVD.
Eagle Creek Pk.
FORT BENJAMIN HARRISION
Highland Country Club
Crows Nest
56TH ST.
Eagle Creek Reservoir
LAFAYETTE RD.
Little Eagle Creek
GUION RD.
KESSLER BLVD.
Broadmoor Country Club
Highwoods
Rocky Ripple
State School For The Deaf
Lawrence
46TH ST.
PENDLETON PIKE
Bob Shank Airport
Spring Hills
Butler University
42ND ST.
State Fairgrounds
W. 38TH ST.
Wynnedale
Woodstock
38TH ST.
Eagle Creek Airport
Clermont
34TH ST.
Moller Rd.
GEORGETOWN RD.
Marian College
30TH ST.
FALL CR. PKWY.
Big Eagle Creek
Coffin G. C.
South Grove G. C.
MASSACHUSETTS AVE.
SHADELAND AVE.
Indianapolis Motor Speedway Museum
Indianapolis Motor Speedway
Indianapolis Country Club
25TH ST.
21ST ST.
TIBBS RD.
Benjamin Harrison Memorial
16TH ST.
Speedway
Bush Stadium
SHERMAN DR.
EMERSON AVE.
ARLINGTON AVE.
RURAL ST.
10TH ST.
War Memorial
Pleasant Run G. C.
Warren Park
I.U.P.U.I.
INDIANAPOLIS
Indiana Univ. Medical Center
War Memorial
P. O.
MICHIGAN ST.
NEW YORK ST.
WASHINGTON ST.
State Capitol
ROCKVILLE RD.
Tremont
HENDRICKS COUNTY
MARION COUNTY
Hoosier Dome
ENGLISH AVE.
SOUTHEASTERN AVE.
GIRLS SCHOOL RD.
Ben Davis
Mickeyville
MORRIS ST.
PROSPECT ST.
MINNESOTA
POST RD.
WASHINGTON ST.
Bridgeport
Union Stock Yards
W. RAYMOND ST.
E. RAYMOND ST.
KEYSTONE AVE.
Indianapolis International Airport
LYNHURST DR.
HOLT RD.
Mars Hill
Maywood
Sarah Shank G. C.
TROY AVE.
Beech Grove
Five Points
FISHER RD.
HANNA AVE.
Hanna
White River
Univ. of Indianapolis
MICHIGAN RD.
Wanamaker
BRUSHWOOD
MANN RD.
HARDING ST.
THOMPSON RD.
EPLER RD.
Edgewood
SHELBYVILLE RD.
FLYNN RD.
BLUFF RD.
EAST ST.
MADISON AVE.
EDGEWOOD
Homecroft
Camby
CAMBY RD.
MOORESVILLE RD.
Antrim
SOUTH PORT RD.
Southport
REYNOLDS RD.
SOUTHPORT RD.
West Newton
STOP 11 RD.
Glenns Valley
MORGANTOWN RD.
MARION COUNTY
Scale of Miles
0 .5 1 2 3
© C.S.C.

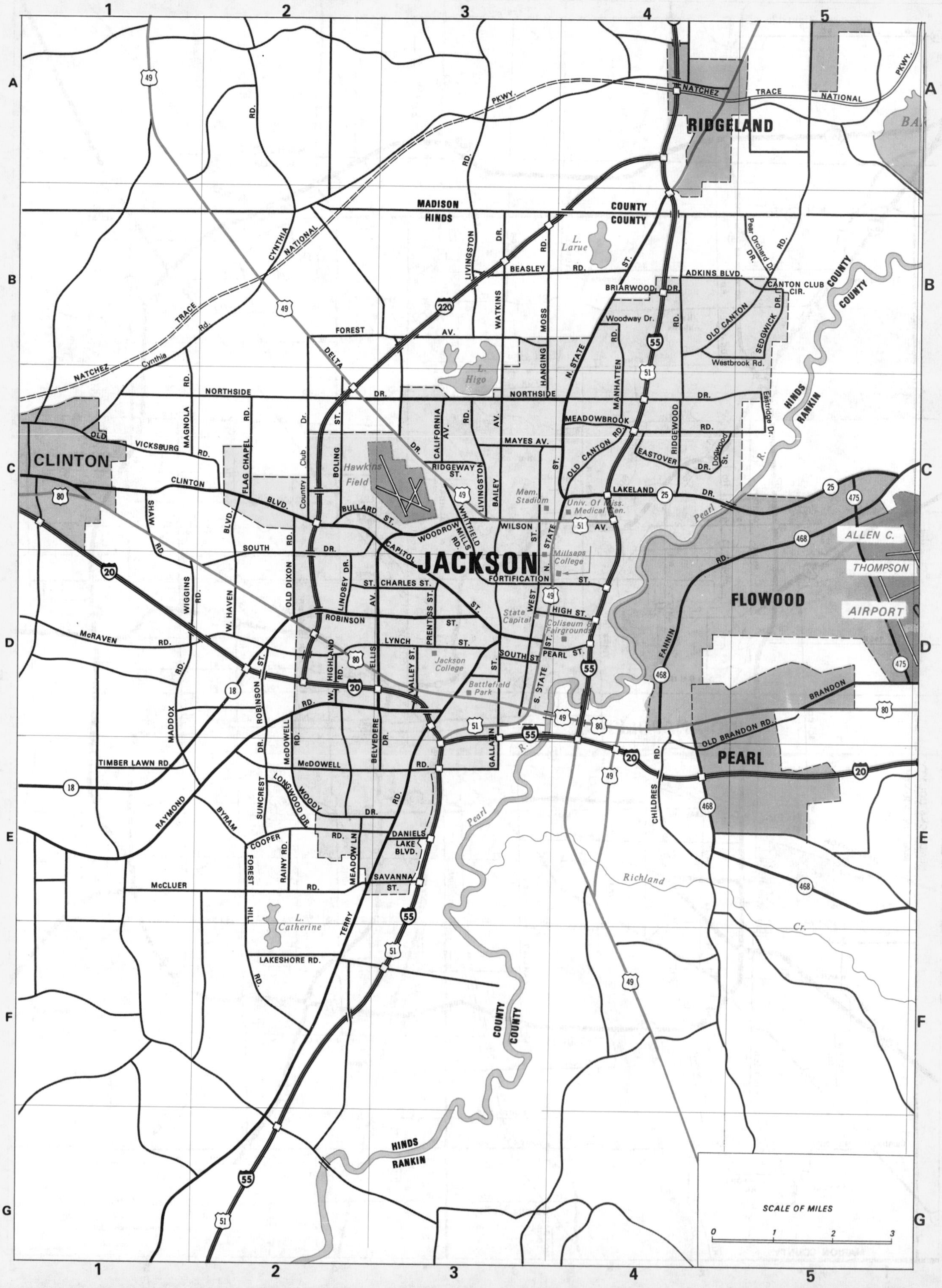

N

1 2 3 4 5

A B C D E F G

NATCHEZ TRACE NATIONAL PKWY.
RIDGELAND
MADISON
HINDS
COUNTY
COUNTY
L. Larue
BEASLEY RD.
BRIARWOOD DR.
ADKINS BLVD.
CANTON CLUB CIR.
Pear Orchard Dr.
OLD CANTON RD.
SEDGWICK DR.
Woodway Dr.
Westbrook Rd.
LIVINGSTON RD.
WATKINS DR.
MOSS RD.
CYNTHIA RD.
Cynthia Rd.
FOREST AV.
DELTA DR.
L. Higo
HANGING MOSS RD.
N. STATE ST.
MANHATTEN RD.
NORTHSIDE DR.
MEADOWBROOK RD.
RIDGEWOOD RD.
Eastridge Dr.
HINDS
RANKIN
MAGNOLIA RD.
FLAG CHAPEL RD.
Country Club Dr.
BOLING ST.
CALIFORNIA AV.
MAYES AV.
EASTOVER DR.
Dogwood St.
CLINTON
OLD VICKSBURG RD.
CLINTON BLVD.
Hawkins Field
RIDGEWAY ST.
BAILEY AV.
Mem. Stadium
Univ. Of Miss. Medical Cen.
LAKELAND DR.
Pearl R.
BULLARD ST.
WOODROW WILSON AV.
MILLS ST.
WHITFIELD
SHAW RD.
SOUTH DR.
CAPITOL ST.
JACKSON
Millsaps College
FORTIFICATION ST.
ALLEN C. THOMPSON AIRPORT
FLOWOOD
WIGGINS RD.
W. HAVEN BLVD.
OLD DIXON RD.
LINDSEY DR.
ST. CHARLES ST.
State Capitol
WEST ST.
HIGH ST.
Coliseum & Fairgrounds
McRAVEN RD.
ROBINSON ST.
LYNCH ST.
PRENTISS ST.
Jackson College
SOUTH ST.
PEARL ST.
FANNIN
W. HIGHLAND RD.
ELLIS AV.
VALLEY ST.
Battlefield Park
S. STATE ST.
ROBINSON RD.
MADDOX RD.
McDOWELL RD.
BELVEDERE DR.
GALLATIN ST.
BRANDON
OLD BRANDON RD.
PEARL
TIMBER LAWN RD.
McDOWELL RD.
CHILDRES RD.
RAYMOND
BYRAM
SUNCREST
LONGWOOD DR.
WOODY DR.
COOPER RD.
DANIELS
LAKE BLVD.
Pearl
Richland Cr.
FOREST HILL RD.
RAINY RD.
MEADOW LN.
McCLUER RD.
SAVANNA ST.
L. Catherine
TERRY
LAKESHORE RD.
COUNTY
COUNTY
HINDS
RANKIN

SCALE OF MILES
0 1 2 3

49 220 55 51 25 475 468 80 20 18

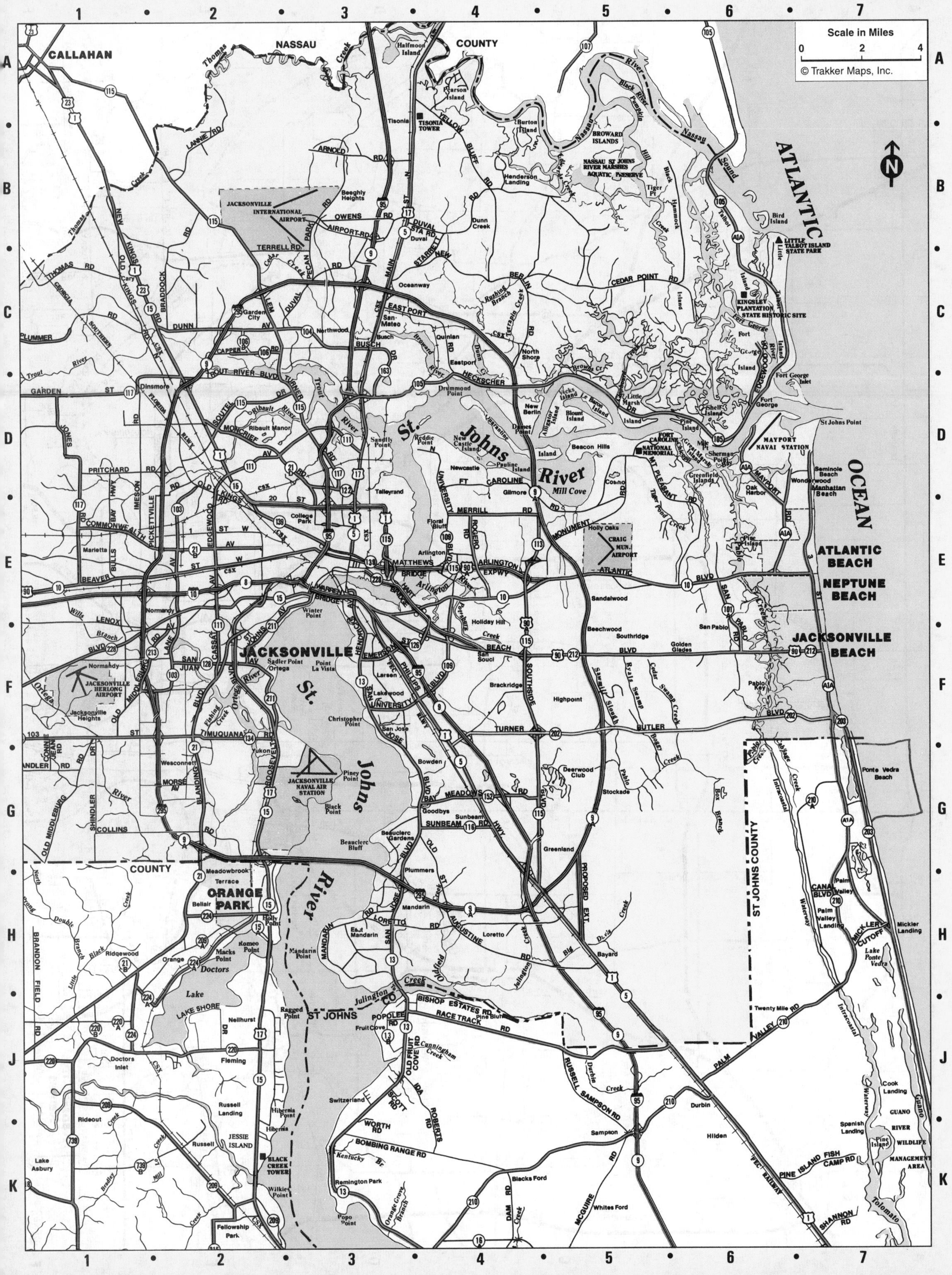

Scale in Miles
0 2 4

1 2 3 4 5 6 7

A B C D E F G H J K

CALLAHAN
NASSAU COUNTY
ATLANTIC OCEAN
JACKSONVILLE INTERNATIONAL AIRPORT
LITTLE TALBOT ISLAND STATE PARK
KINGSLEY PLANTATION STATE HISTORIC SITE
NASSAU ST JOHNS RIVER MARSHES AQUATIC PRESERVE
FORT CAROLINE NATIONAL MEMORIAL
MAYPORT NAVAL STATION
St. Johns River
CRAIG MUN. AIRPORT
ATLANTIC BEACH
NEPTUNE BEACH
JACKSONVILLE BEACH
JACKSONVILLE
JACKSONVILLE HERLONG AIRPORT
JACKSONVILLE NAVAL AIR STATION
ORANGE PARK
Doctors Lake
ST JOHNS COUNTY
Ponte Vedra Beach
GUANO RIVER WILDLIFE MANAGEMENT AREA
BLACK CREEK TOWER
TISONIA TOWER
CLAY COUNTY

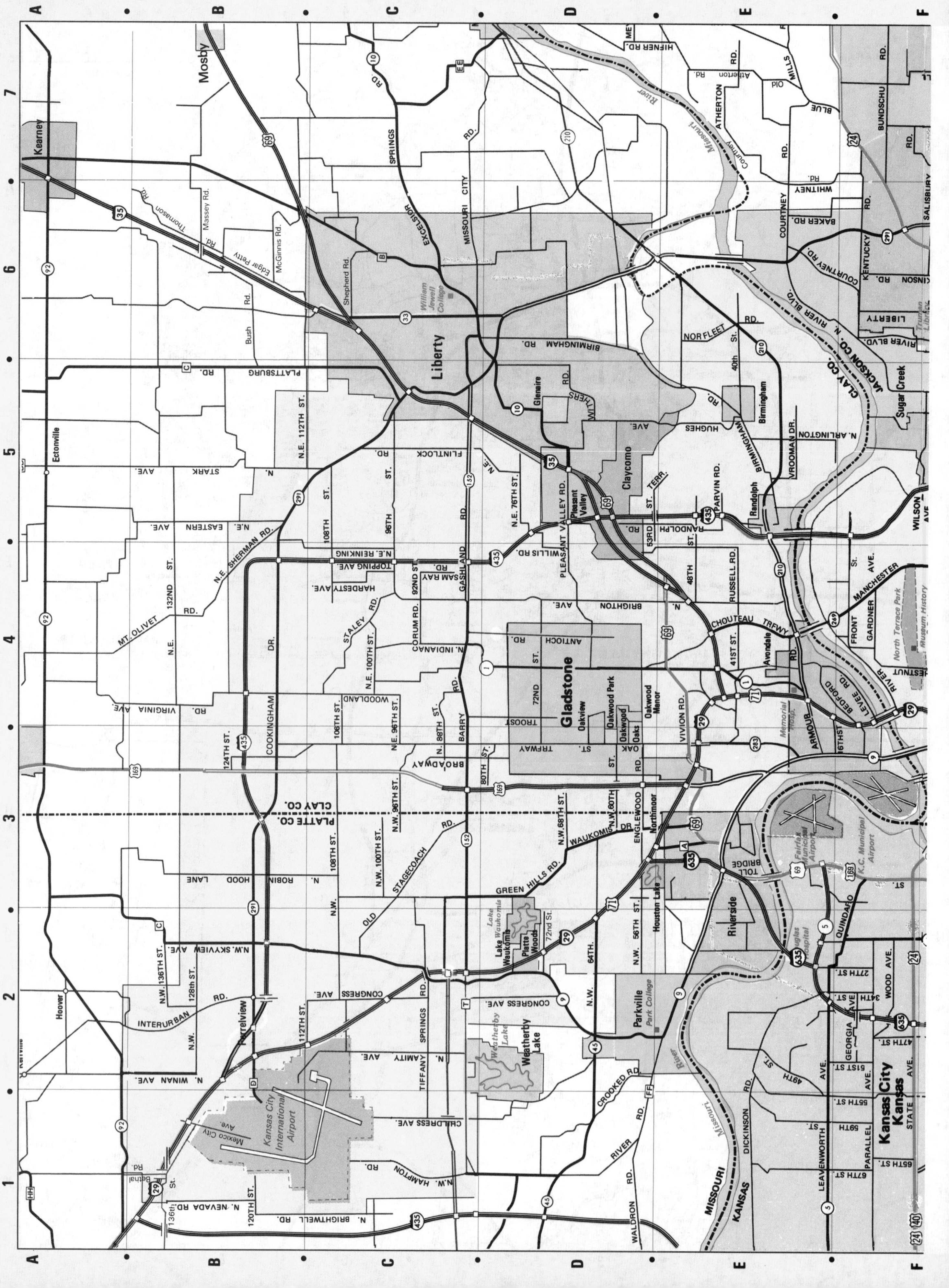

Kearney
Mosby
Liberty
William Jewell College
Claycomo
Pleasant Valley
Gladstone
Oakview
Oakwood Park
Oakwood
Oaks
Oakwood Manor
Avondale
Randolph
Birmingham
Sugar Creek
North Terrace Park
Northmoor
Houston Lake
Riverside
Lake Waukomis
Platte Woods
Parkville
Park College
Weatherby Lake
Kansas City International Airport
Fairfax Municipal Airport
K.C. Municipal Airport
Kansas City Kansas
Ectonville
Hoover
Ferrelview
CLAY CO.
PLATTE CO.
JACKSON CO.
MISSOURI
KANSAS
Missouri River

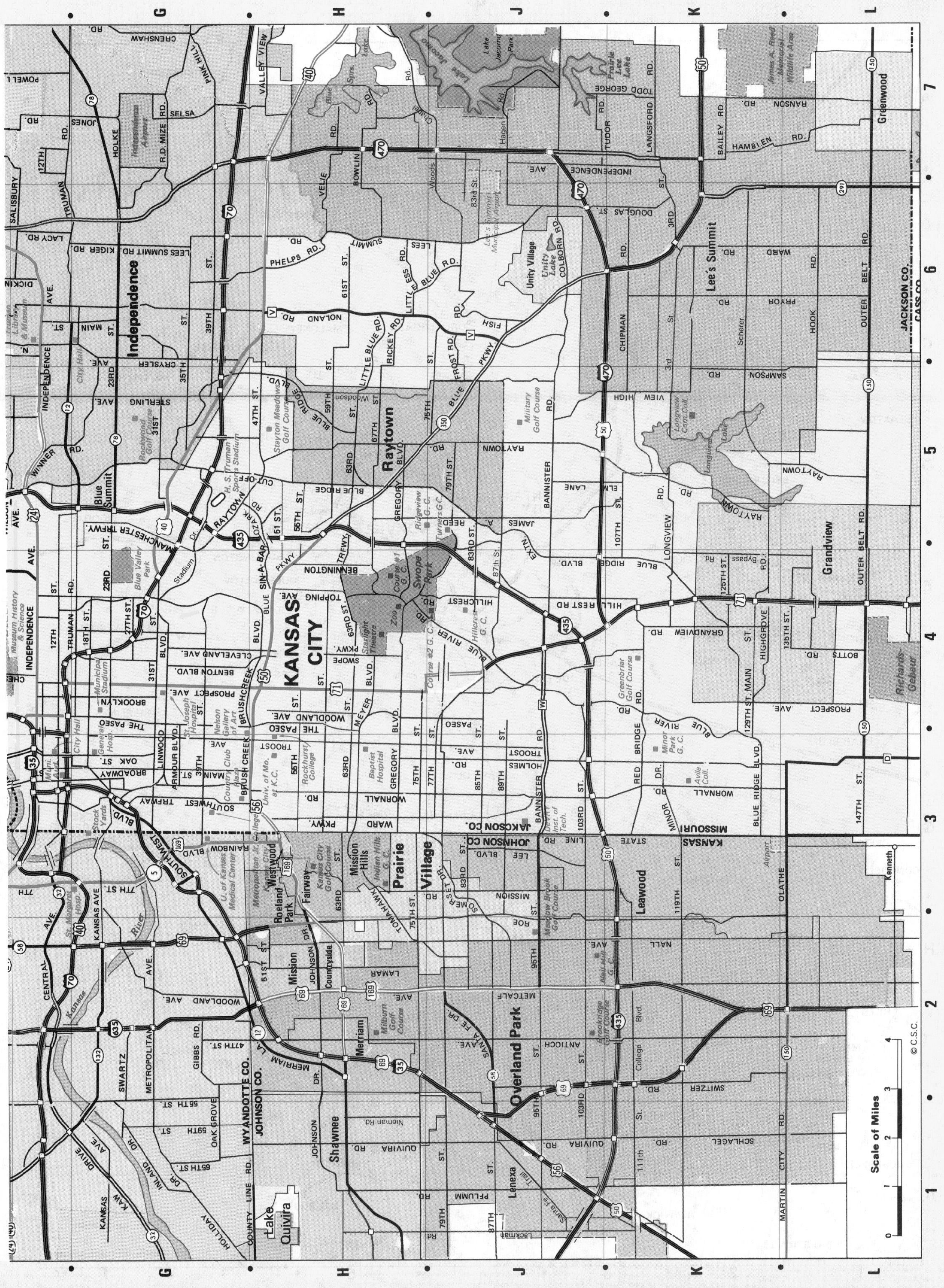

KANSAS CITY

Independence

Raytown

Lee's Summit

Grandview

Blue Summit

Prairie Village

Mission Hills

Westwood

Roeland Park

Fairway

Mission

Merriam

Overland Park

Leawood

Shawnee

Lenexa

Lake Quivira

JACKSON CO.

JOHNSON CO.

WYANDOTTE CO.

CASS CO.

MISSOURI

KANSAS

Scale of Miles

© C.S.C.

Knoxville, TN

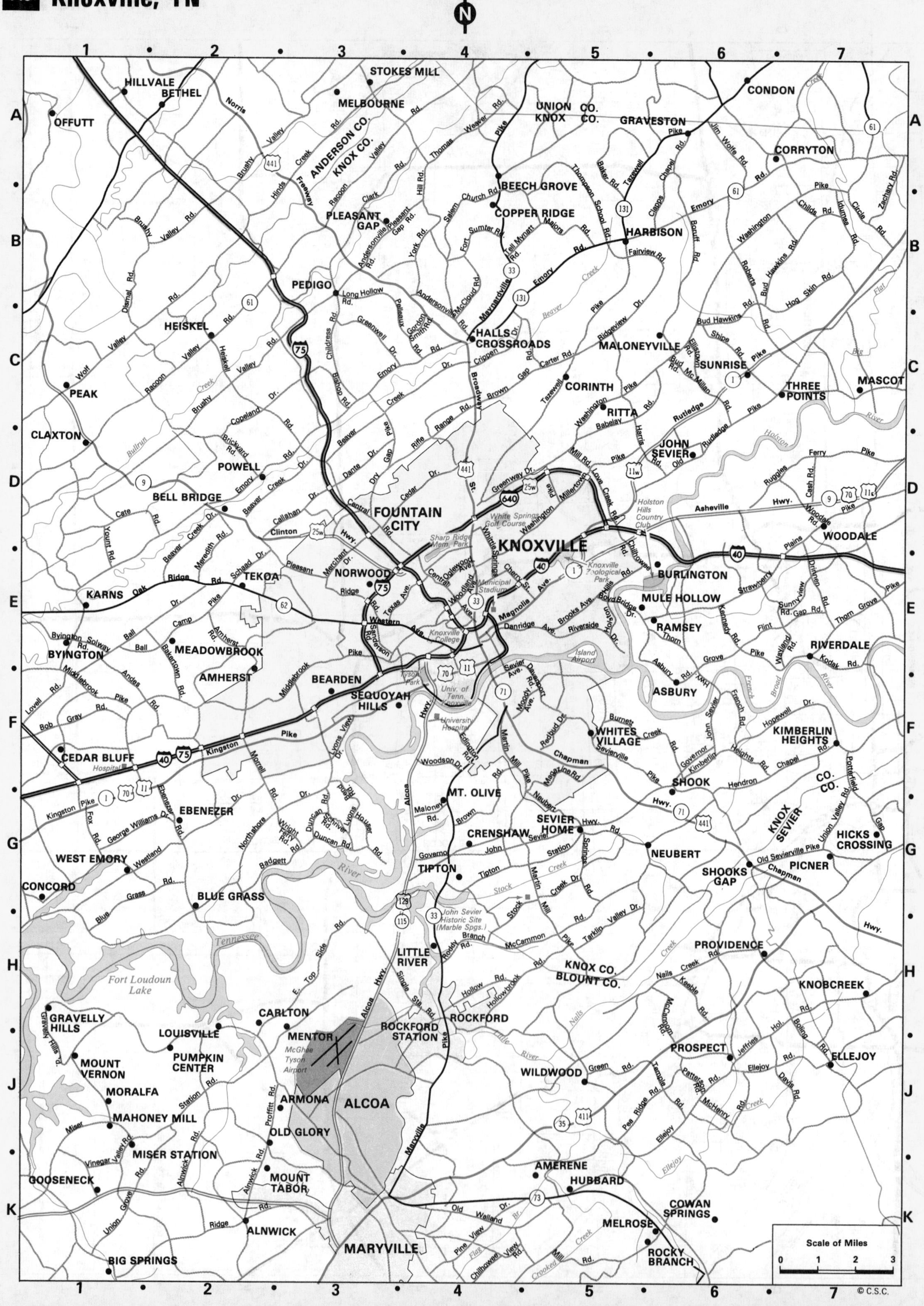

N

HILLVALE
BETHEL
OFFUTT
STOKES MILL
MELBOURNE
ANDERSON CO.
KNOX CO.
UNION CO.
KNOX CO.
GRAVESTON
CONDON
CORRYTON
BEECH GROVE
COPPER RIDGE
PLEASANT GAP
HARBISON
PEDIGO
HEISKEL
HALLS CROSSROADS
MALONEYVILLE
SUNRISE
MASCOT
PEAK
CORINTH
THREE POINTS
RITTA
CLAXTON
JOHN SEVIER
POWELL
BELL BRIDGE
FOUNTAIN CITY
KNOXVILLE
WOODALE
TEKOA
NORWOOD
BURLINGTON
KARNS
MULE HOLLOW
RAMSEY
BYINGTON
MEADOWBROOK
RIVERDALE
AMHERST
BEARDEN
SEQUOYAH HILLS
ASBURY
WHITES VILLAGE
KIMBERLIN HEIGHTS
CEDAR BLUFF
SHOOK
MT. OLIVE
EBENEZER
SEVIER HOME
CRENSHAW
HICKS CROSSING
KNOX CO.
SEVIER CO.
WEST EMORY
NEUBERT
PICNER
TIPTON
SHOOKS GAP
CONCORD
BLUE GRASS
PROVIDENCE
LITTLE RIVER
KNOX CO.
BLOUNT CO.
Fort Loudoun Lake
KNOBCREEK
GRAVELLY HILLS
CARLTON
LOUISVILLE
MENTOR
ROCKFORD STATION
ROCKFORD
PUMPKIN CENTER
MOUNT VERNON
PROSPECT
ELLEJOY
McGhee Tyson Airport
WILDWOOD
MORALFA
ARMONA
ALCOA
MAHONEY MILL
OLD GLORY
MISER STATION
AMERENE
HUBBARD
GOOSENECK
MOUNT TABOR
COWAN SPRINGS
MELROSE
ALNWICK
MARYVILLE
ROCKY BRANCH
BIG SPRINGS

Scale of Miles
0 1 2 3

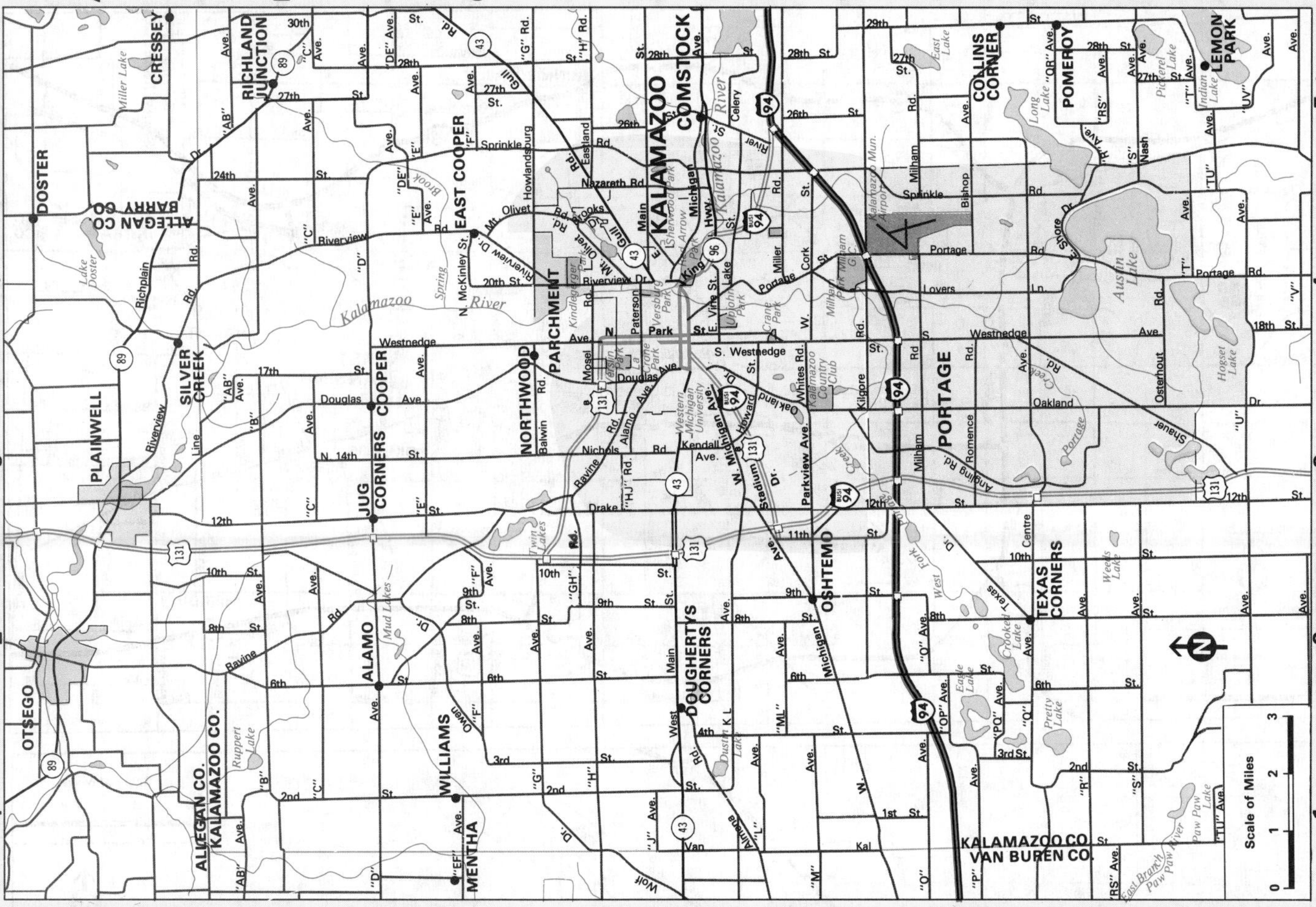

KALAMAZOO
COMSTOCK
PARCHMENT
NORTHWOOD
PORTAGE
OSHTEMO
TEXAS CORNERS
DOUGHERTYS CORNERS
ALAMO
JUG CORNERS
COOPER
EAST COOPER
RICHLAND JUNCTION
CRESSEY
DOSTER
PLAINWELL
SILVER CREEK
OTSEGO
WILLIAMS
MENTHA
COLLINS CORNER
POMEROY
LEMON PARK
ALLEGAN CO.
BARRY CO.
KALAMAZOO CO.
VAN BUREN CO.
Scale of Miles 0 1 2 3

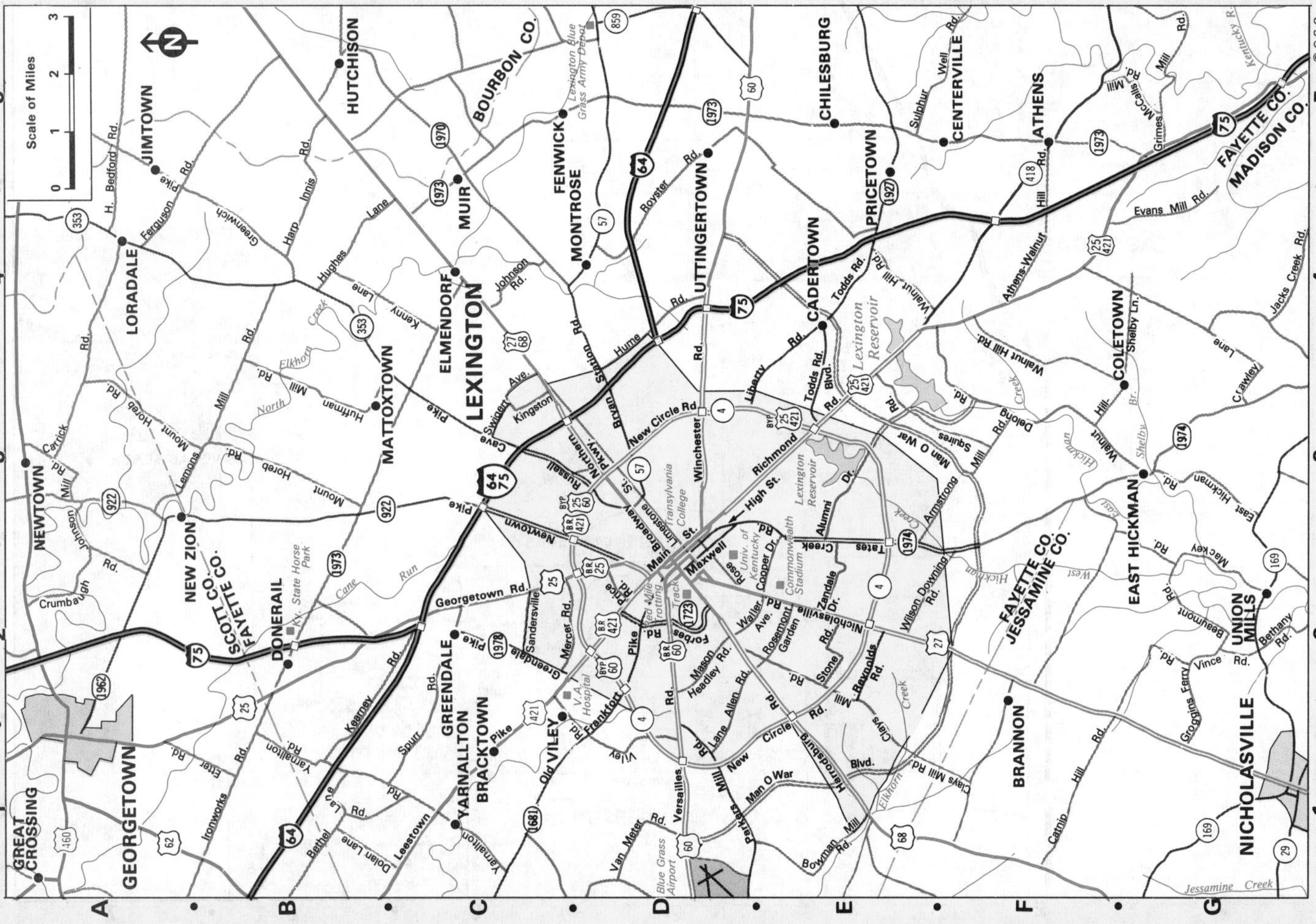

LEXINGTON
GEORGETOWN
GREAT CROSSING
NEWTOWN
NEW ZION
DONERAIL
LORADALE
JIMTOWN
HUTCHISON
MATTOXTOWN
ELMENDORF
MUIR
MONTROSE
FENWICK
UTTINGERTOWN
CADERTOWN
CHILESBURG
PRICETOWN
CENTERVILLE
ATHENS
COLETOWN
EAST HICKMAN
UNION MILLS
NICHOLASVILLE
BRANNON
GREENDALE
YARNALLTON
BRACKTOWN
VILEY
SCOTT CO.
FAYETTE CO.
BOURBON CO.
MADISON CO.
JESSAMINE CO.
Scale of Miles 0 1 2 3

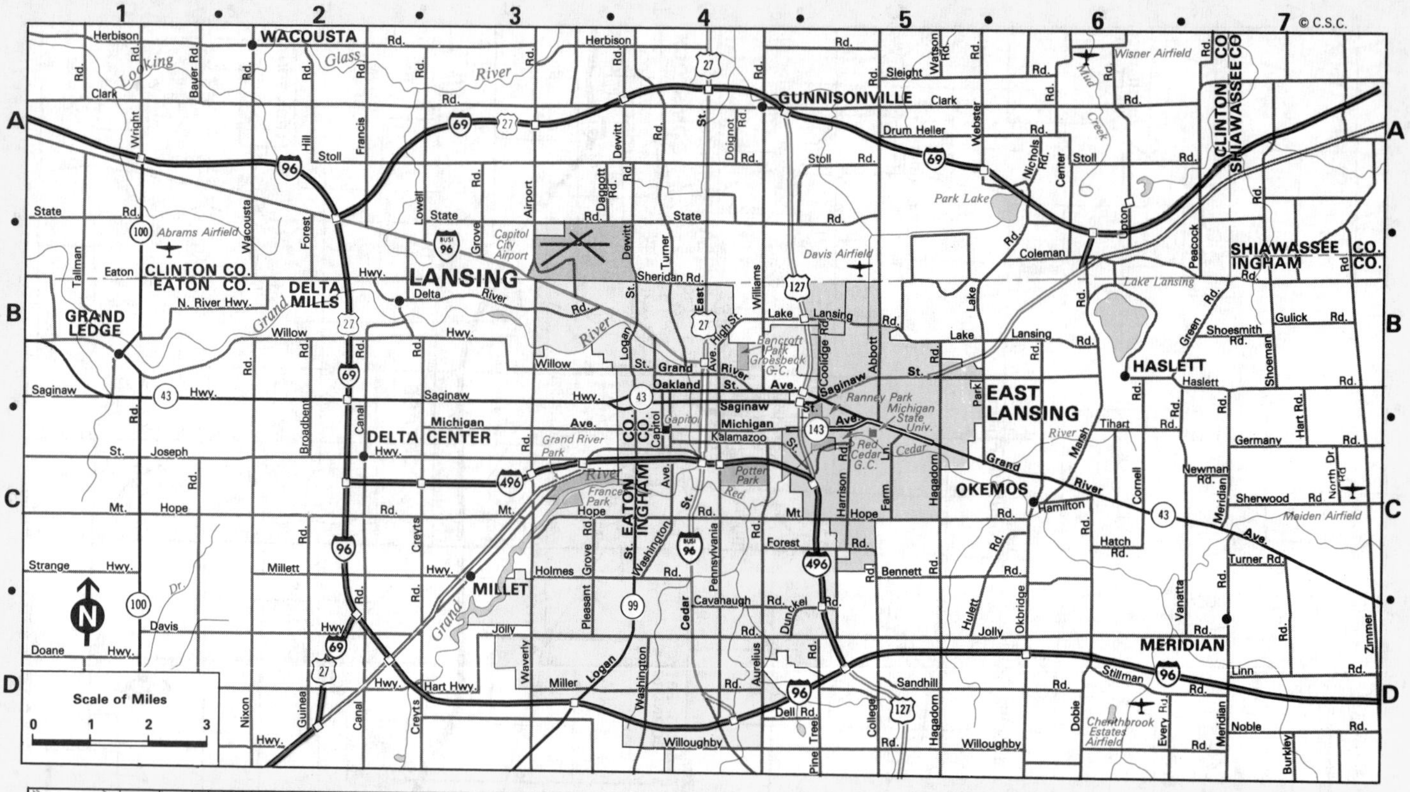

Lansing, MI
Las Vegas, NV

© C.S.C.

WACOUSTA
GUNNISONVILLE
Wisner Airfield
CLINTON CO.
SHIAWASSEE CO.
Park Lake
Abrams Airfield
Capitol City Airport
Davis Airfield
LANSING
CLINTON CO.
EATON CO.
DELTA MILLS
GRAND LEDGE
SHIAWASSEE CO.
INGHAM CO.
Lake Lansing
HASLETT
Bancroft Park
Groesbeck G.C.
Ranney Park
Michigan State Univ.
EAST LANSING
DELTA CENTER
Grand River Park
Potter Park
France Park
Red Cedar G.C.
OKEMOS
EATON CO.
INGHAM CO.
Maiden Airfield
MILLET
MERIDIAN
Chertthbrook Estates Airfield
Scale of Miles
0 1 2 3

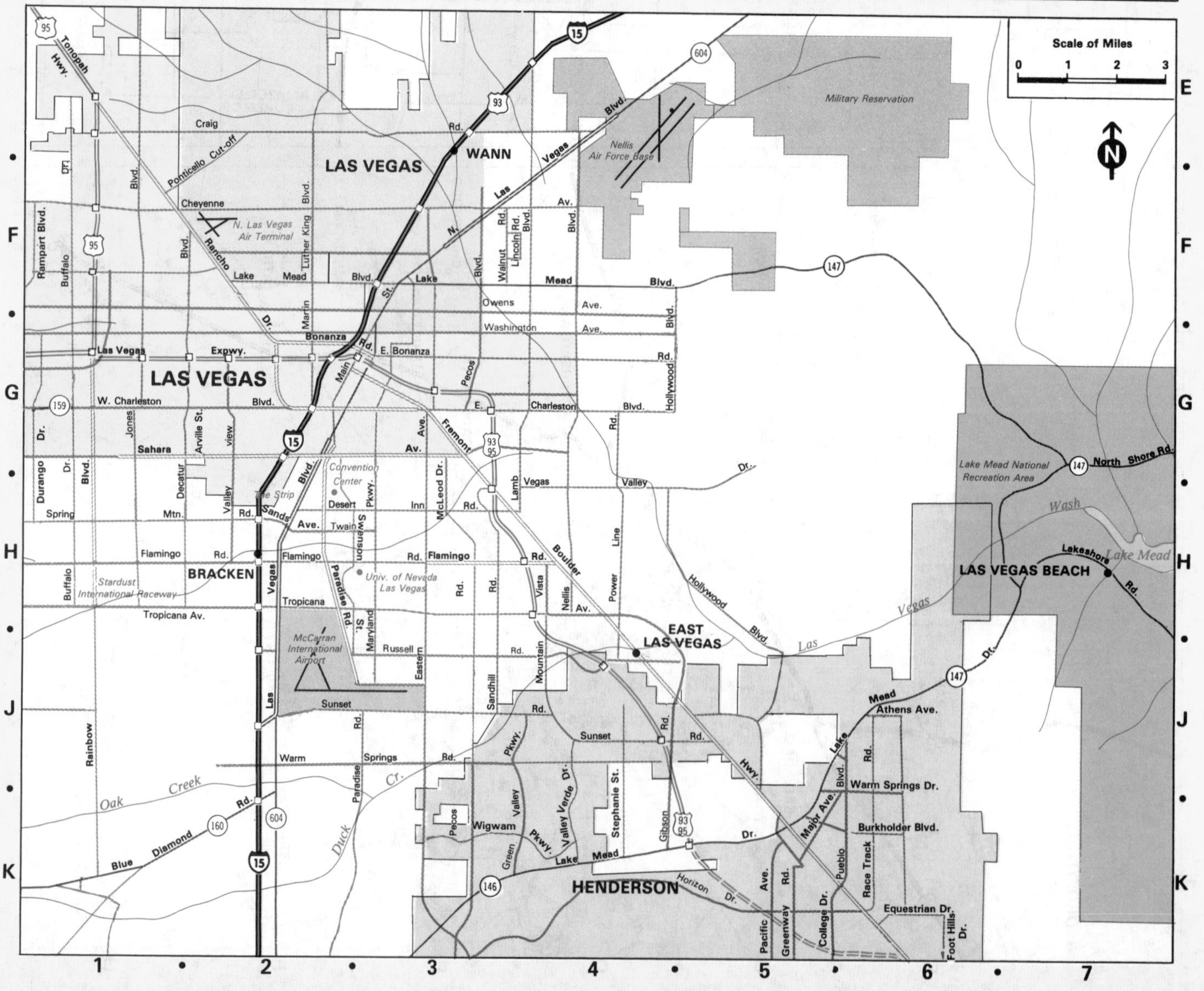

LAS VEGAS
WANN
Military Reservation
Nellis Air Force Base
N. Las Vegas Air Terminal
Scale of Miles
0 1 2 3
LAS VEGAS
Lake Mead National Recreation Area
Convention Center
The Strip
BRACKEN
Stardust International Raceway
Univ. of Nevada Las Vegas
LAS VEGAS BEACH
Lake Mead
EAST LAS VEGAS
McCarran International Airport
HENDERSON

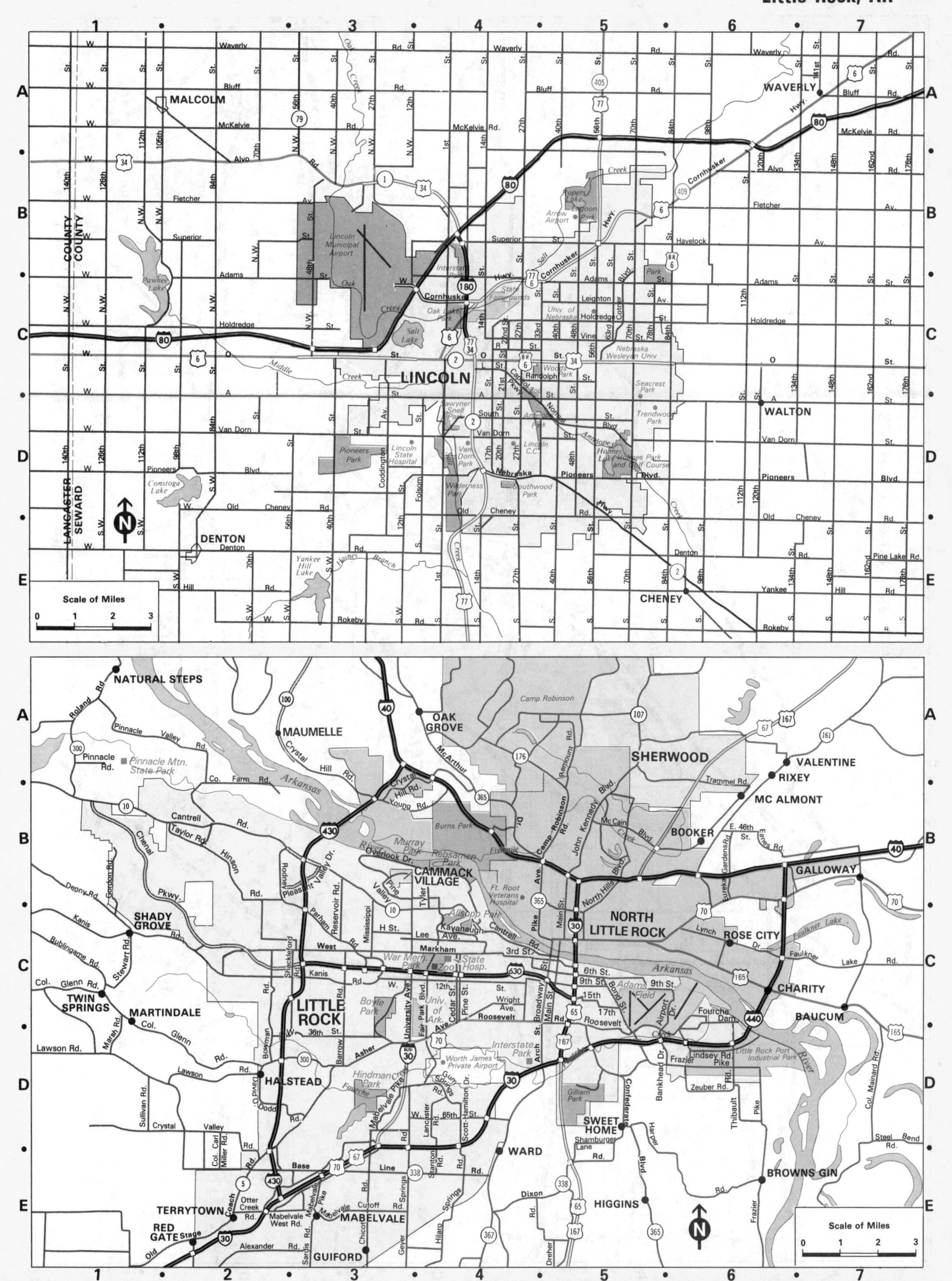
MALCOLM
WAVERLY
LINCOLN
WALTON
DENTON
CHENEY
LANCASTER COUNTY
SEWARD COUNTY
Lincoln Municipal Airport
Arrow Airport
Popers Lake
Lagoon Park
Pawnee Lake
Conestoga Lake
Yankee Hill Lake
Salt Lake
Oak Lake Park
State Fairgrounds
Univ. of Nebraska
Nebraska Wesleyan Univ.
Seacrest Park
Trendwood Park
Pioneers Park
Lincoln State Hospital
Wilderness Park
Southwood Park
Holmes Lake
Holmes Park and Golf Course
Antelope Park
Lincoln C.C.
Scale of Miles
0 1 2 3
NATURAL STEPS
MAUMELLE
OAK GROVE
Camp Robinson
SHERWOOD
VALENTINE
RIXEY
MC ALMONT
BOOKER
GALLOWAY
Pinnacle Mtn. State Park
Burns Park
Murray Park
Rebsamen Park
CAMMACK VILLAGE
Ft. Root Veterans Hospital
NORTH LITTLE ROCK
ROSE CITY
Faulkner Lake
SHADY GROVE
TWIN SPRINGS
MARTINDALE
War Mem. Park
Zoo
State Hosp.
LITTLE ROCK
Boyle Park
Univ. of Ark.
Adams Field
CHARITY
BAUCUM
Interstate Park
Worth James Private Airport
Little Rock Port Industrial Park
HALSTEAD
Hindman Park
Gilliam Park
SWEET HOME
WARD
BROWNS GIN
HIGGINS
TERRYTOWN
RED GATE
MABELVALE
GUIFORD
Arkansas River
Scale of Miles
0 1 2 3

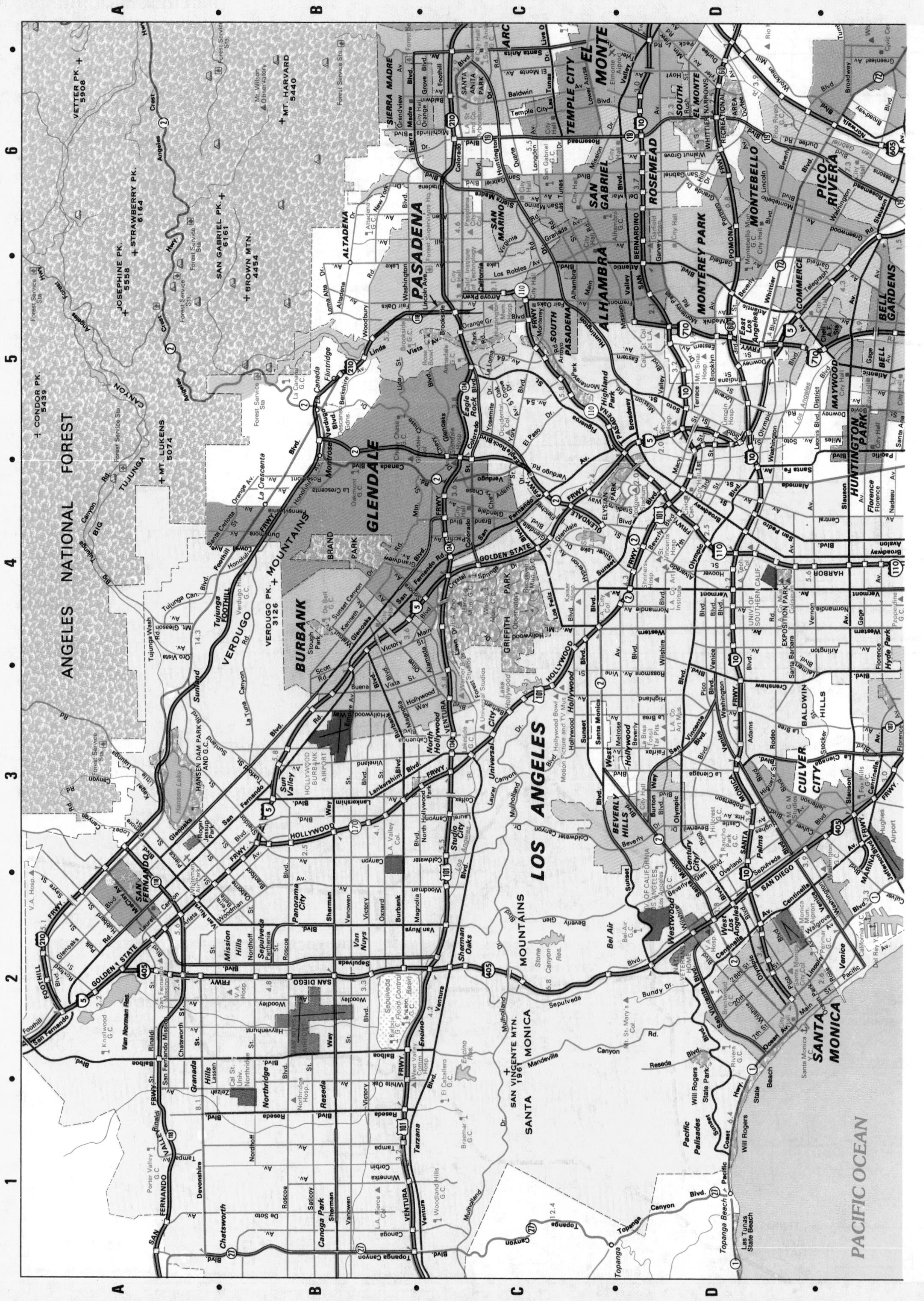
ANGELES NATIONAL FOREST
VERDUGO MOUNTAINS
SANTA MONICA MOUNTAINS
LOS ANGELES
PASADENA
GLENDALE
BURBANK
ALHAMBRA
SAN GABRIEL
SAN MARINO
SOUTH PASADENA
ALTADENA
SIERRA MADRE
ARCADIA
TEMPLE CITY
EL MONTE
SOUTH EL MONTE
ROSEMEAD
MONTEREY PARK
MONTEBELLO
PICO RIVERA
COMMERCE
BELL GARDENS
BELL
MAYWOOD
HUNTINGTON PARK
EAST LOS ANGELES
BEVERLY HILLS
WEST HOLLYWOOD
CULVER CITY
SANTA MONICA
SAN FERNANDO
Chatsworth
Canoga Park
Northridge
Reseda
Tarzana
Encino
Van Nuys
Sherman Oaks
Studio City
North Hollywood
Sun Valley
Panorama City
Sepulveda
Mission Hills
Granada Hills
Tujunga
La Crescenta
Montrose
La Canada Flintridge
Highland Park
Bel Air
Westwood
West Los Angeles
Venice
Pacific Palisades
Topanga
Topanga Beach
PACIFIC OCEAN
JOSEPHINE PK. 5558
STRAWBERRY PK. 6164
SAN GABRIEL PK. 6161
BROWN MTN. 4454
MT. HARVARD 5440
VETTER PK. 5908
CONDOR PK. 5439
MT. LUKENS 5074
VERDUGO PK. 3126
SAN VICENTE MTN. 1961
HOLLYWOOD BURBANK AIRPORT
GRIFFITH PARK
BALDWIN HILLS
EXPOSITION PARK
HANSEN DAM PARK
UNIV. OF SOUTHERN CALIF.
GOLDEN STATE FRWY.
SAN DIEGO FRWY.
VENTURA FRWY.
HOLLYWOOD FRWY.
FOOTHILL FRWY.
SANTA MONICA FRWY.
HARBOR FRWY.
PASADENA FRWY.
SAN BERNARDINO FRWY.
POMONA FRWY.
Will Rogers State Park
Las Tunas State Beach

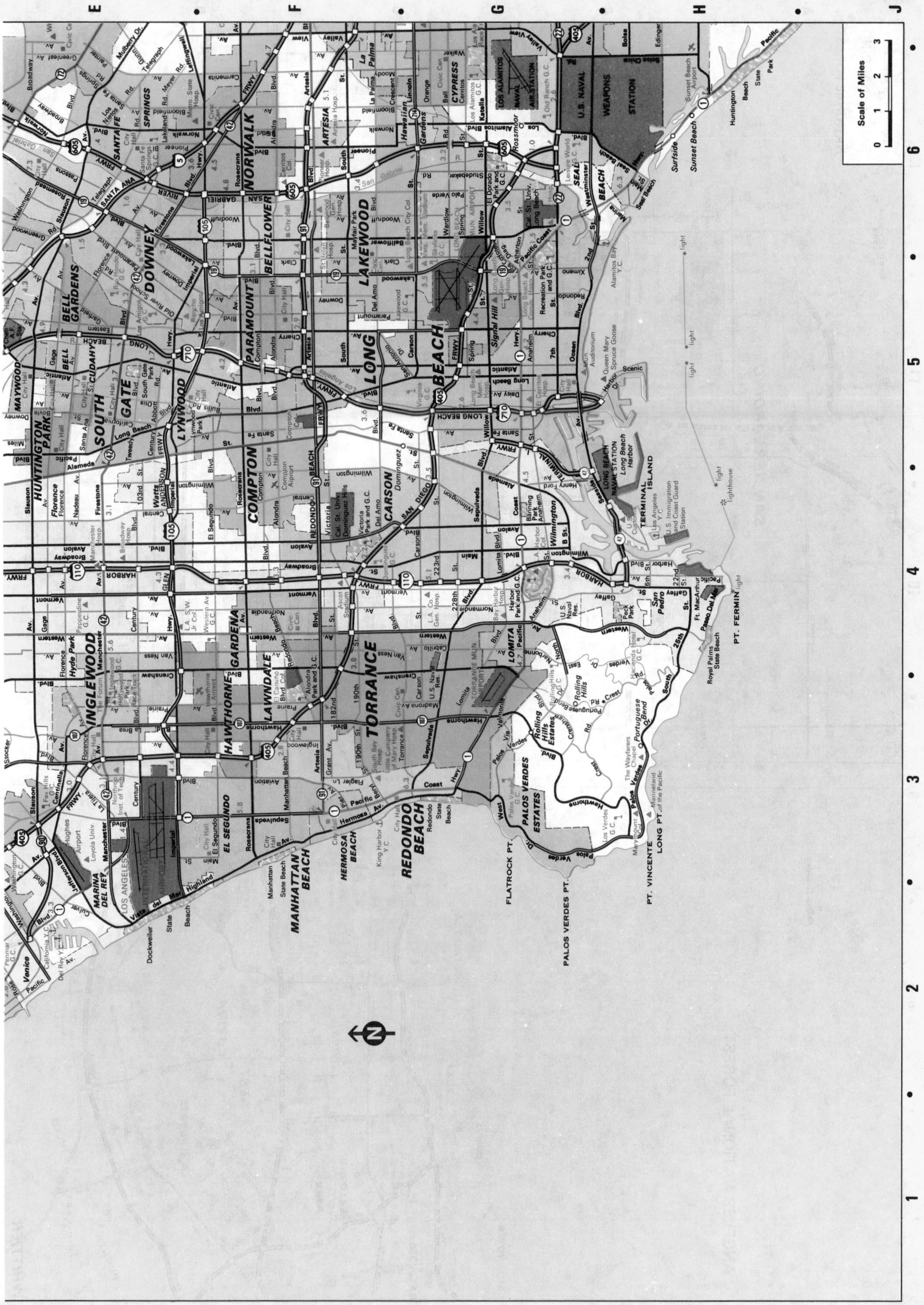

HUNTINGTON PARK
MAYWOOD
BELL
BELL GARDENS
CUDAHY
SOUTH GATE
LYNWOOD
DOWNEY
SANTA FE SPRINGS
NORWALK
ARTESIA
La Palma
CYPRESS
PARAMOUNT
BELLFLOWER
LAKEWOOD
LONG BEACH
Signal Hill
Rossmoor
SEAL BEACH
Surfside
Sunset Beach
COMPTON
CARSON
Wilmington
TERMINAL ISLAND
San Pedro
PT. FERMIN
INGLEWOOD
Hyde Park
Florence
Watts
HAWTHORNE
GARDENA
LAWNDALE
TORRANCE
LOMITA
Rolling Hills
Rolling Hills Estates
Portuguese Bend
PALOS VERDES ESTATES
LONG PT.
PT. VICENTE
PALOS VERDES PT.
FLATROCK PT.
REDONDO BEACH
HERMOSA BEACH
MANHATTAN BEACH
EL SEGUNDO
MARINA DEL REY
Venice
LOS ANGELES INTERNATIONAL AIRPORT
TORRANCE MUN. AIRPORT
LONG BEACH MUN. AIRPORT
LOS ALAMITOS NAVAL AIR STATION
U.S. NAVAL WEAPONS STATION
LONG BEACH NAVAL STATION
Long Beach Harbor
Scale of Miles
0 1 2 3

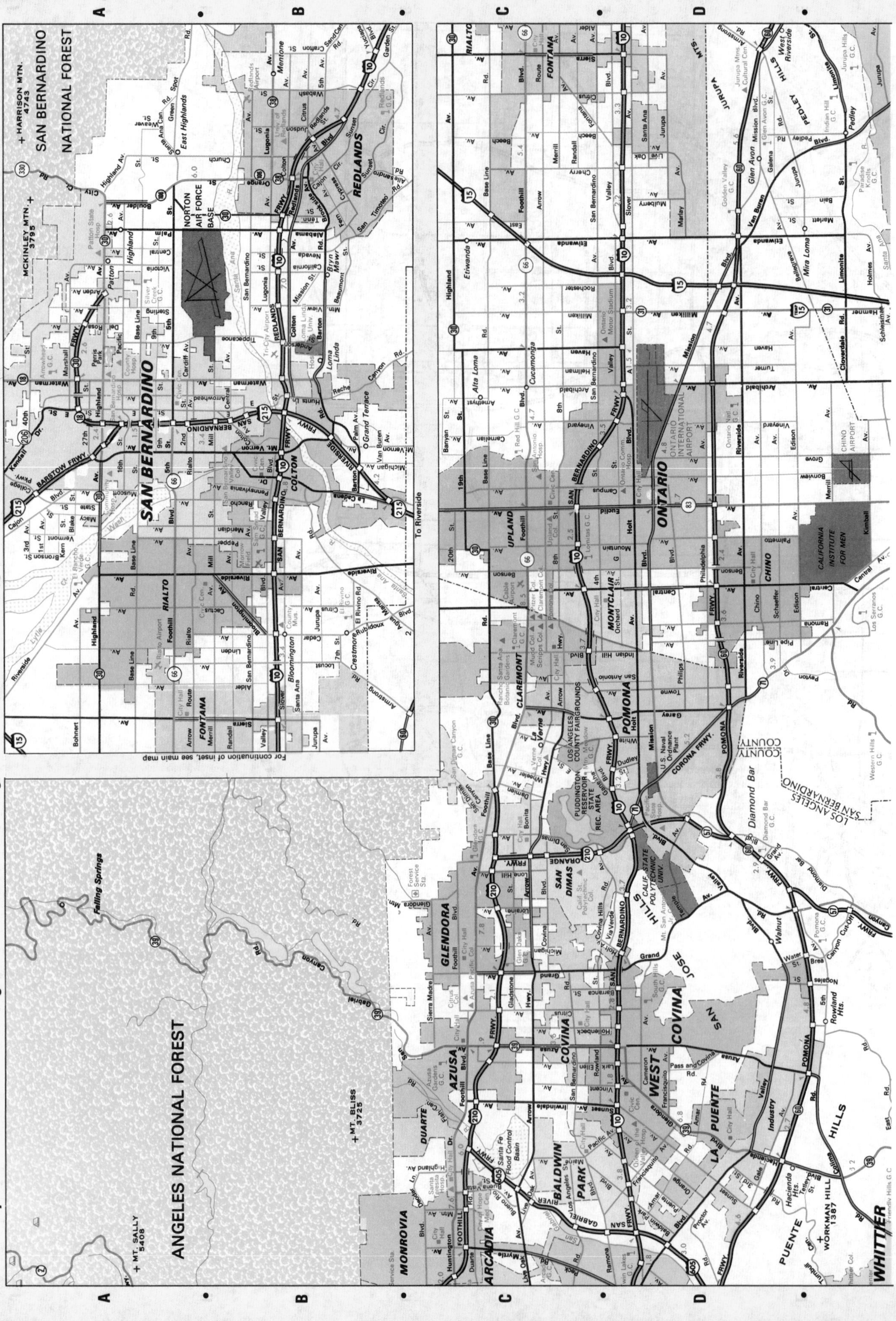

ANGELES NATIONAL FOREST

SAN BERNARDINO NATIONAL FOREST

For continuation of inset, see main map

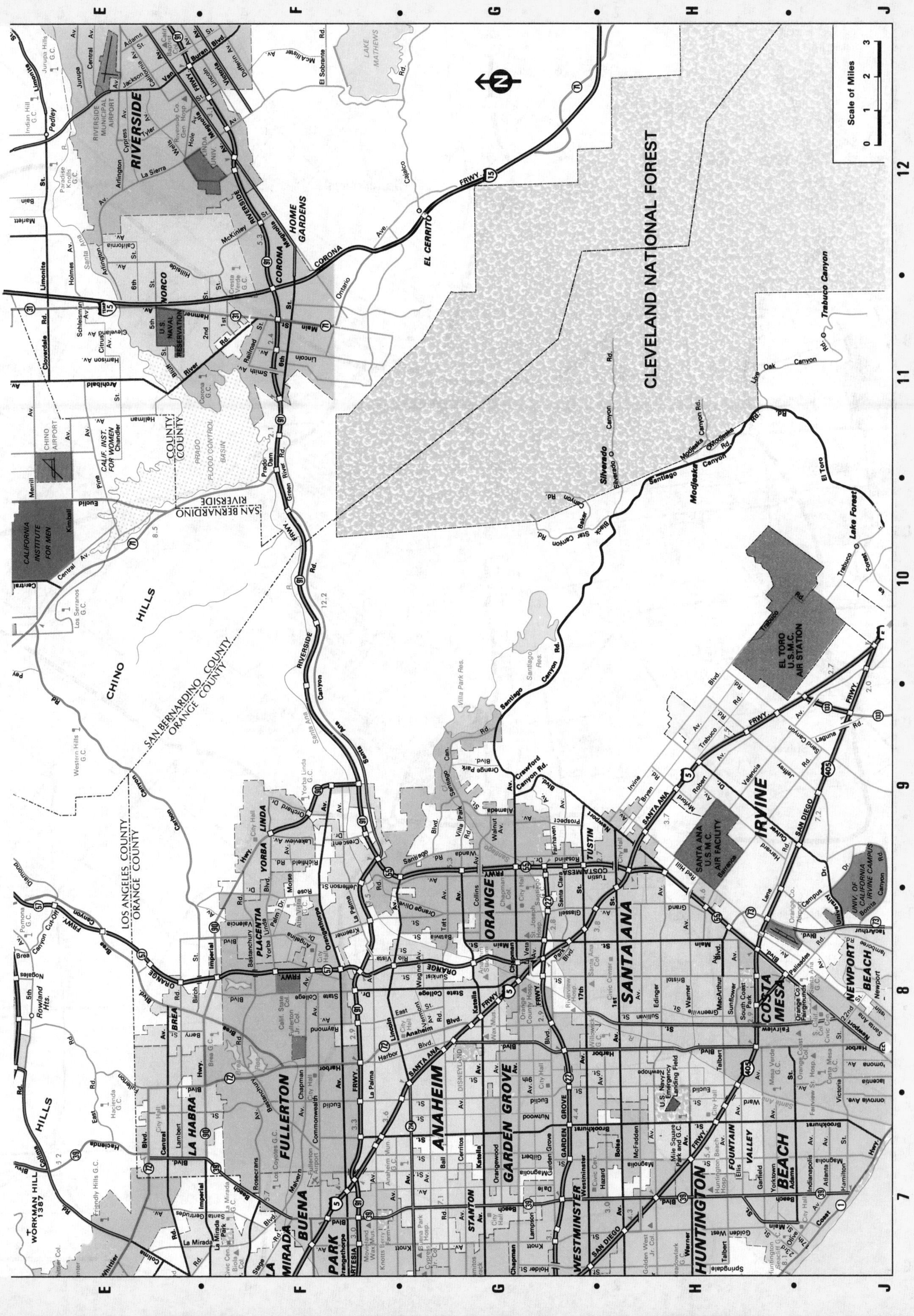

CLEVELAND NATIONAL FOREST

RIVERSIDE

CORONA

NORCO

HOME GARDENS

EL CERRITO

CHINO HILLS

ANAHEIM

FULLERTON

LA HABRA

BREA

PLACENTIA

YORBA LINDA

ORANGE

SANTA ANA

TUSTIN

IRVINE

COSTA MESA

NEWPORT BEACH

HUNTINGTON BEACH

FOUNTAIN VALLEY

WESTMINSTER

GARDEN GROVE

STANTON

BUENA PARK

LA MIRADA

SAN BERNARDINO COUNTY / ORANGE COUNTY

LOS ANGELES COUNTY / ORANGE COUNTY

SAN BERNARDINO / RIVERSIDE

EL TORO U.S.M.C. AIR STATION

SANTA ANA U.S.M.C. AIR FACILITY

CALIFORNIA INSTITUTE FOR MEN

U.S. NAVAL RESERVATION

UNIV. OF CALIFORNIA IRVINE CAMPUS

Scale of Miles 0 1 2 3

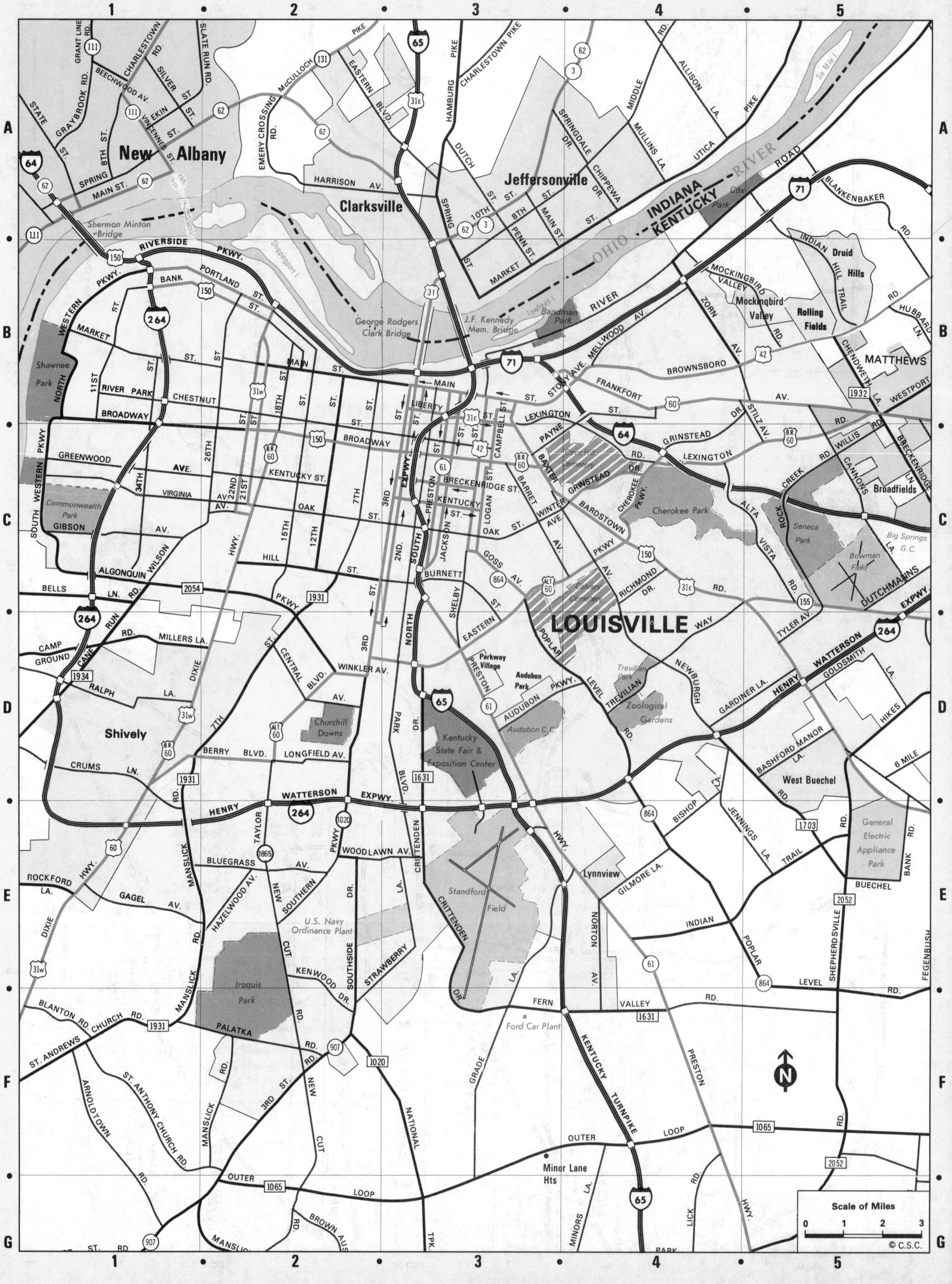
New Albany
Clarksville
Jeffersonville
INDIANA
KENTUCKY
OHIO RIVER
Sherman Minton Bridge
Shippingport I.
George Rodgers Clark Bridge
J.F. Kennedy Mem. Bridge
Towhead I.
Bandman Park
Cox Park
Six Mile I.
Shawnee Park
Commonwealth Park
Druid Hills
Mockingbird Valley
Rolling Fields
MATTHEWS
Broadfields
Cave Hill Cemetery
Cherokee Park
Seneca Park
Big Springs G.C.
Bowman Field
Calvary Cemetery
LOUISVILLE
Parkway Village
Audubon Park
Audubon C.C.
Trevilian Park
Zoological Gardens
Churchill Downs
Kentucky State Fair & Exposition Center
Shively
West Buechel
General Electric Appliance Park
Standford Field
Lynnview
U.S. Navy Ordinance Plant
Iroquis Park
Ford Car Plant
Minor Lane Hts
RIVERSIDE PKWY.
PORTLAND
BANK
MAIN
MARKET
BROADWAY
CHESTNUT
RIVER PARK
GREENWOOD AVE.
VIRGINIA AV.
GIBSON AV.
ALGONQUIN PKWY
BELLS LN.
CAMP GROUND RD.
MILLERS LA.
RALPH LA.
CRUMS LN.
BERRY BLVD.
LONGFIELD AV.
HENRY WATTERSON EXPWY.
WINKLER AV.
CENTRAL AV.
WOODLAWN AV.
BLUEGRASS AV.
ROCKFORD LA.
GAGEL AV.
DIXIE HWY.
MANSLICK RD.
KENWOOD DR.
STRAWBERRY LA.
SOUTHSIDE DR.
NEW CUT RD.
NATIONAL TPK.
CRITTENDEN DR.
GRADE LA.
FERN VALLEY RD.
KENTUCKY TURNPIKE
PRESTON HWY.
OUTER LOOP
BLANTON RD.
ST. ANDREWS CHURCH RD.
ST. ANTHONY CHURCH RD.
ARNOLDTOWN RD.
BROWN AUS
MINORS LA.
LICK RD.
PARK
NORTON AV.
GILMORE LA.
INDIAN TRAIL
POPLAR LEVEL RD.
BISHOP LA.
JENNINGS LA.
BASHFORD MANOR LA.
HIKES LA.
6 MILE LA.
BUECHEL BANK RD.
SHEPHERDSVILLE RD.
FEGENBUSH
GOLDSMITH LA.
GARDINER LA.
NEWBURGH RD.
TYLER AV.
DUTCHMANNS LA.
RICHMOND DR.
EASTERN PKWY
GOSS AV.
BARDSTOWN RD.
BAXTER AV.
BARRET AV.
WINTER AVE
CHEROKEE PKWY.
GRINSTEAD DR.
LEXINGTON RD.
FRANKFORT AV.
BROWNSBORO RD.
MELLWOOD AV.
RIVER RD.
STONY AVE.
ZORN AV.
STILZ AV.
ALTA VISTA RD.
ROCK CREEK DR.
CANNONS LA.
WILLIS
BRECKENRIDGE LN.
CHENOWETH LA.
WESTPORT RD.
HUBBARDS LN.
BLANKENBAKER RD.
INDIAN HILL TRAIL
MOCKINGBIRD VALLEY RD.
HAMBURG PIKE
CHARLESTOWN PIKE
UTICA PIKE
ALLISON LA.
MIDDLE RD.
MULLINS LA.
SPRINGDALE DR.
CHIPPEWA DR.
DUTCH LA.
SPRING ST.
MAIN ST.
PENN ST.
MARKET ST.
HARRISON AV.
EMERY CROSSING RD.
McCULLOCH PIKE
EASTERN BLVD.
SLATE RUN RD.
CHARLESTOWN RD.
GRANT LINE RD.
GRAYBROOK RD.
STATE ST.
BEECHWOOD AV.
VINCENNES ST.
EKIN
SILVER ST.
SPRING ST.
MAIN ST.
WESTERN PKWY
SOUTH WESTERN PKWY
NORTH
WILSON RD.
CANE RUN RD.
TAYLOR BLVD.
HAZELWOOD AV.
SOUTHERN PKWY
PALATKA RD.
3RD ST.
4TH
7TH ST.
12TH
15TH
18TH
22ND
21ST
26TH
34TH
11ST
2ND
3RD
EXPWY.
SOUTH
PRESTON
JACKSON
LOGAN
SHELBY
CAMPBELL ST
BRECKENRIDGE ST.
KENTUCKY
OAK ST.
HILL ST.
BURNETT
LIBERTY
KENTUCKY ST.
WINKLER AV.
BLVD.
EASTERN ST.
POPLAR LEVEL
AUDUBON PKWY.
TREVILIAN WAY
Scale of Miles
0 1 2 3
© C.S.C.
N

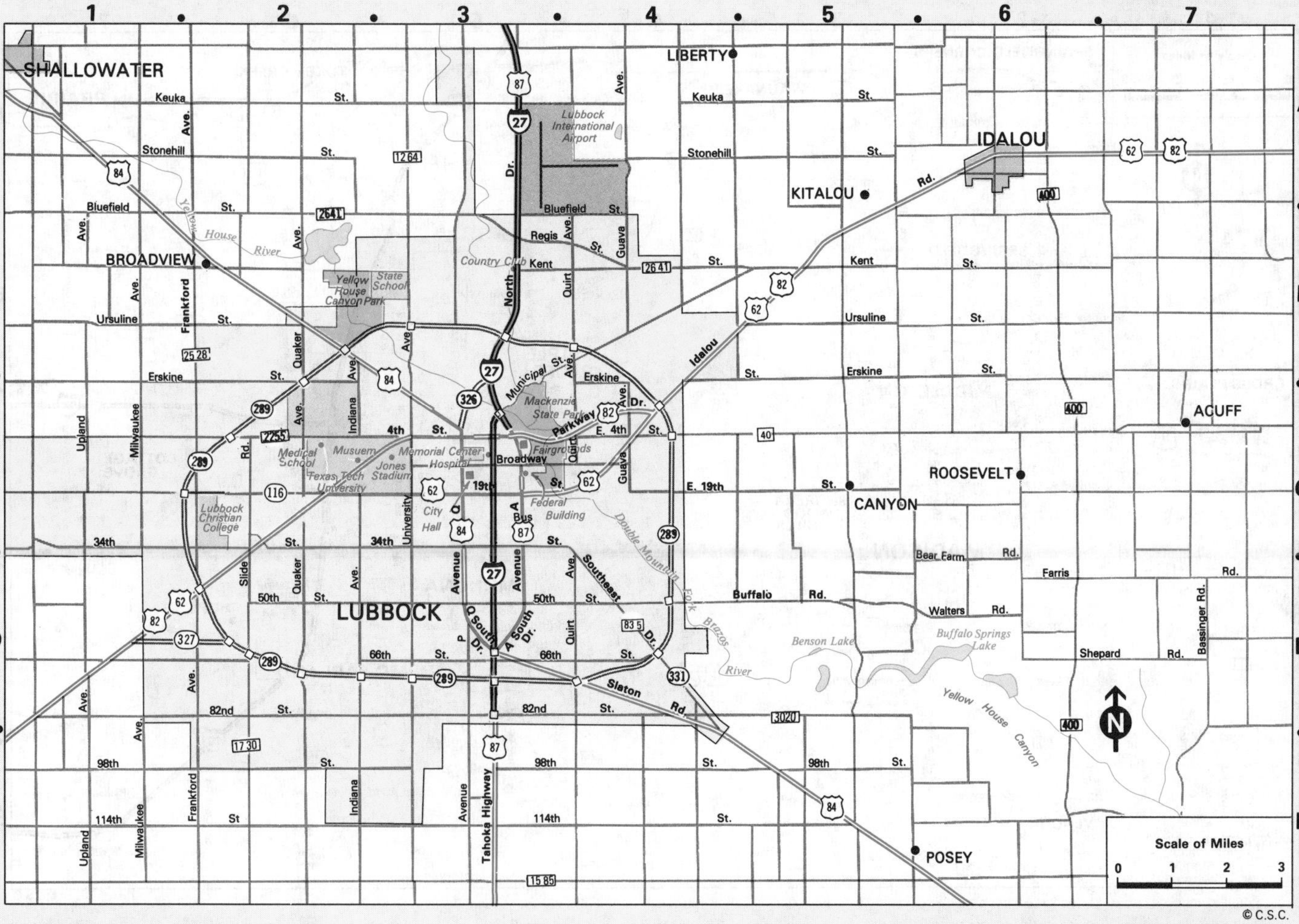

Lubbock, TX (map)

Scale of Miles 0 1 2 3

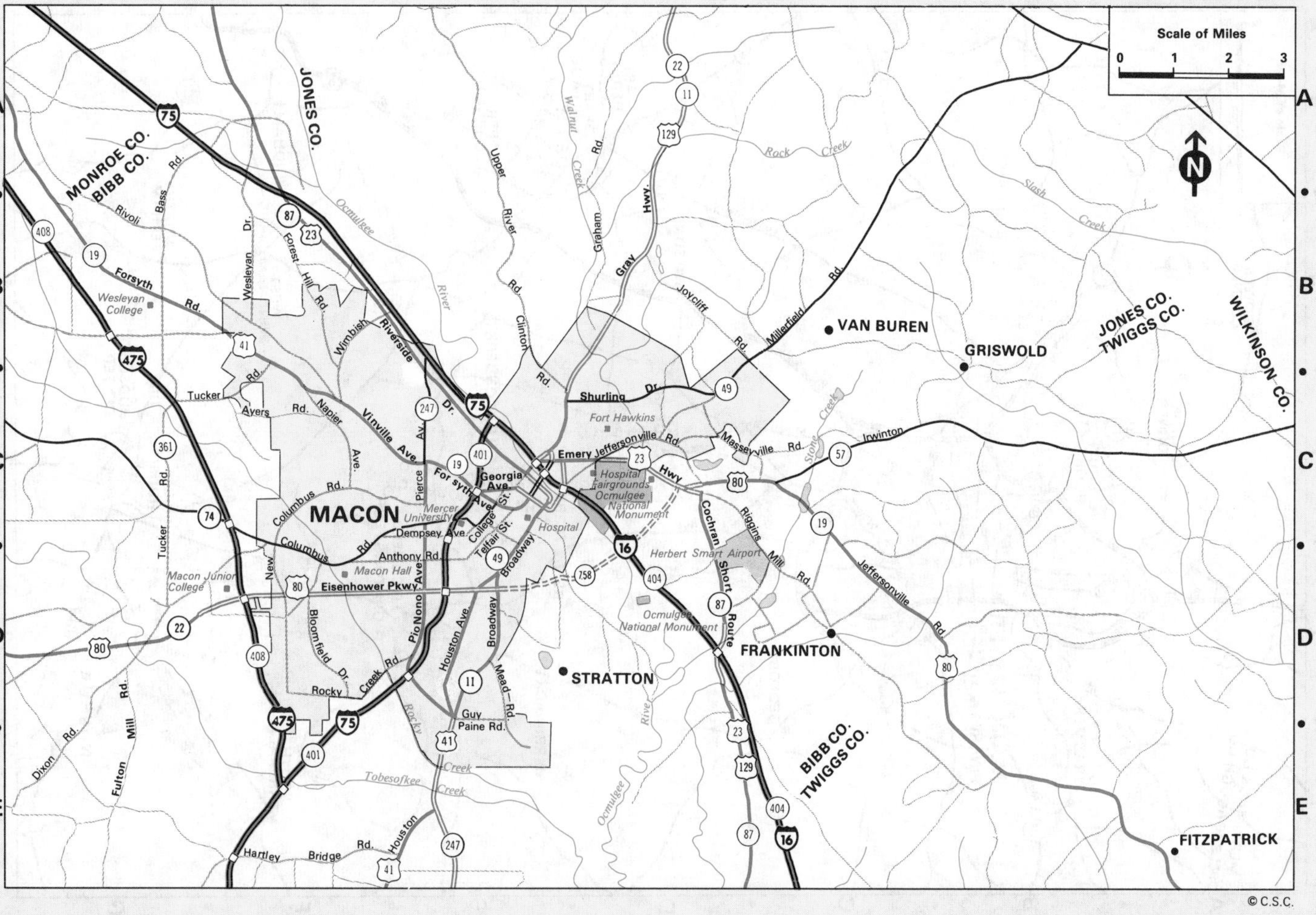

© C.S.C.

Macon, GA (map)

Scale of Miles 0 1 2 3

© C.S.C.

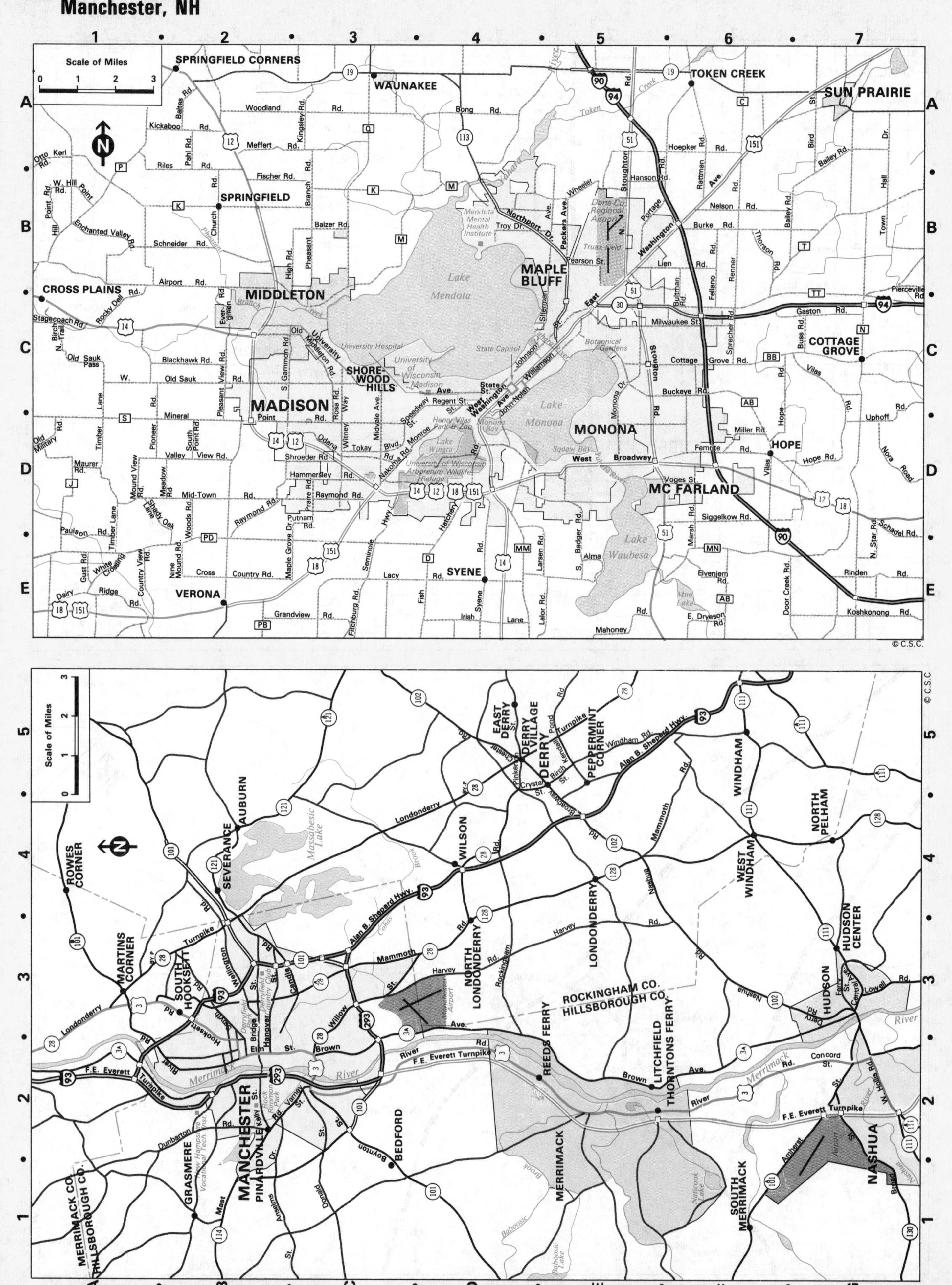
SPRINGFIELD CORNERS
WAUNAKEE
TOKEN CREEK
SUN PRAIRIE
SPRINGFIELD
MAPLE BLUFF
CROSS PLAINS
MIDDLETON
Lake Mendota
COTTAGE GROVE
SHOREWOOD HILLS
MADISON
Lake Monona
MONONA
HOPE
MC FARLAND
Lake Waubesa
SYENE
VERONA
Scale of Miles
©C.S.C.
AUBURN
ROWES CORNER
SEVERANCE
Massabesic Lake
WILSON
DERRY VILLAGE
EAST DERRY
DERRY
PEPPERMINT CORNER
WINDHAM
NORTH PELHAM
WEST WINDHAM
MARTINS CORNER
SOUTH HOOKSETT
NORTH LONDONDERRY
LONDONDERRY
HUDSON CENTER
HUDSON
ROCKINGHAM CO.
HILLSBOROUGH CO.
REEDS FERRY
LITCHFIELD
THORNTONS FERRY
MANCHESTER
PINARDVILLE
BEDFORD
GRASMERE
MERRIMACK
SOUTH MERRIMACK
NASHUA
MERRIMACK CO.
HILLSBOROUGH CO.

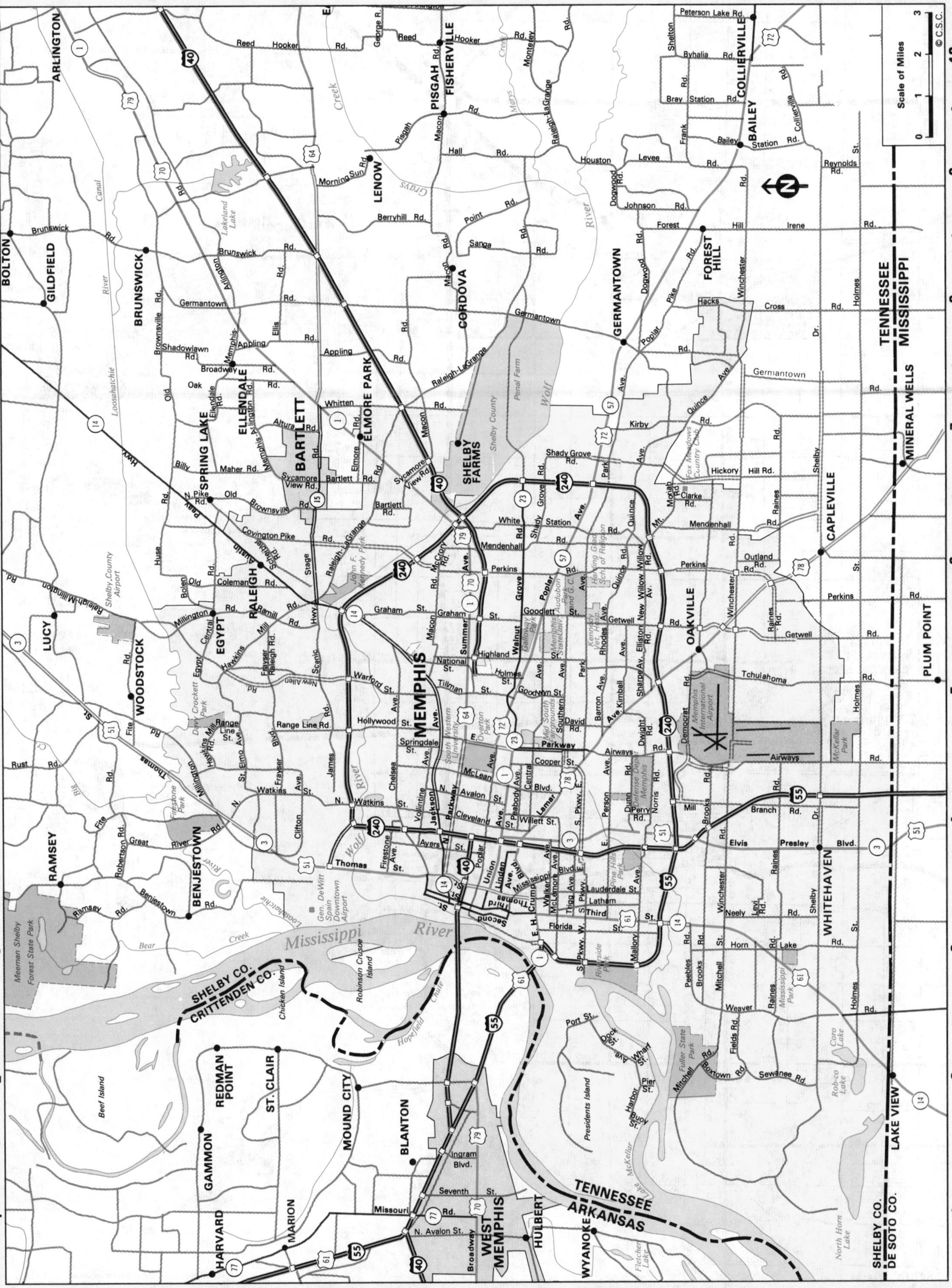

ARLINGTON
GILDFIELD
BOLTON
BRUNSWICK
LUCY
WOODSTOCK
RAMSEY
BENJESTOWN
SPRING LAKE
ELLENDALE
BARTLETT
ELMORE PARK
RALEIGH
EGYPT
LENOW
PISGAH
FISHERVILLE
CORDOVA
SHELBY FARMS
MEMPHIS
GERMANTOWN
FOREST HILL
COLLIERVILLE
BAILEY
OAKVILLE
CAPLEVILLE
MINERAL WELLS
PLUM POINT
WHITEHAVEN
LAKE VIEW
TENNESSEE
MISSISSIPPI
SHELBY CO.
DE SOTO CO.
CRITTENDEN CO.
ARKANSAS
GAMMON
REDMAN POINT
ST. CLAIR
MOUND CITY
BLANTON
MARION
HARVARD
WEST MEMPHIS
HULBERT
WYANOKE
Mississippi River
Meeman Shelby Forest State Park
Shelby County Airport
Memphis International Airport
Gen. DeWitt Spain Downtown Airport
Scale of Miles
0 1 2 3
© C.S.C.

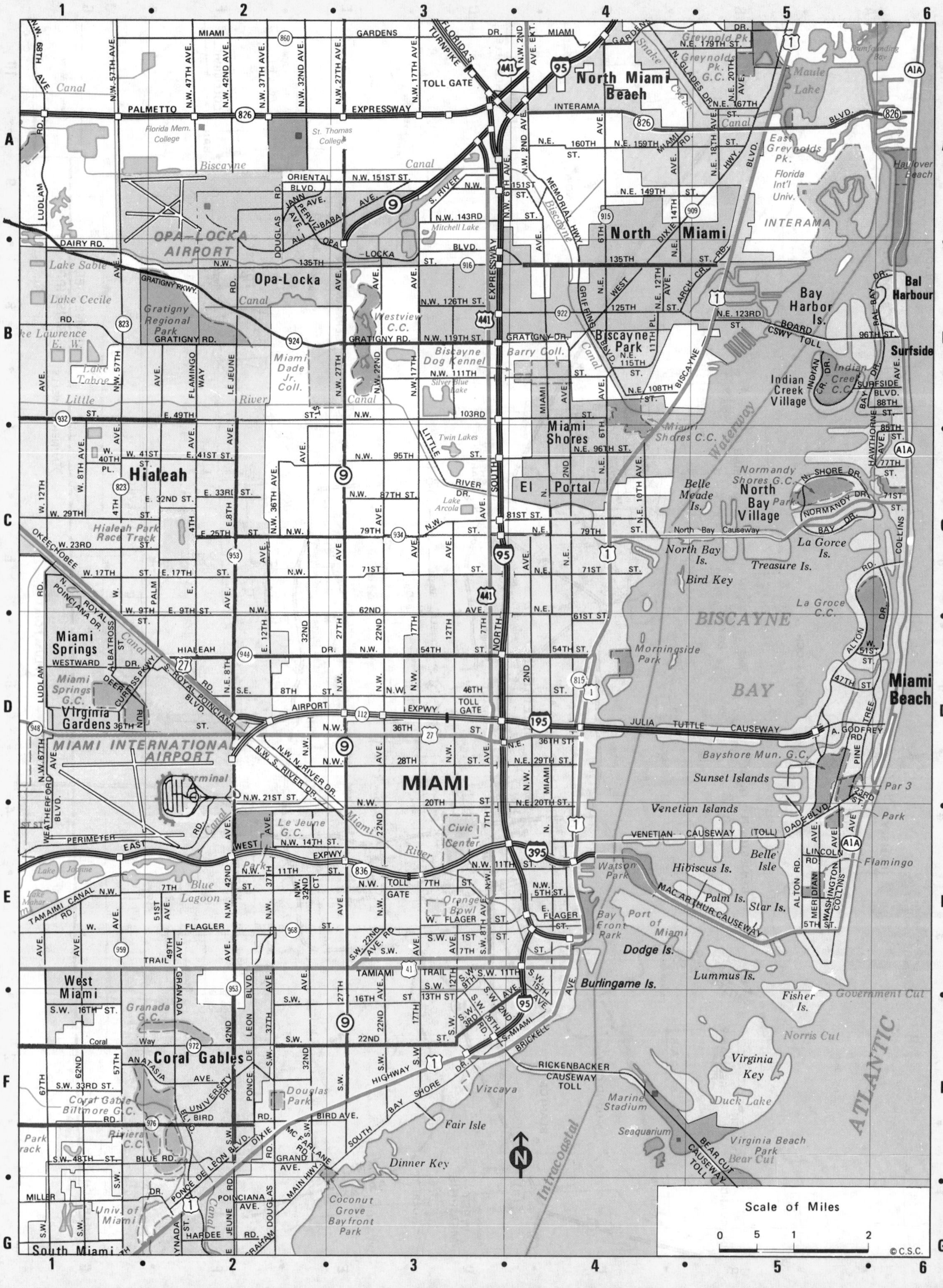

North Miami Beach
North Miami
Opa-Locka
Opa-Locka Airport
Hialeah
Hialeah Park Race Track
Miami Springs
Miami Springs G.C.
Virginia Gardens
Miami International Airport
Terminal
West Miami
Coral Gables
Coral Gables Biltmore G.C.
Riviera C.C.
Granada G.C.
Univ. of Miami
South Miami
Miami Shores
Miami Shores C.C.
El Portal
Biscayne Park
MIAMI
Civic Center
Orange Bowl
Bay Front Park
Port of Miami
Dodge Is.
Lummus Is.
Burlingame Is.
Watson Park
Hibiscus Is.
Palm Is.
Star Is.
Belle Isle
Venetian Islands
Sunset Islands
Bayshore Mun. G.C.
Miami Beach
La Gorce Is.
La Gorce C.C.
Treasure Is.
North Bay Is.
Bird Key
Morningside Park
Normandy Shores G.C.
North Bay Village
Belle Meade Is.
Indian Creek Village
Indian Creek C.C.
Bay Harbor Is.
Bal Harbour
Surfside
Haulover Beach
Florida Int'l Univ.
Interama
Greynolds Pk.
Greynolds Pk. G.C.
East Greynolds Pk.
Maule Lake
Dumfoundling Bay
Florida Mem. College
St. Thomas College
Mitchell Lake
Westview C.C.
Miami Dade Jr. Coll.
Biscayne Dog Kennel
Barry Coll.
Silver Blue Lake
Twin Lakes
Lake Arcola
Le Jeune G.C.
Blue Lagoon
Lake Sable
Lake Cecile
Lake Lawrence
Lake Tahoe
Gratigny Regional Park
Vizcaya
Fair Isle
Dinner Key
Coconut Grove Bayfront Park
Douglas Park
Marine Stadium
Seaquarium
Virginia Key
Duck Lake
Virginia Beach Park
Bear Cut
Norris Cut
Fisher Is.
Government Cut
Flamingo
Par 3
Park
BISCAYNE BAY
ATLANTIC
Intracoastal Waterway
Little River
Miami River
Biscayne Canal
Tamiami Canal
Palmetto Expressway
Airport Expwy
East West Expwy
Julia Tuttle Causeway
Venetian Causeway (Toll)
MacArthur Causeway
Rickenbacker Causeway Toll
Bear Cut Causeway Toll
Broad Cswy Toll
North Bay Causeway
Florida's Turnpike
Toll Gate
North South Expressway

Scale of Miles
0 .5 1 2

©C.S.C.

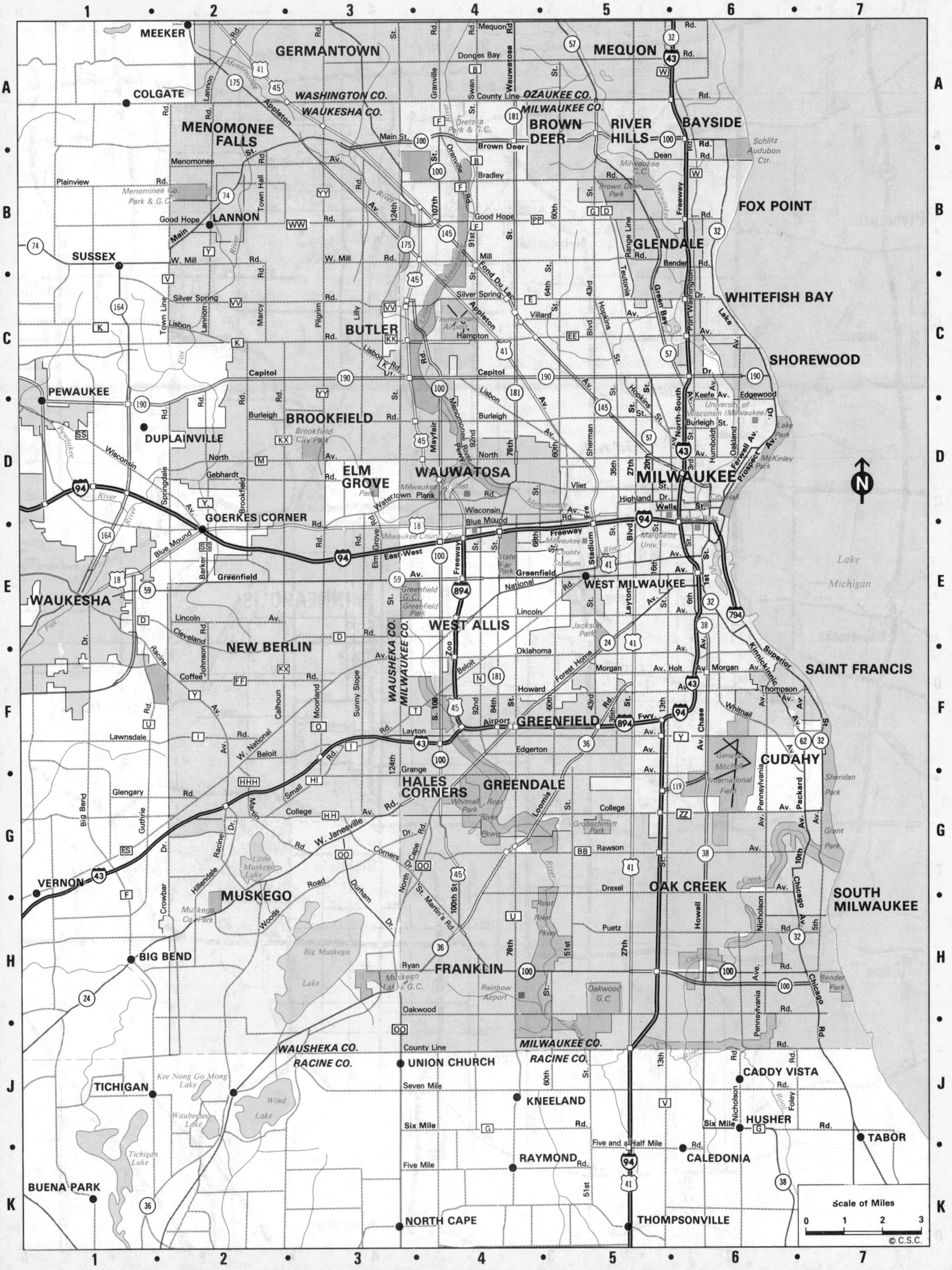
MEEKER
GERMANTOWN
MEQUON
COLGATE
WASHINGTON CO.
WAUKESHA CO.
OZAUKEE CO.
MILWAUKEE CO.
MENOMONEE FALLS
BROWN DEER
RIVER HILLS
BAYSIDE
Schlitz Audubon Ctr.
LANNON
FOX POINT
SUSSEX
GLENDALE
WHITEFISH BAY
BUTLER
SHOREWOOD
PEWAUKEE
BROOKFIELD
Brookfield City Park
DUPLAINVILLE
ELM GROVE
WAUWATOSA
MILWAUKEE
GOERKES CORNER
WAUKESHA
WEST MILWAUKEE
WEST ALLIS
NEW BERLIN
Lake Michigan
SAINT FRANCIS
GREENFIELD
CUDAHY
HALES CORNERS
GREENDALE
General Mitchell International Field
VERNON
MUSKEGO
OAK CREEK
SOUTH MILWAUKEE
BIG BEND
Big Muskego Lake
FRANKLIN
WAUSHEKA CO.
RACINE CO.
UNION CHURCH
CADDY VISTA
TICHIGAN
KNEELAND
HUSHER
TABOR
RAYMOND
CALEDONIA
BUENA PARK
NORTH CAPE
THOMPSONVILLE
Scale of Miles
0 1 2 3
© C.S.C.

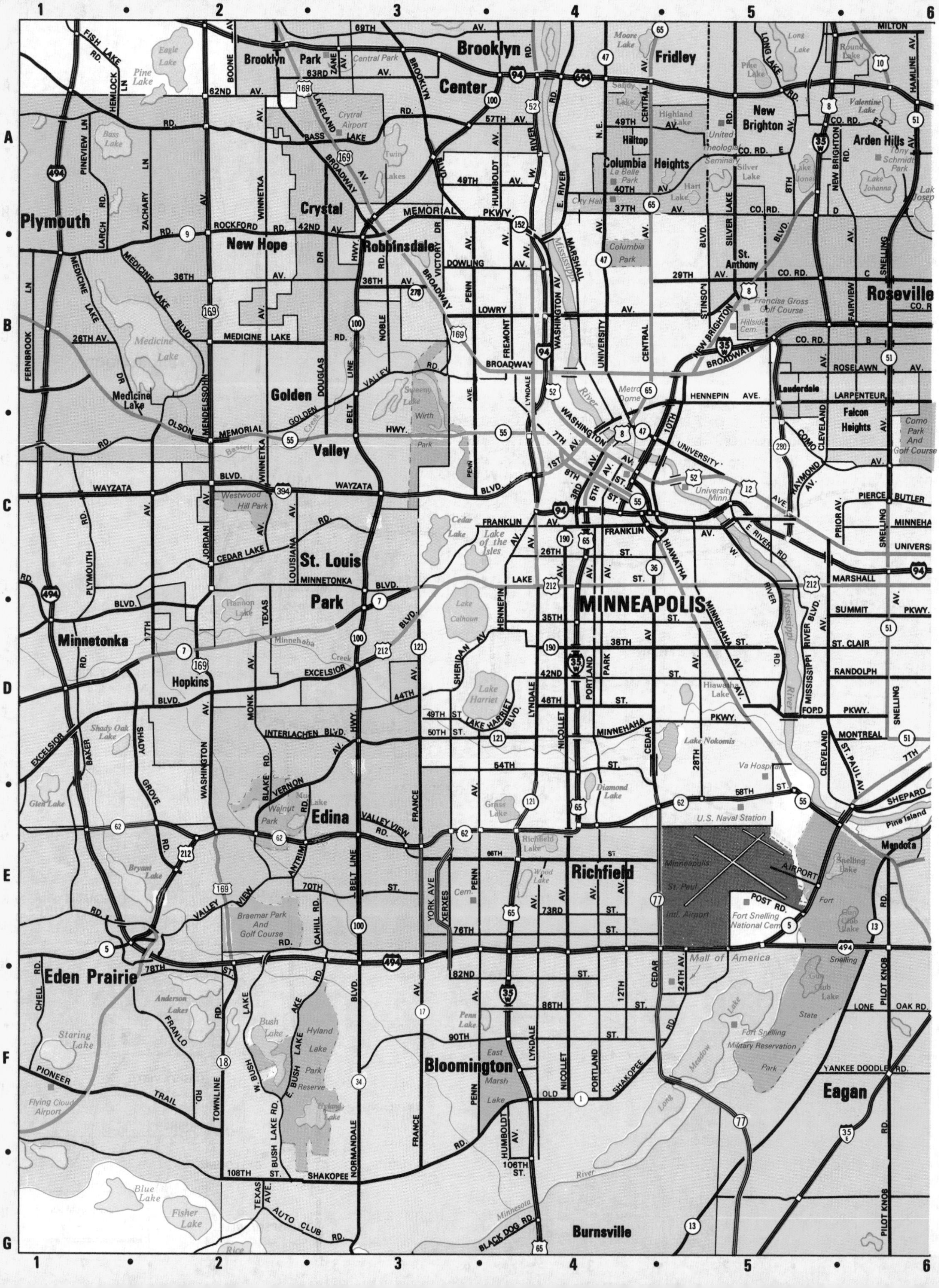

1 2 3 4 5 6

A B C D E F G

Brooklyn Park
Brooklyn Center
Fridley
New Brighton
Arden Hills
Crystal
Columbia Heights
Hilltop
Plymouth
New Hope
Robbinsdale
St. Anthony
Roseville
Medicine Lake
Golden Valley
Lauderdale
Falcon Heights
St. Louis Park
MINNEAPOLIS
Minnetonka
Hopkins
Edina
Richfield
Eden Prairie
Bloomington
Mendota
Eagan
Burnsville

Eagle Lake
Pine Lake
Bass Lake
Crytral Airport
Twin Lakes
Moore Lake
Long Lake
Round Lake
Pike Lake
Sandy Lake
Highland Lake
Valentine Lake
United Theological Seminary
Silver Lake
La Belle Park
Hart Lake
Lake Jones
Tony Schmidt Park
Lake Johanna
Lake Josephine
City Hall
Columbia Park
Francisa Gross Golf Course
Hillside Cem.
Metro Dome
Sweeny Lake
Wirth Park
Bassett Creek
Como Park And Golf Course
University of Minn.
Westwood Hill Park
Cedar Lake
Lake of the Isles
Lake Calhoun
Hannon Lake
Minnehaba Creek
Lake Harriet
Hiawatha Lake
Lake Nokomis
Va Hospital
Diamond Lake
Grass Lake
Shady Oak Lake
Glen Lake
Mud Lake
Walnut Park
Bryant Lake
Richfield Lake
Wood Lake
Cem.
Braemar Park And Golf Course
Minneapolis St. Paul Intl. Airport
U.S. Naval Station
Fort Snelling National Cem.
Snelling Lake
Gun Club Lake
Fort Snelling State Park
Fort Snelling Military Reservation
Mall of America
Long Meadow Lake
Anderson Lakes
Staring Lake
Flying Cloud Airport
Bush Lake
Hyland Lake Park Reserve
Penn Lake
East Marsh Lake
Blue Lake
Fisher Lake
Rice Lake
Mississippi River
Minnesota River
Pine Island

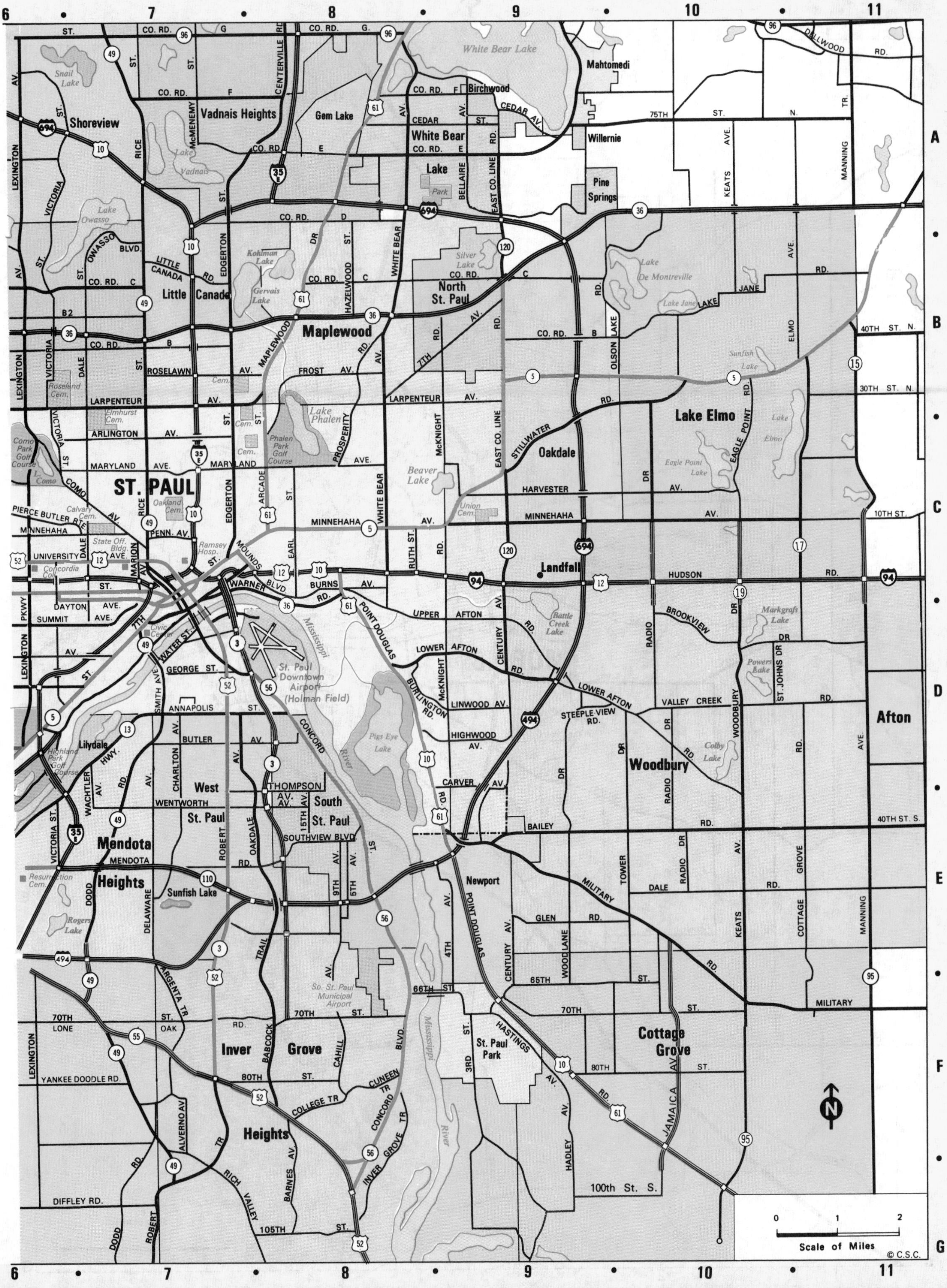

6 • 7 • 8 • 9 • 10 • 11

A • B • C • D • E • F • G

Shoreview
Vadnais Heights
Gem Lake
White Bear Lake
Birchwood
Mahtomedi
Willernie
Pine Springs
Little Canada
North St. Paul
Maplewood
Lake Elmo
Oakdale
ST. PAUL
Landfall
Woodbury
Afton
Lilydale
West St. Paul
South St. Paul
Mendota Heights
Sunfish Lake
Newport
St. Paul Park
Cottage Grove
Inver Grove Heights

White Bear Lake
Snail Lake
Lake Vadnais
Lake Owasso
Kohlman Lake
Gervais Lake
Silver Lake
Lake Phalen
Beaver Lake
Lake De Montreville
Lake Jane
Sunfish Lake
Lake Elmo
Eagle Point Lake
Battle Creek Lake
Markgrafs Lake
Powers Lake
Colby Lake
Pigs Eye Lake
Rogers Lake
L. Como
Mississippi River

Roseland Cem.
Elmhurst Cem.
Como Park Golf Course
Phalen Park Golf Course
Oakland Cem.
Calvary Cem.
Union Cem.
State Off. Bldg.
Ramsey Hosp.
Concordia Coll.
Civic Center
St. Paul Downtown Airport (Holman Field)
Highland Park Golf Course
Resurrection Cem.
So. St. Paul Municipal Airport

LEXINGTON AV.
VICTORIA ST.
RICE ST.
EDGERTON ST.
CENTERVILLE RD.
WHITE BEAR AV.
McKNIGHT RD.
CENTURY AV.
EAST CO. LINE RD.
KEATS AV.
MANNING AV.
CO. RD. G
CO. RD. F
CO. RD. E
CO. RD. D
CO. RD. C
CO. RD. B
75TH ST. N.
40TH ST. N.
30TH ST. N.
LARPENTEUR AV.
MARYLAND AVE.
MINNEHAHA AV.
UNIVERSITY AVE.
HUDSON RD.
UPPER AFTON RD.
LOWER AFTON RD.
STILLWATER RD.
BAILEY RD.
MILITARY RD.
70TH ST.
80TH ST.
100th St. S.
105TH ST.
DIFFLEY RD.
YANKEE DOODLE RD.
ROBERT ST.
DODD RD.
CONCORD ST.
POINT DOUGLAS RD.
HASTINGS AV.
JAMAICA AV.
RICH VALLEY BLVD.

0 1 2
Scale of Miles

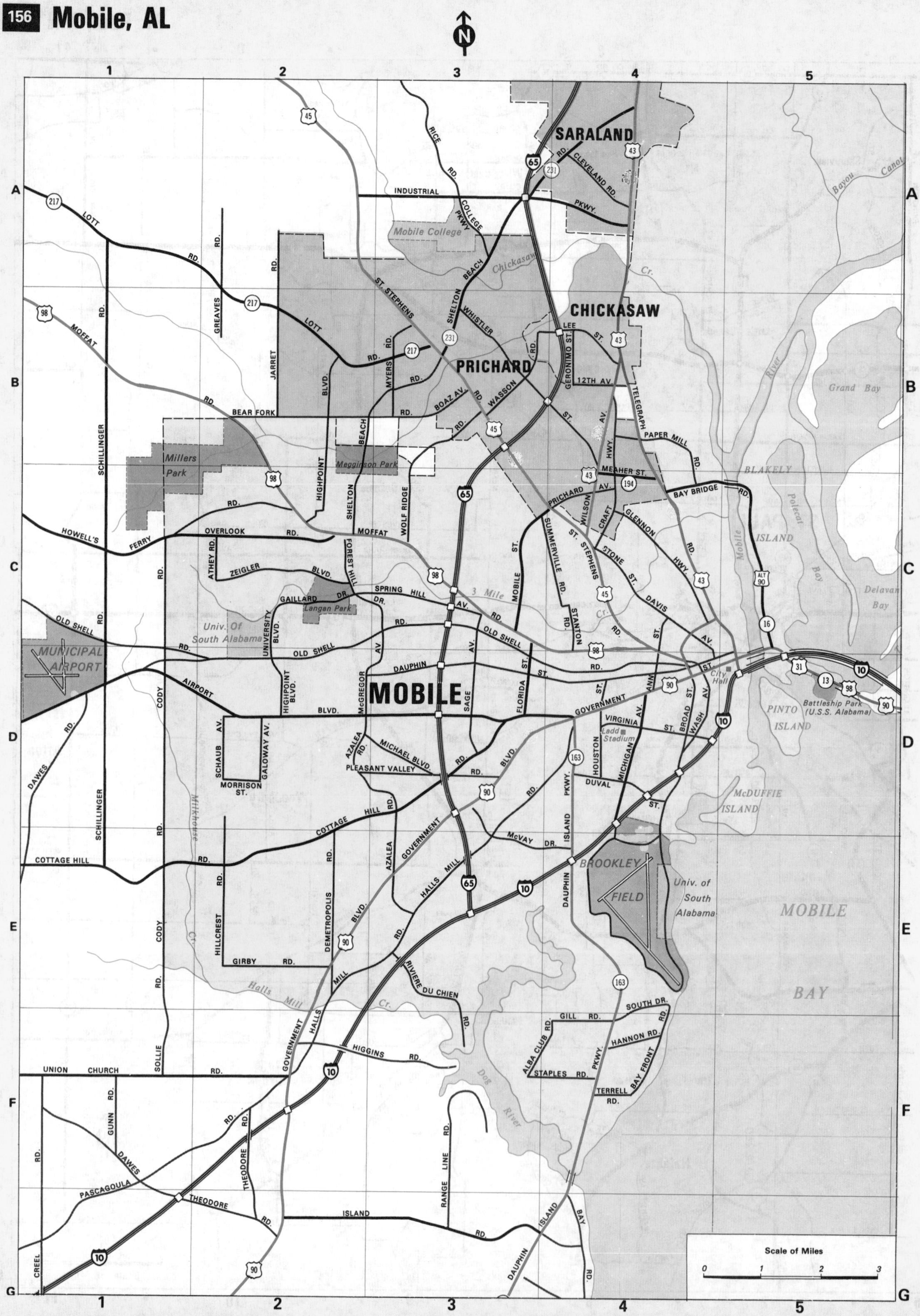

Mobile, AL

SARALAND
CHICKASAW
PRICHARD
MOBILE
Mobile College
Millers Park
Megginson Park
Langan Park
Univ. Of South Alabama
MUNICIPAL AIRPORT
City Hall
Ladd Stadium
BROOKLEY FIELD
Univ. of South Alabama
Battleship Park (U.S.S. Alabama)
BLAKELY ISLAND
PINTO ISLAND
McDUFFIE ISLAND
Grand Bay
Polecat Bay
Delavan Bay
MOBILE BAY
Mobile River
Dog River
Halls Mill Cr.
Chickasaw Cr.
3 Mile Cr.
Milkhouse Cr.
Bayou Canot

Scale of Miles
0 1 2 3

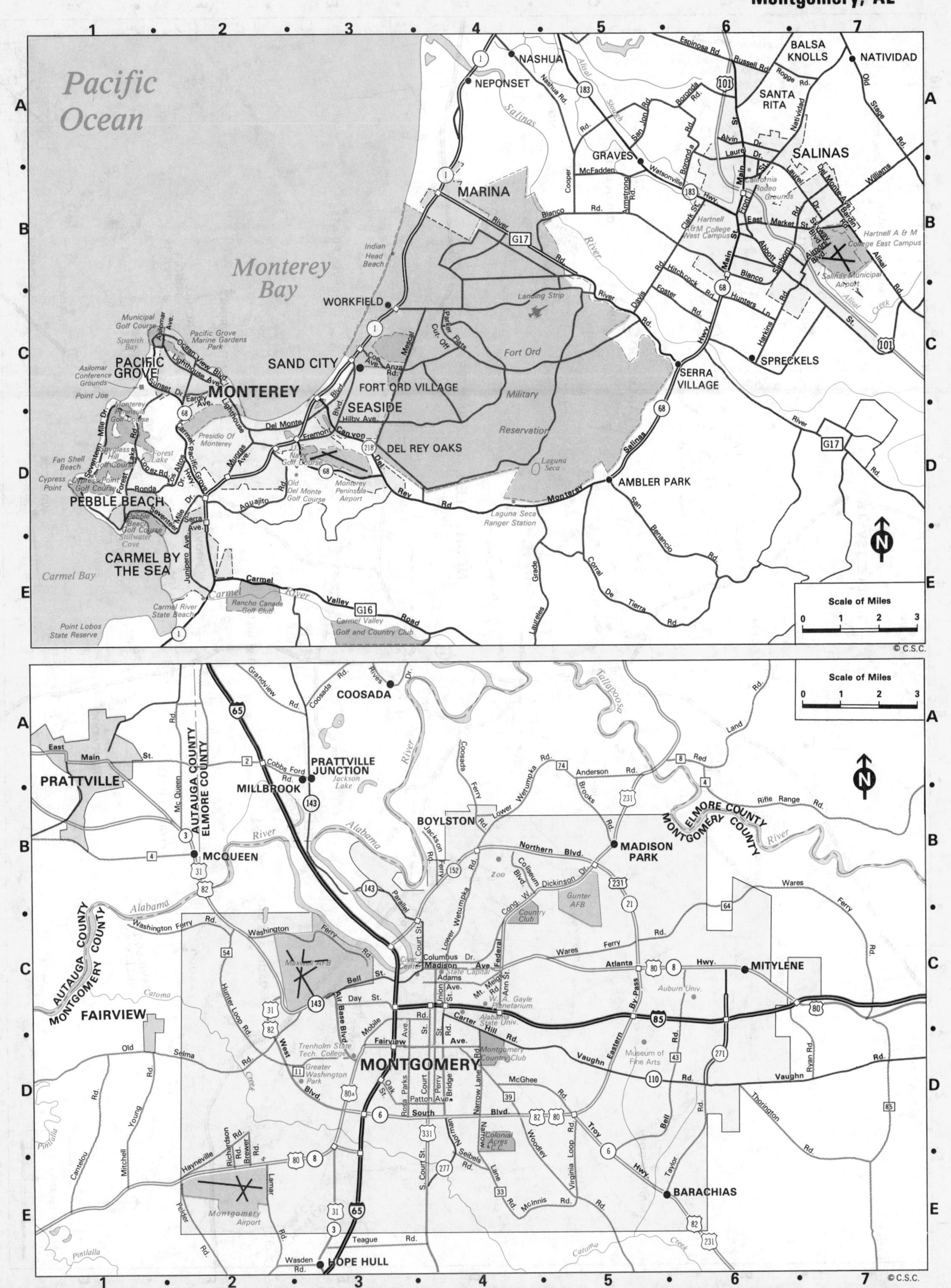
Pacific Ocean
Monterey Bay
MARINA
SALINAS
MONTEREY
PACIFIC GROVE
SAND CITY
SEASIDE
FORT ORD VILLAGE
DEL REY OAKS
PEBBLE BEACH
CARMEL BY THE SEA
NASHUA
NEPONSET
GRAVES
BALSA KNOLLS
SANTA RITA
NATIVIDAD
WORKFIELD
SPRECKELS
SERRA VILLAGE
AMBLER PARK
Fort Ord Military Reservation
Carmel Bay
Scale of Miles
© C.S.C.
PRATTVILLE
PRATTVILLE JUNCTION
MILLBROOK
COOSADA
BOYLSTON
MADISON PARK
MCQUEEN
FAIRVIEW
MONTGOMERY
MITYLENE
BARACHIAS
HOPE HULL
AUTAUGA COUNTY
ELMORE COUNTY
MONTGOMERY COUNTY
Maxwell AFB
Gunter AFB
Montgomery Airport
Scale of Miles
© C.S.C.

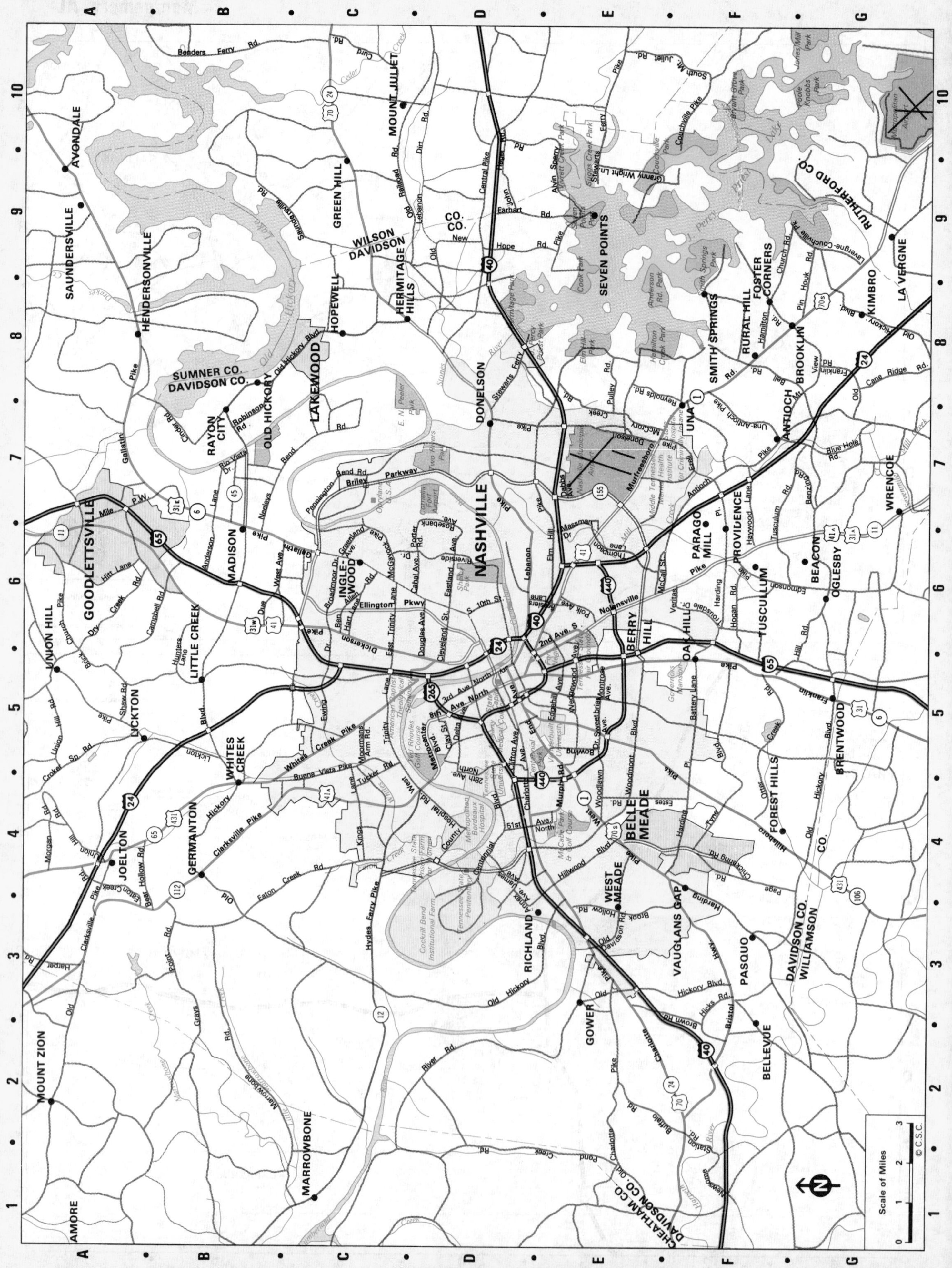
NASHVILLE
AMORE
MOUNT ZION
MARROWBONE
JOELTON
GERMANTON
WHITES CREEK
LICKTON
LITTLE CREEK
UNION HILL
GOODLETTSVILLE
MADISON
RAYON CITY
OLD HICKORY
LAKEWOOD
HENDERSONVILLE
SAUNDERSVILLE
AVONDALE
HOPEWELL
GREEN HILL
MOUNT JULIET
HERMITAGE HILLS
WILSON CO.
DAVIDSON CO.
SUMNER CO.
DAVIDSON CO.
DONELSON
SEVEN POINTS
INGLEWOOD
BERRY HILL
OAK HILL
BELLE MEADE
WEST MEADE
RICHLAND
VAUGLANS GAP
GOWER
BELLEVUE
PASQUO
FOREST HILLS
BRENTWOOD
TUSCULUM
OGLESBY
BEACON
PROVIDENCE
PARAGO MILL
UNA
ANTIOCH
SMITH SPRINGS
RURAL HILL
FOSTER CORNERS
BROOKLIN
KIMBRO
LA VERGNE
WRENCOE
RUTHERFORD CO.
DAVIDSON CO.
WILLIAMSON CO.
CHEATHAM CO.
DAVIDSON CO.
J. Percy Priest Lake
Old Hickory Lake
Scale of Miles
©C.S.C.

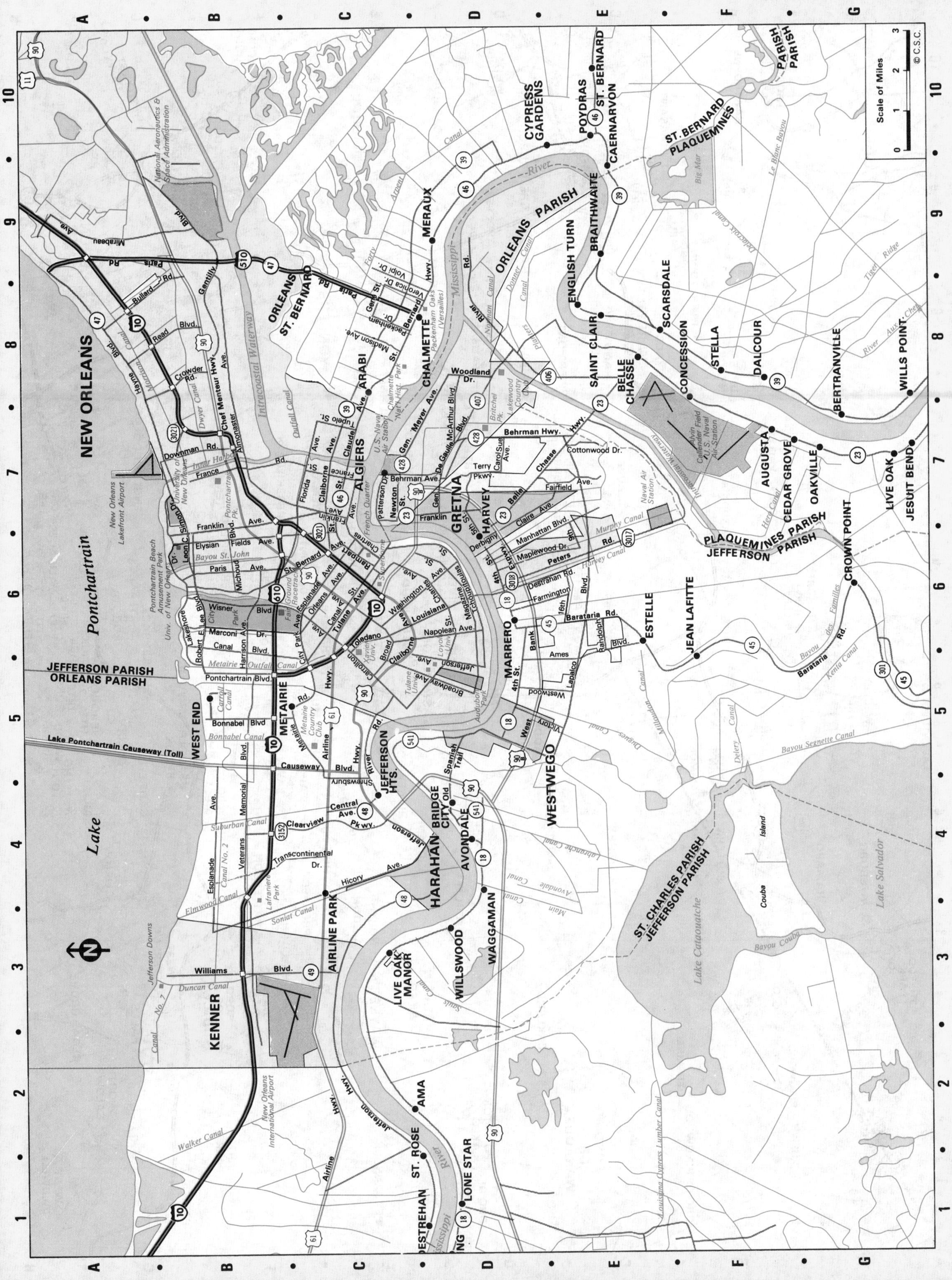
NEW ORLEANS
Lake Pontchartrain
JEFFERSON PARISH
ORLEANS PARISH
WEST END
METAIRIE
KENNER
New Orleans International Airport
New Orleans Lakefront Airport
Lake Pontchartrain Causeway (Toll)
AIRLINE PARK
JEFFERSON HTS.
HARAHAN
BRIDGE CITY
AVONDALE
WAGGAMAN
WILLSWOOD
LIVE OAK MANOR
AMA
ST. ROSE
DESTREHAN
LONE STAR
WESTWEGO
MARRERO
HARVEY
GRETNA
ALGIERS
ARABI
CHALMETTE
MERAUX
ORLEANS
ST. BERNARD
ORLEANS PARISH
CYPRESS GARDENS
POYDRAS
ST. BERNARD
CAERNARVON
ST. BERNARD
PLAQUEMINES
ENGLISH TURN
BRAITHWAITE
SCARSDALE
SAINT CLAIR
BELLE CHASSE
CONCESSION
STELLA
DALCOUR
BERTRANVILLE
WILLS POINT
AUGUSTA
CEDAR GROVE
OAKVILLE
LIVE OAK
JESUIT BEND
CROWN POINT
PLAQUEMINES PARISH
JEFFERSON PARISH
ESTELLE
JEAN LAFITTE
ST. CHARLES PARISH
JEFFERSON PARISH
Lake Cataouatche
Lake Salvador
Mississippi River
Intracoastal Waterway
Scale of Miles
0 1 2 3
© C.S.C.

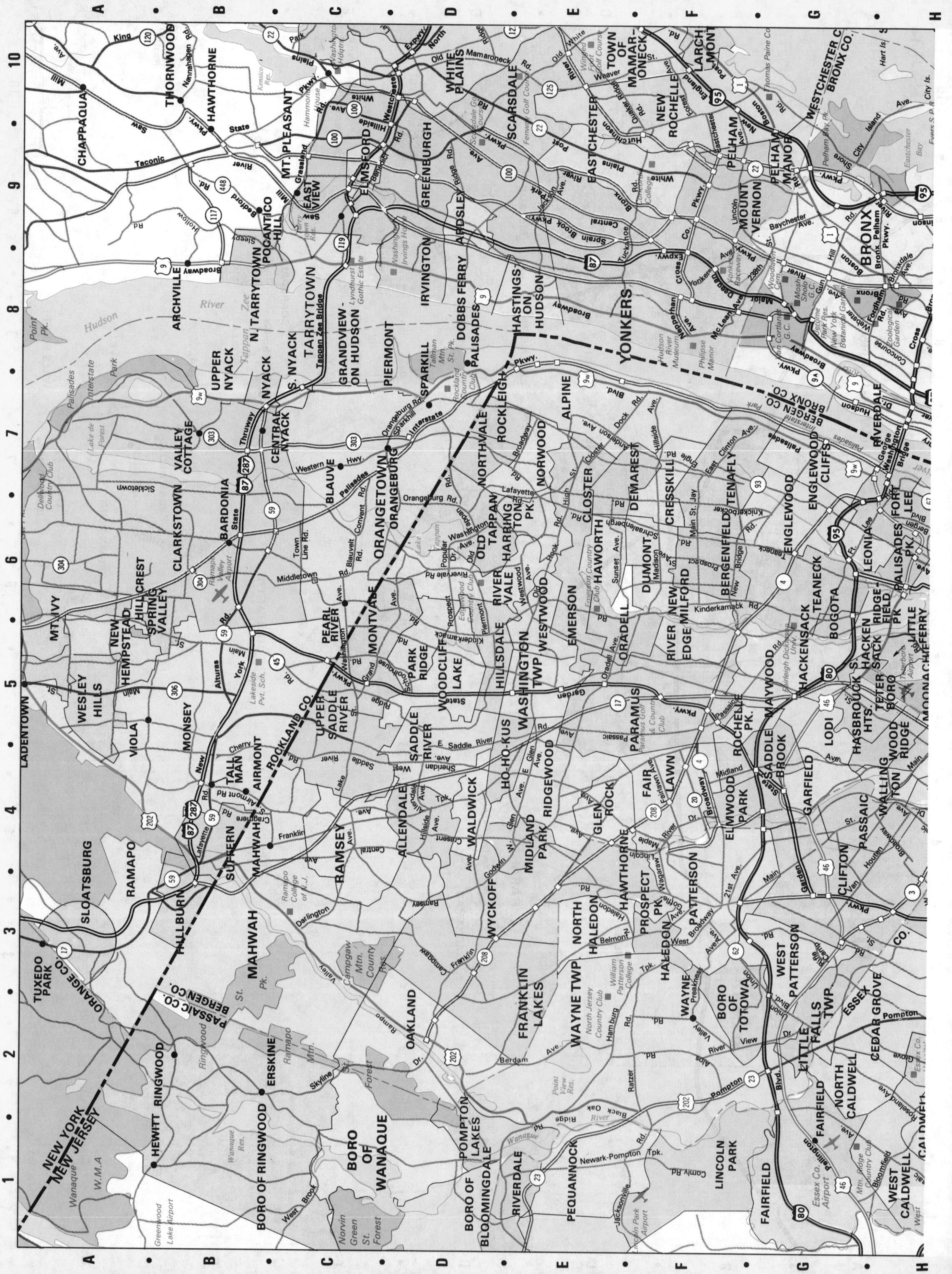

CHAPPAQUA
THORNWOOD
HAWTHORNE
MT. PLEASANT
EAST VIEW
POCANTICO HILL
ARCHVILLE
N. TARRYTOWN
TARRYTOWN
UPPER NYACK
NYACK
S. NYACK
CENTRAL NYACK
VALLEY COTTAGE
GRANDVIEW - ON HUDSON
PIERMONT
SPARKILL
PALISADES
IRVINGTON
DOBBS FERRY
HASTINGS - ON HUDSON
ELMSFORD
GREENBURGH
ARDSLEY
WHITE PLAINS
SCARSDALE
EASTCHESTER
TOWN OF MAMARONECK
NEW ROCHELLE
LARCHMONT
PELHAM
PELHAM MANOR
MOUNT VERNON
YONKERS
BRONX
WESTCHESTER CO.
BRONX CO.
BERGEN CO.
RIVERDALE
ALPINE
ROCKLEIGH
NORTHVALE
NORWOOD
CLOSTER
DEMAREST
CRESSKILL
TENAFLY
ENGLEWOOD CLIFFS
ENGLEWOOD
FORT LEE
LEONIA
PALISADES PK.
TEANECK
BOGOTA
RIDGEFIELD PK.
LITTLE FERRY
HACKENSACK
SOUTH HACKENSACK
TETERBORO
MOONACHIE
HASBROUCK HTS.
WOOD RIDGE
WALLINGTON
LODI
GARFIELD
SADDLE BROOK
ROCHELLE PK.
MAYWOOD
PARAMUS
RIVER EDGE
NEW MILFORD
BERGENFIELD
DUMONT
HAWORTH
EMERSON
ORADELL
WESTWOOD
RIVER VALE
OLD TAPPAN
HARRINGTON PK.
ORANGETOWN
ORANGEBURG
BLAUVELT
PEARL RIVER
MONTVALE
PARK RIDGE
WOODCLIFF LAKE
HILLSDALE
WASHINGTON TWP
UPPER SADDLE RIVER
SADDLE RIVER
HO-HO-KUS
RIDGEWOOD
GLEN ROCK
FAIR LAWN
ELMWOOD PARK
ALLENDALE
WALDWICK
MIDLAND PARK
RAMSEY
MAHWAH
SUFFERN
AIRMONT
TALLMAN
MONSEY
VIOLA
WESLEY HILLS
NEW HEMPSTEAD
SPRING VALLEY
HILLCREST
MT. IVY
CLARKSTOWN
BARDONIA
ROCKLAND CO.
LADENTOWN
SLOATSBURG
RAMAPO
HILLBURN
TUXEDO PARK
ORANGE CO.
PASSAIC CO.
NEW YORK
NEW JERSEY
HEWITT
RINGWOOD
BORO OF RINGWOOD
ERSKINE
BORO OF WANAQUE
OAKLAND
POMPTON LAKES
BORO OF BLOOMINGDALE
RIVERDALE
PEQUANNOCK
FRANKLIN LAKES
WYCKOFF
WAYNE TWP
WAYNE
NORTH HALEDON
HALEDON
HAWTHORNE
PROSPECT PK.
PATTERSON
BORO OF TOTOWA
WEST PATTERSON
LINCOLN PARK
FAIRFIELD
LITTLE FALLS TWP.
CEDAR GROVE
ESSEX CO.
NORTH CALDWELL
WEST CALDWELL
CLIFTON
PASSAIC

MANHATTAN
QUEENS
BROOKLYN
STATEN ISLAND
NEW YORK
JERSEY CITY
NEWARK
ELIZABETH
BAYONNE
HOBOKEN
UNION CITY
WEEHAWKEN
SECAUCUS
KEARNY
HARRISON
IRVINGTON
EAST ORANGE
ORANGE
MAPLEWOOD TWP.
MILBURN
LINDEN
RAHWAY
CARTERET
PERTH AMBOY
WOODBRIDGE
FLUSHING
JAMAICA
ASTORIA
LONG ISLAND CITY
FLATBUSH
CANARSIE
CONEY ISLAND
BAY RIDGE
Atlantic Ocean
Upper Bay
Lower Bay
Jamaica Bay
John F. Kennedy International Airport
Newark International Airport
Statue of Liberty National Monument
Gateway National Recreational Area
Scale of Miles
© C.S.C.

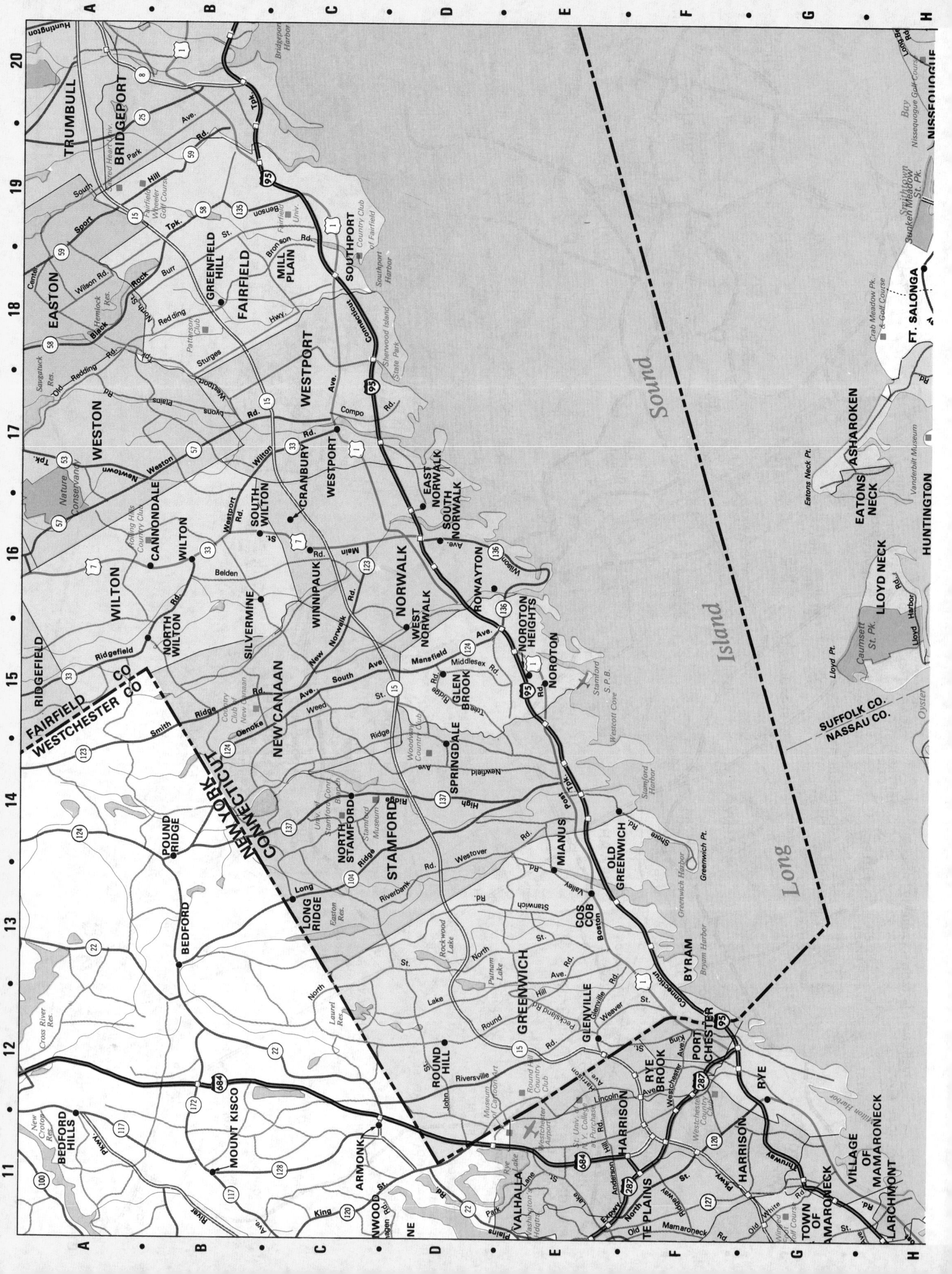

A B C D E F G H

11 12 13 14 15 16 17 18 19 20

TRUMBULL
BRIDGEPORT
EASTON
FAIRFIELD
GREENFIELD HILL
MILL PLAIN
SOUTHPORT
WESTPORT
WESTON
CANNONDALE
WILTON
SOUTH WILTON
NORTH WILTON
RIDGEFIELD
FAIRFIELD CO
WESTCHESTER CO
NEW YORK
CONNECTICUT
SILVERMINE
WINNIPAUK
CRANBURY
EAST NORWALK
SOUTH NORWALK
NORWALK
WEST NORWALK
ROWAYTON
NOROTON HEIGHTS
NOROTON
NEW CANAAN
GLEN BROOK
SPRINGDALE
POUND RIDGE
NORTH STAMFORD
STAMFORD
LONG RIDGE
BEDFORD
MIANUS
OLD GREENWICH
COS COB
GREENWICH
BYRAM
ROUND HILL
GLENVILLE
PORT CHESTER
RYE BROOK
RYE
HARRISON
MOUNT KISCO
BEDFORD HILLS
ARMONK
VALHALLA
WHITE PLAINS
TOWN OF MAMARONECK
VILLAGE OF MAMARONECK
LARCHMONT
Long Island Sound
SUFFOLK CO.
NASSAU CO.
LLOYD NECK
EATONS NECK
ASHAROKEN
HUNTINGTON
FT. SALONGA
NISSEQUOGUE
Caumsett St. Pk.
Sunken Meadow St. Pk.
Sherwood Island State Park
Stamford Harbor
Greenwich Harbor
Bryam Harbor
Greenwich Pt.
Lloyd Pt.
Eatons Neck Pt.
Westchester Airport
Kensico
Nature Conservancy
Easton Res.
Rockwood Lake
Putnam Lake
Laurel Res.
Cross River Res.
New Croton Res.

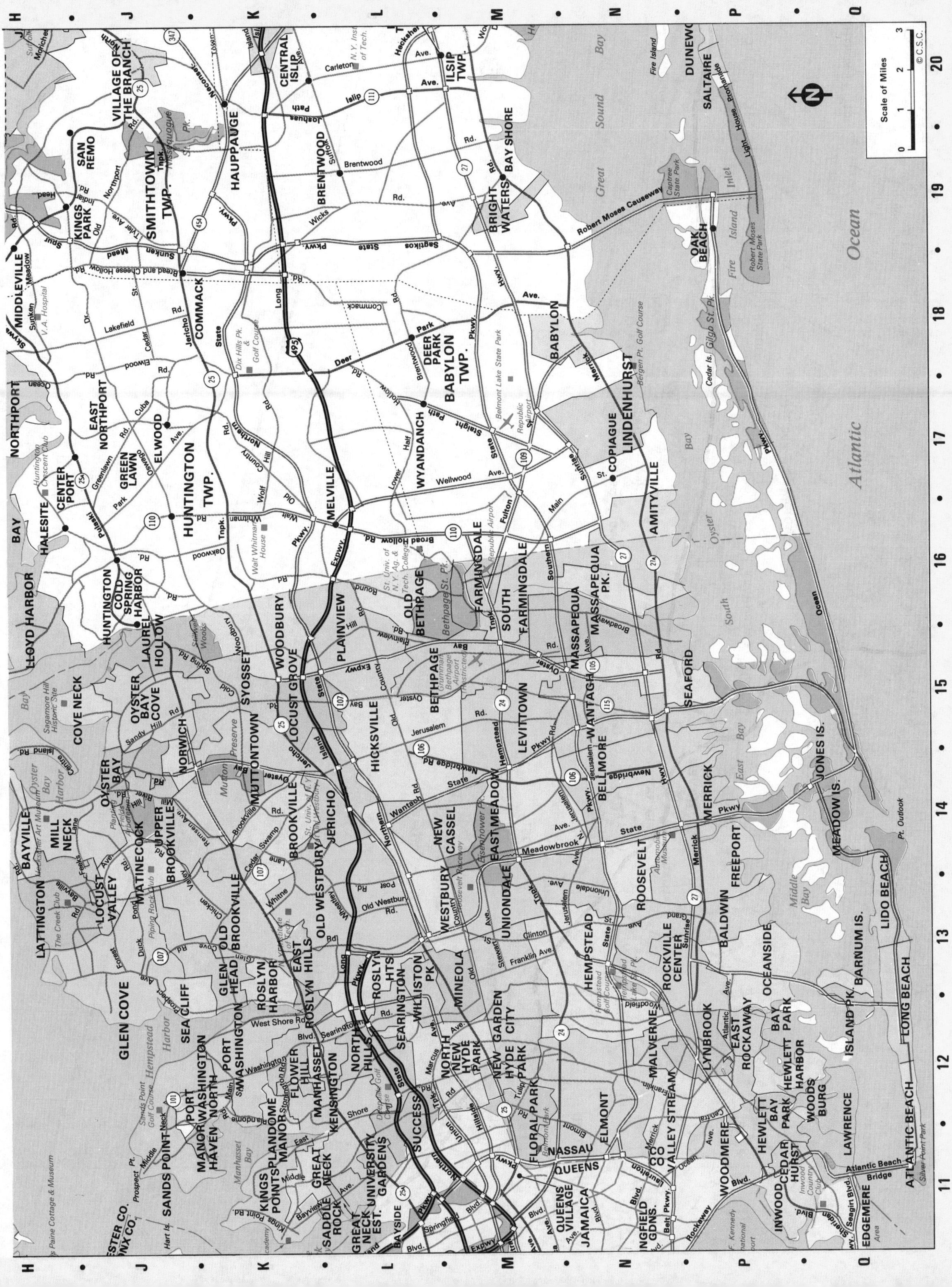

Scale of Miles 0 1 2 3

©C.S.C.

Atlantic Ocean

Great South Bay

South Oyster Bay

HUNTINGTON TWP.
SMITHTOWN TWP.
BABYLON TWP.
ISLIP TWP.

NORTHPORT
EAST NORTHPORT
COMMACK
HAUPPAUGE
BRENTWOOD
CENTRAL ISLIP
BAY SHORE
BRIGHTWATERS
DEER PARK
WYANDANCH
MELVILLE
FARMINGDALE
SOUTH FARMINGDALE
BABYLON
LINDENHURST
COPIAGUE
AMITYVILLE
MASSAPEQUA
MASSAPEQUA PK.
SEAFORD
WANTAGH
BELLMORE
MERRICK
FREEPORT
BALDWIN
ROOSEVELT
HEMPSTEAD
UNIONDALE
EAST MEADOW
LEVITTOWN
HICKSVILLE
BETHPAGE
OLD BETHPAGE
PLAINVIEW
WOODBURY
SYOSSET
JERICHO
WESTBURY
NEW CASSEL
MINEOLA
GARDEN CITY
ROCKVILLE CENTRE
OCEANSIDE
LONG BEACH
LIDO BEACH
JONES IS.
OAK BEACH
SALTAIRE
GLEN COVE
SEA CLIFF
OYSTER BAY
LOCUST VALLEY
ROSLYN
MANHASSET
GREAT NECK
QUEENS
NASSAU
ELMONT
VALLEY STREAM
LYNBROOK
JAMAICA
FLORAL PARK
ATLANTIC BEACH
LAWRENCE
WOODMERE

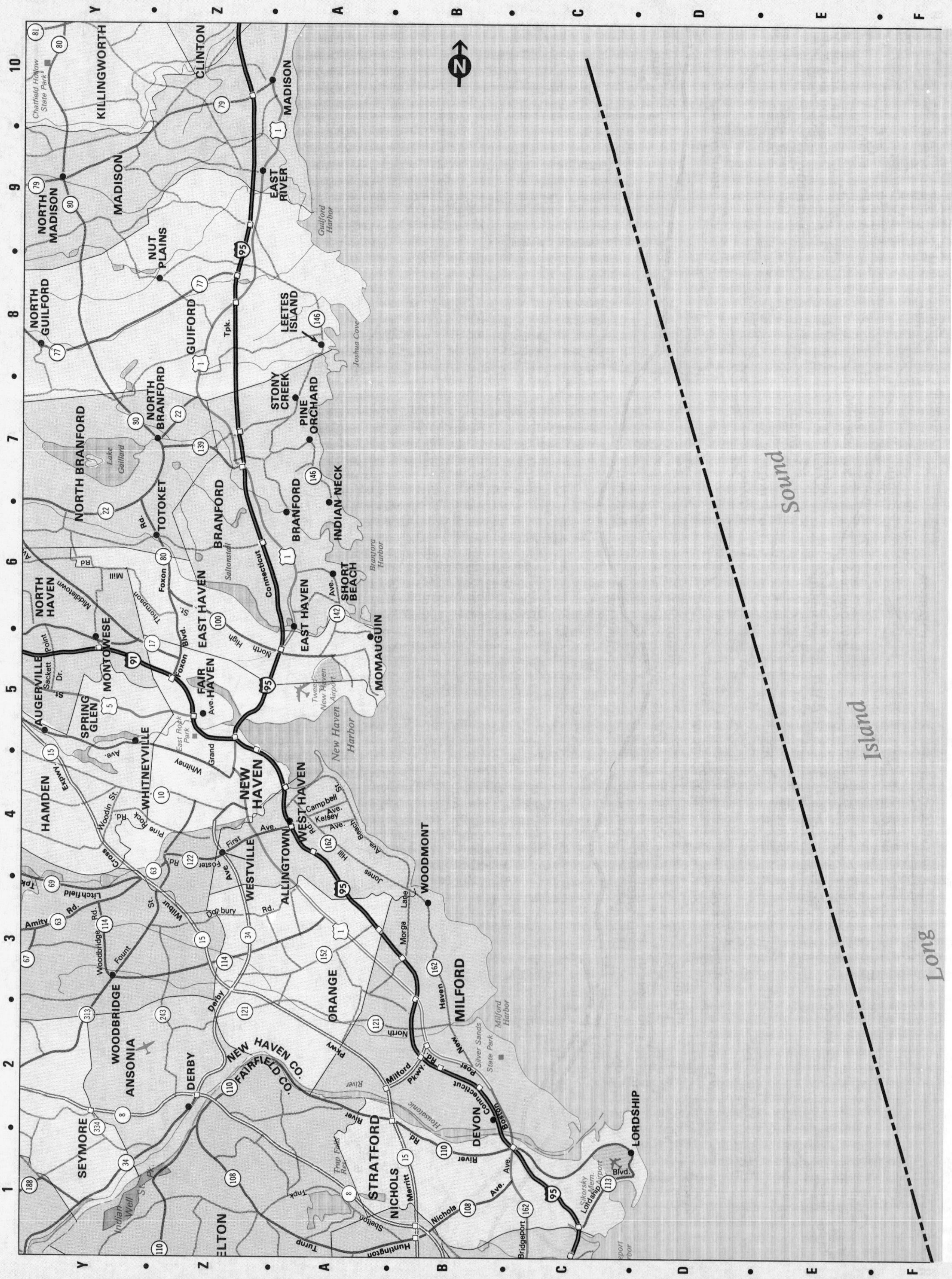

SEYMORE
ANSONIA
DERBY
WOODBRIDGE
ORANGE
MILFORD
DEVON
LORDSHIP
STRATFORD
NICHOLS
NEW HAVEN CO.
FAIRFIELD CO.
HAMDEN
WHITNEYVILLE
SPRING GLEN
AUGERVILLE
MONTOWESE
NORTH HAVEN
NEW HAVEN
WESTVILLE
ALLINGTOWN
WEST HAVEN
WOODMONT
FAIR HAVEN
EAST HAVEN
MOMAUGUIN
SHORT BEACH
BRANFORD
INDIAN NECK
TOTOKET
NORTH BRANFORD
PINE ORCHARD
STONY CREEK
LEETES ISLAND
GUIFORD
NORTH GUILFORD
NUT PLAINS
NORTH MADISON
MADISON
EAST RIVER
KILLINGWORTH
CLINTON
New Haven Harbor
Tweed-New Haven Airport
Branford Harbor
Guilford Harbor
Joshua Cove
Milford Harbor
Silver Sands State Park
Sikorsky Memorial Airport
Chatfield Hollow State Park
Lake Gaillard
Long Island Sound

Long
Atlantic
Ocean
Great
Sound
Bay
Moriches
Bay
Shinnecock
Inlet
NORTHVILLE
RIVERHEAD
RIVERHEAD TWP.
FLANDERS
WESTHAMPTON
QUOGUE
EASTPORT
SPEONK
EAST MORICHES
CENTER MORICHES
MANORVILLE
WADING RIVER
SHOREHAM
ROCKY POINT
SOUND BEACH
MILLER PLACE
MOUNT SINAI
PORT JEFFERSON
BELLE TERRE
POQUOTT
OLD FIELD
SETAUKET
STONY BROOK
ST. JAMES
SOUTH SETAUKET
PORT JEFFERSON STATION
CORAM
SELDEN
GORDON HTS.
MIDDLE ISLAND
RIDGE
YAPHANK
BROOKHAVEN TWP.
FARMINGVILLE
MEDFORD
HOLBROOK
RONKONKOMA
LAKE GROVE
CENTEREACH
VILLAGE OF THE BRANCH
HAUPPAUGE
VILLAGE OF ISLANDIA
BOHEMIA
SAYVILLE
WEST SAYVILLE
OAKDALE
GREAT RIVER
ISLIP TERRACE
BAYPORT
BLUE POINT
PATCHOGUE
HAGERMAN
NORTH BELLPORT
BELLPORT
BROOKHAVEN
SOUTH HAVEN
SHIRLEY
MASTIC
MASTIC BEACH
MASTIC MORICHES
N.Y.S. Conservation Area
Brookhaven St. Pk.
Brookhaven National Laboratory
Wildwood State Park
Grumman Bethpage Airport (Restricted)
Suffolk Co. Airport
Quogue Bird Sanctuary
Brookhaven Airport
Edwards Airport
Islip MacArthur Airport
Connetquot River St. Pk.
Heckscher St. Pk.
Wertheim National Wildlife Refuge
South Haven Pk.
Lake Ronkonkoma
Stony Brook Univ.
Scale of Miles
©C.S.C.

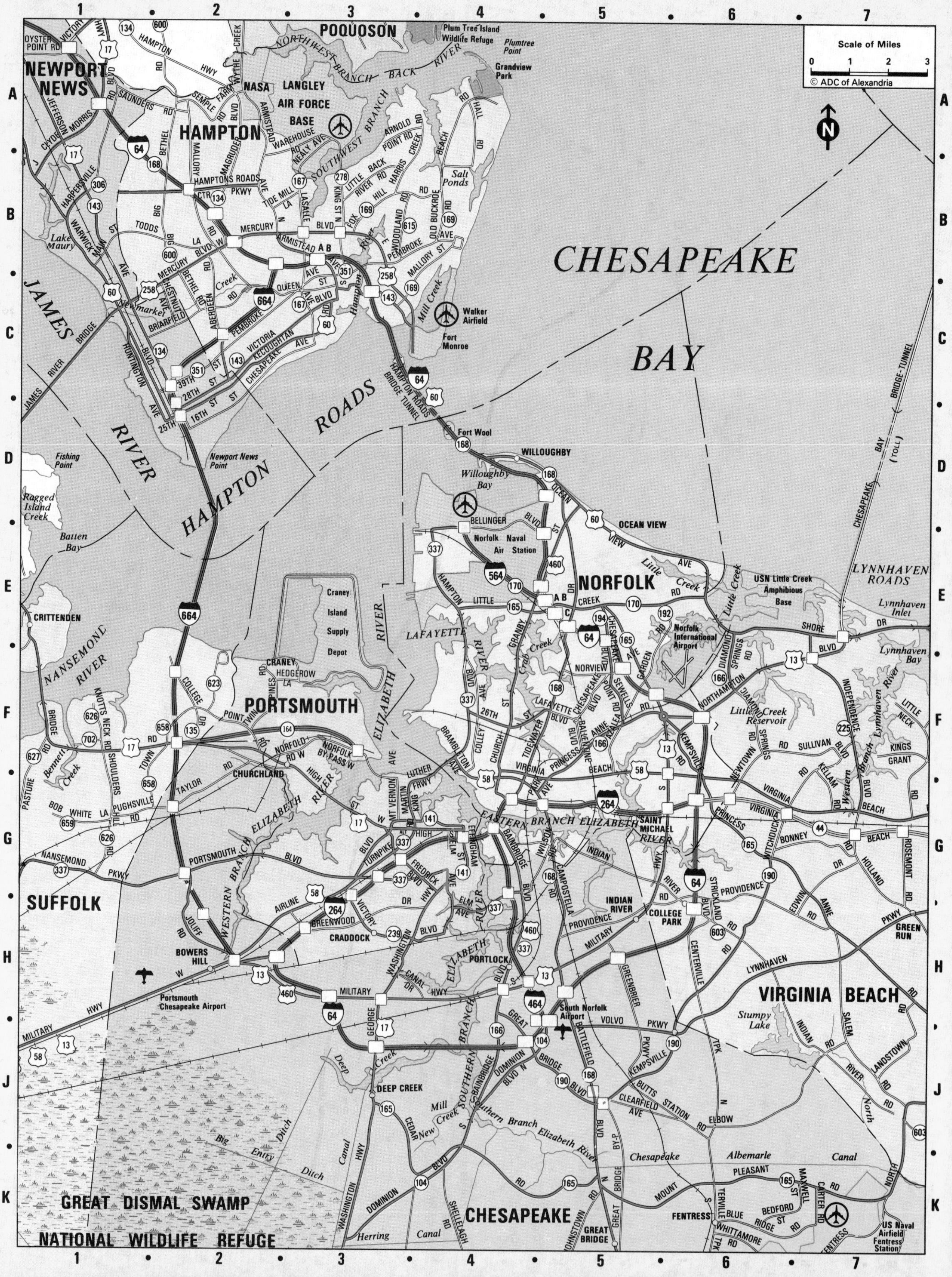

Scale of Miles
0 1 2 3
© ADC of Alexandria

NEWPORT NEWS
HAMPTON
POQUOSON
NASA
LANGLEY AIR FORCE BASE
Plum Tree Island Wildlife Refuge
Plumtree Point
Grandview Park
Salt Ponds
Walker Airfield
Fort Monroe
Fort Wool
Newport News Point
Lake Maury
Newmarket
CHESAPEAKE BAY
JAMES RIVER
HAMPTON ROADS
HAMPTON ROADS BRIDGE TUNNEL
Fishing Point
Ragged Island Creek
Batten Bay
WILLOUGHBY
Willoughby Bay
OCEAN VIEW
Norfolk Naval Air Station
NORFOLK
USN Little Creek Amphibious Base
LYNNHAVEN ROADS
Lynnhaven Inlet
Lynnhaven Bay
Norfolk International Airport
Little Creek Reservoir
CHESAPEAKE BAY BRIDGE-TUNNEL (TOLL)
CRITTENDEN
NANSEMOND RIVER
Craney Island Supply Depot
CRANEY
PORTSMOUTH
CHURCHLAND
LAFAYETTE RIVER
ELIZABETH RIVER
EASTERN BRANCH ELIZABETH RIVER
WESTERN BRANCH
SUFFOLK
BOWERS HILL
CRADDOCK
PORTLOCK
INDIAN RIVER
COLLEGE PARK
GREEN RUN
VIRGINIA BEACH
Stumpy Lake
Portsmouth Chesapeake Airport
South Norfolk Airport
DEEP CREEK
Southern Branch Elizabeth River
Chesapeake Albemarle Canal
FENTRESS
US Naval Airfield Fentress Station
GREAT BRIDGE
CHESAPEAKE
GREAT DISMAL SWAMP NATIONAL WILDLIFE REFUGE

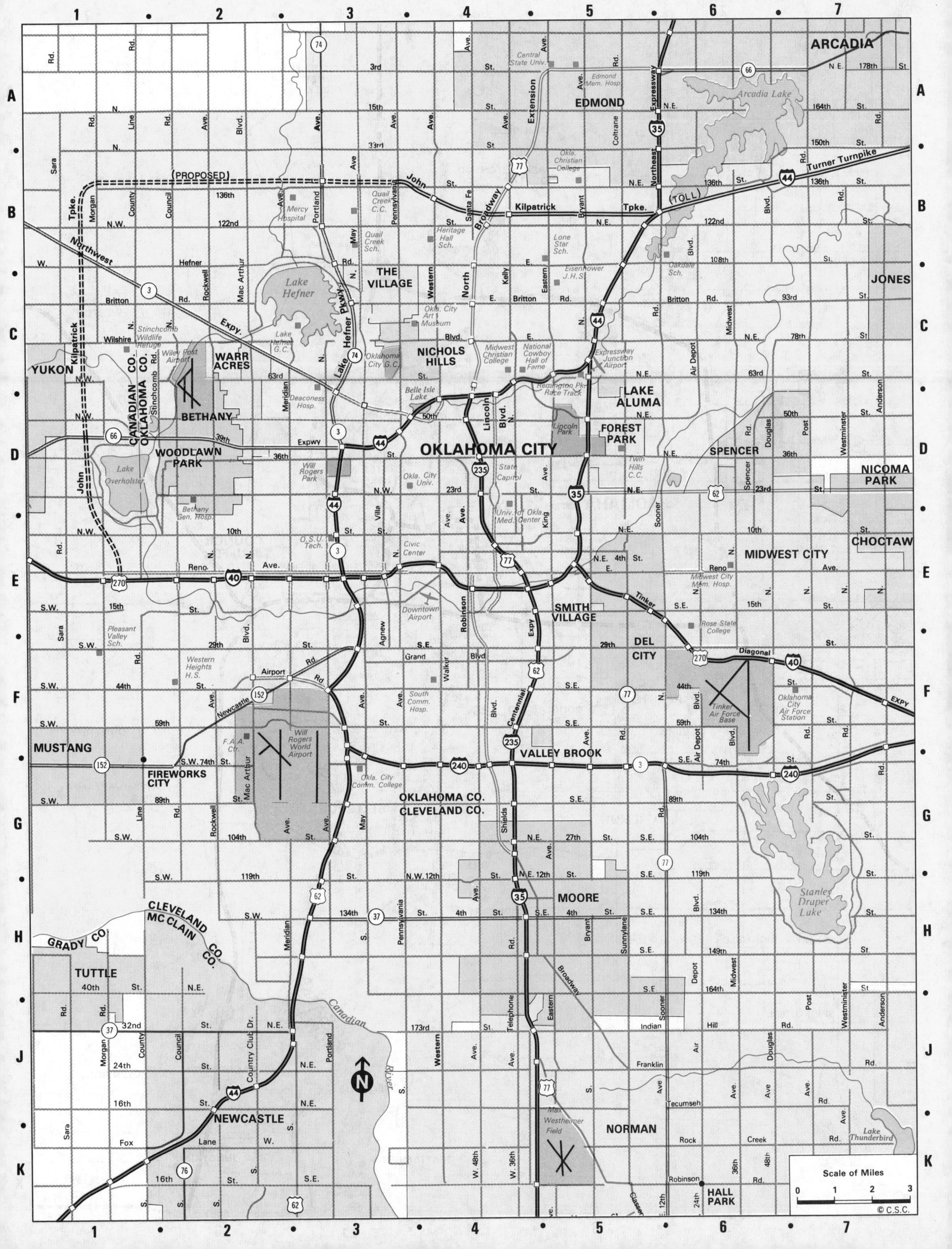

ARCADIA
Arcadia Lake
EDMOND
Central State Univ.
Edmond Mem. Hosp.
Okla. Christian College
Turner Turnpike
(TOLL)
(PROPOSED)
John Kilpatrick Tpke.
Mercy Hospital
Quail Creek C.C.
Quail Creek Sch.
Heritage Hall Sch.
Lone Star Sch.
Oakdale Sch.
Eisenhower J.H.S.
Northwest Expy.
Lake Hefner
THE VILLAGE
JONES
Stinchcomb Wildlife Refuge
Wiley Post Airport
WARR ACRES
Lake Hefner G.C.
Oklahoma City G.C.
Okla. City Art Museum
NICHOLS HILLS
Midwest Christian College
National Cowboy Hall of Fame
Expressway Junction Airport
Remington Pk. Race Track
LAKE ALUMA
YUKON
CANADIAN CO.
OKLAHOMA CO.
BETHANY
Deaconess Hosp.
Belle Isle Lake
Lincoln Park
FOREST PARK
WOODLAWN PARK
Lake Overholser
Will Rogers Park
OKLAHOMA CITY
SPENCER
Twin Hills C.C.
NICOMA PARK
Okla. City Univ.
State Capitol
Univ. of Okla. Med. Center
Bethany Gen. Hosp.
O.S.U. Tech.
Civic Center
CHOCTAW
MIDWEST CITY
Midwest City Mem. Hosp.
Downtown Airport
SMITH VILLAGE
Rose State College
Pleasant Valley Sch.
Western Heights H.S.
DEL CITY
South Comm. Hosp.
Tinker Air Force Base
Oklahoma City Air Force Station
MUSTANG
F.A.A. Ctr.
Will Rogers World Airport
VALLEY BROOK
FIREWORKS CITY
Okla. City Comm. College
OKLAHOMA CO.
CLEVELAND CO.
Stanley Draper Lake
MOORE
CLEVELAND CO.
MC CLAIN CO.
GRADY CO.
TUTTLE
Canadian River
NEWCASTLE
Max Westheimer Field
NORMAN
Lake Thunderbird
HALL PARK

Scale of Miles
0 1 2 3

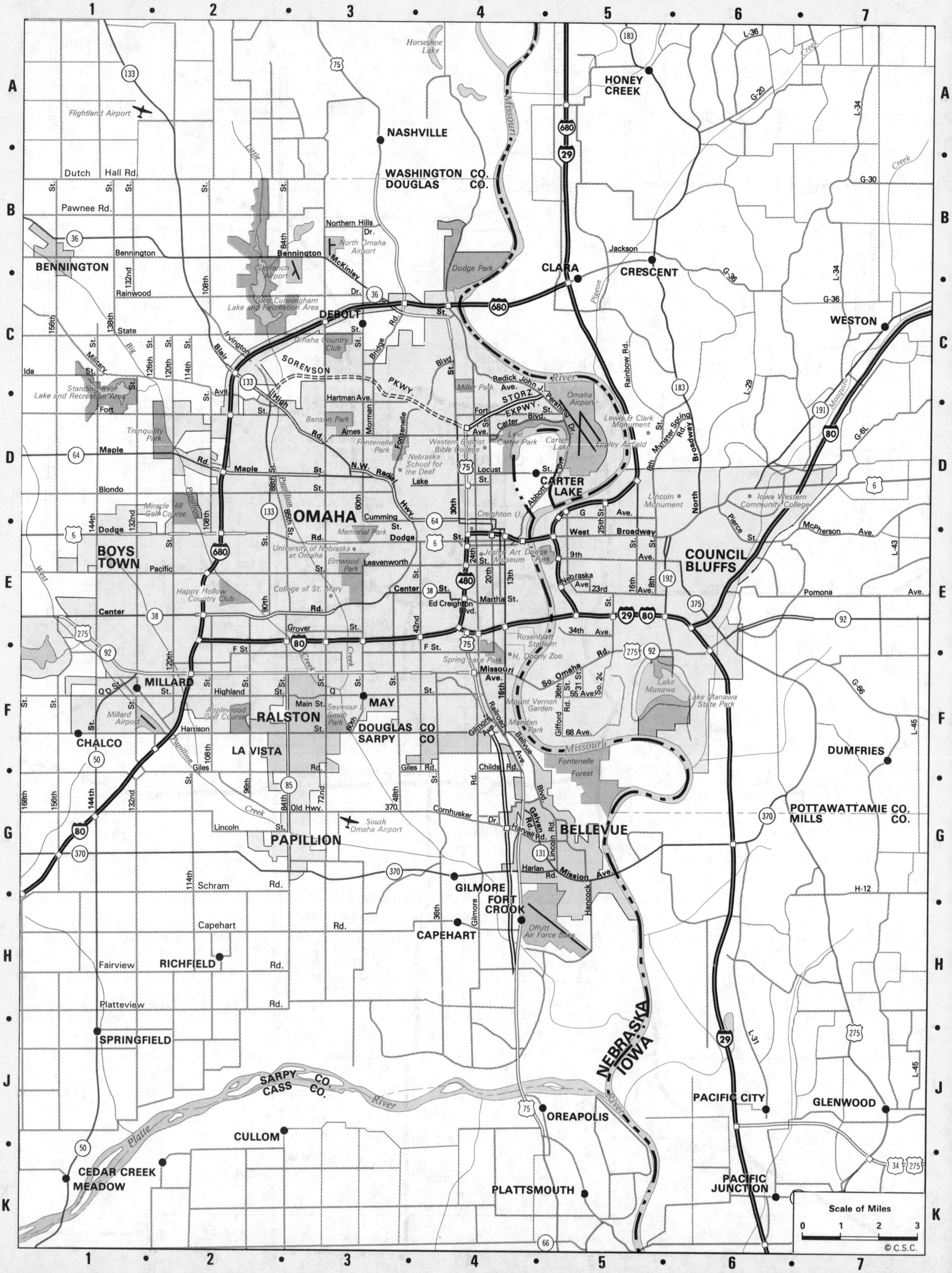

HONEY CREEK
NASHVILLE
WASHINGTON CO.
DOUGLAS CO.
Flightland Airport
Horseshoe Lake
Missouri River
BENNINGTON
North Omaha Airport
Skyranch Airport
Glen Cunningham Lake and Recreation Area
Dodge Park
CLARA
CRESCENT
WESTON
DEBOLT
Omaha Country Club
SORENSON PKWY.
Standing Bear Lake and Recreation Area
Tranquility Park
Miller Park
Benson Park
Omaha Airport
Lewis & Clark Monument
Levi Carter Park
Carter Lake
Eppley Airfield
Fontenelle Park
Western Baptist Bible College
Nebraska School for the Deaf
CARTER LAKE
Lincoln Monument
Iowa Western Community College
Miracle Hill Golf Course
OMAHA
Memorial Park
Creighton U.
University of Nebraska at Omaha
Joslyn Art Museum
Dodge Park
BOYS TOWN
Elmwood Park
COUNCIL BLUFFS
Happy Hollow Country Club
College of St. Mary
Rosenblatt Stadium
H. Doorly Zoo
Spring Lake Park
Lake Manawa
Lake Manawa State Park
MILLARD
Millard Airport
Applewood Golf Course
RALSTON
Seymour Smith Park
MAY
Mount Vernon Garden
CHALCO
DOUGLAS CO
SARPY CO
LA VISTA
Mandan Park
Fontenelle Forest
DUMFRIES
South Omaha Airport
BELLEVUE
PAPILLION
POTTAWATTAMIE CO.
MILLS CO.
GILMORE
FORT CROOK
CAPEHART
Offutt Air Force Base
RICHFIELD
SPRINGFIELD
NEBRASKA
IOWA
SARPY CO.
CASS CO.
Platte River
PACIFIC CITY
GLENWOOD
OREAPOLIS
CULLOM
CEDAR CREEK
MEADOW
PLATTSMOUTH
PACIFIC JUNCTION

Scale of Miles
0 1 2 3

1 2 3 4 5 6 7

A B C D E F G H J K

ZELLWOOD
POKAN
Lake Grassmere
YOTHERS
WAKIWA SPRINGS STATE PARK
Prevatt Lake
WEKIWA
SANLANDO SPRINGS
Lake Brantley
THE SPRINGS
MARKHAM
West Lake
LONGWOOD
ORLANDO
CHASE
MIDWAY PARK
Marl Bed Flats
Eureka Hammock
Moses E. Levy Grant
SANFORD-OVIEDO RD
GARDENIA
WINTER SPRINGS
SEMINOLE PROPOSED
MCDONALD
Wilkins Lake
Lake Francis
Lake Standish
Lake Alden
Lake Coroni
Lake McCoy
PLYMOUTH
SEMINOLE COUNTY
Fairy Lake
Kathryn Lake
Triplet Lake
Little Howell Lake
Philip R. Young Grant
Marshall Lake
Lake Alma
APOPKA
ALTAMONTE SPRINGS
ZELLWOOD FARMS RD
Lake Fuller
Semmes Lake
SOUTH APOPKA
Bear Lake
FOREST CITY
Spring Lake
SEMORAN
Lake Orienta
CASSELBERRY
Red Bug Lake
RED BUG LAKE RD
HOOPER FARMS
LAKE COUNTY
LAKE APOPKA
KEENE RD
Trout Lake
Marden
Lake Lotus
Lake Hill
ORANGE
Lake Howell
Howell
Lake Alpharetta
Trout Lake
Lake Sims
Long Lake
Lockhart Lake
Lake Lovely
Lake Destiny
Minnehaha
MAITLAND
Lake Maitland
WELL BRANCH RD
Bear Gully Lake
JAMES
KENNEDY BLVD
EATONVILLE
SEMINOLE COUNTY
Crown Point
BEN WHITE RACEWAY
Lake Wekiva
Crooked Lake
Lake Prairie
Lake Stanley
Lake Fairview
Lake Killarney
Lake Osceola
WINTER PARK
Lake Virginia
Lake Mizell
Lake Berry
O.N.T.C. HOSP.
MEAD BOTANICAL GARDENS
Lake Sue
Lake Waunatta
Lake Pearl
Lake Irma
Lake Phillips
DEAN
EXPRESSWAY
Gator Island
WINTER GARDEN
PLANT ST
OCOEE
Spring Lake
Starke Lake
Olympia
SILVER STAR RD
HIAWASSEE RD
GOOD HOMES RD
MERCY MEDICAL CENTER
Lake Larne
EDGEWATER DR
ORANGE
MILLS
ORLANDO NAVAL TRAINING CTR
Lake Corrine
ORLANDO SPORTS STADIUM
GOLDENROD
EASTERN
Hull Island
OAKLAND
CSX
COLONIAL DR
Montverde
Clermont
WESTERN EXTENSION PROPOSED
ROBINSON AV
SOUTH ST
EXECUTIVE AIRPORT
Barton Lake
LAKE UNDERHILL DR
Deer Is.
Johns Lake
Lake Rose
Lake Steer
Lake Hiawassa
Lake Mann
CITRUS BOWL
CHURCH STREET STATION
Lake Arnold
Black Lake
WINTER GARDEN
Lake Beulah
Whitney Lake
MAGUIRE RD
Lake Olivia
Clear Lake
ORLANDO
MEDICAL CENTER
Lake Roberts
CURRY FORD RD
SEMORIAN
Lake Porter
Lake Frederica
ORANGE COUNTY
Lake Avalon
TILDEN RD
Lake Roper
Lake Crescent
Lake Down
Turkey Lake
Lake Catherine
Lake Holden
Lake Pinelock
Lake Barber
CONWAY
TERRE BLVD
BLOSSOM TR
CONROY WINDERMERE RD
Lake Butler
WINDERMERE
Lake Gatlin
Little Lake Conway
PROPOSED
BELTWAY
EDGEWOOD
Lake Jessamine
Lake Hartley
Lake Speer
Cypress Lake
Lake Universal
UNIVERSAL STUDIOS
HOFFNER RD
MCKINNEY RD
Lake Lake Sawyer
Little Lake Sawyer
BAY HILL COUNTRY CLUB
OAK RIDGE RD
Lake Conway
BELLE ISLE
NARCOOSSEE
Hancock
JUDGE
Rattle Snake Lake
PINE CASTLE
Bay Lake
Lake Warren
Open Pond
Lake Tibet
Spring Lake
ORLANDO CENTRAL PARK
LAKE
SAND LAKE RD
EXPRESSWAY
Lake Needham
Little Sand Lake
INTERNATIONAL DR
BEELINE
Lake Reams
Lake Sharpe
Huckleberry Lake
REPUBLIC DR
MARTIN MARIETTA CORP.
JOHN YOUNG PKWY
LANDSTREET RD
TAFT
ORLANDO INTERNATIONAL AIRPORT
Lake Mabel
Big Sand Lake
Lake Aldrich
Tub Lake
Lake Sheen
Bull Slough
SEIDEL RD
Bay Lake
South Lake
Pocket Lake
Lake Crowell
Big Sand Lake
TAFT-VINELAND RD
BOGGY
Reedy Lake
Seven Seas Lagoon
Little Fish Lake
Lake Nona
Red Lake
Buck Lake
HARTZOG
CENTRAL FLORIDA PKWY
WALT DISNEY WORLD
Lake Ruby
Lake Willis
SEA HARBOR DR
FLEMINGS RD
LAKE BUENA VISTA
WETHERBEE RD
CREEK
Mud Lake
Barton Lake
Reedy Cr
Boggy Creek Swamp
Lake Bryan
MARRIOTT WORLD RESORT
GATORLAND
Shingle
ORANGE
OSCEOLA
CO CO
SPACE
OAK ISLAND RD
COAST
ENTRY PT BLVD
LIVINGSTON RD
BUNKER HILL RD
SAND HILL RD
PRINCESS WY
L. Cecile
SCOTT BLVD
SEVEN DWARFS LA
DIAMOND MINE RD
BERMUDA
CARROLL
JACKSON
MILL SLOUGH
DONEGAN AV
MICHIGAN
BUENAVENTURA BLVD
BUENAVENTURA LAKES
Fells Cove
Davenport
OSCEOLA COUNTY
OAK ISLAND RD
LEDERER LA
GOODMAN
Creek Swamp
MIDIEVIL TIMES
OREN BROWN RD
COLUMBIA AV
VINE
OAK
MAIN ST
FORTUNE RD
Hillard Isle
EAST LAKE TOHOPEKALIGA
LLOYD LA
VICKIE CT
KEVIN WY
PARTIN
SETTLEMENT RD
POINCIANA
ROME WY
BABB RD
EMMETT ST
NEPTUNE
BROADWAY AV
SHADY LA
KISSIMMEE MUNICIPAL AIRPORT
Grass Island
THACHER AV
PLEASANT HILL
Cypress Cove
KISSIMMEE
Paradise Island
Wilson Cove
Fish Lake
ST. CLOUD
Runnymeade Lake
CSX
DIXIE
ORANGE BLOSSOM
HARRIS BLVD
SKYLINE DR
HENRY PARTIN RD
KINGS HWY
MACY ISLAND RD
KINGS POINT BLVD
LAKESHORE BLVD
Tampa
To Haines City
INDUSTRIAL
INTERCESSION CITY
CAMPBELL
Makinson Island
St. Cloud Canal C-31
TURNPIKE
VERMONT
MISSOURI
FLORIDA
MICHIGAN
ORANGE AV
PKWY
ST. CLOUD AIRPARK
HICKORY TREE RD
Live Oak

Scale of Miles
0 1 2 3
©TRAKKER MAPS INC.

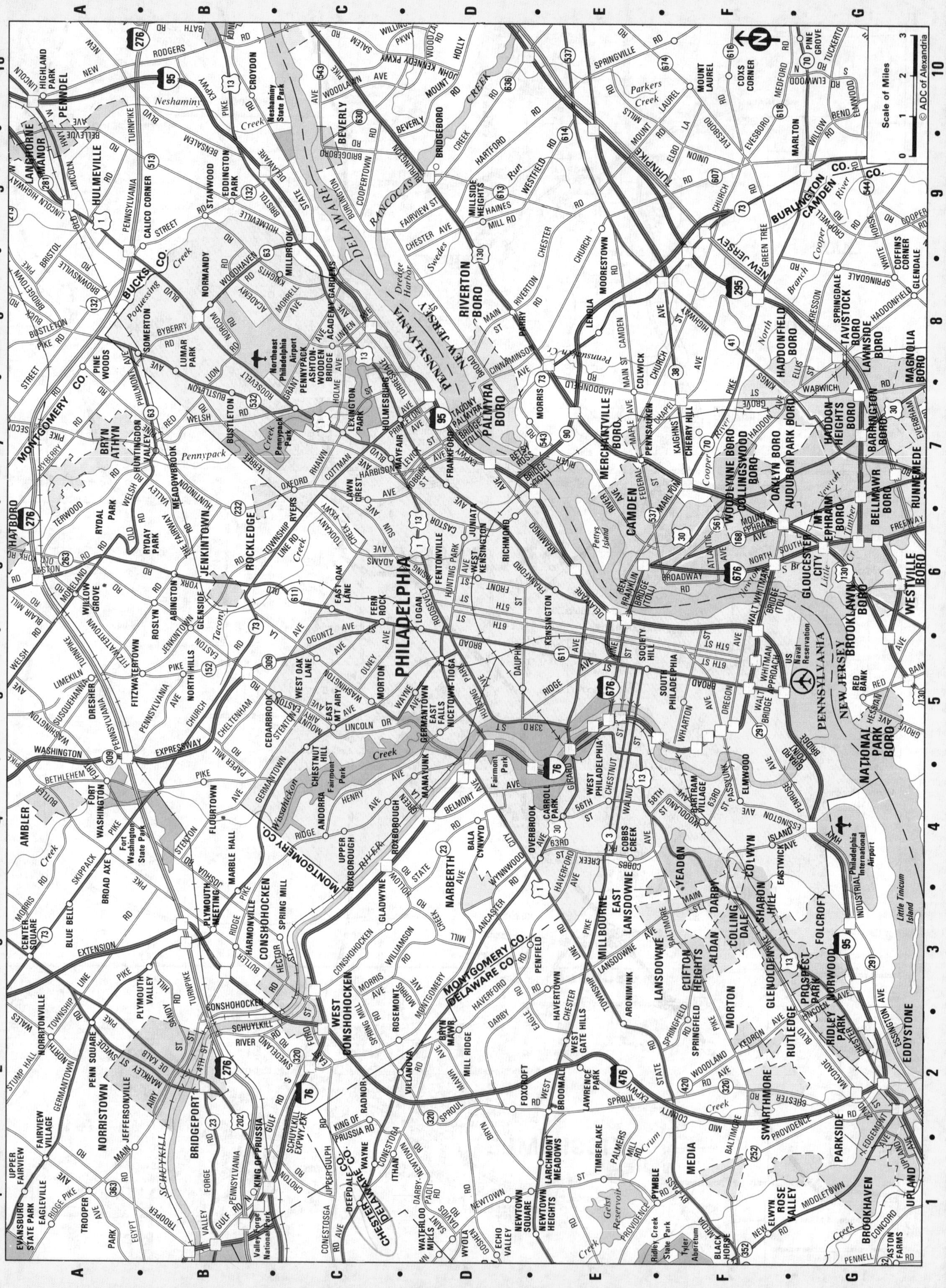
PHILADELPHIA
PENNSYLVANIA
NEW JERSEY
DELAWARE
CAMDEN
BUCKS CO.
MONTGOMERY CO.
DELAWARE CO.
BURLINGTON CO.
CAMDEN CO.
Northeast Philadelphia Airport
Philadelphia International Airport
Fairmount Park
Pennypack Park
Neshaminy State Park
Valley Forge National Park
Fort Washington State Park
Ridley Creek State Park
Tyler Arboretum
NORRISTOWN
CONSHOHOCKEN
WEST CONSHOHOCKEN
KING OF PRUSSIA
BRIDGEPORT
JENKINTOWN
ROCKLEDGE
CHESTNUT HILL
GERMANTOWN
MANAYUNK
ROXBOROUGH
NARBERTH
BRYN MAWR
VILLANOVA
WAYNE
RADNOR
MEDIA
SWARTHMORE
LANSDOWNE
YEADON
DARBY
GLOUCESTER CITY
BROOKLAWN BORO
WESTVILLE BORO
NATIONAL PARK BORO
RIVERTON BORO
PALMYRA BORO
BEVERLY
HADDONFIELD BORO
COLLINGSWOOD BORO
CHERRY HILL
MERCHANTVILLE BORO
PENNSAUKEN
MARLTON
Scale of Miles
© ADC of Alexandria

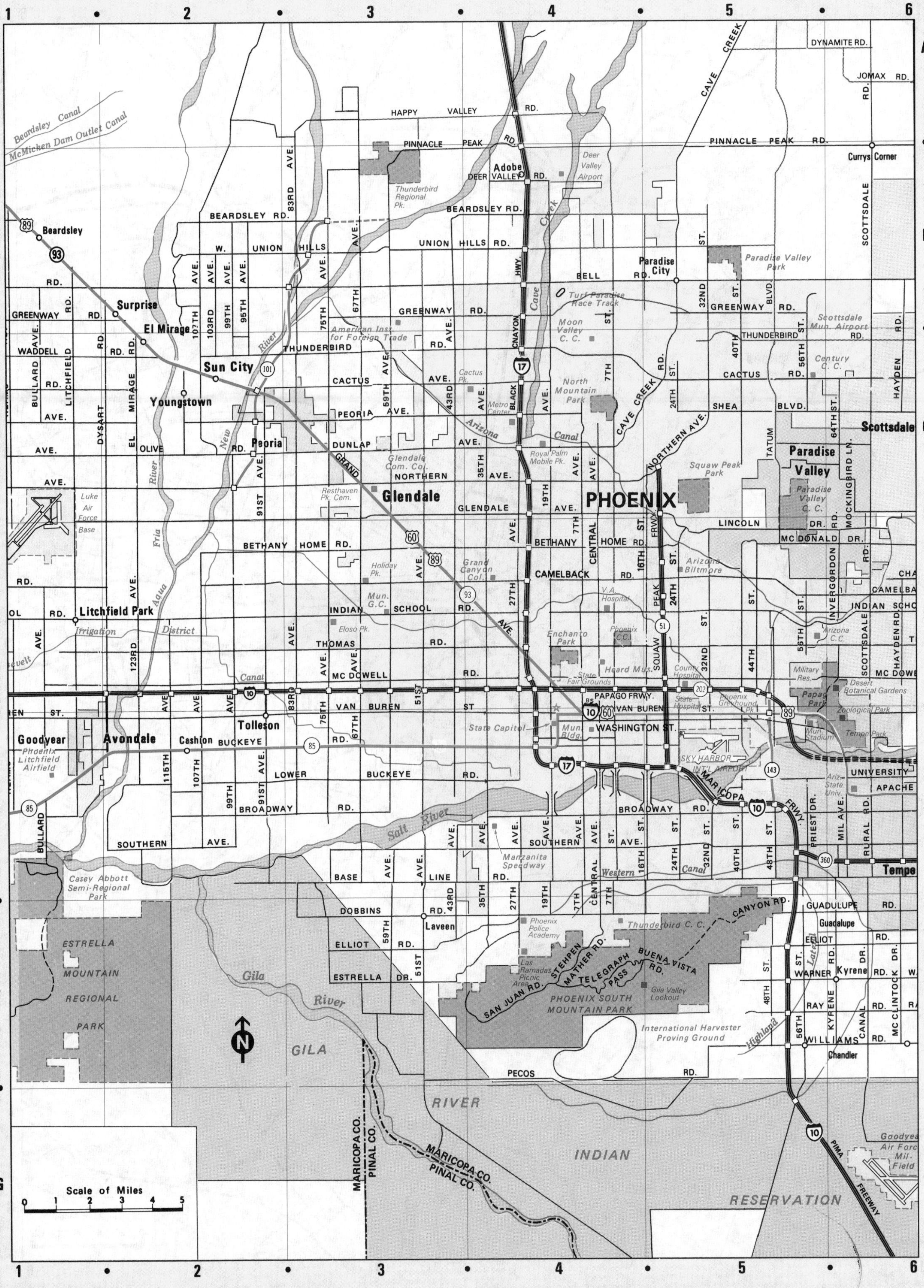
PHOENIX
Glendale
Scottsdale
Paradise Valley
Tempe
Sun City
Youngstown
Peoria
Surprise
El Mirage
Beardsley
Litchfield Park
Goodyear
Avondale
Tolleson
Cashion
Laveen
Guadalupe
Kyrene
Chandler
Adobe
Paradise City
Currys Corner
Beardsley Canal
McMicken Dam Outlet Canal
New River
Agua Fria River
Salt River
Gila River
Cave Creek
Arizona Canal
Western Canal
Highland Canal
Irrigation District
Luke Air Force Base
Phoenix Litchfield Airfield
Sky Harbor Int'l Airport
Deer Valley Airport
Scottsdale Mun. Airport
Goodyear Air Force Mil. Field
Thunderbird Regional Pk.
Turf Paradise Race Track
Moon Valley C. C.
North Mountain Park
Squaw Peak Park
Paradise Valley Park
Paradise Valley C. C.
Century C. C.
Arizona Biltmore
Arizona C. C.
Encanto Park
Phoenix C. C.
Heard Mus.
State Fair Grounds
State Capitol
Mun. Bldg.
County Hospital
State Hospital
Phoenix Greyhound Pk.
Papago Park
Desert Botanical Gardens
Zoological Park
Tempe Park
Mun. Stadium
Ariz. State Univ.
Military Res.
American Inst. for Foreign Trade
Glendale Com. Col.
Grand Canyon Col.
Resthaven Pk. Cem.
Holiday Pk.
Mun. G.C.
Eloso Pk.
Metro Center
Cactus Pk.
Royal Palm Mobile Pk.
V.A. Hospital
Manzanita Speedway
Phoenix Police Academy
Thunderbird C. C.
Las Ramadas Picnic Area
Gila Valley Lookout
PHOENIX SOUTH MOUNTAIN PARK
International Harvester Proving Ground
Casey Abbott Semi-Regional Park
ESTRELLA MOUNTAIN REGIONAL PARK
GILA RIVER INDIAN RESERVATION
MARICOPA CO.
PINAL CO.
PIMA FREEWAY
PAPAGO FRWY.
MARICOPA FRWY.
SQUAW PEAK FRWY.
BLACK CANYON HWY.
GRAND AVE.
HAPPY VALLEY RD.
PINNACLE PEAK RD.
DYNAMITE RD.
JOMAX RD.
DEER VALLEY RD.
BEARDSLEY RD.
UNION HILLS RD.
BELL RD.
GREENWAY RD.
THUNDERBIRD RD.
CACTUS RD.
SHEA BLVD.
PEORIA AVE.
DUNLAP AVE.
OLIVE AVE.
NORTHERN AVE.
GLENDALE AVE.
LINCOLN DR.
BETHANY HOME RD.
MC DONALD DR.
CAMELBACK RD.
INDIAN SCHOOL RD.
THOMAS RD.
MC DOWELL RD.
VAN BUREN ST.
WASHINGTON ST.
BUCKEYE RD.
LOWER BUCKEYE RD.
BROADWAY RD.
SOUTHERN AVE.
UNIVERSITY
APACHE
BASE LINE RD.
DOBBINS RD.
ELLIOT RD.
ESTRELLA DR.
GUADULUPE RD.
WARNER RD.
RAY RD.
WILLIAMS
PECOS RD.
CANYON RD.
SAN JUAN RD.
STEHPEN
MATHER RD.
TELEGRAPH PASS
BUENA VISTA RD.
BULLARD AVE.
LITCHFIELD RD.
DYSART RD.
EL MIRAGE RD.
123RD AVE.
115TH AVE.
107TH AVE.
103RD AVE.
99TH AVE.
95TH AVE.
91ST AVE.
83RD AVE.
75TH AVE.
67TH AVE.
59TH AVE.
51ST AVE.
43RD AVE.
35TH AVE.
27TH AVE.
19TH AVE.
7TH AVE.
CENTRAL AVE.
7TH ST.
16TH ST.
24TH ST.
32ND ST.
40TH ST.
44TH ST.
48TH ST.
56TH ST.
64TH ST.
TATUM BLVD.
MOCKINGBIRD LN.
INVERGORDON RD.
SCOTTSDALE RD.
HAYDEN RD.
PRIEST DR.
MIL AVE.
RURAL RD.
KYRENE RD.
MC CLINTOCK DR.
CAVE CREEK RD.
Scale of Miles
0 1 2 3 4 5
1 2 3 4 5 6
A B C D E F G
10 17 51 60 85 89 93 101 143 202 360

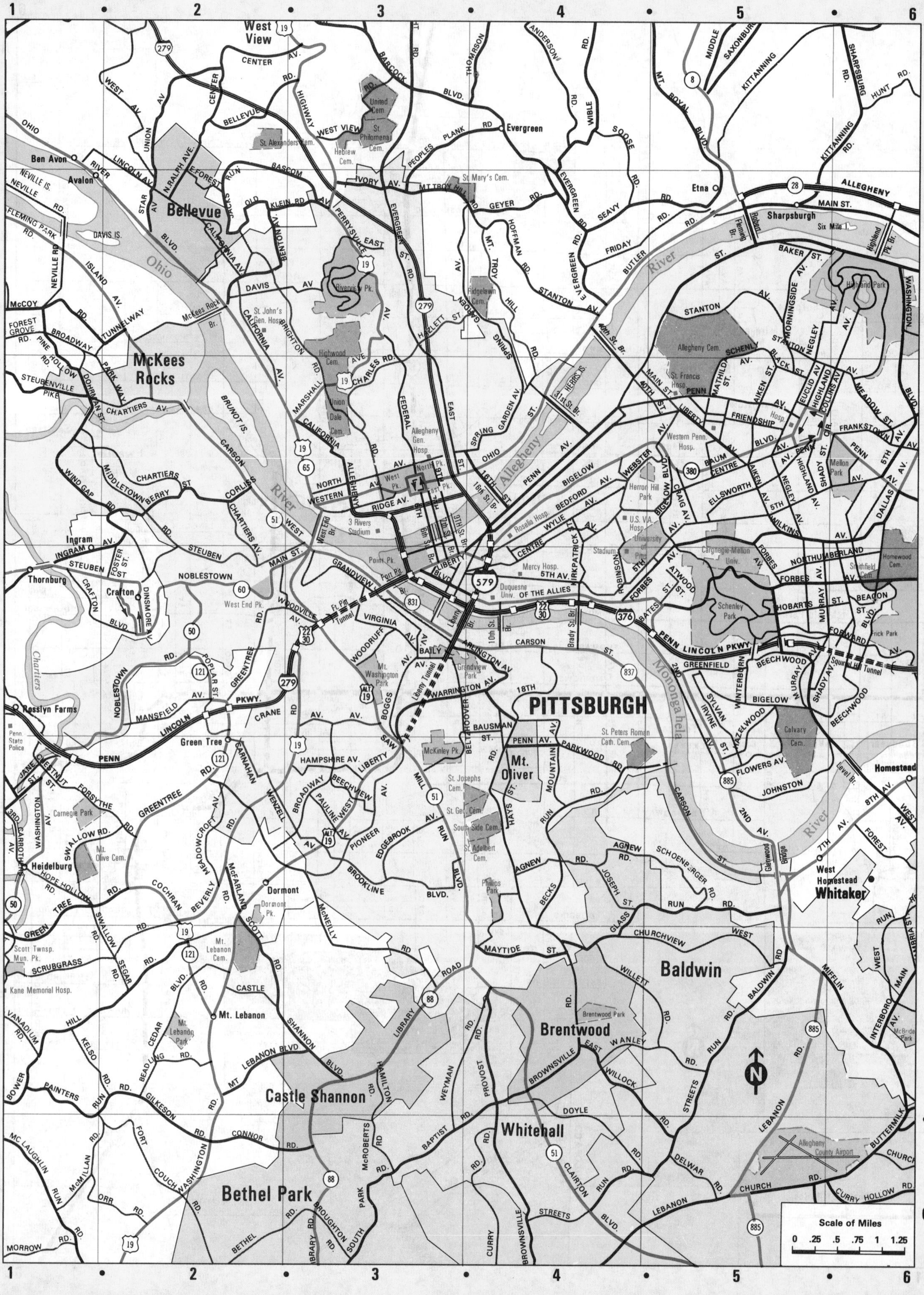
West View
Ben Avon
Avalon
Bellevue
McKees Rocks
Evergreen
Etna
Sharpsburgh
Ingram
Thornburg
Crafton
Rosslyn Farms
Green Tree
Heidelburg
Dormont
Mt. Lebanon
Castle Shannon
Bethel Park
Whitehall
Brentwood
Baldwin
Mt. Oliver
PITTSBURGH
Homestead
West Homestead
Whitaker
Ohio River
Allegheny River
Monongahela River
Chartiers
Allegheny Cem.
Homewood Cem.
Calvary Cem.
Schenley Park
Frick Park
Highland Park
Riverview Pk.
Mellon Park
West End Pk.
Point Pk.
McKinley Pk.
Carnegie Park
Brentwood Park
Allegheny County Airport
3 Rivers Stadium
Duquesne Univ.
University of Pitt.
Carnegie-Mellon Univ.
Ft. Pitt Tunnel
Liberty Tunnel
Squirrel Hill Tunnel
PENN LINCOLN PKWY.
Scale of Miles
0 .25 .5 .75 1 1.25

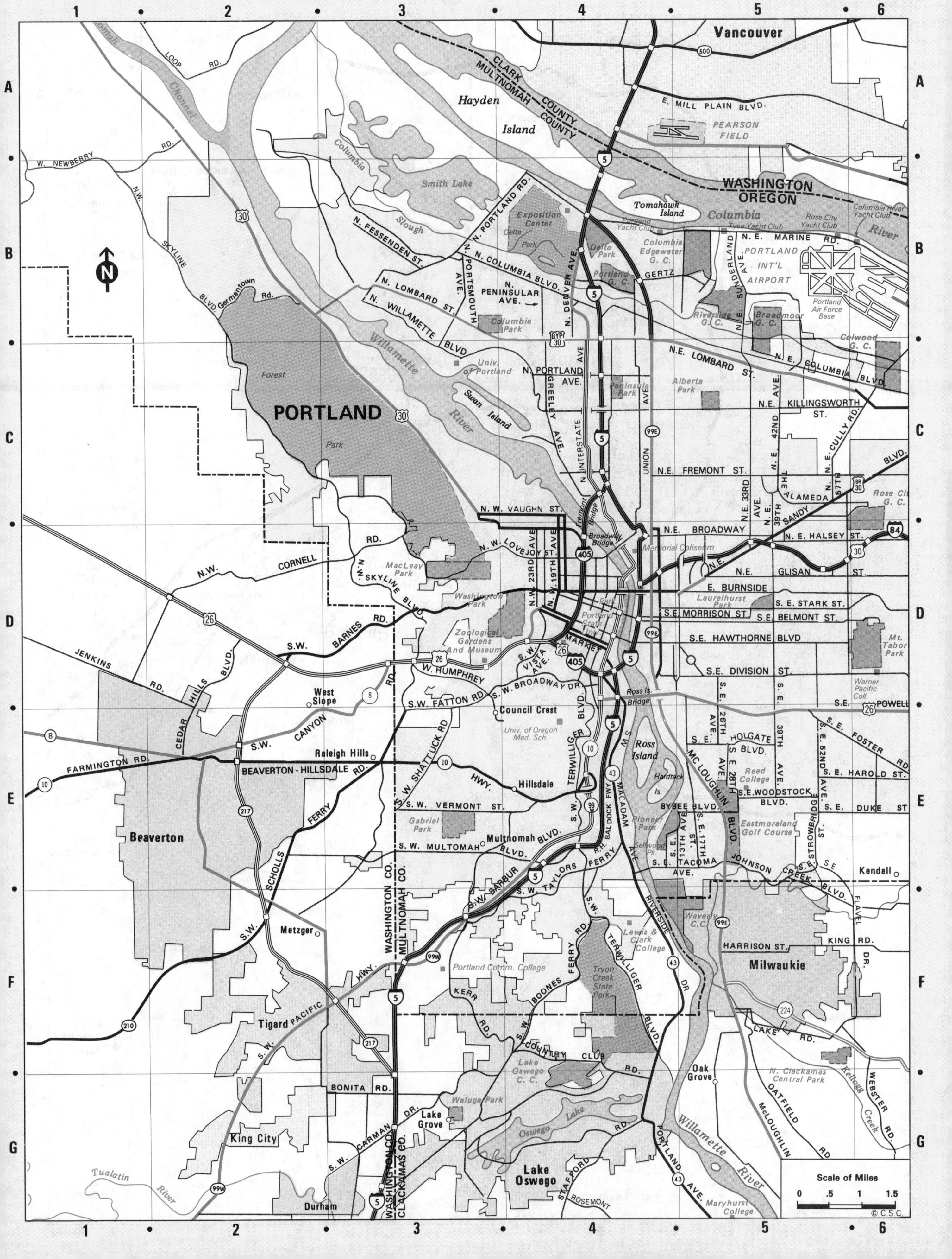

1 2 3 4 5 6

A B C D E F G

Vancouver
E. MILL PLAIN BLVD.
PEARSON FIELD
CLARK COUNTY
MULTNOMAH COUNTY
Hayden Island
WASHINGTON
OREGON
Columbia River
Columbia River Yacht Club
Rose City Yacht Club
Tyee Yacht Club
Tomahawk Island
Portland Yacht Club
Columbia Edgewater G. C.
N. E. MARINE RD.
PORTLAND INT'L AIRPORT
Portland Air Force Base
Riverside G.C.
Broadmoor G. C.
Colwood G. C.
N. E. SUNDERLAND AVE.
GERTZ
Portland G. C.
Delta Park
Exposition Center
Smith Lake
Columbia Slough
N. FESSENDEN ST.
N. PORTLAND RD.
N. COLUMBIA BLVD.
N. PORTSMOUTH AVE.
N. PENINSULAR AVE.
N. DENVER AVE
N. LOMBARD ST.
N. WILLAMETTE BLVD
Columbia Park
BYP 30
Univ. of Portland
N. PORTLAND AVE.
Peninsula Park
Alberta Park
N.E. LOMBARD ST.
N.E. COLUMBIA BLVD
N.E. KILLINGSWORTH ST.
N.E. 42ND AVE.
N.E. CULLY RD.
GREELEY AVE.
N. INTERSTATE AVE
UNION AVE
N.E. FREMONT ST.
THE ALAMEDA
N.E. 33RD AVE.
39TH
57TH
SANDY BLVD.
Rose City G. C.
Multnomah Channel
LOOP RD.
W. NEWBERRY RD.
N.W. SKYLINE BLVD
Germantown Rd.
Forest Park
PORTLAND
Willamette River
Swan Island
N. W. VAUGHN ST.
Fremont Bridge
Broadway Bridge
N.E. BROADWAY
Memorial Coliseum
N. E. HALSEY ST.
N.E. GLISAN ST.
E. BURNSIDE
Laurelhurst Park
S. E. STARK ST.
S.E. MORRISON ST.
S.E. BELMONT ST.
S.E. HAWTHORNE BLVD
Mt. Tabor Park
S.E. DIVISION ST.
Warner Pacific Coll.
S.E. POWELL
N.W. LOVEJOY ST.
N.W. 23RD AVE
N.W. 19TH AVE
N.W. CORNELL RD.
MacLeay Park
N.W. SKYLINE BLVD.
Washington Park
Portland State Univ.
MARKET
Zoological Gardens And Museum
S.W. BARNES RD.
JENKINS RD.
CEDAR HILLS BLVD.
W. HUMPHREY
S.W. VISTA AVE.
S. W. BROADWAY DR.
West Slope
S.W. FATTON RD
Council Crest
Ross Is. Bridge
S.W. CANYON RD
Univ. of Oregon Med. Sch.
S.W. TERWILLIGER BLVD.
Ross Island
S.E. 26TH AVE.
S.E. HOLGATE BLVD.
39TH AVE.
S.E. FOSTER RD
S.E. 52ND AVE.
Raleigh Hills
FARMINGTON RD.
BEAVERTON - HILLSDALE HWY.
S. W. SHATTUCK RD
Hillsdale
Hardtack Is.
MC LOUGHLIN BLVD.
S. E. 28TH AVE.
Reed College
S.E. WOODSTOCK BLVD.
S.E. HAROLD ST.
S. W. VERMONT ST.
BYBEE BLVD.
S.E. DUKE ST
Beaverton
FERRY
Gabriel Park
Pioneer Park
Sellwood Pk.
Eastmoreland Golf Course
STROWBRIDGE ST.
S.W. MULTOMAH BLVD.
Multnomah
R.H. BALDOCK FWY
MACADAM AVE.
S.E. 13TH AVE.
S. E. 17TH ST.
S.E. TACOMA AVE.
JOHNSON CREEK BLVD.
Kendall
SCHOLLS FERRY
S.W. TAYLORS FERRY
S.W. BARBUR
WASHINGTON CO.
MULTNOMAH CO.
FLAVEL DR.
Metzger
S.W. FERRY RD.
RIVERSIDE DR
Lewis & Clark College
Waverly C.C.
HARRISON ST.
KING RD.
Milwaukie
HWY. PACIFIC
Portland Comm. College
KERR RD.
BOONES FERRY RD.
Tryon Creek State Park
TERWILLIGER BLVD.
Tigard
S.W. PACIFIC
LAKE RD.
S.W. COUNTRY CLUB RD.
Lake Oswego C. C.
Oak Grove
N. Clackamas Central Park
Kellogg Creek
WEBSTER RD.
OATFIELD RD
McLOUGHLIN
BONITA RD.
Waluga Park
Lake Grove
Oswego Lake
PORTLAND AVE.
Willamette River
King City
S. W. CARMAN DR.
WASHINGTON CO.
CLACKAMAS CO.
Lake Oswego
STAFFORD RD.
ROSEMONT
Tualatin River
Durham
Maryhurst College

Scale of Miles
0 .5 1 1.5

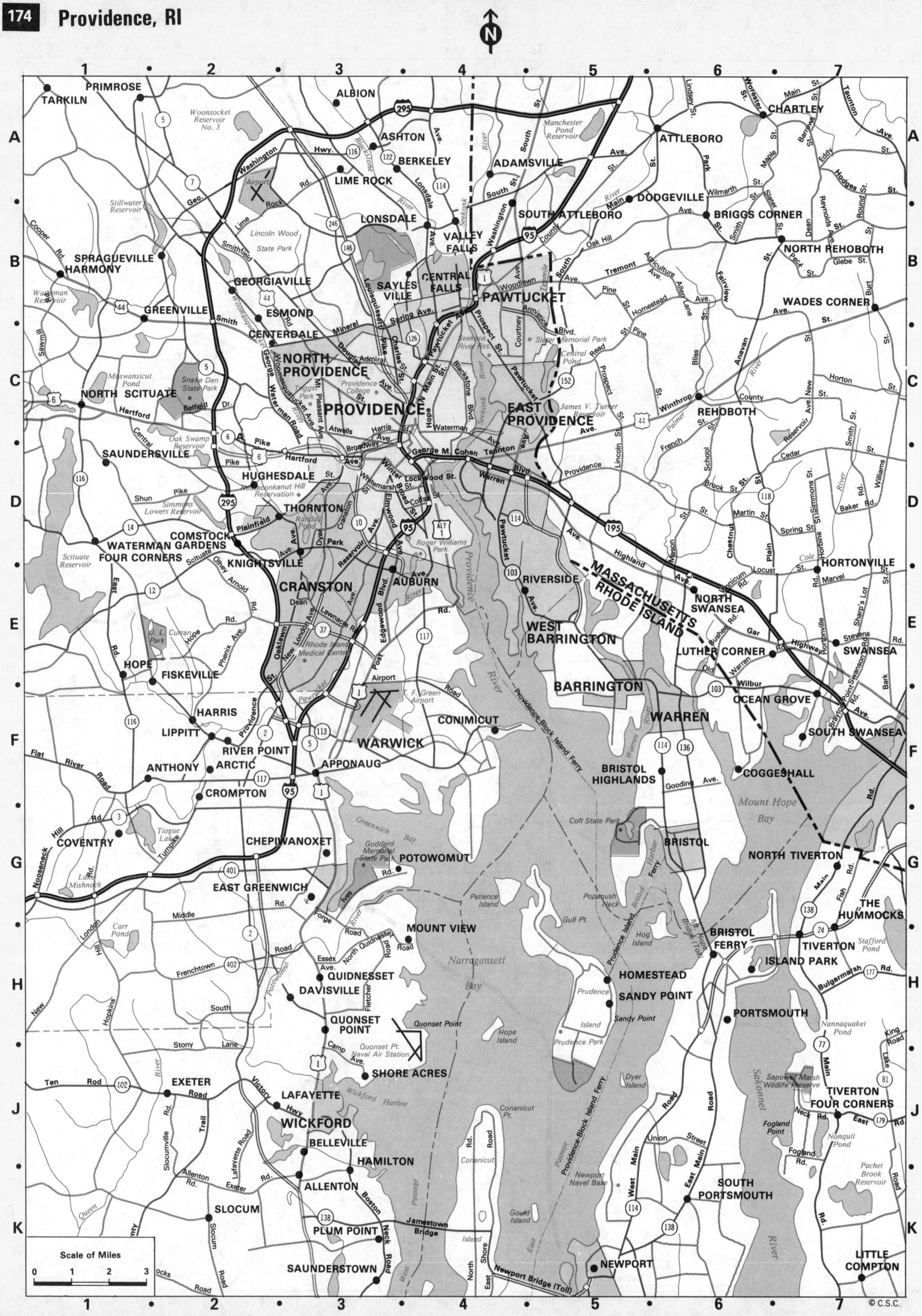

TARKILN
PRIMROSE
ALBION
ASHTON
BERKELEY
ADAMSVILLE
ATTLEBORO
CHARTLEY
LIME ROCK
DODGEVILLE
SOUTH ATTLEBORO
BRIGGS CORNER
LONSDALE
VALLEY FALLS
NORTH REHOBOTH
SPRAGUEVILLE
HARMONY
GEORGIAVILLE
SAYLES VILLE
CENTRAL FALLS
PAWTUCKET
WADES CORNER
GREENVILLE
ESMOND
CENTERDALE
NORTH PROVIDENCE
NORTH SCITUATE
PROVIDENCE
EAST PROVIDENCE
REHOBOTH
SAUNDERSVILLE
HUGHESDALE
THORNTON
COMSTOCK
WATERMAN GARDENS
FOUR CORNERS
KNIGHTSVILLE
CRANSTON
AUBURN
RIVERSIDE
HORTONVILLE
MASSACHUSETTS
RHODE ISLAND
NORTH SWANSEA
WEST BARRINGTON
LUTHER CORNER
SWANSEA
HOPE
FISKEVILLE
BARRINGTON
OCEAN GROVE
HARRIS
LIPPITT
RIVER POINT
CONIMICUT
WARREN
SOUTH SWANSEA
WARWICK
ANTHONY
ARCTIC
APPONAUG
BRISTOL HIGHLANDS
COGGESHALL
CROMPTON
Mount Hope Bay
COVENTRY
CHEPIWANOXET
BRISTOL
NORTH TIVERTON
POTOWOMUT
EAST GREENWICH
THE HUMMOCKS
MOUNT VIEW
BRISTOL FERRY
TIVERTON
ISLAND PARK
Narragansett Bay
QUIDNESSET
HOMESTEAD
DAVISVILLE
SANDY POINT
PORTSMOUTH
QUONSET POINT
Quonset Pt. Naval Air Station
SHORE ACRES
EXETER
LAFAYETTE
TIVERTON FOUR CORNERS
WICKFORD
BELLEVILLE
HAMILTON
ALLENTON
SOUTH PORTSMOUTH
SLOCUM
PLUM POINT
Newport Navel Base
Jamestown Bridge
SAUNDERSTOWN
NEWPORT
Newport Bridge (Toll)
LITTLE COMPTON

Scale of Miles
0 1 2 3

© C.S.C.

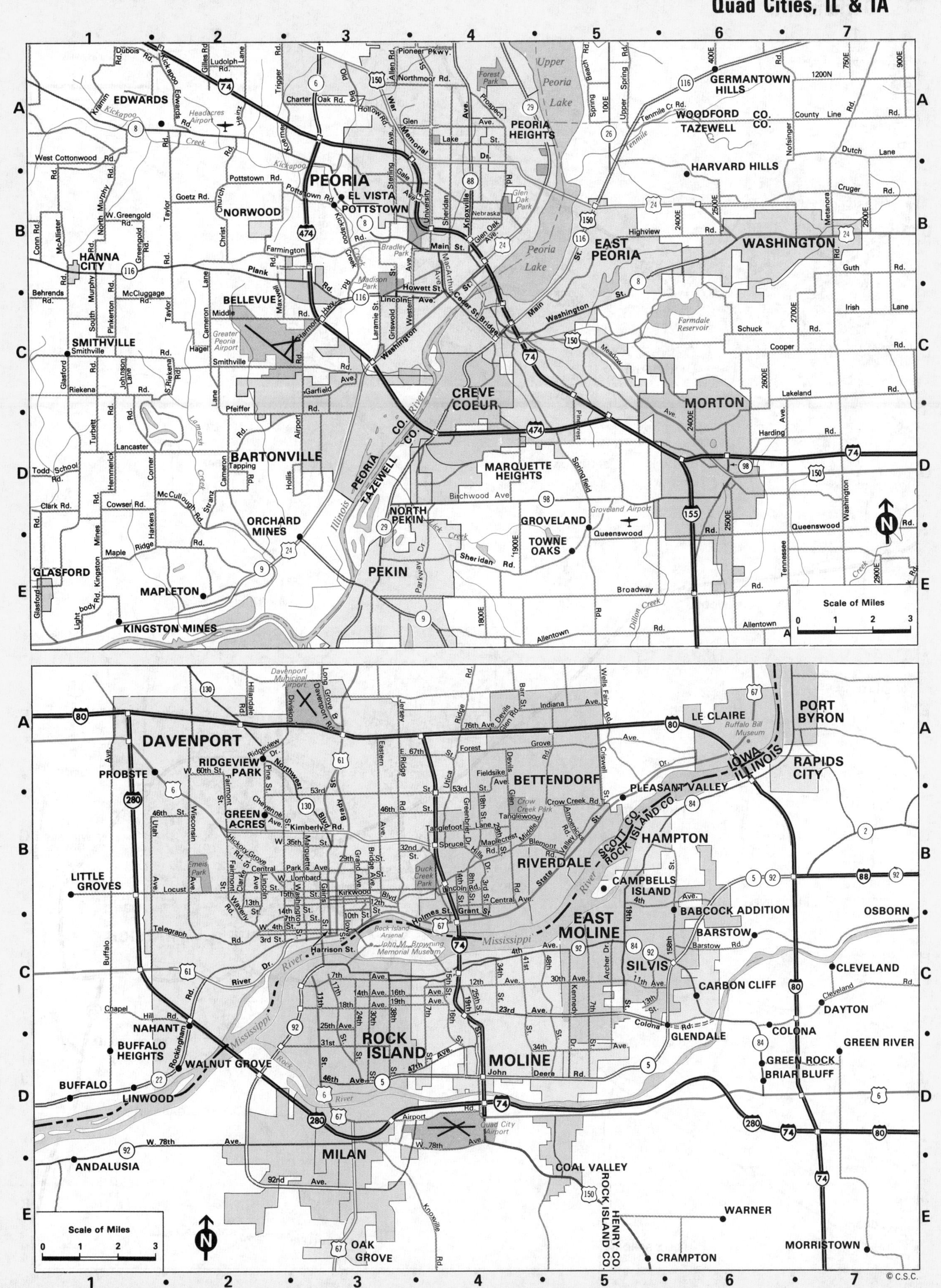
EDWARDS
PEORIA
EL VISTA
POTTSTOWN
NORWOOD
HANNA CITY
BELLEVUE
SMITHVILLE
BARTONVILLE
ORCHARD MINES
GLASFORD
MAPLETON
KINGSTON MINES
PEORIA HEIGHTS
EAST PEORIA
CREVE COEUR
MARQUETTE HEIGHTS
NORTH PEKIN
PEKIN
GROVELAND
TOWNE OAKS
MORTON
WASHINGTON
GERMANTOWN HILLS
HARVARD HILLS
WOODFORD CO.
TAZEWELL CO.
PEORIA CO.
Upper Peoria Lake
Peoria Lake
Illinois River
Greater Peoria Airport
Farmdale Reservoir
Groveland Airport
Scale of Miles
0
1
2
3
DAVENPORT
PROBSTE
RIDGEVIEW PARK
GREEN ACRES
LITTLE GROVES
BETTENDORF
RIVERDALE
LE CLAIRE
PORT BYRON
RAPIDS CITY
PLEASANT VALLEY
HAMPTON
CAMPBELLS ISLAND
BABCOCK ADDITION
EAST MOLINE
BARSTOW
OSBORN
SILVIS
CLEVELAND
CARBON CLIFF
DAYTON
GLENDALE
COLONA
GREEN RIVER
GREEN ROCK
BRIAR BLUFF
ROCK ISLAND
MOLINE
NAHANT
BUFFALO HEIGHTS
WALNUT GROVE
BUFFALO
LINWOOD
ANDALUSIA
MILAN
COAL VALLEY
WARNER
CRAMPTON
MORRISTOWN
OAK GROVE
IOWA
ILLINOIS
SCOTT CO.
ROCK ISLAND CO.
HENRY CO.
Mississippi River
Rock Island Arsenal
John M. Browning Memorial Museum
Davenport Municipal Airport
Quad City Airport
Buffalo Bill Museum
Scale of Miles
0
1
2
3
© C.S.C.

Raleigh, NC
Durham, NC

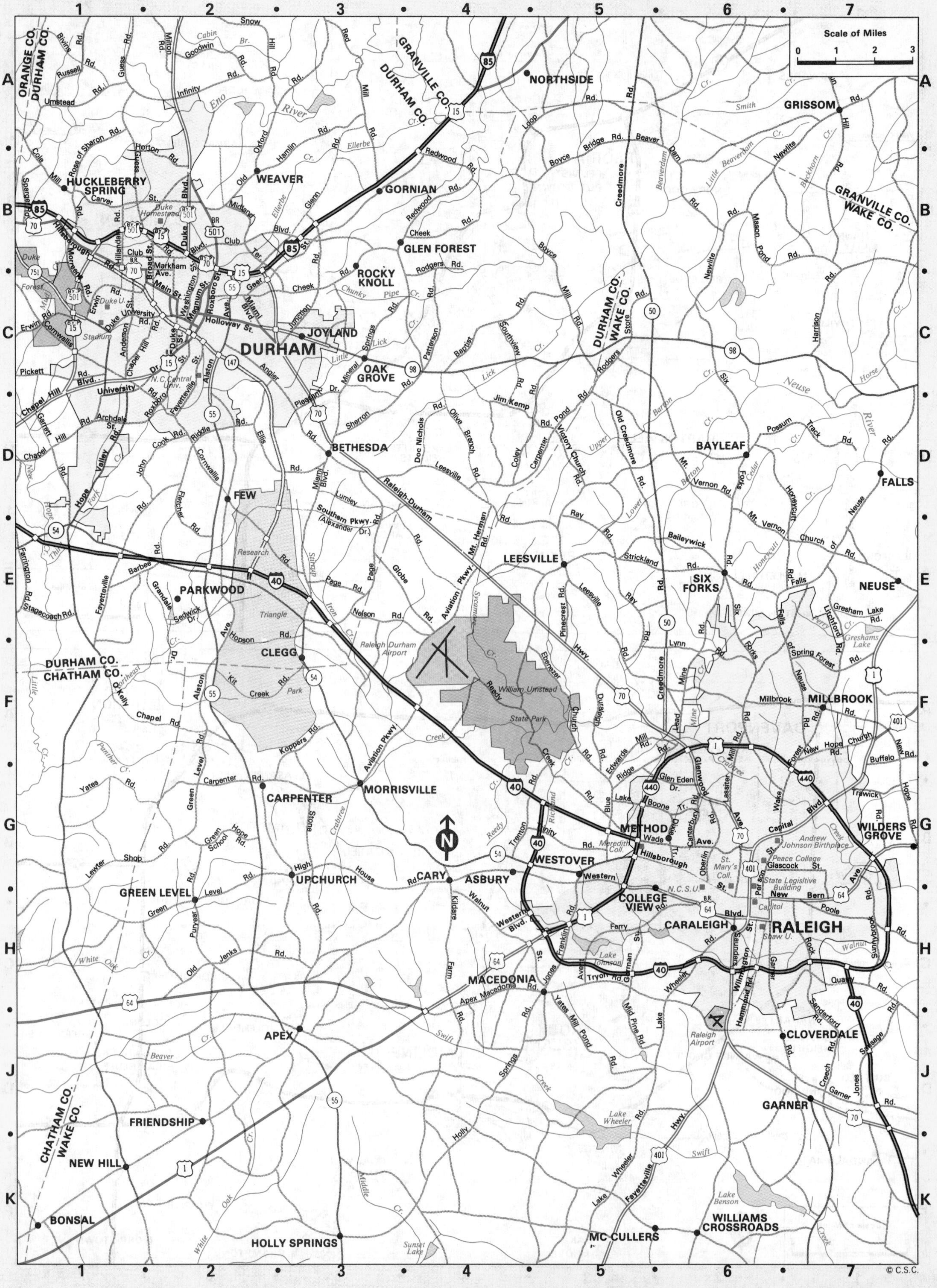

Scale of Miles
0 1 2 3

ORANGE CO.
DURHAM CO.
GRANVILLE CO.
DURHAM CO.
GRANVILLE CO.
WAKE CO.
DURHAM CO.
WAKE CO.
DURHAM CO.
CHATHAM CO.
CHATHAM CO.
WAKE CO.

NORTHSIDE
GRISSOM
HUCKLEBERRY SPRING
WEAVER
GORNIAN
GLEN FOREST
ROCKY KNOLL
JOYLAND
DURHAM
OAK GROVE
BETHESDA
BAYLEAF
FALLS
FEW
LEESVILLE
SIX FORKS
NEUSE
PARKWOOD
CLEGG
MILLBROOK
MORRISVILLE
CARPENTER
METHOD
WILDERS GROVE
WESTOVER
CARY
ASBURY
UPCHURCH
GREEN LEVEL
COLLEGE VIEW
CARALEIGH
RALEIGH
MACEDONIA
APEX
CLOVERDALE
GARNER
FRIENDSHIP
NEW HILL
BONSAL
HOLLY SPRINGS
MC CULLERS
WILLIAMS CROSSROADS

Raleigh Durham Airport
William Umstead State Park
Raleigh Airport
Lake Wheeler
Lake Benson
Lake Johnson
Sunset Lake
Research Triangle Park
N.C. Central Univ.
Duke U.
Duke Forest
N.C.S.U.
Meredith Coll.
St. Mary's Coll.
Peace College
Shaw U.
State Legislative Building
Capitol
Andrew Johnson Birthplace
Duke Homestead

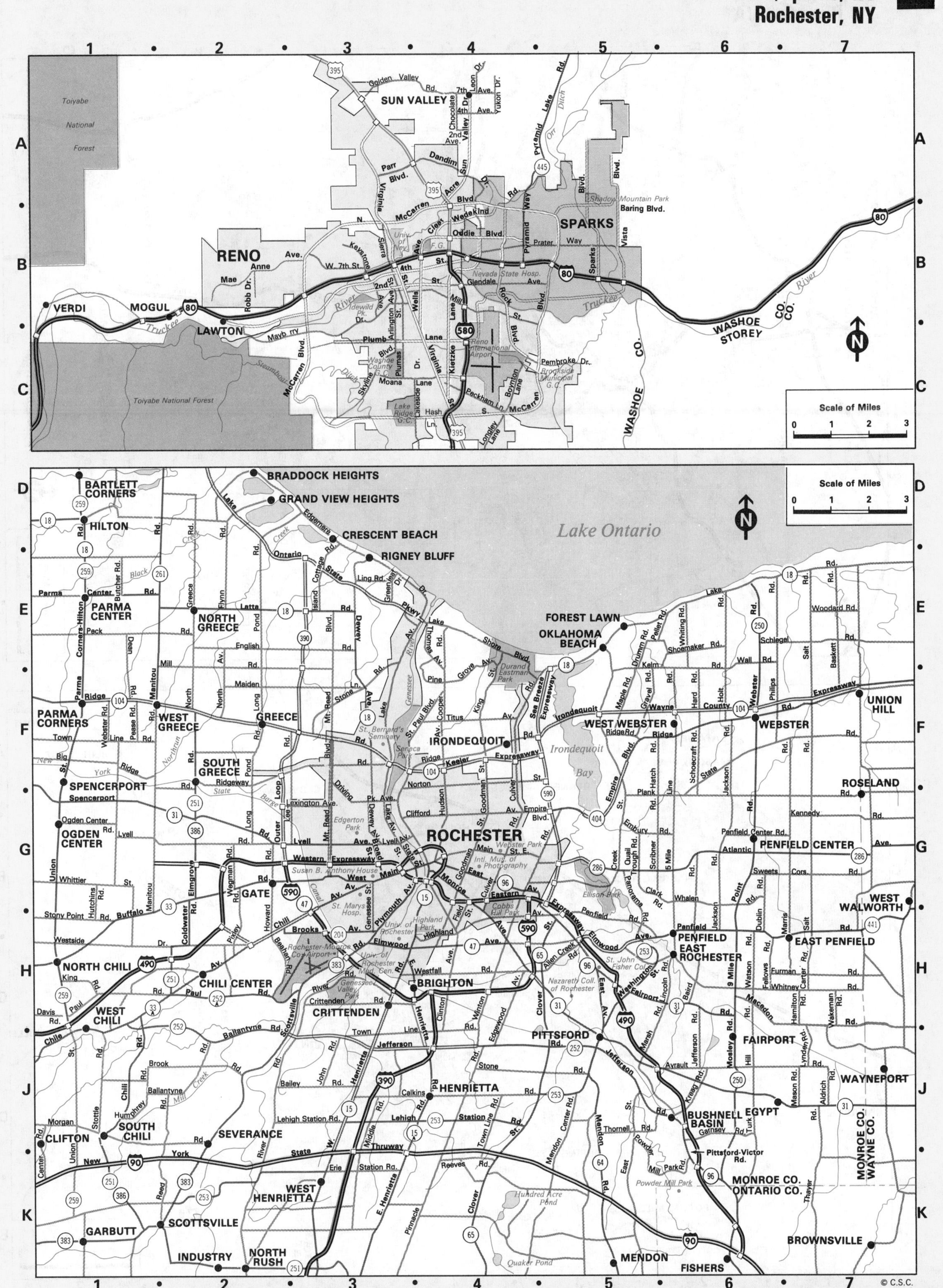
RENO
SPARKS
SUN VALLEY
VERDI
MOGUL
LAWTON
Toiyabe National Forest
Scale of Miles
WASHOE CO.
STOREY CO.
Lake Ontario
ROCHESTER
BRADDOCK HEIGHTS
GRAND VIEW HEIGHTS
CRESCENT BEACH
RIGNEY BLUFF
BARTLETT CORNERS
HILTON
PARMA CENTER
NORTH GREECE
FOREST LAWN
OKLAHOMA BEACH
PARMA CORNERS
WEST GREECE
GREECE
IRONDEQUOIT
WEST WEBSTER
WEBSTER
UNION HILL
SOUTH GREECE
SPENCERPORT
ROSELAND
OGDEN CENTER
PENFIELD CENTER
GATES
WEST WALWORTH
NORTH CHILI
CHILI CENTER
BRIGHTON
PENFIELD
EAST ROCHESTER
EAST PENFIELD
WEST CHILI
CRITTENDEN
PITTSFORD
FAIRPORT
HENRIETTA
WAYNEPORT
EGYPT
BUSHNELL BASIN
SOUTH CHILI
CLIFTON
SEVERANCE
WEST HENRIETTA
MONROE CO.
ONTARIO CO.
WAYNE CO.
GARBUTT
SCOTTSVILLE
INDUSTRY
NORTH RUSH
MENDON
FISHERS
BROWNSVILLE
© C.S.C.

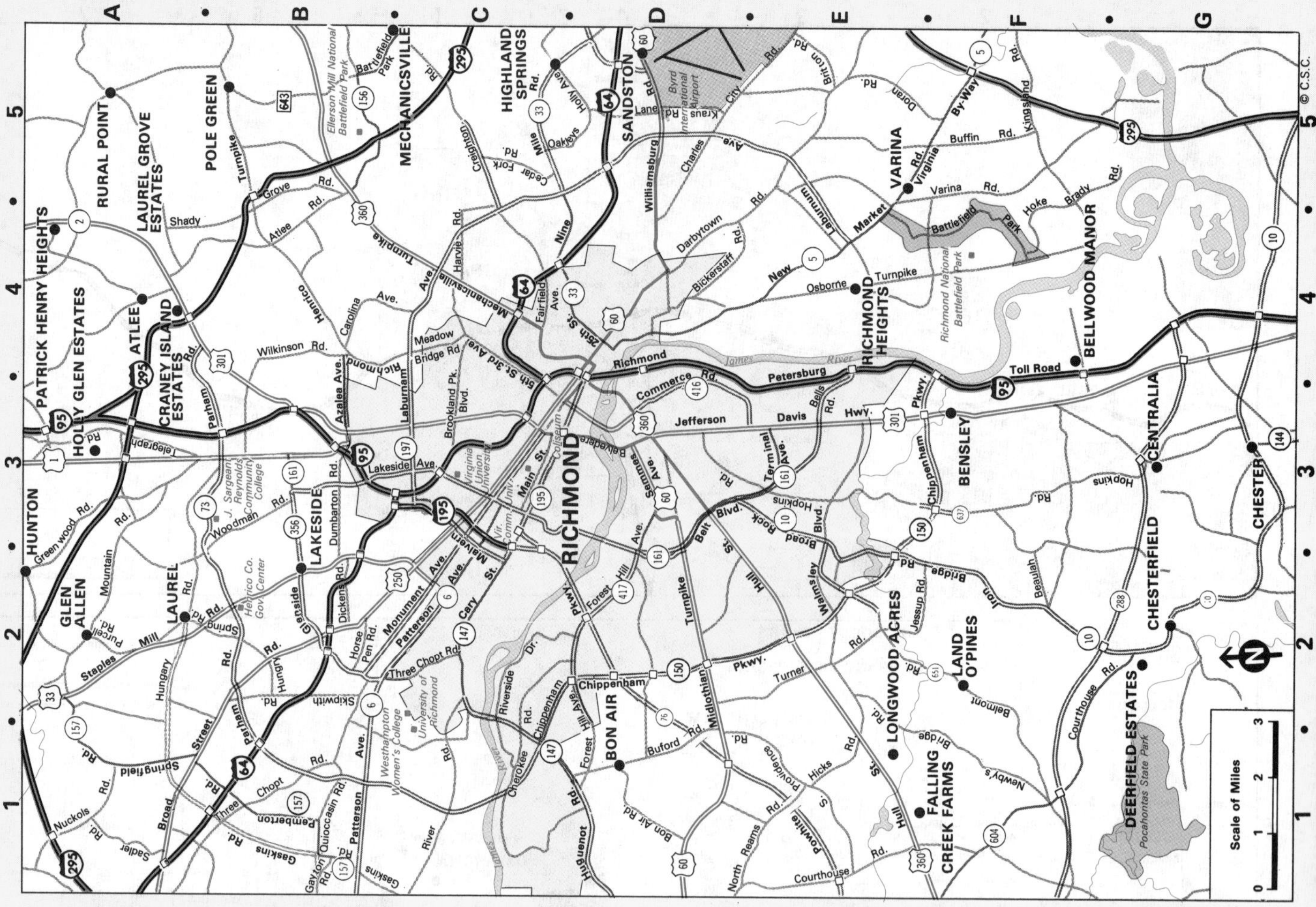

Richmond, VA
Roanoke, VA

Richmond, VA

RURAL POINT
POLE GREEN
LAUREL GROVE ESTATES
MECHANICSVILLE
HIGHLAND SPRINGS
SANDSTON
Byrd International Airport
Ellerson Mill National Battlefield Park
PATRICK HENRY HEIGHTS
HOLLY GLEN ESTATES
ATLEE
CRANEY ISLAND ESTATES
VARINA
Richmond National Battlefield Park
RICHMOND HEIGHTS
BELLWOOD MANOR
RICHMOND
James River
BENSLEY
CENTRALIA
CHESTER
HUNTON
GLEN ALLEN
LAUREL
LAKESIDE
J. Sargeant Reynolds Community College
Henrico Co. Gov't Center
Virginia Union University
Vir. Commonwealth Univ.
University of Richmond
Westhampton Women's College
BON AIR
LONGWOOD ACRES
LAND O'PINES
CHESTERFIELD
DEERFIELD ESTATES
Pocahontas State Park
FALLING CREEK FARMS

Scale of Miles
0 1 2 3

© C.S.C.

Roanoke, VA

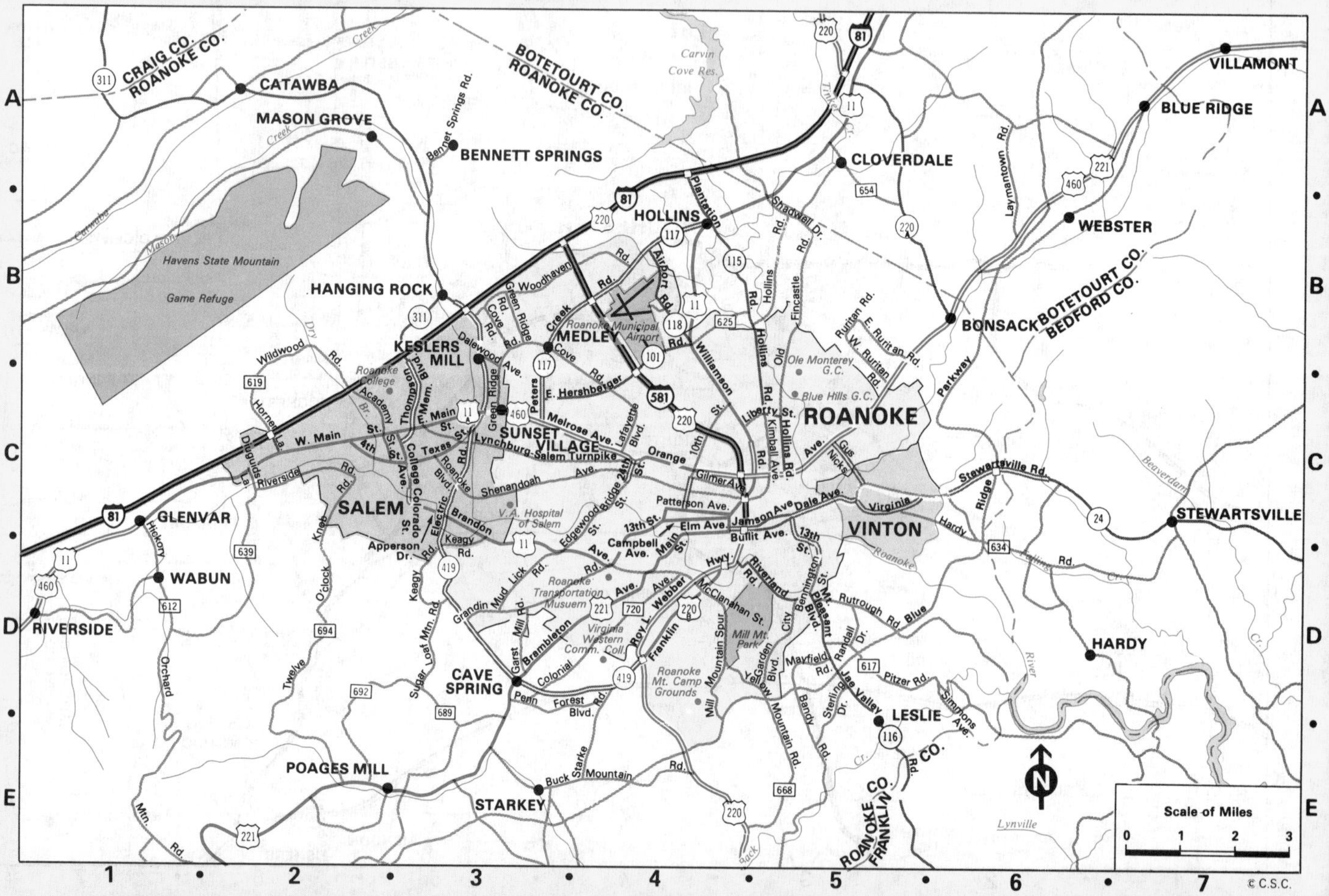

CRAIG CO.
ROANOKE CO.
CATAWBA
BOTETOURT CO.
ROANOKE CO.
Carvin Cove Res.
VILLAMONT
BLUE RIDGE
MASON GROVE
BENNETT SPRINGS
CLOVERDALE
HOLLINS
WEBSTER
Havens State Mountain Game Refuge
HANGING ROCK
Roanoke Municipal Airport
MEDLEY
KESLERS MILL
Roanoke College
BONSACK
BOTETOURT CO.
BEDFORD CO.
Ole Monterey G.C.
Blue Hills G.C.
ROANOKE
SUNSET VILLAGE
SALEM
V.A. Hospital of Salem
GLENVAR
VINTON
STEWARTSVILLE
WABUN
RIVERSIDE
Roanoke Transportation Museum
Virginia Western Comm. Coll.
Mill Mt. Park
Roanoke Mt. Camp Grounds
CAVE SPRING
HARDY
LESLIE
POAGES MILL
STARKEY
ROANOKE CO.
FRANKLIN CO.

Scale of Miles
0 1 2 3

© C.S.C.

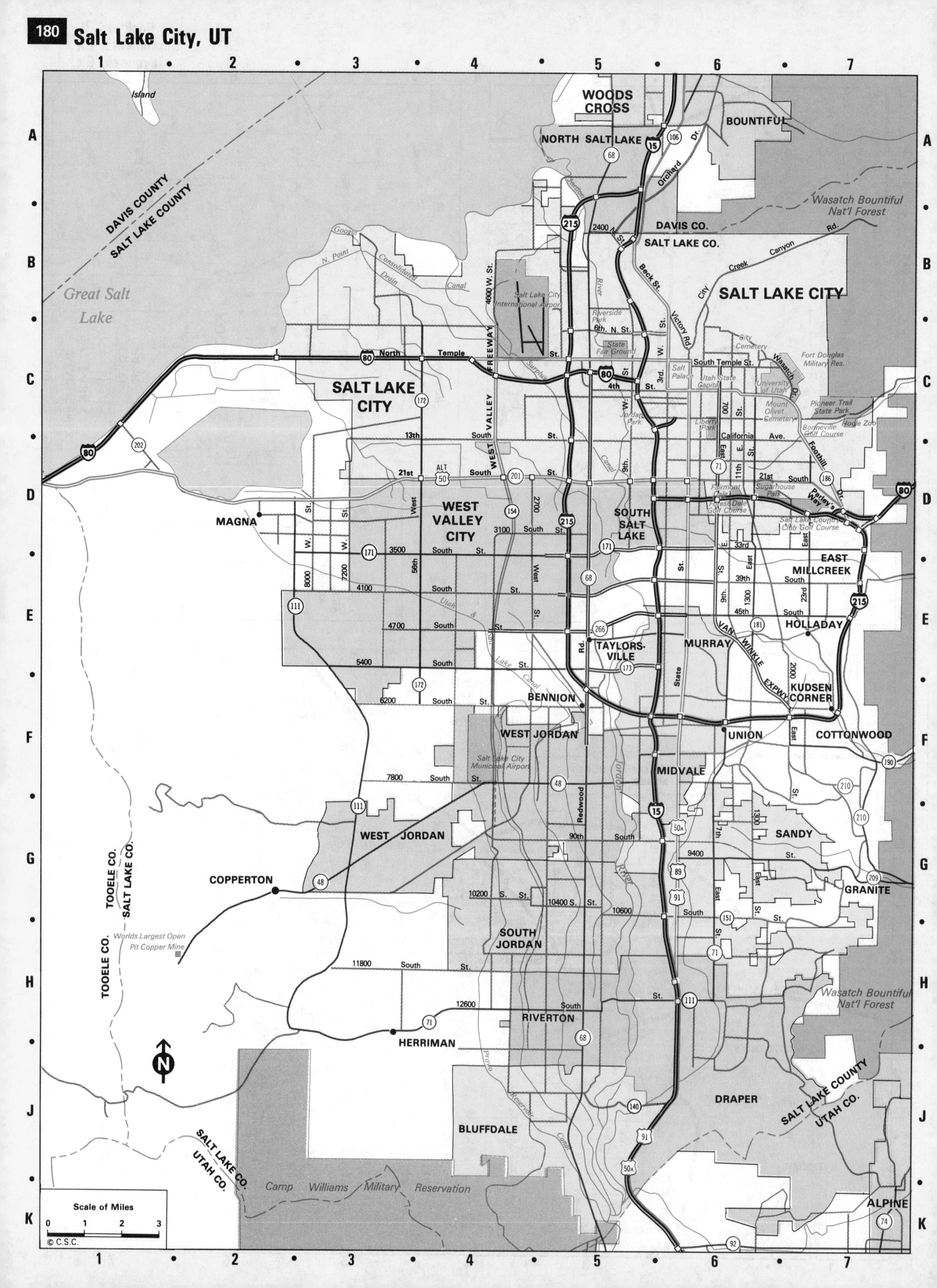
Island
Great Salt Lake
DAVIS COUNTY
SALT LAKE COUNTY
WOODS CROSS
BOUNTIFUL
NORTH SALT LAKE
Orchard Dr.
Wasatch Bountiful Nat'l Forest
DAVIS CO.
SALT LAKE CO.
2400 N. St.
Beck St.
Victory Rd.
City Creek Canyon Rd.
SALT LAKE CITY
Salt Lake City International Airport
4000 W. St.
N. Point Consolidated Drain
Canal
Riverside Park
6th. N. St.
State Fair Ground
City Cemetery
Fort Douglas Military Res.
North Temple St.
South Temple St.
Salt Palace
Utah State Capitol
University of Utah
Wasatch Dr.
SALT LAKE CITY
4th St.
Jordan Park
Liberty Park
Mount Olivet Cemetery
Pioneer Trail State Park
Bonneville Golf Course
Hogle Zoo
Foothill Dr.
13th South St.
California Ave.
21st South St.
21st South
Sugarhouse Park
Fairmont Park
Forest Dale Golf Course
Parley's Way
Salt Lake Country Club Golf Course
MAGNA
WEST VALLEY CITY
WEST VALLEY FREEWAY
SOUTH SALT LAKE
3100 South St.
3500 South St.
33rd
39th South
EAST MILLCREEK
4100 South St.
45th South
4700 South St.
HOLLADAY
TAYLORSVILLE
MURRAY
VAN WINKLE EXPWY
5400 South St.
KUDSEN CORNER
6200 South St.
BENNION
WEST JORDAN
UNION
COTTONWOOD
Salt Lake City Municipal Airport
MIDVALE
7800 South St.
Redwood Rd.
SANDY
WEST JORDAN
90th South
9400 St.
GRANITE
COPPERTON
10200 S. St.
10400 S. St.
10600 South St.
TOOELE CO.
SALT LAKE CO.
Worlds Largest Open Pit Copper Mine
SOUTH JORDAN
11800 South St.
Wasatch Bountiful Nat'l Forest
12600 South St.
RIVERTON
HERRIMAN
Jordan River
Provo Reservoir Canal
DRAPER
SALT LAKE COUNTY
UTAH CO.
BLUFFDALE
SALT LAKE CO.
UTAH CO.
Camp Williams Military Reservation
ALPINE
Scale of Miles
0 1 2 3
© C.S.C.

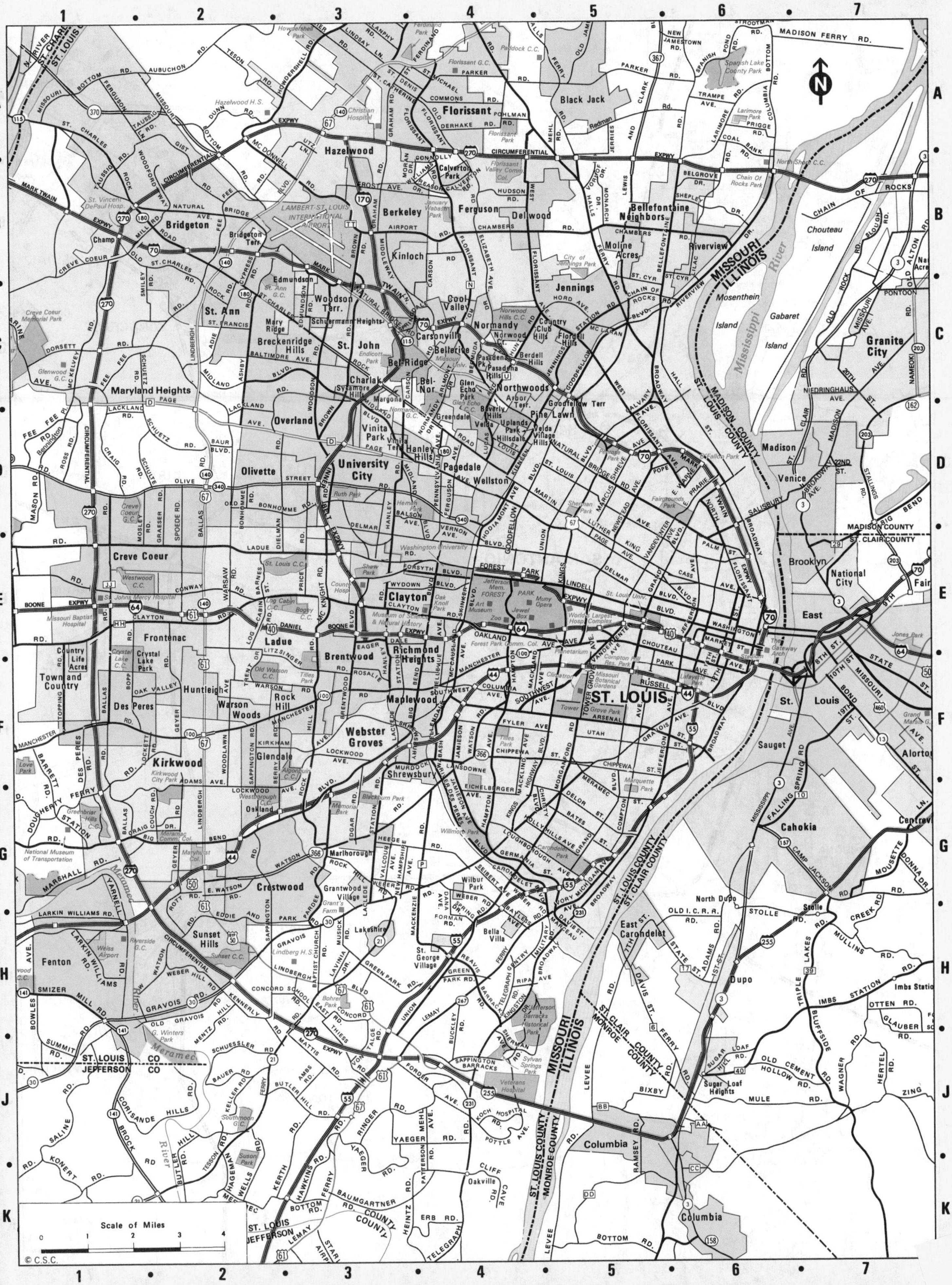

St. Louis
Florissant
Black Jack
Hazelwood
Berkeley
Ferguson
Dellwood
Bellefontaine Neighbors
Bridgeton
Kinloch
Moline Acres
Riverview
Jennings
Edmundson
Woodson Terr.
St. Ann
Cool Valley
Normandy
Carsonville
St. John
Breckenridge Hills
Bel-Ridge
Bellerive
Northwoods
Pine Lawn
Charlack
Bel-Nor
Maryland Heights
Overland
Vinita Park
Hanley Hills
Pagedale
Wellston
University City
Olivette
Creve Coeur
Clayton
Richmond Heights
Brentwood
Ladue
Frontenac
Country Life Acres
Town and Country
Des Peres
Huntleigh
Warson Woods
Rock Hill
Webster Groves
Maplewood
Shrewsbury
Glendale
Kirkwood
Oakland
Marlborough
Crestwood
Grantwood Village
Lakeshire
St. George Village
Bella Villa
Sunset Hills
Fenton
Granite City
Madison
Venice
Brooklyn
National City
East St. Louis
Sauget
Alorton
Cahokia
Centreville
East Carondelet
North Dupo
Dupo
Stolle
Sugar Loaf Heights
Columbia
Oakville
Mississippi River
Missouri River
Meramec River
Missouri / Illinois
St. Louis County / Madison County
Madison County / St. Clair County
St. Louis County / St. Clair County
St. Louis County / Monroe County
St. Clair County / Monroe County
St. Louis Co / Jefferson Co

Scale of Miles
0 1 2 3 4

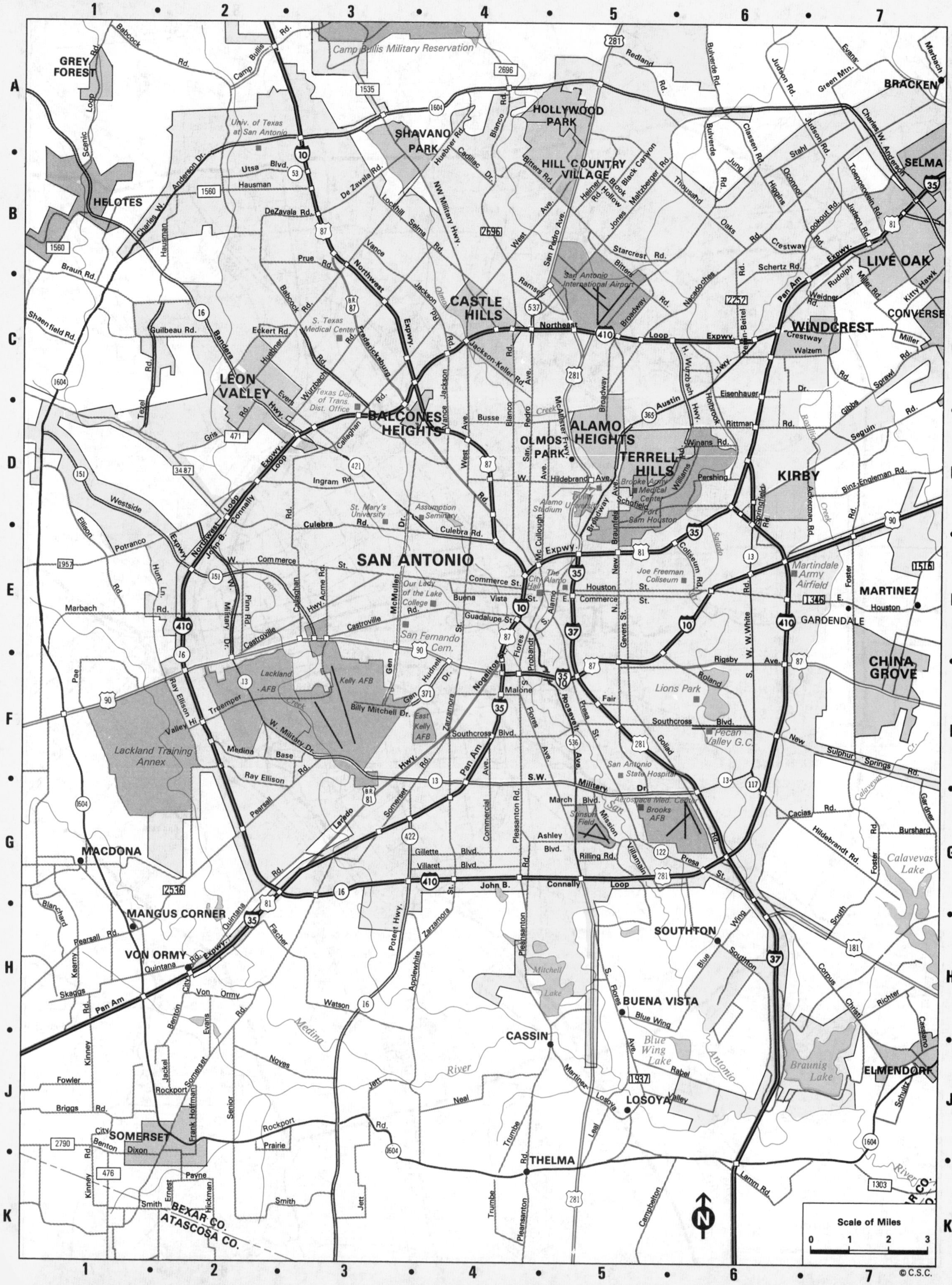

San Antonio, TX

GREY FOREST
HELOTES
SHAVANO PARK
HOLLYWOOD PARK
HILL COUNTRY VILLAGE
BRACKEN
SELMA
LIVE OAK
CONVERSE
WINDCREST
CASTLE HILLS
LEON VALLEY
BALCONES HEIGHTS
ALAMO HEIGHTS
OLMOS PARK
TERRELL HILLS
KIRBY
SAN ANTONIO
MARTINEZ
GARDENDALE
CHINA GROVE
MACDONA
MANGUS CORNER
VON ORMY
SOUTHTON
BUENA VISTA
CASSIN
LOSOYA
ELMENDORF
THELMA
SOMERSET
Camp Bullis Military Reservation
Univ. of Texas at San Antonio
San Antonio International Airport
S. Texas Medical Center
Texas Dept. of Trans. Dist. Office
Trinity University
Alamo Stadium
Brooke Army Medical Center
Fort Sam Houston
St. Mary's University
Assumption Seminary
Our Lady of the Lake College
City Hall
The Alamo
Joe Freeman Coliseum
Martindale Army Airfield
San Fernando Cem.
Lackland AFB
Kelly AFB
East Kelly AFB
Lackland Training Annex
Lions Park
Pecan Valley G.C.
San Antonio State Hospital
Aerospace Med. Center
Brooks AFB
Stinson Field
Mitchell Lake
Blue Wing Lake
Braunig Lake
Calaveras Lake
Medina River
San Antonio River
BEXAR CO.
ATASCOSA CO.
Scale of Miles
0 1 2 3

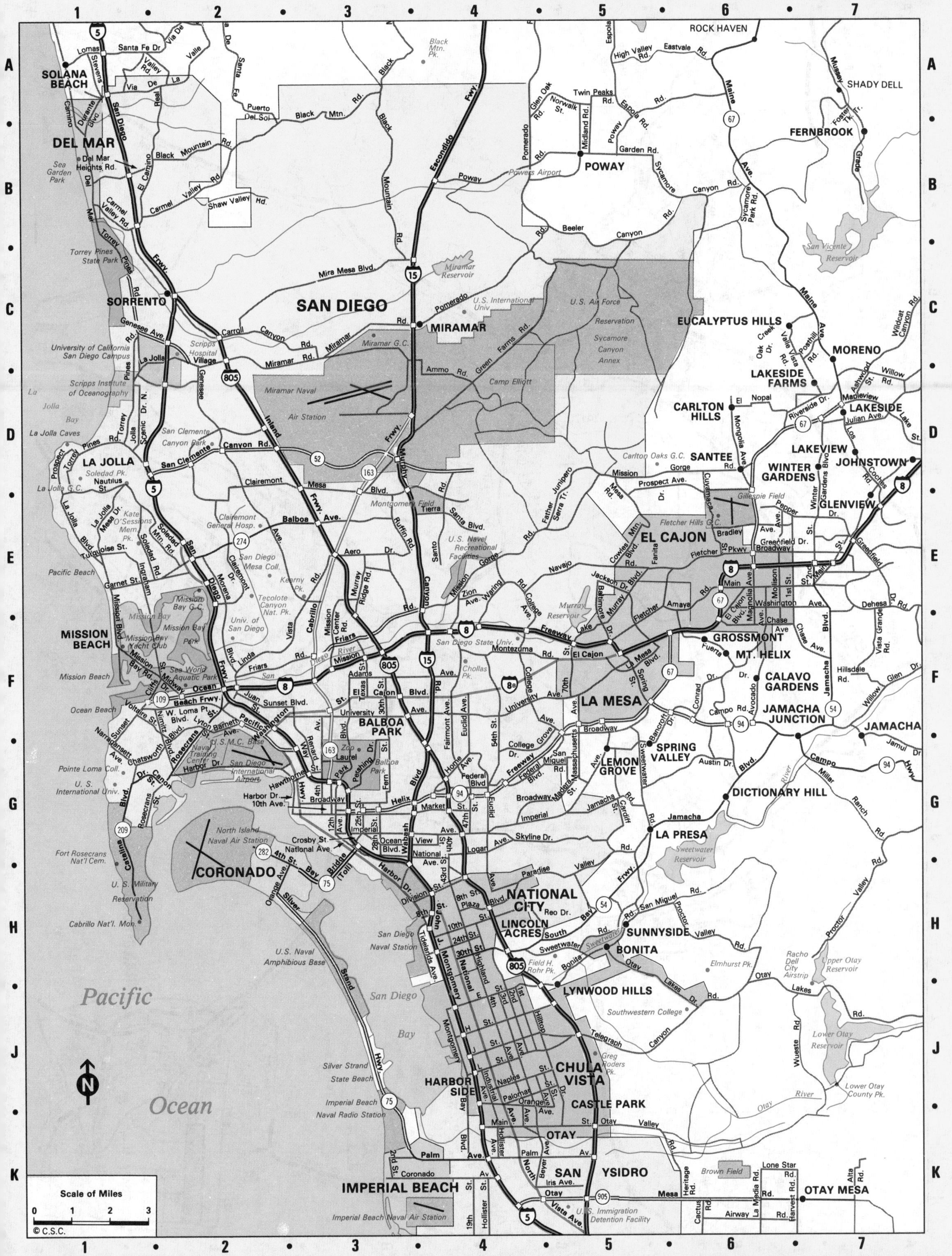
SOLANA BEACH
DEL MAR
Sea Garden Park
Torrey Pines State Park
SORRENTO
SAN DIEGO
MIRAMAR
University of California San Diego Campus
Scripps Hospital
Scripps Institute of Oceanography
La Jolla Bay
La Jolla Caves
LA JOLLA
Soledad Pk.
La Jolla G.C.
Pacific Beach
MISSION BEACH
Mission Bay
Mission Bay G.C.
Mission Bay Park
Mission Bay Yacht Club
Mission Beach
Sea World Aquatic Park
Ocean Beach
Pointe Loma Coll.
U. S. International Univ.
Fort Rosecrans Nat'l Cem.
U. S. Military Reservation
Cabrillo Nat'l. Mon.
North Island Naval Air Station
CORONADO
San Diego Naval Station
U.S. Naval Amphibious Base
Pacific
Ocean
San Diego Bay
Silver Strand State Beach
Imperial Beach Naval Radio Station
IMPERIAL BEACH
Imperial Beach Naval Air Station
Miramar Reservoir
Miramar G.C.
Miramar Naval Air Station
U.S. International Univ.
U.S. Air Force Reservation
Sycamore Canyon Annex
Camp Elliott
San Clemente Canyon Park
Clairemont General Hosp.
Kate O'Sessions Mem. Pk.
San Diego Mesa Coll.
Kearny Pk.
Tecolote Canyon Nat. Pk.
Univ. of San Diego
Montgomery Field
U.S. Naval Recreational Facilities
San Diego State Univ.
Murray Reservoir
Chollas Pk.
BALBOA PARK
Zoo
U.S.M.C. Base
Naval Training Center
San Diego International Airport
POWAY
Powers Airport
ROCK HAVEN
SHADY DELL
FERNBROOK
San Vicente Reservoir
EUCALYPTUS HILLS
MORENO
LAKESIDE FARMS
LAKESIDE
CARLTON HILLS
SANTEE
Carlton Oaks G.C.
LAKEVIEW
WINTER GARDENS
JOHNSTOWN
GLENVIEW
Gillespie Field
Fletcher Hills G.C.
EL CAJON
GROSSMONT
MT. HELIX
CALAVO GARDENS
LA MESA
JAMACHA JUNCTION
JAMACHA
SPRING VALLEY
LEMON GROVE
DICTIONARY HILL
LA PRESA
Sweetwater Reservoir
NATIONAL CITY
LINCOLN ACRES
SUNNYSIDE
BONITA
Field H. Rohr Pk.
LYNWOOD HILLS
Southwestern College
Elmhurst Pk.
Racho Dell City Airstrip
Upper Otay Reservoir
Lower Otay Reservoir
Lower Otay County Pk.
Otay River
Greg Roders Pk.
CHULA VISTA
HARBOR SIDE
CASTLE PARK
OTAY
SAN YSIDRO
U.S. Immigration Detention Facility
Brown Field
OTAY MESA
Scale of Miles
0 1 2 3
© C.S.C.

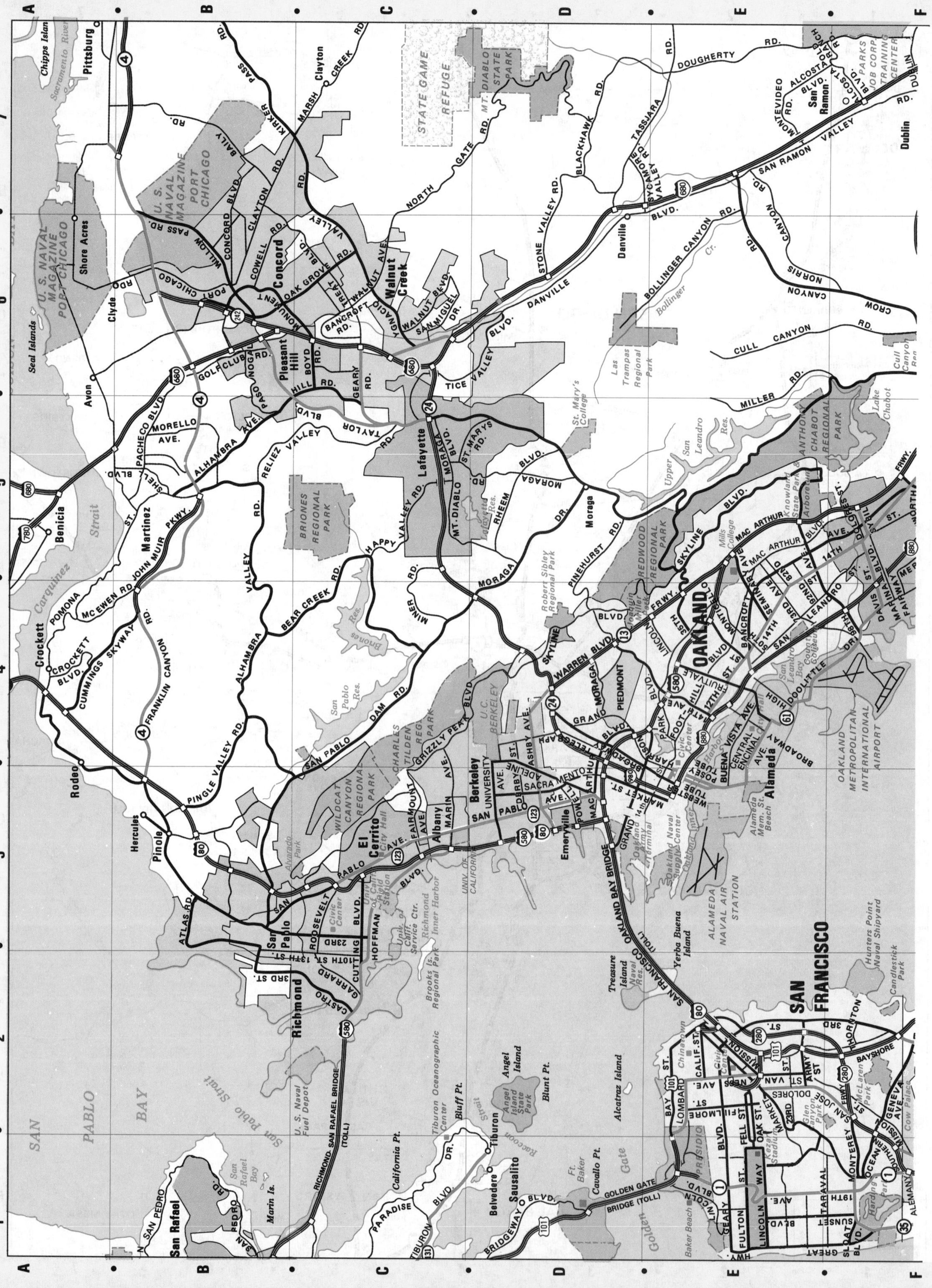

A B C D E F

1 2 3 4 5 6 7

Pittsburg
Chipps Island
Sacramento River
Clayton
STATE GAME REFUGE
MT. DIABLO STATE PARK
U. S. NAVAL MAGAZINE PORT CHICAGO
Shore Acres
Clyde
Concord
Walnut Creek
Pleasant Hill
Seal Islands
Avon
Danville
San Ramon
Dublin
Las Trampas Regional Park
Bollinger Cr.
Lafayette
St. Mary's College
BRIONES REGIONAL PARK
Benicia
Carquinez Strait
Martinez
Moraga
Upper San Leandro Res.
ANTHONY CHABOT REGIONAL PARK
Lake Chabot
Crockett
Robert Sibley Regional Park
REDWOOD REGIONAL PARK
Mills College
Knowland State Park
Arboretum
Briones Res.
San Pablo Res.
Rodeo
Hercules
Pinole
CHARLES TILDEN REG. PARK
WILDCAT CANYON REGIONAL PARK
U.C. BERKELEY
Berkeley
Albany
El Cerrito
Emeryville
Piedmont
OAKLAND
Alameda
Civic Center
San Leandro Bay
Coliseum
OAKLAND METROPOLITAN INTERNATIONAL AIRPORT
ALAMEDA NAVAL AIR STATION
Oakland Naval Supply Center
Alameda Mem. St. Beach
San Pablo
Richmond
Civic Center
Richmond Inner Harbor
Brooks Is. Regional Park
Yerba Buena Island
Treasure Island Naval Res.
SAN FRANCISCO
Hunters Point Naval Shipyard
Candlestick Park
SAN PABLO BAY
San Pablo Strait
U. S. Naval Fuel Depot
RICHMOND SAN RAFAEL BRIDGE (TOLL)
San Rafael
San Rafael Bay
Marin Is.
California Pt.
Tiburon Oceanographic Center
Bluff Pt.
Tiburon
Raccoon Strait
Angel Island
Angel Island State Park
Blunt Pt.
Alcatraz Island
Belvedere
Sausalito
Ft. Baker
Cavallo Pt.
GOLDEN GATE BRIDGE (TOLL)
Golden Gate
PRESIDIO
Baker Beach
Chinatown
Civic Center
Kezar Stadium
Glen Canyon Park
McLaren Park
Harding Park
Cow Palace
SAN FRANCISCO OAKLAND BAY BRIDGE (TOLL)

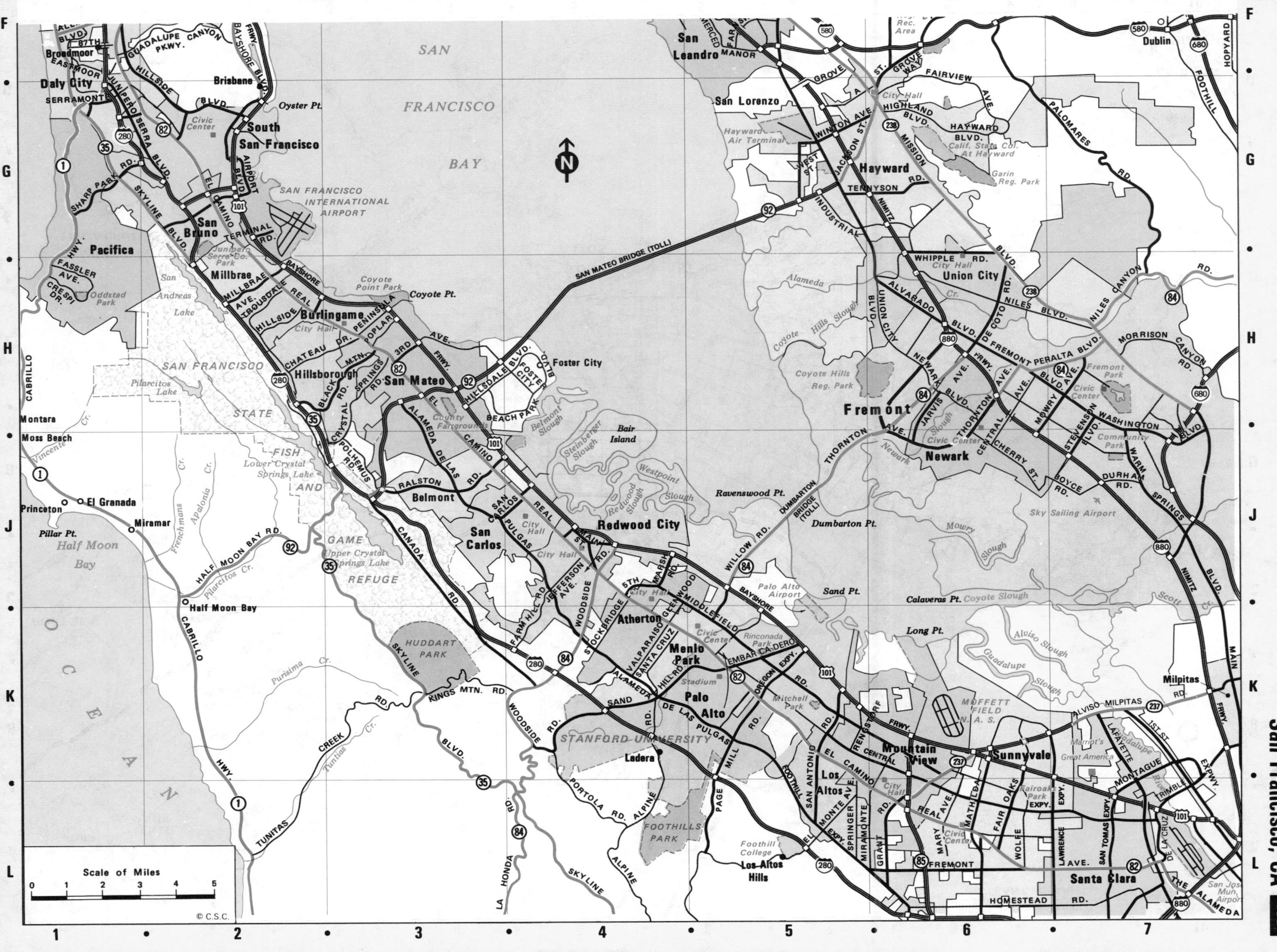
SAN FRANCISCO BAY
O C E A N
Dublin
San Leandro
San Lorenzo
Hayward
Union City
Fremont
Newark
Milpitas
Santa Clara
Sunnyvale
Mountain View
Los Altos
Los Altos Hills
Palo Alto
Menlo Park
Atherton
Ladera
Redwood City
Foster City
San Mateo
San Carlos
Belmont
Burlingame
Hillsborough
Millbrae
San Bruno
South San Francisco
Brisbane
Daly City
Broadmoor
Pacifica
Montara
Moss Beach
Princeton
El Granada
Miramar
Half Moon Bay
San Francisco International Airport
Stanford University
San Mateo Bridge (Toll)
Dumbarton Bridge (Toll)
Scale of Miles
© C.S.C.

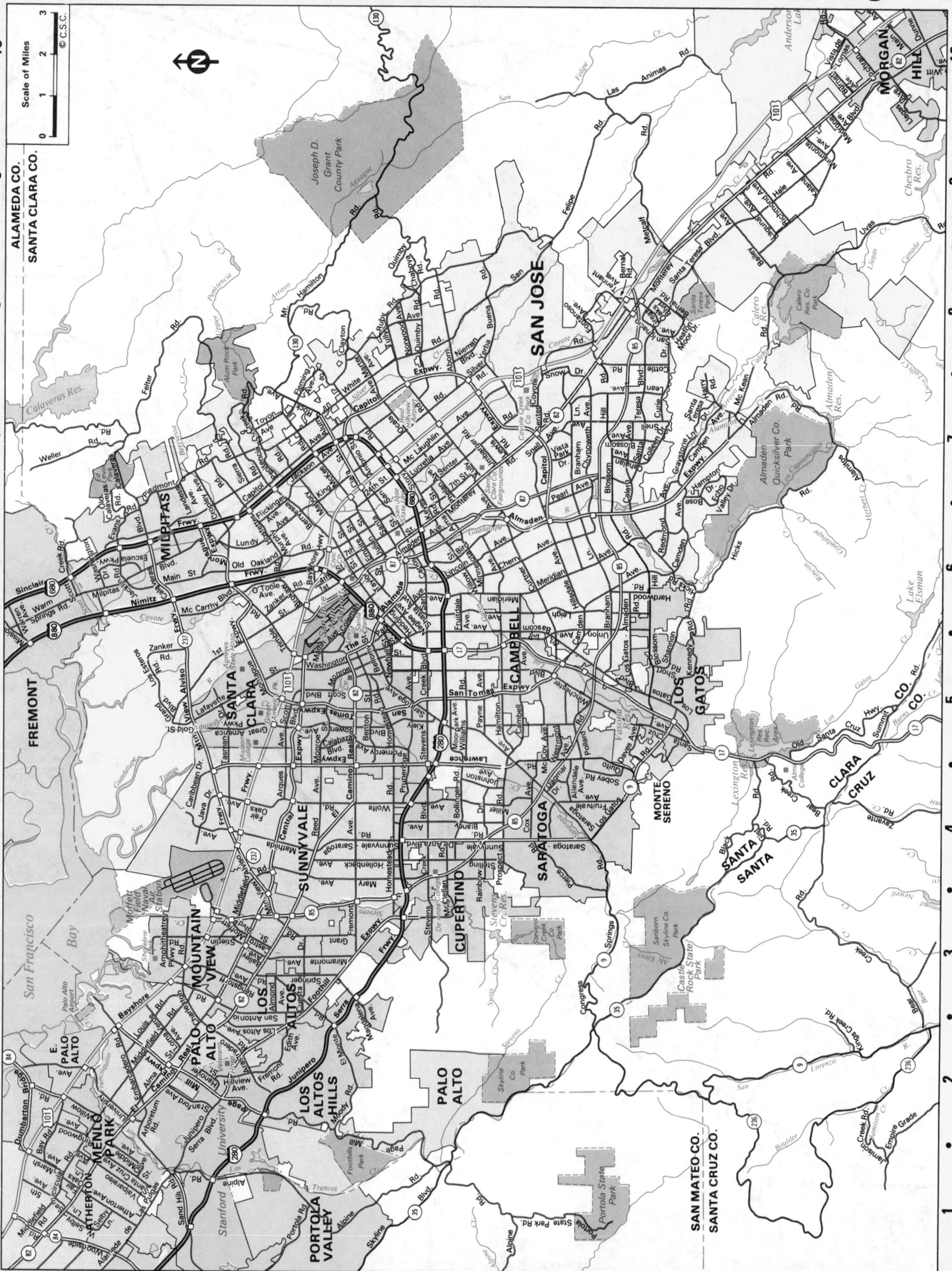

Scale of Miles
0 1 2 3
© C.S.C.

ALAMEDA CO.
SANTA CLARA CO.

SAN MATEO CO.
SANTA CRUZ CO.

SANTA CLARA CO.
SANTA CRUZ CO.

SAN JOSE
MILPITAS
FREMONT
SANTA CLARA
SUNNYVALE
CAMPBELL
LOS GATOS
MONTE SERENO
SARATOGA
CUPERTINO
MOUNTAIN VIEW
LOS ALTOS
LOS ALTOS HILLS
PALO ALTO
E. PALO ALTO
MENLO PARK
ATHERTON
PORTOLA VALLEY
MORGAN HILL

Joseph D. Grant County Park
Alum Rock Park
Calaveras Res.
Almaden Quicksilver Co. Park
Calero Res. Co. Park
Santa Teresa Park
Sanborn Skyline Co. Park
Castle Rock State Park
Portola State Park
Foothills Park
Stevens Creek Co. Park
Skyline Co. Park
Moffett Field Naval Air Station
Stanford University
San Francisco Bay
Lexington Res.
Almaden Res.
Lake Elsman
Anderson Lake
Chesbro Res.

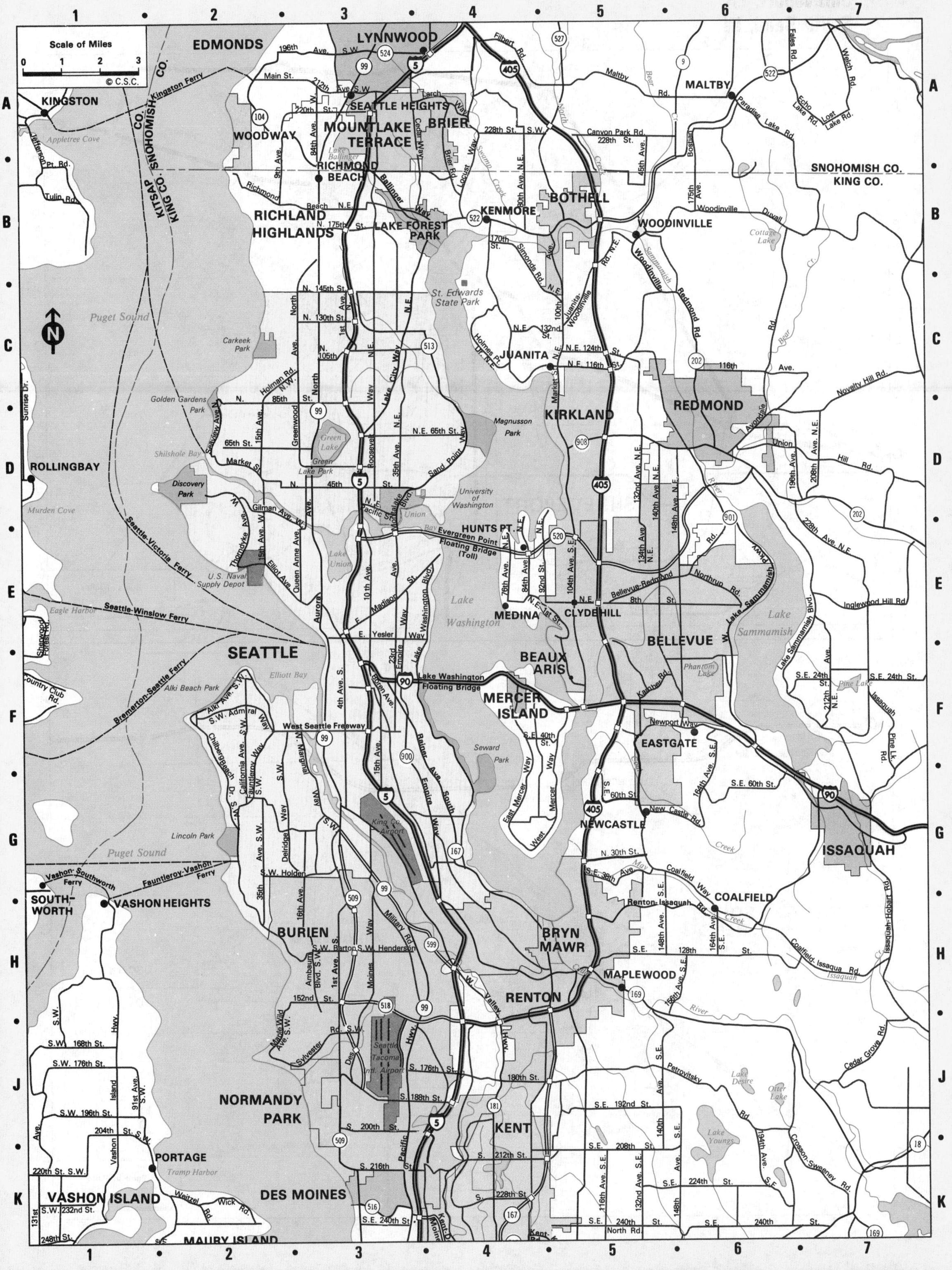
Scale of Miles
0 1 2 3
© C.S.C.
EDMONDS
LYNNWOOD
KINGSTON
Kingston Ferry
CO. SNOHOMISH
KITSAP CO.
KING CO.
WOODWAY
SEATTLE HEIGHTS
MOUNTLAKE TERRACE
BRIER
RICHMOND BEACH
Lake Ballinger
Appletree Cove
Jefferson Pt. Rd.
Tulin Rd.
RICHLAND HIGHLANDS
LAKE FOREST PARK
KENMORE
BOTHELL
MALTBY
WOODINVILLE
Cottage Lake
SNOHOMISH CO.
KING CO.
Filbert Rd.
Maltby Rd.
Canyon Park Rd.
228th St. S.W.
Paradise Lake Rd.
Echo Lake Rd.
Lost Lake Rd.
Welch Rd.
Fales Rd.
Woodinville Duvall
Puget Sound
Carkeek Park
St. Edwards State Park
N. 145th St.
N. 130th St.
N. 105th
Lake City Way
JUANITA
N.E. 124th St.
N.E. 116th St.
Holmes Pt. Dr. N.E.
Juanita Woodinville
Redmond Rd.
116th Ave.
Novelty Hill Rd.
REDMOND
Avondale
Union Hill Rd.
Golden Gardens Park
Holman Rd.
N. 85th St.
Shilshole Bay
Seaview Ave. N.W.
Green Lake
Magnusson Park
N.E. 65th St.
Sand Point Way
KIRKLAND
ROLLINGBAY
Sunrise Dr.
Discovery Park
Market St.
N. 45th St.
Murden Cove
University of Washington
Union Bay
Gilman Ave. W.
Seattle-Victoria Ferry
Thorndyke Ave. W.
15th Ave. W.
Elliott Ave.
Queen Anne Ave.
Aurora
Lake Union
Evergreen Point Floating Bridge (Toll)
HUNTS PT.
Lake Washington
MEDINA
CLYDE HILL
Bellevue-Redmond
N.E. 8th St.
Northrup Way
Inglewood Hill Rd.
Lake Sammamish
W. Lake Sammamish Pkwy.
BELLEVUE
U.S. Naval Supply Depot
Eagle Harbor
Seattle-Winslow Ferry
Sherwood Forest Rd.
SEATTLE
E. Yesler Way
E. Madison St.
Lake Washington Blvd.
BEAUX ARIS
Elliott Bay
Lake Washington Floating Bridge
Country Club Rd.
Bremerton-Seattle Ferry
Alki Beach Park
Alki Ave. S.W.
S.W. Admiral Way
West Seattle Freeway
MERCER ISLAND
S.E. 40th St.
Phantom Lake
S.E. 24th St.
Pine Lake
Lake Sammamish Blvd.
Issaquah Pine Lk. Rd.
Newport Way
EASTGATE
S.E. 60th St.
Seward Park
East Mercer Way
West Mercer Way
Rainier Ave. South
Empire Way
King Co. Airport
NEWCASTLE
New Castle Rd.
Lincoln Park
Puget Sound
N. 30th St.
ISSAQUAH
Coalfield Way
COALFIELD
Renton-Issaquah Rd.
S.W. Holden
Vashon-Southworth Ferry
Fauntleroy-Vashon Ferry
SOUTHWORTH
VASHON HEIGHTS
BURIEN
Military Rd.
BRYN MAWR
S.W. Barton
S.W. Henderson
MAPLEWOOD
S.E. 128th St.
Coalfield-Issaqua Rd.
Issaquah-Hobart Rd.
RENTON
S.W. 168th St.
S.W. 176th St.
S.W. 196th St.
152nd St.
Sylvester Rd. S.W.
Seattle Tacoma Intl. Airport
S. 176th St.
S. 188th St.
180th St.
Petrovitsky Rd.
Lake Desire
Otter Lake
Cedar Grove Rd.
NORMANDY PARK
S. 200th St.
KENT
S.E. 192nd St.
S.E. 208th St.
Lake Youngs
Crosson-Sweeney Rd.
204th St. S.W.
PORTAGE
Tramp Harbor
220th St. S.W.
S. 212th St.
S. 216th St.
DES MOINES
S. 228th St.
S.E. 224th St.
VASHON ISLAND
S.W. 232nd St.
Weitzel Rd.
Wick Rd.
S.E. 240th St.
North Rd.
MAURY ISLAND
248th St.
Kent-Des Moines

Shreveport, LA
South Bend, IN

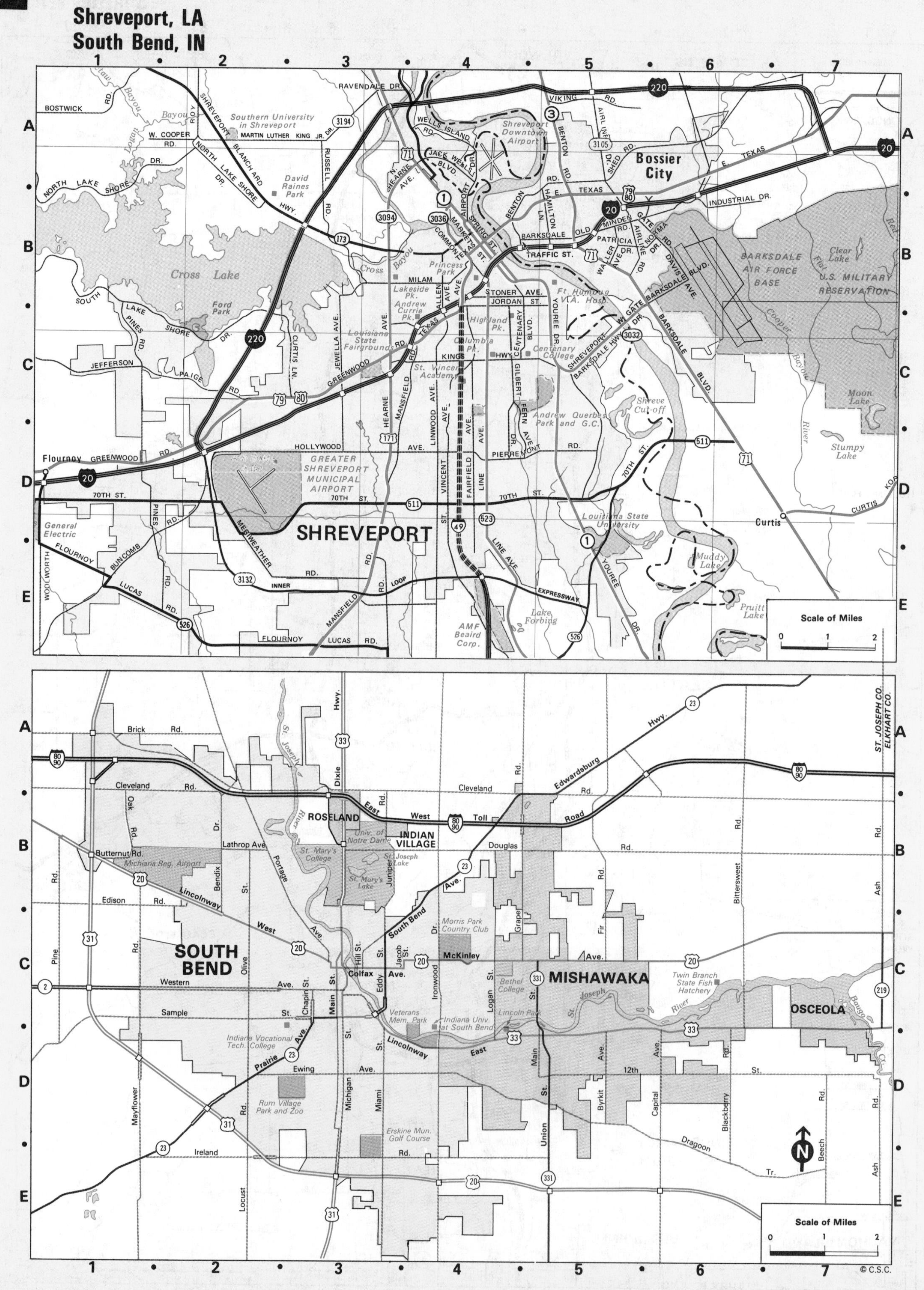

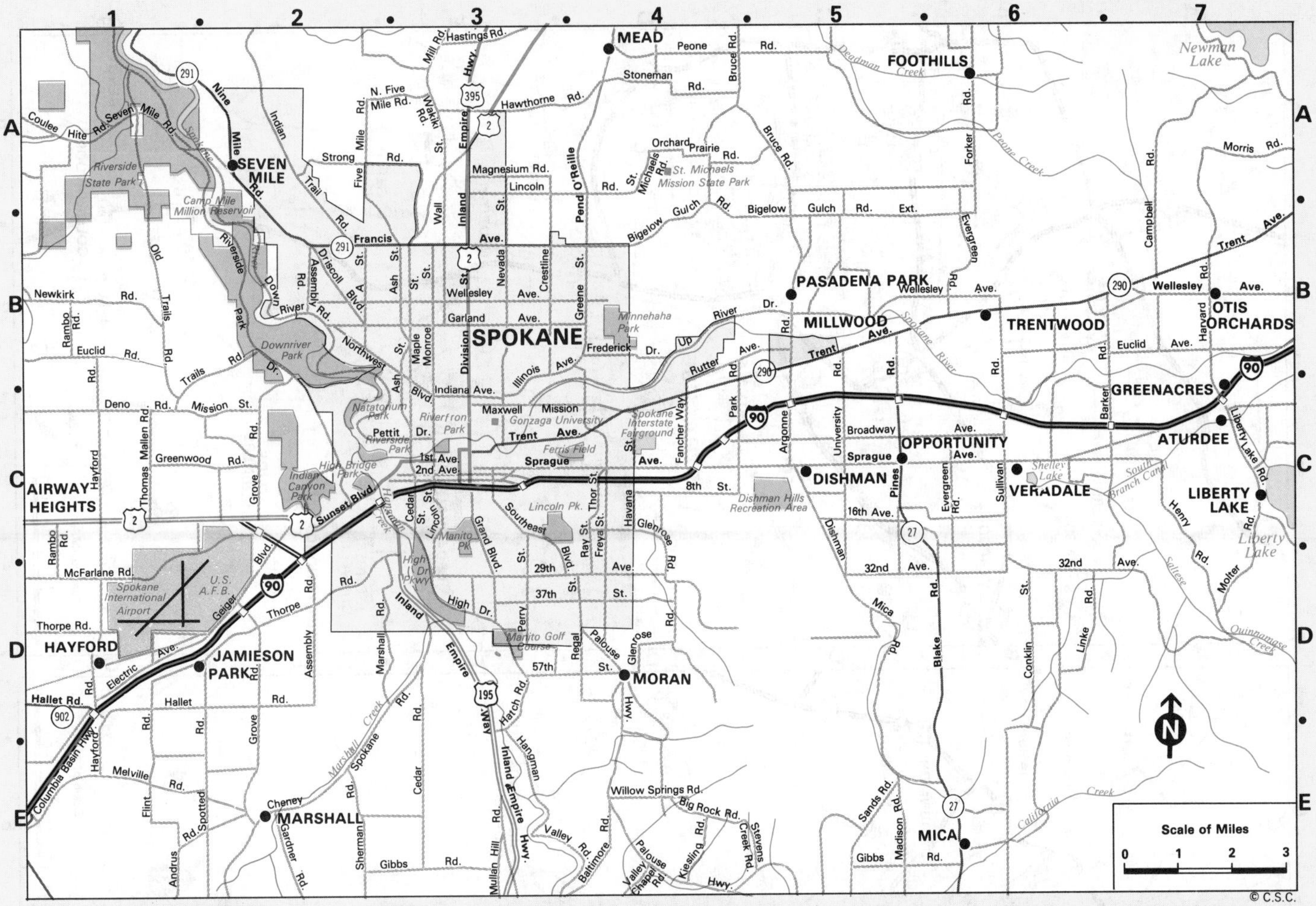

1 2 3 4 5 6 7

A B C D E

MEAD
FOOTHILLS
Newman Lake
SEVEN MILE
Riverside State Park
Camp Nine Mile Reservoir
St. Michaels Mission State Park
PASADENA PARK
MILLWOOD
TRENTWOOD
OTIS ORCHARDS
SPOKANE
Minnehaha Park
Downriver Park
GREENACRES
Natatorium Park
Riverfront Park
Gonzaga University
Spokane Interstate Fairground
Riverside Park
Ferris Field
OPPORTUNITY
ATURDEE
High Bridge Park
Indian Canyon Park
DISHMAN
VERADALE
Shelley Lake
AIRWAY HEIGHTS
LIBERTY LAKE
Liberty Lake
Dishman Hills Recreation Area
Lincoln Pk.
Manito Pk.
High Drive Pkwy
U.S. A.F.B.
Spokane International Airport
Manito Golf Course
HAYFORD
JAMIESON PARK
MORAN
MARSHALL
MICA

Scale of Miles
0 1 2 3

© C.S.C.

SOUTH HADLEY FALLS
HAMPSHIRE CO.
HAMPDEN CO.
LUDLOW CITY
Westover Air Force Base
Springfield Reservoir
BANDSVILLE
HOLYOKE
ROCK VALLEY
LUDLOW CENTER
WHIPPLES
EAST FARMS
Pequot Pond
Chicopee State Park
THORNDIKE
PALMER CENTER
Ashley Pond
LUDLOW
THREE RIVERS
Barnes Municipal Airport
PALMER
CHICOPEE
NORTH WILBRAHAM
Elms College
WESTFIELD
SPRINGFIELD
NORTH MONSON
WEST SPRINGFIELD
S. A. Museum
WILBRAHAM
Hospital
Springfield College
MONSON
SOUTH MONSON
NORTH AGAWAM
WEST AGAWAM
EAST LONGMEADOW
HOSMER CORNER
AGAWAM
HAMPDEN
Bowles-Agawam Airport
LONGMEADOW
MASSACHUSETTS
STATE LINE
CONNECTICUT
RISING CORNER
THOMPSONVILLE

Scale of Miles
0 1 2 3

© C.S.C.

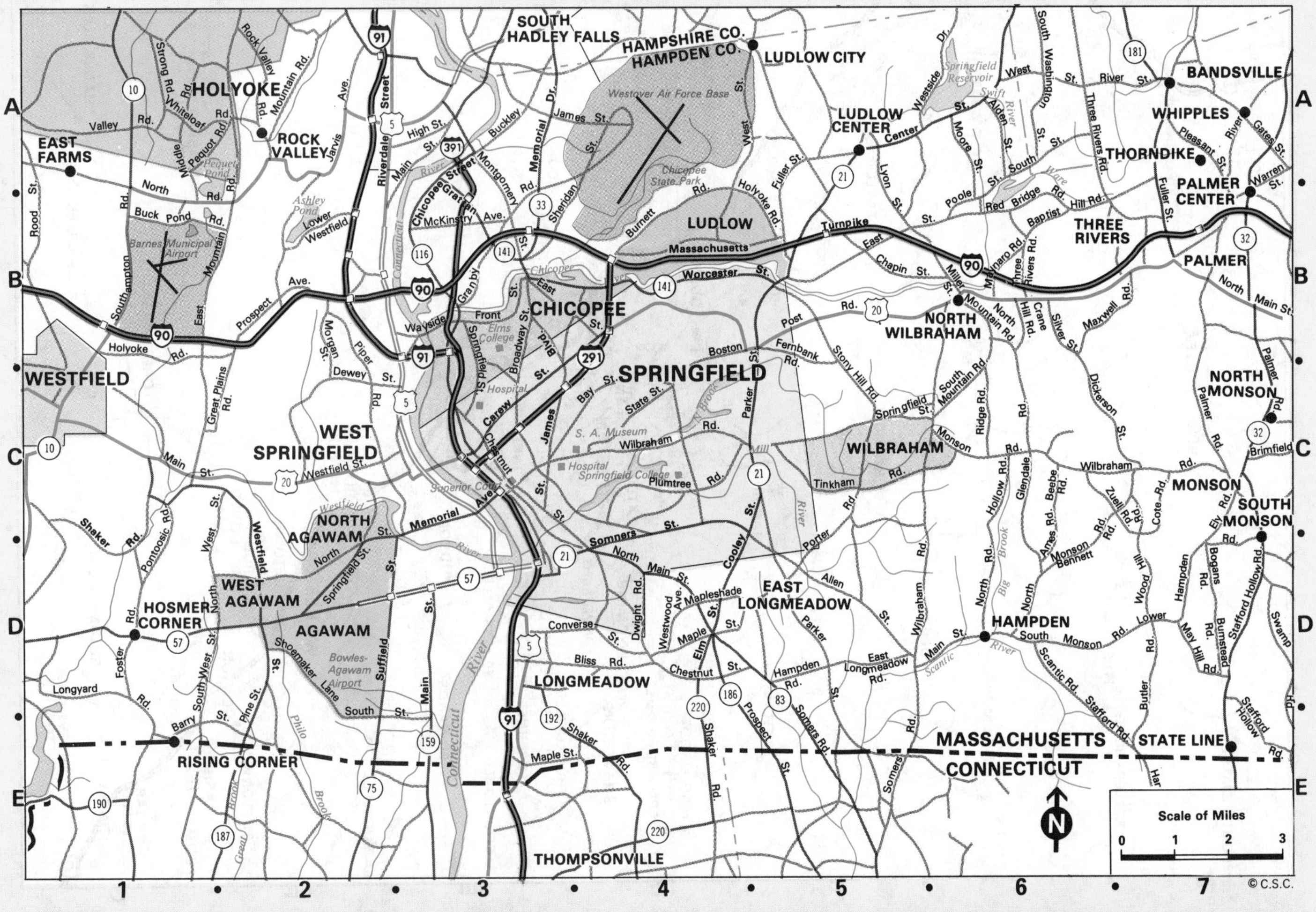

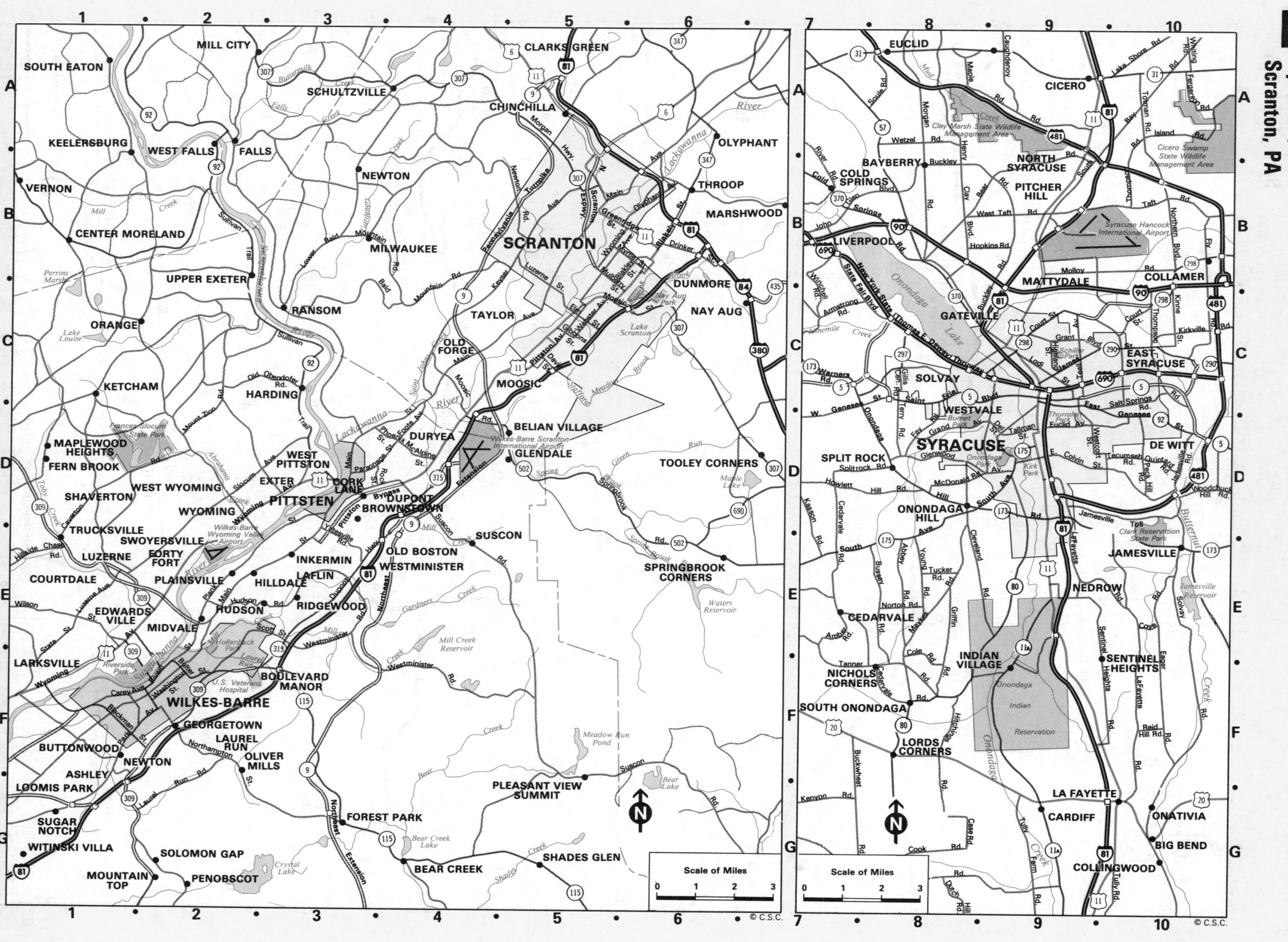
SCRANTON
CLARKS GREEN
CHINCHILLA
MILL CITY
SOUTH EATON
SCHULTZVILLE
KEELERSBURG
WEST FALLS
FALLS
NEWTON
OLYPHANT
THROOP
MARSHWOOD
VERNON
CENTER MORELAND
MILWAUKEE
UPPER EXETER
DUNMORE
NAY AUG
RANSOM
ORANGE
TAYLOR
OLD FORGE
MOOSIC
KETCHAM
HARDING
BELIAN VILLAGE
GLENDALE
DURYEA
MAPLEWOOD HEIGHTS
FERN BROOK
WEST PITTSTON
EXETER
TOOLEY CORNERS
SHAVERTON
WEST WYOMING
PITTSTON
CORK LANE
DUPONT
BROWNSTOWN
WYOMING
TRUCKSVILLE
SWOYERSVILLE
SUSCON
LUZERNE
FORTY FORT
INKERMIN
OLD BOSTON
WESTMINISTER
SPRINGBROOK CORNERS
COURTDALE
PLAINSVILLE
LAFLIN
HILLDALE
RIDGEWOOD
EDWARDSVILLE
HUDSON
MIDVALE
LARKSVILLE
BOULEVARD MANOR
WILKES-BARRE
GEORGETOWN
LAUREL RUN
BUTTONWOOD
NEWTON
OLIVER MILLS
ASHLEY
LOOMIS PARK
PLEASANT VIEW SUMMIT
SUGAR NOTCH
FOREST PARK
WITINSKI VILLA
SOLOMON GAP
SHADES GLEN
MOUNTAIN TOP
PENOBSCOT
BEAR CREEK
Frances Slocum State Park
Wilkes-Barre Scranton International Airport
Wilkes-Barre Wyoming Valley Airport
Hollenback Park
Riverside Park
U.S. Veterans Hospital
Nay Aug Park
Lake Scranton
Mill Creek Reservoir
Waters Reservoir
Meadow Run Pond
Maple Lake
Bear Lake
Bear Creek Lake
Crystal Lake
Lake Louise
Perrins Marsh
Lackawanna River
Susquehanna River
Scale of Miles
0 1 2 3
© C.S.C.
EUCLID
CICERO
Clay Marsh State Wildlife Management Area
Cicero Swamp State Wildlife Management Area
BAYBERRY
NORTH SYRACUSE
COLD SPRINGS
PITCHER HILL
Syracuse Hancock International Airport
LIVERPOOL
MATTYDALE
COLLAMER
GATEVILLE
Onondaga Lake
EAST SYRACUSE
SOLVAY
WESTVALE
SYRACUSE
DE WITT
SPLIT ROCK
ONONDAGA HILL
Clark Reservation State Park
JAMESVILLE
Jamesville Reservoir
NEDROW
CEDARVALE
INDIAN VILLAGE
SENTINEL HEIGHTS
NICHOLS CORNERS
Onondaga Indian Reservation
SOUTH ONONDAGA
LORDS CORNERS
LA FAYETTE
CARDIFF
ONATIVIA
BIG BEND
COLLINGWOOD
Thornden Park
Schiller Park
Burnet Park
Onondaga Park
Kirk Park
New York State (Thomas E. Dewey) Thruway
Scale of Miles
0 1 2 3
© C.S.C.

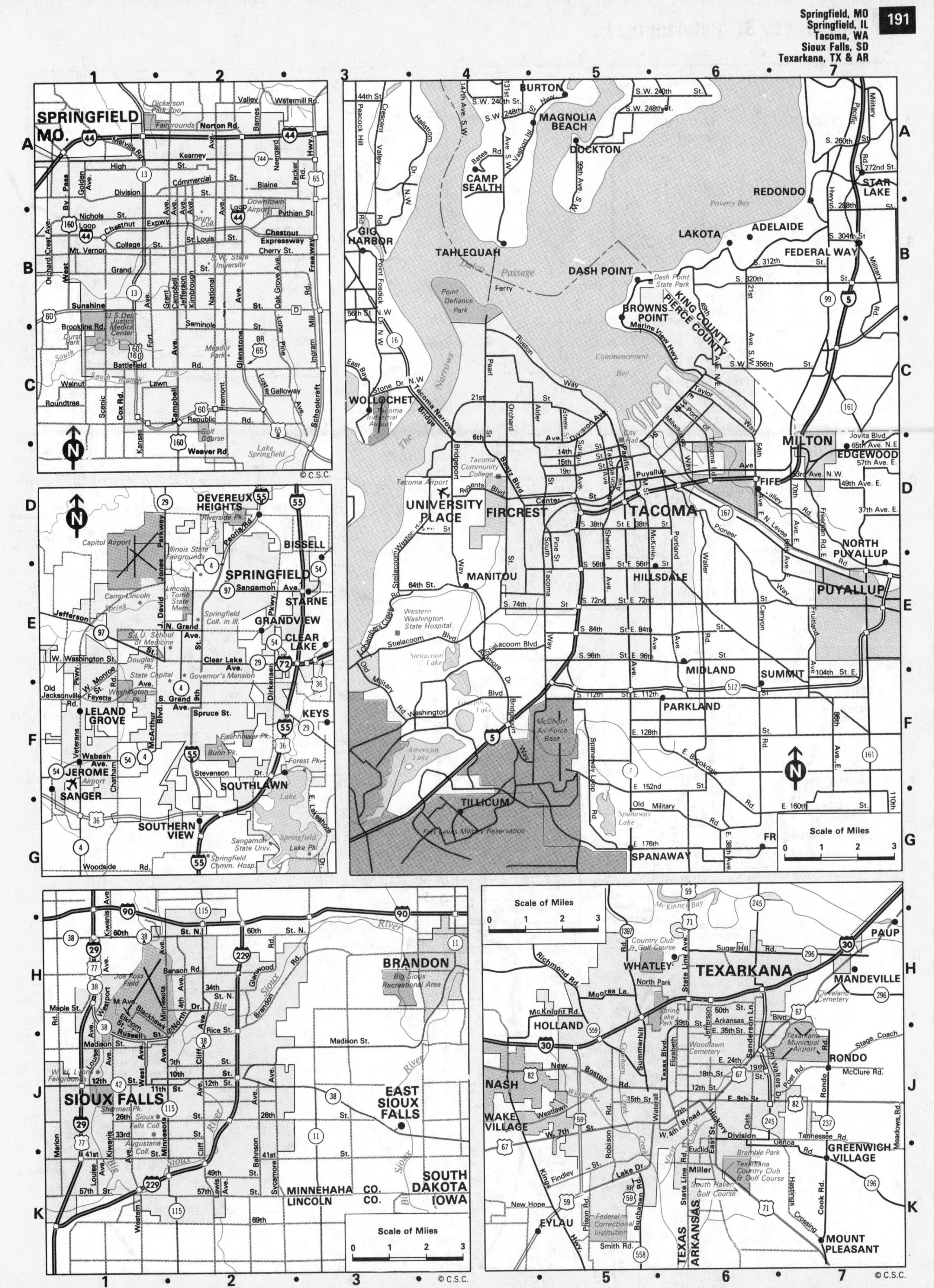
SPRINGFIELD MO.
Dickerson Park Zoo
Fairgrounds
Norton Rd.
Valley
Watermill Rd.
Kearney
High St.
Commercial St.
Division St.
Blaine
Nichols St.
Chestnut Expwy
Chestnut Expressway
Pythian St.
College St.
St. Louis St.
Mt. Vernon
Cherry St.
Grand St.
S.W. State University
Drury Coll.
Downtown Airport
Sunshine St.
U.S. Dept. Justice Medical Center
Brookline Rd.
Seminole St.
Battlefield Rd.
Meador Park
Walnut
Lawn
Galloway
Roundtree
Republic Rd.
Golf Course
Weaver Rd.
Lake Springfield
Glenstone
National
Campbell
Kansas
Scenic
Cox Rd.
Fremont
Schoolcraft
Ingram Mill
Oak Grove Ave
Fort Ave.
© C.S.C.
SPRINGFIELD
DEVEREUX HEIGHTS
Riverside Pk.
BISSELL
Capitol Airport
Illinois State Fairgrounds
Sangamon Ave.
STARNE
Camp Lincoln
Lincoln Tomb State Mem.
Springfield Coll. in Ill.
GRANDVIEW
CLEAR LAKE
Jefferson
S.I.U. School of Medicine
W. Washington St.
Douglas Pk.
Clear Lake Ave.
State Capital
Governor's Mansion
Old Jacksonville Rd.
W. Monroe St.
Fayette
Washington Pk.
S. Grand Ave.
LELAND GROVE
Spruce St.
KEYS
Eisenhower Pk.
Bunn Pk.
Wabash Ave.
JEROME
Airport
SANGER
Stevenson Dr.
Forest Pk.
SOUTHLAWN
Lake
SOUTHERN VIEW
Sangamon State Univ.
Springfield Lake Pk.
Springfield Comm. Hosp.
Woodside Rd.
BURTON
MAGNOLIA BEACH
DOCKTON
CAMP SEALTH
S.W. 240th St.
S.W. 248th St.
REDONDO
STAR LAKE
Poverty Bay
LAKOTA
ADELAIDE
FEDERAL WAY
GIG HARBOR
TAHLEQUAH
Dalco Passage
DASH POINT
Dash Point State Park
Point Defiance Park
Ferry
BROWNS POINT
KING COUNTY
PIERCE COUNTY
Marine View Hwy.
Commencement Bay
The Narrows
Tacoma Narrows Bridge
WOLLOCHET
Tacoma Industrial Airport
Ruston Way
City Hall
MILTON
EDGEWOOD
FIFE
Tacoma Airport
Tacoma Community College
UNIVERSITY PLACE
FIRCREST
TACOMA
NORTH PUYALLUP
PUYALLUP
MANITOU
HILLSDALE
Western Washington State Hospital
Steilacoom Lake
MIDLAND
SUMMIT
PARKLAND
McChord Air Force Base
American Lake
TILLICUM
Fort Lewis Military Reservation
Spanaway Lake
SPANAWAY
FR
Scale of Miles
BRANDON
Big Sioux Recreational Area
Joe Foss Field
Big Sioux River
Madison St.
W.H. Lyons Fairgrounds
SIOUX FALLS
Sherman Pk.
Sioux Falls Coll.
Augustana Coll.
EAST SIOUX FALLS
MINNEHAHA CO.
LINCOLN CO.
SOUTH DAKOTA
IOWA
Scale of Miles
© C.S.C.
McKinney Bay
Country Club & Golf Course
WHATLEY
TEXARKANA
PAUP
MANDEVILLE
Cleveland Cemetery
Richmond Rd.
Moores La.
North Park
McKnight Rd.
HOLLAND
Spring Lake Park
Woodlawn Cemetery
Texarkana Municipal Airport
RONDO
Stage Coach
McClure Rd.
NASH
WAKE VILLAGE
Westlawn
Cowhorn Creek
Days Creek
Tennessee Rd.
Bramble Park
GREENWICH VILLAGE
Texarkana Country Club & Golf Course
South Haven Golf Course
New Hope
EYLAU
Federal Correctional Institution
Smith Rd.
TEXAS
ARKANSAS
MOUNT PLEASANT
© C.S.C.

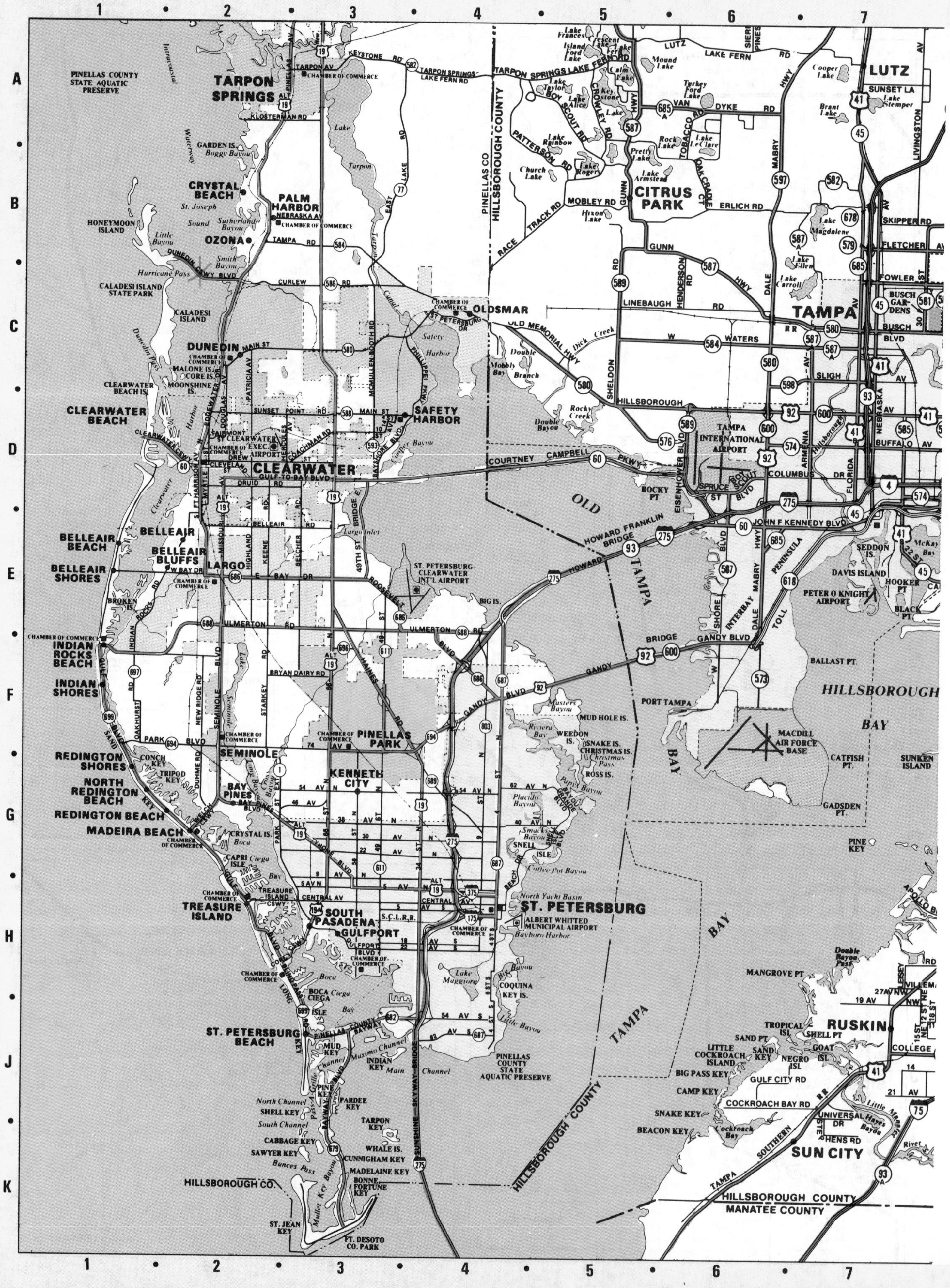

Tampa, FL/St. Petersburg, FL

1 2 3 4 5 6 7

A B C D E F G H J K

PINELLAS COUNTY STATE AQUATIC PRESERVE
TARPON SPRINGS
LUTZ
CRYSTAL BEACH
PALM HARBOR
CITRUS PARK
HONEYMOON ISLAND
OZONA
CALADESI ISLAND STATE PARK
CALADESI ISLAND
OLDSMAR
TAMPA
DUNEDIN
CLEARWATER BEACH IS.
CLEARWATER BEACH
SAFETY HARBOR
TAMPA INTERNATIONAL AIRPORT
CLEARWATER
OLD TAMPA BAY
HOWARD FRANKLIN BRIDGE
BELLEAIR
BELLEAIR BEACH
BELLEAIR BLUFFS
BELLEAIR SHORES
LARGO
ST. PETERSBURG-CLEARWATER INT'L AIRPORT
DAVIS ISLAND
PETER O KNIGHT AIRPORT
INDIAN ROCKS BEACH
INDIAN SHORES
HILLSBOROUGH BAY
PORT TAMPA
MACDILL AIR FORCE BASE
PINELLAS PARK
REDINGTON SHORES
NORTH REDINGTON BEACH
REDINGTON BEACH
SEMINOLE
KENNETH CITY
BAY PINES
MADEIRA BEACH
TREASURE ISLAND
SOUTH PASADENA
GULFPORT
ST. PETERSBURG
ALBERT WHITTED MUNICIPAL AIRPORT
TAMPA BAY
ST. PETERSBURG BEACH
RUSKIN
PINELLAS COUNTY STATE AQUATIC PRESERVE
SUNSHINE SKYWAY BRIDGE
HILLSBOROUGH CO.
HILLSBOROUGH COUNTY
SUN CITY
HILLSBOROUGH COUNTY
MANATEE COUNTY
FT. DESOTO CO. PARK

1 2 3 4 5 6 7

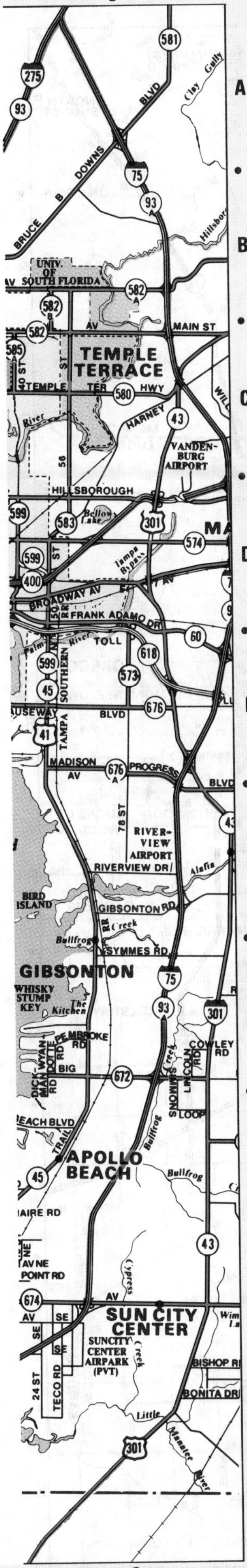
TEMPLE TERRACE
UNIV. OF SOUTH FLORIDA
VANDENBURG AIRPORT
HILLSBOROUGH
BROADWAY AV
FRANK ADAMO DR
RIVERVIEW AIRPORT
RIVERVIEW DR
BIRD ISLAND
GIBSONTON RD
SYMMES RD
GIBSONTON
WHISKY STUMP KEY
BIG
APOLLO BEACH
SUN CITY CENTER
SUNCITY CENTER AIRPARK (PVT)
BISHOP RD
BONITA DR
MAIN ST
MADISON AV
PROGRESS BLVD
BLVD
BEACH BLVD
PEMBROKE RD
COWLEY RD
LOOP
24 ST
TECO RD

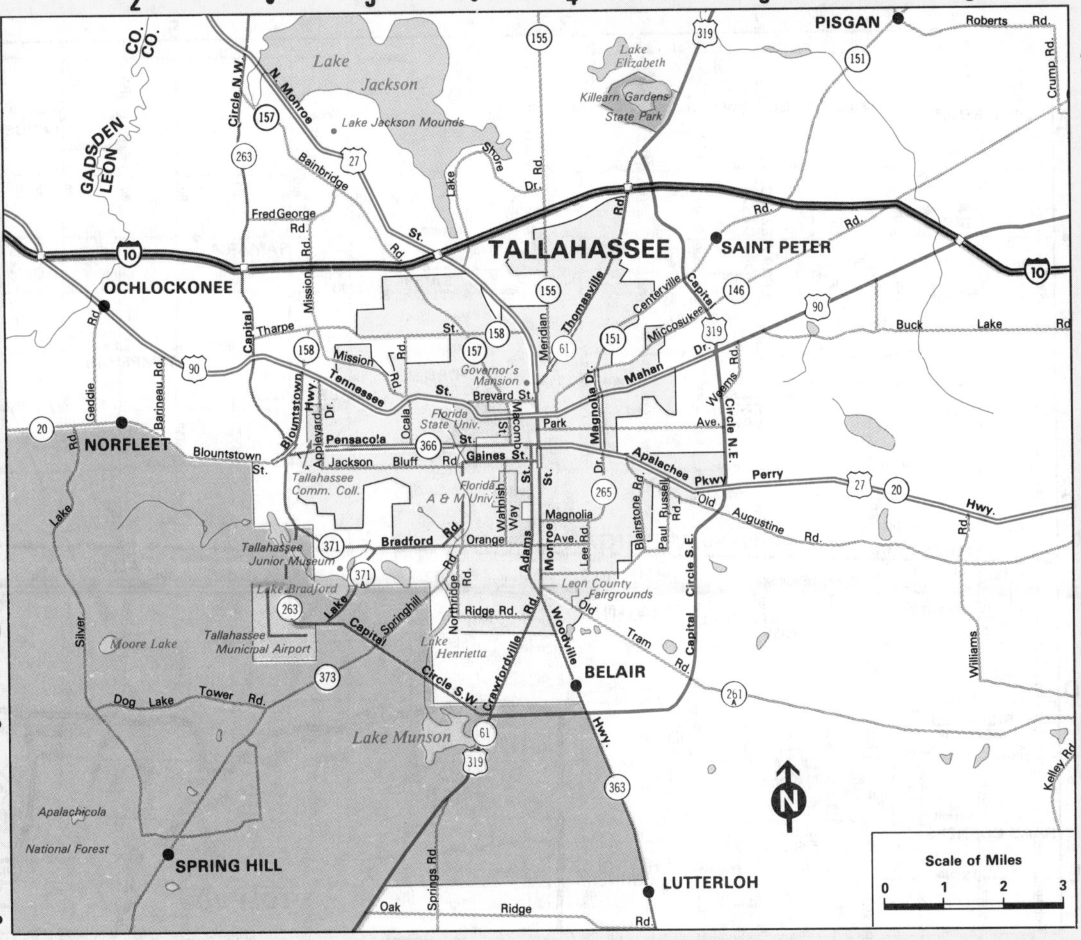
GADSDEN CO.
LEON CO.
Lake Jackson
Lake Jackson Mounds
Lake Elizabeth
Killearn Gardens State Park
PISGAN
Roberts Rd.
TALLAHASSEE
SAINT PETER
OCHLOCKONEE
NORFLEET
Governor's Mansion
Florida State Univ.
Tallahassee Comm. Coll.
Florida A & M Univ.
Tallahassee Junior Museum
Lake Bradford
Leon County Fairgrounds
Moore Lake
Tallahassee Municipal Airport
Lake Henrietta
BELAIR
Lake Munson
Apalachicola National Forest
SPRING HILL
LUTTERLOH
Oak Ridge Rd.
Scale of Miles
0 1 2 3

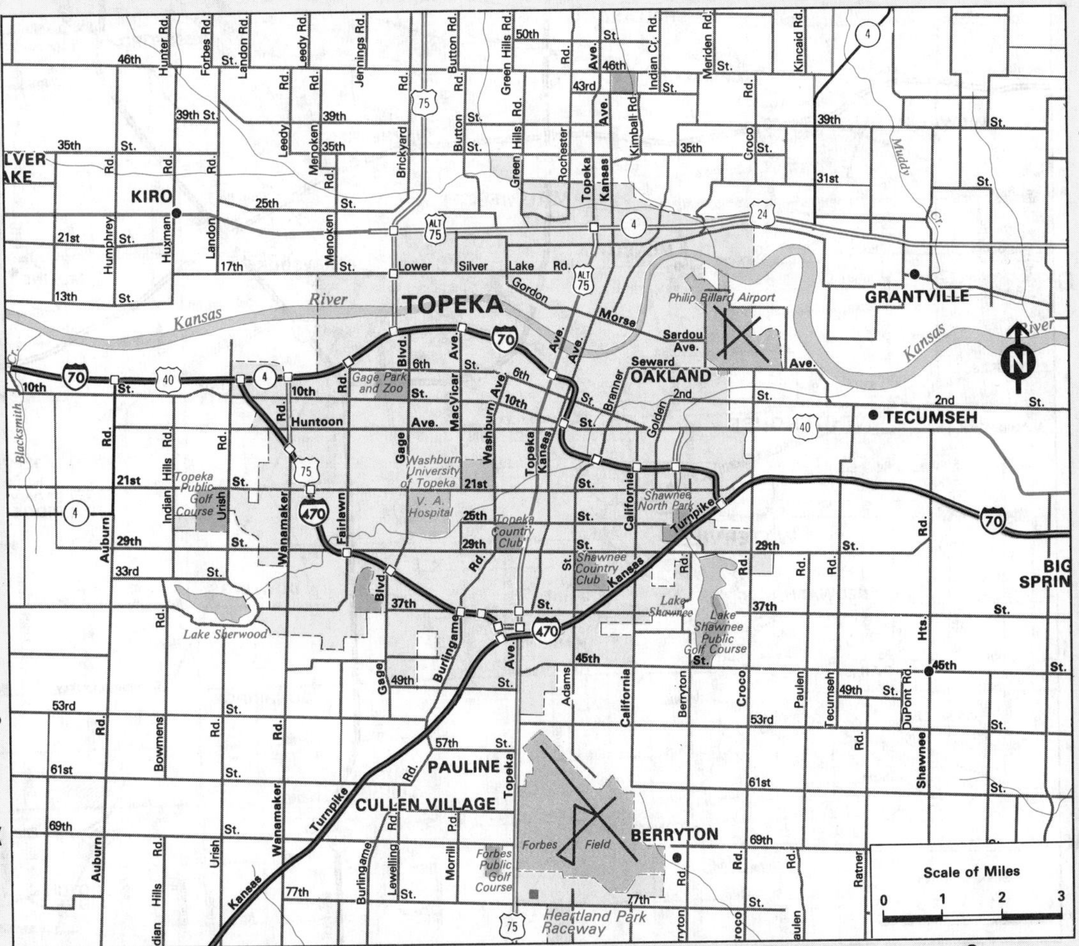
KIRO
TOPEKA
Kansas River
Philip Billard Airport
GRANTVILLE
OAKLAND
TECUMSEH
Gage Park and Zoo
Washburn University of Topeka
V. A. Hospital
Topeka Public Golf Course
Topeka Country Club
Shawnee Country Club
Shawnee North Park
Lake Shawnee
Lake Shawnee Public Golf Course
Lake Sherwood
BIG SPRIN
PAULINE
CULLEN VILLAGE
Forbes Field
BERRYTON
Forbes Public Golf Course
Heartland Park Raceway
Scale of Miles
0 1 2 3

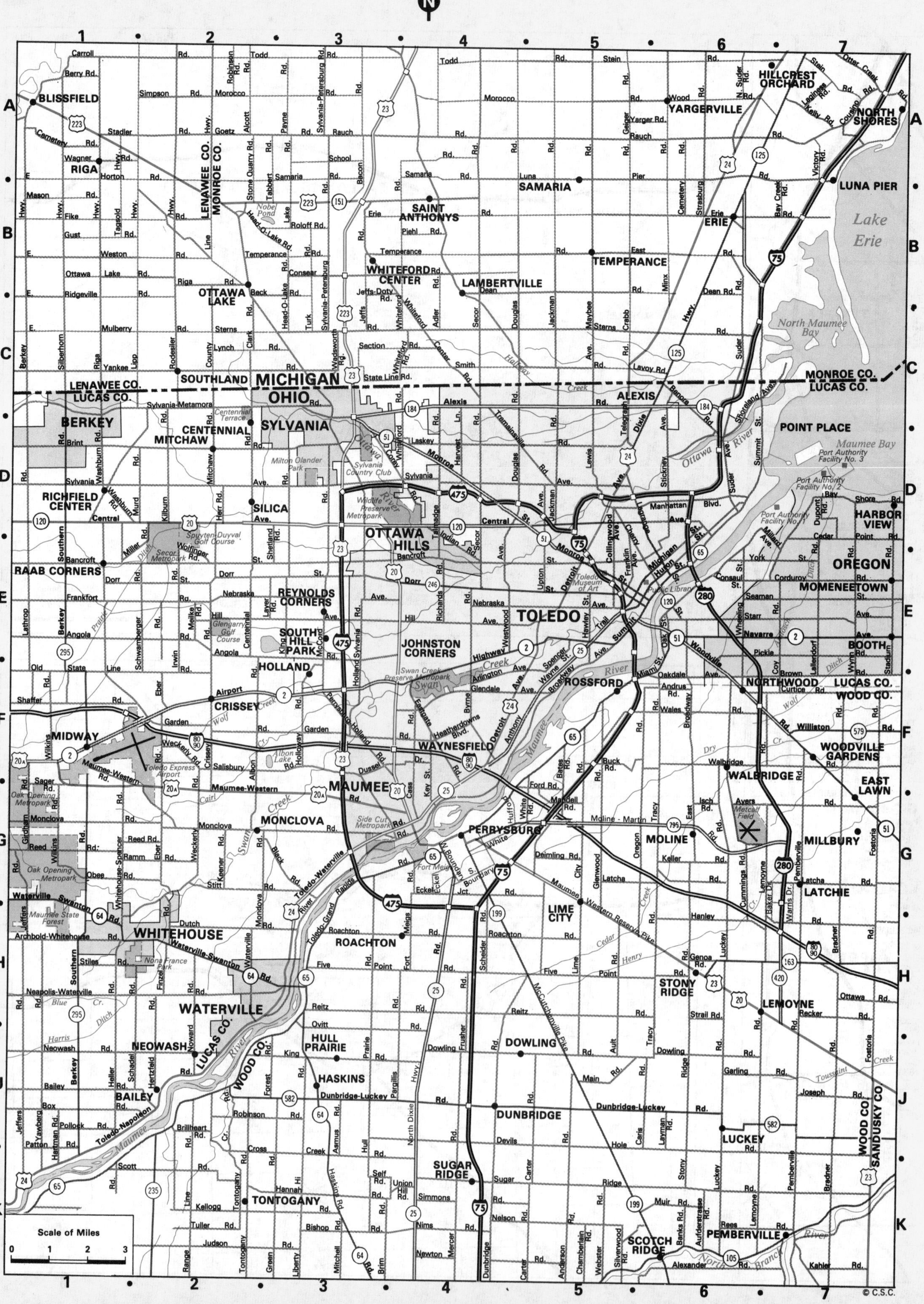

N

BLISSFIELD
RIGA
LENAWEE CO.
MONROE CO.
SAINT ANTHONYS
SAMARIA
YARGERVILLE
HILLCREST ORCHARD
NORTH SHORES
LUNA PIER
ERIE
Lake Erie
TEMPERANCE
WHITEFORD CENTER
LAMBERTVILLE
OTTAWA LAKE
North Maumee Bay
SOUTHLAND
MICHIGAN
OHIO
LUCAS CO.
ALEXIS
BERKEY
SYLVANIA
CENTENNIAL
MITCHAW
POINT PLACE
Maumee Bay
Port Authority Facility No. 3
Port Authority Facility No. 2
Port Authority Facility No. 1
RICHFIELD CENTER
SILICA
OTTAWA HILLS
HARBOR VIEW
RAAB CORNERS
OREGON
MOMENEETOWN
REYNOLDS CORNERS
TOLEDO
Toledo Museum of Art
SOUTH HILL PARK
JOHNSTON CORNERS
HOLLAND
BOOTH
CRISSEY
ROSSFORD
NORTHWOOD
WOOD CO.
MIDWAY
WAYNESFIELD
Toledo Express Airport
WOODVILLE GARDENS
MAUMEE
WALBRIDGE
EAST LAWN
MONCLOVA
PERRYSBURG
MOLINE
Metcalf Field
MILLBURY
LATCHIE
LIME CITY
WHITEHOUSE
ROACHTON
Oak Openings Metropark
Maumee State Forest
STONY RIDGE
WATERVILLE
LEMOYNE
NEOWASH
HULL PRAIRIE
DOWLING
BAILEY
HASKINS
DUNBRIDGE
LUCKEY
SANDUSKY CO.
SUGAR RIDGE
TONTOGANY
SCOTCH RIDGE
PEMBERVILLE

Scale of Miles
0 1 2 3

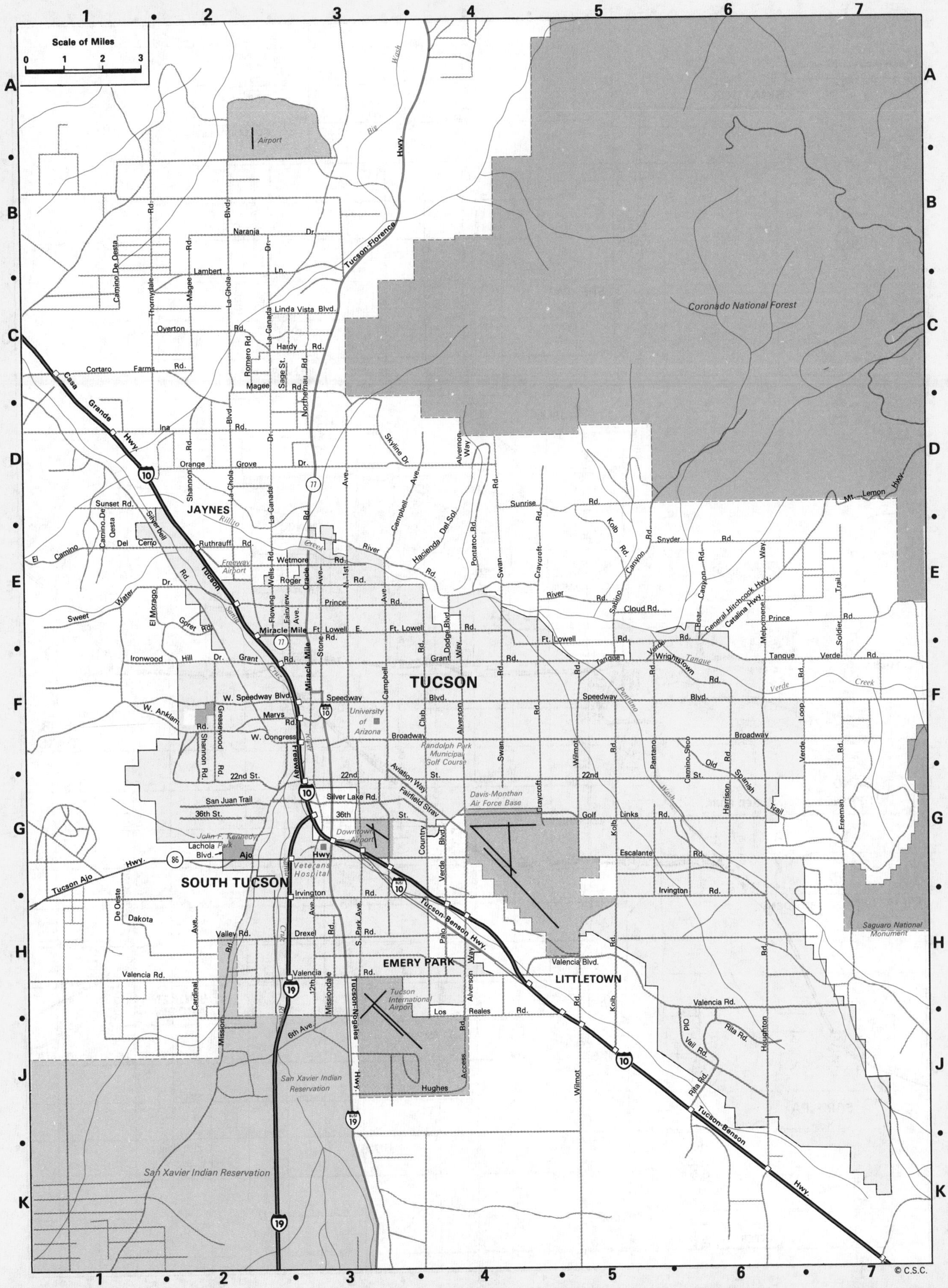
Scale of Miles
0 1 2 3
TUCSON
SOUTH TUCSON
JAYNES
EMERY PARK
LITTLETOWN
Coronado National Forest
Saguaro National Monument
San Xavier Indian Reservation
Davis-Monthan Air Force Base
Tucson International Airport
Downtown Airport
Freeway Airport
Airport
University of Arizona
Randolph Park Municipal Golf Course
Veterans Hospital
John F. Kennedy Park
Casa Grande Hwy.
Tucson-Benson Hwy
Tucson Florence Hwy.
Tucson Ajo Hwy.
Tucson-Nogales Hwy.
Mt. Lemon Hwy.
General Hitchcock Hwy.
Catalina Hwy.
Speedway Blvd.
Broadway
22nd St.
Grant Rd.
Ft. Lowell Rd.
River Rd.
Ina Rd.
Orange Grove Dr.
Valencia Rd.
Irvington Rd.
Golf Links Rd.
Old Spanish Trail
Santa Cruz River
Rillito Creek
Tanque Verde Creek
Pantano Wash
Big Wash
Miracle Mile
Oracle Rd.
Stone Ave.
Campbell Ave.
Swan Rd.
Craycroft Rd.
Wilmot Rd.
Kolb Rd.
Pantano Rd.
Harrison Rd.
Houghton Rd.
Freeman Rd.
Sabino Canyon Rd.
Aviation Way
Alvernon Way
10
19
77
86

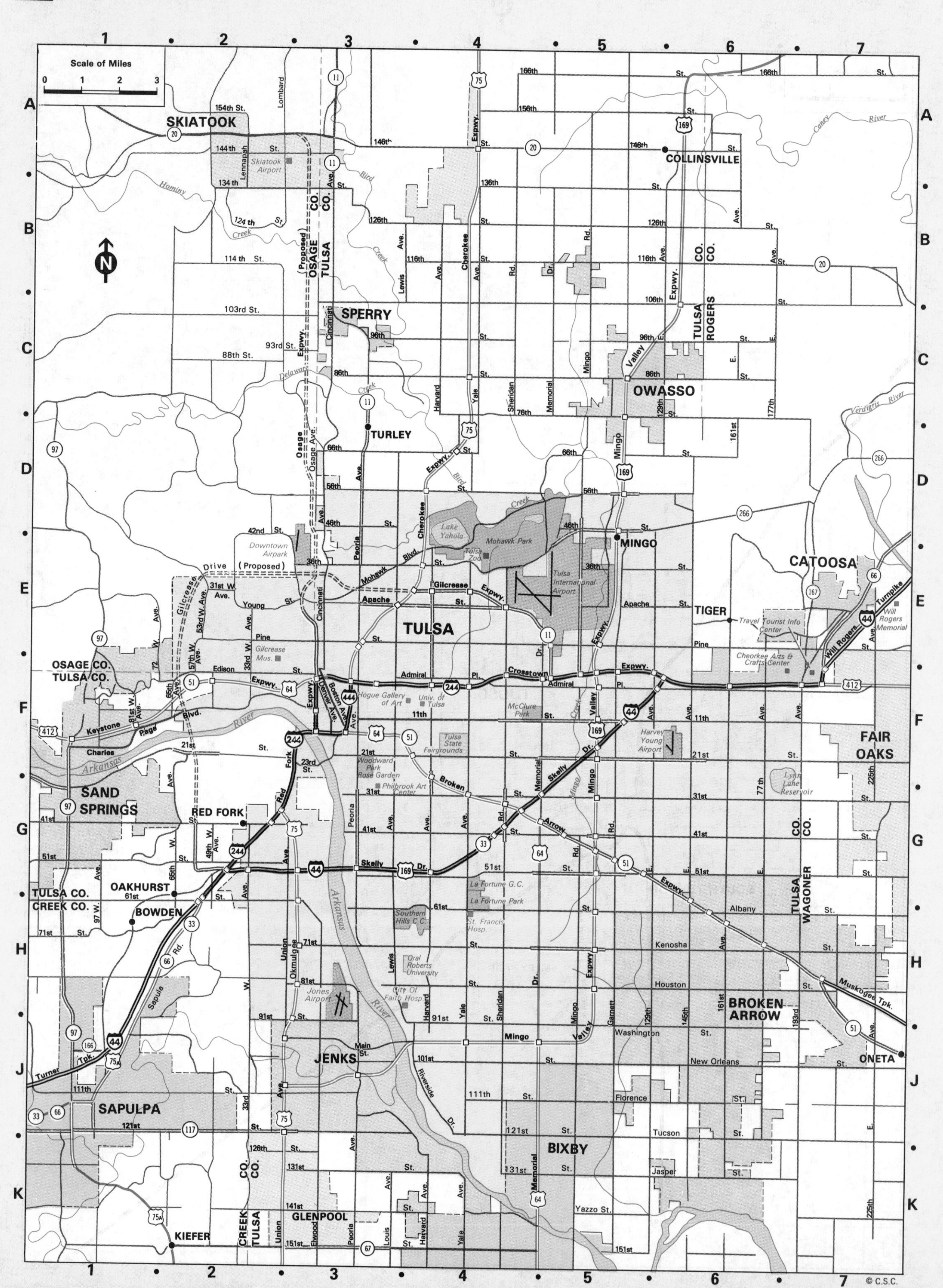
Scale of Miles
SKIATOOK
COLLINSVILLE
SPERRY
OWASSO
TURLEY
MINGO
CATOOSA
TULSA
TIGER
FAIR OAKS
SAND SPRINGS
RED FORK
OAKHURST
BOWDEN
JENKS
SAPULPA
BIXBY
BROKEN ARROW
ONETA
GLENPOOL
KIEFER
Tulsa International Airport
Mohawk Park
Lake Yahola
Arkansas River
Verdigris River
Caney River
Muskogee Tpk.
Will Rogers Turnpike
Oral Roberts University
© C.S.C.

1
2
3
4
5
6
7
A
B
C
D
E
Scale of Miles
0
1
2
3
PENNINGTON
WASHINGTON CROSSING
Washington Crossing State Park
LAWRENCEVILLE
EWINGVILLE
Rider College
Mercer County Airport
Trenton State College
WRIGHTSTOWN
YARDLEY
NEWTOWN
TRENTON
Trenton Speedway
MERCERVILLE
Mercer County Central Park
Merceville
ROBBINSVILLE
NEW SHARON
WHITEHORSE
ALLENTOWN
PENNDEL
Van Sciver Lake
PENNSYLVANIA
NEW JERSEY
FLORENCE
ROEBLING
CHESTERFIELD
ARNEYTOWN
BRISTOL
HEDDING
Delaware River
Assunpink Cr.
Crosswicks Cr.
Neshaminy Cr.
Washington Crossing Pennington Rd.
Lawrenceville Princeton Rd.
Wrightstown Taylorsville Rd.
Mt. Eyre Rd.
Old Lincoln Hwy.
Trenton Ave.
Bridge St.
Oxford Valley Rd.
Levittown Parkway
Maple Rd.
W. Lincoln Hwy.
New Rodgers Rd.
Edgewood Rd.
Yardley Morrisville Rd.
Nottingham Way
Hamilton Ave.
Klockner Rd.
Sharon Rd.
Yardville-Allentown Rd.
Allentown Red Valley Rd.
Allentown Davis Station Rd.
New Jersey Turnpike
Walnford Rd.
29
31
295
1
532
206
95
332
32
33
130
1A
195
413
13
276
132
511
68
ERATH
WACO-MADISON COOPER
Madison Cooper Airfield
NORTHCREST
James Connally Airfield
LACY LAKEVIEW
BELLMEAD
BOSQUEVILLE
McLennan Community College
Airfield
Lake Waco
Bosque Park
Cameron Park
WACO
Brazos River
Bosque River
SPEEGLEVILLE
Speegleville Park
WILLOW GROVE
Baylor Univ.
BEVERLY HILLS
Midway Park
Veterans Hospital
WOODWAY
McGregor Municipal Airfield
Stratton Lake
Tradinghouse Creek Reservoir
Tehuacana Creek
HALLSBURG
HARRISON
ROBINSON
DOWNSVILLE
HEWITT
Lake Creek Lake
Minos Creek
Castleman Creek
McLENNAN CO.
FALLS CO.
ROSENTHAL
ASA
Old Steinbeck
Steinbeck Rd.
Dripping Springs
Gholson Rd.
Orchard Ln.
Kendall Ln.
Chappel Hill Rd.
Lake Shore Dr.
Park Lake Dr.
Mac Arthur
Herring
Hillcrest Dr.
Valley Mills Dr.
Fish Pond Rd.
Sanger Ave.
Franklin Ave.
Dutton
La Salle
Prime Rose
Curley Ln.
Garden Dr.
Tinsley Dr.
Newland Dr.
Old Robinson Rd.
McGregor Rd.
Old Hewitt Rd.
Estates Dr.
Jewell Dr.
Old Temple Rd.
Pantherway
Warren
Greig Dr.
Moonlight
Chapel
Speegleville Rd.
Yankie Rd.
Tree Lake Rd.
Flat Rock Rd.
Street Eason Rd.
Rock Cr. Rd.
Meyers Ln.
Connally
185
412
6
84
77
81
35
88
434
164
1695
2113
©C.S.C.

Scale of Miles

0 1 2 3

© ADC of Alexandria

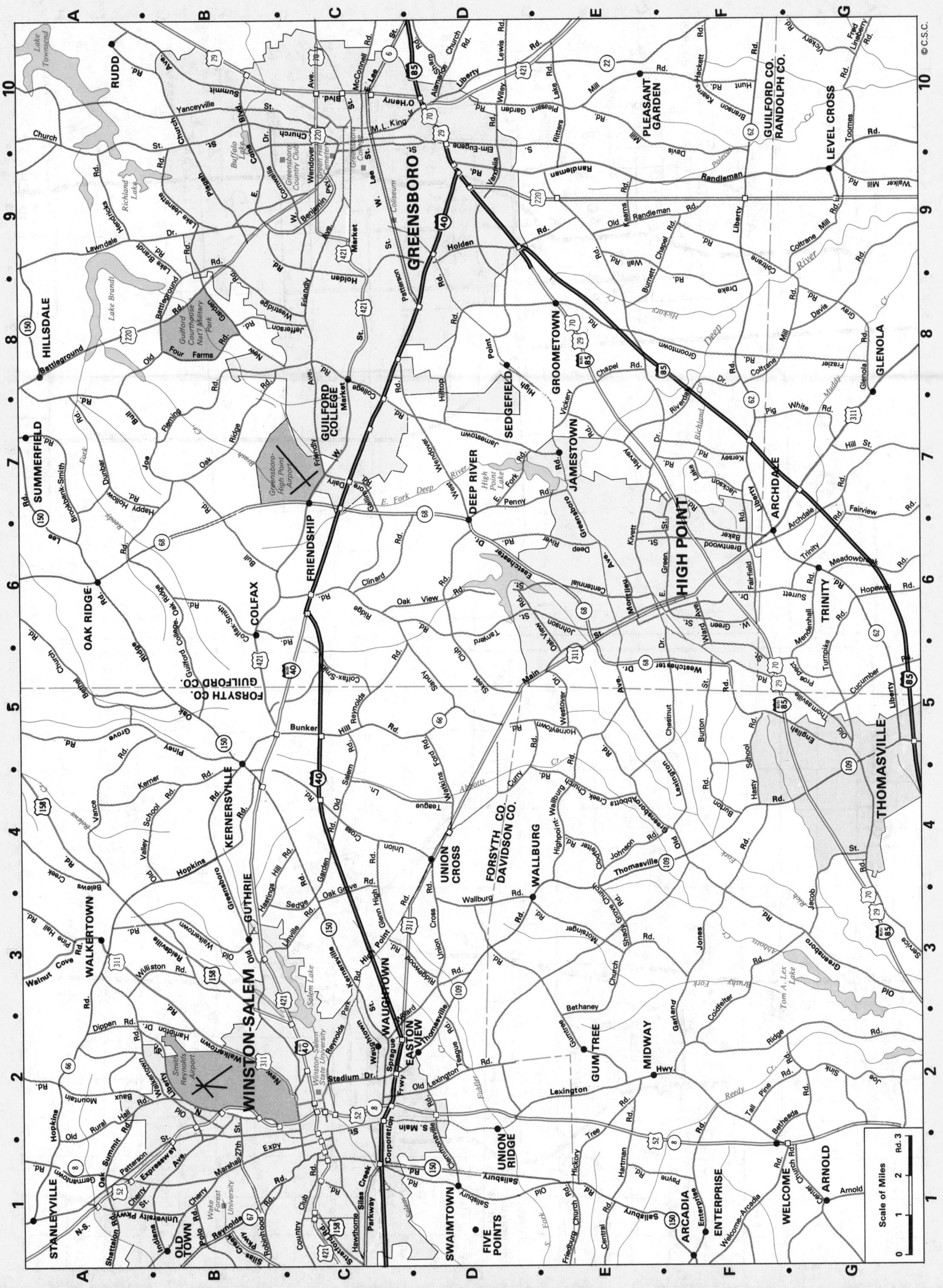

WINSTON-SALEM
GREENSBORO
HIGH POINT
THOMASVILLE
RUDD
HILLSDALE
SUMMERFIELD
OAK RIDGE
COLFAX
KERNERSVILLE
GUTHRIE
WALKERTOWN
STANLEYVILLE
OLD TOWN
FIVE POINTS
SWAIMTOWN
UNION RIDGE
WAUGHTOWN
EASTON VIEW
UNION CROSS
WALLBURG
GUM TREE
MIDWAY
ARCADIA
ENTERPRISE
WELCOME
ARNOLD
FRIENDSHIP
GUILFORD COLLEGE
DEEP RIVER
SEDGEFIELD
GROOMETOWN
JAMESTOWN
PLEASANT GARDEN
LEVEL CROSS
GLENOLA
ARCHDALE
TRINITY

FORSYTH CO.
GUILFORD CO.
FORSYTH CO.
DAVIDSON CO.
GUILFORD CO.
RANDOLPH CO.

Scale of Miles
0 1 2 3

West Palm Beach, FL
Wichita, KS

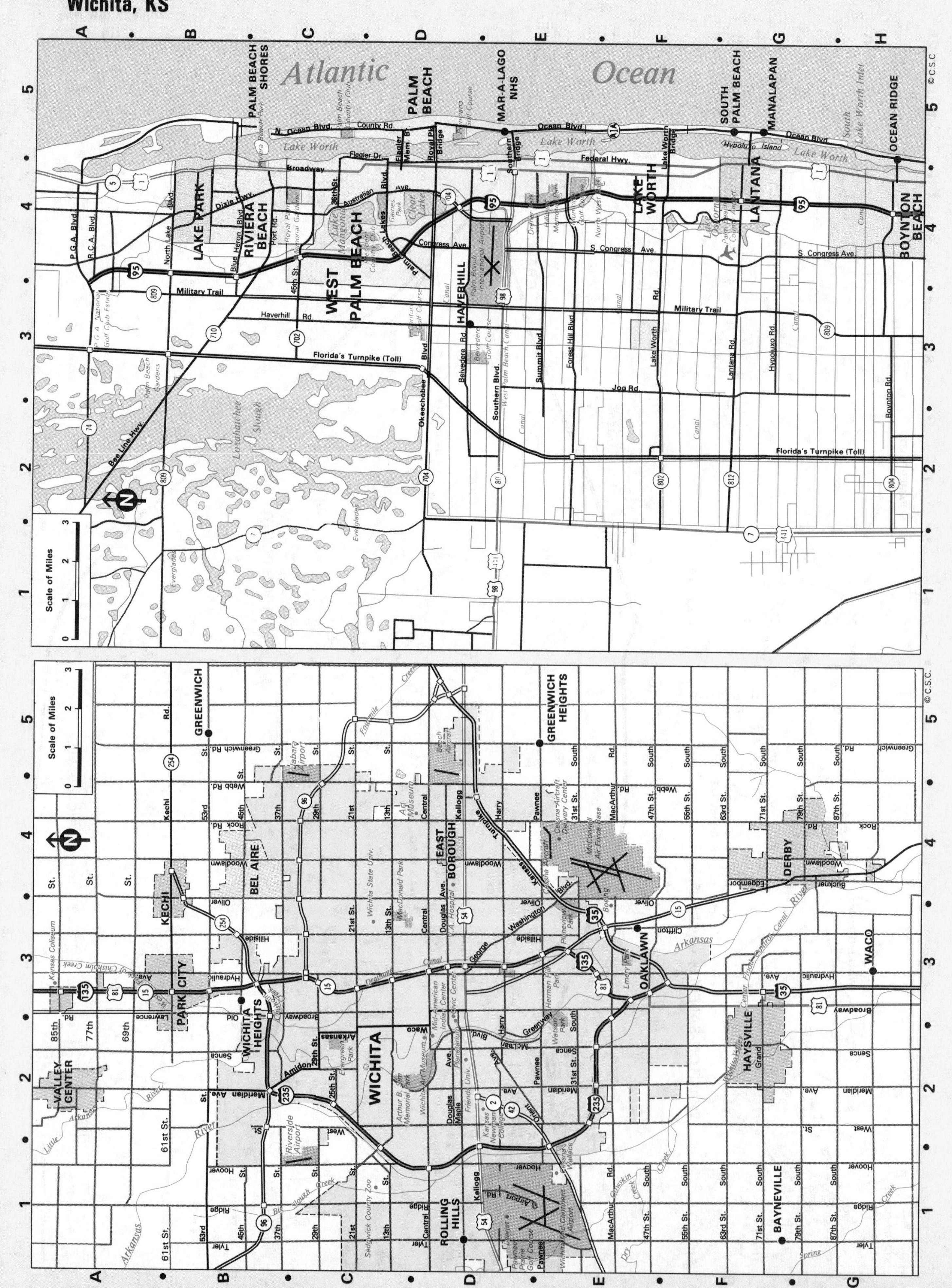

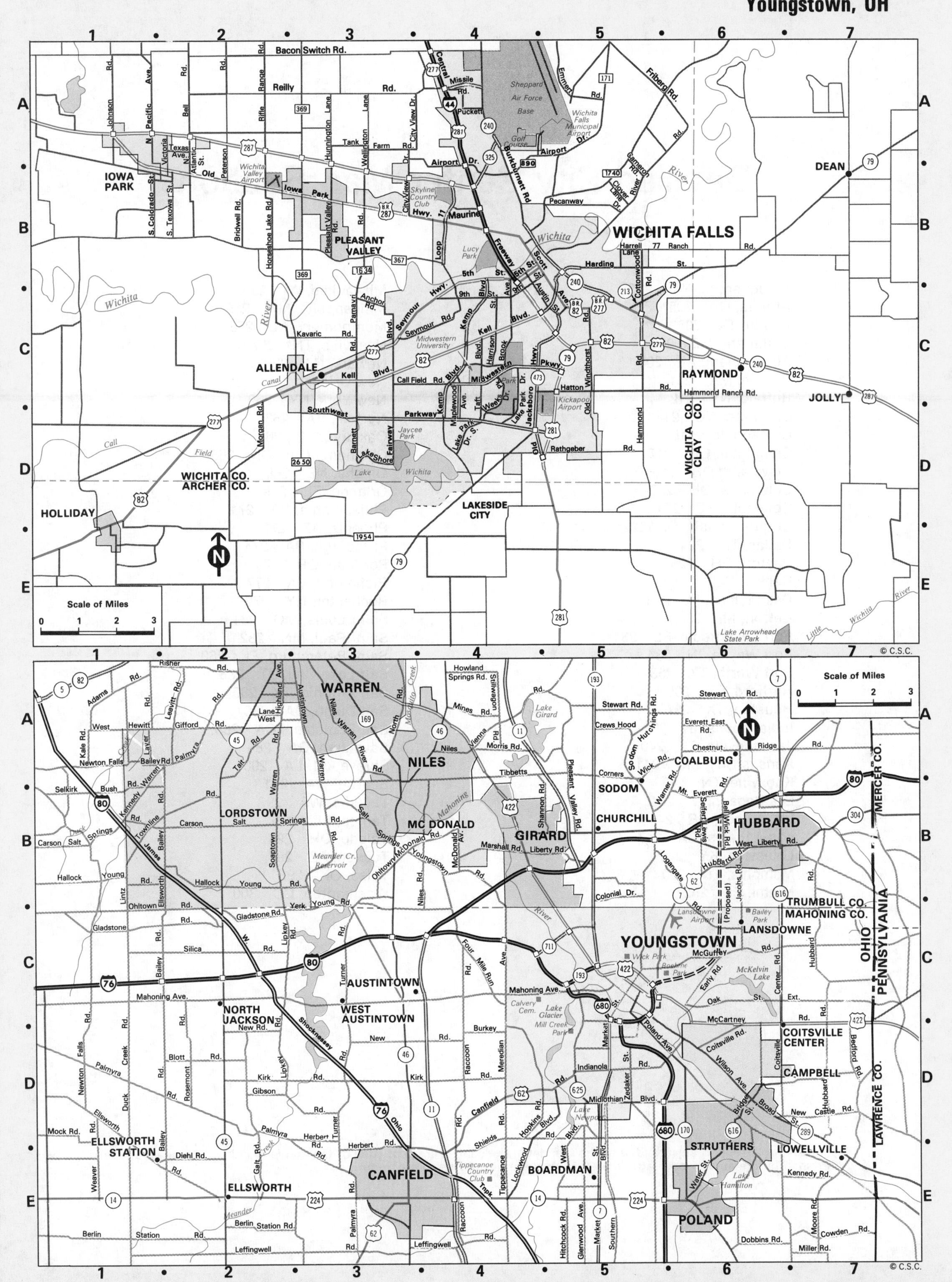
WICHITA FALLS
IOWA PARK
PLEASANT VALLEY
ALLENDALE
RAYMOND
DEAN
JOLLY
HOLLIDAY
LAKESIDE CITY
WICHITA CO.
ARCHER CO.
WICHITA CO.
CLAY CO.
Sheppard Air Force Base
Wichita Falls Municipal Airport
Midwestern University
Kickapoo Airport
Lake Wichita
Lake Arrowhead State Park
Scale of Miles
© C.S.C.
WARREN
NILES
LORDSTOWN
MC DONALD
GIRARD
SODOM
CHURCHILL
COALBURG
HUBBARD
YOUNGSTOWN
LANSDOWNE
AUSTINTOWN
WEST AUSTINTOWN
NORTH JACKSON
COITSVILLE CENTER
CAMPBELL
STRUTHERS
LOWELLVILLE
BOARDMAN
CANFIELD
POLAND
ELLSWORTH STATION
ELLSWORTH
TRUMBULL CO.
MAHONING CO.
MERCER CO.
LAWRENCE CO.
OHIO
PENNSYLVANIA
Meander Cr. Reservoir
Lake Glacier
Lake Hamilton
Lake Girard
Scale of Miles
© C.S.C.

Detailed City Center Map Section

The maps in this section are intended to show in detail the central areas of each of the above 68 cities. The scale of each map was determined by the area of coverage, therefore they are not drawn to a common scale.

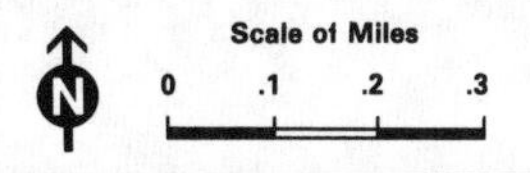
N
Scale of Miles
0
.1
.2
.3

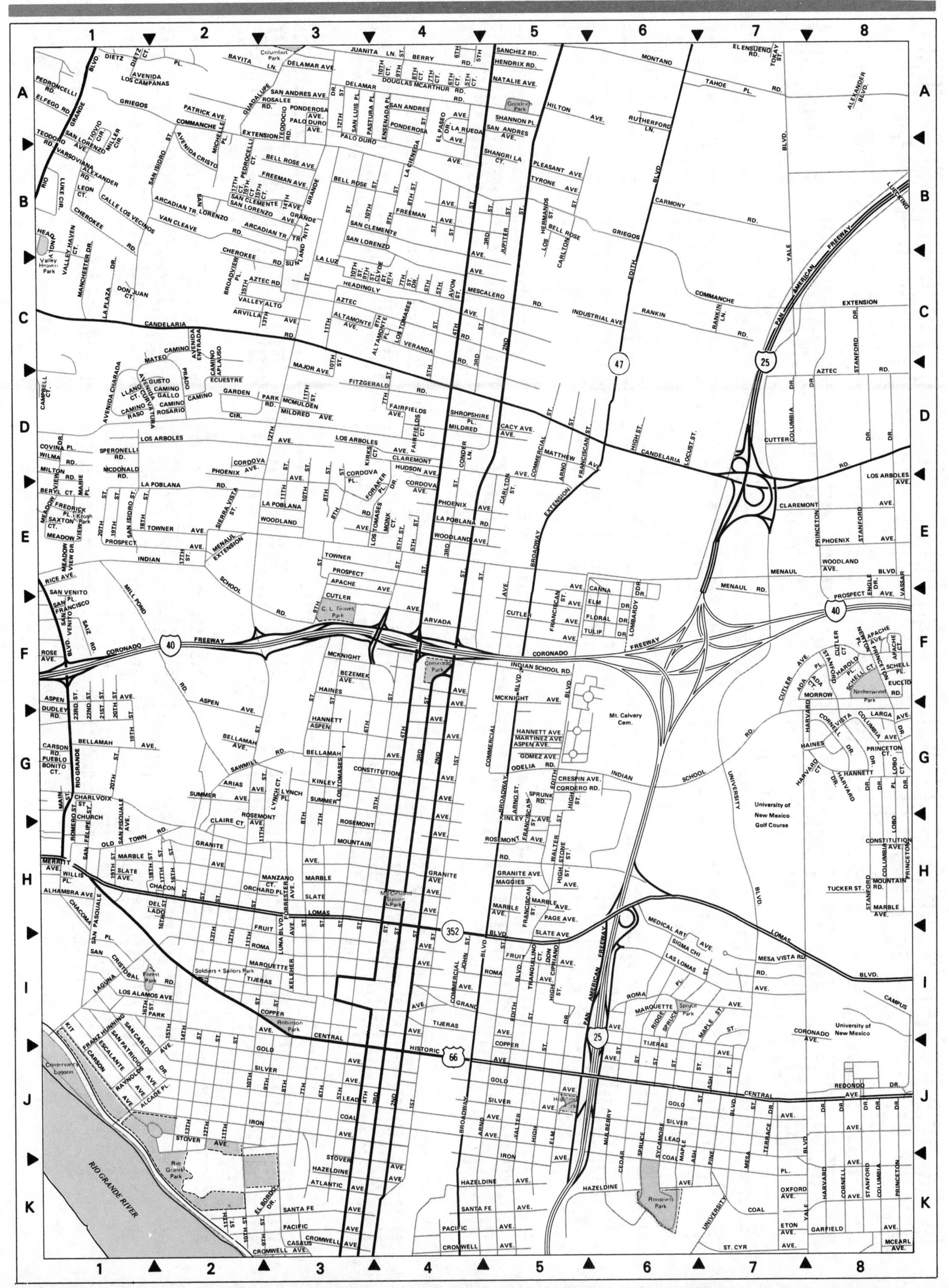
1
2
3
4
5
6
7
8
A
B
C
D
E
F
G
H
I
J
K
RIO GRANDE BLVD.
CANDELARIA RD.
MONTANO
COMMANCHE
MENAUL
CORONADO FREEWAY
40
25
47
352
66
PAN AMERICAN FREEWAY
INDIAN SCHOOL RD.
LOMAS BLVD.
CENTRAL
HISTORIC
BROADWAY
EDITH BLVD.
UNIVERSITY BLVD.
YALE BLVD.
GRIEGOS
LOS ARBOLES
MOUNTAIN
RIO GRANDE RIVER
Conservancy Lagoon
Rio Grande Park
Roosevelt Park
Robinson Park
Forest Park
Coronado Park
C. L. Graves Park
Valley Heaven Park
Goodrich Park
Netherwood Park
Mt. Calvary Cem.
University of New Mexico Golf Course
University of New Mexico
Soldiers + Sailors Park
Columbus Park
Kirk Krogh Park

ALBUQUERQUE

Ada Ct. F-8
Ada Pl. F-7
Alcade Pl. J-1
Alexander Blvd. A-8
Alexander Rd. B-1
Alhambra Ave. H-1
Altamonte Ave. C-3
Altamonte Pl. C-4
Apache Ave. E-3,F-8
Apache Ct. F-8
Arcadian Tr. B-2,B-3
Arias Ave. G-2
Arno St. D-5,G-5,J-4
Arvada Ave. F-4
Arvilla Ave. C-2
Ash St. J-7,K-6
Aspen Ave. F-1, F-2,G-3,G-5
Atlantic Ave. K-3
Avenida Charada D-1
Avenida Cristo B-2
Avenida Curvatura D-1
Avenida Entrada C-2
Avenida Los Campanas A-1
Avon St. C-4
Aztec Rd. C-3,D-8
Bayita Ln. A-2
Bellamah Ave. G-1-3
Bell Rose Ave. B-3,B-5
Berry Rd. A-4
Beryl Ct. E-1
Bezemek Ave. F-3
Broadview Pl. C-2
Broadway Blvd. G-5,J-4
Broadway Ext. E-5
Cacy Ave. D-5
Calle Los Vecinos B-1
Camino C-2
Camino Aplauso C-2
Camino Ecuestre D-2
Camino Gallo D-2
Camino Raso D-1
Camino Rosario D-2
Campbell Ct. D-1
Campus I-8
Candelaria Rd. C-2,D-6
Canna Dr. F-6
Carlton St. B-5,E-5
Carmony Rd. B-6
Carson Rd. G-1
Casaus Ave. K-3
Cedar St. K-6
Central Ave. I-3,J-7
Chacoma Pl. H-1
Chacon H-2
Charlvoix St. G-1
Cherokee Rd. B-1,C-2
Church St. G-1
Claire Ct. H-2
Claremont Ave. D-4,E-7
Clyde St. C-4
Coal Ave. J-3,K-7
Coal Pl. K-6
Columbia Dr. D-7, G-8,H-8,K-8
Commercial Ave. G-4,I-4
Commercial St. D-5
Conder Ln. D-4
Constitution Ave. G-4,H-8
Copper Ave. I-3,J-5
Cordero Rd. G-5
Cordova Ave. E-4
Cordova Pl. D-3
Cornell Dr. G-8,K-8
Coronado Ave. I-8
Coronado Frwy. F-1,F-5
Covina Pl. D-1
Crespin Ave. G-5
Cromwell Ave. K-3,K-4
Cutler Ave. F-3,F-5,F-7
Cutler Ct. F-8
Cutter D-7
Delmar Ave. A-3
Delmar Rd. A-3
Dietz Ct. A-1
Dietz Pl. A-1
Don Cipriano I-5
Don Juan Ct. C-1
Douglas McArthur Rd. A-4
Dudley Rd. G-1
Edith Blvd. G-5,I-5
El Bordo Dr. K-3
El Ensueno Rd. A-7
Elfego Rd. A-1
Elm Dr. F-6,J-5
El Paseo Ave. A-4
El Paseo Dr. A-4
Engle Dr. F-8
Ensenada Pl. A-4
Escalante Ave. J-1
Eton Ave. K-7
Euclid F-8
Fairfields Ave. D-4
Fairfields Ct. D-4
Fitzgerald Rd. D-3
Floral Dr. F-6
Foraker Pl. E-4
Forest Park I-1
Forrester Ave. H-3
Franciscan St. D-5,F-5,H-5
Franz Hunning I-1
Fredrick Pl. E-1
Freeman Ave. B-3,B-4
Fruit Ave. H-3,I-5
Garden Cir. D-2
Garfield Ave. K-8
Gold Ave. J-3,J-5,J-6
Gomez Ave. G-5
Grand Ave. I-4
Grande Ct. B-3
Grande Dr. B-3
Granite Ave. H-2,H-4,H-5
Griegos Commanche Ext. A-1,B-6
Guadalupe A-2
Gusto D-2
Haines G-8
Haines Ave. F-3
Hannett Ave. G-3,G-5
Hannett Dr. G-8
Harold Pl. F-8
Harvard Ct. G-8
Harvard Dr. G-8,K-8
Hazeldine Ave. K-3,K-4,K-6
Headingly B-1
Headingly Ave. C-3
Hendrix Rd. A-5
High St. D-6,G-5, I-5,J-5
High Stone St. H-5
Hilton Ave. A-5
Hudson Ave. D-4
Indian School Rd. E-2,F-5,G-6
Industrial Ave. C-6
Iron Ave. J-2,K-5
John St. I-4
Juanita Ln. A-3
Jupiter St. B-5
Keleher Ave. I-3
Kinley Ave. G-3,H-5
Kit Carson Ave. I-1,J-1
La Cienega St. B-4
Laguna I-1
La Luz C-3
La Plaza Dr. C-1
La Poblana Rd. E-2,E-3,E-4
Larga Ave. G-8
La Rueda A-4
Las Lomas Ave. I-6
Lead Ave. J-3,J-6
Leon Ct. B-1
Llano Ct. D-1
Lobo Ct. G-8
Lobo Pl. H-8
Locust St. D-6
Lomas Blvd. H-3,I-7
Lombardy Dr. F-6
Los Alamos Ave. I-1
Los Arboles Ave. D-2,D-3,E-8
Los Hermanos St. B-5
Los Tomases Dr. C-4,E-4
Luecking B-8
Luke Cir. B-1
Luna Blvd. I-3
Lynch Ct. G-3
Lynch Pl. G-3
Maggies Ave. H-5
Main St. G-1
Major Ave. D-3
Manchester Dr. C-1
Manzano Ct. H-3
Maple St. I-7,J-6
Marble Ave. H-1, H-3,H-5,H-8
Marie Pl. E-1
Marquette Ave. I-2,I-6
Martinez Ave. G-5
Mateo C-2
Matthew Ave. D-5
McDonald Rd. D-1
Mcearl Ave. K-8
McKnight Ave. F-3,F-5
McMulden D-3
Meadow View Dr. E-1
Medical Art Ave. I-6
Menaul Blvd. E-7
Menaul Ext. E-2
Menual Rd. F-7
Merritt Ave. H-1
Mesa St. K-7
Mesa Vista Rd. I-7
Mescalero Rd. C-5
Michelle Pl. A-2
Mildred Ave. D-3,D-4
Mill Pond Rd. F-1
Miller Cir. A-1
Milton Rd. D-1
Monk Ct. E-4
Montano Rd. A-6
Morrow Rd. F-8
Mountain Rd. H-3,H-8
Mulberry St. J-6
Natalie Ave. A-5
Newton Pl. F-8
Odelia Rd. G-5
Old Town Rd. H-1
Orchard Pl. E-2
Oxford Ave. K-7
Pacific Ave. K-3,K-4
Page Ave. H-5
Palo Duro Ave. A-3,A-4
Pan American Frwy. C-7,I-6
Park Ave. I-2
Park Rd. D-3
Pastura Pl. A-4
Patrick Ave. A-2
Pedrocelli Ct. B-2
Pedroncelli Rd. A-1
Phoenix Ave. D-2,E-4,E-8
Pine St. K-7
Pleasant Ave. B-5
Ponderosa Ave. A-3,A-4
Princeton Dr. E-8,F-8, G-8,H-8,K-8
Prospect Ave. E-1,E-3,F-8
Pueblo Bonito Ct. G-1
Rankin Ln. C-7
Rankin Rd. C-6
Raynolds Ave. J-1
Rice Ave. E-1
Ridge Pl. I-6
Roma Ave. I-2,I-5,I-6
Romero St. H-1
Rosalee Rd. A-3
Rose Ave. F-1
Rosemont Ave. G-2, H-3,H-5
Rutherford Ln. A-6
St. Cyr Ave. K-7
Saiz Rd. F-1
San Andres Ave. A-3,A-4,A-5
San Carlos Dr. I-1
Sanchez Rd. A-5
San Clemente Ave. B-2,B-4
San Cristobal Rd. I-1
San Felipe St. H-1
San Francisco F-1
San Isidro St. B-2,E-1
San Lorenzo Ave. A-1,B-2,B-3
San Luis Pl. A-3
San Patricio Ave. I-1
San Pisquale Ave. H-1
Santa Fe Ave. K-3,K-4
San Venito Pl. E-1,F-1
Sawmill Rd. G-2
Saxton Ct. E-1
Schell Ct. F-8
Schell Pl. F-8
Shangri La B-5
Shannon Pl. A-5
Shropshire Pl. D-4
Sierra Vista St. E-2
Sigma Chi Rd. I-6
Silver Ave. J-3,J-5,J-6
Slate Ave. H-1,H-3,I-5
Speronelli Rd. D-1
Spruce St. I-6,J-6
Sprunk Rd. G-5
Stanford Dr. C-8,D-8, E-8,F-8,H-8
Stover Ave. J-2,K-3
Summer Ave. G-2,G-3
Sunland C-3
Sycamore St. J-6
Tahoe Pl. A-7
Teodocio Rd. A-3
Terrace Dr. J-7
Tijeras Ave. I-2,I-4,J-6
Tiovio Cir. A-1
Toedoro Rd. A-1
Tokay St. A-7
Towner Ave. E-2,E-3
Tranquilino Ct. I-5
Trinity B-3
Tucker St. H-8
Tulip Dr. F-6
Tyrone Ave. B-5
University Blvd. G-7,K-7
Valley Alto C-2
Valley Haven Ct. B-1
Valley Heaven Park C-1
Van Cleave Rd. B-2
Varsoviana B-1
Vassar Dr. E-8
Veranda Rd. C-4
Vista Columbia Ave. G-8
Walter St. H-5,J-5
Willis Pl. H-1
Wilma Rd. D-1
Woodland Ave. E-8
Woodland Rd. E-3,E-5
Yale Blvd. K-8
1st St. G-4,J-4
2nd St. C-5,G-4,J-4
3rd St. B-5,C-4, E-4,G-4,J-4
4th St. C-4,G-4,J-3
5th Ct. A-4
5th St. A-4,C-4, D-4,E-4,G-4,J-3
6th Ct. A-4
6th St. A-4,C-4, E-4,G-3,J-3
7th Ct. A-4
7th St. C-4,D-4,H-3,J-3
8th Ct. A-4
8th St. B-4,C-4, E-3, H-3, J-3
9th St. A-4,B-4,C-3, E-3,F-3,J-3,K-2
10th Ct. A-4
10th St. B-4,C-3, E-3,J-2,K-2
11th St. C-3,E-3, I-2,J-2,K-2
12th St. A-3,D-3,I-1,J-2
13th St. C-3,I-2,J-2
14th Ct. B-3
14th St. I-2
15th Ct. B-3
15th St. C-2,I-2
16th Ct. B-2
16th St. H-2,I-1
17th Ct. B-2
17th St. E-2,H-2
18th St. E-1,H-1
19th St. E-1,G-1,H-1
20th St. E-1,G-1
21st St. G-1
22nd St. G-1
23rd St. G-1

POINTS OF INTEREST

C. L. Graves Park F-3
Columbus Park A-3
Coronado Park F-4
Goodrich Park A-5
Highland Park J-5
McClelland Square Park H-4
Netherwood Park F-8
Rio Grande Park K-2
Robinson Park I-3
Roosevelt Park K-6
Soldiers & Sailors Park I-2
Spruce Park I-6

ATLANTA

Abbott E-1
Adair Ave. A-7,K-4
Adrian B-7
Aiken K-3
Alabama E-4
Alamo A-1
Alaska D-7
Albion Av. C-8
Alice F-3
Allen Ave. I-1
Allene Ave. I-1
Alexander St. C-3
Alloway H-8
Alta Ave. D-8
Amal Dr. K-3
Angier Ave. C-5,C-6
Angier Pl. C-6
Anne J-5
Argonne Ave. B-5
Arlington Pl. A-8
Arnold C-6
Ashby B-1,D-1,G-1
Ashby Cir. D-1
Ashby Pl. E-1
Ashland, W. D-7
Ashland Dr. D-7
Ashley B-7
Athas K-8
Athens Ave. J-2
Atlanta Ave. H-4,H-6
Atlantic A-3
Atlantis C-7
Auburn Ave. D-4,D-6
Augusta G-5
Austin Ave. C-8,D-8
Avon Ave. I-1
Ashwood Ave. K-2
Avondale Ave. I-7
Bailey E-2
Baker D-4
Bankhead Ave. B-1
Barker Pl. D-6
Barnett Pl. C-7
Barnett St. B-7
Bass H-5
Bass St. G-3,G-4,G-5
Battery Pl. D-8
Battle E-1
Battle Ct. F-8
Beatie Ave. J-1
Beauregard I-7
Beckwith St. E-1
Bedford St. B-1
Beechwood Ave. I-1
Belfast J-1
Belgrade C-7
Bell E-5
Belmont Ave. J-2
Bender G-2
Benjamin G-2
Benteen I-6,J-6
Benteen Way J-6
Berean Ave. F-6
Berne St. G-6,G-7
Bernina C-7
Beryl H-2
Biglin St. I-1
Bird St. K-5
Bisbee Ave. J-4
Bishop C-5
Bishop Al. C-5
Blashfield J-5
Blossom J-1
Blue Ridge Ave. B-8
Blue Ridge Ct. B-7
Bluff G-2
Boley C-1
Bonar E-1
Bonaventure B-7
Bonn K-4
Bonnie Brae H-1
Booker I-3,J-3
Booker Wash. Dr. D-1
Boulevard C-6,D-6, G-6,I-6
Bouldvard Dr. E-8
Boulevard Pl. C-6
Bowen K-5
Bowen Cir. J-3
Box Al. C-5
Boynton I-4
Bradberry E-3
Bradley D-5,D-6,E-6
Brady Ave. A-1
Brantley D-8
Briarcliff Pl. A-8
Brewer K-1
Broad St. E-3
Brookline St. H-1
Brotherton St. E-3
Brown Ave. J-4
Bruce Cir. J-7
Bryan St. F-6
Buckeye K-1
Buena Vista H-3
Bulloch St. J-4
Burchill J-1
Burns J-1
Burns Dr. H-8
Burroughs St. J-6
Burton K-5
Butler St. C-5,E-4,E-5
Byron J-1
Caldwell K-3
Cahoon St. K-1
Calhoun St. A-2
Capitol Ave. H-4
Carmel C-8
Carnigie D-4
Carroll E-6
Carter St. D-1
Cassanova St. J-6
Casplan K-2
Casplan Ct. K-1
Casplan St. K-1
Castleberry St. E-3
Catherine St. H-1
Central Ave. F-3,G-3
Chamberlain E-6
Chapel F-2
Chapel St. E-2
Charles Allen Dr. B-6
Charleston K-5
Charlton Pl. J-2
Chastain F-7
Cherokee Ave. G-6,H-6
Chester Ave. F-7
Chestnut Pl. D-1
Chestnut St. C-2,D-1
Christman G-2
Circle K-3
Claire Dr. K-1,K-4
Clemont Dr. A-7
Cleveland St. E-8
Clifton C-5
Climax I-5
Clinton J-2
Clover Al. C-5
Cogins Dr. H-7
Cohen H-2
Coleman St. I-2
Colquitt Ave. C-8
Columbia A-4
Cone D-4
Confederate H-8
Confederate Ct. H-7
Connally St. G-5,H-5
Cooledge Ave. A-6
Cooper St. G-3
Copenhill C-7
Corley D-6
Cornelia E-6
Courtland St. C-4,E-4
Crew St. H-4
Crogman K-4
Crumley St. G-3
Curran A-2
Currier St. C-4
Custer Ave. J-7
Cummings E-8
Cypress B-4
Dale Dr. J-7
Dallas St. C-6
Dalney St. A-3
Dalton St. I-5
D'Alvigney B-3
Daniel E-6
Danner I-8
David C-2
David Ct. C-7
Davage J-4
Davis St. C-2
De Kalb Ave. E-7
De Leon St. A-7
Deckner J-2
Deckner Ave. J-1
Degress D-8
Delaware Av. H-8
Delbridge St. D-2
Delevan St. H-2
Delmar Ave. H-7
Desoto D-1
Desoto Ave. J-1
Dewey H-2
Dill Ave. J-1
Division C-1
Dixie D-7
Doane St. H-3
Dodd G-3
Dora F-2
Drewry St. A-7,A-8
Drud Cir. D-7
Drummond E-1
Dunbar I-2
Dupont Pl. J-7
Durant Pl. B-5
Dunlap D-5
Earle D-3
East Ave. C-6
Eastwood H-8
Echo St. B-1
Echo St., W. B-1
Edgewood Ave. E-4,E-6
Edie Ave. I-7
Edith C-7
Eden Ave. H-8
Elbert St. H-1
Electric D-2
Elijah D-1
Elizabeth St. C-7
Elleby Rd. J-7
Elliot St. D-3
Elm St. D-2,E-2
Eloise G-7
Eloise Ct. G-7
Elvira St. J-5
Emerson Av. G-8
Emmett C-2
Englewood Ave. I-5
English Ave. C-1
Erin I-2
Erin Ave. I-1
Essie H-8
Esten E-8
Estoria St. F-7
Euclid C-8
Euclid Ave. C-8,D-7
Eugenia F-3
Euhrlee E-1
Evans St. G-1
Everhart St. J-1
Fair St. E-2,E-3
Fairbanks St. J-1
Faith Ave. F-8
Farrington H-5
Federal J-7
Federal Ter. J-6
Fedler Dr. I-2
Felton Dr. C-5
Fern St. H-4
Field E-7
Fisher Rd. J-7
Fitzgerald E-6
Flat Shoals Ave. E-8
Fletcher St. I-2,I-3
Florence D-7
Florida F-8
Ford Pl. B-7
Forest Ave. C-5,C-6
Formwalt St. G-3
Fort St. D-5,E-5
Fortress Ave. I-3
Fortune C-6
Forsyth St. E-3
Foundry St. D-1,D-2
Fowler B-3
Fowler St. A-3
Fox St. B-1
Francis Ave. I-6
Frank F-1
Franklin St. H-8
Fraser St. F-4,H-4
Fuller D-3
Fulton F-3
Fulton St. F-4
Fulton Terr. E-7
Funton J-7
Gammon I-4
Gardner H-2
Garibaldi St. G-3,I-3
Garland St. J-4
Garnett St. E-3
Gartrell E-5
Gault J-6,I-6
Genesee Ave. J-1
Geneva Dr. J-8
Georgia Ave. G-3
Gerard A-2
Giben Rd. K-3
Gibson St. F-8
Gift F-8
Gilbert H-8,I-8
Gillette H-1
Glen Pl. C-6,G-4
Glendale Pl. A-1
Glendale Ter. B-5
Glenn St. G-1,G-3
Glenwood Ave. F-5,F-7,F-8
Glenwood Pl. G-8
Gordon G-1
Gould K-5
Graham St. J-1
Grant Cir. H-5
Grant Pl. C-4
Grant St. F-5,G-5,J-5
Grant Park Pl. H-5
Grape D-6,I-4
Gray C-3
Gray St. C-2
Green Field A-3
Greens Ferry Ave. F-1
Greenwood B-6,B-7
Gress H-7
Griffin C-1,D-1
Hale St. D-7
Hall Ave. G-8
Hamilton Ave. H-8
Hammond F-1
Hampton St. A-2
Hannah J-5
Hanover E-8
Hansell G-6
Hardee H-2
Harden K-2
Hardwick K-4
Harold E-7
Harriet St. J-5
Harris St. D-4
Harte Dr. K-1
Hartford Ave. J-1
Hartford Pl. I-1
Hatcher H-4
Hayden D-3
Haynes St. D-3
Hawthorne E-8
Hazelrig Dr. K-8
Hemlock Cir. G-8
Hemphill A-2
Hendrix A-1
Highland Ave. B-8,D-6
Highland View A-7
Hill I-5
Hill St. F-5
Hillard St. D-5,E-5
Hills F-2
Hillside Dr. J-2
Hilltop Cir. D-1
Hobart Ave. I-6
Hogue St. D-6
Holiday D-8
Holtzclaw F-5
Home Ave. H-6
Hobson H-2
Hopkins K-5
Houston D-6
Houston St. D-4
Howell E-5
Howell Pl. C-3
Howell St. D-6
Hubbard St. I-2
Hudson Dr. A-8
Hugh G-3
Hulsey St. D-3
Humphries St. G-2
Hunnicutt C-3
Hunter St. E-2,E-3,F-5
Hurt St. D-8
Inman B-5
Ira St. G-3,I-3
Iris Dr. C-6
Irwin St. D-5
Iswald F-7
Ivy St. D-4,E-4
Jackson E-6
Jackson Pkwy. D-6
Jefferson St. B-1
Jenkins K-4
Jett St. C-1
John St. C-2
Johnson C-3
Jones Ave. C-2
Joyce E-1
Joyland J-3
Joyland St. K-3
Julian St. B-2
Juniper St. B-5
Kelly St. F-5
Kendrick Ave. H-5
Kennedy St. C-1
Kenneth G-4
Kent G-5
Kenyon E-7
King E-5
Killian G-7
Kirkwood E-7
Knott I-8
Lake Ave. D-7
Lakewood Ave. I-5
Lampkin D-6
Lansing St. J-4
Larkin St. F-2
Latimer D-3
Lawshe E-1
Lee St. I-1
Lethea St. K-5
Lester H-7
Lexington Ave. I-1
Lillian H-1
Linam H-4
Lincoln Pl. K-3
Linden Ave. B-4
Lindsay St. C-1
Linwood Ave. C-8
Little St. G-4,G-5
Livermore K-5
Logan F-5
Loneta I-6
Longview C-5
Loring St. G-8
Love St. G-3,G-4,G-5
Lovejoy C-3
Lowndes H-1
Luckie St. D-3,D-4
Lucy E-6
Lynch A-2
Lyndale H-8
Lynnhaven J-2
Lynwood St. G-7
Lytle F-8
Maderira C-7
Magnolia St. D-2
Maiden Ln. B-7
Main St. A-1
Manford Rd. I-2,I-3
Mangum St. D-3
Manigault Ave. E-8
Maple St. D-2
Marcus E-7
Margaret J-5
Marietta D-4
Marietta St. A-1,B-3
Marion G-7,J-7
Markham E-1
Martin I-4,I-5,J-4
Martin St. F-4,H-4,I-5
Mary H-2
Maryland A-8,I-2
Mayes D-3
Mayfield A-2
Mayland I-2
McCreary I-4
McCullough D-1
McDaniels St. F-2,I-2
McDonald F-7
McDonald Dr. I-5
McDonough Blvd. I-5,I-7
McMillan B-2
Mead St. H-6
Means St. B-2
Meldon Ave. J-4
Meldrum St. C-1
Mellview J-2
Memorial Dr. F-4
Mercer Pl. G-7
Mercer St. G-7,G-8
Merritts C-3
Merritts Ave. C-4
Michigan D-1
Middleton J-4
Mildred E-1
Milledge Ave. G-5
Miller J-5
Milton Ave. I-4
Milton St. E-1
Milton Ter. I-5
Mitchell St. E-1,E-2,E-4
Monroe B-6
Monroe Cir. B-6
Moore E-5
Moreland Ave. C-8,H-8
Moreland Dr. K-7
Morely I-7
Moton I-3
Moury Ave. J-4
Murphy Ave. G-1
Murray J-4
Myrtle B-5
Neal St. C-1
Nelson St. E-3
New Cir. K-7
Newcastle E-1
Newport St. C-1
Newton D-3
Napier I-8
Narrow St. F-7
Nolan St. I-5
North Ave. B-1,B-3, B-6,B-8
Northern E-8
Northside Dr. A-2
Nuting C-5
Oak Knoll Cir. K-5
Oakhill Ave. H-2
Oakland F-5
Oakland Ave. G-6
Ogden E-1
Old Flat F-7
Old Wheat D-5,D-6
Olive K-4
Oliver St. C-1
Orlean G-5
Ormewood Ave. G-6,G-8
Ormewood Ter. H-7
Ormond St. H-3,H-5
Ozone E-1
Park Ave. G-6, H-6,J-4,J-6
Park Rd. J-7
Park St. F-1
Parsons K-5
Parsons St. E-1
Pavillion G-5
Paynes C-1
Peachtree Pl. A-4
Peachtree St. C-4,E-3
Peachtree St., W. C-4
Pear St. I-4
Pearce St. H-1
Pearl St. F-7
Pelham St. B-1
Penn Ave. B-5
Pickett E-7
Piedmont E-4
Piedmont Rd. B-5,D-5
Pine C-6
Pine St. C-4
Play Lane D-1
Plum A-3
Ponce De Leon Ave. B-4,B-7
Ponce De Leon Ct. B-6
Ponce De Leon Ter. A-7
Pontiac K-8
Pontiac Pl. J-8
Poole Pl. G-1
Poplar D-4
Poplar Cir. D-8
Portland G-8
Porter Pl. C-4
Pratt St. E-5
Prescott C-4
Primrose Cir. I-5
Primrose St. H-5
Proctor C-1
Prospect H-8
Pryor Circle K-3
Pryor Rd. I-3,K-3
Pryor St. E-4,G-3
Pullman St. H-4
Pylant A-7
Rankin Pl. C-6
Rankin St. C-6
Rawlins I-5
Rawlins St. H-5
Rawson F-2
Raymond St. E-1
Reed G-4
Reed Ave. J-5
Reinhard E-6
Renfroe E-8
Rhodes Ave. K-4
Rhodes St. D-2
Ridge Ave. H-4
Richardson St. F-3
Richmond G-4,K-6
Richmond Cir. K-6
Roach F-2
Roberts Dr. I-6
Robins G-2
Robinson Ave. H-6
Rock A-1-3
Rockwell St. H-2
Rosalie St. G-6
Rose Circle H-1
Rosedale G-6
Rosedale Dr. A-8
Roy St. I-2
St. Charles B-6,B-8
St. Charles Ave. B-7
St. Louis Pl. B-8
St. Paul F-5
Sampson St. D-7
Sanders F-8
Savannah E-6
Sawtell Ave. K-5
Schuyler I-6
Sciple Ter. D-1
Scott C-4
Seaboard Ave. D-8
Seal Pl. A-6
Selman E-7
Seminole Ave. C-8
Shannon J-2
Shaw St. J-4
Shelby Pl. I-8
Shelton G-1
Shoals Rd. F-7
Short E-7
Siloam Ave. G-8
Simpson St. C-2,C-3
Sims St. I-2
Sinclair Ave. C-8
Sloan Cir. I-7
Smith St. I-3
Somerset Ter. B-7
Soloman F-4
South Ave. H-5
Spelman F-2
Spencer St. D-1,D-2
Spring St. C-4,E-3
Stephens G-2,G-3
Stonewall St. E-2
Stovall St. F-8
Strong C-2
Sunset Ave. C-2
Sydney F-5,F-6
Sylvan Pl. K-1
Sylvan Rd. I-1
Taft A-6,K-3
Taliaferro St. D-1
Tech Pkwy. B-2
Techwood Dr. B-4,C-4,E-3
Tenelle St. E-6
Terry G-4
Terry St. F-4
Thayer J-4
Thirkeld Ave. J-4
Thomas Dr. K-8
Thomasville Blvd. K-7
Thorton j-3
Thurmond St. D-1,D-2
Tift G-1,H-2
Todd A-8
Trammel St. C-1
Travis B-2
Trenholm F-2
Trinity Ave. E-3
Troup St. K-4
Trusco G-2
Tudor H-3
Turner D-1
Turpin Ave. I-7
Tuskegee H-5
Twiggs J-4
Tyler C-2,C-3
Underwood Ave. H-7
University Ave. I-2,I-4
Vanra Ave. H-4
Vaud C-8
Vedado Way B-6
Venable St. C-3
Vernon C-5,G-7
Vickers St. I-8
Victoria C-3
Victory Dr. K-1
Vine St. D-2,E-2
Violet H-4
Virgil D-7
Virginia A-8
Virginia Ave. A-7
Virginia Cir. A-7
Wade D-8
Waddel D-7
Wall St. D-3
Walker Ave. H-7
Walker St. F-2
Walnut C-2
Walnut St. E-2
Walthan St. E-8
Walton St. D-3
Warner St. I-1
Warren B-2
Warwick G-8
Washita Av. C-8
Washington St. F-4,G-4,H-4
Waverly Way D-8
Wayne J-2
Welch St. H-2,K-7
Wells St. G-2
Wellswood Dr. J-7
West Ave. I-3
West End Ave. F-1
Western Ave. C-2
Weyman I-4
Whatley K-5
Wheeler St. B-1
White St. H-1
Whitehall St. F-2
Whittaker St. B-1
Wilbur F-8
Wilcox J-4
Williams St. B-4,C-4
Williams Mill Rd. C-7
Willow St. B-5
Wilson A-1
Windsor St. G-3,I-3
Winton Ter. C-6
Wood F-6
Woodall St. G-6
Woodbourne Dr. K-1
Woodland Ave. G-8,I-8
Woodlawn Ave. J-8
Woodlawn Cr. J-8
Woodward Ave. F-5
Woodrow Pl. C-5
Woodrow St. I-1
Wylie St. E-7,E-8
Yonge E-6
York Ave. G-1
3rd St. A-2,B-3,B-4
4th St. B-2,B-3,B-5
5th St. B-2,B-3,B-4,B-5
6th St. A-4
7th St. A-4,A-5
8th St. A-3,A-4,A-5
9th St. A-1,A-2,A-5
10th St. A-1,A-2,A-3,A-6
11th St. A-3,A-4

POINTS OF INTEREST

Adair Park I-1
Atlanta Tech J-2
Benteen Park J-7
Chosewood Park I-6
Clark College E-1
Couch Park A-2
Emman Milligan Park J-2
Grant Park H-6
Home Park A-3
Joyland Park J-3
Key Park G-7
Morehouse College E-1
Oaklane Cemetery E-6
Parkerson Park J-1
Piedmont Park Golf Course A-5
Pittman Park H-3
Roosevelt H.S. G-7
Smith H.S. G-5
U.S. Penitentiary J-6

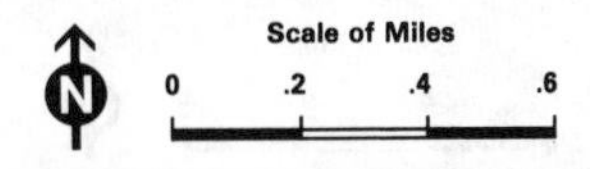
N
Scale of Miles
0 .2 .4 .6

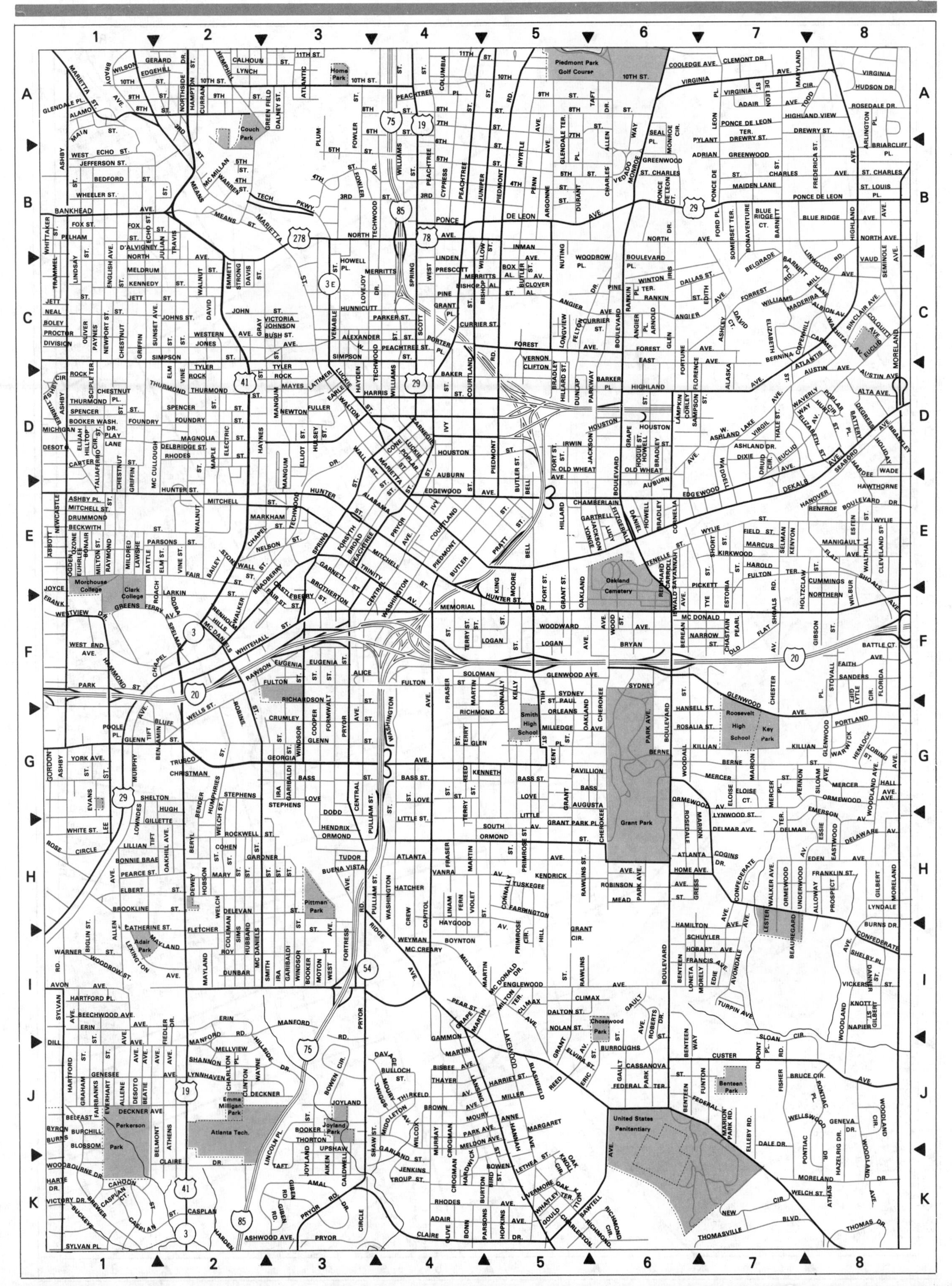
1 2 3 4 5 6 7 8
A B C D E F G H I J K
Piedmont Park Golf Course
Home Park
Couch Park
Morehouse College
Clark College
Oakland Cemetery
Smith High School
Roosevelt High School
Key Park
Grant Park
Pittman Park
Adair Park
Chosewood Park
Benteen Park
Emma Milligan Park
Atlanta Tech.
Joyland Park
Perkerson Park
United States Penitantiary

Austin, TX

Scale of Miles

0 .2 .4 .6

N

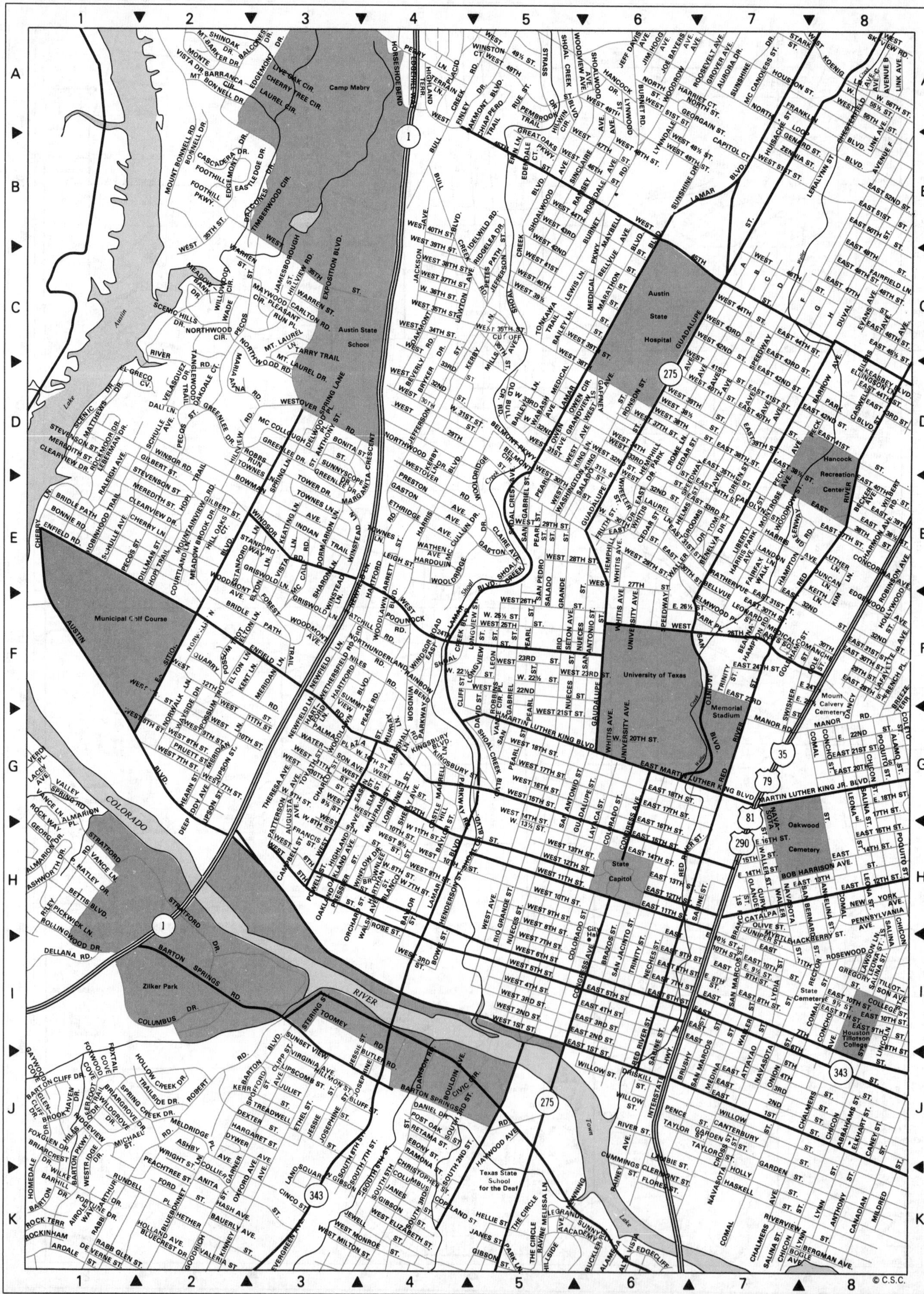

Camp Mabry

Austin State School

Austin State Hospital

Hancock Recreation Center

Municipal Golf Course

University of Texas

Memorial Stadium

Mount Calvary Cemetery

Oakwood Cemetery

State Capitol

State Cemetery

Houston Tillotson College

Zilker Park

Texas State School for the Deaf

COLORADO RIVER

Town Lake

Lake Austin

AUSTIN

Abrahams St.........J-8
Academy Ln.....K-5,K-6
Airole Way..........K-1
Alameda.............K-6
Alamo St............G-8
Almarion Dr.........H-1
Almarion Pl.........G-1
Alta Vista..........K-6
Angelina St.........H-8
Anita St............K-2
Anthony St..........K-8
Ardale St...........K-1
Arthur Rd...........K-1
Ashby Ave...........J-2
Ashworth Dr.........H-1
Attayao St..........J-7
Augusta Ave....G-3,H-3
Aurora Dr...........A-7
Austin Blvd......F-1,G-2
Ave. A...........C-7,D-6
Ave. B......A-8,C-7,D-7
Ave. C......A-8,C-7,D-7
Ave. D...........C-7,D-7
Ave. F.......B-8,C-7,D-7
Ave. G...........C-8,D-7
Ave. H...........C-8,D-7
Bailey Ln........C-5,D-5
Balcones Dr.....A-2,A-3
Barhill Dr..........K-1
Barranca Cir........A-2
Barrow Ave..........D-8
Barton Blvd.......I-3,J-3
Barton Cliff Dr.......J-1
Barton Dr.........J-1,K-1
Barton Pkwy.......J-1,K-1
Barton Springs Rd.
.................I-2,J-4
Baverly Ave.........K-2
Baylor St...........H-4
Beanna..............F-7
Becker Ave..........E-8
Bellvue Ave......B-6,C-6
.................E-7,F-7
Belmont Cir.........D-5
Belmont Pkwy........D-5
Benelva Dr..........E-7
Bergman Ave.........K-8
Bettis Blvd.........H-1
Beverly Rd.......C-4,D-4
Blanco St...........H-4
Bluebonnet Ln.......K-2
Bluecrest Dr........K-2
Bluff St.........J-3,J-4
Bob Harrison Ave.
.................H-7,H-8
Bogle Ave........K-7,K-8
Bonita St...........D-3
Bonnell Dr..........B-2
Bonnie Rd...........E-1
Bouldin Ave.........J-4
Bowie St............I-4
Bowman Ave.......E-2,E-3
Branch St...........H-7
Brazos St...........I-6
Breeze Terr......F-8,G-8
Briarcrest Dr.......J-1
Briargrove Dr.......J-1
Bridle Path...E-1,F-2,F-3
Brookhaven Dr.......J-1
Brownlee St.........H-4
Brushy St...........J-6
Bryker Dr........C-4,D-4
Buckler Rd..........K-6
Bull Creek Rd.......A-4
...........B-4,C-4,C-5
Burnet Rd........A-6,B-6
Butler Rd...........J-4
Campbell St.........H-3
Canadian St.........K-8
Caney St............J-8
Canterbury St.......J-7
Capitol Ct..........B-7
Carlton Rd..........C-3
Carolyn St..........E-7
Cascadera Dr........B-2
Castle Hill.........G-4
Caswell Ave......C-8,D-8
Catalpa.............H-7
Cedar St.........D-6,E-6
Chalmers Ave........I-8
.................J-8,K-7
Charlot St..........G-3
Cherry Ln........E-1,E-2
Cherry Tree Cir.....A-3
Chesterfield Ave.
.................A-8,B-8
Chiappero Trail.....A-5
Chicon St........G-8,H-8
...........I-8,J-8,K-7
Christopher St......K-4
Cinco St............K-4
Circle, The.........K-5
Civic Cir...........J-4
Claire Ave..........E-5
Clearview Dr.....D-1,E-2
Clermont St.........K-6
Cliff St............F-4
Clipp St............J-3
College St..........I-8
Cole St.............F-8
Coleto St...........G-8
Collier St.......K-2,K-3
Colorado St..G-6,H-6,I-5
Columbus Dr.........I-2
Columbus St.........K-4
Comal St.........G-8,H-8
..........I-8,J-7,K-7
Comanche.........F-7,F-8
Concho St...........G-8
Concordia Ave.......E-8
Congress Ave.G-6,H-6,I-6
Copeland St.........K-4
Courtland Ln........E-2
Cross St............J-7
Cullun Dr........E-4,E-5
Cummings St.........J-6
Curve Orlander St....H-7
Dalcones Dr......B-2,B-3
Dali Ln.............D-2
Daniel Dr...........J-4
Dancy............F-8,G-8
David St............G-5
Dawson Rd...........J-4
Deep Eddy Ave..G-2,H-2
Deerfoot Trail......J-1
Dellana Rd..........I-1
Delwood Pl..........D-3
DeVerne St..........E-4
Dexter St........J-2,J-3
Dillman St..........E-2
Dorm Arion Ln.......E-3
Dressier St.........H-3
Driskill St.........J-6
Duncan Ln...........E-8
Duval St.........B-8,C-8
Dywer Ave...........J-2
Eason St.........G-3,G-4
East Dr.............E-6
Ebony St............J-4
Edendale Ct.........B-5
Edgecliff...........K-6
Edgehill Way........A-4
Edgemont Dr....A-3,B-2
Edgewood Ave........F-8
Eilers Ave.......C-8,D-8
Elizabeth St., West..K-4
El Greco Cv......D-1,D-2
Elkhart St..........J-8
Ellingson Ln........D-8
Elm St..............G-4
Elmwood Pl..........F-7
Elton Ln.....E-3,F-2,F-3
Enfield Rd...E-1,F-3,G-3
Erin Ln.............B-5
Ethel St............J-3
Ethridge Ave........E-4
Evans Ave...........C-8
Evergreen Ave.......K-3
Exposition St.......B-3
.............C-3,E-2,F-2
Fairfaild Ln........C-8
Fairfax Walk........F-7
Finley Dr........A-4,A-5
Flores St...........K-6
Foothill Eastledge Dr.
.................B-2,B-3
Foothill Pkwy.......B-2
Ford St.............K-2
Forest Trail........F-3
Fortune Dr..........K-1
Foxglen Dr..........J-1
Foxtail Cove........J-1
Foxwood Cove........J-1
Francis St..........H-3
Franklin Blvd....A-7,B-8
French Pl...........F-8
Fruth St............E-6
Gaffney St..........D-6
Garden St........J-7,K-7
Garner Ave...J-2,J-3,K-2
Gaston Ave.......E-4,E-5
Gaywood Cove........J-1
Genard St...........B-7
Georgain St..A-6,A-7,B-7
George P. Hatley Dr..H-1
Gibson St........K-4,K-5
Gibson St., W.......K-3
Gilbert St.......D-2,E-2
Glencliff Dr........J-1
Goldham St..........F-7
Goodrich St.........K-2
Grande St........E-5,F-5
Grandview........D-5,D-6
Great Oaks Pkwy.....B-5
Greenlee Dr.........D-2
.................D-3,E-3
Green St.........D-7,E-7
Greenway............E-7
Gregory St..........I-8
Griswold Ln......E-3,F-3
Grooms Ave.......E-6,E-7
Grover Ave..........A-7
Guadalupe St.....B-7,C-6
...........E-6,F-6,G-6
Hampton Rd..E-7,E-8,F-7
Hancock Dr..........A-6
Hardouin............E-4
Hargaret St......J-2,J-3
Harmon Ave..........E-8
Harriet Ct.......A-6,A-7
Harris Ave.......E-7,E-8
Harris Blvd......D-4,E-4
Harris Park Ave.....E-7
Hartford Ln.........G-3
Hartford Rd.........E-4
.................F-3,F-4
Harthan St..........H-4
Hash Ave............K-2
Haskell St..........K-7
Haywood Ave.........J-5
Hearn St............G-2
Hellie St...........K-5
Helms St............E-6
Hemphill Dr.........E-6
Hemphill Park.......D-6
Henderson St........H-4
Hether St...........K-2
Highland Ave........H-3
Highland Terr.......A-4
Hillmon St..........J-3
Hill Oaks Ct........E-2
Hills Dr............J-1
Hillside Ave........K-5
Hillview Rd......C-3,D-2
Hilwin Cir..........A-5
Holland Ave.........K-2
Hollow Creek Dr.....J-2
Holly St............K-7
Hollywood St........F-8
Homedale.........J-1,K-1
Hopi Trail.......D-2,E-2
Horseshoe Bend......A-4
Houston St..........A-7
Huisache St......A-7,B-7
Indelwild Rd.....B-5,C-4
Indian Trail........E-3
Interstate Hwy. 35
.................I-6,J-6
Inwood Pl...........F-2
Jackson Ave......B-4,C-4
Jamesborough St.
.................B-3,C-3
Janes St.........K-4,K-5
Jarrett St..........E-4
Jeff Davis Ave......A-6
Jefferson St........B-5
.............C-4,C-5,D-4
Jessie St........I-4,J-3
Jewell...........K-3,K-4
Jim Hogg Ave........A-6
Joe Bayers Ave......A-6
Josephine St.....J-3,J-4
Juliet St...........J-3
Juniper St..........I-7
Keasbey Blvd........D-8
Keating Ln..........E-3
Keith Ln.........E-8,F-8
Kent Ln.............F-2
Kerby Ln.........C-5,D-4
Kerr St.............J-2
Kim Ln...........E-8,F-8
Kingsbury Blvd......G-4
Kingsbury St........G-4
King St.............D-6
King Ln.,West.......D-5
.................D-6,E-5
Kinney Ave.......K-2,K-3
Koenig Ln.,West.....A-8
Lacey Ave...........G-1
Lafayette Ave.......F-8
Lamar Blvd.......B-6,B-7
.....C-6,D-5,E-5,F-4,H-4
Lambie St...........J-6
Land Square.........K-3
Laurel Cir..........A-3
Laurel Ln...........E-6
Lawson Ave..........I-8
Lawton Ave..........C-4
Layaca St........G-6,H-6
Leberman Dr.........D-1
Legrande.........K-5,K-6
Leigh St............E-4
Leonard St..........F-7
Leona St.....G-8,H-8,I-8
Leon St.............F-5
Leralynn St.........B-8
Lewis Ln............C-5
Liberty St..........E-7
Lincoln St..........I-8
Link Ave.........A-8,B-8
Lipscomb St.........J-3
Live Oak Cir........A-3
London Ln...........E-7
Longview St.........F-5
Loop Blvd.,North
...........A-7,B-7,B-8
Lorraine St.........G-4
Luther Ln...........E-8
Lydia St............I-7
Lynndale.........A-6,B-6
Lynn St.............K-8
Lynn St., West...G-3,H-3
Lynwood St..........A-6
Manor Rd.........G-7,G-8
Marathon Blvd...B-6,C-6
Marganita Crescent
...........D-4,E-3,E-4
Maria Anna Rd...D-2,D-3
Marrell Ln..........G-4
Marshall Ln.........G-4
Martin Luther King
Blvd................G-5
Martin Luther King
Blvd. East..........G-6
Martin Luther King, Jr.
Blvd.............G-7,G-8
Mattews Dr..........D-1
Maufraisi St.....G-4,H-4
Maybelle............B-6
Maywood Cir.........C-3
McCall Rd........E-3,F-3
McCanoless St.......A-7
McCollough St.......D-3
Meadowbank Dr.......C-2
Meadowbrook Dr......E-2
Medical Arts Sq. St...F-7
Medical Pkwy........C-6
Medina St...........J-7
Meldridge Pl........J-2
Meredith St......D-1,E-2
Meridan Ln.......F-3,G-2
Michael St..........J-1
Mildred St..........K-8
Mills Ave........C-5,D-5
Milton St., West....K-3
Monroe St., WestK-3,K-4
Monte Vista Dr......A-2
Montrose Ave........E-7
Mountainview Rd.....E-2
M.. Barker Dr.......A-2
Mt. Bonnell Dr......A-2
Mount Bonnell Rd.
.................B-1,B-2
Mt. Laurel Dr.......C-3
Mt. Laurel Ln.......C-3
Murray Ln...........G-4
Myrtle Hackberry St.
.................I-7,I-8
Navasota St.........G-7
...........H-7,J-7,K-7
Neches St...........I-6
Newfield Ln.........E-4
.................F-3,G-3
Newning..........K-5,K-6
New York Ave........H-8
Niles Rd.........F-3,F-4
North St.........A-6,A-7
Northunderland Rd.
.................F-3,F-4
Northwood Cir.......C-2
Northwood Dr........D-4
Northwood Rd.....C-2,D-3
Norwalk Ln.......F-2,G-2
Nueces St....F-5,F-6,H-5
Oakdale Ct..........D-2
Oakland Ave.........H-3
Oakmont Blvd........A-5
.................B-4,C-4
Old Bull Cr. Rd.....D-5
Olive St............H-7
Onion St............J-7
Orchard St..........H-3
Owen Ave............D-5
Owen Cir.........D-5,D-6
Oxford Ave..........K-2
Palma Plaza.........G-3
Park Blvd...........D-8
Park Ln.............K-5
Park Pl.............F-7
Parkway Rd..........G-4
Patterson Ave....G-3,H-3
Peachtree St.....J-2,K-2
Pearl Ave........D-5,E-5
Pearl St.....E-5,F-5,G-5
Pease Rd............G-4
Peck Ave.....D-7,D-8,E-7
Pecos St.....C-2,D-2,E-1
Pembrook Trail......A-5
Pence St............J-6
Pennsylvania Ave....H-8
Perry Ln............A-4
Petes Path..........C-5
Pickwick Ln.........H-1
Placid..............A-4
Pleasant Run Pl.....C-3
Poquito St.......G-8,H-8
Poqunock Rd.........F-4
Possum Trot......F-2,G-2
Post Oak St.........J-4
Powell St...........H-3
Preston Ave.........E-4
Pruett St...........G-2
Quarry Rd...........F-2
Rabb Glen St........K-1
Rabb Rd.............K-1
Rainbow Bend........F-4
Rainey Ave.......J-6,K-6
Raleigh Ave......D-1,E-1
Ramona St...........J-4
Ramsey Ave.A-6,B-5,B-6
Rathervue........E-7,F-7
Ravine Melissa Ln...K-5
Rector St...........I-8
Red River St........D-8
.........E-8,E-7,F-7,G-7
...........I-6,J-6,H-6
Retama St...........J-4
Ridgelea Dr......B-5,C-5
Ridgeview...........J-1
Ridge Dr., West..J-1,K-1
Riley Rd............H-1
Rio Grande St.......H-5
River Rd.........C-2,D-2
River St............J-6
Riverview St........K-7
Robbins Pl..........F-5
Robbs Run...........D-3
Robert Rd...........J-2
Robinhood Trail.....E-1
Robinson Ave.....E-8,F-8
Rockinham...........K-1
Rockmoor Dr.........D-1
Rock Terr...........K-1
Rockway..........G-1,H-1
Rollingwood Dr.
.................H-1,I-1
Rome Ln.............D-6
Ronson St...........D-6
Roosevelt Ave...A-6,A-7
Rose St.............H-4
Rosewood............I-8
Rosindale Ave.......B-6
Rue St..............A-5
Rundell Pl.......K-1,K-2
Sabine St....H-7,I-6,J-6
St. Anthony St......D-3
Salado St........D-6,E-5
.................E-6,F-5
Salina St........G-8,H-8
.................I-8,K-7
San Antonio St...F-6,G-5
San Bernard St......H-8
San Gabriel St......E-5
.................F-5,G-5
San Jacinto St......I-6
San Lacinto Blvd.
.................F-7,G-7
San Marcos St....I-7,J-7
San Pedro St.....E-5,F-5
Scenic Dr...........D-1
Scenic Hills Dr.....C-2
Schulle Ave......D-2,E-1
Seton Ave...........F-5
Sharon Ln........E-3,F-3
Shelly Ave..........G-4
Shinoak Dr..........A-2
Shoal Creek Blvd....A-5
........B-5,C-5,E-5,F-4
..............G-5,H-4,H-5
Shoal Crest Ave..D-5,E-5
Shoalwood Ave..A-6,B-5
Sinclaire Ave.......A-6
.................B-5,B-6
Skyview Rd., West..A-8
Speedway.........C-7,D-7
.................E-6,F-6
Spring Creek Dr..J-1,J-2
Spring Ln........D-3,E-3
Spufford St.........J-3
Stanford Ln......E-2,E-3
Stanford Way........E-3
Stark St............A-7
Stering St..........I-3
Stevenson St.....D-1,E-2
Strass Dr...........C-2
Stratford Dr.....H-1,H-2
Summit View.........F-3
Sunny Ln............K-6
Sunnyscope Dr.
.................D-4,E-4
Sunsetview..........I-3
Sunshine Dr.....A-7,.B-6
Swisher St.......F-7,G-7
Tanglewood Trail....D-2
Tarry Trail.........C-3
Taylor St........J-6,J-7
Terrian Ln..........A-4
Theresa Ave.........G-3
Tillotson Ave.......I-8
Timberwood Cir......B-3
Tonkawa Trail.......C-5
Toomey Rd...........I-3
Tower Dr............E-3
Townes Ln...D-2,E-3,E-4
Toyath St...........G-3
Trailside Dr........J-2
Treadwell St........J-3
Trinity St.......F-7,I-6
University Ave...F-6,G-6
Upson St............G-2
Valasquez Dr........D-2
Valeria St..........K-2
Valley Springs Rd...G-1
Vance Cir........F-5,G-5
Vance Ln.........G-1,H-1
Verdi Pl............G-1
Virginia Ave........J-3
Vista Ln............E-3
Wabash Medical Ave.D-5
Wade Cir............C-2
Waller Dr........E-6,E-7
Waller St........H-7,J-7
Walsh Ave...........H-4
Warren St........C-2,C-3
Washington..........E-5
Watchill.........F-3,F-4
Water-Son Ave.......G-3
Wathen Ave..........E-4
Wayside Dr.......F-2,G-2
West Ave.....D-5,D-6,H-5
Westover Rd.........D-3
.................D-4,E-4
Wethersfield Rd.....F-3
Wheeler St..........E-6
Whitis Ave...E-6,F-6,G-6
Whitis, East........E-6
Wilbert Rd..........E-8
Wildgrove Dr........J-1
Willowood Cir.......C-2
Winsor Rd........F-4,G-4
Windsor Rd. East....F-4
Winflow Dr.......H-3,H-4
Winsor Rd...........D-2
Winstead Ln......E-3,F-3
Winston Ct..........A-5
Wilke Dr............K-1
Willow St........J-6,J-7
Windsor Rd..........E-3
Woodlawn Blvd..F-4,G-3
Woodmont Ave....E-2,F-3
Woodrow Ave.........A-6
Woodrow St..........E-7
Woodview Ave........A-5
Woolridge........D-5,E-4
Wright St...........J-2
Zennia St...........B-7
1st St. East.......I-6J-7
1st St. West........I-5
2nd St. East.....I-6,J-7
2nd St. South....J-4,K-4
2nd St. West........I-5
3rd St. East.....I-6,J-7
3rd St. South....J-4,K-4
3rd St. West.....I-4,I-5
4th St. East....I-6,I-7,J-7
4th St. West........I-5
5th St. East....I-6,I-7,J-7
5th St. South....J-4,K-4
5th St. West.....H-3,I-5
6th St. East.....I-6,I-7
6th St. South....J-4,K-3
6th West.........H-3,I-5
7th St. East.....I-6,I-7
7th St. South....J-4,K-3
7th St. West...G-2,H-4,I-5
8th St. East.....I-6,I-7
8th St. South....J-3,K-3
8th St. West.....G-1,G-2
.........G-3,H-3,H-4,H-5
9th St. East....I-6,I-7,I-8
9th St. West.....G-2,G-3
.............H-3,H-4,H-5
9½ St. E.........I-7,I-8
9½ St. West......G-3,H-4
10th St. East...I-6,I-7,I-8
10th St. West....F-1,F-2
.....G-2,G-3,G-4,H-4,H-5
10½ St. East........I-7
11th St. East....H-6,I-7
11th St. West....F-2,G-3
.................H-4,H-5
12th St. East....H-6,H-8
12th St. West....F-2,G-3
.............G-4,H-5,H-6
13th St. East....H-6,H-8
13th St. West....G-4,H-5
13½ St. W...........H-5
14th St. East.H-6,H-7,H-8
14th St. W.......G-4,G-5
15th St. East.......H-6
.................H-7,H-8
15th St. West.......G-5
16th St. East....H-6,H-7
16th St. West.......G-5
17th St. East....G-6,G-8
17th St. West.......G-5
18th St. East....G-6,G-8
18th St. West.......G-5
20th St. East.......G-8
20th St. W..........G-6
21st St. East.......G-8
21st St. West.......G-5
22nd St. E..........G-8
22nd St. West.......F-5
22½ St. W........F-4,F-5
23rd St. East.......F-7
23rd St. West....F-5,F-6
24th St. East....F-7,F-8
24th St. West....F-4,F-5
25½ St. W...........F-5
26th St. West.F-5,F-6,F-7
26½ St. E...........F-6
27th St. West.......E-6
28th St. E..........F-6
28th St. West....E-5,E-6
29th St.............F-8
29th WestD-4,D-5,E-5,E-6
30th St. East....E-6,E-7
.................F-7,F-8
30th St. West.D-4,E-5,E-6
31st St. East.......E-7
.............E-6,F-7,F-8
31st St. W...D-4,D-5,E-6
31½ St...........D-6,E-6
32nd St.............F-8
32nd St. East....F-7,F-8
32nd St. West.......D-4
.............D-5,D-6,E-6
33rd St. East....E-6,E-7
33rd St. West.......C-4
.............D-4,D-5,D-6
34th St. East.......E-7
34th St. West....C-4,D-6
35th St..........C-5,D-5
35th St. East....D-7,E-7
35th St. West.......B-2
....B-3,C-2,C-3,C-4,C-5
36th St. W..........C-4
37th St. East.......E-8
37th West........C-4,D-6
38th St. East....D-7,E-8
38th St. WestC-4,C-5,D-6
38½ St. East.D-7,E-7,E-8
38½ St. West.....D-6,D-7
39th St. East....D-7,E-8
39th St. West.......B-4
....C-4,C-5,C-6,D-6,D-7
39½ St. West.....C-5,C-6
40th St. East....D-7,E-8
40th St. West.......B-4
.........C-5,C-6,D-6,D-7
41st St. East....D-7,D-8
41st St. WestC-5,C-6,D-7
42nd St. East....D-7,D-8
42nd St. West.......B-5
.................C-6,C-7
43rd St. East.......C-7
.................D-7,D-8

BALTIMORE

Abell...............A-5
Abbotson St.........A-8
Abbott St...........E-8
Aiken St............D-7
Aisquity St.........B-7
...........C-7,E-7,G-7
Albemarle...........H-7
Aliceanna...........I-7
Alluvion St.........J-3
Amity St............H-2
Annapolis Rd........K-3
Ann St..............I-8
Arclay St...........D-6
Argeant St..........I-2
Argyle Ave..........E-2
Arlington Ave.......H-2
Art Museum Dr.......A-4
Ashland Ave.........F-7
Atkinson St.........B-4
Baker St............D-1
Baltimore St.....G-7,H-4
Bank St.............H-7
Barclay St.......B-5,D-6
Barnes St...........E-8
Barney St........K-5,K-6
Barre St.....I-3,I-4,I-5
Bartlett Ave........B-6
Battery Ave......I-6,K-6
Bayard Rd...........J-3
Bayard St...........J-2
Beason St...........K-8
Belt St.............K-6
Belvidere St........D-6
Bethel St........D-8,I-8
Bevan...............J-5
Biddle St........E-6,F-3
Block St............I-7
Bloom St............D-2
Bolton Pl...........E-4
Bolton St...........C-3
Bonaparte Ave.......B-7
Bond St..........F-8,G-7
Boone St.........B-6,C-6
Booth St.........H-1,H-2
Boyle St............K-7
Bradley.............F-3
Brantley............F-2
Brentwood Ave..B-5,E-5
Broadway St.....D-8,H-8
Brookfield Ave......C-2
Brooks Lane.........C-2
Brunt St............E-2
Bush St.............J-2
Byrd St.............K-6
Calender St.........H-2
Calhoun St..........H-1
Callow Ave..........C-2
Calvert St.......B-5,H-5
Camden St...........H-4
Carey St.........D-1,I-2
Carlton St.......G-2,H-2
Caroline St.........I-7
Carroll St.......I-3,J-2
Carrollton St....G-2,H-2
Carswell St.........A-8
Cathedral St........E-4
Cecil Ave........B-7,C-6
Center St...........F-4
Central St..........F-7
Chapel St........D-8,F-8
Charles St.......H-5,J-5
Chase St............E-4
Chauncey Ave........C-2
Chestnut Ave........A-3
Clarkson St.........K-5
Clay St.............G-4
Clement St...J-5,J-6,K-7
Clendenin Ave.......D-2
Cleveland St........J-2
Cliftview Ave.......B-7
Cloverdale Rd.......C-1
Cokesbury Ave.......C-6
Colvin St...........F-6
Commerce St.........H-6
Constitution St.....F-6
Conway St.....I-3,I-5,J-3
Cooksie St..........K-8
Covington St........J-6
Cresmont............A-4
Cresmont Ave........A-4
Cross St.........I-2,J-4
Cuba................K-8
Cumberland..........D-1
Curtain Ave.........C-6
Dallas St........F-8,G-8
Darley Ave.......B-7,C-8
Davis St............G-5
Decatur.............K-8
Dexter St...........I-3
Division St.........D-1
Dock St.............I-7
Dolphin St..........E-3
Dorn St.............F-3
Druid...............C-1
Druid Hill Ave...C-1,E-3
Ducatel St..........C-2
Durham St........D-8,I-8
Eager St............E-5
East St.............F-6
Eastern Ave.........H-7
Eden St.............E-7
Edmonson Ave........F-1
Edyth St............F-7
Eislen St...........I-3
Ellerslie Ave.......A-6
Ellsworth St........D-8
Etting St........D-1,E-3
Eutah Pl.........C-1,E-4
Eutaw St.....H-4,J-4,K-3
Exeter St........F-6,H-7
Exeter Hall Ave.....A-6
Fairmont Ave........G-4
.................G-7,H-1
Faith La............E-8
Falls Rd............B-3
Fallscliff Rd.......A-2
Fawcett.............C-4
Fawn St.............H-7
Fayette St.......G-1,G-7
Federal St.......D-5,D-7
Fenwick Ave.........A-8
Fillmore St.........A-7
Fleet St............H-7
Fort Ave.........K-5,K-7
Francis St..........C-1
Franklin St.........G-4
Frederick St........H-6
Fremont Ave......E-2,G-3
Friendship..........B-7
Frisby St...........A-6
Front St.........F-6,G-6
Garrett Ave.........B-7
Gay St...........F-7,H-6
George St...........F-3
Germania Ave........C-7
Gilbert St..........F-2
Gittings St......J-5,J-6
Glyndon St..........I-2
Gold St.............D-1
Gorsuch Ave.........A-7
Gough St............H-7
Granby St...........H-6
Graves St...........F-6
Greene St...........H-4
Greenmount Ave......E-6
Guilford Ave.....B-5,F-5
Gutman Ave..........B-6
Haines St...........K-3
Hamburg St.......I-3,J-5
Hamilton St.........F-4
Hampden.............C-4
Hampden Ave.........B-4
Hanover St.......H-5,K-5
Harford Ave......E-7,F-6
Harford Rd..........B-8
Harlem Ave..........F-1
Harvey St...........K-7
Haubert St..........K-8
Heath St.........K-5,K-6
Henneman Ave........E-7
Henrietta...........I-5
Herkimer St.........I-1
High St..........F-6,H-7
Hill St.............I-5
Hillen St...........F-6
Hillman St..........E-6
Hoffman St..........D-6
.................E-4,F-3
Holbrook St.........E-6
Holiday.............G-6
Hollins St..........A-1
Holly Cross La......C-8
Homestead St........A-7
Homewood Ave........A-7
.................C-6,E-6
Hope St.............D-7
Hopkins Pl..........H-4
Howard St....B-4,H-4,J-4
Hughs St............I-5
Hull St.............K-8
Hunter Ave..........F-5
Hunter St...........E-5
Huntingdon..........C-4
Huntingdon Ave......A-4
Ilchester Ave.......A-5
Independence St.....A-6
Irvine Pl...........G-8
Jackson St..........K-6
James St............I-2
Jasper St...........F-4
Jefferson St........F-8
John St.............D-4
Johnson St..........K-6
Jones Fall Expwy....A-2
Josephine St........G-3
Kennedy.............B-7
Kennedy Ave.........C-6
Kensett.............C-1
Keswick Rd..........A-3
Key Hwy.....I-5,J-6,.K-7
Kirk Ave.........B-7,C-6
Lafayette Ave.......D-4
.............D-8,E-3,F-1
Lakeview Ave........C-2
Lamont Ave..........E-8
Lamont St...........D-7
Lancaster...........I-7
Lanvale St...D-5,E-3,F-1
Latrobe St..........D-5
Laurens St.......E-1,E-2
Lawrence St.........K-7
Leadenhall St.......J-4
Lee St..............I-5
Lemmon ST...........H-1
Lennox St...........C-3
Lexington St.....G-4,G-6
Liberty St..........G-5
Light St.........H-5,K-5
Linden Ave...C-2,E-3,F-4
Llewelyn Ave........D-8
Lloyd St............G-7
Loch Raven Rd.......B-6
Lombard St.......H-1,H-4
Lorraine Ave.....B-4,B-5
Lovegrove St.....A-5,C-5
Low St..............G-6
Lowman..............K-8
Ludlow St...........K-7
McAllister St.......D-5
McColluh St......C-1,E-3
McDonough St........F-8
McElderry St........F-8
McHenry St..........H-1
McKim St............F-6
McMechan St.........E-2
Madison Ave......C-1,E-3
Madison St.......E-3,F-4
Market Pl...........H-6
Market St...........A-5
Marriott............K-8
Mary St.............F-3
Maryland Ave.....B-4,E-4
May St..............G-7
Miles Ave...........B-4
Miller St...........F-8
Mill Rd.............A-3
Milliman St.........F-8
Monroe St...........J-1
Montgomery St....I-4,I-6
Montpelier St....A-6,B-7
Monument St.........F-7
Mosher St.....D-3,F-1
Mount St............I-1
Mt. Clare St........H-2
Mt. Royal Ave.......D-4
Mt. Royal Terr......C-3
Mt. Vernon Pl.......F-5
Mulberry St.........G-4
Mullikin St.........G-8
Mura St.............E-8
Myrtle Ave.......F-2,G-3
Nanticoke St........J-2
Newington Ave.......C-2
Normal Ave..........C-8
Norris St...........I-1
Oler St.............K-3
Oliver St........D-4,D-7
Olive St.........J-5,J-6
Orchard St..........F-3
Orleans St..........G-7
Ostend St.....I-2,J-3,J-5
Oxford..............F-3
Paca St..........I-3,J-3
Pacific St..........A-3
Park Ave............E-4
Parkin St...........H-3
Park St.............F-4
Park Lake Dr........C-2
Parrish St.......F-1,I-1
Patapsco St......J-5,K-5
Pearl St............G-3
Penn St.............H-3
Pennsylvania Ave.
.................C-1,F-3
Perry St............H-5
Philpot St..........I-7
Pierce St...........G-3
Pine St.............G-3
Pitcher.............E-1
Pleasant............G-4
Polk St.............B-8
Poppleton St.....G-2,H-2
Portland St.........H-3
Poultney St.........J-5
Pratt St.....H-4,H-7,H-8
President St........I-7
Presstman St........D-2
Preston St.......E-6,F-3
Proctor St..........E-6
Race St.............J-5
Ramsay St...........I-3
Randall St..........K-5
Read St..........E-4,F-5
Redwood St.......H-3,H-5
Reese St.........A-6,B-6
Regester St......D-8,I-8
Remington Ave.......A-3
Reservoir St........C-3
Retreat.............D-1
Richardson St.......K-8
Ridgely..........I-3,K-3
Riggs Ave...........E-1
River St............J-6
Riverside Ave.......K-6
Robert St...........E-2
Russell St..........K-3
Rutland Ave......D-8,F-8
Ryan St.............I-2
St. Ann's...........B-6
St. James St........F-7
St. Lo Dr...........C-8
St. Paul Pl.........G-5
St. Paul St.........E-5
Sarah Ann St.....G-2,G-3
Saratoga St.........G-4
School St...........E-1
Schroeder St........H-2
Scott St............J-3
Severn St...........J-2
Shakespeare St......I-8
Sharp St.........H-4,J-4
Shields Pl..........E-2
Sinclair La.........C-8
Sisson St........B-3,C-4
Smithson............E-2
Somerset St.........E-7
South St............H-5
Spring St........D-7,I-7
Stanford Pl.........D-1
Sterrett St.........I-3
Stevenson St........K-7
Stiles St...........H-6
Stirling St.........F-7
Stockton St......E-1,H-2
Stricker St......F-1,I-1
Suther St...........F-8
Thames St...........I-8
The Alameda.........A-8
The Fallsway........G-6
Tilden Dr...........A-3
Tivoly Ave..........A-8
Towson St...........K-8
Truck...............H-3
Tyler...............B-7
Tyson St.........F-4,G-4
Upton St............F-2
Valley St...........E-6
Vincent St.......F-1,I-1
Vine St..........G-1,G-3
Vineyard Lane.......A-5
Wards St............J-2
Warner St...........K-3
Warren Ave..........I-5
Washington Blvd.
.................I-3,K-1
Washington St.......D-8
Water St............H-6
Watson St...........G-7
Webb St.............E-7
Webster St..........K-6
Welcome.............I-5
West................J-4
What Coat St........F-1
Wheeling............J-5
Whitelock St........C-2
Whitridge...........B-5
Wicomico St.........J-2
William St..........K-6
Wills St............I-7
Wilmer Ct...........E-2
Wilson..............E-2
Winchester St.......E-1
Wirton St...........E-6
Wolffe St........D-8,I-8
Woodall.............K-7
Woodbrook Ave.......C-1
Woodyear St......E-1,I-1
Worchester St.......J-3
Wyeth............H-3,I-3
Wyman Park Dr.......A-3
Wyman Pl............A-4
York St.............I-5
20th St..........C-4,C-8
21st................C-4
22nd St.............C-4
23rd St.............C-4
24th St.............C-4
25th St..........B-4,C-8
26th St..........B-4,B-6
27th St.............B-4
28th St......A-4,B-3,B-6
29th St......A-5,A-8,B-3
30th St......A-4,A-5,A-8
31st St......A-3,A-5,A-8
32nd................A-3

POINTS OF INTEREST

Carroll Mansion
Park................J-1
Clifton Park........B-8
Druid Hill Park.....A-1
Federal Hill........I-6
Riverside Park......K-6
Wyman Park..........B-3

Scale of Miles

0 .1 .2 .3 .4

N

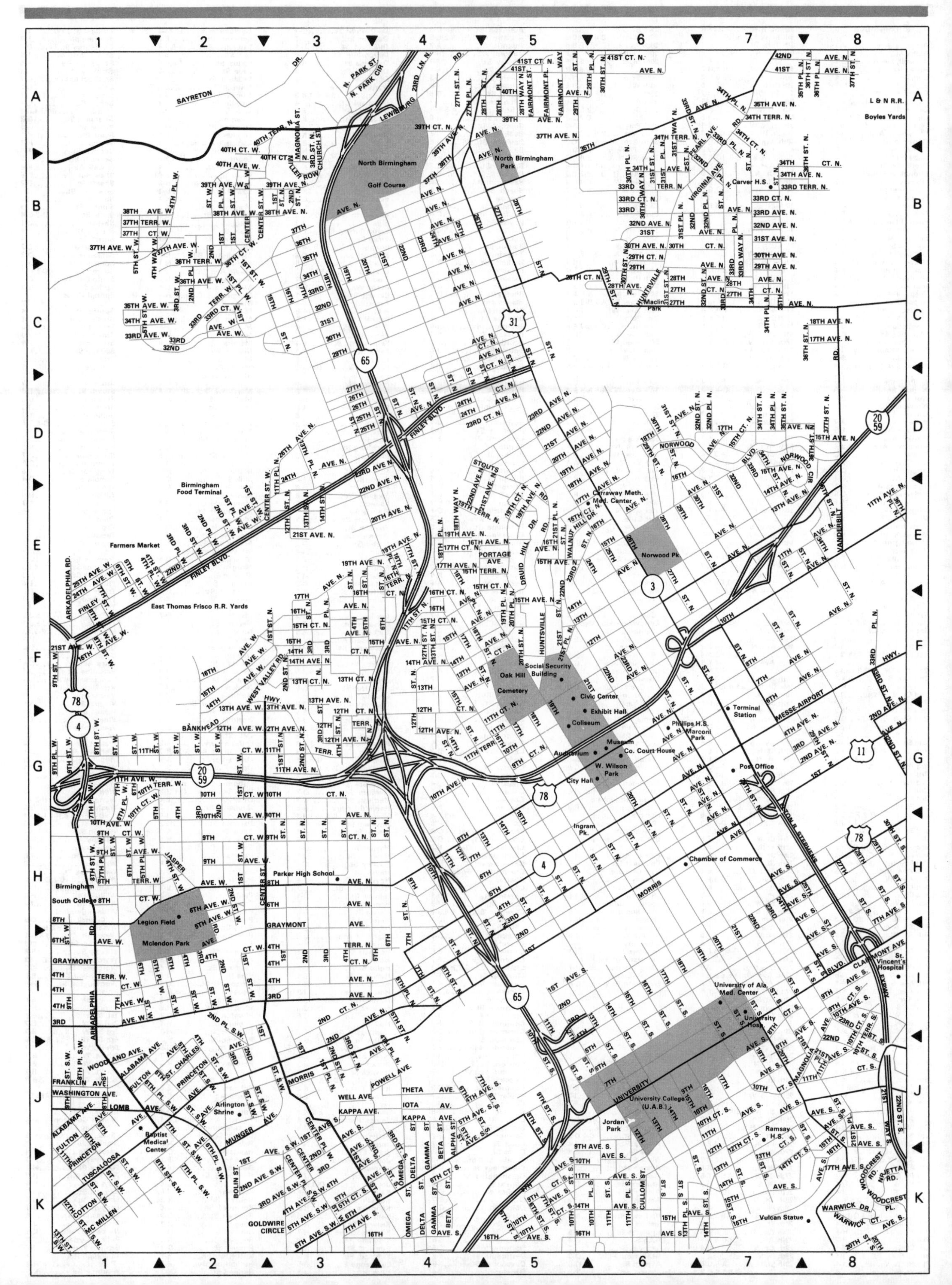
Scale of Miles
0 .1 .2 .3 .4 .5
North Birmingham Golf Course
North Birmingham Park
Carver H.S.
Maclin Park
Birmingham Food Terminal
Farmers Market
East Thomas Frisco R.R. Yards
Norwood Pk
Carraway Meth. Med. Center
Oak Hill Cemetery
Social Security Building
Civic Center
Exhibit Hall
Coliseum
Museum
Auditorium
Co. Court House
W. Wilson Park
City Hall
Phillips H.S.
Marconi Park
Terminal Station
Post Office
Ingram Pk.
Chamber of Commerce
Parker High School
Birmingham South College
Legion Field
Mclendon Park
University of Ala. Med. Center
University Hosp.
St. Vincent's Hospital
University College (U.A.B.)
Jordan Park
Ramsay H.S.
Vulcan Statue
Arlington Shrine
Baptist Medical Center
Boyles Yards
L & N R.R.
Goldwire Circle

BIRMINGHAM

Alabama Ave. J-1
Alpha St. J-4
Arkadelphia Rd. E-1,I-1
Bankhead Hwy. G-2
Beta St. K-4
Bolin St. K-2
Center Pl. S.W. K-3
Center Pl. W. B-2,J-3
Center St. H-2
Center St. W. B-2,E-3,K-3
Church St. B-3
Claremont Ave. I-8
Cotton Ave. K-1
Cullom St. K-8
Delta St. K-4
Druid Hill Dr. E-5
Elton B. Stephens Expwy. H-7
Fairmont Pl. A-5
Fairmont St. A-5
Fairmont Way A-5
Finley Ave. F-1
Finley Blvd. D-4,E-2
Franklin Ave. J-1
Fulton Ave. J-1
Gamma St. K-4
Goldwire Circle K-3
Graymont Ave. H-3,I-1
Henrietta Rd. K-8
Huntsville Rd. C-6,F-5
Iota Ave. J-4
Jasper Rd. H-2
Kappa Ave. J-3,J-4
Lewisburg Rd. A-4
Lomb Ave. J-1
Magnolia Ave. J-8
Magnolia St. A-3
McMillen Ave. K-1
Messe Airport Hwy. G-7
Miller Row B-3
Morris Ave. H-6,J-3
Munger Ave. J-2
N. Park Cir. A-3
N. Park St. A-3
Norwood D-6
Norwood Cir. D-7
Omega St. K-4
Pearl Ave. A-7
Portage Ave. E-5
Powell Ave. J-4
Princeton Ave. J-1,J-2
Sayreton Dr. A-2
St. Charles J-2
Stouts Rd. D-5
Theta Ave. J-4
Tuscaloosa Ave. K-1
University Blvd. J-6
Vanderbilt Rd. E-8
Virginia Ave. B-7
Walnut Hill Dr. E-5
Warwick Ct. K-8
Warwick Dr. K-8
Washington Ave. J-1
Well Ave. J-3
West Valley Rd. F-2
Woodcrest Pl. K-8
Woodcrest Rd. K-8
Woodland Ave. J-1
1st Ave. N. G-8,I-5
1st Ave. S. I-5,J-3
1st Ave. S.W. K-2
1st Pl. N. J-3
1st Pl. W. B-2,C-2,E-2
1st St. N. B-3,F-3,G-3,I-3
1st St. S. K-3
1st St. S.W. I-2
1st St. W. B-2,C-2,-2,G-2 H-2,I-2
2nd Ave. N. G-8,I-3,I-5
2nd Ave. S. I-5,K-3
2nd Ave. S.W. K-2
2nd Ct. N. I-3
2nd Pl. S.W. I-2
2nd Pl. W. C-2,E-2
2nd St. N. B-3,F-3,G-3 I-3,J-3
2nd St. S. J-3
2nd St. S.W. J-2
2nd St. W. B-2,E-2,G-2 H-1,I-2
3rd Ave. N. G-7,H-5,I-3
3rd Ave. S. I-5,K-3
3rd Ave. S.W. K-3
3rd Ave. W. I-1
3rd Pl. N. F-3
3rd Pl. W. E-2
3rd St. N. B-2,F-3,G-3 I-3,J-3
3rd St. S. J-4
3rd St. S.W. J-2
3rd St. W. C-2,E-2,G-2,I-2
4th Ave. N. G-7,H-5,I-3
4th Ave. S. I-5,J-4,K-3
4th Ave. S.W. K-3
4th Ave. W. I-1
4th Ct. N. I-3
4th Ct. W. I-1,I-2
4th Pl. N. J-4
4th Pl. W. B-2,E-1
4th St. N. F-3,G-3,I-3
4th St. S.W. J-2
4th St. W. E-1,G-2 H-2,I-2
4th Terr. N. I-3
4th Terr. W. I-1
4th Way W. C-1
5th Ave. N. H-5
5th Ave. S. J-6,K-3
5th Ave. S.W. K-3
5th Ave. W. H-2
5th Ct. S. K-3
5th Pl. W. H-1,I-2
5th St. N. F-3,G-3,I-3,I-4
5th St. S. J-4,K-5
5th St. S.W. J-2
5th St. W. B-1,C-1,E-1 G-1,I-2
6th Ave. N. F-7,H-3,H-5
6th Ave. S. J-6,K-3
6th Ave. S.W. K-3
6th Ave. W. H-2,I-1
6th Ct. S. K-4
6th Pl. N. I-4
6th Pl. S.W. J-1,J-2
6th Pl. W. G-1
6th St. N. F-4,I-4
6th St. S. J-4,K-5
6th St. S.W. J-2
6th St. W. E-1,G-1,H-1,I-1
7th Ave. N. F-7,H-4
7th Ave. S. H-8,J-6,K-3
7th Pl. S.W. K-2
7th Pl. W. G-1,H-1
7th St. N. I-4
7th St. S. J-5,K-5
7th St. S.W. J-2
7th St. W. F-1,G-1,I-1
8th Ave. N. F-7,H-3,H-4
8th Ave. W. H-1
8th Ct. S. I-7
8th Ct. W. H-1
8th Pl. S.W. J-1
8th St. N. I-4
8th St. S. J-5
8th St. S.W. K-2
8th St. W. F-1,G-1,H-1
8th Terr. W. H-1
9th Ave. S. I-8,J-5,J-6 J-7,K-5
9th Ave. W. H-1,H-2
9th Ct. N. G-5,H-3
9th Ct. S. I-8,K-5
9th Ct. W. H-1,H-2
9th Pl. W. G-1
9th St. N. H-4
9th St. S. J-5
9th St. S.W. J-1
9th St. W. F-1,G-1,I-1
10th Ave. N. FG-7,G-3 G-4,G-5
10th Ave. S. I-8,J-7,K-5
10th Ave. W. G-2,H-1
10th Ct. N. G-3
10th Ct. S. J-7,J-8,K-5
10th Ct. W. G-1,G-2
10th St. N. H-4
10th St. S. J-5,K-5
10th St. S.W. J-1
10th Terr. S. J-8
10th Terr. W. G-1
11th Ave. N. E-7,E-8 G-3,G-5
11th Ave. W. G-1
11th Terr. N. G-3
11th Ct. W. G-1
11th Ct. N. G-5
11th Terr. N. G-4
11th Ave. S. J-7,J-8,K-5
11th Ct. S. J-8
11th St. N. F-4,H-4
11th St. S. I-5,K-6
11th Pl. S. K-6
11th St. S.W. K-1
12th Ave. N. F-6,G-3,G-4
12th Ave. S. J-7
12th Ave. W. G-2
12th Ct. N. G-3,G-4
12th Ct. S. J-7
12th St. N. E-3,F-4 G-4,H-4
12th St. S. I-5,J-6
12th St. S.W. K-1
12th Terr. N. G-3
13th Ave. N. E-7,F-3 F-4,F-5
13th Ave. S. K-7
13th Ave. W. G-2
13th Ct. N. F-3,F-4
13th Ct. S. K-7
13th Pl. N. D-3
13th Pl. S. K-6
13th St. N. E-3,F-4,H-5
13th St. S. I-6,J-6
13th St. S.W. K-1
14th Ave. N. E-7,F-3 F-4,F-5
14th Ave. S. K-5,K-7
14th Ave. W. F-2
14th Ct. N. F-3
14th Ct. S. K-7
14th St. N. E-3,G-4,H-5
14th St. S. I-6,J-6,K-7
15th Ave. N. D-7,D-8,E-6 E-5,F-3,F-4,F-5
15th Ave. S. J-8,K-6,K-7
15th Ave. W. F-2
15th Ct. N. D-7,E-5 F-3,F-4
15th St. N. C-3,F-4,H-5
15th St. S. I-6,J-6
15th Terr. N. E-4
16th Ave. N. D-6,E-4,E-5 E-6,F-3,F-4
16th Ave. S. K-4,K-5 K-7,K-8
16th Ave. W. F-1,F-2
16th Ct. N. E-3,E-4,E-5
16th Pl. N. E-4
16th St. N. C-3,G-5
16th St. S. I-6,J-7
16th Terr. N. E-4
17th Ave. N. C-8,D-7 E-4,E-5,F-3
17th Ave. S. K-8
17th Ct. N. E-4
17th St. N. C-3,E-4 F-4,G-5
17th St. S. I-6,J-7
18th Ave. N. C-8,D-6,E-5
18th Pl. N. E-4
18th St. N. C-3,F-4,G-5
18th St. S. I-6
18th Way N. E-4
19th Ave. N. D-5,E-3 E-4,E-5
19th Ct. N. E-5
19th Pl. N. F-5
19th St. N. C-3,G-5
19th St. S. I-7,J-7
19th Terr. N. E-4
20th Ave. N. D-5,E-4
20th Ave. S. K-8
20th Pl. N. F-5
20th St. N. C-3,F-5,G-6
20th St. S. I-7,J-7,K-8
21st Ave. N. D-5,E-3,E-5
21st Ave. W. F-1
21st Pl. N. E-5,F-5
21st Pl. S. J-8
21st St. N. C-4,F-5,F-6
21st St. S. I-7,J-7
21st Way S. J-8
22nd Ave. N. D-5,E-3,E-4
22nd Ave. W. E-2
22nd St. N. B-4,E-5,F-6
22nd St. S. H-7,J-8
23rd Ave. N. D-3,D-5
23rd Ct. N. D-4
23rd Ln. N. A-4
23rd St. N. B-4,E-5,F-6
23rd St. S. H-7,I-8
24th Ave. N. D-3,D-4
24th Ave. W. E-1
24th Ct. N. D-4
24th St. N. B-4,E-6
24th St. S. H-7
25th Ave. N. D-3
25th Ave. W. E-1
25th Ct. N. D-3
25th St. N. B-4,E-6,G-7
25th St. S. H-8
26th Ave. N. D-3
26th Ct. N. D-3
26th St. N. B-4,E-6
27th Ave. N. C-6,D-3
27th Ct. N. C-6,C-7
27th Pl. N. A-5
27th St. N. A-4,B-5,E-6
27th St. S. H-8
28th Ave. N. C-6,C-7
28th Ct. N. C-5
28th Pl. N. A-5
28th St. N. A-5,B-5,E-6
28th St. S. H-8
28th Way N. A-5
29th Ave. N. C-3,C-6,C-7
29th Ct. N. C-6
29th Pl. N. A-5
29th St. N. A-5,C-6 D-6,E-6,G-8
29th St. S. H-8
30th Ave. N. B-6,B-7,C-3
30th Ct. N. B-6
30th Pl. N. B-6
30th St. N. A-6,C-6 D-6,E-7
30th St. S. H-8
30th Way N. B-6
31st Ave. N. B-6,B-7,C-3
31st Pl. N. B-6
31st St. N. B-6,C-6 D-6,E-7
31st Way N. A-6
32nd Ave. N. B-6,B-7,C-3
32nd Ave. W. C-2
32nd Pl. N. B-7,D-7
32nd St. N. B-6,C-7 D-6,E-7,G-8
33rd Ave. N. B-6,B-7,C-3
33rd Ave. W. C-1,C-2
33rd Ct. N. B-6,B-7
33rd Ct. W. C-2
33rd Pl. N. B-7,C-7,F-8
33rd St. N. A-6,C-7 D-7,F-8
33rd Terr. N. B-6,B-7
33rd Terr. W. C-2
33rd Way N. B-7
34th Ave. N. B-6,B-7,C-3
34th Ave. W. C-1
34th Ct. N. A-7,B-7
34th Pl. N. A-7,C-7,D-7
34th St. N. C-7,D-7,E-8
34th Terr. N. A-6,A-7
35th Ave. N. A-7,B-3,B-5
35th Ave. W. C-1
35th Pl. N. A-7
35th St. N. C-7,D-7,E-8
36th Ave. N. B-3
36th Ave. W. C-2
36th Ct. W. B-2
36th Pl. N. A-8,E-8
36th St. N. A-8,B-8 C-8,D-8
36th Terr. W. C-2
37th Ave. N. A-5,B-3,B-4
37th Ave. W. B-1,B-2
37th Ct. W. B-1
37th St. N. A-8,D-8
37th Terr. W. B-1
38th Ave. N. B-3,B-4
38th Ave. W. B-1,B-2
39th Ave. N. A-5,B-3,B-4
39th Ave. W. B-2
39th Ct. N. A-4
40th Ave. N. A-5
40th Ave. W. B-2
40th Ct. N. B-3
40th Ct. W. B-2
40th Terr. N. A-3
41st Ave. N. A-5,A-7
41st Ct. N. A-5,A-6
42nd Ave. N. A-7

POINTS OF INTEREST

Arlington Shrine J-2
Auditorium G-5
Baptist Medical Center J-1
Birmingham Food Terminal E-2
Birmingham South College H-1
Carraway Meth. Med. Center E-6
Carver H.S. B-7
Chamber of Commerce H-7
City Hall G-5
Civic Center F-6
Coliseum G-5
Co. Court House G-6
East Thomas Frisco R.R. Yards F-2
Exhibit Hall G-6
Farmers Market E-1
Ingram Pk. H-5
Jordan Park J-6
Legion Field H-2
L. & N. R.R. Boyles Yards A-8
Maclin Park C-6
Marcorn Park G-7
Mclendon Park I-2
Museum G-6
North Birmingham Golf Course B-4
North Birmingham Park B-5
Norwood Pk. E-6
Oak Hill Cemetery F-5
Parker H.S. H-3
Phillips H.S. G-6
Post Office G-7
Ramsay H.S. J-7
St. Vincent's Hospital I-8
Social Security Building F-5
Terminal Station G-7
University College (U.A.B.) J-6
University Hospital I-7
University of Ala. Med. Center I-7
Vulcan Statue K-7
W. Wilson Park G-6

BOSTON

A St. B-6
Abbotsford Rd. K-7
Aberdeen St. I-7
Acorn St. J-5
Adams St. K-7
Adams St. D-1
Adrian St. I-1
Albany St. C-6,D-8
Albany St. I-4
Alger St. B-8
Allen St. I-1
Allston St. J-5
Ames St. G-4
Amherst St. G-4
Amory St. J-7
Amory St. I-3
Anderson St. E-3
Andrews St. K-5
Antrim St. J-3
Appleton St. E-6
Arch St. C-4
Arundel St. I-7
Ashburton Pl. D-3
Asylum St. D-6
Athens St. K-3
Atkinson St. D-8
Atlantic Av. B-2,C-5
Atlantic St. A-7
Auburn St. J-4
Audrey St. I-5
Austin St. E-1
Austin St. J-3,I-3
Autumn St. I-8
Avery St. D-5
Avon St. C-4
B St. A-5,B-7
Babbitt St. I-6
Babcock St. K-7
Back St. I-6
Baldwin St. B-6
Ball E-8
Banks St. K-3
Battery St. C-2,C-3
Bay St. K-3
Bay State Rd. I-6
Beach St. C-5
Beacon St. K-8
Beacon St. D-4,F-5
Bedford St. C-4
Beech Rd. J-8
Belvidere St. G-6
Bent St. G-3
Benton St. F-8
Berkeley St. E-5
Berkeley St., E. D-6
Berkshire St. H-3
Bigelow St. J-3
Billerical St. D-2
Binney St. G-3,H-3
Blackstone St. K-4
Blagden St. E-6
Blanche St. I-4
Blandford St. I-6
Blossom St. E-3
Boardman St. I-3
Boden St. J-4
Bolton St. I-2
Bolyston St. F-6 G-7,H-7
Bonner Av. I-1
Border St. B-1
Borland St. J-7
Boston Univ. Bridge J-6
Bowdoin J-1
Bowdoin St. D-3
Braddock Pl. E-7
Bradford St. D-7
Branch St. E-4
Bremen St. A-1
Brimmer St. E-4
Bristol St. C-6
Bristol St. H-3
Broad St. C-3
Broadway D-5,H-3
Bromfield St. D-4
Brookline Ave. I-8
Brookline St. J-5
Brookline St., E. D-7
Brookline St., W. E-7
Brownie St. K-7
Bryant St. K-1
Buckingham St. E-6,J-2
Buick St. K-6
Bullock St. A-6
Burbank St. G-7
Burke St. F-8
Burlington St. H-7
Buswell St. I-7
Byron St. E-4
C St. B-7
Cabot St. F-8
Calahan Tunnel B-2
Calender St. K-4
Calvin St. J-1
Cambridge St. D-3,H-2
Camden St. E-8
Canal St. D-2
Canton St. E-6
Canton St., E. D-7
Canton St., W. E-7
Carleton St. G-4
Carlisle St. I-2
Carlton St. F-7,J-1,J-7
Carver St. D-5,K-1
Causeway St. D-2
Cedar St. E-4
Central St. C-3
Centre St. J-3
Chalk St. J-4
Chandler St. E-6
Chapel St. J-8
Chaplin St. A-6
Charles St. E-3,E-4 G-2,K-8
Charlesgate East G-6
Charlesgate West G-6
Charles River Dam E-2
Charlestown Ave. F-2
Charlestown St. H-1
Charlestown Bridge D-2
Charter St. C-2
Chatham St. C-3,J-3,J-8
Chelsea St. A-1,D-1
Cherry St. I-3
Chestnut St. E-4,J-5
Chilton St. J-7
Church J-1
Church St. D-5
Churchill St. J-7
Claremont St. E-5
Claremont Pk. F-7
Clark St. C-2,H-3,I-1
Clearway St. F-7
Cleveland St. K-3
Clinton St. C-3,J-3
Colchester St. J-8
Columbia St. I-2-4
Columbus Av. I-1
Commercial Av. F-3
Commercial St. C-2,C-3
Common St. D-5
Commonwealth Av. F-6
Concord Av. I-1
Concord St., E. D-8
Concord St., W. E-7
Concord Sq. E-7
Congress St. B-5,C-3
Copley St. K-7
Cordis St. D-1
Corning St. D-6
Cortes St. E-6
Cottage St. A-2,J-4
Cottage Farm Rd. J-7
Court St. C-3
Cross St. C-3
Cumberland St. F-7
Cummings St. C-8
Cummington St. I-7
Cunard St. F-8
Cypher St. B-6
D St. B-7
Dalton St. G-6
Dana St. K-3
Dane Av. J-1
Dane St. J-1
Darnell St. B-8
Dartmouth Pl. E-6
Dartmouth St. F-6
Davenport St. F-8
Decatur St. J-4
Dedham St., E. D-7
Dedham St., W. E-7
Deerfield St. H-6
Devonshire St. C-4
Dewolee St. K-3
Dilworth St. F-8
Dimick St. J-2
Dorchester Ave. B-5,B-7,B-8
Douglass Ct. I-4
Drapers Ln. E-7
Dummer St. K-7
Durham St. F-7,J-1
Dwight St. D-6,K-8
E St. B-7
Earle St. H-1
Edgerly Rd. G-7
Egmont St. K-7
Eighth St. G-2
Elba St. K-7
Eliot St. K-1
Eliot Kneeland St. D-5
Ellery St. K-3
Ellsworth Ave. J-2
Elm St. I-3
Embankment Rd. E-4
Emily St. I-5
Emmons St. A-1
Endicott St. C-2
Erie St. J-5
Essex St. D-5,I-3,J-7
Euston St. J-7
Evans Way H-8
Everett St. I-1
Exeter St. F-5
F St. A-8
Fairfield F-5
Fairmont St. J-4
Fargo St. A-5
Farnsworth St. B-4
Farrar St. K-1
Fay St. D-6
Fayette St. D-5,J-3
Federal St. C-4
Fellows St. E-8
Felton St. K-2
Fenway H-8
Fiarmont Av. K-5
Field St. G-8
Fifth St. G-3
First St. F-3
Fisk Pl. I-3
Fitchburg St. G-1
Flagg St. K-3
Fleet St. C-2
Florence St. K-5
Follen St. F-6
Forsyth St. G-8
Forsyth Way G-8
Foundry St. C-6
Francis Av. K-1
Frankfort St. A-1
Franklin St. C-4,J-3
Freeman St. K-7
Friend St. D-2
Front I-4
Fruit St. E-3
Fulkerson St. H-3
Fullerton St. H-7
Fulton St. C-2
Gaffney St. K-5
Gainsborough St. G-7
Garden St. D-3
Gardner St. I-2
Garrison St. F-6
Gates St. A-8
Glenwood Av. J-5
Gloucester St. G-5
Gore St. A-1,G-2
Granby St. I-6
Granite St. B-6,J-5
Green St. I-4,K-3,K-8
Greenleaf St. G-8
Greenough Av. J-2
Greenwich St. F-8
Grove St. E-3
Hamilton St. J-5
Hammond St. F-8,J-2
Hampshire St. H-3
Hancock St. D-3,J-3
Hanover C-3
Hanson St. D-6,J-1
Harcourt St. F-6
Harding St. H-2
Hardwick St. H-2
Harrison Av. D-7
Harrison St. K-1
Harvard Bridge G-5
Harvard St. D-5,H-3,J-3
Haviland St. G-6
Havre St. A-1
Hawes Pl. J-8
Hawes St. J-8
Hawkins St. I-1
Hawley St. C-4
Hayes St. K-3
Hayward St. G-4
Hemenway St. G-7
Henley St. D-1
Henry St. J-5
Herald St. D-6
Hereford St. G-6
High St. C-4
Highland Ave. J-2
Hingham St. K-4
Hinsdale St. I-6
Holden St. K-1
Hollis St. C-5
Holyoke St. E-6
Horrace St. H-1
Houghton St. I-2
Hovey Ave. J-2
Howard St. D-1,J-4
Hudson St. C-5
Hull St. C-2
Huntington Ave. F-7
Hurley St. G-2
Ilavado St. K-1
India St. C-3
Inman St. J-3
Ipswich St. H-7
Irving St. D-3,K-1,K-2
Irving Terr. K-2
Isabella E-6
Ivy St. J-7
James St. E-8,K-8
James Storrow H-6
Jay St. J-4
Jefferson St. H-7
Jersey St. H-7
John St. K-8
John Fitzgerald Expwy. C-7
Joseph St. I-1
Joy St. D-3,H-1
Kelly Rd. J-4
Kendall St. F-8
Kenmore Sq. H-6
Kent St. J-8
Keswick St. I-7
Kilmarnock St. H-7
Kingman Rd. I-2
Kinnaird St. K-3
Kirkland Pl. K-2
La Grange St. D-5
Lake St. J-1
Lansdowne St. H-7,I-4
Lattmore Ct. E-8
Lawrence St. E-6,J-4
Lee St. J-3
Lenox St. F-8,J-7
Leon St. G-8,J-2
Leonard Av. J-2
Lime St. E-4
Lincoln Pkwy. J-1
Lincoln St. C-5,I-1,I-2
Line St. J-1
Liverpool B-1
Linwood St. H-1
London St. B-1
Longfellow Bridge F-3
Longwood Av. J-8
Lopez Av. F-2
Lopez St. J-4
Louis St. B-6
Louis Pasteur Av. H-8
Lowell St. D-2
Magee St. K-4
Magnolia Ave. K-2
Magazine St. J-5
Magnus J-1
Main St. H-4,I-4
Maitland St. I-7
Malden St. D-7
Manchester Rd. K-7
Mansfield St. H-1
Maple Ave. J-2
Marcella St. H-2
March Oliver St. C-4
Marginal A-1
Marie Av. J-2
Marion St. H-2,J-2
Market St. C-3,H-3
Marlborough St. F-5
Marney St. H-2
Marshall St. J-8
Martha Way E-2
Marys St. I-7
Mason St. D-4,J-7
Massachusetts Ave. F-7,G-6,H-4
Matthews St. C-4
Maverick St. A-1
Melcher St. B-5
Melrose St. D-5
Memorial Dr. G-5,H-5
Mercer St. A-8
Merchants Row C-3
Medford St. D-2
Merriam St. H-1,I-5
Midway St. B-6
Milford St. D-6
Milk St. C-4
Mill St. A-1
Miner St. I-7
Monmouth Ct. I-7
Monmouth St. I-7
Montgomery St. E-7
Monument Ave. D-1
Mountfort St. I-7
Mt. Vernon St. E-3
Mt. Washington St. B-5
Moore St. C-8
Morgan St. K-1
Munroe St. G-3
Munson St. I-7
Museum Rd. G-8
Museum St. K-1
Myrtle St. E-4
Mystic St. D-7
Nashua St. E-2
Necco St. B-5
New St. B-1
Newbury St. F-6,H-6
Newton J-5
Newton St. I-1
Newton St., E. D-8
Newton St., W. F-7
Norfolk St. I-2,I-3
North St. C-3
Northampton St. E-8
Northern Av. B-4
Northfield St. F-8
Norway St. G-7
Oak St. D-5,I-2
O'Brien Hwy. F-2
Old Colony Av. B-7
Old Harbor St. A-8
Opera Pl. G-7
Orleans St. A-1
Osborn St. H-4,K-7
Otis St. C-4,G-2
Otter St. E-4
Overland St. H-7
Pacific St. I-4
Palermo St. H-2
Paris St. A-1
Park Dr. H-8
Park St. D-1,D-4,J-3,K-1
Parkdale J-1
Parkman St. K-8
Parker J-1
Parker St. G-8
Parkman St. E-3
Paul St. K-7
Pearl St. C-4,J-5
Pelham St. G-4
Pembroke St. E-7
Perry St. J-1,J-4
Peterborough St. H-7
Peters I-3
Phillips St. E-3
Pilgrim Rd. I-8
Pinckney St. E-4
Pine St. I-4
Pittsburgh St. B-5
Pleasant St. J-5,K-8
Plymouth H-3
Plymouth St. I-8
Plympton St. D-7
Poplar St. H-1
Portland St. D-2,H-3
Portsmouth H-3
Potter St. G-3
Powell St. J-7
Prescott St. J-7,K-3
Prince St. C-2
Prospect St. I-3
Providence St. E-5
Province St. C-4
Puchase St. C-4
Purrington I-4
Putnam Av. J-5,K-3,K-4
Queensbury St. H-7
Quincy St. J-1,K-3
Raleigh St. H-6
Randolph St. D-7
Reed St. E-8
Revere St. E-4
Richmond St. C-2
River St. E-4
Riverway I-8
Roberts Rd. K-2
Rockwell St. J-4
Rodgers St. G-3
Rose J-1
Rossmore H-1
Ruggles St. G-8
Russell St. D-3
Rutherford E-1
Rutherford Av. E-1
Rutland St. E-7
Rutland St., W. F-7
St. Botolph F-7
St. George St. D-7
St. Germain St. F-6
St. James Av. E-5
St. Mary St. I-2
St. Stephens St. G-7
Salem C-2
Sanborn I-1
School St. C-4 E-1,I-4,J-1
Sciarappa St. G-2
Scotia St. G-6
Scott St. K-1
Seckel H-3
Second St. F-3
Seventh St. G-2
Sewall Av. K-8
Shawmut Av. D-6
Sheafe St. C-2
Sherborn St. I-6
Short St. I-8
Sidney St. I-5
Sixth St. G-3,G-4
Skehan St. J-1
Sleeper St. B-4
Smart St. H-4
Snowhill St. C-2
Somerset St. D-3
Somerville Av. J-1
South St. C-5,H-2
South Bay Av. D-8
Spring St. G-2
Springfield St. F-7,I-2
Spruce St. D-4
Stanhope St. E-6
State St. C-3,I-4
Stearns Rd. J-8
Sterling St. F-8
Stetson St. K-7
Still St. J-7
Stillings St. A-5
Stillman St. C-2
Stone St. I-1
Stuart St. E-6
Sudbury St. D-3
Suffolk St. I-3
Summer Rd. K-2
Summer St. B-5,C-4
Sumner St. A-1
Sumner Tunnel B-2
Surrey St. K-3
Symphony Rd. G-7
Tavern Rd. G-8
Telegraph St. A-8
Temple Pl. D-4
Temple St. D-3,J-3
Tetlow St. H-8
Thatcher St. K-7
Thayer St. C-6
Third St. G-2,G-3
Thorndike St. G-2
Tileston C-2
Topeka St. D-8
Traveler St. D-6
Traverse St. D-3
Tremont St. E-7,F-8,I-3
Tremont St. I-3
Trinity Pl. E-6
Trotter Ct. E-8
Trowbridge K-3
Tudor St. I-4
Tufts St. C-5,J-5
Tyler St. C-5,J-1
Union St. E-1,I-3
Union Park St. D-7
Unity St. C-2
Upton St. J-4
Utica St. C-5
Valentine St. J-4
Vancouver St. G-8
Van Ness St. H-7
Vassar St. I-5
Viaduct St. A-5
Village St. J-1
Vine St. K-1
Wadsworth St. G-4
Waldo St. J-2
Walnut St. D-4
Waltham St. D-6
Wapole St. F-8
Ward St. H-1
Ware St. K-3
Wareham St. D-7
Warren Av. E-6
Warren St. E-1,H-2,I-1
Washington St. D-2 D-6,E-1,E-8,H-3,I-3,J-1
Water St. C-3
Watson St. F-7,J-4
Waverly St. I-5
Webster Av. H-3
Webster St. A-1,K-8
Well Rd. J-8
Wellington St. F-7
West St. B-6,D-4,J-3
Westland Av. G-7
Willow St. H-2
William St. J-4
William Cardinal O'Connell Way D-3
Windsor St. F-8,H-2,I-3
Winter St. D-4,G-2
Winthrop St. D-1
Winton St. B-8
Worcester St. E-8,I-3
Worcester Sq. E-8
Wormwood St. B-5
Worthington St. H-8
Wyatt J-1
Yarmouth St. E-6
York St. H-2
1st St. W. B-6
2nd St. W. B-7
3rd St. W. B-7
4th St. W. B-7,C-6
5th St. W. B-7
6th St. W. B-7
7th St. W. B-7
8th St. W. B-7
9th St. W. B-8

POINTS OF INTEREST

Ahern Field G-2
Boston Common D-4
Black Bay Fens Park H-8
Blackstone Sq. E-7
Central Sq. J-4
Donnelly Field H-2
Fenway Park H-7
Franklin Sq. D-7
Hoyt Field K-4
Knyvet Sq. J-7
Lincoln Park J-1
Longwood Sq. J-8
Madison Park F-8
Mason Sq. J-7
Nickerson Field K-6
Public Garden E-5
Sennott Park I-3
Telegraph Hill Park A-8
Winthrop Sq. K-7

Scale of Miles

0 .1 .2 .3 .4

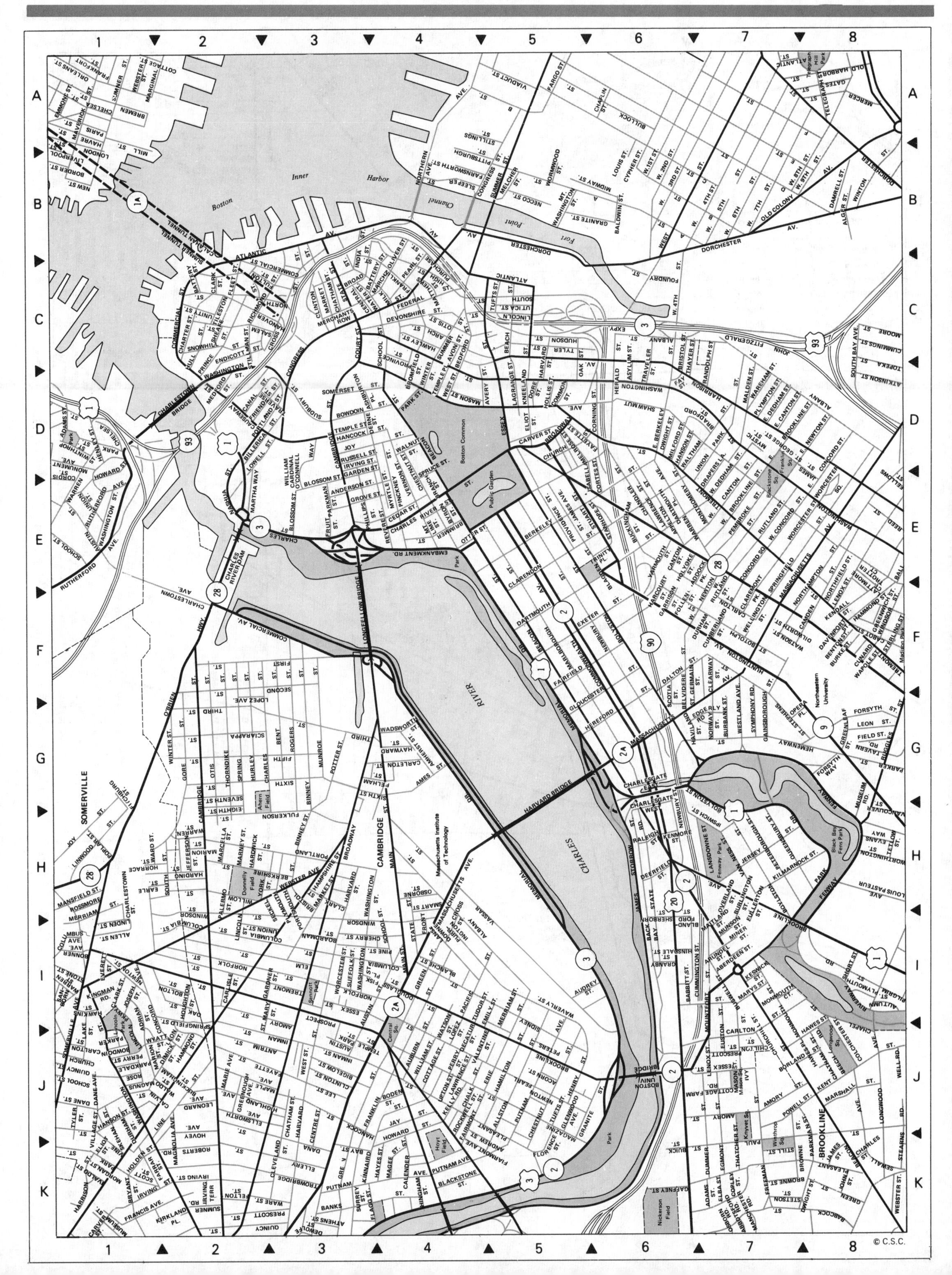

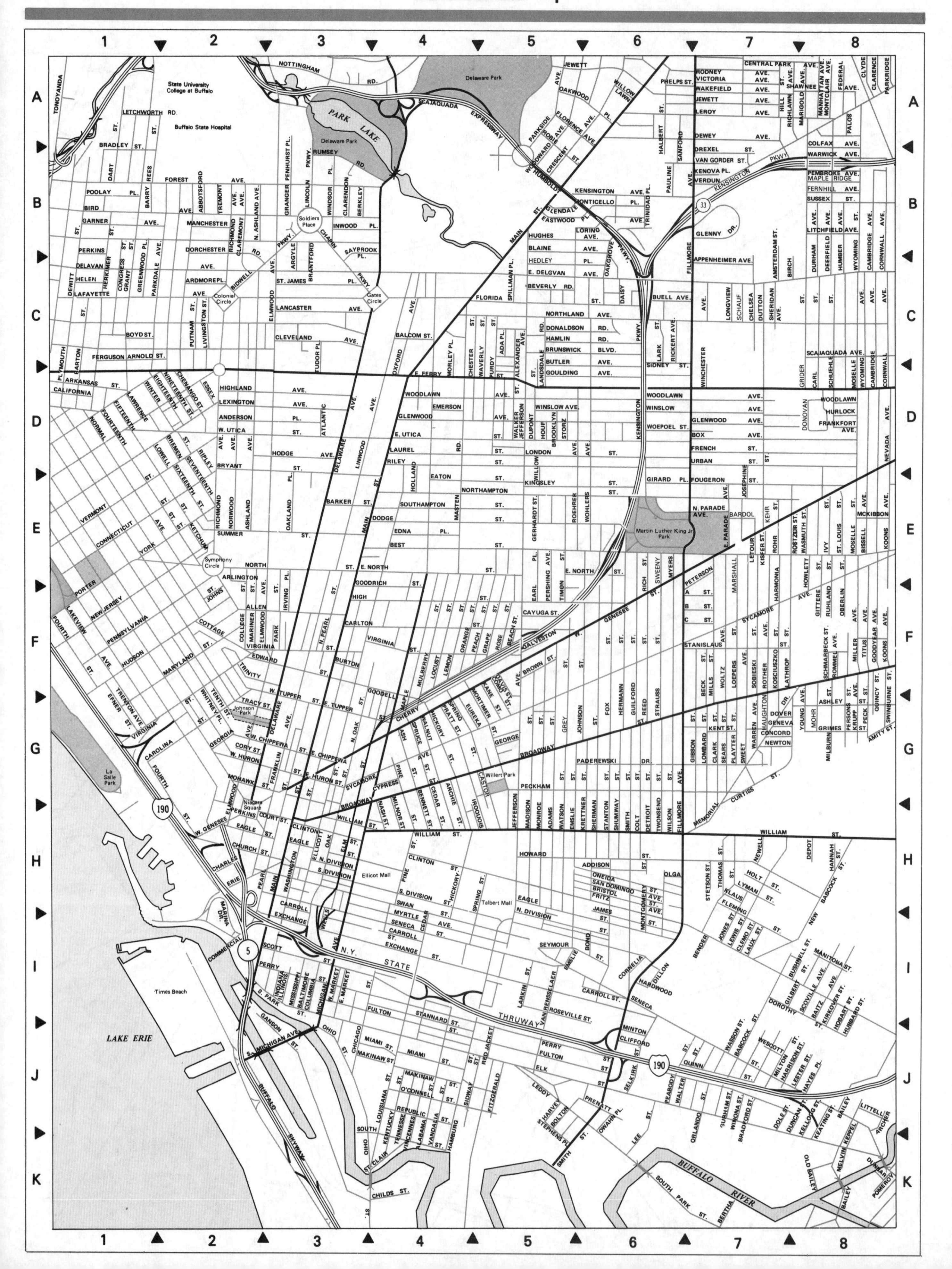
Scale of Miles
0 .1 .2 .3 .4 .5
N
State University College at Buffalo
Buffalo State Hospital
Delaware Park
PARK LAKE
Soldiers Place
Gates Circle
Colonial Circle
Symphony Circle
Martin Luther King Jr Park
Johnson Park
Niagara Square
La Salle Park
Willert Park
Ellicot Mall
Talbert Mall
Times Beach
LAKE ERIE
BUFFALO RIVER
N.Y. STATE THRUWAY
SCAJAQUADA EXPRESSWAY
KENSINGTON PKWY.
BUFFALO SKYWAY
190
33
5

BUFFALO

A St. F-7
Abbotsford B-2
Adams St. H-5
Ada Pl. C-5
Addison H-6
Alabama St. K-4
Alexander Ave. C-5
Amity St. G-8
Amsterdam St. C-7
Anderson Pl. D-2
Appenheimer Ave. C-7
Archer J-8
Archie St. G-4
Ardmore Pl. C-2
Argyle C-3
Arkansas St. D-1
Arlington E-2
Ashland Ave. E-2
Ashland Ave., N. .. B-2,E-2
Ashley St. G-8
Ash St. G-4
Atlantic D-3
B St. F-7
Babcock St. J-7
Bailey J-8,K-8
Baitz Ave. I-8
Balcom St. C-4
Baltimore I-3
Barker St. E-3
Barrymes St. B-1
Barton St. C-1
Beach St. F-5
Beck St. F-7
Bender I-7
Bennett St. G-4
Berkley B-3
Bertha K-7
Best St. E-4
Beverly Rd. C-5
Bidwell Pkwy. C-2
Birch C-7
Bird Ave. B-1
Bissell Ave. E-8
Blaine Ave. B-5
Bolton J-5
Bond I-6
Box Ave. D-7
Boyd St. C-1
Bradford St. J-7
Bradley St. A-1
Brantford C-3
Bremen D-2
Bristol St. H-6
Broadway St. G-5,H-3
Brooklyn D-5
Brown St. F-5
Brunswick Blvd. C-5
Bryant St. D-2
Buell Ave. C-6
Buffalo Skyway .. J-2,K-3
Burton F-3
Bushnell St. I-8
Butler Ave. C-5
C St. F-7
California D-1
Cambridge Ave. .. B-8,D-8
Carl St. D-8
Carlton St. F-5
Carolina St. G-1
Carroll St. H-3,I-4,I-6
Cayuga St. F-5
Cedar I-4
Cedar St. G-4
Central Park Ave. A-7
Chapin Pkwy. B-3
Charles H-2
Chelsea C-7
Chenango St. D-2
Cherry G-4
Chester St. C-4,D-4
Chicago St. J-3
Childs St. K-4
Chippewa St., E. G-3
Chippewa St., W. G-2
Church St. H-2
Claremont Ave. B-2
Clarence A-8
Clarendon B-3
Clark St. G-7
Clemo St. I-7
Cleveland Ave. C-3
Clifford J-6
Clinton St. H-3,H-4
Clyde A-8
Colfax Ave. B-8
College St. F-2
Colt St. H-6
Columbia I-3
Commercial I-2
Como St. F-5
Concord G-7
Congress St. C-1
Connecticut St. E-1
Cornelia I-6
Cornwall Ave. B-8,D-8
Cory St. G-2
Cottage F-2
Court St. H-3
Crescent Ave. B-5
Curtiss St. G-7
Cypress G-4
Daisy C-6
Dart St. B-1
Davis St. F-5
Deerfield Ave. B-8
Delaware Ave. D-3,G-3
Delavan Ave. C-1
Delavan Ave., E. C-5
Depot H-8
Detroit St. H-6
Dewey Ave. A-7
Dillon I-6
Division St., N. .. H-3,H-5
Division St., S. .. H-3,H-4
Dodge St. E-4
Dole St. J-7
Donaldson Rd. C-5
Dorchester Rd. B-2
Dorothy St. I-7
Dover G-7
Drexel St. B-7
Dunbar K-8
Duncan St. J-8
Dupont St. D-5
Durham Ave. B-8,C-8
Dutton C-7
Eagle St. H-2,H-5
Earl Pl. F-5
Eastwood Pl. B-5
Easton St. E-4
Edna Pl. E-4
Edward St. F-2
Efner St. G-1
Eighteenth St. D-2
Elk St. J-5
Ellicott H-3
Elm St. H-3
Elmwood Ave. C-3
.... F-2,G-2
Emerson D-4
Emslie St. H-5,I-5
Erie H-2
Essex D-2
Eureka G-4
Exchange St. I-3,I-4
Farnhill Ave. B-8
Federal A-8
Ferguson Arnold
St. C-1
Ferry St., E. D-4
Fifteenth St. D-1
Fillmore Ave. B-7,C-7,H-6
Fitzgerald J-5
Fleming St. H-7
Florence Ave. A-5
Florida St. C-5
Forest Ave. B-2
Fougeron St. E-7
Fourteenth St. D-1
Fourth St. F-1,G-2
Fox St. G-6
Frankfort Ave. D-8
Franklin Ave. G-3
French St. D-7
Fritz Ave. H-6
Fulton St. I-4,J-5
Galveston F-5
Ganson St. J-3
Garner Ave. B-1
Genesee St., W. .. F-6,H-2
Geneva G-7
George G-5
Georgia G-2
Gerhardt St. E-5
Gibson St. G-7
Gilbert St. I-8
Girard Pl. E-6
Gittere St. E-8
Glendale B-5
Glenny Dr. B-7
Glenwood Ave. .. D-4,D-7
Goodell F-4
Goodrich St. F-4
Goodyear Ave. F-8
Goulding Ave. D-5
Granger Penhurst
Pl. B-3
Grant St. C-1
Grape St. F-5
Gray St. G-5
Grimes St. G-8
Greenwood Pl. C-1
Gridar St. D-8
Guilford St. G-6
Gurham St. J-7
Hadley Pl. C-5
Halbert St. A-6
Hamburg K-4
Hamlin Rd. C-5
Hannah St. H-8
Hardwood I-6
Harmonia St. F-7
Harrison St. J-7
Harvey J-8
Hayes Pl. J-8
Helen C-1
Herkimer St. C-1
Hermann St. G-6
Hickory St. G-4,H-4
Highland Ave. D-2
High St. F-3
Hill St. A-7
Hobart St. I-8
Hodge Ave. D-3
Holland E-8
Holt St. H-7
Houf D-5
Houghton G-7
Howard St. H-5
Howlett I-7
Hubbard St. J-8
Hudson F-1
Hughes Ave. B-5
Humber Ave. B-8,C-8
Humboldt Pkwy. B-5
Hurlock B-8
Huron St., E. G-3
Huron St., W. G-2
Illinois I-3
Indiana I-3
Iroquois H-6
Irving Pl. F-3
Ivy St. E-8
James St. H-6
Jefferson Ave. .. D-5,H-5
Jewett Ave. A-5,A-7
Johnson St. G-6
Jones St. I-7
Josephine E-7
Kahr St. E-7
Kane St. F-5
Keating St. J-8
Kellogg St. J-8
Kenova Pl. B-7
Kensington Ave. B-6
Kensington Pkwy. D-6
Kent St. G-7
Kentucky St. J-4
Ketchum E-2
Kingsley St. E-5
Kirkover St. I-8
Kisfer St. E-7
Klaus H-7
Koons Ave. E-8,F-8
Kosciuszko St. F-7
Krettner St. H-6
Krupp Ave. G-8
Lafayette Ave. C-1
Lakeview Ave. F-1
Lancaster Ave. C-3
Landsdale Rd. D-5
Larkin St. I-5
Lark St. F-6
Latchworth St. A-1
Lathrop St. F-7
Laurel St. D-4
Laux St. I-7
Lowell D-2
Lawrence D-1
Lee St. K-6
Leddy J-5
Lemon St. F-4
Leroy Ave. A-7
Lester St. J-8
Letour E-7
Lewis St. I-7
Lexington Ave. D-2
Lincoln Pkwy. B-3
Linwood Ave. D-3
Litchfield Ave. B-8
Littell I-7
Livingston St. C-2
Locust St. F-4
Loepers St. F-7
Lombard St. G-7
London St. D-5
Longview Ave. C-7
Loring Ave. B-6
Louisiana St. J-4
Lyman St. H-7
Madison St. H-5
Main St. B-5,E-3,H-3
Makinaw St. J-3,J-4
Manchester B-2
Manhattan Ave. A-8
Manitoba St. I-8
Maple G-4
Marigold Ave. A-8
Marine Dr. I-2
Mariner St. F-2
Market, E. I-3
Market, W. I-3
Marshall E-7
Maryland St. F-2
Masten Rd. E-4
McKibbon E-8
Melvin Keppel K-8
Memorial Dr. H-7
Miami St. J-4
Michigan Ave., S. .. I-3,J-2
Milburn St. G-8
Miller Ave. F-8
Mills St. F-7
Milnor St. G-4
Milton St. J-7
Minton J-6
Mississippi I-3
Mohawk St. G-2
Monroe St. H-5
Montclair Ave. A-8
Montgomery St. .. H-6,I-6
Monticello Pl. B-6
Morley Pl. C-4
Mortimer St. G-5
Moselle St. D-8,E-8
Mulberry St. F-4
Myers E-6
Myrtle Ave. I-4
Nash St. H-4
New Babcock St. .. H-8,I-8
Newell H-7
New Jersey F-1
Newton G-7
Nevada Ave. B-8
Nineteenth St. D-2
Normal Ave. D-1
North St., E. E-4,E-6
Northampton St. E-4
Northland Ave. C-5
North St. E-3
Nottingham Rd. A-3
Norwood Ave. E-2
Oak H-3
Oak St., N. G-3
Oak Grove C-6
Oakland Pl. E-3
Oakwood Pl. A-5
Oberlin St. F-8
O'Connell St. J-4
Ohio St. J-3,K-3
Old Bailey K-8
Olga H-6
Oneida St. H-6
Orange St. G-4
Orlando St. J-7
Owahn Pl. J-6
Oxford Ave. C-4
Paderewski Dr. G-6
Palos A-8
Parade Ave., E. E-7
Parade Ave., N. E-7
Park St. F-3
Parkdale Ave. C-1
Parkridge A-8
Parkside Ave. A-5
Pauline B-6
Peabody St. J-6
Peach St. F-5
Pearl St. H-2
Pearl St., N. F-3
Peck St. G-8
Peckham G-5
Pembroke Ave. B-8
Pennsylvania F-1
Perkins B-1
Perry St. I-2,J-4
Pershing Ave. F-5
Persons St. G-8
Peterson E-7
Phelps St. K-8
Pine St. G-4,H-4
Playter St. G-7
Plymouth Dewitt
St. C-1
Pomeroy K-8
Pooley Pl. B-1
Porter E-1,F-1
Pratt St. G-4
Prenatt J-8
Purdy St. C-5
Putnam St. C-2
Quincy St. G-8
Quinn St. J-7
Red Jacket J-5
Reed St. G-6
Republic St. J-4
Richlawn Ave. A-7
Rich St. E-5
Richmond Ave. .. B-2,E-2
Rickert Ave. C-6
Riley St. D-4
Ripley D-2
Robie St. A-5
Rodney Ave. A-7
Roehrer Ave. E-5
Rohr St. E-7
Rommel Ave. F-8
Rose St. F-5
Roseville St. I-5
Rostzor St. E-8
Rother Ave. F-7
Ruhland St. F-8
Rumsey Rd. B-3
St. Clair K-8
St. James Pl. C-3
St. Johns F-2
St. Louis St. E-8
San Domingo Ave. H-6
Sanford B-6
Saybrook Pl. B-3
Scajaquada Ave. C-8
Scajaquada Expwy. A-4
Schmarbeck St. F-8
Schuehle St. D-8
Scott St. I-3
Scoville Ave. I-8
Sears St. G-7
Seneca St. I-4,I-6
Seventeenth St. D-2
Seymour St. I-5
Shawnee Ave. A-8
Sheridan Ave. F-7
Sherman H-6
Shumway St. H-6
Sidney St. C-6
Sidway St. J-4
Sixteenth St. D-2
Smith St. H-6,K-5
Sobieski St. F-7
Southampton St. E-4
South St. J-3
South Park St. I-3,K-6
Spillman Pl. C-5
Spring St. G-4,H-5
Spruce St. G-4
Stanislaus St. F-7
Stannard St. I-4
Stanton St. H-6
Stephens Pl. J-5
Stetson St. H-7
Storz D-5
Strauss St. G-6
Summer St. E-2
Sussex St. B-8
Swan St. H-1
Sweet Ave. G-7
Swinburne St. G-8
Sycamore Ave. F-7,G-3
Tennessee J-4
Tenth St. F-2
Thomas St. H-7
Timon St. F-5
Titus Ave. F-8
Tonoyanda A-1
Tracy St. G-2
Tremont B-2
Trenton Ave. G-1
Trinidad Pl. B-6
Trinity F-2
Tudor Pl. C-3
Tupper St., E. G-3
Tupper St., W. F-3
Townsend St. H-6
Utical St., E. D-4
Utica St., W. D-2
Urban St. D-7
Vandalia St. K-4
Vangarder St. B-7
Van Rensselaer I-5
Verdun B-7
Vermont St. E-1
Victoria Ave. A-7
Vincennes F-4
Virginia St. F-2
.... F-4,G-1
Wakefield Ave. A-7
Walker St. D-5
Walnut G-4
Walter St. D-5
Warren Ave. G-7
Warwick Ave. B-8
Washington St. H-3
Wasmuth St. E-8
Wasson St. J-7
Watson St. G-5
Wells I-3
Wescott J-7
Whitney Pl. F-2
William St. .. H-3,H-4,H-7
Willow D-5,E-5
Willow Lawn A-8
Wilson St. H-6
Winchester Ave. D-7
Windsor Pl. B-3
Winona St. J-7
Winslow Ave. D-5,D-6
Winter St. D-1
Woepel St. D-6
Wohlers Ave. E-6
Woltz Ave. F-7
Woodlawn Ave. D-4
.... D-6,D-8
Woodward AVe. B-5
Wyoming St. B-8,D-8
York St. E-1
Young Ave. G-8

POINTS OF INTEREST

Colonial Circle C-2
Delaware Park .. A-3,A-5
Ellicott Mall H-3
Gates Circle C-4
Johnson Park G-2
La Salle Park G-1
Martin Luther King,
Jr. Park E-6
Niagra Square H-2
Soldiers Place B-3
Symphony Circle E-2
Talbert Mall H-5
Times Beach I-2
Willert Park G-5

CHARLOTTE

A Ave. A-3
Abbey Pl. K-3
Ablewood B-2
Academy D-7
Academy St. D-7
Addison Dr. K-8
Admiral D-8
Alabama B-1
Alberto H-6
Alexander, N. D-6,F-4
Alexander, S. F-4
Alice Ave. D-1
Alma Ct. B-4
Altondale Ave. H-5
Ambassador D-1
Amber Dr. D-1
Ameron Dr. B-6
Amble Dr. A-6
Amherst G-5
Anderson D-7
Anderson St. D-8
Andover Rd. J-6
Andrill D-3
Anthony Cir. J-7
Ansley J-6
Anson St. I-1
Applegate J-1
Arcadia K-4
Arden St. C-4
Ardmore E-5
Ardsley H-4
Argyle B-7
Arlington F-3
Armour D-4
Armoury Dr. F-5
Arnold Dr. E-3
Arosa Ave. G-3
Arty Ave. F-1
Asbury Ave. B-5
Ashby D-5
Ashcraft Ln. J-2
Ashland Ave. F-7
Ashworth Rd. I-7
Atando Ave. B-5,C-6
Atmore St. D-8
Attaberry E-7
Auburn Ave. I-2
Audubon Rd. K-6
Augusta B-3
Auten Rd. A-4,D-2
Avant St. G-4
Aylesfora J-6
Azalea B-5
B Ave. A-3
Bacon D-2
Badger C-4
Baldwin G-5
Baltimore H-1
Bancroft St. C-5
.... D-4,D-5
Bank St. H-1
Barkley Rd. K-5
Barnette G-5
Barnhill Dr. G-7
Barringer Dr. G-1
Barry St. E-6
Bartow E-5
Barwick J-7
Bascom St. G-6
Basin G-1
Baxter St. G-4
Bay St. G-6
Baylor Dr. K-1
Beal St. F-1
Bearwood Ave. C-8
Bellaire B-2
Belleterre E-6
Belmont Ave. E-5
Belor Ave. G-4
Belot A-5
Belrose Ln. J-2
Belton St. H-1
Belvedere Ave. E-6,F-7
Benard B-6
Bennett C-6
Benson C-2
Berkeley Ave. G-3
Berkshire H-2
Berryhill E-1
Bertonley Ave. J-7
Bethel Rd. F-1
Beverly Dr. I-5
Billingsley Rd. I-6
Bingham C-8
Black B-5
Blackthorne Rd. I-1
Blackwood D-8
Blairmore F-6
Bland, W. F-2
Blanton G-1
Blazer C-3
Blythe Blvd. H-4
Bobby Ln. I-7
Booker C-2
Bolling Rd. I-5
Bonwood Dr. J-6
Boone St. C-3
Botany St. B-2
Boyd C-1
Bradbury A-2
Brandon Cir. J-5
Brandon Rd. I-4
Brandywine Ave. K-3
Brantham Rd. J-8
Brentwood Dr. F-1
Brevard St., N. F-4
Brevard St., S. F-3
Briabend J-1
Briarcliff Pl. J-4
Briar Creek Rd. G-7
Brice St. D-1
Bridlepath Ln. J-8
Brighton Pl. H-8
Bromley Rd. H-4
Brook Hill Rd. G-1
Brookhurst H-8
Brookhurst Dr. I-7
Brookridge K-6
Brookshire Frwy. C-2,D-3
Brookwood K-5
Browning Sprague F-6
Brownstone B-3
Bruns Ave. E-2
Brunswick Ave. H-4
Bryant E-2
Buchanan St. H-3
Bucknell I-4,J-4
Buena Vista C-2
Bulford Ave. H-7
Bungalow C-1
Burbank Rd. B-2
Burch Dr. A-4
Burklane I-7
Burleigh K-7
Burrough B-8
Burton D-3
Butterfield C-7
C Ave. A-3
Caldwell St., N. F-4
Caldwell St., S. F-4
Callahan D-4
Calvert E-2
Cameron G-6
Campus C-2,D-2
Cambridge Rd. I-2
Camden G-2
Candy Stick B-8
Canterbury, N. J-7
Canterbury S. J-6
Card D-4
Carmine St. B-4
Carol D-1
Carolyn Dr. G-8
Carowil J-3
Carter St. H-8
Cartier Way F-1
Cassamia Pl. J-5
Castleton Rd. J-8
Castlewood J-2
Caswell Rd. G-5
Catalina Ave. C-5
Catawba E-6
Cedar, N. E-3
Cedar St. S. E-3
Celia B-2,C-2
Cemetery C-2
Central Ave. F-6
Centre B-1
Chamberland D-2
Chandler Pl. K-7
Charles Ave. D-6
Charmapeg Ave. K-8
Chase St. G-5
Chatman Ave. F-7
Cheddington Dr. J-8
Cherokee Pl. H-5
Cherokee Rd. I-5
Cherry G-4
Chesterfield Ave. G-6
Chestnut Ave. F-6
Cheswick Rd. J-7
Chicago AVe. G-1
Chillingworth K-7
Chilton Pl. J-4
Chipley Ave. H-7
Chippendale Rd. I-8
Chelmsford Rd. J-6
Church St., N. D-5,E-4
Church, S. F-3
Churchill Rd. I-6
Cinderella B-8
Clarice Ave. G-6
Clark J-5
Clarkson, N. E-3
Clay D-1
Clayton Dr. I-3
Clement Ave. F-6,G-5
Clemson D-7
Cleveland G-2
Clifford F-2
Cloudman A-1
Clover K-6
Club Rd. F-7
Clyde Dr. C-1
Clydesdale Rd. D-1
Coates E-8
Cochran E-6
Coddington Pl. J-6
Coker D-1
Colfax B-2
Collanade Dr. E-7
College St., N. E-4
College St., S. F-3
Collingwood J-1
Colliseum Dr. H-7
Colonial Ave. G-5
Colony Rd. K-4
Columbus Pl. H-7
Colville Rd. H-5,I-5
Colwick J-7
Commerce F-2
Commercial D-8
Commodore I-8
Commonwealth Ave.
.... G-6,H-8
Concordia C-5
Concordia Ave. C-5
Condon St. C-3
Coniston Pl. H-4
Connecticut C-8
Connecting Rd. I-7
Conway Ave. I-1
Cooper Dr. K-1
Cornelius C-6
Cornel Ave. J-6
Cornet Way C-1
Corporation Circle A-1
Cosby Pl. F-8
Cotes St. E-3
Cottage Pl. H-5
Cottonwood A-6
Country Club E-7
Country Club Rd. E-7
Coventry Rd. J-6
Coxe Ave. D-2
Craghead Rd., W. B-6
Craig Ave. H-8
Cranbrook Ln. H-6
Creek D-3
Creger St. B-1
Creighton B-2
Crest Dr., E. G-7
Crestdale A-3
Crestview C-6
Crestway C-2
Cricketeer Dr. A-2
Crosby K-6
Croydon Rd. J-4
Crystal G-8
Cullman Ave. C-5
Cumberland Ave. I-3
Cummings Ave. B-3
Cunningham D-1
Currituck Dr. K-1
Curtiswood C-8
Cushman St. B-7
Custer B-3
Cyrus Dr. G-8
Dakota A-1
De Lane Ave. I-6
Dalehurst Dr. D-8
Dalton Ave. D-4
Darby Ave. E-8
Dare Dr. B-7
Darsey D-3
Dartmouth H-5
Davidson St., N.
.... C-7,D-6,F-4
Davidson St., S. F-4
Dawson C-2
Dean St. C-3
Dearborn C-4
De Armon Dr. F-7
Dellwood K-3
De Paul E-3
Devon J-3
Diana Dr. H-7
Dickinson I-4
Dilworth Rd. H-3
Dilworth, E. G-3
Dinglewood Ave. C-8
Distribution St. G-2
Dixon D-2
Doggett G-2
Dogwood Ave. C-5
Domino Ct. H-7
Donna D-8
Doris Ave. I-8
Dorothy H-2
Dotger Ave., N. H-6
Dotger, S. H-6
Double St. C-4
Dover I-1
Downs Ave. D-7
Dresden Dr., W. H-8
Drexel Pl. K-3
Drexmore Ave. K-3
Druid Cir. C-4
Drum E-1
Drummond Ave. E-6
Dry Dr. B-7
Duckworth D-2
Dunavant St. H-2
Dunbar St. E-2
Dundeen St. C-2
Dunkirk F-2
Dunlavin Way F-8
Dunn St. H-7
Dupree B-1
Durham G-5
Durwood Dr. G-4
Dwelle B-1
Earle D-4
East Boulevard H-3
Eastover Rd. H-5
Eastview I-8
Eastwood E-8
Echo Glen Rd. A-8
Edgehill Rd. N. H-4
Edgehill Rd. S. H-4
Edgerton B-8
Edison St. C-4
Edwin C-4
Effingham D-1
Elder Ave. E-3
Eldridge F-3
Ellington I-7
Ellison G-4
Ellsworth I-7
Elmhurst Rd. H-2
Elizabeth Ave. F-4
Elkwood D-8
Ella St. A-6
Ellen Ave. E-8
Elm E-4
Emerson H-6
Emory Ln. J-8
Englehardt St. B-2
English Dr. B-2
Ennis Ave. C-5
Enterprise St. C-6
Erie St. B-3
Erskine Dr. B-2
Eastway Dr. D-8
Essex D-8
Estelle St. C-2
Euclid G-3,H-2
Eureka C-4
Everett E-7
Evergreen E-1
Ewing Ave. H-3
Exeter J-6
Fairmont C-3
Falson D-6
Fannie Cir. H-6
Farmcrest A-7
Farmington Ln. F-8
Faulkner I-7
Fenton Pl. H-5
Fenwick K-2
Ferguson F-8
Ferncliff K-6
Ferncliff Rd. K-5
Fieldbrook K-3
Fireside B-8
Firth F-7
Firwood Ln. J-1
Flamingo E-8
Fleetwood Rd. E-1
Flint D-2
Floral Ave. H-3
Florida Ave. E-7
Flynwood F-6
Fontana Ave. D-4
Ford Rd. B-3
Ford Rd., E. E-7
Fordham Rd. F-1
Forest Dr. K-5
Fort St. E-7
Fortune C-3
Foster H-1
Fox G-4
Franklin C-5
Frazier D-3
Freeland I-1
French D-3
Frew Rd. C-6
Friendly A-8
Fugate Ave. H-7
Fulton F-7
Furman Pl. K-1
Galax C-8
Gardenter H-4
Gardner Ave. D-2
Garnette B-3
Gay St. F-1
Gaynor Rd. K-7
Gene Ave. H-6
Geneva Ct. I-3
Gentry Pl. K-2
Georgia Ave. E-7
Gibbs St. C-3
Gilbert St. B-3
Glendale K-4
Glenn St. G-3
Glory St. B-7
Gondola B-7
Goodwin Ave. H-8
Gordon St. E-6
Goshen B-1
Grace E-6
Graham St., N. .. A-6,D-4
Graham, S. F-2
Graham St., S. E-3
Grandin Rd. E-2
Grant G-1
Granville Rd. H-4
Graybark Ave. F-8
Greenleaf E-2
Green Oaks G-7
Greensboro St. C-7
Greenway Ave. G-5
Greenwich Rd. J-7
Greenwood Cliff G-4
Gregg Ave. D-2
Greystone Rd. I-1
Grier St. C-3
Griffith St. H-1
Grimes St. C-5,D-5
Guilford I-2
Gunn C-3
Habersham I-2
Haines B-3
Haleworth J-8
Hall St. B-8
Halsey C-1
Halstead K-2
Hamilton St. D-4
Hammet St. I-8
Hampshire D-8
Hampton Ave. I-5
Hanover G-6
Hanson Dr. H-8
Hardianway K-7
Harding Pl. G-4
Hardwick Rd. K-7
Harris Rd. J-1
Harrison St. C-1
Hartford I-4
Hartford Ave. I-1,J-1
Hartley St. B-5
Harvard H-5
Harvey K-3
Hassell Pl. K-3
Hasting J-3
Hatteras Ave. B-3
Haven Dr. J-3
Havilon Ct. K-8
Hawkins G-2
Hawkins St. G-2
Hawthorne Lane F-6
Heathcliff E-2
Heather Ln. J-2
Heathwood Rd. J-7
Hedgemore K-3
Heflin H-6
Heil C-3
Hempstead Pl. H-5,I-5
Henley Pl. H-4
Hermitage H-5
Hermitage Rd. H-5
Herrin D-7,E-8
Hershey St. B-8
Hickory A-4
Hickorynut St. H-8
Hidden Brook Dr. H-8
Hidden Valley Rd. B-8
Hildebrand B-2
Hill St., E. F-3
Hill St., W. E-2,F-3
Hillard E-8
Hillcrest St. B-7
Hillsdale J-4
Hillside Ave. J-3
Hilo Dr. C-7
Hobbs D-5
Hodgson K-7
Holland B-6
Hollis Rd. I-1
Holly St. B-3
Hollyday Ct. K-2
Holroyd J-6
Holt Ct. D-7
Homorton Pl. F-6
Honeywood Ave. B-1
Hope Dale H-4
Horne Dr. A-7
Hoskins Rd. A-1
Hough Rd. J-3
Howie Cir. C-8
Hudson D-7
Hughes A-8
Hungerford Pl. I-6
Hunter Ln. K-7
Huntley Pl. H-5
Hyde Dr. B-7
Ideal Way H-2
Idlewood Cir. I-3
Independence Blvd.
.... F-1,G-6
Industry Rd. I-1
Inwood Dr. K-1
Irby J-3
Iris Dr. F-7
Irma B-3
Ironwood St. B-7
Irwin E-3
Isenhour B-4
Ison F-1
Iverson Way H-2
Ivey Dr. G-7
Jackson F-3
Jameston Dr. I-3
Jasper F-3
Jay St. E-1
Jeff St. E-8
Jefferson Davis F-2
Jenkins D-2
Jennings B-3
Jersey J-2
Jessie Ct. B-4
Jewel St. H-6
Joe St. C-7
Johnson D-4
Johnston Rd. B-6
Jones St. C-1
Jonquil St. I-7
Jordan Pl. D-6
Judith J-7
Judson C-1
Julia Ave. B-4
Julien St. G-6
June H-8
Justice Ave. B-4
Kasaw A-5
Kay C-3
Keller Ave. B-3
Kendall A-2
Kenilworth H-3
Kenmore G-5
Kennesaw Dr. B-3
Kenney St. C-3
Kennon St. E-6
Kensington Dr. F-6
Kentbrook B-8
Kentucky B-1
Kenwood F-6
Keswick D-5
Kings Dr., S. G-4,H-4
Kingsbury Dr. G-6
Kingston Ave., E. G-3
Kingston Ave., W. F-2
Kirkland D-1
Kirkwood G-3
Knollwood K-6
Kohler C-4
Kotonah D-2
Laburnum Ave. G-6
Lake D-1
Lakeview C-1
Lakewood C-1
Lamar Ave. G-5
Lambeth Dr. B-8
Lander St. C-1
La Salle C-1
La Salle St. B-2
Laurel Ave. G-6,H-5
Leafmore C-8
Leigh Ave. D-6
Lenox Ave. G-3
Leroy St. D-2
Lexington G-3
Liberty F-3
Libeth St. E-8
Liddell St., W. E-4
Lilac Rd. I-3
Lilington Ave. G-6
Lincoln E-2
Lincrest K-8
Linda Ln. J-7
Linden Ln. E-5
Linganore G-3
Linwood Ave. B-1
Litchfield I-8
Little Hope J-2
Lloyd D-4
Lockhart Dr. H-3
Lockridge Dr. I-5
Log Cabin Rd. A-8
Logie Ave. F-7
Lola Ave. E-6
Loma Ln. E-3
Lomax Ave. J-8
Lombardy Cir. H-6
Long D-1
Long Branch F-5
Longfellow G-8
Longwood Dr. E-2
Lorene Ave. J-4
Louise Ave. F-5
Lucena St. C-5
Ludlow B-2
Lunsford Pl. D-6
Lyndhurst Ave. G-2
Lynnwood J-3
Lyon F-7
Lyttleton K-3
Macathur Ave. B-4
Madison D-3
Madrid Rd. B-3
Magnolia Ave. G-2
Manhasset I-1
Maple Grove A-2
Marguerite E-7
Maribe Ave. B-2
Marlwood Ter. I-3
Marsh Rd. H-2
Marshall H-2
Martin St. D-3
Marvin Rd. I-7
Mavista J-3
Maryland Ave. I-4
Masonic F-7
Matheson Ave. .. D-6,E-7
Mathis D-1
Mattoon C-2
May St. H-1
Mayfield I-2
May View Dr. H-1
Maywood E-8
McAlway I-7,J-7
Mc Call St. D-4
Mc Cauley St. I-8
Mc Clintock Rd. G-6
Mc Donald B-2,H-2
Mc Dowell St. E-5
Mc Dowell St., N.
.... D-7,F-4
Mc Dowell St., S. F-4
Mc Millan St. D-8
Mc Quay D-1
Meacham St. H-2
Meadow Brook I-6
Meadow Ln. D-8
Melbourne Ct. H-2
Melchor Ave. I-7
Melita C-2
Mellow Dr. B-8
Mercury St. C-7
Merlane Dr. A-7
Merriman Ave. F-2
Merry Oaks G-8
Merwick Cir. J-7
Metals Dr. A-6
Meyer, N. D-6
Middleton Dr. H-5
Midfield B-7
Miles Ct. C-3
Millbrook Rd. I-7
Millerton Ave. E-1
Mimosa Ave. E-6
Mint St. F-3
Mitchell St. B-3
Mockingbird Ln. K-3
Mohigan H-1
Mona K-3
Montclair K-7
Monterey H-8
Montford Dr. K-2
Montgomery D-2
Montrose Ct. H-6
Mooney D-7
Moorehead St., E. F-3
Moorehead St., W. E-2
Moravian Ln. H-5
Moretz Ave. C-4
Morningside Dr. G-7
Morrow F-4
Morson D-1
Moss St. E-1
Moultrie St. I-1
Mt. Kisco Dr. A-8
Mt. Vernon G-3
Mulberry C-3
Murdoch Rd. D-7
Murray Hill Rd. .. J-1,K-1
Museum Dr. I-5
Museum Pl. H-6
Myers St. F-4
Myrica A-8
Myrtle G-3
Nandina St. F-7
Nancy Dr. I-8
Nassau Blvd. F-6
Nelson Ave. B-1
New Castle C-3
New Hope H-3
Newland C-3
Newland Rd. B-3
Norcross E-7
Norfolk H-1
Norris Ave. C-4
Northaven Dr. B-7
Northcliffe A-6
Northcrest B-7
Northerly Rd. B-4
North Gate I-1
Northmore C-7
Norton Rd. I-4
Norwood Dr. C-1
Nottingham K-8
Oaklawn Ave. D-4
Oakmont Ave. G-8
Oaks Rd. C-4
Oakwood Ave. E-7
October C-1
Olando Ave. C-5
Onyx C-2
Opal St. D-1
Orange St. H-6
Ordermore H-3
Oregon St. E-5
Orvis St. G-3
Osborne G-5
Osmond F-1
Otts St. F-5
Overhill Rd. K-5
Oxford Pl. I-5
Paddock J-2
Palm E-5
Palmer F-2,F-3
Palmisto St. F-6
Park Ave. G-3
Park Ave., E. F-2
Park Ave., W. F-3
Park Dr. F-5
Park Rd. J-3,H-3
Parkside Dr. D-1
Parkway Dr. D-1
Parkwood Ave. E-6
Patch B-2
Patterson C-7
Patton Ave. B-8
Paxton E-3
Pearson St. E-6
Pebble St. A-6
Pecan Ave. F-6,G-5
Pelton St. H-1
Pembroke I-4
Pender J-4
Penman E-2
Pennsylvania B-1
Perrin Pl. H-5
Phifer E-4
Philemon C-7
Phoenix Pl. K-6
Picaroy J-4
Pickens F-8
Pierce St. H-3
Pine St. E-3
Pine St., N. D-5
Pinecrest Ave. G-3
Pine Grove Ave. B-8
Pinehurst K-3
Pineville Rd. K-1
Pinewood Cir. C-1
Pinkney Ave. D-6
Pitts Dr. C-2
Placid I-6
Plainwood A-1
Plantation I-3
Planters C-1
Plum Hearty K-7
Plymouth Ave. B-1
Poindexter Dr. H-2
Polk St. D-4
Pompano St. A-1
Pondella Dr. A-8
Poplar St., N. D-5,E-3
Poplar, S. E-3
Portland I-5
Post St. F-2
Prospect F-5
Prince St. B-1
Princeton Ave. J-3
Providence I-5
Providence Rd. H-5,I-5
Pryor St. D-1
Purnell Ct. J-5
Queens I-4
Queens Rd. H-4
Queens Rd., E. J-4
Queens Rd., W. I-4
Quentin Pl. B-2
Quincy F-2
Rachel St. C-5
Radcliffe Ave. I-4
Radio C-2
Rainbow B-6
Raleigh St. C-7,C-8
Ramona C-1
Rampart G-2
Randall F-6
Randolph Rd. I-6
Ravencroft Dr. D-1
Rayon St. B-1
Redbud C-2
Redwood Ave. C-8
Reece J-3
Renner C-3
Rennsselar G-3
Rensford Ave. I-5
Remington B-2
Remus G-1
Reynolds Dr. J-2
Richland Dr. I-8
Ridgecrest F-7
Ridgedale A-7
Ridgeway Ave. G-5
Ridgewood J-3
Ringwood D-1
Ritch Ave. C-6
Roanoke Ave. H-8
Robin Rd. K-7
Robinhood Rd. K-8
Robinson Ave. C-6
Rockbrook Dr. K-4
Rockford J-1
Rocklyn J-3
Rockway Dr. G-7
Rodman I-6
Roland F-7
Rolling Hill Dr. B-8
Rollins Ave. H-8
Romany Rd. G-3
Rose D-1
Rosetta C-2
Rosewood B-8
Roslyn Ave. C-2,D-2
Roswell Ave. J-4
Rothwood Dr. K-6
Royal Ct. F-3
Rozzelles Ferry
Rd. D-2
Runnymede K-4
Rush D-1
Russell Ave. C-3
Rutgers Ave. B-7
Rutledge K-7
St. Andrews Ln. F-7
St. Bernard C-6
St. George St. G-7
St. John B-3
St. Luke B-3
St. Mark B-3
St. Paul B-3
Samuel B-4
Sandalwood Rd. H-6
Sanders C-2
Sandridge K-1
Salem Dr. I-3
Scaleybark I-1
School St. F-6
Scotland Ave. H-6
Scott Ave. H-3
Sedgefield I-2
Sedgewood J-6
Seldon Dr. C-2
Selwyn Ave. K-3,I-4
Senior B-2
September Ln. D-1
Service St. B-5
Shamrock E-8
Sharon Ave. K-5
Sharon Lane K-6
Shawee Dr. I-2
Shenandoah Ave. E-6
Sherwood I-3
Shoreham Dr. K-6
Silabert Ave. I-8
Simplicity B-5
Simpson Dr. E-7
Skyland Ave. H-6
Skyview Rd. F-1
Sloan St. I-1
Smallwood C-2
Smith St. E-4
Snow White Ln. B-8
Sofley Rd. B-1
Solomon D-2
Somerset Dr. I-3
South Blvd. G-2,H-2
Southwest Blvd. B-2
Southwood F-2
Sparrow Mill Rd. D-7
Spencer D-7
Spratt St. D-4
Spring St. D-4
Springdale H-2
Springview Rd. B-8
Spruce St. F-2
Squirrel Hill Rd. A-8
Stafford K-3
Stancill Pl. I-6
Stanley Ave. H-7
Stanford Pl. I-4
Starita Rd. A-4
State St. D-2
Statesville Ave. .. B-4,D-4
Sterling Rd. J-3
Stevens Seigle F-5
Stoneybrook Rd. E-7
Stonewall, E. F-3
Stonewall, W. F-3
Stratford E-7
Suffolk Pl. J-6
Sugar Creek C-7
Sugar Creek Rd. C-8
Summitt Ave. D-3,E-2
Summit Ave., W. E-2
Sumter D-2
Sunderland J-5
Sunnyside Ave. F-5
Sunset Dr. I-2
Suttle Ave. F-1
Sutton E-2
Sycamore E-3
Sylvania D-5
Syracuse Dr. E-2
Tangle Dr. J-8
Tanglewood Ln. J-8
Teppan Pl. D-7
Tate C-2
Taylor Ave. B-3
Television Ln. H-7
Templeton G-3
Tennessee B-1
Terrybrook Ln. D-8
Thackery G-8
The Plaza D-7
Thera Dr. B-7
Thomas F-6
Thornwood Dr. A-8
Thrift Rd. E-1
Thrush Ln. A-6
Tippan Ave. F-6
Toal St. B-5
Tom Hunter Rd. A-8
Toomey Ave. G-1
Topaz St. B-6
Topsfield Rd. I-8
Torrence St. G-4
Townes Rd. J-3
Trade St. D-2,F-4
Trade St., E. F-4
Trade St., W. E-3
Tranquil J-3
Tremont G-2,G-3
Trentwood A-3
Trinity Pl. K-7
Truman Rd. F-7
Tryclan I-1
Tryon St., N. C-7,E-4
Tryon St., S. F-3,H-1
Tuckoseegee Rd. E-2
Turner Ave. D-2
Twiford Pl. I-5
Twinfield A-2
Tyngway J-6
Tyson St. I-1
University J-1
Unstead St. E-6
Van Buren D-3
Van Every St. F-5
Van Ness H-5
Vance, E. F-3
Vance, W. F-3
Vancouver A-8
Vanderbuilt C-4,F-5
Vane Ct. B-5
Vanizer D-1
Vernon Dr. J-6
Victoria E-3
Vine D-2
Vinewood H-7
Vinton C-3
Virginia Ave. E-7
Waco St. G-4
Waddell St. C-3
Wadsworth E-5
Wainwright B-4
Wakefield K-3
Wales J-3
Walnut Ave. E-2
Walter Oak Rd. J-7
Ware St. B-5
Warp St. C-7
Washburn Ave. H-7
Washington St. C-3
Washington, W. G-2
Waterbury J-2
Webster Pl. I-1
Weddington G-6
Wedgewood Dr. K-2
Weldon D-8
Welker G-4
Wellesley Ave. I-4
Wellingford St. B-8
Wells St. C-5
Wembley Dr. G-7
Wendover Rd. J-5
Wendwood Ln. I-6
Wenworth Pl. K-2
Wesley D-7
Westbrook E-3
Westbury Rd. K-6
Westfield Rd. J-3
Westminster J-4
Westmoreland St. G-7
Westover St. G-6
Wheatley Ave. I-6
Whisnant D-3
Whitby Ln. K-7
Whitehaven D-2
White Oak K-1
White Plains Rd. A-8
Whiting D-7
Wickford G-2
Willow Oak Rd. J-3
Wilmore Dr. F-2
Wilson Ave. A-3
Wilson Ln. A-7
Windemere Ln. J-8
Windingwood K-4
Windsor Ave. I-3
Windsor Dr. K-4
Winnifred F-2
Winston C-5
Winston Dr. D-7
Winter St. F-7
Wintergreen Pl. K-6
Winthrop H-2
Wolfberry D-5
Wolfe F-7
Wonderwood Dr. K-7
Woodcrest G-2
Woodhaven Ln. K-5
Woodland G-8
Woodlawn Rd. K-1,K-3
Woodlark Ln. J-7
Woodruff Pl. E-2
Woodside Ave. D-6
Woodvale D-2
Wood Valley A-2
Woodward Ave. C-4
Worthington Ave. G-3
Wray St. C-4
Wren F-1
Wrenwood J-6
Wright Ave. J-8
Wriston I-2
Wyanoke G-6
Yale Pl. I-3
Yellowstone D-1
Yonkers B-6
Yorkshire I-1
Youngblood H-1
Yuma St. A-8,B-8
1st G-4
1st, E. F-3
1st St., W. E-2,F-3
2nd St., E. F-3
2nd St., W. E-2,F-3
3rd St., E. F-3
3rd, W. E-3
4th St., E. F-4
4th St., W. E-2,E-3
4th St. Ext. E-2
5th St., E. F-4,G-5
5th, W. E-3
6th St., E. F-4
6th, W. E-3
7th St. G-5
7th St., E. E-4
7th, W. E-3
8th St., E. E-4,G-5
8th St., W. E-4
9th St., E. E-4,G-6
9th St., W. E-4
10th St. F-5
10th St., E. E-4
10th St., W. E-4
11th St., E. E-4
11th St., W. D-3,E-4
12th St. E-4
12th St., W. D-3
13th St., E. E-5
13th St. E-5
15th St., E. E-5
16th St., E. E-5
17th St., E. E-5
18th St., E. E-6
19th, E. E-6
19th St., E. E-6
20th, E. E-6
20th St., E. E-6
21st, E. E-6
22nd, E. E-6
23rd, E. D-5
24th, E. E-6
24th St., W. D-5
25th, E. D-6
26th, E. D-6
26th St., E. E-6
27th, E. D-6
27th, W. D-5
28th St., E. D-6
28th, W. C-6
30th St., W. C-6
31st, W. C-6
32nd St. D-6
32nd St., W. C-6
33rd St., E. D-6
34th St., E. D-6
35th St., E. D-7
36th St., E. D-7
37th St., E. D-7

POINTS OF INTEREST

Independence Park G-5
Revolution Park G-1

Scale of Miles

0 .2 .4 .6 .8

N

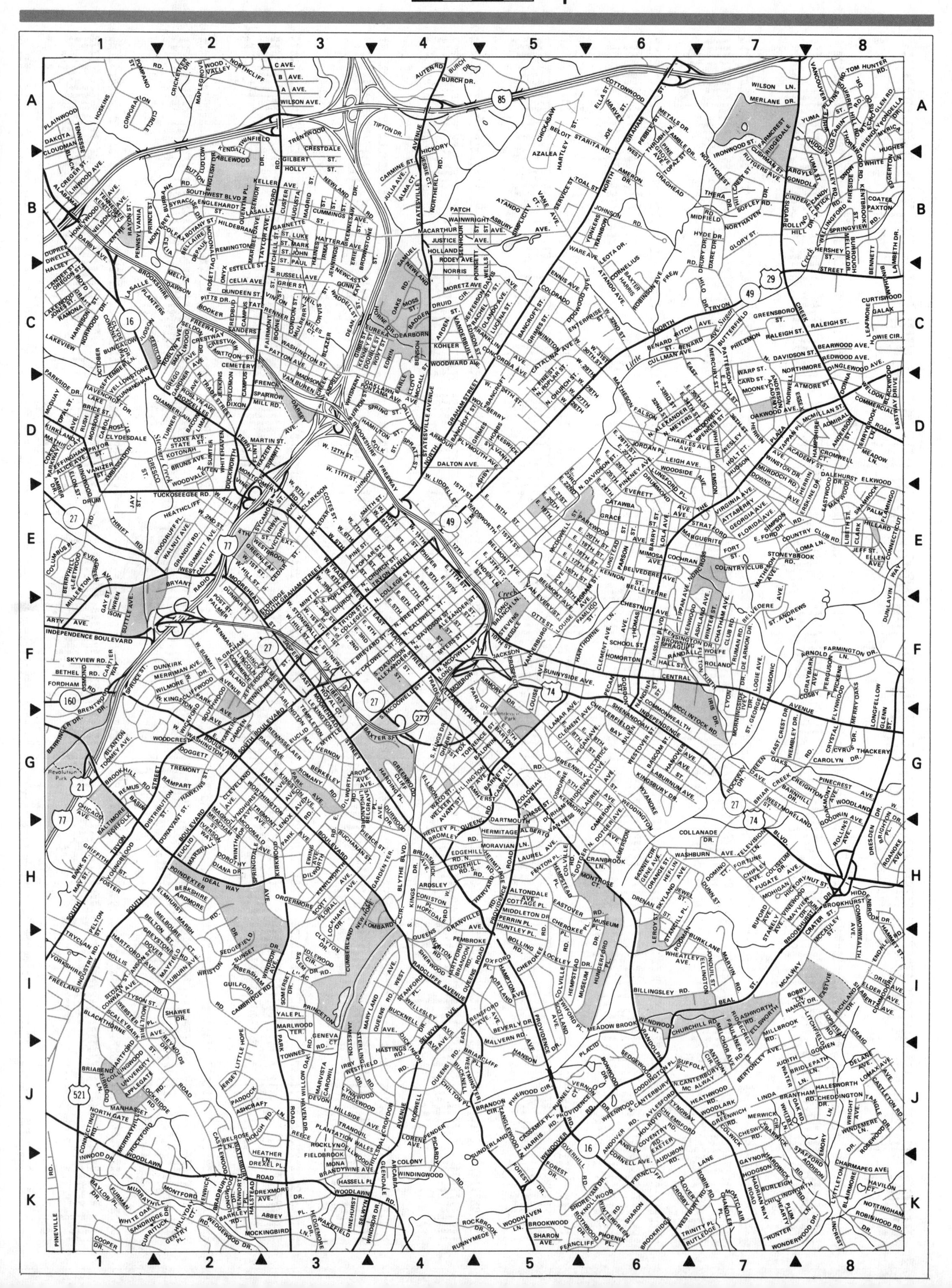

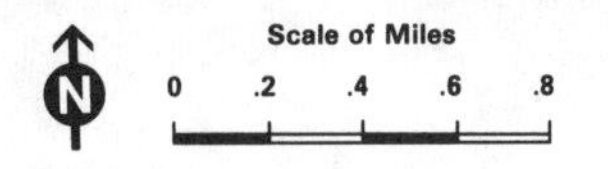
N
Scale of Miles
0
.2
.4
.6
.8

1
2
3
4
A
B
C
D
E
LAKE
MICHIGAN
DIVERSEY
FULLERTON
ARMITAGE
NORTH
DIVISION
CHICAGO
GRAND
RANDOLPH
WASHINGTON
MADISON
JACKSON
ADAMS
EISENHOWER
EXPRESSWAY
ROOSEVELT
CERMAK
JOHN F. KENNEDY
EXPWY
ADLAI E. STEVENSON EXPWY
DAN RYAN EXPWY
WESTERN
DAMEN
ASHLAND
HALSTED
STATE
MICHIGAN
LAKE SHORE DR.
OGDEN
BLUE ISLAND
ARCHER
ELSTON
CLARK
WACKER DR.
HUMBOLDT PARK
DOUGLAS PARK
UNION PARK
HARRISON PARK
GRANT PARK
NAVY PIER
FILTRATION PLANT
ADLER PLANETARIUM
SHEDD AQUARIUM
BUCKINGHAM FOUNTAIN
SOLDIER FIELD
MERCHANDISE MART
COOK COUNTY HOSPITAL
CHICAGO STADIUM
MEIGS FIELD (CGX)
NORTHERLY ISLAND
Burnham Park
MC CORMICK PLACE COMPLEX (CHICAGO EXPOSITION CENTER)
Diversey Yacht Club
Columbia Yacht Club
Chicago Yacht Club
Burnham Park Yacht Club
Chicago Harbor
Outer Harbor
NORTH AVENUE BEACH
OAK STREET BEACH
12TH ST. BEACH
31 ST. STREET BEACH
FULLERTON BEACH
COAST GUARD STATION
BREAKWATER
SHORE ARM BREAKWATER
SOUTHERLY BREAKWATER
COMISKEY PARK
CHINATOWN
ILLINOIS INSTITUTE OF TECHNOLOGY
UNIVERSITY OF ILLINOIS (CHICAGO CAMPUS)
NORTHWESTERN UNIVERSITY (CHICAGO CAMPUS)
DE PAUL UNIVERSITY
CHICAGO HISTORICAL SOCIETY
MICHAEL REESE MEDICAL CEN.
MERCY HOSPITAL
FUTURE SITE CHICAGO PRODUCE MARKET
HOUSE OF CORRECTION
BURNHAM
90
94
290
55
41
14
64

CHICAGO

Aberdeen B-E2
Ada A-E2
Adams C1-3
Alexander D3
Allport D2
Almond C1
Altgeld A1-3
Anacona B2
Anson Pl. B1
Arbour Pl. C2
Arcade Pl. C2
Arch E2
Archer E2
Arlington Pl. A3
Armitage A1,2
Armour B2
Artesian A-E1
Arthington C1-3
Ashland A-E2
Astor B3
Attrill A1
Augusta Blvd. B1,2
Avondale A1,2
Balbo Dr. C3,4
Banks B3
Barber D3
Bauwans B2
Beach B1,2
Beaubien Ct. C3
Belden A1,2
Bell A-E1
Bellevue Pl. B3
Benson E2
Besly Ct. A2
Bingham A1
Bishop B-E2
Bissell A2,3
Blackhawk B2,3
Bliss B2
Bloomingdale A1,2
Blue Island D1,2
Bonaparte E2
Bonfield E2
Bosworth A,B2
Boulevard Way E1
Bowler C1
Brenock Dr. B1
Broad E2
Bross E1
Browning E4
Burling A3
Burton Pl. B3
Cabrini C2,3
California A-E1
Calumet D,E4
Cambridge A,B3
Campbell A-E1
Canal C-E3
Canalport D2,3
Cannon Dr. A3
Carpenter B-E2
Carroll C1-3
Caton A1
Cedar B3
Chanay A1
Charleston A1
Cherry B2
Chestnut B2,3
Chicago B1-4
Churchill A1
Claremont A-E1
Clark A-D3
Cleaver B2
Cleveland B3
Clifton A2
Clinton C,D3
Clybourn A2
Commonwealth A3
Concord Pl. A1-3
Congress Pkwy. C1-3
Corbett D,E2
Cortez B1,2
Cortland A1,2
Cottage Grove D,E4
Cottage Pl. C2
Couch C3
Court Pl. C3
Crilly A3
Crosby B3
Crowell E2
Crystal B1,2
Cullerton D1-3
Damen A-E2
Dayton A2
Dean B2
Dearborn B-E3
Dekoven C3
Delaware Pl. B3
Deming Pl. A1,3
Denvir C1
Des Plaines B-D3
Dewitt Pl. B3
Dickens A1-3
Diversey Dr. A3
Diversey Pkwy. A2
Division B1-3
Dominick A2
Dr. M. L. King Dr. D-E4
Draper A2
Drummond Pl. A1-3
Eastman B2,3
Eberhart E4
Edward Ct. A3
Eleanor E2
Elias Ct. E2
Elizabeth B-E2
Elk Grove A,B2
Ellen B2
Ellis E4
Elm B3
Elston A1,B2
Emerald E3
Erie B1-3
Ernst Ct. B3
Eugenie A3
Evergreen B1-3
Fair Pl. B3
Fairbanks Ct. B3
Fairfield A-E1
Farrell E2
Federal C-E3
Felton Ct. B3
Ferdinand B1
Fern Ct. A3
Financial C3
Flournoy C1,2
Ford D3
Francis Pl. A1
Francisco A-E1
Franklin C3
Fremont A,B2
Fry B2
Ft. Dearborn Dr. E4
Fuller E2
Fullerton A1-3
Fulton C1-3
Garland Ct. C3
Geneva Ter. A3
Germania Pl. B3
Giles E4
Gladys C1-3
Goethe B3
Grady Ct. E2
Grand B1-3
Grant Pl. A3
Gratten E2
Green C-E3
Greenview A,B2
Grenshaw D1-3
Grove D3
Haddon B1,2
Haines B3
Halsted A-E3
Hamilton A-E1
Hampden Ct. A3
Harrison C1-3
Hartland Ct. B2
Hastings D1,2
Haynes Ct. E2
Heath D1
Henry Ct. A1
Hermitage A-E2
Hickory B2
Hill B3
Hillock E2
Hirsch B1
Hobbie B3
Hobson A2
Hoey E2
Holly A2
Homer A1,2
Hooker B2
Horner A1
Howe A,B3
Hoyt E1
Hubbard B1-3
Hudson A,B3
Huguelet Pl. B3
Humboldt Blvd. A1
Humboldt Dr. B1
Huron B1-4
Illinois B3
Indiana D,E3
Institute Pl. B3
Iowa B1,2
Iron E2
Jackson Blvd. C1-3
Jackson Dr. C3,4
Janssen A2
Jasper Pl. E2
Jefferson C,D3
Jessie Ct. B1
Jones A1
Jourdan Ct. D3
Julia Ct. A1
Julian B2
Justine C-E2
Keeley E2
Kemper Pl. A3
Kenmore A2
Kingsbury A-C2
Kinzie C1-3
LaSalle B-E3
Laflin C-E2
Lake C1-3
Lake Park E4
Lake Shore Dr. A,B3,C-E4
Lake View A3
Lakewood A2
Larrabee A,B3
Le Moyne B1,2
Leavitt A-E1
Lee Pl. B1,2
Lehmann Ct. A3
Lessing B2
Levee E2
Lexington C1-3
Liberty D2,3
Lill A2
Lincoln A3
Lincoln Park W. A3
Lister A2
Lituanica E2
Lloyd E2
Lock E2
Locust B3
Logan Blvd. A1
Loomis C-E2
Loomis Pl. E2
Louis Munoz Marin Dr B1
Lowe D,E3
Lumber D2,3
Luther Pl. E1
Lyman E2
Lyndale A1
Lytle C,D2
Madison A1,3
Magnolia A2
Mansfield D2
Maple B3
Maplewood A-E1
Marcey A2
Marion Ct. B1
Marshall Blvd. D1
Marshfield A-E2
Mary E2
Maud A2
Mautene Ct. B2
Maxwell D2,3
May B-E2
Maypole A1-2
McClurg Ct. B4
McLean A1,2
Medill A1,2
Meis Van Der Rohe B3
Mendell A2
Menominee A3
Meyer A3
Michigan B-E3
Miller C,D2
Milwaukee B2
Moffat A1,2
Mohawk A3
Monroe C1-3
Monroe Dr. C3,4
Montana A1,2
Moore B2
Moorman B2
Morgan C-E2
Mozart A-E1
Nada B2
New B,C4
Newberry D3
Newburg D3
Noble A,B2
Normal D,E3
North A1-3
North Branch B2
North Park A,B3
North Water B4
Norton C2
Nursery A2
O'Brien D3
Oak Pl. B3
Oakley Blvd. B1
Ogden Blvd. D1,C2
Ohio B1-4
Ontario B1-4
Orchard A,B3
Orleans A,B3
Oswego B,C2
Palmer A1
Park E4
Parnell E3
Paulina A-E2
Pearl Ct. A3
Pearson B3
Peoria B-E3
Peshtigo Ct. B4
Pitney Ct. E2
Plymouth Ct. C,D3
Poe A2
Point A1
Polk C1-3
Poplar E2
Potomac B1,2
Prairie D,E3
Princeton E3
Prindiville A1
Quincy C1-3
Quinn E2,3
Race B1,2
Racine A-E2
Randolph Dr. C1-4
Rhodes E4
Rice B1,2
Richmond A-E1
Ridge Dr. A3
Ritchie Ct. B3
River E. C4
Robinson E2
Rockwell A-E1
Roosevelt Rd. D1-3
Roslyn Pl. A3
Ruble D3
Rundell Pl. C2
Rush B3
Sacramento A-E1
Sacramento Dr. D1
Sacramento Sq. B1
Sangamon B-E2
Schick Pl. B3
Schiller B1-3
Scott B3
Sedgwick B3
Seeley A-E2
Seminary A2
Seneca B3
Senour E2
Shakespeare A1-3
Sheffield A2
Shelby Ct. D2
Sherman C3
Shields E3
Short E2
Shubert A1-3
Siebens Pl. B3
Southport A2
St. Clair B3
St. Georges Ct. A1
St. Helen A1
St. James Pl. A3
St. Mary A1
St. Michaels Ct. A3
St. Paul A1-3
Stark E2
State B-E3
Stave A1
Stetson C3
Stewart D,E3
Stockton Dr. A3
Stone B3
Streeter Dr. B,C4
Sullivan B3
Surrey Ct. A2
Talman A-E1
Taylor C1-3
Terra Cotta Pl. A2
Thomas B2
Throop A-E2
Union B-E3
Van Buren C1-4
Vernon E4
Vernon Pk. Pl. C2,3
Vine A3
Wabansia A1,2
Wabash B-E3
Wacker Dr. C3
Wallace D,E3
Walnut C1-3
Walton B1-3
Warren Blvd. C1-3
Washburne D1,2
Washington Blvd. C1-3
Washtenaw A-E1
Water North C4
Water South C3
Wayman A1-3
Wayne A2
Webster A1-3
Weed B2,3
Wells A-E3
Wendell B3
Western A-E1
Wicker Park B2
Wieland B3
Wilcox C1
Willard Ct. B,C2
Willets Ct. A1
Willow A2
Wilmot A1
Wilton A2
Winchester A-E2
Winnebago A1
Wisconsin A2,3
Wolcott A-E2
Wolfram A2
Wood A-E2
Wrightwood A1-3
8th C3
9th C3
11th C2,3
11th Pl. D3
12th Pl. D1-3
13th D1-3
13th Pl. D1-3
14th D1-3
14th Blvd. D4
14th Pl. D1-3
15th D1-3
15th Pl. D1-3
16th D1-3
17th D1-3
17th Pl. D2,3
18th D1-3
18th Pl. D1-3
19th D1-3
19th Pl. D2,3
20th Pl. D2,3
21st D1-4
21st Pl. D1,2
22nd D3
22nd Pl. D1-3
23rd D1-4
23rd Pl. D1-3
24th D1-4
24th Blvd. E1
24th Pl. E1-3
25th E1-4
25th Pl. E1-3
26th E1-4
26th Pl. E3
27th E1-4
28th E1-3
28th Pl. E3
29th E1-4
29th Pl. E3,4
30th E1-4
30th Pl. E3
31st E1-4
31st Pl. E2-4
32nd E1-4
32nd Pl. E2-4
33rd E1-4
33rd Pl. E2-4
34th E1-3
34th Pl. E1-3
35th E1-4
35th Pl. E1-3
36th E1-4

CINCINNATI

Adela K1
Adler E1
Agnes A1
Alabama C2
Alaska Ave. B7
Alaska Ct. A7
Albion F6
Alfred E2
Alpine F8
Altamont J3
Alter A8
Altoona D8
Alvin F6
Amor F1
Ann H4,K8
Armory G4
Ash J2
Atkinson E5
Auburn F6
Auburncrest E6
Audrey F6
Augusta St. J5
Avon D2
Back G5
Bader D3
Bains G8
Baltimore D1
Bank F3
Barker A7
Barnard F3
Bates C2
Bathgate D7
Bauer G4
Baumer H7
Baymiller F3
Beech I8
Beecher D8
Beekman E1
Beekman St. C1
Beldare B5
Bella Vista F7
Bellevue E6,J3
Belsaw Pl. A4
Bethesda C6
Betts H4
Biddle B5
Bigelow F6
Bishop C5
Blackwell K5
Blaine E1
Blair C8
Blair Conn. C8
Bleeker D1
Boal G6
Bodmann F6
Bogart B8
Bond J3
Boone E7
Borden A1
Borrman D7
Bosley E5
Boston H1
Bowdle B3
Bowman D7,G1
Branch F4
Braxton C8
Brent Spence Bridge J4
Broadway I6
Brookline B5
Bryant B4
Buck E3
Budd I3
Buenavista D8
Burbank E7
Burgoyne A2
Burlington D2
Burnett D7,F6
Burns I1
Burton A8
Butler I7,J1
Byron F4
Calhoun E5
Calumet F5
Camden A7
Canyon A6
Carll D1
Carmalt F6
Carneal J2
Carney H7
Carplin C7
Catherine C7
Celestial H7
Central E3,F4,K7
Central Bridge J7
Central Pkwy. B2,H5
Chalfonte B8
Channing G6
Chappel D8
Charles H5
Charlotte G3
Charlton E6
Chestnut H5,K7
Chickasaw E4
Church K1
Clark H4,K3
Claypole G1
Clearwater F3
Cleveland C8
Cleveland Ave. C8
Cliff E5,G8
Clinton G4
Clifton B4
Clifton Crest B3
Clifton Hills B3
Clifton Hills Terr. B3
Clifton View C3
Cloister C5
Clybourn E3
C & O Bridge J5
Colfrain D3,F3
Columbia K8
Concord E7
Concordia D7
Conroy F3
Convention Way I5
Cook E3
Coon E4
Copelin E8
Cornell B4
Cormany C3
Corry E6
Cortlandt G3
Court H4,H5,J6
Craig K5
Crescent K4
Crescent, N. A8
Crescent, S. A8
Crestmont B5
Crooked Stone E3
Cross E8
Crown E7
Culvert H6
Cumber G6
Cumberland E6
Cummins D2
Curtis E8
Cutter H4
Cypress Garden A4
Dalton H3,I3
Dandridge G6
Daniels D6
Daves J1
David G4
Dawson A1
Dayton F4,J8
Deerfield F7
Delleva A5
Dellway C8
Dempsey D1
Denman G3
Dennis E5
Depot H1
Detzel E5
Deverhill K1
Devotie D3,D4
Dick B6
Digby D4
Dirr A1
Division K3
Dixmyth C3
Donahue D6
Dorsey G6
Dover E7
Draper E2
Dreman A1
Drexel C8
Duke D3
Duluth B5
Dunkirk A6
Dunlap F4
Dunore B3
Dury C6
Earnshaw F6
Eden B6,E6
Edgehill A8
Edinburg F6
Edmunds D7
Eggleston H6
Ehrman B6
Eighth K7
Eighth St. H5
Eighth St. Viaduct H2
Elam C2,D2
Eleanor F5
Eleventh K8
Elgin G7
Eliza E3
Elizabeth H4
Elland C6
Elland Cir. C6
Elliot H6
Elm J1,K1, K5,K7
Elmore Ct. A2
Elmore St. A1
Elsinore G7
Elysian F5
Emma B1
Emming F4
English St. I1
Erkenbrecker C6
Ernst F2
Esmond F1
Essex E7
Estelle F6
Ethan C2
Euclid C6,D6,J1
Eupeka F5
Eutaw A6
Evans H1
Evanswood B4
Evening Star C3
Eyre A3
Fairmont E1
Fairview E3
Fairview Pl. F3
Faulkner G3
Fifth J7,K5
Fifth St. H7,I3,I5
Filson H7
Findlay G3,G5
Fitzpatrick G1
Flora E4
Florence E8
Flint G3
Follett A1
Forest J1,K3
Forest Park B7
Foretaker D8
Fortune E4
Fort View H7
Fosdick D6
Foulke D3
Fourth J7,K5,K6
Fourth St. I5
Forest Ave. B6
Fourteenth H5
Fort Washington Way I6
Fox B1
Francisco F8
Frank G6
Fredonia D8
Freeman F3,J3
Freeman Ave. I1
Fricke B1
Front St. I7
Fuller H7
Fulton F8
Gage F5
Gano C4
Garrard D2,K6
George I5
Genessee H5
Gerard D6
Gest H1,I5
Gholson B7
Gilbert C8,H6
Glencoe F6
Glendora C5,D6
Gleneste A6
Glenmary B5
Glenridge A6
Glenwood J1
Glenwood Ave. A6,B7
Gilbert D8
Goeth G6
Goodman D6
Graham F5
Grand St. I5
Greendale A5
Greenhill A5
Greenup K6
Greenwood A8
Gregory G7
Guido H7
Hale C7
Hallwood C7
Halmar D3
Halstead E3
Hamer G5
Hammond H6
Harriet I3
Harrison F3
Hartford C8
Hartshorn E5
Hastings F4
Hatch H8
Hathaway J3
Hatmaker I1
Haven B6
Hazen K2
Hearne C6
Hebron A1
Hedgegrow B4
Helen E6,K1
Hemlock E7
Henry F5
Henshaw C2,D3
Herman E4
Heywood D3
Hickman C7
Hickory B7
High K3
Highland E6,G6
Hill H7,K3
Hillcrest J3
Hilton E6
Hollister E5
Hooper J1
Hopkins H1,H3
Hopple Ct. D1
Hopple St. Viaduct D1
Hosea C5
Howard K1
Howard Taft E6
Howell C3
Hughes G5
Hulbert G3
Hull D4
Huntington F6
Hutchins B8
Highway Ave. K3
Ida G7
Iowa E7
Imperial D5
Intervine B5
Interwood Ave. B5
Inwood E6
Iroquis E1
Irving B6
Isabella K7
Jackson H5
James E8
Jay C7
Jean F1
Jefferson C5,D5
Jerome H7
Jessamine D2
John G4,K3,K8
John St. I5
Johnson K5
Joselin D4
Josephine G6
Juergens A5
Juliaann C3
June E7
Justis F5
Kasota C7
Kemper E8,F8
Kenner G3,J1
Kenton E7
Kerper D8
Keturah K7
Kilgour H7
Kindel F3
Kinsey E6
Klotter F3,F4
Koebel F3
Kottman E3
Knob C1
Knott C7
Knox E2
Lafayette A3,A4
Lafayette Cir. A5
Lafayette Ln. A5
Lake K1
Lakewood D5
Lakes F8
Landon C7
Lane F1
Lange G5
Laredo E8
Larona B6
Latta J1
Lauk A5
Laurel K1
Lawson Ct. I6
Leblie C2
Lee B8
Lehman G1
Leroy F6
Lexington B8,I8,J3
Liberty G3,G4,K7
Liberty Hill G6
Library H5
Lidell E1
Lillie B1
Lincoln D8
Lincoln Park Dr. H4
Lincoln Pl. D8
Linden D1,J1,K1
Lindsey K7
Linton D7
Lionel F1
Livingston G3,J3
Llewellyn A1
Lloyd E3
Lock St. H7
Lockhurst G3
Locust J2
Logan G4
Lorraine B5
Lossing B6
Loude G7
Louis C6
Lowell C3,K7
L & N Bridge I7
Ludford K1
Ludlow J2
Ludlow Ave. B3
Luna E4
Luray F8
Lyleburn B4
Lyon E5
MacAuley E6
Madison K6
Malvern F6
Main G5,K5
Maiseldeckebach D3
Manchester F4
Mantiou F7
Mann B8
Manor Hill B3
Mansfield G6
Maple C7
Maplewood F6
Marmet C5
Marquis F7
Marshall D3
Martin G8
Maryland I1
Mason F5,K2
Massachusetts D3
Mathers D8
May St. E7
McAlpin B3
McAlpin Pl. B3
McCormick B5
McFarland I5
McGregor F7
McGregory F7
McLean F3
McMicken E3,F3
McMillan E4
Meadow A5
Meeker D2
Mehring J5
Mehring Way I3
Melbourne C8
Melish D6
Melrose E8
Memorial I1
Mentor D8
Mercer G5
Mill H4
Mill Creek Expwy. C3
Millvale Ct. B1
Milton C8,G6
Minnesota E7
Mistletoe G1
Moerlein E5
Mohawk Pl. F4
Mohawk St. F4
Montclair J2
Monfort D8
Monmouth C2,J8
Monroe F7,J8
Montrose J2
Moore G5
Moosewood C1
Morgan E7
Morrison B3
Mound St. H4
Mulberry F5
Muriel E4
Myrtle E8
Nassau F8
Neave I1
Nelson I8
Nevada H1
New St. H6
Neyer B1
Ninth K7
Ninth St. H5
Nixon C5
Northeast Expwy. I5
Northern B6
Northern Blvd. C1
Norway A7
Norwich Ln. A6
Oak D6,E8,E7,K1,J1
Odeon G5
Old Ludlow B3
Oliver G4
Omaha D7
Orchard K8
Oregon St. H7
Ormond C4
Osterfield F1
Overton J8
Oxford C5
Paradome G7
Parchman A6
Paris E6
Park E8,J8,K1
Parker F4
Parkside G7
Parson H8
Patchen I7
Patterson F3
Pavillion H7
Peete F5
Peggie I1
Pendleton H6
Perkins C8
Perr I5
Philadelphia K5
Piedmont D6
Pike K6
Pike St. I7
Pinetree E1
Pleasant G5
Plum H5
Polk E5
Poplar G3,J1
Post J2
Postal I6
Powers Ave. A1
Powers Pl. A1
Powers St. A2
Probasco D4
Produce St. J5
Prospect C7
Providence G4
Prior H5
Pulte D1
Purdue B6
Putnam K8
Queen E1
Race G5
Rachel D3
Ralston B1
Randolph B5
Rankin E1
Ravine F4
Rawson Woods Cir. A4
Rawson Woods Ln. B4
Reading E7,H6
Reedy H6
Renner F4
Renner St. F4
Renshaw E5
Republic G5
Resor Pl. B4
Revel A7
Rice F5
Richmond St. H4
Riddle D3
Riddle Crest C4
Riddleview D4
Ridge K3
Ridgeway C7
Ringgold G6,J1
River Rd. J2
Riverside I8,J6
Robert J8
Rochelle D6
Rockdale B6
Rogers E8
Rohs E6
Roll A1
Rose St. J4
Rural A5
Russell K5
Ruther C5
Ryan F7
St. Clair D5
St. Clair Ext. C4
St. James F8
St. Paul St. H7
Sanford K6
Saratoga G1,J8
Sassafras C2
Sauer E5
School D7
Scioto D5,E5
Scott K6
Second I8,J6,J7
Second St. J5
Seegar D1
Seitz F5
Selim F1
Seminole D6
Senator B5
Sentinel H6
Seventh K6,K7
Seventh St. I4
Severn A5
Shadewell Clara F1
Shields C6
Shelby G1
Sherlock C4
Sherman G3
Shiloh B4
Sidney C3,D3
Sinton F8
Sixth K6,K7
Sixth St. Expwy. I3
Slack G6
Sloo I3
Sohn F4
Somerset J1
South H1
Southern F6
South Gate J7
Southview E4
Spring G6,K3
Springs A7
Springhouse A7
Stokesay K1
Stadium Dr. J6
Stanton E8
Storrs I1
Staedler I1
Stark F4
State H1
Sterrett G1
Stetson D6
Stewart C7
Stonewall F4
Stratford D4,E4
Straight E3
Summer H1
Sunset K3
Suspension Bridge J6
Sutter D1
Swain J4
Sycamore H6
Sylvan A1
Symmes E7
Syracuse D8
Tafel E3
Tallant B6
Taylor C3
Telford B4
Tennis D7
Tenth K8
Terrace C3
Thill F5
Third I8,J7, K5,K6
Third St. I6
Thirteenth G6
Thrall B4
Tillotson E4
Tower B5
Townsend St. A1
Townsend Pl. A1
Township D3
Tremont E1
Trevor D1
Twelfth St. H6
Union D7
University D4,D5
Uphill K2
Valencia F5
Valeroy F6
Valley D2,E2
Valley View E4
Vanantwerp Pl. B7
Van Buren D7
Van Lear E5
Van Meter H7
Vaughn D6
Vernon D7
Vestry E4
Victor E4
Victoria J1
View J3
View Ct. J3
Vine A6,E6,G5
Vinecrest B5
Vinton E1
Volker E4
Wade G3,G4
Wagner D4
Walker G6
Wall A6
Wallace G1
Walnut G5
Walter D8
Wareham Dr. H7
Warner E4
Warren A4
Washington B7,J8
Waverly E1
Wayne E7
Weber A1
Webman B1
Wehrman D8
Wellington E6
Wentworth C5
Wesley St. H4
West B5
Western F3,K4
Western Hills Viaduct E2
Westwood E1
Whatley H1
Wheeler E4
Whitefield B4,C4
Whiteman F4
Whittaker H8
Whittier C7
Wilder H1
Wilkinson F7
Wm. Howard Taft E8
Willow D2
Wilstach G3
Winchell F3,G3
Windham C8
Windsor F8
Winkler F5
Wirham C3
Wood Ave. B4
Woodrow H1
Woodside D5
Woodward G6
Woolper B5
Wright K4
Yale E8
York F3,G3,K8
Young G6

POINTS OF INTEREST

Burnet Woods C5
Coy Field & Playground D4
Devon Park K2
Dury Ave. Play Area B6
Eden Park G7
Fleischmann Gardens B7
Goeber Park K5
Haven Tot Lot B6
Millvale Play Area B1
Mt. Storm Park A3
Rawson Woods B4
Shirley Bonne Camp A6
St. Clair Heights Park E1
University of Cincinnati D5
Windham Park C8
Zoological Gardens B6

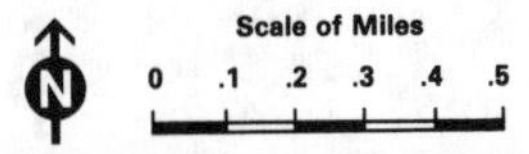
N
Scale of Miles
0 .1 .2 .3 .4 .5

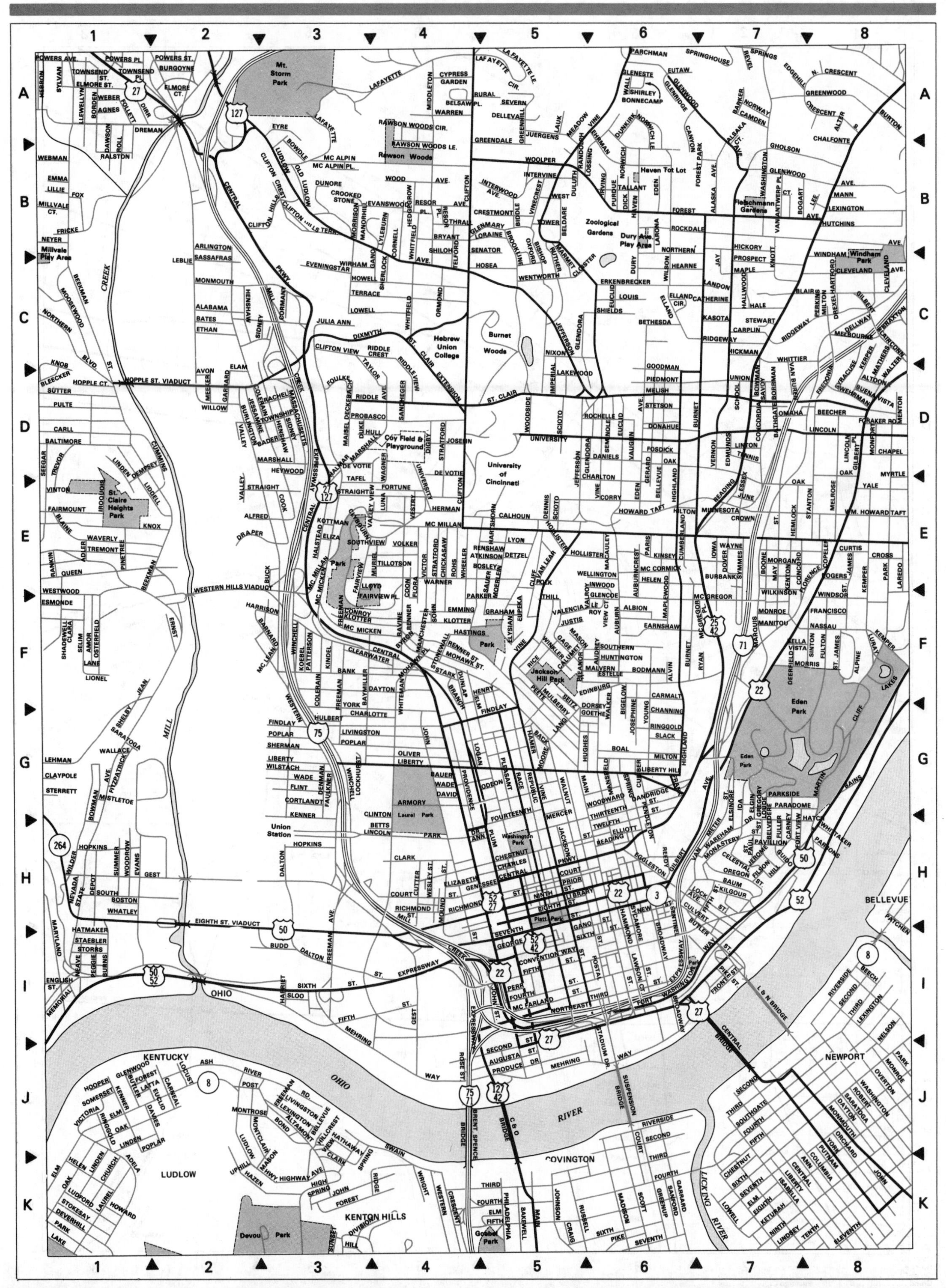
1 2 3 4 5 6 7 8
A B C D E F G H I J K
Mt. Storm Park
Rawson Woods
Haven Tot Lot
Zoological Gardens
Dury Ave. Play Area
Fleischmann Gardens
Windham Park
Millvale Play Area
Hebrew Union College
Burnet Woods
Coy Field & Playground
University of Cincinnati
St. Claire Heights Park
Park
Hastings Park
Jackson Hill Park
Eden Park
Armory
Laurel Park
Union Station
Washington Park
Piatt Park
HOPPLE ST. VIADUCT
WESTERN HILLS VIADUCT
EIGHTH ST. VIADUCT
KENTUCKY
OHIO
RIVER
LUDLOW
COVINGTON
NEWPORT
BELLEVUE
KENTON HILLS
Devou Park
Goebel Park
BRENT SPENCE BRIDGE
C & O BRIDGE
SUSPENSION BRIDGE
CENTRAL BRIDGE
L & N BRIDGE
LICKING RIVER
MILL CREEK

Scale of Miles
0 .1 .2 .3 .4

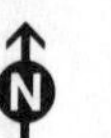

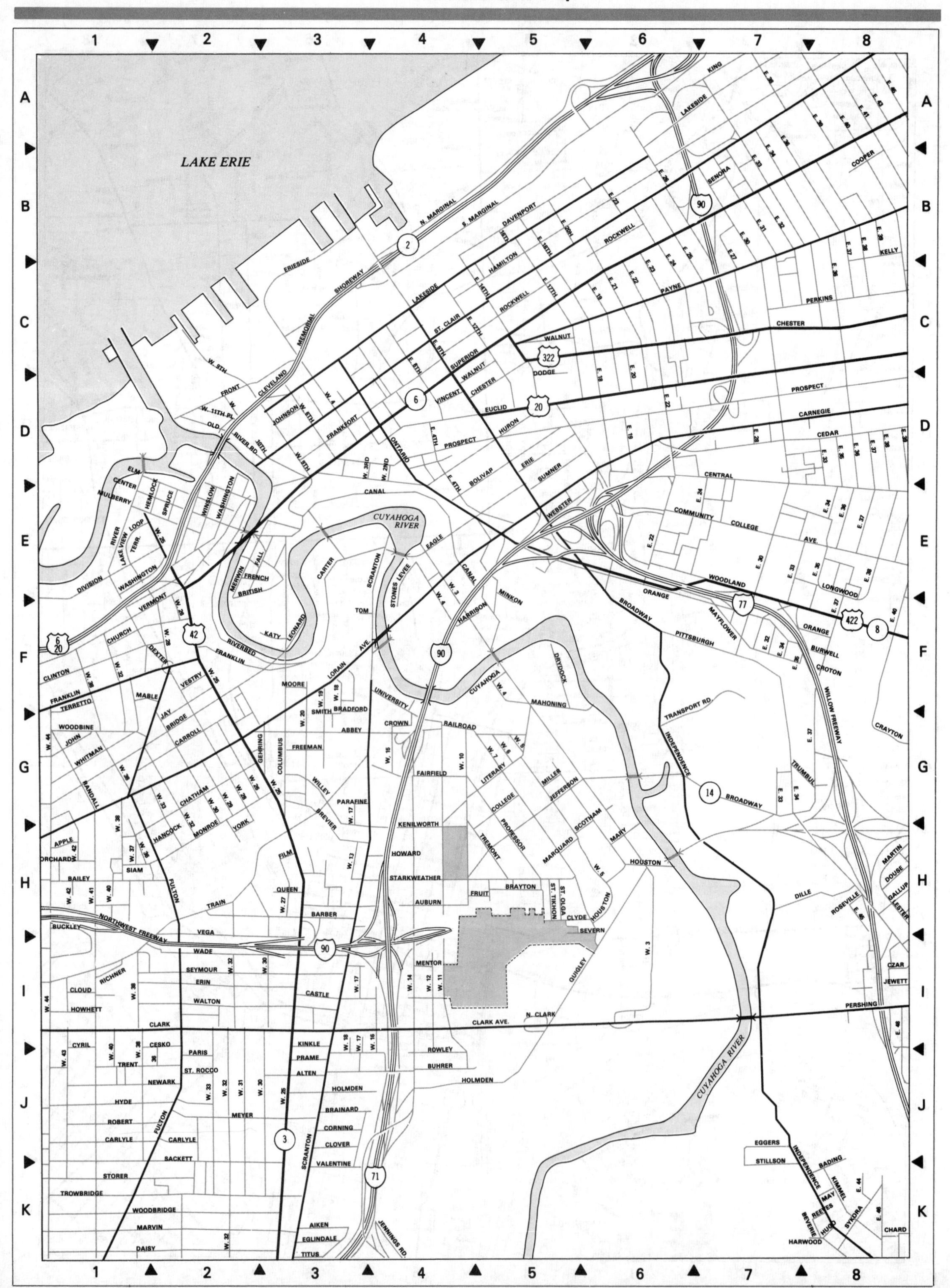

CLEVELAND

Abbey ... G-3
Aiken ... K-3
Alten ... J-3
Apple ... H-1
Auburn ... H-4
Bading ... K-8
Bailey ... H-1
Barber ... H-3
Beverie ... K-8
Bolivar ... D-5
Bradford ... F-3
Brainard ... J-3
Brayton ... H-5
Brevier ... G-3
Bridge ... G-2
British ... E-2
Broadway ... F-6,G-7
Buckley ... H-1
Buhrer ... J-4
Burwell ... F-8
Canal ... E-4
Carlyle ... J-1,J-2
Carnegie ... D-8
Carroll ... G-2
Carter ... E-3
Castle ... I-3
Cedar ... D-8
Center ... E-1
Central ... D-7
Cesko ... I-2
Chard ... K-8
Chatham ... G-2
Chester ... C-7,D-5
Church ... F-1
Clark ... I-2
Clark, N. ... I-5
Clark Ave. ... I-5
Cleveland Memorial Shoreway ... C-3,D-3
Clinton ... F-1
Cloud ... I-1
Clover ... J-3
Clyde ... H-5
College ... G-5
Columbus ... G-3
Community College Ave. ... E-7,E-8
Cooper ... B-8
Corning ... J-3
Crayton ... G-8
Croton ... F-8
Crown ... G-4
Cuyahoga ... F-5
Cyril ... I-1
Czar ... I-8
Daisy ... K-1
Davenport ... B-5
Dexter ... F-2
Dille ... H-7,H-8
Division ... E-1
Dodge ... C-5
Douse ... H-8
Drylock ... F-5
Eagle ... E-4
Eggers ... J-7
Eglindale ... K-3
Elm ... D-1
Erie ... D-5
Erieside ... B-3,C-3
Erin ... I-2
Euclid ... D-5
Fairfield ... G-4
Fall ... E-3
Film ... H-3
Frankfort ... D-3
Franklin ... F-1,F-2
Freeman ... G-3
French ... E-2
Front ... D-2
Fruit ... H-5
Fulton ... H-2,J-1
Gallup ... H-8
Gehring ... G-3
Hamilton ... B-5
Hancock ... H-2
Harrison ... F-5
Harwood ... K-8
Hemlock ... E-2
Holmden ... J-3
Houston ... H-6
Howard ... H-4
Howhett ... I-1
Hugo ... K-8
Huron ... D-5
Hyde ... J-1
Independence ... G-6,K-8
Jefferson ... G-5
Jennings Rd. ... K-4
Jewett ... I-8
John ... G-1
Johnson ... D-3
Katy ... F-3
Kelly ... B-8
Kenilworth ... H-4
Kimmel ... K-8
King ... A-7
Kinkle ... I-3
Lakeside ... A-6,A-7,C-4
Lakeview Loop Terr. ... E-1
Leonard ... F-3
Literary ... G-5
Longwood ... E-8
Mable ... F-2
Mahoning ... F-5
Marginal, N. ... B-4
Marginal, S. ... B-5
Marquard ... H-5
Martin ... H-8
Marvin ... K-1
Mary ... H-6
May ... K-8
Mayflower ... F-7
Mentor ... I-4
Merwin ... E-2
Meyer ... J-2
Miller ... G-5
Minkon ... E-5
Monroe ... H-2
Moore ... F-3
Mulberry ... E-1
Newark ... J-2
Northwest Frwy. ... H-1
Old River Rd. ... D-2
Ontario ... D-4
Orange ... E-6,F-8
Orchard ... H-1
Parafine ... G-3
Paris ... J-2
Payne ... C-6
Perkins ... C-8
Pershing ... I-8
Pittsburgh ... F-6
Prame ... J-3
Professor ... H-5
Prospect ... D-4,D-8
Quigley ... I-5
Railroad ... G-4
Randall ... G-1
Reeves ... K-8
Robert ... J-1
Rockwell ... B-6,C-5
Roseville ... H-8
Rowley ... J-4
Richner ... I-1
River ... E-1
Riverbed ... F-2
Sackett ... J-2
St. Clair ... C-4
St. Olga ... H-5
St. Rocco ... J-2
St. Tikhon ... H-5
Scotham ... H-6
Scranton ... E-4,K-3
Senora ... B-7
Severn ... H-6
Seymour ... I-2
Siam ... H-1
Smith ... F-3
Spruce ... E-2
Starkweather ... H-4
Stillson ... K-7
Stones Levee ... E-4
Storer ... K-1
Summer ... D-5
Superior ... C-4
Sykora ... K-8
Terretto ... F-1
Titus ... K-3
Tom ... F-3
Train ... H-2
Transport Rd. ... F-6,F-7
Tremont ... H-5
Trent ... J-1
Trowbridge ... K-1
Trumbul ... G-7,G-8
University ... F-4
Valentine ... K-3
Vega ... H-2
Vermont ... F-1
Vestry ... F-2
Vincent ... D-4
York ... H-2
Wade ... I-2
Walnut ... C-4,C-5
Walton ... I-2
Washington ... E-1,E-2
Webster ... E-5
Whitman ... G-1
Willey ... G-3
Willow Freeway ... G-8
Winslow ... E-2
Woodbine ... G-1
Woodland ... E-7
Woodbridge ... K-1
2nd West ... D-4
3 West ... E-4,I-6
3rd West ... D-4
4th East ... D-4,E-5
4 West ... E-4
4th West ... F-3
5 West ... G-5,H-6
6th East ... C-4
6th West ... D-3,G-5
7 West ... G-5
9th East ... C-4
9th West ... C-2,D-3
10th West ... D-2,G-4
11 West ... I-4
11th Pl. West ... D-2
12th East ... C-5
12 West ... I-4
14th East ... C-5
14 West ... I-4
15 West ... G-4
16th ... B-5
16 West ... I-4
17th East ... C-5
17 West ... G-3,I-3
18th East ... B-5,C-6
18 West ... F-3,H-3,I-3
19 East ... C-6,D-6
19 West ... F-5
20 East ... C-6
20th East ... B-5
20 West ... G-3
21 East ... C-6
22 East ... C-6,D-6,E-6
23 East ... B-6,C-6
24 East ... E-7
24th East ... C-6
25 East ... B-6
25 West ... E-2,F-2,G-3,J-3
26 East ... B-6
26 West ... G-2
27 East ... B-7
27 West ... H-3
28 East ... D-7
28 West ... F-2,G-2
29 West ... G-2
30 East ... B-7,E-7
30 West ... G-2,I-3,J-3
31 East ... B-7
31 West ... J-2
32 East ... B-7,F-7
32 West ... F-1,G-2,I-2,J-2,K-2
33 East ... B-7,D-8,E-7,G-7
33 West ... G-2,J-2
34 East ... B-7,E-8,F-7,G-7
35 East ... D-8,E-8,F-7
36 East ... A-7,C-8,D-8,E-8
36 West ... H-1,J-2
37 East ... B-8,D-8,E-8,F-8,G-7
37 West ... H-1
38 East ... A-7,B-8,D-8,E-8
38 West ... F-1,H-1,I-1,G-1,J-1
39 East ... A-8,B-8,D-7
40 East ... A-8,F-8
40 West ... H-1,J-1
41 East ... A-8
41 West ... H-1
42 West ... H-1
43 East ... A-8
43 West ... J-1
44 East ... K-8
44 West ... G-1,I-1
45 East ... A-8,H-8
46 East ... K-8
48 East ... I-8

COLUMBUS

Aberdeen Ave. ... B-5,B-6
Abner ... B-5
Ackerman Rd. ... C-2
Action Rd. ... A-3
Adams Ave. ... C-3
Advance ... J-7
Afton ... B-1
Agawam Cir. ... A-7
Agler ... B-8
Agler Rd. ... B-6
Akola Ave. ... B-4
Alamo Ave. ... B-4
Albert Ave. ... A-6
Alcott ... K-5
Alden ... C-4
Allegheny ... E-8
Allen ... G-5
Alum ... H-8
Alum Creek Dr. ... I-8
Alvason ... C-6
Amazon ... A-2
Amberly ... A-1
Amherst Ave. ... I-1
Arcadia ... C-3
Arden Rd. ... A-3
Ardmore Rd. ... F-8
Argyle ... D-7
Arlington ... B-4
Arlington Ave. ... B-5
Armstrong ... F-4
Ashbourne ... F-8
Ashdowne Rd. ... D-1
Ashland Ave. ... E-1
Astor ... G-8
Atcheson St. ... F-5
Atwood ... A-5,C-5
Atwood Ter. ... B-5
Auburn ... F-6
Audrey ... A-6
Audubon Rd. ... B-4
Avalon ... E-7,E-8
Avondale ... F-1,G-3
Avon Lea ... A-8
Azelda Ave. ... B-4
Baitz ... J-5
Bancroft St. ... C-6
Bar Harbor ... C-7
Bartha ... A-8
Barthman Ave. ... J-5
Bassett ... E-6
Baughman Ave. ... B-7
Beaumont Rd. ... D-1
Beautyview Ct. ... A-1
Beck St. ... H-5
Becker ... J-5
Bedford ... G-6
Belle St. ... G-4
Bellevue ... J-5
Bellevue Ave. ... J-8
Bellows Ave. ... H-2
Bellwood ... E-8
Belmont Ave. ... D-3
Belvidere ... I-1
Benfield Ave. ... K-5
Bernard ... B-1
Berrel ... C-6
Berrel Ave. ... B-6
Bethesda ... C-7
Beulah Rd. ... A-4
Bevis Rd. ... B-2
Bexford ... F-8
Bexley Park Rd. ... G-8
Birchmont ... A-1
Blake Ave. ... C-3,C-6
Blenkner ... H-4
Bluff Ave. ... F-1
Bonham ... D-4,D-5
Boston ... F-8
Boyleston ... A-7
Bradford ... B-1
Brehl Ave. ... H-2
Bremen ... A-5
Bremen St. ... A-5
Brentnell ... C-7
Brentnell Ave. ... D-7
Brentnell Blvd. ... B-7
Brentford Dr. ... E-8
Brentwood Rd. ... G-8
Bretton ... D-6
Brevoort ... A-3
Briarwood Ave. ... B-5
Brickel ... F-4
Bricker ... B-1
Bridgeview Dr. ... A-7
Bridgewalk ... A-7
Brighton ... B-2,B-4
Broad St. ... G-5
Broadbelt ... F-3
Brockton ... C-7
Brookcliff ... D-7
Brooksgerald ... I-2
Brown Rd. ... I-2
Brown Leaf Rd. ... J-2
Bruck ... H-5,J-5
Brunson ... F-7
Bryden Rd. ... G-5,G-8
Buckeye Park Rd. ... J-6
Buckingham ... F-4
Bulen Ave. ... K-7
Burley Dr. ... J-7
Burnbrae ... C-2
Burr ... F-2
Burrell ... F-2
Butler ... I-1
Buttles Ave. ... F-3
Cable ... G-2
Caldwell Ct. ... F-5
California Ave. ... B-3
Camaro ... I-8
Camden Ave. ... E-5
Campbell Ave. ... H-2
Caniff ... B-1
Capital St. ... G-5
Caralee Dr. ... B-8
Cardiff Rd. ... D-1
Carlisle Ave. ... A-4
Caroline ... A-5,E-8
Cassady ... A-8,F-8
Cassady Ave. ... K-8
Cassady Pl. ... B-8
Cassingham Rd. ... E-8,G-8
Castle ... K-5
Catherine ... H-2
Central ... D-2
Century Dr. ... B-7
Chambers Ave. ... E-4
Champion Ave. ... H-6
Charles ... G-8
Chauncy ... D-6
Chelford Dr. ... A-8
Chesapeake Ave. ... E-1
Chester ... E-7
Chestnut St. ... F-4
Cherry St. ... G-6
Chicago ... G-2
Chilcote ... C-4
Chittenden ... D-5
Chittenden Ave. ... D-4
Civic Center Dr. ... G-4
Claredon ... I-1
Clarendon Ave. ... H-1
Clark ... E-3,I-1
Clearview ... B-1
Cleveland ... A-6
Cleveland Ave. ... D-5,F-5
Clifton Ave. ... F-6,F-7
Clinton ... C-4,C-5
Clinton Heights Ave. ... B-3
Cochrane ... J-5
Cole St. ... G-6
Colerain Ave. ... A-3
College Ave. ... G-8
College Hill Dr. ... C-1
Collins Ave. ... F-3
Columbia Ave. ... E-8
Columbian ... I-1
Columbus ... E-8,H-4
Columbus St. ... H-7
Commonwealth Park N. ... F-7
Commonwealth Park S. ... F-7
Como Ave. ... B-4
Como St. ... B-2
Coolidge ... H-3
Copeland ... F-2
Cordell ... C-6
Corr Rd. ... K-6
Corrugated ... E-4
Corvair ... I-8
Corwin ... E-6
Courtland ... E-4
Crestview Rd. ... B-3
Criswell Dr. ... A-1
Cullman ... K-5
Cumberland ... E-7
Curtis Ave. ... F-5
Cypress ... G-2
Dakota ... G-2
Dale ... F-8
Dale Ave. ... F-8
Dana Ave. ... H-2
Dartmouth ... F-6
Daugherty ... E-7
Davis Ave. ... G-3
Dawnlight Ave. ... B-7
Dawson ... E-8,F-8
Dayton ... C-3
Deckebach Rd. ... I-4
Delamere ... B-1
Deibert ... C-6,D-6
Delhi Ave. ... B-2
Delmar ... E-8
Delno ... B-3
Delray Rd. ... K-5
Delta, N. ... A-2
Deming Ave. ... C-4
Dennison ... F-4
Denune Ave. ... B-7
Denver ... F-8
Deporres ... D-7
Deshler Ave. ... I-5,I-7
Devonsh ... C-7
Dewberry Rd. ... K-7
Dewey ... D-6
Dick Ave. ... E-5
Dillward ... A-3
Dodridge St. ... C-2
Doone Dr. ... D-1
Doren ... H-1
Doris Ave. ... B-2
Doten ... E-2
Douglas ... G-8
Dover ... E-2
Dow ... E-4
Drake ... C-8
Dresden ... A-5
Dresden St. ... A-5
Drexel ... E-8
Drexel Ave. ... E-8
Drummond Island ... A-1
Dublin Ave. ... G-3
Dublin Rd. ... F-2
Dunbar Rd. ... A-6
Duncan ... C-3
Dunedin ... A-4,A-6
Dunedin Rd. ... A-3
Dunham ... I-1
Dunning ... C-7
DuPont Ave. ... E-5
Durham Dr. ... H-1
Duxberry Ave. ... C-4
Eagle Ave. ... K-6
Eakin Rd. ... I-1
Earl Ave. ... B-8
Earncliff ... A-8
East Ave. ... B-3,C-3
Eastfield Dr. N. ... J-1
Eastview ... E-2
Eastwood ... F-8
Eaton ... I-4
Eddyston ... A-4
Eddystone Ave. ... A-6
Edenburch ... C-7
Edgar ... B-4
Edgefield ... C-1
Edgehill Rd. ... E-2
Edsel ... K-7
Edwards St. ... F-5
Eisenhower Rd. ... A-4
Elda St. ... F-6
Eldorn ... I-7
Elm ... B-1
Elm St. ... G-4
Elmwood Ave. ... E-1
Elsmere ... H-5
Elwood ... J-6
Emerald ... F-7
Emerson ... H-1
Enderly ... C-6
Engadine Ave. ... C-6
Engler St. ... G-5
Erie ... A-2
Essex ... E-5
Essex Rd. ... D-1
Euclaire Ave. ... G-8
Euclid ... E-4
Eulalia ... A-5
Eureka ... J-1
Evergreen Rd. ... K-7
Faber Ave. ... K-7
Faculty ... B-1
Fair ... G-6
Fair Ave. ... G-8
Fairbank Rd. ... K-7
Fairfield ... E-6
Fairmont Ave. ... H-1
Fairview Ave. ... F-1
Fairwood ... G-7
Fairwood Ave. ... H-7
Fairwood Rd. ... K-7
Fallis Rd. ... A-3
Feddern ... K-2
Fern Ave. ... B-6
Ferndale ... H-8
Fields ... E-4
Filmore ... I-4
Findley Ave. ... C-3
Finland Ave. ... K-3
Floral Ave. ... H-1
Forest ... J-2
Forest St. ... H-5,H-7
Fornoff Rd. ... K-5
Fradena Ave. ... C-4
Frambes ... D-3
Framington Dr. ... A-7
Francis Ave. ... G-8
Frank Rd. ... K-3
Frankfort St. ... H-7
Franklin ... J-1
Franklin Ave. ... G-6
Franklin Rd. ... H-8
Franklin Park S. ... G-7
Franklin Park W. ... F-7
Franllin ... G-5
Frebis Ave. ... I-5
Frey Rd. ... A-8
Friar ... B-1
Front St. ... G-4
Fulton St. ... G-6,H-5
Gantz Rd. ... K-1
Gardendale ... D-7
Garfield Ave. ... F-5
Garling Ave. ... I-2
Garrett ... A-1
Gate Rd. ... B-1
Gates St. ... I-5
Gault St. ... H-7
Gay ... G-3
Gay St. ... F-6,G-4
Geers ... H-7
Genessee Ave. ... B-5,B-6
Geneva ... I-1
Gerbert ... A-5,C-5
Gerbert Rd. ... A-5
Gerrard ... E-2
Gibbard ... E-5
Gibbard Ave. ... E-6
Gift St. ... G-3
Gilbert St. ... H-6
Gladden ... F-2
Glencoe Rd. ... A-3
Glendower ... K-6
Glendower Ave. ... K-6
Glen Echo ... C-4
Glenmawr ... B-4
Glenn Ave. ... E-1
Glenwood ... H-2
Goodale Blvd. ... F-1,F-2
Governor ... F-6
Grafton ... B-1
Graham ... F-6
Granden ... A-3
Grant ... E-5
Grant Ave. ... G-5
Granville ... F-6
Grasmere ... C-5
Gray ... E-5
Greenlawn Ave. ... I-3
Greenleaf Rd. ... J-2
Greenway ... F-7
Greenwich ... A-6
Greenwich St. ... B-6
Greenwood ... E-3
Grenoble Rd. ... D-1
Griggs Ave. ... H-3
Grogan ... D-5
Grove ... F-5
Groveport Rd. ... K-6
Grovewood ... K-7
Grubb St. ... G-3
Guilford Rd. ... D-1
Halstead ... C-1
Hamilton ... F-5
Hamlet St. ... E-4
Hanford St. ... I-5,I-7
Hanna Dr. ... B-7
Hanover ... G-3
Hardy Pkwy. ... K-2
Harley Dr. ... B-2
Harmon Ave. ... J-4
Harrison ... F-3
Hart Rd. ... J-2
Hartford ... G-3
Harvard ... F-6
Haskell ... A-8
Havendale ... B-3
Hawkes ... G-3
Hawthorne ... F-6
Hayden ... G-2
Heiman ... K-3
Helen ... H-2
Helena Ave. ... D-6
Hendrix Dr. ... K-2
Hennepin ... A-3
Hess ... E-2
Heyl Ave. ... H-6
Hiawatha ... A-4,C-4
Hickory St. ... G-4
Higgs ... F-2
High St. ... B-3,G-4
Highland ... A-1,E-3,H-1
Hildreth Ave. ... F-6
Hilltonia ... I-1
Hingham Ln. ... A-7
Hinkle ... J-5
Hinman Ave. ... J-5
Holburn ... I-7
Holloway ... I-1
Hollywood ... E-1
Holt Ave. ... D-7
Holton Ave. ... H-2
Holtzman ... G-7
Homecroft ... A-4
Homestead ... A-5
Homewood ... I-1
Hope ... F-2
Hopkins ... J-3
Hopkins Ave. ... J-2
Hosack ... J-5
Hoster ... H-4
Hosack ... J-5
Howard St. ... E-5
Howey ... C-5
Hubbard ... F-3
Hudson ... C-3,C-5
Humphrey Ave. ... I-1
Hunt Ave. ... I-1
Hunter ... D-3
Hunter Ave. ... E-3
Huntington Rd. ... K-8
Ida Ave. ... E-1
Independence ... E-1
Indianola ... E-4
Indianola Ave. ... D-4
Ingleside ... F-3
Innis Ave. ... J-7
Innis Pl. ... J-5
Innis Rd. ... A-6
Integrity Dr. N. ... I-8
Integrity Dr. S. ... I-8
Ipswick Cir. ... A-7
Isabell ... D-5
Iuka ... D-4
Jackson Rd. ... I-2
Jade ... K-1
Jaeger St. ... H-5
Jane ... D-7
Jefferson ... C-5,F-5
Jefferson Pl. ... F-5
Jenkins Ave. ... I-5
Jeri ... C-7
Jermain ... B-7
John Ave. ... G-2
Jonathan ... J-7
Joyce Ave. ... B-6,D-6
Kail ... K-5
Karon ... D-7
Kelenard ... D-8
Kelso Rd. ... B-3
Kelton ... H-7,I-7
Kemper ... D-6
Kendall ... G-6
Kenlawn ... A-6
Kenley ... A-1
Kenmore Rd. ... C-6
Kennedy ... H-5
Kenny ... E-2
Kenny Rd. ... C-1
Kenraydor ... A-8
Kent St. ... H-7
Kenton ... G-8
Kenwood ... A-1
Kenworth Rd. ... A-4
Kermit ... I-7
Kessler ... E-5
Kettering Rd. ... B-2
Keywest ... D-6
Kian Ave. ... J-5
Kilbourne ... F-4
Kimball ... H-6
Kimberley Dr. ... A-4
King Ave. ... E-1
Kingston ... J-5
Kinnear ... C-3
Kinnear Rd. ... D-1
Kitchner ... J-7
Koebel Ave. ... K-6,K-8
Koebel Rd. ... K-7
Kohr Pl. ... C-6
Korbel ... D-4
Kossuth ... H-4
Kossuth St. ... H-5,H-7
Kramer ... F-2
Kutchins Pl. ... G-6
Lafayette St. ... G-4
Lakeview Ave. ... B-2
Lambeth ... B-1
Lancashire ... C-7
Lane Ave. ... D-1,D-3
lansing St. ... H-5
Lansmere ... A-1
Larcomb ... A-1
Latham Ct. ... A-1
Laurel ... I-2
Lawn ... J-5
Lawndale ... J-6
Lawndale Ave. ... J-8
Lawrence Dr. ... J-6
Lazelle St. ... G-4
Lear St. ... H-5
Lechner ... H-1
Lee ... D-6
Lenore ... A-5
Leon ... B-8
Leona Ave. ... E-5
Leonard Ave. ... E-6,F-5
Lexington ... D-5
Lilley ... I-7
Lincoln St. ... F-4
Linden ... C-6
Linnet ... I-2
Linview ... B-6
Linwood ... J-6
Linwood Ave. ... H-6
Liscomb ... I-7
Liston Ave. ... J-8
Little Ave. ... J-2
Livingston Ave. ... H-5,H-7
Lockbourne Rd. ... H-6
Locust St. ... G-4
Loew ... E-5
Long St. ... G-4
Longview ... B-3,B-4
Longwood Ave. ... K-2
Loretta ... C-6
Loretta Ave. ... C-5
Louis ... D-6
Loxley Dr. ... K-5
Lucas St. ... G-3
Lynn St. ... G-4
Mabel ... C-3
Madison Ave. ... G-6
Main St. ... G-5,G-7
Maize ... A-4
Manchester Ave. ... B-5
Manning ... G-7
Maple St. ... F-3
Marburn Dr. ... A-1
Marcia Dr. ... B-7
Marconi ... G-4
Margaret ... E-7,E-8
Margaret Pl. ... E-7
Marina-Toni ... C-7
Marion Rd. ... J-6
Markison Ave. ... I-6,J-5
Marsdale ... J-1,K-2
Marston ... C-7
Martin ... G-3
Martha ... B-8
Maryland St. ... E-8
May ... G-3
Mayfield ... H-8
Maynard ... C-3,C-5
McAllister ...
McClain ... F-2
McClelland ... D-5
McCoy ... F-5
McDowell ... G-3
McGuffy ... C-5
McKinley ... G-2
McKinley Ave. ... G-1
McMillen ... E-2,E-3
Meade ... I-1
Meadow ... E-2
Meadowdale ... D-7
Mecca Rd. ... A-6
Medary Ave. ... C-3
Medbrook ... A-2
Medhurst ... A-1
Medina ... A-5
Medinah Ave. ... A-5
Melrose ... B-4,B-5
Melrose Ave. ... B-7
Memory Ln. ... H-8
Menlo Pl. ... F-6
Mennesota Ave. ... B-5,B-6
Meredith ... D-7
Merrimac ... G-2
Merryhill ... F-7
Miami ... F-6
Michigan ... F-3
Middlehurst ... C-7
Midgard ... B-4
Midland ... I-1,J-1
Midland Ave. ... H-1
Midway ... K-7
Milford Ave. ... B-4,B-5,B-7
Millcreek ... A-1
Miller Ave. ... G-7
Milton Ave. ... B-2
Miner ... I-3
Mithoff St. ... I-5
Mohawk ... H-5
Moler St. ... I-5,I-7
Monroe Ave. ... F-5,G-6
Montrose ... A-2
Montrose Ave. ... G-8
Mooberry ... G-7
Moon Rd. ... A-4
Morning ... E-1
Morrill Ave. ... J-5
Morrison ... G-7
Mound ... H-3
Mound St. ... G-5,G-6
Mt. Pleasant ... E-4
Mt. Vernon Ave. ... F-5
Mura ... B-4
Murray Ave. ... E-7
Myrtle Ave. ... B-5,B-6
Nace Ave. ... H-2
Naghten ... F-5
Nashoba ... I-1
Nason ... I-7
Neff Ave. ... K-5
Neil ... C-3
Neil Ave. ... B-3
Neilston ... F-4
Nelson Rd. ... G-7
Newton Ave. ... H-6
Niantic Dr. ... A-1
North St. ... C-3
North Broadway ... A-3
Northglen ... A-7
Northmoor ... A-2
North Star Ave. ... C-1
Northview ... D-8
Northwest Blvd. ... E-2
Northwood Ave. ... C-3,C-4
Norton Ave. ... E-2
Norway ... B-1
Norwich ... C-3,C-4
Norwood ... A-5
Norwood St. ... A-5
Nuway ... K-7
Oak St. ... G-5,G-6
Oakland Ave. ... C-3,F-1
Oakland Park Ave. ... A-3,A-5
Oaklawn ... A-6
Oaklawn St. ... A-6
Oakwood Ave. ... H-6
Ohio Ave. ... H-6
Ohlen ... D-4
Olentangy ... A-2
Olentangy St. ... B-3
Oletangy River Rd. ... B-2
Olmstead Ave. ... E-5
Olpp ... J-5
Omar Dr. ... J-7
Ontario ... A-5
Ontario St. ... A-5
Orchard Ln. ... A-2
Oregon Ave. ... E-3
Oriole ... D-7
Osborn Dr. ... D-1
Osceola ... A-4
Osceola Ave. ... B-4
Overlook Dr. ... A-1
Oxford ... D-5
Oxley ... E-2,F-2
Pacemount Rd. ... B-2
Palmer Ave. ... F-2
Pamella ... K-7
Pannell ... I-7
Park Dr. ... F-7
Park Leigh ... A-1
Parkview Ave. ... E-8
Parkway N. ... E-2
Parkwood ... C-6,D-6
Patterson Ave. ... C-3
Pauline ... A-5
Pauline Ave. ... A-5
Payne ... G-8
Peabody ... J-1
Pearl ... E-4
Pearl St. ... G-4
Peasley ... D-3
Pegg ... B-1
Pembrooke ... F-7
Pennsylvania Ave. ... E-3
Penny ... E-4
Perry Ave. ... E-3
Peters ... E-5
Pickwick ... C-1
Piedmont ... A-4,A-6
Piedmont Rd. ... A-3
Pierce Dr. ... H-3
Pleasant Ridge ... G-8
Pontiac ... C-4
Pontiac Ave. ... B-4
Poplar ... F-8
Powell ... G-3
Presidential ... E-1
Preston Rd. ... F-8
Price ... E-4
Princeton ... G-2
Princeton Ave. ... H-2
Progress ... I-8
Purdue Ave. ... B-7
Puritan ... D-6
Quay ... F-2
Rainbow Park ... H-7
Rankin ... C-7
Ransburg Ave. ... I-2
Raynor ... A-8
Rea Ave. ... I-1
Red Rock Blvd. ... K-1
Reeb ... A-8
Reeb Ave. ... J-5,J-6
Refugee Rd. ... J-7
Reinhard Ave. ... H-5
Remington ... F-8
Renick St. ... H-3
Republic ... C-6
Republic Ave. ... C-5
Reynolds ... E-5
Rhoads Ave. ... H-7,K-7
Rhoda ... E-2
Rich ... H-2
Rich St. ... G-5-7, H-3
Richards Rd. ... A-3
Richmond ... F-7
Richmond Rd. ... J-2
Richter Rd. ... J-2
Ridge ... C-8
Ridge St. ... F-1
Rightmire Blvd. ... B-1
Riverside Dr. ... B-2
Riverview Dr. ... B-2
Robert ... B-6
Roberts ... I-7
Rodgers ... G-2
Rose ... E-7
Rosedale Ave. ... J-1
Rosemont ... I-1
Rosewood ... D-7
Ross ... E-7
Rudwick ... A-8
Ruhl ... E-8
Ruhl Ave. ... E-8
Russell ... F-4
Ryan ... H-2
Ryder ... F-2
Safford Ave. ... I-1
Sagamore ... C-7
St. Clair ... D-5,F-5
St. Clair Ave. ... F-5
Sampson ... D-6
Sandlin ... B-5
Saugus Cir. ... A-7
Say Ave. ... E-4
Schenley ... C-7
Schultz ... G-2
Scioto ... I-4
Scott Rd. ... I-2
Scott St. ... G-3
Sells ... E-2
Seneca Park Pl. ... F-8
Seymour Ave. ... H-7
Shady Ln. ... J-2
Shady Hill ... C-1
Shadywood ... C-1
Sharon ... A-3
Shattuck ... A-1
Shattuck Ave. ... B-1
Sheldon ... I-6
Sheldon Ave. ... I-5
Sherbourne ... F-7
Sheridan Ave. ... G-8
Sherwood Rd. ... G-8
Shoemaker Ave. ... E-5
Short St. ... H-4
Sidney ... E-5
Siebert St. ... H-5
Sigsbee ... D-6
Silver Dr. ... A-4
Skidmore St. ... G-3
Smith ... E-1
Smith Rd. ... I-6,I-7
Somersworth ... D-7
Souder Ave. ... G-3
South Ave. ... E-2
South Ln. ... H-5
Southard ... K-5
Southwood Ave. ... J-5
Spring St. ... F-5,G-3
Springmont ... I-1
Spruce St. ... F-3
Stambaugh ... J-6
Stanbery ... F-8
Stanhope ... B-1
Stanley Ave. ... H-5
Starling ... G-4
Starr ... E-4,E-5
State St. ... G-5
Steelwood ... E-1
Stevens ... G-2
Stewart Ave. ... I-5
Stimmel Rd. ... I-3
Stinchcomb Dr. ... B-2
Stoddart ... G-7
Stonington Ave. ... K-1
Stratford ... F-7
Studer Ave. ... H-6
Sullivant Ave. ... H-3
Summit ... D-4,E-4
Sunbury ... B-8
Sunbury Dr. ... D-7
Suncrest ... I-4
Sunny Hill ... C-1
Swan St. ... F-5
Sycamore ... H-4,H-5
Sycamore St. ... H-7
Talmadge St. ... F-5
Taylor ... D-6,E-6
Thelma ... I-2
Thomas ... F-2,H-2
Thomas Ave. ... H-3
Thomas Ln. ... B-1
Thorn ... F-6
Thornwood ... E-1
Thurber ... F-3
Thurman Ave. ... E-3
Tibet ... B-4
Tibet Rd. ... B-3
Tilbur Ave. ... E-3
Tompkins ... C-3
Toronto ... F-6
Torrence Rd. ... A-3
Town St. ... G-5
Townsend ... H-2
Trentwood ... B-1
Tulane ... B-3,B-4
Tulane Rd. ... B-3
Tuller St. ... C-3
Tuttle Park Pl. ... D-3
Twin Rivers ... G-3
Union ... H-2,I-1
Universal Rd. ... J-7
Urana ... A-4
Valcon ... I-7
Valley ... B-8
Varsity ... B-1
Vassar Dr. ... C-1
Vaughn ... I-1
Velma Ave. ... C-5
Vendome ... C-7
Vernon ... F-6
Vine St. ... F-4
Virginia ... E-2
Waldeck ... C-3
Walhalla Rd. ... B-3
Wall St. ... G-4
Walmar Dr. ... A-5
Walnut St. ... G-5
Walsh ... H-2
Walters ... E-5
Wareham ... C-1
Warren St. ... F-4
Washington ... E-5,J-5
Washington Ave. ... G-5
Watkins ... K-8
Watkins Rd. ... K-7
Weber Rd. ... B-3,B-5
Webster Park ... A-2
Wedge ... B-7
Weiler ... I-7
Welch Ave. ... J-5
Weldon Ave. ... B-4,B-5
Wellesley Dr. ... C-1
Wendell ... G-2
Wentworth ... D-6
Westfield Dr. N. ... J-1
Westland Ave. ... F-8
Westminster Dr. ... C-1
Weston ... A-2
Westwood Ave. ... E-1
Wheatland ... H-1
Whitehead ... I-1
Whitehorn ... J-1
Whitehorne ... I-1
Whitehorne Ave. ... H-1
Whittier St. ... H-5,H-7
Wildwood ... E-7
Willamont ... E-7
Willard ... F-2
Williams ... F-2
Willow ... H-5
Wilnis Ave. ... C-4
Wilson Ave. ... H-6,K-6
Windsor Ave. ... D-6
Winslow Dr. ... J-8
Winthrop ... A-2
Wisconsin ... G-2
Withers Ave. ... I-2
Wood Rd. ... C-2
Woodbury ... H-1
Woodcroft Rd. ... J-1
Woodford Ave. ... D-6
Woodland ... F-7
Woodland Ave. ... B-7,D-7
Woodmere ... A-1
Woodnell ... D-7
Woodrow Ave. ... J-5,J-6
Woodruff Ave. ... D-3
Woodward ... D-7
Worthington ... E-3
Wrexham ... I-1
Wyandotte ... C-4
Wyandotte Ave. ... E-1
Yale Ave. ... H-2
Yolanda ... J-7
Yorkcliff ... A-8
Zebulon Ave. ... A-5
Zimpfer St. ... I-5
1st Ave. ... E-4,F-1,F-3
2nd ... E-2
2nd Ave. ... E-3,E-4
3rd ... E-4
3rd Ave. ... E-1,E-3,E-5
3rd St. ... G-4,I-5
4th ... E-6,E-8
4th Ave. ... E-3,E-4,E-5
4th St. ... E-4,G-4,I-5
5th Ave. ... E-1,E-3,E-5,E-7
5th St. ... E-4,G-5
6th ... E-3,E-4,E-8,F-4
6th Ave. ... E-1
6th St. ... E-4,G-5
7th ... E-3,E-4,E-8
7th Ave. ... E-1
8th ... D-8,E-5
8th Ave. ... E-3
9th ... D-2,D-8
9th Ave. ... E-3
9th St. ... G-5
10th ... D-2,D-8
10th Ave ... D-3
11th ... D-2,D-8
11th Ave. ... D-3,D-4
12th ... D-4,D-6,D-8
12th Ave. ... D-5
13th ... D-4,D-8
13th Ave. ... D-5
14th ... D-4,D-5
14th Ave. ... D-8
15th ... D-4
15th Ave. ... D-5
16th ... D-4,D-6
16th Ave. ... D-5
17th ... D-3,H-6,I-6
17th Ave. ... D-4,D-5
17th St. ... F-5,G-6
18th ... D-3,D-4,H-6,I-6
18th Ave. ... D-5
18th St. ... F-6,G-6
19th ... D-4,E-6,I-6
19th Ave. ... D-5
20th ... D-4,E-6,G-6
20th Ave. ... D-5
20th St. ... F-6,I-6
21st Ave. ... C-5
21st St. ... E-6,F-6,G-6
22nd ... E-6,G-6
22nd Ave. ... C-5
22nd St. ... F-6,H-6
23rd ... C-6,E-6
23rd Ave. ... C-5
24th ... C-6
24th Ave. ... C-5
25th Ave. ... C-5
26th Ave. ... C-5

POINTS OF INTEREST

Fort Hays ... F-5
Franklin Park ... F-7
Goodale Park ... F-4
Lincoln Park ... I-6
Schiller Park ... H-5
Southview Park ... I-4
Sunshine Park ... H-3

Scale of Miles

0 .2 .4 .6 .8 1

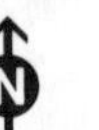

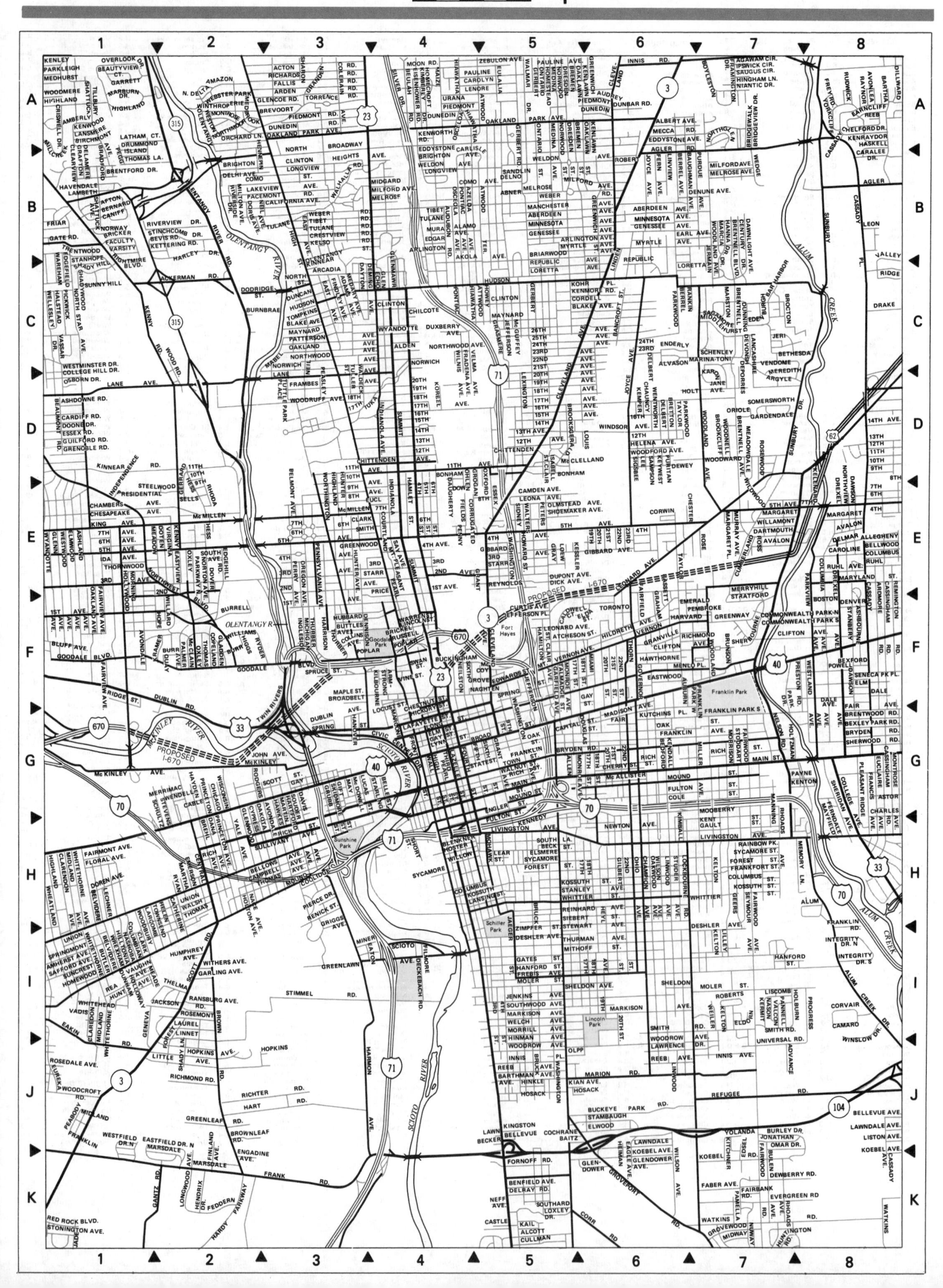

Scale of Miles
0 .1 .2 .3 .4

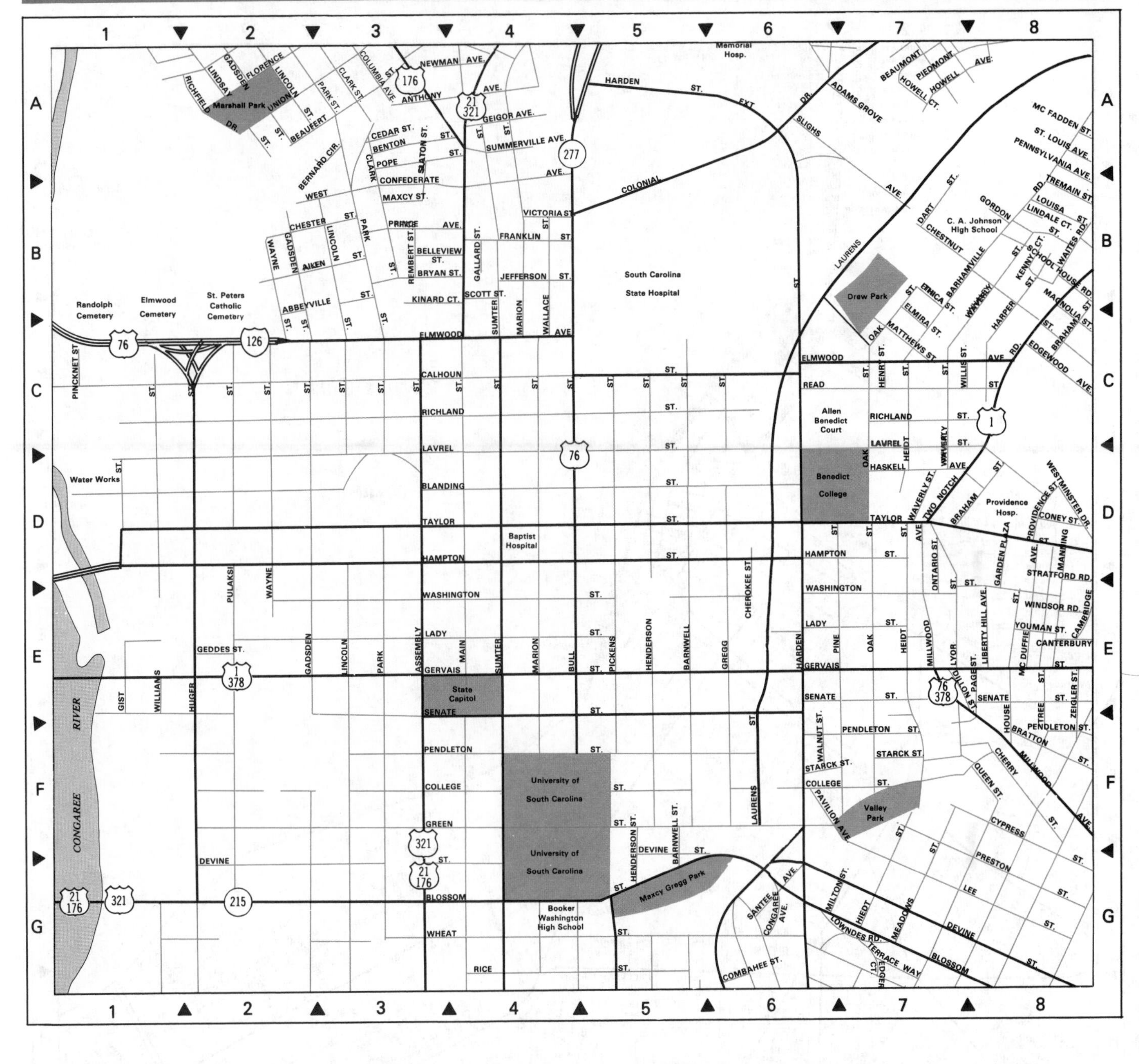

COLUMBIA

Abbeyville St. B-2
Adams Grove. A-7
Allen St. B-3
Anthony Ave. A-3
Assembly St. E-3
Barhamville Rd. B-8
Barnwell St. E-5
. . . . F-5, G-5
Beaufert St. A-2
Beaumont Ave. A-7
Belleview St. B-3
Benton St. A-3
Bernard Cir. A-3
Blanding St. D-3
Blossom St. G-3, G-7
Braham St. C-8, D-7
Bratton St. F-8
Bryan St. B-3
Bull St. E-4
Calhoun St. C-3
Cambridge E-8
Canterbury E-8
Carver St. A-7
Cedar St. A-3
Cherokee St. E-6
Cherry St. F-8
Chester St. B-2
Chestnut St. B-7
Clark St. A-3
College St. F-3, F-6
Colonial Dr. B-5
Columbia St. A-3
Combahee St. G-6
Coney St. D-8
Confederate Ave. B-3
Congaree Ave. G-6
Cypress St. F-8
Dart St. B-7
Devine St. F-5
. . . . G-2, G-7
Dillon St. E-7
Edgerton Ct. G-7
Edgewood Ave. C-8
Elmira St. C-7
Elmwood Ave. . . . C-3, C-6
Florence St. A-2
Franklin St. B-4
Gadsden St. A-2
. . . . B-2, E-2
Gallard St. B-4
Garden Plaza D-8
Geddes St. E-2
Gervais St. E-3, E-6
Geigor Ave. A-4
Gist St. E-1
Gordon St. B-8
Green St. F-3
Gregg St. E-6
Grove St. A-6
Hampton St. D-3, D-6
Harden St. E-6
Harden St. Ext. A-5
Harper St. B-8
Haskell Ave. D-7
Heidt St. D-7, E-7, G-7
Henderson St. . . . E-5, G-5
Henry St. C-7
House St. F-8
Howell Ave. A-7
Howell Ct. A-7
Huger St. E-2
Ithica St. B-7
Jefferson St. B-4
Kenney Ct. B-8
Kinard Ct. B-3
Lady St. E-3, E-6
Laurel St. C-3, C-7
Laurens St. F-6
Lee St. G-7
Liberty Hill Ave. E-8
Lincoln St. . . . A-2, B-3, E-3
Lindale Ct. B-8
Lindsay St. A-2
Louisa St. B-8
Lowndes Rd. G-7
Lyor St. E-7
Magnolia St. B-8
Main St. E-4
Manning D-8
Mapion St. C-4
Marion St. E-4
Matthews St. C-7
Maxcy St. B-3
McDuffie Ave. E-8
McFadden St. A-8
Meadows St. G-7
Millwood Ave. . . . E-7, F-8
Milton St. G-6
Newman Ave. A-3
Oak St. C-7, D-7, E-7
Ontario St. D-7
Page St. E-8
Park St. A-2, B-3, E-3
Pavilion Ave. F-6
Pendicton St. F-3
. . . . F-7, F-8
Pennsylvania Ave. . . . A-8
Pickens St. E-5
Piedmont Ave. A-7
Pincknet St. C-1
Pine St. E-6
Pope St. A-3
Preston St. G-8
Prince Ave. B-3
Providence St. D-8
Pulaksi St. D-2
Queen St. F-8
Read St. C-6
Rembert St. B-3
Rice St. G-4
Richfield Dr. A-2
Richland St. C-3, C-7
St. Louis Ave. A-8
Santee Ave. G-6
School House Rd. . . . B-8
Scott St. B-4
Senate St. E-3
. . . . E-6, E-8
Slaton St. A-3
Slighs Ave. A-6
Starck St. F-6, F-7
Stratford Rd. D-8
Summerville Ave. . . . A-4
Sumter St. C-4, E-4
Taylor St. D-3, D-7
Terrace Way G-7
Tree St. E-8
Tremain St. B-8
Two Notch Rd. D-7
Union St. A-2
Victoria St. B-4
Waites Rd. B-8
Wallace St. C-4
Walnut St. F-6
Washington St. . . E-3, E-6
Waverly St. B-8, D-7
Wayne St. B-2, D-2
West B-3
Westminster Dr. D-8
Wheat St. G-3
Williams St. E-1, G-1
Willis St. C-7
Windsor Rd. E-8
Youman St. E-8
Zeigler St. E-8

POINTS OF INTEREST

Allen Benedict Court C-6
Baptist Hospital D-4
Benedict College D-6
Booker Washington H.S. G-4
C. A. Johnson H.S. . . . B-7
Drew Park B-7
Elmwood Cemetery . . B-1
Marshall Park A-2
Maxcy Gregg Park . . . G-5
Providence Hosp. D-8
Randolph Cemetery . B-1
Richland Memorial Hosp. A-6
St. Peters Catholic Cemetery B-2
South Carolina State Hospital B-5
State Capitol E-4
University of South Carolina . . F-4, G-4
Valley Park F-7

Corpus Christi, TX

Scale of Miles
0 .2 .4 .6

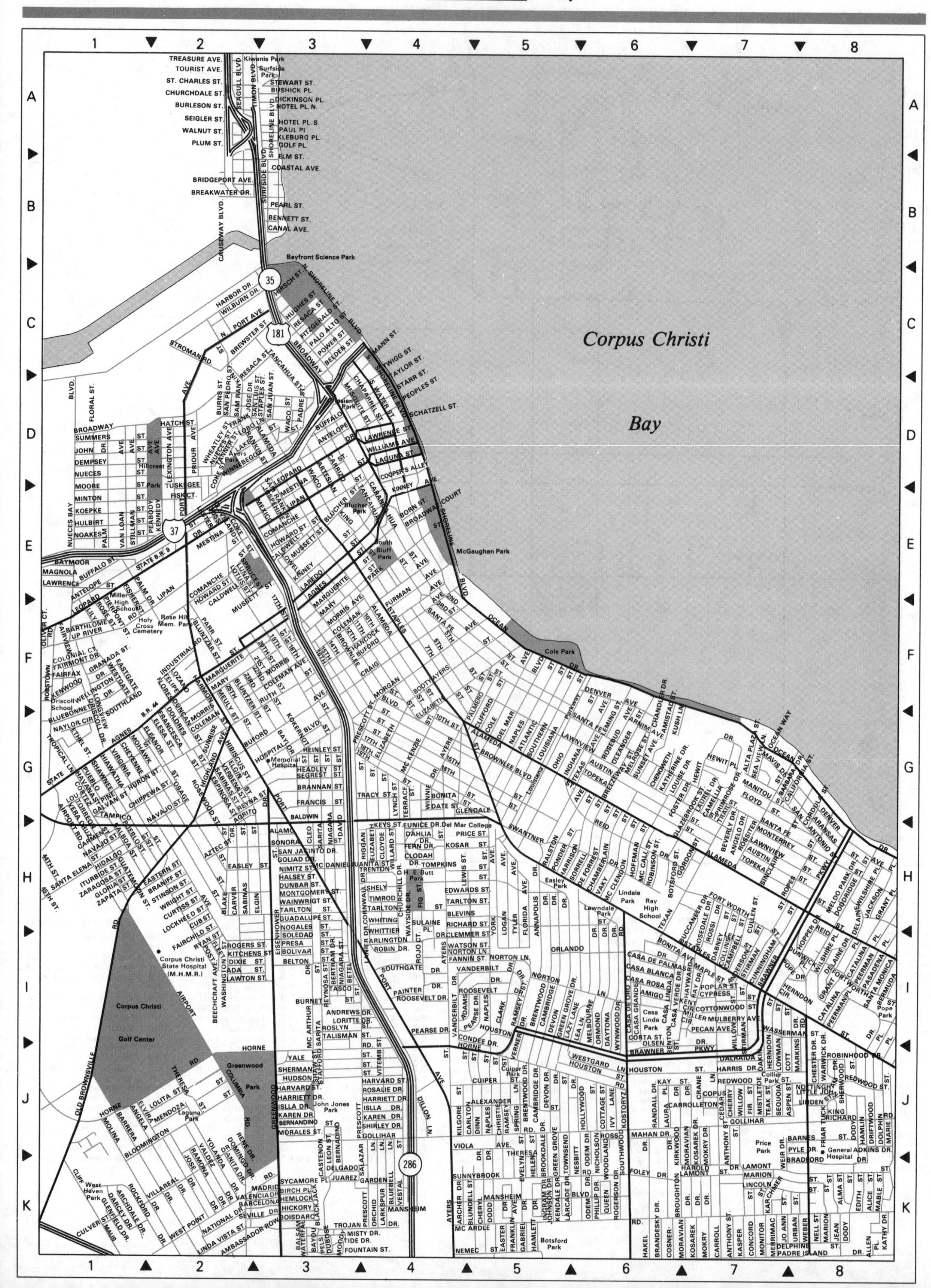

CORPUS CHRISTI

Ada St. I-2
Adams I-4
Adkins Dr. J-8
Agnes St. E-3,G-1
Agrito G-2
Airport Rd. G-1,H-1,I-2
Alame[illegible] D-3,F-4
Alamo H-3
Alexander St. J-5
Alice St. K-8
Allen Pl. K-8
Alma St. K-8
Alameda St. G-5,H-7
Alta Plaza St. G-7
Ambassador Row K-2
Amigo I-6
Amistad St. G-6
Andrews Dr. I-3
Angela Dr. J-1
Angelo Dr. H-7
Anita Dr. K-2
Annapolis Dr. H-5
Antelope St. D-3,E-1
Anthony St. J-7K-7
Aransas St. H-8
Archdale Dr. K-1
Archer Dr. K-4
Arlington Dr. I-4
Arnim Ave. G-6
Artesian D-3
Aspen St. J-7
Atlantic St. G-5
Austin G-6
Austin St. H-7
Ayers G-4,I-4,K-4
Aztec St. H-2
Balsam K-3
Barcelona K-3
Barnes St. J-8
Barrera Dr. G-7,J-1
Barthlome Rd. F-1
Baylor G-3
Baymoor E-1
Bayou K-3
Beechcraft Ave. I-2
Belden St. C-3
Belton St. I-3
Bennett St. B-3
Benton St. I-6
Bermuda I-8
Bernanding St. J-3
Bernadino J-3
Betel St. I-3
Bertram Dr. J-3
Beverly Dr. H-7
Birch Pl. K-3
Blackjack K-3
Blake St. H-2
Blevins St. H-4
Bloomington St. J-2
Blucher St. G-3
Bluebell Ln. K-4
Bluebonnet Dr. G-1
Blundell St. K-5
Bluntzer St. F-2
Boisdarc K-3
Bolivar St. I-3
Bonita Ave. I-6
Bonita St. G-4
Born St. E-4
Botsford St. H-6
Bradford Dr. J-8
Brandesky Dr. K-6
Braniff St. G-3
Brannan St. G-3
Brawner Pkwy. I-7,J-6
Breakwater Dr. B-2
Breezeway St. G-6
Brentwood Dr. I-5,J-5
Brewster St. C-2
Bridgeport Ave. B-2
Broadway C-3,D-1
Broadway Court E-4
Brookdale Dr. J-5
Broughton K-6
Brownlee F-3
Brownlee Blvd. G-5
Buccaneer Dr. H-7
Buffalo St. D-3,E-1
Buford St. F-4,G-2
Burleson St. A-2
Burnet St. I-3
Burns Dr. D-2
Bushick Pl. A-3
Caldwell St. E-2,E-3
California St. G-7
Cambridge Dr. I-5,J-5
Camellia G-3
Campbell St. I-7
Canal Ave. B-3
Caranca Hua E-4
Carlton St. J-5
Carmen H-1
Carrizo D-3
Carroll Dr. K-7
Carrolleton J-7
Carver Dr. H-2
Casa Blanca I-6
Casa De Oro Rd. I-6
Casa De Palmas I-6
Casa Grande I-6
Casa Linda I-6
Casa Rosa I-6
Casa Verde Dr. I-6
Castenon J-3
Catalina Pl. I-8
Causeway Blvd. B-2
Cedar J-7
Chamberlain St. H-6
Chandler Dr. G-6
Chaparral St. D-4
Chennoweth G-6
Cherry J-7
Cheryl Dr. K-5
Chester Dr. J-8
Cheyenne St. G-1
Chippewa St. G-1
Christie St. J-5
Churchdale St. A-2
Churchhill Dr. H-4
Clark Dr. K-5
Clemmer St. I-4
Cleo H-3
Cleveland E-2
Clifford St. G-5
Clodah Dr. H-4
Cloyde St. H-4
Coastal Ave. B-3
Coke St. D-2,E-2
Cole Ave. G-5
Coleman Ave. F-3,G-2
Collins St. I-7
Colonial Ct. F-1
Comanche St. E-2,E-3
Concord St. K-5
Condee Dr. I-5
Cooper's Al. D-4
Copus J-7
Cordula St. G-8
Cornwall Dr. H-4
Corta St. I-6
Cortez G-1
Cosner Dr. K-6
Cott St. J-7
Cottage St. J-6
Cottonwood St. I-7
Craig F-4
Crane J-7
Cub St. H-2
Cuiper St. J-5
Cullen St. H-7
Culver St. K-1
Cunningham St. I-7
Curtiss St. H-2
Cypress St. I-7
Dabney St. I-7
Dahlia Dr. H-4
Dakin J-7
Dalraida J-7
Darcey Dr. K-1
Date St. G-4
David St. H-3
Daytona Dr. I-6
De Forest St. H-6
DeLaine Dr. I-8
Delaine Wilshire Pl. H-8
Delgado K-3
Del Mar Blvd. G-5
Delphine St. K-8
Dempsey St. D-1
Denver Ave. F-6,G-8
Derson St. I-7
Devon Dr. I-5,J-5
Dickinson Pl. A-3
Dillon Ln. J-4
Dinn St. J-5
Dixie St. I-2
Dodd Dr. K-5
Doddridge St. H-7
Dody St. J-8,K-8
Dolphin Pl. J-8
Dolores St. G-2
Domingo Dr. J-2
Doss E-2
Driftwood Pl. J-8
Driscoll Dr. G-1
Dubose K-3
Dunbar St. H-3
Duncan St. G-2
Dunwick Pl. I-8
Easley St. H-2
Easter Dr. K-5
Eastern St. H-2
Eastgate F-1
Edith St. K-8
Edwards St. H-4
Eisenhower St. I-3
Eleanor St. G-2
Elesa St. G-2
Elgin St. G-2,H-2
Elizabeth St. G-4,H-4
Elm St. B-3
Elvira Dr. J-2
Espinosa St. G-2
Ethel St. G-1
Eunice Dr. H-4
Evelyn St. K-5
Fairchild St. I-2
Fairfax F-1
Fairmont Dr. F-1
Fairview I-4
Fannin St. J-5
Felipe F-2
Fern Dr. H-4
Fig St. H-4
Fir St. J-7
Fisher St. E-1
Fitch St. E-2
Fitzgerald St. C-3
Fleet I-2
Floral St. D-1
Flores F-2
Florida Ave. H-5
Floyd St. G-7
Foley Dr. K-6
Fort Worth St. H-7,I-8
Foster Dr. G-6
Fountain St. K-4
Francesca St. G-2
Francis St. G-3
Frank D-2
Franklin Ave. K-5
Friar Tuck J-8
Furman Ave. F-1
Gabriel Dr. K-5
Galvan St. G-1
Garden K-4
Glazebrook St. G-7
Glendale G-4
Glenfield Dr. K-1
Gloria St. H-1
Golf Pl. A-3
Goliad Dr. H-3
Gollihar J-4,J-7
Gordon St. H-6
Granada St. F-1
Grant Pl. H-8,I-8
Green Grove Dr. I-5,J-5
Greenwood Dr. J-3
Guadalupe St. H-3
Guatemozin St. H-1
Hakel Dr. K-6
Halsey St. H-3
Hamlett Dr. K-5
Hamlin Dr. J-8
Hancock Ave. F-4
Harbor Dr. C-2
Harmon St. F-2
Harold Dr. K-7
Harriett Dr. J-3,J-4
Harris Dr. J-7
Harrison St. H-5
Harvard St. J-3,J-4
Hatch St. D-2
Hudson St. J-3
Hughes St. C-3
Hulbirt St. E-1
Huron St. G-1
Havana St. G-1
Hayward I-7
Headley St. G-3
Heinley St. G-3
Helen St. K-5
Hemlock K-3
Herndon St. J-7
Hewit Pl. G-7
Hiawatha G-1
Hibiscus St. G-2
Hickory K-3
Hidalgo St. H-1
Highland Ave. G-2
Hirsch St. C-3
Hoffman St. H-6
Hollywood J-6
Horne J-8
Hopper Dr. I-8
Horne Rd. J-1,J-3
Hospital Blvd. G-3
Hotel Pl. N. A-3
Hotel Pl. S. A-3
Houston St. I-5,J-6
Howard St. E-2,E-3
Indiana J-5
Industrial Rd. F-2
Ingram Dr. K-5
Isla Dr. J-3,J-4
Iturbide St. H-1
Ivy Lane J-6
Jackson Pl. H-8
Jean St. I-8
Jo Ann St. K-7
John St. G-1
Johnson Dr. K-5
Jose Dr. D-2,K-2
Josephine E-3
Juarez K-3
Karchmer K-7
Karen Dr. J-3,J-4
Kasper St. K-7
Katherine Dr. G-6
Kathy Dr. K-8
Kay St. J-6
Kendale Dr. K-5
Kennedy Ave. E-2,E-4
Kent Cir. I-7
Kenwood F-1
Keys St. H-4
Kilgore St. J-4
King Richard Rd. J-8
Kinney St. E-3
Kirkwood Dr. J-6
Kitchens St. I-2
Kleburg Pl. A-3
Koepke St. I-7
Kokernot G-3
Kosar St. H-4
Kosarek Dr. J-7,K-7
Kostoryz J-6
Kush Ln. G-6
Laguna St. D-4
Lake St. D-2
Lamont St. K-7
Lanier Dr. I-4
Laredo St. E-2
Larkspur La. K-4
Larcade Dr. K-5
Laura Pl. J-6
Laurel Dr. G-7
Lawnview St. G-5,H-7
Lawrence St. D-4,E-1
Lawton St. I-2
Lazy Lane I-5
Lily F-1
Lincoln St. K-7
Linda Vista St. K-4
Linden St. J-8
Lipan E-2,E-3
Little John J-8
Lobo Ln. D-2
Lockhead St. H-2
Logan Ave. H-5
Lolita St. J-2
Loritte Dr. I-3
Lotus St. E-2
Lou St. G-1
Louise Dr. G-6
Louisiana St. G-5
Lowell G-3
Lowman St. J-7
Lozano F-2
Lea Ln. I-6
Lee St. I-7
Leming St. G-6
Leon St. J-3
Leopard Dr. D-3,F-1
Lester E-2
Lewis St. H-4
Lexington Ave. D-2
Luna St. E-2
Lynch St. G-4
Mable St. K-8
Madrid K-3
Magnola E-1
Mahan Dr. J-6
Manitou St. G-7
Mann St. C-4
Manshum Blvd. K-4,K-5
Maple St. I-7
Marguerite St. E-3,F-2
Marie St. J-8
Marion St. K-7, K-8
Markins Dr. J-8
Mavis Dr. G-7
Mary F-2,F-3
McArdle Rd. K-5
McArthur Dr. I-3
McCall St. H-6
McClendon St. H-6
McDaniel Juanita St. H-3
McGaughan Park E-5
McKenzie St. G-4
Melbourne I-6
Meldo Park Dr. H-8
Melrose St. G-6
Mendoza St. J-2
Merrimac K-5
Mestina St. E-2,E-3
Mesquite St. D-3
Meuly St. E-3
Mexico F-2
Miller Cir. I-7
Minton St. E-1
Mistletoe J-7
Misty Dr. K-3
Mohawk St. G-1
Motry Dr. J-7,K-7
Molina Dr. J-1
Monitor St. K-7
Monterrey St. H-7
Montgomery St. H-3
Moody K-3
Moore St. E-1
Morales St. F-3
Moravian Dr. J-7,K-6
Morgan St. F-4
Morris Ave. F-3,G-2
Mulberry Ave. I-7
Mussett St. E-3,F-2
Naples St. G-5,I-5,J-5
Naterfal K-3
National Dr. K-2
Navajo St. G-2,H-1
Naylor Circle G-1
Nell St. K-8
Nemec St. K-5
Nesbitt Dr. J-6
Niagara St. H-3,I-3
Nicholson St. J-6
Nimitz St. H-3
Noakes St. E-1
Nogales St. H-3
Norton Ln. I-4,I-5
Nottingham J-8
Nueces St. D-1,D-2
Nueces Bay Blvd. E-1
Ocean Dr. F-5,G-7
Ocean Way G-7
Odem Dr. J-6,K-6
Ohio St. G-5
Old Brownsville Rd. F-1,J-1
Oleander St. G-6
Oliver Ct. F-1
Olsen Dr. I-6
Orchid Ln. K-4
Orlando I-5
Ormond Dr. J-8
Osage G-2
Padre St. D-3
Padre Island Dr. K-8
Painter Dr. I-4
Palboa St. H-1
Palermo St. F-5
Palm Dr. E-1
Palo Alto St. C-3
Park St. E-4
Parr St. F-2
Pasadena I-8
Paul Pl. A-3
Peabody Ave. F-2
Pearl St. B-3
Pearse Dr. I-4,I-5
Pecan Ave. I-7
Peerman Pl. I-8
Pel's K-3
Peoples St. D-4
Phillip Dr. K-6
Pierpont St. F-1
Pine G-1
Plum St. A-2
Ponder St. H-5
Pope Dr. I-7
Poplar St. I-7
Port Ave. E-2,G-3,I-4
Port Avenue N. C-2
Post I-2
Power St. C-3
Pueblo St. G-1,G-5
Presa St. J-3
Prescott St. G-3,J-3,K-4
Price St. H-3
Primrose Dr. G-7
Priour Ave. D-2
Pyle Dr. J-8
Queen K-6
Ralston Ave. H-5
Ramona Dr. K-2
Ramsey St. I-5,I-6
Randall Dr. J-6
Rankin St. D-2
Ray Dr. I-7
Redwood St. J-7,J-8
Reid Dr. H-6,H-8
Rene Dr. I-6
Resaca St. C-2,C-3
Reyna St. G-2
Reynosa St. I-3
Richard St. H-4
Riggan St. H-4
Robin Dr. I-4
Robinhood Dr. I-8
Robinson St. H-6
Rockford Dr. I-6
Rogers St. J-7
Rogerson Dr. K-6
Rojo Ct. I-4
Roosevelt Dr. I-4,I-5
Ropes St. H-7
Roslie Dr. J-7
Rose St. I-7
Rosebud Ave. G-6
Rosedale Dr. H-7
Rosewood St. G-2
Roslyn St. I-3
Ross J-6
Rossi Dr. H-7
Rossiter St. H-7
Ruth F-3
Ryan St. I-2
St. Charles St. A-2
Sabinas St. G-2,H-2
Salazar K-4
Sam Rankin St. D-2
San Antonio St. H-8
Santa Elena St. H-1
Santa Fe St. G-6,H-6
San Jacinto Dr. H-3
San Juan St. D-3
San Luis St. D-2
Santa Monica I-8
San Pedro St. D-2
San Rankin E-3
Sarita St. H-3,I-3
Schatzell St. D-4
Sea Gull Blvd. A-2
Sea View Ln. G-7
Segrest St. G-3
Seigler St. A-2
Sequoia J-7
Seville Dr. K-3
Shaw St. J-3
Shawnee St. G-1
Shely St. H-4
Sherman St. J-3
Sherwood J-8
Shirley Dr. J-4
Shoreline Blvd. A-3,E-4
S. Shoreline Blvd. D-4
Sinclair St. H-7
Southgate Dr. I-4
Soledad St. I-3
Sonora H-3
Sorrell St. H-6
Southern St. G-5
Southland F-1
Southwood J-6
Spring Ln. J-5
Spruce St. E-2
Stafford J-3
Staples St. D-3,F-4,I-7
Starr St. D-4
Stewart St. A-3
Stillman Ave. E-1
Stinson St. H-2
Stirman St. I-7
Stromanro C-2
Sulaine Pl. H-4
Summers St. D-1
Sunnybrook Rd. K-5
Sunrise J-2
Sunset Ave. G-6
Surfside Blvd. B-3
Swantner H-5
Sycamore Pl. K-3
Talisman St. I-3
Tampico G-1
Tancahua St. C-3,E-4
Tanner St. D-2
Tarlton St. H-3,H-4
Tasco I-3
Taylor St. C-4
Teak St. J-7
Terrace St. G-4
Texan H-2
Texas Ave. G-6
Theresa St. J-2,J-5
Tide Dr. A-3
Timon Blvd. A-2
Timrod H-4
Tompkins St. H-4
Topeka G-6,H-7
Torreon G-1
Tourist Ave. A-2
Townsend St. J-5
Tracy St. G-4
Trail G-7
Treasure Ave. A-2
Trenton St. H-4
Trojan Dr. K-3
Tropical Ln. G-3
Tuskegee E-2
Twigg St. C-4
Tyler Ave. H-5
Up River F-1
Urban St. K-8
Vaky St. H-6
Valdez Dr. J-2
Valencia Dr. K-3
Vanderbilt Dr. I-4,I-5
Van Loan Ave. E-1
Vera Cruz G-1
Verbena St. G-2
Verner Dr. J-5
Vestal St. K-4
Villareal Dr. K-2
Viola Ave. J-4
Virginia G-1
Vitemb St. J-4
Waco St. D-3
Wainwright St. H-3
Walnut St. J-7
Warwick Dr. J-8
Washington St. I-2
Wasserman St. I-7
Water St., S. D-4
Watson St. I-4
Wayside Dr. H-4
Weber Rd. K-8
Weir Dr. K-7
Wellington F-1
Westgard Ln. J-6
Westgate F-1
West Point Rd. K-2
Whiting St. D-2
Whittier H-4
Wilburn Dr. C-2
Willard St. H-4
Williams Ave. D-4
Willow St. I-7,J-7
Wilshire Pl. I-8
Winnebago St. D-2
Winnie Dr. G-4
Woodland K-6
Wright St. H-7
Wynwood Dr. I-6
Yale St. J-3
Yolanda Dr. J-2
York Ave. H-5
Zapata St. H-1
Zaragosa St. H-1
2nd St. F-4
3rd St. F-4
5th St. F-4
6th St. F-4
7th St. F-4
10th St. G-4
11th St. F-3
12th St. F-3
14th St. F-3
15th St. F-3,G-4
16th St. F-3,G-4
17th St. F-3,G-4
18th St. F-3,G-4
19th St. F-3,G-4
20th St. F-3
21st St. F-3
22nd St. F-3
23rd St. F-2
25th St. F-2
44th St. H-1
46th St. H-1

POINTS OF INTEREST

Artesian Park D-3
Ayers Park D-2
Bayfront Science Park B-3
Blucher Park E-3
Botsford Park K-5
Casa Linda Park I-6
Cole Park F-5
Collier Park J-7
Corpus Christi Golf Center I-1
Corpus Christi State Hospital (M.H.M.R.) I-2
Cuiper Park J-5
Del Mar College H-4
Driscoll School F-1
Easley Park H-5
General Hospital J-8
Greenwood Park J-2
H. E. Butt Park H-4
Hillcrest Park D-1
Holy Cross Cemetery F-1
John Jones Park J-3
Kiwanis Park A-2
Laguna Park J-2
Lawndale Park H-6
Lindale Park H-6
Louisiana Parkway G-5
Memorial Hospital G-3
Miller H.S. F-1
Pope Park I-8
Price Park J-7
Ray H.S. H-6
Rose Hill Memorial Park F-2
South Bluff Park E-4
Surfside Park A-3
West Haven Park K-1

DALLAS

Abbott B-6
Abbott Ave. D-6
Acorn D-1
Adair G-7
Addison J-4
Adolph G-6
Afton E-3
Airfreight B-1
Air Line Rd. A-7,B-7
Akard H-5
Akard St. I-6
Albright K-6
Alcott D-6,E-6
Alexander J-6
Algiers F-1
Alice Cir. C-5
Allen G-6
Allison J-2
Amelia E-2
Amesbury A-8
Amonette H-3
Andrews B-5
Angelina G-1
Anita B-7,B-8
Ann H-2
Annex F-7
Apple G-7
Arcady C-3,C-4
Armstrong D-6
Armstrong Pkwy. B-4
Arnold J-6
Arroyo E-3
Asbury B-6
Ash I-7
Ash Ln. H-8
Ashland G-4
Athens Ave. A-6
Atlanta I-7
Atoka E-6
Atwater C-6
Atwell C-2
Aubrey C-2
Auburndale B-6
Austin H-5,I-5
Bank G-8
Bataan A-7
Baylor H-7
Bayonne I-2
Bayside G-1
Beatrice I-3
Beaver B-7
Beckley Ave. K-3
Beeville I-2
Belclaire C-3,C-4
Belfort C-4
Bell I-6
Belleau J-1
Bellview I-6
Belmont Ave. E-7
Belmont Rd. D-7
Bengal E-2
Bennett Ave. E-7
Benson H-8
Beverly C-3,C-6
Bickers G-1,H-2
Binkley B-6
Birch A-8,G-8
Bird G-7
Birmingham Ave. J-8
Bishop K-3
Blakney I-6
Blaylock Dr. K-4
Bluebonnet A-2
Boaz A-3
Bobbie K-5
Boedecker A-7
Boll G-6
Bomar D-6
Bonita D-7
Bookhout F-6
Bordeaux B-3,C-4
Borger H-1
Bowen St. E-4
Bowser C-5
Bowser Ave. C-3
Bradford D-2
Bradley J-1
Brantley H-2
Bremen C-8
Bristol A-2
Brook H-3
Broom H-4
Browder I-6
Brown E-3,F-4
Bryan E-8,G-6
Bryan Pkwy. E-8
Buena Vista E-6
Buffalo J-6
Burgess F-1
Burlew F-7
Burr I-2
Butler E-2
Byron C-6
Cabell Dr. F-6
Caddo F-6
Cadis St. I-6
Cadiz H-6,I-5
Cadiz St. Viaduct J-4
Cagemi B-1
Caillet A-3
Calvary F-5
Cambrick E-5
Cambridge C-6
Camden Ave. B-8
Campbell F-6
Canada Dr. G-1
Cantegral G-6
Canterbury Ct. J-1
Canton H-6,H-7
Canty St. K-2
Carnation E-1
Carrol Ave. F-7
Capitol Ave. E-6
Carlson D-2
Cass E-2
Cedar D-3,F-5
Cedar Crest Blvd. K-7
Cedar Hill Rd. J-2,K-2
Cedar Plaza Ln. D-3
Cedar Springs Rd. C-2,E-4
Central H-6
Central Expwy. E-6 G-6,I-7
Chambers E-7
Chattanooga D-1
Chemical F-2
Cherrywood Ave. D-2
Chicago H-1
Chihuahua H-1
Clara E-3
Clark F-5
Cleburne A-7
Clyde F-5
Clarence J-7
Cleaves K-5
Clinton K-1
Cobb G-6
Cochran F-6
Cockrell I-6
Colby G-5
Cole F-5,G-3
Cole Ave. E-5
Coleman K-7
College B-7
Colonial K-6
Colorado Blvd. J-1,J-3
Colson G-8
Comal K-4
Commerce St. H-5,H-8
Commerce St. Viaduct I-3
Conant F-2
Concho D-8
Concord B-1
Congress E-4
Conklin H-2,I-2
Conner F-5
Connerly B-5
Conrail B-1
Conroe H-2
Continental Ave. H-4
Continental Viaduct H-3
Convent G-7
Converse F-2
Coombs I-7
Cooper J-8,K-7
Corbin H-5
Corinth St. J-6
Cornell C-6
Coronet H-2
Corsicana I-6
Cortland B-1
Cowan A-2
Cragmont D-5
Crampton F-2
Crescent D-6
Crest D-3
Cresthaven A-2
Crockett G-5
Crossman H-1
Crowdus H-7
Crutcher G-8
Culcourt A-2
Cullen E-7
Curtis St. H-7
Dallas-Ft. Worth Tpk. I-2
Daniels A-6
Dartmouth C-6
Dathe J-8
Dawson H-7,I-7
Deere F-6
Delano E-7
Delmar Dr. E-8
Delta H-1
Denley K-5
Denton D-2
Denton Dr. C-1
Dickason E-4
Dorothy St. C-3
Douglas E-3,E-4
Douglas Ave. A-4,C-4
Dragon G-4
Drane Dr. B-3
Drexel B-6
Driskell A-5
Druid A-5
Druid Ln. A-3
Dublin A-7
Duluth I-2
Dumas F-8
Dumont F-8
Durham B-7
Dyer B-7
Eads K-4
Eastern E-3
Eastside H-8
Eastus Dr. J-2
Edgefield I-1
Edgefield Dr. J-1
Edgewater D-5
Edgewood J-8
Edison G-4
Edmondson C-3
Edwards B-2
El Benito I-1
El Dorado K-3
Elihu H-8
Elizabeth E-6
Ellis F-5
Ellsworth C-7
Elm St. H-5,H-7
Elsbeth I-5
Emerson A-3,A-4
Englewood Rd. K-3
Enid F-4
Ervay St. H-5,I-6
Eton C-7
Euclid C-6,E-7
Eugene K-8
Evergreen H-6,J-2
Ewing Ave. K-4
Exposition H-8
Express G-2
Fairfax C-3
Fairfield E-5
Fairfield Ave. C-5
Fairley C-3
Fairmount F-4
Fairway D-4
Famous I-2
Farrington F-1,G-3
Federal H-5
Ferris I-7
Fielder Ct. C-1
Fitzhugh Ave. D-6,E-6
Flanders I-1
Fleetwood D-2
Fletcher G-8
Flora G-5
Floride G-7
Floyd G-6
Folsom I-2
Fondren A-7
Fordyce G-3
Forest J-7,J-8
Ft. Worth Ave. J-1
Fox I-2
Fuqua E-7
Gallagher I-2
Gano I-6
Garrett E-7
Gaston Ave. G-7
Gilbert C-3,D-4
Gilmer I-2
Glass G-3
Glencoe D-7
Glenwick Ln. A-3,A-4
Glenwood Ave. D-5
Golf B-5
Goliad D-8
Good Latimer G-6
Goodwin D-7,D-8
Gordon G-7
Granada B-6
Grand Ave. I-8,J-7
Grassmere A-4,A-5
Green K-2
Greenbriar J-2,J-3
Greenville Ave. D-8
Greenway B-4
Greenway, W. A-3
Greer K-8
Gregg D-2
Gretna F-1
Griffin H-5
Griffin, N. I-5
Griffin, S. I-6
Guest I-2
Guilden H-3
Guillot G-5
Gunter H-8
Guymon I-1
Haggar C-2
Haines J-3,K-2
Hall D-1,D-3,E-4
Hall St. G-6,H-7
Hamburg K-7
Handley J-3
Hardwick I-3
Harlem F-5
Harry Hines Blvd. E-2,G-4
Harston G-1,I-1
Hart K-5
Hartford E-4,F-4
Harvard C-6
Harwood G-4,G-5 H-6,I-6
Haskell Ave. F-7
Haskill E-5
Haslet I-2
Harvester G-3
Hawes C-1
Hawkins G-6,H-6
Hawthorne E-3
Hawthorne Ave. D-4
Haynie A-6
Hearne I-1
Hedgerow Dr. D-2
Henderson Ave. E-7
Henning J-6
Henson F-4
Herbert H-2
Herschel D-4
Hester D-6
Hickman K-8
Hickory I-8,I-7
Highland C-5
High School B-6
Hillcrest G-5
Hill G-7,K-1
Hillcrest Ave. B-6
Holland D-4
Holland Ave. C-3
Homeland D-3
Homer D-7,E-6
Hondo E-3
Hood F-4
Hope E-8
Hopkins Ave. A-2
Hord H-5
Horrace F-6
Hoskins E-8
Houston I-4
Houston St. Viaduct J-4
Howell F-6,H-3
Hubert E-8
Hudnall St. D-2
Hudson E-2
Hugo F-5
Hunt G-4
Hunters Glen Rd. A-5
Hursey A-6
Iberia F-1
Indiana H-1
Industrial Blvd. F-3,J-5
Inspiration G-4
Inwood Rd. B-3,D-2
Irving D-5
Irving Blvd. F-1
Ivan G-4
Jackson H-5
Jefferson Blvd. Viaduct J-4
Jeffries H-7,I-8
Jewette F-6
Johnson F-6
Jordan J-8
Julian J-6
Junior J-2,J-3
Junius F-8,G-7
Kay G-2
Kaywood Dr. A-3
Kelly J-7
Kelton Dr. A-2
Kenwell B-3
Kenwood C-8
Kessler Ave. J-2
Kessler Pkwy. J-1,J-2
Key B-6
Kidd Springs K-1
Kimsey Dr. E-2
Kings Hwy. K-1
Kings Rd. E-3
Knight F-3
Knott Pl. J-2
Knox D-6
Kyle K-1
La Clede F-5
Ladd H-1
Lafayette F-6,F-7
Lafoy Blvd. C-3
Lagoon E-1
Lahoma D-3
Lake E-3
Lake Cliff Dr. K-3
Lakeside C-5
Lamar J-7
Lamar St. H-4
Lamesa J-6
Lancaster K-4
Lane I-7
Lanier D-7
Langford I-3
Lansing K-4
Larchmont B-4
Lark Dr. B-3
La Salle AVe. D-4
Latimer I-7
La Vista E-8
La Vista Ct. E-8
Lear I-6
Leath G-1
Leatrice J-2
Lee Ave. D-6
Lee Pkwy. E-5
Lemmon C-3
Lemmon Ave. A-1,E-4
Lenway J-8,K-7
Levee St. G-3
Lewis E-7
Liberty G-8
Lincoln I-7
Lindell E-8
Lindenwood C-6
Linnet Ln. B-2
Live Oak St. F-7,G-6
Livingston E-3
Llano D-8
Lockheed B-2
Lofland E-2
Logan B-2
Lomo Alto A-4,C-4
Longview C-7
Loomis I-6
Lorraine Ln. C-3
Louise I-7
Lovers Ln. A-2,A-6
Lucas Ave. E-3
Lupo E-1
Mabel B-2
Mac Arthur D-3
Macatee E-2
Mack E-2
Madison K-3
Magnolia H-5
Mahanna Springs D-3
Main E-4
Main St. H-5
Malt H-8
Manett E-6
Manett Ave. D-7
Manor E-6
Manor Wy. C-1
Manufacturing E-3
Maple E-3
Maple Ave. C-1,F-4
Maplewood B-5
March A-2
Marilla I-5
Market H-5
Marquita C-8
Marsalis Ave. K-4
Martel C-7
Mary E-8
Mary Cliff K-1
Matalee C-7
Matilda D-8
Matilda St. A-8
Matton F-5
May I-3
Mayo F-6
McBroom H-1,H-2
McCommas Blvd. C-7
McCoy G-6
McDonald K-8
McFarlin Blvd. A-4,B-7
McKee I-6,J-6
McKell F-7
McKinney G-5
McKinney Ave. E-6
McKinnon F-4
McMillan D-7
Meadow I-8
Medical Center E-1
Melrose E-7
Mercedes C-7
Merrimac C-7
Meyers I-8
Milam D-6
Miles C-3
Milton A-6,A-8
Mississippi F-1
Mobile I-2
Mockingbird C-1
Mockingbird Ln. B-4
Monarch E-7
Monitor F-2
Montclair K-1
Montclair Rd. J-1
Montgomery J-6
Monticello C-7,C-8
Montrose B-3
Moody G-5
Moore F-8
Moreland F-7
Morgan J-3
Morningside C-7
Morton A-3
Moser E-7
Motor E-2,F-1
Mt. Vernon B-6
Muncie I-2
Munger F-8,G-5,H-6
Murphey H-5
Myrtle I-8
Nakoma B-3
Nash C-2
Navaro H-1,I-1,J-1
Neal I-1,J-1
Neches K-2
Neola B-3
Newmore A-2
New Orleans D-2
Newton Ave. K-2
Nomas H-2
Nob Hill J-2
Noble E-5,F-5
Normandy B-4,B-5
Northway Dr. A-8
Nusbaume C-7
Oak G-6,H-8,I-7,J-2
Oakenwald J-3
Oak Grove F-5
Oakland H-7
Oakland Ave. H-7,I-8
Oak Lawn G-3
Oak Lawn Ave. E-4
Obenchain H-2
Olive G-5
Oliver D-8
Olympia J-1
Oram E-8
Oregon J-3
Orleans Ave. I-7
Ormsbey I-2
Overhill D-5
Ownby B-6
Oxford Ave. B-7,C-6
Pacific Ave. H-6
Palo Pinto D-8
Park A-5,H-6,J-8,I-7
Park Rd. D-1
Paker J-6
Parkhouse H-3
Parkland D-2
Parkmont E-8
Parvia H-2
Pastor I-2
Patterson H-5
Payne G-4,H-3
Peabody J-7
Peabody Ave. I-8
Peak St. F-7
Pear K-8
Pearl F-4,G-5,H-6
Pearlstone H-7
Pecan K-5
Pecos G-6
Pennsylvania J-7,K-7
Penrose D-8
Pershing D-6
Pine K-8
Pittman I-2
Pittsburg H-3
Plowman J-3,K-3
Poe E-6
Poplar K-8
Potomac B-5
Potomac, W. B-4
Powhattan I-5
Poydras H-5
Prairie View E-1
Prentice D-4
Prescott D-4
Preston B-5
Preston Rd. C-5
Princeton C-6
Prospect E-8
Prosper B-2
Pueblo H-1
Quaker F-1
Rainbow J-2
Raleigh D-3
Ralston C-2
Rampart E-2
Randall D-4
Ranier K-1
Rankin A-6
Rawlins C-3,E-4
Reagan E-4,F-3
Record I-5
Redfield D-2
Revere A-6
Rheims Pl. D-4
Richard D-7
Richardson I-7
Richmond Ave. D-7
Ridgedale C-7
Rio Vista J-3
River H-4
Roberts Pkwy. B-8
Robin Rd. A-3
Rock Creek D-5
Rock Island Rd. J-5
Roe J-5
Roland Ave. D-4
Romine K-3
Roper B-2
Rosedale A-6
Rosewood E-3
Ross Ave. E-8,F-6
Roswell E-4
Routh F-4,G-5
Rowena E-7
Rusk E-6
Rutz H-1
Sadler Cir. D-2
St. Johns C-6
St. Joseph G-7
St. Louis H-7
St. Paul G-5
Sale E-5
Salmon K-2
Sam Dealy I-2
San Carlos B-4
Sandhurst Ln. B-8
Sanger J-6
San Juacinto G-6,H-5
Santa Fe H-8
Savage B-2
Savannah I-6
Scurry F-7
Seale I-2
Sears E-8
Seegar J-6
Sewanee C-6
Shannon B-5
Shaw H-1
Shelby F-4
Shenandoah B-4,B-5
Shreveport G-7
Simpson G-7
Singleton Blvd. H-1
Skiles G-6
Slocum G-3,G-4
Sneed F-5
Snider Plaza A-6
Somerville B-8
South J-7
Spann J-3
Sparks K-5
Spence K-8
Springbrook D-5
Springs Blvd. D-3
Stacey G-6
Stanhope B-4
Starling Cir. B-3
Starr K-3
State F-6,G-5
Stemmons I-4
Stemmons Frwy. N. G-4
Stigall Dr. A-2
Stone Bridge D-5
Stoneman K-8
Stonleigh D-2
Stratford C-5
Stutz Dr. D-1
Sullivan I-6
Sulphur I-3
Summit E-7
Swiss G-7
Sycamore F-7
Sylvan Ave. H-2
Sylvester E-3,F-4
Sylvi K-2
Taft K-1
Taos A-2
Taylor H-7,I-6
Texas G-6
Tex Oak E-2
Thackeray B-6
Thedford B-2
Thomas F-6,G-5
Thornton Frwy. K-4
Throckmorton D-5,E-4
Thrush B-3
Tilden J-3
Timbergrove K-1
Topeka H-2,I-2
Toronto H-1
Travis E-5
Treadway D-1
Tremont F-8
Tully E-2
Turner Ct. J-2,K-2
Turtle Creek G-3
Turtle Creek Ln. A-5
Tyler K-2
Tyree B-2
University B-8
University Blvd. A-2,A-5,A-7
Vagas E-3
Valdina F-2
Vanelia D-3
Vanderbilt D-7,D-8
Vandilia D-3
Vantage F-2
Vassar A-5
Velasco D-8
Versailles, N. C-4
Versailles, S. C-4
Viaduct K-5
Vickery Blvd. D-7
Vicksburg F-1
Victor G-7
Victoria B-2
View D-3
Vilbig Rd. H-1,J-1
Villars G-6
Virginia F-7
Wade G-6
Wake H-2
Wales I-3
Walkway I-2
Wall I-6
Walton H-7
Wanetka B-3
Warren J-7,J-8
Washington F-6,H-8
Wateka Rd. B-2,B-3
Watt F-6
Webb C-6
Webster B-2
Welborn E-4
Welbourn G-4
Weldon F-6
Wendelkin K-8
Wenoah B-3
Westchester A-4
West Main I-2
Westminster A-6
West Side C-4
Westward I-2
Westway D-4
Westwick Rd. B-5
Wheeler D-3
Wichita G-4,H-4
Wickford K-2
Willard A-7
Williams Dr. B-5
Williams Pkwy. B-5
Willis D-7
Willow G-8,H-8
Windomere J-1,K-1
Windsor B-4
Windsor Pkwy. B-4
Winnetak H-2,I-1,K-1
Winton B-8,C-7
Wolf H-5
Wood H-5
Woodall Rogers Frwy. G-5
Woodlawn K-2
Woodside E-5
Worcola B-8,D-7
Word G-6
Worth G-7,G-8
Worthington F-5
Wren B-3
Wycliff D-4
Wycliff Ave. E-3,F-2
Yeargan F-5
Young H-5,H-6
Yuma I-2
Zang J-4
Zangs Blv.d K-3
1st H-7,H-8
2nd H-7,H-8
3rd H-8
4th H-8
5th St. K-2
6th St. K-2

POINTS OF INTEREST

Buckner Park G-8
Cole Park E-6
Exall Park G-6
Greenbelt Park G-2,K-5
Greenway Park B-4
Griggs Park F-6
Highland Park C-6
Kidd Springs Park K-2
Lake Cliff Park K-3
Lee Park E-5
Love Field Municipal Airport B-1
Reverchon Park F-4
University Park B-6

Scale of Miles
0 .2 .4 .6 .8

N

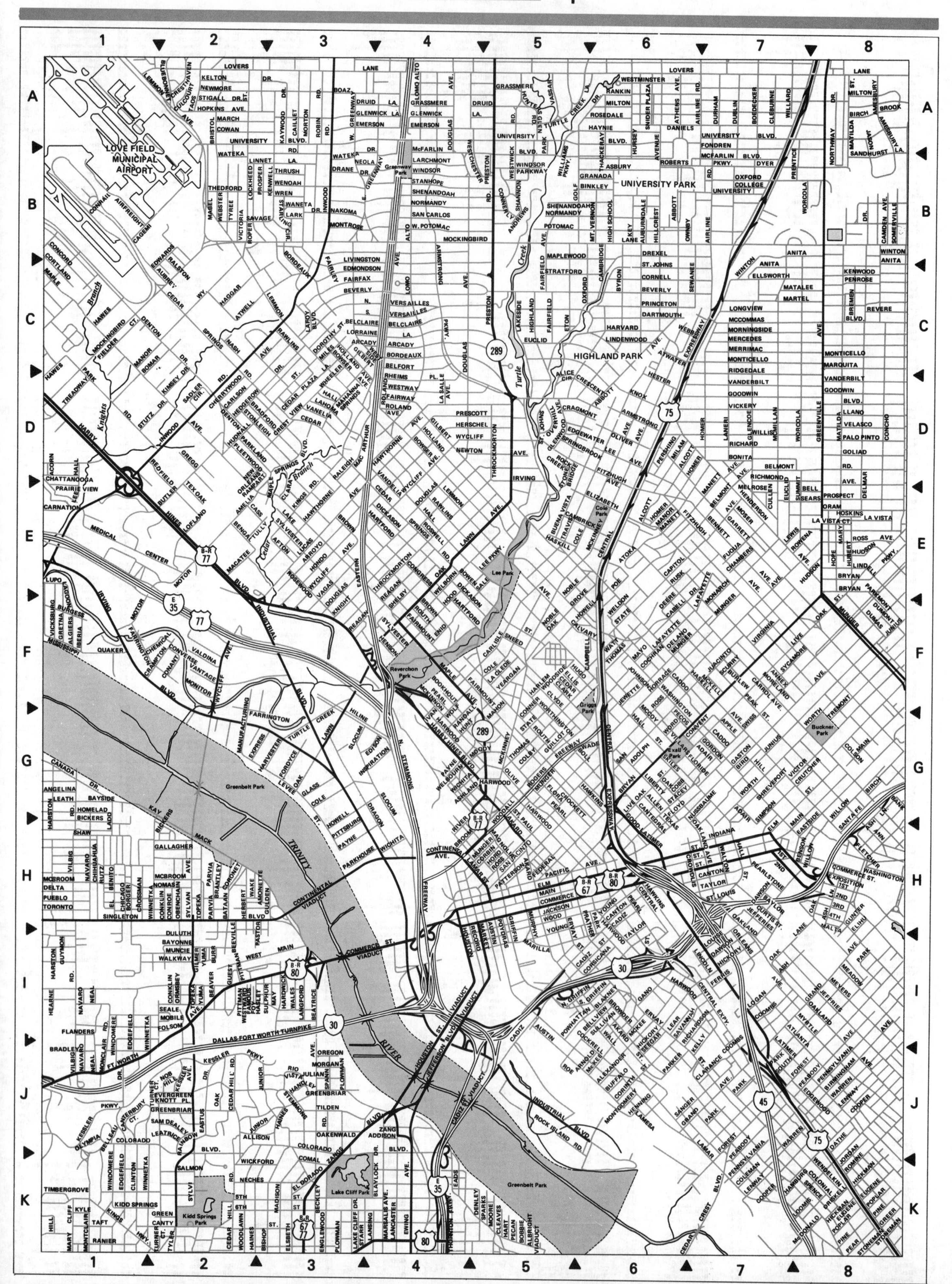

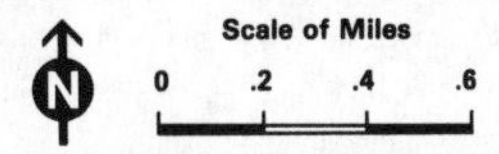
N
Scale of Miles
0
.2
.4
.6

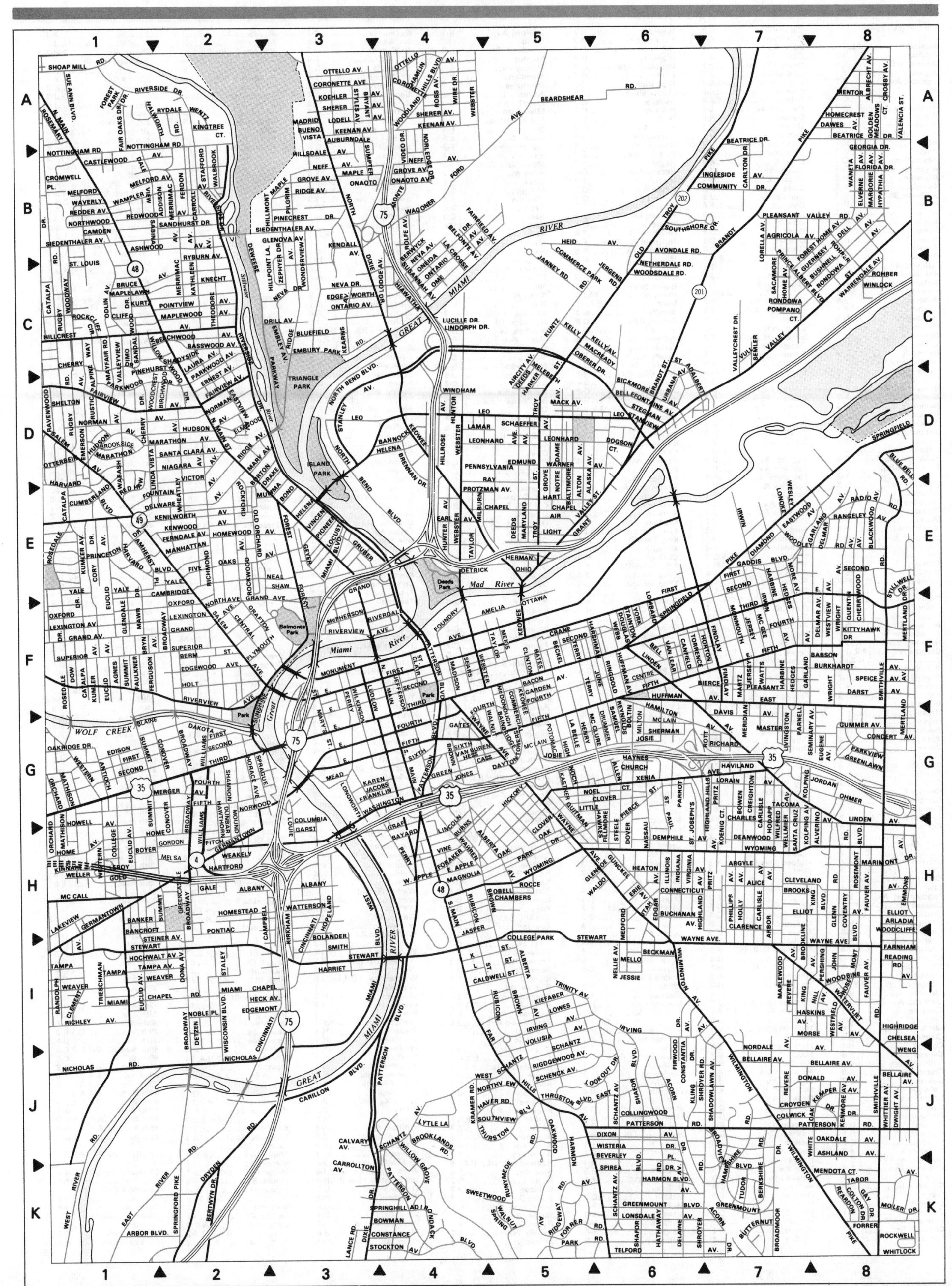
1
2
3
4
5
6
7
8
A
B
C
D
E
F
G
H
I
J
K
GREAT MIAMI RIVER
Stillwater
Mad River
Miami River
WOLF CREEK
TRIANGLE PARK
ISLAND PARK
Deeds Park
Belmonte Park
75
35
48
49
4
201
202
SHOAP MILL RD.
RIVERSIDE DR.
N. MAIN ST.
NOTTINGHAM RD.
SIEDENTHALER AV.
FOREST
KEOWEE
TROY
SPRINGFIELD
LEO
HUFFMAN
XENIA
STEWART
WAYNE AVE.
WOODCLIFFE
NICHOLAS RD.
GERMANTOWN
CARILLON BLVD.
PATTERSON
IRVING
WILMINGTON
SMITHVILLE
STROOP
CATALPA
RIVERVIEW
MONUMENT
SECOND
THIRD
FOURTH
FIFTH
HARSHMAN
BRANDT PIKE
VALLEY

DAYTON

Acorn Dr.....J-6,K-7
Addison.....B-2
Adirondack.....K-4
Agnes.....F-1
Agricola Av.....B-7
Air.....E-5
Aircity Av.....C-5
Alaska Av.....D-5
Albany.....H-2,H-3
Alberta Ave.....H-4,I-5
Albrecht Av.....A-8
Alice.....H-7
Allen.....G-6
Alpine Way.....C-1
Alton v.....D-5
Alverino.....H-8
Amelia.....F-5
Amherst Blvd.....E-1
Antioch.....G-1
Apple,E.....H-4
Apple,W.....H-4
Arbor.....H-7
Arbor Blvd.....K-1
Argyle.....H-7
Arladia.....H-8
Ashland Av.....J-8
Ashwood Av.....B-1
Auburndale.....A-3
Avondale Rd.....B-6
Babson.....F-8
Bacon.....F-5
Bainbridge.....G-5
Baltimore.....D-5
Bancroft.....H-1
Banker.....H-1
Bannock.....D-4
Basswood Av.....C-2
Bates.....F-5
Bayard.....H-4
Beardshear Rd.....A-5
Beatrice Dr.....A-7
Beckel.....F-5
Beckman.....I-6
Beechwood Av.....C-2
Belfonte Av.....B-4
Bell.....F-6
Bellaire Av.....J-7,J-8
Bellefontaine Av.....D-6
Berkshire Rd.....K-7
Berm St.....F-2
Bertwyn Dr.....K-2
Berwick.....B-4
Beverley Pl.....J-6
Bickmore Av.....D-6
Bierce.....F-7
Blackwood.....E-8
Blue Bell Rd.....D-8
Bluefield.....C-3
Bolander.....H-3
Boltin.....G-6
Bond.....E-3
Bowen.....G-7
Bowman.....K-4
Boyer.....H-1
Brandt St.....D-6
Brandt Pike.....B-7
Brennan Dr.....D-4
Broadmoor Dr.....K-7
Broadvie Blvd.....J-7
Broadway.....F-2,G-2
.....H-2,I-2
Brooklands Rd.....J-4
Brookline.....H-7
Brooks.....H-7
Brown.....G-4,H-5
Bruce Av.....C-1
Bryan Av.....F-1
Bryant.....A-3
Buchanan Av.....H-6
Burkhardt Av.....F-8
Burns.....H-4
Burton.....D-2
Bushnell Av.....C-8
Butternut.....K-7
Caldwell St.....I-5
Calvary Av.....J-3
Cambridge.....E-2
Camden.....B-1
Campbell.....H-2
Canfield.....F-6
Carillon Blvd.....J-2
Carl Av.....E-4
Carlisle Av.....G-7,H-7
Carlton Dr.....B-4
Carroll.....B-2
Carrollton Av.....K-3
Cass.....G-4
Castlewood Av.....B-1
Catalpa Dr.....C-1,E-1,F-1
Central.....F-2
Chambers.....H-5
Chapel.....E-5
Charles.....G-7
Chelsea.....I-8
Cherrwood Av.....F-8
Cherry Dr.....C-1,D-1
Church.....G-6
Cincinnati.....H-3,I-2
Clarence.....H-7
Clement Av.....I-1
Cleveland.....H-7
Cliff.....C-1
Clinton.....F-5
Clover.....G-6
College.....H-1
College Park.....I-5
Collingwood.....J-6
Colton Dr.....K-8
Columbia.....G-3
Colwick Dr.....J-7
Commerce Park.....B-5
Commercial.....F-5
Community Dr.....B-7
Condert Av.....G-8
Connecticut.....H-6
Conover.....G-2
Constance.....K-4
Constantia Av.....J-6
Coronett Ave.....A-3,A-4
Cory Dr.....E-1
Coventry Rd.....H-8
Crane.....F-5
Creighton.....G-7
Cromwell Pl.....B-1
Crosby Av.....A-8
Croyden Dr.....J-7
Cumberland.....E-1
Dakota.....G-2
Dale View.....B-1
Darst Av.....F-8
Davis Av.....G-7
Dayton Towers.....G-5
Deanwood.....H-7
Deeds Ave.....D-5,E-5
Delaine Ave.....K-6
Delmar Av.....E-8,F-8
Delware Av.....E-2
Demphile.....H-6
Detrick.....E-4
Detzen.....I-2
Deweese Pkwy.....B-3
Diamond.....E-7
Dixie Dr.....K-3
Dixon Av.....J-6
Dogson Ct.....D-6
Dona Av.....I-2
Donald Av.....J-8
Douglas.....F-6
Dover.....H-6
Dow.....F-1
Drake.....D-3
Drill Av.....C-3
Drummer.....G-6
Dryden Rd.....J-2,K-1
Dunbar.....G-2
Dwight Av.....J-8
East River Rd.
.....J-2,K-1
East View Av.....D-2
Eastwood Av.....E-8
Edgar.....H-6
Edgemont.....I-2
Edgewood Ave.....F-2
Edgeworth.....C-3
Edison.....G-1
Edmund.....D-5
Elliot.....H-7,H-8
Elverine Av.....B-8
Embley Av.....C-3
Embury Park.....C-3
Emerson Av.....D-1
Emmons.....H-8
Erie.....H-6
Ernest Av.....D-2
Euclid Av.....E-1,F-1
.....H-1,I-1
Eugene Av.....G-8
Fairfield Av.....B-4
Fair Oaks Dr.....A-1
Fairview Av.....D-1,D-2
Faulkner.....F-1
Fauver Av.....H-8,I-8
Far Hills Av.....J-5
Farnham.....I-8
Ferdon Rd.....B-2
Ferguson.....F-1
Ferndale Ave.....E-2
Fifth.....F-6,G-2
Fifth, E.....F-7,G-3
Filmore.....G-6
Findlay.....F-7
First.....E-6,E-7
.....F-4,G-1,G-2
Firwood Dr.....J-6
Fitch.....G-1
Five Oaks Av.....E-2
Florida Dr.....B-8
Foraker.....H-4
Forest.....E-3
Forest Home Av.....B-8
Forest Park Dr.....A-1
Forrer Rd.....K-5,K-8
Foundry.....F-4
Fountain Av.....E-2
Fourth.....F-5,F-7,G-2,G-4
Fourth, E.....G-3
Franklin.....G-3
Gaddis Blvd.....E-7
Gale.....H-2
Garden Av.....F-5
Garland Av.....E-8,F-7
Garst.....H-3
Gates.....G-4
Gay Dr.....K-8
Georgia Dr.....B-8
Germantown.....H-1
Geyer.....E-3
Glencoe.....H-5
Glendale.....F-1
Glenn Rd.....H-8
Glenoua Av.....B-3
Gold.....H-1
Golden Meadows Ct.A-8
Graf.....H-4
Grafton.....F-2
Grand Ave.....E-2,E-3
.....F-1,F-2
Grant.....E-5
Greenlawn.....G-8
Greenmount Blvd.....K-6
Grove Av.....D-5
Gruber.....E-3
Guernsey Dell Av.....B-8
Gummer Av.....G-8
Gunckel.....H-6
Halworth.....A-2
Hamilton.....G-6
Hamlin.....A-4
Hampshire Rd.....J-7
Harbine Av.....E-7,F-7
Harker.....D-5
Harmon Blvd.....J-5,K-6
Harriet.....I-3
Harshman.....F-6
Hart.....E-5
Hartford.....H-2
Harvard Blvd.....D-1,E-1
Haskins Av.....I-7
Hathaway Rd.....K-6
Haver Rd.....J-5
Haviland.....G-7
Hawkes.....G-6
Hawthorn.....G-2
Haynes.....G-6
Heaton.....H-6
Heck Av.....I-2
Hedges.....E-7,F-7
Heid Av.....B-5
Helena.....D-4,E-3
Henry.....G-5
Herman.....E-5
Hess.....G-4
Hiawatha.....C-4
Hickory.....G-5
Hillcrest Av.....C-1
Hilldale Av.....B-3
Hillmont.....B-3
Hill Point Ln.....B-3
Hill Rose Av.....D-4
Highland.....H-6
Highland Hills.....G-7
High Noch.....G-5
High Ridge.....I-8
Hochwalt Av.....I-1
Hodapp.....G-7
Holly Av.....H-7
Holt.....F-2
Home Av.....H-1
Homecreast.....A-8
Homestead.....H-2
Homewood Av.....E-2
Hopeland.....H-3
Horace Ave.....G-8
Horton.....F-7
Howe Av.....C-7
Howell Av.....G-1
Hudson Av.....D-1,D-2
Hunter Av.....E-4
Huntor.....D-4
Huffman Av.....F-6
Huffman Av. East.....F-6
Hypathia Av.....B-8
Illinois.....H-6
Imo.....C-1
Indiana.....H-6
Ingleside Av.....B-7
Irving Av.....I-5,I-6
Irwin.....E-7,F-7
Jacobs.....G-3
Janney Rd.....C-5
Jasper.....H-4
Jefferson.....F-4
Jergens Rd.....C-6
Jersey.....F-7
Jessie.....I-6
John Glen Rd.....I-8
Jones.....G-4
Jordan.....G-8
Joseph's.....G-6
Josie.....G-5,G-6
June.....F-6
K St.....I-5
Karen.....G-3
Kathleen Av.....C-2
Kearns.....C-3
Keenan Av.....A-3,A-4
Kelly Ave.....C-5,C-6
Kendall Av.....B-3
Kenilworth Av.....E-2
Kenimore Av.....J-8
Kenwood Av.....E-1
Keowee.....D-4,G-5
Kiefaber.....I-5
King Av.....H-8,I-7
King Tree Ct.....A-2
Kinnard.....H-1
Kirkham.....H-3
Kitty Hawk Dr.....F-8
Kling Av.....J-6
Kling Dr.....J-6
Knecht.....C-2
Koehler Av.....A-3
Koeing Ct.....H-7
Kolping Av.....G-7,H-7
Kramer Rd.....J-4
Kumler Av.....E-1,G-1
Kuniz Rd.....C-5
Kurtz.....C-1
L St.....I-5
La Belle.....G-5
La Crosse.....B-4
Lakeview.....H-1
Lamar.....D-4
Lance Rd.....K-3
Laura Av.....C-2
Leo.....D-3,D-4,D-5
Leonhard.....D-4,D-5
Leroy.....H-1
Lexington Av.
.....F-1,F-2
Light.....E-5
Lincoln.....H-4
Linda Vista.....D-1
Linden Av.....F-6,H-8
Lindorph Dr.....C-4
Little.....G-5
Livingston.....G-7
Locust.....E-3
Lodell Av.....A-3
Lodge Av.....C-4
Lombard.....F-6
Longworth.....G-3
Lonoke.....E-7
Lonsdale Av.....K-6
Lookout Dr.....J-5
Lorain Av.....G-7
Lorella Av.....B-7
Louie.....G-3
Lowes.....I-5
Lucille Dr.....C-4
Ludlow.....F-3
Lytle Ln.....J-4
Mack Av.....D-5
Macready.....C-6
Madrid Bueno
Vista.....A-3
Madson.....F-4
Magnolia.....H-4
Main St., N.....A-1
.....D-2,F-4
Main, S.....G-4,H-4
Manhattan.....E-2
Martz.....F-7
Maple Grove Av.....B-3
Maple Lawn.....C-1
Maplewood Av.....C-2,I-7
Marathon Av.....D-1
Margoire Av.....B-8
Marimont Dr.....H-8
Mary Av.....D-2
Maryland Av.....E-5
Master.....G-7
Mathison.....G-1,H-1
Mawr Dr.....F-1
Mayfair Rd.....C-1
McCall.....H-1
McClure.....G-5
McDonough.....F-5
McGee.....F-7
McLain.....G-5,G-6
McPherson.....F-3
Mead.....G-3
Medford.....H-6
Meigs.....F-5
Melbeth.....C-5
Melford Av.....B-1
Mello.....I-6
Mendota Ct.....K-8
Mentor.....A-8
Merger Av.....G-2
Meridian.....G-7
Merrimac Av.....B-2,C-2
Mertland Dr.....F-8
Miami Blvd.....E-3
Miami Blvd.,W.....I-3
Miami Chapel.....I-1
Milburn.....E-4
Moler Dr.....K-8
Monmouth.....F-7
Monte Video Dr.....B-4
Monument Ave.....F-3
More Av.....E-7
Morse.....I-7
Mound.....G-2
Mumma.....E-3
Nassau.....H-6
Neal.....E-3
Neff Av.....B-3,B-4
Nellie Av.....I-6
Netherdale Rd.....C-6
Neva Av.....B-4
Neva Dr.....C-3
Niagara Av.....D-2
Nicholas Rd.....J-1,J-2
Nill Av.....I-8
Noble Pl.....I-2
Noel.....G-5
Nordale Av.....J-7
Norledge Dr.....B-4
Norman Av.....D-1,D-2
North Ave.....E-2
North Bend Blvd.....D-3
North Dixie.....B-3
Northwood.....B-1
Northview.....J-5
Norwood.....G-2
Notre Dame Av.....D-5
Nottingham Rd.....A-1
Oak.....H-5
Oak Dale Av.....J-8
Oak Kemper Av.....J-8
Oakridge Dr.....G-1
Oakwood Av.....J-5
Obell.....H-5
Oberer Dr.....C-5
Odlin Av.....C-1
Ohio.....E-5
Ohmer.....G-8
Old Orchard.....E-2
Old Troy Pike.....B-6
Ome Av.....C-4
Onaoto Av.....B-3
Oneida Dr.....B-4
Ontario Av.....C-3,C-4
Orchard.....G-1,H-1
Ottawa.....F-5
Ottello Av.....A-3
Otterbelli.....D-1
Oxford.....E-2,F-1
Park Dr.....H-5
Park Rd.....K-5
Parkview.....G-8
Parkway.....C-3
Parkwood Av.....C-2
Parkwood Dr.....C-1
Parnell.....G-7
Parrot.....G-6
Patterson Blvd.
.....F-4,G-4,I-4
.....J-6,J-8,K-4
Pennsylvania.....D-4
Perry.....F-3,H-4
Pershing Blvd.....I-8
Phillips Av.....H-7
Pierce St.....G-6
Pilgrim.....B-3
Pioneer.....E-3
Pinecrest Dr.....B-3
Pinehurst.....C-1
Pleasant.....F-7
Pleasant Valley Rd.....B-7
Plymouth.....F-2
Pointview Av.....C-2
Pompano Ct.....C-7
Pontiac.....H-2
Prince Albert Blvd.....C-7
Princeton Dr.....E-1
Pritz.....G-7,H-7
Protzman Av.....E-4
Quentin Av.....F-8
Quitman.....G-5
Radio Rd.....E-8
Randolph.....I-1
Rangeley Av.....E-8
Ravenwood.....D-1
Ray.....D-4
Reading Rd.....I-8
Reardon.....K-8
Reddar Av.....B-1
Red Haw.....D-1
Redwood Av.....B-1
Revere Av.....I-7,J-7
Reynolds.....G-6
Richard.....G-7
Richley Av.....I-1
Richmond.....E-2
Ridge Av.....C-3,D-2
Ridgeway Rd.....K-5
Ridgewood Ave.....J-5
Ringgold.....F-6
Riverdale.....F-3
Riverside Dr.....A-2
.....B-2,C-2
Riverview Ave.....F-2,F-3
Rocce.....H-5
Rockcliff Cir.....C-1
Rockford.....E-2
Rockwell.....K-8
Rockwood Av.....E-2
Rohrer.....B-8,C-8
Rolfe Av.....B-4
Rondowa.....C-7,C-8
Rosedale.....E-1,F-1
Rosemary.....A-1
Rosemont Blvd.....H-8,I-8
Ross Av.....A-4
Rott.....G-7
Rubicon.....H-4,I-5
Rugby Rd.....C-1,D-1
Runymede Rd.....K-5
Rustic Av.....D-1
Ryburn Av.....B-2
Rydale Wentz.....A-2
Sabina.....B-2
Sacamore.....C-7
St. Adalbert.....C-6
St. Brown.....I-5
St. Clair.....F-4
St. Louis.....B-1
St. Mary's St.....F-3
St. Paul.....G-6
Salem Av.....D-1,F-2
Samuel.....G-6
Sandal.....C-1
Sandhurst Dr.....B-2
Santa Clara Av.....D-2
Santa Cruz.....H-7
Schaefer.....D-5
Schantz.....J-4,J-5,J-6
Schenck Av.....J-5
Sears.....F-4
Second.....E-1,E-8,F-1
.....F-5,G-1,G-2
Seminary Av.....G-8
Semler.....C-7
Shaddy Side.....C-2
Shadowlawn Av.....J-7
Shafor Blvd.....J-6,K-6
Shannon.....G-2
Shaw.....E-3
Shelton.....D-1
Sherer Av.....A-3,A-4
Sherman.....G-6
Shoapmill Rd.....A-1
Shroyer Rd.....J-6,K-6
Siedenthaler Av.....B-1
Sixth.....G-4
Smith.....I-3
Smithville Rd.....E-8,F-8
.....H-8,I-8,J-8
Southshore Dr.....B-6
Southview.....J-5
Speice Av.....F-8
Spirea Dr.....K-6
Sprague.....G-2
Springfield Pike
.....D-8,E-7,F-6
Springford Pike.....K-2
Spring Hill.....K-4
Stafford.....B-2
Stainton.....F-6
Staley.....I-2
Stanley Av.....D-3
Stanview.....D-6
Steele.....G-6
Stegman.....D-6
Stewart.....I-3,I-1
Stewart Wayne Ave.
.....I-5,I-6
Stillwell Dr.....E-8
Stockton Av.....K-4
Styles Av.....A-3
Sue Ann Blvd.....A-1
Summit.....F-1,G-1,H-2
Sumter.....A-
Superior.....F-1,G-2
Susannah Av.....C-4
Sweetwood.....K-5
Tabor Av.....K-8
Tacoma.....G-7
Tampa.....I-1
Taylor.....E-4,F-5
Telford Av.....K-6
Terry.....F-5
Theodore Av.....C-2
Third.....F-4,G-1,G-2
Third, E.....F-7
Thruston Blvd., E.....J-5
Thurston Blvd.....J-5
Trieschman.....I-1
Trinity Av.....I-5
Torrence.....F-6
Troy St.....D-5,E-5
Tudor.....K-7
Utah Av.....H-6
Urbana Av.....D-6
Valencia St.....A-8
Valley St.....C-7,E-5
Valleycrest Dr.....C-7
Valleyview.....C-1
Van Buren.....G-4
Van Lear.....F-6
Victor Av.....D-2
Vincent.....E-3
Virginia.....H-6
Volusia Av.....I-5
Vull.....C-7
Wabash Av.....D-1
Wagner Ford Rd.....B-4
Walbrook.....B-2
Waldo.....H-5
Walnut.....G-5
Walnut Spring.....K-5
Wampler Av.....B-1
Waneta Av.....B-8
Warner Av.....D-5
Warren.....H-4
Warrendale Av.....C-8
Washington.....G-3
Watervliet Av.....I-8
Watterson.....H-3
Watts.....F-7
Wayne Ave.....G-4,H-5,I-8
Waverly.....B-1
Weakely.....H-2
Weaver.....I-1
Webb.....F-6
Webster.....A-4,D-4,E-4
Weller.....H-1
Wellmier.....H-7
Wesley.....E-7
Western.....G-1,H-1
Westfield Av.....I-8
West River Rd.....J-1,K-1
West Schantz.....J-4
Westview Av.....F-8
Wheatley Av.....E-2
White.....J-8
Whitlock.....K-8
Whittier Av.....J-8
Wilford.....G-7
Wilkinson.....F-3
Williams.....G-2
Willow Grove.....J-4
Wilmington Av.....I-6
Wilmington Pike.J-7,K-7
Willow Wood Dr.....C-2
Windham.....D-4
Winlock.....C-8
Wire Dr.....A-4
Wisconsin Blvd.....I-2
Wisteria Dr.....J-6
Wood Dr.....C-1
Woodbine Av.....I-8
Woodcliffe.....I-8
Woodland Hills Blvd.
.....A-4
Woodley Rd.....E-7
Woodsdale Rd.....C-6
Woodway.....C-1
Wonderview.....B-3
Wright Av.....F-8
Wyoming.....H-5,H-7
Xenia Ave.....G-6
Yale.....E-1
York.....F-6
Zephyer Dr.....B-3

DENVER

Acoma St.....A-4,H-4,K-4
Alcott St.....B-1,F-1,K-1
Allcott Way.....H-1
Arapahoe St.....F-3
Archer Pl.....K-3
Argyle Pl.....D-1
Arkins Ct.....D-4
Baldwin Ct.....A-6
Bannock St.....H-4,J-4
Barberry.....I-1
Bassett St.....E-2
Bayaud Ave.....K-3
Beach St.....A-1
Blake St.....D-4,F-3
Boulder St.....D-1
Bryant St.....D-1,F-1,I-1
Bryant Way.....K-1
Brighton Blvd.....A-7,C-5
Broadway.....A-4,G-4,J-4
Bryon Pl.....E-1
Bunker Pl.....E-1
Cahita.....B-4
California St.....G-4
Cedar Ave.....K-1,K-8
Chaffee.....B-2
Champa St.....F-3
Cherokee St.B-4,H-4,K-4
Cherry Creek North
Dr.....K-8
Cherry Creek South
Dr.....K-8
Chestnut Pl.....C-5
Circle Dr.....J-7
Clarkson St.....H-5,K-5
Claude.....B-7
Clayton St.....B-8,H-8,J-8
Colfax Ave.....G-1,G-8
Columbine St.....B-8
.....H-8,J-8
Corona St.....H-6,K-6
Court Pl.....F-5,G-4
Crescent Dr.....F-1
Curtis St.....F-3,G-2
Delaware St.....B-4,H-4,J-4
Delgany St.....C-5,D-4,F-2
Denargo St.....D-4
Detroit St.....I-8
Dixie Pl.....A-2
Downing St.....H-6,K-6
Elati St.....A-4
.....C-4,H-4,J-4
Elgin Pl.....A-5
Elizabeth St.....E-8,I-8
Elk Pl.....A-1,A-3
Ellsworth Ave.....K-3
Emerson St.....H-5,K-5
Erie St.....D-2
Explande.....G-8
Fairway Ave.....H-1
Fife Ct.....D-1
Fillmore St.....A-8
.....D-8,I-8
Firth Ct.....E-1
Fox St.....A-3,C-3,E-3,K-3
Franklin St.....A-6
.....H-6,J-6
Gaithness Pl.....D-1
Galapago St.....C-3
.....K-3,H-3
Gaylord St.....A-7
.....H-7,J-7
Gilpin St.....E-7,H-6,J-7
Glenarm Pl.....E-5
.....F-5,G-4
Globeville Rd.....C-4
Grant St.....A-5,H-5,J-5
Grinell Ct.....E-2
Hawthorne Pl.....J-7
High St.....A-7,H-7,J-7
Holden Pl.....H-1
Humboldt St.....A-6
.....D-6,H-6
Inca St.....C-3,E-3,K-3
Irvington Pl.....K-2
Jason St.....C-3
Lawrence St.....F-3,G-2
Leaf Ct.....A-4
Lincoln St.....A-4,H-4,J-4
Lipan St.....D-3,H-3
Logan St.....A-5,H-5,J-5
Lyle St.....D-1
Josephine St.....D-7
.....H-7,J-7
Kalamath St.....A-3
.....C-3,H-3,K-3
Kensing Ct.....E-2
Lafayette St.....H-6
Larimer St.....F-3
Maple Ave.....K-2,K-4
Marion St.....H-6
Mariposa St.....D-2,H-2,J-2
Market St.....F-3
Milwaukee St.....B-8
.....D-1,I-8
Mulberry Pl.....I-1
Navajo St.....D-2,I-2
Ogden St.....H-6,K-5
Osage St.....D-2,H-2,J-2
Park Ave.....F-5
Park Pl.....F-7
Pearl St.....A-5,H-5,J-5
Pecos St.....D-2,H-2
Pennsylvania St.
.....A-5,H-5,J-5
Quieto Ct.....B-2
Quivas St.....A-2,B-2
.....D-2,J-2
Race St.....A-7,H-7,J-7
Raritan.....A-2,B-2
Raritan Way.....J-2
Ringsby Ct.....C-5
Rio Ct.....H-2
River Dr.....F-1
St. Paul St.....B-8,D-8,I-8
Santa Fe St.....H-3,K-3
Scott Pl.....B-1
Sheer Blvd.....H-4
Sherman St.....A-4,H-4,J-4
Shoshone St.....D-2,H-2
Steavenson Pl.....A-7
Steele St.....B-8,D-8,I-8
Stout St.....G-3
Tejon St.....B-1,I-1,K-1
Thompson Ct.....B-8
Tremont Pl.....F-5,G-4
Umatilla St.....C-1,E-1,I-1
Vallejo St.....B-1,D-1,J-1
Viaduct.....G-1
Vine St.....A-7,H-7,J-7
Warner Pl.....A-3
Washington St.....H-5,K-5
Water St.....F-1
Wazee St.....C-5,F-2
Welton St.....G-4
Westwood Dr.....J-7
Wewatta St.....F-2
Williams St.....A-7
.....D-7,J-7
Wyandot St.....B-1,D-1,J-1
Wynkoop St.....C-5
Xinca St.....H-1
York Ln.....A-7
York St.....B-7,D-7,H-7
Yuma St.....K-1
Zuni St.....B-1,H-1,K-1
1st Ave.....K-6
2nd Ave.....J-3,J-8
3rd Ave.....J-3,J-6
4th Ave.....J-3,J-6
5th Ave.....J-3,J-6
5th St.....G-2
6th Ave.....I-2,I-6
6th St.....G-2
7th Ave.....I-2,I-7
7th St.....G-2
8th Ave.....I-2,I-6
8th St.....G-2
9th Ave.....I-2,I-6
9th St.....G-2
10th Ave.....H-2,H-8
11th Ave.....H-2,H-7
12th Ave.....H-2,H-6
12th Pl.....H-1
12th St.....F-2
13th Ave.....H-1,H-3,H-6
13th St.....F-3,G-3
14th Ave.....H-3,H-6
14th St.....G-3
15th St.....E-2
16th Ave.....G-6
16th St.....E-2
17th Ave.....G-6,G-8
17th St.....D-2,F-3
18th Ave.....G-6
18th St.....D-2,E-3
19th Ave.....F-6
19th St.....E-3
20th Ave.....F-6
20th St.....E-3
21st Ave.....F-6
21st St.....E-4
22nd Ave.....F-6
22nd St.....E-4
23rd Ave.....F-6
23rd St.....E-4
24th Ave.....F-6
24th St.....E-4
25th Ave.....E-6
25th St.....E-4
26th Ave.....E-6
26th St.....E-5
27th Ave.....E-6
27th St.....E-5
28th Ave.....E-6
29th Ave.....E-3,E-6
29th St.....D-5
30th Ave.....D-3,E-6
30th St.....D-5
31st Ave.....D-3,D-6
31st St.....D-5
32nd Ave.....D-1,D-6
32nd St.....D-5
33rd Ave.....D-1,D-6
33rd St.....D-5
34th Ave.....D-1,D-6
34th St.....C-5
35th Ave.....D-1,D-6
35th St.....C-6
36th Ave.....C-1,C-6
36th St.....C-6
37th Ave.....C-1,C-6
38th Ave.....C-1
38th St.....B-5
39th Ave.....C-1,C-3
39th St.....B-5
40th Ave.....C-1,C-3,C-6
40th St.....C-6
41st Ave.....B-1,B-3,B-8
42nd Ave.....B-1,B-3,B-8
43rd Ave.....B-1,B-4,B-8
44th Ave.....B-1,B-4,B-8
45th Ave.....B-1,B-4,B-8
46th Ave.....A-1
47th Ave.....A-1,A-5,A-7
48th Ave.....A-4,A-7,A-8
49th Ave.....A-4,A-7,A-8

POINTS OF INTEREST

Chessman Park.....H-6,H-7
City Park.....F-8
City Park Golf
Course.....F-8
Congress Park.....I-8
Denver Botanical
Gardens.....I-7
Lincoln Park.....H-2
Sunken Gardens.....H-3

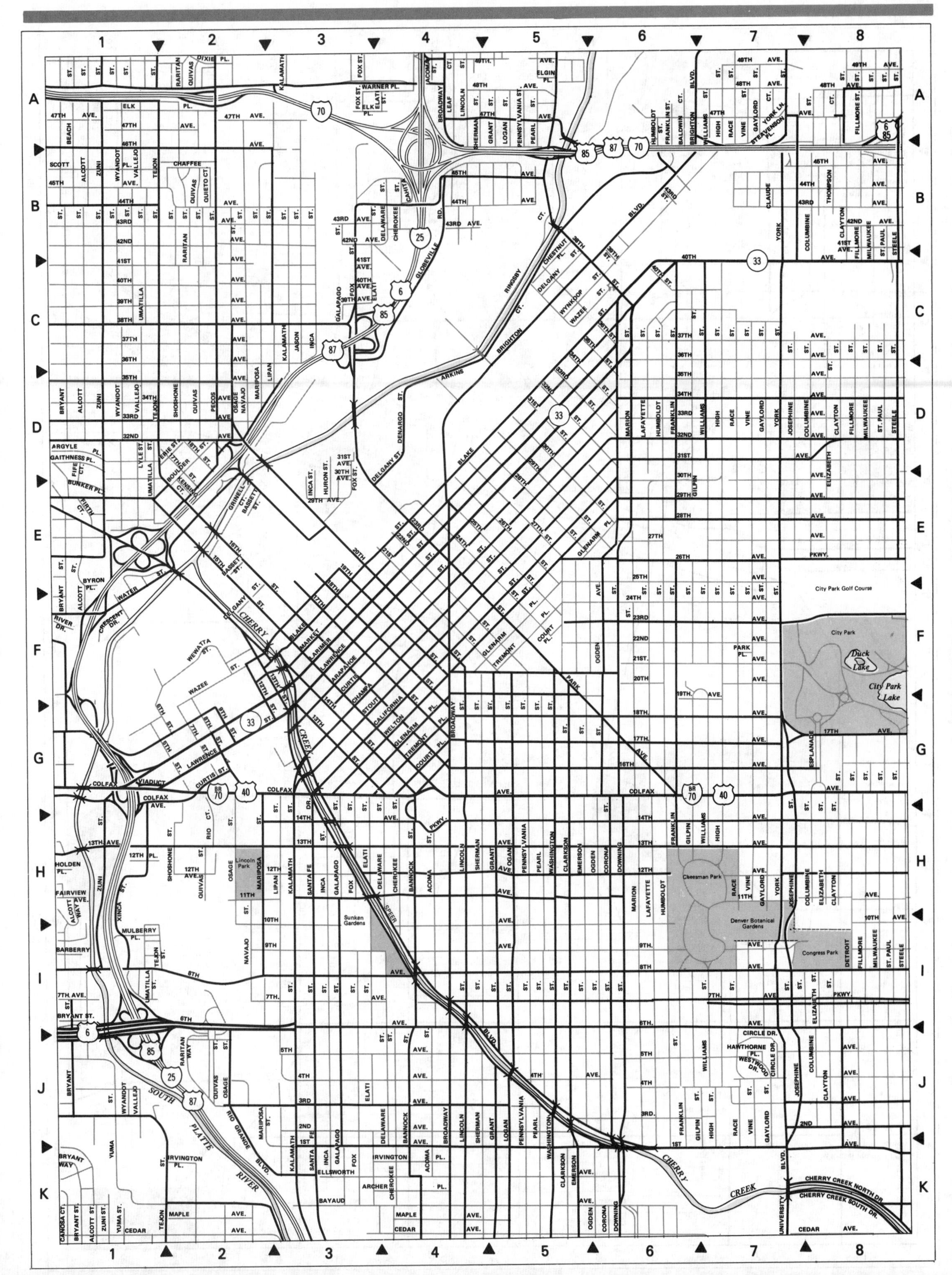
Scale of Miles
0 .1 .2 .3 .4 .5
N
City Park Golf Course
City Park
Duck Lake
City Park Lake
Cheesman Park
Denver Botanical Gardens
Congress Park
Lincoln Park
Sunken Gardens
CHERRY CREEK
SOUTH PLATTE RIVER
SPEER BLVD.
COLFAX AVE.
BRIGHTON BLVD.
CHERRY CREEK NORTH DR.
CHERRY CREEK SOUTH DR.
RIO GRANDE BLVD.
UNIVERSITY BLVD.
BROADWAY
DOWNING
YORK
COLORADO
FEDERAL
PARK AVE.
6TH AVE.
8TH AVE.
17TH AVE.
26TH AVE.
40TH AVE.
47TH AVE.

Des Moines, IA

Scale of Miles
0 .1 .2 .3 .4 .5

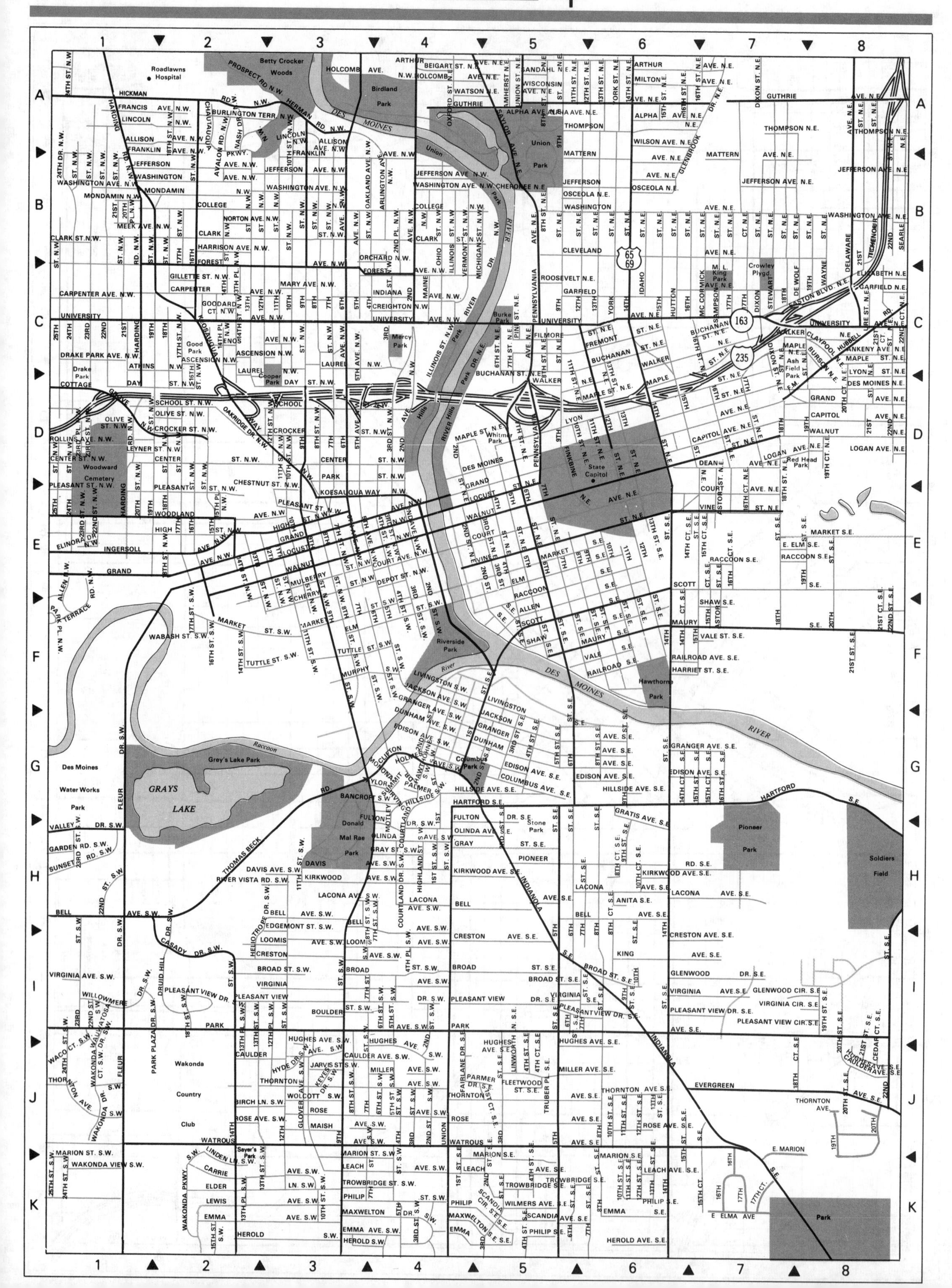

DES MOINES

Allen N.W. E-1,F-1
Allen S.W. E-6,F-5
Allison Ave. N.W. A-1
.... A-2,A-3
Alpha Ave. N.E. A-5,A-6
Amherst N.E. A-5
Anita S.E. H-6
Ankeny Ave. N.E. C-8
Arlington Ave. N.W. B-4
Arthur Ave. N.E. A-4
.... A-5,A-6,A-7
Ascension N.W. C-2,C-3
Astor St. N.E. D-7,E-7
Astor St. S.E. E-7,F-7
Athins N.W. C-1,C-2
Avalon Rd. N.W. A-2,B-2
Bancroft S.W. G-3,G-4
Bell Ave. S.E. H-4
.... H-5,H-6
Bell Ave. S.W. H-1
.... H-3,H-4
Biegart St. N.E. A-4
Birch Ln. S.W. J-2,J-3
Boulder St. S.W. I-3
Broad St. S.E. I-4,I-5,I-6
Broad St. S.W. I-3,I-4
Buchanan St. N.E.
.... C-6,C-7,D-5
Burlington Terr. N.W.
.... A-2,A-3
Burson N.E. C-8
Capitol Ave. N.E.
.... D-6,D-7,D-8
Carpenter Ave. N.W.
.... C-1,C-2,C-3
Carrie Ave. S.W. K-2,K-3
Casady Dr. S.W. I-2
Caulder Ave. S.E. J-8
Caulder Ave. S.W.
.... J-2,J-3,J-4
Cedar Ct. S.E. I-8,J-8
Center St. N.W.
.... D-1,D-2,D-3,D-4
Chautauqua A-2
Cherokee N.E. B-5
Cherry St. N.W. E-3,F-3
Chestnut St. N.W.
.... E-2,E-3
Clark St.N.W. B-1,B-2
.... B-3,B-4
Claypool N.E. C-8
Cleveland Ave. N.E.
.... B-5,B-6,B-7
Clifton Ave. S.W. G-4
College N.W. B-2,B-3,B-4
Columbus Ave. S.E. G-5
Columbus Park G-4,G-5
Cottage Grove N.W. D-1
Court Ave. N.E. E-4
.... E-5,E-6,E-7
Court Ave. N.W. E-4
Courtland Dr. S.W.
.... H-4,G-4
Creighton N.W. C-4
Creston Ave. S.E.
.... I-4,I-5,I-6,I-7
Creston Ave. S.W.
.... I-3,I-4
Crocker St. N.W.
.... D-2,D-3,D-4
Davis Ave. S.W.
.... H-2,H-3,H-4
Day St. N.W. D-1,D-2,D-3
Dean Ave. N.E. D-7
Delware Ave. N.E.
.... A-8,B-8,C-8
Depot St. N.W. E-4
Des Moines D-4,D-5
.... D-7,D-8
DeWolf St. N.E. B-7,C-7
Dixon St. N.E. A-7
.... B-7,C-7,D-7
Drake Park Ave. N.W.
.... C-1
Druid Hill Dr. S.W.
.... H-2,I-2
Dunham Ave. S.E.
.... G-4,G-5,G-6
Dunham Ave. S.W. G-4
Edgemont St. S.W. H-3
Edison Ave. S.E.
.... G-5,G-6,G-7
Edison Ave. S.W. G-4
Elder Ln. S.W. K-2,K-3
Elinora Dr. N.W. E-1
Elizabeth N.E. C-8
Elm St. S.W. E-5,E-6
.... E-7,E-8,F-3,F-4
Emma S.E. K-4,K-5,K-6
Emma Ave. S.W. K-2
.... K-3,K-4
Enos Ave. N.W. C-2,C-3
Evergreen Ave. S.E.
.... J-7,J-8
Fairlane Dr. S.E.
.... I-4,J-4
Faston Blvd. N.E.
.... C-7,C-8
Filmore St. N.E.
.... C-5,C-6
Finkbine N.E. D-5,E-5
Fleetwood St. S.E. J-5
Fleur Dr. S.W. G-1
.... H-1,I-1,J-1
Forest Ave. N.W. C-2
.... C-3,C-4
Francis Ave. N.W.
.... A-1,A-2
Franklin Ave. N.W. B-1
.... B-2,B-3,B-4
Fremont St. N.E. C-6
Fulton Dr. S.E. H-4,H-5
Fulton Dr. S.W. H-3,H-4
Garden Rd. S.W. H-1
Garfield Ave. N.E.
.... C-5,C-6,C-7,C-8
Gillette St. N.W. C-2
Glenbrook Dr. N.E.
.... A-6,A-7,B-6
Glenwood Cir. S.E.
.... I-7,I-8
Glenwood Dr. S.E.
.... I-6,I-7
Glover Ave. S.W. J-3
Goodard Ct. N.W. C-2
Grand Ave. N.E. D-7
.... D-8,E-5
Grand Ave. N.W. E-1
.... E-2,E-3,E-4
Granger Ave. S.E.
.... G-5,G-6,G-7
Granger Ave. S.W.
.... F-4,G-4
Gratis Ave. S.E.
.... G-6,H-6
Gray St. S.E. H-4,H-5
Gray St. S.W. H-4
Guthrie Ave. N.E. A-4
.... A-5,A-6,A-7,A-8
Harding Rd. N.W. A-1
.... B-1,C-1,D-1,E-1
Harriet St. S.E. F-6,F-7
Harrison Ave. N.W. B-2
Hartford S.E. G-4
.... G-5,G-7,G-8
Hawthorne G-4
Heliotrope Dr. S.W.
.... H-3,I-2
Herman Rd. N.W. A-1
.... A-2,A-3
Herold Ave. S.E. K-6
Herold S.W. K-2,K-3,K-4
High St. N.W. E-2,E-3
Highland St. S.W. H-4
Hillside Ave. S.E.
.... G-4,G-5,G-6
Hillside S.W. G-4
Holcomb Ave. N.E.
.... A-4,A-5
Holcomb Ave. N.W.
.... A-3,A-4
Holmes G-4
Hughes Ave. S.E. I-5
.... I-6,I-8
Hughes Ave. S.W. I-3
.... I-4,J-4
Hyde Dr. S.W. J-3
Idaho St. N.E. B-6,C-6
Illinois St. N.W.
.... B-4,C-4,D-4
Indiana Ave. N.W. C-4
Indianola S.E. H-5
.... I-5,J-6,K-7
Ingersoll Ave. N.W.
.... E-1,E-2
Irving H-2
Jackson Ave. S.E.
.... G-5,G-6
Jackson Ave. S.W. F-4
Jarvis St. S.W. J-3
Jefferson Ave. N.E.
.... B-5,B-6,B-7,B-8
Jefferson Ave. N.W.
.... B-1,B-2,B-3,B-4
Keosauqua Way N.W.
.... C-2,D-2,D-3,E-3,E-4
Keyes Dr. S.W. J-3
King Ave. S.E. I-6,I-7
Kirkwood Ave. S.E.
.... H-4,H-5,H-6,H-7
Kirkwood Ave. S.W.
.... H-3,H-4
Lacona Ave. S.E.
.... H-6,H-7
Lacona Ave. S.W.
.... H-3,H-4
Laurel N.W. C-2,C-3,C-4
Leach Ave. S.E. K-4
.... K-5,K-6,K-7
Leach Ave. S.W.
.... K-3,K-4
Lewis Ave. S.W. K-2,K-3
Leyner St. N.W. D-1,D-2
Lincoln N.W. A-1,A-2,A-3
Linden Ln. S.W.
.... J-2,K-2
Linworth Ln. S.E.
.... I-5,J-5
Livingston S.E. F-5,G-5
Livingston S.W. F-4
Locust St. N.E. E-4,E-5
Locust N.W. E-3,E-4
Logan Ave. N.E. D-7,D-8
Loomis Ave. S.W. I-3,I-4
Lyon St. N.E. C-8,D-5,D-7
Maine N.W. B-4,C-4
Maish Ave. S.W. J-3,J-4
Maple St. N.E. C-7
.... C-8,D-4,D-5,D-6,D-7
Marion S.E. J-5,J-6
Marion St. S.W. J-1
.... J-3,J-4
Market S.E. E-5,E-6,E-8
Market St. S.W. F-2
.... F-3,F-4
Mary Ave. N.W. C-3
Mattern Ave. N.E.
.... B-5,B-6,B-7
Maury S.E. F-6,F-7,F-8
Maxwelton S.E. K-4,K-5
Maxwelton Dr. S.W.
.... K-3,K-4
McCormick St. N.E.
.... B-7,C-7
Meek Ave. N.W. B-1,B-2
Michigan St. N.W.
.... B-4,C-4
Miller Ave. S.E.
.... J-5,J-6
Miller Ave. S.W. J-4
Milton Ave.N.E. A-6,A-7
Mondamin N.W. B-1,B-2
Monona G-4
Morton Ave. N.W.
.... B-2,B-3
Motley G-4,H-4
Mulberry St. N.W. E-3
Murphy S.W. F-3,F-4
Nash Dr. N.W. A-2
Oak Bridge Dr. N.W.
.... D-2,D-3
Oakland Ave. N.W.
.... A-3,B-3
Ohio St. N.W. B-4,C-4
Olinda Ave. S.E. H-4,H-5
Olinda Ave. S.W. H-4
Olive St. N.W. D-1,D-2
Orchard N.W. B-3,B-4
Osceola N.E. B-5,B-6
Oxford St. N.E. A-4
Palmer St. S.W. G-4
Park Ave. S.E. I-4,I-6,I-7
Park Ave. S.W. I-2,I-4
Park Pl. N.W. F-1
Park Plaza Dr. S.W.
.... I-1,J-1
Park St. N.W. D-3,D-4
Parmer Dr. S.E.
.... J-4,J-5
Pennsylvania Ave. N.E.
.... B-5,C-5,D-5
Philip S.E. K-4,K-5,K-6
Philip St. S.W. K-3,K-4
Pine St. N.E. C-5
Pioneer Rd. S.E. H-5
.... H-6,H-7
Pkwy. B-2
Pleasant St. N.W.
.... E-1,E-2,E-3
Pleasantview Cir. S.E.
.... I-7,I-8
Pleasantview Dr. S.E.
.... I-4,I-5,I-6,I-7
Pleasantview Dr. S.W.
.... I-2,I-3,I-4
Prospect Rd. N.W.
.... A-2,A-3
Pl. N.W. B-1
Raccoon S.E. E-5
.... E-6,E-7,E-8
Railroad Ave. S.E.
.... F-6,F-7
River Dr. N.E. C-4,D-4
River Dr. N.W. B-5
.... C-4,C-5
River Vista Rd. S.W.
.... H-2,H-3
Rollins Ave. N.W. D-1
Roosevelt N.E. C-5
Rose Ave. S.E. J-4
.... J-5,J-6
Rose Ave. S.W. J-2
.... J-3,J-4
Sampson St. N.E. B-7,C-7
Sandahl N.E. A-5
Saylor Ave. N.E. A-5,B-5
Scandia Ave. S.E. K-5,K-6
Scandia Cir. S.E. K-5
School St. N.W. D-2
.... D-3,D-4
Scott S.E. F-5,F-6
.... E-6,E-8
Searle N.E. A-8,B-8
Shaw S.E. F-5,F-7
Stewart St. N.E. B-7,C-7
Summit G-4
Sunset Rd. S.W. H-1
Taylor G-3,G-4
Terrace Rd. N.W.
.... E-1,F-1
Thomas Beck Rd.
.... G-3,H-2
Thompson A-5,A-6
.... A-7,A-8
Thornton Ave. S.E.
.... J-4,J-5,J-6
Thornton Ave. S.W.
.... J-1,J-3,J-4
Tichendur St. N.E.
.... A-8,B-8
Trowbridge S.E.
.... K-5,K-6
Trowbridge St. S.W.
.... K-3,K-4
Truber Pl. S.E. J-5
Tuttle St. S.W. F-2,F-3,F-4
Union St. N.E. A-5
Union St. S.W. H-4,J-4
University N.E. C-5
.... C-6,C-7,C-8
University Ave. N.W.
.... C-1,C-2,C-3,C-4
URE St. N.E. C-8
Vale St. S.E. F-6,F-7
Valley Dr. S.W. H-1
Vermont St. N.W.
.... B-4,C-4
Vine St. N.E. E-4
.... E-5,E-6,E-7
Virginia Ave. S.E.
.... I-5,I-6,I-7
Virginia Ave. S.W.
.... I-1,I-2,I-3,I-4
Virginia Cir. S.E.
.... I-7,I-8
Wabash St. S.W. F-1,F-2
Waco Ct. S.W. I-1,J-1
Wakonda Ct. S.W. J-1
Wakonda Dr. S.W. J-1
Wakonda Pkwy. S.W.
.... K-2,J-2
Wakonda View S.W. K-1
Walker N.E. C-6,C-7
.... C-8,D-5,D-6
Walnut St. N.E. D-7
.... D-8,E-4,E-5
Walnut N.W. E-3,E-4
Washington Ave. N.E.
.... B-5,B-6,B-7,B-8
Washington Ave. N.W.
.... B-1,B-2,B-3,B-4,B-5
Watrous Ave. S.E. J-4
.... J-5,J-6
Watrous Ave. S.W. J-2
.... J-3,J-4
Watson N.E. A-4,A-5
Wauwatosa Dr. S.W.
.... I-1,J-1
Wayne St. N.E. B-8,C-8
Willowmere Dr. S.W. I-1
Wilmers Ave. S.E. K-5
Wilson Ave. N.E. A-6
Wisconsin Ave. N.E. A-5
Wolcott S.W. J-3
Woodland Ave. N.W.
.... E-2,E-3
York St. N.E. A-6,B-6,C-6
1st Ct. S.E. J-5
1st St. S.E. F-5,G-4,K-4
1st St. S.W. G-4,H-4
2nd Ave. N.W. B-4,C-4
.... D-4,E-4
2nd Pl. N.W. B-4
2nd St. N.E. D-4,E-4
2nd St. S.E. E-5,F-5
.... G-5,J-5,J-6
2nd St. S.W. E-4,F-4
.... G-4,I-4,J-4
3rd St. N.E. E-5
3rd St.N.W. B-4,C-4
.... D-4,E-4
3rd St. S.E. E-5,F-5
.... G-5,H-5,J-5,K-5
3rd St. S.W. E-4,F-4
.... J-4,K-4
4th Ct. S.E. I-5,J-5
4th Pl. S.W. I-4
4th St. N.E. E-5
4th St. N.W. B-3,C-3,E-4
4th St. S.E. E-5,F-5
.... G-5,I-5,J-5,K-5
4th St. S.W. E-4,F-4,J-4
5th St.N.E. E-5
5th Ave. N.W. B-3
.... C-3,D-3,E-3
5th St. S.E. E-5
.... F-5,G-5,H-5,I-5,J-5
5th St. S.W. F-4
.... I-4,J-4,K-4
6th St. N.E. C-5,E-5
6th Ave. N.W. B-3,C-3
.... D-3,E-3
6th St. S.E. E-5,F-5
.... G-5,H-5,I-5,J-5,K-5
6th St. S.W. F-4,I-4,J-4
7th St. N.E. C-5,D-5,E-5
7th St. N.W. B-3,C-3
.... D-3,E-3
7th St. S.E. E-5,F-5
.... H-5,I-5,K-5
7th St. S.W. F-3,F-4
.... H-4,I-3,J-3,K-4
8th Ct. S.E. H-6,I-6
8th St.N.E. A-5,B-5
.... C-5,D-5
8th St. N.W. B-3,C-3,
.... D-3,E-3
8th St. S.E. E-5,F-6
.... G-6,H-6,I-6,J-6,K-6
8th St. S.W. F-3,H-3
.... I-3,J-3
9th St. N.E. A-5,B-5
.... C-5,D-5,E-5
9th St. N.W. B-3
.... C-3,D-3,E-3
9th St. S.E. E-6,G-6
.... H-6,I-6
9th St. S.W. F-3,I-3,J-3
10th Ct. S.E. H-6
10th St. N.E. D-5,E-6
10th St. N.W. A-3,B-3
.... C-3,D-3,E-3,F-3
10th St. S.E. E-6,F-6
.... H-6,I-6,J-6,K-6
10th St. S.W. K-3
11th St. N.E. A-5,B-5
.... C-5,D-6
11th St. N.W. B-3,C-3
.... D-3,E-3
11th St. S.E. E-6
.... F-6,J-6,K-6
11th St. S.W. F-3,H-3
12th Pl. S.W. I-3
12th St. N.E. A-5
.... B-5,C-5,D-6
12th St. N.W. B-2,C-2
.... D-3,E-3,F-3
12th St. S.E. E-6
.... F-6,J-6,K-6
12th St. S.W. I-3,J-3
13th Pl. N.W. B-2,C-2
13th Pl. S.W. I-2,J-2,K-2
13th St. N.E. A-6
.... B-6,C-6,D-6
13th St. N.W. B-2
.... C-2,E-2,F-3
13th St. S.E. E-6,F-6
.... I-6,J-6,K-6
13th St. S.W. I-2,J-2,K-3
14th Ct. S.E. E-6,F-6,G-6
14th Pl. N.W. C-2
14th St. N.E. A-6
.... B-6,C-6,D-6
14th St. N.W. B-2,C-2,E-2
14th St. S.E. E-6,F-6
.... G-6,H-6,I-6,K-6
14th St. S.W. F-2
.... J-2,I-2
15th Ct. S.E. E-7,F-7
.... G-7,J-7,K-7
15th Pl. N.W. E-2
15th St. N.E. A-6,B-6
.... C-6,D-6,D-7
15th St. N.W. D-2,E-2
15th St. S.E. E-6,F-6
.... G-6,J-6,K-6
15th St. S.W. K-2
16th Ct. N.E. D-7,E-7
16th Ct. S.E. E-7,G-7
16th St. N.E. A-6
.... B-6,C-6,D-7
16th St. N.W. B-2
.... C-2,D-2,E-2
16th St. S.W. F-2
17th Ct. N.E. B-7,C-7
17th St. N.E. B-7,C-7,D-7
17th St. N.W. B-2
.... C-2,D-2,E-2
17th St. S.W. E-2,F-2
18th St. S.E. J-7
18th St. N.E. B-7
.... C-7,D-7,E-7
18th St. N.W. A-2,B-2
.... C-2,D-2,E-2
18th St. S.E. E-7,F-7
18th St. S.W. I-2
19th Ct. N.E. D-8
19th St. N.E. B-8
.... C-7,D-7
19th St. N.W. B-1
.... C-1,D-1,E-1
19th St. S.E. E-7,I-8
20th Ct. N.E. D-8
20th St. N.W. B-1,D-1,E-1
20th St. S.E. E-8
.... F-8,I-8,J-8
21st Ct. C-8,E-8,F-8
21st St. N.E. A-8
.... B-8,C-8,D-8
21st St. N.W. B-1,C-1
21st St. S.E. F-8,I-8,J-8
22nd Ct. N.E. C-8
22nd St. N.E. A-8
.... B-8,C-8,D-8
22nd St. N.W. B-1,C-1,E-1
22nd St. S.E. E-8,F-8
.... I-8,J-8
22nd St. S.W. H-1,I-1
23rd Pl. N.W. D-1
23rd St. N.W. B-1,C-1
.... D-1,E-1
23rd St. S.W. H-1,I-1
24th Dr. N.W. A-1,B-1
24th St. N.W. A-1,B-1
.... C-1,D-1,E-1
24th St. S.W. I-1,J-1,K-1
25th St. N.W. B-1
.... C-1,D-1,E-1
25th St. S.W. K-1

POINTS OF INTEREST

Ash Field Park
.... C-7,D-7
Bayer's Park J-3,K-3
Betty Crocker Woods
.... A-3
Birdland Park A-4
Burke Park C-5
Cooper Park D-3
Crowley Plygd. C-7
Des Moines River A-3
.... A-4,B-5,F-5,F-6,G-7
Des Moines Waterworks
Park G-1
Donald Mal Rae
Park H-3
Drake Park C-1,D-1
Good Park C-2
Gray's Lake G-1,G-2
Grey's Lake Park G-2
Hawthorne Park F-6
Hills Park C-4,D-4
Mercy Park C-4
M. L. King Park C-7
Park K-8
Pioneer Park H-7
Raccoon River G-2
.... G-3,F-4
Red Head Park D-7
Riverside Park F-4
Roadlawns Hospital A-1
Soldiers Field H-8
State Capitol D-6
Stone Park H-5
Union Park A-5
.... B-4,B-5
Wakonda Country Club
.... J-2
Whitmer Park D-5
Woodard Cemetery D-1

DETROIT

Abbott J-6
Ackley C-8
Adams E. I-7
Adams W. I-6
Adelaide H-7,I-6
Adele D-6
Alexandrine H-4
Alexandrine E. F-7,G-6
Alfred H-7,H-8
Alger C-5,D-4
Alice C-6
Alpena A-7
Amsterdam F-4
Andrus C-6
Anna C-5
Antietam I-7
Antoinette G-2
Arden Park C-3
Ash I-3,I-4,J-2
Atkinson C-3,D-1
Atlas A-4
Atwater E. J-8
Austin K-4
Avalon A-1,B-1
Avery F-3,G-3
Avery Ter. G-3
Bagley J-5,J-6,K-2
Baltimore E-4
Bangor H-1
Bates J-7
Beaubien J-7
Beech J-6
Belmont A-5,B-3
Benham C-7
Bernard B-5
Berres B-6
Bethune D-4,E-3
Blaine D-3,E-2
Boston Blvd. E. C-3
Boston Blvd. W. D-1
Brainard I-4
Brandon K-1
Brewster G-8,H-6
Broadway I-7
Brockton A-7
Brooklyn G-4,K-6
Brush A-3,D-4,J-7
Bryanston H-8
Bryant G-3
Buchanan H-3,I-1
Buena Vista A-1
Buffalo A-7
Burger A-5
Burlingame C-2
Burroughs F-4
Butler B-7
Butternut I-3,J-2
Byron D-2,E-3
California A-3
Calumet H-3
Calvert C-2
Cameron B-4
Campbell K-1
Canfield E. F-7
Canfield W. G-5,H-3
Caniff A-5,B-4
Canton C-8
Cardoni B-4
Carrie B-8,C-8
Carter E-1
Casmer A-5
Cass I-6,J-6
Centre I-7
Chandler D-4
Charest A-6
Charlotte I-5
Chateaufor Pl. I-8
Chene F-7
Cherboneau Pl. I-8
Chicago Blvd. D-1
Chipman K-4
Christopher B-7
Church J-5
Churchill E-2
Civic Center Dr. K-6
Clairmount D-1,D-3
Clark K-2
Clarkdale K-2
Clay C-6,D-5
Clifford I-6
Clinton I-7
Cochiane I-4
Collingwood C-2
Colorado A-3
Columbia E. I-7
Columbia W. I-6
Columbus F-1
Commonwealth F-3,G-4
Commor A-4
Comstock A-7,B-6
Conant C-7
Concord B-8,C-8
Congress E. J-7
Connecticut B-3
Cortland B-2
Cromwell K-4
Custer E-4
Cymbal C-7
Dallas C-5
Dalzelle J-4
Dane D-7
Danforth C-6
Davenport H-5
Davison Expwy. A-1
Deleware E-2,E-3
Delmar B-4
Deming K-2
Denton C-6
Dequindre G-7
Dodge B-7
Domine B-7
Doremus A-6,A-7
Dorothy B-7
Dubois B-5,C-6
Dunderin F-2
Dunn Rd. C-7
Dwyer B-7
Dyar A-4
Eastern Pl. G-1
East Grand Blvd.
.... D-6,D-8
Edison C-3,D-1
Edsel Ford Frwy.
.... E-6,G-3
Edwin A-6,A-7,B-5
Elijah McCoy Dr. F-3
Eliot H-6
Elizabeth J-5
Elizabeth E. I-7
Elizabeth W. I-6
Ellery A-6,D-8,E-8
Elm I-4
Elmhurst B-2,C-1
Elmwood F-8
Endicott E-4
Englewood B-3
Erskine G-8,H-6
Euclid E. D-4
Euclid W. D-3,E-2
Evaline A-5,A-7
Faber B-6
Farmer I-6
Farnsworth E-7,F-5
Farr C-8
Farrand A-3
Ferry E. E-7,F-5
Ferry Park F-2
Field C-8
Filer B-8
Finley D-6
Fisher Frwy I-6,J-3
Fleming A-5
Florian B-5
Ford A-1
Fordyce C-6
Forest Ct. G-4
Forest E. F-7,G-5
Forest W. G-5,H-3
Foster B-8
Fort W. K-6
Franklin J-8
Frederick E-7,F-5
Frontenac C-8
Gage F-3
Gallagher A-6
Garfield F-7,G-5
Geimer B-6
Georgia B-7,B-8
Gibson H-4
Gillett C-5
Girardin B-8
Gladstone D-3,E-1
Glendale A-4,B-1
Glynn Ct. C-2
Goldner J-1
Goodson B-6
Goodwin B-4,C-4
Grand A-1
Grand Haven A-4
Grand River G-1,I-6
Grandy C-6,F-8
Gratiot G-8,I-7
Grayling C-6
Greenley C-5
Griffin C-7
Grinnell A-8
Griswold J-7
Guthrie A-8
Hague C-5,D-4
Hale G-8
Hamilton B-2
Hammond K-1
Hancock H-3,I-1
Hancock E. F-7
Hancock W. G-5
Hanley B-6
Hanover F-2
Harmon B-3
Harper E-5
Harrison I-4
Hartwick C-5
Hastings E-5
Hawthorne B-4
Hazelwood D-1,D-3
Heck Pl. E-8
Hecla F-3,G-3
Hedge A-7
Heintz B-7
Helen B-8,C-8
Hendricks H-8
Hendrie E-7
Henry I-5
Hewitt B-6
Highland B-2
Hindle B-4
Hobart F-4
Hobson I-5
Hogarth F-1
Holborn D-8
Holbrook B-6,C-4
Holmes A-5
Hooker G-2
Hope Pl. G-1
Horatio I-1
Horton D-6,E-4
Howard J-6,K-4
Hubbard K-2
Huber A-8
Hudson G-2
Hughes Ter. F-2
Humboldt I-3
Hunt G-8
Huron I-3
Hyde H-8
Illinois G-6,G-8
Jacob B-6
Jay H-8
Jefferson W. K-6
Jeffries I-5
Jeffries Frwy. I-2
John C. Lodge
Service Dr. D-3,J-6
John R. E-4,I-6
Joliet Pl. I-8
Jos Campau A-6,F-8
Josephine C-4
Junction K-1
Kanter D-7
Kenilworth C-4
King C-4
Kingsley D-4
Kipling E-3
Kirby F-5,F-6,G-4
Kirby E. E-7,F-5
Kirby W. H-1
Klein C-7
Knox C-5
Konkel K-1
Kopernick J-1
LaBelle A-1
LaBrosse J-5
Lafayette E. I-8
Lafayette W. J-6
Lambert D-8
Lambie Pl. K-3
Lamothe F-2
Lanman I-1
Larned E. J-7
La Salle E-1,F-2
La Salle Gds. N. F-1
La Salle Gds. S. F-1
Lawrence C-2
Lawton E-1,I-3
Ledyard I-5
Lee Pl. E-2
Legrand C-7
Lehman B-6
Leicester Ct. C-4
Leland G-6,G-8
Leverette J-5
Liberty C-6
Lincoln B-1,F-3,G-4
Linwood E-1,G-2
Lockwood J-1
Lodi D-8
Longfellow C-3,D-1
Loraine G-2
Lothorp E-3,F-1
Lovett I-1
Lumpkin A-5,C-5
Lyman D-6
Lynn B-4
Lysander H-3
McDougall A-6,F-8
McGraw G-2
McGregor K-1
McKinley I-2
McLean A-3
Mack G-6,H-8
Macomb I-7
Madison I-7
Magnolia I-2,I-3,J-1
Manhattan B-7
Manson K-1
Mansur E-5
Manuel B-7
Maple I-7
Marantette J-4
Marcus B-8
Marjorie A-8
Mark G-3
Marquette G-2
Marston D-4,D-6
Massachusetts A-3
Mechanic I-7
Medbury D-8,E-7
Melbourne D-4
Melrose D-5
Merrick G-4,H-1
Merrill E-3
Merritt K-1
Michigan J-2,J-6
Middle J-6
Milford H-1
Miller B-7,B-8,C-6
Milwaukee D-7,E-4
Missouri G-2
Mitchell A-5,A-6,F-3
Monning Ct. E-6
Monroe I-7
Montclam E. I-7
Monterey B-2
Montgomery F-1
Moore Pl. G-1
Moran E-8
Morrow C-5
Mt. Elliott B-7,D-8
Mt. Vernon D-4
Mulberry I-3
Myrtle H-5,I-3,J-2
Nagle B-8
Nall I-1
Nebraska F-2
Newark K-1,K-4
Newhall B-7
Newton C-6
Nicolet Pl. I-8
Northwestern F-1
Norwalk A-6,A-7,B-5
Oakland A-3
Oliver B-7
Orleans D-6,G-7,I-8
Osborne Pl. E-6
Otis J-1
Owen C-4
Pallister E-2,E-3
Palmer E. E-7,F-5
Palmer W. F-4
Palms K-2
Park Dr. H-8
Parks H-3
Parsons H-5
Pasadena A-1
Pease B-8
Perry I-5,J-3
Peterboro I-5
Philadelphia E. D-4
Philadelphia W.
.... D-3,E-2
Pierce G-8
Pine I-5
Pingree D-3,E-2
Piquette D-7,E-5,E-6
Plaza Dr. J-6
Plum J-5
Plumer K-1
Poe E-3
Poland B-5
Poplar I-2,I-3
Porter K-5
Prentis G-5
Preston G-8
Pulaski A-5
Putnam G-5
Randall K-3
Randolph J-7
Reed Pl. G-4
Rhode Island A-3
Rich I-1
Richardson B-8
Richton B-2
Riopelle C-5,D-6
.... F-7,H-7,J-8
Risdon J-2
Rivard H-7,J-8
Roby E-6
Rockwood B-7
Roland A-8
Roosevlet B-6,G-1,I-2
Rosa G-3
Rosa Parks Blvd. B-1
Rose J-3
Rosedale B-3
Rugg B-8
Ruskin K-3
Russell F-7
Saginaw D-8
St. Antone D-5,F-6
St. Aubin A-4,C-5,F-7,I-8
St. Cyril B-8
St. Hedwig J-1
St. Joseph G-8
St. Thomas C-8
Sallan B-7
Sampson K-4
Sargent D-7
Schewitzer Pl. J-8
Scott G-8
Scotten I-1,K-2
Scovel Pl. G-1
Second B-2,E-4
Selden H-4,H-5
Selkirk B-7
Service H-8
Seward E-2,E-3
Sheehan A-8
Shelby J-7
Sherwood B-8
Sibley I-6
Sloman B-4
Smith C-6,D-4
Sproat I-6
Spruce I-5
Stanley G-2
Sterling F-3
Stimson H-6
Stroh Dr. I-7
Strong C-8
Sturtevant B-1
Superior F-7,G-6
Sycamore I-4
Taylor D-1,D-3
Temple I-5
Tennyson B-3
Theodore F-5,F-7
Third B-2,E-3
Times Square J-6
Toledo K-1
Torrey J-1
Torrey Ct. I-1
Trombly D-6
Trowbridge A-5,B-3
Trumbull A-1,F-3
.... G-4,K-5
Tuscola H-5
Tuxedo B-2
Tyler A-1
Varney C-7
Vermont I-4,K-4
Vernor Hwy. E. H-8
Vernor Hwy. W. K-2
Vicksburg F-1
Vincen B-7
Vinewood K-2
Virginia Park D-3,E-2
Walter P. Chrysler
Frwy. B-5,F-6
Warren E. F-7
Warren W. G-4,H-1
Washington J-7
Washington Blvd. J-6
Watson G-8,H-6
Waverly A-1
Webb B-2
Weitzel Ct. D-8
Wellington C-5
West Grand Blvd.
.... F-1,G-1,J-2
Westminster C-3
Whalin A-5
Whitney F-1
Widman Pl. D-6
Wildemere E-1
Wilkins G-8,H-7
Willis G-6,H-5
Willis E. F-7
Winder H-7,I-6
Winfield B-8
Wing Pl. K-4
Winkleman B-7
Winona A-1
Witherell I-6
Woodbridge J-8
Woodland B-3
Woodrow Wilson
.... B-1,D-2,E-2
Woodward B-2,F-5,J-7
Wreford G-2
Wyandotte B-6
Yemans A-5
Zinow A-5
2nd I-5,J-6
2nd Blvd. E-4
3rd G-5,J-6
4th H-4,J-5
5th J-5
6th K-6
8th H-4,J-5,K-5
10th K-5
14th E-2,I-3,K-4
15th I-3
16th I-3,K-4
17th I-3,K-4
18th I-3,K-4
20th K-3
21st K-3
22nd I-2
23rd I-2
24th I-2
25th I-2,K-3
26th I-2
28th I-1
29th I-1
30th I-1,J-1
31st J-1
32nd J-1

POINTS OF INTEREST

Cass Park I-5
Northwestern Field G-1
Roosevelt Field C-1

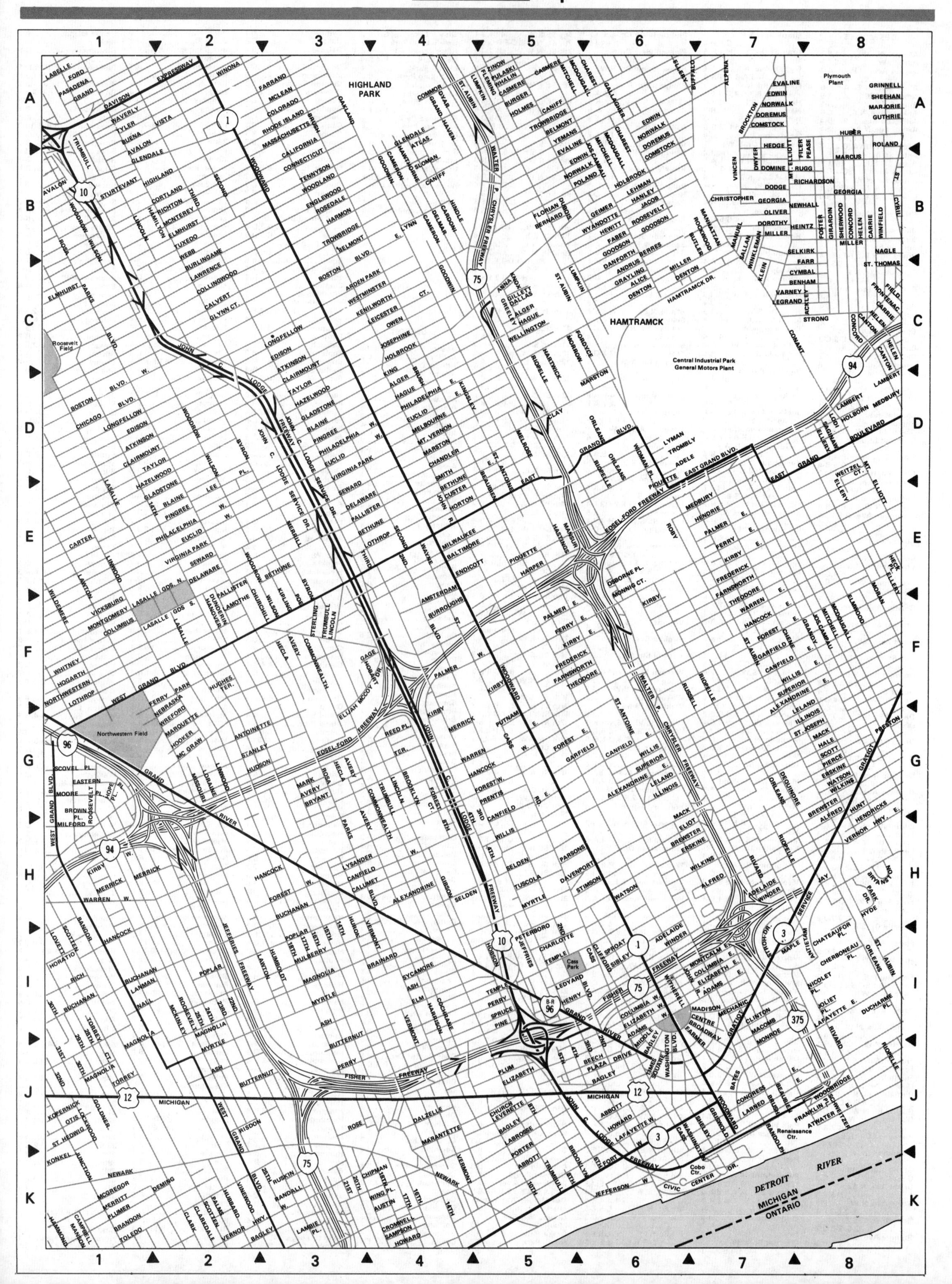
Scale of Miles
0 .2 .4 .6
N
HIGHLAND PARK
HAMTRAMCK
Central Industrial Park
General Motors Plant
Plymouth Plant
Roosevelt Field
Northwestern Field
Cass Park
Cobo Ctr.
Renaissance Ctr.
DETROIT RIVER
MICHIGAN
ONTARIO
DAVISON EXPRESSWAY
WOODWARD
JOHN C. LODGE FREEWAY
CHRYSLER FREEWAY
EDSEL FORD FREEWAY
FISHER FREEWAY
JEFFRIES FREEWAY
GRAND RIVER
GRAND BLVD.
WEST GRAND BLVD.
EAST GRAND BLVD.
MICHIGAN
GRATIOT
JEFFERSON
CIVIC CENTER DR.
HAMTRAMCK DR.
CONANT
TRUMBULL
LINWOOD
DEXTER
OAKLAND
ST. AUBIN
CASS
SECOND
THIRD
JOSEPH CAMPAU
WARREN
FOREST
CANFIELD
MACK
VERNOR HWY.
FORT
LAFAYETTE
1
2
3
4
5
6
7
8
A
B
C
D
E
F
G
H
I
J
K

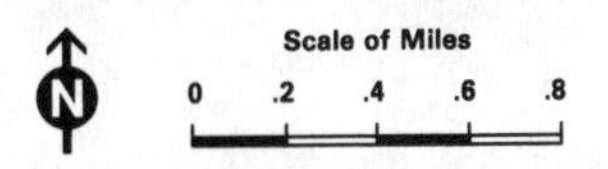
N
Scale of Miles
0 .2 .4 .6 .8

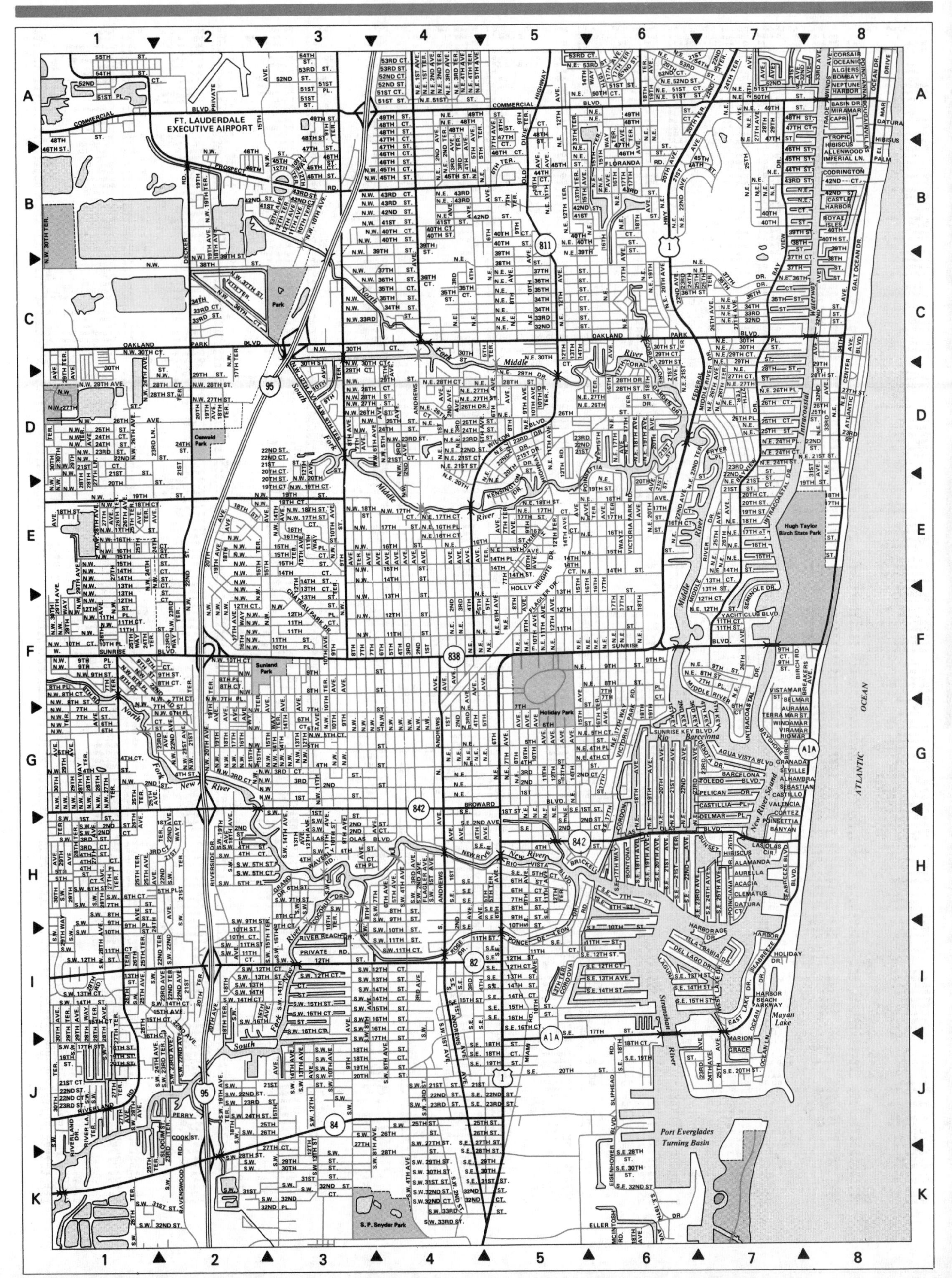
FT. LAUDERDALE EXECUTIVE AIRPORT
Park
Oaswald Park
Sunland Park
Holiday Park
Hugh Taylor Birch State Park
S. P. Snyder Park
Port Everglades Turning Basin
ATLANTIC OCEAN
Mayan Lake
Intracoastal Waterway
New River Sound
Stranahan River
Middle River
North Fork
South Fork
New River
COMMERCIAL BLVD.
OAKLAND PARK BLVD.
SUNRISE BLVD.
BROWARD BLVD.
LAS OLAS BLVD.
FEDERAL HWY
GALT OCEAN DR.
95
811
838
842
82
84
A1A
1

FT. LAUDERDALE

Acacia ... H-7
Agua Vista Blvd. ... G-7
Alamanda ... H-7
Algiers ... A-8
Alhambra ... G-7
Allenwood ... B-8
Andrews Ave. ... D-4
Andrews Ave., N. ... G-4
Andrews Ave., S. ... H-4,I-4
Atlantic Blvd. ... D-8
Aurama ... G-7
Aurelia ... H-7
Banyan ... H-7
Barcelona ... G-7
Basin Dr. ... A-8
Bayshore ... G-7
Bayview Dr. ... C-7,E-7
Belmar ... G-7
Birch Rd. ... F-7,G-7
Bombay ... A-8
Bontona ... H-6
Breakers Ave. ... F-8
Brickell ... H-5
Broward Blvd. ... G-4
Capri ... A-8
Castle Harbor ... B-8
Castilla Pl. ... G-7
Castillo ... G-7
Center Ave., N.E. ... D-8
Chateau Park Dr. ... F-3
Clematus ... H-7
Coconut Dr. ... H-3
Codrington ... B-8
Commercial Blvd.
... A-1,A-5
Cook St. ... J-2
Coral Gardens Dr.
... C-6,D-6
Coral Shore Dr.
... C-6,D-6
Cordon ... H-6
Cordova ... I-5
Corsair ... A-8
Cortez ... H-7
Datura ... A-8
Datura Ct. ... H-7
Decker Rd. ... B-2
Del Lago Dr. ... I-6
Delmar Pl. ... H-7
Desota Dr. ... G-7
Dixie Hwy. ... E-5
East Lake Dr. ... I-7
Eisenhower Blvd. ... K-6
Eller Dr. ... K-6
El Mar Dr. ... B-8
Federal Hwy. ... D-7
Flagler Ave. ... H-4
Flagler Dr. ... F-5
Floranda Rd. ... B-6
Fryer Pl. ... D-7
Galt Ocean Dr. ... C-8
Garden Dr. ... D-8
Grace ... J-7
Granada ... G-7
Grand Dr. ... H-3
Harbor ... A-8,I-7
Harborage Dr. ... I-6,I-7
Harbor Beach Pkwy. ... I-7
Hibiscus ... B-8,H-7
Holiday Dr. ... I-7
Holly Heights Dr. ... E-5
Imperial Ln. ... B-8
Intracoastal Dr.
... E-7,G-7
Isle Babia Dr. ... I-6
Kensington Dr. ... E-5
Kensington St. ... E-5
Laguna ... I-6
Las Olas Blvd. ... H-3,H-6
Las Olas Cir. ... H-7
Marion ... J-7
McIntosh Rd. ... K-6
Miami Rd. ... J-5
Middle River Dr. ... D-7,F-6
Miramar ... A-8
Neptune ... A-8
New River ... H-4
Oakland Park Blvd.
... C-1,C-2,C-6,C-7
Ocean Dr. ... A-8,I-7
Oceanic ... A-8
Ocean Ln. ... J-7
Old Dixie Hwy. ... A-5,B-5
Palm ... B-8
Pelican Dr. ... G-7
Perry ... J-2
Poinciana
Bougainvillas ... A-8
Poinciana Dr. ... H-7
Poinsettia ... H-7
Poinsettia St. ... D-6,E-5
Ponce De Leon Dr. ... I-5
Private ... A-2
Private Rd. ... I-3
Prospect Rd. ... B-2
Ravenswood Rd. ... K-2
Riomar ... G-7
Rio Vista Blvd. ... H-5
Riverland Dr. ... J-1
Riverland Rd. ... J-1
Riverland Ter. ... J-1
River Reach Dr. ... I-5
Riverside Dr. ... H-6
Rose Dr. ... I-4
Royal Isles ... B-8
Seabreeze Blvd.
... H-7,I-7
Sebastian ... G-7
Seminole Dr. ... F-7
Seville ... G-7
Sliphead Rd. ... J-6
Slocum St. ... J-2
Sunrise Blvd. ... F-6
Sunrise Blvd, W. ... F-1
Sunrise Key Blvd. ... G-6
Sunset ... H-7
Terra Mar St. ... G-7
Toledo Blvd. ... G-7
Tradewinds Ave. ... A-8
Tropic ... A-8
Victoria Park Rd.
... E-6,G-6
Vistamar St. ... F-7
Waverly Rd. ... H-3
West Lake Dr. ... I-7
Wilton Blvd. ... D-5
Windamar ... G-7
Yacht Club Blvd. ... F-7
1st Ave. N.E. ... A-4,D-4,G-4
1st Ave. N.W. ... F-4,G-4
1st Ave. S.E. ... H-4,I-4
1st Ave. S.W. ... H-4,I-4
1st Ct. ... H-2
1st Key ... G-6
1st St. S.E. ... G-4,G-5
1st St. S.W. ... G-1,G-3
1st St. N.E. ... G-4
1st St. N.W. ... G-2,G-3
1st Ter. N.E. ... A-4
2nd Ave. N.E. ... A-4,B-4
... D-4,F-4,G-4
2nd Ave. N.W. ... D-4
... E-4,G-4
2nd Ave. S.E. ... H-4,I-4
2nd Ave. S.W. ... H-4
2nd Ct. ... G-5
2nd Ct. S.E. ... H-5
2nd Ct. S.W. ... H-1,H-3
2nd Key ... G-6
2nd St. N.E. ... G-4
2nd St. N.W. ... G-1,G-2,G-3
2nd St. S.E. ... H-5
2nd St. S.W. ... H-1
... H-2,H-3,K-4
2nd Ter. N.E. ... A-4,B-4
3rd Ave. N.E. ... A-4,B-4
... C-4,D-4,F-4,G-4
3rd Ave. N.W. ... D-4,G-4
3rd Ave. S.E. ... I-4
3rd Ave. S.W. ... H-4,I-4
3rd Ct. ... H-1
3rd Ct. N.W. ... G-2,G-3
3rd Key ... G-6
3rd St. N.E. ... G-4
3rd St. N.W. ... G-3
3rd St. S.W. ... H-1,J-4
3rd Ter. N.E. ... A-4,B-4
4th Ave. N.E. ... A-4,B-4
... C-4,F-4,G-4
4th Ave. N.W. ... F-4,G-4
4th Ave. S.E. ... H-4,I-4
4th Ave. S.W. ... H-4,K-4
4th Ct. ... G-1,H-1
4th Ct. N.E. ... G-5
4th Ct. S.W. ... H-2,H-3
4th Key ... G-7
4th Pl. ... H-1,H-3
4th Pl. N.E. ... G-5
4th St. ... G-1,G-2,H-1
4th St. N.E. ... G-4
4th St. N.W. ... G-1,G-2
4th St. S.E. ... H-4,H-5
4th St. S.W. ... H-2,H-3
4th Ter. N.E. ... A-4
5th ... H-5
5th Ave. ... D-4,H-4
5th Ave. N.E. ... A-5,B-4
... D-5,F-4,G-5
5th Ave. N.W. ... D-4,G-4
5th Ct. ... H-1
5th Ct. N.E. ... G-5
5th Ct. N.W. ... G-3
5th Ct. S.E. ... H-5
5th Ct. S.W. ... H-2
5th Key ... G-7
5th Pl. ... H-2
5th Pl. S.W. ... H-1
5th St. ... H-1
5th St. N.E. ... G-4
5th St. N.W. ... G-1,G-2
5th St. S.W. ... H-2
5th Ter. ... C-5,H-5
5th Ter. N.E. ... B-5,C-5
6th Ave. N.E. ... B-5,C-5
... D-5,F-5,G-5
6th Ave. N.W. ... D-4,F-4,G-4
6th Ave. S.E. ... H-5,I-5
6th Ave. S.W. ... H-4,J-4
6th Ct. ... G-3,G-7
6th Ct. N.W. ... G-1,G-2
6th Ct. S.E. ... H-4
6th Ct. S.W. ... H-1
6th Pl. N.W. ... G-2
6th Pl. S.W. ... H-1
6th St. ... G-3
6th St. N.E. ... G-4
6th St. N.W. ... G-1,G-2
6th St. S.E. ... H-4
6th St. S.W. ... H-1
6th Ter. ... B-5
7th S.W. ... H-3
7th Ave. ... D-3
7th Ter. N.W. ... F-3,G-3
7th Ave. N.E. ... A-5
... F-5,G-5
7th Ter. ... B-5,E-5
7th Ave. N.W. ... F-3,G-3
7th Ct. ... F-6
7th Ct. N.W. ... F-1,F-2
7th Pl. ... F-6,F-7
7th Pl. N.W. ... F-2
7th St. N.E. ... G-4
7th St. N.W. ... F-1
... G-1,G-2,G-3
7th St. S.E. ... H-4,H-6
7th St. S.W. ... H-1,H-3
8th Ave. ... D-3
8th Ave. N.E. ... A-5
... C-5,D-5,F-5,G-5
8th Ave. N.W. ... F-3,G-3
8th Ave. S.W. ... H-3,I-3,K-3
8th Ct. ... F-1,F-2,H-3
8th Ct. N.W. ... F-1
8th Pl. ... F-1,F-2
8th Pl. N.W. ... F-1
8th Rd. ... F-1
8th St. N.E. ... F-6,F-7
8th St. N.W. ... F-1,F-2
8th St. S.E. ... H-5,H-6
8th St. S.W. ... H-1,H-3
9th ... F-3
9th Ave. ... H-3
9th Ave. N.E. ... B-5,D-5
... E-5,F-5,G-5
9th Ave. N.W. ... G-3
9th Ave. S.E. ... H-5
9th Ave. S.W. ... J-3
9th Ct. ... F-7,H-3
9th Ct. N.W. ... F-1,F-2
9th Pl. ... F-6
9th Pl. N.W. ... F-1
9th Pl. S.W. ... H-1
9th St. ... F-7
9th St. N.E. ... F-5,F-7
9th St. N.W. ... E-3,F-3
9th St. S.E. ... I-5
9th St. S.W. ... H-1,H-2,H-4
9th Ter. ... A-5
9th Ter. N.W. ... A-3,D-3
10th Ave. N.E. ... B-5,C-5
... D-5,E-5,G-5
10th Ave. N.W. ... B-3
... D-3,E-3,F-3,G-3
10th Ave. S.E. ... I-5
10th Ave. S.W. ... J-3
10th Ct. ... I-2
10th Ct. N.W. ... F-1,F-2
10th Pl. ... F-1
10th Pl. N.W. ... F-2,F-3
10th St. ... B-3
10th St. S.E. ... I-4,I-6
10th St. S.W. ... I-1,I-2,I-4
10th Ter. ... B-3,D-5,I-3
10th Ter. N.W. ... F-3,G-3
10th Way ... B-3
11th Ave. ... B-3
11th Ave. N.E. ... B-5
... D-5,F-6,G-5
11th Ave. N.W. ... G-3
11th Ave. S.E. ... H-5
11th Ave. S.W. ... H-3
11th Ct. ... F-7
11th Ct. N.W. ... F-1,F-2,F-3
11th Ct. S.E. ... I-5
11th St. S.W. ... I-1,I-2,I-4
11th Pl. N.W. ... F-1,F-2,F-3
11th St. ... F-7,I-4
11th St. N.E. ... F-5
11th St. N.W. ... F-1
... F-2,F-3,F-4
11th St. S.E. ... I-5,I-6
11th St. S.W. ... I-2,I-4
11th Way ... E-3
12th ... C-5,K-3
12th Ave. ... B-3,D-3,E-3
12th Ave. N.E. ... E-6
12th Ave. N.W. ... D-3,G-3
12th Ave. S.W. ... H-3,J-3
12th Ct. ... F-7
12th Ct. N.W. ... F-1,F-2,F-3
12th Ct. S.E. ... I-5,I-6
12th Rd. ... D-5
12th St. N.E. ... F-4,F-6,F-7
12th St. N.W. ... F-1
... F-2,F-3,F-4
12th St. S.E. ... I-5,I-6
12th St. S.W. ... I-1,I-3,K-3
12th Ter. ... E-3,H-5
12th Ter. N.E. ... B-5,E-5
13th ... C-5
13th Ave. ... B-3
13th Ave. N.E. ... G-5
13th Ave. N.W. ... G-3
13th Ave. S.E. ... I-6
13th Ave. S.W. ... H-3,J-3
13th Ct. ... E-7
13th Ct. N.W.
... E-1,E-2,E-3
13th Ct. S.W. ... I-1,I-2
13th St. ... F-7
13th St. N.E. ... F-4,F-5
13th St. N.W. ... E-1,E-3
13th St. S.E. ... I-5,I-6
13th St. S.W. ... I-1,I-2,I-3
13th Ter. ... A-5,B-5,I-5
14th ... C-5,E-5
14th Ave. N.E. ... E-5,G-5
14th Ave. N.W. ... E-3,G-3
14th Ave. S.W. ... H-3,I-3,J-3
14th Ct. ... E-5
14th Ct. N.W. ... E-1,E-2,E-3
14th Ct. S.E. ... I-5
14th Ct. S.W. ... I-1,I-2,I-3
14th Pl. ... E-5
14th St. ... E-5
14th St. N.E. ... E-6,E-7
14th St. N.W. ... E-1
... E-2,E-3,E-4
14th St. S.E. ... I-5,I-6
14th St. S.W. ... I-1,I-2,I-3
14th Ter. N.W. ... G-3
14th Way ... A-6
14th Way N.W. ... G-3
15th S.W. ... I-3
15th Ave. ... I-1
15th Ave. N.E. ... B-6
... E-5,F-5,G-5
15th Ave. N.W. ... A-3
... B-3,E-3,G-3
15th Ave. S.W. ... J-3
15th Ct. ... I-1
15th Ct. N.W. ... E-1,E-2,E-3
15th Ct. S.W. ... I-3
15th Pl. N.W. ... E-2,E-3
15th St. N.E. ... E-6,E-7,E-8
15th St. N.W. ... E-1
... E-2,E-3,E-4
15th St. S.E. ... I-4,I-5,I-6
15th Ter. N.E. ... B-6,D-6
15th Ter. N.W. ... E-2,G-2
15th Ter. S.W. ... I-2
15th Way ... G-3
15th Way N.E. ... B-6
16th ... E-3
16th Ave. ... I-2
16th Ave. N.E. ... C-6,D-6
16th Ave. N.W. ... G-2
16th Ct. ... I-1
16th Ct. N.E. ... E-4,E-5,E-6
16th Ct. N.W. ... E-1,E-2
16th Ct. S.E. ... I-5
16th Ct. S.W. ... I-3
16th Pl. N.E. ... E-4
16th St. N.E. ... E-4,E-5,E-6
16th St. N.W. ... E-1
... E-2,E-3,E-4
16th St. S.E. ... I-5
16th St. S.W. ... I-2,I-3
16th Ter. N.E. ... B-6
... F-6,G-6
16th Way ... F-2
17th ... E-8
17th S.E. ... H-6
17th Ave. ... D-6,G-6
17th Ave. N.E. ... A-6
... B-6,C-6,D-6,F-6
17th Ave. N.W. ... F-2,G-2
17th Ct. ... E-6
17th Ct. N.E. ... E-4,E-5
17th Ct. N.W. ... E-3,E-4
17th St. ... E-5
17th St. N.E. ... E-5,E-6,E-7
17th St. N.W. ... E-1,E-3,E-4
17th St. S.E. ... I-5
17th St. S.W. ... J-1,J-3
17th Ter. N.E. ... A-6
... B-6,D-6,F-6
17th Ter. N.W. ... C-2
17th Way N.E. ... F-6,G-6
17th Way S.E. ... H-6
18th ... D-6,E-8,G-6
18th Ave. ... B-2
18th Ave. N.E. ... B-6,C-6
... D-6,E-5,E-6,F-6,G-6
18th Ave. N.W. ... E-2,G-2
18th Ave. S.E. ... H-6
... J-6,K-6
18th Ave. S.W. ... H-2,I-2
18th Ct. N.W. ... E-1,E-2,E-3
18th Ct. S.E. ... J-4,J-6
18th Ct. S.W. ... J-3
18th St. ... E-2
18th St. N.E. ... E-5,E-7
18th St. N.W. ... E-3,E-4
18th St. S.E. ... J-4
18th St. S.W. ... J-1,J-3
18th Ter. ... D-2,I-2
18th Ter. N.E. ... A-6
18th Ter. S.W. ... J-2
19th Ave. N.E. ... A-6
... C-6,D-6,G-6
19th Ave. N.W. ... B-2
... D-2,E-2,G-2
19th Ave. S.E. ... H-6,K-6
19th Ave. S.W. ... H-2,J-2
19th Ct. N.W. ... E-3
19th St. ... E-6,E-7
19th St. N.W. ... E-1,E-2,E-3
19th St. S.E. ... J-4,J-6
19th St. S.W. ... J-3
19th Ter. N.W. ... B-2,G-2
20th ... D-2
20th N.E. ... D-4
20th Ave. ... I-2
20th Ave. N.E. ... B-6
... C-6,E-6,G-6
20th Ave. N.W. ... E-2
20th Ave. S.E. ... H-6
20th Ct. ... D-3,E-7
20th Dr. N.E. ... D-5
20th St. ... D-3,E-7,J-1
20th St. N.E. ... D-5
20th St. N.W. ... D-1,D-3,E-1
20th St. S.E. ... J-5,J-7
20th St. S.W. ... J-3
20th Ter. ... A-7,I-2
21st ... A-7,D-5
21st Ave. ... D-6
21st Ave. N.E. ... B-6,G-6
21st Ave. N.W. ... G-2
21st Ave. S.E. ... H-6
21st Ct. ... C-1
21st Ct. N.E. ... D-4,E-7
21st Ct. N.W. ... D-1,D-4
21st Dr. ... D-5
21st St. ... D-3,D-6
21st St. N.E. ... D-4,D-8,E-7
21st St. N.W. ... D-1,D-3
21st St. S.W. ... J-3,J-4
21st Ter. ... D-2
21st Ter. N.E. ... D-6
21st Ter. N.W. ... G-2
21st Ter. S.W. ... H-2,I-2
21st Way ... I-2
22nd ... D-5
22nd Ave. ... I-2
22nd Ave. N.E. ... B-6
... C-6,E-6,G-6
22nd Ave. N.W. ... G-2
22nd Ave. S.E. ... H-6,J-7
22nd Ave. S.W. ... I-2,J-2
22nd Ct. ... D-3,J-1
22nd Dr. N.E. ... D-5
22nd Rd. N.W. ... F-1
22nd St. ... D-3,D-8,E-7,J-1
22nd St. N.E. ... D-4
22nd St. N.W. ... D-1,E-2
22nd St. S.E. ... J-4
22nd St. S.W. ... J-2,J-4
22nd Ter. ... H-2,I-2
22nd Ter. N.E. ... A-7,E-6
23rd ... C-6,D-5,G-7
23rd Ave. N.E. ... G-6
23rd Ave. S.W. ... J-2
... I-2,J-2
23rd Dr. N.E. ... D-5
23rd Ln. ... D-1
23rd St. ... D-7,J-1
23rd St. N.E. ... D-4,D-8
23rd St. N.W. ... D-1,D-4
23rd St. S.E. ... J-4
23rd St. S.W. ... J-2,J-4
23rd Ter. N.W. ... F-2
23rd Ter. S.W. ... J-2,K-2
23rd Way ... F-2
24th ... C-6
24th Ave. N.W. ... D-1,E-1
24th Ave. S.E. ... H-7,J-7
24th Ave. S.W. ... J-1
24th Ct. ... D-7
24th Ct. N.E. ... D-7
24th Ct. N.W. ... D-1
24th Ct. S.W. ... J-1
24th St. N.E. ... D-4,D-8
24th St. N.W. ... D-1,D-3,D-4
24th St. S.W. ... J-2
24th Ter. ... F-1
24th Ter. N.E. ... A-7
24th Ter. N.W. ... E-1
25th ... C-7
25th Ave. ... E-1,G-1,H-1,I-1
25th Ave. N.W. ... B-7,E-7
25th Ave. S.E. ... H-7,J-7
25th Ct. N.E. ... D-7
25th Pl. ... D-7
25th St. ... D-8
25th St. N.E. ... D-4,D-7
25th St. N.W. ... D-1,D-3,D-4
25th St. S.W. ... J-2,J-4
25th Ter. ... C-7,E-1
... G-1,H-1,I-1,K-1
25th Way ... F-1
26th ... D-8,H-7
26th Ave. N.E. ... A-7
... C-7,D-7,E-7,F-7
26th Ave. N.W. ... D-1,E-1
26th Ave. S.W. ... I-1
26th Ct. N.E. ... D-7
26th Dr. N.E. ... D-4,D-5,D-6
26th Pl. N.E. ... D-7
26th St. N.E. ... D-4,D-5
26th St. N.W. ... D-3
26th St. S.W. ... J-2,J-3,J-4
26th Ter. ... E-1,I-1
26th Ter. N.E. ... D-7
26th Ter. S.W. ... K-1
27th ... J-3
27th Ave. ... C-7
27th Ave. N.E. ... A-7,B-7
27th Ave. N.W. ... E-1
27th Ave. S.W. ... H-1,J-1
27th Ct. ... D-7,J-2
27th Dr. N.E. ... D-4,D-5
27th Ln. ... D-1
27th St. N.E. ... D-4
... D-5,D-6,D-7
27th St. N.W. ... D-1,D-2,D-3
27th St. S.E. ... J-4
27th St. S.W. ... J-4
27th Ter. ... D-7,H-1
... I-1,J-1
27th Ter.N.W. ... G-1
28th Ave. N.E. ... A-7,B-7
28th Ave. N.W. ... D-1
... E-1,F-1,G-1
28th Ave. S.W. ... H-1
... I-1,J-1
28th Ct. ... D-2
28th St. N.W. ... D-3,D-4
28th Rd. ... I-1
28th St. N.E.
... D-6,D-7,D-8
28th Dr. ... D-6
28th St. N.W. ... D-1,D-2
... D-3,D-4
28th St. S.E. ... K-4,K-6
28th St. S.W. ... K-2,K-3
28th Ter. N.W. ... D-1,G-1
28th Ter. S.W. ... H-1,I-1,J-1
28th Way N.W. ... G-1
28th Way S.W. ... I-1,J-1
29th S.E. ... K-4
29th Ave. ... D-1,F-1,G-1
29th Ave. N.E. ... A-7,B-7
29th Ave. N.W. ... D-1
29th Ave. S.W. ... H-1,I-1
29th Ct. N.E. ... C-7
29th Dr. N.E. ... D-4,D-5
29th Pl. N.E. ... D-6
29th St. N.E. ... C-7,D-8
29th St. N.W. ... D-2,D-3,D-4
29th St. S.W. ... K-2,K-4
29th Ter. N.W. ... D-1,G-1
29th Ter. S.W. ... H-1,I-1
29th Way ... F-1,I-1
30th Ave. N.W. ... D-1
... F-1,G-1
30th Ave. S.W. ... H-1,I-1
30th Ct. N.E. ... C-5
30th Ct. N.W. ... C-1,C-3
30th Pl. N.E. ... C-7
30th St. ... C-1,C-3
... C-4,C-6
30th St. N.E. ... C-7
30th St. S.E. ... K-4,K-6
30th St. S.W. ... K-2,K-3,K-4
30th Ter. N.W. ... B-1,E-1
30th Ter. S.W. ... I-1
31st ... A-7
31st Ave. N.E. ... B-8
31st Ave. ... A-7,E-8
31st St. S.W. ... K-1
... K-2,K-3,K-4
32nd Ave. ... A-8,C-8
32nd Ave. N.E. ... E-8
32nd Ct. N.E. ... C-8
32nd Ct. S.E. ... K-4
32nd Ct. S.W. ... K-2,K-4
32nd Pl. ... K-2
32nd St. N.E. ... C-5,C-7
32nd St. S.E. ... K-4,K-6
32nd St. S.W. ... K-1,K-3,K-4
33rd Ave. ... C-8
33rd Ave. N.E. ... D-8
33rd Ct. ... C-2
33rd St. ... C-2
33rd St. N.E. ... C-5,C-7
33rd St. N.W. ... C-3,C-4
33rd St. S.E. ... K-4
33rd St. S.W. ... K-4
34th ... C-2
34th Ave. N.E. ... C-8,D-8
34th Ct. N.E. ... C-5
34th St. N.E. ... C-5,C-7
34th St. N.W. ... C-3,C-4
35th ... C-4
35th Ct. N.E. ... C-7
35th Dr. N.E. ... C-6
35th St. N.E. ... C-5,C-7
35th St. N.W. ... C-3,C-4
36th Ct. ... C-2
36th Ct. N.W. ... C-2
36th St. ... C-5
36th St. N.E. ... C-5,C-7,C-8
36th St. N.W. ... C-3,C-4
37th Ct. ... C-7
37th Dr. N.E. ... C-7
37th St. ... C-8
37th St. N.E. ... C-5
37th St. N.W. ... C-2,C-3,C-4
38th ... C-8
38th St. ... B-7
38th St. N.W. ... C-1
... C-2,C-4,C-5
39th ... B-8
39th St. ... B-4,B-5,B-6
39th St. N.E. ... B-5,B-6
39th St. N.W. ... B-2,B-3,B-4
40th ... B-8
40th St. ... B-4,B-5
... B-6,B-7
40th Ct. N.W. ... B-4
40th St. ... B-5,B-6
... B-7,B-8
40th St. N.E. ... B-5,B-6
40th St. N.W. ... B-3,B-4
41st St. ... B-3,B-4
... B-5,B-6
41st St. N.W. ... B-3,B-4
42nd Ct. ... B-3,B-8
42nd St. ... B-2,B-3,B-5
42nd St. N.E. ... B-4
... B-5,B-8
42nd St. N.W. ... B-3,B-4
43rd Ct. N.E. ... B-4,B-5
43rd Ct. N.W. ... B-3,B-4
43rd St. ... B-3
43rd St. N.E. ... B-4,B-6,B-7
43rd St. N.W. ... B-3,B-4
44th Ct. ... B-5
44th St. ... B-6
44th St. N.E. ... B-7
45th Ct. ... B-3
45th Ct. N.W. ... B-4
45th St. ... B-3,B-4
... B-5,B-7,B-8
45th St. N.W. ... B-4
46th Ct. ... B-4,B-5
46th St. ... A-1,A-2
... B-2,B-4,B-7
46th St. N.E. ... B-5,B-6
46th St. N.W. ... B-2
47th Ct. ... A-4,A-5,A-7
47th St. ... B-3,B-4
47th St. N.E. ... A-7,B-6
48th Ct. ... A-5
48th Ct. N.E. ... A-4,A-5,A-6
48th St. ... A-1,A-3
... A-4,A-5,A-7
49th St. ... A-3,A-4
49th St. N.E. ... A-4,A-6,A-7
50th Ct. N.E. ... A-5,A-6,A-7
51st Ct. ... A-4
51st Ct. N.E. ... A-6
51st Pl. ... A-1,A-3
51st St. ... A-3,A-4,A-6
51st St. N.E. ... A-4
... A-6,A-7
52nd Ct. ... A-1,A-4,A-6
52nd St. ... A-3,A-4
... A-6,A-7
52nd St. N.E. ... A-6
53rd Ct. ... A-4,A-5
53rd St. ... A-3,A-4
53rd St. N.E. ... A-5
... A-6,A-7
54th St. ... A-1,A-3
54th St. N.E. ... A-6,A-7
55th St. ... A-1

POINTS OF INTEREST

Ft. Lauderdale
Executive Airport ... A-2
Holiday Park ... F-5,G-5
Hugh Taylor Birch
State Park ... E-7,E-8
... F-7,F-8
Osswald Park ... D-2
S.P. Snyder Park
... K-3,K-4
Sunland Park ... F-2,F-3

FORT WORTH

A Ave. ... G-7
Abney ... J-5
Ada St. ... H-7
Adams St. ... G-3,I-3,K-3
Adolph ... E-2
Adrian ... E-1
Akers ... E-5
Alabama ... F-3
Alcannon St. ... K-7
Alice St. ... K-5
Allen Ave. ... H-3,H-5
Alloway ... J-6
Alston Ave. ... G-3
Alta View ... A-6
Alvin ... K-7
Alvord ... E-6
Amont ... D-4
Ampspoker ... K-5
Amy ... H-6
Andy Ave. ... F-8
Angle ... B-1
Ann ... G-6
Anna St. ... J-6
Annabell ... J-6
Annglenn ... J-6
Annie St. ... G-4,G-5
Arch St. ... G-8,H-8
Arch Adams ... E-1
Arizona ... G-4,H-4
Arlington ... H-3,H-5,H-6
Armour ... C-4
Arnold ... D-7
Arthur ... E-2
Ash Crescent ... G-6,I-6
Aster ... B-5,B-6
Atkins St. ... J-5
Aurline ... E-5
Austin ... D-6,E-2
Aval ... F-8
Avenue ... B-3
Avondale ... H-1
Ayers Ave. ... F-8,G-8
Azalea ... E-1
B Ave. ... G-6,G-7
Bailey ... E-1
Baker Lane ... B-8
Baker St. ... I-1
Baldwin St. ... J-3,K-3
Ballinger ... F-3
Balsam ... B-5
Baltimore Ave.
... H-3,H-5
Barbella ... B-6
Barclay ... C-6
Barden ... E-1
Baylor St. ... E-1
Beach St. ... A-6
... D-6,D-7,E-7,G-7
Beckham ... G-2
Bedsoe ... F-1
Belford Ct. ... F-8
Belknap ... A-8,D-5,E-3
Belmead St. ... K-4
Belmont ... E-1
Belzise ... G-6
Belzise Terr. ... I-6
Benbrook Dr. ... J-1,J-2
Benhal Ct. ... H-2
Benjamin ... E-2
Berke Rd. ... K-5
Berner ... A-5
Bernice ... B-7
Berry St. ... I-7,I-8
... J-3,J-4
Berryhill ... H-8
Bessie ... F-5
Betty ... C-7
Beverly Ave. ... H-5
Bewick ... J-1,J-3
Biddison ... J-1,J-2,J-4
Bideker Ave. ... H-7
Binkley St. ... G-7,H-7,I-7
Bird ... C-5
Birdville ... C-8
Bishop ... I-8
Blanch ... E-1
Blandin ... B-5,C-5
Blalock Ave. ... K-4,K-5
Blevins ... D-6
Blitton ... K-5
Blodgett Ave. ... K-5
Blue ... C-5
Bluebird ... B-4
Blue Grove ... D-6
Blue Smoke Ct. ... F-6
Boland ... E-1
Bolt ... K-1
Bomar ... F-6,F-7
Bonnett ... B-5,C-5
Bonnie Brae ... B-6,C-6
Booker ... E-5
Boston Ave. ... F-8,G-8
Bowie St. ... I-1,I-3,I-4
Boyd Ave. ... I-1,K-1
Brady ... K-1
Brandies ... E-6
Bransford ... G-5
Braswell ... A-4
Brenning ... B-3
Brents ... D-4
Bright ... I-8
Bristol ... E-4
Brittian ... B-6
Broadway ... F-2,F-3
... F-5,F-6
Brookshire ... E-4
Brown ... D-6
Bruce ... A-5
Bryan Ave. ... G-4,I-4,K-4
Buck St. ... G-2
Bundie St. ... K-7
Burchill ... H-7
Burchill Ct. ... H-7
Burnett ... E-3,F-3
Burson ... J-4
Burton St. ... I-8
Buster Ct. ... C-6
Butler ... J-1,J-2
... K-2,K-4,K-5
Buxton ... G-5
Calhoun ... B-2,D-3
... E-3,F-4
Calvert ... E-2
Calvin ... I-6
Camilla ... G-6
Campbell St. ... G-8
... H-8,I-8
Canberra ... H-6
Cannon ... G-3,G-5
Cantey ... I-1,I-3
... I-4,I-5
Canyon Ct. ... G-8
Capps St. ... I-3
Cardinal ... B-4
Carlock St. ... H-3
Carnation ... B-5,B-6
Carnes ... K-5
Carroll ... E-2
Carter ... F-6
Carter Park Dr. ... K-5
Carver ... C-4
Casablanca ... E-1
Castleman ... I-8
Catalpa ... B-8,C-7
C Ave. ... G-6,G-7
Cedar ... F-5
Central ... C-2
Chambers ... E-4
Chandler ... C-5,E-6
Chandler Dr. ... B-6
Chenault ... D-5
Cherry ... F-3
Cheryl ... B-6
Chase Ct. ... H-3
Chester ... B-4,F-6
Chester Boyer Rd. ... A-6
Chestnut ... B-1
Chicago Ave. ... F-8,G-8
Chickasaw Ave. ... J-8
Childress ... J-7
Christin ... H-7
Circle ... D-7
Circle Park Blvd. ... C-2
Circle Pkwy. ... H-1
Clairemont Ave. ... F-8,G-8
Clara St. ... G-1
Clarence ... C-7,D-7
Clary ... C-6
Clearview ... K-6
Cleckler ... C-6
Clement ... G-6,G-7
Cleveland ... F-4
Cleyburn ... J-2
Clinton ... A-2,B-2,C-2
Coer ... K-1
Cobb ... F-3
Cobb Park Dr. ... H-6
Cockerell Ave. ... I-1,J-1,K-1
Cole St. ... K-4
Cole Spring Rd.
... C-4,D-4
Collard ... I-7
College St. ... G-3
... I-3,J-3
Collier ... F-3
Collin ... J-7,J-8
Collins ... H-1
Colonial ... D-6
Columbus ... B-1
Colvin Ave. ... I-4,I-5
Comanche ... I-7
Commerce ... B-2
Commercial ... E-2
Concord ... J-6,K-4
Congress ... E-2
Conkling ... A-7
Conway ... D-6,D-7
Cook ... D-4
Cooper ... G-4
Corner Ave. ... G-7
Corpus Ct. ... J-3
Cosgrove Ct. ... A-8
Cottonwood ... D-6
Court ... G-7
Covella ... D-6
Coventry ... E-1
Crawford St. ... G-4
Creach Rd. ... B-7
Crenshaw St. ... H-7
Crest ... J-6
Crockett ... E-1
Cromwell ... F-6
Crump ... E-4
Cullen ... D-2
Cummings ... E-3
Currie ... E-1
Cutter St. ... K-7
Cypress ... F-5
D Ave. ... G-6,G-7
Daggett ... F-2-4
Dakota ... E-2
Daisy La. ... B-5
Dalford ... C-5
Dallas-Ft. Worth
Tpk. ... F-5
Damon ... D-6
Darby ... H-8
Darcy ... E-1
Dashwood ... G-3,G-4
David ... B-6
Davis Blvd. ... H-5
Dayton ... B-8,C-5
Debbie St. ... K-5
Decatur ... A-3
Decosta ... D-6,E-6
Degar ... G-1
Delaware St. ... H-5
Delga ... C-4
Dell ... B-7,C-5
Denair ... E-6
Denman St. ... K-7
Denver ... C-1
Devitt ... J-2,J-4,J-5
Dewey ... A-3,A-4
Dickson ... K-1,K-2,K-4
Donalee St. ... H-8,I-8
Donna St. ... J-4
Dooling ... A-5
Dowell ... J-8
Drew St. ... K-1,K-2,K-4
Duell St. ... H-6
Dundee ... B-4
Dunford ... I-6
Dunlap dr. ... K-6
E Ave. ... G-6,G-7
E Dr. ... F-1
Eagle ... B-6,C-6
Earl ... B-6
Eastland ... I-7
Eastline Dr. ... J-6,K-6
East Ridge ... A-7,B-7
Eden Ave. ... B-8
Edith ... A-7
Edwin ... H-2
Elinor ... A-6
Elizabeth Blvd. ... I-3
Ella ... E-4
Elliot ... D-7
Ellis ... A-2,B-2,C-2
Elm ... A-2,E-4
Elmwood Ave. ... H-5,H-6
El Paso ... F-3
El Rancho Rd. ... K-6
Elsie ... G-7
Elton ... B-8
Elva Warren St. ... K-5
Embry ... C-5
Emerson ... I-7
Enneca ... F-2
Ennis ... E-6
Eoch ... E-1
Erath St. ... K-8
Ernst ... G-6
Essex ... G-6
Esthill ... I-7
Etsie ... C-8
Eugene ... K-5
Evans St. ... H-4,I-4,J-4
Exchange ... B-2
Exeter ... F-5,G-5
Faett Ct. ... J-6
Fain ... C-7
Fairfax ... K-8
Fairline ... K-8
Fairmont Ave. ... I-3
Fair Park Blvd. ... K-5
Fair View ... A-6,B-6
... C-6,D-6
Fairway ... J-7
Fairway Rd. ... J-7
Felcher-Andrews ... A-8
Field ... A-8
Fincher ... B-8
Finley ... E-5
First Ave. East ... D-5
First St. East ... D-7
Fisher ... E-6
Fitzhugh ... I-7,I-8
Flint ... K-1,K-2
Florence ... F-3
Foard ... K-7
Fogg ... K-3,K-4
Forbes ... I-8
Forby Ave. ... F-8
Forest Blvd. ... H-2,I-2
Forest Park Blvd.
... F-2
Fossil ... A-8
Fouriner ... F-2
Fox ... C-7
Frazier Ave. ... I-2,K-2
Freddie ... I-8,J-8
Freman ... G-5
Fry St. ... K-4
Fulton ... F-3
G Ave. ... H-7,H-8
Gage ... B-8
Galves ... D-5,D-6
Galveston ... H-4
Gardina St. ... K-7
Garvey ... D-4
Gibson ... C-7,E-5
Gilcrest ... C-6
Gillis ... D-5
Gilmore ... D-7
Glen ... J-6
Glencoe Terr. ... H-2
Glenda ... A-7
Glendale ... B-3
Glengarden Dr. ... I-4
Glenhaven ... B-8
Glenmore ... C-4
Goldenrod ... B-5
Goldie ... B-7
Goddard ... D-7
Gorden Ave. ... I-2,K-2
Gould ... B-1,C-1
Gounah ... D-5
Grace ... B-5
Grainge ... G-3
Grand ... D-1,D-2
Grant ... D-4
Grapewood ... D-5
Grayson ... J-7
Greene Ave ... I-1,J-1
Greenfield ... D-4
Greenleaf ... E-2
Greer ... D-3
Gross ... D-7
Grover ... A-4,B-4
Grove St. ... C-2,E-4
... J-4,K-4
Gunther ... B-3
H Ave. ... H-7,H-8
Hale ... A-4,B-3
Hall ... G-6
Haltom Rd. ... A-8,B-8
Hampshire ... G-8
Hampton ... A-3,D-4,E-4
Hanger Ave. ... H-7
Harding ... A-3,D-4,E-4
Hardy ... A-3
Harper ... C-7
Harrington ... D-1
Harris ... G-2
Harris Lane ... B-7
Harrison ... G-2
Harrold ... E-2
Harrow ... A-8
Harvey ... H-5
Harwent Terr. ... J-1
Harwood St. ... F-7
Hathcox ... K-5
Hattie ... G-5
Hawthorne ... H-2,H-3
Haynes Ave. ... F-8,G-8
Hazeline ... G-8
Heath Ct. E. ... E-5
Hedrick ... A-6
Hemlock ... C-8
Hemphill St. ... G-3,I-3,K-3
Henderson St. ... F-3
... G-3,J-3
Hendricks ... G-6
Hickory ... C-8
Higgens ... B-7
Highcrest ... A-5
Highland ... C-1
High Point Rd. ... J-6
Hill Dr. ... I-2
Hill Place ... C-1
Hillview ... J-6
Hoffman ... E-5
Hollis St. ... B-7
Holmes ... G-6
Holtzer ... H-1
Homan ... D-1
Honey Suckle ... B-5,B-6
Houston ... B-2,C-2
... D-2,D-3,F-3
Howard ... A-8,J-7
Hudgins ... D-6
Humboldt ... G-2,G-4
Hunting Dr. ... K-7
Huntington ... H-2
Hutchinson ... A-3
I Ave. ... H-7,H-8
Illinois ... F-5,J-5
Inderly ... H-2
Inderly Pl. ... H-2
Industrial Blvd. ... F-4
Irion ... B-4,E-7
Irma ... G-4
Irwin ... G-2
Ivey ... B-5,B-7
J Ave. ... H-7,H-8
Jackson ... J-7
Jamaica ... F-6,G-8
James Ave. ... J-2,K-2
Jane Ln. ... K-7
Janis St. ... F-3
Jarvis St. ... F-2
Jeanette ... J-1,K-1
Jefferies ... A-7
Jefferson Ave. ... H-3,H-5
Jennings Ave. ... F-3,G-3
... H-3,I-3
Jerome ... F-2,G-2
Jessamine St. ... H-3,H-5
Jones St. ... C-2,D-2
... I-4,K-4
Joplin ... C-7
Juanita ... D-5
Judd St. ... I-4,I-6
Juniper ... B-8
K Ave. ... H-7,H-8
Kansas ... E-2
Karnes ... C-6,D-6
Kearby ... B-7
Kellis ... K-8
Kennedy St. ... F-5
Kent ... I-1
Kentucky Ave. ... G-4
Kerry ... A-8
Killian ... J-8
Kimble St. ... H-6
Kimbo Ct. ... A-6
King ... B-8,C-7,D-7
King Oaks ... D-7
Kingsdale ... J-6
Knoll ... B-7
Knox Dr. ... K-8
L Ave. ... H-7,H-8
Lagonia ... C-1
Lake ... F-3,G-3
Lakeland ... A-6
Lames ... J-1
Lancaster Ave. ... F-1
... F-3,F-4,F-6
Lane ... A-7
Lasalle ... E-5
Laughton ... J-3
Lawnwood ... E-6
Layton ... B-7
Lee ... B-1,C-2
Leming ... A-4
Leona ... B-7
Leon Ave. ... G-5
Leota ... C-4
Leslie ... G-2
Leuda ... G-3,G-5
Lewis St. ... F-8
Lexington ... E-3,F-3
Lilac St. ... H-3
Lillian ... C-5
Lincoln ... B-1,C-2
Linda Ln. ... K-6
Lipscomb St. ... G-3,I-3,K-3
Lisbury ... B-4
Little St. ... H-8,I-8
Little John St. ... H-7,H-8
Live ... D-4
Livingston ... I-2,J-2,K-2
Locust ... D-3
Loeton Terr. ... H-1
Logan ... H-6
Lois St. ... I-8
Lola ... B-6
Lollita Ct. ... J-6
Lomita ... J-7
Lomo ... K-4
Loney St. ... G-5
Loraine ... A-1,A-4,A-7
Lotus ... B-5
Louisiana ... H-4
Loving ... B-1
Lowden ... I-1,I-3,I-4
Lowe St. ... J-3
Lower Birdville
Rd. ... B-8,C-7,E-7
Lowriemore ... I-6
Lucinda ... K-7
Lucy ... G-4
Ludell ... F-6
Lulu ... A-3
Lynch ... D-7
Lynfield ... G-8
M Ave. ... H-7,H-8
Macon ... F-3
Maddox Ave. ... H-3,H-5
Magnolia Ave. ... G-2-4
Main St. ... D-3,E-3,J-4,K-4
Malone ... B-3
Malta St. ... K-3,K-4
Malvern ... I-6
Mandell ... F-7
Mansfield Ave. ... G-5
Maple ... D-4
Maple Leaf ... B-5
Marigold ... C-6
Marion Ave. ... I-5
Market ... B-1
Marsalis ... C-7
Marshall ... D-5
Martin St. ... K-8
Mason St. ... K-3,K-4
Maurice ... C-6
Maxine St. ... C-7
May St. ... H-3,J-3
... J-3,K-4
Mayfield ... D-3
McCart ... I-2,K-2
McComas ... H-6
McCurdy ... G-6,H-6
McIvey ... E-5
McKenzie St. ... G-7,H-7
McLean St. ... F-7
McLemore Ave. ... C-5
McLennon ... B-3
McNutt ... C-7
McPherson Ave. ... I-1,I-2
Meadowbrook Dr. ... F-7
Mecca St. ... J-7
Medford Ct. ... H-1
Menzer ... E-8
Menzer Ct. ... E-8
Mercedes ... E-1
Mercury ... B-7
Merida Ave. ... I-1,J-1
Meriwether ... K-5
Merrimac ... E-1
Mesquite ... A-6
Midland ... G-7
Millet Ave. ... H-7
Mills ... E-3
Milton ... F-6
Minden ... J-2,J-5
Minnie ... B-7
Mission ... J-1
Mississippi Ave. ... H-5
Missouri Ave. ... G-4,H-4
Mistletoe ... G-1
Mistletoe Ave. ... G-2
Mistletow Blvd. ... G-2
Mitchell Blvd. ... H-6,I-7
Mitchell St. ... H-3
Moberly St. ... K-8
Moline ... B-8
Monda St. ... J-2
Montague ... J-7
Moore ... B-4
Moreby ... I-6
Morgan ... E-4
Morning Glory ... B-4
Morningside Dr.
... I-3,I-4
Morphy ... G-3,G-5
Morrison ... D-3
Morton ... F-1
Mt. Vernon St. ... F-7,F-8
Mt. View ... F-8
Mulke St. ... I-3,I-5
Murphy ... C-5
Myrtle St. ... H-3,H-5
N Ave. ... H-7,H-8
Nashville ... G-7
Neal ... B-3
Nebraska ... E-2
Nelson ... B-5
Newman ... A-7
Newton ... B-8
New York ... F-5,J-4
New York Ave. ... H-5,I-4
Nichols ... A-3,E-4
Nixon ... C-4
Noble ... D-5,D-6
Noe St. ... I-8
Nolan St. ... K-8
Normandy ... F-8
North ... C-2
North Glen Dr. ... J-6
Northside Dr. ... C-2
Norwood ... E-1
Oak ... C-7,D-4
Oak Grove St. ... G-4
Oak Hurst Scenic
Dr. ... C-4,E-5
Oak Ridge ... B-8
Oak View ... C-7
Oakwood ... B-7
Odessa Ave. ... J-1
Old Mansfield Rd. ... J-6
Oleander ... G-3
Ollie St. ... K-7
Orange St. ... J-2,J-3
Orr ... G-2
Oscar ... A-3
Otto St. ... H-8
Owens ... B-8
Oxford ... B-4
Pafford St. ... K-1,K-2,K-3
Page St. ... I-3
Panola ... G-8
Paradise ... E-5
Park ... C-1,C-3,I-1
Parkdale ... G-7
Park Dr. ... D-1
Parkins St. ... F-8
Park Place Ave. ... H-3
Park Place Dr. ... H-2
Park Ridge Blvd. ... J-1
Parrish ... C-7
Patton Ct. ... H-2
Pavillion ... C-3
Peach ... E-3
Peak ... B-3
Pearl ... B-1
Pecan St. ... A-2,E-4,J-4
Pecos ... K-6-8
Pembroke Dr. ... H-2
Penn ... F-2
Pennsylvania Ave. ... G-3
Perry ... A-7
Petersmith ... F-2,F-3
Pharr ... D-4
Photo St. ... G-1
Pine ... F-5
Pioneer ... K-7,K-8
Pittsburgh ... D-5
Plumwood ... D-5
Poindexter ... C-3
Poplar ... F-4
Porter St. ... G-6
Portland ... C-4
Post ... B-7
Powell Ave. ... H-3,H-5
Prairie ... B-1
Pratcher ... K-6
Premier ... A-6
Presidio ... F-3,F-5
Prince ... J-6
Primrose ... C-5
Prospect ... B-1
Pruitt ... G-2,G-3
Pulaski ... G-3,G-4
Quentin Ct. ... A-4
Quinn ... H-7
Race ... C-5
Ramey ... H-8,I-8
Ramsey ... H-4,I-6
Ranch Ter. ... J-6
Rapurt ... A-4
Rattikin ... I-6
Ravin ... G-7
Ray ... K-7
Raynor ... D-5
Reaford ... A-7
Refugid ... B-1
Retta ... E-5
Riardale Rd. ... J-6
Richmond ... H-3,H-5,H-6
Ridgeview ... J-6
Riley St. ... J-5
Rio Grand ... F-3
Ripy St. ... J-2,J-3,J-4,J-5
Riverridge ... C-6
Riverside Dr. ... G-6,I-5,K-6
Roberts ... I-2-5
Robert Burns Dr. ... J-6
Robinwood ... B-6
Rodeo St. ... K-7
Rogers Ave. ... H-1,I-1
Rolling Hills Dr. ... J-6
Roosevelt ... B-1
Rosedale St. ... G-3,G-5,G-7
Ross ... A-2,B-2
Rouse ... D-5
Rufus ... I-8
Runnels ... A-4
Rupert ... E-2
Rusty ... B-6
Rutan ... K-8
Ryan Ave. ... J-2,K-2
Ryan Place Dr. ... I-3
St. Louis Ave. ... J-4,K-4
St. Louis St. ... G-4
Samuels ... C-3
Sanborn ... F-7
Sandage ... I-1
Sanders ... B-7
Sanderson St. ... F-8
Sandgate ... I-6
Santa Rosa ... B-8
Sarah ... K-6
Sargent ... E-7
Seaman ... C-6,D-6
Seaman Ave. ... D-6
Selk ... B-6
Selma ... B-5,B-6
Scenery Hill Ct. ... E-8
Scenic Ct. ... C-5
Schadt ... A-3
Schwartz ... A-3
Scott ... E-4,F-6
Shackelford ... K-8
Shadow ... C-7
Shamrock ... D-2
Shane ... A-7
Shaw St. ... J-2-5
Shirley Way ... I-1
Shotts ... E-1
Shropshire ... I-6
Simpson ... F-5
Smilay ... C-4
South Freeway ... I-4
South Hill ... J-1
Southland St. ... G-2
Spiller ... I-6
Spring St. ... G-1
Springdale Rd. ... B-6
Stadium Dr. ... I-1,J-1
Stanley ... H-2,I-2
... J-2,K-2
Stephen Lee ... J-6
Stardust ... J-6
Stayton ... E-2
Stearnes ... G-6
Stella ... F-5
Stephenson ... F-4
Stratford Dr. ... F-8
Strong ... I-7
Stuart Dr. ... I-4,K-4
Summit ... F-2
Sunday ... A-7
Sunset ... F-2
Surrey ... K-1
Swayne ... E-6
Swift ... C-4
Sycamore ... G-6
Sydney ... I-8
Sylvania ... D-5
Tallman St. ... K-8
Taylor ... E-3
Templeton ... E-1
Tennessee ... G-5
Terrace ... D-1
Terrell ... G-2,G-3,G-4
Terry ... A-2,E-4
Thanntson ... I-7
Thockmorton ... D-3,E-3
Thrall St. ... G-7,H-7,I-7
Tillar ... E-1
Timberline Dr. ... K-5,K-6
Todd Ave. ... J-5
Tom Ellen ... A-6
Townshend Dr. ... I-2,K-2
Transcontinental
Cutoff ... A-5
Travis St. ... G-3,I-3,K-3
Troost ... G-5
Trueland Dr. ... J-7,K-7
Tucker St. ... G-5
Ubbock Ave. ... I-1,J-1,K-1
University Dr. ... F-1
... I-1,J-1
Uvalde ... H-6
Vacek ... E-2
Valley ... E-3
Vance ... C-7
Van Horn ... E-5
Vaughn Blvd. ... I-7
Vaughn St. ... J-8
Vera Cruz ... A-4
Verben E. ... G-4
Vickery Blvd. ... F-3,F-6
Vicki Ln. ... I-4
View ... F-7
Vinetta ... K-8
Viola ... E-2
Virginia ... C-5,G-5
Vogt St. ... H-8
W Dr. ... F-1
Wabash ... I-1
Wabash Ave. ... J-1
Waggoman St. ... K-1
... K-3,K-4
Waits Ave. ... I-1,J-1
Waldemar ... B-7,B-8
Walker ... B-4
Wall ... D-4
Wallace St. ... E-7,H-8
Ward Pkwy. ... H-2
Ward St. ... F-7
Warner ... H-2
Warwick ... B-4
Washington Ave. ... G-3
Watauga Rd. ... B-5
Waterman ... D-4
Wathal ... A-8
Wayne ... B-7,C-7,D-7
Wayside Dr. ... I-2,K-2
Weatherbee St. ... H-2
Weatherford ... E-3
Weber ... A-3
Weisenberger St.
... E-1,J-5
Wesley ... A-6
Wesleyan Dr. ... G-7,H-7
Westbrook ... C-5
Westchester ... B-8
Wheeler ... C-7
White Ave. ... G-5
White Settlement
Rd. ... E-1
Whitmore ... E-1
Wichita St. ... K-7
Wilbarger St. ... J-7
Wilkinson ... F-7
William ... F-4
Williams ... A-6
Williams Pl. ... B-6
Willing Ave. ... J-2,K-2
Wilshire ... H-2
Wimberly ... E-1
Windham ... F-6
Windmill ... C-4
Winfield ... K-1
Wingate ... E-1
Winslow Pl. ... H-2
Winston Rd. ... J-1
Winston Terr. ... H-1
Withers St. ... G-6
Woodland ... J-3
Woodrow St. ... G-6
Woods ... B-8
Woodward ... E-2
Worth St. ... G-2
Worthills ... J-1
Wyatt Ct. ... J-6
Wynne ... C-6
Young ... F-7
Yucca ... C-6
Yuma St. ... H-5,I-5,J-5
Zelma ... B-8
1st St. ... D-5,E-3
2nd ... E-3
3rd ... E-3
4th St. ... D-3,D-5,D-6
... E-1,E-3
5th St. ... D-3,E-1-3
5th Ave. ... D-3,E-1-3,I-3,K-3
6th ... D-3,E-1-3
6th Ave. ... I-3,K-3
7th ... D-3,E-1,E-3
8th ... E-3
8th Ave. ... H-2,J-2,K-2
9th ... D-3,E-3
10th ... C-2,C-3
11th St. ... C-3,F-2,G-2
12th St. ... E-5,G-2
13th St. ... C-3,G-2
14th St. ... C-1,F-2,F-4
15th ... C-1,F-4
16th ... C-1,C-3,F-4
17th ... C-3,F-4
18th ... C-1,C-3,F-5
19th St. ... C-1,F-4
20th ... C-1,C-2
21st ... C-1,C-2
22nd ... B-1,B-2
23rd ... B-1,B-2
24th ... B-1
25th ... B-1
26th ... B-1
27th ... B-1,B-2
28th ... A-1,A-2,A-5
29th ... A-1,A-3,A-6
30th ... A-1,A-2,A-3
31st ... A-1,A-2,A-3
32nd ... A-1,A-2,A-3

Scale of Miles
0. .1 .2 .3 .4 .5

N

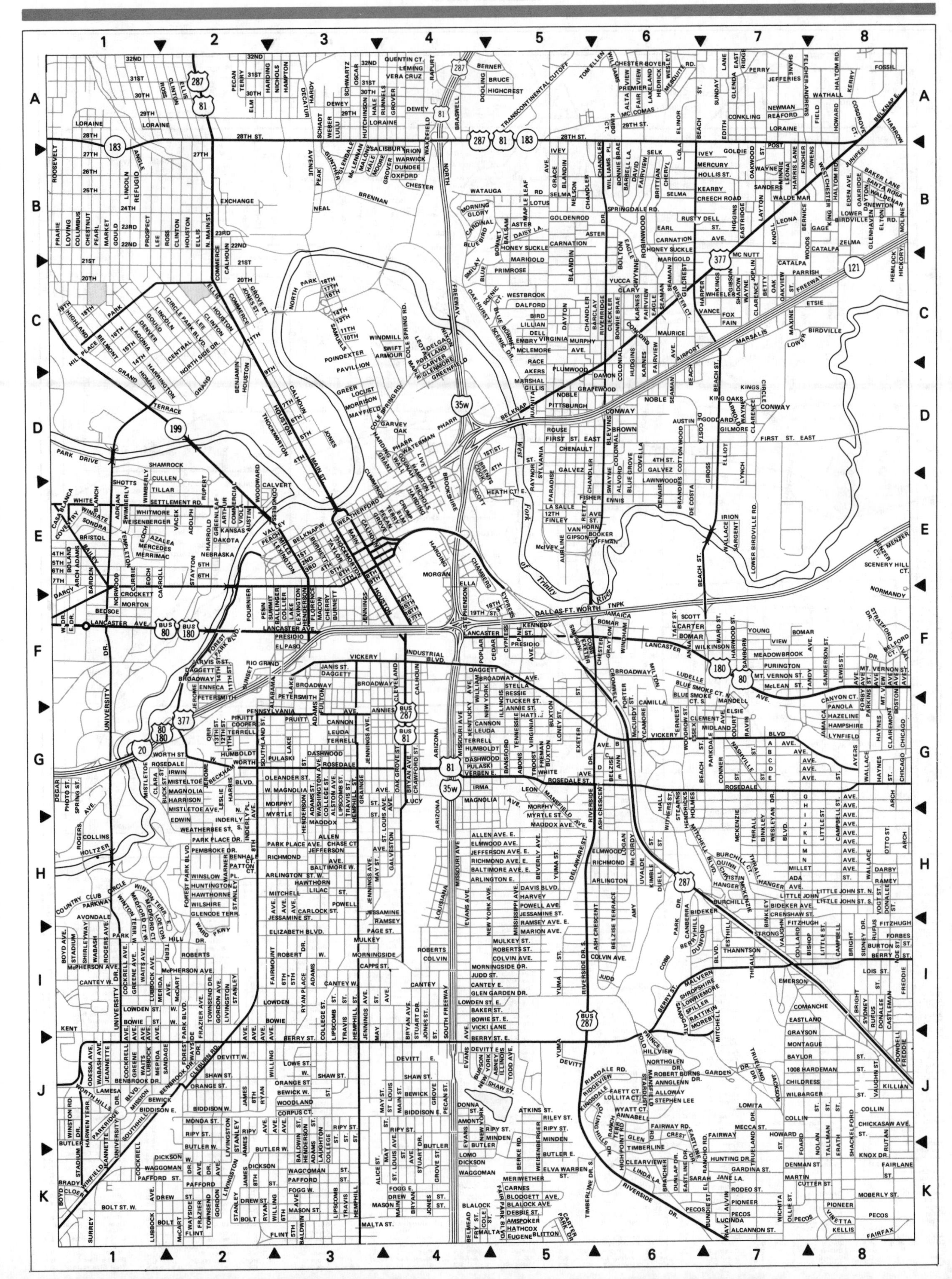

Fort Wayne, IN

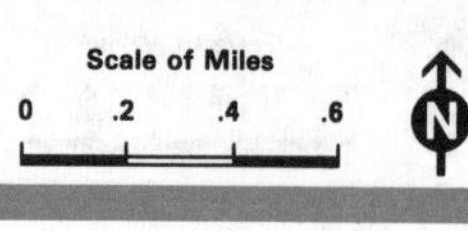

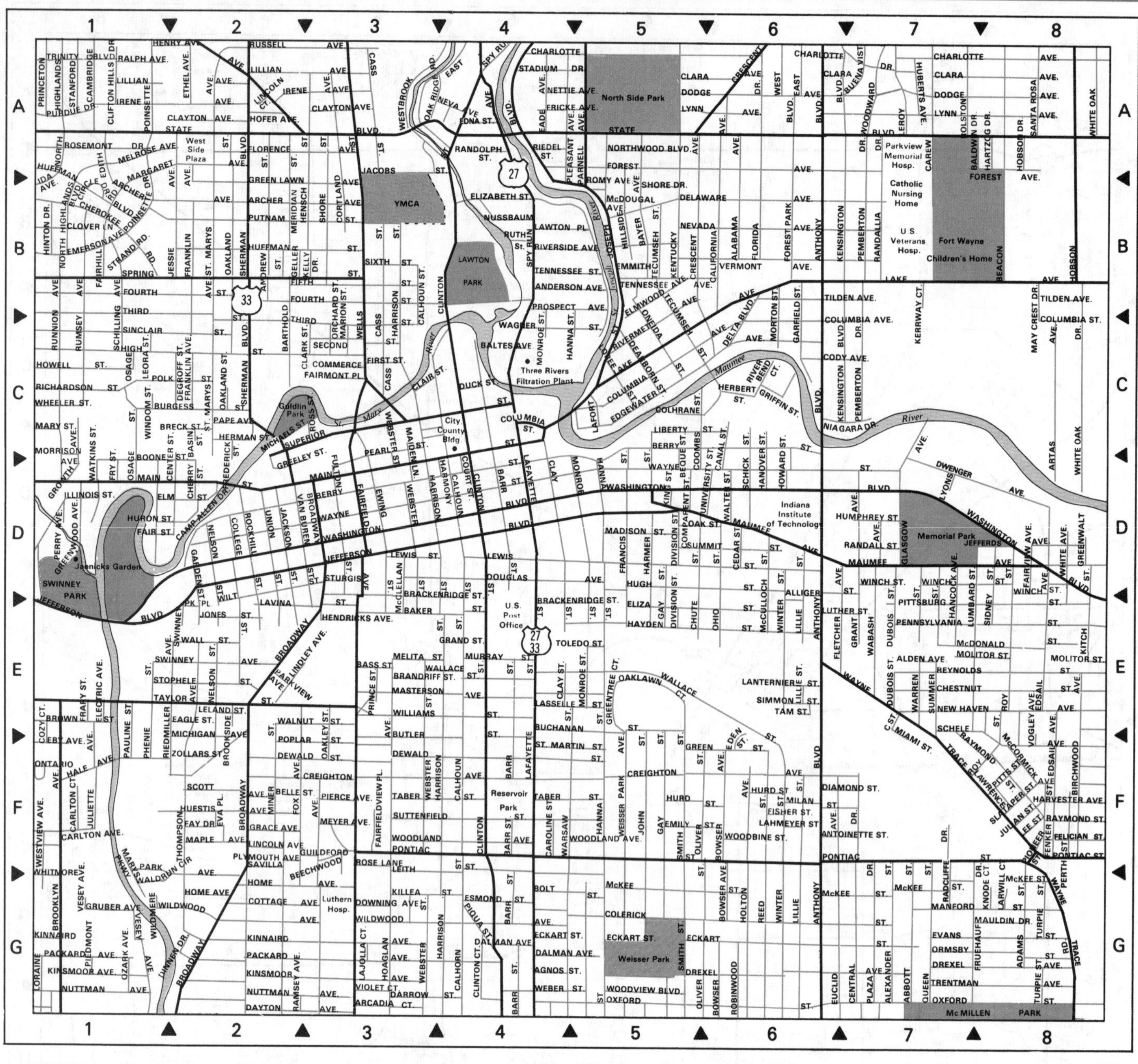

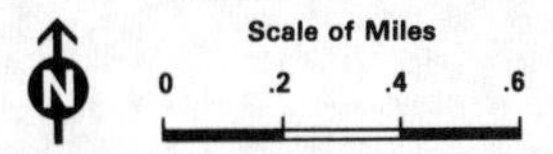
N
Scale of Miles
0 .2 .4 .6

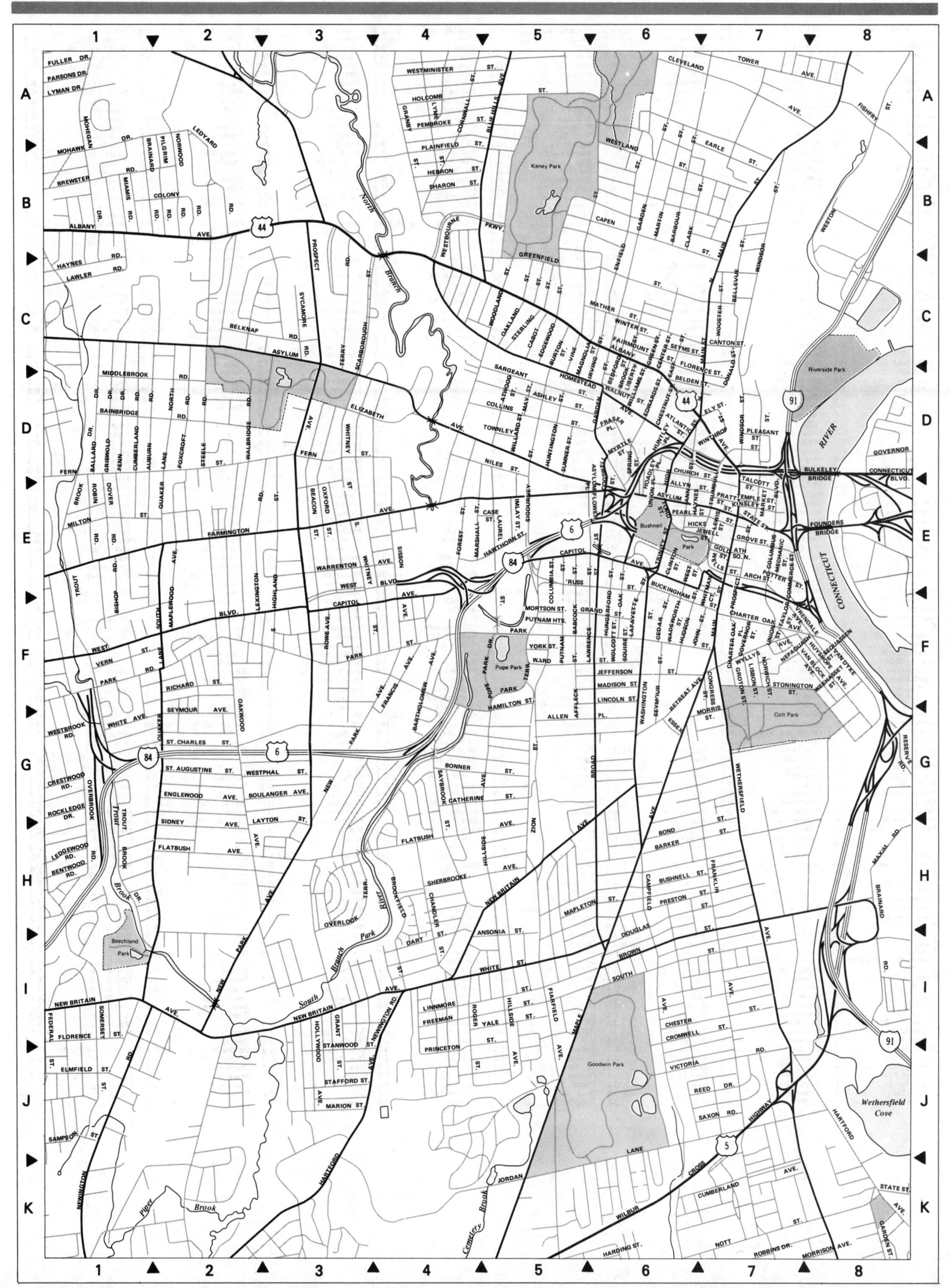
1 2 3 4 5 6 7 8
A B C D E F G H I J K
Keney Park
Riverside Park
Bushnell Park
Pope Park
Colt Park
Goodwin Park
Beechland Park
Wethersfield Cove
CONNECTICUT RIVER
North Branch
South Branch Park River
Trout Brook
Piper Brook
Cemetery Brook
FULLER DR.
PARSONS DR.
LYMAN DR.
MOHAWK DR.
BREWSTER RD.
COLONY RD.
ALBANY AVE.
HAYNES RD.
LAWLER RD.
BELKNAP RD.
ASYLUM AVE.
MIDDLEBROOK RD.
BAINBRIDGE RD.
ELIZABETH ST.
FARMINGTON AVE.
WARRENTON AVE.
WEST BLVD.
CAPITOL AVE.
PARK ST.
NEW BRITAIN AVE.
PROSPECT AVE.
WESTMINISTER ST.
HOLCOMB ST.
PEMBROKE ST.
PLAINFIELD ST.
HEBRON ST.
SHARON ST.
GREENFIELD ST.
WESTLAND ST.
CLEVELAND AVE.
TOWER AVE.
EARLE ST.
CAPEN ST.
MATHER ST.
WESTON ST.
BULKELEY BRIDGE
FOUNDERS BRIDGE
CHARTER OAK AVE.
WETHERSFIELD AVE.
MAIN ST.
WASHINGTON ST.
BROAD ST.
MAPLE AVE.
FLATBUSH AVE.
NEW BRITAIN AVE.
HARTFORD RD.
NEWINGTON RD.
JORDAN LANE
WILBUR CROSS HIGHWAY
HARDING ST.
NOTT ST.
ROBBINS DR.
MORRISON AVE.
STATE ST.
GARDEN ST.
MAXIM RD.
BRAINARD RD.
RESERVE RD.
HARTFORD AVE.
FISHFRY ST.
44
6
84
91
5

HARTFORD

Affleck St........F-5
Albany Ave........B-1,C-6
Allen Pl........G-5
Allyn St........E-6
Ann St........E-6
Arch St........E-7
Ashley St........D-5
Asylum Ave........C-3
Asylum Pl........D-6
Asylum St........E-6
Ath Sq. N........E-7
Atlantic St........D-6
Atwood St........D-5
Auburn Rd........D-1
Babock St........F-5
Bainbridge Rd........D-1
Ballard Dr........D-1
Barbour St........B-6
Barker St........H-6
Bartholomew Ave.
........F-5,G-4
Beacon St........E-3
Bedford St........C-6,D-6
Beldeno St........D-6
Belknap Rd........C-2
Bellevue St........C-7
Bishop Rd........F-1
Blue Hills Ave........A-5
Bond St........H-6
Bonner St........G-4
Boulanger Ave........G-2
Brainard Rd........A-1
Brainard Rd.————H-8
Brentwood Rd........H-1
Brewster Rd........B-1
Broad St........G-6
Brookfield St........H-4
Brook St........D-6
Brown St........I-6
Buckingham St........E-6
Burton St........C-5
Bushnell St........H-6
Cabot St........C-5
Campfield Ave........H-6
Canton St........C-7
Capen St........B-6
Capitol Ave........E-5,F-3
Case St........E-5
Catherine St........G-4
Cedar St........F-6
Center St........C-6
Chandler St........H-4
Charter Oak Ave........F-7
Charter Oak Pl........F-7
Chester St........I-6
Chestnut St........D-6
Church St........D-6
Clark St........B-6
Cleveland Ave........A-6
Clinton St........E-6
Cogswell St........D-6
Collins St........D-5
Colony Rd........B-2
Columbia St........E-5
Columbus Blvd........E-7
Commerce St........E-7
Congress St........F-7
Connecticut Blvd........D-8
Cornwall St........A-4
Crestwood Rd........G-1
Cromwell St........I-6
Cumberland Ave........K-7
Cumberland Rd........D-1
Dart St........I-4
Donald St........D-7
Douglas St........I-6
Dover Rd........E-1
Earle St........B-7
East St........C-6
Edgewood St........C-5
Edwards St........D-6
Elizabeth........D-3
Elmfield St........J-1
Ely St........D-7
Enfield St........C-6
Englewood Ave........G-2
Essex........G-6
Fairfield Ave........I-5
Fairmount St........C-6
Farmington Ave........E-2
Federal St........I-1
Fern St........D-1,D-3
Fishery........A-8
Flatbush Ave........H-2,H-4
Florence St........C-6,I-1
Flower St........E-6
Ford St........E-6
Forest St........E-4
Foxcroft Rd........D-2
Francis Ave........F-4
Franklin Ave........H-7
Fraser Pl........D-6
Freeman St........I-4
Fuller Dr........A-1
Garden St........B-6,D-6,K-8
Gold St........E-7
Governor St........F-7
Granby St........A-4
Grand St........F-6
Grant St........I-3
Green St........C-6
Greenfield St........B-5
Griswold Dr........D-1
Groton St........F-7
Grove St........E-7
Hamilton St........F-5
Harding St........K-6
Hartford Ave........J-8,K-3
Hawthorn St........E-5
Haynes Rd........C-1
Haynes St........E-6
Hebron St........B-4
Hicks........E-6
High St........D-6
Highland St., S........F-3
Hillside Ave........I-5
Hoadley Pl........D-6
Holcomb St........A-4
Hollywood Ave........I-3
Homestead Ave........D-5
Hudson St........F-6
Hungerford St........F-6
Huntington St........D-5
Huntley Pl........D-6
Huyshope Ave........F-8
Imlay St........E-5
Irving St........D-6
Jefferson St........F-6
Jewell St........E-7
John St........F-6
Jordan Ln........K-5
Keney Park........B-5
Kinsley St........E-7
Lafayette St........F-6
Laurel St........E-5
Lawler Rd........C-1
Lawrence St........F-6
Laxton St........G-2
Ledgewood Rd........H-1
Ledyard Rd........A-2
Lewis St........E-7
Lexington Rd........F-2
Liberty........D-6
Lincoln St........F-6
Linnmore St........I-4
Lisbon St........F-7
Lyman Dr........A-1
Lyme St........A-4
Madison St........F-6
Magnolia St........C-5
Main St., N........B-7,D-7
Maple Ave........I-5
Mapleton St........H-5
Maplewood Ave........F-2
Marion St........J-3
Market St........E-7
Martin St........B-6
Marshall St........E-4
Mather St........C-6
Maxim Rd........H-8
May St........D-5
Mechanic St........E-7
Miamis Rd........B-1
Middlebrook Rd........D-1
Milton St........E-1
Mohawk Dr........B-1
Mohegan Dr........A-1
Morrison Ave........K-8
Morris St........G-7
Mortson St........F-5
Myrtle St........D-6
Nepaquash........F-7
New Britain Ave.
........H-5,I-1,I-3
Newington Rd........K-1
New Park Ave........G-3,I-2
Niles St........D-5
Norwich St........F-7
Norwood Rd........B-2
Nott St........K-7
Oak St........F-6
Oakland St........C-5
Oakwood Ave........G-2
Overbrook Rd........G-1
Overlook Terr........H-3
Oxford St........E-3
Park Rd........F-1
Park St........F-3,F-5
Park Terr........F-5
Parsons Dr........A-1
Pearl St........E-6,E-7
Pembroke St........A-4
Penn Dr........D-1
Pilgrim Rd........B-2
Plainfield St........A-4
Pleasant St........D-7
Pope Park Dr........F-5
Potter St........E-7
Pratt St........E-7
Preston St........H-6
Princeton St........J-4
Prospect Ave........B-3
Prospect St........E-7,F-7
Putnam Hts........F-5
Putnam St........F-5
Quaker Lane, N........D-2,E-2
Quaker Lane, S........F-2,G-2
Reed Dr........J-7
Reserve Rd........G-8
Retreat Ave........F-6
Richard St........F-2
Robbins Dr........K-7
Robin Rd........E-1
Rockledge Dr........G-1
Roger St........I-4
Rowe Ave........F-3
Russ St........E-5
St. Augustine St........G-2
St. Charles St........G-2
Sampson St........J-1
Sargeant St........D-5
Saxon Rd........J-7
Saybrook St........G-4
Scarborough St........C-3
Sequassen St........F-8
Seymour St........F-6,G-6
Seyms St........C-6
Sharon St........B-4
Sherbrook Ave........H-4
Sidney Ave........G-2
Sigourney St........E-5
Sisson Ave........E-4
Somerset St........I-1
South St........I-6
Spring St........D-6
Squire St........F-6
Stafford St........J-3
State St........E-7,K-8
Steele Rd........F-2
Sterling St........C-5
Stonnington St........F-7
Sumner St........D-5
Sycamore Rd........C-3
Talcott St........E-7
Taylor St........F-7
Temple St........E-7
Terry Rd........C-3
Tower Ave........A-7
Townley........D-5
Trinity St........E-6
Trout Brook Dr........E-1,H-1
Trumbull St........E-7
Union Pl........E-6
Union St........F-7
Van Block Ave........F-8
Van Dyke........F-8
Vern St........F-1
Victoria Rd........J-6
Vine........C-5
Vrendale Ave........F-7
Wadsworth St........F-6
Walbridge Rd........D-2
Walnut St........D-6
Ward St........F-5
Warrenton Ave........E-3
Washington St........F-6,G-6
Weehasset St........F-8
Wells St........E-7
West Blvd........E-3,F-1
West St........E-6
Westbourne Pkwy........B-4
Westbrook Rd........G-1
Westland St........A-6
Westminster St........A-4
Weston St........B-8
Westphal St........G-2
Wethersfield Ave........G-7
White Ave........G-1
White St........I-5
Whitney St........D-3
Whitney St., S........E-3
Wilbur Cross Hwy........K-6
Willard St........D-5
Williams St........D-6
Windsor St........C-7,D-7
Winter St........C-6
Winthrop St........D-7
Wolcott St........F-6
Woodland St........C-5
Wooster St........C-7
Wyllys Ave........F-7
Yale St........I-5
York St........F-5
Zion St........H-5

POINTS OF INTEREST

Beechland Park........I-1
Bulkeley Bridge........D-8
Bushnell Park........E-6
Coh Park........G-7
Founders Bridge........E-8
Goodwin Park........J-6
Pope Park........F-5
Riverside Park........C-8

HOUSTON

Alabama St........J-5,J-7
Alabama St. W........H-1
Alamo........C-4
Albany........G-3
Alber St........A-7
Allen........D-1
Allen Parkway........E-2
Allston St........B-1
Alpha St........F-1
Andrews St........F-3
Angella........F-1
Anita St........H-4,I-7
Ann St........F-8
Arbor St........J-4,K-6
Arch........E-7
Archer........A-4
Arlington St........B-2
Arthur St........F-4
Ashby........J-1
Attucks........J-7
Audobon........H-3
Austin........J-3
Austin St........G-5,J-3
Autrey........I-2
Averill........A-6
Avondale St........G-3
Baer........D-8
Bagby St........F-5,G-4
Bailey........G-3
Baker........E-6
Baldwin........F-4
Banks........I-2
Banks St........J-1
Barbee St........J-5
Bardwell........A-7
Barkdull........J-2
Barnes........D-1
Bass St........D-1
Bastrop St........G-6
........H-6,J-5
Bayard........J-2
Bayland St........B-3
Beach........D-5
Bell........H-8
Bell, W........F-3
Bell Ave........G-6
Bering........F-8
Berry........I-5
Berthea........J-2
Beverly St........B-2
Bigelow St........A-7
Billingsley St........A-6
Bingham St........D-4
Binz St........J-3,K-4
Bishop........C-5
Bissonnet St........J-1
Blair St........A-1
Blodgett St........J-5,K-6
Bolsover........K-1
Bomar........G-2
Bonnie Brae........I-1
Boone........B-8
Booth St........B-6
Boswell St........A-7
Boundary........B-4
Bradley........B-4
Brailsfort St........I-6
Branard St........I-1
Brandt........I-3
Brazos St........F-5
Bremond........G-4
Bremond St........I-7
Briley St........I-6
Brooks Pl........C-8
Brooks St........D-6
Bruce........A-4
Buckner St........F-4
Buel........G-3
Buffalo Terrace........E-3
Bunton St........A-7
Burkett........J-6
Burkett St........K-6
Burnett St........D-6
Bute St........I-3
Butler St........E-2
Calendar........A-5
Calhoun Ave........G-5
California........H-2
Calle........J-8
Calumet St........J-3
Campbell St........C-7
Canal........F-8
Canfield St........I-7
Capitol........E-4
Capitol Ave........F-5
Carl........C-5
Caroline St........G-5,J-3
Carr........A-8
Castle Ct........I-2
Catherine........A-5
Cetti........B-6
Chapman........C-7
Chartres........K-3
Chase St........C-7
Chelsea........I-3
Chenevert St........G-6,J-4
Cherokee........J-1
Cherry........A-8
Churchill........B-5
Clay, W........F-3
Clay Ave........G-6
Clay St........H-7
Cleburne St........J-5
Cleveland........F-3
Cline St........E-8
Clinton Dr........E-7
Cobb St........K-6
Cochran St........B-6
Collingsworth St........A-7
Collins........D-6
Colorado St........D-4
Colquitt........I-1
Columbia St........B-2
Columbus........F-2
Commerce St........E-6
Common........B-6
Common St........C-6
Congress Ave........E-6
Conti........D-8
Cook........A-4
Cordell........A-4
Cordier........F-7
Cortland St........B-2
Cottage........A-3
Court St........D-2
Courtland Pl........H-3
Coyle St........H-7
Cranberry........C-2
Crawford........J-4
Crawford St........F-6
Crocker St........F-3
Crockett........D-4
Crosby St........F-4
Cullen St........K-7
Cushing St........F-4
Dallas, W........F-2
Dallas Ave........F-5
Daly........D-6
D'Amico St........F-1
Damon........G-2
Dart St........D-4
Davis St........C-8
Day........I-3
Decatur........E-4
De George........A-4
DeLano St........H-7,I-6
Dell........A-4
Dennis Ave........H-5
Denver........H-8
De Pel........C-4
Dewey........D-4
Diesel........D-3
Dora........J-2
Douglas St........I-7
Dowling St........G-7,J-4
Drew........G-4,H-5
Drew St........G-4
Driscoll St........G-1,I-1
Dunlavy St........H-1
Dunstan........K-1
Eagle St........J-5
Eberhard........F-2
Edison St........A-6
Edmundson St........I-8
Edwards St........D-4
Elder........D-4,D-5,E-5
Elgin........H-4
Eli St........D-1
Elmen St........G-1
Elser........B-6
Elysian........C-7,E-7
Embry........B-6
Emerson........H-3
Engeike St........F-8
Enid........A-4
Ennis, N........F-8
Ennis St........H-7,J-6
Erin St........A-7
Euclid St........B-3
Eunice........B-5
Evella........A-8
Ewing St........J-3
Fairview St........G-1
Fannin St........G-5,I-3
Fargo St........G-3
Fletcher St........C-6
Flora........H-3
Florence St........B-4
Flynn........F-8
Foote........E-8
Fowler........E-1
Fox........F-8
Francis........H-4,J-7
Franklin Ave........E-6
Frasier........C-2
Freund........E-8
Fugate St........A-3
Fulton St........A-5,B-6
Gano........C-7
Gardner, W........A-3
Gargan........C-5
Garrott........I-3
Garrow St........G-8
Garvin Ct........D-4
Genese St........G-3
Gilette........G-3
Gillespie........E-8
Givens........D-2
Gladys........C-4
Glaser........D-6
Goldenrod........B-5
Golf Links........F-3
Goliad St........D-5
Grant........G-2
Gravstark........J-2
Gray Ave. W........F-3
Gray St........G-4,G-5
Grayson........E-8
Greeley St........I-3
Greenle........C-4
Grigsby........B-8
Gross........F-1
Gulf Freeway........I-8
Haddon Ave........G-1
Hadley Ave........G-4
Hain........A-5
Hamilton St........G-6,H-5
Hammock........B-6
Hardy........C-7
Harold St........H-1
Harrington St........D-6
Harrisburg........G-8
Harvard St........B-2,D-2
Hathaway St........H-3
Hawkins........G-8
Hawthorne St........H-2
Hays St........B-6
Hazard St........H-1
Hazel........F-2
Heights Blvd........B-2,D-2
Helen St........B-4
Helena........G-3
Hemphill St........E-4
Henderson St........E-4
Herkimer St........B-1
Hermann St........K-3
Hickory........D-5
Hiensley........D-5
Highland St........B-3
Hill St........E-1
Hogan St........C-6
Hogg........C-8
Holman Ave........H-4,J-7
Holy........D-5
Home........D-2
Honsin........D-2
Hopkins St........G-3
Hopson........F-3
Houston St........D-5
Howe........F-4
Hussion St........I-8
Hutcheson St........H-8
Hutchins St........G-6,J-4
Hyacinth........B-5
Hyde St........G-3
Ideal........B-5
Indiana Ave........G-1
Ingborg........I-8
Institute........J-2
Isabella St........J-5
Jack St........I-3
Jackson........J-4
Jackson St........F-6
James St........C-6
Jasmine........F-1
Jefferson Ave........G-5
Jensen........E-8
Jensen St........C-8
Jessamine........A-5
Joe Anne St........F-2
Johnson St........D-4
Jones........B-8
Joseph St........H-8
Julian........C-3
Kane........E-4
Karnes........A-5
Keating St........H-8
Keene........C-6
Kennedy........F-8
Kent........J-1
Key St........A-3
Kipling St........H-1
Koehler St........D-1
Kolb St........C-1
Kuester........H-1
Kyle........I-2
La Branch........J-4
La Branch St........F-6
Lamar........F-2
Lamar, W........F-3
Lamar Ave........F-5
Larkin........D-2,E-2
La Rue........F-2
Lee St........C-7
Leek St........J-8
Leeland Ave........G-6
........H-8,I-8
Leona........D-8
Leonidas St........F-1
Leverkuhn........E-4
Lewis St........D-6
Lexington........I-1
Lincoln........F-2
Live Oak, N........F-8
Live Oak St........G-7,K-4
Loretto St........I-2
Lorraine St........C-7
Lottmann St........E-8
Louisiana St........F-5,H-4
Lovett St........H-2
Lubbock........E-4
Lucinda St........J-7
Luzon St........B-7
Lyle........E-7
Lyons Ave........D-7,D-8
MacGregor........K-4
Maggie........B-6
Main, N........A-4
Main St........G-5,I-3
Main St. W........I-1
Mandell........H-1,J-1
Marconi St........F-2
Marie St........C-5
Marigold........B-5
Marina........C-1
Marshall St........H-1
Marstow........F-1
Mary St........C-8
Maryland Ave........G-1
Mason St........G-3
Matthews........F-4
Maury St........C-7
Maverick........A-5
Maylor........D-6
McGowen........G-4
McGrower St........I-7
McKee........C-7
McKenney, W........F-3
McKinney Ave........F-5
McLihenney........G-4
McLihenney Ave........I-7
McMillan St........E-2
McNeil St........B-7
Melwood St........A-3
Menefee........C-1
Merrill St........B-3
Michaux........B-3
Michigan Ave........G-1
Milam St........F-5
Milby St........I-8
Milford St........J-1
Miller St........I-8
Mills........C-8
Mills St........C-7
Mirimar........I-2
Missouri........G-2
Montana St........D-1
Montrose........I-2
Moore........B-6
Mop........C-7
Morgan........G-3
Morris........B-6,B-8
Morrison St........B-4
Morse St........G-1
Moss St........A-4
Mt. Vernon St........I-2
Mulberry St........H-2
Nagle Alley........F-8
Nagle St........G-7,I-6
Nagle St. N........F-8
Nance St........D-7
Napoleon St........J-7
Nettletor........J-6
Nevada........G-1
Newhoff........B-8
Newhouse........F-1
Nicholson........B-1
Noble St........C-7
Norfolk........I-1
Norhill........A-3
North........B-3
North Blvd........J-1
Northwood........B-4
Oak Ct........G-4
Oak Dr........C-4
Oakdale........J-3
Oakley........I-3
Oak Ridge St........B-3
Odin........D-8
Olive St........D-1
Omar St........B-3
O'Neil St........F-4
Opelousas St........D-7
Orr St........B-7
Oxford........B-2
Ovid........C-4
Pacific........G-3
Paige........F-8
Paige St........H-7,I-6
Palm........J-3
Palmer........H-7,K-5
Palmer, N........F-8
Panama St........B-6
Park........B-6,G-1
Parkview........B-5
Paschall........C-6
Patterson........E-1
Payne........B-5
Pease........H-7
Pease Ave........G-5
Peden Ave........F-2
Peveto........F-2
Pierce, W........F-3
Pierre Ave........G-5
Pinckney St........C-6
Pinedale........J-3
Pineridge........B-3
Polk........F-5
Polk, W........F-3
Polk Ave........G-6
Portland........J-3
Portsmouth........I-1
Prairie Ave........E-6
Preston Ave........E-6,G-8
Prospect St........J-3
Providence........D-7
Quinn........A-4
Quitman........C-6
Race........E-7
Railey........A-4
Rains........E-7
Ralph........H-1
Raymond........D-2
Reagan........B-4
Redan........B-4
Reeves........J-7
Rein........B-6
Reisner........E-5
Renfro........F-1
Renner Ct........E-4
Reynolds........B-6
Rice Blvd........K-1
Richmond Ave........I-1
Ridge St........B-4
Ridgewood........G-1
Riverside Dr........K-4
Roanoke........E-7
Roberts........H-8
Robertson........B-6
Robila St........F-3
Robin St........F-3
Rochow........F-1
Rockwood St........K-8
Rosalie St........H-4,I-7
Rose........D-1
Rosedale........J-3,K-5
Roseland........I-3
Rosewood........J-5
Rosine........F-2
Rossmoyne St........I-2
Ruiz St........E-7
Rusk Ave........F-5
Ruth St........J-5,K-7
Ruthven........F-3
Rutland Pl........C-1
Rylis........F-2
Ryon St........B-7
Sabine St........C-4,E-4
Sachs St........E-2
St. Charles St........G-7,I-6
St. Charles St. N.
........F-8
St. Emmanuel St.
........G-6,J-4
Saltus........F-8
Sampson........H-8,K-5
Sampson St........K-6
Sanders........J-7
San Jacinto St........G-5,I-3
Sauer St........I-5,K-5
Saulnier, W........F-3
Sawyer St........D-3
Scott........K-7
Scott St........J-7
Schulan St........D-1
Schwartz........D-8,E-8
Searle........A-5
Sellers........D-2
Semmes St........C-7
Shaw........F-5
Shearn St........D-4
Shelby St........A-6
Shelley........A-4
Sherman St........F-8
Shiloh........E-8
Silver St........E-4
Simmons........J-7
Sledge........C-3
Smith St........F-5
Snover........E-1
South Blvd........J-1
Southmore St........J-3
Southwest Freeway
........I-1
Spencer........C-1
Spring........C-4
Spruce St........E-7
Stalker........D-2
Stanford St........I-3
State St........E-4
Stevens St........C-8
Stout........B-7
Stuart........H-4,J-7
Stude........C-3
Studenmont........D-3
Studewewood........B-3
Sul Ross St........H-1
Summer........D-4
Sumpter St........C-7
Sun Ct........F-7
Sunset Blvd........K-2
Sutton St........G-3
Sweetwood........F-1
Sydnor........D-8
Tabor........A-4
Tackaberry........B-6
Taft St........G-3
Taylor St........E-3
Temple........A-4
Temple St. W........A-3
Terrell........F-1
Terry........C-7
Texas Ave........E-6,G-8
Thelma........C-4
Thomas........C-5
Thompson........D-1
Threlkeld........C-3
Tierwester St........K-6
Top........E-7
Travis........I-3
Travis St........G-5,I-3
Tretham........D-6
Trimble........A-5
Trinity........E-4
Trulley........H-6
Truxillo........J-5
Tuam St........H-4,I-5
Tulane St........B-1
Union St........D-3
Usener Blvd........C-3
Valentine St........F-4
Van Buren........F-2
Varsity........K-8
Vassai........I-2
Velasco........G-8,K-5
Vermont Ave........G-1
Vick........E-1
Victor........F-3
Vincent........A-4
Violet........F-1
Voight........C-3
Wagner St........D-2,E-2
Walker Ave........F-5
Walker, W........F-3
Walton........A-4
Watson........A-3
Waugh Dr........F-2
Waverly Ct........J-2
Waverly St........B-1
Weber........C-4
Weber Ave........G-4
Webster........G-4,H-6
Welch Ave........G-1
Wendel........C-3
Wentworth St........J-4,K-6
West Blvd........J-2
West St........C-8
Westmoreland........H-3
Wheeler Ave........J-5,K-7
White St........E-4
White Oak Dr........C-2
Whitney........H-3
Wichita St........J-3
Wilkenson St........F-2
Wilkes........B-6
Willard Ave........G-2
Wilson........G-4
Winbern........J-5
Winnie........C-5
Woodhead St........H-1
Woodland St........B-3
Wood Leigh........H-8
Wood Row........I-3
Wrightwood........A-5
Yale St........B-1,D-1
Yoakum........G-2,J-2
York........H-8
Yupon St........G-2,H-2
2nd........D-2
4th W........C-2
4th St., E........C-2
4½ E........C-2
5th, E........C-2
5th, W........C-2
6th St., W........C-2
6½ St., E........B-2
7th St., E........B-2
7½ St., E........B-2
8th St., E........B-2
8th St., W........B-2
8½ St. E........B-2
9th St. E........B-2
9th St. W........B-2
9½ E........B-2
10th, E........B-2
10th St., W........B-2
10½ St., E........A-2
10½ St., W........A-1
11th St., E........A-2
11th St., W........A-2
11½ St., E........A-2
12th St., E........A-2
12th St., W........A-1
13th St., E........A-2
13th St., W........A-1
13½ St., E........A-2
14th St., E........A-2
14th St., W........A-1

POINTS OF INTEREST

Hermann Park........K-2,K-3
Moody Park........A-6
Stude Park........C-3
Woodlawn Park........C-5

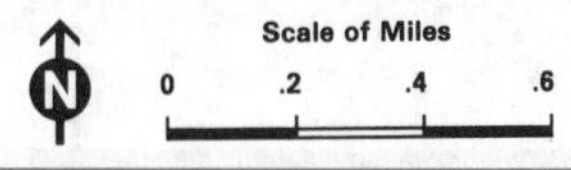
N
Scale of Miles
0
.2
.4
.6

Moody Park
Woodlawn Park
Stude Park
White Oak Bayou
Buffalo Bayou
Hermann Park
ALLEN PARKWAY
SOUTHWEST FREEWAY
GULF FREEWAY
WHITE OAK DR.
N. MAIN
HOUSTON
MEMORIAL
WASHINGTON
10
45
59
75
90
CONGRESS
PRESTON
PRAIRIE
TEXAS
CAPITOL
RUSK
WALKER
McKINNEY
LAMAR
DALLAS
LOUISIANA
MILAM
TRAVIS
MAIN
FANNIN
SAN JACINTO
CAROLINE
AUSTIN
LA BRANCH
CRAWFORD
JACKSON
CHENEVERT
HAMILTON
JEFFERSON
CALHOUN
PIERCE
GRAY
WEBER
McGOWEN
ELGIN
HOLMAN
ALABAMA
WESTHEIMER
BISSONNET
DUNLAVY
MONTROSE
SHEPHERD
YALE
HEIGHTS BLVD.
STUDEWOOD
CULLEN
SCOTT
HOLMAN AVE.
WHEELER
BLODGETT
MacGREGOR
ROCKWOOD ST.
VARSITY
1
2
3
4
5
6
7
8
A
B
C
D
E
F
G
H
I
J
K

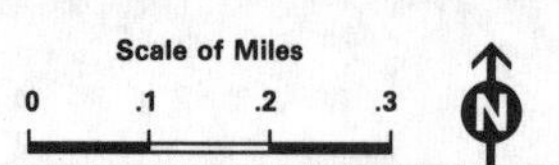

Scale of Miles
0
.1
.2
.3
N

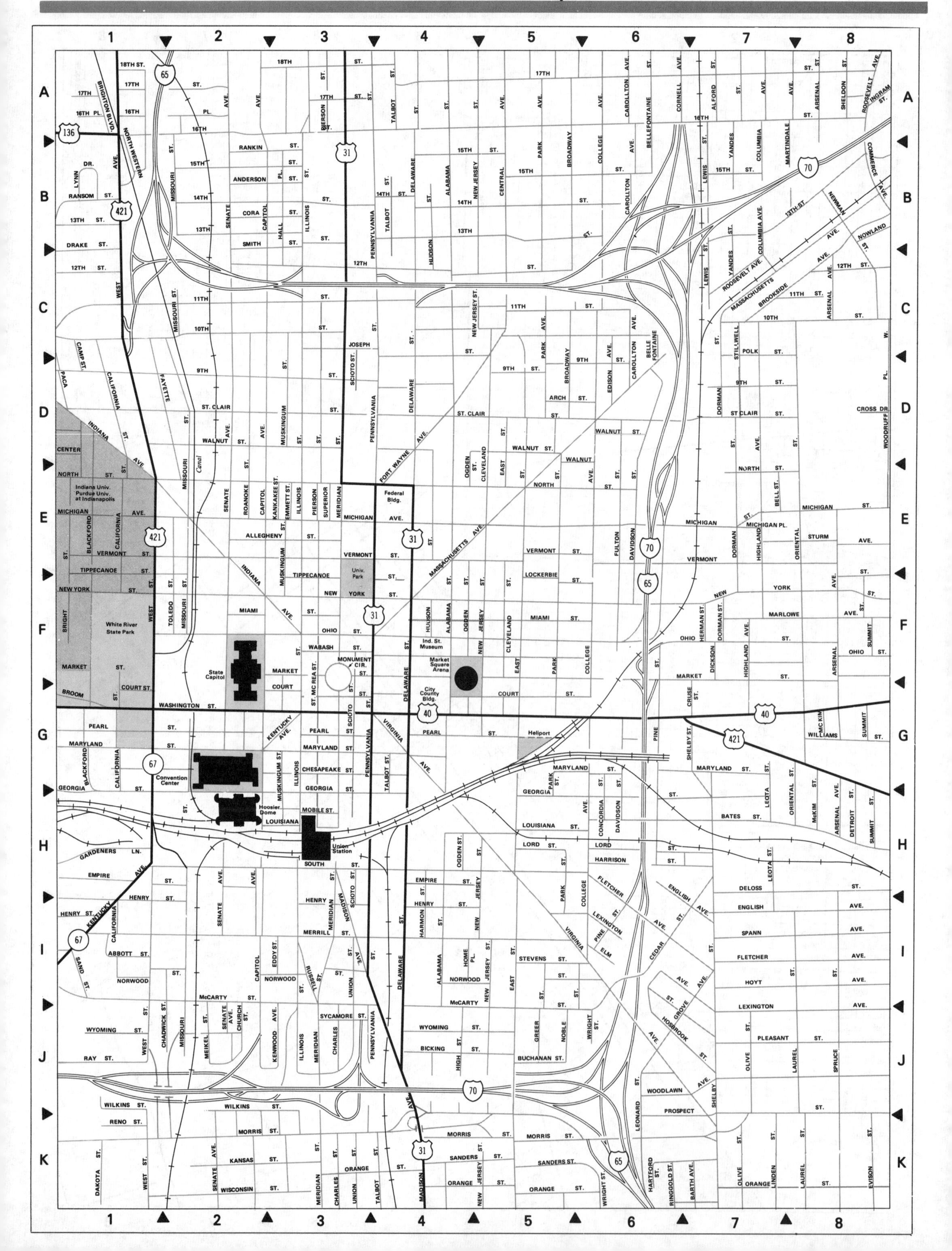

1
2
3
4
5
6
7
8
A
B
C
D
E
F
G
H
I
J
K
Indiana Univ. Purdue Univ. at Indianapolis
White River State Park
State Capitol
Convention Center
Hoosier Dome
Union Station
Univ. Park
MONUMENT CIR.
Federal Bldg.
Ind. St. Museum
Market Square Arena
City County Bldg.
Heliport
Canal
18TH ST.
17TH
16TH
16TH PL.
15TH
14TH
13TH
12TH
11TH
10TH
9TH
ST. CLAIR
WALNUT
NORTH
MICHIGAN
VERMONT
TIPPECANOE
NEW YORK
MIAMI
OHIO
WABASH
MARKET
COURT
WASHINGTON ST.
PEARL
MARYLAND
GEORGIA
LOUISIANA
SOUTH
MOBILE ST.
CHESAPEAKE
GARDENERS LN.
EMPIRE
HENRY
MERRILL
ABBOTT ST.
NORWOOD
McCARTY
WYOMING
RAY ST.
WILKINS
RENO ST.
MORRIS
KANSAS
WISCONSIN
SANDERS
ORANGE
BRIGHTON BLVD.
NORTH WESTERN AVE.
WEST
MISSOURI
SENATE
CAPITOL
ILLINOIS
MERIDIAN
PENNSYLVANIA
DELAWARE
ALABAMA
NEW JERSEY
CENTRAL
PARK
BROADWAY
COLLEGE
CAROLLTON AVE.
BELLEFONTAINE
CORNELL AVE.
ALFORD
YANDES
COLUMBIA
MARTINDALE
ARSENAL
SHELDON
ROOSEVELT AVE.
INGRAM
COMMERCE AVE.
NEWMAN
NOWLAND
MASSACHUSETTS AVE.
BROOKSIDE
FORT WAYNE AVE.
INDIANA AVE.
KENTUCKY AVE.
VIRGINIA AVE.
ENGLISH AVE.
SHELBY ST.
FLETCHER
LEXINGTON
PLEASANT
WOODLAWN
PROSPECT
DELOSS
SPANN
HOYT
ORIENTAL
HIGHLAND
DORMAN
STILLWELL
WOODRUFF PL.
CROSS DR.
SUMMIT
LAUREL
SPRUCE
OLIVE
LINDEN
EVISON
BARTH AVE.
RINGGOLD ST.
HARTFORD ST.
LEONARD
WRIGHT ST.
HOSBROOK
GROVE
CEDAR
MADISON
TALBOT
UNION
CHARLES
RUSSELL
KENWOOD
MEIKEL
CHADWICK ST.
DAKOTA ST.
CALIFORNIA
BLACKFORD
65
136
421
31
70
40
67

INDIANAPOLIS

Alabama St...B-4,F-4,I-4
Allegheny St.........E-2
Alford St............A-7
Anderson St.........B-2
Arch St............D-5
Arsenal St...A-8,F-8,H-8
Barth Ave..........K-7
Bates St.............H-7
Bell St...............E-7
Bellefontaine St.....A-6
Bicking St...........J-4
Blackford E-1,G-1
Brighton Blvd.......A-1
Broadway.......B-5,D-5
Brookside Ave.......C-7
Broom..............G-1
Buchanar St.........J-5
California D-1,G-1
Capitol Ave...B-2,F-2,I-2
Carollton Ave....A-6,D-6
Cedar St.............I-6
Center..............D-1
Central Ave..........B-5
Chadwick St.........J-1
Charles St...........K-3
Chesapeake St......G-3
Church St...........J-2
Cleveland St.........F-5
College Ave......B-6,F-6
Columbia Ave....B-7,C-7
Commerce Ave......B-8
Cora St..............B-2
Cornell Ave..........A-6
Court St.............G-4
Cruse St.............G-7
Dakota St...........K-1
Davidson St.........E-6
Delaware St..B-4,G-4,I-4
DeLoss St...........H-7
Detroit St...........H-8
Dickson............F-7
Dorman St.......D-7,E-7
Drake St............B-1
East St...........F-5,I-5
Edison Ave..........D-6
Elm St..............I-6
Empire St........H-1,I-4
English Ave......H-6,I-7
Evison St...........K-8
Fayette D-2
Fletcher Ave......H-6,I-7
Fort Wayne Ave......E-4
Fulton St............E-6
Gardeners Ln........H-1
Georgia St.......G-3,H-5
Greer St............J-5
Grove Ave...........J-6
Hall Pl..............B-3
Harrison St..........H-6
Hartford St..........K-6
Henry...........H-1,I-3
Highland Ave.....E-7,F-7
High St..............J-4
Home Pl.............I-4
Hosbrook St.........J-6
Hoyt Ave............I-7
Hudson St.......C-4,F-4
Illinois St....B-3,E-3,J-3
Indiana Ave..........E-2
Ingram St...........A-8
Joseph St...........C-3
Kansas St...........K-2
Kentucky Ave....G-3,I-1
Kenwood Ave........J-3
Laurel St.........J-8,K-8
Leonard St..........K-6
Leota St.............H-7
Lexington Ave....I-6,J-7
Lewis St.........B-7,C-7
Linden St...........K-7
Lockerbie St.........F-5
Lord St..........H-5,H-6
Louisiana St.....H-3,H-5
Lynn Dr.............B-1
McCarty St........I-2,I-4
McKim St...........H-8
Madison St..........K-4
Market St........F-3,F-7
Marlowe Ave........F-7
Martindale Ave......B-7
Maryland St.G-3,G-5,G-7
Massachusetts Ave. C-7,E-4
Meikel St............J-2
Meridian St......E-3,J-3
Merrill St............I-3
Miami St............F-2
Michigan Ave........E-3
Michigan Pl..........E-7
Missouri St. B-2,F-2,J-2
Mobile St...........H-2
Monument Cir.......F-3
Morris St........K-2,K-5
Muskigum St....E-3,G-3
New Jersey St. B-5,F-5,I-5
Newman St..........B-8
New York St......F-3,F-7
Nobel St............J-5
North St.........Ep3,E-7
Northwestern A-1
Norwood St.......I-1,I-4
Ogden St.........E-4,F-4
Ohio St...........F-3,F-7
Olive St..........J-7,K-7
Orange St....K-2,K-5,K-7
Oriental St.......E-8,H-8
Paca................D-1
Park Ave........B-5,C-5 ....F-5,G-5,I-5
Pearl St...........G-1,G-3
Pennsylvania St. B-4,G-3,J-4
Pierson St...........E-3
Pine St...........G-6,I-6
Pleasant St..........J-7
Polk St..............C-7
Prospect St..........J-6
Rankin St...........B-2
Ransom St...........B-1
Ray St...............J-1
Reno St.............K-1
Ringgold St..........K-6
Roanoke St..........E-2
Roosevelt.......A-8,B-7
St. Clair St.......D-2,D-7
Sanders St..........K-4
Senate Ave.......B-2,F-2 ....I-2,K-2
Shelby St........G-7,J-7
Sheldon St..........A-8
Smith St............B-2
South St............H-3
Spann Ave...........I-7
Spruce St............J-8
Stevens St...........I-5
Stillwell.............C-7
Sturm Ave...........E-8
Summit St.......F-8,H-8
Sycamore St.........J-3
Talbot St....A-4,G-4,K-4
Terrace Ave.........K-8
Toledo St............F-2
Union St..........I-3,K-3
Vermont St......E-3,E-7
Virginia Ave.........G-4
Wabash St...........F-3
Walnut St.......D-2,D-5
Washington St.......G-2
West Ave............C-1
West St..........F-1,K-1
Wilkins St........J-1,J-2
Williams St..........G-8
Wisconsin St........K-2
Woodlawn Ave......J-6
Woodruff Pl.........D-8
Wright St........J-6,K-6
Wyoming St.........J-4
Yandes St.......B-7,C-7
9th St.......D-2,D-5,D-7
10th St..........C-2,C-7
11th St.......C-2,C-5,C-8
12th St.......C-1,C-3,C-8
13th St.......B-1.B-2,B-4
14th St..........B-2,B-4
15th St...B-2,B-4,B-5,B-7
16th St......A-1,A-2,A-7
17th St..........A-1,A-5
18th St..........A-1,A-3

POINTS OF INTEREST

Military Park........F-1
Univ. Park...........E-3

JACKSONVILLE

Acosta Bridge.......H-3
Adams St....F-1,F-3,G-5
Alvarez St...........J-5
Arch St.............A-3
Ashley St.....E-4,F-6,G-8
Bay St.......F-1,G-3,G-6
Beaver St....D-1,E-4,F-6
Blanche St..........D-1
Broad St.....B-4,D-4,F-4
Bugbee St...........J-6
Catherine St.........H-7
Cemetery St..........F-8
Chelsea St...........H-1
Church St....E-2,F-5,G-8
Clay St...........B-5,F-4
Cleveland St.....D-2,F-1
Crothe St............B-1
Dante Pl.............K-4
Davis St.........C-3,F-3
Dewdrop North......E-4
Dewitt St............E-1
Dora St.............H-1
Duval St......F-2,F-5,G-8
Eaverson St......B-1,D-1
Edison Ave...........I-1
Elm St..............G-1
Flagler St........J-5,K-5
Forest St............H-1
Forsyth St....F-1,G-4,G-6
Francis St...........C-2
Fuller Warren Bridge J-2
Gary St.............K-4
Gilmore St...........J-1
Gulf Life Dr...........I-5
Hart St..............C-1
Hendricks Ave.......K-6
Hogan St........E-5,G-4
Home St.............J-6
Illinois..............B-4
Ionia St..........B-8,D-8
Jackson St...........H-1
Jefferson St.....D-4,F-3
Jessie St............E-8
Johnson St......C-2,F-2
Julia St..........E-5,G-4
Kings Ave...........K-6
Kings Rd............D-2
Kipp St..............J-6
Laura St.........B-6,F-5
Lee St...........C-2,F-2
Leila St.............H-2
Liberty St........B-7,G-7
Lisbon St............K-5
Louisa St............K-6
Louisiana B-3
Madison St..........F-3
Magnolia St.........H-1
Main St..........B-6,F-5
Main St. Bridge......I-5
Market St....B-7,E-7,G-6
Mars A-3
Mary St.............J-5
May St..............H-1
McConihe St........A-1
McCoy St...........H-3
Minnie St...........D-1
Monroe St.......F-3,G-6
Mt. Herman St.......C-2
Myrtle Ave.......B-1,E-1
Nadia...............B-3
Newman St.........G-6
Nira St..............K-4
Oak St..............H-1
Ocean St.........H-5,F-6
Odessa St...........E-8
Orange St...........E-5
Palm St.............K-4
Park St..............H-1
Pearl St......B-5,E-4,G-4
Phelps St........E-6,E-8
Pippin St............F-8
Price St.............H-1
Prudential Dr........J-4
Reiman A-2
Riverside Ave....H-2,J-1
Roselle St............I-1
San Marco Blvd......J-5
Schofield St.........E-7
Spruce St...........H-1
State St..........E-4,E-6
Steele St............B-1
Stonewall St........H-2
Stuart St.............F-2
Union St..D-1,E-4,F-6,F-8
Venus A-3
Walnut St.........B-8,E-7
Washington St.......G-7
Water St............G-3
Wilcox St........B-1,D-1
1st St..........D-5,D-7
2nd St..........D-5,D-7
3rd St.......C-1,C-5,D-7
4th St.......C-1,C-5,C-7
5th St.......B-1,C-5,C-7
6th St.......B-1,B-5,B-7
7th St....B-1,B-3,B-5,B-7
8th St.......A-1,A-5,B-7
9th St.......A-1,A-5,A-7
10th St......A-1,A-5,A-7

POINTS OF INTEREST

Baptist Hospital.....J-4
Cemetery E-7
St. Lukes Hospital...A-4
Springfield Park C-4

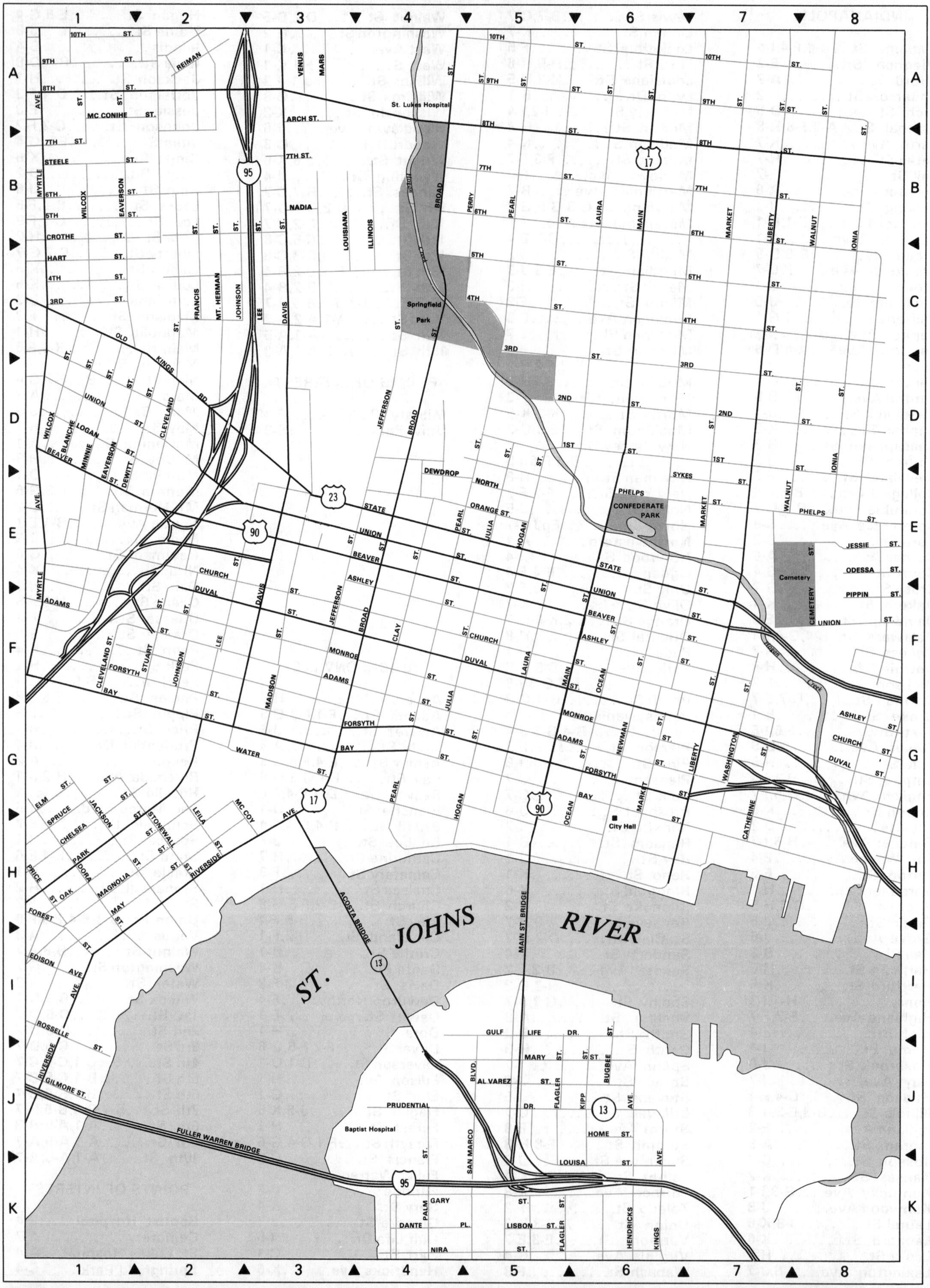

Jacksonville, FL

Scale of Miles
0 .1 .2 .3
N

1 2 3 4 5 6 7 8
A B C D E F G H I J K

10TH ST.
9TH ST.
8TH ST.
MC CONIHE ST.
7TH ST.
STEELE ST.
6TH. ST.
5TH ST.
CROTHE ST.
HART ST.
4TH ST.
3RD ST.
REIMAN
MYRTLE AVE.
WILCOX
EAVERSON
FRANCIS
MT. HERMAN
JOHNSON
LEE
DAVIS
VENUS
MARS
ARCH ST.
7TH ST.
NADIA
LOUISIANA
ILLINOIS
St. Lukes Hospital
Springfield Park
BROAD ST.
PERRY
PEARL
LAURA
MAIN
MARKET
LIBERTY
WALNUT
IONIA
OLD KINGS RD
UNION ST.
LOGAN ST.
BLANCHE
MINNIE
CLEVELAND
BEAVER ST.
DEWITT
JEFFERSON
2ND ST.
1ST ST.
SYKES
PHELPS
DEWDROP
NORTH
ORANGE ST.
STATE
JULIA
HOGAN
CONFEDERATE PARK
Cemetery
CEMETERY
JESSIE ST.
ODESSA ST.
PIPPIN ST.
CHURCH
DUVAL
ADAMS
ASHLEY
MONROE
FORSYTH
BAY
WATER
CLEVELAND ST.
STUART
MADISON
CLAY
OCEAN
NEWMAN
WASHINGTON
CATHERINE
City Hall
ELM
SPRUCE
CHELSEA
JACKSON
PARK
DORA
PRICE
OAK
MAGNOLIA
STONEWALL
LEILA
MC COY
RIVERSIDE AVE.
FOREST
MAY ST.
EDISON AVE
ROSSELLE ST.
RIVERSIDE
GILMORE ST.
FULLER WARREN BRIDGE
ACOSTA BRIDGE
MAIN ST. BRIDGE
ST. JOHNS RIVER
Hogan Creek
GULF LIFE DR.
MARY ST.
AL VAREZ ST.
BLVD.
FLAGLER
KIPP
BUGBEE AVE.
PRUDENTIAL DR.
Baptist Hospital
HOME ST.
LOUISA ST.
SAN MARCO
PALM
GARY ST.
DANTE PL.
LISBON ST.
NIRA ST.
HENDRICKS AVE.
KINGS
95
17
23
90
1 90
1 17
13

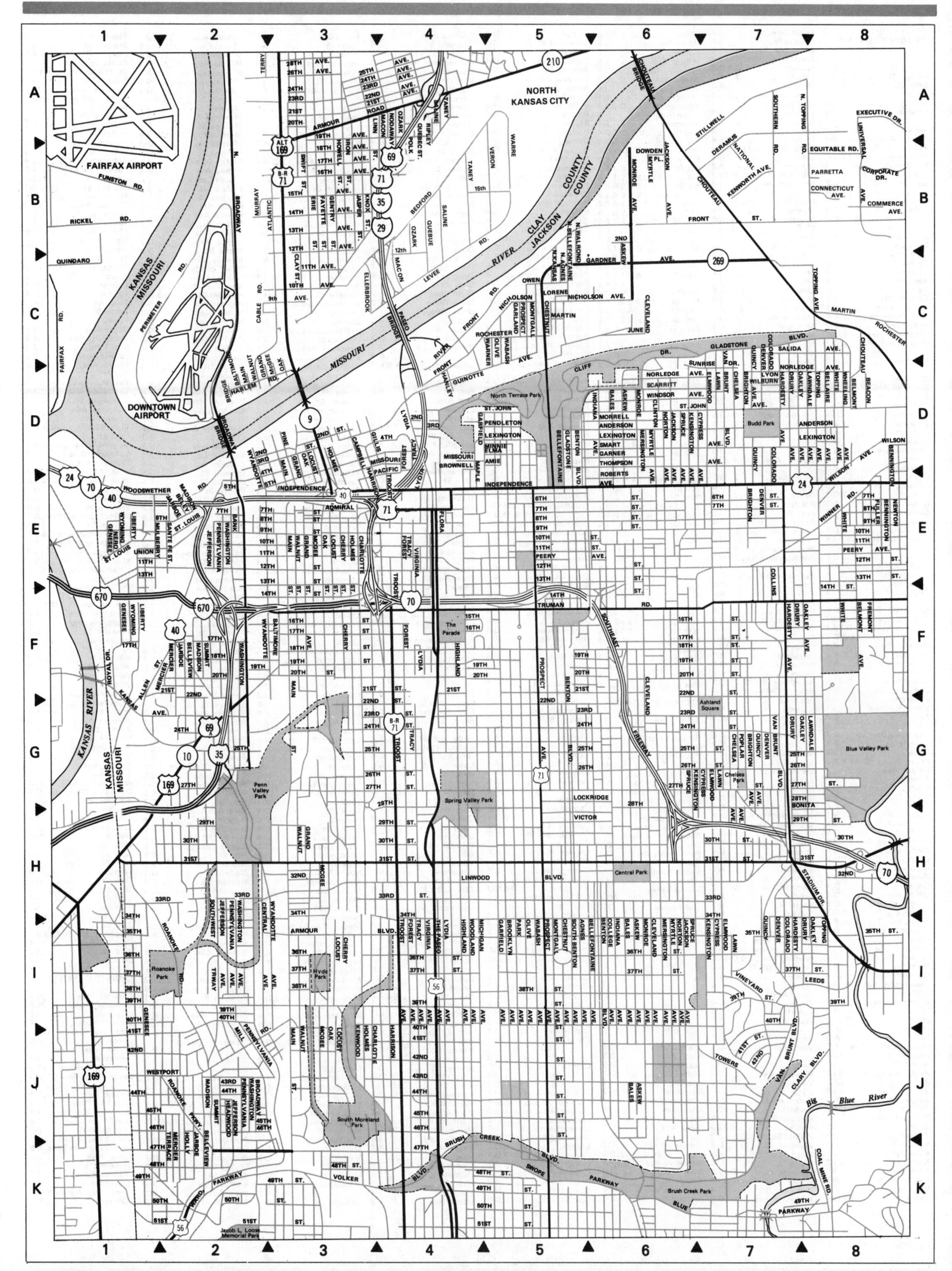
Scale of Miles
0 .2 .4 .6 .8
N
FAIRFAX AIRPORT
DOWNTOWN AIRPORT
NORTH KANSAS CITY
KANSAS
MISSOURI
CLAY COUNTY
JACKSON COUNTY
MISSOURI RIVER
KANSAS RIVER
Big Blue River
North Terrace Park
Budd Park
The Parade
Penn Valley Park
Spring Valley Park
Ashland Square
Chelsea Park
Blue Valley Park
Central Park
Roanoke Park
Hyde Park
South Moreland Park
Brush Creek Park
Jacob L. Loose Memorial Park
INDEPENDENCE AVE.
TRUMAN RD.
LINWOOD BLVD.
ARMOUR BLVD.
BRUSH CREEK BLVD.
SWOPE PARKWAY
BLUE PARKWAY
WARD PARKWAY
SOUTHEAST FREEWAY
STADIUM DR.
PASEO BRIDGE
BROADWAY BRIDGE
CHOUTEAU BRIDGE
FRONT ST.
GARDNER AVE.
COAL MINE RD.
TOPPING AVE.
PROSPECT AVE.
TROOST AVE.
MAIN
BROADWAY
WESTPORT
RICKEL RD.
QUINDARO
FUNSTON RD.
WOODSWETHER RD.
GLADSTONE BLVD.
CLIFF DR.

KANSAS CITY, MO.

Admiral....E-3
Agnes, N....C-5
Allen St....F-1
Amie....D-5
Anderson....D-6,D-8
Armour Blvd....I-3
Armour Rd....A-3
Askew....B-6,D-6
Askew Ave....I-6,J-6
Bales....D-6
Bales Ave....I-6,J-6
Baltimore....F-3
Bank....E-2
Beacon....D-8
Beley....E-2
Bellaire Ave....D-8
Bellefontaine....D-5
Bellefontaine, N....B-5
Bellefontaine Ave....I-6
Belleview....F-2
Bellview Ave....J-2
Belmont....D-8
Belmont Ave....F-8
Bennington....D-8
Bennington....E-8
Benton Blvd....D-5
Benton Blvd....F-5
Benton Blvd....I-6
Blue Parkway....K-6
Broadway, N....B-2
Broadway Ave....J-2
Broadway Bridge....D-2
Brooklyn Ave....I-5
Brownell....D-4
Brighton....D-7
Brighton....E-7
Brighton Ave....G-7
Brush Creek Pkwy....J-4
Campbell....D-3
Central Ave....H-2
Charlotte....E-3
Charlotte....I-3
Chelsea....D-7
Chelsea Ave....G-7
Cherry....F-3,I-3
Cherry St....E-5
Chestnut....C-5
Chestnut Ave....I-5
Chouteau....B-7,C-8
Chouteau Bridge....A-6
Clary Blvd....J-7
Clay St....B-3
Cleveland....C-6
Cleveland Ave....F-6,I-6
Cliff Dr....D-5
Clinton....D-6
Coal Mine Rd....K-8
College Ave....I-6
Colorado....C-7,D-7
Colorado Ave....I-7
Collins....E-7
Corporate Dr....B-8
Cypress....G-7
Cypress Ave....D-7,I-7
Denver....C-7,E-7,G-7
Denver Ave....I-7
Deramus....A-7
Dowden Pl....B-6
Drury....F-8,G-7
Drury Ave....D-7,I-8
Elma....D-5
Elmwood....D-7,G-7
Elmwood Ave....I-7
Equitable Rd....B-8
Erie St....B-3
Executive Dr....A-8
Fayette St....B-3
Flora....E-4
Forest....D-4,E-4,F-4
Forest Ave....I-4
Fremont....F-8
Front....D-4
Front St....B-6
Fuller....E-8
Funston Rd....B-1
Gardner Ave....C-6
Garfield....D-5
Garfield Ave....I-5
Garland....C-5
Garner....D-6
Genesee....E-1,F-1,I-1
Gentry St....B-3
Gillis....D-4
Gladstone....D-5
Gladstone Blvd....C-7
Grand....D-3,H-3
Grand St....E-3
Guinotte Ave....D-4
Hanley....D-4
Hardesty Ave.D-7,F-8,I-7
Harlem Rd....D-2
Harrison....D-3
Harrison Ave....J-4
Headwood Ave....J-2
Highland....F-4
Highland Ave....I-4
Holly Ave....J-2
Holmes....D-3,I-3
Holmes St....E-3
Howell St....A-3
Independence Ave....E-5
Indiana Ave....D-6,I-6
Iron St....A-3
Jackson Ave..B-6,D-6,I-6
Jarboe....E-2,F-2
Jarboe Ave....J-2
Jasper St....B-3
Jefferson....E-2
Jefferson Ave....H-2,J-2
June....C-6
Kansas, N....C-5
Kansas Ave....F-1
Kensington....G-6
Kensington Ave..D-6,I-7
Kenwood....I-3
Kenworth Ave....B-7
Knox St....B-3
Lawn....D-7,G-7
Lawn Ave....I-7
Lawndale....G-8
Lawndale Ave....D-8
Leeds Ave....I-8
Lexington....D-5,D-6,D-8
Liberty....E-1,F-1
Linn St....A-4
Linwood Blvd....H-4
Lockridge....G-5
Locust....D-3,I-3
Locust St....E-3
Lorene....C-5
Lydia....D-4,F-4
Lydia Ave....I-4
Lyon....D-7
Macon....A-4
Madison....E-2,F-2
Madison Ave....J-2
Main....D-3
Main St....E-3,F-3,J-3
Maple....D-4
Martin....C-5
McGee....E-3,H-3,I-3
Mercier....F-2
Mercier Ave....J-2
Mersington....D-6
Mersington Ave....I-6
Michigan Ave....I-4
Mill....I-2
Minnie....D-5
Missouri....D-4
Monroe....D-6
Monroe Ave....B-6,I-6
Montgall....C-5
Montgall Ave....F-5,I-5
Morrell....D-6
Mulberry....E-2
Myrtle....B-6
Myrtle Ave....D-6,I-6
Nero....E-1
Newton....E-8
Nodaway....A-4
Norledge Ave....D-6,D-7
Norton Ave....D-6,I-6
Nicholson Ave....C-5
Oak....D-3,I-3
Oak St....E-3
Oakley....G-8
Oakley Ave....D-8,F-8,I-8
Olive....C-5
Olive Ave....I-5
Owen....C-5
Ozark....A-7
Pacific....D-4
Park Ave....I-5
Paseo Bridge....C-4
Peery Ave....E-5,E-8
Pendleton....D-5
Pennsylvania....E-2,I-2
Pennsylvania Ave....H-2
....J-2
Perimeter Rd....C-1
Pine....D-3
Polk....A-4
Poplar Ave....G-7
Prospect....C-5
Prospect Ave....I-5
Quebec St....A-4
Quincy....C-7,D-7
Quincy Ave....G-7,I-7
Quindaro....B-1
Rickel Rd....B-1
Ripley....A-4
River Front Rd....C-4
Roanoke Pkwy....J-2
Roanoke Rd....I-2
Roberts....D-6
Rochester....C-5
Royal Dr....F-1
St. John....D-5
St. John Ave....D-7
St. Louis....E-2
St. Louis....E-2
Salida Ave....C-7
Saline....A-4
Sante Fe St....E-2
Scarritt....D-6
Smart Ave....D-6
South Benton Ave....I-5
Southeast Freeway..F-6
Southern Rd....D-5
Southwest Trway....H-2
Spruce....G-6
Spruce Ave....D-6,I-6
Stillwell....A-7
Summit....F-2
Summit Ave....J-2
Sunrise Dr....C-7
Swift St....B-3
Swope Pkwy....K-5
Taney....A-4
Terrace Ave....J-2
The Paseo Ave....I-4
Thompson Ave....D-6
Topping Ave..C-8,D-8,I-8
Topping Rd., N....A-8
Towers....J-7
Tracy....D-4,E-4,G-4
Tracy Ave....I-4
Troost....E-4,G-4
Troost Ave....I-4
Truman Rd....F-5
Union....E-1
Universal Ave....B-8
Van Brunt Blvd....C-7
....G-7,J-7
Victor....H-5
Vineyard Ave....I-7
Virginia....E-4
Virginia Ave....I-4
Volker Blvd....K-3
Wabash....C-5
Wabash Ave....I-5
Walnut....H-3,I-3
Walnut St....E-3
Walrond, N....B-5
Ward Pkwy....K-2
Warner....C-5
Washington....E-2,F-2
Washington Ave.H-2,J-2
Westport Rd....J-1
Wheeling Ave....D-8
White....E-8,F-8
White Ave....D-8
Wilburn....D-7
Wilson Ave....E-8
Windsor Ave....D-6
Winner Rd....E-8
Woodland Ave....I-4
Woodswether Rd....E-1
Wyandotte....D-2,F-3
Wyandotte Ave....H-3
Wyoming....E-1,F-1
2nd St....B-6,D-3,D-4
3rd St....D-3,D-4
4th....D-3,D-4
5th....D-3,E-2
6th St....E-5,E-7
7th....E-2,E-3,E-5,E-7,E-8
8th St....E-2,E-3,E-5,E-8
9th St....E-3,E-5,E-8
10th....C-3,E-3,E-5,E-8
11th..C-3,E-1,E-3,E-5,E-8
12th St....B-3,E-3,E-5,E-8
13th St....B-3,E-1,E-3
....E-5,E-8
14th St....B-3,E-1,E-3
....F-5,F-8
15th....B-3,F-4
16th St....B-3,F-1,F-3
....F-4,F-6
17th St....B-3,F-1-3, F-6
18th St....A-3,F-2,F-3,F-6
13th St....A-3,F-2-6
20th St....A-3,F-2-6
21st St....A-3,A-4
....F-2-5
22nd St....A-4,F-2,F-3
....F-6,G-5
23rd St....A-3,A-4,G-3
....G-5,G-6
24th St....A-3,A-4,G-2
....G-3,G-5,G-6
25th St....A-4,G-2,G-3
....G-5,G-6,G-7
26th St....A-3,G-3
....G-5-7
27th St....G-2,G-3
....G-4,G-6,G-8
28th St....A-3,G-2
....G-4,G-6,G-7
29th St....H-2,H-4,H-7
30th St..H-2,H-4,H-7,H-8
31st St....H-2-8
32nd St....H-3,H-8
33rd St..H-1,H-2,H-4,H-7
34th St....H-1,H-3,H-4-7
35th St....I-1,I-7,I-8
36th St....I-1,I-3,I-4,I-6
37th St....I-1,I-3,I-4
....I-6,I-7
38th St....I-1,I-3,I-5
39th St....I-1-8
40th St....I-1,I-2,I-4,I-7
41st St....I-1,I-4,I-7
42nd St....J-1,J-4,J-7
43rd St....J-2,J-4
44th St....J-1,J-2,J-4
45th St....J-1,J-2,J-4
46th St....J-1,J-2,J-4
47th St....K-1,K-4
48th St....K-1,K-3,K-4
49th St....K-1,K-3
....K-4,K-7
50th St....K-1,K-2,K-4
51st St....K-1,K-2,K-4

POINTS OF INTEREST

Ashland Sq....G-7
Blue Valley Park....G-8
Broadway Bridge....D-2
Brush Creek Park....K-6
Budd Park....D-7
Central Park....H-6
Chelsea Park....G-7
Chouteau Bridge....A-6
Downtown Airport..D-1
Fairfax Airport....B-1
Hyde Park....I-3
Jacob L. Loose Memorial Park....K-2
North Terrace Park....D-5
Penn Valley Park....G-2
Roanoke Park....I-1
South Moreland Park....J-3
Spring Valley Park....G-4
The Parade....F-4

KNOXVILLE

Acker St....B-6
Adair Ave....B-6
Adams Ave....D-8
Adcock St....D-1
Agee St....I-7
Ailor Ave....H-4,I-2
Alabama Ave....E-2
Albert Ave....B-6
Alexander Ave....E-5
Alpha Ave....H-3
Amber St....B-8
Ambrister St....H-2
Ambrose St....D-2
Anderson Ave....E-5,F-4
Andy Holt Ave....J-4
Arcade St....C-8
Armstead St....E-5
Armstrong Ave....D-6
Arthur St....H-4
Ashwood St....C-5
Atchley St....I-7
Atlantic Ave....C-4
Atlantic St....C-3
Augusta Ave....J-6
Austin St....F-8
Avondale Ave....A-7
Bales St....F-5
Bank Ave....C-4
Barber St....I-8
Barton St....B-6
Battery Ave....J-1
Baxter Ave...F-4,F-5,G-3
Bearden Pl....F-5
Beaumont Ave....F-2
Belleaire Ave....F-3
Bell St....G-7
Bellevue St....A-6
Belmont Hts....G-1
Belvoir....B-6
Bernard St....F-5
Bertrand St....E-8
....F-8,G-8
Bethel Ave....G-8
Blount Ave....I-7,K-6
Bluff St....D-6
Boggs Ave....I-8
Bond St....D-3
Bonnyman Dr....F-2
Booker St....G-3
Bookwalter Dr....B-1
Boone St....C-8
Booth St....I-1
Boright Ave....A-7
Boright Place....A-7
Boruff St....E-7
Bowling Ave....G-1
Boxwood Ln....B-7
Boyd St....F-3,G-4
Bradley Ave....G-5
Bragg St....E-1
Brandau St....G-3
Branner St....E-4
Branson Ave....C-6
Breda Dr....A-3
Brice St....B-7
Bridalwood Dr....A-8
Bridge Ave....H-5
Brigham Ave....G-7
Brisco St....F-7
Britton Hts....J-8
Broadview Rd....A-1
Broadway....E-6,G-6
....H-6,I-6,J-6
Brock Ave....I-2
Brookside Ave...F-3,G-3
Brown Ave....D-7
Browning Ave....H-1
Brown St....D-6
Bruhin Rd....A-1
Brunswick St....A-7
Buchanan Ave....A-5
Buck St....G-1
Buffalo Ave....H-3
Buffat Mill Rd....B-7,B-8
Burge Dr....G-8
Burgess Ave....G-4
Burnside St....E-1
Burwell Ave....D-3
Byrd Ave....B-4
Caldwell Ave....D-4
Caledonia St....J-4
Callaway Ave....G-4
Cambridge St....J-6
Cammichael St....I-7
Campbell Ave....G-7
Canary St....K-8
Cansler Ave....G-4
Capella....A-5
Carrick St....H-4
Carson St....C-8
Carus Rd....B-2
Cary St....J-1
Caswell Ave....E-6
Cecil....C-8,D-6
Cedar St....C-2
Central....C-2,D-3
....E-5,H-6
Century St....H-2
Chamberlain Blvd....K-7
Chapman St....J-7
Charles Pl....H-6
Cherry St....C-8
Chicago St....C-6
Chickamauga Ave.
....B-4,C-3
Chilhowee Ave....F-7
Chipman....D-7
Church Ave....H-6,H-8
Churchwell....D-4
Citico St....H-3
Citrus St....C-8
Claiborne Pl....B-5
Clancy St....J-6
Clark St....G-4
Claude St....I-8
Claudius....B-2
Clay St....C-8
Clearview St....A-6
Cleveland St....F-5
Clifford St....G-8
Clinch Ave....I-4
Clyde St....H-3
Coker Ave....C-6
College St....G-3
Colonial Ave....B-5
Colter St....I-8
Columbia St....D-4
Commerce Ave....H-5
Common Ave....G-6
Cooper St....F-5
Concord St....I-2
Connecticut Ave....E-1
Copeland St....B-6
Coram....C-3
Cornelia....D-3,D-4
Coster St....A-1
Cottage Pl....D-6
Council Pl....I-7
Cowan St....H-4
Cox St....D-5
Cristine Ave....I-8
Crockett St....C-8
Cruze St....E-8,F-8
Cumberland Ave....J-3
Dale Ave....H-4
Dameron Ave....F-4
Dandridge Ave....G-8
Davanna St....E-4
Davenport Rd....I-8
Davison Ave....J-6
Dawn St....G-1
Dawson St....I-7
Deadrick Ave....G-4
Deery St....F-6
Delaware Ave....E-1
Delden Dr....A-4
Dell St....G-3
Dempster St....B-6
Denson Ave....H-1
Depot Ave....G-6
Derieux St....B-7
Dill St....C-2
Dinwiddie St....D-1
Division St....I-1
Dixie St....I-8
Dodson Ave....D-8
Dora St....G-4
Douglas Ave....G-4
Dow Dr....K-8
Drive One....F-2
Drive Three....F-2
Drive Two....F-2
Druid Dr....K-8
Dunbar St....G-4
Dundee St....D-7
Durbin....B-1
Dutch Rd....A-2
Edgewood St....B-6,B-7
Edmonds Ave....I-8
Eleavor Ave....F-6
Ellis St....K-8
Elm Ave....E-3
Ely Ave....E-1
Emerson St....I-6
Emerald St....E-4,F-3
Emmett St....I-1
Emoriland Blvd....A-6
Eubanks Ave....H-2
Euclid Ave....H-3,H-4
Exeter Ave....G-3
Ezell St....C-7
Fairfax....A-4
Fairmont Blvd....B-5
Fairview St....B-6
Fairwood....B-6
Farragut....B-3
Felts St....C-4
Fillmore Ave....H-1
Fine Ave....C-7
Flagg Ave....E-8
Flamingo St....K-8
Flemming St....C-7
Folsom Ave....D-5,E-5
Fontana....A-7
Forest Ave....H-4,I-3
Forestdale Ave....A-8
Forsythe St....B-4
Fort Ave....J-7
Fort Dickerson Rd....K-7
Fort Hill Rd....J-8
Fox St....C-2
Francis St....J-4
Fraternity Park Dr....J-5,K-5
Frazier St....F-6
Freemason St....B-4
Fremont Ave....D-5
Front Ave....I-6
Fulton Pl....F-5
Galba Rd....B-2
Galbraith St....D-1
Gale St....C-8
Galway St....D-8
Gap Rd....E-1
Gaston Ave....B-8
Gay St....G-5,H-6
Georgia St....F-6,G-7
Geraldine St....E-4
Gerson Dr....K-8
Gertrude Ave....I-8
Geta Rd....B-2
Gibbsons St....F-7
Gill....F-5
Gillespie Ave....D-7
Gladstone....B-4
Glen Ave....H-1
Glen Oaks St....A-1
Glenview Dr....C-6
Glenwood St....C-4
....D-4,E-6,E-7
Glider Ave....D-7
Godfrey St....B-6
Goldfinch Ave....K-8
Goins Dr....F-1
Gordon St....F-2
Grace St....B-5
Grainger....D-6
Grand Ave....H-4,I-3
Gratz St....E-6,F-6
Greenfield Ln....C-7
Griffin St....G-2
Grove St....B-3
Hague, The....A-4
Hall St....J-8
Hancock St....C-4
Hannah Ave....H-4,H-5
Hanover ST....B-4
Hardin Hill Rd....B-8
Harp St....D-8
Harriet Tubman St....F-7
Harris Ave....H-1
Hart Ave....I-3
Hartford St....J-7
Harvey St....C-4,D-4
Hatton Ave....F-4
Haun Dr....A-8
Hawkins St....F-1
Hawthorne St....J-6
Haynes Pl....E-6
Hays Dr....A-1
Hazel Pl....F-5
Hazen St....G-8
Heins Ct....A-2
Heiskell....D-1
Helen St....C-5
Helland Ave....E-5
Henegar St....C-5
Henley St....H-6
Henrietta Ave....F-8
Hickley Pl....H-5
Hickman St....I-8
High Ave....J-8
Highland Ave....I-3,I-4
High School St....A-1
Hill Ave....H-7
Hill St....I-6
Hiwassee Ave....B-4
Hiwassee St....C-3
Hoitt Ave....D-7
Holland Dr....A-4
Holly St....F-7
Hooker St....F-2
Hooray Ln....C-2
Howard St....C-6
Humes St....G-6
Huron St....B-4,D-5
Hutchinson Ave....C-6
Hutton Dr....A-1
Iredell Ave....G-3
Irene Ave....I-8
Irwin St....E-5
Isabella Cir....I-8
Islington Ave....F-8
Jackson Ave....G-6
James Ave....G-4
Jefferson Ave....E-7
Jenkins St....D-1
Jennings Ave....F-5
Jersey Ave....J-2
Jessamine St....F-7
Johnston St....D-1,D-2
Jonathan Ave....J-8
Jones St....I-7
Jordan St....G-2
Jourolman Ave....F-2
Kanterbury Dr....B-8
Karns Ave....A-5
Katherine Ave....D-2
Keith Ave....G-1
Keller St....F-8
Kenilworth Ln....B-7,B-8
Kennington Rd....B-8
Kentucky St....F-7
Kenyon St....C-4,D-5
Kern Pl....E-5
Keystone Ave....B-7
King St....G-6
Kingland Ave....J-8
Krohn St....C-5
Kuhlman St....A-5
Kyle St....F-8
Lake Ave....J-4
Lake Loudoun Blvd....J-5
Lamar St....F-6
Langford Ave....I-8
Langland St....F-8
Larkwood Ln....G-1
Laurel Ave....I-3,I-4
Laurens Ave....G-8
Lawson Ave....C-6
Lee St....F-4
Ledgerwood Ave....C-5
Leinard Ln....K-5
Leland St....C-4
Lemon St....D-8
Leonard Pl....D-6
Leslie Ave....H-3
Lincoln St....I-8
Linden Ave....F-7,F-8
Lindsay St....J-2
Link St....F-4
Lippincott St....J-8
Locust St....H-6
Logan Ave....H-4
Lonsdale Ave....E-2
Loraine Ave....H-1
Louisiana Ave....E-1
Lovenia Ave....F-6
Ludlow....A-4
Luck St....D-6
Luttrell....D-6,F-6
Lynn Ave....G-4
Magnolia....F-8,G-6
Major Ave....G-2
Main Ave....I-6
Mall St....H-1
Maria Ave....G-3
Marietta Ave....A-5
Marion St....E-3,F-4
Market St....H-6
Martin Mill....J-7
Maryland....F-1
Massachusetts Ave.
....E-2,F-1
Matthews Pl....E-5
Maxwell St....A-7
May St., N.W....F-5
Maynard....B-4
McCalla Ave....F-7
McClellan St....F-1
McCroskey Ct....C-7,D-6
McElroy Ave....J-1
McGhee Ave....G-4
McGuffey St....D-5
McKinley St....A-5
McMillan....D-3
McMinn St....C-7
McMurry St....B-2
McNutt St....A-7
McPherson St....F-1
McSpadden St....F-3
McTeer St....G-1
Mead Ave....E-3
Melbourn St....C-5
Melrose Ave....I-5,J-4
Melrose Pl....J-4
Melrose St....J-4
Mercer St....F-3
Metler St....C-3
Miami St....A-6
Middle St....G-4
Middlebrook Pike Ave.
....H-1,H-2
Midway St....D-1
Millers Ave....J-7
Milton St....A-5
Mimosa St....J-7
Mingle Ave....H-1
Minnesota Ave....E-1
Mississippi Ave....G-2
Mitchell Ave....D-7
Mitchell St....A-1,E-7
Money Pl....B-7
Monmouth....B-5
Monroe St....E-7
Monticello....B-5
Morelia St....D-3
Morgan St....F-6,G-6
Morris Ave....H-1
Moses Ave....G-4
Mount Castle St....J-4
Mulvan St....G-7
Murphy St....E-3
Myers St....G-2
Mynatt Ave....J-1
Myrtle St....F-7
Nadine Ave....C-6
Nelson Ave....G-7
Nerva Rd....B-2
New St., West....G-7
New College Hill Dr..G-2
Newman St....C-7
New York Ave....E-1
Neyland Dr....J-3,K-5
Nickerson Ave....B-6
Nolan Ave....G-3
North Ave....G-4
Oak St....G-4,G-5
Oakhill Ave....E-3
Oakhill St....E-4
Oakota Ave....D-1
Ocoee Trail....B-5
Oglewood St....D-4
Ohio St....E-1
Oldham Ave....E-4
Oldham, West....F-3
Old Vine Ave....G-8
Olive St....E-8
Oklahoma Ave....E-5
O'Neal St....D-1
Orange Ave....H-2
Orchard Dr....A-8
Orlando St....A-6
Oswald St....B-5
Overton Pl....E-6
Pacific St....B-5
Paige St....C-5
Painter Ave....J-2
Palmer St....J-6
Park Circle....J-5
Patton St....G-7
Pedigo Ave....I-8
Pembrokelinciad....A-5
Pershing St....C-4
Pertinax Dr....B-2
Phillips Ave....I-8
Pickett Ave....G-3
Pike St....B-7,E-2
Pitner Pl....K-6
Polk St....E-8
Poplar St....I-6
Portland St....I-1
Potter St....C-8
Powers Ave....A-5,A-6
Pratt St....E-6,E-7
Probus Rd....B-2
Pruitt St....F-5
Quincy Ave....D-4
Rader Ave....E-5
Radford Pl....C-3,C-4
Raleigh Ave....A-4,B-5
Ramona Ave....H-1
Randolph St....F-6
Rector St....D-1
Redwine St....K-6
Reed St....F-3
Reverse Curve Ave....I-8
Reynolds St....G-2
Richmond Ave....G-1
Rider Ave....B-5
Rim St....C-7
Riverside Ave....H-7
Rogers St....C-6
Rosedale Ave....G-8
Rosemond Dr....C-6
Ross St....C-6
Roy St....E-5
Ruckart Ln....B-1
Rudy St....D-1
Russell St....G-3
St. Mary St....D-5
St. Paul St....J-6
Salem St....I-1
Salvus St....H-7
Sam Houston St....J-7
Sandra St....A-6
Sanford St....C-5
Schade Ave....G-8
Schofield St....G-1
Scott Ave....E-5
Scott, West....F-3
Scottish Pike....K-5
Seaman St....I-1
Sevier Ave....I-7
Seymour Ave....B-6
Shamrock Ave....C-4
Sharp Ln....A-1
Sharps Ridge Park Rd....B-2,B-3
Shaw Dr....A-6
Shawmut Ave....I-2
Shea St....G-3
Shepard Pl....E-5
Sheridan St....E-2
Sherman St....E-1
Sherrod St....J-7
Silver St....E-5,F-5
Southwood Dr....K-8
Spencer St....E-8
Spring Ave....E-8
Springs Rd....B-7
Springdale Ave....D-3
Spruce St....E-8
Stadium Dr....J-5
Stair Ave....F-3
State St....G-6,H-6
Stephenson St....K-4
Sterchi St....G-1
Stewart St....E-5
Stone St....F-5
Stonewall St....E-1,F-2
Sunrise St....F-3
Susanne Ave....I-8
Sutherland Ave....I-2
Sullins St....J-2
Tecoma Dr....A-6
Teeple St....B-7
Tellico St....H-3
Temple St....J-4
Tennessee....F-1
Terrace Ave....J-4
Testerman St....F-2
Texas....E-1
Thomas St....E-1
Thompson Pl....D-6
Tiberius Rd....B-2
Tillery St....C-7
Tims Rd....A-2
Tindell St....J-6
Tipton Ave....K-8
Titus Way....B-2
Toms St....G-2
Topaz St....B-8
Town View Dr....G-7
Truman St....G-1
Tulip Ave....H-4
Tulip St....H-5
Tyson St....F-5
Unaka St....H-3
Union Ave....H-5
University Ave....H-3
U of T Rd....J-4
Upland Ave....A-6
Valley Ave....I-8
Valley Rd....A-2
Val St....F-4
Vandeventer Ave....J-8
VanGilder St....F-8
Vermont Ave....F-2
Vincent St....B-7
Vine Ave....G-8,H-6
Vine St....G-7
Virginia Ave....F-2
Volunteer Blvd....J-4
Walker Blvd....A-5
Walker St....F-3
Wall Ave....H-5
Wall St....H-5
Wallace St....G-4
Walnut St....H-6
Walnut Grove, West.A-3
Warren St....D-4
Washington Ave.C-6,E-7
Washington Pike....A-8
Watauga Ave....B-3
Watson Ave....G-4
Waycross St....H-1
Weaver....A-5
Webster Ave....H-3
Wells Ave....E-5,E-7
Western Ave....G-1,H-3
Westover Ave....A-4
Westview Ave....H-2
Wheeler St....D-7
White Ave....I-4
Whitney Pl....A-7
Whittle St....A-7
Wilkins St....G-3
Willow Ave....G-7
Winona St....F-7
Woodbine Ave....E-7
Woodland....E-4,F-3
Woodlawn Dr....A-1,K-7
Woodmont....A-8
Worth St....E-4
Wray St....F-5
Wynderse Ave....F-1
Yale Ave....J-5
Yarnell Ave....J-8
3rd Ave....F-6
4th Ave...D-6,E-6,F-6,G-4
5th St....F-6,F-7,G-4,G-6
6th, North....D-6
7th, North....D-7
8th, North....D-7
9th, North....D-7
10th St....H-5
11th St....H-5
12th St....H-5
13th St....H-5
14th St....I-5
15th St....H-5
16th St....H-4
17th St....I-4
18th St....I-4
19th St....I-4,J-4
20th St....I-4
21st St....H-3,I-3,J-4
22nd St....I-3,I-4
23rd St....I-2
24th St....H-2

POINTS OF INTEREST

Caswell Park....E-7
Chevans Park....B-6
City Hall....H-5
Court House....I-6
East Tenn. Baptist Hospital....I-6
Ft. Saunders Presb. Hospital....I-4
Knoxville College....G-3
Leslie Street Park....H-2
Linden Park....F-8
Lonsdale Park....E-1
Morning Side Park....G-8
Post Office....I-6
Preston Park....F-8
Sharps Ridge Park....B-3
Tom Dickerson Park.K-6
Tyson Park....J-3
University of Tennessee....I-5
University of Tennessee Agriculture Center....K-2

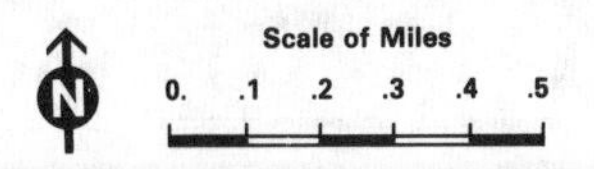

N
Scale of Miles
0. .1 .2 .3 .4 .5

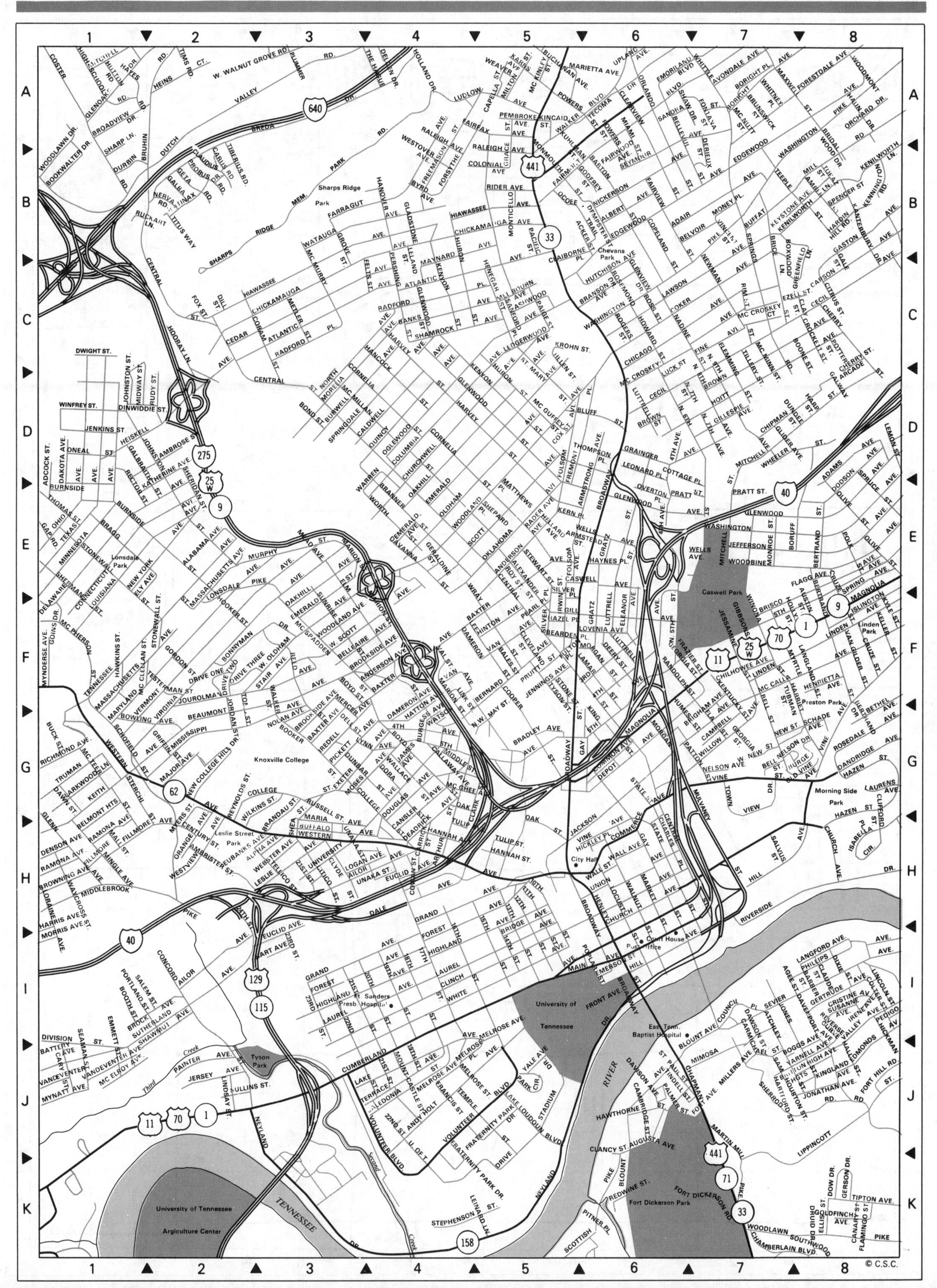

1 2 3 4 5 6 7 8
A B C D E F G H I J K
Sharps Ridge Park
Chevans Park
Lonsdale Park
Caswell Park
Linden Park
Preston Park
Knoxville College
Leslie Street Park
Morning Side Park
City Hall
Court House
Post Office
University of Tennessee
Ft. Sanders Presb. Hospital
East Tenn. Baptist Hospital
Tyson Park
Fort Dickerson Park
University of Tennessee Argiculture Center
TENNESSEE
RIVER
Third Creek
Second Creek
640
441
33
275
25 W
9
40
62
129
115
11
70
1
71
158
CENTRAL
BROADWAY
MAGNOLIA
CUMBERLAND
WESTERN AVE.
CHAPMAN
NEYLAND
KINGSTON PIKE
© C.S.C.

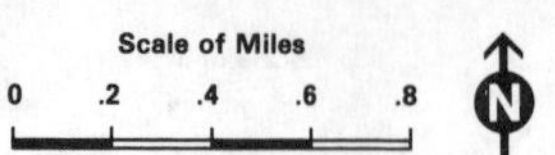
Scale of Miles
0 .2 .4 .6 .8
N

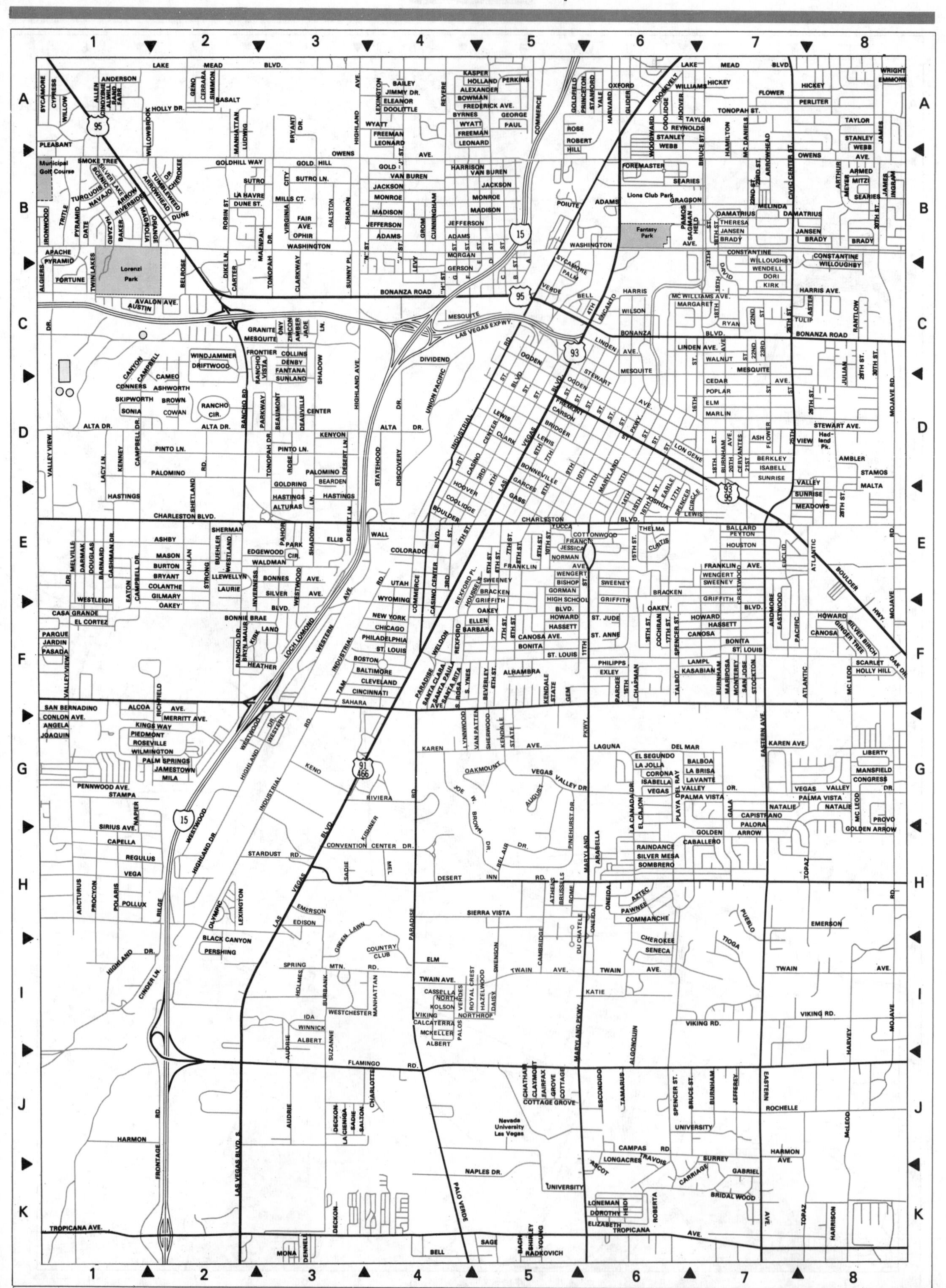
1 2 3 4 5 6 7 8
A B C D E F G H I J K
LAKE MEAD BLVD.
95
15
93
93 95
91 466
Municipal Golf Course
Lorenzi Park
Lions Club Park
Fantasy Park
Nevada University Las Vegas
COUNTRY CLUB
Had-land Pk.
OWENS AVE.
WASHINGTON
BONANZA ROAD
LAS VEGAS EXPWY.
UNION PACIFIC
ALTA DR.
CHARLESTON BLVD.
SAHARA
DESERT INN RD.
FLAMINGO RD.
TROPICANA AVE.
LAS VEGAS BLVD. S.
LAS VEGAS BLVD.
RANCHO RD.
HIGHLAND AVE.
INDUSTRIAL
WESTERN AVE.
PARADISE
MARYLAND PKWY.
EASTERN AVE.
PALO VERDE
BOULDER HWY.
MOJAVE RD.
NELLIS
TWAIN AVE.
SPRING MTN. RD.
CONVENTION CENTER DR.
VEGAS VALLEY DR.
KAREN AVE.
STEWART AVE.
FREMONT
HARMON AVE.
UNIVERSITY
NAPLES DR.
ROCHELLE
VIKING RD.
SIERRA VISTA
STARDUST RD.
RIVIERA
FRONTAGE RD.
HARRISON
BELL
SAGE
RADKOVICH

LAS VEGAS

A St. B-5
Adams B-4,B-6
Albert I-3,I-4
Alcoa Ave. F-1
Alexander A-5
Algiers C-1
Algonquin I-6
Alhambra F-5
Allen A-1
Alta Dr. D-1,D-2,D-4
Alturas E-3
Alwill A-1
Amber C-3
Ambler D-8
Anderson A-1
Angela G-1
Apache B-1
Arabella H-6
Arcturus H-1
Ardmore F-7
Armed B-8
Arrow B-1
Arrowhead B-2,B-7
Arthur B-8
Ascot K-6
Ash D-7
Ashby E-2
Ashworth D-2
Aster C-8
Athens H-5
Atlantic E-8,F-8
Audrie J-3
August G-5
Austin C-1
Avalon C-1
Aztec H-6
B St. C-5
Bach K-5
Bailey A-4
Baker B-1
Balboa G-6
Ballard E-7
Baltimore F-3
Barnard E-1
Barbara F-4
Basalt A-2
Beardee D-3
Beaumont D-3
Bel Air Dr. H-5
Bell C-5,K-4
Belrose C-2
Berkley D-7
Beverly F-5
Bishop E-5
Black Canyon I-2
Bonanza Blvd. C-6
Bonanza Rd. C-4,C-8
Bonita F-5,F-7
Bonnie Brae F-2
Bonnes Ave. E-3
Bonneville D-5
Boston F-3
Boulder E-4
Boulder Hwy. E-8
Bowman A-5
Bracken E-5,E-6
Brady B-7,B-8
Bridal Wood K-7
Bridger D-5
Brown D-2
Brown, Joe W. Dr. G-4
Bruce St. A-7,J-6
Brussels H-5
Bryant E-2
Bryant Dr. A-3
Burbank I-3
Bryn Mawr F-2
Buehler E-2
Burnham F-7,J-7
Burnham Ave. D-7
Burton E-2
C St. C-5
Caballero H-7
Cahlan E-2
Calcaterra I-4
Campbell C-1
Campbell Dr. D-1,E-1
Cameo C-2
Campas Rd. J-6
Canosa F-7,F-8
Canosa Ave. F-5
Canyon C-1
Capella H-1
Capistrano G-7
Carrara A-2
Carriage K-6
Carson D-5
Carter C-2
Casa Grande F-1
Cashman Dr. E-1
Casino Center Blvd. D-5,E-4
Cassella I-4
Cedar Ave. D-7
Center D-3
Cervantes D-7
Chapman F-6
Charleston Blvd. E-2,E-5
Charlotte J-4
Chatham B-1
Cherokee B-2,I-6
Chicago F-4
Cincinnati F-3
Cinder Ln. I-1
Cindysue A-1
Circle E-6
Clark D-5
Claymont J-5
Cleveland F-3
Cochran F-6
Colanthe E-2
Collins C-3
Colorado E-4
Commanche H-6
Commerce A-5,F-4
Congress G-8
Conlon Ave. G-1
Conners D-1
Constantine B-7,B-8
Convention Center Dr. H-3
Coolidge A-6,E-4
Corona G-6
Cottage J-5
Cottage Grove J-5
Cottonwood E-5
Country Club I-4
Cowon D-2
Crestwood E-7
Cunningham B-4
Curtis E-6
Cypress A-1
D St. B-5
Daisy I-5
Dale B-1
Damatrius B-7
Darmak E-1
David C-7
Deauville D-3
Deckon J-3,K-3
Del Mar G-6
Denby C-3
Dennell K-3
Desert Inn Rd. H-5
Desert Ln. D-3,E-3
Dike Ln. C-2
Discovery Dr. D-4
Dividend C-4
Doolittle A-4
Dori C-7
Dorothy K-6
Douglas E-1
Dune B-2
Dune St. B-2
E St. B-5
Earle E-6
Eastern Ave. G-7,J-7
Eastwood F-7
Eaton E-1
Edgewood E-3
Edison H-3
El Cajon H-6
El Cortez F-1
El Segundo G-6
Eleanor A-4
Elizabeth K-6
Ellen F-4
Ellis E-3
Elm D-7,I-4
Emerson H-3,H-8
Emmone A-8
Encanto C-6
Escondido J-6
Euclid E-7
Exley F-6
F St. C-4
Fair Ave. B-3
Fairfax J-5
Farr A-1
Flamingo Rd. J-3
Flower A-7
Flower St. D-7
Fontana C-3
Foremaster B-6
Fortune C-1
Francis E-5
Franklin Ave. E-5,E-7
Frederick Ave. A-5
Freeman A-4,A-5
Fremont St. D-5
Frontage Rd. K-2
Frontier C-2
G St. C-4
Gabriel K-7
Gala G-7
Garces D-5
Gass E-5
Gem F-5
Geno A-2
George A-5
Gerson C-4
Gilmary E-2
Glider A-6
Gold B-4
Gold Hill B-3
Golden Arrow H-7,H-8
Goldfield A-5
Goldhill Way B-2
Goldring D-3
Gorman E-5
Gragson B-6
Granite C-2
Green Lawn I-3
Griffith F-5--7
Grom B-4
Grove J-5
H St. C-4
Hamilton A-7
Harmon J-1
Harmon Ave. J-7
Harris C-6
Harris Ave. C-8
Harrison B-4,K-8
Harvard A-6
Harvey J-8
Hassett F-5,F-7
Hastings E-1,E-3
Hazard B-1
Hazelwood I-5
Heather F-2
Heidi K-6
Held B-7
High School F-5
Highland Ave. A-3,D-3
Highland Dr. G-2,H-2,I-1
Hickey A-7,A-8
Hill A-5
Holly Dr. A-2
Holly Hill F-8
Holmes I-3
Hoover A-6,E-4
Houssels E-4
Houston E-7
Howard F-5,F-7,F-8
Ida I-3
Industrial Rd. D-4,F-3,G-3
Ingram B-8
Inverness E-2
Ironwood B-1
Isabell D-7
Isabella G-6
J St. B-4
Jade C-3
Jackson B-4,B-5
James A-8,B-8
Jamestown G-2
Jansen B-7,B-8
Jefferson B-4
Jeffery J-7
Jessica E-5
Jimmy Dr. A-4
Joaquin G-1
Joshua E-6
Julian C-8
Karen G-4
Karen Ave. G-4,G-7
Kasabian F-6
Kasper A-4
Katie I-5
Kendale F-5,G-5
Kenney D-1
Keno G-3
Kenyon D-3
Kings Way G-1
Kirk C-7
Kirkland F-2
Kishner H-3
Kolson I-4
La Brisa G-6
La Canada Dr. H-6
Lacy Ln. D-1
La Cienega J-3
Laguna G-6
La Havre B-2
La Jolla G-6
Lake Mead Blvd. A-2,A-6
Lampl F-7
Las Vegas Blvd. D-5,H-3
Las Vegas Blvd. S. K-2
Las Vegas Expwy. C-4
Laurie E-2
Lavante G-6
Leonard A-4,A-5
Levy C-4
Lewis D-5,E-6
Lexington A-4,H-2
Liberty G-8
Linden Ave. C-6,C-7
Llewellyn E-2
Loch Lomond F-3
Lon Gene D-6
Loneman K-6
Longacres J-6
Ludwig A-2
Lynnwood G-4
Madison B-4,B-5
Maenpah B-3
Magnolia B-1
Malta E-8
Manhattan A-2,I-4
Mansfield G-8
Margaret C-6
Mariposa F-7
Marlin D-7
Maryland Pkwy. D-6,H-6,J-5
Mason E-2
McDaniels A-7
McKeller I-4
McLeod F-8,G-8,J-8
McWilliams Ave. C-6
Meadows E-8
Mel H-4
Melinda B-7
Melville E-1
Merritt Ave. G-2
Mesquite C-2,C-4,C-6,C-7
Meyer B-8
Mila G-2
Mills Ct. B-3
Mitzi B-8
Mojave Rd. D-8,F-8,I-8
Mona K-3
Monroe B-4,B-5
Monterey F-7
Morgan B-4
N St. B-4
Napier G-1
Naples Dr. K-4
Natalie G-7,G-8
Navajo B-1
New York F-4
Norman E-5
North I-4
Northrop I-4
Oak Dr. F-8
Oakey Blvd. F-2,F-5,F-6
Oakmont G-4
Ogden C-5,D-5
Olympic H-2
Oneida H-6
Ony C-3
Ophir B-3
Orange B-2
Owens Ave. A-3,B-8
Oxford A-6
Pacific F-7
Pahor E-3
Pamos B-6
Palm C-5
Palm Springs G-1
Palma Vista G-6,G-8
Palo Verde K-4
Palomino D-2,D-3
Palora H-7
Palos Verdes I-4
Paradise F-4
Paradise Rd. I-4
Pardee F-6
Park Cir. E-3
Parkway D-3
Parque F-1
Pasada F-1
Paul A-5
Pawnee H-6
Pennwood Ave. G-1
Perkins A-5
Perliter A-8
Pershing I-2
Peyton E-7
Philadelphia F-4
Phillips F-6
Piedmont G-1
Pinehurst Dr. H-5
Pinto Ln. D-2,D-3
Playa Del Ray G-6
Pleasant A-1
Poiute B-5
Polaris H-1
Pollux H-1
Poplar D-7
Princeton A-6
Procyon H-1
Provo G-8
Pueblo H-7
Pyramid B-1
Radkovich K-5
Raindance H-6
Ralston B-3
Rancho C-3
Rancho Cir. D-2
Rancho Dr. F-2
Rancho Rd. D-2
Rand A-1
Rantlow C-8
Regullis H-1
Revere A-4
Rexford F-4
Rexford Pl. E-4
Reynolds A-6
Richfield F-2
Rilge H-2
Riverside Dr. B-1
Riviera G-4
Robert A-5
Roberta K-6
Robin St. B-2
Rochelle J-7
Rome H-5
Roosevelt A-6
Rosa F-4
Rosa, S. F-4
Rose A-5,D-3
Roseville G-1
Royal Crest I-4
Ryan C-7
Sadie H-3,J-3
Sage K-5
Sagman B-6
Sahara Ave. F-3
St. Anne F-6
St. Jude F-6
St. Louis F-4,F-5,F-7
Salton J-3
San Bernadino F-1
San Jose F-7
Santa Clara F-4
Santa Paula F-4
Santa Rita F-4
Scarlet F-8
Scenic B-1
Searles B-6,B-8
Seneca I-6
Shadow Ln. C-3,E-3
Sharon B-3
Sherman E-2
Sherwood G-5
Shetland Rd. E-2
Shirley K-5
Sierra Vista H-4
Silver B-1
Silver Ave. E-3
Silver Mesa H-6
Simmon A-2
Sirius Ave. H-1
Skipworth D-1
Smoke Tree B-1
Sombrero H-6
Sonia D-1
Spencer E-6
Spencer St. F-6,J-6
Spring Mtn. Rd. I-3
Stamos D-8
Stampa G-1
Stanford A-6
Stanley A-6
Stardust Rd. H-2
State F-5,G-5
Statehood D-4
Stewart Ave. D-6,D-8
Stockton F-7
Strong E-2
Sunny Pl. C-3
Sunrise D-7,E-8
Surrey J-7
Sutro B-2
Sutro Ln. B-3
Suzanne J-3
Sweeney E-5-7
Swenson I-5
Sycamore A-1,C-5
Talbot F-6
Tam F-3
Tamarus J-6
Taylor A-6
Thelma E-6
Theresa B-7
Tioga I-7
Tonapah Dr. C-3,D-3
Tonopah St. A-7
Topaz H-8,K-8
Tritle B-1
Tropicana Ave. K-1,K-6
Tulip C-8
Tumbleweed B-2
Turquois B-1
Twain Ave. I-4-7
Twin Lakes B-1
Union Pacific D-4
University J-6,K-5
Utah E-4
Valley D-8
Valley View D-1
Valley View Dr. F-1
Van Buren B-4,B-5
Van Patten G-5
Vega H-1
Vegas Valley Dr. G-5,G-6,G-8
Verde C-5
View D-7
Viking I-4
Viking Rd. I-6,I-8
Virginia City B-3
Vista C-3
Waldman E-3
Wall E-4
Walnut C-1
Washington B-3
Washington AVe. B-6
Webb A-6,A-8
Weldon F-4
Wendell C-7
Wengert E-5,E-7
Westchester I-3
Western F-3,G-2
Westland E-2
Westleigh E-1
Westwood E-3,G-2,H-2
Williams A-6
Willoughby B-7,C-8
Willow H-1
Willow Brook A-2
Wilmington G-1
Wilson C-6
Windjammer C-2
Winnick I-3
Woodward A-6
Wright A-8
Wyatt A-4,A-5
Wyoming E-4
Yucca E-5
Ynes, S. F-4
Young K-5
Zircon C-3
1st St. D-4
3rd St. D-5,E-4
4th St. C-6,D-5,E-4
5th St. E-5
6th St. D-5,E-5,F-5
7th St. D-5,E-5,F-5
8th St. E-5,F-5
9th St. D-5,E-5
10th St. D-6,E-5
11th St. D-6,F-6
13th St. D-6
14th St. E-6
15th E-6,F-6
16th St. D-7,E-6,F-6
17th E-6
17th St. B-7,F-6
18th C-7
18th St. B-7,D-7
19th C-7
20th Ave. D-7
21st St. D-7
22nd C-7
22nd St. B-7
23rd St. B-7,C-7
25th St. C-7,D-7
26th St. D-8
28th St. E-8
29th St. C-8
30th St. B-8,C-8

POINTS OF INTEREST

Fantasy Park B-6
Hadland Pk. D-8
Lions Club Park B-6
Lorenzi Park C-1
Municipal Golf Course B-1
Nevada University Las Vegas J-5

LITTLE ROCK

Allen B-6
Allis H-1,I-1,J-1
Appianway G-1
Arch H-6,J-5
Augusta B-4
Barber I-8,K-8
Barton F-2
Battery H-3,J-3
Berry E-1
Bishop H-4,K-4
Booker G-1,H-1,J-1
Bragg J-8
Branch Dr. E-2
Broadway E-8
Broadway H-6,J-6
Brookwood Dr. C-1
Cantrell Rd. D-1,D-2
Cantrell, N. E-2,E-3
Capitol, W. F-1,G-5
Cedar Hill Dr. C-1
Centennial Park H-3
Center H-6,J-6
Chandler B-6
Charles E-1
Chester H-5,J-5
Coates G-2
College Cir. A-6
College Pk. A-6
Collins G-8
Commerce G-7,H-7,J-7
Coolwood Dr. E-2
Cottondale La. D-1,D-2
Crest C-4
Cross H-4,J-4
Crutcher B-4
Cumberland H-7,J-7
Cypress B-8,D-8
Denison F-2,H-2,J-2
Division B-5,C-5
Dunbar Park I-4,I-5
Fairfax E-1
Fern F-1
Ferry G-8
Flora B-4
Fort Roberts Dr. A-3
Fountain F-2
Frank B-4,C-4
Franklin B-5,C-5
Gaines H-5,J-5
Garland F-3
Gill E-3
Grove F-2
Gum B-6,E-6
Harold B-3
High H-4,J-4
Howard J-2
Izard H-5,J-5
Jessie Rd. D-2
Johnson G-1,H-1,J-1
Jones G-2,H-2
Julian B-3
Karrot E-6
Lamar G-1
Linden F-1
Little E-1
Lloyd Ct. E-3
Locust D-8
Long I7th, W. B-5
Louise E-1
Louisiana H-6,J-6
Magnolia B-7,D-7
Main B-7,D-7,H-6,J-6
Maple B-7,D-7
Marion B-6
Markham, E. F-7,F-8
Markham, W. F-1,F-2,F-5
Marshall H-3,J-3
Maryland G-2
McAlmont I-8,J-8
McGowan H-8,I-8
Melrose Cir. C-6,C-7
Melrose Dr. C-6,D-6
Midland E-1,F-1
Moss B-5
Nanette B-3
Nona B-4
North E-3,E-4,E-5,F-5
Oak A-2
Oak D-6,E-6
Olive B-8,D-8
Orange B-7,D-7
Ozark E-1
Park F-3,J-3,H-2
Park Dr. C-4
Parker B-5,C-5
Park La. J-7
Pearl F-1
Phyllis B-3
Pike Ave. A-5,B-5C-5
Pine B-8,D-8,E-8
Poplar B-7,D-7
Pulaski H-4,J-4
Recton G-8
Rice F-2,H-2,J-2,K-2
Ringo H-5,J-4
River C-3,D-4
Riverdale Rd. B-1
Riverfront Dr. C-1
Riverside E-4
Riverview E-3
Rock H-7,J-7
Roosevelt Rd., E. K-7
Roosevelt Rd., W. K-4,K-5
Rosetta F-1
Schaer B-5
Schiller F-3,H-3,J-3
Scott H-7,J-6,J-7
Sherman G-8,K-7
Short 17th, W. B-5
Spring H-6,J-6
State H-5,J-5
Summit F-3,H-3,J-3
Summit Ave. E-3
Sycamore B-6
Terrace Dr. E-2
Thayer F-2,H-2,J-2
Turtle Creek La. B-1
Tuxedo Ct. K-7
Vance J-8
Vernon F-1
Vestal C-4
Victory G-4,H-4
Vine B-8,D-8,E-8
Water Works Rd. D-2,E-2
Willow B-6,D-6
Wolfe H-3,J-3
Wood Ln. G-4
Woodrow E-1,F-1,G-1,H-1,K-1
Wright Ave. I-5
W. Wright Ave. I-2,I-3
2nd A-2
2nd, E. E-8
2nd, E. F-7
2nd, W. D-5,E-6
2nd, W. F-3,F-5
3rd A-2
3rd, E. G-7
3rd, W. D-5,D-6
3rd, W. F-1,F-5
4th A-2
4th, E. D-8
4th, E. G-7
4th, W. D-5,D-6
4th, W. F-1,F-5
5th, E. D-8
5th, E. G-7
5th, W. D-7
6th, E. D-8
6th, E. G-7
6th, W. D-7
6th, W. G-1,G-5
7th, E. D-8
7th, E. G-7
7th, W. D-5,D-7
7th, W. G-1,G-5
8th, E. D-8
8th, E. H-7
8th, W. D-5,D-7
8th, W. G-5
9th, E. C-7,C-8
9th, E. H-7
9th, W. C-5,C-7
9th, W. G-5
10th, E. C-7,C-8
10th, E. H-7
10th, W. C-5
10th, W. G-2,H-5
11th, E. C-7
11th, E. H-7
11th, W. C-5
11th, W. H-2,H-5
12th, E. C-7
12th, E. H-7
12th, W. C-5,C-6
12th, W. H-2,H-4,H-5
13th, E. C-7
13th, E. H-7
13th, W. C-5,C-6
13th, W. H-2,H-4,H-5
14th, E. B-7
14th, E. I-7
14th, W. B-6
14th, W. H-2,H-4,H-5
15th, E. B-7
15th, E. I-7
15th, W. B-5
15th, W. H-4,I-5
16th, E. B-7
16th, E. I-7
16th, W. B-5
16th, W. I-2,I-4
17th, E. B-7
17th, E. I-7
17th, W. B-6
17th, W. I-2,I-4
18th, E. B-7
18th, W. A-5
18th, W. I-2,I-4
19th, E. A-7
19th, E. I-7,J-7
19th, W. A-5
19th, W. I-4
20th, E. A-7
20th, E. J-7
20th, W. A-5,A-6
20th, W. J-4
21st, E. A-7
21st, E. J-7
21st, W. A-5,A-6
21st, W. J-4
22nd, E. A-7
22nd, E. J-7
22nd, W. A-5
22nd, W. J-4
23rd, W. J-4
24th, E. J-7
24th, W. J-4
25th, W. J-1
26th, E. K-7
26th, W. J-1,K-4
27th, E. K-7
27th, W. J-1,K-4
28th, E. K-7
28th, W. K-4
32nd, W. K-1

POINTS OF INTEREST

Ark. Baptist College I-4
Arkansas Livestock Show Grounds K-2
Arkansas River D-3,E-5
Arkansas School for the Blind and the Deaf E-2
Barton Park J-2
Calvary Cemetery I-1,J-1
Central Baptist Hospital H-3
City Hall E-7,F-6
County Hospital K-1
Court House F-6
David D. Terry Lake E-6,E-7,F-6,F-7
Fletcher Park H-1
Fort Roots Veterans Hospital A-2
Little Rock Central H.S. H-2
MacArthur Park H-8
Missouri Pacific Hospital E-4
Mt. Holly Cemetery H-5,H-6
Pettaway Park J-7
Philander Smith College H-5
Riverview Park B-2,B-3
Roselawn Cemetery J-1,J-2
Shorter College D-8
State Capitol F-4
Vestal Park B-4
Veterans Hospital K-7

Little Rock, AR

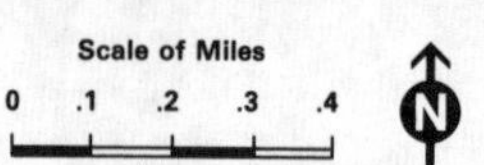

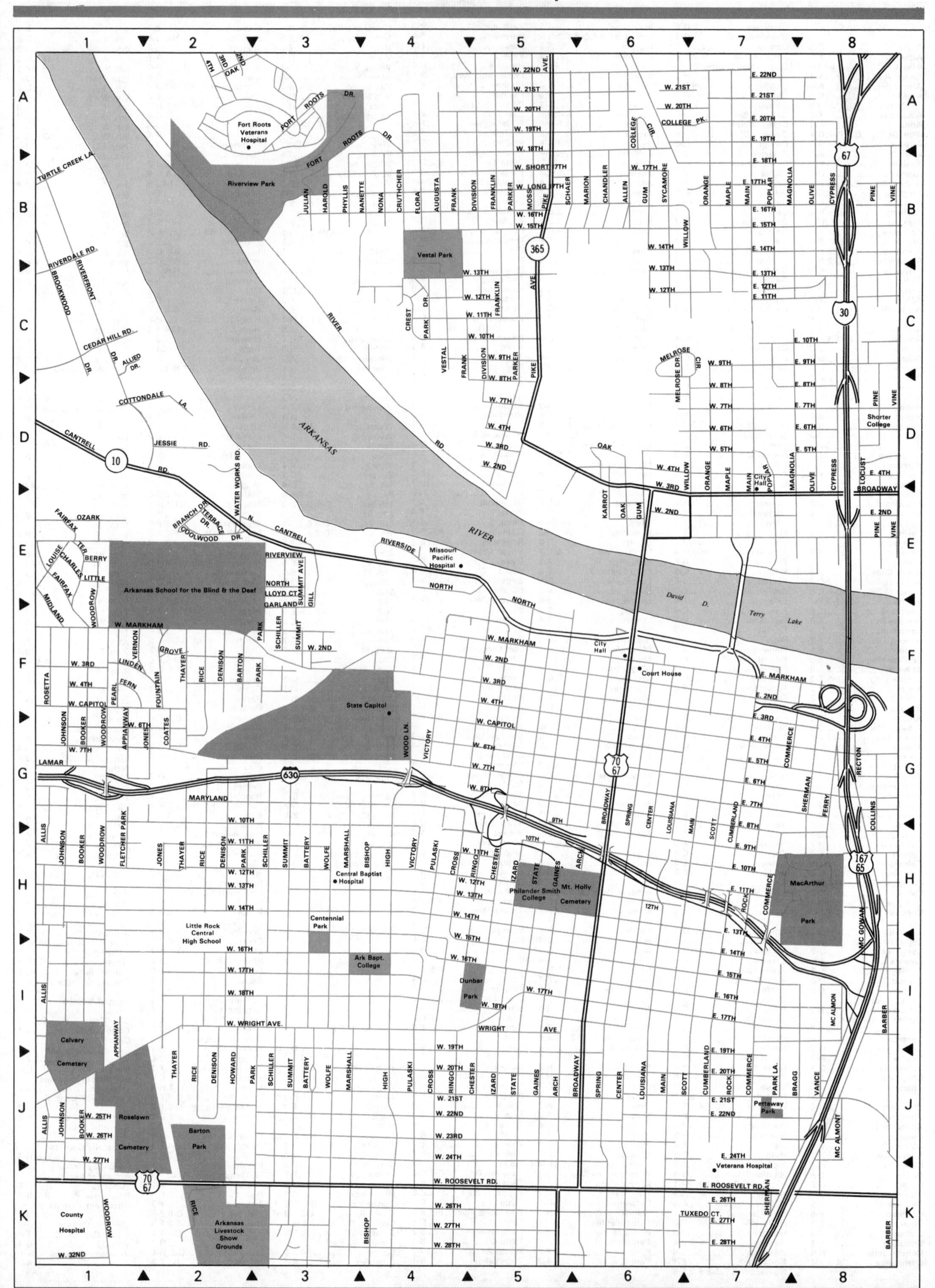

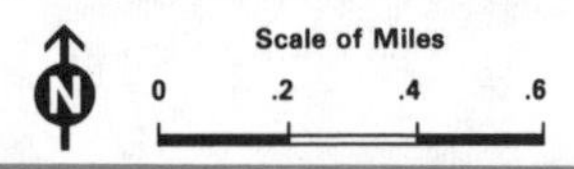
N
Scale of Miles
0
.2
.4
.6

1
2
3
4
5
6
7
8
A
B
C
D
E
F
G
H
I
J
K
Barnsdale Park
SILVER LAKE RES.
LOS ANGELES RIVER
Elysian Park
Dodger Stadium
Bellevue Park
ECHO PARK LAKE
Lafayette Park
Mac Arthur Park
Rosedale Cemetery
Terr Park
St. James Park
HOLLYWOOD
SUNSET
SANTA MONICA
BEVERLY
WILSHIRE
OLYMPIC
PICO
VENICE
WASHINGTON
ADAMS
JEFFERSON
HOLLYWOOD FRWY.
SANTA ANA FRWY
SANTA MONICA
HARBOR FRWY
PASADENA FRWY.
101
110
10
ALAMEDA
CENTRAL
FIGUEROA
BROADWAY
VERMONT
HOOVER
NORMANDIE
ALVARADO
GLENDALE
ELYSIAN PARK
STADIUM WAY
ACADEMY
SHORELAND DR.
BROOKS AVE.
SUNSET BLVD.
BEVERLY BLVD.
TEMPLE
FIRST
WESTERN

LOS ANGELES

Aaron St. C-6
Academy D-7
Adair St. K-6
Adams Blvd. J-1,J-3
Agatha St. I-7
Alameda St. H-8,K-8
Albany St. I-4
Alexandria Ave. B-2
. . . . C-2,E-2
Allesandro St. . . . C-5,D-5
Allison Ave. D-6
Alpine St. F-8
Alvarado St. . D-5,F-4,H-3
Altman St. B-8
Angelina St. F-6
Angelus Ave. C-5
Apex Ave. B-6
Arapahoe St. H-3,I-3
Ardmore Ave. . . . B-1,C-1
. . . . E-1,H-1,I-1
Ashmore Pl. C-6
Avon Terr. B-7
Avon Pk. Ter. B-7
Azusa H-8
Banning St. H-8
Bard St. B-7
Bates Ave. A-6
Baxter St. B-6
Bay St. J-8
Beaudry Ave. E-7,F-6,G-6
Bellevue Ave. D-3,E-6
Belmont F-5
Benedict St. A-7
Berendo St. B-2
. . . . E-2,H-2,I-2
Benton Way. C-5,D-4,F-3
Berkeley Ave. C-5
Beverly Blvd. D-1,E-4
Bimini Pl. E-2
Birch St. J-7
Birkdale B-8
Bixel St. G-5,H-5
Blackstone Ct. I-6
Blake Ave. A-7
Blimp St. A-8
Bonniebrae . . . E-5,H-4,I-3
Bonsallo Ave. J-4
Boston St. F-7
Boyd St. H-7
Boylston Dr. D-7
Boylston St. F-6,G-6
Branden St. C-6
Brighton Ave. J-1
Broadway G-7,I-6,K-4
Broadway, N. F-8
Brooks Ave. D-8
Bruce Ct. D-6
Budlong Ave. . I-2,J-2,K-2
Burlington Ave. F-5
. . . . H-4,I-3
Burns Ave. C-3
Byram H-5
Cabot B-8
Calumet Ave. E-6
Cambridge St. H-1
Cambria St. G-5
Camero Ave. A-3
Cameron Ln. I-5
Carroll Ave. E-6
Carondelet St. . . . E-4,G-3
Catalina St. A-2,E-2
. . . . H-2,J-2,K-2
Catsby Ln. I-5
Cecelia St. I-7
Centennial St. E-7
Central Ave. . . I-8,K-6,K-7
Ceres Ave. I-7
Cerro Gordo St. B-7
Cherry I-4
Clayton Ave. A-3
Cleveland F-8
Clifford St. C-6
Clinton St. . . C-1,C-2,D-5
College St. E-7
Colton St. F-6
Columbia Ave. G-5
Colyton St. I-8
Commercial G-8
Commonwealth Ave.
. . . . B-3,F-3
Congress Ave. J-1
Connecticut St. H-4
Constance St. H-4
Cordova St. I-2
Coronado St. C-5
. . . . D-5,F-3
Corralitas Dr. A-6
Cortez St. E-6
Cottage Pl. H-5
Council St. D-1-3,5
Court St. E-5,F-6
Cove Ave. B-6
Crandall St. E-4
Crestmont St. B-4
Crocker St. I-7,J-7
Crownhill Ave. F-5
Cumberland Ave. A-3
Curran St. B-7
Dahlia B-4
Dallas B-8
Dalton Ave. J-1
Dawson F-5
Deacon Ave. H-4
Delong St. I-4
Delongpre Ave. . . A-1,A-2
Denby A-7
Denker Ave. K-1
Descanso C-4
Dewey Ave. H-2
Diamond St. F-6
Dillon St. B-5,D-4,E-3
Donaldson St. B-7
Douglas St. D-7,F-6
Dorris Pl. B-8
Duane St. B-6
Ducommun G-8
Eads B-8
Earl St. B-6
East Edgeware E-6
Echo Park Ave. . . C-7,D-6
Edgecliffe Dr. B-4,C-4
Edgemont Ave. C-2
Edgemont St. E-2
Effie St. C-6
Ellendale J-3
Ellwood St. K-8
Elsinore D-5
Elyian Park Dr.
. . . . C-7,C-8,D-7
Essex St. J-7
Everett St. E-7
Ewing St. B-6
Factory Pl. I-8
Fairbanks Pl. D-6
Fargo St. B-6,B-7
Fedora St. H-2
Figueroa Pl. E-7
Figueroa St. I-5,K-3
Figueroa Ter. F-7
Firmin St. F-6
Florida St. H-5
Flower St. H-5,I-5
Forney B-8
Fountain Ave. B-1,B-3
Francis Ave. G-3
Francisco H-5
Frank Ct. H-7
Fremont Ave. F-7
Gail B-8
Garey St. H-8
Garland Ave. H-5
Gateway Ave. B-3
Georgia St. H-5,I-4
Glassell St. E-4
Gladys Ave. I-8
Glendale B-6
Glendale Blvd. A-6
Glover Pl. B-8
Golden Gate Ave. B-4
Grand Ave. H-6,K-4
Grandview St. G-3
Grattan St. H-4
Green Ave. H-4
Griffith Ave. K-6
Griffith Park Blvd.
. . . . A-4
Halldale Ave. J-1
Harbor Frwy. H-5
Harlem Pl. H-7
Harold Way A-1
Hartford Ave. G-5
Harvard B-1
Harvard Blvd. C-1
. . . . D-1,E-1,I-1,J-1
Heliotrope D-2
Hemlock St. J-7,J-8
Hewitt St. H-8,I-8
Hidalgo Ave. A-6
Hill St. F-8,I-6,K-4
Hobart Blvd. A-1
. . . . B-1,E-1,I-1
Hobart Pl. D-1
Hollywood Blvd.
. . . . A-1,A-2,A-3
Hooper Ave. K-7
Hoover St. A-3,F-3
. . . . H-3,K-3
Hope St. G-6,I-5
. . . . J-4,J-5,K-4
Hunter St. J-8
Huntley Dr. F-6
Hyans St. E-4
Hyperion Ave. A-4,C-3
Imogen Ave. C-3
Industrial St. I-8
Ingraham St. G-5
Irolo St. H-1
Jaunita Ave. C-2
Jefferson Blvd. K-1-3
Juliet St. J-2
Kellam Ave. E-6
Kenilworth Ave. A-5
Kenmore Ave. B-2
. . . . E-2,H-2
Kensington Rd., E. E-7
Kensington Rd., W. . . . E-6
Kent St. D-4,D-5
Kenwood Ave. J-1
Kinglsey Dr. B-1,C-1
. . . . E-1,H-1,I-1
Kingswell Ave. A-2
Knox A-7
Kodak Dr. C-3
Kohler St. I-7
Lafayette Park Pl. F-3
Lake St. F-4,G-3
Lakeshore Ave. . . . C-6,D-6
Lake View Ave. A-6
La Mirada Ave. B-2
Landa St. B-4,B-7
Larissa Dr. C-4
Lasalle Ave. I-1,J-1
Laveta Terr. D-6,E-6
Lawrence St. K-8
Lebanon St. H-6,I-5
Leeward Ave. F-2
Lemoyne Ave. B-7
Lemoyne St. C-6,D-6
Lexington Ave. B-2
Lilac Ter. E-8
Lily Crest Ave. C-2
Lindley Pl. H-7
Little St. G-4
Lockwood Ave. B-3
Logan St. D-6
Loma Dr. G-5
Loma Pl. F-5
Loma Vista Pl. B-6
London St. D-3,D-4
Long Beach Ave. K-8
Los Angeles St.
. . . . I-7,J-5,J-6
Lucas Ave. G-5
Lucille Ave. A-4,C-3
Lucretia C-6
Macy St. G-8
Madison Ave. D-2,E-3
Magnolia Ave. I-3,J-3
Main St. H-6,I-6,K-5
Maltman Ave. C-3
Malvern Ave. H-3
Manzanita St. B-3
Maple Ave. I-7,J-6,K-5
Mar Ave. B-3
Marathon St. C-3,C-4
Mariposa Ave. C-1
. . . . E-1,H-1,I-1
Margo I-5
Marview Ave. E-7
Maryland St. F-4,G-6
Mayberry St. C-5
McClintock Ave. K-3
McCollum Pl. C-5
McCready Ave. A-6
McGarry St. K-8
Meadow A-6
Melrose Ave. C-1,C-3
Melrose Hill C-1
Menlo Ave. H-2
Merwin St. D-5
Micheltonena B-4,C-4
Middlebury St. D-3
Mignonette Ct. F-6
Miramar St. F-4
Modjeska St. A-7
Mohawk St. C-5
Moore St. B-6
Monroe Ave. C-2
Montana St. C-5
Morton Ave. C-6
Montrose St. D-5
Mountainview Ave.
. . . . E-5,F-5
Myra Ave. A-3,B-3
Myrtle St. J-6
Naomi Ave. J-7,K-7
New England St. I-3
New Hampshire Ave.
. . . . B-2,E-2,H-2
New High F-8
Newell St. A-7
Newton St. K-7
Normal Ave. C-3
Normandie Ave. C-1
. . . . E-1,H-1,K-1
Normandie Pl. D-1
North Main St. G-8
North Spring F-8
Norwood I-4
Oak St. H-4,I-4
Oakwood Ave. D-1,D-3
Occidental Blvd.
. . . . C-5,D-4,E-3
Oceanview Ave. F-4
Olive St. H-6,J-5,K-4
Olympic Blvd. G-2
. . . . H-4,J-7
Omar H-8
Orchard Ave. I-3,K-3
Ord St. F-8
Oxford Ave. G-1,H-1
Palm Dr. J-4
Palmetto St. I-8
Paloma St. J-7
Panorama A-4
Park Dr. C-7
Park Pl. D-4
Parkman Ave. D-4
Parkview St. G-3
Park Grove Ave. J-4
Patton St. E-6
Pembroke Ln. I-5
Pico Blvd. H-1,H-2,I-4
Pirtle St. A-7
Plata St. D-3,D-4
Portland St. J-3
Preston Ave. C-6
Princeton St. B-7
Prospect Ave. A-2,A-3
Queen St. A-7
Quintero St. D-7
Rampart Blvd. D-4,F-3
Raymond Ave. I-2,J-1
Reno St. E-3
Reservoir St. C-5
Rich St. A-7
Ridgeway E-6
Riverdale B-8
Robinson St. D-3,D-4
Rockwood St. F-5
Rosalia Rd. A-3
Rose St. H-8
Roselake Ave. E-4
Rosemont Ave. D-5,E-4
Roosevelt Ave. I-1
Sacramento St. J-8
St. James Pk. J-4
St. Paul Ave. G-6
St. Paul Pl. G-5
Sanborn Ave. A-4,B-3
San Julian St. I-6
. . . . I-7,J-6
San Marino St. G-1,G-2
San Pedro I-7,K-6
Santa Ana Frwy. . F-7,G-8
Santa Monica Blvd. . . B-1
Santa Monica Frwy.
. . . . I-2,I-3,I-5,K-7,K-8
Santa Ynez St. D-5
Santee St. I-6
Sargent Pl. D-6
Scarff St. J-3
Scott Ave. C-5,D-6
Scott Pl. C-5
Scout Way E-5
Seaton St. I-8
Sentous St. H-4
Serrano Ave. . B-1,E-1,H-1
Severance St. J-3
Shatto Pl. F-2
Shatto St. G-5
Shoreland Dr. D-8
Shrine Pl. K-3
Silverlake D-4
Silver Ridge Ave. A-6
Spring St. H-6
Stadium Way D-7,E-7,E-8
Stanford Ave. I-7
. . . . I-8,J-7,K-6
Staunton K-8
Sunbury Ct. H-5
Sunflower A-7
Sunset Dr. A-3
Sunset Blvd. A-1
. . . . C-4,D-6,F-7
Sunset Pl. F-3
Sutherland St. D-6
Sunvue Pl. E-7
Swansea Pl. B-1
Talmadge St. A-3
Tarleton K-7
Teed St. F-8
Temple St. G-7
Terminal St. J-8
Teviot St. A-6
Toberman St. I-4
Toluca St. F-6
Towne Ave. I-7,J-7
Traction H-8
Trenton St. I-5
Trinity St. K-5
Tularosa Dr. C-3,C-4
Turner St. H-8
Union F-5
Union Ave. G-4
Union Dr. G-5
Union Pl. F-5
University Ave. K-3
Valencia St. I-4,H-4
Valentine B-7
Van Buren Pl. J-2
Vendome D-4,E-3
Venice Blvd. H-2,I-4
Vermont Ave. H-2,K-2
Vestal Ave. B-7,C-6
View St. E-4
Virgil Ave. C-3,F-3
Virgil Pl. A-3
Virginia B-3
Wall St. I-6,J-6,K-5
Wallace E-6
Walnut St. K-7
Walton Ave. I-2,K-2
Warehouse St. J-8
Washington Blvd.
. . . . I-1,I-2,K-5
Waterloo St. B-6,D-5
Welcome F-6
Weller H-8
Werdin Pl. H-7,I-5
West Edgeware E-6
Westmoreland Ave.
. . . . B-3,D-3,F-3
Westmoreland Blvd. . . . I-1
Westerly Ter. C-4
Westlake Ave. F-5
Westlake St. H-3
West Silver Lake
Dr. A-5,B-5
White Knoll Dr. E-7
Whitmore Av. B-7
Wilde St. I-8
Willowbrook Ave. B-2
Wilshire Blvd. F-2,G-4
Wilshire Pl. F-3
Wilson K-8
Winona Blvd. A-1
Winslow Dr. C-4
Winston St. H-7
Witmer St. G-5
Wright St. I-4
1st St. D-1,D-2,G-7
2nd St. E-1,E-3,E-4
3rd St. E-1,E-3,F-5,G-7
4th Pl. H-8
4th St. E-1,E-3,F-4
. . . . F-5,G-6,H-7,H-8
5th St. E-1,E-3,F-4,G-6
6th St. E-1,F-4,H-6,I-8
7th Pl. J-8
7th St. F-1,F-2
. . . . G-4,H-6,J-8
8th St. F-2,G-4,H-6,J-8
9th Pl. I-7
9th St. G-1,G-4,H-6
10th Pl. H-5
10th St. G-3,I-7,J-8
11th Pl. H-4
11th St. G-3,I-6,J-7
12th St. H-4,I-5,J-7
14th Pl. I-5,J-6,J-7
14th St. H-1,H-3
. . . . I-4,I-5,J-7,K-8
15th St. H-1,H-3
. . . . I-5,J-6,K-7
16th St. J-6,K-8
17th St. H-2,I-3
18th St. I-3,I-4
20th St. I-1,I-3
. . . . J-4,J-6,K-7
21st St. I-3,J-4,J-5,K-7
22nd Pl. I-2
22nd St. I-1,I-3
J - 4, K - 6, K - 7
23rd St. I-1,I-3,J-3,J-5
24th St. I-1,J-3,J-5
25th Pl. J-3
25th St. J-1,J-2,J-3,K-5
27th St. J-1,J-3
. . . . J-4,K-4,K-5
28th St. . . . J-2,J-3,K-4,K-5
29th St. J-1,J-2
. . . . J-3,K-4,K-5
30th Pl. K-2
30th St. J-1,J-3,K-5
31st St. K-2,K-4
32nd St. K-3,K-4
34th St. K-3
35th Pl. K-1,K-2
36th St. K-1,K-2
36th Pl. K-1,K-2
36th St. K-1
37th St. K-1

POINTS OF INTEREST

Barnsdale Park A-2
Bellevue Park C-3
Dodger Stadium E-8
Lafayette Park F-3
MacArthur Park F-4
Rosedale Cemetery
. . . . I-1,I-2
St. James Park J-4
Terr Park H-3

LOUISVILLE

Abraham B-6
Adams St. A-8
Adelia Ave. F-2
Adair St. I-6
Airview Dr. J-8
Algonquin Pkwy. E-2
Alley A-3
Alleycourt Pl. B-5
Alliston Ave. E-1
Alma Ave. J-2
Almond Ave. J-5
Anna Lane J-1
Ann Ct. F-2
Appleton Ct. G-1
Appleton Ln. G-1
Arcade Ave. G-3
Ardmore Dr. F-7
Argonne Ave. G-1
Arling Ave. K-3
Arling Ct. K-2
Armory B-5
Ash St. E-7
Ashbury Rd. K-4
Ashland Ave. I-3,I-5
Ashton Ave. I-6
Auburndale Ave. K-4
Audobon Parkway G-7
Auery Ave. E-5
Bachman Dr. H-1
Badger Ave. J-5
Ballard Ct. B-7
Bancroft St. C-6
Bank St. A-4
Barbee Ave. F-6
Barbee Way F-6
Barkwood Rd. H-2
Baroness St. D-7
Barret Ave. C-7
Bardstown C-8
Barret C-8
Barrowdale Dr. K-1
Barton Ave. I-7
Bates Ct. D-8
Baxter St. B-8
Beacher St. I-3
Becker Ct. F-1
Beech Ave. I-2
Beech St. E-1
Beecher St. I-3
Beeler K-4
Bedgravia Ct. E-5
Bellevue J-4
Bellevue Ave. I-4
Belmar Ave. H-8
Bergman St. E-6
Berkley Sq. J-8
Bernheim Ave. E-2
Berry Blvd. H-2
Bicknell Ave. J-3
Bloom E-5
Blue Grass J-3
Bobolink Rd. F-8
Bohannon St. G-3
Bolling Ave. D-2
Bourban Ave. H-8
Boxley Ave. G-6
Boyle St. D-7
Brandies F-5
Breckinridge St. . . C-4,C-7
Brent St. D-8
Brentwood Ave. H-3
Briden Ave. I-8
Broadway B-1,C-6
Brook C-6,F-5,I-5
Brookline Ave. J-3
Buchanan A-7
Buckner Ave. K-4
Burnett Ave. . D-2,E-5,E-7
Burwell E-1
Byrne Ave. G-5
Cable St. A-8
Caldwell C-7
Camden Ave. I-2
Campbell St. C-7
Camp St. D-6
Cardinal Dr. H-8
Carl Ct. I-1
Carlisle Ave. H-2
Carrico I-3
Cassin Ave. E-5
Castlevale Dr. E-8
Castlewood Ave. D-8
Catalpa St. D-1
Cawthan St. C-4
Cayuga I-2
Cedar St. A-2
Center Cliff I-4
Central Ave. G-3
Chalmer F-2
Chapel St. A-4
Charles St. D-7
Cherokee C-8
Chester Rd. H-1
Chestnut B-1,B-4,B-7
Cheyenne Ave. H-6
Chickadee Rd. G-8
Chicopee I-5
Christopher Pl. J-4
Churchman Ave. J-2
Clara Ave. H-2
Clarks Lane F-7
Clay C-7
Clay St. B-6
Cliff Ave. J-4
Clover Hills Dr. J-1
College St. C-5
Colgan St. B-3
Collins Ct. H-5
Colorado Ave. G-3
Columbia St. A-2
Compton St. F-4
Concord Dr. G-6
Conestoga Ave. E-1
Congess A-2,A-5
Conlin Ave. F-7
Conn St. H-2
Conrad C-6
Cooper St. B-8
Cornette Way J-1
Craig Ave. I-2
Crittenden Dr. G-6,J-6
Crop St. A-2
Crossbill Rd. G-8
Crown St. D-8
Cumberland Ave. K-4
Curry Dr. I-7
Cypress St. D-2,E-1
Dahlia Dr. E-8
Dakota Ave. H-6
Dale Ave. G-3,K-4
Dana Dr. I-1
Dandridge D-7
Date St. C-2
Davies Ave. E-4
Dearcy H-3
Dellwood Dr. J-1
Del Mar Ave. K-1
Delmar Ct. J-1
Delmar Ln. J-1
Delor Ave. F-7
Del Park B-1
Dena Dr. I-1
Denmark Ave. H-3
Dixdale Ave. E-1
Dixie St. F-8
Dixie Hwy. E-2,H-1
Dixon St. E-7
Doerr Dr. H-1
Dove Ln. G-8
Dove Rd. G-8
Dresden Ave. H-3
Dubourg Ave. G-1
Dumesnil D-2
Durrett Lane I-8
Duvall G-6
Eagan Ave. J-6
Eagle Pass F-8
Earl Ave. G-3
Eastern Pkwy. F-7
Edward St. C-8
Eicher Rd. H-1
Eigelbach Ave. E-8
Elderwood Way I-1
Elliot Ave. B-2
Ellison Ave. D-8
Elm St. A-8
Emil Ave. G-6
English Ave. E-8
Estate Dr. K-1
Euclid F-4
Euclid Ave. F-4
Evelyn Ave. H-4
Expressway St. B-6
Fairmount Ave. H-5
Falcon Dr. G-8
Farmdale Ave. I-8
Farmington Ave. H-5
Farnsley Rd. H-1
Fayette Ave. H-8
Faywood Way G-2
Federal Pl. B-5
Fern Dr. G-8
Finzer St. C-6
Fischer Ave. F-2
Fitzgerald Dr. G-1
Flexner Way B-8
Floral Terr. D-5
Florence Ave. I-4,I-5
Floyd C-6,G-5
Fontaine Ave. H-7
Forest Ave. E-7
Forum Ave. K-4
Foster St. G-6
Fountain Ct. E-5
Francis Ave. J-5
Franklin Ave. H-8
Franklin St. A-8
Fust Ave. F-2
Gagel Ave. K-1
Garden Row J-8
Garey Ln. G-2
Garland Ave. C-2,C-4
Garrett St. H-4
Garvey Ct. J-1
Garvey Dr. J-1
Garvin Pl. D-5
Gaulbert Ave. D-2,E-5
Geiger St. A-7
Georgetown Ave. H-2
Gillette Ave. G-1,I-5
Glenafton Ln. G-7
Glendale Ave. H-2
Glenrock Rd. K-1
Glenview Rd. F-2
Goddard Ave. D-8
Goss St. D-7
Grand Ave. C-1,C-2
Grant H-5
Gray B-6
Greenleaf Rd. G-7
Green St. A-2
Greenup F-7
Greenwood Ave. C-2
Gregg St. E-2
Guthrie St. B-5
Hale C-2
Hamilton Ave. A-7
Hancock St. C-6,E-6
Hannah Ln. I-8
Hardesty Ave. F-1
Harding Ave. I-6
Harold Ave. F-2
Harrison Ave. F-6
Hartwell K-5
Harwell G-6
Haskin Ave. K-3
Hathaway Ave. H-3
Hawes Ln. I-6
Hawthorne E-8
Hazel St. D-1
Hazelwood Ave. J-2
Hazelwood Ct. K-2
Hdertz Ave. E-7
Helm St. G-6
Hemlock St. D-1
Hepburn Ave. C-8
Herbert Ave. I-1
Herr F-4
Hess Lane G-8
Heywood Ave. G-4
Hickory St. E-7
Highland Ave. C-8
Highway H-8
High Pine Dr. K-1
Hill Ave. D-2,E-4
Hill Top Ct. I-4
Hobart Dr. I-2
Homeview Dr. F-3,G-4
Hopkins Ave. F-4
Howard St. C-2
Hull St. B-8
Huntoon Ave. K-3
Huron Dr. H-5
Innis Ct. C-8
Inverness K-4
Iowa Ave. G-4
Iroquois Ave. K-3,K-4
Ivy Ct. A-1
Jackson St. C-6,E-6
Jacob St. C-5,C-6
Jefferson A-1,A-4
Jefferson Ct. G-7
Johnson St. B-7
Jordan E-4
Joseph St. I-1
Julia Ave. D-8
Kahlert Ave. I-3
Kathleen Ave. I-4
Kelland Way H-1
Keller Ave. I-7
Kennedy Dr. F-2
Kenton St. H-5
Kentucky St. C-1,C-4,C-6
Kenwood Way J-4
Keswick F-7
Kings K-2
Kingston K-4
Knight Ave. J-2
Knight Rd. K-1
Krieger St. E-7
Lafayette H-2
Lammers D-8
Lampton St. C-6
Lancaster Ave. K-2
Lance Dr. J-1
Lansing Ave. I-5
Larchmont Ave. G-3
Larue Ave. H-8
La Salle Ave. J-2
Lawrence H-2
Lawrie Ln. K-2
Lee Dr. J-8
Lee St. E-5,E-6
Lehigh Ave. J-3
Lennox St. H-5
Lentz Ave. I-3
Leroy Ave. H-1
Lester I-3
Lewis Ave. B-2
Lexington Rd. B-8
Libertybell H-2
Liberty St. B-5
Lillian Ave. F-3
Lily Ave. F-6
Lindberg G-2
Lindberg Dr. F-3
Lincoln Ave. F-4
Linnet Rd. G-8
Linwood Ave. E-1
Livingston Ave. I-4
Locust Lane G-7
Logan St. C-7
Lone Oak Ave. K-4
Loney Ln. F-2,G-2
Longfield H-2
Loretta Ave. H-2
Lou Gene Way J-1
Louisville J-6
Lucas Ave. J-8
Lukin Dr. H-1
Lupino Ct. H-8
Lydia St. E-7
Lynnhurst Ave. J-3
Lynn St. E-5
Lyttle A-4
Madelon St. B-1
Madison St. B-1,B-6
Magazine B-1
Magnolia Ave. D-2,D-4
Main A-1,A-4
Malcom Rd. J-1
Malcome Pl. J-2
Manitau Ave. I-3
Mann Ave. J-2
Manning Rd. H-7
Manor Rd. H-7
Manslock Ct. I-1
Maple B-2
Maple Ct. I-4
Mapleton I-3
Market St. A-1,A-4
Marrett Ave. D-6
Mary St. D-7
Mary Catherine
Dr. H-1
Mason Ave. C-7
Mathias Ln. H-1
Mayer Ave. E-8
May Lawn Ave. G-6
McAtee B-1
McCloske Ave. F-1
McCoy G-3
McKinley Ave. F-8
Meade St. F-8
Meadowlark G-8
Mellwood Ave. B-8
Merhoff St. G-6
Merriweather St. E-6
Merwin G-6
Mill St. A-7
Millers Lane F-1
Milton Ave. E-7
Mission Dr. H-7
Mitscher Ave. K-4
Mix Ave. E-4
Model Dr. F-2
Mohawk Ave. I-5
Molter Ave. G-7
Montana Ave. G-3
Morgan Ave. E-7
Morgan St. H-8
Morrison Ave. K-3
Morton Ave. C-8
Mulberry Ave. E-7
Myrtle St. D-4
Nancy Lee Dr. J-8
Naneen Dr. J-1
Narraganset Dr. E-1
Navaho Ave. I-3
Nelson Ave. F-1
New High St. G-6
Nichols Dr. H-2
Nobel H-1
Nobel Pl. H-1
North-South St. B-6
Oak St. C-2,D-5,D-7
Oakwood Ave. J-3
Oehrle Dr. H-1
Ohio St. A-8
Oldham St. D-4
Oleanda Ave. G-3,H-3
Oleanda Ct. G-3
Oleanda Ave. G-1
Olive St. D-1
Oneida Ct. I-2
Oregon Ave. E-1
Oriole Dr. G-8
Ormsby Ave. D-3,D-6
Osage Ave. C-2
Ottawa Ave. I-5
Packard Ave. G-7
Park Ave. D-5,H-1
Park Rd. H-1
Parkway Dr. F-7
Parthenina I-2
Parthenina Ave. I-2
Paul Ave. H-3
Payne St. B-8
Peachtree H-3
Peachtree St. I-3
Peck Ave. I-3
Peerless Ct. D-3
Penquin G-8
Penway Ave. E-1
Phillips Lane I-6
Phyllis Ave. G-3
Picadilly Ave. J-3
Pidgeon Ave. G-8
Pikeview Rd. H-7
Pindell Ave. F-7
Pine St. B-8
Piper Ct. G-7
Place St. B-5
Plantation Dr. F-1
Plover F-8
Plymouth Ct. B-2
Pocahantis A-7
Poplar Level Rd. F-8
Powell Ave. I-2
Preston C-6,G-7
Prosperity Ct. B-1
Pylon Ct. I-7
Queen Ave. G-3
Quincy St. A-8
Racine Ave. G-4
Ralph Ave. G-1
Rammers D-7
Ramser Ave. G-1
Ratcliffe Ave. E-2
Rawlings Ave. E-6
Reasor Ave. E-7
Reutlinger St. D-7
Rhonda Dr. J-1
Rhonda Way J-1
Rice Ave. H-6
River Rd. A-6
Robin Dr. H-1
Robin Rd. G-7
Rochester Dr. K-4
Rodman St. G-4
Rondean Dr. J-1
Roosevelt H-8
Rosa Ter. H-1
Roselane Ave. C-6
Rosemary Dr. G-7
Rowan A-2,A-4
Royal Ave. E-8
Rubel Ave. C-8
Ruffer Ave. D-8
Rutland I-3
Sadie Ave. J-1
St. Bertand St. D-5
St. Catherine C-4,D-7
St. James Ct. E-5
St. Louis Ave. D-2
St. Michael D-7
St. Paul St. D-6
Sale Ave. H-3
Samuel St. D-7
Sanders Ln. K-1
Scanlon St. G-5
Schaffner F-3
Schiley St. J-2
Schiller C-7
Schneiter Ave. G-3
School K-2
School Way K-4
Seelbach Ave. I-2,I-3
Seneca I-5
Sharp Ave. F-3
Shelby St. C-7
Sherry Dr. G-6
Shingo Ave. K-3
Short H-8
Sioux Ave. H-6
Skyway Dr. J-8
Sonne Rd. F-2,F-3
Southcrest Dr. J-4
Southern Ave. D-1
Southern Pkwy. K-3
Southern Heights
. . . . I-2,I-4
Southgate Ave. H-3
Southland Blvd. K-3
South St. F-3
Speckert Ct. C-7
Spratt St. E-7
Spring St. B-8
Springdale I-8
Squires Dr. I-3
Standard Ave. D-2
Stanford Ave. J-8
Stanford Lane J-8
Stanley Ave. I-3
Stephan Lane K-1
Story Ave. B-7
Strader I-2
Strader Ave. I-2
Stratter Alley C-6
Summit Ave. E-8
Swako Ln. K-1
Swan St. C-7
Sylvia Ave. E-7
Tallulah Ave. I-6
Talmadge Way J-1
Taylor Ave. J-3
Teal Ave. G-8
Tennessee Ave. F-3
Tenny Ave. J-4
Terry K-5
Texas Blvd. F-7
Theresa K-5
Theresa Ave. H-1
Thornberry Ave. G-4
Thrush Rd. G-8
Thruston Dr. F-7
Tokay Ave. G-6
Towne Way J-5
Tuberose Ave. K-8
Tuscarora Way J-3
Utah G-3
Valley Forge H-2
Valley Rd. D-8
Valley View Dr. I-1
Vine C-7
Virginia Ave. C-1
Vorster Ave. E-2
Wabasso Ave. J-6
Wagner Ave. E-7
Wainwright F-6
Wallie F-2
Walnut St. A-1,B-4
Walter Ave. J-3
Wampum Ave. I-5
Warnock F-5
Warren Ave. H-4,I-3
Washington St. A-7
Wathens Ln. F-2
Watkins Ave. C-8
Wawa I-5
Weaver Ct. D-1
Webster A-8
Weyler G-3
Weller Ave. F-3
Wellington Ave. J-5
Wenatchee E-1
Wenona Ave. I-6
Wenzel St. B-7
Wetterau E-7
Wheeler J-2
Wheeler Ave. I-2
Whitewood Dr. I-6
Whitney Ave. H-5
Wilson E-1
Wilson Ave. D-3
Wingfield F-1
Wingfield Ave. E-3
Winkler Ave. F-4
Winter St. D-8
Wirth G-7
Wolfe Ave. H-8
Woodbine D-5
Woodlawn Ave. D-1
. . . . D-2,J-4,J-6
Woodruff I-3
Woody Ave. J-3
Wren Rd. G-7
Wue St. F-2
Wurtele Ave. Fp3,G-1
Wyandotte Ave. E-1
Yolanda Dr. J-1
York St. C-4
Youngland Ave. F-1
Zane St. C-5
Ziegler E-6
1st St. C-5,E-5,J-5
2nd St. C-5,E-5,J-5
3rd St. C-5,G-5,J-4
4th St. C-5,G-4,J-4
5th St. C-5,G-4,I-4,J-4
6th St. C-5,E-5,G-4,J-4
7th St. C-5,G-2
8th St. C-4,E-4,I-4
9th St. C-4,E-4,G-4
10th St. B-4,E-4
11th St. D-4,E-4
13th St. B-4,E-3
14th St. B-4,E-3
15th St. B-4,E-3
16th St. B-3,D-3,E-3
17th St. B-3,D-3,E-3
18th St. B-3
19th St. B-3
20th St. B-3,E-2
22nd C-2,E-2
23rd A-2,B-2,C-2,E-2
24th St. . . . A-2,B-2,C-2,E-2
25th St. A-2,C-2,E-2
27th St. B-2
28th St. A-2,C-1,D-1
29th St. A-2,B-1,E-1
30th St. A-1,B-1,C-1
31st St. B-1,D-1
32nd D-1
33rd St. C-1
34th St. C-1
35th St. B-1
36th St. A-1

POINTS OF INTEREST

Standiford Field
Airport J-6
Algonquin Park E-1
Boone Park A-3
Central Park D-5
Elliot Park B-1
George Rogers Clark
Park F-8
Iroquois Park K-3
Shelby Park D-6
Sheppard Park B-3
South Central
Park F-3
Victory Park C-2
Westonia Park A-1
Wyandotte Park I-3

N

Scale of Miles
0 .2 .4 .6

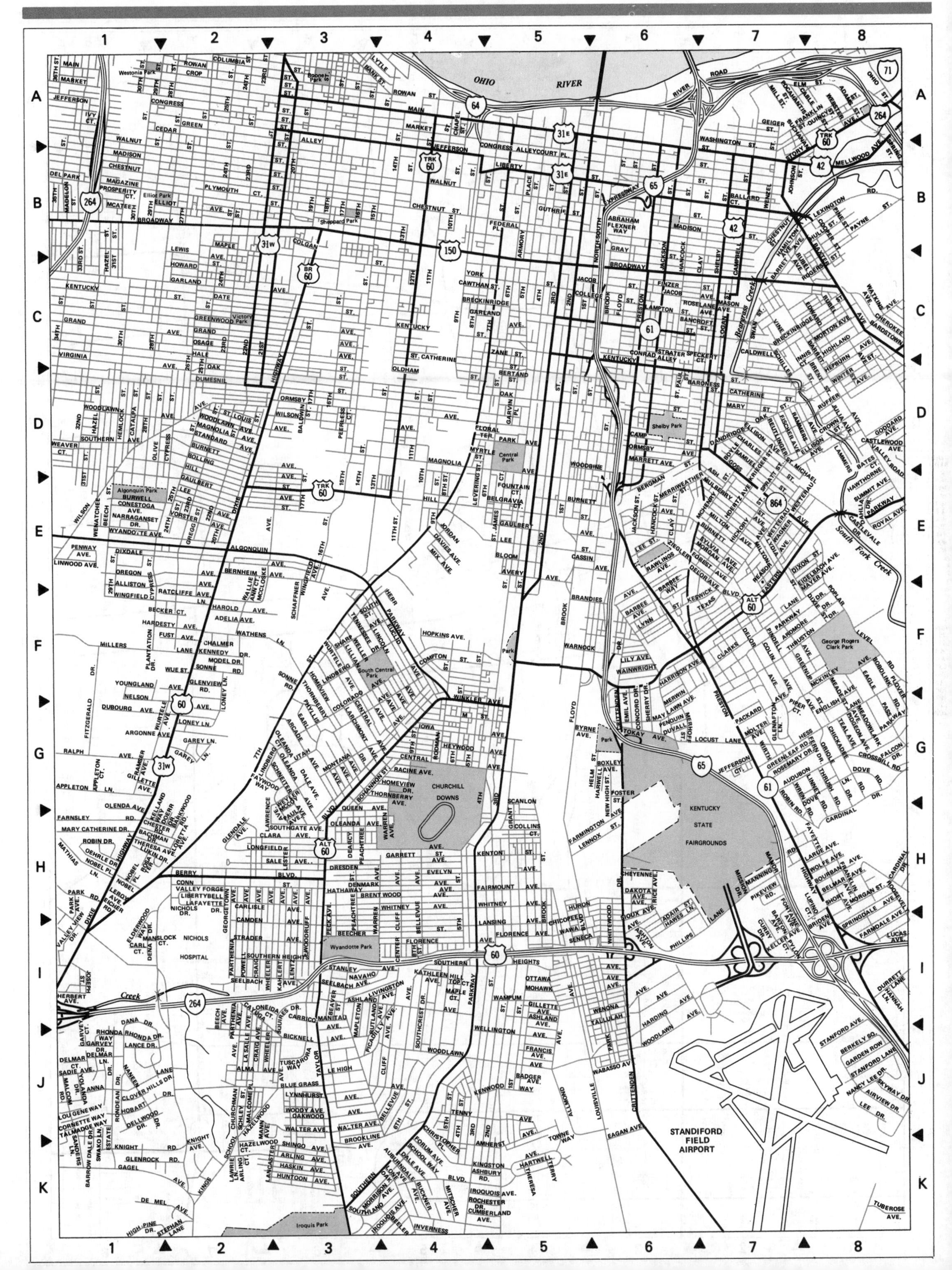

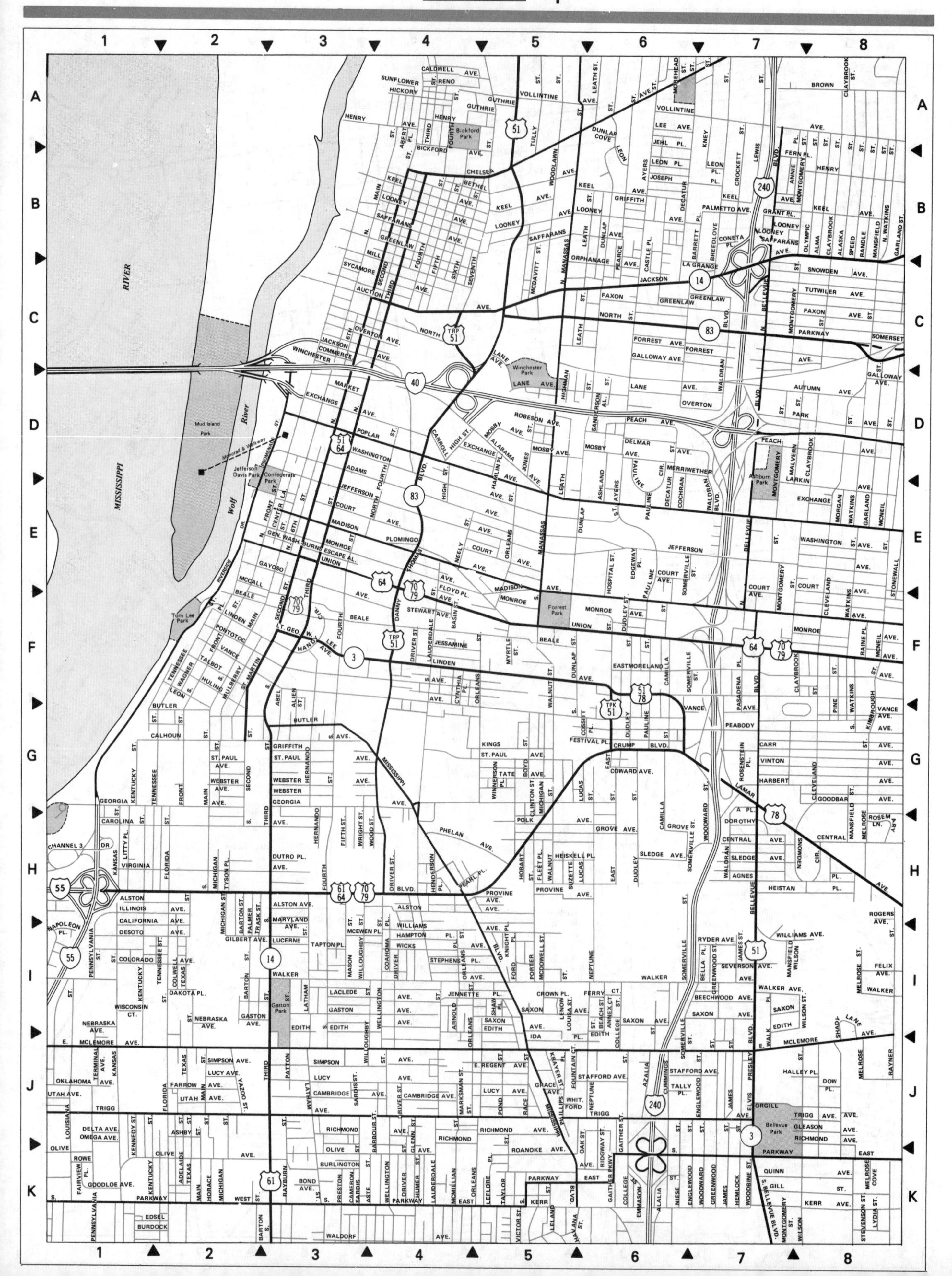
Scale of Miles
0 .1 .2 .3 .4
N
MISSISSIPPI
RIVER
Wolf River
Mud Island Park
Jefferson Davis Park
Confederate Park
Tom Lee Park
Winchester Park
Forrest Park
Bickford Park
Ashburn Park
Gaston Park
Bellevue Park

MEMPHIS

A Pl....G-7
Abel....F-3
Abert Pl....A-4
Adams Ave....D-3
Adolph....J-3
Agnes Pl....H-7
Allen St....F-3
Alma St....B-8
Alston Ave....H-1,H-4
Alabama Ave....D-5
Alalia St....K-6
Alaska St....B-8
Annex Ct....I-6
Annie Pl....B-7
Arkansas Riverside Dr....E-2
Arnold Pl....I-4
Ashburn Park....D-7
Ashby....J-2
Ashland St....D-6
Aste St....K-3
Auction Ave....C-3
Autumn Ave....D-8
Ayers St....B-6,E-6
Azalia St....J-6
Barboro....E-3
Barbour St....J-4
Barrett Pl....B-7
Barton St....H-2,K-2
Basin St....F-4
Beach St....I-6
Beachwood Ave....I-7
Beale....F-2,F-3,F-5
Bellevue Blvd....H-7
Bellevue Blvd., N....C-7,E-7
Bellevue Blvd., S....K-7
Bethel Ave....B-4
Bickford Ave....A-4
Bond Ave....K-3
Boyd St....G-5
Breedlove St....B-7
Brown Ave....A-8
Burdock....K-1
Burlington Ave....K-3,K-5
Butler....G-1,G-3
Caldwell Ave....A-4
Calhoun Ave....G-1
California Ave....H-1
Cambridge Ave....J-3
Cameron....K-3
Camilla St....F-6,H-6
Carolina Ave....H-1
Carr Ave....G-7
Carroll....D-4
Castle Pl....B-6
Center La....E-3
Central Ave....H-7,H-8
Channel 3 Dr....H-1
Chelsea Ave....A-6,B-4
Claybrook St....A-8
....D-8,F-7
Cleveland St....F-8
Clinton St....G-5
Coahoma Pl....I-4
Cochran....E-6
College St....I-6,K-6
Colorado Ave....I-1
Colwell....I-2
Commerce....C-3
Coneta Pl....B-7
Cossitt Pl....G-6
Court Ave....E-3,E-6,E-8
Coward Ave....G-6
Crockett St....A-7
Crown Pl....I-5
Crump Blvd....G-6,H-1
Cummings....J-6
Cynthia Pl....F-4
Dakota Pl....I-2
Danny Thomas Blvd....E-4
Decatur St....B-6,E-6
Delmar Ave....D-6
Delta Ave....J-1
Desota Ave....I-1
Dorothy....H-7
Dow Pl....J-8
Driver St....F-4
....H-4,J-4,K-4
Dudley St....F-6,H-6
Dunlap Cove....A-6
Dunlap St....B-6,E-5
Dunlap St., S....F-5
Dutro Pl....H-3
East St....G-6
Eastmoreland Ave....F-6
Edgeway Pl....E-6
Edith Ave....I-3,I-7,J-6
Edsel....K-1
Elvis Presley Blvd....J-7
Emmason St....K-6
Englewood St....J-7,K-7
Exchange Ave....D-3
....D-5,E-8
Fairview Pl....K-1
Farrow Ave....J-2
Faxon Ave....C-6,C-8
Felix Ave....I-8
Fern Pl....B-7
Ferry Ct....I-6
Festival Pl....G-6
Fifth St....C-4,H-3
Fleet Pl....H-5
Florida St....H-2
Floyd Pl....E-4
Ford Pl....I-5
Forrest Ave....C-6
Fountian Ct....J-5
Fourth St....A-4,B-4
Fourth St., N....E-4
Fourth St., S....F-3,H-3
Front St....G-2
Front St., N....E-3
Front St., S....F-2
Gaither Pkwy....K-6
Gaither St....J-6
Galloway Ave....C-6,D-8
Garland St....B-8,E-8
Gaston Ave....I-2,I-3
Gayoso Ave....E-3
Gen Wash.-Burns Escape Al....E-3
Georgia Ave....G-1,G-3
Gilbert Ave....I-2
Gill St....K-7
Gleason Ave....J-8
Glenn St....K-4
Goodbar Ave....G-8
Goodloe Ave....K-1
Grace Ave....J-5
Grant Pl....B-7
Greenlaw....B-4,C-6
Greenwood St....I-7,K-7
Griffith....B-6,G-3
Grove Ave....H-6
Guthrie Ave....A-4
Halley Pl....J-7,J-8
Hamlin Pl....D-5,E-5
Hampton Pl....I-4
Handy Cir....F-3
Hanover....K-1
Harbert Ave....G-7
Havana St....K-5
Heiskell Pl....H-5
Heistan Pl....H-7
Hemlock St....K-7
Henderson Pl....H-4
Henry Ave....A-3,B-8
Hernando....G-3,H-3
Hickory Ave....A-4
Highman....D-5
High St....D-4,E-4
Hobart....H-5
Horace St....K-2
Hospital St....E-6
Huling....F-2
Humer St....K-4
Ida Pl....J-5
Illinois Ave....H-1
Jackson Ave....C-3,C-6
James St....I-7,K-7
Jefferson Ave....E-3,E-6
Jehl Pl....A-6
Jennette Pl....I-4
Jessamine....F-4
Jones....D-5
Joseph Pl....B-6
Kansas St....H-1,J-1
Keel Ave....B-4,B-8
Kennedy St....K-1
Kentucky St....G-1,I-1,K-1
Kerr Ave....K-5,K-8
Kings....G-5
Kiney St....A-7
Krayer St....J-5
Laclede Ave....I-3
La Grange....C-6,C-7
Lamar Ave....G-7
Lane Ave....C-5,D-6
Larkin Ave....E-8
Latham St....I-3,J-3
Lauderdale....K-4
Lauderdale St., S....F-4
Leath St....A-6,B-6,E-5
Lee Ave....A-6
Leflore Pl....K-5
Leland St....K-5
Lenow St....I-5
Lewis St....B-7
Leon....F-2
Leon Pl....B-6,B-7
Leon St....B-6
Lt. Geo. W. Lee Ave....F-3
Linden Ave....F-2,F-4
Litty Pl....H-1
Looney Ave....B-4,B-7
Lucerne....I-3
Lucy Ave....J-2,J-3,J-5
Lydia St....K-8
Madison Ave....E-3,E-5
Main St....J-2,K-2
Main St., N....B-4
Main St., S....F-2,G-2
Malvern St....D-7
Manassas St., N....C-5
Manassas St., S....E-5
Mansfield St....B-8
....H-8,I-7
Market....D-3
Marksman St....J-4
Mars Hall Ave....E-4,F-5
Maryland Ave....H-3
Mason....I-3
McCall....E-2
McDavitt St....C-5
McDowell St....I-5
Mcewen Pl....I-3
McLemore Ave., E....I-1,J-8
McMillian St....K-4
McNeil St....E-8,F-8
Melrose Cove....K-8
Melrose St....H-8,J-8
Merriwether....D-6
Michigan....G-5
Michigan St....H-2,K-2
Mill....B-4
Mississippi Blvd....G-4,J-5
Monroe Ave....E-3,F-5,F-8
Montgomery St....B-7
....C-7,E-7,K-7
Morehead St....A-6
Morgan....E-8
Mosby Ave....D-5
Mulberry....F-2
Myrtle St....F-5
Napoleon Pl....I-1
Nebraska Ave....I-1,I-2
Neeley St....E-4
Neptune St....I-6,J-6
Niese St....K-6
North....C-4
Oklahoma Ave....J-1
Olive Ave....K-1-3
Olympic St....B-8
Omega Ave....J-1
Orgill Ave....J-7
Orleans St....E-5,F-5
....J-4,K-4
Orphan Ave....A-4
Orphanage Ave....B-5,C-6
Overton Ave....C-3,D-6
Palmer....I-2
Palmetto Ave....B-7
Park Ave....D-7
Parkway....C-6
Parkway East, S....K-4,K-8
Parkway West, S....K-1
Pasadena Pl....F-7
Patton St....J-3
Pauline Cir....D-6
Pauline St., F-6,G-6
Peabody Ave....G-7
Peach Ave....D-6,D-7
Pearce St....C-6
Pearl Pl....H-4
Pennsylvania St....I-1,K-1
Phelan Ave....H-4
Phillips Pl....J-5
Pine St....G-8
Plomingo....E-4
Polk Ave....H-5
Pond St....J-5
Pontotoc Ave....F-2
Poplar Ave....D-3,E-8
Porter St....I-5
Preston....K-3
Promenade St....C-3
Provine Ave....H-5
Quinn Ave....K-7
Race St....J-5
Raine Pl....F-8
Randle St....B-8
Rayburn St....K-3
Rayner St....J-8
Regent Ave., E....J-5
Reno Ave....A-4
Richmond Ave....J-3
....J-5,J-8
Ridgway St....K-6
Roanoke Ave....K-5
Robeson Ave....D-5
Rogers Ave....H-8
Rosemary Ln....H-8
Rosenstein Pl....G-7
Rowe....K-1
Ryder Ave....I-7
Louisa St....I-5
Louisiana St....J-1
Lucas St....G-5,H-5
Saffarans Ave....B-4,B-5,B-7
St. Kimbrough St....G-8
St. Martin....F-2
St. Paul Ave....G-2,G-5
Sanderson Al....D-6
Sardis St....J-3,K-3
Saxon Ave....I-5,I-7
Second St....F-3
Second St., N....C-4
Second St., S....G-2
Seventh St....C-4
Severson Ave....I-7
Shady Lane....I-8
Shaw Pl....I-5
Simpson....J-2,J-3
Sixth St....C-4
Sledge Ave....H-6,H-7
Snowden Ave....C-8
Snowden Cir....H-8
Somerset....C-8
Somerville St....E-6
....F-6,H-6,I-6,J-6
Speed St....B-8
Stafford Ave....J-6
Stephens Pl....I-4
Stevenson St....K-8
Stewart Ave....F-4
Stonewall....E-8
Sunflower Ave....A-4
Suzette....H-5
Sycamore Ave....C-3
Talbot....F-2
Tally Pl....J-6
Tapton Pl....I-3
Tate Ave....G-5
Taylor St....K-5
Tennessee St....F-2
....G-1,I-2
Terminal Ave....J-1
Texas St....I-2,K-2
Third St., N....A-4,C-4
Third St., S....E-3
....H-3,J-3
Trask St....I-2
Trigg Ave....J-1,J-8
Tully St....A-5
Tutwiler Ave....C-8
Tyson Pl....H-2
Union Ave....E-3,F-5
Utah Ave....J-1,J-2
Vance Ave....F-2,G-6,G-8
Victor St....K-5
Vinton Ave....G-7
Virginia Ave....H-1
Vollintine Ave....A-5,A-6
Wagner Pl....F-2
Waldorf Ave....K-3
Waldran Blvd....D-7
....E-7,H-7
Walker Ave....I-3,I-8
Walk Pl....I-7,J-7
Walnut St....F-5,H-5
Washington Ave....D-3,D-4,E-8
Watkins....E-8,F-8
Watkins St., N....B-8
Watkins, S....F-8,G-8
Webster Ave....G-2,G-3
Wellington St....I-4,K-4
Whitford....J-5
Wicks Ave....I-4
Williams Ave....I-4,I-7
Willoughby St....J-3
Wilson St....I-7,I-8,K-8
Winchester....C-3
Winnerson Pl....G-5
Wisconsin Ct....I-1
Woodbine St....K-7
Woodlawn St....B-5
Wood St....H-4
Woodward St....H-7,K-7
Wright St....H-3
Yazoo St....J-2
6th St., N....C-3,E-3

POINTS OF INTEREST

Ashburn Park....D-7
Bellevue Park....J-7
Bickford Pk....A-4
Confederate Park....D-3
Forrest Park....F-5
Gaston Park....I-3
Jefferson Davis Park....D-2
Tom Lee Park....F-2
Winchester Park....C-5

MIAMI

Airport Expwy....A-1
Bay Shore Dr., N....F-7,G-7
Biscayne Blvd....A-6
....B-6,H-6
Chopin Plaza....J-7
East-West Expwy....G-2,G-5
Flagler St. E....I-6
Flagler St. W....I-2,I-5
Flagler Ter....J-1
Herald Plaza....F-7,G-7
Julia Tuttle Causeway....B-8
McArthur Causeway....G-7,G-8
Miami Ave. N....C-5,H-5
Miami Ave. S....K-5
Palm....A-7
River Dr. N.W....H-2
River Dr. S....H-2
Sabal....A-7
Venetian Causeway....F-7,F-8
1st Ave. N.E....A-6,H-6
1st Ave. N.W....C-5,H-5
1st Ave. S.E....J-6
1st Ave. S.W....K-5
1st Ct. N.W....E-5,F-5
1st Pl. N.W....E-5,F-5
1st St. N.E....I-6
1st St. N.W....I-2,I-5
1st St. S.E....J-6
1st St. S.W....J-2
2nd Ave. N.E.A-6,D-6,H-6
2nd Ave. N.W....C-4,H-5
2nd Ave. S.E....K-6
2nd Ave. S.W....K-5
2nd St. N.E....I-6
2nd St. N.W....I-2,I-4,I-5
2nd St. S.E....J-6
2nd St. S.W....J-2
3rd Ave. N.W....H-4
3rd Ave. S.E....J-6
3rd Ave. S.W....K-4
3rd St. N.E....I-6
3rd St. N.W....I-2,I-4,I-5
3rd St. S.E....J-6
3rd St. S.W....J-2
4th Ave. N.W....C-4,F-4
....G-4,H-4
4th Ave. S.W....J-4
4th St. N.E....I-6
4th St. N.W....I-2,I-4,I-5
4th St. S.E....J-6
4th St. S.W....J-2,J-5
5th Ave. N.W.C-4,F-4,H-4
5th Ave. S.W....J-4
5th Pl. N.W....E-3
5th St. N.E....H-6
5th St. N.W....H-3,H-5,I-2
5th St. S.E....K-6
5th St. S.W....J-2
6th Ave. N.W....C-3,I-3
6th Ave. S.W....J-3
6th St. N.W....H-2,H-3,H-5
6th St. S.E....K-6
6th St. S.W....K-2,K-4
7th Ct....C-3,D-3
7th Ave. N.W.C-3,H-3,I-3
7th Ave. S.W....J-3
7th Pl....C-3,D-3
7th Ct. S.W....J-3
7th St. N.E....H-6
7th St. N.W....H-2
....H-3,H-5
7th St. S.E....K-6
7th St. S.W....K-2,K-5
8th Ave. N.W....A-2
....C-3,F-3,I-3
8th Ave. S.W....J-3
8th Ct. S.W....J-3,K-3
8th St. N.E....H-6
8th St. N.W....H-2,H-3,H-5
8th St. S.E....K-6
8th St. S.W....K-2,K-5
8th St. Rd. N.W....G-2
9th Ave....H-3
9th Ave. N.W....B-2,F-2,I-2
9th Ave. S.W....J-2
9th Ct. N.W....H-2
9th St. N.E....H-6
9th St. N.W....H-5
9th St. S.E....K-6
9th St. S.W....K-2,K-5
10th Ave....H-2
10th Ave. N.W....A-2
....C-2,G-2,I-2
10th Ave. S.W....J-2
10th Ct....E-2,H-2
10th St. N.W....G-3,G-5
11th Ave....G-2
11th Ave. N.W....C-2,I-2
11th Ave. S.W....J-2
11th Ct....C-2,E-2,H-2
11th Pl....C-2
11th St. N.E....G-6
11th St. N.W....G-2
....G-3,G-5
11th St. Rd. N.W....G-3
12th Ave. N.W....B-1
....C-1,I-1
12th Ave. S.W....J-1
12th Ct....H-1
12th St. N.E....G-6
12th St. N.W....G-3
....G-4,G-5
13th Ave. N.W....A-1
....C-1,I-1
13th Ave. S.W....J-1
13th Ct. N.W....F-1,H-1
13th St. N.E....G-6
13th St. N.W....G-3,G-5
14th Ave. N.W....A-1
....C-1,F-1,I-1
14th Ave. S.W....J-1
14th Ct. N.W....H-1
14th St. N.E....G-6
14th St. N.W....G-1,G-5
14th Ter....F-1,F-2,F-4
15th Ave. N.W....C-1,I-1
15th Ave. S.W....J-1
15th St....F-1
15th St. N.E....F-6
15th St. N.W....F-3,F-5
15th St. Rd....E-1
15th Ter....F-6
16th St....F-1,F-2
16th St. N.E....F-6
16th St. N.W....F-5
17th St. N.E....F-6
17th St. N.W....F-3-5
17th Ter. N.E....F-6
18th St. N.E....E-6
18th St. N.W....F-3,F-5
18th Ter....H-1
18th Ter. N.W....E-3
19th St. N.E....E-6
19th St. N.W....E-1,E-3-5
19th Ter. N.E....E-6
20th St. N.E....E-6
20th St. N.W....E-2,E-4
20th Ter. N.E....E-6
21st St. N.E....E-6
21st St. N.W....E-2,E-5
22nd Ln. N.W....D-4
22nd St. N.E....D-6
22nd St. N.W....D-2,E-4
22nd Ter. N.W....D-2,D-4
23rd St. N.E....D-7
23rd St. N.W....D-2,D-4
24th St. N.E....D-6
24th St. N.W....D-2,D-4
25th St. N.E....D-6
25th St. N.W....D-2,D-4
26th St. N.E....D-6
26th St. N.W....D-2,D-4
27th St. N.E....D-6
27th St. N.W....D-2
27th Ter....D-4
28th St. N.E....C-6
28th St. N.W....C-2,C-4
29th St. N.E....C-6
29th St. N.W....C-2,C-4
29th Ter. N.W....C-2
30th St. N.E....C-6
30th St. N.W....C-2-4
31st St. N.E....C-6
31st St. N.W....C-2,C-4
32nd St. N.E....C-6
32nd St. N.W.C-1,C-3,C-4
33rd St. N.E....B-7
33rd St. N.W....B-1,C-4
34th St. N.E....B-7
34th St. N.W....B-1,B-3-5
34th Ter....B-5
35th St. N.E....B-6
35th St. N.W....B-1,B-3-4
35th Ter. N.E....B-6
36th St. N.E....B-6
36th St. N.W....B-2,B-4
37th St. N.W....B-2,B-4
38th St. N.E....B-6,B-7
38th St. N.W....B-2,B-5
39th St. N.E....A-6,A-7
39th St. N.W....A-2,A-4
40th St. N.E....A-6
40th St. N.W....A-2,A-4
41st St. N.E....A-6
41st St. N.W....A-2,A-4
42nd St. N.E....A-6
42nd St. N.W....A-2,A-4
43rd St. N.E....A-6
43rd St. N.W....A-2,A-4
44th St. N.E....A-6
44th St. N.W....A-2,A-4

POINTS OF INTEREST

Bay Front Park....I-7
Moore Park....B-2

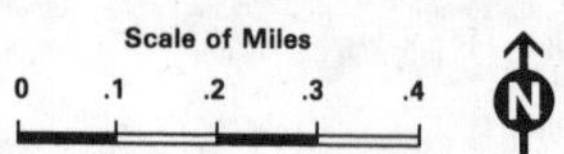
Scale of Miles
0
.1
.2
.3
.4
N

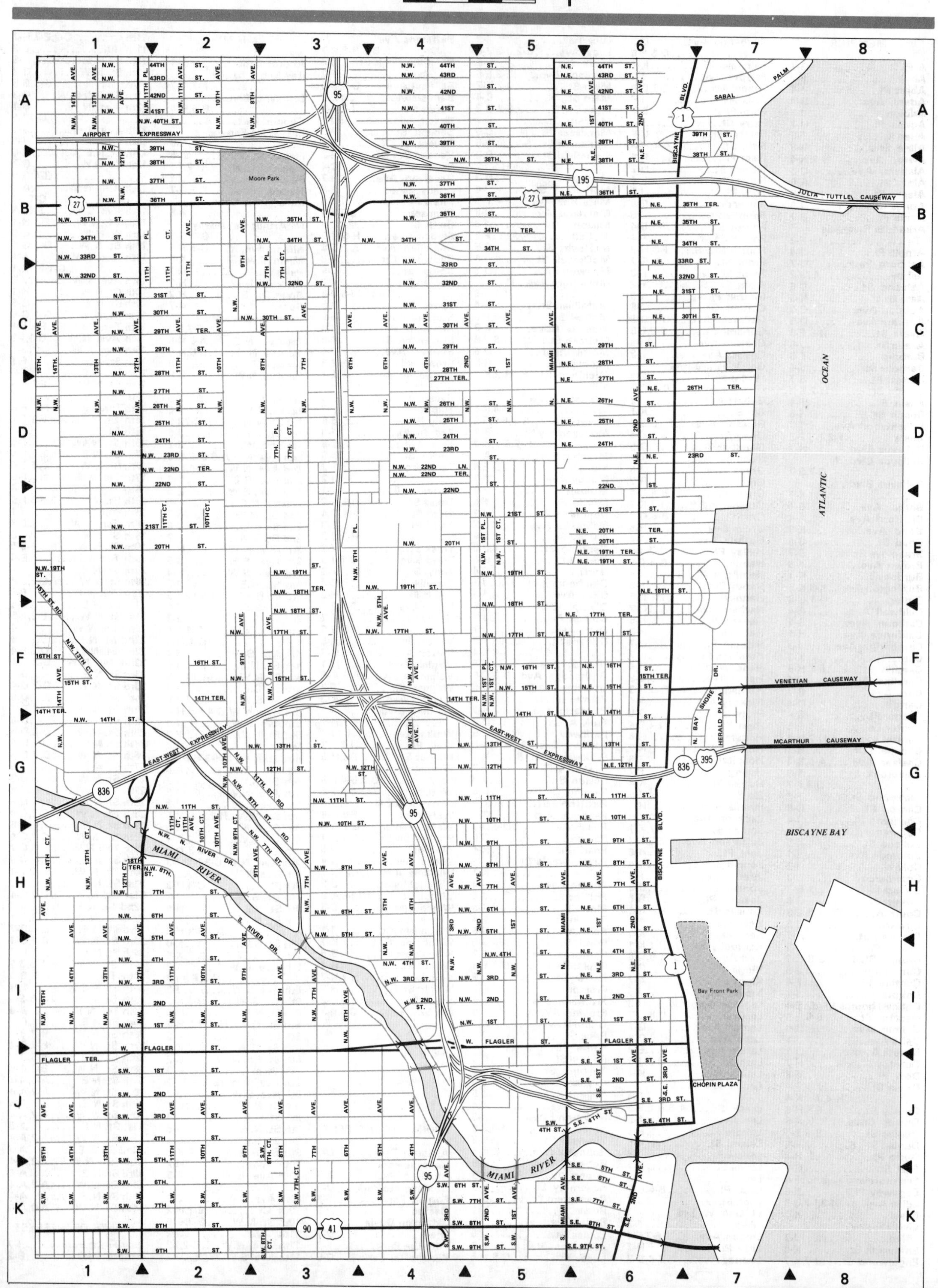
Moore Park
AIRPORT EXPRESSWAY
EAST-WEST EXPRESSWAY
JULIA TUTTLE CAUSEWAY
VENETIAN CAUSEWAY
MCARTHUR CAUSEWAY
SABAL PALM
BISCAYNE BLVD.
MIAMI AVE.
ATLANTIC OCEAN
BISCAYNE BAY
MIAMI RIVER
N.W. N. RIVER DR.
S. RIVER DR.
15TH ST. RD
N.W. 13TH CT.
N.W. 14TH ST. RD
N.W. 8TH ST. RD
N.W. 7TH ST. RD
N. BAY SHORE DR.
HERALD PLAZA
Bay Front Park
CHOPIN PLAZA
FLAGLER ST.
FLAGLER TER.
95
195
395
836
27
1
90
41

Scale of Miles
0 .2 .4 .6 .8

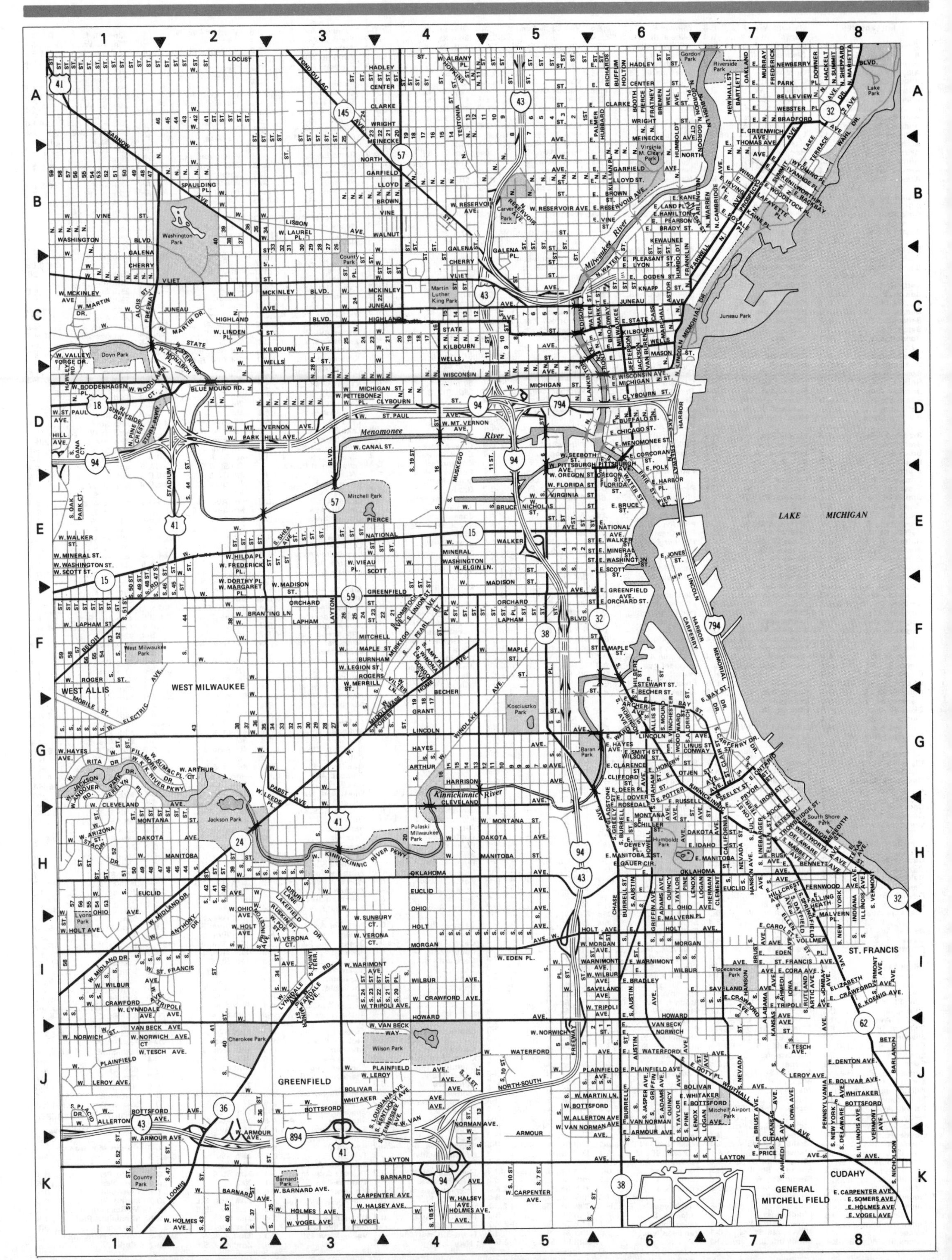

MILWAUKEE

Adams Ave., S...H-6,J-6
Ahmedi Ave., S...I-7,K-7
Alabama Ave., S.......I-7
Albany Pl., W........A-4
Aldrich St., S.........G-6
Allerton Ave., W..J-1,J-5
Allis St., S..........G-6
Alois St...........C-1
Amy Pl., S..........F-4
Andover Rd., W.......G-1
Anthony Dr., W.......I-2
Archer Ave., E.......G-6
Arizona St., W........H-1
Arlington B-7
Armour Ave., E.......K-6
Armour Ave., W.
...........J-1,K-1,K-5
Arthur Ave., W.......G-4
Arthur Ct., W........G-2
Artic Ave...........I-8
Astor St., N.........C-6
Austin St., S.
...........H-6,I-6,J-6
Backbay, E..........B-8
Barland Ave., S.......J-8
Barnard Ave., W.
...........K-2,K-3,K-4
Bartlett, N..........A-7
Bay St., E.........G-6,G-7
Becher St., E.........G-6
Becher St., W....F-4,G-4
Belleview, E..........A-7
Beloit Rd...........F-1
Bennett Ave., E.......H-8
Betz J-8
Blue Mound Rd., W..D-2
Boddenhagen Pl., W.D-1
Bolivar Ave., E...J-6,J-8
Bolivar Ave., W.......J-3
Bombay Ave., S.......I-8
Booth St., N..........A-6
Bottsford, E.......J-6,J-8
Bottsford Ave., W.
...........J-1,J-3,J-5
Bradford Ave., E.......A-7
Bradley Ave., E.......I-6
Brady St., E..........B-6
Branting Ln., W.......F-2
Bremen St., N.........A-6
Broadway St., N.......C-6
Brown St., E..........B-6
Brown St., W..........B-4
Bruce St., E..........E-6
Bruce St., W..........E-5
Brust Ave., S......I-7,J-7
Buffalo St., E.........D-6
Buffum St., N.........A-6
Burnham St., W.......F-3
Burrell St., S......H-6,J-6
Bush Ln., N...........A-7
California St., S......H-7
Cambridge, N..........B-7
Canal St., W..........D-3
Carferry Dr., S...F-6,G-7
Carol St., E...........I-7
Carpenter Ave., E.....K-8
Carpenter Ave., W.
...............K-3,K-5
Cass, N..............C-6
Center St., E..........A-6
Center St., W.........A-4
Chase Ave., S.........H-6
Cherry St., W....C-1,C-4
Chicago St., E.........D-6
Clarence St., E.......G-6
Clark St., E...........A-6
Clarke St., W.........A-4
Clement Ave., S.......H-7
Cleveland Ave., W.
...............G-1,H-4
Clifford St., E.........G-6
Clybourn St., E.......D-6
Clybourn St., W.......D-4
Comstock Ave., S.....F-4
Congo Ave...........F-4
Conway St., E.........G-6
Cora Ave., E..........I-7
Corcoran St., E.......D-6
Crawford Ave., E.
...............I-7,I-8
Crawlford Ave., W.
...............I-1,I-4
Cudahy Ave., E..K-6,K-7
Dakota Ave., W..H-1,H-5
Dakota St., E..........H-7
Dana Ct., S...........D-1
Dayfield St., S.........I-7
Deer Pl., E............G-6
Delaware Ave., S.
.......H-7,H-8,J-8,K-8
Denton Ave., E.........J-8
Dewey Pl., E..........H-6
Dorothy Pl., W........E-2
Doty Pl., E............J-7
Dover St., E..........G-6
Downer, N............A-8
Drury Lane............H-3
Eden Pl., E............I-7
Eden Pl., W...........I-5
Edison, N............C-5
Electric Ave., W......G-1
Elgin Ln., W..........E-4
Elizabeth I-8
Ellen St., S.........H-7,I-7
Erie St., E............E-6
Estes St., E...........H-7
Euclid Ave., E.....H-6,H-7
Euclid Ave., W...H-1,H-4
Falling Heath, E.......H-8
Fardale Ave., W.......I-3
Farwell Ave., N.......C-7
Fernwood Ave., E.....H-8
Fillmore Dr., W.......G-1
Florida St., E.........E-6
Florida St., W........E-5
Fond Du Lac Ave.....A-3
Forest Home Ave., W.
...............F-4,G-4
Franklin, N...........C-6
Fratney St., N........A-6
Frederick, N..........A-7
Frederick Pl., W......E-2
Front St., N...........D-6
Fulton St., S..........H-7
Galena Pl.............C-5
Galena St., W....B-1,B-4
Garfield Ave., E.......B-6
Garfield Ave., W......B-4
Gauer Cir., E.........H-6
Gladstone, S..........H-6
Gordon Ct., N.........A-7
Gordon Pl., N....A-7,B-7
Graham St., S...D-6,G-6
Grant St., W..........G-4
Greeley St., S.........H-6
Greenfield Ave., E.....F-6
Greenfield Ave., W....F-4
Greenwich Ave., E....A-7
Griffin Ave., S....I-6,J-6
Hackelt, N............A-8
Hadley St., E.........A-6
Hadley St., W.........A-4
Halsey Ave., W...K-3,K-4
Hamilton, E...........B-6
Hansen Ave., S........H-7
Harbor, N............D-6
Harbor Dr., S.........F-7
Harbor Pl., E.........E-6
Harrison Ave., W.....G-4
Hawley Rd............D-1
Hayes Ave., E........G-6
Hayes Ave., W...G-1,G-4
Herman St., S.........H-7
Highland Blvd,....C-2,C-4
Hilbert St., S.........F-6
Hilda Pl., W..........E-2
Hill Ave..............D-1
Hillcrest Ave., E......H-7
Holmes Ave., E.......K-8
Holmes Ave., W.
...........K-2,K-3,K-4
Holt Ave., W..........I-1
...........I-2,I-4,I-6
Holton St., N.........A-6
Homer St., E..........G-6
Hopkins, W...........A-4
Howard Ave., E........I-6
Howell Ave., S........H-6
Howard Ave., W........I-4
Hubbard St., N........A-6
Humboldt Blvd, N.
...........A-6,B-6,C-6
Illinois Ave., S.......K-8
Indiana Ave., S........I-8
Iowa Ave., S......I-7,J-7
Iron St., E............H-7
Irving Ave., E.........B-7
Ivanhoe Pl............B-8
Jackson St., N........C-6
Jackson Park Dr., W.
....................G-1
Jasper Ave., S........J-6
Jefferson St., N.......C-6
Jerelyn Pl............G-1
Jones St., E...........E-6
Juneau Ave., E.......C-6
Juneau Ave., W..C-2,C-4
Kaine Pl., E..........B-7
Kane, E..............B-6
Kansas Ave., S........J-7
Kenilworth Pl., E.
...............B-7,B-8
Kentucky Ave., S.....J-4
Kewaunee B-6
Kilbourn Ave., E......C-6
Kilbourn Ave., W.C-3,C-4
Kinnickinnic River
Pkwy., W........G-1,H-3
Kinnickinnic Ave., S.
...............G-7,H-7
Knapp St., E..........C-6
Koenig Ave., E........I-8
La Fayette Pl., E......B-7
Lake Dr., N...........B-8
Lake Freeway.........D-6
Lakefield Dr., W......H-3
Land Pl., E............B-6
Lapham St., W.
...........F-1,F-3,F-5
Laurel Pl., W.........B-3
Layton Ave., E........K-7
Layton Blvd,..........F-3
Leeds Pl., W..........G-2
Legion St., W.........F-3
Lenox St., S......H-6,J-6
Leroy Ave., E..........J-7
Le Roy Ave., W...J-1,J-3
Lincoln Ave., E.......G-6
Lincoln Ave., W.......G-4
Lincoln Memorial Dr.
N.....................C-6
Lincoln Memorial Dr.
S.....................F-7
Linden Pl., W.........C-2
Linebarger Terr., S...H-7
Linus St., E..........G-6
Lisbon Ave., W.......B-3
Lloyd Ave., W.........B-4
Lloyd St., E...........B-6
Locust St., W.........A-2
Logan Ave., S....H-7,J-7
Loomis Rd............K-2
Louisiana Ave., S......J-4
Lynndale Ave., W..I-1,I-3
Lyon St., E............C-6
Mabbett Ave., S.H-7,H-8
Madison St., W.......F-5
Malvern Pl., E.....I-6,I-8
Manitoba St., E..H-6,H-7
Manitoba St., W.H-2,H-5
Maple St., E...........F-6
Mapel St., W.....F-3,F-5
Marietta, N...........A-8
Market St., N.........C-6
Marshall St., N.......C-6
Martin Dr., W.....C-1,C-2
Martin Ln., W.........J-5
Mason St., E..........C-6
McKinley Blvd., W.
...............C-3,C-4
McKinley Ave., w.....C-1
Meinecke Ave., E.....B-6
Meinecke Ave., W.....A-4
Meinecke St., E.......A-6
Menomonee St., E...D-6
Meredith St., E........H-8
Merrill St., W.........F-3
Michigan St., E.......D-6
Michigan St., W.
...............D-3,D-5
Middlemass, W........G-4
Midland Dr., W....H-2,I-1
Milwaukee St., N......C-6
Miner St., S...........I-3
Mineral St., E.........E-6
Mineral St., W.....E-1,E-4
Mitchell St...........F-3
Mobile St............G-1
Monarch Pl.....W.....C-2
Montana St., E........H-6
Montana St., W..H-1,H-5
Morgan Ave., W...I-4,I-6
Mound, S.............G-6
Mt. Vernon Ave., W.
...............D-3,D-4
Murray, N............A-7
Muskego Ave., S.
...............E-4,F-4
National Ave.........E-4
National Ave., E......E-6
New Hall St., N.......A-7
Newberry Blvd., E...A-7
New York, S.......I-8,J-8
Nevada St., S.....H-7,J-7
Nicholas St., W.......E-5
Nicholson, S..........K-8
Nock St., E...........H-7
North Ave., E.........B-6
North Ave., W.........B-4
North-South Frwy....J-5
Norwich Ave., W.......J-1
Norwich Ct., W........J-1
Norwich St., E.........J-6
Norwich St., W...J-1,J-5
Oak Park Ct., S.......E-1
Oakland, N............A-7
Ogden St., E..........C-6
Ohio Ave., E...........I-6
Ohio Ave., W.
...........H-1,H-2,H-4
Oklahoma Ave., E.....H-6
Oklahoma Ave., W....H-4
Ontario St., E.........G-7
Orchard St., E.........F-6
Orchard St., W....F-3,F-5
Oregon St., E..........E-6
Oregon St., W.........E-5
Otjen St., E...........G-6
Pabst Ave., S..........G-3
Palmer St., N.........A-6
Park Pl., E............A-7
Park Hill Ave., W.....D-2
Pearl St., S...........F-4
Pearson, E............B-6
Pennsylvania Ave., S.
....................J-8
Petibone Pl., W.......D-3
Pier St...............E-6
Pierce St.............E-4
Pierce St., N..........A-6
Pine Ave., S......H-6,J-6
Pine Crest, N..........D-1
Pittsburgh Ave., W.
...............D-5,D-6
Placid Dr., S..........J-1
Plainfield Ave., E.....J-6
Plainfield Ave., W.
...........J-1,J-3,J-4,J-5
Plankinton, N.........D-6
Pleasant St., E........C-6
Poe St., W............I-3
Point Terr., S.........I-3
Polk, E...............E-6
Potter Ave., E........G-6
Price Ave., E..........K-7
Princeton Ave., S.....I-2
Prospect Ave., N.....B-7
Pryor St..............H-7
Pryor St., E...........G-7
Pulaski St., N.........B-6
Quincy Ave., S...H-6,J-6
Reservoir Ave., E.....B-6
Reservoir Ave., N.....B-5
Reservoir Ave., W.....B-4
Richards St., N.......A-6
Rita Dr., W...........G-1
Robinson, S...........G-6
Rogers St., W.....F-1,F-3
Rosedale, E...........H-6
Royale Pl., E..........B-7
Rusk Ave., E..........H-7
Ruskin St.............I-3
Ruskin St., W.........H-3
Russell, E.............H-6
Rutland, S............I-8
St. Clair St., S........G-7
St. Francis Ave.......I-7
St. Francis Ave., W.
...............I-1,I-2
St. Paul Ave., W.D-1,D-4
Sarnow Ave...........A-1
Saveland Ave., E......I-7
Saveland Ave., W......I-6
Schiller St., E.........H-6
Scott St., E...........E-6
Scott St., W.......E-1,E-4
Seeboth St., W.......D-5
Seeley St., E..........G-7
Shea Ave., S..........E-3
Sheppard, N...........A-8
Smith St., E...........G-6
Somers Ave., E........K-8
Spaulding Pl..........B-2
Springfield, S.........I-8
Stadium Frwy....C-1,E-2
Stark Dr., W...........H-1
State St., E...........C-6
State St., W......C-2,C-4
Sterling Pl., W.......C-2
Stewart St., E.........F-6
Story Pkwy...........D-1
Sumac Pl., W.........G-2
Summit Ave., N..A-8,B-7
Sunbury Ct., W........I-3
Sunnyside Dr., W.....D-1
Superior St., S........H-8
Taylor Ave., S....H-6,J-6
Tennessee Ave., S.....J-4
Terrace Ave., N.......B-8
Tesch Ave., E..........J-7
Tesch Ave., W.........J-1
Teutonia St., N.......A-4
Texas Ave., E..........H-8
Thomas Ave., E........B-7
Tripoli Ave., E.........I-7
Tripoli Ave., W.
...........I-1,I-3,I-6
Trowbridge St., E.....H-7
Union St., S...........F-4
Valley Forge Dr., W..C-1
Van Beck Ave., E......J-6
Van Beck Ave., W.
...............J-1,J-3
Van Buren St., N......C-6
Van Norman Ave., E..J-6
Van Norman Ave., W.
...............J-4,J-5
Vermont, S.....H-8,I-8,J-8
Verona Ct., W.........I-3
Vieau Pl., W..........E-3
Vilter Ln., W..........F-4
Virginia St., W........E-5
Vine St., E...........B-6
Vine St., W.......B-1,B-4
Vliet St., W......C-2,C-4
Vogel Ave., E.........K-8
Vogel Ave., W.........K-3
Vollmer, E............I-8
Wahl Dr..............B-8
Walker St., E..........E-6
Walker St., W....E-1,E-5
Walnut St., W.........B-4
Ward St..............G-6
Warnimont Ave., E....I-6
Warnimont Ave., W.
...............I-3,I-6
Warren N.............B-7
Washington Blvd., W.B-1
Washington St., E.....E-6
Washington St., W.
...............E-1,E-4
Water St., N..........C-6
Water St., S..........E-6
Waterford Ave., E....J-6
Waterford Ave., W....J-5
Webster Pl., E........A-7
Wells St., E..........C-6
Wells St., N..........A-6
Wells St., W.....C-3,C-4
Wentworth Ave., S..H-8
Whitaker Ave., E.
...............J-6,J-8
Whithall Ave., S.......J-7
Wilbur Ave., E.........I-6
Wilbur Ave., W.
...........I-1,I-4,I-6
Wilson St............G-6
Winchester, S.........G-6
Windlake Ave., W.....G-4
Windsor, E...........B-7
Winona Ln., S.........F-4
Wisconsin Ave., E....D-6
Woodlawn Ct., W.....D-1
Woodstock Pl., E.....B-7
Woodward St., S.....G-6
Wright St., E..........A-6
Wright St., W.........A-4
Wyoming Pl...........B-8
1st Pl., S.............J-6
1st St., N............A-6
1st St., S............J-6
2nd St., N.......A-5,D-5
2nd St., S.....E-5,J-6,K-5
3rd St., N.........A-5,C-4
3rd St., S.........E-5,J-5
4th St., N.........A-5,C-5
4th St., S.............E-5
5th Pl., S.........G-5,J-5
5th St., N.........A-5,C-5
5th St., S.........G-5,J-5
6th St., N.........A-5,C-5
6th St., S.........D-5,G-5
7th St., N.........A-5,C-5
7th St., S.........G-5,K-5
8th St., N.........A-5,C-5
8th St., S.............G-5
9th Pl., S.............G-5
9th St., N.........A-5,C-5
9th St., S.............G-5
10th St., N.......A-5,C-5
10th St., S.......G-5,K-5
11th Ln., N............A-5
11th St., N.......A-5,C-5
11th St., S.......D-5,G-5
12th Ln., N............A-4
12th St., N............C-4
12th St., S............G-5
13th St., N.......A-4,C-4
13th St., S.......G-4,J-4
14th St., N.......A-4,C-4
14th St., S.......G-4,J-4
15th St., N.......A-4,C-4
15th St., S............G-4
16th St., N.......A-4,C-4
16th St., S............G-4
17th St., N.......A-4,C-4
17th St., S............G-4
18th St., N.......A-4,C-4
18th St., S.......G-4,K-4
19th St., N.......A-4,C-4
19th St., S.......D-4,G-4
20th St., N.......A-4,C-4
21st St., N.......A-4,C-4
21st St., S.......F-4,I-4
22nd St., N.......A-4,C-4
22nd St., S.......F-4,I-3
23rd St., N.......A-4,C-3
23rd St., S.......F-3,I-3
24th Pl., N............C-3
24th St., N.......A-3,C-3
24th St., S.......F-3,I-3
25th St., N.......A-3,C-3
25th St., S............F-3
26th St., N............B-3
26th St., S............F-3
27th St., N............B-3
28th Pl., N............B-3
28th St., N............B-3
28th St., S............G-3
29th St., N............B-3
29th St., S............G-3
30th St., N............B-3
30th St., S............G-3
31st St., N............B-3
31st St., S............G-3
32nd St., N............B-3
32nd St., S............G-3
33rd St., N............B-3
33rd St., S............G-3
34th St., N............B-3
34th St., S.......G-3,I-3
35th St., N............B-2
35th St., S.......G-2,K-2
36th St., N............B-2
36th St., S.......G-2,J-2
37th St., N............B-2
37th St., S.......G-2,K-2
38th St., N............B-2
38th St., S.......F-2,G-2
39th St., N...........B-2
39th St., S...........H-2
40th St., N...........B-2
40th St., S....H-2,J-2,K-2
41st St., N...........A-2
41st St., S.........H-2,J-2
42nd St., N...........A-2
42nd St., S...........H-2
43rd St., N...........A-2
43rd St., S.........G-2,K-2
44th St., N...........A-2
44th St., S.....E-2,F-2,H-2
45th St., N...........A-2
45th St., S.........E-2,H-2
46th St., N...........A-2
46th St., S.........E-2,H-2
47th St., N...........B-1
47th St., S.....E-1,H-1,K-1
48th St., N...........B-1
48th St., S...........E-1
49th St., N...........B-1
49th St., S.........E-1,H-1
50th St., N...........B-1
50th St., S.........E-1,H-1
51st St., N...........B-1
51st St., S.....F-1,H-1,K-1
52nd St., N...........B-1
52nd St., S....F-1,H-1,K-1
53rd St., N...........B-1
53rd St., S.........F-1,H-1
54th St., N...........B-1
54th St., S.........F-1,H-1
55th St., N...........B-1
55th St., S.........F-1,H-1
56th St., N...........B-1
56th St., S.........F-1,H-1
57th St., N...........B-1
57th St., S.........F-1,H-1
58th St., N...........B-1
58th St., S...........F-1,I-1
59th St., N...........B-1
59th St., S...........F-1

POINTS OF INTEREST

Baran Park......G-5,G-6
Barnard Park..........K-3
Carver Park...........B-5
Cherokee Park........J-2
County Park...........C-3
County Park...........K-5
Doyn Park.............C-1
Grodon Park...........A-6
Humboldt Park........H-6
Jackson Park..........H-2
Juneau Park...........C-7
Kosciuszko Park......G-5
Lake Park.............A-8
Lyons Park............I-1
Martin Luther King
Park..................C-4
Mitchell Airport
Park..................J-7
Mitchell Park.........E-3
Pulaski Milwaukee
Park..................H-4
Riverside Park........A-7
South Shore Park.....H-8
Tippecanoe Park.......I-7
Virginia M. Cleary
Park..................B-6
Washington Park......B-2
West Milwaukee
Park..................F-1
Wilson Park...........J-4

MINNEAPOLIS

Aldrich................B-3
Aldrich Ave. S....E-3,I-3
Arthur Ave............D-8
Bedford St............D-8
Belmont...............I-3
Blaisdell Ave......E-3,H-3
Bloomington..........E-5
Bloomington Ave.......J-5
Border Ave............B-3
Bossen Ter............K-7
Bradford..............B-3
Bryant Ave. N........B-2
Bryant Ave. S.....D-2,J-2
Calhoun Pkwy..........G-2
Cecil St...............D-8
Cedar Ave.........F-6,J-6
Cedar Lake Rd.........C-1
Central Ave...........B-5
Chateau Pl............J-5
Chestnut Ave.....C-1,C-3
Chicago...............C-4
Chicago Ave...........E-5
...............H-4,K-4
Clarence..............D-8
Clifton...............D-3
Clinton Ave...........D-4
...............H-4,K-4
Colfax................C-2
Colfax Ave. S....D-2,J-2
Columbus Ave....E-4,H-4
Como Ave..............B-7
Currie Ave............B-3
Delaware C-7
Diamond Lake Rd......J-3
Dight Ave.............F-7
Dorman Ave............E-8
Douglas Ave...........B-2
Dupont Ave. N.........B-2
Dupont Ave. S....D-2,J-2
Edgewater Blvd........J-5
Elm Ave...............B-7
Elmwood Pl. W.........I-3
Emerson Ave. N.......B-2
Emerson Ave. S..D-2,J-2
Erie St. S.E...........D-7
Essex St..............C-7
Fairmount Ave.........B-7
Farwell Ave...........A-1
Floyd B. Olson Mem.
Hwy...................B-2
Franklin Ave. E......D-5
Franklin Ave. W......D-3
Fremont Ave...........F-2
Fremont Ave. N.......B-2
Fremont Ave. S.......D-2
...............F-2,J-2
Fulton St.............C-7
Garfield Ave......E-3,I-3
Girard Ave............F-2
Girard Ave. S.........J-2
Girard Ter............B-2
Glenwood Ave..........B-2
Grand Ave...E-3,H-3,K-3
Grant St..............C-3
Grass Lake............K-2
Groveland Ave........D-3
Groveland Ter........C-2
Harding...............A-7
Harmon Pl.............C-3
Harriet Ave.........E-3,I-3
Harvard...............C-7
Hawthorne Ave..C-1,C-3
Hennepin Ave..........B-4
...............E-2,F-2
Hennepin Ave. E.
...............A-7,B-5
Hiawatha Ave..........G-7
Holmes................F-2
Hoover St.............A-8
Humboldt Ave. N.....B-2
Humboldt Ave. S.
...............F-2,K-2
Irving Ave. N....A-2,C-2
Irving Ave. S.....F-2,J-2
James Ave. N.....A-2,C-2
James Ave. S......F-2,J-2
Kasota Ave............B-8
Keewaydin Pl..........I-7
Kennedy St............A-7
Kenwood...............D-1
Kenwood Pkwy..C-2,D-1
Knox Ave. N...........A-2
Knox Ave. S...........J-2
Lagoon Ave............E-2
Lake Pl...............E-2
Lake St. E............F-5
Lake St. W............F-2
Lake Harriet Pkwy.
E.....................I-2
Lake Harriet Pkwy.
W.....................H-1
Lakeside Ave..........B-3
Lasalle Ave...........C-3
Laurel Ave.......C-1,C-3
Lincoln Ave...........D-2
Linden Ave............C-3
Linden Hill Blvd......H-1
Logan Ave. N..........A-2
Logan Ave. S.....D-2,J-2
Longfellow Ave........G-6
Luverne Ave...........J-4
Lyndale AVe. S.
...............E-3,I-3
Main St. S.E..........B-5
Malcom Ave.......C-8,D-8
Marquette Ave.........C-4
Melbourne Ave.........D-8
Minneapolis Ave.......D-7
Minnehaha Ave..D-6,F-7
Minnehaha Pkwy......I-5
Mondamin St..........J-7
Morgan Ave. N.........A-1
Morgan Ave. S.........J-1
Mt. Curve Ave.........D-2
Newton Ave. N.........A-1
Newton Ave. S...D-1,J-1
Nicollet Ave...........B-4
...............E-3,H-3
Nokomis Ave...........H-7
Nokomis Ct............K-5
Nokomis Ln............K-5
Nokomis Pkwy..........J-5
Oak St................C-7
Oak Grove St..........D-3
Oakland Ave......E-4,H-4
Oak Park..............A-1
Oliver Ave. N.........A-1
Oliver Ave. S....D-1,J-1
Ontario St.S.E........D-7
Orlin Ave.............D-8
Park Ave.....C-4,E-4,H-4
Park Dr...............I-2
Penn Ave. N...........A-1
Penn Ave. S.......D-1,J-1
Pickfield.............G-1
Pillsbury Ave.....D-3,H-3
Pleasant Ave......E-3,H-3
Pleasant St...........C-6
Plymouth Ave. N......A-2
Portland Ave..........C-4
...........E-4,H-4,K-4
Prospect Ave..........I-3
Queen Ave. N..........A-1
Queen Ave. S.....D-1,J-1
River Rd..............D-8
River Ter.............D-8
Riverside Ave.........D-6
Roosevelt.............A-7
Royalston.............B-3
Russell Ave. N........A-1
Russell Ave. S........K-1
Rustic Lodge Ter......I-3
St. Mary's............C-8
Sander................K-7
Seabury Ave...........E-8
Seymour...............D-8
Sharon Ave............D-8
Sheridan Ave. N.......A-1
Sheridan Ave. S.
...............D-1,K-1
Shore View Ave........J-6
Snelling..............D-6
Snelling Ave..........F-7
Spring St.........A-5,A-8
Stevens Ave......D-4,H-4
Summit Ave............D-2
Sunrise Dr............K-2
Taft St...............A-7
Talmadge Ave..........A-7
Thomas Ave. N.........A-1
Thomas Ave. S...D-1,J-1
University Ave........B-5
University Ave. N.E.
......................A-5
Upton Ave. N..........A-1
Upton Ave. S.....E-1,J-1
Valley View Pl........J-3
Vincent Ave. N.
...............A-1,C-1
Vincent Ave. S........J-1
Walnut................C-7
Warwick St............D-8
Washburn Ave. N.
...............A-1,C-1
Washburn Ave. S......J-1
Washington Ave. N.
......................A-3
Washington Ave. S.E.
......................C-6
Wayzata...............C-2
Weeks.................B-8
Weenonan Pl...........I-7
Wentworth.........H-3,K-3
Williams..............D-8
Wilson St.............A-7
Winter St.........A-6,A-8
Woodlawn Blvd.
...............I-7,J-6
Yale Pl...............C-3
Yndale Ave. N.........B-3
1st Ave. N............B-4
1st Ave. N.E..........A-5
1st Ave. S....E-3,H-3,K-3
1st St. N.............A-4
1st St. S.............B-5
2nd...................F-2
2nd Ave. N....B-1,B-2,B-4
2nd St. N.E...........A-5
2nd Ave. S........C-4,D-4
2nd Ave. S.E..........B-5
2nd St. N.........A-3,A-4
2nd St. S.............B-4
2nd St. S.E...........B-5
3rd Ave. N....B-1,B-3,B-4
3rd Ave. N.E..........A-5
3rd Ave. S...C-4,D-4,H-4
3rd Ave. S.E..........B-5
3rd St................A-3
3rd St. N.............A-3
3rd St. S.............C-5
4th...................C-5
4th Ave. N........B-1,B-2
4th Ave. N.E..........A-5
4th Ave. S....C-4,E-4,H-4
4th Ave. S.E..........B-5
4th St. N.........A-3,B-3
4th St. N.E...........A-5
4th St. S.........C-5,C-6
4th St. S.E.......B-5,C-8
5th...................C-5
5th Ave. N............B-1-3
5th Ave. N.E..........A-4
5th Ave. S....C-4,E-4,H-4
5th Ave. S.E..........B-6
5th St. N.........A-3,B-3
5th St. N.E...........A-5
5th St. S.........C-4,C-6
5th St. S.E...........B-5
6th...............B-3,C-5
6th Ave. N........B-1,B-3
6th Ave. N.E..........A-4
6th Ave. S.E..........B-6
6th St. N.........A-3,B-4
6th St. S.........C-4,D-6
6th St. S.E...........A-5
7th Ave. N............B-3
7th Ave. N.E..........A-4
7th Ave. S.E..........B-6
7th St. N.............A-3
7th St. S.........C-4,D-6
7th St. S.E...........B-6
8th...................C-5
8th Ave. N............A-1
8th Ave. N.E..........A-4
8th Ave. S.E..........B-6
8th St. S.........C-4,D-7
8th St. S.E...........A-6
9th...................A-3
9th Ave. S............C-5
9th St. S.............C-4
9th St. S.E...........A-6
10th Ave..............E-5
10th Ave. N...........B-3
10th Ave. S...........C-5
10th Ave. S.E.........B-6
10th St. S............C-4
11th Ave..............E-5
11th Ave. S...........C-5
11th Ave. S.E.........B-6
11th St. S............C-4
12th Ave..............E-5
12th Ave. N...........A-2
12th Ave. S...........C-5
12th Ave. S.E.........B-6
12th St. S............C-4
13th..................C-4
13th Ave..............E-5
13th Ave. S...........D-5
13th Ave. S.E.........B-6
14th Ave. N...........A-2
14th Ave. S.......E-5,H-5
14th Ave. S.E.........B-7
14th St. E............C-4
14th St. W............C-3
15th..................D-5
15th Ave. N...........A-2
15th Ave. S...........D-5
15th Ave. S.E.........B-7
15th St. E............D-4
16th..............B-7,C-5
16th Ave. N...........A-2
16th Ave. S...........D-5
16th Ave. S.E.........A-7
16th St. E............D-4
16th St. W............D-3
17th..................C-7
17th Ave. S...........E-5
17th Ave. S.E.........B-7
17th St. E............D-4
18th..................C-7
18th Ave. S.......E-5,K-5
18th Ave. S.E.........B-7
18th St. E........D-4,D-5
19th..................C-7
19th Ave. S.......D-6,F-6
19th Ave. S.E.........B-7
19th St. E............D-5
20th..................K-6
20th Ave. S.......D-6,F-6
20th Ave. S.E.........B-7
21st Ave. S.E.........B-7
21st St. E............D-5
21st Ave. S.......D-6,F-6
22nd Ave. S.......D-6,F-6
22nd Ave. S.E.........B-7
22nd St. E.......D-5,D-7
22nd St. W............D-2
23rd Ave. S.......D-6,F-6
23rd Ave. S.E.........B-8
24th..................B-8
24th Ave. S...D-6,F-6,K-6
24th Ave.S.E..........B-8
24th St. E........D-7,E-5
24th St. W........D-1,D-2
25th..................C-7
25th Ave. S.......D-6,F-6
25th Ave. S.E.........B-8
25th St. E........E-5,E-7
25th St. W............E-2
26th..................C-7
26th Ave. S.......D-6,G-6
26th Ave. S.E.........B-8
26th St. E........E-5,E-7
26th St. W............E-2
27th Ave. S...D-7,G-7,K-7
27th Ave. S.E.....B-8,D-7
27th St. E........E-5,E-7
27th St. W............E-2
28th Ave. S...D-7,G-7,K-7
28th St. E........E-5,E-7
28th St. W............E-2
29th..................B-8,C8
29th Ave. S.......F-7,G-7
29th Ave. S.E.........B-8
29th St. E............E-5
29th St. W............E-2
30th..................B-8,C-8
30th Ave. S.......F-7,G-7
30th Ave. S.E.........B-8
31st Ave. S.......F-7,H-7
31st St. E............F-5
31st St. W............F-2
32nd Ave. S.......F-7,H-7
32nd St. E............F-5
32nd St. W............F-2
33rd Ave. S.......F-7,H-7
33rd St. E............F-5
33rd St. W............F-2
34th Ave. S.......F-7,H-7
34th St. E............F-5
34th St. W............F-2
35th Ave. S.......F-7,H-7
35th St. E............F-5
35th St. W............F-2
36th Ave. S...........F-8
36th St. E............G-5
36th St. W............G-2
36½ St. E.............G-6
37th Ave. S...........F-8
37th St. E............G-5
37th St. W............G-3
38th St. E............G-5
38th St. W........G-1,G-3
38th St. S........F-8,J-8
39th Ave. S...........F-8
39th St. E............G-5
39th St. W........G-1,G-3
40th Ave. S...........F-8
40th St. E............G-5
40th St. W........G-1,G-3
41st..................H-5
41st Ave. S...........F-8
41st St. W........H-1,H-3
42nd E................H-5
42nd Ave. S.......F-8,J-8
42nd St. W........H-1,H-3
43rd W................H-1
43rd Ave. S...........H-8
43rd St. E............H-5
43rd St. W........H-1,K-3
44th W................H-1
44th Ave. S...........H-8
44th St. E........H-5,H-8
44th St. W............H-2
45th W................H-1
45th Ave. S...........H-8
45th St. E........H-5,H-8
45th St. W............H-2
46th W................H-1
46th Ave. S.......H-8,J-8
46th St. E........H-5,H-8
46th St. W............H-2
47th W................I-1
47th St. E........I-5,I-7
47th St. W............I-2
48th W................I-1
48th St. E........I-5,I-7
48th St. W............I-2
49th W................I-1
49th St. E........I-5,I-7
49th St. W............I-2
50th St. E........I-4,I-7
50th St. W............I-2
51st St. E........I-5,I-7
51st St. W........I-1,I-2
52nd W................J-3
52nd St. E........J-5,J-7
52nd St. W........J-1,J-3
53rd St. E........J-5,J-7
53rd St. W........J-1,J-2
54th St. E........J-5,J-7
54th St. W............J-3
55th St. E........J-5,J-7
55th St. W............J-3
56th St. E........J-5,J-7
56th St. W............J-3
57th St. E........J-5,J-7
57th St. W............J-2
58th St. E........K-5,K-8
58th St. W............K-2
59th St. E....K-4,K-6,K-8
59th St. W............K-2
59½ St. E.............K-7
60th St. E............K-4
60th St. W........K-1,K-3
61st St. E............K-4
61st St. W........K-1,K-3

POINTS OF INTEREST

Bryn Mawr Meadows
......................C-2
Dorilus Morrison
Park..................E-4
Dr. Martin Luther
King Park.............H-3
Elliot Park...........C-5
Kenwood Park..........D-1
Loring Park...........C-3
McRae Park............I-5
Minnehaha Park........I-8
Pearl Park............J-4
Powder Horn Lake
Park..................F-5

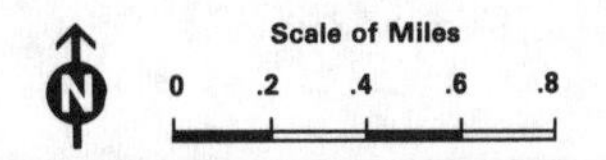
N
Scale of Miles
0 .2 .4 .6 .8

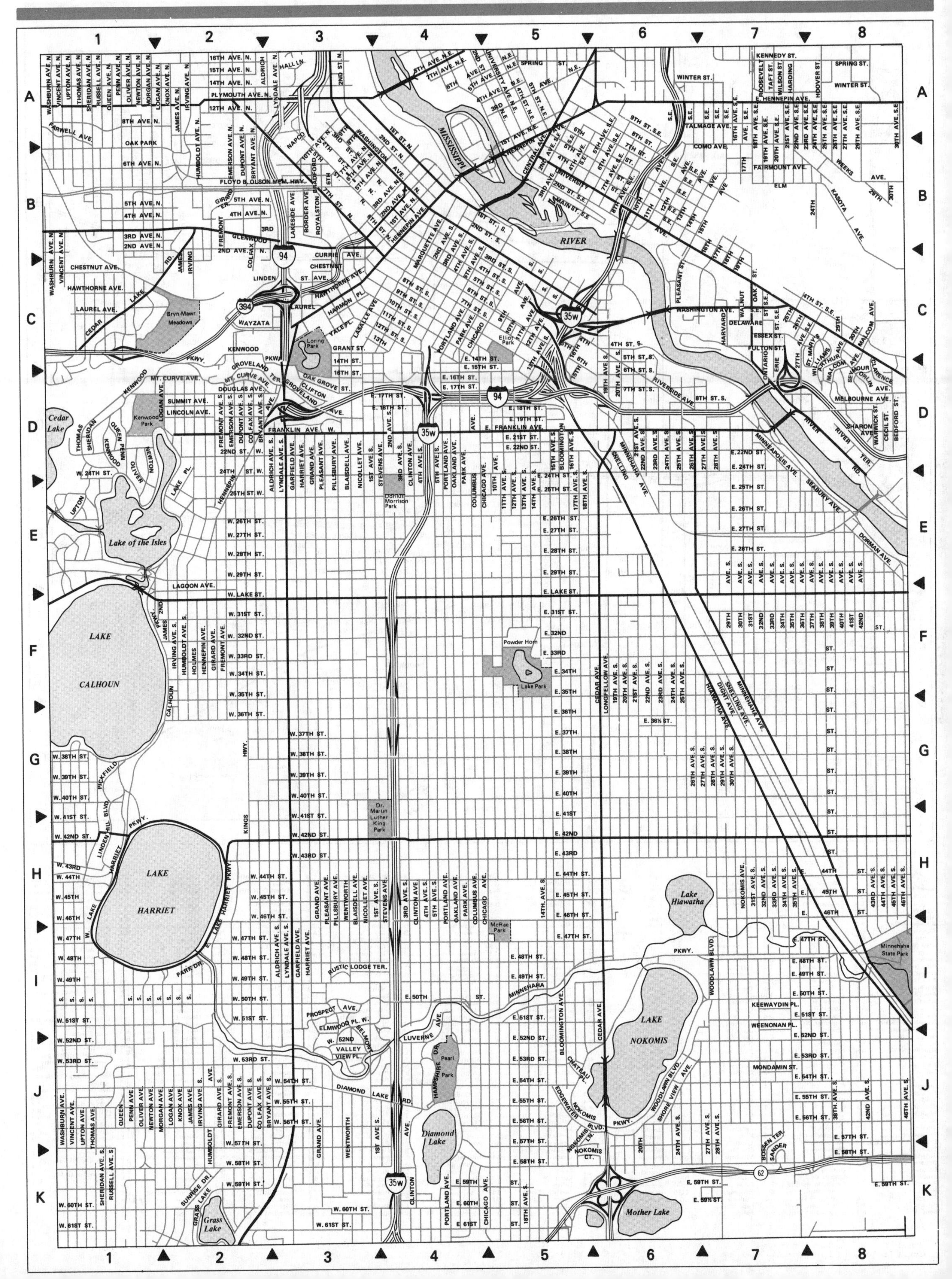
1 2 3 4 5 6 7 8
A B C D E F G H I J K
MISSISSIPPI
RIVER
LAKE CALHOUN
LAKE HARRIET
Lake of the Isles
Cedar Lake
Lake Hiawatha
LAKE NOKOMIS
Mother Lake
Diamond Lake
Grass Lake
Loring Park
Kenwood Park
Bryn-Mawr Meadows
Elliot Park
Powder Horn Lake Park
Dorilus Morrison Park
Dr. Martin Luther King Park
McRae Park
Pearl Park
Minnehaha State Park
94
394
35w
62
HENNEPIN AVE.
NICOLLET AVE.
LYNDALE AVE. S.
FRANKLIN AVE.
W. LAKE ST.
E. LAKE ST.
CEDAR AVE.
HIAWATHA AVE.
MINNEHAHA AVE.
UNIVERSITY AVE.
WASHINGTON AVE.
RIVERSIDE AVE.
E. HENNEPIN AVE.
PORTLAND AVE.
CHICAGO AVE.
BLOOMINGTON AVE.
W. 42ND ST.
E. 42ND ST.
E. 50TH ST.
MINNEHAHA PKWY.
DIAMOND LAKE RD.
W. 61ST ST.

Scale of Miles
0. .1 .2 .3 .4 .5

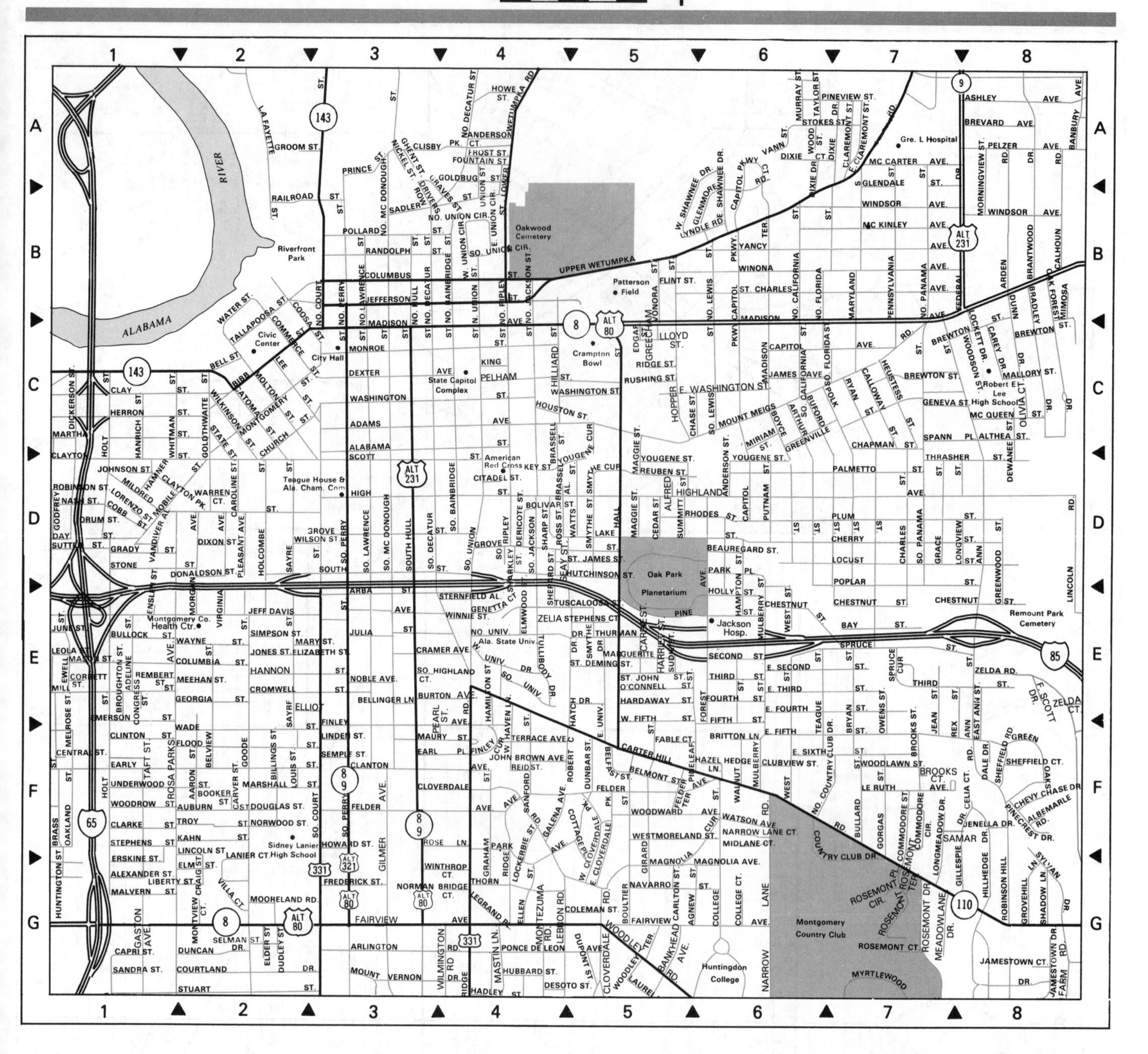

MONTGOMERY

POINTS OF INTEREST

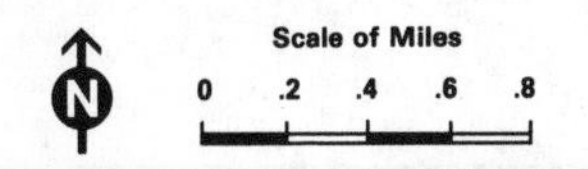
N
Scale of Miles
0 .2 .4 .6 .8

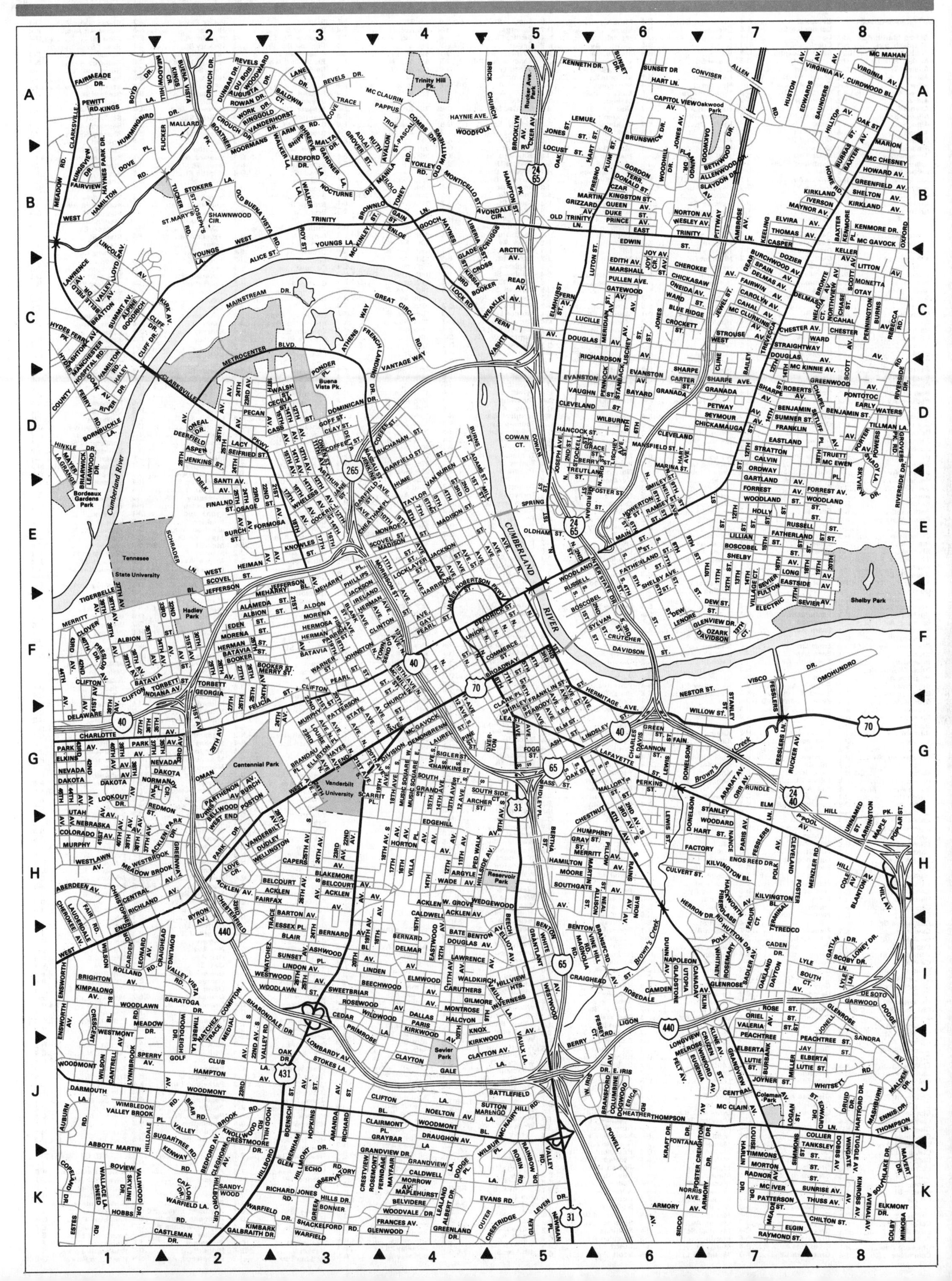
1
2
3
4
5
6
7
8
A
B
C
D
E
F
G
H
I
J
K
Cumberland River
CUMBERLAND RIVER
Tennessee State University
Hadley Park
Bordeaux Gardens Park
Centennial Park
Vanderbilt University
Shelby Park
Trinity Hill Pk.
Rucker Ave. Park
Oakwood Park
Buena Vista Pk.
Reservoir Park
Sevier Park
Coleman Park
Brown's Creek
24 65
265
40
70
65
31
440
431
24 40
CLARKSVILLE PK.
TRINITY LN.
METROCENTER BLVD.
MAINSTREAM DR.
GREAT CIRCLE RD.
VANTAGE WAY
DOMINICAN DR.
JEFFERSON ST.
CHARLOTTE AV.
BROADWAY
WEST END AV.
HILLSBORO RD.
WOODMONT BL.
THOMPSON LN.
GALLATIN RD.
MAIN ST.
WOODLAND ST.
SHELBY AV.
DAVIDSON ST.
HERMITAGE AVE.
LAFAYETTE
NOLENSVILLE
MURFREESBORO
CHESTNUT
CRAIGHEAD
HARDING PL.
BATTLEFIELD
GRANNY WHITE
BELMONT BL.
BLAKEMORE
ACKLEN AV.
WEDGEWOOD
SWEETBRIAR
GLENROSE
PEACHTREE ST.
ELBERTA
WHITSETT
MC MAHAN
VIRGINIA AV.
CAPITOL VIEW
OMOHUNDRO
VISCO
FESSLERS LN.
RUCKER AV.
SOUTHGATE
CLAYTON AV.
KIMBARK
GALBRAITH DR.
CASTLEMAN DR.
ABBOTT MARTIN
WIMBLEDON
VALLEY BROOK
GLENWOOD
GREENLAND DR.
RAYMOND ST.
ELGIN

NASHVILLE

Abbott Martin Rd....J-1
Aberdeen Ave.......H-1
Acklen Ave.........H-2-4
Acklen Park Dr......H-1
Adams St.............D-4
Alameda St...........F-2
Albert Dr.............K-4
Albion St.........F-1,F-2
Aldon..................F-3
Allen Rd...............A-7
Allenwood Dr.........B-7
Alice St...............B-2
Allison................h-5
Alpine..................C-1
Amanda Av............J-3
Ambrose Av...........B-7
Ararat Ave............G-6
Archer St..............G-4
Arctic Av..............B-5
Argyle..................H-4
Arm Rd.................A-3
Armory Ave...........K-6
Arrington..............H-8
Arthur Ave............D-3
Ash St..................G-5
Ashton Av.............C-1
Ashwood Av............I-3
Aspen...................D-2
Athens Way...........C-3
Auburn La.............J-1
Augusta Dr............A-2
Avalon St.............B-4
Avenal Ave............K-8
Avondale Cir..........B-5
Bailey..................C-7
Baldwin Ct............A-3
Barton Ave............H-3
Bass St.................G-5
Batavia.................F-3
Batavia St.........F-1,F-2
Bate.....................I-4
Battlefield.............J-5
Baxter..................B-8
Baxgter Av............B-8
Bayard..................D-6
Bear Rd................J-2
Bedford Av............K-2
Beechwood Av.........I-3
Belcourt Ave..........H-3
Belle Field Dr.........C-1
Bellwood..............H-2
Belvidere..............K-4
Benham Av............K-3
Benjamin..............D-7
Benton..................I-5
Benton Av.............I-4
Bernard Av............I-3
Berry Rd................J-5
Berry St................D-5
Bertha St..............H-5
Bethwood Dr..........B-7
Beuna Vista Pk.......A-2
Blair.....................I-3
Blakemore.............H-3
Blank St................F-3
Blanton.................H-8
Blue Ridge............C-6
Boatner.................A-2
Boensch St............J-3
Booker..............C-4,F-2
Booker St..............F-2
Bornbuckle Ln.........D-1
Boscobel St............E-7
Boview..................K-1
Bowling Av............I-2
Boyd Dr.................A-1
Brandau Pl.............G-3
Bransford Av.....I-6,J-6
Bratton.................C-1
Briarwick..............D-1
Brick Church Pk.......A-5
Brighton Av............I-1
Broadway..............F-5
Brooklyn Av...........A-5
Bronte Av..............C-8
Brownlon...............B-3
Brunswick Dr..........A-6
Buchanan St...........D-4
Burbank Av............J-7
Burch...................G-2
Burch St................E-2
Burchwood Av.........B-7
Burns Av...............G-2
Burns St..........C-8,D-4
Burras St...............B-8
Byron Ave.......H-2,H-6
Caden Dr...............I-1
Cahal Av.......—C-7,C-8
Caldwell...............H-4
Calvin...................D-7
Camden.................I-6
Canaday................I-6
Cannon St..............G-6
Cantrell Av.............J-1
Capers Ave.............H-3
Capitol View Av.......A-6
Carden Av..............I-7
Carolyn Av.............C-7
Carter St................D-6
Caruthers...............I-4
Casper..................B-7
Cass.....................D-3
Castleman Dr..........K-1
Caylor Dr...............K-2
Cecilia St...............D-3
Cedar Ln................I-3
Centennial Blvd........F-1
Central..................J-7
Central Ave............H-1
Chapel Av..............D-8
Charlotte Av...........G-1
Chase St................C-8
Cheatham Pl............E-3
Cherokee Av.......C-6,H-1
Chester.................C-8
Chesterfield...........H-2
Chestnut St............G-6
Chester Av.............C-7
Chickamauga..........D-7
Chickasaw..............C-6
Chilton St..............K-8
Christopher............H-1
Church St...............F-4
Cladwell Ln.............K-4
Clairmont Pl............J-3
Clay St..................D-3
Clayton..................J-4
Clayton Av.............J-4
Cleghorn Av...........K-2
Cleveland..............D-6
Cleveland Ave.........H-7
Cleveland St...........D-5
Cliff Dr..................C-1
Clifton..............F-1,F-3
Clifton Ln...............J-3
Cline.....................C-7
Clinton St...............F-3
Clover...................E-1
Cockrill St..............E-3
Coffee St..........D-3,D-4
Colby....................K-8
Cole Ave...............H-8
Collier...................J-8
Colorado Ave..........H-1
Columbine..............J-6
Combs Dr..............A-4
Commerce St..........F-5
Compton................I-2
Copeland Dr...........K-1
County Hospital Rd..D-1
Cowan St...............D-5
Craighead..............I-1
Craighead St...........I-5
Crescent................I-1
Crestmore Dr..........J-2
Crestridge Dr..........K-5
Crestview...............K-3
Crockett St..............C-6
Cross.....................C-4
Crouch Dr...............A-2
Crutcher St.............F-6
Cruzen St...............J-7
Culvert St...............H-6
Curdwood Blvd........A-8
Czar......................B-6
Dakota Av...............G-1
Dakota Pl................G-1
Dallas....................I-4
Darmouth...............J-1
Davidson................F-7
Davidson St.............F-6
Dayton...................I-7
Deadrick St.............F-4
Deerfield.................D-2
Delaware...............G-1
Delk Av..................E-2
Delmar...................I-4
Delmas...................C-7
Delmas Av..............C-7
Delta......................D-3
Demonbrenn St.........G-4
Dew.......................F-6
Dew St....................F-7
Division..................G-4
Doak Av..................D-1
Dobbs Av................J-8
Dodge.....................I-8
Dodge Pl.................K-4
Dogwood.................J-6
Dominican Dr...........D-3
Donald St................B-6
Donelson................G-6
Douglas Av......C-5,C-7,I-4
Dove Pl...................B-1
Dozier....................B-7
Draughon Av............J-4
Druid Dr..................J-8
DuBois....................A-2
Dudley....................H-2
Duke St...................B-6
Dunbar Dr................A-2
Dunn Av...................I-6
Early.......................D-8
Eastland Av..............D-7
Eastside Av..............E-7
East Trinity Ln..........B-6
Eastwood.................I-4
Eden St....................F-2
Edith Av...................C-6
Edwards Av..............A-8
Edwin St..................B-6
Elberta....................J-7
Eletric Av.................F-7
Elgin.......................K-7
Elkins Av..................G-1
Elkmont Dr...............K-8
Elliot Av...................I-5
Elliston Pl................G-3
Elm St......................G-5
Elm Hill....................G-6
Elmhurst St...............C-5
Elmwood..................I-2
Elvira Av..................B-7
Enloe.......................B-4
Ennis Dr...................J-8
Ensworth..................I-1
Ensworth Av..............I-1
Essex Pl....................H-3
Estes Rd...................K-1
Eugenia Av................J-5
Evans Rd..................K-5
Evanston Av.......D-5,D-6
Factory.....................H-6
Fadur Ct...................H-7
Fain.........................G-6
Fairfax Ave................H-2
Fairmeade Dr.............A-1
Fair Rd......................H-1
Fairview....................B-1
Fairwin Av.................C-7
Fatherland St......E-6,E-7
Felicia St...................G-2
Felt Av......................J-6
Fern Av.....................C-5
Ferndale....................K-3
Fessey Ct...................I-6
Fesslers.....................F-7
Fiberglass Rd..............H-7
Frances Ave................K-4
Finland St...................E-2
Fisk St.......................F-3
Flamingo Dr................B-6
Flicker.......................A-2
Forest Av............E-7,E-8
Formosa.....................E-2
Foss St......................G-5
Foster Ave..................H-7
Foster St....................E-6
Foster Creighton
Dr..............................K-6
Franklin Av..................D-7
Franklin St..................F-5
French Landing Dr.......C-3
Fresno........................B-6
Fulton.........................F-7
Gain St.......................A-4
Galbraith Dr................K-2
Gale Ln.......................J-4
Gardner Ln..................B-3
Garfield St...........D-4,E-3
Gartland Av.................D-7
Gatewood Av...............C-6
Gatlin Dr.....................I-8
Gay............................F-4
Gear...........................C-7
Georgia.......................F-2
Gilmore.......................I-4
Glade.........................B-4
Gladstone....................I-6
Glen Echo Rd...............K-3
Glen Leven Dr..............K-5
Glenrose Av...........I-7,I-8
Glenview Dr.................F-7
Glenwood....................K-3
Goff St........................D-3
Golf Club Ln.................J-2
Gooch.........................B-4
Goodrich.....................C-1
Gordon Terr.................B-6
Grace..........................D-6
Gray St........................H-5
Granada......................B-4
Granada Av.................D-7
Grand St.....................G-4
Grandview Av..............J-7
Grandview Dr...............K-3
Grantland....................I-5
Graybar Ln...................J-3
Great Circle Rd.............C-4
Green St......................I-5
Greenfield Av...............B-8
Green Hills Dr...............K-3
Greenland Dr...............K-4
Greenway....................H-2
Greenwood Av..............D-8
Grizzard Av..................B-5
Grove, W.....................H-4
Grover St.....................A-3
Grovers Pk. Rd.............D-8
Hackworth...................H-7
Halcyon.......................I-4
Hamilton Ave................H-5
Hamilton Rd.................C-1
Hampton Av.................J-2
Hampton St..................B-5
Harlin Dr.....................K-7
Harrison St..................E-4
Hart Ave......................D-6
Hart Ln........................A-6
Hart St...............B-6,H-6
Hartford Dr..................J-1
Hawkins St..................G-4
Haynes Park Dr.............B-1
Haynes St............B-4,G-3
Herman.......................I-4
Herman St...........F-2,F-3
Hermitage Ave..............F-6
Hermosa......................F-3
Herron Dr.....................H-6
Hill Ave........................H-8
Hillboro.......................K-2
Hillboro Cir..................K-2
Hillmont Dr...................K-3
Hillside Ave..................H-4
Hilltop Av......................A-8
Hillview Hts...................K-3
Hobbs Rd.......................K-1
Holly St..........................E-7
Home Rd........................B-8
Home St.........................D-4
Hoodhill Rd.....................J-2
Hopkins St......................J-3
Horton Ave......................H-4
Houston Av.....................A-7
Howard Av......................B-8
Howerton St....................E-6
Hummingbird Dr...............A-1
Humphrey.......................H-5
Hutton Dr........................I-7
Hydes Ferry Pk.................C-1
Hydes Ferry Rd.................C-1
Indiana Av.......................F-1
Interstate Dr....................E-6
Inverness.........................I-5
Ireland St.........................F-3
Iris Dr., W.......................J-5
Iris Rd., E.......................J-6
Iverson............................B-8
Jackson St...............E-3,E-4
James Robertson
Pkwy...............................F-4
Jane St.............................E-3
Jefferson..........................E-2
Jefferson Blvd...................E-3
Jenkins Ave......................D-3
Jenkins St.........................D-2
Jewel St...........................C-7
Johnston Ave....................F-3
Jones Av....................A-6,C-6
Jones St...........................A-5
Joseph Ave.......................B-5
Joy Av..............................B-6
Joy Cir..............................C-6
Joyner St...........................J-7
Keeling.............................B-7
Kenmore...........................B-8
Keller Av...........................B-8
Kenmore Dr.......................B-8
Kenneth Dr........................A-5
Kenway..............................B-4
Kilvington Blvd...................H-7
Kimbark.............................K-2
Kimpalong Ave.....................I-1
Kings Cr..............................A-2
Kings Ln..............................A-1
Kingston St..........................B-6
Kingsview Dr........................B-1
Kinross Ave..........................C-2
Kirk Av...............................A-3
Kirkland.............................B-8
Kirkland Av..........................B-8
Kirkwood Av...............I-4,J-4
Kissia..................................C-4
Klin......................................I-7
Kline Av...............................I-7
Knollwood.............................J-2
Knowles................................E-3
Knox.....................................I-3
Kraft Rd.................................K-6
Lacy......................................D-2
Lafayette St...........................G-6
Lane.......................................K-3
Lauderdale..............................H-1
Laurel St..................................G-4
Lawrence..................................I-4
Lawrence Av..............................I-1
Le St...G-5
Lealand Ln.................................K-4
Leawood Dr................................D-1
Ledford Dr.................................B-3
Lemuel Rd..................................A-5
Lenore St...................................F-6
Leonard Av.................................I-1
Lewis St......................................G-6
Liberia St.....................................B-4
Ligon...I-6
Lillian St......................................E-7
Lincoln Av.....................................B-1
Linden Av......................................I-5
Lindon Av......................................I-3
Lindsley St.....................................G-5
Lischey Ave.....................................D-6
Litton Av...C-7
Lloyd Av..C-1
Lock Rd...C-7
Locklayer St......................................E-4
Locust St..B-5
Logan St...J-7
Lombardy Av......................................J-3
Loney Dr...I-8
Long Av...E-7
Longview...J-6
Lookout Dr...G-1
Louise Ave..G-3
Louise Dr...J-7
Love Cir...H-2
Lucille..C-5
Lutie St..J-7
Luton St...C-6
Lyle Ln..I-7,I-8
Lynnbrook Rd.......................................I-1
Madison St...E-4
Magnolia Rd...I-5
Main St...E-6
Malden Dr..J-8
Mallory St..G-6
Malta Dr...A-3
Manchester...C-1
Manila..B-4
Mansfield St..D-6
Maple...H-8
Maplehurst..K-4
Marina St..D-6
Marion...A-8
Marion Av...J-4
Marshall St..C-6
Martin..B-5
Martin St...H-5
Mashburn..J-8
Mavert Dr...K-8
Mayer..D-1
Mayfair..K-4
Maynor Av..B-7
McChesney...B-8
McClain..J-7
McClurkins Av.......................................C-7
McEwen...D-8
McGavock...B-8
McGavock St...G-4
McIver..K-7
McKinley St..B-3
McKinnie Av..C-7
McMahan..A-8
McMillin..F-4
McNairy..K-6
Mead Ave..K-7
Meadow Dr..I-1
Medow Rd...B-1
Meadowbrook..H-1
Meadow Hill..A-2
Medial..I-2
Meharry..F-2
Meharry Pl..E-3
Melrose..J-6
Menzler Rd...H-8
Meridian St....................................C-6,E-5
Merritt Ave...H-5
Merry St...F-2
Mill St...E-4
Miller St...J-7
Mimosa..K-8
Monticello St...B-4
Montrose..J-3
Moore..H-5
Morena..F-2
Morena St...F-2
Morrison St...E-4
Morrow Av..K-4
Mortons St..K-7
Murphy Rd..H-1
Murphy St...G-3
Music Square E......................................G-4
Music Square W.....................................G-4
Napoleon..I-6
Nassau St...D-3
Matchez Trace.......................................I-2
Neal..H-6
Nebraska..A-1
Neldia Ct..C-8
Nestor St..F-6
Nevada...G-1
Nevada Av..G-1
Newman Pl..K-5
Niel Ave...E-6
Nocturne Dr..B-3
Noelton Av...J-4
Norton Av..B-6
Northview Av...C-8
Normandy Cir..G-1
Oak Dr..J-3
Oak St...................................A-8,B-5,G-5
Oakland..I-7
Oakwood Av..A-7
Observatory...K-3
Old Buena Vista
Rd...B-2
Oldham St...E-5
Old Matthews...B-4
Old Trinity Ln...B-5
Oman...G-2
Omohundro...F-8
Oneal Dr...D-2
Oneida Av...C-6
Ordway Pl...D-7
Oriel St..I-7
Orr Ave...G-6
Osage...E-2
Otay...C-8
Overton...G-5
Owen St..D-3
Oxford..B-8
Ozark...F-7
Paris...I-4
Paris Ave..H-7
Parish St...F-3
Park Av...G-1
Parthenon Av...G-2
Patterson St..................................G-3,I-6
Peachtree...I-7
Peachtree St...J-8
Peabody St...G-5
Pearl St...F-3,F-4
Pecan...D-2
Pennington...C-8
Pennock..D-6
Perkins St..G-6
Petway Av...D-7
Pewitt Rd...A-1
Phillips St...E-3
Pillow St...H-6
Pine St..G-4
Pittway...B-6
Plum St...B-8
Polk Ave...H-7
Ponder Pl..C-3
Pontotoc..D-8
Poplar St...H-8
Porter Rd...D-8
Poston...G-2
Powell Av...J-6
Powers...D-8
P'Pool Ave...H-1
Preslor Dr..F-1
Primrose..I-3
Prince Av..B-6
Pullen Ave...C-6
Queen Av..B-6
Radnor St..K-7
Rainbow...D-6
Rains Ave...H-5
Ramsey St..E-6
Raymond St..G-5
Read Av..C-5
Rebecca Rd...C-8
Redmon St..G-1
Red Walk..H-4
Revels Dr..A-2
Richard St...J-3
Richard Jones Rd..K-5
Richardson Av..C-5
Richland Ave..H-1
Ridley Pl...G-5
Ringgold...A-2
River Dr...D-1
Riverside Dr...D-8,E-8
Robin Rd...K-5
Roberts...F-2
Rolland Rd..I-1
Roscobel St...F-5
Rose St..I-7
Rosedale Av...I-6
Rosemary..I-7
Rosemont..K-3
Rosewood Av..I-3
Rowan Dr..A-2
Roy St..B-3
Rucker Ave..A-5,G-6
Russell St..E-5,E-7
Ruth..A-3
Sadler Av...I-7
St. Edward Dr...J-8
Santi Av...D-2
Saratoga...I-2
Saunders Av...A-3
Scarrt Pl..G-3
Scott Ave..C-8
Scovel St..E-2,E-4
Scruggs..B-5
Setliff Pl...D-8
Sevier...E-7,F-7
Seymour Av...D-7
Shackelford Rd...K-3
Shady Ln..D-8
Sharondale Dr..I-2
Sharpe Av..D-6,D-7
Shelby Ave..E-6,E-7
Shelton Av..B-8
Shifried St...D-2
Shipp..A-3
Shirley St..G-5
Shreeve Ln..A-3
Sidco Dr..K-6
Sigler St..G-4
Simmons St..K-7
Skyline Dr..K-1
Skyview Dr...D-8
Slaydon Dr...B-7
Smiley St...E-6
Sneed Rd...K-1
South Ct...I-7
South St...G-4
Southgate Ave...H-5
Southlake Dr...K-8
South Side Ct...G-4
Spain Av..C-7
Sperry...J-1
Spring St...E-5
Stanback Av...D-6
Stanley St..F-7
State St...E-3
Stockell St...D-5
Stokers Ln...B-3
Stratton...D-7
Strokes Ln..J-3
Strouse Av...C-7
Sugartree Rd..J-1
Summit Av...C-1
Sumner St..D-7
Sunrise Ave..K-8
Sunset Dr...A-6
Sunset Pl..I-3
Sutton Hill Rd...J-5
Sweetbriar Av..I-3
Sylvan St..F-6
Tanksley..J-7
Taylor St...E-4
Terminal Blvd..H-7
Thomas Av...B-7
Thomson Ln...J-6,J-8
Thuss Ave..K-8
Tigerbelle..F-1
Tillmania...D-8
Timber Ln...I-2
Timons St..K-7
Toney Rd...B-4
Torbett..F-2
Torbett St...F-1
Town Send Dr..F-4
Trace..H-2
Tredco..I-7
Treutland St...D-5
Trevecca...C-7
Truett...D-8
Tucker Rd...B-2
Tuggle Av...J-8
Unamed St..H-8
Union St..F-4
Utah Av..G-1
Utopia..I-6
Valeria St...I-7
Valiwood Dr..K-1
Valley Av...C-1
Valley Brook Pl...J-1
Valley Brook Rd...J-2
Valley Vista Rd...I-2
Van Buren St...D-4
Vanderbilt...H-2
Vanderhorst Dr...A-2
Vantage Way..C-5
Vashti...C-5
Vaughn St..D-5
Vaulx Ln...I-5,J-5
Venture Cir..C-4
Villa Pl...H-4
Village Ct..F-7
Vine Hill Rd..I-5
Virginia Av..A-8
Visco Dr..F-7
Wade Ave..H-4
Waldkirch...I-4
Walker Ln...A-3,B-3
Wallace Ln...K-1
Walsh St...D-3
Ward..C-8
Ward St..C-6
Warfield..K-3
Warfield Dr..K-2
Warfield Ln..K-2
Warner St...F-3
Waters..D-8
Weakley Av..C-5
Wedgewood..H-4
Westmont..I-1
Wellington..H-2
West...C-7
Westend...H-2
West End Ave...G-3
West End Dr..I-1
West Hamilton Rd...B-1
West Heiman St...E-2
Westlawn Av..H-1,I-2
Westley Av..B-6
West Trinity Ln..B-3
Westwood Av..I-2
Westwood...I-5
Wharf Ave...G-6
White Av...I-5
Whitney..I-7
Whitsett Rd...J-8
Wilbur Pl...K-4
Wilburn..D-6
Wildwood..I-3
Willow St...G-6
Wilson Bl..I-1,J-1
Wimbleton Rd..J-1
Winford Av...J-6
Wingate Ave...J-8
Woodfolk...A-4
Woodhill Dr..B-8
Woodland St..E-5,E-7,E-8
Woodlawn Dr...I-1
Woodleigh Dr..I-2
Woodmont Blvd.
....J-1,J-2,J-4
Woodvale Dr...K-4
Woodward...A-2,H-7
Work Dr..A-2
Yokley Rd..B-4
Youngs Ln...B-2,B-3
1st Ave. N..E-4
1st Ave. S...F-5
1st St..F-5
1st St. N..E-5
2nd Ave. N...E-4
2nd Ave. S...F-5,G-6
2nd St..F-5
2nd St. N..D-5,E-5
3rd..C-7
3rd Ave..E-4
3rd Ave. S..F-5,G-6
3rd St. S...F-6
4th Ave. N..E-4
4th Ave. S...F-5,G-6
4th St. S..........E-6
5th Ave. N..........E-4
5th Ave. S..........F-5
5th St..........D-6
5th St. S..........E-6
6th Ave. N..........E-4
6th Ave. S..........F-5
6th St..........D-6
6th St. N..........E-6
6th St. S..........E-6
7th Ave. N..........E-4
7th Ave. S..........F-5
7th St. S..........E-6
8th Ave. N..........E-4,E-6
8th Ave. S..........F-5
8th St..........E-6
9th Av..........I-4
9th Ave. N..........E-4,E-6
9th St..........E-6
10th Av..........I-4
10th Ave. N..........E-4
10th St..........E-7
11th Av..........D-3,H-4,I-4
11th Ave. N..........E-4
11th St..........E-7
12th Ave..........D-3,E-3,H-4
12th Ave. S..........G-4
12th St..........C-7,D-7,E-7
13th Av..........E-3
13th Ave. S..........G-4
13th Ct..........F-7
13th St..........E-7
14th Ave..........D-3,E-3,H-4
14th St..........D-7,E-7
14th Ave. S..........G-4
15th Av..........D-3,E-3,H-4
15th Ave. N..........F-4
15th Ave. S..........G-4
15th St..........E-7
16th Ave..........E-3,H-4
16th St..........E-7
17th Ave..........E-3,H-4
17th St..........F-7
18th Av..........D-3,E-3,J-3
18th Ave. N..........F-3,F-4
18th Ave. S..........G-4,H-4
18th St..........D-8,E-8
19th Ave..........H-3
19th Ave. N..........F-3
19th Ave. S..........G-4,H-4
19th St..........E-8
20th Ave..........I-3
20th Ave. N..........F-3
20th Ave. S..........G-3
20th St..........E-8
21st Ave..........E-3,F-3
21st Ave. N..........F-3
21st Ave. S..........G-3
22nd Av..........E-2,H-3
22nd Ave. N..........F-3
22nd Ave. S..........J-2
23rd Av..........D-2,E-2,H-3
23rd Ave. N..........G-3
23rd Ave. S..........J-2
24th..........D-2,F-2
24th Ave..........D-2,E-2,G-3
24th Ave. S..........G-3,H-3
25th..........F-2
25th Ave..........D-2,E-2
..........F-2,H-3,I-3
26th..........F-2,G-2
26th Ave..........D-2,H-3,I-3
27th..........F-2,G-2
28th..........F-2
28th Ave..........D-2,G-2,G-3
29th..........F-2
30th..........F-2
31st Ave..........F-2,G-2
32nd Ave..........F-2
33rd Ave..........F-2
34th Ave..........F-1
35th Ave..........F-1
35th..........G-1
36th..........G-1
37th Ave..........E-1,G-1,H-1
38th Ave..........F-1,G-1,H-1
39th..........F-1,G-1,H-1
40th Ave..........F-1,G-1,H-1
41st Ave..........G-1,H-1
42nd Ave..........F-1,G-1
43rd Ave..........F-1,G-1
44th Ave..........F-1,G-1
45th Ave..........G-1

NEWARK

Abington Ave., E..........B-6
Abington Ave., W..........A-4
Academy St..........F-5
Adams St..........H-7
Afton..........C-8
Alexander St..........E-1
Alpine St..........J-1,K-1
Alpine St., E..........I-4
Alpine St., W..........I-4
Algea..........H-8
Amherst St..........C-1
Ampere Pkwy..........B-4
Ann St..........H-8
Argyle..........G-1
Arlington..........G-6
Arlington, N..........B-3
Arlington Ave., S..........D-2
Ashland..........B-2
Astor St..........I-5
Austin St..........I-5
Avenue A..........I-5
Avenue B..........I-6
Avenue C..........I-6
Avon Ave..........H-2
Badger Ave..........I-3
Baldwin Ave..........H-3
Ballantine..........B-6
Bank St..........F-6
Barclay St..........H-5
Bayview Ave..........J-1
Beacon St..........G-5
Beardsley Ave..........A-4
Bedford..........C-1,G-4
Beech St..........C-2
Belgrove Dr..........D-8
Belmont..........A-5
Bergen Ave..........C-8
Bergen St..........F-8,G-4,K-2
Berkeley..........G-1
Berkeley Ave..........A-5
Berwyn..........B-1
Beverley St..........H-1
Bigelow St., E..........I-4
Bigelow St., W..........I-3
Bleeker St..........F-6
Bloomfield Ave..........C-6
Bock Ave..........I-2
Boston St..........F-5
Boyd St..........H-4
Boyden St..........E-6
Boylan..........E-1
Bragaw Ave..........I-1
Branford..........I-4
Branford Pl..........G-6
Breckenridge..........F-1
Brenner St..........G-3
Bridge St..........E-8
Brighton..........C-8, G-1
Briley Ave..........K-1
Broad St..........D-7,G-6
Broadway..........C-7
Brockside..........J-1
Brookwood St..........E-1
Broone St..........G-5
Bruce St..........F-4
Bruen St..........G-7
Brunswick St..........H-5
Burnet, N..........B-2
Burnet St., S..........C-1
Cabinet St..........E-4
Calumet..........I-7
Camden..........G-4
Camp St..........H-6
Carlton St..........B-2
Carnegie Ave..........C-1
Central Ave..C-1,E-4,E-8
Center St..........A-1,F-7
Chadwick Ave..........J-3
Chancellor Ave..........F-5
Charlton St..........H-4
Chelsea Pl..........A-1
Chester Ave..........B-7
Chestnut..........C-2
Chestnut St..........H-6
Clark Ave..........D-8,K-1
Clark St..........D-7
Clay St..........E-7
Clifford St..........I-7
Clifton Ave..........D-6
Clifton St..........I-4
Clinton Ave..........H-2,H-4
Clinton Pl..........J-1
Clinton St..........F-6
Clinton St., N..........B-2
Clinton St., S..........C-1
Clover St..........G-8
Columbia St..........J-7
Commerce St..........F-6
Concord St..........B-2
Congress St..........H-7
Conklin Ave..........K-1
Court..........G-5
Crane St..........D-6
Crawford..........C-1
Crawford St..........H-5
Crescent Ave..........J-1
Cross St..........E-8
Custer Ave..........J-2
Cutler St..........D-6
Cypress St..........H-1
Davenport Ave..........A-5
Davis..........D-8
Dawson St..........I-6
DeLancy St..........I-7
Delavan Ave..........A-7
Delmar..........G-1
Devon..........D-8
Dewey St..........J-1
Dickerson St..........E-4
Division St..........E-7
Dorer Ave..........K-1
Downing St..........G-8
Duryea..........F-5
Eagles..........E-6
Earl St..........I-5
Eastwood..........A-2
Edgar St..........D-1
Edgerton Terr..........A-3
Edison..........A-3,A-5
Edison Pl..........G-8
Elizabeth Ave..........I-4,K-2
Ellington St..........B-4
Elliott..........A-5
Ellis Ave..........G-1,H-1
Elm St..........A-1,G-6
Elmwood Ave..........D-1
Elwood Ave., E..........A-7
Emmet St..........I-5
Empire..........J-4
Essex St..........F-8
Evergreen..........C-1
Fabyan Pl..........I-1
Fairmount..........G-4
Fairmount Terr..........E-1
Fairview Ave..........G-4
Farley Ave..........H-3
Ferguson St..........H-8
Ferry St..........G-7
Franklyn..........G-6
Freeman Ave..........E-1
Frelinghuysen Ave..........K-4
Fuller Pl..........H-7
Fulton..........F-7
Garden..........H-7
Garden State Pkwy.
..........C-3,D-2,E-2
Garrison St..........I-8
Garside St..........D-6
Gillett..........H-5
Glenwood Ave..........B-1
Goble St..........I-6
Goldsmith Ave..........J-1
Goodwin Ave..........J-2
Gotthart St..........I-8
Gould Ave..........D-4
Gouverneur..........D-7
Grafton Ave..........A-7
Grand Ave..........E-1
Grant..........E-7
Grant Ave..........D-8
Gray St..........D-4
Green St..........G-6
Greenwood Ave..........C-3
Grove St..........F-2,H-11
Grove St., N..........A-4,C-3
Grove St., S..........D-3
Grove Terr..........F-1
Grumman Ave..........K-1
Halleck St..........A-8
Halsey St..........G-6
Halstead St..........C-1
Hamilton Ave..........A-2
Hamilton St..........G-7
Hampton Terr..........B-1
Harding Terr..........J-2
Harrison Ave..........E-8
Harrison Pl..........H-1
Harrison St., S..........C-1
Harper St..........I-6
Hartford St..........E-5
Harvey St..........C-7
Hawthorne..........D-2
Hawthorne Ave...I-1,J-4
Hayes St..........F-4
Heckel St..........D-1
Hecker St..........E-5
Hedden Terr..........I-3
Heller Pkwy..........A-7
Hensler..........H-8
Herbert Pl..........B-7
Hickory St..........B-1
Highland..........G-1
Highland Ave....C-6,D-8
High St.....D-6,F-6,H-5
Hill St..........G-6
Hillside Ave..........I-4
Hinsdale Pl..........B-7
Hobson St..........J-1
Hoffman Blvd..........A-4
Holland St..........G-3
Hollywood..........D-3
Hopkins Pl..........H-1
Hose..........D-8
Houston St..........I-8
Howard St..........G-5
Hoyt St..........D-8
Hudson St..........K-5
Humboldt..........D-4
Hunter St..........I-4
Hunterdon St..........J-3
Huntington Terr..........J-2
Irving..........J-2
Irving Turner
Blvd..........H-4
Isabella Ave..........F-1
Jabez St..........I-8
Jacob St..........G-3
James St..........E-6
Jay St..........E-5
Jefferson St..........H-7
Jelliff Ave..........I-3
Jersey St..........E-8
John St..........E-8
Johnston Ave..........D-8,I-4
Johnson St..........H-6
Jones St..........F-5
Kearney Ave..........D-8
Kearney St..........C-6
Keer Ave..........J-1
Kellor St..........F-7
Kent St..........G-3
Kinney St., E..........H-6
Kinney St., W..........G-4
Lafayette St..........G-6,H-7
La France..........A-4
Lake St..........C-6
Lang St..........H-8
Lehigh Ave..........J-2
Lenox Ave..........C-2
Leslie St..........B-4,J-1
Liberty St..........G-7
Lincoln Ave..........B-7
Lincoln St..........B-2,G-5
Linden Ave..........D-1,G-1
Littleton Ave..........G-3
Livingston St..........H-4
Lock St..........F-5
Lombardy..........F-7
Longworth..........H-5
Lyons Ave..........J-1
Madison..........G-8
Madison Ave..........G-1
Magnolia St..........G-3
Main St..........B-1,C-3
Malvern St..........H-7
Mapes Ave..........J-2
Maple Ave..........H-1,K-1
Maple Ave., N..........A-4,C-3
Maple Ave., S..........D-3
Maple St..........D-8
Market St..........F-6
Market St., W..........E-4,F-5
Marshall..........D-8
May St..........B-7
McCarter Hwy..........D-7,H-6
McNary..........D-8
McWhorter St..........H-7
Meeker Ave..........J-3
Melmore Gardens..........B-2
Melrose..........F-1
Mercer St..........G-5
Merchant St..........H-8
Middlesex St..........F-8
Milford Ave..........I-4
Miller St..........B-1,I-5
Millington Ave..........I-2
Milton St..........E-8
Mohammad Ali Ave..H-5
Monroe..........G-8
Montgomery..........G-5
Montgomery Ave..........H-5
Montrose..........F-2
Morris..........A-3,E-5,F-4
Morton..........G-5
Mountainview Ave..........E-1
Mt. Prospect AVe..........D-6
Mulberry St..........H-6
Munn Ave., N..........C-3
Munn Ave., S..........E-2
Murray St..........H-5
Myrtle..........D-4
Myrtle Ave..........F-1
Nairin Pl..........I-2
Napoleon St..........I-8
Nesbitt Ave..........E-6
Nevada St..........G-6
New St..........A-1,E-5,F-6
Newfield St..........B-4
New York Ave..........H-8
Newark St..........F-5
Nobel..........K-4
Norfolk St..........F-5
Norwood St..........D-1
Nursery..........C-7
Nye Ave..........H-1
Oak Ave..........F-2
Oak St..........D-1
Oakwood Ave..........B-1
Oakwood Pl..........B-1
Oliver St..........H-7
Orange St....D-4,E-6,F-5
Orange Ave., S.
..........E-1,F-4
Oraton St..........B-7
Oriental St..........C-7
Osborne Terr..........J-2
Park Ave..........A-1
Park Pl..........A-1,F-7
Park St..........A-1
Parker St..........C-6
Parkhurst St..........H-6
Parkview Terr..........K-1
Patterson..........H-8
Peabody..........B-8
Peddle St., E..........J-4
Peddle St., W..........I-3
Pennington St..........H-6
Peshine Ave..........J-3
Poinier St..........I-5
Polk St..........H-8
Pomona Ave..........J-1
Prince St..........H-4
Prospect St..........B-2,H-7
Pulaski St..........I-8
Quitman St..........H-5
Randolph..........I-5
Rankin..........G-5
Raymond Blvd...F-6,G-8
Raymond Plaza E....G-7
Raymond Plaza W....G-7
Renner Ave..........J-2
Richmond..........F-5
Ridge St..........D-6
Ridgewood Ave..........I-4
Rhode Island Ave....D-1
Roosevelt Ave..........A-4
Rose St..........G-3,H-4
Rose Terr..........H-3
Roseville Ave..........D-4
Rutledge Ave..........A-3
Runyon St., E..........I-4
Runyon St. W..........I-3
St. Agnes Lane..........D-2
Sanford St..........C-1
Saybrook..........F-7
Scheerer Ave..........J-2
Schley St..........J-1
Seymour Ave..........J-2
Shanley Ave..........H-3
Shepard..........D-1
Shephard Ave..........J-2
Sherman Ave..........D-8,I-5
Shipman..........G-5
Snyder..........A-1
Somerset St..........F-8,H-5
South St..........H-6
Springdale Ave..........A-2
Springfield Ave..........G-2
Spruce St..........G-4
Standard..........G-2
Stanton St..........I-5
State St..........A-1,E-6
Stengel..........J-2
Sterling St..........F-5
Stockton..........C-3
Stockton Pl..........F-1
Stone St..........D-6
Stratford Pl..........H-4
Stueben..........D-3
Summit..........B-2
Summit St..........F-6
Sunnyside..........E-1
Sunset Ave..........E-1
Sussex..........E-8
Sussex Ave..........D-4
Taylor..........C-6
Telford St..........D-1
Thomas St..........H-6
Tichenor..........F-1
Tichenor St..........H-6
Tillinghast St..........I-2
Treacy Ave..........H-3
Tremont..........D-1,G-1
Union St..........G-7
University..........F-6
University Ave..........G-6
Van Buren St..........H-8
Vanderpool St..........I-5
Vanness..........I-2
Vassar Ave..........J-1
Vernon..........H-2
Vernon Terr..........C-3
Vermont Ave..........F-1
Victoria..........J-4
Wainwright St..........J-1
Wall..........H-8
Wallace..........A-1
Walnut St..........G-6
Walnut St., N..........A-3,B-2
Ward St..........A-1
Warren St..........E-8,F-5
Warrington Pl..........B-3
Warwick St..........B-4
Washington St..........A-1
..........F-6,G-6
Washington Terr..........B-2
Watson..........J-3
Watson Ave..........D-2
Waverly Ave..........G-3
Webster Pl..........B-1
Webster St..........D-6
Webster Pl..........B-1
White Terr..........I-2
Wickliffe St..........H-5
William St.....B-1,E-8,G-6
Williamson Ave..........K-1
Willow St..........J-1
Wilsey St..........F-5
Wilson..........H-8
Wilson Ave..........C-8
Winans Ave..........G-3,K-1
Winans St..........C-2
Woodland..........G-2
Woodland Ave..........A-2,D-8
Woodside..........B-7,C-7
Wright St..........I-5
Yates Ave..........I-3
1st Ave. W..........B-5
1st St..........D-5,F8
2nd Ave. E..........C-6
2nd Ave. W..........B-5
2nd St..........D-5,F-8
2nd St., N..........E-8
3rd Ave. E..........C-6
3rd Ave., W..........B-4
3rd St..........B-6,D-5,F-8
4th Ave..........B-3
4th Ave., E..........D-6
4th St..........C-5,D-5,F-8
5th St..........F-8
5th St., N..........C-5
6th Ave., E..........D-6
6th Ave., W..........C-3
6th St., N..........B-5
7th Ave., E..........D-6
7th Ave., W..........D-4
7th St., N..........B-5
8th..........A-5
8th Ave..........E-6,E-7
9th..........A-5
9th Ave..........D-3
10th St., N..........B-5
10th St., N..........B-3
11th Ave..........E-4
12th Ave..........E-4
12th St., N..........B-4
13th Ave..........E-2,E-3,E-5
14th Ave..........F-3
14th St., S..........D-4,H-2
15th Ave..........F-3
15th St., N..........A-5,C-4
15th St., S..........H-2
16th Ave..........F-3
16th St., S..........H-2
16th St., N..........A-5,C-4
17th Ave..........F-2,G-4
17th St..........H-2
17th St., N..........A-4,C-4
18th Ave..........G-3
18th St..........F-1
18th St., N..........C-4
19th Ave..........G-2
19th St., N..........C-3
20th St..........B-4
21st St..........B-4,H-1
22nd St..........B-4,G-1

POINTS OF INTEREST

Branch Brook Park..A-6
Hayes Park North.....C-7
Independence Park..H-7
Lincoln Park..........H-6
Newark International
Airport..........K-7
Riverbank Prk..........G-8
Stickle Bridge..........E-7
Vailsburg Park..........E-1
Wahington Park..........F-7
Weequahil Park..........K-3
West Side Park..........G-2

N

Scale of Miles

0. .1 .2 .3 .4 .5

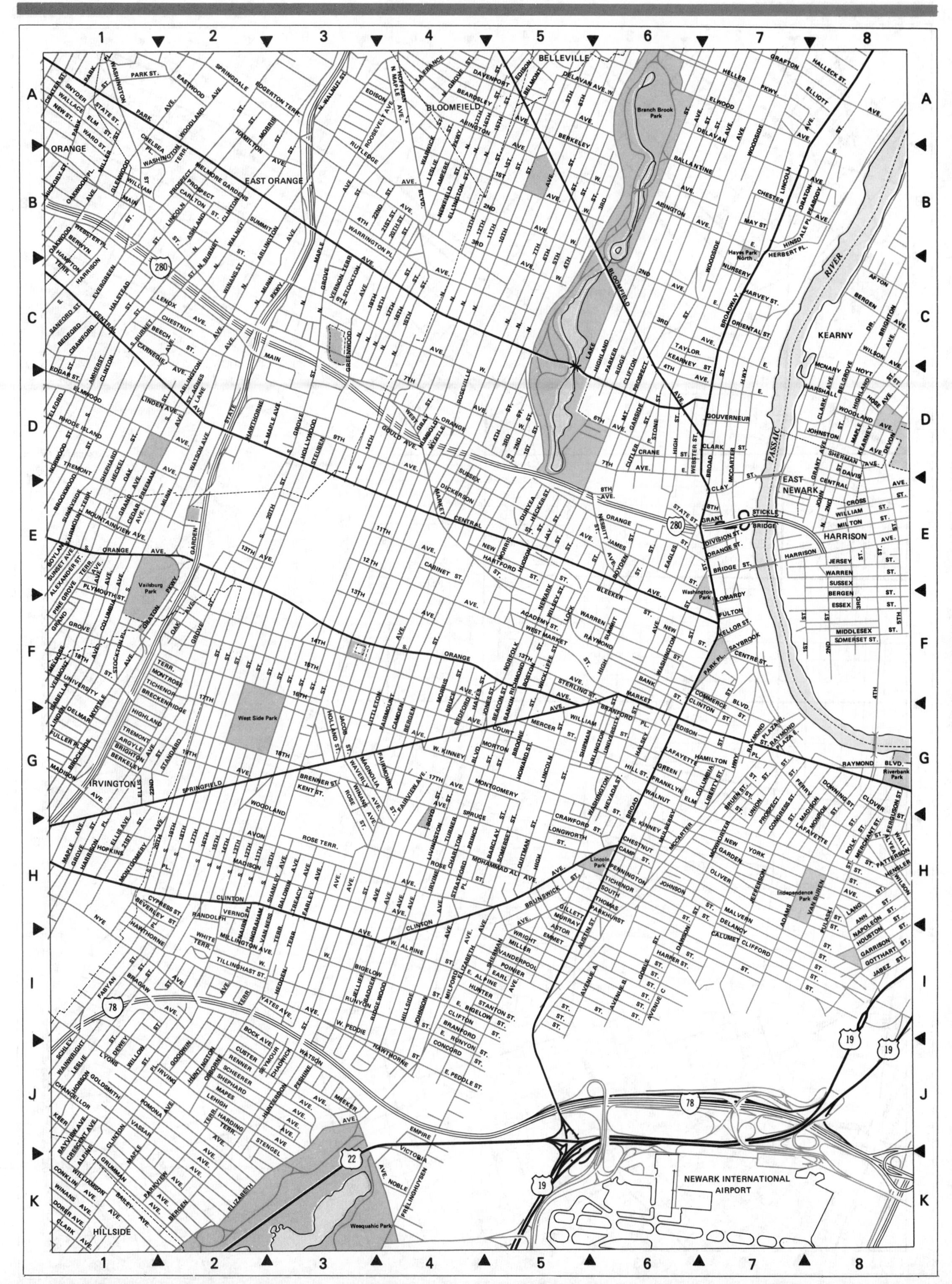

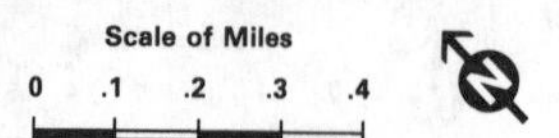
Scale of Miles
0 .1 .2 .3 .4

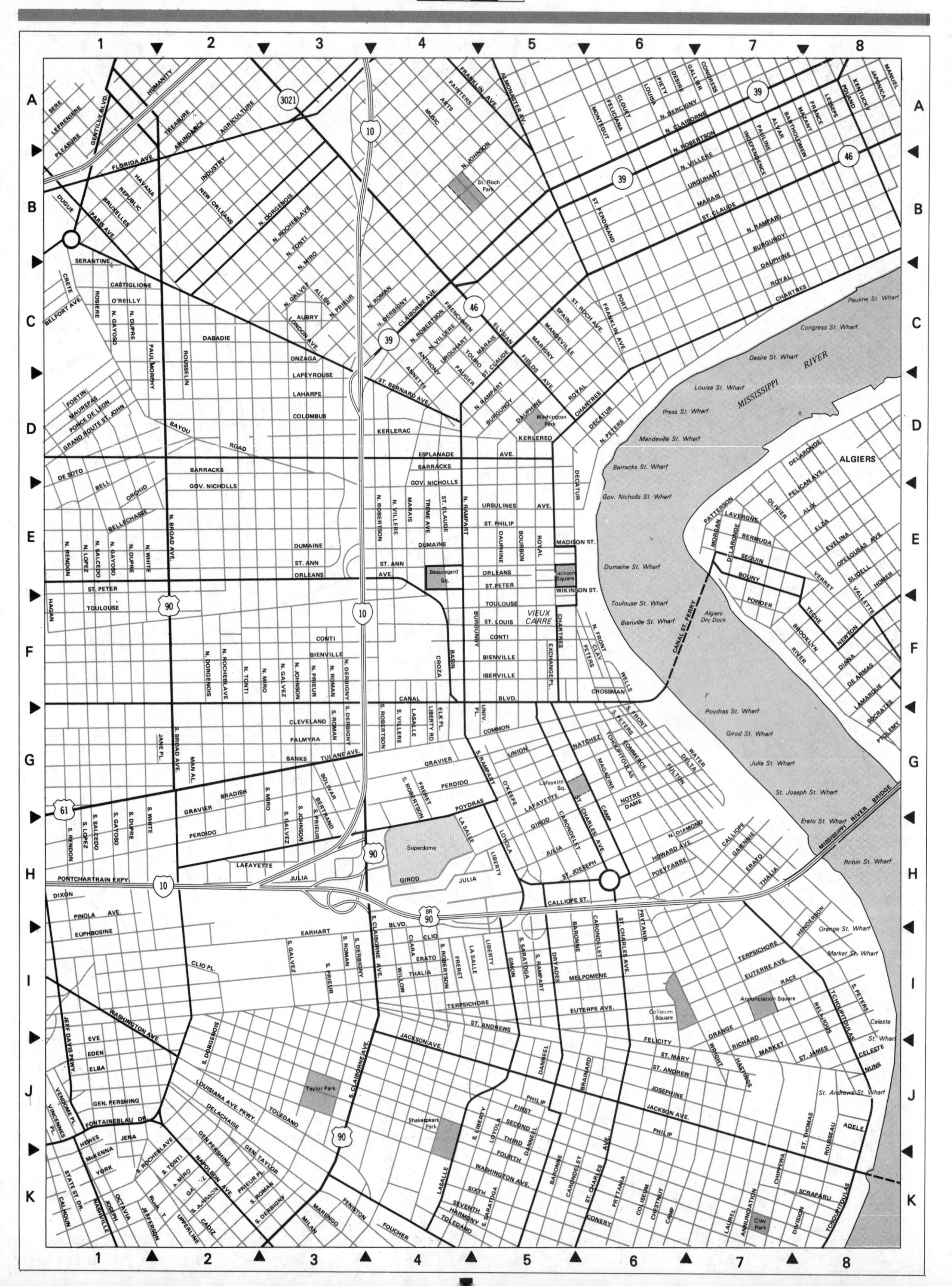
1
2
3
4
5
6
7
8
A
B
C
D
E
F
G
H
I
J
K
SERE
LEFRENIERE
PLEASURE
GENTILLY BLVD.
HUMANITY
TREASURE
ABUNDANCE
AGRICULTURE
3021
FLORIDA AVE.
HAVANA
REPUBLIC
BRUXELLES
INDUSTRY
NEW ORLEANS
DUGUE
PARIS AVE.
N. DORGENOIS
N. ROCHEBLAVE
N. TONTI
N. MIRO
N. GALVEZ
ALLEN
N. PRIEUR
AUBRY
LONDON AVE.
SERANTINE
CASTIGLIONE
O'REILLY
CRETE
BELFORT AVE.
ROBIERE
N. GAYOSO
N. DUPRE
DABADIE
PAUL MORPHY
ROUSSELIN
ONZAGA
LAPEYROUSE
LAHARPE
COLUMBUS
FORTIN
MAUREPAS
PONCE DE LEON
GRAND ROUTE ST. JOHN
BAYOU
ROAD
KERLERAC
ESPLANADE
BARRACKS
GOV. NICHOLLS
DE SOTO
BELL
ORCHID
BELLECHASSE
N. BROAD AVE.
N. RENDON
N. LOPEZ
N. SALCEDO
N. WHITE
DUMAINE
ST. ANN
ORLEANS
AVE.
ST. PETER
TOULOUSE
HAGAN
10
90
FRANKLIN AVE.
PAINTERS
ARTS
MUSIC
N. JOHNSON
St. Roch Park
ALMONASTER AV.
ROMAN
N. DERBIGNY
CLAIBORNE AVE.
N. ROBERTSON
FRENCHMEN
N. VILLERE
ANTHONY
URQUHART
TOURO
MARAIS
ST. CLAUDE
ANNETTE
PAUGER
ELYSIAN
FIELDS AVE.
MARIGNY
MANDEVILLE
SPAIN
ST. ROCH AVE.
PORT
FRANKLIN AVE.
ROYAL
CHARTRES
DECATUR
N. PETERS
N. RAMPART
BURGUNDY
DAUPHINE
Washington Park
KERLEREC
ST. BERNARD AVE.
39
46
MONTEGUT
FELICIANA
CLOUET
LOUISA
PIETY
DESIRE
GALLIER
CONGRESS
N. DERBIGNY
N. CLAIBORNE
N. ROBERTSON
N. VILLERE
URQUHART
MARAIS
ST. CLAUDE
ST. FERDINAND
INDEPENDENCE
PAULINE
ALVAR
BARTHOLOMEW
MAZANT
FRANCE
LESSEPS
POLAND
KENTUCKY
JAPONICA
MANUEL
BURGUNDY
DAUPHINE
ROYAL
CHARTRES
Pauline St. Wharf
Congress St. Wharf
Desire St. Wharf
MISSISSIPPI RIVER
Louisa St. Wharf
Press St. Wharf
Mandeville St. Wharf
Barracks St. Wharf
Gov. Nicholls St. Wharf
Dumaine St. Wharf
Toulouse St. Wharf
Bienville St. Wharf
Algiers Dry Dock
CANAL ST. FERRY
ALGIERS
DELARONDE
PELICAN AVE.
OLIVIER
ALIX
ELZA
EVELINA
OPELOUSAS AVE.
PATTERSON
LAVERGNE
MORGAN
BERMUDA
SEGUIN
BOUNY
POWDER
VERRET
SLIDELL
HOMER
VALLETTE
TECHE
BROOKLYN
NEWTON
RIVER
DIANA
DE ARMAS
LAMARQUE
SOCRATES
PTOLEMY
N. ROBERTSON
N. VILLERE
MARAIS
TREME AVE.
ST. CLAUDE
N. RAMPART
URSULINES
ST. PHILIP
DAUPHINE
BOURBON
ROYAL
MADISON ST.
Beauregard Sq.
Jackson Square
WIKINSON ST.
VIEUX CARRE
ST. LOUIS
CONTI
BIENVILLE
IBERVILLE
BURGUNDY
BASIN
CROZA
EXCHANGE PL.
CHARTRES
N. FRONT
CLAY
PETERS
WELLS
CROSSMAN
CANAL
BLVD.
N. DORGENOIS
N. ROCHEBLAVE
N. TONTI
N. MIRO
N. GALVEZ
N. JOHNSON
N. PRIEUR
N. ROMAN
N. DERBIGNY
S. ROBERTSON
S. VILLERE
LASALLE
LIBERTY RD
ELK PL.
UNIV. PL.
COMMON
CLEVELAND
PALMYRA
S. ROMAN
S. DERBIGNY
BANKS
TULANE AVE.
JANE PL.
S. BROAD AVE.
MAN AL.
S. FRONT
S. PETERS
TCHOUPITOULAS
COMMERCE
NATCHEZ
MAGAZINE
WATER
DELTA
FULTON
Poydras St. Wharf
Girod St. Wharf
Julia St. Wharf
St. Joseph St. Wharf
Erato St. Wharf
Robin St. Wharf
MISSISSIPPI RIVER BRIDGE
61
S. RENDON
S. LOPEZ
S. SALCEDO
S. GAYOSO
S. DUPRE
S. WHITE
GRAVIER
BRADISH
PERDIDO
S. MIRO
S. GALVEZ
S. JOHNSON
S. PRIEUR
BERTRAND
BOLIVAR
S. ROBERTSON
FRERET
POYDRAS
S. RAMPART
UNION
O'KEEFE
LAFAYETTE
Lafayette Sq.
GIROD
CARONDELET
ST. CHARLES AVE.
CAMP
NOTRE DAME
N. DIAMOND
HOWARD AVE.
POEYFARRE
CALLIOPE
GAIENNIE
ERATO
THALIA
LA SALLE
Superdome
LOYOLA
LIBERTY
JULIA
ST. JOSEPH
LAFAYETTE
PONTCHARTRAIN EXPY
DIXON
PINOLA AVE.
EUPHROSINE
EARHART
BLVD.
CLIO
BR 90
CALLIOPE ST.
S. GALVEZ
S. ROMAN
S. DERBIGNY
S. PRIEUR
S. CLAIBORNE AVE.
CLARA
ERATO
WILLOW
THALIA
S. ROBERTSON
FRERET
LA SALLE
LIBERTY
SIMON
S. SARATOGA
S. RAMPART
DRYADES
BARONNE
CARONDELET
ST. CHARLES AVE.
PRYTANIA
MELPOMENE
CLIO PL.
TERPSICHORE
EUTERPE AVE.
ST. ANDREWS
Coliseum Square
HENDERSON
Orange St. Wharf
Market St. Wharf
TERPSICHORE
EUTERPE AVE.
RACE
Annunciation Square
S. PETERS
RELIGIOUS
TCHOUPITOULAS
Celeste St. Wharf
ORANGE
RICHARD
MARKET
ST. JAMES
CELESTE
NUNS
WRIGHT
HASTINGS
WASHINGTON AVE.
JEFF DAVIS PKWY.
EVE
EDEN
ELBA
S. DORGENOIS
LOUISIANA AVE. PKWY.
DELACHAISE
TOLEDANO
Taylor Park
S. CLAIBORNE AVE.
JACKSON AVE.
FELICITY
ST. MARY
ST. ANDREW
JOSEPHINE
JACKSON AVE.
PHILIP
DANNEEL
BRAINARD
PHILIP
FIRST
SECOND
THIRD
FOURTH
Shakespeare Park
S. LIBERTY
LOYOLA
DANNEEL
St. Andrews St. Wharf
ST. THOMAS
ROUSSEAU
ADELE
GEN. PERSHING
VENDOME PL.
VINCENNES PL.
FONTAINEBLEAU DR.
HEWES
JENA
McKENNA
YORK
STATE ST. DR.
CALHOUN
NASHVILLE
JOSEPH
OCTAVIA
S. ROCHEBLAVE
S. TONTI
S. MIRO
GEN PERSHING
NAPOLEON AVE.
GEN. TAYLOR
PRIEUR PL.
S. ROMAN
S. DERBIGNY
JEFFERSON
UPPERLINE
CADIZ
MARENGO
MILAN
PENISTON
FOUCHER
LASALLE
WASHINGTON AVE.
SIXTH
SEVENTH
HARMONY
TOLEDANO
S. SARATOGA
BARONNE
CARONDELET
ST. CHARLES AVE.
PRYTANIA
CONERY
COLISEUM
CHESTNUT
CAMP
LAUREL
ANNUNCIATION
CHIPPEWA
Clay Park
DIVISION
SCRAPARU
TCHOUPITOULAS

NEW ORLEANS

Abundance A-2
Adele J-8
Agriculture A-2
Alix E-8
Allen C-3
Almonaster Ave. A-5
Alvar A-7
Annette C-4
Annunciation K-7
Anthony C-4
Arts A-4
Aubry C-3
Banks G-3
Baronne K-5
Barracks D-2,D-4
Bartholomew A-7
Basin F-4
Bayou Rd. D-2
Belfort Ave. C-1
Bell E-1
Bellechase E-1
Bermuda E-7
Bertrand G-3
Bienville F-3,F-5
Bolivar G-3
Bouny E-7
Bourbon E-5
Bradish G-2
Brainard J-6
Broad Ave., N. E-2
Broad AVe., S. G-2
Brooklyn F-8
Bruxelles B-1
Burgundy B-7,D-5,F-5
Cadiz K-2
Calhoun K-1
Calliope H-7
Callipest H-5
Camp K-6,G-6
Canal Blvd. F-4
Carondelet H-5,H-6,K-5
Castiglione C-1
Celeste J-8
Chartres C-7,D-6,F-5
Chestnut K-6
Chippewa K-7
Claiborne, N. A-6
Claiborne Ave. C-4
Claiborne Ave., S.
H-4,J-3
Clara I-4
Clay F-6
Cleveland G-3
Clio I-4
Clio Pl. I-2
Clouet A-6
Coliseum K-6
Colombus D-3
Commerce G-6
Common G-5
Conery K-6
Congress A-7
Conti F-3,F-5
Crete C-1
Crossman F-6
Croza F-4
Dabadie C-2
Danneel J-5
Dauphine C-7,D-5,E-5
De Armas F-8
Decatur D-5,D-6
DeLachaise J-2
DeLaronde D-7,E-7
Delta G-6
Derbigny,N. A-6,C-4,F-3
Derbigny,S. G-3,I-3,K-2
Diamond,N. H-6
Diamond,S. H-6
De Soto D-1
Diana F-8
Division K-8
Dixon H-1
Dorgenois,N. B-3,F-2
Dorgenois,S. J-2
Dryades I-5
Dugue B-1
Dumaine E-3,E-4
Dupre,N. C-1,E-1
Dupre,S. G-1
Earhart Blvd. I-3
Eden J-1
Elba J-1
Elk Pl. G-4
Elysian Fields Ave. C-5
Elza E-8
Erato H-7,I-4
Esplanade Ave. D-4
Euterre Ave. I-5,I-7
Euphbosine I-1
Eve J-1
Evelina E-8
Exchange Pl. F-5
Felciana A-6
Felicity J-6
First J-5
Florida AVe. B-1
Fontaineblau Dr. J-1
Fortin D-1
Foucher K-4
Fourth K-5
France A-8
Franklin Ave. A-4,C-6
Frenchmen C-4
Freret G-4
Front, N. F-6
Front, S. G-6
Fulton G-6
Gaiennie H-7
Gallier A-7
Galvez, N. C-3,F-3
Galvez, S. H-3,I-3,K-2
Gayoso, N. C-1,E-1
Gayoso, S. G-1
Gen. Pershing J-1,J-2
Gen. Taylor K-2
Gentilly Blvd. A-1
Girod H-5
Gov. Nicholls E-2,E-4
Grand Route St.
John D-1
Gravier G-2,G-4
Hagan F-1
Harmony K-4
Hastings J-7
Havana B-1
Henderson I-8
Hewes J-1
Homer E-8
Howard Ave. H-6
Humanity A-2
Iberville F-5
Industry B-2
Independence A-7
Jackson Ave. I-4,J-6
Jane Pl. G-2
Japonia A-8
Jeff Davis Pkwy. I-1
Jefferson K-1
Jena J-1
Johnson, N. B-3,F-3
Johnson, S. H-3,K-2
Joseph K-1
Josephine J-6
Julia H-3,H-5
Kentucky A-8
Kerlerac D-4,D-5
LaFayette G-5,H-2
Laharpe D-3
Lamaroue F-8
Lapeyrouse D-3
Lasalle G-4,K-4
Laurel K-7
Lavergne E-7
Lefreniere A-1
Lesseps A-8
Liberty, N. E-4
Liberty, S. H-4,J-5
London Ave. C-3
Lopez, N. E-1
Lopez, S. H-1
Louisa A-6
Louisiana Ave. Pkwy.
J-2
Loyola J-5
McKenna K-1
Madison St. E-5
Magazine G-6
Man Pl. G-2
Mandeville C-5
Manuel A-8
Marais B-7,C-5,E-4
Marengo K-3
Marigny C-5
Market J-7
Maurepas D-1
Mazant A-8
Melpomene I-5
Milan K-3
Miro, N. B-3,F-3
Miro, S. G-3,K-2
Mississippi River
Bridge H-8
Morgan E-7
Montegut A-6
Mujsic A-4
Napoleon Ave. K-2
Nashville K-1
Natchez G-6
Newton F-8
New Orleans B-2
Notre Dame G-6
Nuns J-8
Octavia K-1
O'Keefe G-5
Oliver E-7
Onzaga C-3
Opelousas Ave. E-8
Orange I-7
Orchid E-1
O'Reilly C-1
Orleans Ave. E-3,E-5
Painters A-4
Palmetto H-1
Palmyra G-3
Paris AVe. B-1
Patterson E-7
Pauger C-4
Pauline A-7
Paulmorphy C-1
Pelican Ave. E-7
Peniston K-3
Perdido G-4,H-2
Peters F-6
Peters, N. D-6
Peters, S. G-6,I-8
Philip J-5,J-6
Piety A-6
Pleasure A-1
Poeyfarre H-6
Poland A-8
Ponce De Leon D-1
Pontchartrain Expy. H-1
Port C-6
Powder F-7
Poydas G-4
Prieur Pl. K-2
Prieur, N. C-3,F-3
Prieur, S. H-3,I-3
Prytania H-6,K-6
Ptolemy G-8
Race I-7
Rampart, N. B-7
D-5,E-4
Rampart, S. G-5,I-5
Religious I-8
Rendon, N. E-1
Rendon, S. H-1
Republic B-1
Richard J-7
River F-8
Robert K-2
Robertson, N. A-7
C-4,E-4
Robertson, S. G-4,I-4
Rocheblave, N. B-3,F-2
Rocheblave, S. K-2
Romain, N. C-4,F-3
Romain, S. G-3,I-3,K-2
Rosiere C-1
Rousseau J-8
Rousselin C-2
Royal C-7,D-5,E-5
St. Andrews I-5,J-6
St. Ann E-3,E-4
St. Bernard Ave. D-4
St. Charles Ave. G-5
I-6,K-6
St. Claude B-7,C-5
St. Ferdinand B-6
St. James J-8
St. Joseph H-5
St. Louis F-5
St. Mary J-6
St. Peter E-1,E-5
St. Philip E-5
St. Roch Ave. C-5
St. Thomas J-8
Salcedo, N. E-1
Salcedo, S. G-1
Saratoga G-4
Saratoga, S. I-5,K-5
Scraparu K-8
Second J-5
Seguin E-7
Serantine B-1
Sere A-1
Seventh K-4
Simon I-5
Sixth K-5
Slidell E-8
Socrates G-8
Spain C-5
State St. Dr. K-1
Tchoupitoulas
G-6,I-8,K-8
Teche F-8
Terpsichore I-4,I-7
Thalia H-7,I-4
Third J-5
Toledano J-3,K-4
Tonti, N. B-3,F-2
Tonti, S. K-2
Toulouse F-1,F-5
Touro C-4
Treasure A-2
Tulane Ave. G-3
Union G-5
University Pl. G-5
Upper Line K-2
Urquhart B-7,C-4
Ursulines Ave. E-5
Vallette F-8
Vendome Pl. J-1
Verret E-8
Villere, N. B-7,C-4,E-4
Villere, S. G-4
Vincennes Pl. J-1
Washington K-5
Washington Ave. I-1
Water G-7
Wells F-6
White, N. E-1
White, S. G-1
Willow I-4
Wilkinson St. E-5
Wright J-7
York K-1

POINTS OF INTEREST

Annunciation Sq. I-7
Beauregard Sq. E-4
Canal St. Ferry F-6
Clay Park K-7
Coliseum Sq. I-6
Jackson Sq. E-5
Lafayette Sq. G-5
Mississippi River
Bridge G-8
St. Roch Park B-5
Shakespere Park J-4
Taylor Park J-3
Washington Park D-5

NEW YORK

Adams St. A-1
Albany I-1
Allen H-3
Amsterdam Ave.
B-4,C-3
Ann H-2
Anslie St. H-6
Ash F-5
Ashland Pl. J-4
Astor G-3
Astoria Blvd. C-7,C-8
Atlantic Ave.
J-2,K-4,K-6
Attorney H-3
Auburn Pl. J-4
Avenue of the
Americas D-3
Ave. A G-3
Ave. B G-3
Ave. C G-3
Ave. D G-4
Bainbridge St. K-7
Baltic St. K-3
Bank St. F-2
Banker St. G-5
Barclay St. H-1
Barrow G-2
Baruch H-4
Bayard St. H-6
Baxter H-2
Beach H-1
Beadle St. H-6
Beaver I-1,I-6
Bedford Ave. H-5,I-4
I-5,J-5
Beekman E-4,I-2
Bergen St. K-3
Bergenline Ave. A-1
Berginine A-1
Bergenwood A-1
Berry I-4
Bethune St. F-2
Bleecker St. G-2,J-7
Bonn C-1
Borden Ave. F-5
Bowery H-2
Bridge St. J-3
Broad I-1
Broadway A-2,A-4
B-1,D-3,D-7,E-3,E-8
H-3,I-4,I-6,J-7,K-8
Broadway St. G-2
Bromme G-2
Brooklyn Bridge I-2
Brooklyn Queens
Connecting Hwy. J-4
Broome G-2
Buchanan B-1
Bushwick Ave. J-7,I-6
Butler St. K-3
Calyer G-5
Cambridge K-5
Canal H-2
Cardinal H-2
Carlisle I-1
Carlton Ave. J-4
Carmine G-2
Carroll St. K-2
Cedar I-1
Central C-3,J-8
Centre H-2
Central Park South D-3
Central Park West C-4
Chambers St. H-1
Charles F-2
Charlton G-2
Chauncey K-7,K-8
Cherry H-3,I-2
Christopher St. F-2
Chrystie H-2
Clark St. J-2
Clarkson G-2
Classon Ave. J-5
Clay St. F-5
Clermont Ave. J-4
Clifton C-1
Clinton H-3,I-4
Clinton Ave. J-4
Clinton St. K-2
Clifton Pl. J-5
Clymer I-5
Coenties I-1
Columbia C-1,H-3
Columbus B-4,C-3
Commercial F-5
Conimer St. G-5
Conselyea St. H-6
Cook St. I-6
Cooper C-1
Cooper St. K-8
Cornelia G-2
Cornelia St. J-8
Cortlandt H-1,I-1
Cour St. K-2
Covert St. K-8
Cranberry St. I-2
Crescent St. C-7,E-6
Crosby St. G-2
Cuylers I-1
Dean St. K-4
De Bevoise St. I-6
Decatur St. K-6,K-8
Degraw St. K-2,K-3
De Kalb St. J-5,J-7
Delancey St. H-3
Des Brosses G-2
Devoe St. H-6
Dey H-1
Diamond St. G-6
Ditmars Blvd. B-8
Division H-2,I-4
Division Pl. H-6
Dobbin St. G-5
Dominick G-2
Douglass St. K-3
Doughty I-2
Downing G-2
Driggs Ave. G-6
H-5,I-5
Duane H-1,H-2
Duer Pl. C-1
Durham A-1
Eckford St. G-6
Eldert St. K-8
Elizabeth H-2
Eldorado C-1
Eldridge H-3
Emerson J-5
Entrance St. E-4
Engert Ave. H-6
Essex St. H-3
Evergreen St. J-7
Fairview I-8
Flatbush Ave. J-3
Fletcher I-2
Forsyth H-2
Franklin H-1,H-2
Franklin Ave. J-5
Franklin St. F-5
Franklin Roosevelt
Dr. F-4,H-4
Freedom Pl. C-3
Front I-2
Front St. I-3
Frost H-6
Frost St. H-5
Fulton C-1
Fulton St. H-1
I-2,J-3,K-5
Gansevoort St. F-2
Gardner G-7
Gates Ave. J-6,J-8,K-4
George I-7
Gerry I-6
Gould I-2
Gouveneur H-3
Graham Ave. H-6
Grand G-2
Grand Ave. K-5
Grand St. H-5,H-6,H-7
Grandview I-8
Green St. K-4
Greene G-2
Greene Ave. J-7,K-7
Greenpoint Ave.
F-6,G-5
Greenwich F-2,G-1
Grove G-2
Grove St. J-7
Guernsey St. G-5
Hall St. J-4
Halsey St. J-8,K-5,K-7
Hamill H-2
Hamilton C-1
Hancock St. J-8
K-5,K-7
Hanson Pl. K-4
Hansman G-6
Harman St. J-7
Harrison A-1,H-1
Harrison Ave. I-5
Hart St. I-7,J-5
Havemeyer H-5
Henry H-2
Henry St. J-2,K-2
Henry Hudson C-3
Hester H-3
Hewes I-5
Hewes St. I-5
Heyward St. I-5
Hicks St. J-2
Highland C-1
Himrod St. J-7
Hobart Ave. D-8
Hooper St. I-5
Hope St. H-5
Hopkinson Ave. K-8
Honeywell St. E-7
Horatio St. F-2
Houston St. G-2,H-3
Howard H-2
Howard Ave. K-7
Hoyte Ave. C-7
Hubert H-1
Hudson C-1
Hudson Ave. A-2,B-1
Hudson Blvd. B-1
Hudson Blvd. E. B-1
Hull St. K-8
Humboldt St. G-6,H-6
Hunter St. E-6
Huron St. G-5
Imlay St. K-1
India St. G-5
Irving K-5
Irving Ave. I-7
Irving St. K-2
Jackson H-3
Jackson Ave. E-6
Jackson St. H-6
James H-2
Jane St. F-2
Java St. G-5
Jay St. H-1
Jefferson H-3
Jefferson Ave.
I-7,J-8,K-5,K-7
Jefferson St. A-1
Jewel St. G-6
John I-2
Johnson Ave. I-6,I-7
John St. I-3
Jones G-2
Kane St. J-2
Keap St. I-5
Kent I-4
Kent Ave. H-4,J-5
Kent St. G-5
King C-1
King St. G-2,K-1
Kings Island Ave. G-6
Knickerbocker Ave.
I-7,J-7
Kosciusko St. J-5
Lafayette H-2
Lafayette Ave.
J-6,K-4
Laight G-1
Lawrence J-3
Lee Ave. I-5
Leonard H-2
Leonard St. G-5,H-6
LeRoy St. G-2
Lewis H-3
Lewis Ave. J-5
Lexington Ave.
C-5,F-3,J-6,K-5
Liberty C-1,I-1
Lincoln Tunnel D-1
Linden St. J-7
Lispenard H-2
Livingston St. J-3
Lombardy St. G-6
Lonimer St. I-5,I-6
Ludlow St. H-3
Lynch St. I-5
Macon St. K-6
MacDougal St. G-2,K-8
Madison H-2,I-8
Madison Ave. C-5,F-3,K-7
Madison St. J-8,K-5
Mangin H-4
Manhatten Ave.
F-5,H-6
Manhattan Bridge I-3
Marcy H-5
Marcy Ave. I-5,J-5
Marion St. K-7
Marshall I-3
Maspeth Ave. H-6
Mavjer St. I-6
McKibbin St. I-6
Meadowview A-1
Meeker Ave. G-6
Melrose St. I-7
Menahan St. J-7
Mercer St. G-2
Messerole H-7,I-6
Messerole Ave. G-6
Metropolitan Ave.
H-4,H-6,H-8
Middagh I-3
Miller Hwy. D-2,G-1
Milton St. G-5
Mitchell E-4
Monitor St. G-6
Montague St. J-2
Montgomery H-3
Mt. Guiness Blvd. F-5
Monroe B-1,H-2
Monroe St. K-5,K-7
Montrose Ave. I-6
Moore St., N. H-1
Moore St. I-6,I-7
Morgan Ave. G-6,H-6
Morgan I-7
Morton G-2,I-4
Mott H-2
Moultrie G-6
Mulberry H-2
Murray St. H-1
Myrtle Ave. J-5,J-6,J-7
Nassau I-1
Nassau Ave. G-5
Navy St. J-3
New I-1
Newel St. G-6
Newkirk A-1
Newton Ave. C-7
Noble St. G-5
Norfolk St. H-3
Norman Ave. G-5
North Henry St. G-6
Northern Blvd. E-7
Nostrand Ave. I-5
Oak St. G-5
Old I-2
Onderdonk I-8
Pacific St. J-3,K-4
Page Pl H-8
Paidge Ave. F-5
Palisade Ave. A-2,B-1
Palmetto St. J-8
Park H-3
Park Ave. A-2,B-1
C-5,D-4,F-3
Park St. H-1
Park Row H-2
Park Row East H-2
Patchen Ave. J-7
Pearl H-2,I-2
Peck I-2
Pell H-2
Penn St. I-5
Perry St. F-2
Pierrpont St. J-2
Pilling K-8
Pitt St. H-3
Platt I-2
Pleasent C-1
Pleasant Ave. A-7
Plymouth St. I-3
Poplar I-3
Porter G-6,I-7
Powers St. H-6
President St. K-3
Prince St. G-2
Prospect I-3
Provost St. F-6
Pulaski St. J-5,J-6
Putman Ave. J-8,K-5
Putnam Ave. K-7
Quay St. G-5
Queens Blvd. E-6,E-7
Queens Midtown
Tunnel E-5
Quincy St. J-6,K-5
Ralph Ave. J-7
Randolph St. I-7
Rector St. I-1
Reid Ave. J-7
Remsen St. I-3
Renwick G-2
Review Ave. F-6
Richardson St. H-6
Ridge H-3
River H-4
Riverside Dr. A-4,B-3
Rivington H-3
Roebling H-5
Rockaway K-8
Ross St. I-5
Rutgers H-3
Rutledge St. I-5
Russell St. G-6
Sackett St. K-2,K-3
St. James Pl. J-4
St. Marks Ave. K-4
St. Nicholas Ave. I-8
Sanford J-5
Sands I-3
Saratoga Ave. K-8
Seigel St. I-6
Seneca Ave. I-8
Schermephorne St. J-3
Scholes H-7,I-6
Shaefer St. K-8
Skillman Ave. E-7
F-6,H-6
Skillman St. J-5
Slip H-3
Smith Ave. A-1,K-2
Somers ST. K-8
South I-2
South Ave. E-4
Spencer St. J-5
Spring G-2
Spruce H-2
Stagg St. H-7,I-6
Stanhope St. J-7
Stanton H-3
Starr St. I-7
Sterling C-1
Stewart Ave. I-7
Stockholm St. I-7
Stone K-8
Stuyvesant Ave. J-7
Suffolk H-3
Sullivan St. G-2,K-1
Summitt St. K-2
Sumner Ave. J-6,J-7
Sumpter St. K-8
Sutton Pl. D-5
Sutton St. G-6
Suydam St. I-7
Taaffe Ave. J-5
Taylor St. I-4
Thomas St. H-2
Thompson St. G-2
Thomson Ave. E-6
Throop Ave. I-6,J-6,K-6
Tillary J-3
Tompkins Ave. J-6
Tonsor H-8
Trinity I-1
Troutman St. I-7,I-8
Tudor City Pl. E-4
Tunnel Exit St. E-4
Union I-5
Union St. K-2
University G-2
Van Buren J-6
Vandam G-2
Van Dam St. F-7
Vanderbilt E-4
Vanderbilt Ave. J-4
Van Brunt St. K-1
Vandervoort H-6
Varet St. I-6
Varick H-2
Varick Ave. H-7
Vernon Blvd. D-6,E-5
Vernon Ave. J-5
Vestry G-2
Vessey St. H-1
Vine I-2
Wall I-1
Walker H-2
Wallabout I-5,I-6
Walton St. I-5
Walworth Pl. J-5
Warren St. H-1,K-3
Warsoff Pl. J-5
Washington G-1,J-3
Washington Ave. I-4
Water I-2
Water St. I-3
Watts G-2
Waverly I-4
Weirfield St. J-8
West A-1
West St. F-4,G-2
Westevd C-3
White St. H-2
White Hall I-1
Willoughby St. I-7
Willett H-3
William I-2
Willoughby Ave. J-4
Williamsburg
Bridge H-4
Wilson Ave. J-8
Wolcott K-1
Woodbine I-8
Woodbine St. J-8
Woodside E-8
Wooster St. G-2
Woodward Ave. I-8
Worth St. H-2
Wyckoff Ave. I-7
Wyckoff St. K-3
Wythe H-4
York I-3
York Ave. C-6,D-5
Zold Pl., S. G-4
1st Ave. C-5,F-4,G-3
1st Pl. K-2
1st St. C-6,H-6
1st St. H-5
2nd Ave. C-5,F-3,G-3
2nd Pl. K-2
2nd St. C-6,F-5,H-5
2nd St., E. G-3
3rd Ave. C-5,E-4,G-3
3rd Pl. K-2
3rd St. C-7,H-5
3rd St., E. G-3
3rd St., W. G-2
4th Ave. G-3
4th Pl. K-2
4th St. H-4,H-5
4th St., E. G-3
5th Ave. B-5
5th St. E-5,H-4,H-5
5th St., E. G-3
6th Ave. K-4
6th St. H-5
6th St., E. G-3
7th Ave. F-2,K-2
7th St. H-5
7th St., E. G-3
8th I-4
8th Ave. D-3,F-2
8th St. H-5
8th St., E. G-3
8th St., W. F-2
9th Ave. E-2
9th St. C-7,D-6,H-5
9th St., E. G-3
9th St., W. F-2
10th Ave. E-2
10th St. D-6,E-5,E-6,H-5
10th St., E. G-3
10th St., W. F-2
11th Ave. E-2
11th St. D-6,E-5,E-6,H-5
11th St., E. G-3
11th St., W. F-2
12th St. D-6,E-5,G-6
12th St., E. G-3
12th St., W. F-2
13th St. E-6
13th St., W. F-2
14th St. C-7,D-6
14th St., E. F-3
14th St., W. F-2
15th St. H-5
15th St., E. F-3
15th St., W. F-2
16th St., E. F-3
16th St., W. F-2
17th St., E. F-3
17th St., W. F-2
18th St., E. G-3
18th St., W. F-2
19th St. C-7
19th St., E. F-3
20th Ave. B-8
20th St. B-8
20th St., E. F-3
20th St., W. F-2
21st Ave. B-8
21st St. B-8,C-7
D-7,E-6,F-6
21st St., E. F-3
21st St. W. F-2
22nd St. E-6
22nd Rd. B-8
22nd St., E. F-3
23rd St. C-7,D-7,E-6
23rd St., E. F-3
23rd St., W. E-2
24th St. C-7,E-6
25th Ave. D-8
25th Rd. C-7
25th Rd., W. E-2
25th St., E. F-3
25th St., W. F-2
26th Ave. C-7
26th St. C-8
26th St., E. F-3
26th St., W. E-2
27th Ave. C-7
27th St. C-8,E-6
27th St., E. F-3
27th St., W. E-2
28th St. C-8,E-6
28th St., W. E-2
29th Ave. C-7,D-8
29th St. C-8,E-6
29th St., W. E-2
30th Ave. D-8
30th Dr. C-7
30th Pl. F-6
30th Rd. C-7
30th St. C-7,D-7,E-6,F-6
30th St., E. F-3
30th St., W. E-2
31st Ave. C-7
31st Dr. D-7
31st Pl. F-6
31st St. E-6
31st St., E. F-3
31st St., W. E-2
32nd St. D-7,E-7
32nd Pl. F-7
32nd St., E. F-3
32nd St., W. E-3
33rd St. C-8,D-7,E-6,F-7
33rd St., E. F-3
33rd St., W. E-2
34th Ave. D-7
34th St. F-7
34th St., E. F-4
34th St., W. E-2
35th Ave. D-6
35th St. C-1,C-8,E-7,F-7
35th St., E. E-4
35th St., W. E-2
36th Ave. D-6
36th St. C-1,C-8,E-7,F-7
36th St., E. E-4
36th St., W. E-2
37th Ave. D-6,E-8
37th St. C-1,C-8,E-8
F-7,G-6
37th St., E. E-4
37th St., W. E-2
38th E-8
38th Ave. D-6
38th St. C-8,E-8,F-7
38th St., E. E-4
38th St., W. E-3
39th Ave. E-6,E-8
39th Pl. F-7
39th St. C-1,F-7
39th St., E. E-4
39th St., W. D-3
40th Ave. E-6
40th St. F-7
40th St., E. E-4
40th St., W. D-3
41st Dr. E-8
41st St. C-1,C-8,E-7,F-7
41st St., E. E-4
41st St., W. D-3
42nd Pl. F-7
42nd St. C-8,E-7,F-7
42nd St., E. F-4
42nd St., W. D-3
43rd Ave. E-5,E-7
43rd Rd. E-5
43rd St. C-1,E-8,F-7,G-7
43rd St., E. F-4
43rd St., W. D-3
44th Ave. E-5
44th Dr. E-5
44th Rd. E-5
44th St. E-8,F-7,G-7
44th St., E. E-4
44th St., W. D-3
45th B-1
45th Rd. E-5
45th St. E-8,F-7
45th St., E. E-4
45th St., W. D-7
46th Ave. E-5
46th Rd. E-5
46th St. C-1,E-8,F-7
46th St., E. E-4
46th St., W. D-3
47th Ave. E-5,F-6
47th St. C-1,E-8,F-8,H-7
47th Rd. E-5
47th St., E. E-4
47th St., W. D-3
48th Ave. F-5,F-6
48th St. B-1,E-8,F-8,G-7
48th St., E. E-4
48th St., W. D-3
49th Ave. F-5
49th St. B-1,D-8,E-8,F-8
49th St., E. D-4
49th St., W. D-3
50th B-1
50th Ave. D-8,E-8,F-5
50th St. B-1,E-8,F-8,G-8
50th St., E. D-4
50th St., W. D-3
51st B-1
51st Ave. F-5
51st St. B-1,F-8
51st St., E. D-4
51st St., W. D-3
52nd Ave. F-8
52nd Rd. F-8
52nd St. B-1,E-8
52nd St., E. D-4
52nd St., W. D-3
53rd H-8
53rd St. B-1,F-8
53rd St., E. D-4
53rd St., W. D-3
54th Ave. F-5,G-7,G-8
54th Dr. G-7
54th Pl. H-8
54th Rd. G-7
54th St. B-1,E-8,F-8,H-8
54th St., E. D-4
54th St., W. D-3
55th Ave. F-5,G-7
55th Dr. G-7
55th St. B-1,E-8,F-8,H-8
55th St., E. D-4
55th St., W. D-3
56th Rd. G-7,G-8
56th St. B-1,E-8,F-8,H-8
56th St., E. D-4
56th St., W. D-3
57th Ave. G-7
57th Dr. H-8
57th St. B-1,E-8,F-8
57th Rd. G-8
57th St., E. D-4
57th St., W. C-3
58th Ave. H-8
58th Ln. F-8
58th Rd. H-8
58th St. B-1,E-8,F-8
58th St., E. D-4
58th St., W. C-3
59th St. A-1,G-8
59th Dr. H-8
59th Pl. H-8
59th St., E. D-4
59th St., W. C-3
60th St. A-1
60th St., E. D-4
60th St., W. C-3
61st St. A-1
61st St., E. D-4
61st St., W. C-3
62nd St. A-1
62nd St., E. D-4
62nd St., W. C-3
63rd St. A-1
63rd St., E. D-4
63rd St., W. C-3
64th St. A-1
64th St., E. D-4
64th St., W. C-3
65th St. A-1
65th St., E. D-4
65th St., W. C-3
66th St. A-1
66th St., E. D-4
66th St., W. C-3
67th St. A-1
67th St., E. C-4
67th St., W. C-3
68th St., E. C-4
68th St., W. C-3
69th St. A-1,C-4
69th St., E. C-4
69th St., W. C-4
70th St. A-1
70th St., E. C-4
71st St. A-2,C-3
71st St., E. C-4
72nd St. A-2,B-3
72nd St., E. C-5
73rd St. A-2,B-3
73rd St., E. C-5
74th St. A-2,B-3
74th St., E. C-5
75th St. A-2,B-3
75th St., E. C-5
76th St. A-2
76th St., E. C-5
76th St., W. B-4
77th St. A-2,B-3
78th St. A-2,B-3,C-5
79th St. A-2,B-3,C-5
80th St. B-3,C-5
81st St. B-3,C-5
82nd St. B-3,C-5
83rd St. B-3,C-5
84th St. B-4,C-5
85th St. C-4,C-5
86th St. B-4,B-5
87th St. A-4,B-4,B-5
89th St. A-4,B-5
90th St. A-4,B-5
91st St. A-4,B-5
92nd St. A-4,B-5
93rd St. A-4,B-5
94th St. A-4,B-5
95th St. A-4,B-4
96th St. A-4,B-5
97th St. A-4,B-5
98th St. A-4,B-5
99th St. A-4,B-6
100th St. A-4,B-6
101st St. B-6
102nd St. A-4,A-5,B-6
103rd St. A-4,A-6
104th St. A-4,A-6
105th St. A-4,A-6
106th St. A-6
107th St. A-6
108th St. A-6
109th St. A-6
110th St. A-6
111th St. A-6
112th St. A-6
113th St. A-6
114th St. A-6
115th St. A-6
116th St. A-6
117th St. A-6
118th A-6
119th St. A-6

New York City, NY

Scale of Miles
0 .2 .4 .6 .8
N

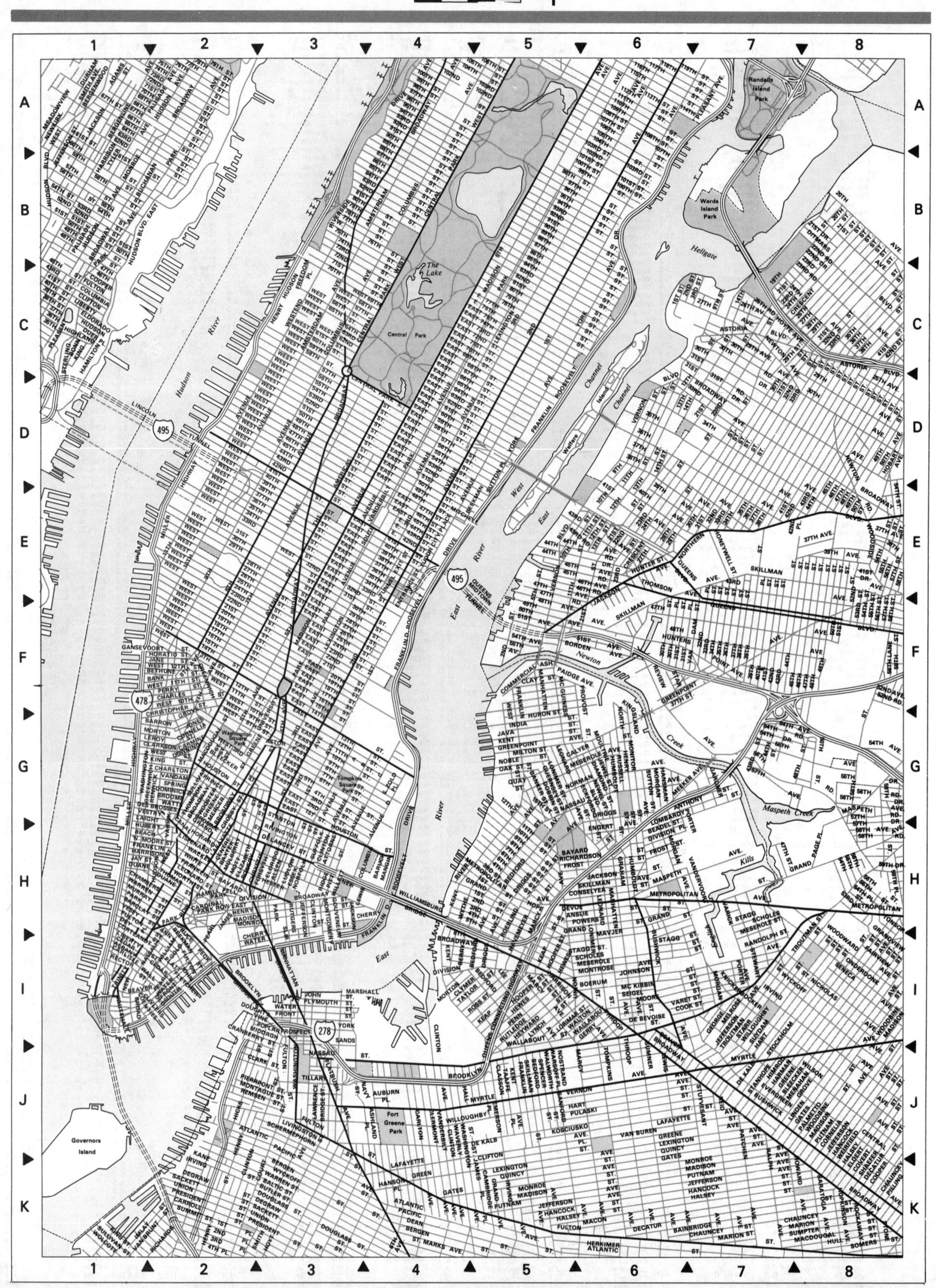

Scale of Miles
0 .1 .2 .3 .4 .5

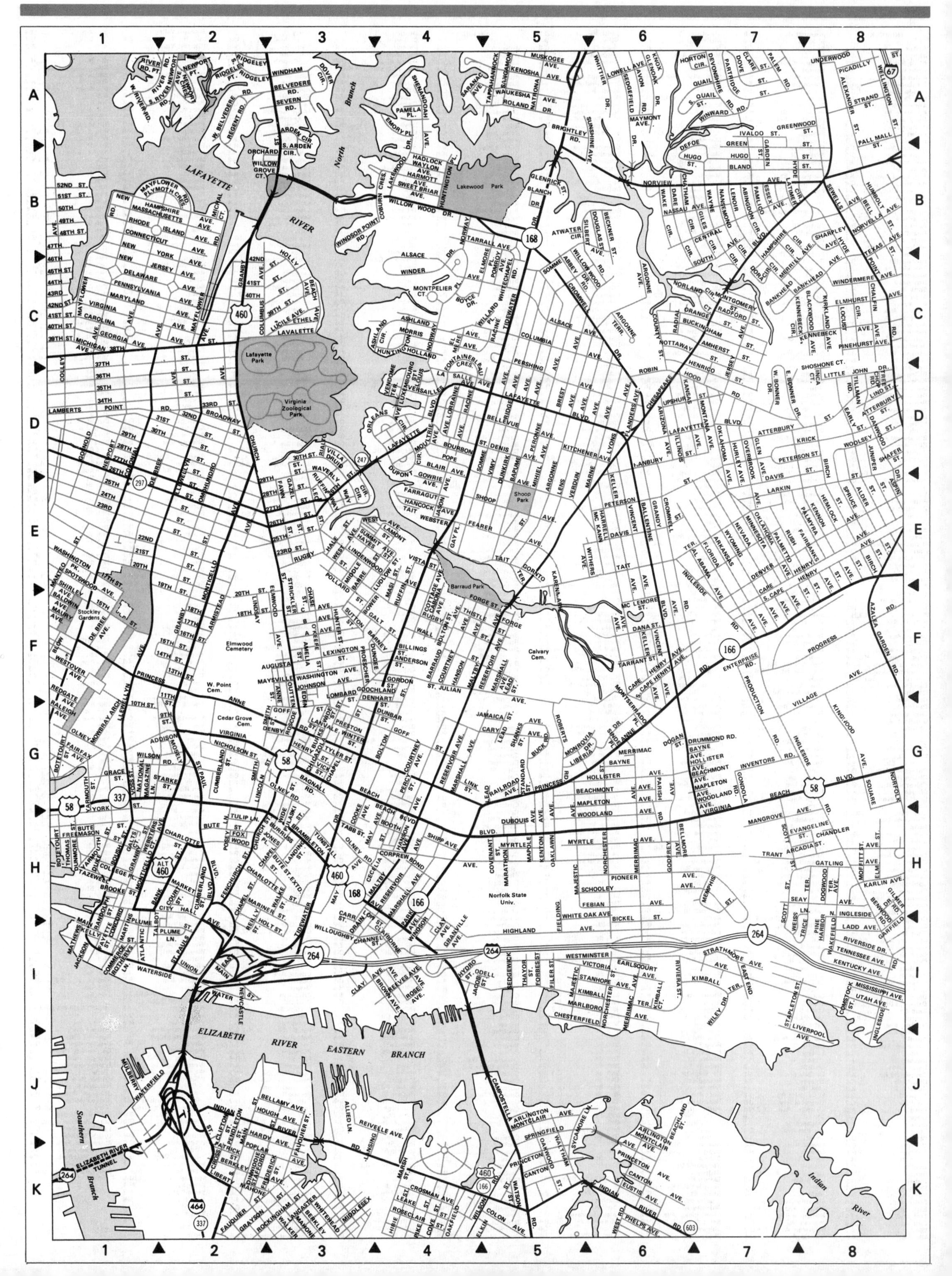

NORFOLK

A Ave....F-3
Abbey Rd....C-5
Abingdon Cir....B-7
Alabama Ave....E-7
Alden St....E-8
Alexander St....A-8
Alsace Ave....B-4,C-5
Amelia St....F-3
Amherst St....C-7
Anderson St....F-4
Anne St....F-3
Arcadia St....H-8
Arden Cir....A-3
Arden Cir., S....A-3
Argonne Ave....C-6,E-5
Arizona Ave....D-6
Arkansas Ave....E-5
Arlington....J-5,J-8
Armistead Ave....F-2
Ashland Ave....C-4
Ashland Cir....C-4
Ashton Ave....J-3
Aspin....E-8
Atlantic....I-1
Atterbury St....D-7,D-8
Atwater Cir....B-5
Augusta....F-3
Avon Rd....A-6
Azalea Garden Rd....F-8
B Ave....F-3
Bagnall Rd....G-3
Baldwin Ave....F-1
Ballentine Blvd....E-6
Bankhead Ave....C-7,C-8
Bank St....H-2
Bapume Ave....D-5
Barney....F-4
Barraud Ave....F-4
Barre St....E-3
Bayne Ave....G-6,G-7
Beach Ave....C-3
Beachland St....J-6
Beachmont Ave....G-6,G-7
Beckner St....B-6
Bell St....B-8
Bellamy Ave....H-5
Bellmore Ave....H-6
Bellevue Ave....D-5
Belvedere Rd....A-3
Belvedere Rd., W....A-2
Benwood Rd....H-8
Berkley Ave....K-2,K-3
Bickel St....I-6
Billings St....F-4
Birch St....D-8,E-8
Blackwood Ave....C-8
Blair Ave....D-4
Blanch Dr....B-5
Bland St....B-7
Blow....F-1
Botton St....F-4,G-4
Bonner Dr., E....D-7
Bonner Dr., W....D-7
Booth....H-4
Botetourt St....H-1
Botherys Ln....I-1
Bourbon....D-4
Boush St....H-1
Bower....F-3
Boyce Dr....C-4
Brambleton Ave....H-3
Brest Ave....D-4
Brightley Rd....A-5
Broadway....D-2
Brooke St....H-1
Brown Ave....I-4
Buckingham St....C-7
Bucks Rd....G-5
Burke Wide St....G-3
Burruss....H-3
Bute St....H-1,H-2
Bute St. Ext....H-3
C Ave....F-3
Cabona Ave....F-4
Calt St....F-4
Campostella Rd....J-5
Canton Ave....K-5,K-6
Cape Henry Ave....F-6,F-7
Cape Henry Ave., S.
....F-6,F-7
Carolina Ave....C-1
Carr St....H-3
Cary Ave....G-5
Cecelia St....H-4
Central Ave....B-7
Chalfin Ave....C-8
Chandler....H-8
Channel St....I-3
Chapel St....G-3,H-2,H-3
Charlotte St....H-2
Chase St....F-3
Chatham Cir....B-6
Chesapeake Blvd....D-6
Chesterfield....I-5
Chicazola St....G-3
Church St....D-2,H-2
City Hall Ave....H-2
Clare St....A-7
Clay Ave....I-3,I-4
Clayborne Ave....I-4
Clifton St....J-2
Coburn Cres....B-4
College Pl....H-1
Colley Ave....D-1
Columbia Ave....C-5
Columbus Ave....C-3
Colon Ave....K-5
Colonial Ave....D-1
Commerce....I-1
Concord....I-1
Connecticut Ave....B-1
Cooke Ave....H-3
Corprew Ave....H-4
Cottage Ave....F-4
County St....C-8
Cour St....F-2
Courtney Ave....F-4,G-4
Covenant St....H-5
Cove St....K-4
Cromwell Dr....C-5
Cromwell Rd....E-6
Cross St....K-2
Crossman Ave....A-4
Cumberland St....G-2,H-2
Dana St....F-6
Danwood Dr....D-8
Dare Cir....B-6
Davis St....E-6,E-7
De Bere Ave....D-2
De Bree Ave....F-1
Deede....B-6
Delaware Ave....C-1
Denby....G-3
Denhart St....F-3
Denver Ave....E-7
Devonshire Rd....A-7
Dinwiddie....K-2
Dodson St....K-2
Dogan St....G-6
Dora Cir....C-7
Douglass....B-8
Dover Cir....A-3
Dove St....A-7
Drain....I-3
Druid Cir....D-3
Drummond Rd....G-7
Dubious....H-5
Duke St....H-1
Dunbar St....G-4
Dungee....F-4
Dunkirk Ave....D-5
Dunmore St....H-1
DuPont Cir....D-4
Earlscourt....I-6
Early St....D-8
East End....I-7
East St....I-2
Eden....F-3
Elkin St....K-5
Elmhurst Ave....C-8
El Mere....C-4
Elmore Pl....C-5
Elm St....H-8
Elmwood Ave....F-3
Emmett Pl....J-2
Emory Pl....A-4
Enterprise Rd....F-7
Essex Cir....B-7
Ethel....C-3
Eustis Ave....K-6
Evangeline St....H-8
Fairbanks Ave....E-8
Fairfax Ave....G-1
Farragut....E-4
Fauquier St....K-2,K-3
Fawn St....E-3
Fayette St....I-1
Fearer Ave....E-5
Febian Ave....H-6
Fenchurch St....H-2
Fielding....I-5
Filer St....I-5
Flanders Ave....D-6
Florida Ave....E-7
Fontainebleau Cres.
....C-4
Forbes St....I-5
Forge St....F-5
Fox....H-2
Frederick St....K-2
Freemason St....H-1
Galts Ct....H-1
Garden St....B-7
Garfield....I-8
Gatling Ave....H-8
Gay Pl....E-4
Gazel St....E-3
Georgia Ave....C-1
Gilmer Rd....H-8
Glen Ave....D-4
Glenoak Dr....A-6
Glenrick St....B-5
Godfrey Ave....H-6
Goff St....G-3,G-4
Gondola Rd....G-7
Goochland....F-3
Gordon St....F-4
Gornto....E-5
Gosnold Ave....D-1
Gowrie Ave....E-4
Grace St....G-1
Granby St....C-2,F-2,H-1
Grandy Ave....E-6
Granville Ave....I-4
Grayson St....K-2
Green St....A-7
Greenwood St....A-7
Hadlock....B-4
Hale St....E-3
Hampshire Ave....B-7
Hanbury St....D-6
Hancock Ave....E-4
Hanson Ave....F-4,H-4
Hardy Ave....J-2
Harmott Ave....B-4
Harrell Ave....E-6
Hayes St....E-3
Hemlock St....E-8
Henrico St....D-7
Henry St....G-3
Hibie St....K-4
Highland Ave....I-5
Holland....C-4
Hollister Ave....G-6,G-7
Holly Ave....C-3
Holt St....I-3
Hood Rd....D-7
Hope Ave....K-2
Horton Cir....A-7
Hough Ave....J-3
Hugo St....B-6,B-7
Humbolt St....B-8
Hunter St....F-3
Huntington....C-4
Huntington Pl....B-4
Hurley Ave....D-7
Hyde Cir....B-8
Hydro St....I-8
Ingleside Dr., N....H-8
Ingleside Rd....E-6,G-8
Illinois....D-6
Inca Ct....D-8
Indian River Rd.
....J-2,K-6
Inventors Rd....G-7
Ivaloo St....A-7
Ivy St....K-2
Jackson....I-1
Jacob St....I-4
Jamaica Ave....G-5
Jason Ave....E-4
Jersey Ave....C-7
Johnson Ave....F-3
Jones Pl....H-3
Juniper St....D-8
Kansas Ave....D-6
Karlin Ave....H-8
Keller Ave....E-6,F-6
Kelley....I-1
Kennebeck Ave....C-8
Kennebeck Cir....C-8
Kennon Ave....E-8
Kenosha Ave....A-5
Kenton Ave....H-5
Kentucky Ave....I-8
Kimball Ct....I-6
Kimball Ter....I-6,I-7
Kingwood Ave....G-8
Kirkland Ave....C-8
Kitchener Ave....D-5
Knox....A-6
Krick St....D-8
Ladd Ave....I-8
LaFayette Blvd.
....D-4,D-5,D-6
Lakewood Dr....B-4
Lamberts Point
Rd....D-1
Lamont St....E-4
Lancaster....K-3
Landing St....H-3
Lanvale St....G-3
Larkin St....F-3
La Salle Ave....D-4
Latimer....B-7
Laxalette....C-3
Lead St....F-5,G-5
Leake St....A-4
Lenoir Cir....B-7
Lena Ave....E-5
Leo St....E-3
Lexington St....F-3
Liberia Dr....G-5
Liberty St....K-2
Lincoln St....G-2
Lind St....D-8
Lindenwood Ave....E-3
Lindsay....F-2
Link St....G-4
Little John Rd....D-8
Liverpool Ave....J-8
Llewellyn Ave....E-2,G-1
Locust....E-8
Lombard St....F-3
Lorraine Ave....D-4
Lovitt....I-2
Lowell Ave....A-6
Low St....H-3
Lucile Ave....C-3
Ludlow St....E-4
Luxembourg Ave....D-4
Lyons Ave....D-6
Mahone....K-3
Main St....I-1,I-2
Majestic Ave....H-5,I-5
Maltby Ave....F-4,H-4
Mangrove Ave....H-7
Manteo St....E-1
Mapleton Ave....G-6,G-7
Mapole Ave....H-5
Marathon....H-5
Mariner St....H-2
Market....H-2
Marlboro Ave....I-5
Marne Ave....E-6
Marsh St....K-4
Marshall Ave....F-5,G-4,H-4
Martins....I-1
Masi St....E-4
Massachusetts Ave....B-1
Mathews....I-1
Maury Ave....F-1
May Ave....H-3
Mayflower Rd.
....B-1,C-1,C-2
Maryland Ave....C-1
Maymont Ave....A-6
Maysville....F-3
McKann Ave....E-6
McLemore St....F-6
Memphis St....H-7
Merrill Ave....C-7
Merrimac....G-6
Merrimac Ave....H-6,I-6
Michigan Ave....C-1
Middle....E-3
Middlesex....K-3
Middleton St....K-1
Mihiel Ave....D-5
Minnesota Ave....E-7
Mississippi Ave....I-8
Moffitt St....H-8
Monrovia Dr....G-5
Montana Ave....D-7
Montclair Ave....J-5,J-6
Montgomery St....C-7
Monticello Ave.
....E-2,H-1
Montserrado Pl....G-6
Morris Ave....C-4
Moultrie Ave....D-4
Mowbray Arch....G-1
Muskogee Ave....A-5
Myrtle Ave....H-6
Nansemond Cir....B-7
National Magazine
Ln....G-1
Natrona....A-5
New Hampshire
Ave....B-1
New Jersey Ave....B-1
Newport Ave....A-2,D-1
Newport Cres....A-2
Newport Pt....A-2
New York Ave....B-1
Nevada Ave....E-7
Nicholson St....G-2
Norchester St.
....H-6,I-6
Norfolk....G-1
Norland Cir....C-7
Norland Ct....C-7
Norvella Ave....B-8
Norview Ave....B-6
Norway....B-4,C-4
Nottaway St....C-6
Oak Cir....B-6
Oaklawn Ave....H-5
Oakwood St....K-5
Odell....I-4
O'Keefe St....F-3
Oklahoma Ave....D-7,E-7
Olney Rd....G-1,G-2,H-3
Omohundro Ave....E-2
Orange St....C-7
Orchard....B-3
Orleans Cir....D-4
Outten....F-3
Overbrook Ave....D-7
Palem Rd....A-8
Pall Mall St....A-8
Palmetto St....E-7
Palmyra St....E-8
Pamela Pl....A-4
Pamlico Cir....B-7
Parish Rd....G-6
Park Ave....H-5
Partridge St....A-7
Patrick St....K-2
Payne....K-2
Pearl St....J-1
Pendleton St....J-2
Pennsylvania Ave....C-1
Percy St....G-4
Peronne Ave....D-5
Pershing Ave....C-5
Peterson St....D-8,E-8
Phelps Ave....K-5
Picadilly....A-8
Pike St....K-4
Pine Harbor....I-8
Pinehurst Ave....C-8
Pine St....J-1
Pioneer Ave....H-6
Plume Ln....I-2
Plume St....I-1
Plymouth Cres....B-2
Pollard St....E-3
Pomroy Ave....C-5
Pope Ave....D-4
Poplar Ave....D-2
Posey Ln....H-2
Preston....A-2
Princess Anne
Rd....F-1,G-5
Princeton Ave....K-5,K-6
Production Rd....F-7
Proescher....F-3
Progress Rd....F-8
Pulaski....H-3
Quail St....A-7
Quail St., S....A-7
Racine....C-5
Racine Ave....D-4
Radford St....C-7
Radia....G-6
Railroad Ave....G-6
Raleigh Ave....G-1
Randolph St....H-1
Redgate Ave....F-1
Reeves Ave....I-4
Regent Rd....A-2
Reilly St....I-2
Relvelle Ave....J-3
Reservoir Ave.
....F-5,G-4,H-4
Rhode Island Ave....B-1
Ridgeley....A-2
Ridgeley Pt....A-2
Ridgeley, S....A-2
River Rd....A-2
River Rd. Pt....A-1
River Rd., S....A-2
River Rd., W....A-1
Riverside Dr....I-8
Riviera St....I-8
Roberts Rd....B-3
Rockingham St....K-2
Roland Dr....A-5
Roscoe....G-3
Rose Ave....I-4
Roseclaire St....K-4
Ruffin St....E-3
Ruffin Way....E-3
Rugby St....E-3,F-4
Rush St....E-3
Sangamon....A-5
Schooley Ave....H-6
Scott St....H-7
Seay Ave....A-5
Sedgefield Dr....A-6
Sedgewick....I-5
Severn Rd....A-5
Sewells Point Rd....B-8
Shafer St....D-8
Shanks St....G-5
Shapley Ave....B-6
Shenandoah Ave....A-4
Shipp Ave....H-4
Shirley Ave....E-1
Shoop Ave....E-5
Shoshone Ct....D-8
Silbert Rd....B-6
Sinde Pl....G-6
Smith St....G-2,G-3
Somme Ave....C-5,D-5
South Ave....C-7
Spotwood Ave....E-1
Springfield Ave....J-5
Spruce St....E-8
Square....G-1
Stafford St....K-2
St. Denis Ave....D-5
St. Julian Ave....F-4
St. Louis....D-4
St. Pauls Blvd....I-2
Standard St....G-5
Stanhope Ave....I-6
Stansbury Ave....F-2
Stapleton St....I-7
Starke St....G-2
Strand St....A-8
Strathmore Ave....I-7
Strickler St....F-3
Summit Ave....E-3
Sunshine Ave....A-6
Sutton St....F-3
Sweet Briar Ave....B-4
Sycamore Ln....K-5
Tabb St....H-3
Taft Ter....E-6
Tait....E-4,E-5
Talbot....I-1
Tappahannock....A-5
Terrall Ave....B-6
Tarran St....F-6
Thayor St....I-5
Thistle St....F-4
Thomas....H-1
Tennessee Ave....I-8
Texas Ave....B-8
Tidewater Dr....C-5,I-3
Tiller Ave....E-5
Tillman Rd....D-8
Trant Ave....H-7
Tree Chop Rd....D-8
Trice Ter....I-8
Tulip Ln....E-8
Tunstall....H-3
Tyler St....G-3
Underwood....A-8
Upshur St....D-6
Utah Ave....I-8
Vendome Ter....D-4
Verdun Ave....E-5
Versailles Ave....D-4
Vick St....G-3
Victoria Ave....I-6
Villa Cir....D-3
Village Ave....F-8
Vimy Ridge Ave....D-5
Vincent Ave....E-6,F-6
Virginia Ave....C-1
Virginia Beach Blvd.
....G-4,H-7
Vista St....E-4
Voss St....G-7
Wake Cir....B-6
Wakefield Ave....I-8
Walker....K-3
Walker St....H-3
Wall St....F-4
Waltham....K-5
Washington Ave....I-3
Washington Pk....E-1
Water St....I-2
Waterfield St....J-1
Waterfront Dr....I-1
Watson St....K-5
Waukesha Ave....A-5
Waverly Way....B-4
Waylon Ave....B-4
Wayne Cir....B-7
Webster....E-4
Websters Ct....H-1
Weiss Ln....H-7
Wellington St....A-8
West Ave....H-5
West Rd....K-6
Westminster....I-5
Westover Ave....F-1
Whitechapel Rd....C-5
Whitehead....K-3
White Oak Ave....H-6
Whittier Dr....A-6
Wide St....H-3
Wiley Dr....I-7
Willard Pl....C-5
Willoughby....I-3
Willow Grove Ct....B-3
Willow Wood Dr.
....B-4,C-5
Wilson Ave....G-1
Wilson Rd....K-4
Winckley Ave....J-3
Winder Dr....C-4
Windermere Ave....C-3
Windham....A-3
Windsor....I-4
Windsor Point Rd....B-3
Winter St....G-3
Winward Rd....A-7
Withers Ave....E-5
Wood St....H-2
Woodland Ave....G-7,H-6
Woolsey St....D-8
Wyoming Ave....E-7
Yarmouth....H-1
Yarmouth St....G-1
9th St....G-1
10th St....G-1
11th St....F-2
13th St....F-2
14th St....F-2
15th St....F-1,F-2
16th St....F-2
17th....E-1,F-2
18th St....F-2
19th St....F-2
20th St....E-1,F-2
21st St....E-1
22nd St....E-1
23rd St....E-1,E-3
24th St....E-1
25th St....E-1,E-3
26th St....D-1,E-3
27th St....D-1,E-3
28th St....D-1,E-3
29th St....D-1,E-3
30th St....D-2,D-3
31st St....D-2
32nd St....D-2
33rd St....D-2
34th St....D-1
35th St....D-1
36th St....D-1
37th St....D-1
38th St....D-1
39th St....C-1,C-3
40th St....C-1,C-2
41st St....C-1
42nd St....C-1,C-2
43rd....C-1
44th....C-1
45th St....C-1
46th....B-1
47th....B-1
48th St....B-1
49th....B-1
50th....B-1
51st St....B-1
52nd St....B-1

POINTS OF INTEREST

Barraud Park....E-4
Elizabeth River
Tunnel....K-1
Lafayette Park....D-3
Lakewood Pk....B-4
Shoop Park....E-5
Stockley Gardens....F-1

OAKLAND

Abbey....K-8
Abbott....F-8
Acacia....D-7
Action St....A-4
Ada....D-6
Adams St....G-5
Adeline St....C-4
....G-2,H-1
Aqua Vista....K-8
Alcatraz Ave....C-3
Alice St....I-3
Alicia....F-3
Allen Dale....K-8
Allman....I-7
Allston Way....A-3,A-5
Alma....H-6
Alma Pl....H-6
Alpine....D-7
Alpine Rd....G-8
Alta....E-8
Alta Ave....F-7
Alta Vista....G-5
Alvarado....G-6
Alvarado Rd....B-8
Amito....B-8
Amy Dr....E-8
Anthony St....B-2
Apgar....E-4
Arbon....D-7
Arbor Dr....F-6
Arden....A-7
Ardley....I-7
Ardmore....H-7
Arimo Ave....G-6
Arkansas....J-8
Arlington....D-4
Arroyo Pkwy....F-6
Ashby Ave....B-3
Ashmount....H-7
Athens....F-3
Athol Ave....H-5
Atwell....J-7
Auburn....C-6
Avalon....B-7
Avoca Ave....E-8
Avon....D-5
Baker St....B-3
Balfour Ave....G-6
Bancroft....A-7
Barrows....H-7
Bartlett Ave....K-8
Bateman....C-6
Bates Rd....H-7
Bay....K-3
Bay St....C-2
Bayview Ave....I-6
Beacon....H-5
Beaumont....I-7
Beechwood....D-7
Belgrave....D-6
Bell....F-8
Bella Vista....I-6
Bellevue....H-4,H-5
Belmont....G-5
Bequa....G-7
Berlin....J-8
Benvenue Ave....C-6
Birch Ct....D-6
Blair....F-8
Blake Way....A-4
Bonar St....A-3
Bonita Ave....F-7
Boston Ave....J-8
Bowditch....A-6
Boyd Ave....D-5
Brighton....I-7
Brandon....F-6
Broadway....E-5,G-3,H-2
Broadway Terr....E-6
Brook....F-4
Brookdale....K-8
Brookdale Ave....J-7
Brookhurst....F-3
Brooklyn Ave....H-5
Brookside....C-6,D-7
Brookwood....H-6
Browning St....A-3
Brush St....H-2
Bryant....D-6
Buckeye....D-7
Buena Vista....D-7
Buena Vista Ave....K-2
Burnett St....B-3
Burk....H-5
Byron....A-3
Caldwell....E-8
California St....C-4
Calmar Ave....H-6
Cambridge Way....F-6
Campbell St....F-2
Canon....I-8
Caperton....G-7
Capp St....J-8
Cappell....H-6
Carleton St....B-2
Carlston....H-7
Carlton....D-6
Carlton St....B-4
Carmel....F-7,I-8
Carroll St....I-5
Casterline....H-8
Castro....G-3
Cavour St....D-5
Cedar....F-1
Center....G-1
Central State Hwy.
Ave....K-2
Chabot....C-6
Chabot Rd....C-6
Champion St....J-8
Channing Way....A-2
Chapin....K-3
Charing Cross....C-8
Chase St....G-1
Cherry....G-5
Chester St....G-1
Chestnut St....G-2
Chetwood....G-5
Claremont....B-7
Claremont Ave....D-6
Claremont Hill....D-7
Clarewood Dr....E-7
Clarke....E-4
Clay St....H-2
Clemens....H-8
Clement Ave....K-4
Cleveland St....H-5
Clifton St....D-5
Clover....C-6
Cochrane....E-8
Colby....C-5
College Av....C-6
Colma....I-8
Commerce Way....J-6
Contra Costa Rd....D-7
Coolidge....J-8
Cordova....J-7
Coron....E-5
Country Club....D-6
Cox Ave....B-8
Craig....F-7
Creed....H-7
Crescent Clarendon....H-7
Crest Rd....G-8
Crocker....G-7
Crofton....F-4,G-6
Crosby....K-7
Cross Rd....D-6
Curran....J-8
Curtis....G-3
Curtis St....A-3
Custer....K-8
Cypress....G-1
Dale Pl....K-8
Damuth....I-8
Dana....C-5
Dana St....B-5
Dartmouth....B-8
Davis St....J-7,K-8
Deering....K-8
Deleware....J-8
Delmer....I-8
Dennison....K-5
Denoa St....D-4
Derby St....B-4,B-6
Desmond....E-5
Diamond....I-8
Dohr....B-3
Dona St....J-7
Dormidera....F-7
Dover....F-4
Dover St....D-4
Doyle St....D-3
Dracena....F-6
Dulwich....E-7
Duncan Way....E-8
Durant Ave....A-5
Dwight....B-7
Eagle Ave....K-3
East....D-7
Echo Ave....F-6
Edgewood....H-8
Edwards St....A-4
El Camino Real....C-7
El Centro....H-8
Elcerrito Ave....F-6
Eldorado....F-5
Eliot....I-6
Ellis St....C-4
Ellita....H-5
Ellsworth....G-8
Ellsworth Dr....B-5
Elm St....F-4
Elmwood....G-5
Elston....I-7
Embarcadero West....H-1
Emerson....I-6
Emerson St....B-5
Entrada....F-6
Erie St....G-6
Essex St....C-5
Estates....F-8
Estates Dr....H-8
Estrander....D-7
Estrella....F-7
Eton....C-6
Euclid....G-5
Evans Blvd....H-8
Everett....H-7,I-8
Excelsior Ave....I-7
Fairbanks....G-6
Fairlane....D-8
Fairmount....F-5,G-4
Fairview Ave....F-6
Fairview St....C-4
Fallon St....I-3
Farracut....G-7
Filbert St....G-2
Fitzgerald St....E-2
Fleet....H-7
Florence....E-7
Florio St....C-6
Folger....C-2
Foothill....J-6
Foothill Blvd....I-5
Forest....D-5
Forest Ave....B-6
Forest Hill Ave....I-8
Franklin St....I-3
Frisbie....F-4
Fruitvale Ave....I-8,K-7
Fulton....B-8
Fulton Ave....B-8
Fulton St....C-5
Galindo....K-7
Garden St....J-7
Garland....G-4
Gaskill St....D-3
Gelston....B-8
Georgia....J-8
Gilbert St....E-5
Gleason....J-5
Glen....G-8
Glen Ave....F-5
Glenbrook....D-7
Glendora....H-8
Glen Eden....F-6
Glenfield....I-7
Glenview....G-5
Golden Gate Vista....D-7
Grace....D-3
Grand....G-4
Grand St....F-6,K-4
Grandview....C-8
Grand Vista....J-7
Grant St....B-4
Gravatt....C-7,C-8
Gray....K-7
Grayson St....B-2
Greenbank Ave....F-6
Greenwood....H-7
Grove St....C-4,E-4
....F-3,H-2
Grosvenor....H-7
Hadoon....H-5
Hagar....F-7
Hagemann....K-8
Halcyon....C-5
Hampton Rd....G-8
Hanly Rd....I-8
Hanover Ave....H-5
Harbord....E-7
Harboad Dr....F-8
Harding....I-8
Harmon St....C-3
Harper....K-7
Harper St....C-4
Harrison....G-4,G-5
Harrison St....I-3
Harrington Ave....K-7
Harwood Av....C-6
Haskell St....C-3
Haste....A-5
Haven St....D-3
Hawthorne....F-4
Hazel....G-7
Hearst....I-8
Heinz Ave....B-2
Henrietta....K-8
Henry St....G-1
Heron....E-8
Highland....J-6
Highland Ave....F-7
Hillcrest Rd....C-7,C-8
Hillcroft....H-6
Hillegass....B-6
Hillegass Ave....C-6
Hiller Dr....C-8,D-8
Hillgirt Cir....H-6
Holden St....E-2
Hollis St....E-2
Hollywood....H-8
Holman....H-7
Horton St....E-2
Howard....F-6
Howe....B-5
Howe St....F-5
Hubbard....E-2
Hubert Rd....H-7
Hudson St....D-5
Humboldt Ave....J-8
Huntleigh....G-8
Hyde St....K-7
Independence....J-6
Indiana....G-8
Inyo Ave....J-6
Irving St....K-6
Isabella St....G-3
Ivy Dr....I-5
Jack London Sq....I-2
Jackson St....I-3
Jacobus....E-7
James Ave....E-5
Jayne Ave....G-5
Jean Blvd....G-6
Jean St....G-5
Jefferson Ave....A-4
Jefferson St....H-2
Johnston Dr....F-8
Julia St....B-4
Kales....D-6
Keith Ave....D-6
Kempton....F-5
Kenmore....G-6
Kenilworth....B-8
Kenwyn....H-5
King St....C-4,K-5
Kingston....F-6
Kinsley....I-6
Kirkham....G-1
Lacoste St....C-2
La Cresta Ave....I-7,I-8
La Fayette....G-7
La Guna Ave....J-8
Lagunita....H-5
Lake Ave....F-6
Lake Park....H-5
Lake Shore....I-4
Lake Shore Ave....H-6
Lakeside Dr....H-4
Larkspur....H-6
Larkview....G-7
La Salle Ave....G-7
Latham St....F-6
Latimer....E-4
Laurel....J-8
Lawton....D-6
Lawton Ave....D-5
Leach....H-7
Lee St....G-4
Leimert Blvd....H-8
Lenox....G-4
Leo Terrace....E-8
Lester....H-5
Lewis St....G-1
Lewiston....C-6
Lexford Rd....G-8
Lincoln....G-7,I-8
Lincolnshire....D-7
Linda Ave....F-6
Linden....C-6
Linden St....G-2
Linwood....I-7
Livingston....K-5
Locksley Ave....D-5
Logan St....J-7
Longridge Rd....H-6
Los Angeles St....D-3
Lowell St....D-4
Lusk St....E-4
Lyman Rd....I-8
Lynde....J-7,K-8
Lyon Ave....K-8
Mabel....B-3
MacArthur Blvd.
....H-6,I-7,I-8
Madeline....I-8
Madison St....I-3
Magnolia Ave.
....C-6,G-2,G-6
Manchester....D-7
Manila Ave....E-5,F-5
Manoa....C-6
Manor Crest....C-7
Maple Ave....J-8
Margarido....D-6,D-7
Marin Way....J-5
Market....C-4
Market St....E-3,G-2
Mariposa....F-5
Masonic....F-8
Mather....D-6
Mather St....E-5
Mathews....B-3
Max Welton....F-8
McAuley....D-5
McClure....F-4
McDonald....B-8
McGee Ave....B-4
McKillop....J-7
McKinney St....A-4
McMilan....D-6
Mead....F-3
Medocino....D-6
Meadow....K-8
Merritt....H-5
Mesa Ave....F-7
Middle Harbor Rd....H-1
Miles Ave....D-5,D-6
Miller Ave....K-6
Milton....F-3
Milva St....B-5
Mitchell St....K-6
Modoc....E-8
Mondana Blvd....H-6
Monroe....D-6
Montana....J-8
Montana St....I-8
Monte....F-7
Montecello Ave....F-6
Montecito....G-4
Monte Cresta....F-5
Monte Vista Ave....F-5
Montgomery St....E-5
Monticlair....H-5
Monzale....E-8
Moraga....F-6
Moraga Ave....F-8
Morpeth....E-7
Mosswood Rd....A-7
Mountain Ave....F-7
Mountain Blvd....D-8
Muir....G-7
Munson....J-6
Murray St....B-3
Myrtle St....G-2
Nace....F-6
Nason....K-2
Nevil St....K-8
Nicol Ave....J-8
Nimitz Frwy....G-1
Norfolk Rd....C-8
North Hill....C-8
North St....C-5
Northvale....H-6
Nova....F-6
Oak Grove....D-6
Oakland Ave....F-7,G-5
Oakmore....H-8
Oakmont....G-7
Oak Ridge Rd....C-7
Oak St....I-3
Oakvale....C-7
Ocean Ave....C-3
Ocean View Dr....D-6
Octavia....K-8
Olive Ave....F-5
Opal St....F-5
Orange St....G-5
Oregon St....B-4
Overland....C-2
Pacific Ave....F-7,K-2
Pala Ave....F-7
Palm....G-5
Palmetto....I-8
Paloma....H-6
Panoramic....A-7
Panoramic Rd....A-7
Panoramic Way....B-8
Paramount Rd....H-7
Pardee St....B-3
Park Ave....E-3,H-7
Park Blvd....H-7,I-5
Park St....B-3
Parker St....B-2,B-4,B-6
Parkside Dr....C-7
Parkview....G-4,G-6
Patton....C-7
Paxton....K-7
Penniman Ave....K-8
Peralta St....G-1
Perkins....G-5
Pershing....G-8
Perth....C-8
Piedmont....C-6
Piedmont Ave....B-6,F-5
Pine....C-6
Pine St....F-1
Pinewood....E-8
Pleasant St....J-8
Pleasant Valley Ave....E-6
Pleasant Valley
Ct. N....E-6
Pleasant Valley
Ct. S....E-6
Poirer St....C-4
Poplar St....G-2
Portland Ave....H-5
Portsmouth....G-7
Posey Tube....I-2
Potter St....B-2
Powell St....D-2
Prince....C-4
Prince St....C-6
Proctor Ave....E-8
Prosepct....G-7,H-6
Ramona Ave....F-6
Rand....H-6
Randolph....I-7
Randwick....F-4
Ransom Ave....F-6
Raonor....H-5
Redondo....D-5
Regent Av....B-6
Regent St....C-6
Rhoda Ave....J-8
Ricardo....F-6
Richmond....F-4
Rio Vista....F-5
Rispin....B-8
Rockridge Blvd. N....D-6
Rockridge Blvd. S....D-6
Rockwell....C-6
Roden....H-5
Roosevelt Ave....A-4
Rosal....G-6
Ross....C-6
Rose Ave....F-5
Ruby....E-4
Russell St....B-4
Sacramento St....C-4
St. James....H-8
St. Paul....E-8
Salisbury....K-7
San Antonio Way....J-6
San Carlos....F-6
Sandringham....H-8
San Pablo Ave....B-3,E-3
San Sebastian....H-8
Santa Clara Ave.
....G-5,K-2
Santa Ray Ave....H-6
Santa Rita St....K-8
Scenic....F-7,I-8
School St....J-8
Schyler....J-8
Scott St....G-6
Selborne....H-8
Shafter Ave....D-5,E-5
Shatuck Ave....B-5
Sheffield Ave....J-7
Shellmound St....D-2
Sheridan....G-7
Sheridan Rd....E-8
Sherman....K-3
Short....K-8
Sloan....I-8
Soland Way....J-5
Sonia....E-7
Spalding Ave....A-4
Spruce....H-6
State....B-7
Staten....H-5
Stanford Ave....D-3
Stanton....B-3,B-4
Stonewall....B-7
Stow....H-5
Strathmoor....B-8
Stuart St....B-4,B-6
Summit....F-4
Sunnyhills Rd....H-6,H-7
Sunny Side....F-6
Sunnyslope....G-6
Sunset Ave....J-7
Suter St....J-8
Sycamore St....F-3
Sylvan Ave....G-6
Taft....D-6
Telegraph Av....B-5,G-3
Terrace St....E-5
Texas St....J-8
The Plaza Dr....C-7
The Uplands....C-7
Thomas....D-6
Tiffin....H-8
Townsend....H-8
Trestle Glen Rd.
....H-6,H-7
Tunnel State Rd....C-7
Tyler....B-4
Underhills....H-7
Union....G-1
Union St....G-2
Uplands....C-7
Valdez St....G-4
Vallejo St....C-3
Valley St....B-3,G-4
Valley Vista....G-5
Van Buren Ave....G-5
Vermont St....G-6
Vernon....G-5
Vernon St....G-5
Vicente....C-8
View St....E-5
Viola....K-8
Viona....H-6
Virmar Ave....D-6
Vista....F-7,I-8
Wakefield....J-6
Walla Vista....G-6
Waldo....F-7
Walker....G-6
Wallace....E-2
Wallace St....B-3,I-6
Ward....J-8
Ward St....B-4
Warfield Ave....G-6
Warring Av....B-6
Warring St....B-6
Warwick....G-5
Washington St....H-3
Waterhouse....I-8
Watts....E-3
Waverly St....G-4
Wayne....I-5
Webster Ave....E-5
Webster St....C-5
....F-4,I-3,J-2
Webster St. Tube....I-2
Wellington....I-7
Wesley....D-6
Wesley Ave....H-5
West Grand Ave....F-2
West MacArthur
Blvd....F-4
Westminister....D-7
West St....A-3,E-4,F-3
Westview....C-8
Whittle Ave....I-8
Wickson....G-5
Wilbur....I-8
Wildwood Ave....G-7
Wildwood Gardens....G-7
Wildwood Palm Ave.
....G-6
Williams....G-3
Willow St....F-2
Wilson....J-8
Wilson Ave....G-6
Wood....F-8,K-2
Woodland Way....G-7
Woodruff....I-7
Wood St....F-1
Woolsey St....C-5
Wyngarde....H-8
Yerba Buena....E-2
York Dr....F-6
York St....G-6
Yosemite....F-5
1st....I-4
2nd Ave....I-4
2nd St....H-2
3rd Ave....I-4
3rd St....H-1
4th....A-2
4th Ave....I-4
4th St....H-2
5th....A-2
5th Ave....I-4
6th....A-2
6th Ave....I-4
6th St....H-2
7th....A-2
7th Ave....I-4
7th St....G-1,H-2,I-4
7th St., E....I-4
8th....A-3,K-2
8th Ave....I-5
8th St....G-1,H-2
8th St., E....I-4
9th....A-3,K-2
9th Ave....I-5
9th St....G-1,H-2
10th....A-3
10th Ave....I-5
10th St....G-1,G-2,H-2
10th St., E....I-4
11th Ave....J-5
11th St....F-1,H-2
11th St., E....I-4
12th Ave....J-5
12th St....F-1,G-2
12th St., E....I-4,J-5
13th Ave....J-5
13th St....F-1,G-3
14th Ave....I-7,J-5
14th St....F-1,G-3
14th St., E....J-5
15th Ave....J-5
15th St....F-1,G-3
15th St., E....J-5
16th Ave....J-5
16th St....F-1,G-3
17th Ave....J-5
17th St....F-1,G-3
17th St., E....I-5,J-6,K-6
18th Ave....J-5
18th St....F-2,G-3
18th St., E....I-5,J-6
19th Ave....J-6
19th St....F-2,G-3
19th St., E....I-5
20th Ave....J-5
20th St....F-2,G-3
20th St., E....I-5
21st Ave....J-5
21st St....F-2,G-3
21st St., E....I-5
22nd Ave....J-5
22nd St....G-3
22nd St., E....J-6
23rd Ave....K-6
23rd St....G-3
23rd St., E....J-6,K-7
24th Ave....K-6
24th St....F-2,G-3
24th St., E....J-6
25th....G-4
25th Ave....J-7,K-6
25th St....F-3
25th St., E....J-6
26th....G-4
26th Ave....K-6
26th St....F-2,F-3
26th St., E....J-6
27th....G-4
27th Ave....K-6
27th St....F-3
27th St., E....I-6
28th St....F-2,F-3,G-4
28th St., E....J-7
29th St....F-3
29th, E....J-7
30th St....F-3
30th St., E....I-7
31st St....F-3
31st St. E....I-6
32nd....F-3,F-4
32nd St....E-2
32nd St., E....I-6
32nd St., E....I-6
33rd Ave....K-7
33rd St....F-3
33rd St., E....I-6
34th Ave....K-7
34th St....E-3
34th St., E....I-6
35th Ave....K-8
35th St....E-3
36th Ave....K-7
37th....E-3,F-4
38th St., E....I-7
39th....E-3
40th....E-3
40th St....E-4
41st....E-3
41st St....E-4,E-5
42nd....E-3,E-5
43rd....E-3,E-5
43rd St....D-3
44th St....E-3,E-5
45th....D-3
45th St....D-3,E-5
46th....D-4,E-4
47th....D-4,E-4
47th St....D-3
48th....E-5
49th St....E-5
50th St....E-5
51st St....E-5
52nd St....D-4
53rd St....D-3,D-4
54th St....D-3,D-4
55th St....D-4
56th St....D-3,D-4
57th St....D-4,D-5
58th St....D-4,D-5
59th St....C-3,D-3
....D-4,D-5
60th St....C-3,C-4,C-5
61st St....C-2,C-4,C-5
62nd....C-5
62nd St....C-3
63rd....C-3
63rd St....C-2,C-5
64th St....C-2
65th St....C-2,C-4,C-5
66th St....C-2,C-5
67th St....C-2

POINTS OF INTEREST

Adams Park....G-4
Aquatic Park....B-2
Bushrod Park....D-5
Crocker Park....G-7
De Fremery Park....G-2
Dracena Park....F-6
Ernie Raimondi
Park....F-1
Garber Park....B-7
Harrison Square....H-3
Jefferson Square....H-2
Joaquin Miller
Park....H-8
Lafayette Square....H-3
Lakeside Park....H-4
Lakeview Park....G-5
Lynch Square....H-3
Mosswood Park....F-4
Piedmont Park....G-7
Piney Knoll Park....H-4
Regional Park....D-7
San Antonio Park....J-5

Scale of Miles
0 .2 .4 .6 .8

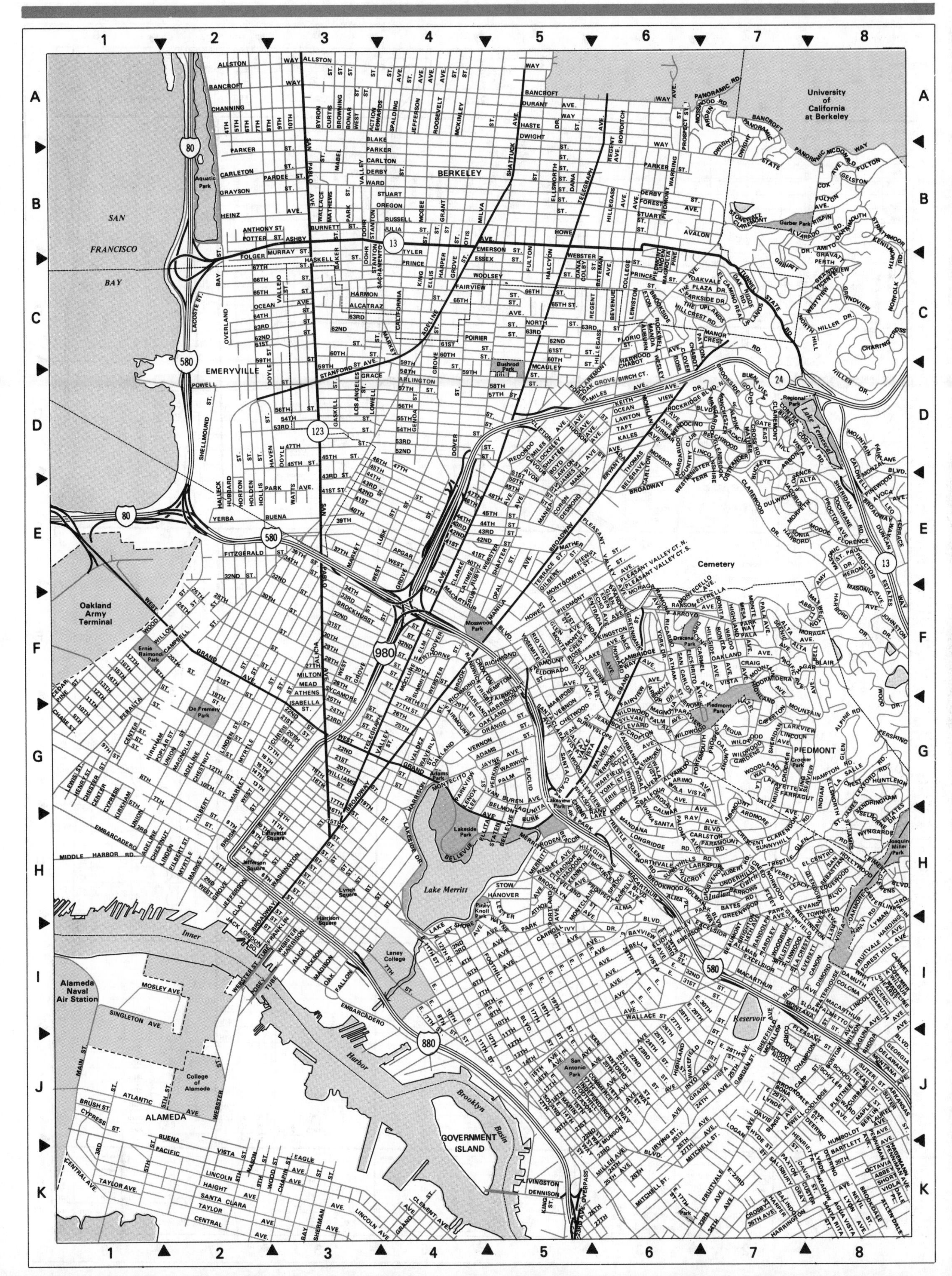

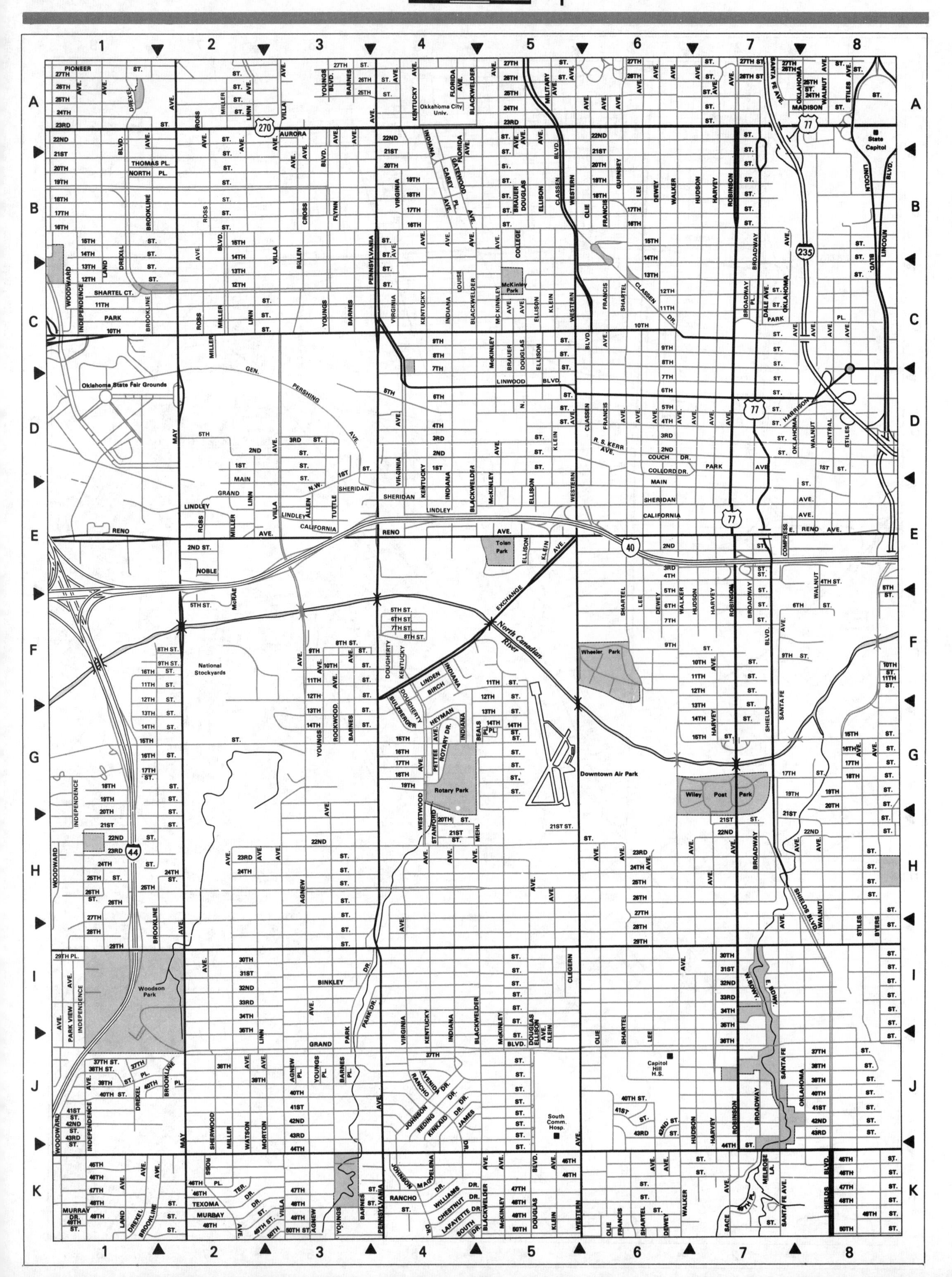

Scale of Miles
0 .2 .4 .6
N
Oklahoma City Univ.
State Capitol
McKinley Park
Oklahoma State Fair Grounds
National Stockyards
Tolan Park
North Canadian River
Wheeler Park
Downtown Air Park
Rotary Park
Wiley Post Park
Woodson Park
Capitol Hill H.S.
South Comm. Hosp.
CLASSEN
LINWOOD BLVD.
GEN. PERSHING AVE.
EXCHANGE
R. S. KERR AVE.
RENO AVE.
SHIELDS BLVD
MAY
PENNSYLVANIA
WESTERN
ROBINSON
BROADWAY
LINCOLN BLVD.

OKLAHOMA CITY

Agnew Ave......H-3,K-3
Agnew Pl...........J-3
Allen..............E-3
Aurora.............A-3
Avenida.............J-4
Barnes Ave......A-3,C-3
...............G-3,K-3
Barnes Pl...........J-3
Beals Pl............G-5
Beatrice Ave.........F-2
Billen Ave...........B-3
Binkley..............I-3
Birch................F-4
Blackwelder Ave.
.........A-5,E-4,I-4,K-5
Brauer Ave......B-5,C-5
Broadway....F-7,H-7,J-7
Broadway, E.........I-7
Broadway, W.........I-7
Broadway Pl.........C-7
Brookline Ave.
........B-1,I-1,J-2,K-1
Byers Ave...........I-8
California Ave.......E-6
Carey Pl............B-4
Cedar Ave...........F-1
Central Ave.........D-8
Chestnut Dr.........K-4
Classen Blvd........D-6
Classen Dr..........C-6
Clegern.............I-5
College.............B-5
Compress............E-7
Concord Dr..........D-6
Couch Dr............D-6
Cross Ave...........B-3
Dale Ave............C-7
Daugherty..........G-4
Dewey Ave...B-6,F-6,K-6
Doffing.............E-2
Douglas Ave.........B-5
...............C-5,I-5
Douglas Ave., N.....D-5
Douglas Blvd........K-5
Drexel Ave.......J-1,K-1
Drexel Blvd.........B-1
Durland.............D-8
Ellison Ave.........B-5
...........C-5,E-5,J-5
Exchange St.........F-1
Florida Ave.........A-4
Flynn Ave...........B-3
Francis Ave.........B-6
...............D-6,K-6
Gatewood Ave.......B-4
Grand Blvd..........J-3
Gurnsey Ave.........B-6
Harrison Ave........D-8
Harvey Ave..........B-7
...........F-7,G-7,J-7
Heyman..............G-4
Hudson Ave..B-7,F-7,J-7
Independence Ave.
...............C-1,J-1
Indiana Ave......A-4,E-4
.............F-4,G-4,I-4
James St............J-4
Johnson Dr......J-4,K-4
Kentucky Ave.......A-4
...............E-4,I-4
Kinkaid Dr..........J-4
Klein Ave.......C-5,D-5
..........E-5,J-5,K-5
Lafayette Dr........K-4
Land Ave.......C-1,K-1
Lee Ave......B-6,F-6,J-6
Lincoln Blvd........B-8
Linden..............F-4
Lindley.............E-2
Linn Ave............A-2
............C-2,E-2,I-2
Linwood Blvd........D-5
Madison St..........A-8
Main St.............E-6
Magdelena...........K-4
May Ave.........D-2,J-2
McKinley Ave.......C-5
...........E-5,J-5,K-5
McRae...............F-2
Mehl................H-5
Melrose Ln..........K-7
Miller Ave....C-2,E-2,J-2
Miller Blvd.........C-2
Miltary Ave..........A-5
Morton Ave..........J-2
Murray Dr.......K-1,K-2
Noble...............E-2
North Pl.......B-1,B-2
Oak St..............G-1
Oklahoma Ave.......A-8
...........C-7,D-8,J-8
Olie Ave.....B-6,J-6,K-6
Park Dr.............I-3
Park Pl.............C-7
Park St.............C-1
Parkview Ave........I-1
Pennsylvania Ave.
...............C-4,K-4
Pettee Ave..........G-4
Pioneer St..........A-1
Pott................F-2
Rancho Dr.......J-4,K-4
Reding Dr...........J-4
Reno Ave.....E-1,E-4,E-8
Reno Ave., E........E-8
Robinson Ave.
...........B-7,F-7,J-7
Ross Ave........A-2,C-2
...............E-2,K-2
Rotary Dr...........G-4
R. S. Kerr Ave.......D-6
Sage Ave............K-7
Santa Fe Ave.
...........A-7,J-7,K-7
Schuneman St...G-1,G-2
Shartel Ave...F-6,J-6,K-6
Shartel Ct..........C-1
Sheridan Ave........E-6
Sherwood Ave........J-2
Shields Blvd.....H-8,K-8
South Dr............K-4
Stanford............H-4
Stiles Ave....A-8,D-8,I-8
Sullivan St.......F-1,F-2
Suzberger...........G-4
Texoma Dr...........K-2
Thomas Pl...........B-1
Tuttle..............E-3
Villa Ave........A-3,B-3
...............E-3,K-3
Virginia Ave.
...........B-4,D-4,I-4
Walker Ave...B-6,F-6,K-6
Walnut Ave......A-8,D-8
...............E-8,H-8
Watson Ave..........J-2
Western Ave.B-5,E-5,K-5
Westwood Ave......H-4
Williams Dr.........K-4
Woodward...........C-1
Woodward Ave..H-1,J-1
Youngs Ave.......G-3,K-3
Youngs Blvd.....A-3,C-3
Youngs Pl...........J-3
1st St.......D-2,D-4,D-8
1st St., N.W........E-3
2nd St.......D-2,D-4,D-6
...............E-2,E-6
3rd St.......D-3,D-4,D-6
...............E-2,E-6
4th St..........D-4,D-6
5th St..........D-4,D-6
6th St..........D-4,D-6
7th St..........C-4,D-6
8th St.......C-4,C-6,F-3
9th St..........C-4,C-6
...............F-3,F-6
10th St.........C-1,C-6
...........F-3,F-7,F-8
11th St.........C-1,C-6
.........F-3,F-5,F-7,F-8
12th St......C-1,C-2,C-6
...........F-3,F-5,F-7
13th St.........B-1,C-2
.......C-6,G-3,G-5,G-7
14th Pl..............G-5
14th St......B-1,B-2,B-6
...........G-3,G-5,G-7
15th St......B-1,B-2,B-6
.......G-1,G-4,G-7,G-8
16th St......B-1,B-4,B-6
...........G-1,G-4,G-8
17th St......B-1,B-4,B-6
...........G-1,G-4,G-8
18th St......B-1,B-4,B-6
...........G-1,G-4,G-8
19th St......B-1,B-4,B-6
...........G-1,G-4,G-8
20th St......B-1,B-4,B-6
...........G-1,G-8,H-4
21st St......A-1,A-4,A-6
...........H-1,H-4,H-7
22nd St......A-1,A-4,A-6
...........H-1,H-3,H-7
23rd St.........A-1,A-5
...........H-1,H-2,H-6
24th St......A-1,A-5,A-8
...........H-1,H-2,H-6
25th St......A-1,A-5,A-8
...............H-1,H-6
26th St......A-1,A-5,A-6
.......A-7,A-8,H-1,H-6
27th St......A-1,A-5,A-6
...........A-7,H-1,H-6
28th St............I-1,I-6
29th St............I-1,I-6
30th St............I-2,I-7
31st St............I-2,I-7
32nd St............I-2,I-7
33rd St............I-2,I-7
34th St............I-2,I-7
35th St............I-2,I-7
36th St..............J-7
37th St.......J-1,J-4,J-8
38th St.......J-1,J-2,J-8
39th Pl..............J-1
39th St..........J-2,J-8
40th St..........J-1,J-3
...............J-6,J-8
41st St...J-1,J-3,J-6,J-8
42nd St...J-1,J-2,J-6,J-8
43rd St..........J-1,J-3
...............J-6,J-8
44th St.........K-3,K-7
45th St.......K-1,K-5,K-8
46th Pl..............K-2
46th St.......K-1,K-5,K-8
46th Terr............K-2
47th St.........K-1,K-3
...............K-5,K-8
48th St.........K-1,K-2
...........K-3,K-5,K-8
49th Pl..............K-7
49th St.........K-1,K-2
...........K-3,K-5,K-8
50th St...K-2,K-3,K-5,K-8

POINTS OF INTEREST

Downtown Air Park..........G-6
McKinley Park......C-5
Oklahoma State Fair Grounds..........D-1
Rotary Park.........G-4
State Capitol........A-8
Tolan Park..........E-5
Wheeler Park.........F-6
Wiley Post Park.....G-7

OMAHA

A St............H-1,H-2
Abbott Dr...........C-7
Arbor St........G-1,G-2
...........G-3,G-5,G-7
Archer Ave..........K-6
Arthur St............I-7
Atlas St.............I-7
B St.........H-1,H-6,I-5
Bancroft St..........G-1
...........G-2,G-5,G-7
Barker Ave..........E-2
Blaine St............I-7
Blake St.............H-8
Blondo..........A-1,A-5
Buckingham Ave..I-4,J-4
Burdette....A-2,A-5,A-6
Burt St......C-1,C-3,C-5
C St..............I-1,I-5
Cady Ave............A-6
Caldwell St......B-1,B-5
California St........C-1
...............C-3,C-6
Capitol Ave......D-1,D-6
Cass St......C-1,C-3,C-5
Castelar St..G-3,G-4,G-6
Center St..F-2,F-5,F-6,F-7
Charles St...B-1,B-4,B-6
Chicago St..........C-1
Clark St........B-6,B-7
Cottage St...........C-4
Culew St.............H-1
Cuming St...C-1,C-2,C-6
D St........I-2,I-3,I-5,I-7
Dahlman Ave.....I-3,J-4
Davenport St.........D-1
..............D-3,D-6
Dayton St............I-2
Decatur.........A-1,B-4
Deer Park Blvd......H-6
Dewey Ave...........D-3
Dodge St...D-1,D-3,D-5
Dorcas St.......G-6,G-8
Douglas St......D-2,D-5
DuPont St...........G-4
E St...........I-2,I-3,I-5
Ed Creighton Ave....G-4
Elm St..........G-4,G-5
Erskine St.......A-1,A-2
F St...........I-1,I-3,I-5
Farnam St...D-1,D-3,D-5
Forest Ave...........F-7
Frances St....F-2,F-3,G-8
Franklin St.......B-1,B-4
Frederick St.........G-1
...........G-2,H-4,H-7
G St...........I-1,I-2,I-5
Garfield St..........I-6
Gibson Rd...........J-7
Gold St.............G-3
Gordon St...........G-3
Grace St............A-6
Grant St.....A-1,A-5,A-6
Grover......H-1,H-3,H-8
H St..........I-2,J-4,J-5
Hamilton St......B-1,B-4
Happly Hollow Blvd.............. B-1
Harney St.......D-3,D-5
Hascall St...H-1,H-3,H-8
Hawthorne St.......B-2
Hickory St...........F-2
...............F-4,F-5,F-6
Hillsdale Ave.........J-2
Hoctor...............I-6
Holmes St...........J-1
Homer St............I-7
Howard St.......D-1,D-5
Hugo St..............I-7
I St..............J-1,J-2
...............J-3,J-5,J-6
Izard........C-1,C-2,C-5
J St..............J-1,J-2
...............J-3,J-4,J-5
Jackson St...........D-1
...........D-3,E-4,.E-6
John Crieghton Blvd............A-3,B-3
Jones St.........E-3,E-6
K St......J-1,J-2,J-4,J-5
Kayah St.............I-7
Krugh Ave...........H-1
L St.......J-1,J-2,J-3,J-5
Lafayette St..........B-1
...............B-2,B-4
Lake St......A-1,A-2,A-5
Leavenworth........ E-1
...............E-3,E-6
M St.........J-1,J-2,J-5
Marcy St.........E-2,E-3
...............E-4,E-5,E-6
Marinda St...........G-2
Martha St...G-3,G-6,G-8
Mason St.E-3,E-4,E-5,E-7
Mayberry St......E-1,E-2
Miller St............I-7
Missouri Ave........J-6
Morton...............F-2
Myrtle Ave...........B-4
N St.............J-2,K-5
Nicholas St..B-1,B-4,B-6
Northwest Military ... Ave.... A-1,A-2,B-2
O St.........K-1,K-3,K-5
Oak St...........G-1,H-5
Ohern St.............K-1
Ontario St...........I-6
Orchard Ave.....J-1,J-2
P St..............K-1,K-5
Pacific St...........E-1
...............E-3,E-7
Park Ave............F-4
Parker..............B-4
Parker St............A-1
Patterson........J-2,J-3
Patrick St.......A-2,A-5
Paul St..............B-5
Phelps St............I-6
Pierce St.....E-1,E-5,E-6
Pine St....F-2,F-3,F-6,F-7
Poppelton Ave........E-2
...............E-4,F-8
Q St.........K-1,K-3,K-5
R St..............K-2,K-4
Radial Hwy......A-2,B-2
Richelieu St..........G-8
Riverview Blvd..G-8,H-8
Rees St..............E-5
S St.........K-1,K-2,K-5
Saddle Creek Rd.
...........B-1,D-2,E-1
St. Mary's Ave.......E-5
Seward St...B-1,B-4,B-7
Shady Ln............H-2
Sherwood Ave.......A-7
Shirley St...........F-2
South Omaha Bridge
...............J-7,K-8
Spring St.........H-1,H-3
.......H-4,H-5,H-6,H-8
Turner Blvd..........E-4
Underwood Ave.....C-1
Valley St............H-2
...........H-3,H-4,H-8
Van Camp St.........I-6
Victor St............A-6
Vinton St............H-2
...........H-3,H-4,H-5
Wakely St............C-2
Walnut St........F-1,F-4
Webster St...C-1,C-3,C-6
Weir St..............K-1
Western St...........B-1
William St........F-2,F-6
Willis...............A-6
Woodland Rd.........H-8
Woolworth Ave.......F-2
...............F-4,F-7
Wright St............G-3
Yates St.............A-6
3rd St...............F-8
4th St............F-8,H-8
5th St........A-8,F-8,G-8
6th St............D-8,G-8
7th St............E-7,G-7
8th St............E-7,F-7
9th St............E-7,F-7
10th St..............F-7
11th St.......C-7,D-7,F-7
12th St.........C-7,D-7
...........E-7,F-7,J-7
13th St.......C-7,F-7,J-7
14th St.......D-7,F-7,J-6
15th St.......D-6,F-6,J-6
16th St...B-6,D-6,F-6,J-6
17th St.........D-6,F-6,J-6
18th Ave..B-6,D-6,F-6,J-6
19th St...B-6,D-6,G-6,J-6
20th St...B-6,D-6,G-6,J-6
21st St..........B-6,C-6
...........F-6,G-6,J-6
22nd St...B-6,C-6,F-6,J-6
23rd St...C-5,E-5,G-5,J-5
24th Ave............D-5
24th St...C-5,F-5,G-5,J-5
25th Ave.........D-5,F-5
25th St..........B-5,D-5
...........F-5,G-5,J-5
26th Ave.............D-5
26th St...B-5,F-5,G-5,J-5
27th St...B-5,F-5,J-5,K-5
28th Ave.............H-5
28th St..............K-4
29th St.......B-4,F-4,K-4
30th Ave.............F-4
30th St.........B-4,D-4
...........E-4,G-4,K-4
31st Ave.........D-4,E-4
31st St.......B-4,D-4,E-4
........F-4,G-4,J-4,K-4
32nd Ave........E-4,G-4
32nd St.........B-4,C-4,
........D-4,F-4,J-4,K-4
33rd St.......B-4,D-4,F-4
........G-4,K-4,I-4,J-4
34th Ave.............A-4
34th St..........B-4,D-4
........F-4,G-4,I-4,J-3
35th Ave.........D-3,F-3
35th St..........B-4,D-4
........F-4,G-4,I-4,J-4
36th Ave.......C-3,I-3,K-3
36th St..........B-3,D-3
...............F-3,I-3,K-3
37th St.......D-3,I-3,K-3
38th Ave.......E-3,G-3,H-3
38th St..........B-3,D-3
...........H-3,I-3,K-3
39th Ave.........I-3,K-3
39th St..........C-3,E-3
........G-3,H-3,I-3,K-3
40th St..........C-3,E-3
........H-3,O-3,J-3,K-2
41st St..........C-2,E-2
...........H-2,I-2,J-2,K-2
42nd Ave.............J-2
42nd St.......C-2,G-2,J-2
43rd St.......G-2,H-2,K-2
44th Ave.........H-2,I-2
44th St.......E-2,G-2,H-2
45th St..........B-2,C-2
........G-2,H-2,J-2,K-2
46th Ave.........G-2,J-1
46th St.......G-2,J-2,K-1
47th St..........G-1,J-1
48th Ave.........A-1,D-1
...............G-1,J-1,K-1
48th St...B-1,D-1,G-1,K-1
49th Ave.A-1,D-1,G-1,K-1
49th St...A-1,D-1,G-1,K-1
50th St...D-1,G-1,J-1,K-1
50th Terr...........K-1
51st Ave.............A-1
51st St.......D-1,F-1,H-1

POINTS OF INTEREST

Clarkson Hospital...D-2
Crieghton University.C-5
Douglas County Hospital...........E-3
Lutheran Hospital...D-5
Graceland Park Cemetery..........J-2
Hanscom Park.......F-4
H. Doorly Zoo.......H-7
Hitchcock Park.......K-2
Holy Sepulchre Cemetery..........D-1
Medical Center.......D-2
Metro Tech South...K-4
Rosenblatt Stadium..I-7
St. Mary Magdalene Cemetery..........K-1
St. Jospeh Hosp.....C-4
Spring Lake Park.....I-6
Union Stockyards...K-4
Veterans Hospital....F-3

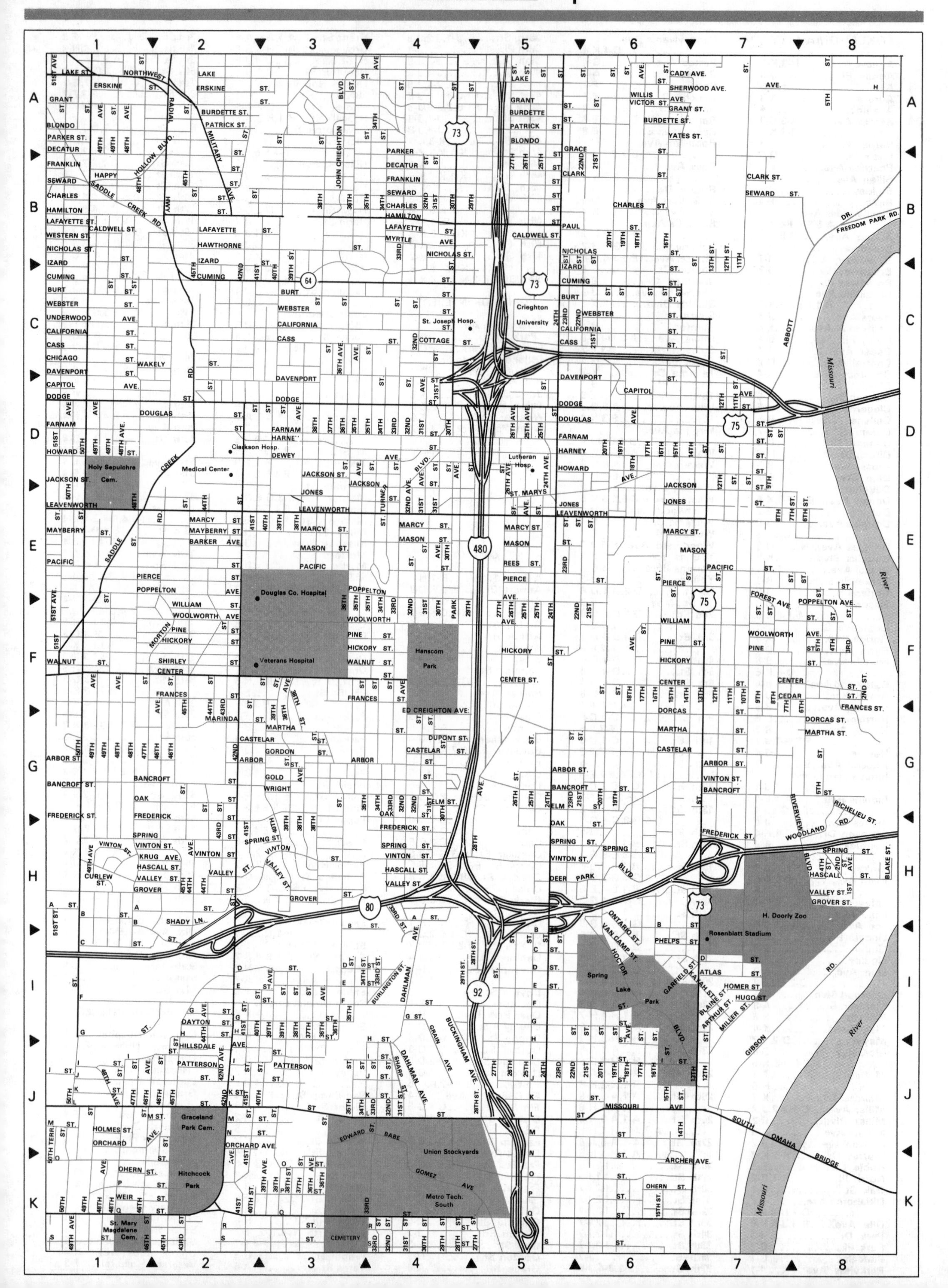
Scale of Miles
0 .2 .4 .6
N
Missouri River
Creighton University
St. Joseph Hosp.
Clarkson Hosp.
Medical Center
Lutheran Hosp.
Holy Sepulchre Cem.
Douglas Co. Hospital
Veterans Hospital
Hanscom Park
H. Doorly Zoo
Rosenblatt Stadium
Spring Lake Park
Graceland Park Cem.
Hitchcock Park
Union Stockyards
Metro Tech. South
St. Mary Magdalene Cem.
CEMETERY
SOUTH OMAHA BRIDGE
FREEDOM PARK RD.
SADDLE CREEK RD.
NORTHWEST RADIAL HWY.
MILITARY AVE.
HAPPY HOLLOW BLVD.
JOHN CRIEGHTON BLVD.
DODGE
DOUGLAS
FARNAM
LEAVENWORTH
PACIFIC
CUMING
CENTER ST.
ED CREIGHTON AVE.
HICKORY
PINE
WOOLWORTH
MARCY
MASON
FREDERICK ST.
SPRING
VINTON ST.
HASCALL ST.
VALLEY ST.
GROVER ST.
DEER PARK BLVD.
DAHLMAN AVE.
BUCKINGHAM AVE.
GRAIN AVE
EDWARD ST. BABE GOMEZ AVE
MISSOURI AVE
ARCHER AVE.
RIVERVIEW BLVD.
GIBSON
ABBOTT
73
75
80
92
480
64

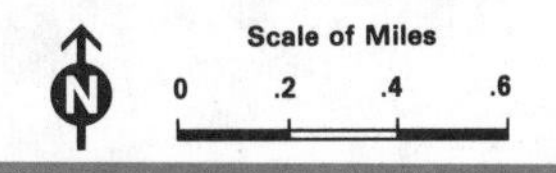
N
Scale of Miles
0 .2 .4 .6

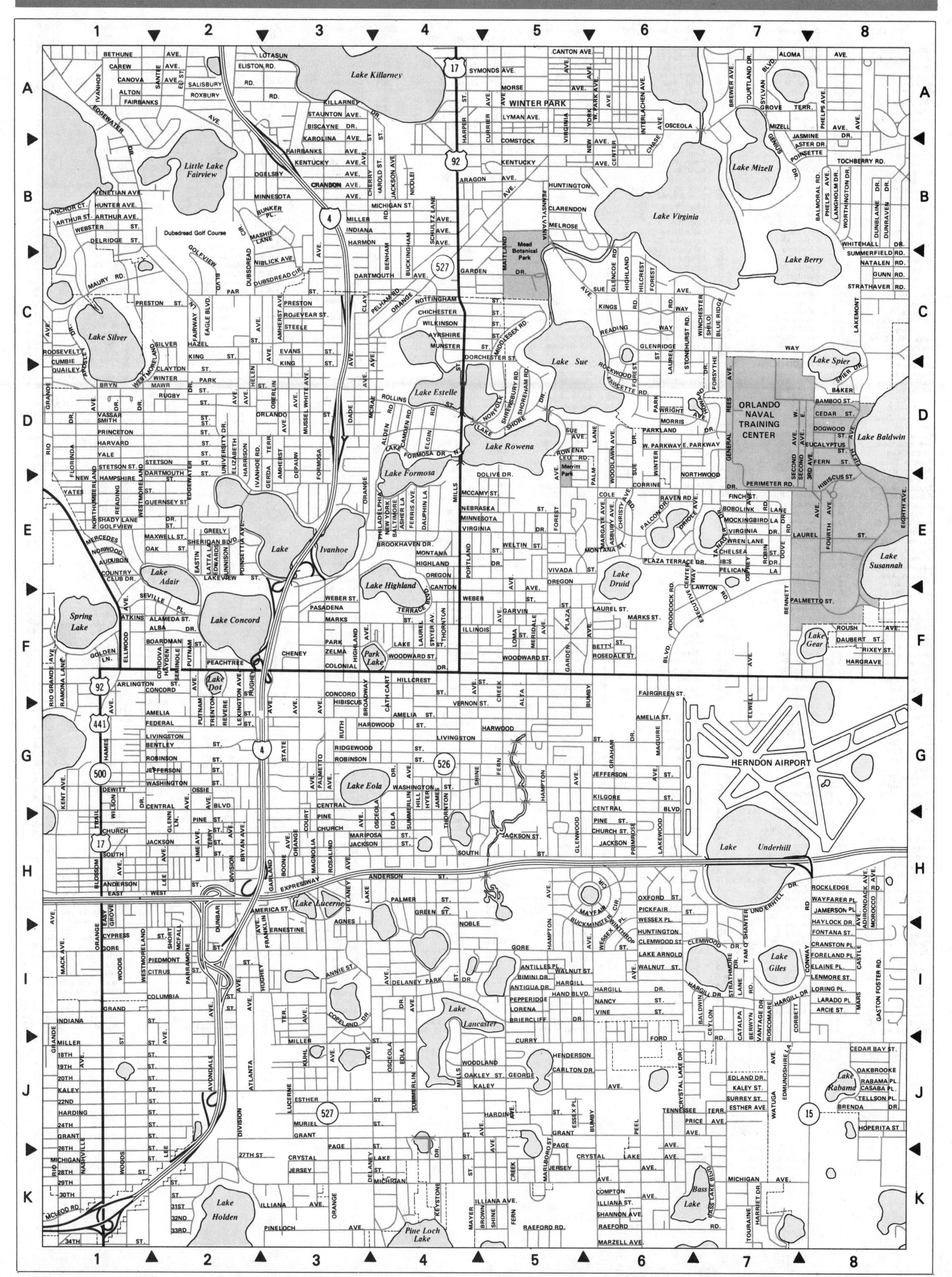
Lake Killarney
WINTER PARK
Little Lake Fairview
Dubsdread Golf Course
Lake Virginia
Lake Mizell
Lake Berry
Mead Botanical Park
Lake Silver
Lake Sue
Lake Spier
Lake Estelle
Lake Rowena
ORLANDO NAVAL TRAINING CENTER
Lake Baldwin
Lake Formosa
Merritt Park
Lake Ivanhoe
Lake Adair
Lake Highland
Lake Druid
Lake Susannah
Spring Lake
Lake Concord
Lake Gear
Park Lake
Lake Dot
HERNDON AIRPORT
Lake Eola
Lake Underhill
Lake Lucerne
Lake Giles
Lake Lancaster
Lake Rabama
Bass Lake
Lake Holden
Pine Loch Lake
EXPRESSWAY
CANTON AVE.
SYMONDS AVE.
MORSE
LYMAN AVE.
COMSTOCK
KENTUCKY
ARAGON
OSCEOLA
ALOMA
FAIRBANKS
MINNESOTA
ORANGE
COLONIAL
CONCORD
LIVINGSTON
ROBINSON
CENTRAL
CHURCH
SOUTH
ANDERSON
GORE
MICHIGAN
ILLIANA AVE.
RAEFORD RD.
MARZELL AVE.
PINELOCH
1 2 3 4 5 6 7 8
A B C D E F G H I J K

ORLANDO

Adanson Ave........A-1
Adirondack Ave.....H-8
Agnes..............H-3
Alameda St.........F-2
Alba Dr............F-2
Alden Rd...........D-4
Alta Loma Ave......F-5
Alton..............A-1
Amelia St....G-1,G-4,G-6
America St.........H-3
Amherst Ave....C-3,D-3
Anchor Ct..........B-1
Anderson St.....H-1,H-4
Annie St...........I-3
Antigua Dr.........I-5
Antilles Pl........I-5
Aragon Ave.........B-4
Ardsley Dr.........C-1
Arlington St.......F-1
Arthur Ave.........B-1
Arthur St..........B-1
Asbury Ave.........E-6
Asher Ln...........E-4
Aster Dr...........A-7
Atkins.............F-1
Atlanta Ave........J-2
Audobon............E-1
Avondale Ave.......J-2
Ayrshire St........C-4
Baker..............D-8
Baldwin............I-7
Balmoral Rd........B-8
Baltimore..........E-4
Bass Lake Blvd.....K-7
Benham Rd..........C-4
Bennett Rd.........F-7
Bentley St.........G-2
Berwyn Rd..........I-7
Bethune Ave........A-1
Betty St...........F-6
Bimini Dr..........I-5
Biscayne Dr........A-3
Blossom Trail......H-1
Blueridge..........C-7
Boardman St........F-2
Bobolink Lane......E-7
Boone Ave..........H-3
Brenda Dr..........J-8
Brewer Ave.........A-7
Briarcliff Dr......I-5
Broadway Ave.......F-3
Brookhaven Dr......E-4
Brown Ave..........K-5
Bryan Ave..........H-2
Bryn Mawr St.......D-1
Buckingham.........C-4
Buckminster Cir....H-5
Bumby Ave.....F-6,J-6
Bunker Pl..........B-3
Camden Rd..........D-4
Canova Ave.........A-1
Canton Ave.........E-4
Carew Ave..........A-1
Carlton Dr.........J-5
Casaba Pl..........J-8
Catalpa Lane.......I-7
Cathcart...........F-4
Cedar Bay St.......J-8
Cedar St...........D-8
Center Ave.........B-6
Central Blvd.G-2,G-3,G-6
Ceylon.............I-7
Chanson Ave........B-3
Chase Ave..........B-6
Chelsea............E-7
Cheny..............F-3
Cherry St..........B-3
Chichester St......C-4
Christy Ave........E-6
Church St...H-1,H-3,H-6
Citrus St..........I-2
Clarendon..........B-5
Clay...............C-3
Clayton St.........D-2
Clemwood Dr........I-7
Clemwood St........I-6
Cole Rd............E-6
Colonial...........F-3
Columbia St........I-2
Compton............K-6
Comstock Ave.......A-5
Concord........F-2,F-3
Conway Rd..........I-8
Copeland Dr........I-3
Corbett............I-7
Cordova............F-2
Corrine Dr.........E-6
Country Club Dr....E-1
Court..............H-3
Cranston Pl........I-8
Crystal Lake St...K-3,K-6
Cumbie.............C-1
Currier Ave........A-5
Curry..............J-5
Cypress............I-1
Dade Ave...........D-3
Dartmouth Ave......C-4
Dartmouth St.......D-2
Daubert St.........F-8
Dauphin Ln.........E-4
Delaney Ave.....H-3,K-3
Delaney Park Dr....I-4
Delridge St........B-1
De Pauw Ave........D-3
Dewitt.............G-1
Division Ave.......J-2
Dogwood St.........D-8
Dolive Dr..........D-5
Dorchester St....C-4,C-5
Dove Dr............E-7
Dubsdread Cir......C-3
Dunbar.............H-2
Dunblaine Dr.......B-8
Dunraven Dr........B-8
East...............H-1
Easy Grove.........H-1
Edgewater..........E-2
Edgewater Dr.......A-1
Edland Dr..........J-7
Edmundshire Ln.....J-7
Edwards............E-2
Eighth Ave.........E-8
Elaine Pl..........I-8
Elgin Rd...........D-4
Elizabeth Ave......D-2
Ellwood Ave........F-1
Eola Dr............H-4
E. Parkway.........D-7
Ernestine..........I-3
Essex Pl...........J-5
Esther St..........J-3
Eucalyptus St......D-8
Evans St...........C-3
Executive Center
Way................F-7
Fairbanks Ave......B-3
Fairgreen St.......F-6
Falcon Dr..........E-6
Fawcette Rd........D-6
Federal St.........G-2
Fern St............D-8
Fern Creek Ave.
...............F-5,G-5
Ferris Ave.........E-4
Fifth Ave..........D-8
Florinda Dr........D-1
Fontana St.........I-8
Ford Rd............J-6
Foreland Pl........I-8
Forest Ave.........E-5
Forest Rd.......C-6,D-6
Formosa St.........D-3
Forsythe...........D-7
Fourth Ave.........E-8
Franklin...........I-3
Garden Dr..........C-4
Garden Plaza.......F-5
Garland............H-3
Garvin St..........F-5
General Reed Ave...D-7
Genius Dr..........B-7
George Ave.........J-5
Gerda Terr.........D-3
Glencoe Rd.........C-6
Glenn Ln...........H-2
Glenridge Way......C-6
Glenwood Ave.......H-5
Golden Lane........F-1
Golfview Blvd......B-2
Golfview St........E-1
Gore...............I-1
Gore St............I-5
Graham.............G-6
Grand St...........I-1
Grant Ave.......J-3,J-5
Grant St........J-1,J-3
Green St...........H-4
Grove Terr.........A-7
Guernsey St........E-2
Gunn Rd............C-8
Hames Ave..........G-1
Hampton Ave.....G-5,I-5
Hand Blvd..........I-5
Harding St......J-1,J-5
Hardwood St.....G-3,G-5
Hargill............I-5
Hargill Dr.........I-6
Hargrave...........F-8
Harmon Ave.........B-3
Harper St..........A-4
Harriet Dr.........K-7
Harrison Ave.......D-2
Harvard St.........D-1
Hayden.............F-2
Haylock Dr.........H-8
Hazel St...........C-2
Helen Ave..........D-2
Henderson..........J-5
Hibiscus...........F-3
Hibiscus St........E-8
Highland.....C-6,E-4,F-3
Hill...............G-4
Hillcrest..........C-6
Hillcrest St.......F-4
Hoperita St........J-8
Howard Dr..........D-7
Hughey.............I-3
Hunter Ave.........B-1
Huntington......B-5,I-6
Hyer...............G-4
Ibis Dr............E-7
Illiana Ave.....K-3,K-5
Illiana St.........K-6
Illinois...........F-4
Indiana Ave........B-3
Indiana St.........I-1
Ivanhoe............A-1
Ivanhoe Rd.........D-2
Jackson St..H-2,H-3,H-5
James..............G-4
Jasmine Dr.........A-8
Jefferson St.
.............G-1,G-2,G-6
Jersey Ave......K-3,K-5
Kaley Ave..........J-5
Kaley St...........J-1
Karolina Ave.......A-3
Kent Ave...........G-1
Kentucky Ave....B-3,B-5
Keystone Dr........K-4
Kilgore St.........G-6
Killarney Dr.......A-3
King St.........C-2,C-3
Kings Way..........C-6
Kuhl Ave...........J-3
Lake Ave...........H-3
Lake St.........F-4,K-4
Lake Arnold........I-6
Lake Formosa Dr. N.
...................D-4
Lakefront Ave......C-8
Lake Shore Dr......D-5
Lakeview St........E-2
Lakewood Ave.......H-6
Langholm Dr........B-8
Lattala............E-2
Laurel.............F-4
Laurel Rd.......C-6,D-6
Laurel St.......E-7,F-6
Lawson Dr..........I-7
Lawton.............E-7
Lawton Rd..........E-7
Lee Ave.........H-2,K-2
Lenmore St.........I-8
Leu Rd.............D-5
Lexington Ave......G-2
Lime Ave...........H-2
Livingston.........G-4
Livingston St.
...............G-2,G-4
Lorena.............I-5
Loring Pl..........I-8
Lucerne Ter........J-3
Mack Ave...........I-1
Magnolia...........H-3
Maguire Blvd.......G-6
Maitland Ave.......C-5
Margate Ave........E-6
Mariposa St........H-3
Marks St........F-3,F-6
Marlboro St........K-5
Mars Castle Rd.....I-8
Marzell Ave.....A-7,K-6
Mashie Lane........B-3
Maxwell St.........E-1
Mayer St...........K-4
McCamy St..........E-4
McFall.............I-2
McRae Ave.......D-3,D-4
Melrose............B-5
Mercedes...........E-1
Meridale Ave.......F-5
Merritt Park.......D-5
Michigan...........K-4
Michigan Ave.......K-7
Miller Ave.........B-3
Miller St.......J-1,J-3
Mills..............E-4
Mills St...........J-4
Minnesota..........E-4
Minnesota Ave......B-3
Mockingbird Ln.....E-7
Montana St......E-4,E-6
Morocco Ave........H-8
Morris Dr..........D-6
Morse Ave..........A-5
Munster St.........C-4
Muriel St..........J-3
Mussel White Ave...D-3
Nancy St...........I-6
Nashville Ave......K-1
Natalen Rd.........C-8
Nebraska...........E-4
New Hampshire St.
...................E-1
New York...........E-4
New York AVe.......A-6
Niblick Ave........C-3
Noble..............H-4
Norfolk............D-5
Northumberland.....E-1
Northwood..........D-6
Norwood............E-1
Nottingham St......C-4
Oak St.............E-2
Oakbrook...........J-8
Oakley St..........J-4
Oberlin Ave........D-3
Ogelsby Ave........B-3
Orange.......C-4,H-3,I-1
Orange Ave.........E-3
Oregon..........E-4,E-5
Oriole Ave.........E-6
Orlando St.........D-3
Osceola.........A-6,H-4
Osceola Ave........J-4
Osceola Dr.........J-4
Osprey Ave.........E-7
Ossie..............G-2
Oxford St..........H-6
Page Ave........J-3,J-5
Palm Ln............D-6
Palmer St..........H-4
Palmetto St.....F-7,G-3
Par................C-2
Park...............F-3
Parkland Dr........D-6
Parramore..........I-2
Pasadena...........F-3
Peachtree..........F-2
Peel Ave...........J-6
Pelican La.........E-7
Pennsylvania.......B-5
Pepperidge.........I-5
Perimeter Rd.......E-7
Phelps Ave......A-8,B-8
Philadelphia.......E-4
Pickfair St........H-6
Piedmont St........I-2
Pine St......H-2,H-3,H-6
Pineloch Ave.......K-3
Plaza Terrace Dr...E-6
Poinsette..........B-7
Poinsettia Ave.....E-2
Portland...........E-4
Preston.........C-1,C-3
Price Ave..........J-6
Primrose Dr........H-6
Princeton St.......D-1
Putnam..........F-2,G-2
Quailey............D-1
Rabama Pl..........J-8
Raeford Rd......K-5,K-6
Ramona Lane........F-1
Raven Rd...........E-6
Reading............E-1
Reading Way........C-6
Revere.............E-4
Ridgewood St.......G-3
Rio Grande Ave.
.............D-1,F-1,K-1
Rixey St...........F-8
Robin Rd...........E-7
Robinson St.....G-1,G-3
Rockledge Rd.......H-8
Rockwood...........D-6
Roosevelt..........C-1
Rosalind...........H-3
Roscomare..........I-7
Rosedale St........F-6
Rosevear St........C-3
Roush Ave..........F-8
Rowena Rd..........D-5
Rugby St...........D-2
Ruth...............G-3
Schultz Lane.......B-4
Second Ave. E......D-7
Second Ave. W......D-7
Seminole...........F-2
Seville Pl.........F-1
Shady Lane Dr......E-1
Shannon Ave........K-6
Sheridan Blvd......E-2
Shilo..............C-7
Shine Ave.......G-4,K-5
Shoreham Rd........D-5
Short..............I-2
Shrewsbury Rd......D-5
Silver.............C-2
Smith St...........D-1
South St........H-1,H-4
Spier Dr...........D-8
Staunton Ave.......A-3
Steele.............C-3
Stetson St......D-1,D-2
Strathaver Rd......C-8
Strathmore Dr......I-7
Summerfield Rd.....C-8
Sue Ave.........C-6,D-5
Sue Dr.............D-6
Summerlin Ave..H-4,J-4
Surrey St..........J-7
Tam O'Shanter......I-7
Tanager Dr.........E-7
Tellson Pl.........J-8
Tennessee Terr.....J-6
Terrace Blvd.......F-4
Thornton...........H-4
Touraine...........K-7
Trenton............G-2
Underhill Dr.......H-7
Vantage Dr.........I-7
University Dr......D-2
Vassar St..........D-1
Venetian Ave.......B-1
Vine St............I-6
Virginia...........E-4
Virginia Ave.......A-5
Virginia Dr........E-7
Vision Ave.........H-2
Vivada St..........E-5
Walnut St.......I-5,I-6
Washington St......G-2
Wayfarer Pl........H-8
Weber St........F-3,F-4
Webster St.........B-1
Weltin St..........E-5
Wessex Pl.......H-6,I-6
West...............H-2
W. Park Ave........A-6
W. Parkway.........D-6
Westmoreland Dr.
...............E-1,I-1
Whitehall Dr.......B-8
Wilkinson St.......C-4
Wilson Ave.........G-1
Winchester.........C-7
Winter Park St.
...............D-2,D-6
Winthrop St........I-6
Woodcock Rd........F-6
Woodland...........J-4
Woodlawn Ave.......D-6
Woods..............I-1
Woods St...........K-1
Woodward St.....F-4,F-5
Worthington Dr.....B-8
Wren Lane..........E-7
Wright Ave.........D-6
Yale St............D-1
Yates St...........E-1
Zelma..............F-3
18th St............J-1
19th St............J-1
20th St............J-1
22nd St............J-1
24th St............J-1
26th St............J-1
27th St.........K-1,K-2
28th...............K-1
29th St............K-1
30th St............K-1
31st...............K-2
32nd...............K-2
33rd...............K-2

POINTS OF INTEREST

Herndon Airport....G-7
Mead Botanical
Park...............B-5
Merritt Park.......D-5
Orlando Naval
Training Center....D-7

PHILADELPHIA

Abigail............A-8
Addison............I-6
Alder..............A-5
Allen..............D-8
Alter..............J-1
Alter St...........J-2
Amber..............A-8
American........F-7,I-7
Annin..............J-1
Annin St........J-1,J-3
Appletree..........G-3
...............G-5,G-6
Arch...............G-1
Arch St.....G-2,G-3,G-6
Arizona............A-8
Arlington..........A-3
Aspen St...........E-2
Bailey St.......C-1,K-1
Bainbridge St......I-2
Bambrey.........D-2,K-2
Bancroft........A-4,I-4
Baring.............F-1
Beach..............E-8
Belgrade...........C-8
Benjamin Franklin
Bridge.............F-7
Benjamin Franklin
Parkway............G-3
Berks...........A-3,A-6
Beulah.............K-6
Bodine.............F-7
Bodine St..........A-7
Bolton.............C-2
Boston.............A-8
Bouvier St......A-3,C-4
...............J-3,K-3
Bowers.............D-4
Brandywine St......E-2
Bread..............G-7
Broad St.....C-4,G-4,I-4
Brown..............D-1
Brown St...........D-4
Bucknell........B-2,E-2
Burns..............G-4
Butler..........H-5,I-5
Buttonwood St...F-3,F-4
Cabot St...........C-3
Cadwallader........C-7
Callowhill St......F-3
Camac...........D-5,K-5
Camac St...........H-5
Cambridge.......D-5,D-7
Cambridge St.......D-3
Cameron............D-3
Canal St...........E-8
Capitol............K-3
Capitol St.........E-3
Carlisle........I-4,K-4
Carlisle St.....A-4,C-4
Carlton............F-3
Carlton St.........F-5
Carpenter St.......J-3
Catharine.......I-1,I-2
Catharine St.......I-7
Chadwick........A-4,I-4
Chancellor..H-2,H-3,H-7
Chancellor St.
...............H-3,H-5
Chang..............D-1
Chelton............F-7
Cherry.............G-1
Cherry St.......G-3,G-6
Chestnut St........H-3
Christian St...J-1,J-3,J-5
Church St..........G-7
Clarion............K-5
Clarion St.........G-5
Clay...............E-3
Clay St............E-5
Cleveland St....J-3,K-3
Clifton......F-5,G-5,I-5
Clinton............H-5
Clover.............G-5
Clymer.............I-3
Colona.............A-5
Colorado...........A-4
Colorado St.....J-4,K-4
Columbia....B-1,B-6,C-8
Commerce St.....G-3,G-5
Cooper.............D-8
Coral..............A-8
Corinthian Ave.....E-3
Crease.............C-8
Croskey.....B-2,G-2,I-2
Cross...........K-2,K-5
Cuthbert........G-4,G-7
Cuthbert St........G-3
Cypress St......H-2,H-3
...............H-4,H-6
Dakota.............A-8
Darien..........I-6,J-6
Dauphin.........A-3,A-8
Day St.............C-8
DeGray..........G-4,G-5
Delancey St........H-2
...............H-3,H-7
Delaware Ave....G-8,I-8
Diamond.........A-3,A-6
Dickinson.......K-1,K-5
Dondill Pl.........C-5
Dorrance St.....J-3,K-3
Dover........A-1,C-1,K-1
Dreer..............A-8
Earl...............C-8
Edgley.............A-2
Edgley St..........A-2
Elfreth's Al.......G-7
Ellsworth..........J-1
Ellsworth St.......J-2
Emerald............A-8
Etting.............K-1
Eyre...............B-1
Fairhill.....A-6,I-6,K-6
Fairmount Ave...E-2,E-5
Fawn...............A-5
Federal St.........J-2
Filbert............G-7
Filbert St.........G-5
Fitzwater St....I-2,I-7
Fletcher........A-1,A-8
Flora..............D-4
Fountain...........A-4
Francis............D-4
Frankford Ave......C-8
Franklin........E-6,K-6
French..........A-1,A-4
Front..............C-8
Front St...........H-7
Fulton St..........I-7
Garnet.............K-3
Gaskill............I-7
George.............D-3
George St..........D-7
Germantown Ave.....B-6
Gerritt......K-1,K-3,K-6
Girard Ave......D-1,D-6
Glenwood........A-2,B-1
Green St........E-3,E-7
Greenwich..........K-2
...............K-5,K-7
Hamilton...........F-1
Hamilton St.....F-3,F-4
Hancock.........B-7,C-7
...............F-7,I-7
Harlan..........C-2,C-3
Harper.............D-5
Harper St..........D-3
Hicks...........F-4,K-4
Hicks St...........I-4
Hollywood..........C-1
Hope...............C-8
Howard..........B-8,K-7
Hutchinson St......A-5
...........D-5,G-5,I-5
Ingersoll.......C-2,C-3
Ingram St..........J-1
Ionic...........H-2,H-6
Iseminger.......A-5,K-5
Jefferson.......C-1,C-6
Jessup.............I-5
John F. Kennedy
Blvd............G-1,G-3
Judson.............B-2
Judson St..........E-2
Juniper.........I-4,K-4
Kater St........I-3,I-4
Kenilworth.........I-4
Kershaw............C-3
Kimball St......J-2,J-5
Lambert.........B-3,K-3
Latimer St......H-2,H-4
Latona..........J-1,J-3
Laurel.............D-7
Lawrence........A-7,C-7
...........E-7,F-7,H-7
League St.......J-3,J-6
Lee................D-8
Leithgow St........A-7
...........D-7,H-7,I-7
Lemon St...........E-5
Leopard............D-8
Letitia............G-7
Locust St..........H-3
Lombard St......I-2,I-7
Ludlow St.......G-3,G-5
Manning St.........H-2
...............H-3,H-4
Manton St.......J-1,J-3
Market St.......G-3,G-6
Marlborough........C-8
Marshall........C-6,J-6
Marston......A-1,C-1,K-1
Martha.............A-8
Mascher.........C-7,G-7
Master..........C-2,C-6
Melon St...........E-4
Melvale............D-8
Meredith St........E-2
Milford.........I-6,J-6
Mole.........F-4,I-4,K-4
Monroe.............I-7
Montgomery......B-1,B-6
Montrose...........J-6
Montrose St........J-3
Moravian........H-3,H-5
Morse..............B-6
Mott...............J-4
Mt. Vernon.........E-2
Moyamensing Ave.
...............J-7,K-7
Myrtle.............D-5
Myrtlewood.........C-1
Nassau.............C-2
Natrona............F-1
Naudian St.........I-2
Nectarine.......E-3,F-4
Nevada.............A-5
New St.............F-7
Newkirk.........A-1,C-1
Newson.............B-6
Nicholas...........B-2
Noble St...........F-4
Norris.......B-3,B-6,B-8
North...........E-3,E-5
N. College Ave.....D-2
Oakford............J-1
Ogden..............D-5
Ogden St...........D-3
Olive St...........E-2
Opal...............K-3
Opal St............I-3
Orianna.........A-7,D-7
...............F-7,G-7
Orkney.............A-6
Oxford.......C-1,C-6,C-8
Page............A-2,A-4
Palmer.............B-8
Palsthrop..........B-7
Panama.......H-2,H-3,H-5
Park...............C-4
Parrish.........D-1,D-5
Passyunk Ave.......K-5
Pearl...........F-1,F-3
Pearl St...........F-5
Peltz..............J-1
Pemberton St....I-2,I-7
Penn St............E-8
Pennock............D-1
Percy...........D-5,I-5
Perkiomen..........D-3
Perot..............E-2
Perth...........H-6,I-6
Philip..........C-7,H-7
Pine St............I-2
Point Breeze Ave.
...............K-2,K-3
Pollard............D-7
Poplar St.......D-2,D-6
Potts..............E-4
Powelton Ave.......F-1
Quarry St.......G-5,G-7
Queen St...........J-7
Quince St..........H-5
Race St.........G-3,G-7
Ramp A.............F-2
Ramp B.............F-3
Ramp C.............F-2
Ramp D.............F-2
Randolph........F-6,I-6
Ranstead St........G-2
...............G-3,G-6
Redner.............C-2
Reed............K-1,K-6
Reese...........A-6,H-6
...............I-6,K-6
Reno...............D-4
Richmond...........D-8
Ridge Ave.......B-2,F-5
Ringgold........B-2,K-2
Ringgold St........E-2
Rodman St..........I-2
...............I-3,I-5
Rosewood........H-4,I-4
St. James St.......H-6
Sansom St..........H-3
Schell.......G-6,H-6,I-6
Schuylkill Ave.....I-1
Schuylkill Expwy.
...............H-1,I-1
Sears..............K-1
Sedgley............B-1
Seybert.........C-2,C-3
Shackamaxon........C-8
Sharswood.......C-2,C-3
Shirley............E-3
Smedley.........H-4,I-4
South St...........I-2
S. College Ave.....D-2
Spring..........F-4,F-5
Spring Green St.
...........E-4,E-5,F-1
Spruce St.......H-2,H-7
Stamper............I-7
Starr Garden.......I-6
Stewart............C-2
Stiles St..........D-3
Stilman............K-2
Stockbeck..........H-3
Strawberry......G-7,H-7
Summer..........F-1,F-5
Summer St..........F-4
SusquehannaA-3,A-6,A-8
Swain..............E-2
Sydenham St........A-4
...............C-4,H-4
Taney St.....E-1,I-2,K-1
Tasker..........K-2,K-5
Taylor St....B-2,E-2,K-2
Thompson........C-2,C-8
Titan..............K-1
Titan St...........K-3
Tulip..............B-8
Turner.............C-2
Uber St............B-3
Van Horn...........D-7
Van Pelt........G-3,I-3
Van Pelt St........H-3
Vine St.........F-2,F-4
Walden.............G-3
Wallace............E-6
Wallace St.........E-3
Walnut St.......H-3,H-7
Warnock.........H-5,I-5
Washington Ave..J-1,J-5
Waterloo...........B-7
Watts...........A-4,E-4
...........G-4,I-4,K-4
Waverly St......I-2,I-5
Webster St......I-1,I-3
Westmont...........A-4
Wharton St......K-1,K-6
Wilcox St..........E-3
Wilder..........K-1,K-3
...............K-6,K-7
Wildey..........D-7,D-8
Willington.........C-4
Willow St..........F-6
Winter..........F-1,F-5
Winter St..........F-3
Wood............F-3,F-6
Wood St............F-5
Woodstock.......B-3,K-3
York...............A-8
2nd St..........B-7,I-7
3rd St..........B-7,I-7
4th St..........C-7,I-7
5th St..........C-6,I-6
6th St..........C-6,I-6
7th St..........C-6,I-6
8th St..........C-6,I-6
9th St..........C-6,I-6
10th St.........C-5,I-5
11th St.........C-5,I-5
12th St.........C-5,I-5
13th St.........C-5,I-5
15th St.........B-4,I-4
16th St.........B-4,I-4
17th St.........B-4,I-4
18th St.........B-3,I-3
19th St.........B-3,I-3
20th St.........B-3,I-3
21st St.........B-3,I-3
22nd St.........B-2,I-2
23rd St.........C-2,E-2
...............H-2,J-2
24th St.........C-2,E-2
...........F-2,H-2,J-2
25th St......C-2,I-2,J-2
26th St.........C-1,I-2
...............J-1,K-1
27th St.........A-1,C-1
...............I-2,K-1
28th St......A-1,C-1,K-1
29th............C-1,K-1
30th St.........B-1,C-1
...............H-1,K-1
31st St............H-1
32nd............G-1,H-1
33rd St.........G-1,H-1

POINTS OF INTEREST

Fairmount Park.....E-1
Franklin Field.....H-1
Franklin Square....F-6
Independence
Square.............H-6
Independece Square
Mall...............G-6
Jefferson Square...J-7
Logan Circle.......F-3
Norris Square......A-7
Passyunk Square....K-5
Rittenhouse
Square.............H-3
Vine St. Bridge....F-2
Washington Square..H-6
Wharton Square.....K-2

N

Scale of Miles
0 .1 .2 .3 .4

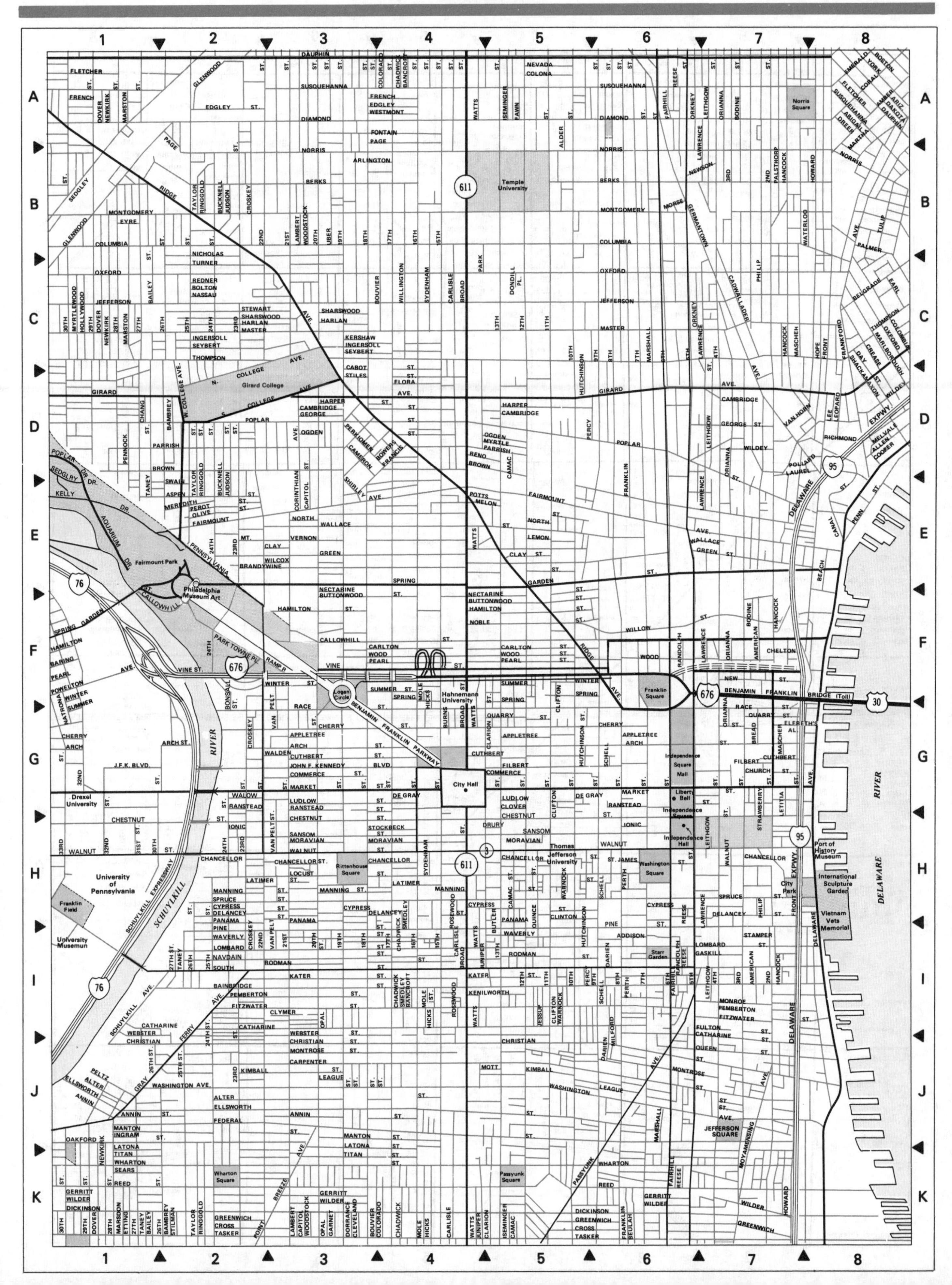

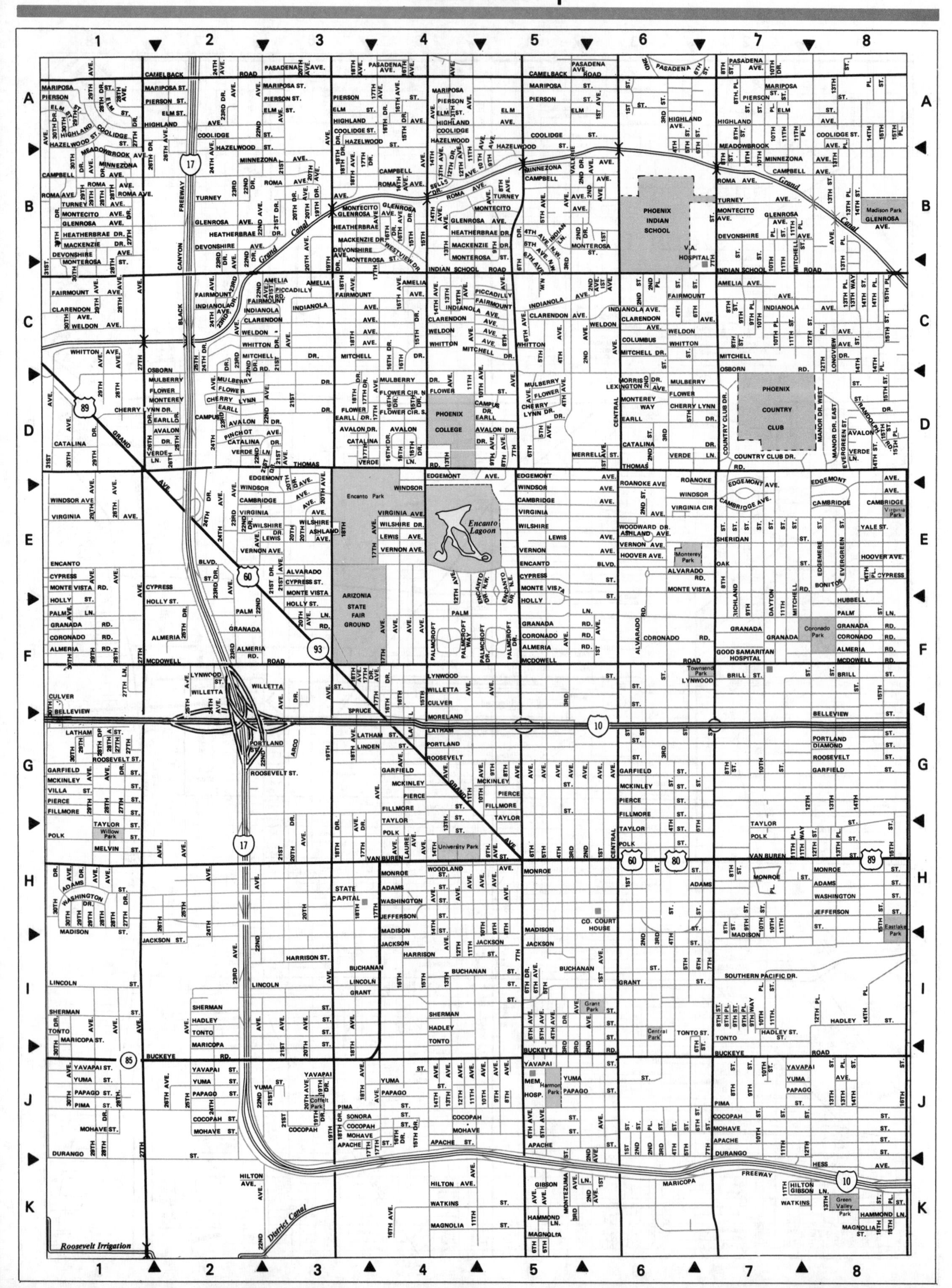
Scale of Miles
0 .2 .4 .6
N
PHOENIX INDIAN SCHOOL
V.A. HOSPITAL
Madison Park
PHOENIX COLLEGE
PHOENIX COUNTRY CLUB
Encanto Park
Encanto Lagoon
ARIZONA STATE FAIR GROUND
Monterey Park
Coronado Park
GOOD SAMARITAN HOSPITAL
Townsend Park
Willow Park
University Park
STATE CAPITAL
CO. COURT HOUSE
Eastlake Park
Grant Park
Central Park
Coffelt Park
MEM. HOSP.
Harmon Park
Green Valley Park
Grand Canal
District Canal
Roosevelt Irrigation
BLACK CANYON FREEWAY
CAMELBACK ROAD
INDIAN SCHOOL ROAD
THOMAS
MCDOWELL ROAD
VAN BUREN
BUCKEYE RD.
MARICOPA FREEWAY
GRAND AVE.
17
60
93
89
10
80
85

PITTSBURGH

Abstract J-5
Adams C-1
Addison D-7
Admiral A-6
Agnew K-7,K-8
Ajax C-8
Albany G-1
Alcor D-4
Alice I-6
Allegheny C-3
Allen G-6
Allequippa E-7
Alplaus J-7
Alries K-7
Altaview K-4
Alton I-3
Alturia J-3
Alverado I-2
Amabell F-3
Amanda H-6,J-6
Anderson D-5
Anthony I-7
Arcena D-7
Arch C-4,D-4
Arion H-4
Arlington G-5,H-7
Armandale B-3
Arthur D-6
Atmore B-1
Attica E-1
Augusta F-2
Avery C-5
Badger A-6
Bailey G-5
Bank D-4
Banksville H-2,J-1
Bark A-4
Barker D-5
Barlton J-7
Basin B-6
Bausman I-6
Bayonne I-2
Bealty J-3
Beaufort K-3,K-4
Beaver C-1
Becks Run J-8
Bedford D-6
Beech C-3
Beechview I-2
Beelen E-7
Behan C-3
Belasco I-2
Belinda D-7
Belham I-2
Belle Isle K-3
Bellingham J-1
Belmont D-2
Belonda G-3
Beltzhoover G-5
Bensonia K-2
Bentley E-7
Berg H-8
Berwin K-3
Bexley I-1
Biatto A-6
Bidwell C-2
Bigelow D-6,E-5
Bigger I-4
Biggs A-4
Bigham F-3
Bingham F-6
Birmingham J-7
Birtley K-3
Bloom I-4
Bluff F-6
Boggs G-4,I-4
Boggston H-4
Blvd. of the Alles E-5,F-6
Bon Air I-5
Bonifay I-8
Boustead J-2
Bowmore J-2
Boyd E-6
Boyle B-4
Brabec B-6
Brackenridge D-8
Bradley E-2
Brady F-8
Brahm B-5
Branchport B-1
Breed G-7
Brenham E-7
Brent H-8
Brereton C-7
Brighton B-2,C-3
Brighton Pl. B-3
Brightridge B-3
Broadway J-2,K-2
Brook J-6
Brookside I-3
Brosville G-6
Brownsville J-6
Bruner J-7
Buchanan B-6
Buena Vista B-3
Buente A-5
Burgess A-3
Burham G-8
Burrows E-7
Butler A-8
Cadet I-4
Cagwin J-2
Cairo C-3
Calhoun J-6
California B-1,B-2
Calle J-5
Campfield J-5
Candace J-2
Canton I-2
Caperton I-5
Capital J-3
Carl I-6
Carnahan I-1
Carrie B-4
Carrington B-3
Carson F-6
Casement A-1
Cassatt D-6
Cathedral I-7
Catoma B-4
Cecl Pl. E-4
Cedar C-4
Cedarhurst H-5
Cedricton J-6
Celtic K-6
Centre D-7,D-8,E-6
Chalfont H-5
Chancery E-4
Chaplotte A-8
Chappel I-1
Charles A-3,B-3,H-5
Chateau C-2
Chauncey C-8,D-8
Chautaugua B-3
Cherryhill, E. K-7
Cherryhill, W. K-6
Cherry Way E-5
Chester A-3
Church I-7
Claim A-6
Clarance F-2
Clark E-6
Clayton B-3
Cliff D-6
Clifferty D-3
Climax H-5
Clinton G-7
Clover H-8
Cloverdale J-6
Coast J-3
Cobden H-8
Coffey E-5
Cohasset F-3
Colerain K-6
Colfax B-2
Cologne H-8
Colorado B-1
Columbus C-2
Colwell E-6
Commonwealth E-4
Compton I-1
Concordia K-7
Conkling C-8
Connecticut K-1
Conniston J-5
Copperfield K-6
County J-5
Cowan G-4
Cowley A-6
Craighead G-5
Crailo K-7
Crane H-3
Crawford E-6
Creswell I-8
Crispen A-3
Crosby J-2
Crysler J-4
Curtin H-5
Curtis H-4
Cushman G-4
Cuthbert G-4
Dagmar I-3
Daleland I-1
Dalemount J-1
Damas A-6
Danbury A-2
Dasher D-4
Davenport D-7
Dawes H-7
Dawn I-4
Daytona I-6
Dell Ave. K-2
Dellrose K-6
Delray E-4
Denham B-4
Devilliers D-7
Diamond C-4
Dickson A-1
Dilworth G-3
Dinwiddie E-7
Divinity B-3
Dobson B-8
Dodds K-5
Donora A-5
Dowling K-7
Drum B-3
Drycove J-5
Duff D-8
Dunster J-4
Durham J-1
East G-6
Eathan K-3
Edenvale B-4
Edgebrook K-4
Edgemont G-5
Edith F-2
Edle D-7
Eccles H-8
Elba D-8
Eldora I-5
Eleanor H-8
Elkton D-1
Ellers E-7
Elliott E-1
Elmbank K-4
Elmira A-4
Elmore D-7
Elsdon A-3
Engstler I-8
Enoch D-7
Erie I-7
Erin D-7
Estella G-5
Etna D-5
Eureka G-5
Eutaw G-4
Fairacres J-3
Fairview E-1
Fallowfield I-2
Federal B-4
Ferncliff K-4
Fernhill K-4
Fiat K-5
Fifth E-5,E-6
Fingal E-2
First E-5
Firth B-6
Fisher I-8
Fitler H-8
Flach H-8
Fontella C-2
Forbes E-5,E-6
Fordyce I-5
Foreland C-5
Forsythe A-1
Fort Duquesne Blvd. E-4
Fort Pitt Blvd. E-4
Fort Pitt Tunnel F-2
Fountain B-4
Fourth E-5
Francis C-8
Franklin C-2
Fredell J-7
Freeland H-5
Freemont K-2
Freyburg G-6
Frontier A-5
Fulton C-2,D-2
Galveston C-3
Garden A-6
Gardner B-6
Garrison D-5
Gearing H-5
General D-4
Georgia I-6
Gibson E-1,E-6
Giffin I-6,I-7
Gifford A-1
Giller J-6
Gist E-7
Glade K-6
Gladys H-2
Glenrose A-4
Goebel D-2
Goering B-5
Goettmann B-6
Goldbach H-8
Goldstrom I-2
Grace G-3
Grandview F-4
Granite C-8
Grant E-5
Grape I-6
Gray G-4
Graymore I-2
Green D-7
Greenbush G-4
Greenside I-1
Greenleaf Pl. F-2
Greentree F-1
Greenleaf E-1,E-2
Griffin G-4
Grimes H-6
Grogan I-6
Grove D-7
Haberman G-5
Hackstown G-7
Hallett D-8
Hallock F-3,G-2
Hallowell K-5
Halsey B-1
Hampshire I-2
Hanover G-6
Harbor A-6
Harlen A-3
Harmony K-5
Harpster A-6
Hartford G-6
Hartranft K-4
Harwood G-5
Haslage B-5
Haug A-6
Hays I-6
Hazelton A-4
Hechkleman A-7
Heinz C-6
Hemans D-7
Hemlock B-4
Henderson B-4
Henger I-8
Herman B-6
Herndon E-1
Herron B-8
Hestor G-1
Hetzel A-6
Hibbs H-3
High B-6
Highwood A-1
Hodgkiss A-1
Holbrook A-1
Hollace C-8
Holyoke A-3
Homer B-6
Hooper E-6
Hornaday K-7
Horner E-2
Horton C-8
Howard, East A-4
Iberia G-5
Independence F-1
Industry H-5
Ingham A-1
Institute I-5
Irwin B-3
Isabella D-4
Island B-1,B-2
Itan B-5
Jacksonia C-3
Jacob I-7
James C-4
Jane G-7
Jasper H-4
Jefferson B-3
Jessie J-1
Jonquil H-7
Josephine G-8
Jucunda H-5
Jumonville E-7
Juniata C-2,C-8
Kambach G-5
Kathleen G-5
Kenberma I-3
Kenwood A-3
Kingsboro H-5
Kiraley J-3
Kirkbridge B-2
Kirkpatrick D-8
Kirsopp I-1
Knapp A-1
Knoll B-5
Knox B-1,H-6
Kunkle B-2
Kuth G-4
Labella F-3
Lacock D-4
Lacona J-7
Lafayette A-4
Lafferty H-4
Lager B-6
Lampe G-1
La Marido K-4
La Moine K-4
Lanark B-4
Lander E-1
Langley B-3
Laplace D-8
Lappe A-5
La Rose J-4
Laughlin J-6,J-7
Lawson D-7
Leavitte J-4
Leech A-8
Leister B-5
Lelia H-4
Leolyn K-7
Leticoe G-8
Letsche B-4
Liberty C-7
Liberty, West K-3
Lighthill D-2
Ligonier A-8
Lilian H-5
Lincoln, N. C-3
Linda J-4
Linial I-4
Linneview J-7
Linton D-7
List A-5
Lithgow B-4
Liverpool C-2
Lockhart C-5
Locust E-7
Lofink A-7
Loleta K-6
Lombard E-7
Lonergan J-3
Longmore I-3
Los Angeles K-2
Louisiana K-1
Lovelace E-1
Lowen G-4
Lowenhill H-3
Lowrie B-6
Lupton F-2
Lyceum A-4
Lynnbrook K-4
Lyzell A-4
Mackinaw K-2
Maddock D-5
Maginn A-3
Mahon D-8
Main E-1
Manhattan C-2
Manilla D-6
Marena D-1
Marengo H-7
Margaret I-6,I-7
Marion E-7
Market St. E-4
Marloff K-4
Marlow D-1
Marquis C-3
Marvista B-2
Mary G-7
Maryland I-6
Mathews I-6
Mathias A-5
May I-3
Maydell J-1
Mayville K-4
Maywood A-3
McClintock B-3
McCullough B-3
McIntyre A-3
McKean F-5
McKeever A-2
McKinley I-6
McKinney Ln. G-8
McLain G-6
Meadville B-4
Medhurst I-1
Melrose B-3
Mercer D-6
Merchant C-4
Meredith K-7
Meridan F-3
Merrimac F-3
Merriman F-7
Meta F-3
Metcalf B-3
Methyl I-2
Metropolitan C-1
Meyer, E. J-7
Meyers, W. K-6
Michigan H-5
Millbridge G-6
Miller E-7
Minooka J-6
Minsinger H-4
Mission G-8
Mohawk E-7
Monastery G-7
Monterey C-3
Moore I-6
Moredale K-5
Morgan C-8
Morrison B-2
Moultrie E-7
Mt. Joseph K-6
Mt. Oliver H-6
Mountain I-8
Mountford A-4
Moye H-7
Mullins A-1
Muriel F-6
Napoleon I-2
Natchez G-4
Neeld J-2
Neff G-4
Neidell E-1
Neptune E-1
Neville B-8
New Hampshire B-2
Newton G-6
New York G-1
Niggel A-7
Nixon B-1
Nobles J-6,K-6
Noblestown F-1
Norman B-3
North C-2
Northcrest I-4
North Shore Dr. D-4
Norton G-3
Noster A-6
Oakhurst K-6
Oakville H-1
O'Hern B-3
Ohio, E. B-6
Ohio, W. C-4
Ohio River Blvd. D-3
Oliver E-5
Olympia F-3
Oneida F-3
Onyz I-6
Orangewood I-3
Orchard H-5
Oriana B-2
Ormsby I-7,I-8
Orr E-7
Osgood A-4
Ottillia I-7
Overbeck A-5
Page C-2
Palisade A-2
Palm Beach K-2
Palo Alto C-3
Parallel J-7
Parkhurst C-4
Park Low I-5
Parkwood I-8
Pasadena G-5
Patterson H-8
Paul H-4
Pauline J-2
Paulowna B-8
Pawnee F-3
Peekskill A-4
Penn C-7,D-5,I-7,J-8
Penn-Lincoln Parkway F-6
Pennsylvania C-2
Peralta C-5
Perrilyn K-3
Perry D-7
Perrysville B-3
Petunia K-5
Phineas C-5
Piermont F-3
Pioneer I-4,K-3
Pitler A-1
Pius G-7
P. J. McArdle F-4
Plainview K-3
Platt I-3
Plough A-1
Plunkett G-2
Plymouth F-3
Pocono Dr. H-1
Popargrove K-7
Potomac Ave. K-1
Preble C-1
Pressley C-5
Pride E-6
Princess I-3
Progress C-5
Prospect G-4
Protectory D-6
Province B-6
Pynchon A-4
Quarry H-6
Queen B-5
Quincy J-6
Ravine H-4
Rectenwald I-7
Reddour C-4
Redlyn I-6
Redrose K-7
Reed D-8
Reedsdale D-2
Reese E-2
Regina G-6
Reifert I-6
Republic F-2
Resaca C-3
Rescue A-6
Rhine A-6
Rhode Island G-1
Ridge D-2
Ridgemont G-1
Ridgeway C-7,D-6
Ridgewood B-3
Ringwalt J-3
Risby H-2
Rising Main B-4
River C-6
River Blvd. D-4
Robinson D-5,D-8
Rochelle I-6
Rockland I-3
Rockledge A-6
Romanhoff A-5
Rose E-7
Roseanne J-1
Roseberry I-1
Rosegarden J-1
Roseton I-5
Ross E-5
Rothman H-8
Royal A-5
Rubicon G-3
Rugraff H-6
Russell A-3
Ruth G-4
Rutherford I-2
Rutledge E-2
Ruxton G-5
St. Ives B-2
St. Joseph I-7
St. Lucas I-7
St. Paul G-7
St. Thomas G-7
Salerna G-1
Salisbury H-8
Sampsonia C-3
Sanctus G-1
Sarah G-7
Saranac K-2
Saw Mill Run H-3,J-5
Schuler I-8
Scott Pl. D-5
Schafer H-8
Scripway E-5
Seabright A-4
Sebring I-3
Sedgewick C-2
Seneca E-7
Seventh E-5
Seward E-2
Shadycrest G-2
Shaler F-2
Sharon G-6
Sheffield C-2
Sherman C-4
Shiras J-2
Shulze H-7
Sidney F-7,F-8
Sigel B-2
Simms H-4
Sixth E-5
Smallman C-7
Smith Field E-5
Soffel H-4
Soho D-8
Somers D-7
Sophia B-5
Sorrell A-1
Southcrest J-4
Southern H-4
Spahrgrove G-3
Spring B-6,H-8
Springfield F-1,G-1
Spiral A-4
Sprucewood K-7
Square D-7
Stamm H-7
Stanwik E-4
Stayton A-2
Steil I-8
Stetson J-3
Steuben E-1
Stevenson E-6
Stockholm C-8
Stockton D-4
Strachan K-1
Stranmore B-2
Strauss A-2
Stromberg G-8
Success B-2
Suffolk A-4
Suismon C-5
Sulgrove F-3
Summer G-8
Suncrest I-6
Sunderman A-7
Superior A-1
Sweetbriar F-2,G-3
Sycamore F-2,F-3
Sylvania H-4,H-5
Taft H-4
Tarragunna J-6
Taylor C-3
Terrace E-8
Texdale J-3
The Boulevard J-7
Thelma A-1
Thielman K-6
Third E-5
Timberland J-5
Tinsbury A-6
Tippet J-1
Tonapah J-2
Topeka I-8
Transverse I-7,J-7
Traymore I-4
Trent D-7
Triana J-7
Tripoli C-5
Tropical H-2
Trost J-6
Troy Hill B-6
Tucker G-1
Tumbo A-1
Tuscola G-4
Tyrone C-7
Ulysses F-4
Valonia D-1
Van Bram E-7
Varley A-5
Vaughn E-2
Venango A-5
Vincent H-5
Vickroy E-6
Vine E-7
Vinial B-6
Virginia F-3
Vista B-5
Vodeli K-2
Voskamp B-6
Wabash E-1
Wade G-4
Wadsworth E-7
Wagner J-7
Walde I-7
Walden G-4
Walnut H-7
Walter H-6,I-7
Wandless C-8
Warden F-1
Warfield C-5
Warning D-8
Warren B-4
Warrington H-4
Washington E-6
Watson E-6
Watt C-8
Weaver F-1
Webster D-7
Well F-2
Wenke H-8
Wentworth I-2
Wenzell K-2
Western D-2
Westfield I-3
Westmont K-6
Westwood G-4
Wharton F-7,F-8
White C-8
Whited K-5
Whiteside D-7
Wick E-7
Wieland G-7
Wiggins B-8
Wilber J-6
Wilbert G-3
Wilhelm E-1
William G-4
William Penn E-5
Wilmar F-2
Wilmerding G-2
Wilson A-3
Winfield B-2
Wingate J-7
Winton G-5
Wisdom G-6
Woessner K-3
Wolford K-3
Wood E-5
Woodcover J-1
Woodland A-1
Woodruff G-2,G-3
Woodward K-3
Woodville F-1
Wooster D-7
Wurzell A-4
Wyandotte E-7
Wylie D-7,E-5
Wynoka J-7
Wyola F-2
Wyoming G-4
Yetta B-5
Younger J-1
Zara H-5
Zimmerman K-5
1st St. K-8
1st St., S. F-5
3rd, S. F-5
4th, S. F-5
5th, S. F-6
6th D-5
6th, S. F-6
7th D-5
7th, S. G-6
8th D-5
8th, S. F-6
9th D-5
9th, S. F-6
10th D-5
10th, S. G-6
11th, S. G-6
12th D-6
12th, S. G-6
13th D-6
13th, S. G-7
14th D-6
14th, S. G-7
15th D-6
15th, S. G-7
16th D-6
16th, S. G-7
17th C-6
17th, S. G-7
18th C-6,F-7
18th, S. F-7,G-7,H-7
19th C-6,F-7
19th, S. F-7
20th C-6,F-7
20th, S. F-7
21st C-6
21st, S. G-8
22nd C-6
22nd, S. G-8
23rd C-6
24th C-7
24th, S. G-8
25th C-7
25th, S. G-8
26th B-7
26th, S. G-8
27th B-7
28th B-7
29th B-7
30th B-8
31st St. B-8
32nd B-8
33rd A-8
34th A-8
35th A-8
36th A-8
38th A-8

POINTS OF INTEREST

Brady St. Bridge F-8
East Park C-5
Fort Duquesne Bridge D-4
Fort Pitt Bridge E-4
Grandview Park G-5
Herr's Island A-7
Liberty Bridge F-5
McKinley Park I-5
Melon Square Park E-5
Mt. Washington Park G-3
Ninth St. Bridge D-5
North Park C-4
Olympia Park F-3
Point State Park E-4
Roberto Clemente Memorial Park D-3
Seventh St. Bridge D-5
Sixteenth St. Bridge C-6
Sixth St. Bridge D-4
Smithfield St. Bridge F-5
Southside Park H-7
Tenth St. Bridge F-6
West Park C-3,C-4
Westend Bridge D-2
31st St. Bridge A-7

PHOENIX

Adams H-7
Adams St. H-1-8
Almeria F-2
Almeria Rd. F-1-8
Alvarado E-3
Alvarado Rd. E-6,F-6
Amelia C-3,C-4
Amelia Ave. C-7
Apache St. J-3-7
Arco Dr. G-3
Ashland E-3
Ashland Ave. E-6
Avalon D-2,D-4,D-8
Avalon Dr. D-2-5
Belleview St. F-1,G-8
Black Canyon Frwy. B-2-C-2
Bonitos E-8
Brill St. F-7,F-8
Buchanan I-3
Buchanan St. I-4,I-5
Buckeye Rd. J-1-7
Cambridge E-8
Cambridge Ave. E-2-7
Camelback Rd. A-2,A-5
Campbell Ave. B-1-7
Campus D-2
Campus Dr. D-5
Catalina Dr. D-1-6
Central Ave. D-6-H-6
Cherry Lynn D-2
Cherry Lynn Dr. D-1-8
Clarendon Ave. C-1-6
Cocopah J-3
Cocopah St. J-2-7
Columbus Ave. C-6
Coolidge A-4
Coolidge St. A-1-8
Coronado Rd. F-1-8
Country Club Dr. D-7
Culver St. F-1,F-4,G-3
Cypress E-1,E-8
Cypress St. E-1,E-5
Dayton St. F-7
Devonshire Ave. B-1-7
Diamond St. G-8
Durango St. J-1-7
Earll D-2
Earll Dr. D-2-6
Edgemere St. E-8
Edgemont D-2,D-3
Edgemont Ave. D-4-8
Elm St. A-1-8
Encanto Blvd. E-1,E-5
Encanto Dr. N.E. E-5
Encanto Dr. N.W. E-5
Evergreen St. E-8,E-8
Fairmount C-2-5
Fairmount Ave. C-1-7
Fillmore St. G-1-6
Flower D-2-6
Flower cir. N. D-4
Flower Cir. S. D-4
Flower St. D-4
Garfield G-4
Garfield St. G-1-8
Glenrosa B-4
Glenrosa Ave. B-1-8
Gibson Ln. K-5-7
Granada F-7
Granada Rd. F-1-8
Grand Ave. D-1,G-4
Grant St. I-3-6
Hadley St. I-2-8
Hammond Ln. K-5,K-8
Harrison St. I-3,I-4
Hazelwood A-4
Hazelwood St. A-1,A-4
Hazelwood St. A-2,A-5
Heatherbrae B-3
Heatherbrae Dr. B-1-4
Hess Ave. K-8
Highland A-1
Highland Ave. A-2-7
Hilton K-7
Hilton Ave. K-2,K-3
Holly St. E-1,F-3,F-5
Hoover Ave. E-6,E-8
Hubbell F-8
Indian Ln. B-5
Indianola C-2,C-3
Indianola Ave. C-2-7
Indian School Rd. C-4-7
Jackson I-4,I-5
Jackson St. I-1-5
Jefferson St. H-4-8
Latham St. G-1-4
Laurel Ave. G-4,H-4
Lewis Ave. E-3-5
Lexington Ave. D-6
Lincoln I-3
Lincoln St. I-1-3
Linden St. G-3,G-4
Longview Ave. C-8
Lynwood F-6
Lynwood St. F-2-4
Mackenzie Dr. B-1-4
Madison H-7
Madison St. H-1-5
Magnolia K-5
Magnolia St. K-4,K-8
Manor Dr. East D-8
Manor Dr. West D-8
Maricopa Frwy. K-6
Maricopa St. I-1,I-2
Mariposa A-1,A-4
Mariposa St. A-2,A-5
McDowell Rd. F-2-8
McKinley G-4
McKinley St. G-1-6
Meadowbrook Ave. A-1,A-7
Melvin St. H-1
Merrell St. D-5,D-6
Minnezona B-1
Minnezona Ave. B-2-7
Mitchell Dr. C-2-7
Mitchell St. B-7,F-7
Mohave J-3
Mohave St. J-1-7
Monroe St. H-4-8
Montecito B-3
Montecito Ave. B-1-7
Monterey D-2
Monterey Way D-6
Monterosa St. B-1-6
Monte Vista E-3,E-5
Monte Vista Rd. E-1,E-6
Montezuma St. K-5
Moreland G-3
Moreland St. G-1-8
Morris Dr. D-6
Mulberry D-2,D-5
Mulberry Dr. D-2,D-4
Mulberry St. D-6
Oak St. E-7
Osborn Rd. C-2-7
Palm Ln. F-1-8
Palmcroft F-4
Palmcroft Dr. F-5
Palmcroft Way F-4
Papago J-4,J-7
Papago St. J-1-5
Pasadena A-6
Pasadena Ave. A-3-7
Piccadilly C-5
Piccadilly Rd. C-3
Pierce G-4,G-5
Pierce St. G-1,G-6
Pierson A-4
Pierson St. A-1-8
Pima St. J-1-7
Pinchot Ave. D-2
Polk St. H-1-7
Portland St. G-2-8
Randolph Rd. D-8
Richland St. F-7
Roanoke D-6
Roanoke Ave. D-6
Roma Ave. B-1-7
Roosevelt St. G-1-8
Sells Dr. B-4
Sheridan St. E-7
Sherman St. I-1-4
Sonora St. J-3
Southern Pacific Dr. I-7
Spruce F-3
Taylor G-5
Taylor St. G-1,G-4
H-6,H-7
Thomas Rd. D-3-6
Tonto I-1-4
Tonto St. I-2-7
Turney Ave. B-1-7
Valerie B-5
Van Buren St. H-4-8
Verde Ln. D-2-8
Vernon Ave. E-2-6
Villa St. G-1
Virginia Ave. E-1-5
Virginia Cir. E-6
Washington Dr. H-1
Washington St. H-4-8
Watkins St. K-4-7
Weldon Ave. C-1-6
Westview Dr. B-4
Whitton C-5,C-6
Whitton Ave. C-1-4
Willetta F-2
Willetta St. F-2-4
Wilshire E-3,E-5
Wilshire Dr. E-2,E-4
Windsor Ave. E-1-6
Woodland Ave. H-4
Woodward Dr. E-6
Yale St. E-8
Yavapai J-7
Yavapai St. J-1-5
Yuma Ave. J-7
Yuma St. J-1-5
1st Ave. A-5-K-5
1st St. A-6-J-6
2nd Ave. B-6-K-6
2nd Dr. B-5,B-6
2nd Pl. C-6,J-6
2nd St. C-6-J-6
3rd Ave. B-5-K-5
3rd Dr. I-5
3rd St. A-6-J-6
4th A-6
4th Ave. C-5-I-5
4th Ave. N.W. B-5
4th St. C-6-J-6
5th C-5,I-5
5th Ave. B-5-K-5
5th Ave. N.W. B-5
5th St. A-6-J-6
6th Ave. C-5-K-5
6th Ave. N.W. B-5,C-5
6th Dr. B-5,I-5
6th St. A-6-J-6
7th Ave. D-5-I-5
7th St. B-7-J-7
8th G-5
8th Ave. B-5-J-5
8th Pl. A-7,I-7
8th St. A-7-J-7
9th Ave. B-5-J-5
9th Pl. C-7,I-7
9th St. B-7-J-7
9th Way I-7
10th Ave. B-5-J-5
10th Dr. A-7
10th Pl. A-7-I-7
10th St. B-7-J-7
11th B-7,K-8
11th Ave. B-4-K-4
11th Pl. A-7,H-7
11th St. A-7-J-7
11th Way H-8
12th Ave. B-4-J-4
12th Dr. B-4
12th Pl. C-8,I-8
12th St. C-8,J-8
13th C-4,K-8
13th Ave. A-4-J-4
13th Pl. B-8,C-8,H-8,J-7
13th St. A-8-J-8
13th Way C-8
14th H-4
14th Ave. B-4-J-4
14th Pl. A-8-I-8
14th St. B-8-J-8
15th Ave. B-4-I-4
15th Dr. A-4-J-4
15th Pl. A-8,D-8,K-8
15th St. A-8-K-8
16th C-4,D-4
16th Ave. A-4-K-4
16th Dr. A-4-J-4
16th St. J-8
17th J-3
17th Ave. A-4-J-4
17th Dr. B-3-H-3
18th Ave. A-3-J-3
18th Dr. B-3-J-3
19th Ave. B-3-J-3
19th Dr. B-3,J-3
20th E-3
20th Ave. A-3-J-3
20th Dr. B-3-H-3
21st C-3,J-3
21st Ave. A-3-J-3
21st Dr. B-3-E-3
22nd B-2-E-2
22nd Ave. A-2-K-2
23rd Ave. B-2-I-2
23rd Dr. A-2-E-2
24th J-2
24th Ave. A-2-H-2
24th Dr. C-2,E-2
25th C-2
25th Ave. F-2-J-2
25th Dr. D-2-F-2
26th D-2
26th Ave. B-2,H-2,J-2
26th Dr. B-1,D-1
27th G-1
27th Ave. D-1-J-1
27th Dr. A-1,G-1,H-1
27th Ln. F-1,G-1
28th B-1,C-1,H-1
28th Ave. A-1-J-1
28th Dr. A-1-J-1
29th A-1-G-1
29th Ave. A-1-J-1
29th Dr. B-1-H-1
30th A-1-G-1
30th Ave. D-1-J-1
30th Dr. A-1,B-1
31st H-1,I-1
31st Ave. B-1-D-1

POINTS OF INTEREST

Central Park I-6
Coffelt Park J-3
Coronado Park F-8
Eastlake Park H-8
Encanto Park E-3
Grant Park I-6
Green Valley Park K-8
Harmon Park J-5
Madison Park B-8
Monterey Park E-6
Townsend Park G-7
University Park H-4
Virginia Park E-8
Willow Park H-1

Scale of Miles
0 .1 .2

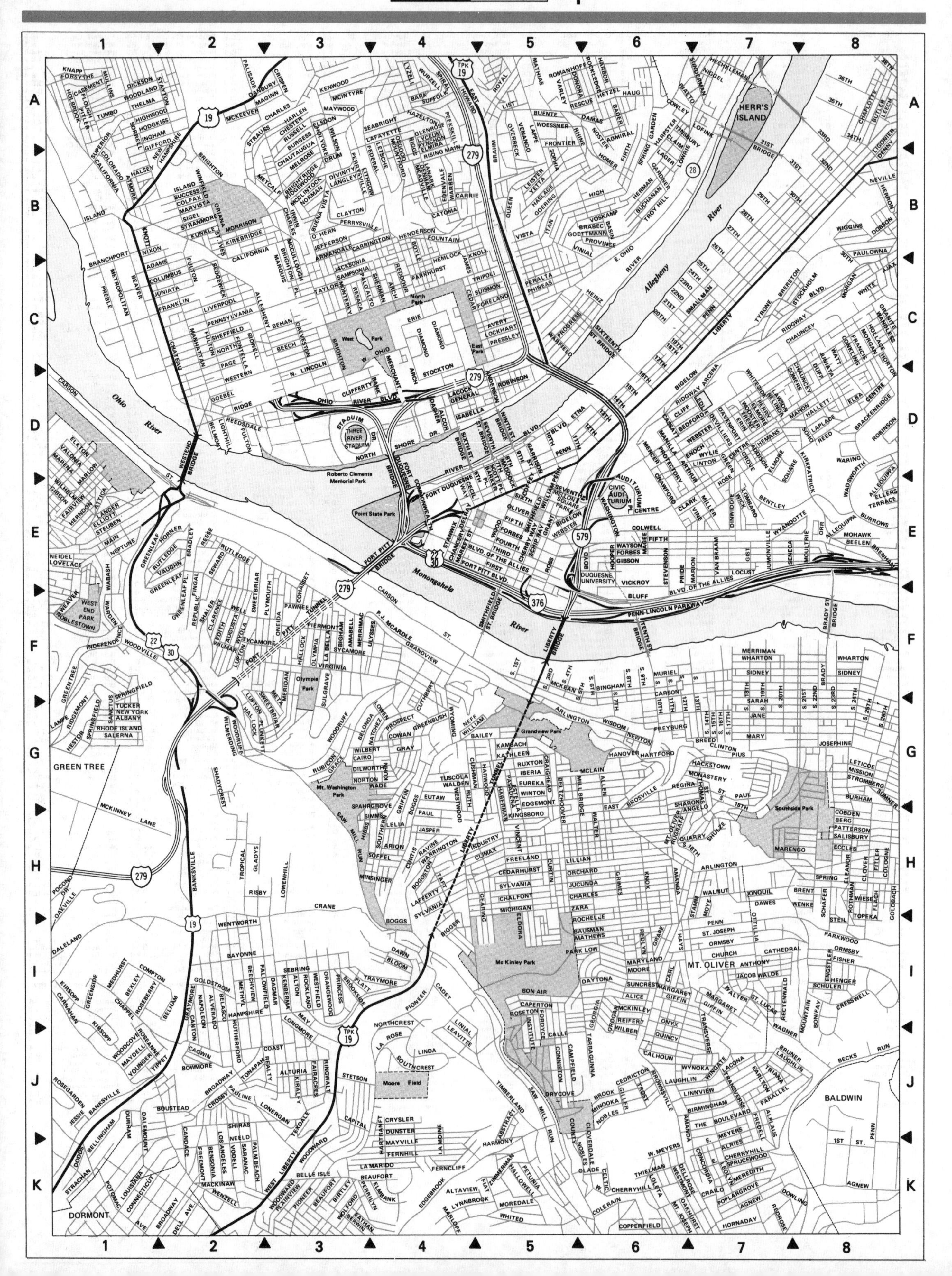

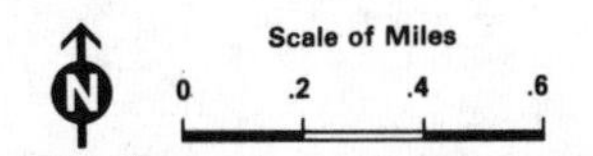
N
Scale of Miles
0
.2
.4
.6

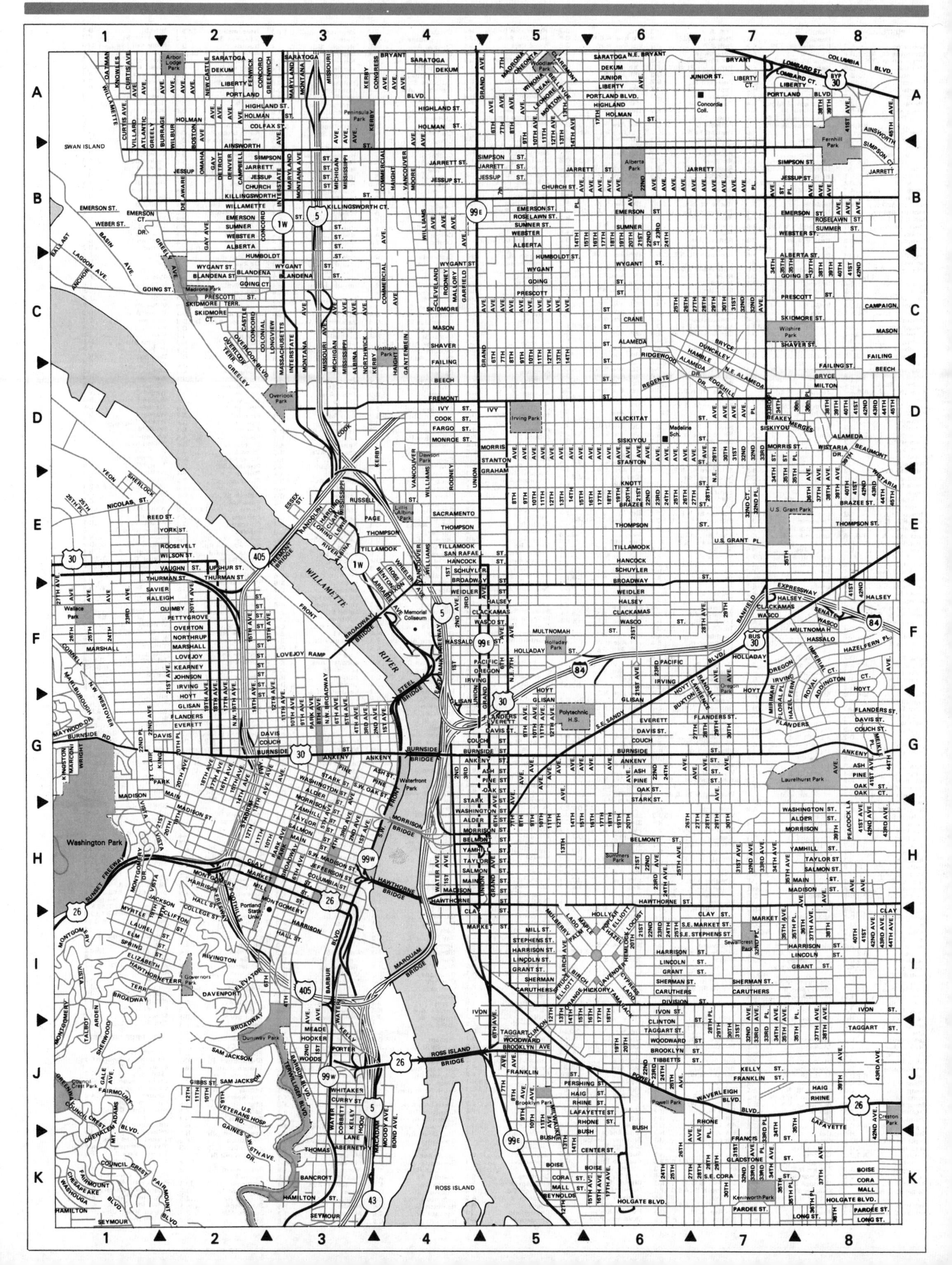

PORTLAND

Abernethy....K-3
Addington Ct. N.E...G-8
Ainsworth St. N.E...A-2
Alameda....D-8
Alameda N.E....D-7
Alberta....B-5
Alberta St. N....B-2
Alberta St. N.E....C-7
Albina....E-3
Albina Ave. N....B-3,D-3
Alder St. S.E.
....H-4,H-8
Alder St. S.W....G-3
Anchor....C-1
Ankeny....G-3
Ankeny St. S.E.
....G-4,G-8
Arden S.W....J-1
Ash St. S.E....G-4,G-8
Ash St. S.W....G-4
Atlantic Ave. N....A-1
Ballast....B-1
Bancroft S.W....K-3
Banfield Expwy....F-7
Barbur Blvd. S.W....J-3
Basin Ave. N....B-1
Beakey....D-7
Beaumont....D-8
Beech....D-8
Beech St....D-4
Bellevue....A-5
Belmont St. S.E.
....H-4,H-5,H-6
Benton....E-4
Birch....I-5
Blandena....C-2
Blandena St. N...C-2,C-3
Boise S.E....K-5,K-8
Bondvale Ave. S.W..K-4
Boston Ave. N....A-2
Brazee St. N.E....E-6,E-8
Broadway....I-1
Broadway N.W....G-3
Broadway S.W...H-3,J-2
Broadway Bridge....F-3
Broadway St. N.E.
....E-4,F-6
Brooklyn St. S.E....J-6
Bryant N.E...A-4,A-6,A-7
Bryce....D-8
Bryce, N.E....C-7
Burnside Bridge....G-4
Burnside Rd....G-1
Burnside St....G-3
Burnside St. E...G-4,G-6
Burrage Ave. N....A-2
Bush S.E....K-5
Buxton....G-6
Campaign....C-8
Campbell Ave. N....B-2
Caruthers S.E....I-5,I-7
Castle N....C-2
Center St. S.E....K-5
Chehalem Blvd....K-1
Chesapeake Blvd....K-1
Church St....B-2
Church St. N.E....B-5
Clackamas....F-7
Clackamas, N.E...F-5,F-6
Claremont....A-5
Clark....E-3
Clay St. S.E....I-4,I-7,I-8
Clay St. S.W....H-2
Cleveland Ave....C-4
Clifton....I-2
Clinton St. S.E....J-6
Colfax St. N....A-2
College St....H-2
Columbia Blvd. N.E..A-8
Columbia St. S.W....H-3
Commercial Ave. N.
....B-4,C-4
Concord N....A-2,C-2
Congress N....A-4
Cook....D-3
Cook St. N.E....D-4
Cora St. S.E....K-5,K-8
Corbett....J-3
Cornell N.W....F-1
Couch N.W....G-3
Couch St. N.E...G-4,G-8
Council Crest....J-1,K-1
Curry St....J-3
Curtis Ave. N....A-1
Cypress....I-6
Davenport....I-2
Davis....G-2
Davis N.W....G-3
Davis St. N.E.
....G-5,G-6,G-8
Dean....A-5
Dekum....A-4
Dekum N....A-2
Dekum N.E....A-6
Delaware Ave. N....B-2
Denver Ave. N....B-2
Detroit Ave. N....B-2
Division St....I-6
Dixon....F-4
Dunckley, N.E....C-7
East Bank Frwy....F-4
Edgehill Pl....D-7
Elevator....I-2
Elizabeth....I-1
Elliott....I-5,I-6
Elm St. S.W....I-1
Emerson St. N....B-2
Emerson St. N.E.B-5,B-8
Everett....G-2,G-5
Failing....C-8
Failing St....D-8
Failing St. N....D-4
Fairmount....J-1
Fairmount Blvd..K-1,K-2
Fargo St. N.E....D-4
Fenwick....A-2
Flanders....G-2,G-5
Flanders St. N.E..G-6,G-8
Floral Pl....G-7
Foothills Frwy....H-2
Francis St. S.E....K-7
Franklin St. S.E...J-5,J-7
Freemont Bridge....E-3
Fremont St....D-4
Front N.W....F-3
Front S.W....H-4
Gaines S.W....J-2
Gale Ave....J-1
Gantenbein N....D-4
Garfield Ave....C-4
Gay Ave. N....B-2
Gibbs St. S.W....J-2
Gladstone St. S.E....K-7
Glisan....G-2,G-4
Glisan N.E....G-5,G-6
Going Ct....C-2
Going St....C-1
Going St. N.E....C-5
Graham....E-5
Grand....G-4
Grand Ave. N.E..A-5,C-5
Grand Ave. S.E....H-5
Grant St. S.E....I-5,I-8
Greeley....D-2
Greely Ave. N....A-1,B-2
Greenway....J-1
Greenwich....A-3
Haig....J-8
Haig St. S.E....J-5
Haight Ave. N.
....B-4,D-4
Halsey....F-8
Hall St. S.W....H-2,I-3
Halsey N.E....F-5,F-6
Hamble N.E....C-7
Hamilton....K-1
Hamilton St. S.W....K-3
Hancock....E-6
Hancock St. N.E....E-4
Harding....E-3
Harrison St. S.E.
....I-5,I-8
Harrison S.W....H-2
Hassalo....F-8
Hawthorne Bridge....H-4
Hawthorne St....H-4
Hawthorne St. S.E....H-6
Hawthorne Terr....I-1
Hazel....I-6
Hazelfern....G-7
Hazelfern Pl....F-8
Hemlock....I-6
Hickory....I-6
Highland....A-6
Highland St. N.
....A-2,A-4
Holgate Blvd. S.E.
....K-6,K-8
Holladay....F-7
Holly S.E....I-6
Holman N....A-2,A-4
Holman St. N.E....A-6
Hood....J-3
Hooker S.W....J-3
Hoyt....G-5,G-6
Hoyt N.E....F-8
Hoyt St. N.W....G-2
Humboldt St....C-5
Humboldt St. N....C-2
Imperial....F-8
Interstate Ave. N.
....B-3,D-3
Irving N.E....F-4,F-6,F-7
Irving St. N.W....F-2
Ivon....I-4
Ivon St. S.E....I-6,I-8
Ivy....D-5
Ivy St. N.E....D-4
Jackson....H-1
Jarrett....B-7
Jarrett St....B-2
Jefferson St. S.W....H-3
Jessup N....B-2
Jessup St....B-2
Jessup St. N.E.
....B-4,B-6,B-8
Johnson St. N.W....F-2
Junior N.E....A-6
Junior St....A-7
Kearney St. N.W....F-2
Kelly....J-3
Kelly St. S.E....J-7
Kerby N....A-4,D-4
Killingsworth N.E.
....B-2
Killingsworth Ct....B-3
King....G-2
Kingston....G-1
Klickitat St....D-6
Knott St. N.E....E-6
Knowles N....A-1
Ladd S.E....I-5,I-6
Lafayette....J-8
Lafayette St. S.E....J-5
Lagoon Ave. N....C-1
Lane....K-3
Larch Ave....I-5
Laurel St. S.W....F-1
Lavender....I-6
Lawarence....F-7
Leonore....A-5
Lewis....E-3
Liberty N....A-2
Liberty N.E....A-6,A-7
Lincoln St. S.E....I-5,I-8
Locust....I-6
Lombard Ct....A-7
Lombard St. N.E....A-8
Long St....K-8
Longview N....C-3
Loring....E-3
Lovejoy Ramp....F-3
Lovejoy St. N.W....F-2
Macadam S.W....K-4
Madison S.W....G-1
Madison St....H-4
Madison St. S.E....H-8
Madison St. S.W.
....H-2,H-3
Madrona....A-5
Main....G-2
Main St. S.E....H-4,H-8
Main St. S.E....H-3
Mallory Ave....C-4
Mall St. S.E....K-5,K-8
Maple S.E....I-5
Marconi....G-1
Market St. S.E....I-4,I-7
Market St. S.W....H-2
Marlborough....F-1
Marquam Bridge....I-4
Marshall N.W....F-1,F-2
Maryland N....A-3
Mason N.E....C-8
Mason St....C-4
Massachusetts N....C-3
Maywood Dr....G-1
Meade S.W....J-3
Meilkie Pl....G-8
Merges....D-8
Michigan Ave. N.
....B-3,C-3
Mill St. S.E....I-5
Mill St. S.W....H-2
Milton....D-8
Milwaukee....J-5
Mirimar....G-7
Mississippi Ave. N.
....B-3,C-3
Missouri Ave. N..A-3,C-3
Monroe St. N.E....D-4
Montana Ave. N.
....A-3,B-3,C-3
Montgomery....J-1
Montgomery S.W.
....H-2,I-1
Montgomery Dr....H-1
Moody Ave. S.W....K-4
Moore Ave. N....B-4
Morris....D-5
Morris St....D-7
Morrison Bridge....H-4
Morrison St. S.E.
....H-4,H-8
Morrison St. S.W....G-3
Morton....A-5
Mt. Adams....K-1
Mulberry....I-5
Multnomar N.E....F-8
Myrtle St. S.W....H-1
Newcastle....A-2
Nicolas St. N.W....E-1
Northrop St. N.W....F-2
Northwick Ave. N....D-3
Oak Ct. S.E....G-8
Oak St. S.E....G-4
....G-6,G-8
Oak St. S.W....G-3
Oatman....A-1
Omaha Ave....B-2
Oneonta....A-5
Orange....I-5
Oregon N.E....F-4,F-7
Overlook Blvd....C-2
Overlook Terr....C-2
Overton St. N.W....F-2
Pacific N.E....F-4,F-6
Page N....E-3
Palm....I-5
Pardee St....K-7,K-8
Park Ave....H-3
Park Ave. N.W....G-3
Park S.W....G-2
Peacock La....H-8
Pershing St. S.E....J-5
Pettygrove St. N.W...F-2
Pine S.W....G-3
Pine St. S.E....G-4,G-8
Poplar....I-6
Porter....J-3
Portland Blvd. N.E.
....A-2,A-6,A-7
Powell Blvd. S.E....J-6
Prescott St....C-5
Prescott St. N.E....C-7
Quimby St. N.W....F-2
Raleigh....F-1
Randall....F-7
Randolph....E-3
Reed St. N.W....E-1
Regents Dr....D-6
Reynolds....K-5
Rhine....J-8
Rhine St. S.E....J-5
Rhone St. S.E....J-6,J-7
Ridgewood Dr....C-6
River....E-3
Rivington....I-1
Rodney....E-4
Rodney Ave....C-4
Roosevelt N.W....E-2
Roselawn St. N.E....B-5
Ross....E-4
Ross Island Bridge...J-4
Royal Ct. N.E....G-8
Russell....E-5
St. Claire....G-1
Sacramento....E-4
Salmon St. S.E...H-4,H-5
Salmon St. S.W....H-3
Sam Jackson....J-2
Sandy Blvd. S.E....G-6
San Rafael St. N.E....E-4
Saratoga....A-2,A-4
Saratoga N.E....A-6
Savier....F-1
Schuyler....E-4,E-6
Senate....F-8
Seymour....K-1,K-3
Shaver St....C-4
Shaver St. N.E....C-7
Sherlock....E-1
Sherman St. S.E.
....I-6,I-7
Sherwood....J-1
Simpson St. N....B-2
Simpson St. N.E....B-7
Siskiyou....D-7
Siskiyou St....D-6
Skidmore Ct....C-3
Skidmore St....C-4
Skidmore St. N.E....C-7
Skidmore Terr....C-2
Spring St. S.W....I-1
Spruce....I-5
Stadium Fwy....H-2
Stanton....D-5
Stanton St. N.E....D-6
Stark St. S.E....G-4,G-6
Stark St. S.W....G-3
Steel Bridge....F-4
Stephens St. S.E...I-5,I-7
Sumner....B-6
Sumner St. N....B-2
Sumner St. N.E....B-5
Sunset Fwy....H-1
Swan Island....A-1
Taggart....J-5
Taggart St. S.E...J-6,J-8
Talbot....J-1
Tamarack....I-6
Taylor St. S.E....H-4,H-8
Taylor St. S.W....H-3
Terr. S.W....I-1
Thomas....K-3
Thompson N....E-4
Thompson St. N.E.
....E-6,E-8
Thurman St....E-2
Tibbetts St. S.E....J-6
Tillamook....E-4,E-6
Union....E-4
Union N.E....G-4
Union S.E....H-5
Upshur St. N.W....E-2
U.S. Grant Pl....E-7
U.S. Veterans
Hosp. Rd....J-2
Vancouver N....B-4,E-4
Vaugn St. N.W....E-2
Villard Ave. N....A-1
Vista....H-1,I-1
Wasco....F-7,F-8
Wasco St. N.E...F-5,F-6
Washington St. S.E.
....H-4,H-8
Washington St. S.W.
....G-3
Washouga....K-1
Water....I-3,J-3
Water Ave. S.E....H-4
Waverleigh Blvd....J-7
Webster....B-5
Webster St. N....B-2
Webster St. N.E....B-8
Weidler St. N.E.
....F-4,F-6
Westover N.W....G-1
Whitaker S.W....J-3
Wilbur Ave.N....A-2
Willamette N....A-1,B-2
Williams N....B-4,E-4
Wilson St. N.W....D-1
Winona....A-5
Wistaria....E-8
Wistaria Dr....D-8
Woods....J-3
Woodward....J-5
Woodward St. S.E....J-6
Wright....G-1
Wygant....C-5
Wygant St. N....C-2,C-3
Wygant St. N.E..C-4,C-6
Yamhill St. S.E.
....H-4,H-8
Yamhill St. S.W.
....H-3
Yeon N.W....D-1
York St. N.W....E-2
1st Ave....F-4,H-3,J-3
1st Ave. N.W....G-4
1st Ave. S.E....H-4
2nd S.E....G-4
2nd Ave....H-3,J-3
2nd Ave. N.E....F-4
2nd Ave. N.W....G-4
3rd S.E....G-4
3rd Ave....H-3
3rd Ave. N.E....F-4
3rd Ave. N.W....G-3
4th S.W....I-3
4th Ave....H-3
4th Ave. N.W....G-3
5th Ave....H-3
5th Ave. N.W....G-3
6th S.W....I-2
6th Ave....H-3
6th Ave. N.E..A-5,C-5,F-5
6th Ave. N.W....G-3
6th Ave. S.E....H-5,J-5
6th Ave. S.W....K-2
7th Ave. N.E..A-5,C-5,F-5
7th Ave. S.E....H-5,J-5
8th Ave. N.E..A-5,C-5,E-5
8th Ave. N.W....G-3
8th Ave. S.E....H-5,J-5
9th....J-2
9th Ave. N.E.A-5,C-5,G-5
9th Ave. N.W....G-3
9th Ave. S.E....H-5,J-5
10th....J-2
10th S.W....H-3
10th Ave. N.E.
....A-5,C-5,G-5
10th Ave. N.W....G-3
10th Ave. S.E....H-5,J-5
11th....J-2
11th Ave. N.E.
....A-5,C-5,G-5
11th Ave. N.W....G-3
11th Ave. S.E....H-5,J-5
11th Ave. S.W....H-2
12th....J-2
12th Ave. N.E.
....A-5,C-5,G-5
12th Ave. N.W....G-3
12th Ave. S.E....H-5,K-5
12th S.W....H-2
13th Ave. N.E.
....A-5,C-5,E-5
13th Ave. N.W....F-3
13th Ave. S.E....H-5,K-5
13th S.W....H-2
14th Ave. N.E.
....A-5,C-5,E-5
14th Ave. N.W....G-2
14th Ave. S.E....H-5,K-5
14th Ave. S.W....G-2
15th Ave. N.E....B-6,E-6
15th Ave. N.W....F-2
15th Ave. S.E....H-5,K-6
15th Ave. S.W....G-2
16th Ave. N.E....B-6,E-6
16th Ave. N.W....G-2
16th Ave. S.E....H-6,K-6
16th Ave. S.W....G-2
17th Ave. N.E.
....A-6,B-6,E-6
17th Ave. N.W....G-2
17th Ave. S.E....H-6,K-6
17th Ave. S.W....G-2
18th Ave....H-6,I-6
18th Ave. N.E....B-6,E-6
18th Ave. N.W....G-2
18th Ave. S.W....G-2
19th Ave. N.E....B-6,E-6
19th Ave. N.W....G-2
19th Ave. S.E....H-6,J-6
20th....I-6
20th Ave....G-2
20th Ave. N.E....B-6,E-6
20th Ave.N.W....F-2
20th Ave. S.E....H-6,J-6
20th Pl....G-2
21st....H-6,I-6
21st Ave....F-6
21st Ave.N.E....B-6,E-6
21st Ave. N.W....F-2
22nd....G-6,J-6
22nd Ave....B-6
22nd Ave. N.E....B-6,E-6
22nd Ave. N.W....G-1
22nd Pl....G-1
23rd....F-6
23rd Ave....F-1,H-6,J-6
23rd Ave. N.E....B-6,E-6
24th....G-6,J-6
24th S.E....K-6
24th Ave. N.E....B-6,E-6
24th Ave. N.W....F-1
25th....E-1
25th Ave....H-6,J-6
25th Ave. N.E....C-6,E-6
25th Ave. N.W....F-1
25th Ave. S.E....K-6
25th Pl....E-1
26th Ave....F-1
26th Ave. N.E....C-6,E-6
26th Ave. S.E....H-6,K-6
27th....G-7
27th Ave. N.E....C-7,E-7
27th Ave. N.W....F-1
27th Ave. S.E....H-7,K-7
28th Ave. N.E.
....C-7,E-7,G-7
28th Ave. S.E....H-7,K-7
28th Pl....J-7
29th....F-7,J-7
29th Ave. N.E.
....C-7,D-7,G-7
29th Ave. S.E....H-7,K-7
30th Ave. N.E.
....C-7,D-7,G-7
30th Ave. S.E....H-7,K-7
31st S.E....K-7
31st Ave....H-7,J-7
31st Ave. N.E.
....C-7,D-7
32nd S.E....K-7
32nd Ave....H-7,J-7
32nd Ave. N.E....C-7,D-7
32nd Ct. N.E....E-7
32nd Pl....I-7
32nd Pl. N.E.
....C-7,D-7,E-7
33rd Ave....H-7,J-7
33rd Ave. N.E....D-7
33rd Ave. S.E....K-7
33rd Pl....D-7,K-7
33rd Pl. S.E....K-7
34th Ave....H-7,J-7
34th Ave. N.E....C-7,E-7
34th Ave. S.E....K-7
35th S.E....K-7
35th Ave....I-7,J-7
35th Pl....C-7,E-8,K-7
35th St. N.E....C-7,E-7
36th Ave....I-8
36th Ave. N.E....E-8
36th Pl....K-8
37th Ave. N.E....C-8,E-8
37th Ave. S.E....I-8,K-8
38th....K-8
38th Ave. N.E.
....A-8,C-8,E-8
38th Ave. S.E....I-8
39th....D-8
39th Ave. N.E.
....A-8,C-8,E-8
39th Ave. S.E....H-8,I-8
40th Ave....I-8
40th Ave. N.E....C-8,E-8
41st Ave....F-8,H-8
41st Ave. N.E.
....A-8,C-8,E-8
41st Ave. S.E....G-8,I-8
42nd Ave....F-8,H-8
42nd Ave. N.E....C-8,E-8
42nd Ave. S.E....I-8,K-8
43rd N.E....E-8
43rd Ave....H-8,I-8,J-8
44th N.E....E-8
44th Ave. S.E....G-8,I-8
45th N.E....E-8
45th Ave. N.E....A-8,D-8

POINTS OF INTEREST

Alberta Park....B-6
Arbor Lodge Park....A-2
Broadway Bridge....F-3
Brooklyn Park....J-5
Buckman Field....G-5
Burnside Bridge....G-4
Council Crest
Park....J-1
Creston Park....J-8
Dawson Park....D-4
Duniway Park....J-2
Fernhill Park....A-8
Fremont Bridge....E-3
Governors Park....I-2
Hawthorne Bridge..H-4
Holladay Park....F-5
Irving Park....D-5
Kenilworth Park....K-7
Laurelhurst Park....G-7
Lillis Albina
Park....E-4
Madrona Park....C-2
Marquam Bridge....I-4
Morrison Bridge....H-4
Oregon Park....F-7
Overlook Park....D-3
Penninsula Park....A-3
Powell Park....J-6
Ross Island
Bridge....J-4
Sewallcrest Park....I-7
South West Park....H-3
Steel Bridge....F-4
Summers Park....H-6
Unthank Park....C-4
U.S. Grant Park....E-7
Wallace Park....F-1
Washington Park....H-1
Wilshire Park....C-7
Woodlawn Park....A-5

RICHMOND

Abington....G-5
Accommodation St..F-7
Adams St....F-5
Addison St....F-3
Admiral St....E-5
Akron St....A-5
Albany Ave....J-5
Alder....E-7
Allen Ave....F-3
Allison St....E-3
Alpine....C-7
Althea....D-7
Altmont....D-3
Alton....B-6
Alvis Ave....C-7
Ambler St....G-7
Ambrose....E-8
Amelia....G-3
Amherst Ave....A-4
Ancester Rd....D-5
Anniston St....E-8
Antrim Ave....C-1
Appomattox St....G-3
Arch St....G-5
Archer....I-2
Arizona Dr....K-1
Arizona Pl....K-2
Arlington Rd....C-3
Arnold....C-7
Auburn Ave....E-2
Augusta Ave....C-1
Avondale Ave....A-4
....B-3,B-4
Bacon St....E-5,E-6
Bainbridge....J-4
Baker St....F-6
Balding St....F-7
Bancroft....C-7
Bates....F-6
Barry St....J-4
Bartee Rd....K-1
Barton Ave....D-6
Bassett Ave....J-1
Bath St....D-4
Battery....D-6
Beattie St....J-3
Beaumont Ave....E-1
Bedford....I-6
Beech St....F-4
Beleville St....C-2
Bellevue Ave....A-4
Belmont Ave....E-2,F-1
Belt Blvd....K-1
Belvidere St....G-5
Benton....B-8
Berkley....G-2
Bermuda Ave....K-1
Bethell....E-8
Bethlehem....A-1
Birch St....F-4
Bishop Rd....B-1
Blacker St....C-1
Blair St....G-3
Blake Ln....J-3
Bland St....J-2
Blanton Ave....F-2
Blue Shingles Ln....G-1
Booker St....A-5
Boscobel....I-2
Boston Ave....J-5
Botetour St....D-4
Boulevard....C-3,E-2
Boulevard Bridge....H-1
Bowe St....E-4
Bowen....K-6
Bowling Green....E-7
Boyd....F-4
Brandon Ln....F-2
Brandon Rd....K-1
Bragg....G-5
Brang....F-8
Bravers Ln....F-8
Brenton St....B-6
Brinser....K-5
Broad St....B-1,D-3
....F-6,H-8
Brook Rd....B-5,E-5,F-5
Brookfield....E-6
Brookland Pkwy....C-5
Brookside....J-2
Bruce St....K-6
Brunswick St....F-4
Bryan....F-7,F-8
Burfoot St....J-3
Burns St....C-8
Burton....G-7
Bute Lake....F-1
Byrd St....G-5
Byron St....B-8,C-8
Caledonia Rd....H-1
Calhoun....E-6
Call St....E-7
Cameron St....D-1
Canal St....G-5
Carillon Ct....G-1
Carlton....C-3
Carmine St....E-8
Carolina Ave....C-8
Carrington St....G-7
Carroll St....E-8
Carrolton St....F-1
Carson....J-2
Carter St....H-3
Cary St....E-1,E-2
....G-5,H-7
Catalra....B-3
Cathedral Pl....F-4
Catherine....E-5
Carver St....F-8
Cecil....D-4
Cedar Ln....I-1
Cersley St....J-4
Chamberlayne...B-5,E-5
Chambers....K-5
Chantilly St....B-1
Chapel Dr....K-3
Charity....F-6
Charlotte....E-1
Chatham Dr. E....A-6
Chatham Rd....B-4
Cheatwood Ave....A-6
Cherry St....G-4
Chesterfield....I-3
Chesterman....K-5
Chestnut....E-7
Chevy Chase....A-4
Chicago Ave....J-5
China St....G-4
Chuckatuck Ave....K-1
Church St....G-5
Circle Wood....K-1
Claiborne St....G-3
Claremont Ave....A-4
Clarence....I-1
Clay St....C-2,F-6,H-8
Cleveland St....E-2
Cliff Ave....B-6,C-7
Clifton....D-7
Clinton Ave....A-4
Clopton St....J-4
Clover Rd....A-1
Coalter St....F-7
College St....G-6
Colonial....D-2,F-2
Colorado St....G-3
Columbia St....K-5
Commerce Rd....I-6,K-6
Commonwealth Ave.C-1
Condie St....F-2
Confederate Ave....B-4
Conrad....E-8
Conway St....B-7
Corbin....A-5
Coutts....E-6
Cowardin Ave....I-4
Crafton La....D-7
Craig Ave....A-7
Craigie....C-7
Crawford St....B-7,C-6
Crenshaw....E-1
Crestwood....A-4
Crump St....E-8
Crutchfield St....J-1
Culpepper St....B-6
Cummings St....C-4
Cumberland....F-4
Custer St....D-8
Cutshaw Ave....C-1,D-2
Cypress St....D-7
Dabney Rd....A-2
Dakota Ave....G-3
Daniel....E-7
Davis Ave....F-2
Dawn St....D-6
Decatur St....I-5,K-3
Deforest St....E-8
Delaware Ave....B-8
Delmont St....A-6,B-7
Denny St....I-8
Detroit Ave....D-8
Devonshire Rd....I-1
Dinneen....D-4
Dinwidde Ave....J-5
Dixon....K-1
D.M.U. Dr....D-3
Dobson St....G-4
Dock St....H-7
Donnan....B-7
Dooley Ave....E-2
Douglasdale Ave....F-1
Dove St....D-6
Drewery....K-6
Dubois Ave....D-4
Dundee....J-3
Dunn Ave....D-7
Dunston Ave....I-1,J-2
Dupont Cir....C-5
Durham St....D-4
Duval St....F-6
Dwyer St....K-8
Edgehill....D-5
Edgewood Ave....B-6
Edwards St....J-5
Eggleston St....G-3
Elizabeth....E-4
Ellen Rd....C-3
Ellwood St....E-1
Elmdale....K-6
Elmsmere Ave....B-3
Elton....K-5
Endicott....C-4
Enfield Ave....K-5
Enslow Ave....C-8
Essex St....C-6,C-7
Evarts St....J-4
Evelyn Byrd....H-1
Everett St....I-5
Evergreen....K-2
Exeter Rd....E-1
Fairfax St....J-5
Fairfield....F-7
Fairmont Ave....G-8
Fauquier Ave....A-4
Fayette Ave....A-7
Federal St....E-6
Fells....E-5,E-6
Fendall Ave....B-6,E-6
Fifth....E-7
First Ave....B-7,C-8
Fitz Hugh Ave....B-1
Florida....B-7
Floyd St....D-1,E-2
Fluvanna St....J-4
Ford St....F-7
Forest Cir....I-2
Forest Hill Ave....I-1,J-3
Forestview Dr....I-1
Forqurean Ln....D-7
Fourth....D-7
Foushee St....F-5
Franklin St....C-1
....D-2,E-4,G-5
Freeman Rd....E-1
French....F-1,F-2
Fritz....E-5
Front St....C-8
Garland....B-6,D-6
Garrett St....F-1
Gate St....G-3
Gay St....F-8
Georgia....H-2
Gilbert St....F-3
Giles....C-7
Gilmer St....F-5
Gladstone Ave....C-7
Glenfield....K-6
Glenlea....E-8
Goddin St....J-8
Goodes....J-6
Gordon Ave....K-5
Goshen St....F-5
Grace Ave....C-1
....D-3,F-6,G-7
Graham....D-5,D-6
Granby St....F-3
Grant St....F-1
Graveland....G-2
Grayson Ave....A-7
Grayson St....B-7
Green St....F-4
Green Acres Ave....K-1
Green Alley....F-4
Green Brier Ave....A-7
Greenview Dr....K-8
Greenville Ave....H-2
Greenwood Ave....D-6
Grey Court Ave....A-4
Greystone....K-2
Griffin Ave....B-6
Grove Ave....D-1,E-3
Groveland....D-7
Gun Club Rd....E-1
Halboro St....A-1
Halifax Ave....K-5
Hamilton St....D-1
Hammond Ave....C-5
Hampton St....H-2
Hancock St....E-5
Hanes Ave....C-6
Hanover Ave....D-1,E-3
Harold....D-7
Harrison St....E-4
Harry St....D-4
Harvey Ave....D-6
Harvie St....F-4
Harwood....K-6
Hatcher....J-8
Hawthorne....B-5,D-6
Hazelhurst....B-7
Henrico Blvd....B-6
Henry St....F-5
Herbert St....I-1
Hermitage Rd....A-3,D-4
Hickory St....E-6
High....E-5
Highland St....B-8,C-8
Highland View....C-8
Highpoint....D-2
Hildreth....F-7
Hill St....E-6
Hillcrest....I-2,I-3
Hillmonument Pkwy.B-3
Hills Blvd....I-1
Hilltop Dr....I-1
Holly St....G-4
Holly Springs Ave....K-4
Home St....D-6
Hooper....C-6
Hopkins Rd....K-4
Hospital St....E-6
Howe....G-5
Hull St....I-5,K-3
Hunt Ave....B-7
Huxley St....F-1
Idlewood Ave....E-1
....F-3,G-4
Ingram Ave....K-5
Ivanhoe....B-7
Jackson Ave....B-8
Jackson St....F-6
Jacque....B-2
Jacquelin....F-3
Jasper Ave....A-6
Jefferson Ave....G-7
Jefferson St....F-5
Jefferson Davis Hwy.K-5
Jessanine....G-8
Jeter Ave....B-7
John St....A-5
Joplin Ave....K-5
Joshua St....E-6
Juniper....D-7
Kelly Rd....B-3
Kemper St....F-3
Kensington Ave..D-1,D-2
Kent Rd....D-1
Kern St....K-5
Keswick AVe....K-5
King St....E-6
King William Rd....H-1
Kinney St....E-4
Laburum Ave....B-4,B-8
Ladies Mile Rd....C-6
Lafayett St....D-1
Lakeview St....F-2
Lamb Ave....C-6,D-6
Lamont....A-4,B-4,C-4
Langston Ave....D-4
Lanvale....C-3
Larchmont St....J-2
Laurel St....G-4
Lawson St....K-3
Lebanon....E-7
Leftwich....K-8
Leigh St....C-2,D-3,F-6
Lenmore....K-6
Letcher....C-7
Libby Terrace....I-8
Lincoln Ave....A-6
Lindel St....F-4
Lipscomb....I-5
Littlejohn Rd....C-4
Lodge St....E-4
Lombardy St....E-4
....F-4,G-3
Lone St....K-6
Longdale Ave....K-3
Lordley Rd....K-2
Loudon Ave....B-6
Louisa St....D-4
Louisiana St....I-8
Loumor Ave....B-2
Loxley Rd....C-4
Lunenburg St....D-4
Lynhaven Ave....K-5
M St....G-8
Macarthur....A-4
Mactavish....D-2
Maddox....E-8
Madison St....F-5
Maggie Walker Ave..A-6
Magnolia....D-8
Maiden....G-5
Main St....E-2,G-5,H-7
Malvern....C-1
Manchester Bridge..H-5
Mansion Ave....K-5
Maple....B-3
Maplewood Ave..E-1,F-2
Marcy....J-2
Mardick....K-2
Marshall Rd....D-3
....E-4,F-6,H-8
Marston Ln....E-1
Maryland Ave....C-7
Mason St....KL-5
Massie....A-1
Maury....I-5,J-4
Mayhem St....K-3
Mayo's Bridge....H-6
Maywill St....B-1
McCloy St....F-1
McCoul St....K-8
McDonough St....I-3
McLennan....J-3
McGuire....K-3
McRand St....J-3,K-3
Meadow Bridge....B-7
Meadow St....F-3,H-2
Meade....G-3
Mechanicville Turnpike
....G-7
Melbourne....F-8
Melklenburg....E-7
Melrose....B-5
Menole Ave....A-7
Meredith....C-6
Merry Oaks....K-3
Meyers St....D-3
Middlesex St....D-4
Midland....K-5
Miller Ave....D-6
Milton St....B-7,C-6,C-8
Mineffe....K-6
Missouri Ave....B-7
Mitchell....E-5
Mobile Rd....K-1
Monroe St....F-5
Monterio....E-6
Monticello St....A-4
Montrose Ave...A-6,B-6
Montvale....C-7
Monument Ave..C-1,E-3
Moody Ave....J-3
Moore St....C-2,D-4,E-5
Mordie....B-1
Morris....F-4
Mosside Ave....A-6,B-6
Mottoway Ave....A-4
Mount Vernon....A-4
Mulberry St....F-2
Munford....F-5
Myrtle St....D-7
N St....G-8
Nansemond St...E-1,F-1
Napolean St....C-7
Nevada....G-3
New Kent....H-1,I-2,I-3
Newkirk....K-1
Newport Ave....B-4
New York Ave....H-3
Nicholson St....I-8
Ninemile....G-8
Noble Ave....B-5
Norfolk....C-3
North Ave....B-5,D-6
Northrup....J-2
Northside Ave....C-7
North Umbrella Ave..C-5
North Umbrella Ln...D-5
Norton St....E-4
Norwood St....D-6
O St....G-8
Oak....E-5
Oakdale Ave....B-3
Oak Park....A-5
Old Brook Cir....A-5
Old Brook Rd....A-5
Old Midlothian Tpk...J-1
Ordway Ave....A-5
Orleans St....J-8
Orongo Ave....A-6
Orville Rd....A-1
Osborne Rd....K-8
Overbrook Rd...D-4,D-5
Overlook St....K-5
Oz Ave....I-1
P St....G-8
Palmyria Ave....B-4
Park Ave....C-1,D-2
Park La....A-1
Parkview....B-7
Parkwood Ave...E-2,F-3
Pattick Ave....B-7
Patterson Ave...D-1,D-2
Patton....C-3
Peach St....I-7
Pear St....H-7
Pelham....D-7
Pennsylvania Ave....G-3
Pensacola....B-7,B-8
Perdue Ave....K-4
Perry St....I-5,J-3
Peterborough....H-1
Peyton....J-2,J-3
Phaum St....F-8
Pilkington St....J-4
Pinalto Dr....A-7
Pine St....G-4
Pinetree....A-1
Piney....A-5
Pink....G-8
Plagn St....J-3
Plaza Ave....K-1
Pleasants St....G-7
Plum St....F-4
Plymouth....B-6
Poe....E-6
Pollock St....C-7,C-8
Pope Ave....A-3
Porter....I-4,I-5,J-3
Portland....F-1
Powhaten....G-3
Preble....I-6
Preston St....F-6
Prince Arthur Rd....I-1
Prince Edward....H-1
Prince George....I-1
Pulaksi....D-8
Pulaksi Ave....A-6
Pumphouse Dr...G-1,G-2
Q St....G-8
Queen....E-1
Quckshall....E-6
R St....G-8
Radford Ave....B-1
Randolf St....G-3
Ratcliffe Rd....B-8
Raven....F-7
Redd....F-8
Red Oak Ln....K-1
Redwood....E-8
Reedy Ave....I-1
Rendale....F-1
Reneille St....D-1
Rennie Ave....C-4
Reserve St....C-4
Rex Ave....A-5
Rhoadmiller....D-4
Rice Rd....K-1
Richmond St....F-8
Richmond & Henrico
Trpk....A-7,B-7,D-7
Richmond Petersburgs
Trpk....E-5,J-7
Riverside Dr....H-1,I-2
Roane....E-5
Roanoke....J-2
Robert E. Lee Bridge.H-4
Roberts....D-4,D-5,D-6
Robin Hood Rd....C-4
Robinson St....F-2
Robinview Dr....K-3
Rock Bridge....C-3
Rockingham....E-7
Rogers St....F-8
Rose....E-6
Rosecrest....K-6
Rosedale....A-3,B-3,C-3
Rosedale Blvd. West.C-3
Rosenath Rd....C-2
Rosewood Ave...E-1,F-2
Rothesay Cir....F-1
Rowen....E-7
Rowland St....F-3
Rugby Rd....F-1
Russell....G-8
Ryland....E-4
S St....G-8
St. James....F-6
St. John....E-6
St. Paul....E-6
St. Peter....E-6
St. James....E-6
Savannah Ave...B-7,B-8
Saver Ave....C-1
School St....E-5,E-6
Schuttle Ave....I-1
Second Ave....D-7
Seddon Way....B-3
Seminary Ave....B-5,D-5
Semmes Ave....I-3
Sewell....E-7
Shafer St....F-4
Sharon Ct....I-2
Shelby Dr....K-1
Shenandoah St....C-1
Sheppard St....E-2
Sherwood Ave....C-4
Shields Ave....F-3
Shipley....G-2
Shockoeslip....H-6
Sledd St....E-5
Smith St....F-5
Smithdeal....I-2
Southcliff Rd....H-2
South Hampton....H-3
Southlawn....K-6
Spaine St....K-3
Spotsylvania St.
....F-7,F-8
Spottswood Rd....F-2
Spring St....G-4
Springhill AVe....I-2,I-3
Spruce St....D-7
Stafford Ave....E-3
Stan Hope....A-4
Staples Mill....B-1
Steege....K-6
Sterl St....D-1
Stockton St....I-5,K-3
Stonewall Ave....I-2,I-3
Stony Run....I-8
Strawberry St....E-3
Stuart Ave...D-1,D-2,E-3
Suburban Ave....A-1
Suffolk....B-4
Summit Ave....D-3
Sumpter....G-3
Sunset....F-1
Sunset Ln....G-1
Sussex St....E-8
Swansboro Ln....K-3
Sylvan Rd....I-1
T St....F-8
Taylor Ave....I-1
Tazewell....D-5
Teakwood....A-3
Terminal Rd....D-3
Texas Ave....H-3
Thayer St....K-4
Third Ave....D-7
Thoms....D-4
Thompson....E-1
Thurman St....K-5
Tilden St....E-2
Tobacco....G-4
Tomacee Rd....E-1
Tomlynn St....B-2
Trafton....B-6
Trigg....E-7
Tulip....G-8
Tuskegee....F-8
Tyler....E-6
U St....F-8
Utah Pl....C-8
V St....F-8
Valley Rd....E-7
Vawter Ave....B-8
Veranda....C-7
Victor....D-8
Vine....F-3
Virginia St....H-6
W St....F-8
Waddey....B-7
Walker....G-3
Wall....I-4
Wallace....G-4
Walnut....B-8
Walpol....F-1
Walton Ave....B-5
Waverly....C-8
Wayne....D-2
Weaver Ct....J-2
Wellford....E-6
Wellington....C-7
Welton Ave....K-2
West Ave....E-4
Westbridge Rd....C-4
Westcliffe....B-6
Westmoreland St....B-1
Westove Rd....G-2
Westwood...B-2,B-4,B-6
Whatley....C-8
Whitby Rd....C-5
Whitcomb....E-7
Wickham....D-6
Wilcox Rd....K-1
Williamsburg Ave....I-8
Willis St....K-5
Willow St....D-7
Wilmington Ave....B-4
Winber Dr....G-3,K-2
Windsor Ave....A-4
Winston St....B-7
Wise St....J-4
Wood St....E-7,E-8
Woodbine Rd....I-2
Woodcliff....C-7,D-7
Woodhaven....K-1
Woodrow Ave...B-6,C-7
Wormley....G-4
Wythe Ave....C-1,D-2
X St....F-8
Yancey....E-6
Young....B-8
1st St....G-5,I-6
2nd St....G-54,I-6
3rd St....G-5,I-6
4th St....G-5,H-6
5th St....G-5,I-6
6th St....G-6,H-5
7th St....G-6,H-5
8th St....G-6,H-6
9th St....G-6,H-5
10th St....G-6,I-5
11th St....G-6,I-5,J-6
12th St....G-6,I-5,J-6
13th St....G-6,I-5
14th St....G-6,I-5,J-6
15th St....H-6,I-5
16th St....J-5
17th St....G-7,H-7,J-5
18th St....G-7,J-5
19th St....F-8,G-8
....H-7,I-4,J-5
20th St....F-8,G-8
....H-7,I-4,J-5
21st St....G-7,G-8
....H-7,I-4,J-4
22nd St...G-8,H-7,I-4,J-5
23rd St....G-8,H-7
24th St...G-8,H-7,I-4,J-4
25th St....H-7,I-4
26th St....H-7,I-4
27th St....H-7,I-3,J-4
28th St....H-8,I-3,J-4
29th St....H-8,I-3,J-3,J-4
30th St....H-8,I-3,J-3,J-4
31st St....H-8,I-3,J-3
32nd St...H-8,I-3,J-3,K-3
33rd St....H-8,I-3,J-3
34th St....H-8,I-3
34½ St....I-8
35th St....H-8,I-8,J-3
35½ St....I-8
36th St....I-8,J-2
36½ St....I-8
37th St.I-8,J-237th St.J-2
42nd St....J-2
43rd St....I-2
44th St....J-1
45th St....I-1
46th St....I-1
47th St....I-1
48th St....I-2
49th St....I-1

POINTS OF INTEREST

Atlantic Rural
Expositioin GroundsA-8
Broad St. R.R.
Station....D-3
Capitol Square....G-6
Circle Shopping
Center....K-2
Forest Hill Park....I-2
Forest Lawn Cemetery
....A-6
Hollywood CemeteryG-4
James River Park....I-3
Joseph Bryan Park..A-2
Mavry Cemetery....K-4
Monroe Park....F-5
Mt. Calvary Cemetery
....H-3
Mt. Olivet Cemetery.K-4
Parker Field Arena...C-3
Riverview Cemetery.H-4
Southside Plaza....K-2
Travel Info Center...C-4
Virginia Union Univ..D-5
William Byrd Park...G-1

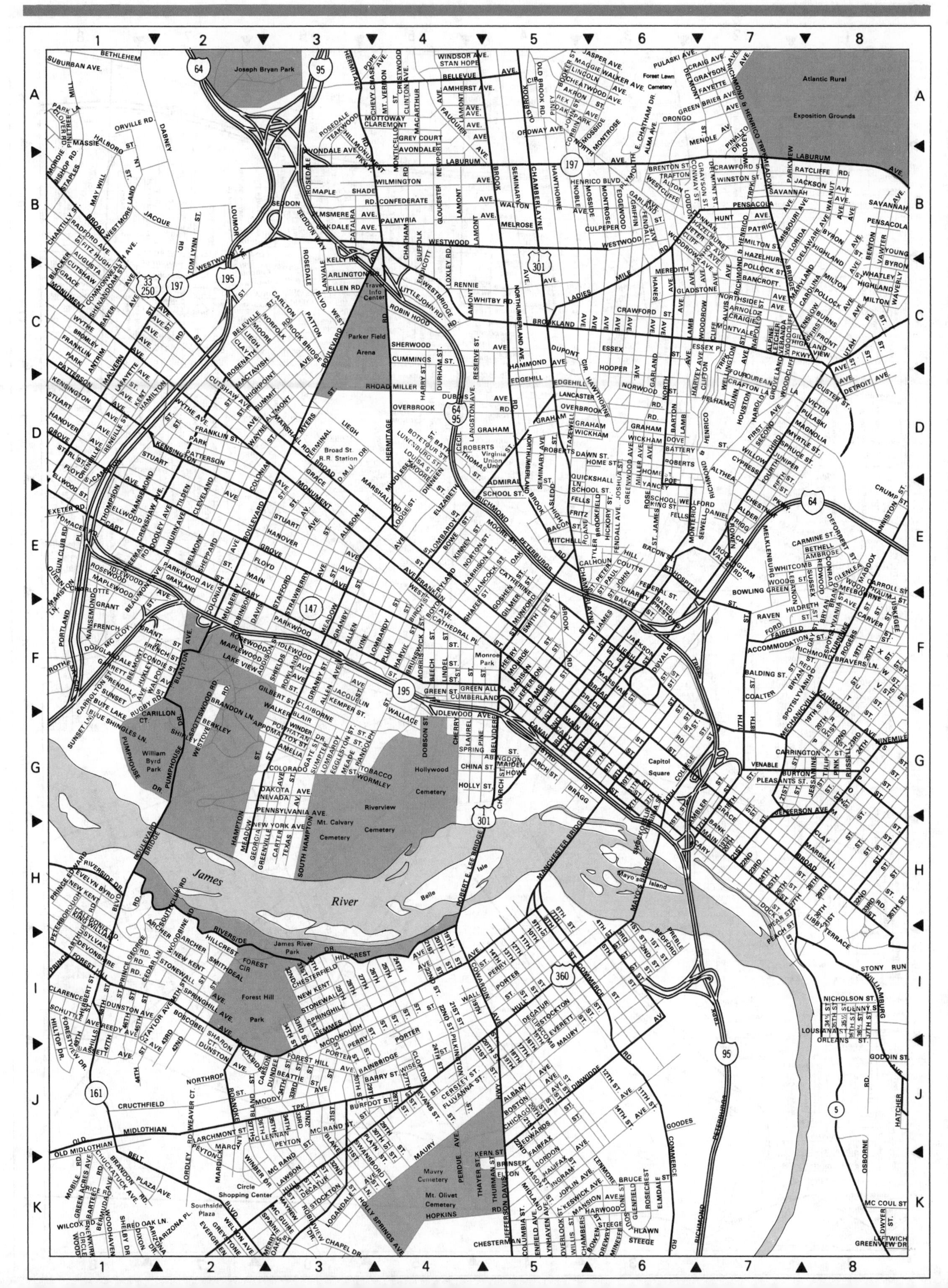
Scale of Miles
0 .2 .4 .6
N
Joseph Bryan Park
Atlantic Rural Exposition Grounds
Forest Lawn Cemetery
Parker Field Arena
Travel Info Center
Broad St. R.R. Station
Virginia Union Univ.
Monroe Park
Capitol Square
William Byrd Park
Hollywood Cemetery
Riverview Cemetery
Mt. Calvary Cemetery
James River
Belle Isle
Mayo's Island
James River Park
Forest Hill Park
Maury Cemetery
Mt. Olivet Cemetery
Circle Shopping Center
Southside Plaza
Robert E. Lee Bridge
Manchester Bridge
Mayo's Bridge
Boulevard Bridge
Laburum
Westwood
Broad
Cary
Main
Monument
Chamberlayne
Hermitage
Boulevard
Midlothian
Hull St.
Commerce
Jefferson Davis Hwy

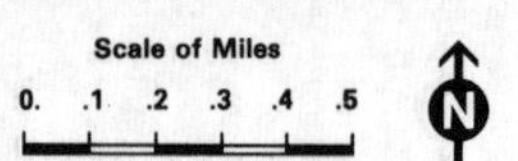
Scale of Miles
0. .1 .2 .3 .4 .5
N

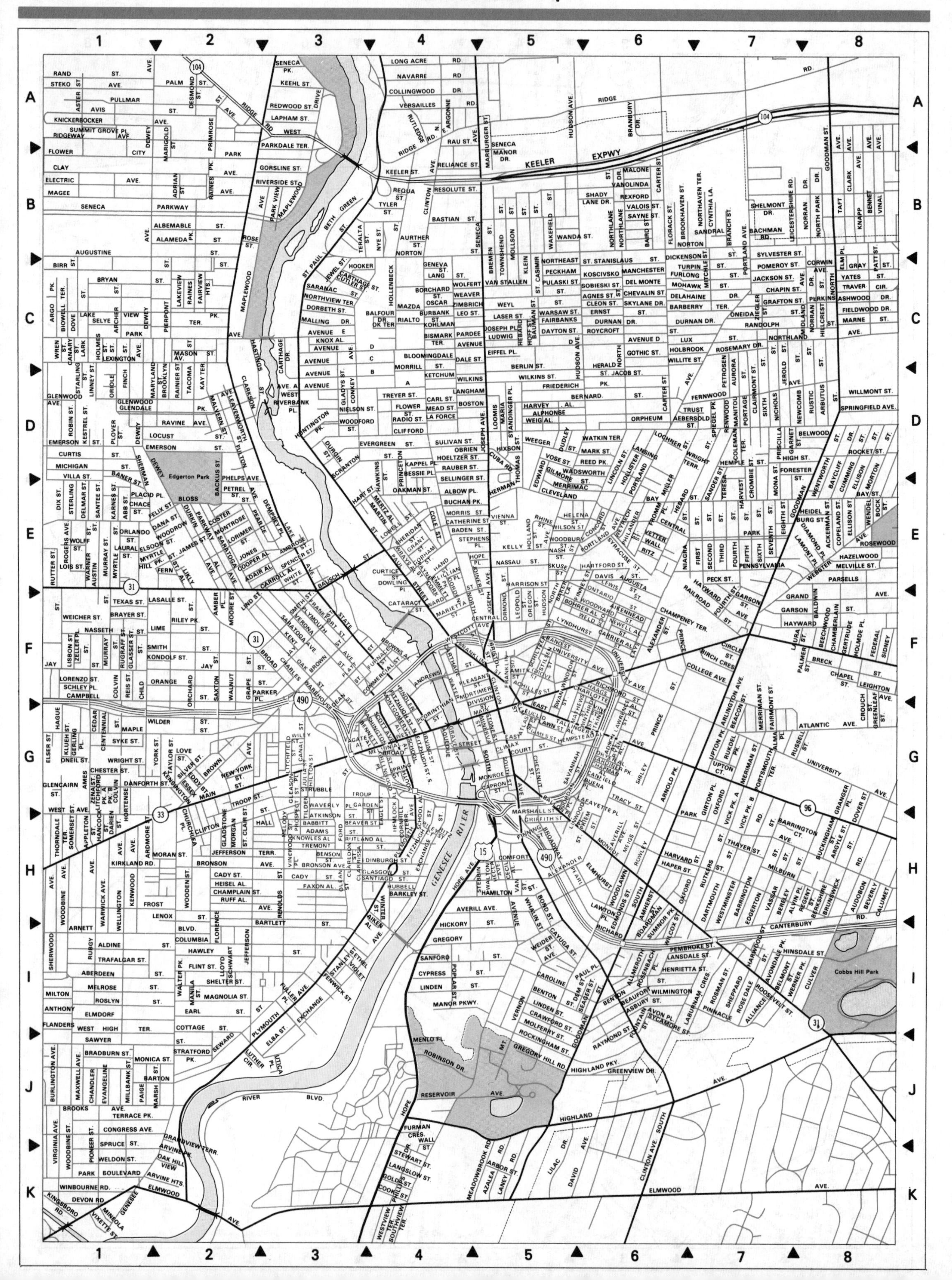
1
2
3
4
5
6
7
8
A
B
C
D
E
F
G
H
I
J
K
KEELER EXPWY
RIDGE RD.
Edgerton Park
Cobbs Hill Park
GENESEE RIVER
104
31
33
490
15
96
LAKE AVE.
ST. PAUL ST.
CLINTON AVE.
HUDSON AVE.
PORTLAND AVE.
NORTON
AVENUE D
AVENUE E
BLOSS
CAMPBELL
LYELL
STATE
PLYMOUTH AVE.
EXCHANGE
MAIN
EAST AVE.
UNIVERSITY AVE.
MONROE AVE.
SOUTH AVE.
MT. HOPE AVE.
ELMWOOD AVE.
HIGHLAND AVE.
CULVER
GOODMAN ST.
WINTON RD.
BRIGHTON
RESERVOIR AVE.
GREGORY HILL RD.
HIGHLAND PKY.
CANTERBURY RD.
GENESEE
ALEXANDER ST.
CHILI AVE.
WEST AVE.
AMES
THURSTON RD.
GENESEE PK. BLVD.
BROAD ST.
BAUSCH ST.
NORTH ST.
JOSEPH AVE.
SENECA AVE.
CARTER ST.
CULVER RD.

ROCHESTER

Abb St.....E-1
Aberdeen St.....I-1
Achilles St.....F-5
Ackerman St.....E-8
Adair Al.....E-2
Adams.....H-3
Adrian St.....B-2
Adson Al.....G-6
Aebersold St.....D-7
Agate Al.....G-3
Agnes St.....C-6
Aiken Al.....H-4
Aikenhead St.....F-6
Alameda St.....B-2
Albemarle St.....B-2
Albow Pl.....E-4
Aldine St.....I-1
Alexander St.....F-6,H-5
Alliance.....I-7
Allmeroth.....I-6
Alma Pl.....G-7
Alphonse.....D-5
Alvin Pl.....H-8
Amber Pl.....F-2
Ames.....G-1
Amherst St.....H-6
Amity St.....F-5
Amrose St.....E-3
Andrew St.....F-4
Anthony.....I-1
Appleton St.....H-1
Aqueduct.....G-4
Arbor St.....K-5
Arbutus St.....D-8
Archer St.....C-1
Ardmore St.....H-1
Argule St.....H-8
Argo Pk.....C-1
Argonine.....A-4
Arlington Ave.....G-7
Arnett Blvd.....I-1
Arnold Pk.....G-6
Arvine Hts.....K-2
Arvine Pk.....K-2
Asbury St.....I-6
Ashwood Dr.....C-8
Aster St.....A-1
Atkinson St.....H-3
Atlantic Ave.....G-8
Atlas St.....G-5
Audobon.....H-8
Augusta St.....E-6
Augustine St.....B-1
Austin St.....E-1
Aurora.....D-7
Aurther St.....B-4
Avenue A.....D-3
Avenue B.....D-3
Avenue C.....C-3
Avenue D.....C-3,C-5,C-6
Avenue E.....C-3
Aderill Ave.....H-5,H-6
Avis.....A-1
Avon Pl.....I-6
Avondale Pk.....I-7
Ayne St.....E-4
Azalea Rd.....K-5
Babbitt.....H-3
Bachman Rd.....B-7
Backus St.....E-2
Baden St.....E-4
Baker Ln.....F-5
Baldwin.....F-8
Balfour Dr.....C-4
Ballon Pl.....G-5
Baner St.....D-1
Barberry Ter.....C-7
Barkley St.....H-4
Barrington Ct.....H-7
Barrington St.....H-7
Bartlett St.....I-3
Barton.....J-2
Bastian St.....B-4
Bauman St.....C-5
Bausch St.....E-3
Bay St.....E-6,E-8
Baycliff Dr.....D-8
Beacon St.....G-7
Beaufort.....I-6
Beaver St.....H-3
Beechwood.....F-8
Belmont St.....I-7
Belwood.....D-8
Bennett Al.....G-4
Benson St.....H-3
Benton St.....I-5
Bereley St.....H-7
Berkshire.....H-8
Berlin St.....C-5
Bernard St.....D-6
Bessie Pl.....D-4
Beth Green.....B-3
Beverly.....H-8
Biowell Ter.....C-1
Birch Crest.....F-7
Birr St.....C-1
Bismark Ter.....C-4
Bloomingdale.....C-4
Bloss St.....E-2
Benton St.....I-6
Boardman.....I-6
Bock St.....E-8
Bohrer Al.....F-6
Bond St.....H-5
Borchard St.....C-4
Boston.....D-4
Bradburn St.....J-1
Branbury Dr.....A-6
Branch St.....B-7
Brayer St.....F-1
Breck St.....F-8
Bremen St.....C-5
Bristol St.....F-5
Broad.....F-3
Broad St., E.....G-4
Broadway.....H-5
Bronson Ave.....H-2,H-3
Brookhaven St.....B-6
Brooklyn.....D-2
Brooks Ave.....J-1
Brown.....G-2
Browns Al.....F-4
Brown St.....F-3
Brunswick St.....H-8
Buchan Pk.....E-4
Buckingham St.....H-8
Buena Pl.....G-6
Burbank St.....C-4
Burlington Ave.....J-1
Cady St.....H-2,H-3
Calumet.....H-8
Campbell St.....F-1
Canal St.....G-3
Canary St.....C-1
Canfield Park Ave.....G-6
Canterbury Rd.....I-7
Capron St.....G-5
Carl St.....D-4
Caroline.....I-5
Carrier Al.....F-6
Carroll Al.....E-3
Carter St.....B-6,D-6
Carthage.....F-4
Carthage Dr.....C-3
Carthage St.....C-3
Cataract St.....F-4
Catherine St.....E-4
Cayuga St.....I-5
Cecil Al.....H-5
Cedar.....G-1
Centennial.....G-1
Central.....E-6
Central Ave.....F-5
Chace St.....E-1
Chamberlian.....F-8
Champerey Ter.....F-6
Champlain St.....H-2
Chandler.....J-1
Chapel St.....F-8
Chapin St.....C-7
Chapman Al.....G-6
Charles.....F-3
Charlotte St.....F-6
Chester St.....G-1
Chestnut St.....G-5
Churchlea Pl.....H-2
Chevalin St.....C-6
Circle St.....F-7
Clairmont St.....D-7
Claredon St.....H-3
Clark Ave.....B-8
Clarkson.....D-2
Clarrissa.....H-3
Clat Ave.....B-1
Cleon St.....C-6
Cleveland.....E-5
Clifford.....D-4
Clinton Ave.....B-4,F-5
Clinton Ave., S.....K-6
Clifton St.....H-2
Climax.....G-5
Cole St.....E-4
Coleman Ter.....D-7
College Ave.....F-7
Collingwood Dr.....A-4
Columbia.....I-2
Colvin.....F-1,G-1
Comfort.....H-5
Commercial St.....F-4
Concord St.....E-6
Congress Ave.....J-1
Cook St.....K-4
Cooper Al.....E-2
Conkey.....D-3
Copeland St.....E-8
Corinthian St.....G-4
Cork.....E-4
Cornhill Terr.....H-4
Corwin Ave.....C-8
Coster.....E-2
Cottage St.....I-2
Court St.....G-4
Crawford St.....I-5
Crombie St.....E-7
Crouch St.....G-8
Cuba Rd.....D-5
Culver.....I-8
Cumming St.....D-8
Curtice St.....E-4
Curtis St.....D-1
Cutler St.....C-3
Cynthia Ave.....B-7
Cypress St.....I-4
Dale St.....C-4
Dana St.....E-2
Danforth St.....G-1
Darien St.....H-1
Dartmouth St.....H-7
David Ave.....K-4
Davis St.....E-6
Dayton St.....C-5
Dean St.....F-3
Delahaine Dr.....C-7
Delmar St.....E-1
Del Monte.....C-6
Dem St.....I-6
Dempsey Pl.....E-3
Desmond St.....A-2
Devon Rd.....K-1
Dewey Ave.
.....C-1,D-1,D-2
Diamond Pl.....E-8
Dickenson St.....B-6
Dinburgh St.....H-4
Division St.....F-5
Dix St.....E-1
Dorbeth St.....C-3
Dove St.....C-1
Dover St.....G-8
Dowling Pl.....E-4
Dryer Al.....G-6
Dudley.....D-5
Durgin St.....D-3
Durkin Al.....E-2
Durnan Dr.....C-6,C-7
Eagle St.....H-4
Earl St.....I-2
East Ave.....G-5
Eddy St.....G-5
Edgerton St.....H-7
Edmonds St.....H-6
Edward.....D-5
Eiffel Pl.....C-5
Eighth St.....E-7
Elba St.....I-3
Electric Ave.....B-1
Ellison St.....D-8,E-8
Elm Pl.....C-8
Elm St.....G-5
Elmdurf.....I-1
Elmhurst.....H-6
Elmwood Ave.....K-2,K-6
Elser St.....G-1
Elsoon St.....E-2
Euclid St.....G-5
Emerson St.....D-1,D-2
Emmett.....E-4
Ernst St.....C-6
Essex.....G-2
Ethel.....I-3
Evangeline.....J-1
Evergreen St.....D-4
Exchange Ave.....I-3
Exchange St.....H-4
Fairbank.....C-5
Fairmont St.....G-7
Fairview Hts.....C-2
Faxon Al.....H-3
Federal.....F-3
Felix St.....E-2
Fenwick St.....I-3
Fern.....E-2
Fernwood Ave.....D-7
Fieldwood Dr.....C-8
Fifth St.....E-7
Finch St.....D-1
Finney St.....F-6
Finning St.....H-5
First St.....E-7
Fitzhugh St.....H-4
Fitzhugh St., N.....G-4
Fitzhugh St., S.....G-4
Flanders.....I-1
Flint St.....I-2
Florack St.....B-6
Florence.....I-2
Flower City Park.....B-1
Flower St.....D-4
Ford St.....H-3
Forester St.....D-8
Fountain St.....I-6
Fourth St.....E-7,F-7
Frankfort St.....F-3
Franklin So.....F-5
Franklin St.....F-5
Friederich Pk.....D-5
Fromm Pl.....E-6
Frost Ave.....H-2
Fuller Pl.....I-3
Fulton Ave.....D-2
Furlong St.....C-7
Furman Cres.....J-4
Furnace St.....F-4
Garden St.....G-4
Garnet St.....D-8
Garson St.....F-7,F-8
Genesee.....K-1
Geneva St.....C-4
Gerling Pl.....G-1
Gertrude.....F-8
Gilmore St.....D-5
Girton Pl.....G-7
Gladstone.....H-2
Gladys St.....D-3
Glasgow St.....H-4
Glasser St.....F-1
Gleason Pl.....G-3
Glencairn St.....G-1
Glendale Pk.....D-1
Glenwood.....D-1
Glenwood Ave.....D-1
Gold St.....K-4
Goodman St.....E-8,I-6
Gorden Pk.....E-4
Gorham St.....E-4
Gorsline St.....B-3
Grafton St.....C-7
Grand Ave.....F-8
Grandview Terr.....J-2
Granger Pl.....G-8
Grant St.....E-4
Grape.....F-2
Graves St.....G-4
Gray St.....C-8
Greenleaf St.....G-8
Greenview Dr.....J-6
Greenwood Pl.....G-4
Gregory.....I-4
Gregory Hill Rd.....J-5
Griffith St.....H-5
Grove St.....F-5
Haags Al.....F-6
Hague.....G-1
Hall.....H-3
Hancock St.....H-1
Hand St.....E-4
Harper St.....H-6
Harrison St.....E-5
Hart St.....E-3
Hartford St.....E-6
Harwood St.....I-7
Harvest.....E-7
Harvey Al.....D-5
Hastings St.....C-2
Hawkins St.....E-4
Hawley St.....I-2
Hayward.....F-8
Hayward Ave.....F-7
Hazelwood.....E-8
Hebard.....C-5
Hebard St.....E-6
Hecla Ave.....H-5
Heidelburg St.....E-8
Heisel Al.....H-2
Helena St.....E-5
Hemlock Al.....H-4
Hemple St.....D-7
Hempstead Al.....G-5
Henrietta St.....I-6
Herman.....E-5
Hickory St.....I-4
High St.....D-8
Highland Ave.....J-6
Highland Pky.....J-6
Hillcrest.....C-8
Hilton St.....G-3
Hinsdale St.....I-8
Hixson.....D-5
Hoeltzer St.....D-4
Hoff St.....C-5
Holbrook.....C-6
Holland St.....E-5
Hollenbeck St.....C-4
Hollister.....D-6
Holmde Pl.....F-8
Holmes St.....C-1
Hooker.....C-3
Hope Ave.....H-4,J-4
Hope Pl.....E-5
Hortense.....G-1
Hubbell.....H-4
Hudson.....F-5
Hudson Ave.....A-5
Huntington Pk.....D-3
Irving St.....G-4
Jackson St.....C-7
James St.....E-3
Jay St.....F-1,F-2,F-3
Jefferson Ave.....I-2
Jefferson Terr.....H-2
Jerold St.....D-7
Joiner St.....E-5
Jones.....E-2
Jordan Pl.....G-5
Joseph Ave.....D-5,F-5
Joseph Pl.....C-5
Julia St.....G-3
Kappel Pl.....D-4
Karnes St.....E-1
Kay Ter.....D-2
Keehl St.....A-3
Keeler St.....B-4
Kelly St.....E-5
Kensington.....G-2
Kent.....F-3
Kenwood Ave.....H-1
Kestrel St.....D-1
Ketchum.....D-4
Kingsboro Rd.....K-1
Kirkland Rd.....H-1
Kislingbury St.....C-1
Klein St.....C-5
Klueh St.....G-1
Knapp Ave.....B-8
Knowles Al.....H-3
Knox Al.....C-3
Kohlman.....C-4
Kondolf St.....F-2
Kosciusko.....C-6
Laburnam Cres.....I-7
Lafayette Pl.....G-6
La Force.....D-4
Lake Ave.....E-3
Lakeview.....C-2
Lake View Pk.....C-1
Lamont Pl.....E-8
Lansing.....D-6
Lansdale St.....I-6
Langslow St.....K-4
Langham.....D-4
Lang St.....C-4
Laney Rd.....K-5
Lapham St.....A-3
Lark.....C-1
La Salle St.....F-2
Laser St.....C-5
Laura St.....E-1
Laura St.....F-8
Lawn St.....G-5
Lawrence St.....G-6
Lawton Pl.....H-6
Leavenworth.....D-2
Leicesteshire Rd.....B-7
Leighton Ave.....F-8
Lenox St.....H-2
Leo St.....C-4
Leopold St.....F-5
Lewis St.....E-6
Likly Al.....E-2
Lilac Dr.....K-5
Lillian Pl.....E-4
Lime St.....F-2
Lincoln St.....D-6
Lind St.....F-2
Linden St.....I-4,I-5
Linney St.....D-1
Linwood Pl.....H-6
Lisbon St.....F-1
Litchfield St.....G-3
Litchford.....G-1
Lloyd.....I-2
Lochner.....D-6
Locust St.....D-2
Lois St.....E-1
Long Acre Rd.....A-4
Loomis.....D-5
Lorenzo St.....F-1
Lorimer St.....E-2
Love St.....G-2
Lowell St.....E-4
Ludwig.....C-5
Luther Cir.....J-2
Lux St.....C-7
Lyndhurst St.....F-5
Magee Ave.....B-1
Magnolia St.....I-2
Main St.....G-2
Malling Dr.....C-3
Malone.....B-6
Malvern St.....D-2
Manchester.....C-6
Manila St.....I-2
Manitou.....D-7
Manor Pkwy.....I-4
Maple St.....G-1
Maplewood Ave.....B-3,C-2
Marburger St.....A-5
Maria.....D-5
Marietta St.....F-4
Marigold St.....A-2
Mark St.....D-6
Marne St.....C-8
Marsh St.....J-2
Marshall St.....G-5
Martin.....E-4
Maryland St.....D-2
Mason St.....C-2
Mathews St.....G-5
Maxwell Ave.....J-1
Mazda.....C-4
Mead St.....D-4
Meadowbrook Rd.....K-5
Meigs St.....H-6
Melody St.....H-3
Melrose St.....I-1
Melville St.....E-8
Melvin Al.....G-4
Menlo Fl.....J-4
Merrimac.....E-5
Merriman St.....G-7
Mertz Al.....E-4
Milbank St.....J-1
Milburn.....H-7
Michigan St.....D-1
Midland Ave.....C-8
Milton.....I-1
Mineola.....K-1
Minerva St.....G-5
Mitchle.....C-7
Mohawk St.....C-7
Molferry St.....I-5
Mollson St.....B-5
Mona St.....E-7
Monica St.....J-1
Monroe.....G-5,H-6
Montgomery Al.....G-4
Moore St.....F-2
Morgan.....H-2
Moran St.....H-2
Morrill St.....C-4
Morris St.....E-4
Mortimer.....F-5
Morton St.....D-8
Montrose.....E-2
Mt. Vernon Ave.....I-5
Muller St.....E-6
Murray St.....E-1,F-1
Myrtle Hill Pk.....E-2
Myrtle St.....E-1
Nash St.....E-5
Nassau St.....E-5
Nasseth St.....F-1
Navarre Rd.....A-4
Newell Al.....F-6
Newcomb St.....D-8
New York St.....G-2
Niagra St.....E-7
Nichols St.....D-7
Nickerbocker Ave.
.....A-1
Nielson St.....D-3
Norran Dr.....B-8,C-8
Norton.....B-7
Norton St.....B-4
North Park Dr.....B-8
North St.....C-6,F-5
Northaven Ter.....B-7
North Clinton.....F-5
Northeast.....C-5
North Goodman St.....B-8
Northland Ave.....C-7
Northlane Dr.....B-6
Northview Ter.....C-3
Nye St.....B-4
Oak St.....F-3
Oak Hill View.....K-2
Oakman St.....E-4
Obrien.....D-4
Ok Ter.....C-4
Olean St.....H-3
Oneida St.....C-7
Oneil St.....G-1
Ontario St.....F-6
Orange St.....F-2
Orchard St.....F-2
Oregon St.....F-5
Oriole St.....D-1
Orlando St.....E-1
Orpheum.....D-6
Ormond St.....F-5
Oscar.....C-4
Oxford.....G-7
Oxford St.....H-7
Paige.....J-1
Palm St.....A-2
Palmer St.....F-8
Pardee.....C-4
Pk. A.....G-1
Park Ave.....C-1
Pk. B.....G-1
Park Blvd.....K-1
Park Dale Ter.....A-3
Parker Pl.....F-3
Parkview.....B-3
Parkway Ave.....E-2
Parsells.....E-8
Patt St.....C-8
Paul Pl.....I-6
Paul St.....E-4
Pearce.....E-3
Pearl.....H-6
Peckham.....C-5
Pembroke St.....I-6
Perkins.....C-8
Petrel St.....E-2
Petrosen.....D-7
Phelps Ave.....D-2
Philander St.....E-6
Phillips Dr.....K-4
Pierpont St.....C-2
Pindler Al.....G-4
Pinnacle.....I-7
Pitkins.....G-6
Placid Pl.....E-1
Platt St.....F-3
Pleasant St.....F-5
Plover St.....D-1
Plymouth.....H-4
Plymouth Ave.....F-3,I-3
Pomeroy St.....C-7
Poplar St.....I-4
Portage St.....D-7
Portland Ave.....C-7,D-6,E-6
Portsmouth Ter.....G-7
Primrose St.....A-2
Prince.....F-6
Prince St.....G-6
Princeton.....D-4
Priscilla Ave.....D-7
Prospect St.....H-3
Pulaski St.....C-5
Pullmar Ave.....A-1
Radio St.....D-4
Railroad St.....F-7
Raines.....C-2
Raines Pk.....B-2
Rainier St.....D-2
Randolph St.....C-7
Rand St.....A-1
Rano St.....F-5
Rau St.....A-4
Rauber St.....D-4
Ravine Ave.....D-2
Raymond St.....J-6
Redwood St.....A-3
Reed Pk.....D-6
Regent St.....H-8
Reis St.....F-1
Reliance St.....B-4
Renwood.....D-7
Reoua St.....B-4
Reservoir Ave.....J-4
Resolute St.....B-4
Rexford.....B-6
Rhine St.....E-5
Rialto.....C-4
Richard.....I-6
Richmond St.....F-6
Ridge Rd.....A-6,B-4
Ridge Rd., W.....A-2
Riley Ct.....H-4
Riley Pk.....F-2
Ritz.....E-6
Ritz St.....H-4
River Blvd.....J-2
Riverside St.....B-3
Robin St.....D-1
Robinson Dr.....J-4
Rocket St.....D-8
Rockingham St.....J-5
Rodgers Ave.....E-1
Roosevlet St.....I-7
Rose St.....B-2
Rosedale.....I-7
Rosemary Dr.....C-7
Rosenbach Pk.....I-6
Rosewood.....E-8
Roslyn St.....I-1
Rosman St.....I-7
Rowley.....H-6
Roycroft.....C-6
Rubgy Ave.....I-1
Ruff Al.....H-2
Rugraff St.....F-1
Rukdel Pk.....G-7
Russell St.....G-8
Rustic St.....D-8
Rutkers St.....H-7
Rutland Al.....H-3
Rutter St.....E-1
Ryan Al.....G-6
St. Casimin St.....C-5
St. Clair St.....H-2
St. James St.....E-2
St. Paul St.....C-3
St. Priem.....H-6
St. Stanislaus St.....C-6
Sander St.....E-7
Sandral.....B-7
Sanford St.....I-4
Santee St.....E-1
Santiago St.....H-4
Saranac St.....C-3
Saratoga.....E-2
Saratoga Ave.....F-3
Sardiner Pk.....G-6
Savannah St.....G-5
Sawyer St.....J-1
Sayne St.....B-6
Schiltzer St.....F-5
Schley Pl.....F-1
School.....G-4
School Al.....H-4
Schwart.....I-2
Scott Ave.....G-4
Scranton St.....D-3
Seager St.....I-6
Second St.....E-7
Selye Ter.....C-1
Seneca.....B-5
Seneca Manor Dr.....A-5
Seneca Pk.....A-3
Seneca Pkwy.....B-1
Seventh St.....E-7
Seward St.....J-2
Shady Lane Dr.....B-6
Shelmont Dr.....B-7
Shelter St.....I-2
Sheppard.....I-7
Sheridan St.....E-4
Sherman.....D-1
Sherwood Ave.....I-1
Sibley.....G-6
Sidney.....F-8
Silver St.....G-2
Sixth St.....D-7,E-7
Skuse St.....E-5
Skylane Dr.....C-6
Smith St.....F-2,F-3
Spencer St.....E-3
Spiegel Pk.....D-7
Spring St.....G-4
Springfield Ave.....D-8
Spruce St.....J-1
Sobieski St.....C-6
Somerset St.....H-1
South Ave.....G-5
South St.....G-5,H-6
Southview Ter.....K-4
Standinger Pl.....D-5
Stanley St.....I-3
Starling St.....D-1
State.....F-3
Stebbin.....H-5
Steko Ave.....A-1
Stephens Pl.....E-5
Sterling.....E-1
Stewart St.....K-4
Stilsen St.....F-5
Stone St.....G-5
Stratford Pk.....J-2
Strubble.....G-3
Sulivan St.....D-4
Summit Grove Pl.....A-1
Sumnor Pk.....I-6
Swan St.....F-5
Sycamore St.....I-6
Syke St.....G-1
Sylvester St.....B-7
Syracuse St.....E-6
Tacoma St.....D-2
Taft Ave.....B-8
Tallinger Al.....G-5
Taylor St.....G-2
Teralta St.....B-3
Teresa.....E-7
Terrace Pk.....J-1
Texas St.....F-1
Thayer St.....H-7
Third St.....E-7
Thomas St.....D-5
Thorndale Ter.....H-1
Tilden.....G-3
Townshend St.....C-5
Tracy St.....G-6
Trafalgar St.....I-1
Trajer Cir.....C-8
Tremont St.....H-3
Treyer St.....D-4
Troop St.....G-2,G-3
Trust.....D-7
Turpin St.....C-7
Tyler St.....B-4
University Ave.
.....F-5,G-8
Upton Ct.....G-7
Upton Pk.....G-7
Utica Pl.....J-3
Valois St.....B-6
Vanolinda.....B-6
Van St.....H-5
Van Stallen.....C-5
Vassar St.....H-7
Versailles Rd.....A-4
Vetter.....E-6
Verona St.....F-3
Vick Pk. A.....G-7
Vick Pk. B.....G-7
Vienna St.....E-5
Villa St.....D-1
Vinal Ave.....B-8
Vine.....G-6
Vinewood Pl.....H-3
Violet.....I-3
Virginia Ave.....J-1
Vixette St.....K-1
Vose St.....D-5
Wadsworth.....D-6
Wakefield St.....B-5
Wall.....E-6
Wall St.....J-4
Walnut.....F-2
Walter.....F-4
Walter Pk.....I-2
Walton.....H-5
Wanda St.....B-5
Ward St.....F-4
Warehouse.....F-3
Warner St.....E-1
Warsaw St.....C-5
Warwick Ave.....H-1
Washington.....G-4
Watkin Ter.....D-6
Waverly Pl.....G-3
Weaver.....C-4
Webster Ave.....E-8
Weeger St.....D-5
Weicher St.....F-1
Weider St.....I-5
Weig Al.....D-5
Weld St.....F-6
Weldon St.....K-1
Wellington Ave.....H-1
Wendel.....E-8
Wentworth St.....D-8
Werner Pk.....I-8
Westcott Ave.....F-4
West High Ter.....I-1
West Main St.....G-4
Westminster Rd.....H-7
West Riverbank Pl.....D-3
Westview Ter.....K-4
Weyl St.....C-5
Weyrech.....E-6
Whalin St.....H-5
White St.....E-3
Windsor St.....F-5
Wiley St.....G-3
Wilcox St.....I-6
Wilkins.....D-4,D-5
Willite St.....C-6
Willmont St.....D-8
Wilmington.....I-6
Wilson St.....E-5
Winbourne Rd.....K-1
Winter St.....H-4
Winthro.....G-6
Wolfert.....C-4
Wolffe St.....E-1
Woodbine Ave.....H-1
Woodbine St.....K-1
Woodbury St.....E-5
Wooden St.....H-2
Woodford St.....D-3
Woodlawn.....H-6
Woodrow.....E-2
Woodward St.....F-6
Wrays Al.....E-4
Wren St.....C-1
Wright St.....G-1
Wright Terr.....D-7
Yates St.....C-8
York St.....G-2
Zeller Pl.....F-1
Zena St.....G-1
Ziegler St.....C-7
Zimbrich.....C-4

POINTS OF INTEREST

Cobbs Hill Park.....I-8
Edgerton Park.....D-2

ST. PETERSBURG

Abington Ave.....G-1
Alcazar Way.....I-4
Alhambra Way.....I-4
All States Ct.....G-5
Almedo.....A-7
Almeria.....I-4
Amber.....K-4
Amelia.....I-6
Anastasia Way.....I-4
Appian.....B-8
Aranda.....B-7
Argon.....I-5
Arlington Ave.....D-5
Asturia Way.....I-5
Auburn.....F-3
Bahama Shores Dr.....J-6
Barcelona.....I-4
Bay Isle.....F-7
Bay St.....B-6,F-6
Bayou Blvd.....H-5
Bayou Cir.....I-5
Bayshore Dr.....D-6
Bayview.....A-8
Beach Dr.....C-7,D-6,F-6
Bellaire.....A-8
Bennett.....A-3
Benson Ave. N.....A-1
Bethel.....G-6
Bluefish.....H-6
Bonita Way.....I-4
Booker.....C-3
Brightwaters.....B-7,B-8
Burlington Ave. N.
.....D-1,D-5
Caesar Way.....I-3
Canton.....J-3
Capri.....A-8
Carlisle.....D-2
Casilla.....I-3
Catalan.....B-8
Catalonia Way.....H-3
Central Ave.....D-1,D-6
Cherry St.....B-7
Cobia.....H-6
Coffee Pot Blvd.....A-6
Columbus Way.....H-3
Coquin Dr.....H-7
Coral.....K-4
Cordova.....A-8,B-8,I-3
Coronada.....H-3
Cortez.....H-3
Country Club Way.....G-3
Dartmouth Ave. N.....C-1
Desota Ave.....H-3
Desota Way.....H-3
Dolphin.....H-7
Dolphin Dr.....H-6
Driftwood Dr.....G-6
Eden Isle Blvd.....A-8
Elkham Blvd.....H-6
Elm St.....B-7
Estado.....B-7
Fairfield Ave.....E-1
Fairway Ave.....G-3,H-4
Florida Ave.....F-6
Friendly.....K-5
Giralda.....B-8
Granada Circle East
.....G-3
Granada Circle West
.....H-3
Green Way.....H-3
Grove St.....C-5,E-5
Hammond.....E-2
Highland St.....G-5,I-5
Hyacinth St.....I-5
Inner Cir.....K-4
Irondale.....G-5
Isle Blvd.....B-8
James Ave.....E-5
Jasmine Way.....I-5
Juanita Way.....I-6
Karlton.....K-4
Kensington.....E-2
Kingston.....D-2,F-2
Lafayette St.....C-2
Laredo.....A-7
Leslee Lake Dr.....B-1
Lake Maggiore.....I-5
Lamparilla.....G-3
Locust.....A-7
Locust St.....B-7
Madrid.....H-3
Manatee Dr.....G-7
Manor Way.....K-4
Maron.....B-7
Melrose Ave.....E-5
Melton.....A-5
Mentiaden.....H-6
Mirrow Lake Ave. N.
.....D-6
Monterey Blvd.....A-7
Morris St.....A-3
Murilla.....G-3
Murok Way.....K-5
Myrtle.....H-5
Naryarez.....H-3
Neptune.....H-6
Newton Ave.....F-5
Nina.....B-8
North Shore Dr.....C-7
Oak St.....B-6
Pampano Dr.....H-6
Pinellas Point Dr.
.....K-4
Poplar St.....B-7
Porpoise.....H-6
Prescott.....F-4
Preston Ave.....F-5
Queen.....E-4,F-4
Queensboro Ave.....F-2
Quincy St.....F-1
Ramon.....A-8
Riv Way.....A-7
Roser Park Dr.....E-5
Seminole.....F-5
Serpentine.....K-4
Seville.....A-8
Snell Blvd.....A-7
Sunfish.....H-7
Sunrise Dr.....F-7,H-6,I-6
Tangerine Ave.....F-4
Taylor.....G-6
Terminal.....D-4
Toldeo.....B-8
Valencia.....G-5
Walnut.....A-7
Walnut St.....B-7
Walton Ct.....F-3
Way, S.....H-3
Whiting.....H-6
Woodlawn Circle.....B-5
Xenia St.....A-3
Yale St.....A-3
Yardley Ave. N.....B-1
1st Ave. N.....D-1,D-5
1st Ave. S.....D-1,D-5
1st St. N.....B-6
1st St. S.....E-6
2nd Ave. N.....D-1
2nd Ave. S.....D-1,D-5
2nd St. N.....B-6
2nd St. S.....E-6,J-6
3rd Ave. N.....C-1,D-5
3rd Ave. S.....D-1,D-5
3rd St. S.....E-6,J-6
4th Ave. N.....C-1,C-6
4th Ave. S.....D-1,D-5
4th St. N.....B-6
4th St. S.....E-6,J-6
5th Ave. N.....C-1,C-5
5th Ave. S.....D-1
5th St. N.....B-6
5th St. S.....E-6
6th Ave. N.....C-1,C-6
6th Ave. S.....D-1
6th St. S.....E-6,H-6
7th Ave. N.....C-1,C-5
7th Ave. S.....E-1,E-5
7th St.....F-5
7th St. N.....B-5
7th St. S.....J-5
8th Ave. N.....C-1,C-6
8th Ave. S.....E-1,E-5
8th St. N.....B-5,E-5
8th St. S.....K-5
9th Ave. N.....C-1,C-5
9th Ave. S.....E-1,E-5
9th St. N.....E-5
9th St. S.....G-5,J-5
10th Ave.....E-5
10th Ave. N.....C-1
10th Ave. S.....E-1,E-3
10th St. S.....G-5
11th Ave. N.....C-1,C-5
11th Ave. S.....E-1,E-4,E-5
11th St. N.....A-5,E-5
11th St. S.....G-5,J-5
12th Ave. N.....C-1
12th Ave. S.....E-1,E-4
12th St. N.....A-5,C-5,E-5
12th St. S.....J-5
13th Ave.....E-1,E-6
13th Ave. N.....B-1
13th St. N.....C-5
13th St. S.....G-5
14th Ave. N.....B-1,C-5
14th Ave. S.....E-2,E-6
14th St. N.....A-5,E-5
14th St. S.....K-4
14th Way.....J-4
15th.....F-4
15th Ave.....E-6
15th Ave. N.....B-1,B-5
15th Ave. S.....E-2,F-5
15th St. N.....A-4,C-4
15th St. S.....J-4
16th Ave.....F-6
16th Ave. N.....B-1,B-5
16th Ave. S.....F-2,F-5
16th St. N.....C-4,E-4
16th St. S.....J-4
16th Way.....J-4
17th Ave. N.....B-1
17th Ave. S.....F-2,F-6
17th St.....F-4
17th St. N.....A-4,D-4
17th St. S.....K-4
18th Ave. N.....B-3,B-5
18th Ave. S.....F-2,F-6
18th St. N.....A-4
18th St. S.....G-4,K-4
19th Ave. S.....F-1,F-6
19th St.....D-4
19th St. N.....B-4
19th St. S.....G-4,J-4
20th Ave. N.....B-3
20th Ave. S.....F-1,F-6
20th St. N.....A-4,D-4
20th St. S.....G-4,J-4
21st Ave. S.....F-1,F-6
21st St. N.....D-4
21st St. S.....G-4,K-4
21st Way.....J-4
22nd Ave.....F-7
22nd Ave. N.....B-1,B-5
22nd Ave. S.....F-1
22nd St. N.....D-4
22nd St. S.....F-4,J-4
22nd Way.....K-4
23rd Ave.....F-1,F-6
23rd Ave. N.....B-1
23rd St.....F-3
23rd St. N.....A-4,D-3
24th Ave.....F-1,F-6
24th Ave. N.....B-1
24th St.....F-3
24th St. N.....D-3
25th Ave. N.....A-1
25th Ave. S.....F-1,F-6
25th St. N.....D-3
25th St. S.....F-3,J-3
26th Ave. N.....A-1,A-5
26th Ave. S.....F-1
26th St. N.....D-3
26th St. S.....F-3
27th Ave.....G-1,G-5
27th Ave. N.....A-1
27th Ave. S.....G-4
27th St. N.....D-3
27th St. S.....F-3
28th Ave.....G-1,G-5
28th Ave. N.....A-1
28th Ave. S.....G-4
28th St. N.....D-3
28th St. S.....F-3,J-3
29th Ave.....G-1,G-5
29th Ave. N.....A-1
29th St.....B-3
29th St. N.....D-3
29th St. S.....F-3,J-3
30th Ave.....G-1
30th Ave. N.....A-1,A-5
30th St.....A-3
30th St. N.....D-3
30th St. S.....F-3,J-3
31st Ave.....A-7
31st Ave. N.....A-1
31st St. N.....D-3
31st St. S.....G-3
31st Ter.....A-7
32nd.....D-2
32nd Ave.....G-5
32nd Ave. N.....A-1
32nd Ave. S.....G-2
32nd St. S.....G-2
33rd Ave. N.....A-1
33rd St.....D-2
34th Ave.....G-5
34th Ave. N.....A-5
34th Ave. S.....G-1
34th St. N.....D-2
34th St. S.....G-2
35th Ave.....G-5
35th Ave. S.....G-1
35th St. N.....D-2
35th St. S.....F-2
36th Ave.....H-5
36th Ave. S.....H-2
36th St.....D-2
36th St. S.....F-2
37th Ave.....H-5,H-6
37th St. N.....D-2
37th St. S.....G-2
38th Ave.....H-5,H-6
38th Ave. S.....H-1
38th St.....D-2,F-2
38th St. S.....H-2
39th.....H-2
39th St.....D-2,F-2,H-1
39th St. N.....D-2
40th Ave.....H-5
40th St.....G-1
40th St. N.....D-1
40th St. S.....F-1,H-1
41st Ave.....H-6
41st St. N.....D-1
41st St. S.....F-1,H-1
42nd.....I-1
42nd Ave.....H-5
42nd Ave. S.....H-1
42nd St. N.....D-1
42nd St. S.....F-1
43rd St. N.....D-1
43rd St. S.....F-1
43rd Terr.....H-6
44th St. S.....F-1
45th Ave.....H-5
45th St.....H-1
45th St. N.....D-1
45th St. S.....F-5
46th Ave.....H-3
46th Ave. S.....H-1
46th St. N.....D-1
46th St. S.....F-1
47th Ave. S.....H-3
47th St. N.....D-1
48th Ave. S.....H-1
49th Ave. S.....I-1
50th Ave. S.....I-1,I-3,I-5
50th Pl.....I-1
51st Ave. S.....I-3,I-5
53rd Ave. S.....I-1,I-5
54th Ave.....I-3
55th Ave.....I-6
56th Ave. S.....I-6
57th Ave. S.....I-6
58th Ave. S.....I-3
58th Ter.....J-3
59th Ave. S.....J-3,J-6
60th Ave. S.....J-1,J-3,J-5
61st Ave. S.....J-3,J-5
62nd Ave.....J-6
62nd Ave. S.....J-1,J-3
62nd Pl.....J-4
62nd Ter.....J-4
63rd.....J-3
63rd Ave. S.....J-4
63rd Ter.....J-4
64th.....J-3
64th Ave. S.....J-5
65th.....J-3
65th Ave. S.....J-5
66th Ave. S.....J-3, J-5
66th Ter.....J-3
67th Ave. S.....J-3
68th Ave. S.....J-3,J-5
68th Ter.....J-4
69th Ave. S.....J-3,J-5
70th Ave.....K-3

POINTS OF INTEREST

Albert Whitted Airport.....E-7
Lake Maggiore Park.....H-4

Saint Petersburg, FL

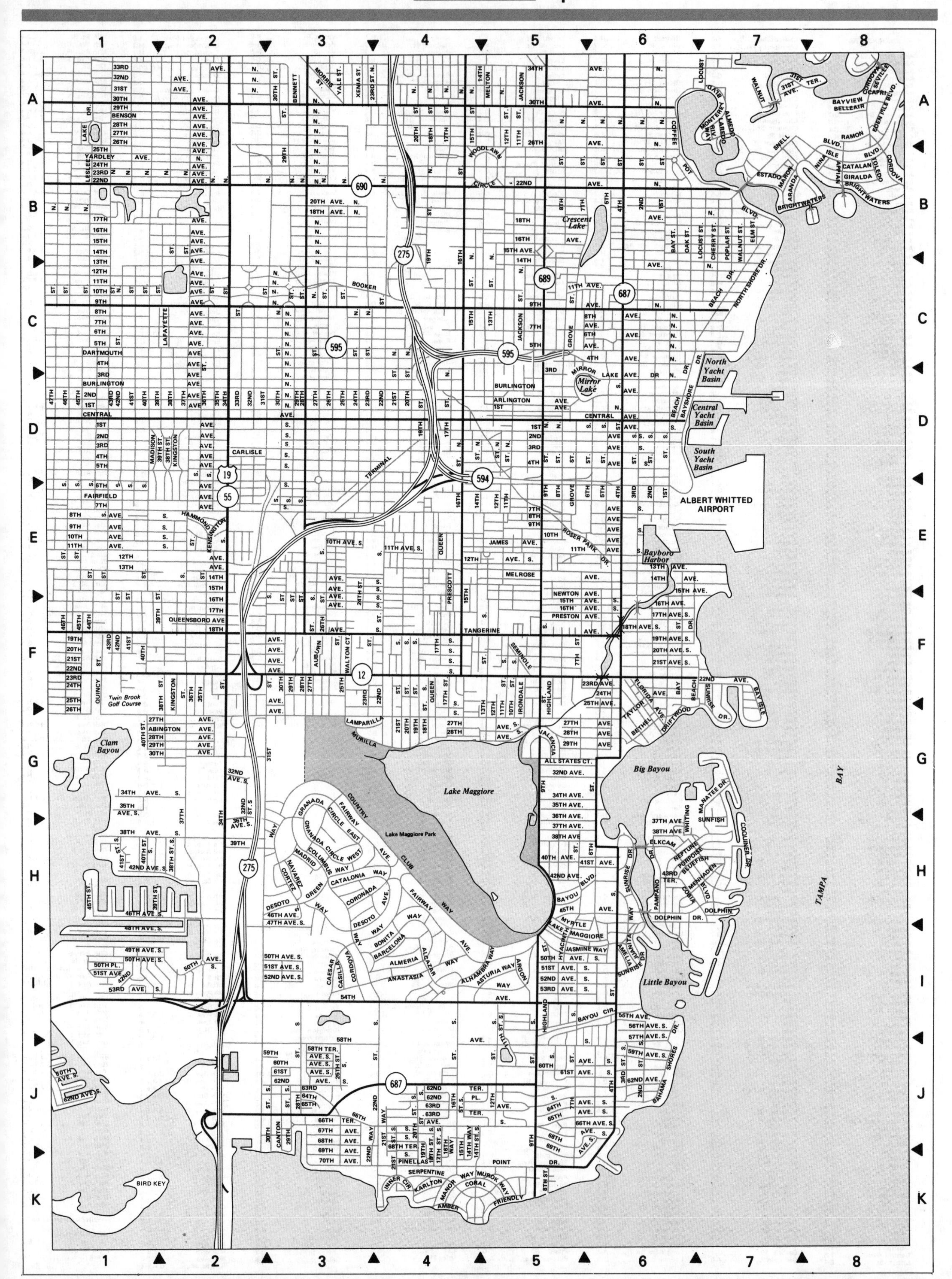

N

Scale of Miles
0 .2 .4 .6

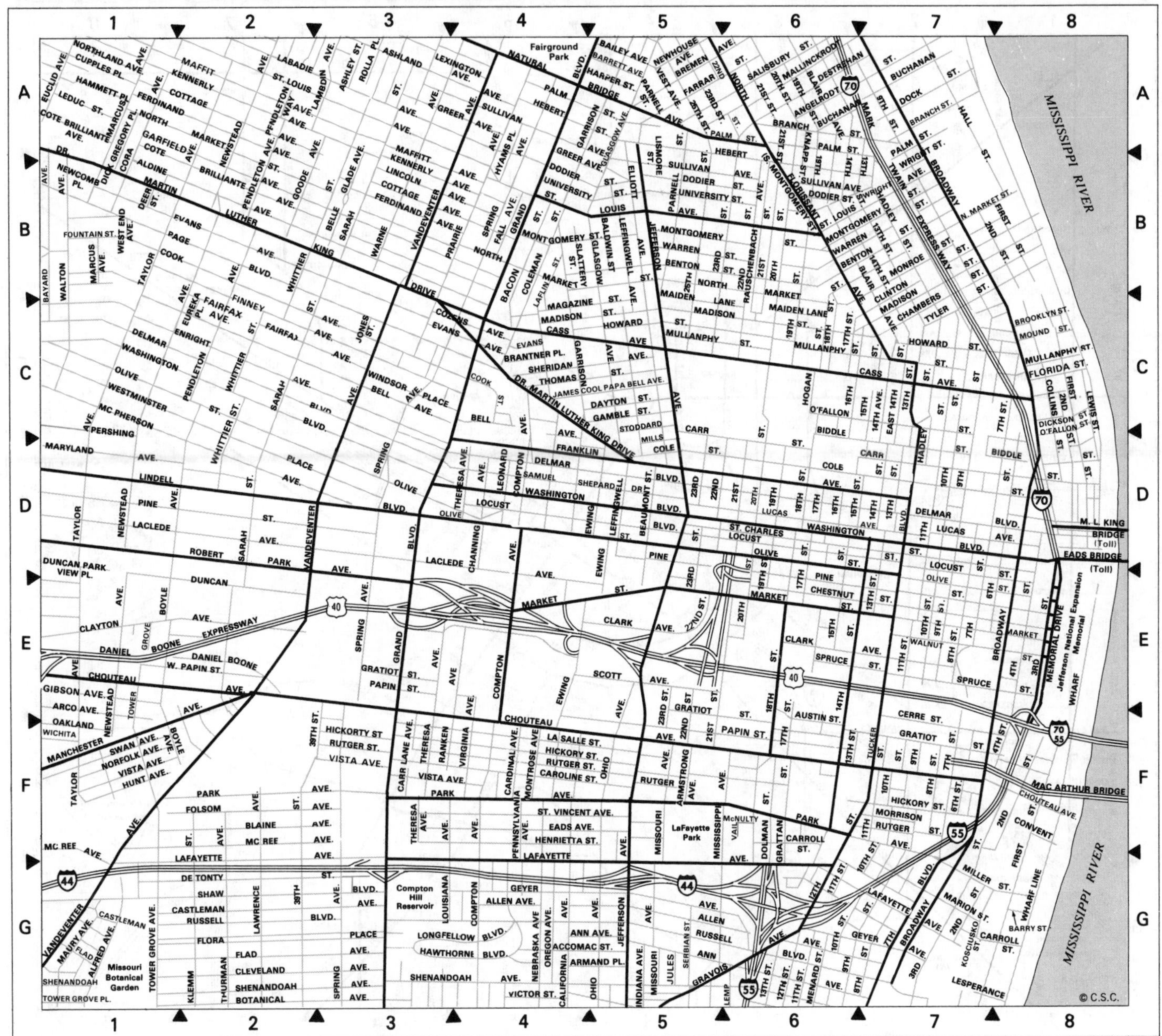

ST. LOUIS

Accomac St. G-4
Aldine Ave. B-1
Alfred Ave. G-1
Allen Ave. G-4,G-5
Angelrodt St. A-6
Ann Ave. G-5,G-6
Arco Ave. E-1
Armand Pl. G-4
Armstrong Ave. F-5
Ashland Ave. A-3
Ashley St. A-3
Austin St. E-6
Bacon St. B-4
Bailey Ave. A-5
Baldwin St. B-5
Beaumont St. D-5
Bell Ave. C-3,C-4
Belle Glade Ave. B-3
Benton St. B-5,B-6
Biddle St. C-6,D-8
Blaine Ave. F-2
Blair Ave. A-6,B-7
Botanical Ave. G-2
Boyle Ave. F-1,G-2
Branch St. A-6
Brantner Pl. C-4
Bremen Ave. A-5
Broadway A-7,E-8,G-7
Buchanan A-6,A-7
California Ave. G-4
Cardinal Ave. F-4
Carr St. C-5,D-7
Carrlane Ave. F-3
Carroll St. F-6,G-8
Cass Ave. C-4,C-7
Castleman Ave. G-2
Cerré St. F-7
Chambers St. C-7
Channing Ave. D-4
Chestnut St. E-6
Chouteau Ave. E-1,F-4
Clark Ave. E-5,E-6
Clayton Ave. E-1
Cleveland Ave. G-2
Clinton St. B-7
Cole St. D-5,D-6
Coleman St. B-4
Collins St. C-8
Compton Ave. D-4, E-4,G-4
Convent F-8
Cook Ave. B-1
Cora Ave. A-1
Cottage Ave. A-2,B-3
Cote Brilliante Ave. A-1,B-2
Cozens Ave. C-3
Cupples Pl. A-1
Daniel Boone E-2
Daniel Boone Expwy. E-1
Dayton St. C-5
Deer St. B-1
Delmar Blvd. C-1, D-4,D-7
Destrehan A-6
Detonty St. G-2
Dickson St. C-4
Dick Gregory Pl. A-1
Dock St. A-7
Dr. Martin Luther King Dr. B-2,C-4
Dodier St. B-4,B-5,B-6
Dolman St. F-6
Duncan E-2
Duncan Park View Pl. D-1
Eads Ave. F-4
Eads Bridge D-8
Elliott Ave. B-5
Enright Ave. C-2
Euclid Ave. A-1
Eureka Pl. C-2
Evans Ave. B-2,C-3
Ewing D-5,E-4
Ewing Ave. D-5
Fairfax Ave. C-2
Fall Ave. B-4
Farrar St. A-5
Ferdinand Ave. A-1,B-3
Finney Ave. B-2
First St. B-8,C-8,G-8
Flad Ave. G-2
Flora G-2
Florida St. C-8
Folsom Ave. F-2
Franklin Ave. D-4
Gamble St. C-5
Garfield Ave. A-1
Garrison Ave. A-4,C-4
Geyer Ave. G-4,G-7
Gibson Ave. E-1
Glasgow Ave. B-5
Goode Ave. B-2
Grand Blvd. B-4,E-3,G-3
Gratiot St. E-3,E-5,F-7
Grattan St. F-6
Gravois Ave. G-5
Greer Ave. A-4
Hadley St. B-7,D-7
Hall St. A-7
Hammett Pl. A-1
Harper St. A-5
Hawthorne Blvd. G-3
Hebert St. A-4,A-6
Henrietta St. F-4
Hickory St. F-3,F-4,F-7
Hogan St. C-6
Howard St. C-5,C-7
Hunt Ave. F-1
Hyams Pl. A-4
Indiana Ave. G-5
Jefferson Ave. B-5,G-5
Jones St. C-3
Jules G-5
Kennerly Ave. A-2,B-3
Klemm St. G-2
Knapp St. A-6
Labadie Ave. A-2
Laclede Ave. D-1,D-3
Lafayette Ave. F-2,F-4,G-7
Lambdin Ave. A-3
LaSalle St. F-4
Lawrence Ave. G-2
Leduc St. A-1
Leffingwell D-5
Leffingwell Ave. B-5
Lenard D-4
Lesperance G-7
Lewis St. C-8
Lexington Ave. A-4
Lincoln Ave. B-3
Lindell Blvd. D-1
Lismore St. A-5
Locust Blvd. D-4
Locust St. D-6,D-7
Longfellow Blvd. F-3
Louisiana Ave. G-3
Lucas Ave. D-4,D-7
MacArthur Bridge F-8
Madison St. C-4,C-5,C-7
Maffitt Ave. A-2,A-3
Magazine St. C-4
Maiden Lane B-5,C-6
Mallinckrodt A-6
Manchester Ave. F-1
Marcus Ave. A-1,B-1
Marion St. G-7
Martin Luther King D-8
Market Ave. A-2
Market St. E-4,E-6
Maryland Ave. D-1
Maury Ave. G-1
McNair Ave. G-5
McPherson Ave. C-1
McRee Ave. F-1,F-2
Memorial Dr. E-8
Menard St. G-6
Miller St. G-7
Mississippi Ave. F-5
Missouri Ave. F-5,G-5
Monroe St. B-7
Montgomery St. B-4, B-5,B-6
Montgomery St., S. B-6
Montrose Ave. F-4
Morrison St. F-7
Mullanphy St. C-5, C-6,C-8
Natural Bridge Ave. A-4
Nebraska Ave. G-4
Newcomb Pl. B-1
Newhouse Ave. A-5
Newstead Ave. A-2, D-1,E-1
Norfolk Ave. F-1
North Florissant Ave. B-6
Northland Ave. A-1
North Market St. A-2,B-4,B-6
Oakland Ave. E-1
O'Fallon St. C-6
Ohio Ave. F-5,G-5
Olive St. C-1,D-3,D-6
Oregon Ave. G-4
Page Ave. B-1
Palm St. A-4,A-6,A-7
Papin St. E-3,F-6
Papin St., W. E-1
Park Ave. F-2,F-3,F-6
Parnell St. A-5,B-5
Pendleton Ave. B-2,C-2
Pendleton Way A-2
Pennsylvania F-4
Pershing C-1
Pine D-5
Pine St. D-1,E-6
Place G-3
Prairie Ave. B-4
Ranken Ave. F-3
Rauschenbach Ave. B-6
Robert Park Ave. D-2
Rolla Pl. A-3
Russell Blvd. G-2,G-5
Rutger St. F-3,F-4,F-5,F-7
St. Charles D-6
St. Louis Ave. A-2, B-5,B-6
St. Vincente Ave. F-4
Salisbury St. A-6
Sarah St. B-3,C-2,D-2
Shaw Blvd. G-2
Shenandoah Ave. G-2,G-3
Sheridan Ave. C-4
Slattery St. B-4
Spring Ave. B-4, D-3,E-3,G-3
Spruce St. E-6,E-7
Sullivan Ave. A-4,B-5,B-6
Scott Ave. E-5
Swan Ave. F-1
Taylor Ave. B-1,D-1,F-1
Theresa Ave. D-4,F-3
Thomas St. C-4
Thurman Ave. G-2
Tower Grove Ave. G-1
Tyler St. C-7
University St. B-4,B-5
Vandeventer Ave. B-3,D-2,G-1
Vest Ave. A-5
Victor St. G-4
Virginia Ave. F-4
Vista Ave. F-1,F-3
Walton Ave. B-1
Warne Ave. B-3
Warren St. B-5,B-6
Washington Blvd. C-1,D-4,D-6
West End Ave. B-1
Westminster Pl. C-1
Wharf Line E-8,G-8
Whittier St. B-2,C-2
Windsor Pl. C-3
Wright St. B-7
2nd St. B-7,C-8,F-8,G-7
3rd St. E-8,G-7
4th St. E-8,F-8
6th St. E-7,F-7
7th Blvd. G-7
7th St. C-8,E-7,F-7
8th St. E-7,F-7,G-6
9th St. A-7,D-7,E-7, F-7,G-6
10th St. D-7,E-7, F-7,G-6,G-7
11th St. D-7,E-7,F-7,G-6
12th St. E-7,G-6
13th St. B-7,C-7,D-7, E-7,F-6,G-6
14th Ave. C-7
14th St. B-6,B-7,D-7,E-6
14th St. E. C-7
15th St. C-7,D-6,E-6
16th D-6
16th St. C-6
17th D-6
17th St. C-6,E-6,F-6
18th St. C-6,D-6,E-6,F-6
19th D-6
19th St. A-6,B-6,C-6,E-6
20th St. A-6,B-6,E-6
21st D-6
21st St. A-6,B-6,F-5
22nd D-5
22nd St. B-6,E-5,F-5
23rd D-5
23rd St. A-5,B-5,E-5
25th St. A-5,B-5
39th St. F-2,G-2

POINTS OF INTEREST

Fairground Park A-4
Jefferson National Expansion Memorial E-8
Missouri Botanical Gardens G-1

Saint Paul, MN

Scale of Miles

0 .1 .2 .3 .4 .5

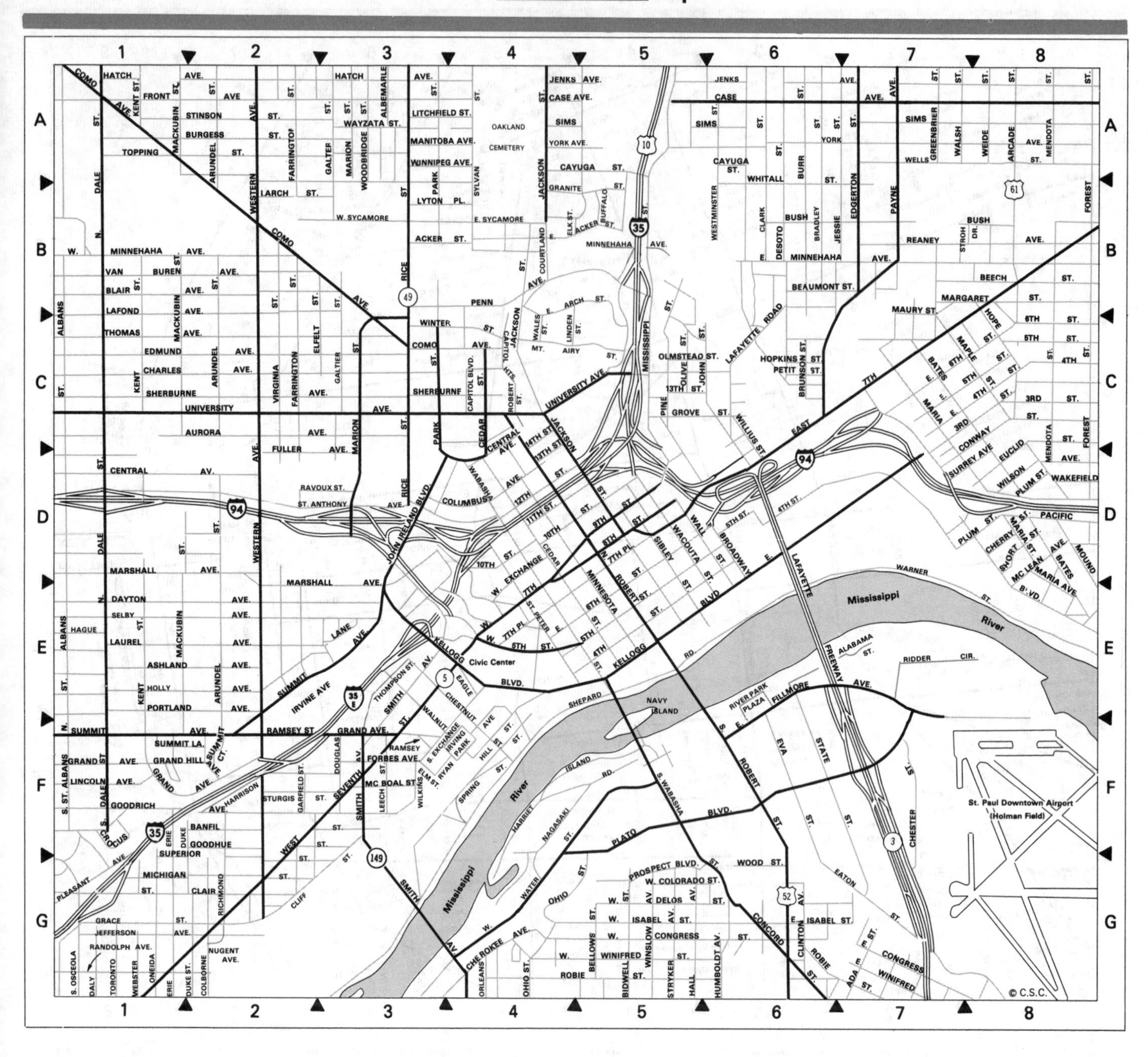

ST. PAUL

Acker St. B-3
Ada St. G-7
Albemarle A-3
Arcade St. A-8
Arundel St. A-2, C-2
Ashland Ave. E-1
Arundel St. E-2
Aurora Ave. C-2
Banfil St. F-2
Bates Ave. C-7, D-8
Beaumont St. B-6
Beech St. B-8
Bidwell St. G-5
Blair Ave. B-1
Bradley St. B-6
Broadway D-6
Brlinson St. C-6
Burgess St. A-2
Burr St. A-6
Bush B-8
Case Ave. A-5, A-6
Cayuga St. A-4, A-6
Cedar St. C-4, D-4
Central Ave. C-4
Charles Ave. C-1
Cherokee Ave. G-4
Cherry St. D-8
Chester St. F-7
Clark St. B-6
Clinton Ave. G-6
Colorado St., W. G-5
Columbus Ave. D-4
Como Ave. . . A-1, B-2, C-3
Concord St. G-6
Congress, E. G-7
Congress St., W. G-5
Conway St. D-7
Crocus F-1
Dale St., N. B-1, E-1
Dale St., S. F-1
Daly St. G-1
Dayton Ave. E-1
Delos St., W. G-5
Desoto St. B-6
Eva St. F-6
Edgerton St. B-7
Edmund Ave. C-1
Elfelt St. C-2
Euclid St. D-8
Exchange, W. E-4
Farrington St. . . . A-2, C-2
Fillmore Ave., E. E-6
Forbes Ave. F-3
Forest St. B-8, C-8
Front Ave. A-1
Fuller Ave. C-2
Galter St. A-3, C-3
Gelby Ave. E-1
Goodhue St. F-2
Goodrich Ave. F-1
Gran St. B-4
Grand Ave. F-1, F-3
Grand Hill Ave. F-1
Greenbrier St. A-7
Grove St. C-5
Hall Ave. G-5
Hatch Ave. A-1, A-3
Holly Ave. E-1
Hope B-8
Hopkins St. C-6
Humboldt Ave. G-5
John Ireland Blvd. D-3
Irvine Ave. E-2
Isabel St., E. G-6
Isabel St., W. G-5
Jackson St. B-4, C-4
Jefferson Ave. G-1
Jenks Ave. A-4
Jessie St. B-6
John St. C-5
Kellogg Blvd. E-3
Kellogg Blvd., E. E-5
Kent St. A-1, C-1, E-1
Lafond Ave. B-1
Lafayette Frwy. D-6
Lafayette Rd. C-6
Larch St. B-2
Laurel Ave. E-1
Lincoln Ave. F-1
Litchfield St. A-3
Lyton Pl. B-3
Mackubin St. . . . . A-1 . . . . C-1, E-1
Maiden Lane E-2
Manitoba Ave. A-3
Maple St. C-7
Margaret St. B-7
Maria Ave. C-7, D-8
Marion St. A-3, C-3
Marshall Ave. D-1, E-2
McBoal St. F-3
McLean Ave. D-8
Maury St. B-7
Michigan St. G-1
Minnehaha Ave., E. . . B-6
Minnehaha Ave., W. . . B-1
Minnesota D-5
Mississippi St. C-5
Ohio St. G-4
Olive St. C-5
Olmstead St. C-5
Pacific D-8
Park St. B-3
Penn Ave. B-4
Petit St. C-6
Pine St. C-5
Plato Blvd. F-5
Plum St. D-7, D-8
Portland Ave. E-1
Prospect Blvd. G-5
Ramsey St. F-2
Reaney Ave. B-7
Rice St. B-3, D-3
Robert, N. D-5
Robert St., S. F-6
Robie St. G-4, G-6
St. Albans C-1
St. Albans, N. F-1
St. Albans, S. F-1
St. Clair G-1
Seventh St., W. F-2
Sherburne Ave. . . C-1, C-3
Short St. D-8
Sibley St. D-5
Sims A-4, A-5, A-7
Smith Ave. . . E-3, G-3, G-3
State St. F-6
Stinson St. A-2
Stryker Ave. G-5
Summit Ave. E-2, F-1
Summit Ct. F-2
Summit Ln. F-1
Superior St. F-1
Surrey Ave. D-7
Thomas Ave. C-1
Topping St. A-1
University Ave. C-2, C-4
Van Buren Ave. B-1
Virginia St. C-2
Wacouta St. D-5
Wakefield D-8
Wall St. D-5
Walsh St. A-7
Wazata A-3
Weide St. A-8
Western Ave. B-2, D-2
Willius St. C-6
Wilson Ave. D-8
Winifred, E. G-7
Winifred St., W. G-5
Winnipeg Ave. A-3
Winslow Ave. G-5
Winter St. C-3
Wood St. G-6
Woodbridge St. A-3
3rd St. C-7, C-8
4th C-8
4th St. E-5
4th St., E. C-7
5th St. C-8, E-4, E-5
5th St., E. C-7
6th St. C-8
6th St., E. C-7, E-4
7th, E. C-6
7th, W. E-4
7th Pl. D-5
7th Pl., W. E-4
8th St. D-5
9th St. D-5
10th St. D-4
11th St. D-4
12th St. D-4
13th St. C-5, D-4
14th St. C-4

POINTS OF INTEREST

Civic Center E-4
St. Paul Downtown Airport (Holman Field) F-8

Scale of Miles
0 .1 .2 .3 .4 .5

N

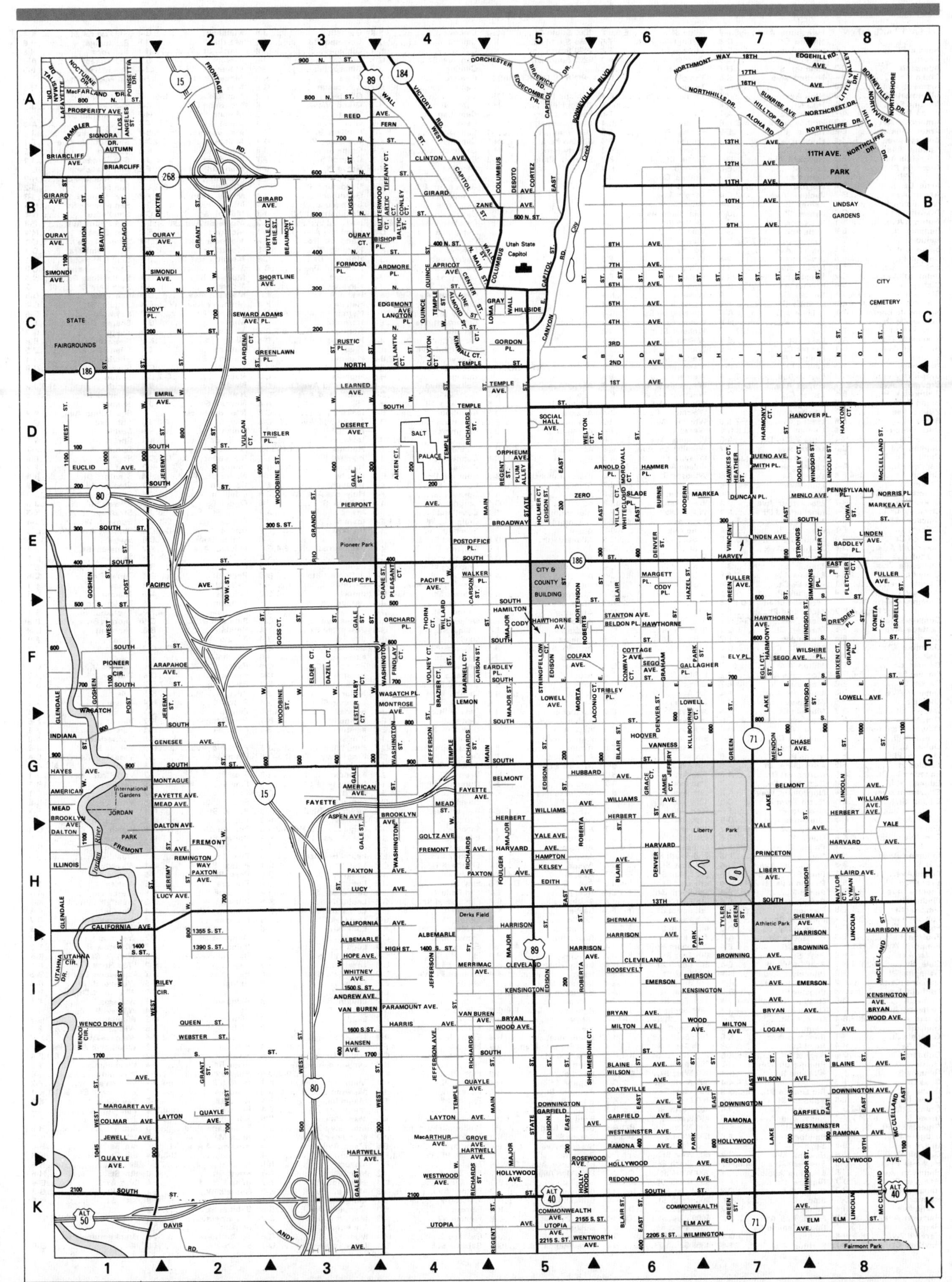

SALT LAKE CITY

Acadia Pl....F-6
Acorn Ct....G-6
Adams Pl....C-3
Aiken Ct....D-4
Alameda Ave....D-6
Albermarle Ave....I-3
Almond St....C-4
Aloha Rd....A-7
American Ave...G-1,G-3
Amos Villa....E-6
Andrew Ave....I-3,I-4
Apricot Ave....C-4
Arapahoe Ave....F-2
Ardmore Pl....C-4
Arnold Pl....D-6
Artic Ct....B-4
Aspen Ave....G-3
Athletic Park....H-7
Atlantic Ct....C-4
Autumn Ave....B-1
Baddley Pl....E-8
Baltic St....B-4
Beaumont Ct....B-3
Beldon Pl....F-6
Bellhaven Ct....F-7
Belmont Ave....G-7
Bishop Pl....B-4
Blaine Ave....J-6,J-8
Blair St....E-6,G-6,K-6
Bonneville Blvd....A-5
Bonneville Dr....A-8
Bothwell St....F-1
Braewick Rd....A-5
Briarcliff Ave....B-1
Brixten Ct....F-8
Broadway....E-5
Brooklyn Ave....G-1,G-4
Browning Ave....I-5,I-7
Brunswick Pl....E-5
Bryan Ave....I-6,I-7
Bueno Ave....D-7
Burns St....E-8
Butterwood Ct....B-4
California Ave....H-1,H-4
Canyon Rd....C-5
Capitol St., E....C-5
Capitol St., W....A-4
Carson St....E-4,F-4
Center St....C-4
Chamberlain Ct....F-5
Chase Ave....G-7
Chesney Ct....E-4
Chicago St....B-1
Citizens Ct....E-3
Clayton Ct....C-4
Cleveland Ave....I-6
Clinton Ave....B-4
Coatsville Ave....J-6
Cody Pl....E-6
Colmar Ave....J-1
Coloney Ct....B-4
Commonwealth Ave.
...K-5,K-6,K-7
Conway Ct....F-6
Cottage Ave....F-6
Crane St....E-4
Dallmore Ct....G-5
Dalton....H-1
Delmar Ct....E-4
Denver St....E-6,G-6
Depot Ct....E-2
Deseret Ave....D-3
Dine Ct....J-5
Dooley Ct....D-7
Dorchester....A-5
Downington Ave.J-5,J-8
Dresden Pl....F-8
Dubec Ct....E-6
Duncan Pl....E-7
Eardley Pl....F-5
Edgecombe Dr....A-5
Edgehill Rd....A-8
Edgemont Ave....C-4
Edison St....E-5
...G-5,I-5,J-5
Edith Ave....H-5
Egli Ct....F-7
Elm Ave....K-6,K-8
Ely Pl....F-7
Emerson Ave....I-6
Erie St....B-3
Euclid Ave....D-1
Exchange Pl....E-5
Fayette Ave....G-2,G-4
Findlay Ct....F-4
Fletcher Ct....E-8
Folsom Ave....D-1,D-4
Formosa Pl....C-3
Foulton Pl....E-3
Fremont Ave....H-1,H-4
Fuller Ave....E-7,E-8
Gale St....D-3,F-3,H-3
Gallagher Pl....F-6
Gardena Ct....C-2
Garfield Ave....J-6,J-8
Gemmill....D-6
Genesee Ave....G-2
Girard Ave...B-1,B-3,B-4
Glen St....F-1
Glendale St....H-1
Goltz Ave....H-4
Goodard Ct....D-5
Gordon Pl....C-5
Goshen St....F-1
Goss Ct....F-3
Grace Ct....G-6
Grand Pl....F-8
Grant St....B-2,J-2
Green St....G-7,H-7,K-7
Greenlawn Pl....C-3
Gregory Pl....F-5
Grove Ave....J-4
Gudgell Ct....F-6
Hammer Pl....D-6
Hampton Ave....H-5
Hanover Pl....D-7
Hansen Ave....I-3
Harmony Ct....D-7
Harris Ave....I-4
Harrison Ave....I-6,I-8
Hartwell Ave....J-3,J-4
Harvard Ave....H-5,H-8
Hawkes Ct....D-7
Hawthorne Ave....F-5,F-7
Haxton Ct....D-8
Hayes Ave....G-1
Hazel St....E-6
Heather St....D-7
Herbert Ave....G-5
...G-6,G-8
High Ave....I-4
Hills Dr....A-8
Hilltop Rd....A-7
Hollywood Ave..K-6,K-8
Holmer Ct....E-5
Hope Ave....I-3
Hoyt Pl....C-1
Hubbard Ave....G-5
Iowa St....E-8
Ivan Ct....E-7
Iverson St....F-6
Jackson Ave....C-2,C-4
James Ct....G-6
James St....J-1
Jefferson Ave....J-4
Jefferson St....G-4
Jeremy St....D-2,F-2,H-2
Jewell Ave....J-1
Juniper Ct....F-6
Kelsey Ave....H-5
Kensington Ave...I-6,I-8
Kilbourne Ct....G-6
Kimball Ct....C-4
Laconio Ct....F-6
Lafayette Dr....A-1
Laird Ave....H-8
Lake St....G-7,J-7
Laker Ct....E-7
Lanark Ct....F-6
Langton Pl....C-4
Laxon Ct....C-1
Layton Ave....J-2,J-4
Lerned Ave....D-1,D-3
Lester Ct....G-3
Lexington Ave....H-1
Liberty Ave....H-7
Linden Ave....E-7,E-8
Lincoln St....D-8,G-8
Little Valley Dr....A-8
Logan Ave....I-7
Long Ave....J-1
Los Angeles St....A-1
Lowell Ave....F-5,F-6
Lucy Ave....H-2,H-4
Lyman Ct....H-8
MacArthur Ave....J-4
MacFarland Dr....A-1
Main St....E-5,J-5
Main St., N....B-4
Major St....F-5,K-5
Margaret Ave....J-1
Margett Pl....E-6
Marion St....B-1
Markea Ave....E-6-8
Markell Ct....F-4
Maryland Pl....E-5
McClelland St....D-8,I-8
McDonald Pl....F-7
Mead Ave....G-2,G-4
Mendon Ct....G-7
Menlo Ave....E-7
Meredith Ct....D-2
Merriman Ave....I-4
Milton Ave....I-6,I-7
Montague Ave....G-2
Montclair Ct....E-7
Montrose Ave....F-4
Mordivall Ct....D-6
Mortem Ct....F-5
Mortenson Ct....F-5
Naylor Ct....H-8
Newman Ct....F-4
Nocturne Dr....A-1
Norris Pl....E-8
Northcliffe Dr....A-8
Northcrest Dr....A-8
Northhills Dr....A-7
Northmont Way....A-6
North Temple St....C-3
Orchard Pl....F-4
Orpheum Ave....D-5
Ouray Ave....B-1,B-2
Owen Pl....E-6
Pacific Ave....E-2,E-4
Pacific Pl....E-3
Paramount Ave....I-4
Park St....E-6,F-6,J-6
Paxton Ave..H-2,H-3,H-4
Penna Pl....E-8
Pershing Pl....E-4
Pierpont....E-3,E-4
Pleasant Ct....E-4
Plum Alley....D-5
Poinsettia Dr....A-1
Poplar Ct....E-4
Post St....E-1
Post Office Pl....E-4
Princeton Ave....H-7
Prosperity Ave....A-1
Pugsley St....B-3
Quayle Ave....J-1,J-4
Queen St....I-2
Quince St....C-4
Ramona Ave....J-6,J-8
Redondo Ave....K-6
Reed Ave....A-3
Regent St....D-5,K-5
Remington Way....H-2
Richards St....D-4
...G-4,J-4,K-4
Rio Grande St...E-3,G-3
Roberta St....I-5
Roosevelt Ave....I-6
Rosewood Ave....K-5
Rustic Pl....C-3
Sandberg Pl....F-5
Sego Ave....F-6,F-7
Seward Ave....C-2
Sherman Ave....H-6,H-7
Shortline Ave....C-3
Signora Dr....A-1
Simondi Ave....C-1,C-2
Slade Pl....E-6
Smith Pl....D-7
Social Hall Ave....D-5
South Temple St.
...D-1,D-4
Stanton Ave....F-6
State St....E-5,J-5
Strongfellow Ct....F-5
Strongs Ct....E-8
Sunrise Ave....A-7
Talisman Dr....A-1
Temple Ave....D-5
Temple St....C-4
Temple St., W....C-4
...D-4,J-4
Thorn Ct....F-4
Tiffany Ct....B-4
Tribley Pl....F-6
Trinidad Pl....F-4
Trisler Pl....D-3
Turtle Ct....B-3
Tyler St....H-7
Utahna Cir....I-1
Utahna Dr....I-1
Utopia Ave....K-5
Van Ct....F-5
Van Buren Ave....I-4
Victory Rd....A-4
Vine St....C-4
Vissing Pl....F-4
Volney Ct....F-4
Voyles Pl....E-8
Vulcan Ct....D-2
Walker Pl....E-4
Wall St....B-5
Wasatch....K-4
Wasatch Pl....F-4
Washington St....G-4
Wayne Ct....E-4
Webster St....I-2
Welton Ct....D-5
Wenco Cir....I-1
Wenco Dr....I-1
Wentworth Ave....K-5
Werner Pl....E-6
Westminster Ave....J-6
Westwood Ave....K-4
Whitecloud Ct....E-6
Whitmore Ct....D-4
Whitney Ave....I-3
Wilard Ct....F-4
Williams Ave....G-5,G-8
Wilmington Ave....K-6
Wilshire Pl....F-8
Wilson Ave....J-6,J-8
Windsor St....D-8
...F-7,H-7,J-7
Winthrop Ct....E-6
Wood Ave....I-5,I-8
Woodbine St....D-3,F-3
Yale Ave....H-5,H-7
Zane Ave....B-5
A St....C-5
B St....C-6
C St....C-6
D St....C-6
E St....C-6
F St....C-6
G St....C-7
H St....C-7
I St....C-7
J St....C-7
K St....C-7
L St....C-7
M St....C-8
N St....C-8
O St....C-8
P St....C-8
Q St....C-8
1st Ave....D-6
2nd Ave....C-6
3rd Ave....C-6
4th Ave....C-6
5th Ave....C-6
6th Ave....C-6
7th Ave....C-6
8th Ave....B-6
9th Ave....B-7
10th Ave....B-7
11th Ave....B-7
12th Ave....B-7
13th Ave....A-7
16th Ave....A-7
17th Ave....A-7
18th Ave....A-7
1st South St....D-1
2nd East St....E-5,J-5
2nd South St....D-1,E-4
3rd East St....E-6,J-6
3rd South St....E-1,E-7
4th East St....E-6,J-6
4th South St....E-1,E-4
5th East St....J-6
5th South St....F-1,F-4
6th East St....J-7
6th South St....F-1,F-4
7th East St....J-7
7th South St....F-1,F-4
8th East St....E-7,J-7
8th South St....G-1,G-4
9th East St....E-8,J-8
9th South St....G-1,G-4
10th East St....J-8
11th East St....J-8
13th South St....H-2,H-6
14th South St....I-1,I-4
15th South St....I-3
16th South St....I-3
17th South St....J-1,J-6
21st South St....K-1,K-6
200 North St....C-3
200 West St....D-4,J-4
300 North St....C-3
300 West St....D-3,J-3
400 East St....K-6
400 North St....B-3
400 West St....D-3
500 North St....B-1,B-4
500 West St....D-3,J-3
600 North St....B-3
600 West St....D-2
700 North St....A-3
700 West St....C-2,J-2
800 North St....A-1,A-3
800 West St....D-2,H-2
900 North St....A-3
900 West St....D-1,J-1
1000 West St....D-1,I-1
1045 West St....J-1
1100 West St....D-1
1355 South St....H-2
1390 South St....I-2
2155 South St....K-5
2170 South St....K-4
2205 South St....K-6

POINTS OF INTEREST

Athletic Park....H-7
Derks Field....H-4
Fairmont Park....K-8
International
Gardens....G-1
Jordan Park....H-1
Liberty Park....H-7
Pioneer Park....E-3

SAN ANTONIO

Abiso Ave....B-6
Ace....K-3
Adams St....K-5
Aganier....A-3,D-3
Agarita Ave., W....E-4
Agnes St....C-4
Alabama St....J-7
Alametos....B-2,B-3
Alamo St....G-6
...H-6,J-4,J-5
Alamo Heights Blvd..B-6
Alamo Plaza....I-5
Albany....A-8
Albert....G-1
Alexander....F-1
Alhambra....A-1
Allen....B-7
Allensworth....D-8
Alling....F-7
Alpen....C-5
Alps....J-6
Alta....A-6
Alta Vista....J-1
Alvarez....K-3
Alvin....D-6
Amestoy....H-2
Ancona....E-6
Andrews....E-8
Angeles....C-1
Angella St....J-1
Apache....J-3
Appler....F-7
Arbor Pl....G-1
Arbutus....B-7
Arcadia....B-7,B-8
Archer....F-6
Arcinega....I-5
Arden Grove....G-6
Argo Ave....A-6
Argyle Ave....B-7
Armour....E-6
Army Blvd....F-7
Arsenal St....I-4
Arthur....H-8
Artillery....F-7
Arkansas Ave....K-7
Ashby Pl....F-3
Ashby Pl., W....E-1
Atlanta....G-5
Aubrey....I-5
Auburn....A-8
Audubon....B-4
August St....G-5
Austin....H-6
Ave. A....F-6,H-5
Ave. B....D-7,H-5
Avery....E-2
Baltimore St....G-5
Barbe....K-5
Barilla....C-8
Barrera....J-5
Basse Rd....A-1
...A-2,A-3
Beacon....B-3
Becky Renee....B-1
Belknap....F-4
Belmont St....I-8
Belvidere Dr....C-5
Belview....D-8
Benton....E-8,G-8
Blaine....I-5
Blanco Rd....E-3
Bluebonnet....H-8
Blue Bonnet Blvd....A-7
Blue Star St....J-4
Blum St....I-6
Boerne....K-6
Booker St....H-6
Bonham St....H-5
Boyer Ave....K-7
Brackenridge Rd....E-7
Brad....C-1,D-1
Brahan Blvd....F-7
Brazos St....H-2
Breeden Ave....B-4
Brittany....B-5
Broadway Ave....E-7,H-6
Brooklyn Ave....G-5
Brooks....G-7
Burbank....K-3
Burleson St....H-6
Burnet St....H-6
Burr Rd....C-8
Bushnell Ave....D-5
Cactus St., S....J-7
Calaveras St., S....G-1
Calhoun....E-8
Callahan....J-5
Calles....K-1
Camada St....G-1
Camargo....J-5
Camaron....H-4
Cambridge....B-7
Camden St....G-5
Camp St....J-4
Canadian St....H-8
Canal....J-5
Canton St....I-8
Capitol....E-2
Cardinal Ave....A-6
Carleton....E-5
Carolina....K-6
Carnanan....D-8
Carson St., E....F-7
Casablanca....G-6
Casanova....J-2
Cash....J-6
Catalina....A-2,C-2
Catapalpa....D-8
Center....I-7,J-8
Ceralvo St....K-1
Cevallos St., E....J-4
Cevallos St., W....J-3
Cherry St....J-6
Chestnut St....I-6
Cincinnati Ave., W....F-1
City St....J-4
Claborne....B-7
Claremont Ave....E-8
Caludia....J-5
Clay St....J-4
Cloverleaf Ave....A-7
Clower....B-2,B-3
Colina St....I-1
College....H-5
College Blvd....A-6
Colonial Dr....C-1
Colorado St....H-3
Columbine....B-7
Columbus St....H-4
Commerce St. E.
...H-1,H-4,I-7
Connell....F-8
Connelly St....J-8
Conrad....K-6
Contour Dr....A-5,B-5
Cooper St....K-5
Cornell....F-3
Corona Ave....A-6
Courtland....F-5
Craig Pl., W....E-1,E-4
Crockett....I-6
Crockett St., E....I-6
Crofton....K-5
Cross Lane....C-8
Crupaderas....H-1
Culebra....F-2
Cunningham Ave....F-7
Cypress St....G-3,G-4
Dakota....J-6
Dakota St....J-6
Dalehurst....A-1
Dallas St....G-5
Davis Ct....C-8
Dawson St....H-6
Dean....B-4
Delaware....K-6,J-6
Delgado....G-1
Delmar St....K-8
Denver Blvd....K-7
Denver Rd....C-6
Devine St....J-6
Dewey....F-5
Dickman Rd....E-8
Dolorosa St....I-4
Donaldson Ave....D-1
Dora St....C-4
Dowdy St....K-4
Dreiss St....J-7
Durango Blvd....I-2
...I-4,J-6
Duval....G-7
Dwyer Ave....I-4
Earl St....C-4
Eaton....B-8
Edison Dr....B-1
Eleanor....E-8
Eleanor Ave....E-8
Elizabeth Rd....C-8
Ellerman....J-3
Elm St....I-6
Elm Endorf....I-1
Elm Endorf St....G-1
Elmhurst....D-8
Elmira, W....G-4
El Monte Blvd...A-1,A-3
Elmwood....C-3
El Paso....I-1
El Prado Dr....C-5
Elsmere Pl....D-4
Elsmere Pl., W....D-2
Elwood....B-8
Encinal....B-5
Encino....B-7
Erie Ave....G-5
Essex St....K-7
Euclid Ave., W....G-4
Evans Ave....A-6
Evergreen....F-4,F-5
Excelsior Dr....C-1
Fest....K-3
Fest St....K-4
Flores, N....H-4
Flores, S....I-4
Florida St....K-6
Forcke....J-5
Forrest....K-3
Fredericksburg Rd....C-1
Fredonia....J-7
French, E....F-5
French Pl., W....E-1,E-3
Fresno Dr....B-1
Frio St., S....I-3
Frio City Rd....K-1
Fulton Ave....D-3
Funston Pl....E-7
Furnish, E....K-3
Furnish, W....K-3
Furr Dr....D-1
Galitzen....H-4
Gardina St....B-1
Garfield....J-5
Geneseo Rd....C-8
Gibbs....I-7,I-8
Gibson....H-8
Gillespe....F-5
Glass....K-3
Gloucester....G-8
Goodwin....K-2
Grand View....B-8
Grant....A-3,C-3,E-3
Grant Ave....B-7
Graham Rd....E-8
Gramercy Pl....D-4
Grape St....J-6
Grayson St....F-7
Grove Pl....B-8
Groveland....D-8
Guadalupe St....I-1
Guenther....J-4
Guilbeau....I-4
Hackberry St., S....J-7
Hall....B-3
Hamilton Ave., N....G-1
Hancock Rd....F-8
Harding Pl....J-8
Harrigan Ct....C-8
Harrison AVe....B-7
Harvard....F-2
Hays St....H-8
Hazel Ave....K-1
Helena St....I-5
Hermine Blvd....A-1,A-3
Hermosa....B-5
Hermosa, W....B-2,B-3
Hermosa Dr., E....B-5
Hess....F-3
Hickman....F-3
Hidalgo St....J-1
High St....K-7
Hildbrand Ave., E....C-2
Hildebrand Ave....D-6
Hill....G-4
Holland....C-5
Hollywood Ave., W..D-4
Hondo St....I-7
Horace....B-4
Houston St...H-1,H-3,I-6
Howard St....E-4
Hubbard....B-8
Huerta....K-1
Huisache....E-2
Huisache Ave., W.E-1,E-4
Indiana St....J-6
Inslee....A-7
Iowa St....J-8
Ira Ave....E-7
Isleta....F-6
Jackson St....F-4,G-4
Jacob....K-6
Jean....J-2
Jefferson St....H-5
Jeffery....F-2
Jerry....C-2
Jessup Rd., N....F-8
Johnson....J-5
Jones Ave....G-6
Josephine St....F-5
Joy St....B-4
Judson....C-5
Kansas....I-6
Katherine Ct....C-8
Kearney....C-4
Keck Ave....K-1
Keller....J-4
Kemper St....J-1
Kendall....C-5,F-5
Kennedy....B-7
Kentucky Ave....F-1
Kinder....A-3
King William....J-5
Kings Hwy...D-1,D-3,D-5
Klein....K-3
Krempkau....F-3
Lachappelle St....K-3
Lagos....C-7
Lamar St....H-6
La Manda Blvd...A-1,A-3
Lambert St....K-3
Laredo....H-4,J-2,J-4
Las Moras....H-2
Laurel....F-2,F-4
Laurel Heights....D-5
Leal St....G-1
Ledge....D-5
Lee....B-2
Leona St., S....I-4
Leopold....K-6
Lewis St....F-4,G-4
Live Oak St....I-6
Lockhart....H-8
Locust....F-5,F-6
Logan....H-8
Loma Vista....J-2
Lombrand....F-2
Lone Star Blvd....K-4
Lotus St....K-6
Louise....E-2
Lovera Blvd....A-3
Lovera Dr....A-3
Lyons....G-1
Lucas....E-8
Lullwood Ave., W.
...D-3,D-4
Luther Dr....C-5
Lynwood Ave., W.
...D-2,D-4
Madero....K-1
Madison....J-5
Magnolia, W....E-1,E-4
Main Ave....E-4
Mandalay....B-5
Mandalay Dr., E....B-5
Mandalay Dr., W....B-3
Mann Ave....I-4
Maple St....G-4
Marcia....B-8
Mardell....A-1
Margaret....E-7
Marine....K-5
Mariposa Dr....B-5
Mariposa Dr., E....B-5
Mariposa Dr., W....B-3
Market St., E....I-5
Mary....C-8
Mary Louise Dr....D-1
Marshall....G-4
Martin....H-8
Martin St., W....H-1
Mason....G-7
Matamoras....I-3
Maverick St....F-4
May....A-7
McCullough St...E-5,G-5
McIlvaine Dr....C-3
Medina St....H-3
Meigs Rd....F-8
Melbourne....B-1
Melrose....C-4,C-5
Mercedes....K-2
Merida St....J-1
Meridian....F-2
Mesquite, N....J-7
Michigan....C-3,E-3
Mickle John....F-2
Milam....G-6
Millard St....C-4
Mistletoe Ave., W....E-1
Mockert....K-3
Montana St....I-7
Montclair....B-8
Monumental St....J-7
Mora....I-1
Morales St....H-1
Moses....I-1
Mulberry Ave....D-1
Mulberry Ave., W....E-4
Muncey St....H-7
Murry....G-1
Myrtle....F-4
Nancy Pl....K-2
Natalen....E-8
Navarro St....I-5
Navidad....G-2
Neer....C-2,D-2
Nelda St....B-1
Nelson....J-8
Nevada St....J-6
New Braunfels
Ave....F-8,J-8
Nolan....H-6,H-8
Noria St....K-1
Normady....A-6
Norwood....C-3,C-4
Nueces....H-2
Neuva St., W....I-4
Obregon....K-1
Odell St....C-4
Ogden La....B-7
Olive St., S....J-7
Olmos Dr....C-1,C-3
Olmos Dr., E....C-5,C-6
Olmos Dam....C-6
Oriental Ave....K-2
Osburn....G-8
Oval....B-7
Pace St....G-1
Palmetto, S....J-8
Paloma....A-5
Panama....K-6
Park Ave....F-4,F-6
Park Hill Dr....C-5
Parklane....B-5
Parland Pl....D-7
Pasadena....C-1,C-3
Paschal....F-5
Paseo....A-5,B-5
Pastores....D-6
Patterson....B-7,C-7
Peacan....H-5
Pecos St....H-4
Penalosa....J-1
Pendleton Ave....K-2
Perez St....G-1
Perry Ct....C-8
Pershing Ave....D-8
Piedmont....J-7
Pierce....F-8
Pine St., S....J-7
Pinecrest....A-8
Pinto....H-2
Poco....C-8
Poplar St....G-1
Popular....G-3
Popular St....G-4,J-8
Porter St....K-7
Post Ave....E-7
Post Rd....F-7
Potosi St....J-1
Presa St., S....I-5
Primera Dr....C-5
Princess....E-5
Princeton....F-3
Probandt St....K-4
Produce....H-4
Queen Anne Ct....D-8
Quitman St....F-7
Ralph....K-3
Rancho Blanco....A-2
Ray....K-2
Red Wood....A-8
Refugio....J-5
Rehman....J-3
Reina Dr....C-1
Rex....B-4
Richter....H-3
Richmond....G-5
Ridgewood, W...C-2,C-4
Rise....B-8
Rivas St....G-2
River Rd....F-6
Road S 2....F-7
Road S 3....E-8,F-7
Road S 4....F-7
Road S 6....F-8
Rogers....F-8,H-8
Roman....H-4
Rose Lane....E-5
Rose Mary Ave....A-7
Rosewood Ave., W.
...D-3,D-4
Rosillo....H-1
Routt St....B-8
Russell Pl., E....E-5
Russell, W....E-4
Rudolph....G-8
Sabinas St....G-2
Sacramento....C-1,C-3
St. Ann....F-2
St. Anthony....K-8
St. Charles....H-8
St. Francis St....K-4
St. James....H-8
St. John....H-8
St. Lukes....A-6
St. Marys....E-5
St. Mary's Rd., N.
...F-5,H-5
St. Mary's Rd., S....J-5
Saltillo....J-1
San Angelo Blvd.
...A-2,A-3
San Antonio....J-5
San Antonio Ave....E-1
Sanders....J-8
San Fernando St....I-1
San Francisco....C-1
San Juacinto St....H-2
San Luis St....I-2
San Marcos....J-3
San Marcos St....H-3
San Pedro....G-4
San Pedro Ave....E-4
San Saba....H-4
Santa Anna....C-1
Santa Barbara....C-1
Santa Monica....C-1
Santa Paula....C-2
Santa Rosa St....I-4
Saunders St....I-2
Schreiner....D-5
Seguin St....G-8
Seymour....G-5
Sharpe Rd....F-8
Sharer....G-7
Sheraton Dr....B-8
Sheridan, E....J-5
Sheridan, W....J-4
Sherman St....G-6
Simon St....K-4
Smith....H-3,J-2
Soledad St....H-5
Sonora....K-3
Southolme Ave....K-2
Spofford....G-8
Spring....I-1
Spring St....G-1
Spruce St....J-7
Stadium....E-6
Staff....F-7
Stanford Dr....B-5
Stanley Rd....F-7
Stark....J-3
Stieren St....J-5
Stribling....K-3
Summit, W....D-4
Summit Ave., W....D-1
Swiss....I-6
Tampico St....J-1
Temple....K-5
Tendick....E-7
Terrell Rd....C-8
Texas....F-1
Thelma....C-5
Thelma Dr....C-5
Thorain Blvd....B-1
Thorain Blvd, W....B-3
Thorman....D-8
Tilden....F-8
Toldeo St....J-7
Tolle....I-5
Torcido Dr....C-7
Torreon....I-2
Trinity St....G-2
Tyler....K-2
University....F-2
Urban Loop....I-4
Valdez....E-5
Valero....F-5
Vanderneck....A-8
Vargas St....J-8
Vassar....C-4
Venice....A-2,A-3
Vera Cruz St....I-1
Verada....B-1
Viendo....C-1
Vitra....K-6
Vollum Ave....E-1
Walnut St....I-6
Walsh....G-3
Warner....C-2,D-2
Warren....G-4
Washington....J-5
Waverly....F-1
Welhausen....J-8
Wesley....E-7
West Ave....A-1
Westfall....K-7
Westgate....E-8
Westover Rd....C-7
Westwood....C-3
Weymouth....F-3
Wheaton Rd....E-8
Wickes St....K-5
Wild Rose Ave....A-7
Wildwood Dr., E....B-5
Wildwood, W....B-2
Wildwood Dr., W....B-3
Williams....F-1
Willow St....H-7
Wilmington....G-5
Wilson....F-8
Wingate Ave....K-2
Woodlawn Ave., W.
...E-1,E-4
Wyckham....B-8
Wyoming....I-8
Yale Ave....F-2
Zarzamora St., S..G-1,K-1
Zavalla....K-3
Zilla St....B-4
3rd....K-2
3rd St....H-5
4th St....H-5
5th St....H-6
6th St....H-6
8th St....H-6
9th St....H-6
10th St....G-6
12th St....G-6
13th St....G-6

POINTS OF INTEREST

Brackenridge Park
...D-6

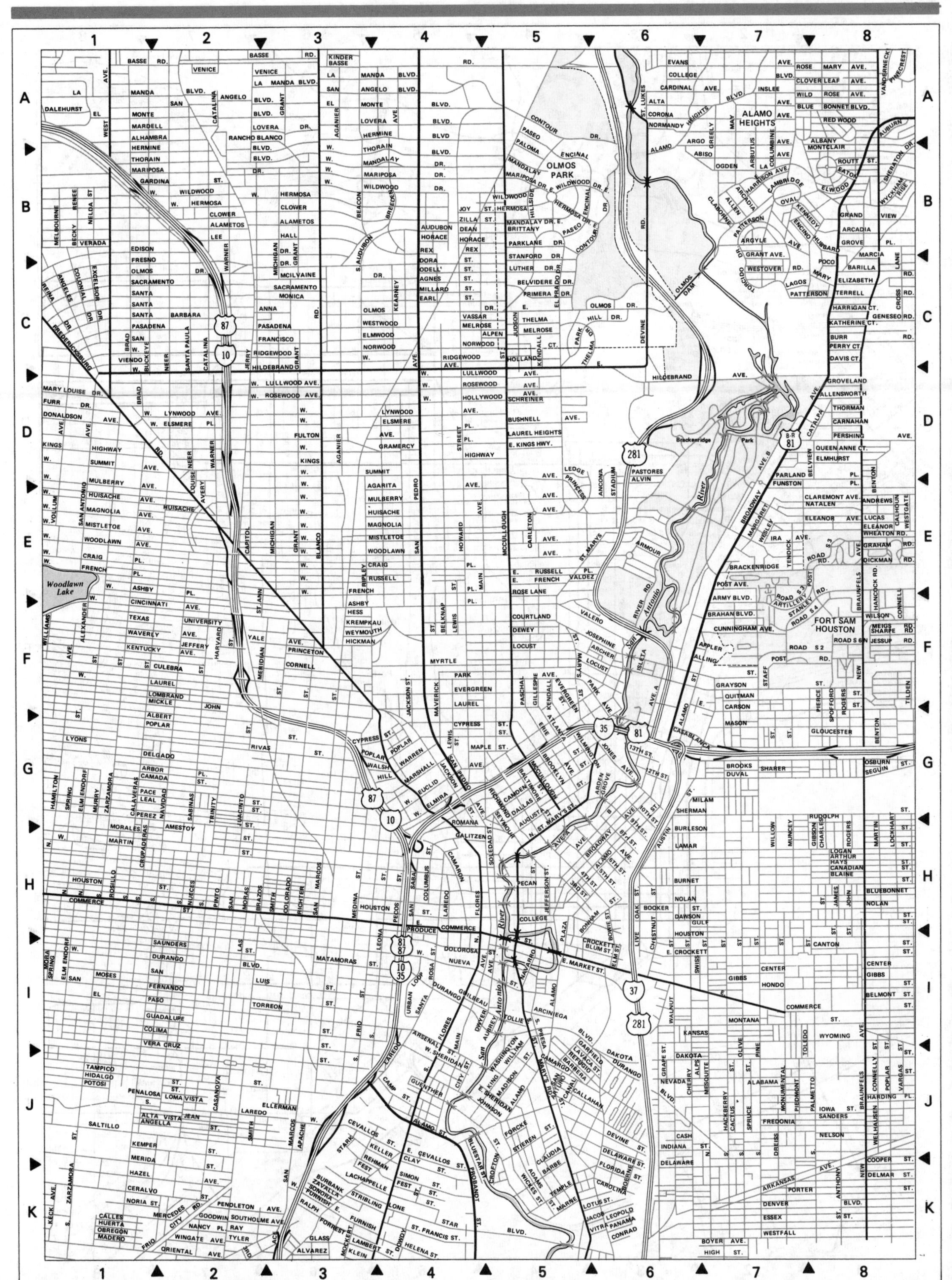
1
2
3
4
5
6
7
8
A
B
C
D
E
F
G
H
I
J
K
ALAMO HEIGHTS
OLMOS PARK
FORT SAM HOUSTON
Woodlawn Lake
Brackenridge Park
San Antonio River
OLMOS DAM
87
10
281
35
81
B-R 81
37
COMMERCE
BROADWAY
HILDEBRAND
FREDERICKSBURG RD.
SAN PEDRO
MCCULLOUGH
NAVARRO
E. MARKET ST.
DURANGO
CULEBRA
BASSE RD.
ZARZAMORA
NEW BRAUNFELS
AUSTIN
ALAMO
S. PRESA
NOGALITOS
SOUTH FLORES
HOUSTON
MARTIN
ST. MARYS
CASTROVILLE RD.
HACKBERRY
PROBANDT
MAIN
DEVINE
JUDSON
AUBURN
WILDWOOD
HERMOSA
OLMOS
ELMIRA
EUCLID
POPLAR
CYPRESS
ROMANA
CAMDEN
SOLEDAD
PECAN
JEFFERSON
ARMOUR
GRAYSON
GROVELAND
CARNAHAN
PERSHING
BURR
MELROSE
WESTWOOD
ELMWOOD
NORWOOD
HOLLAND
LULLWOOD
ROSEWOOD
HOLLYWOOD
SCHREINER
BUSHNELL
LAUREL HEIGHTS
E. KINGS HWY.
ASHBY
CINCINNATI
TEXAS
WAVERLY
KENTUCKY
WOODLAWN
MAGNOLIA
MISTLETOE
HUISACHE
MULBERRY
SUMMIT
FRENCH
CRAIG
SAUNDERS
MATAMOROS
GUADALUPE
COLIMA
VERA CRUZ
TAMPICO
HIDALGO
POTOSI
LAREDO
CEVALLOS
E. CEVALLOS
SIMON
FEST
LONE STAR
ST. FRANCIS ST.
HELENA ST.
KELLER
REHMAN
STEVES

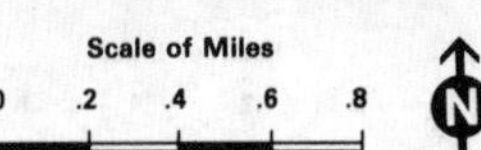
Scale of Miles
0
.2
.4
.6
.8
N

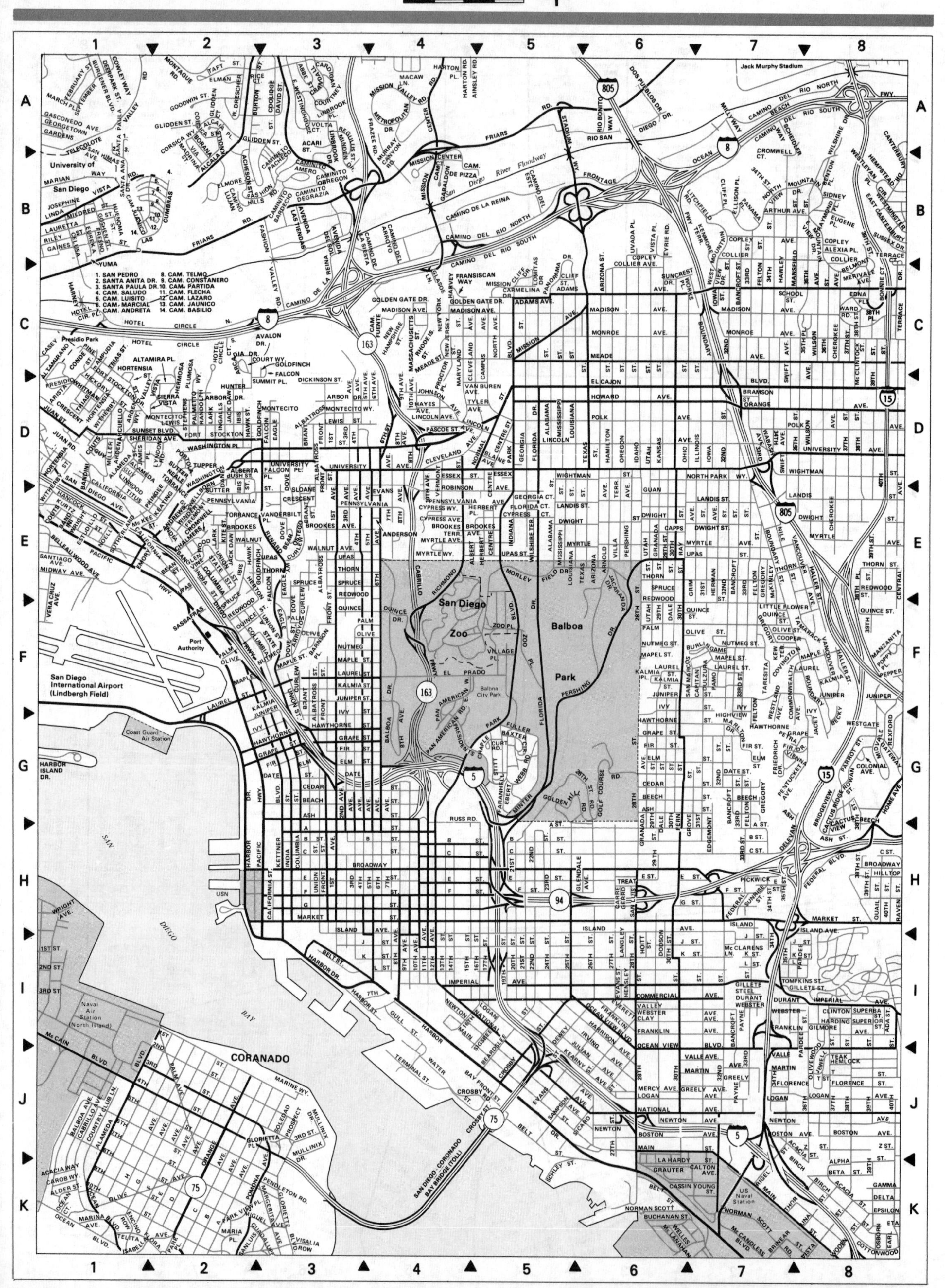
Jack Murphy Stadium
University of San Diego
Presidio Park
San Diego Zoo
Balboa Park
Balboa City Park
Port Authority
San Diego International Airport (Lindbergh Field)
Coast Guard Air Station
Naval Air Station (North Island)
CORANADO
US Naval Station
USN
SAN DIEGO BAY
San Diego River Floodway
CAMINO DEL RIO NORTH
CAMINO DEL RIO SOUTH
FRIARS RD.
MISSION CENTER RD.
HOTEL CIRCLE N.
SAN DIEGO CORONADO BAY BRIDGE (TOLL)
HARBOR DR.
BROADWAY
MARKET
IMPERIAL AVE.
UNIVERSITY AVE.
EL CAJON BLVD.
ADAMS AVE.
FEDERAL BLVD.
1. SAN PEDRO
2. SANTA ANITA DR.
3. SANTA PAULA DR.
4. CAM. SALUDO
5. CAM. LUISITO
6. CAM. MARCIAL
7. CAM. ANDRETA
8. CAM. TELMO
9. CAM. CONSTANERO
10. CAM. PARTIDA
11. CAM. FLECHA
12. CAM. LAZARO
13. CAM. JAUNICO
14. CAM. BASILIO
805
8
163
5
15
94
75
A
B
C
D
E
F
G
H
I
J
K
1
2
3
4
5
6
7
8

SAN DIEGO

A Ave....K-2
A St....H-3,H-4,H-5,H-7
Abbe St....A-3
Acacia St....J-8,K-8
Acacia Way....K-1
Acar St....A-3
Acheson St....B-2
Adams Ave....C-5
Ada St....I-8
Ainsley Rd....A-4
Alabama St....D-5,E-5
Alameda....D-1
Alameda Blvd....J-1
Albatross St..D-3,E-3,F-3
Alcala Ct....B-2
Alder St....K-1
Alexia Pl....B-8
Alpha St....K-8
Altamirano....C-1
Ampudia....C-1
Arbor Dr....D-3
Arcadia Dr....C-2
Arch....C-4
Arden....D-1
Arden Wy....D-1
Arista Ct....C-1
Arista St....D-1
Arizona Ave....E-6
Arnold Ave....E-6
Arroyo....F-3
Arthur Ave....B-7
Ash St....G-3,G-6
Avalah Dr....C-2
B Ave....K-2
B St....H-3,H-4,H-5,H-7
Balboa Ave....J-1
Balboa Dr....G-4
Bancroft St...E-7,G-7,I-7
Barnson Pl....F-3
Baxter Cir....G-5
Bay Front St....J-5
Beach St....G-3,G-6
Bean St....E-2
Beardslee St....J-5
Becky....G-8
Belmont....B-8
Belt St....J-5,K-6
Benton Pl....B-8
Berana....A-2
Beta St....K-8
Birch St....K-8
Blaine Ave....D-5
Bonnie Ct....C-8
Boston Ave....J-6-8
Boundary....F-8
Boundary St....E-7
Bramson St....D-7
Brant St....D-3,E-3,F-3
Brinear St....K-7
Broadway....H-3,H-8
Brookes Ave..E-2,E-3,E-5
Brookes Terr....E-4
Buchanan St....K-6
Burgener Blvd....A-1
Burgener Dr....A-1
Burlingame....F-7
C Ave....K-2
C St....H-3-5,H-7-8
Cabrillo Ave....J-1
California St.
....D-1,E-1,H-3
Calton Ave....K-7
Cam Andreta....B-2
Cam Basilio....B-1
Cam Costanero....B-2
Cam Estoban....B-1
Cam Flecha....B-2
Caminito Amero....B-3
Caminito Bardecio..B-3
Caminito Cantti....B-3
Caminito Degrazia..B-3
Camino de la Reina
....B-4,C-3
Camino Del Arroyo..B-4
Camino de la Siesta..B-4
Camino del Este....B-5
Camino del Rio North
....A-7,A-8,B-4,B-5
Camino Del Rio South
....A-7,A-8,B-5
Caminito Mangana..B-3
Caminito Obregon..B-3
Camineto Pacheco..B-3
Cam Juanico....B-1
Cam Lazaro....B-2
Cam Luisito....B-2
Cam Marcial....B-2
Cam Partido....B-2
Campus Ave....D-5
Cam Saludo....B-2
Cam Talmo....B-2
Canterbury Dr....B-8
Canterbury, E....B-8
Capps....E-6
Captain....F-7
Cardigan....A-3
Cardinal....A-4
Carmelina....C-5
Carob Wy....K-1
Carre Gerro....H-6
Casey....C-1
Cassin Young St....K-7
Cedar St....G-3,G-6
Celusa....B-1
Central Ave....E-8
Centre St....D-5
Chaple....G-5
Cherokee Ave....C-8,E-8
Clay Ave....I-6
Cleveland Ave...C-4,D-4
Cliff Pl....B-7
Cliff St....C-5
Clinton....I-8
Collier Ave..B-7,B-8,C-8
Columbia St..E-2,F-2,H-3
Commercial Ave....I-6
Commonwealth Ave.F-7
Coolidge St....B-2
Cooper....F-7
Copley Ave..B-7,B-8,C-8
Corsica St....A-2
Corsica Wy....A-2
Cottonwood....K-8
Country Club Ln....J-1
Court Way....C-2
Courtney....A-3
Couts....E-1
Covington....F-7
Crescent....E-3
Crescent Dr....D-1
Cromwell Ct....A-7
Crosby St....J-5
Crowell....E-2
Curlew....E-3
Curlow....F-3
Curt Rd....G-5
Cypress Ave....E-4
Cypress Ct....E-5
Cypress Wy....E-4
D Ave....K-2
Dale St....F-6,G-6
Dane....D-2
Date St....G-3,G-7
Delta....K-8
Dewey....I-5
Dickinson St....D-3
Dodson....I-6
Douglas....D-2
Dove....E-3
Drescher St....A-2
Drescher, W....A-2
Duluzura....F-7
Durant....I-7
Dwight St....E-6,E-8
E Ave....K-2
E St....H-3-7
Eagle....D-3,E-3,F-3
Earl....K-8
Edna Pl....C-8
El Cajon Blvd....D-6
Ellison Pl....B-7
Elman....A-2
Elmore St....B-2
Elm St....G-3,G-6,G-7
El Prado....F-4,F-5
Emery....E-1
Encino Row....K-1
Epsilon....K-8
Essex St....D-4
Estudillo....E-1
Eta....K-8
Eugene Pl....B-8
Eureka....B-1
Evans St....E-4,I-6,J-5
Everett....I-6
Eyrie Rd....B-6
F Ave....K-2
F St....H-3-5,H-7
Falcon....D-3
Falcon Pl....D-3
Faranhell Ebert....G-5
Fashion Valley Rd....B-3
February St....A-1
Federal Blvd....H-7,H-8
Felton St....E-7,G-7
Fern St....G-6
Fir Dr....G-7
Fir St....G-3,G-6,G-7
Flora....K-2
Florence St....J-8
Florida Ct....E-5
Florida Dr....D-5,F-5
Fort Stockton Dr.D-1,D-2
Franciscan Wy....C-5
Franklin Ave....I-6,I-7
Frazee Rd....A-4
Fremont....E-2
Fresno....B-1
Friars Rd....A-5,B-2
Friedrich Dr....G-7
Frontage....B-6
Front St....D-3,F-3,H-3
Fuller....G-5
G Ave....K-1
G St....H-3,H-6
Gaines St....B-1
Gamma....K-8
Gardens Ave....A-1
Gasconedo Ave....A-1
Gateway....G-8
Georgetown....A-1
Georgia Ct....E-5
Georgia St....D-5
Gillete St....I-7
Gilmore....I-8
Glasoe Ln....C-4
Glendale Ave....H-6
Glenwood....E-2
Glidden Ct....A-2
Glidden St....A-2
Gloriette Blvd....K-3
Gloriette Pl....J-2,J-3
Golden Gate Dr....C-4
Golden Mill Rd....G-5
Golf Course Rd....G-6
Goodwin St....A-2
Goshawk Ave....A-4
Goshen St....B-1
Granada Ave....E-6,H-6
Grape St....G-3,G-6,G-7
Grauter....K-6
Greenly Ave....J-7
Gregory....E-7,F-7,G-7
Grim Ave....E-7
Grove St....G-7
Guadalupe....K-3
Guan St....E-6
Gull St....I-4
Guy....D-1,E-2
H Ave....K-1
Haller St....F-8
Hamilton St....D-6
Hancock St....E-1
Harbor Dr....H-2,I-4
Harding....I-8
Harrison Ave....I-6
Harton Pl....A-4
Harton Rd....A-4
Harvey Rd....C-4
Hawk....E-2
Hawley....C-7
Hawthorne St.
....G-3,G-6,G-7
Haydon....D-1
Hayes Ave....D-4
Heitt St....I-6
Hemlock....J-8
Hempstead, N....B-8
Hensley....I-6
Herbert Pl....E-5
Herman Ave....E-7
Hertensia St....C-1,D-1
Herton....E-2
Hickory St....D-1
Highview....G-7
Hilldale....B-8
Hilltop....H-8
Hoffman St....C-4
Hortensia....D-1
Hotel Circle Ct....C-2
Hotel Circle N....C-2
Hotel Cir. Pl....C-1
Hotel Circle S....C-2
Howard Ave....D-6
Huenoma St....B-1
Ibis St....D-2,E-2
Idaho St....D-6
Illinois St....D-7
Impasse....D-1
Imperial Ave....I-4,I-8
India St....Ep2,H-2
Ingalls....D-2
Inland Fwy....A-6
Iowa....D-7
Irving Ave....I-6
Isabella Ave....K-1
Island Ave...H-3,H-6,H-8
Ivy St....F-6,F-8,G-3
J St....I-4,I-6,I-7
Jacaranda Dr....E-6
Jacet....G-8
Jack Daw....D-2,E-2
Jewell....J-8
Johnson Ave....D-4
Josephine....B-1
Juan....D-1
Julian Ave....J-6
Juniper St....F-3,F-6,F-8
K St....I-4,I-6-8
Kalmia Pl....F-6
Kalmia St....F-3,F-6,F-8
Kansas St....D-6
Kearny Ave....J-6
Keating....E-2
Kenmore Terr....B-7
Kettner Blvd....H-3
Kew Ter....F-7
Kurtz St....E-1
L St....I-4,I-7,I-8
Las Cumbras....B-1
LaHardy St....K-6
Lake Ct....A-2
Landis St....E-6,E-8
Langley....I-6
Lark....D-2,E-2
Laurel St....F-2,F-3
....F-6,F-7,F-8
Lauretta St....B-1
Lewis St....D-2,D-3
Linbrook Pl....A-3
Lincoln Ave....D-4,D-5
Linda Vista Rd....B-1
Linwood....D-1
Litchfield Rd....B-7
Little Flower....F-7
Logan Ave....I-5,J-6,J-8
Lomitas Dr....C-5
Louisiana St....E-5
Lyndon Rd....D-2
Macaw Ln....A-4
Madison Ave.C-4,C-6,C-7
Main St....I-4,J-6,K-7
Manzanita....F-8
Maple St....F-3,F-6-8
March Pl....A-1
Margerita Ave....K-3
Marian Way....B-1
Marina Ave....K-1
Marine Wy....J-3
Market St....H-3,H-8
Marlton Dr....G-7
Maryland St....C-4
McCain Blvd....I-1,J-1
McCandless Blvd....K-7
McClarens Ln....I-7
McClintock St....C-8
McKee St....E-1
McKinley....E-7
McLanahan....K-6
Meade Ave....C-6
Mercy Ave....J-6
Merivale Ave....C-8
Midvale....G-8
Midway Ave....E-1
Miguel Ave....K-3
Mildred St....B-1
Miller....D-1
Minden Dr....A-3
Minter....G-5
Mission Ave....C-5
Mission Center Ct....B-4
Mission Center Rd.
....A-4,B-4
Mission Cliff Dr....C-5
Mississippi St....E-5
Monroe Ave....C-6,C-7
Montecito....D-2,D-3
Montecito Wy....D-3
Morgha....D-1
Morley Field Dr....E-5
Morse Ct....A-3
Mountain View Dr.,E.B-8
Mountain View Dr., N.
....B-7,B-8
Mullinix Dr....J-3
Murray Canyon Rd....B-4
Myrtle Wy....E-4
Myrtle Ave...E-4,E-7,E-8
National Ave....I-5,J-6
New Hampshire St....C-4
New Jersey....C-4
Newton Ave...I-4,J-6,J-7
New York....C-4
Nile St....E-7
Noell....E-1
Normal St....D-5
Norman Scott Rd.
....K-6,K-7
North Ave....C-5
North Park Wy....D-7
Nutmeg St....F-3,F-6,F-7
Ocean Beach Fwy....A-7
Ocean Blvd....K-1
Ocean Ct....K-1
Ocean View Blvd....I-6
Ohio St....D-6
Old Town....D-1
Olive Ave....K-1
Olive St....F-3,F-7
Olivewood....J-8
Orange Ave.D-7,J-2,ZK-2
Oregon St....D-6
Osborn....K-8
Otis....A-3
Pacific Hwy....E-1,H-3
Palm....F-3,F-6
Palm Ave....J-2
Palmetta....D-2
Pamo....F-7
Panama Pl....B-7
Pan American Rd. E..G-4
Pan American W....F-4
Panorama Dr....C-5
Pardee....I-8
Park Blvd....D-5,G-5
Park Pl....K-2
Parrot St....G-8
Pascoe St....D-4
Payne....I-7,J-7
Pendleton Rd....K-3
Pennsylvania Ave.
....E-2,E-3,E-4
Pentucket Ave....G-7
Pepper....F-8
Pershing Ave....E-6
Pershing Dr....F-5,F-6
Petra....G-7
Pickwick....H-7
Pine St....C-1
Plumosa Wy....D-2
Polk Ave....D-6,D-7
Pomona Ave....K-2
Poplar Pl....D-2
Poppy Pl....F-8
Presidents....G-4
Presidio....D-1
Pringle....E-2
Procter Pl....D-4
Prospect....J-3
Quail St....H-8
Quince Dr....F-3,F-4
Quince St....F-2,F-7,F-8
Quine....A-3
Randolph....D-2
Raven St....H-8
Raymond Pl....B-8
Ray St....E-6
Redwood St..E-3,E-6,E-8
Regulus St....A-3
Renard....E-3
Rexford....G-8
Rhode Island St....C-4
Rice Ct....A-2
Richmond St....E-4
Rigel St....K-7
Riley St....B-1
Robinson Ave....E-4
Rowan St....G-8
Russ Rd....G-4
Sampson St....J-5
San Diego Ave....D-1,E-1
San Luis....H-6,K-2
San Marcos....F-7
Santa Anita Dr....A-1
Santa Paula Dr....A-1
Santiago Ave....E-1
San Pedro St....A-1
San Rimas Ave....B-1
Schley St....K-5
School St....C-7
September....A-1
Sheridan Ave....D-1
Sicard St....J-6
Sidney....B-8
Sigsbee....I-5
Slean....E-3
Soledad....J-3
Spruce....E-3,E-6,F-2
Stadium Wy....A-5
State St....E-2
Steel....I-7
Stephens....D-2
Stitt....G-5
St. James Pl....D-1
Summit Pl....D-2,D-3
Suncrest Dr....C-6
Sunrise St....H-7
Sunset Blvd....D-1
Superba St....I-8
Superior St....I-8
Sussex Dr....B-8
Sutherland....E-1
Sutter St....E-2
Swift Ave....C-7,D-7
Switzer St....J-4
T St....J-8
Taft St....A-2
Tamarack....F-8
Taresita....F-7
Teak....J-8
Telita....K-1
Terminal St....J-4
Terrace Ct....B-8
Terrance....D-2,E-2
Texas St....D-6,E-6
Thorn....E-2,E-3,
....E-6,E-7,E-8
Thorn St....E-8
Thor St....K-8
Titus....E-1
Tompkins St....I-7
Trent....H-6
Trias St....D-1
Tyler Ave....D-5
Una St....K-8
Union St....E-2,F-3,H-3
University Ave....D-3,D-8
Upas St....E-2,E-3,E-7
Utah St....E-6,F-6
Uvada Pl....B-6
Valla....J-7
Valley....I-6
Van Buren Ave....D-5
Vancouver Ave...E-8,F-8
Vanderbilt Pl....E-3
Vermont....E-4
Via Madrina St....A-2
Vienna....G-7
Village Pl....F-5
Villa Terr....E-6
Vine....E-2
Visalia Row....K-3
Vista Pl....B-6
Vista St....K-8
Volta Ct....A-3
Wabash Ave....D-7
Wade Dr....G-5
Water St....I-3,J-4
Wallas....K-6
Walnut Ave....E-2,E-3
Ward Rd....C-8
Washington St....D-2
Webster Ave....I-6,I-7
Wedan St....K-8
Wellborn....E-2
Westgate....G-8
Westinghouse, E....A-3
Westland Ave....G-7
Westlayan Pl....B-8
Westminster....B-8
West Mountain View
Dr....B-7
Wightman St....D-5,D-8
Wilshire Dr....A-8
Wilson....C-8
Winder....E-2
Witherby St....D-1,E-1
Works Pl....C-6
Wright Ave....E-1,H-1
Yuma St....B-1
Z St....J-8
Zoo Dr....F-5
Zoo Pl....F-5
1st Ave....D-3,H-3
1st St....I-1,J-2
2nd Ave....G-3
2nd St....I-1,J-2
3rd Ave....D-3,H-3
3rd St....I-1,J-2,J-3
4th Ave....D-3,E-3,H-3
4th St....J-2
5th Ave....D-3,E-4,H-4
5th St....J-1
6th Ave....D-4,E-4,H-4
6th St....J-1
7th Ave....E-4,H-4
7th St....J-1
8th Ave...D-4,E-4,G-4,I-4
8th St....J-1
9th Ave....D-4,I-4
9th St....K-1
10th Ave....D-4,I-4
10th St....K-1
11th Ave....I-4
12th Ave....I-4
13th St....I-4
14th St....I-4
15th St....I-4
16th....I-4
16th St....I-5
17th St....I-5
19th St....I-5
20th St....I-5
21st St....G-5,H-5,I-5
22nd St....H-5,I-5
23rd St....H-5
24th St....I-5
25th St....I-5
26th St....I-6
27th St....I-6,J-6
28th St....F-6,G-6,I-6,J-6
28th St. Rd....G-6
29th St....F-6,H-6
30th St....E-6,F-6,G-6
....I-6,J-6
31st St....E-7,G-7,J-7
32nd St....E-7,G-7,J-7
33rd St...C-7,E-7,G-7,J-7
34th St...B-7,C-7,H-7,J-7
35th Pl....C-8
35th St....H-7,I-7,J-7
36th St....J-8
37th St....J-8
38th St....B-8,C-8
....G-8,H-8,J-8
39th St....F-8,H-8,J-8
40th St....H-8,J-8

POINTS OF INTEREST

Balboa City Park....F-5
Coast Guard Air Station....G-1
Naval Air Station (North Island)....I-1
San Diego International Airport (Lindbergh Field)....F-2
San Diego Stadium..A-7
U.S. Naval Station....K-7

SAN FRANCISCO

Abbey St....I-2
Alabama St....K-4
Alameda St....H-5
Albion St....I-3
Alpine Ter....H-1
Alvardo....K-1
Ames St....K-2
Annie St....E-6
Anthony St....E-6
Arkansas St....J-5
Ash St....F-2
Austin St....E-2
Balmy St....K-4
Bartlett St....K-3
Battery St....C-6
Bay St....B-2
Beach St....A-3,B-1
Beale St....D-7
Beaver St....I-1
Bedwood Gate Ave...F-3
Belcher St....H-2
Bernice St....H-4
Berry St....G-7
Birch St....G-2
Bird St....I-3
Boardman Pl....G-6
Brady....H-3
Brannon St....G-6
Broadway....C-5,D-2
Brosnan....H-3
Bryant St....F-7,G-6,K-5
Buchanan....D-2,H-2
Buena Vista Ter....I-1
Bush St....E-2,E-4
Caledonia....I-3
California St....D-3
Camp St....I-3
Capp St....K-3
Carolina St....K-6
Casa Way....B-1
Castro....K-1
Cedar St....E-3
Center Pl....F-7
Cervantes Blvd....V-1
Channel....H-6
Chattanooga St....K-2
Chestnut....B-3
Chula Ln....I-2
Church St....K-2
Clara St....F-6
Clarence Pl....F-7
Clay St....D-3
Clementina St....F-5
Cleveland St....G-5
Clinton Park St....H-2
Collingwood St....K-1
Collin P. Kelley Jr. St....F-7
Colton St....H-3
Columbia Square....G-5
Columbus Av....B-4
Connecticut St...J-5,K-6
Croft St....G-5
Cumberland St....J-1
Dagget St....I-6
Dakota....K-7
Davis....D-6
Dearborn St....I-3
De Boom....F-7
Deharo St....K-6
Diamond St....K-1
Divisadero....H-1
Division St....H-5
Dolores St....K-2
Dore St....G-4
Dorland St....I-2
Drumm....D-7
Duboce Ave....H-1
Ecker St....E-6
Eddy St....F-1,F-4
Elgin Park St....H-3
Elizabeth....K-1
Ellis St....E-4
Elm St....F-3
Embarcadero St....C-7
Erie St....H-3
Essex St....E-7
Eureka St....K-1
Fair Oaks St....K-2
Falmouth St....F-5
Federal St....F-7
Fell St....G-2
Fern St....E-2
Filbert St....C-1,C-3
Fillmore St....D-1,G-1
Flint St....I-1
Florida St....K-4
Folsom St....F-6,K-4
Ford St....I-1
Francisco....B-3
Franklin St....D-3
Freelon St....G-6
Fremont St....D-7
Front St....D-6
Fulton St....G-1
Geary....E-2
Germania St....H-1
Gilbert St....G-6
Gough....H-3
Gough St....D-2,G-2
Grace St....G-4
Grant St....D-5
Green St....C-1,C-3
Greenwich St....C-1,C-3
Grove St....G-2
Guerrero St....K-3
Guy....E-7
Haight St....H-1
Hampshire St....K-5
Hampton Pl....F-6
Hancock St....J-1
Harriet St....F-5
Harrison St....G-6,K-4
Hartford St....J-1
Hawthorne St....E-6
Hayes St....G-2
Hemlock St....E-3
Hermann St....H-1
Hickory St....G-3
Hidalgo Ter....I-2
Hill....K-1
Hoff St....I-3
Holly St....C-5
Hooper St....I-6
Howard St....G-4
Hubbell St....I-6
Hunt St....E-6
Hyde St....C-3,F-4
Indiana St....K-7
Irwin St....I-6
Isis St....H-4
Ivy St....G-2
Jackson St....D-1,D-3
James Lick Frwy....K-5
Jefferson St....A-4
Jessie Otis....H-3
Jessie St....F-5
Jones St....C-4,E-4
Julian Ave....I-3
Juniper St....H-4
Kansas St....J-6
Kearny St....C-5
Kenneth St....B-6
King Av....G-7
Kissling St....H-4
Lafayette St....G-3
Laguna....D-2
Landers St....I-2
Lansing....E-7
Lapidge St....J-3
Larch St....F-3
Larkin St....C-3,F-3
Laskie St....G-4
Laussat St....H-1
Leavenworth St..C-4,E-4
Leslie St....C-5
Lexington St....J-3
Liberty St....J-1
Lily St....G-2
Linda St....J-3
Linden St....G-2
Lombard....B-3
Lucerne St....G-6
Lucky St....K-4
Lusk....G-7
McAllister....F-1
McCoppin St....H-3
Main St....D-7
Mallorca Way....B-1
Marina Blvd....A-1
Mariposa St....I-4
Market Ave....I-2
Market St....F-5
Mason St....C-4,E-5
Masset....E-6
Mersey....K-2
Middle St....E-1
Midway....B-5
Minna St....G-4,I-3
Mint Mary....F-5
Mission St....G-4,K-3
Mississippi St....J-5
Missouri St....J-5
Montgomery St..D-6,E-6
Moss St....G-5
Myrtle St....E-3
Natoma St....G-4,I-3
Nellie St....K-2
Noe St....J-1
Norfolk St....H-4
Northpoint St....B-3
Oak St....G-3
Oakwood St....J-2
Octavia St....D-2
O'Farrell St....E-4,F-1
Olive St....E-3
Orosemont St....H-2
Owens St....H-7
Pacific Av....D-1,D-3
Page St....G-1
Paul St....G-5
Pearl....H-3
Pennsylvania Ave....K-7
Pfeiffer....B-5
Pierce St....E-1,H-1
Pine St....D-4,E-2
Plum....H-3
Polk St....C-3,F-3
Pond St....I-1
Post St....E-2,E-4
Potrero Ave....J-5
Powell St....B-4
Prosper St....I-1
Quane St....K-2
Ramona....I-2
Reservoir St....H-2
Retiro Way....B-1
Rhode Island St....K-6
Robert St....B-6
Rondel....I-3
Rose St....G-2
Russ St....F-5
Sacramento St....D-3
San Bruno Ave...J-6,K-5
San Carlos St....J-3
Sanchez....H-1,K-2
San Francisco-Oakland Bay Bridge (Toll)....E-8
San Jose Ave....K-3
Sansome St....C-6
Scott St....H-1
Severn St....K-2
Seymour St....F-1
Sharon St....I-2
Sheridan....H-4
Sherman....G-5
Shirley St....F-6
Shotwell St....K-4
S. Park Ave....F-7
South Van Ness Ave.
....K-4
Spear St....D-7
Stanford St....F-7
States St....I-1
Steiner St....E-1,F-1
Steuart....D-7
Stevenson St....E-5,H-3
Stockton St....D-5
Sutter St....E-2,E-3
Sycamore St....I-3
Taber Pl....F-7
Taylor St....C-4,E-4
Tehama St....F-7
Texas St....J-5,J-7,K-7
Toledo Way....B-1
Townsend St....G-7
Trainor St....H-4
Treat Ave....I-4,K-4
Tubbs St....K-7
Turk St....F-2
Union St....C-1,C-3
Utah St....J-5,K-5
Valencia....K-3
Vallejo St....C-1,C-4
Vandewater St....B-4,
Van Ness Ave....E-3
Varney Pl....F-7
Vermont St....K-5
Verona Pl....F-6
Vicksburg St....K-2
Waller St....H-1
Water St....B-4,H-1
Washburn St....G-4
Washington St....D-3
Webster St....D-1,G-2
Welsh St....G-6
Wiesse St....I-3
Willow St....F-3
Wilmot St....E-1
Wisconsin St....K-6
Woodward St....K-5
York St....K-5
Zoe St....F-7
1st St....E-7
3rd St....F-6,J-5
4th St....F-6
5th St....F-5
6th St....F-5,H-7
7th St....F-7
8th St....G-4
9th St....G-4
10th St....G-4
11th St....G-4
12th St....G-3,H-3
14th St....H-1
15th St....I-1,I-6
16th St....I-2,I-6
17th St....I-1,I-4
18th St....J-1,J-4
19th St....J-1,J-4
20th St....J-1,J-4
21st St....J-1
22nd St....J-6,K-1
23rd St....K-1,K-6
24th St....K-1,K-8
25th St....K-6,K-8

POINTS OF INTEREST

Alamo Sq....G-1
Alta Plaza....D-1
Aquatic Park....A-3
Duboce Park....H-1
Franklin Sq....I-5
Ft. Mason (U.S. Mil. Res.)....B-2
Jackson Park....I-6
Jefferson Sq....F-2
Marine Park....A-1
Mission Dolores Park....J-2
Union Sq....E-5

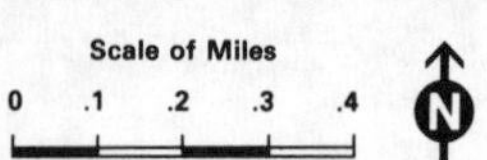
Scale of Miles
0 .1 .2 .3 .4
N

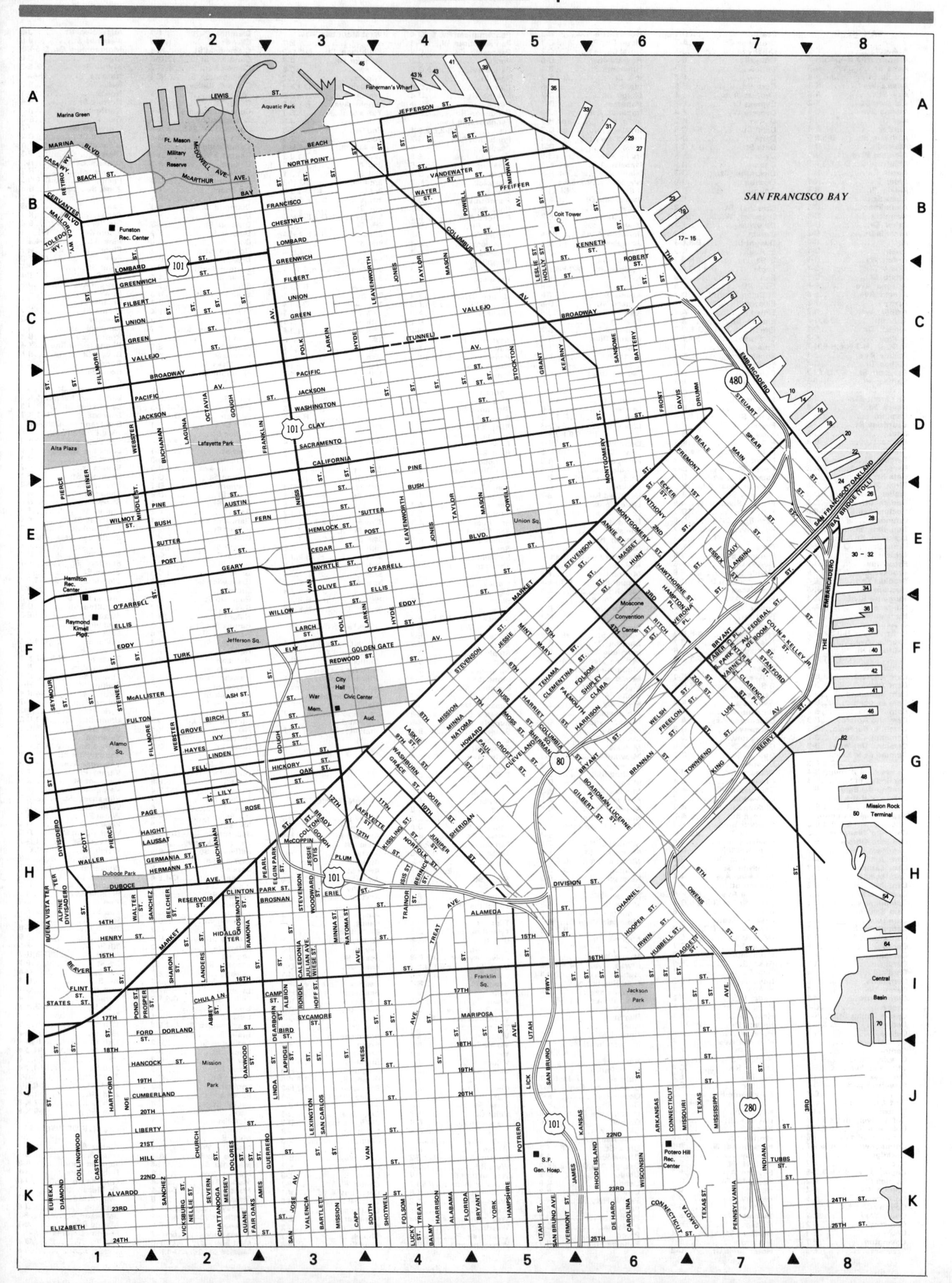
1 2 3 4 5 6 7 8
A B C D E F G H I J K
SAN FRANCISCO BAY
Marina Green
Aquatic Park
Fisherman's Wharf
Ft. Mason Military Reserve
Funston Rec. Center
Coit Tower
Alta Plaza
Lafayette Park
Union Sq.
Hamilton Rec. Center
Raymond Kimell Plgd.
Jefferson Sq.
City Hall
Civic Center
War Mem.
Aud.
Alamo Sq.
Duboce Park
Moscone Convention Center
Franklin Sq.
Jackson Park
Mission Park
S.F. Gen. Hosp.
Potero Hill Rec. Center
Mission Rock Terminal
Central Basin
101
480
80
280
SAN FRANCISCO-OAKLAND BAY BRIDGE (TOLL)
THE EMBARCADERO
MARKET
MISSION
VAN NESS
COLUMBUS AV.
BROADWAY
CALIFORNIA
GEARY
LOMBARD
FRANKLIN
GOUGH
FILLMORE
DIVISADERO
CASTRO
POTRERO
3RD

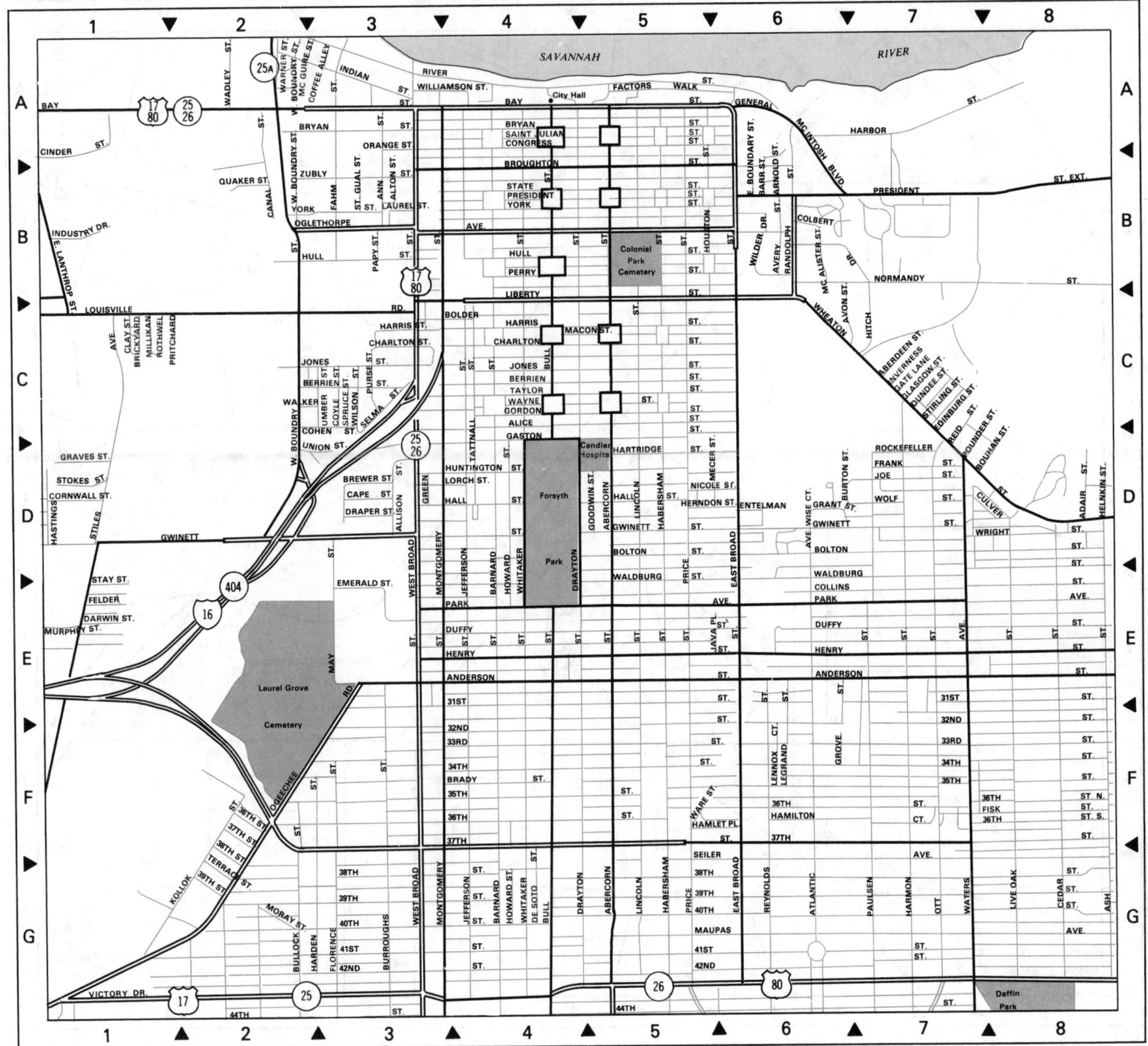

SAVANNAH

Abercorn St.....D-5,G-5
Aberdeen St.....C-7
Adair St.....D-8
Alice St.....C-4
Allison St.....D-3
Alton St.....B-3
Anderson St.....E-4,E-6
Ann St.....B-3
Arnold St.....B-6
Ash St.....G-8
Atlantic St.....G-6
Avery St.....B-6
Ave. Wise Ct.....D-6
Avon St.....C-6
Barnard.....D-4,G-4
Barr St.....B-6
Bay St.....A-1,A-4
Berrien St.....C-3,C-4
Bolder.....C-4
Bolton St.....D-5,D-6
Bouhan St.....D-7
Boundary St., E.....B-6
Boundry St., W.....A-2
.....B-2,D-2
Brady St.....F-4
Brewer St.....D-3
Brickyard.....C-1
Broad St., East..D-6,G-6
Broad St., West.D-3,G-3
Broughton St.....B-4
Bryan St.....A-3,A-4
Bull St.....C-4,G-4
Bullock St.....G-2
Burroughs St.....G-3
Burton St.....D-6
Canal St.....B-2
Cape St.....D-3
Cedar St.....G-8
Charlton St.....C-3,C-4
Cinder St.....A-1
Clay St.....C-1
Coffee Alley.....A-3
Cohen St.....C-3
Colbert.....B-6
Collins St.....E-6
Congress St.....A-4
Corwall St.....D-1
Coyle St.....C-3
Culver St.....D-8
Darwin St.....E-1
DeSoto St.....G-4
Draper St.....D-3
Drayton St.....D-4,G-4
Duffy St.....E-4,E-6
Dundee St.....C-7
Edinburg St.....C-7
Emerald St.....E-3
Entelman St.....D-6
Factors Walk.....A-5
Fahm St.....B-3
Felder.....E-1
Fisk St.....F-8
Florence St.....G-3
Frank St.....D-7
Gaston St.....D-4
Gate Lane.....C-7
General McIntosh Blvd.....A-6
Glasgow St.....C-7
Goodwin St.....D-5
Gordon St.....C-4
Grant St.....D-6
Graves St.....D-1
Green St.....D-3
Grove St.....F-6
Gwinett St.....D-2
.....D-5,D-6
Gual St.....B-3
Habersham St...D-5,G-5
Hall St.....D-4,D-5
Hamilton Ct.....F-6
Hamlet Pl.....F-5
Harbor St.....A-7
Harden St.....G-2
Harmon St.....G-7
Harris St.....C-3,C-4
Hartridge St.....D-5
Hastings.....D-1
Helnkin St.....D-8
Henry St.....E-4,E-6
Herndon St.....D-5
Hitch Dr.....C-7
Houston St.....B-5
Howard St.....D-4,G-4
Hull St.....B-3,B-4
Huntington St.....D-4
Indian St.....A-3
Industry Dr.....B-1
Inverness.....C-7
Java Pl.....E-5
Jefferson St.
.....D-4,G-4
Joe St.....D-7
Jones St.....C-3,C-4
Kollok St.....G-2
Lanthrop St., E.....B-1
Laurel St.....B-3
Legrand St.....F-6
Lennox Ct.....F-6
Liberty St.....C-4
Lincoln St.....D-5,G-5
Live Oak St.....G-8
Lorch St.....D-4
Louisville Rd.....C-1
Macon St.....C-5
Maupas Ave.....G-5
May St.....E-3
McAlister St.....B-6
McGuire St.....A-2
Mecer St.....D-5
Millikan.....C-1
Montgomery St.
.....D-3,G-3
Moray St.....G-2
Murphey St.....E-1
Nicole St.....D-5
Normandy St.....B-7
Ogeechee Rd.....F-2
Oglethorpe Ave.
.....B-3,B-4
Orange St.....A-3
Ott St.....G-7
Papy St.....B-3
Park Ave.....E-4,E-6
Paulsen St.....G-7
Perry St.....B-4
Pounder St.....D-7
President St.....B-4
Price St.....E-5,G-5
Pritchard.....C-1
Purse St.....C-3
Quaker St.....B-2
Randolph St.....B-6
Reid St.....D-7
Resident St. Ext.....B-7
Reynolds St.....G-6
River St.....A-3
Rockefeller.....D-7
Rothwel.....C-1
Saint Julian St.....A-4
Seiler Ave.....G-5
Selma St.....C-3
Stan St.....D-1
State St.....B-4
Stiles Ave.....D-1
Stirling St.....C-7
Stokes St.....D-1
Spruce St.....C-3
Tattnall St.....D-4
Taylor St.....C-4
Terrace St.....G-2
Umber St.....C-3
Union St.....D-3
Victory Dr.....G-1
Wadley St.....A-2
Waldburg St.....E-5,E-6
Walker St.....C-2
Ware St.....F-5
Warner St.....A-2
Waters Ave.....G-7
Wayne St.....C-4
Wheaton St.....C-6
Whitaker St.....D-4,G-4
Wilder Dr.....B-6
Williamson St.
.....A-3,A-4
Wilson St.....C-3
Wolf St.....D-7
Wright St.....D-8
York St.....B-3,B-4
Zubly St.....B-3
31st St.....E-4,E-7
32nd St.....F-4,F-7
33rd St.....F-4,F-7
34th St.....F-4,F-7
35th St.....F-4,F-7
36th St.....F-2,F-4,F-6
36th St. N.....F-8
36th St. S.....F-8
37th St.....F-2,F-4,F-6
38th St.....F-2,G-3,G-5
39th St.....G-2,G-3,G-5
40th St.....G-3,G-5
41st St.....G-3,G-5
42nd St.....G-3,G-5

POINTS OF INTEREST

Candler Hospital.....D-5
City Hall.....A-4
Colonial Park Cemetery.....B-5
Forsyth Park.....D-4
Laurel Grove Cemetery.....E-2

Shreveport, LA

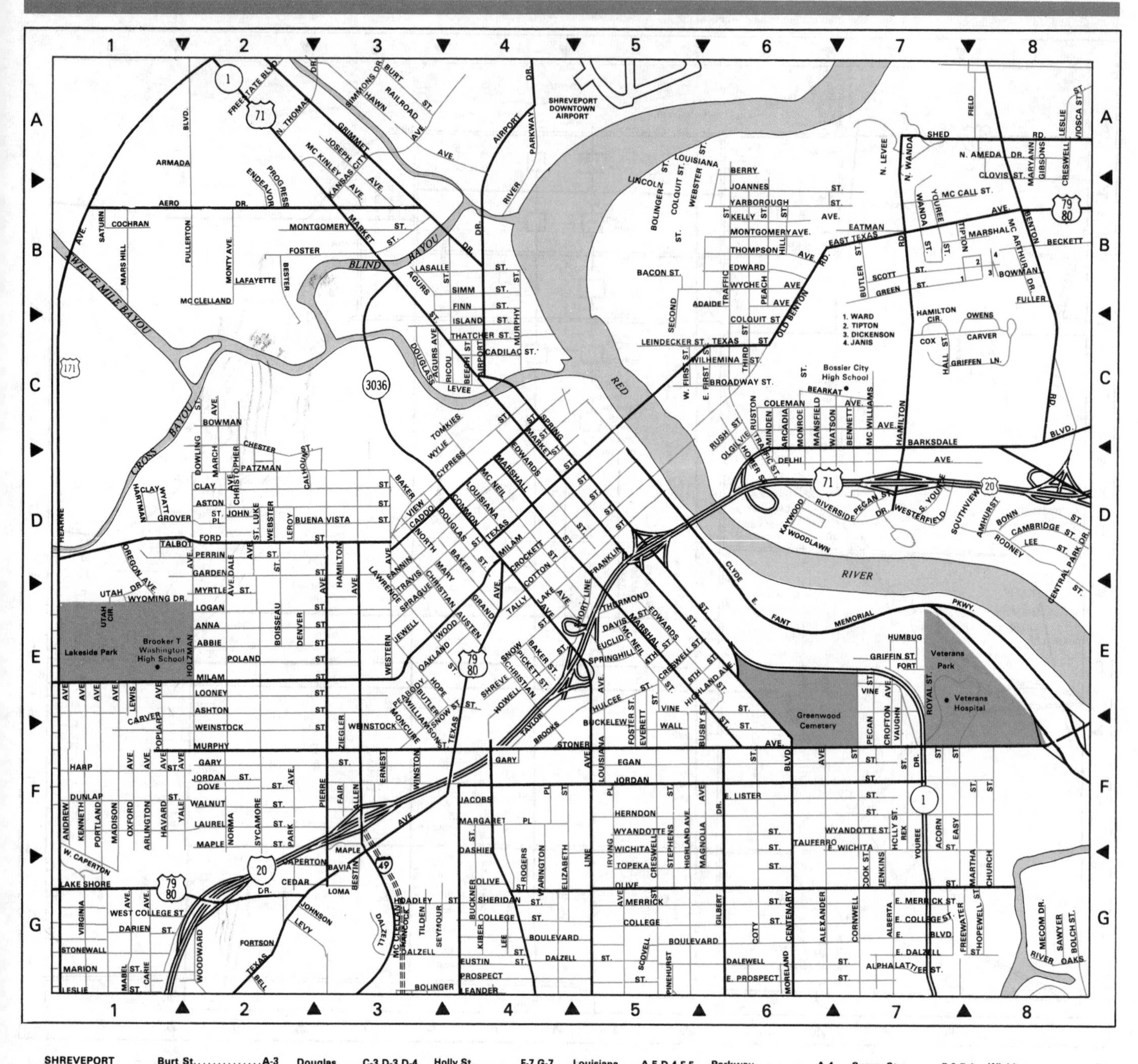

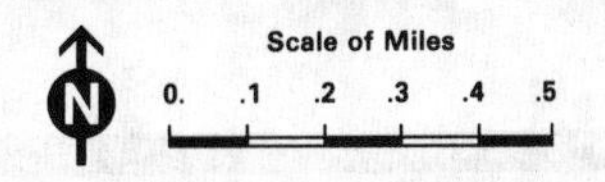
Scale of Miles
N
0. .1 .2 .3 .4 .5

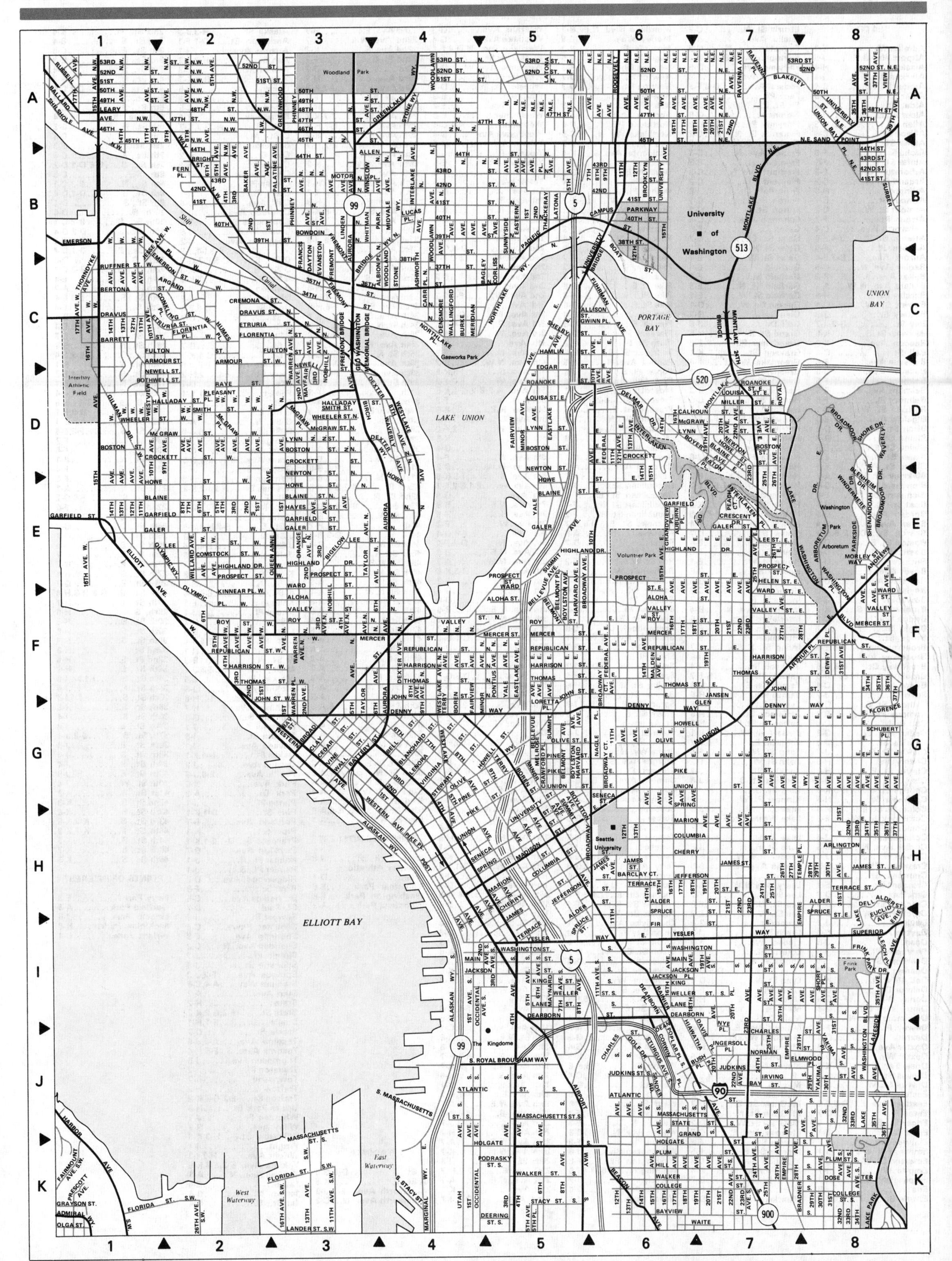
1 2 3 4 5 6 7 8
A B C D E F G H I J K
Woodland Park
University of Washington
UNION BAY
PORTAGE BAY
LAKE UNION
Gasworks Park
Interbay Athletic Field
Volunteer Park
Washington Park Arboretum
Seattle University
ELLIOTT BAY
The Kingdome
Frink Park
East Waterway
West Waterway
Ship Canal
5
99
513
520
90
900
ELLIOTT AVE.
ALASKAN WY.
DENNY WAY
MERCER ST.
YESLER WAY
MADISON
BROADWAY
MONTLAKE BLVD.
AURORA AVE. N.
S. ROYAL BROUGHAM WAY
GEO WASHINGTON MEMORIAL BRIDGE
FREMONT BRIDGE
UNIVERSITY BRIDGE
LAKE WASHINGTON BLVD.
EASTLAKE AVE. E.
WESTLAKE AVE. N.
DEXTER AVE. N.
AIRPORT WAY
RAINIER AVE. S.
S. MASSACHUSETTS

SEATTLE

Admiral Wy..........K-1
Airport Way S........J-5
Alaskan Wy.........H-4
Alaskan Wy. S.........I-4
Albion Pl. N.........C-4
Alder...............H-8
Alder St............H-5
Alder St. E.........H-6
Allen Pl............B-3
Allison St...........C-6
Aloha St............F-5
Aloha St. E..........F-6
Aloha St. N..........F-3
Arboretum Dr........E-8
Argand St. W........C-2
Arlington...........H-8
Armour St...........C-1
Armour St. W........C-2
Arthur Pl............F-7
Ashworth Ave. N....C-4
Atlantic St. S.....J-4,J-6
Auburn Pl...........E-6
Aurora Ave......B-3,E-4
Bagley Ave. N.......C-5
Baker Ave...........B-2
Ballard Ave. N.W....A-1
Barclay Ct..........H-6
Barrett St. W........C-1
Battery St...........G-3
Bay St.........G-3,J-7
Bayview St. S........K-6
Beacon Ave..........K-6
Bell St.............G-4
Bellevue Ave.........F-5
Bellevue Ave. E.......G-5
Belmont.............F-5
Belmont Ave. E.......G-5
Belmont Pl...........F-5
Bertona St. W........C-1
Bigelow Ave.........E-3
Birch St.............D-3
Blaine St............D-7
Blaine St. E.........E-5
Blaine St. N.........E-3
Blaine St. W.........E-1
Blakeley St. N.E.....A-7
Blanchard St.........G-4
Blenheim Dr.........E-8
Boat St.............B-6
Borden Ave..........G-5
Boren Ave. N.........F-4
Boston St...D-1,D-5,D-7
Boston St. E.........D-5
Boston St. N.........D-3
Bothwell St..........D-1
Boyer Ave. E.........D-7
Boylston Ave.........G-5
Boylston Ave. E......G-5
Bowdoin.............B-3
Bradner Ave. S.......K-7
Bridge Wy. N.........C-3
Bright St............B-2
Broad St.............G-3
Broadmoor Dr...D-8,E-8
Broadway Ave.........F-5
Broadway Ct.....F-6,G-6
Brooklyn Ave. N.E...B-6
Burke Ave. N.........C-4
Bush Pl.............J-7
Calhoun St. E........D-7
Campus Pkwy..........B-6
Carr Pl. N...........C-4
Cedar St.............G-3
Charles St...........J-6
Charles St. S........J-7
Cheery St............H-5
Cherry St. E.........H-6
Clay St..............G-3
College St. S.....K-6,K-8
Columbia St..........H-5
Columbia St. E.......H-6
Comstock St. W.......E-2
Conkling Pl..........C-2
Connecticut St. S....J-4
Corliss Ave. N.......C-5
Corwin Pl............I-6
Crawford Pl..........G-5
Cremona St...........C-2
Crescent Dr..........E-7
Crockett.............D-6
Crockett St..........D-3
Crockett St. W.......D-2
Davis Pl.............J-7
Dayton Ave. N........C-3
Dean................J-6
Dearborn Pl..........I-6
Dearborn St..........I-4
Deering St. S........K-5
Dell Alder St........H-8
Delmar Dr............D-6
Denny Way......F-4,G-6
Denny Way E..........G-8
Densmore Ave. N......C-4
Dewey Pl.............F-8
Dexter...............D-4
Dexter Ave. N........F-4
Dexter Wy............D-3
Dose Ter.............K-8
Dravus St. N.........C-2
Dravus St. W.........C-1
Eastern Ave. N.......B-5
Eastlake Ave. E...D-5,F-5
Eaton Pl.............D-7
Edgar St. E..........D-5
Elliot Ave...........E-1
Elmwood Pl...........J-7
Emerson Pl. W........B-1
Emerson St. W........C-2
Empire Wy............I-7
Empire Wy. S.....J-7,K-7
Erie.................I-8
Etruria St...........C-2
Etruria St. N.........C-2
Euclid Ave............I-8
Evanston Ave. N......C-3
Fairmount Ave. S.W...K-1
Fairview Ave. E......D-5
Fairview Ave. N......F-4
Federal Ave. E....D-6,F-6
Fern Pl..............B-2
Fir St. E............I-6
Florence.............G-8
Florentia Pl.........C-2
Florentia St. N......C-2
Florida St. S.W...K-1,K-3
Francis Ave. N.......C-3
Fremont..............B-3
Fremont Ave. N.......C-3
Fremont Pl...........C-3
Frink Park Dr........I-8
Fulton St............C-1
Fulton St. N.........C-3
Fuhrman Ave..........C-6
Galer................E-5
Galer St. E..........E-7
Galer St. N..........E-3
Galer St. W..........E-1
Garfield.............E-6
Garfield St..........E-1
Garfield St. N.......E-3
Garfield St. W.......E-1
Gilman Dr. W.........D-1
Glen.................G-7
Golf Dr..............J-5
Grand St. S..........K-6
Grandview Pl.........E-6
Grayson St...........K-1
Greenlake Wy.........A-4
Greenwood............A-3
Gwinn Pl.............C-6
Halladay St......D-2,D-3
Hamlin St. E.........C-5
Harbor Ave. S.W......J-1
Harrison St. E....F-5,F-6
Harrison St. N.......F-4
Harrison St. W.......F-2
Harvard Ave. E...F-5,G-5
Hawatha Pl...........J-6
Hayes St. N..........E-3
Helen St. E..........E-7
Highland Dr..E-2,E-5,E-7
Highland Dr. N.......E-3
Hill St. S...........K-6
Holgate St. S.....J-4,K-6
Howe.................E-4
Howe St..............D-7
Howe St. E...........E-5
Howe St. N...........E-3
Howe St. W...........E-1
Howell St............G-5
Howell St. E.........G-6
Humes Pl.............C-2
Ingersoll Pl.........J-7
Interlake Ave. N.....B-4
Interlaken Blvd......D-6
Interlaken Dr........E-7
Interlaken Pl........E-7
Irving St. S.........J-7
Jackson Pl...........I-6
Jackson St...........I-6
James St..........H-5-7
James St. E..........H-8
James Wy.............H-6
Jansen E.............F-6
Jefferson St.........H-5
Jefferson St. E......H-6
Jesse Ave. W.........B-1
John St. E........F-5,F-7
John St. N...........F-4
John St. W...........F-2
Judkins St...........J-7
Judkins St. S........J-6
King St. S...........I-5
Kinnear Pl. W........E-2
Lake.................J-8
Lake Washington
Blvd.................E-8
Lake Park............K-8
Lane St. S...........I-5
Lander St. S.W.......K-3
Latona Ave. N.E......B-5
Latona...............A-5
Leary Way N.W........A-1
Lee St. E............E-7
Lee St. N............E-3
Lee St. W............E-2
Lenora St............G-4
Lesch Pl.............I-8
Linden Ave...........B-3
Loretta Pl...........F-5
Louisa St. E.........D-7
Lynn St. E........D-5,D-7
Lynn St. N...........D-3
Lucas Pl.............B-4
Madison St. E........H-5
Main St..............I-4
Main St. S...........I-5
Malden Ave. E........F-6
Marginal Wy. E.......K-4
Marion St............H-5
Marion St. E.........H-6
Massachusetts St.
S...............J-4,J-5
Mayfair Av...........C-3
McGraw Pl............D-2
McGraw St. E.........D-7
McGraw St. N.........D-3
McGraw St. W.........D-2
Melrose Ave. E.......G-5
Mercer St............F-8
Mercer St. E.........F-5
Meridian Ave. N......C-4
Midvale Ave. N.......B-4
Miller St. E.........D-7
Minor Ave.........D-5,G-5
Montlake.............D-7
Montlake Blvd. N.E..B-7
Morley Way...........E-8
Motor N. Pl..........B-3
Nagle Pl.............G-6
Newton Pl............B-3
Newton St.........D-5,D-7
Newton St. N.........D-3
Newell St.........C-3,D-1
Nob Hill.............C-3
Norman St. S.........J-7
Northlake Pl.........C-4
Northlake Wy. N......C-5
Nye Pl...............J-7
Occidental Ave. S....K-4
Olga St..............K-1
Olive St. E..........G-6
Olive Wy.............G-4
Orange Pl............E-3
Olympic Pl. W........F-2
Olympic Wy...........E-2
Pacific St...........B-5
Palatine Ave.........B-3
Park Ave. N..........B-4
Parkside Dr..........E-8
Peach Ct.............E-7
Phinney..........A-3,B-3
Pike St..............H-4
Pike St. E...........G-6
Pike Pl..............H-4
Pike St. E...........G-5
Pine St..............G-4
Pine St. E...........G-5
Pleasant Pl..........D-2
Plum St. S........K-6,K-8
Podrasky St. S.......K-5
Pontius Ave. N.......F-5
Poplar Pl. S.........J-6
Post Ave.............H-4
Prescott Ave.........K-1
Prospect St...E-2,E-5,E-7
Prospect St. E.......E-6
Prospect St. N.......E-3
Queen Anne Ave.
N....................E-3
Rainer Ave...........I-6
Raye St. W...........D-2
Ravenna Ave..........A-7
Ravenna Pl...........A-7
Republican St........F-8
Republican St. E.....F-5
Republican St. N.....F-4
Republican St. W.....F-2
Roanoke St. E...D-5,D-7
Roosevelt............A-6
Royal................D-7
Roy St. N............F-3
Roy St. W............F-2
Ruffner St. W........C-1
Sand Point N.E.......A-8
Sander Rd............J-6
Seneca St.........G-6,H-4
Schubert Pl..........G-8
Shelby St. E.........C-5
Shenandoah Dr........E-8
Shilshole Ave. N.W..A-1
Shore Dr.............D-7
Short Pl.............I-8
Smith St. W.......D-2,D-3
Spring St............H-4
Spring St. E.........H-6
Spruce St............I-4
Spruce St. E......H-6,I-8
Stacy St. S..........K-5
St. Andrews..........E-8
State St. S..........J-6
Stone Wy.............A-4
Stone Wy. N..........C-4
Sturgus Ave. S.......J-6
Stewart St...........G-4
Summit Ave........E-5,G-5
Summit Ave. E........G-5
Sunnyside Ave. N.....B-5
Superior.............I-8
Surber...............B-8
Taylor Ave. N........E-3
Temple Pl............H-7
Terrace..............I-5
Terrace St.E......H-6,H-8
Terry Ave............G-5
Terry Ave. N.........F-4
Thackery Pl. N.E.....B-5
Thomas St. E......F-5,F-6
Thomas St. N.........F-4
Thomas St. W.........F-7
Thorndyke Ave........C-1
Union Bay Pl. N.E....A-8
Union St.............H-4
Union St. E..........G-5
University...........A-8
University St........G-5
University Wy........B-6
Utah Ave. S..........K-4
Valley St.........F-6,F-8
Valley St. E.........F-7
Valley St. N......F-3,F-4
View.................A-8
Vine St..............G-3
Virginia St..........G-4
Waite................K-6
Walker St. S.....K-5,K-6
Wallingford Ave. N...C-4
Wall St..............G-3
Ward.................E-5
Ward St..............F-8
Ward St. E...........F-7
Ward St. N...........E-3
Warren Ave. N........C-3
Warren Pl............F-3
Washington Blvd......J-8
Washington Pl........F-8
Washington St. S..I-4,I-6
Waverly..............D-7
Waverly Ave..........D-4
Wellar St. S.........I-5
Western Ave..........G-3
Westlake Ave.........G-4
Westlake Ave. N.D-4,F-4
W. Mercer St.........F-3
Westview Ave. W......D-1
Wheeler St. N........D-3
Wheeler St. W........D-1
Whitman..............B-3
Willard Ave..........E-2
Windermer............E-8
Woodland.........A-4,C-4
Woodlawn Ave. N......B-4
Yakima Ave...........J-8
Yakima Pl............J-8
Yale Ave.E...........E-5
Yale Ave. N..........F-5
Yesler Way...........I-4
1st Ave..............G-4
1st Ave. N...........F-3
1st Ave. N.E.........B-5
1st Ave. N.W.........B-3
1st Ave. S........J-4,K-4
1st Ave. W........E-2,F-2
2nd Ave.........D-3,G-4
2nd Ave. N........E-3,F-3
2nd Ave. N.E.........B-5
2nd Ave. N.W.........B-2
2nd Ave. S...........I-4
2nd Ave. W........E-2,F-2
3rd Ave..............G-4
3rd Ave. N........C-3,F-3
3rd Ave. S........I-4,K-5
3rd Ave. W........E-2,F-2
4th Ave..............G-4
4th Ave. N...........F-3
4th Ave. N.W.........B-2
4th Ave. S........I-5,K-5
4th Ave. W........E-2,F-2
5th Ave..A-2,B-2,G-4,K-5
5th Ave. N...........F-3
5th Ave. N.E.........B-5
5th Ave. S...........I-5
5th Ave. W........E-2,F-2
5th Pl...............K-5
6th Ave..............G-4
6th Ave. N...........F-4
6th Ave. N.W.........B-2
6th Ave. S........I-5,K-5
6th Ave. W.....E-1,E-2,F-7
7th Ave..............G-4
7th Ave. N.E.........B-6
7th Ave. S...........I-5
7th Ave. W...........E-2
8th Ave...........F-4,G-4
8th Ave. N........D-4,F-4
8th Ave. N.E.........B-6
8th Ave. N.W.........A-2
8th Ave. S........I-5,K-5
8th Ave. W...........E-2
9th Ave..............G-5
9th Ave. N...........D-4
9th Ave. N.E.........B-6
9th Ave. N.W.........A-2
9th Ave. W...........D-2
10th Ave..........C-1,D-6
10th Ave. E..........D-6
10th Ave. W..........D-1
11th Ave.............H-6
11th Ave. E.......D-6,G-6
11th Ave. N.E........B-6
11th Ave. N.W........A-1
11th Ave. S..........I-6
11th Ave. S.W........K-3
11th Ave. W.......C-1,E-1
12th.................B-6
12th Ave. E.......D-6,H-6
12th Ave. N.E........B-6
12th Ave. S..........K-6
12th Ave. W.......C-1,G-1
13th Ave.............D-6
13th Ave. E..........H-6
13th Ave. S..........K-6
13th Ave. S.W........K-3
13th Ave. W.......C-1,E-1
14th.................D-6
14th Ave. E.......F-6,H-6
14th Ave. N.W........A-1
14th St. S...........K-6
14th Ave. W.......C-1,E-1
15th.................D-6
15th Ave.............D-1
15th Ave. E.......E-6,H-6
15th Ave. N.E........B-6
15th Ave. N.W........A-1
15th St. S...........K-6
16th.................D-6
16th Ave. E.......F-6,H-6
16th Ave. N.E........A-6
16th Ave. S..........I-6
16th St. S...........K-6
16th Ave. S.W........K-3
16th Ave.W...........C-1
17th Ave.............A-1
17th Ave. E.......F-6,H-6
17th Ave. S..........K-6
17th Ave. W..........C-1
17th St. N.E.........A-6
18th Ave.............I-6
18th Ave. E.......H-6,H-7
18th Ave. N.E........A-7
18th Ave. S..........K-6
19th Ave.............D-7
19th Ave. E.......F-7,H-7
19th Ave. N.E........A-7
19th Ave. S.......I-6,K-6
20th Ave..........D-7,J-7
20th Ave. E.......F-7,H-7
20th Ave. N.E........A-7
20th Ave. S..........K-7
20th Pl..............I-6
21st Ave. E.......F-7,H-7
21st Ave. N.E........A-7
21st Ave. S..........K-7
22nd Ave..........D-7,J-7
22nd Ave. E..........H-7
22nd Ave. N.E........A-7
22nd Ave. S..........K-7
22nd St. E...........F-7
23rd.................D-7
23rd Ave. E..........H-7
23rd Ave. S.......J-7,K-7
23rd St. E...........F-7
24th Ave. E..........H-7
24th Ave. S.......J-7,K-7
25th Ave. E.......E-7,H-7
25th Ave. S.......J-7,K-7
26th Ave. E.......E-7,H-7
26th Ave. S..........I-7
26th Ave. S.W........K-2
27th Ave. E..........E-7
28th Ave. E..........H-8
28th Ave. S..........K-7
29th Ave. E..........H-8
29th Ave. S.......J-8,K-8
30th Ave. E..........H-8
30th Ave.S........J-8,K-8
31st Ave..........F-8,J-8
31st Ave. E..........H-8
31st Ave. S..........K-8
32nd Ave. E..........H-8
32nd Ave. S.......J-8,K-8
33rd Ave. E..........H-8
33rd Ave. S.......J-8,K-8
34th Ave. E.......F-8,H-8
34th Ave. S..........K-8
34th St. N...........C-3
35th Ave......A-8,I-8,J-8
35th Ave. E..........H-6
35th St. N...........C-3
36th Ave..........A-8,J-8
36th Ave. E.......F-8,H-8
36th St. N...........C-3
37th.................F-8
37th Ave.............A-8
37th Ave. E..........H-8
38th.................C-4
38th St..............B-6
39th Ave.............A-8
39th St. N........B-2,B-4
40th St. N........B-2,B-4
41st St...........B-6,B-8
41st St. N........B-2,B-4
42nd St.......B-2,B-6,B-8
42nd St. N...........B-4
43rd St.......B-2,B-6,B-8
43rd St. N...........B-4
44th.................B-2
44th St...........B-3,B-8
44th St. N...........B-4
45th St. N...........A-3
45th St. N.E.........A-6
45th St. N.W.........A-1
46th St. N...........A-3
46th St. N.W.........A-1
47th.................A-8
47th St..............A-5
47th St. N........A-3,A-5
47th St. N.E.........A-6
47th St. N.W.........A-2
48th St..............A-8
48th St. N...........A-3
48th St. N.W.........A-2
49th St. N...........A-3
49th St. N.W.........A-1
50th St. N...........A-3
50th St. N.E......A-6,A-8
50th St. N.W.........A-1
51st St..............A-3
51st St. N...........A-4
51st St. N.W.........A-1
52nd.................A-8
52nd St..............A-2
52nd St. N........A-4,A-5
52nd St. N.E......A-6,A-8
52nd St. N.W.........A-1
53rd St..............A-7
53rd St.N.........A-4,A-5
53rd St. N.W.........A-1

POINTS OF INTEREST

Frink Park...........I-8
Interbay Athletic
Field................D-1
Volunteer Park.......E-6
Washington Park......E-8
Woodland Park........A-3

TACOMA

A St..............D-6,G-6
A St., E.............K-6
Ainsworth Ave.....E-3,J-3
Ainsworth Ave., N....C-2
Alaska St.........H-3,J-3
Anderson St.......D-1,F-1
Arizona Ave..........I-1
Ash St............H-3,J-2
Asotin St............J-3
B St., E..........G-6,K-7
Baker................C-5
Bell St..............K-6
Broadway.............C-5
Broadway, N..........B-4
C St., E..........G-7,K-7
C St., N.............A-3
California Ave.......I-1
Carr St..............A-2
Center St............H-4
Chandler St..........I-3
Chelan Pl............I-1
Clark Pl.............I-1
Cliff Ave............D-6
Colorado Ave.........J-2
Commerce St..........E-6
Crandall Ln..........J-6
Cushman Ave..........C-2
.............E-3,I-4,J-3
D St., E.......C-6,E-7,G-7
D St., N.............A-3
D St., S.............K-6
Division.............C-4
Division Ln..........I-8
Dock St...........C-6,E-6
E St.................G-7
E St., E..........C-6,K-7
E St., N.............A-3
F St.................G-7
F St., E..........C-6,K-7
Fawcett Ave.......E-5,J-6
Ferry St......E-2,G-2,H-2
Fife.................D-1
Fife St...........I-1,K-1
G St.................G-7
G St., E.............J-8
G St., N.............A-3
G St., S..........E-5,J-5
George St............J-6
Grant Ave., N........C-2
Grant Ave., S.....E-3,G-3
Gregory St...........I-7
Harrison St..........I-8
Hosmer St.....H-2,I-2,K-2
I St., E.............J-8
I St., N.............B-3
Idaho St.............J-1
J St., E.......E-8,H-8,J-8
J St., N.............B-3
J St., S..........E-4,J-4
Jefferson Ave........F-5
Junett St.....D-1,I-1,K-1
K St., E.............J-8
K St., N.............B-3
K St., S..........E-4,J-4
Kellogg..............E-1
Kitsap Pl............J-1
L St., E.............G-8
L St., N.............B-3
L St., S..........E-4,J-4
Lewis Pl.............I-1
M St., N.............B-3
M St., S..........E-3,J-4
Market St............E-5
McKinley Ave.........J-8
McKinley Rd..........H-8
McKinley Way E.......H-7
Molgate St...........G-6
Montana Ave..........J-1
Morton St............I-8
Nevada Ave...........I-1
Oakes St..........D-1,F-1
Oregon Ave...........J-2
Pacific Ave.......E-6,J-6
Park Ave.............K-5
Park Dr..............A-3
Pierce Pl............J-1
Pine St...........D-1,K-1
Portland Ave.........B-7
Prospect.............H-1
Prospect St..D-1,E-1,K-1
Puyallup Ave.........G-7
Rainier Pl., N.......J-1
Rainier Pl., S.......J-1
Ridgewood Ave........D-2
River St.............F-8
St. Helens Ave.......C-5
St. Paul.............D-7
Sawyer St............I-3
Schuster Pkwy........C-5
Sheridan Ave.........E-3
Sheridan Ave., N.....C-2
Skagit Pl............J-1
Spokane St...........J-8
Sprague Ave.......E-2,J-2
Stadium Way.......A-4,C-5
Starr Ave............A-2
State................H-2
State St......E-2,G-2,K-2
Steele............E-1,H-1
Steele St............G-2
Tacoma Ave........A-2,J-5
Tacoma Ave., S..E-5,H-4
Thompson Ave.........J-5
Thurston Pl..........I-2
Townsend.............D-7
Trafton..............H-2
Trafton St....E-2,G-2,K-2
Upper Park St........H-8
Whitman Pl...........J-1
Wiley Ave............G-7
Wilkeson St.......H-3,K-3
Wright...............I-1
Wright Ave........H-7,I-3
Yakima Ave..B-3,E-4,K-5
Yakima Ave., N.......A-1
1st, S...............E-4
1st St., N...........B-4
1st St., S...........B-4
2nd St., N...........B-4
2nd St., S...........B-4
3rd St., E...........C-6
3rd St., N...........B-3
3rd St., S...........C-4
4th St., N...........B-3
4th St., S........C-4,C-5
5th St., N...........B-3
5th St., S...........C-4
6th Ave., S..........C-4
6th St., N...........B-3
7th..................C-5
7th St...............B-3
7th St., E...........C-6
7th St., S.....C-5,D-1,D-2
8th, S...............C-5
8th St., N........B-3,C-1
8th St., S........D-1,D-4
9th St., S...........D-4
10th, S..............D-6
10th St., N.......B-2,C-1
10th St., S.......D-2,D-4
11th St., N.......B-2,C-1
11th St., S.......D-1,D-4
12th, S..............D-6
12th St., N.......B-2,C-1
12th St., S..........E-4
13th St., N..........A-2
13th St., S.......E-1,E-4
14th, S..............E-6
14th St., N..........B-1
14th St., S.......E-1,E-4
15th St., E..........E-7
15th St., N..........B-1
15th St., S.......E-1,E-4
16th St., N..........B-1
16th St., S.......E-4,F-2
17th St., N..........B-1
17th St., S.......F-1,F-4
18th, S..............E-6
18th St., E.......D-8,E-7
18th St., N..........B-1
18th St., S.......F-2,F-4
19th St., E..........F-7
19th St., S..........F-4
19th St., N..........B-1
20th St., S..........F-4
21st St., E..........F-8
21st St., N..........A-1
21st St., S.......F-4,G-2
22nd St..............F-6
23rd St., S...G-2,G-4,G-5
24th St., S..........G-6
25th St., E..........G-7
25th St., S.......G-2,G-4
26th St., E..........G-7
27th St., E..........G-7
27th St., S..........H-4
28th St., E..........G-7
28th St., S...G-6,H-3,H-4
29th St., E.......G-8,H-7
29th, S..............H-6
30th St., S..........H-6
30th St., E..........H-8
30th St., S..........H-4
32nd St..............H-8
32nd St., S..........H-6
34th St., E..........H-8
34th St., S.....I-1,I-4,I-6
35th St., E..........I-8
35th St., S.......I-4,I-6
36th St., E..........I-8
36th St., S..........I-4
37th St., E..........J-8
37th St., S.......J-2,J-4
38th St., E..........J-8
38th St., S..........J-4
39th St., S..........J-4
40th St., E..........J-8
40th St., W..........J-4
41st St., S..........K-3
42nd St., S..........K-3
43rd St., E..........K-8
43rd St., S.......K-3,K-6
44th St., E.......K-7,K-8
44th St., S..........K-3
45th St., E..........K-7
45th St., S..........K-3

POINTS OF INTEREST

Ferry Park...........E-3
Garfield Park........A-3
Lincoln Park.........I-5
McKinley Park........H-8
Wright Park..........C-4

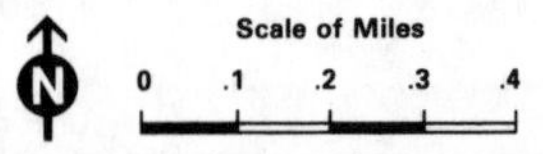
N
Scale of Miles
0 .1 .2 .3 .4

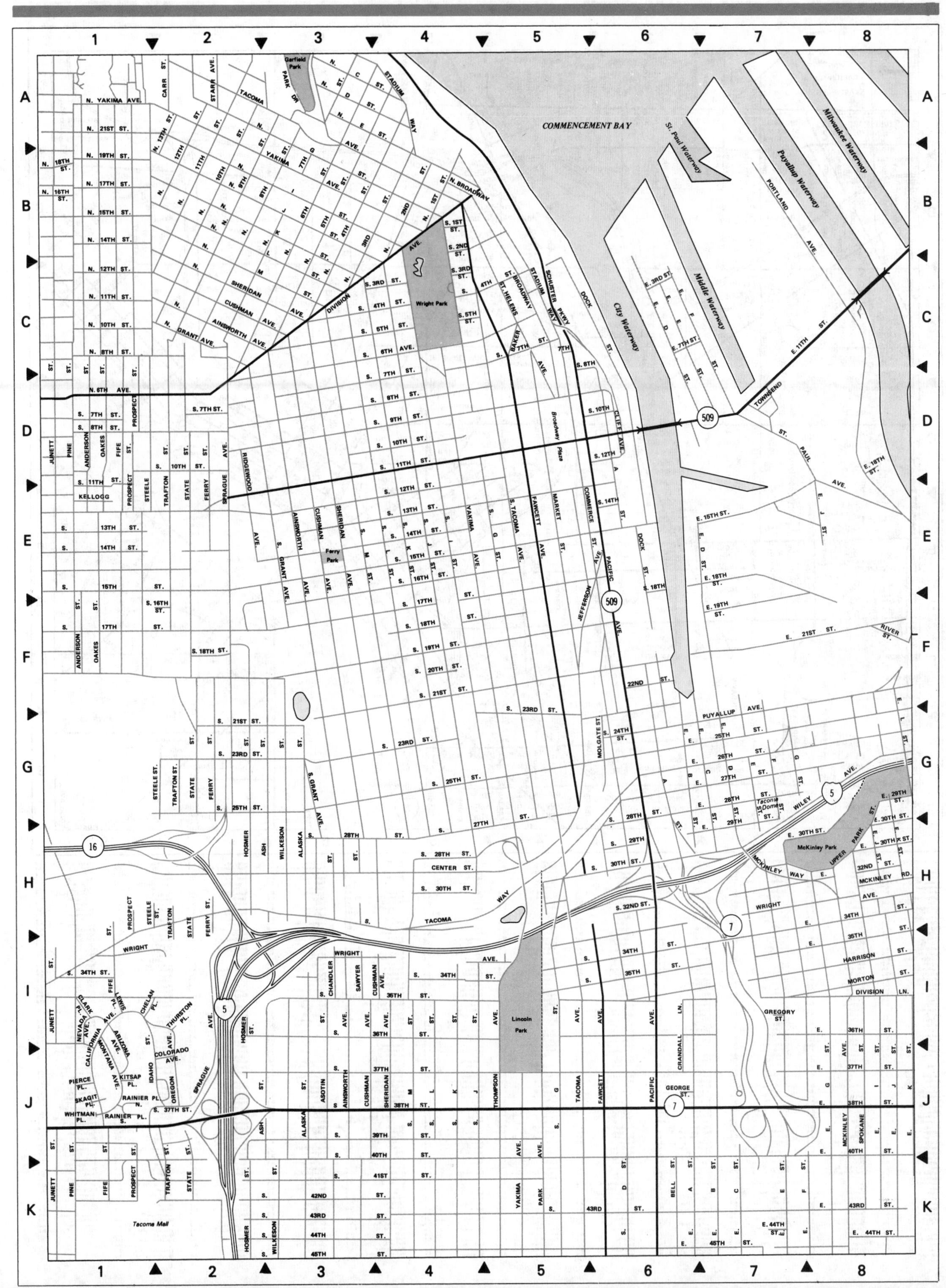
1 2 3 4 5 6 7 8
A B C D E F G H I J K
COMMENCEMENT BAY
St. Paul Waterway
Milwaukee Waterway
Puyallup Waterway
Middle Waterway
City Waterway
Garfield Park
Wright Park
Ferry Park
McKinley Park
Lincoln Park
Tacoma Dome
Tacoma Mall
N. YAKIMA AVE.
N. 21ST ST.
N. 19TH ST.
N. 17TH ST.
N. 15TH ST.
N. 14TH ST.
N. 12TH ST.
N. 11TH ST.
N. 10TH ST.
N. 8TH ST.
N. 6TH AVE.
N. BROADWAY
STADIUM WAY
SHERIDAN
CUSHMAN
AINSWORTH
N. GRANT AVE.
DIVISION AVE.
S. 7TH ST.
KELLOGG
RIDGEWOOD
SPRAGUE
SCHUSTER PKWY
DOCK ST.
Broadway Plaza
CLIFF AVE.
COMMERCE ST.
MARKET ST.
FAWCETT AVE.
S. TACOMA AVE.
YAKIMA AVE.
JEFFERSON
PACIFIC AVE.
PORTLAND AVE.
E. 11TH ST.
TOWNSEND
ST. PAUL AVE.
E. 15TH ST.
E. 21ST ST.
RIVER ST.
PUYALLUP AVE.
S. 23RD ST.
S. 25TH ST.
MOLGATE ST.
WILEY AVE.
MCKINLEY WAY
UPPER PARK ST.
MCKINLEY AVE.
WRIGHT AVE.
HARRISON
MORTON
DIVISION LN.
GREGORY ST.
TACOMA WAY
CENTER ST.
S. 32ND ST.
HOSMER
ASH
WILKESON
ALASKA
THOMPSON
CRANDALL LN.
GEORGE ST.
S. 37TH ST.
S. 38TH ST.
SPOKANE
MCKINLEY
JUNETT
PINE
FIFE
PROSPECT
TRAFTON
STATE
STEELE ST.
FERRY
WRIGHT
S. 34TH ST.
5
7
16
509

Toledo, OH

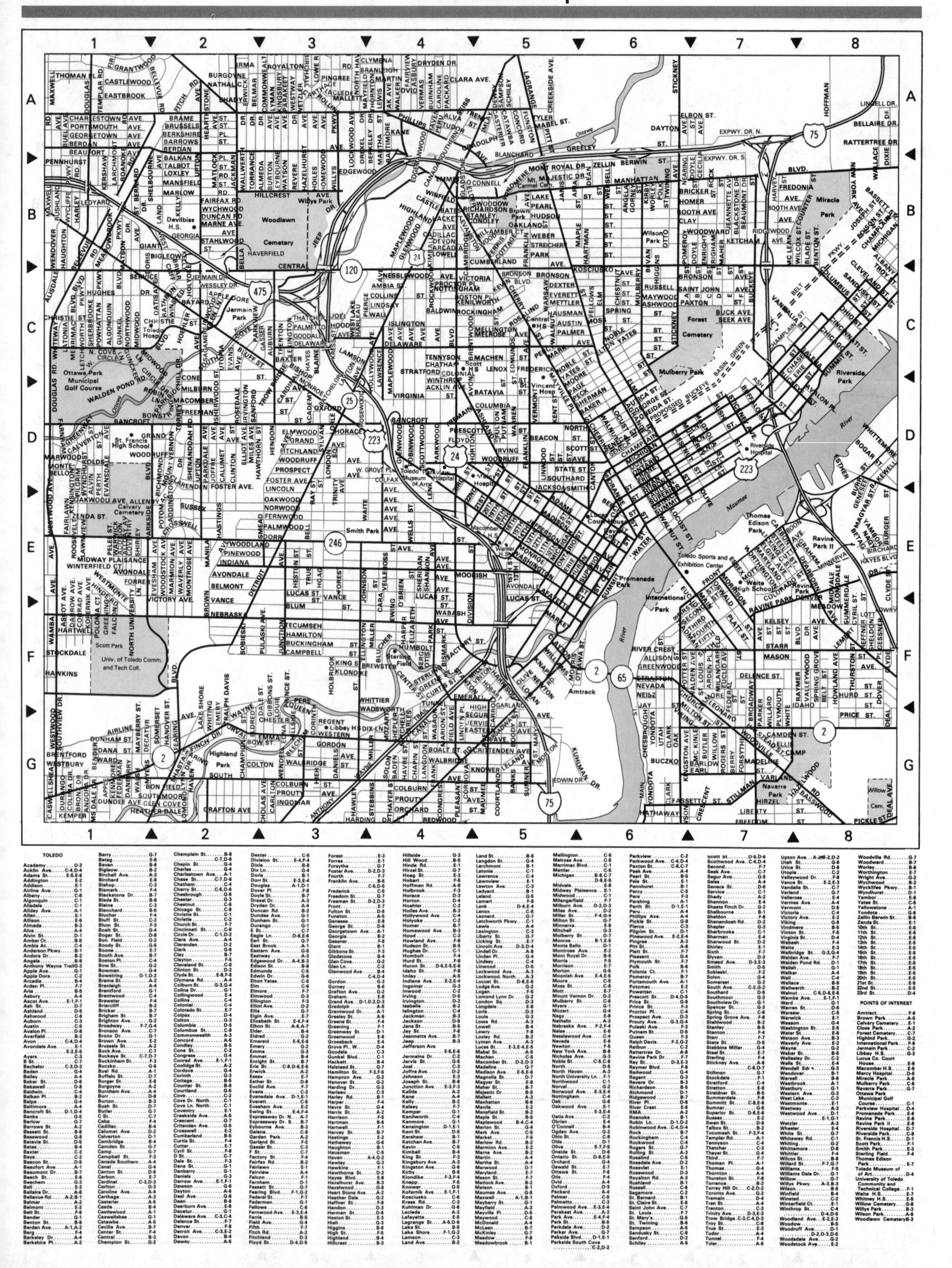
Scale of Miles
0 .2 .4 .6 .8
N
Woodlawn Cemetery
Ottawa Park Municipal Golf Course
Forest Cemetery
Mulberry Park
Riverside Park
Miracle Park
St. Francis High School
Calvary Cemetery
Smith Park
Scott Park
Univ. of Toledo Comm. and Tech Coll.
Highland Park
Thomas Edison Park
Toledo Sports and Exhibition Center
International Park
Promenade Park
Waite High School
Navarre Park
Ravine Park II
Amtrack
Maumee River
TOLEDO
POINTS OF INTEREST

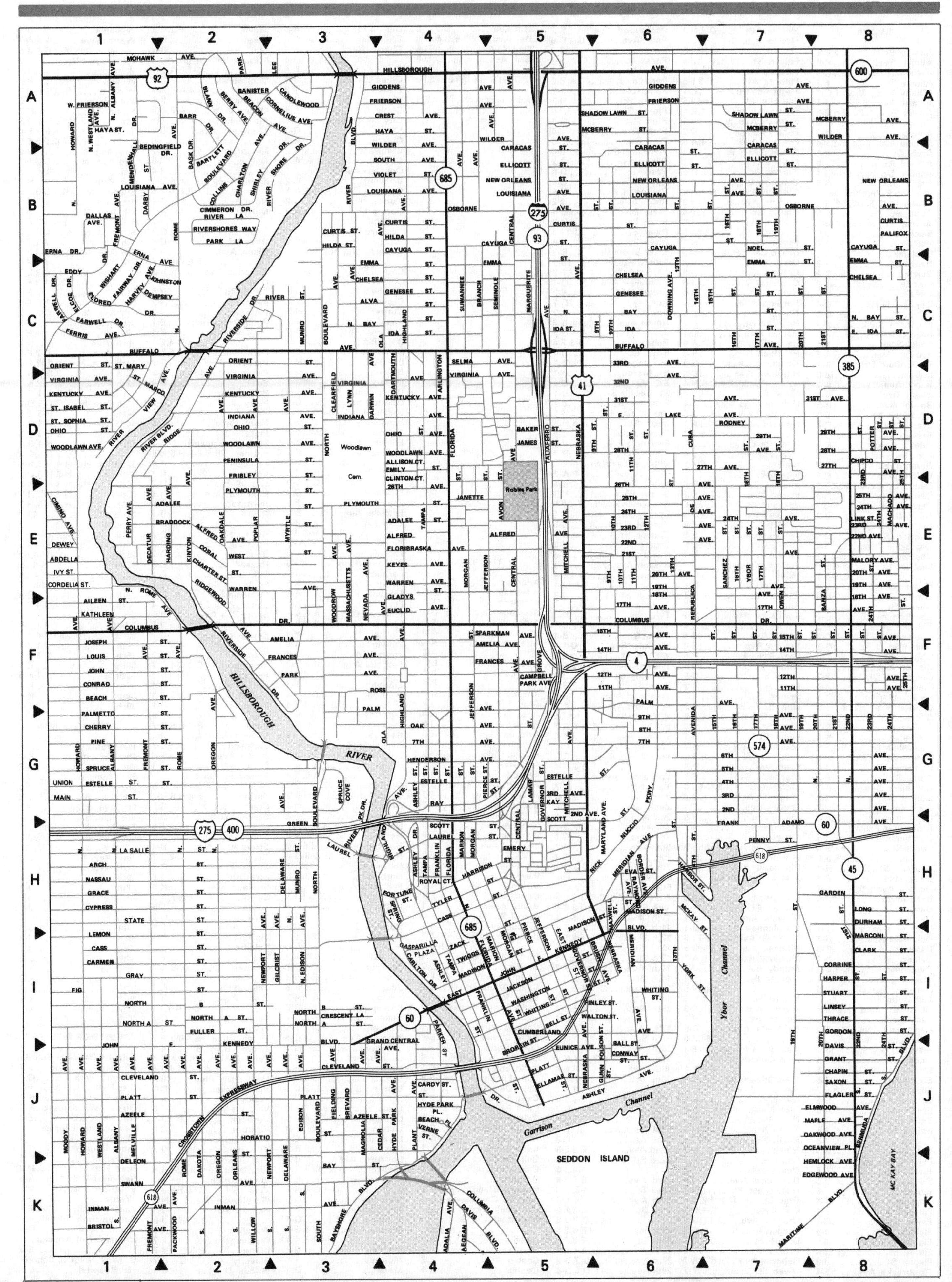
Scale of Miles
0 .1 .2 .3 .4
N
HILLSBOROUGH AVE.
BUFFALO
COLUMBUS
HILLSBOROUGH RIVER
Woodlawn Cem.
Robles Park
SEDDON ISLAND
Ybor Channel
Garrison Channel
MC KAY BAY
CROSSTOWN EXPRESSWAY
BAYSHORE BLVD.
NEBRASKA
FLORIDA
TALIAFERRO
MARGUERITE
CENTRAL
ARMENIA
HOWARD
BOULEVARD
NORTH
RIVER BLVD.
RIVERSIDE DR.
KENNEDY
CLEVELAND
PLATT
SWANN
KEYES
OSBORNE
LOUISIANA
NEW ORLEANS
CHELSEA
CAYUGA
CURTIS
SLIGH
ADAMO
MERIDIAN
NUCCIO PKWY.
MARITIME BLVD.
GREEN
92
600
685
275
93
41
385
4
574
60
618
45
400

TAMPA

Adalee St. ... E-2,E-4
Adalia Ave. ... K-4
Aegean ... K-4
Aileen St. ... F-1
Albany Ave., N. ... A-1,G-1
Albany Ave., S. ... J-1
Alfred Ave. ... E-2,E-4,E-5
Allison Ct. ... D-4
Alva St. ... D-3
Amelia Ave. ... F-3,F-4
Arch St. ... H-1
Ashley Ave. ... J-5
Ashley Dr. ... H-4,I-4
Ashley St. ... G-4
Avenida Republica De Cuba ... D-5,E-6,F-6,G-6
Avon St. ... E-5
Azeele St. ... J-1,J-3
Baker St. ... D-5
Ball St. ... J-6
Banister ... A-2
Banza St. ... F-8
Bartlett ... B-2
Barr Dr. ... A-2
Bay St. ... K-3
Bay St., N. ... C-3,C-4 ... C-6,C-8
Bayshore Blvd. ... K-3
Beach Pl. ... J-4
Beach St. ... F-1
Beacon Ave. ... A-2
Bedingfield Dr. ... A-1
Bell St. ... Ip-5
Berry Ave. ... A-2
Blann Dr. ... A-2
Border Ave. ... H-6
Braddock St. ... E-2
Branch Ave. ... C-4
Brevard ... J-3
Bristol Ave. ... K-1
Brorein St. ... J-5
Brush Ave. ... I-6
Buffalo Ave. ... C-1,C-6
Campbell Grove ... F-5
Candlewood ... A-3
Caracas St. ... B-5-7
Cardy St. ... J-4
Carlton Dr. ... I-4
Carmen St. ... I-1
Casparilla Plaza ... I-4
Cass St. ... H-4,I-1
Cayuga St. ... B-4,B-5 ... B-6,B-8
Cedar Ave. ... J-4
Central Ave. .. B-5;E-5,H-5
Chapin St. ... J-8
Charlton Ave. ... B-2
Charter St. ... E-2
Chelsea St. ... C-3,C-6,C-8
Cherry St. ... G-1
Chipco St. ... D-8
Cimino Ave. ... E-1
Cimmeron Dr. ... B-2
Clark St. ... I-8
Clearfield Ave. ... D-3
Cleveland St. ... J-1,J-3
Clinton Ct. ... D-4
Collins ... B-2
Columbia ... K-4
Columbus Dr. ... F-1,F-6
Conrad St. ... F-1
Conway ... J-6
Coral St. ... E-2
Cornelius Ave. ... A-3
Corrine St. ... I-8
Crescent La. ... I-3
Crest Ave. ... A-4
Crosstown Expwy. ... J-2
Cumberland Ave. ... I-5
Curtis St. ... B-3,B-4 ... B-5,B-8
Cypress St. ... H-1
Dallas Ave. ... B-1
Dakota Ave., S. ... K-2
Dartmouth ... D-4
Davis Blvd. ... K-4
Davis St. ... J-8
Decatur Ave. ... E-1
Delaware Ave., N. ... H-3
Delaware Ave., S. ... K-3
Deleon St. ... K-1
Dempsey ... C-1,C-2
Downing Ave. ... C-6
Durham St. ... I-8
East John F. Kennedy Blvd. ... I-4,I-5
Eddy Dr. ... C-1
Edgewood Ave. ... K-8
Edison Ave., N. ... I-3
Edison Ave., S. ... J-3
Elcoe Dr. ... C-1
Eldred Dr. ... C-1
Ellamae St. ... J-5
Ellicott St. ... B-5-7
Elmwood Ave. ... J-8
Emery ... H-5
Emily St. ... D-4
Emma St. ... C-3-5 ... C-7-8
Erna Dr. ... B-1
Estelle St. ... G-1,G-4,G-5
Euclid Ave. ... F-4
Eunice Ave. ... J-5
Eva St. ... H-6
Fairway Dr. ... C-1
Farwell Dr. ... C-1
Ferris Ave. ... C-1
Fielding Ave. ... J-3
Fig St. ... I-1
Finley St. ... I-6
Flagler St. ... J-8
Floribraska Ave. ... E-4
Florida Ave. ... D-4,H-4
Florida Ave. N. ... I-4,I-5
Folson St. ... J-6
Fortune St. ... H-4
Frances Ave. ... F-3,F-5
Frank Adamo Ave. ... H-7
Franklin St. ... H-4,I-4
Fremont Ave. ... B-1,K-1
Fremont Ave. N. ... G-1
Fribley St. ... D-2
Frierson Ave. ... A-4,A-6
Frierson Ave. W. ... A-1
Fuller St. ... I-2
Garden St. ... H-8
Genesee St. ... C-4,C-6
Giddens Ave. ... A-4,A-6
Gilcrsit Ave. ... I-3
Gladys St. ... F-4
Gordon St. ... I-8
Governor St. ... G-5,I-5
Grace St. ... H-1
Grand Central Ave. ... I-3
Grant St. ... J-8
Green ... H-3
Gunn St. ... J-6
Harbor St. ... H-6
Harding Ave. ... E-1
Harper St. ... I-8
Harrison St. ... H-4
Harvey Ave. ... C-1
Haya St. ... A-1,A-4
Hemlock Ave. ... K-8
Henderson Ave. ... G-4
Highland Ave. ... C-4 ... F-4,G-4,H-4
Hilda St. ... B-3,B-4
Hillsborough Ave. ... A-4
Horatio St. ... J-2
Howard Ave. N. ... A-1 ... G-1,J-1
Hyde Park Ave. ... J-4
Hyde Park Ppl. ... J-4
Ida St. ... C-4-6
Ida St. E. ... C-8
Indiana Ave. ... D-2,D-3
Inman Ave. ... K-1,K-2
Jackson St. ... I-5
James St. ... D-5
Janette ... E-4,E-5
Jefferson St. ... E-5 ... F-5,I-5
John St. ... F-1
John F. Kennedy Blvd. ... J-1-3
Johnston ... C-2
Joseph St. ... F-1
Kay ... G-5
Kentucky Ave. ... D-1 ... D-2,D-4
Keyes Ave. ... E-4
Kinyon Ave. ... E-2
Lake Ave., E. ... D-6
Lamar St. ... G-5
Laurel St. ... H-1-5
Lee ... A-3
Lemon St. ... I-1
Link St. ... E-8
Linsey St. ... I-8
Long St. ... H-8
Louis St. ... F-1
Louisiana Ave. ... B-1 ... B-4-6
Lynn Ave. ... D-3
Machado St. ... E-8
Madison St. ... H-5,H-6,I-4
Magnolia Ave. ... J-3
Malory Ave. ... E-8
Maple Ave. ... J-8
Marconi St. ... I-8
Marguerite ... C-5
Marion St. ... H-5,I-5
Maritime Blvd. ... K-7
Maryland Ave. ... H-6
Massachusetts Ave. ... E-3,F-3
McBerry St. ... A-6-8
McKay St. ... H-6
Melville Ave. ... J-1
Mendenhall Dr. ... B-1
Meridian Ave. ... H-6,I-6
Mitchell Ave. ... E-5,G-5
Mohawk Ave. ... A-1
Moody Ave. ... J-1
Morgan St. ... E-4 ... Hp4,I-5
Munro St. ... C-3,H-3
Myrtle Ave. ... E-3
Nassau St. ... H-1
Nebraska Ave. ... D-5 ... J-5,I-6
Nevada AVe. ... F-3
New Orleans Ave. ... B-6,B-8
New Orleans St. ... B-5
Newport Ave. ... I-2
Newport Ave. S. ... K-2
Nick Nuccio Pkwy. ... H-6
Noel St. ... B-7
North A St. ... I-2,I-3
North B St. ... I-1,I-3
North Blvd. ... C-3,D-3,H-3
Oak Ave. ... G-4
Oakdale Ave. ... E-2
Oakwood Ave. ... J-8
Oceanview Pl. ... J-8
Ohio St. ... D-1,D-2,D-4
Ola Ave. ... C-4,G-4
Oregon Ave. ... G-2
Oregon Ave. N. ... G-2
Orient St. ... C-1,C-2
Orleans Ave. S. ... K-2
Osborne Ave. ... B-4,B-7
Owen St. ... F-7
Packwood Ave. ... K-2
Palm Ave. ... G-3,G-6
Palmetto St. ... G-1
Park ... A-2
Park Ave. ... F-3
Park La. ... B-2
Peninsula St. ... D-2
Penny St. ... H-7
Pierce St. ... I-5
Pine St. ... G-1
Plant Ave. ... J-4
Platt St. ... J-1,J-3,J-5
Plymouth St. ... E-2,E-3
Poplar Ave. ... E-2
Potter ... D-8
Ray St. ... G-4
Raymond Ave. ... H-6
Ridge AVe. ... D-2
River Blvd. ... B-3,D-1
River La. ... B-2
River St. ... C-3,I-4
River Pk. Dr. ... H-3
River Shore Dr. ... B-3
Rivershores Way ... B-2
Riverside Dr. ... C-2,F-2
Riverview Ave. ... O-1
Ridgewood ... E-2
Rodney ... D-7
Rome Ave. ... K-2
Rome Ave. N. ... B-2 ... E-1,G-2
Ross ... F-3,F-4
Royal Ct. ... H-4
St. Isabel St. ... D-1
St. Marco ... D-1
St. Mary ... C-1
St. Sophia St. ... D-1
Sanchez St. ... E-7
Saxon St. ... J-8
Scott St. ... H-4,H-5
Selma Ave. ... C-4
Seminole Ave. ... C-5
Shirley Dr. ... B-2
Shadow Lawn St. ... A-6,A-7
South Ave. ... B-4
South Blvd. ... K-3
Sparkman Ave. ... F-4,F-5
Spring St. ... H-4
Spruce Cove ... G-3
Spruce St. ... G-1
Stuart St. ... I-8
Suwannee Ave. ... C-5
Swann Ave. ... K-1
Tampa St. ... D-4,E-4 ... H-4,I-4
Thrace St. ... I-8
Twiggs St. ... I-4
Tyler St. ... H-4
Verne St. ... J-4
Violet St. ... B-4
Virginia Ave. ... D-1-4
Walton St. ... I-6
Warren Ave. ... E-2,E-4
Washington St. ... I-5
West St. ... E-2
Westland Ave. ... J-1
Westland Ave. N. ... A-1
Whiting St. ... I-5,I-6
Wilder Ave. ... A-4,A-5,A-8
Willow Ave. ... K-2
Wishart ... C-1
Woodlawn Ave. ... D-1 ... D-2,D-4
Woodrow Ave. ... F-3
Ybor St. ... E-7
York St. ... I-6
Zack St. ... I-4
2nd Ave. ... G-7
3rd Ave. ... G-5,G-7
4th Ave. ... G-7
5th Ave. ... G-7
6th Ave. ... G-7
7th Ave. ... G-4,G-6
8th Ave. ... G-6
9th Ave. ... G-6
9th St. ... C-6,D-6,E-6
10th St. ... C-6,E-6
11th Ave. ... F-6,F-7
11th St. ... D-6,E-6
12th Ave. ... F-6,F-7
12th St. ... E-6
13th St. ... C-6,E-6,I-6
14th Ave. ... F-6,F-7
14th St. ... C-7,H-6
15th Ave. ... F-6,F-7
15th St. ... C-7,G-7
16th St. ... B-7,C-7 ... E-7,G-7
17th Ave. ... F-6,F-7
17th St. ... C-7,F-7,G-7
18th Ave. ... F-6,F-8
18th St. ... B-7,D-7,G-7
19th Ave. ... E-6,E-8
19th St. ... B-7,G-7 ... I-7,J-7
20th Ave. ... E-6,E-8
20th St. ... C-7
20th St. N. ... F-8,J-8
21st Ave. ... E-6
21st St. ... C-8
22nd Ave. ... E-6,E-8
22nd St. N. ... G-8 ... H-8,I-8
23rd Ave. ... E-6,E-8
23rd St. ... D-8,G-8
24th Ave. ... E-6-8
24th St. ... E-8,F-8 ... G-8,J-8
25th Ave. ... E-6,E-8
25th St. ... E-8,F-8
26th Ave. ... E-4,E-6
27th Ave. ... D-7,D-8
28th Ave. ... D-6,D-8
29th Ave. ... D-7,D-8
31st Ave. ... D-5,D-6,D-8
32nd Ave. ... D-6
33rd Ave. ... D-6

POINTS OF INTEREST

Robles Park ... E-5

TUCSON

Adams ... C-2,C-6
Adelaide Dr. ... A-5
Alameda ... F-1
Alamogordo ... G-1
Alder Ave. ... E-2
Almartin ... A-8
Alta Vista ... A-4 ... A-6,A-8
Alto Pl. ... D-1
Alturas ... B-1 ... B-2,B-4
Amalia ... J-6
Amigo Ave. ... J-7
Anderson ... E-8
Arcade ... G-1
Arizona Ave. ... E-3,H-3
Armijo ... H-2
Arroyo Chico ... G-7
Ash Ave. ... E-2
Avenida de Palmas ... G-8
Avenidas El Captain ... B-3
Aviation Hwy. ... I-7
Balboa Ave. ... H-8
Barleycorn Ln. ... J-6
Bean Ave. ... F-3,H-3
Benedictine Dr. ... E-8
Bentley Ave. ... A-8 ... D-8,E-8,F-8
Bermuda Ave. ... A-8
Beverly Dr. ... H-7
Birdman ... A-2
Bonita ... F-1
Brady Ave. ... E-1
Branif Pl. ... F-2
Bristol Ave. ... I-7
Broadway E. ... F-3 ... F-5,F-7
Budmoore Terr. ... B-1
Butler ... G-7
Caddie ... F-4
Calle Adelanto ... C-1
Calle Belleza ... F-8
Calle Carlos Arruza ... G-2
Calle Central ... C-1
Calle Claravita ... E-8
Calle Corta ... F-8
Calle de Amistad ... F-8
Calle Ematus ... B-1
Calle Encanto ... F-8
Calle De Felicidad ... F-8
Calle Glorietta ... F-8
Calle Portal ... F-8
Calle Primorosa ... F-8
Calle Progresso ... C-1
Calle Retama ... C-1
Calle Santa Ana ... B-1
Calle Sierra ... C-1
Calle Ventura ... B-1
Camilla ... J-8
Camilla Blvd. ... B-8,D-8
Camino Campestre ... G-8
Campbell, N. ... A-6,E-6
Campbell Ave. S. ... H-6,K-6
Carrillo ... G-2
Carthay Cr. ... F-7
Castro Ave. ... B-2
Central ... G-1
Cerrito Pl. ... A-4
Chauncey ... C-4,C-6
Cherry Ave. ... B-5 ... C-5,D-5,I-5
Chico ... B-4
Chula Vista ... B-8
Church ... F-2
Circle E. ... A-4
Circle W. ... A-4
City Parkway ... I-8
Clark ... G-1
Coleta ... H-2
Congress ... F-2
Conner ... A-6
Contzen Ave. ... E-1
Convent ... G-2
Copper ... A-5,A-7
Corral ... G-2
Cortez Pl. ... A-3
Council ... F-2
Country Club Rd. N. ... B-8,E-8
Country Club Rd. S. ... H-8,I-8
Court ... F-2
Crest ... E-8
Crosby ... I-7
Croyden ... G-7
Curtis ... H-5
Cushing ... G-2
Cushman St. ... B-8
Dahil Rd. ... A-2
Davis ... E-1
Defensa ... H-2
Delong ... D-1
Devon ... G-7
Drachman ... D-2,D-4
Eastbourne ... G-8
Eastland ... H-6
Echols ... E-2
Edison ... B-4,B-6,B-7
Edison Terr. ... B-8
Elberta ... C-5
Elias ... G-2
Elm ... C-1,C-2 ... C-3,C-6
El Paso ... G-2
Espanol ... F-8
Estrella Ave. ... B-2
Euclid Ave. ... A-3,B-3 ... C-3,E-3,H-3,J-3
Exeter ... G-7
Factory ... H-5
Fairland Strav. ... I-6
Farmington Rd. ... I-1
Felisa Ave. ... A-1
Flanwill ... A-8
Florence ... A-1
Florence Dr. ... A-6,A-7
Flores ... B-2,B-3
Floirta ... F-4
Flower ... A-8
Fontana Ave. ... B-3
Forgeus Ave. ... C-7,D-7 ... F-7,H-7,J-7
Franklin ... F-1
Fremont ... A-5,D-4,F-4 ... G-4,H-4,J-4
Freeway ... D-1
Fritz Dr. ... A-2
Geneva ... F-8
Geronimo Ave. ... B-3
Gill Ave. ... A-6
Glenn ... A-1,A-3 ... A-5,A-8
Granada ... F-2
Grant Rd. E. ... B-3,B-7
Green ... H-1
Grossetta ... F-2
Hampton ... B-4,B-6,B-7
Harris ... F-8
Hawthorne ... E-6,E-8
Helen ... D-3,D-6
Henndon ... I-2
Herbert ... E-3,G-3,H-3
Highland Ave. ... D-5 ... G-5,I-5
Hoff ... F-3
Hopi ... A-3
Howard ... J-8
Hughes ... E-1,F-4
Incas ... A-3
Jacinto ... A-1,A-2
Jackson ... G-2
Jacobus Ave. ... F-3,H-3
Juanita ... B-6
Kelso ... A-1,A-2
Kennedy ... G-2
Keswick Cr. ... I-7
Kitchen ... E-1
Kramer Ave. ... C-6
Kroeger Lane ... H-1
La Mirada ... F-6
La Paz ... H-2
Lee ... C-2,C-6
Lester ... C-2,C-3 ... C-4,C-7
Linda ... F-1
Linden ... C-3,C-4,C-7
Lisbon ... F-8
Lord ... E-1
Loretta ... B-8
Los Altos Ave. ... B-8
Mabel ... D-3,D-6
Maciver ... B-1,B-2
Main ... D-1,G-2,H-2
Malvern ... G-5
Manchester ... G-7
Manlove ... G-4
Martin Ave. ... D-6,G-6 ... H-6,I-6,K-6
McFee ... J-7
McKey ... F-4
Mesa ... G-1
Meyer ... F-2,G-2
Miles ... G-4
Mill ... G-4
Miracle Mile Strip ... A-1,B-1,C-1
Miramonte ... J-8
Mitchell ... A-4,A-5,A-7
Montery ... K-6
Monte Vista ... A-7,A-8
Mt. Angel Dr. ... E-8
Mountain Ave. ... B-5 ... D-5,F-5,H-5,J-5
Mountain View ... I-7
Neff ... I-6
New ... I-4
Norris Ave. ... A-6,B-6 ... E-6,H-6,J-6
Norton ... G-7
Norton Ave. ... A-7,C-7 ... F-7,J-7
Ochoa ... G-2
Olive Rd. ... D-4
Olsen Ave. ... B-6 ... E-6,G-6,J-6
Osborne Ave. ... H-2
Oury ... E-1
P.F.E. Rd. ... I-6
Palmera ... H-2
Papago ... I-2
Paris Promenade ... B-1
Park Ave. N. ... B-4 ... D-4,E-4
Park Ave. S. ... H-4,K-4
Parkway Terr. ... G-7
Paseo Redondo ... F-2
Peak ... G-1
Pennington ... F-2
Perry ... D-2
Pima, E. ... C-8
Pinal Vista ... K-6
Plata ... C-2
Plaza del Encanto ... F-8
Plumer .. B-6,E-6,H-6,K-6
Pompeii Cr. ... F-8
Potter ... C-7
Queen Ave. ... D-2
Railroad Ave. ... H-3
Remmoc ... A-6
Rillito ... B-2
Rincon Ave. ... D-5
Roosevelt Pl. ... D-2
Rosales ... G-2
Roskruge ... G-7
Russell ... G-3
Sahuaro ... B-1
St. Marys Rd. ... E-1
Samaniego ... G-2
Santa Rita ... A-4
Santa Rita Ave. ... B-5 ... D-5,F-5,H-5,J-5
Sawtelle Ave. ... D-7 ... F-7,J-7
Scott ... F-2,G-2
Seneca ... C-3,C-4,C-7
Sentinel ... G-1
Sequoyah ... A-4
Shibell ... E-1
Short St. ... H-1
Sierra Vista Dr. ... C-6
Silver ... B-4,B-6,B-7
Silver Lake Rd. ... J-4
Simpson ... G-2
Smith Ave. ... F-7
Soto ... F-1
Speedway Blvd. E. ... D-4,D-8
Spring ... B-4,B-6,B-7
Star ... G-5
Stewart Ave. ... D-8 ... E-8,F-8
Stlovosa ... J-6
Stone Ave. N. ... B-2 ... C-2,E-2
Stone S. ... G-2
Stratford ... G-7,G-8
Sunland Vista ... K-6
Sylvia ... A-7,A-8
Telles ... F-2
Tero ... G-2
Terra Alta ... E-8
Tipton ... A-1
Toledo ... F-8
Toole ... F-3
Towner ... A-6,A-7
Towner Ave. ... A-8
Towner Ln. ... A-7
Treat Ave. ... D-7 ... E-7,H-7,I-7
Tucker Dr. ... C-7
Tucson ... J-7
Tucson Blvd. N. ... B-7,E-7
Tucson Blvd. S. ... H-7
Tyndall Ave. .. A-4,B-4,D-4 ... E-4,F-4,H-4,J-4
University Blvd. ... E-2
Van Alstine ... D-1
Vanover ... A-1
Vera Vista ... K-6
Verdugo ... H-1,I-1
Ventura ... C-2
Via Ecantadora ... A-4
Via Gaudalupe ... G-8
Via Golendrina ... G-8
Via Palos Verdes ... G-8
Via Rotonda ... F-7
Vine Ave. ... D-5,F-5,J-5
Warehouse Ave. ... H-5
Warren Ave. ... D-5 ... H-5,J-5
Warwick V. ... I-7
Washington ... F-2
Water ... B-3,B-4,B-7
Waverly ... C-3,C-6,C-7
Wilcox Dr. ... B-3
Williams ... D-1
Wilson Ave. ... A-6,A-7 ... E-7,G-6,J-6
Winchester V. ... I-7
Winmore ... I-8
Winnie Pl. ... B-4
Winsett ... G-5,G-6,G-7
York Pl. ... A-3
Yuma Rd. ... D-1
1st ... D-2,D-5,D-8
1st Ave. N. ... B-4,E-4
2nd ... E-2,E-5,E-8
2nd Ave. ... H-3
2nd Ave. N. ... B-3,D-3,F-3
3rd ... E-6,E-8
3rd Ave. ... H-3
3rd Ave. N. ... B-3,D-3,F-3
4th ... E-2,E-6,E-8
4th Ave. N. ... C-3,E-3
5th ... E-1,E-2,E-6
5th Ave. N. ... C-3,E-3
6th ... E-3,E-6
6th Ave. N. ... C-3,E-3
7th ... F-3,F-6
7th Ave. N. ... C-2,E-2
8th ... F-3,F-6
9th ... F-7
9th E. ... F-3,F-5
9th Ave. N. ... C-2,E-2
9th Ave. S. ... H-2
10th ... F-3,F-4,F-7
10th Ave. ... H-2
10th Ave. N. ... C-2,E-2
11th Ave. N. ... C-2,E-2
12th ... G-3,G-4,G-6
12th Ave. ... K-1
13th ... G-3,G-4,G-6
13th Ave. N. ... C-1,D-1
14th ... G-5,G-6
14th Ave. ... B-1,D-1
15th ... G-3,G-5,G-6
15th Ave. ... B-1,D-1
16th ... G-3,G-5,G-6
17th ... H-3,H-4,H-5,H-6
18th ... H-3,H-5,H-6
19th ... H-1,H-3,H-4,H-6
20th ... H-1,H-3,H-7
21st ... H-2,H-5,H-7
22nd E. ... I-4,I-7
22nd W. ... I-2
23rd ... I-2,I-4,I-7,I-8
24th ... I-1,I-2,I-4,I-8
25th ... I-2,I-4,I-8
26th ... I-4,I-8
27th ... I-4,I-8
28th ... I-5,I-8
29th ... I-8
30th ... J-4,J-8
31st ... J-4
32nd ... J-4,J-6
33rd ... J-1,J-4,J-6
34th ... J-1,J-6
35th ... K-4,K-6
36th ... K-1
36th E. ... K-4,K-5
37th ... K-4
38th ... K-4
39th ... K-4

SOUTH TUCSON

Herndon ... I-2
Rio Douche ... K-2
Rios ... I-2
2nd Ave. ... J-3
3rd Ave. ... J-3
4th Ave. S. ... J-3
5th Ave. S. ... J-3
6th Ave. S. ... J-3
7th Ave. S. ... J-2
8th Ave. S. ... J-2
9th Ave. S. ... J-2
10th Ave. S. ... J-2
11th Ave. S. ... J-2
26th ... I-2
27th ... I-2
28th ... I-2
29th ... J-2
30th ... J-3
31st ... J-3
32nd ... J-3
33rd ... J-2
34th ... J-2
35th ... K-3
36th E. ... K-2
37th ... K-2
38th ... K-2
39th ... K-2
40th ... K-2

POINTS OF INTEREST

Arizona National Guard Armory ... J-7
Armory Park ... G-3
City Hall ... F-2
Civic Center ... G-2
Downtown Airport ... K-5
Himmel Park ... E-7
Pacific Fruit Express Yards ... J-7
Pima County Hospital ... K-3
Ried Park ... H-8
Salpointe H.S. ... A-5
Southern Pacific Yards ... G-4
Tucson High ... F-3
Tucson High Athletic Field ... G-6
University of Arizona ... E-5
U. of A. Medical School & Hospital ... C-5

Scale of Miles
0 .1 .2 .3 .4 .5

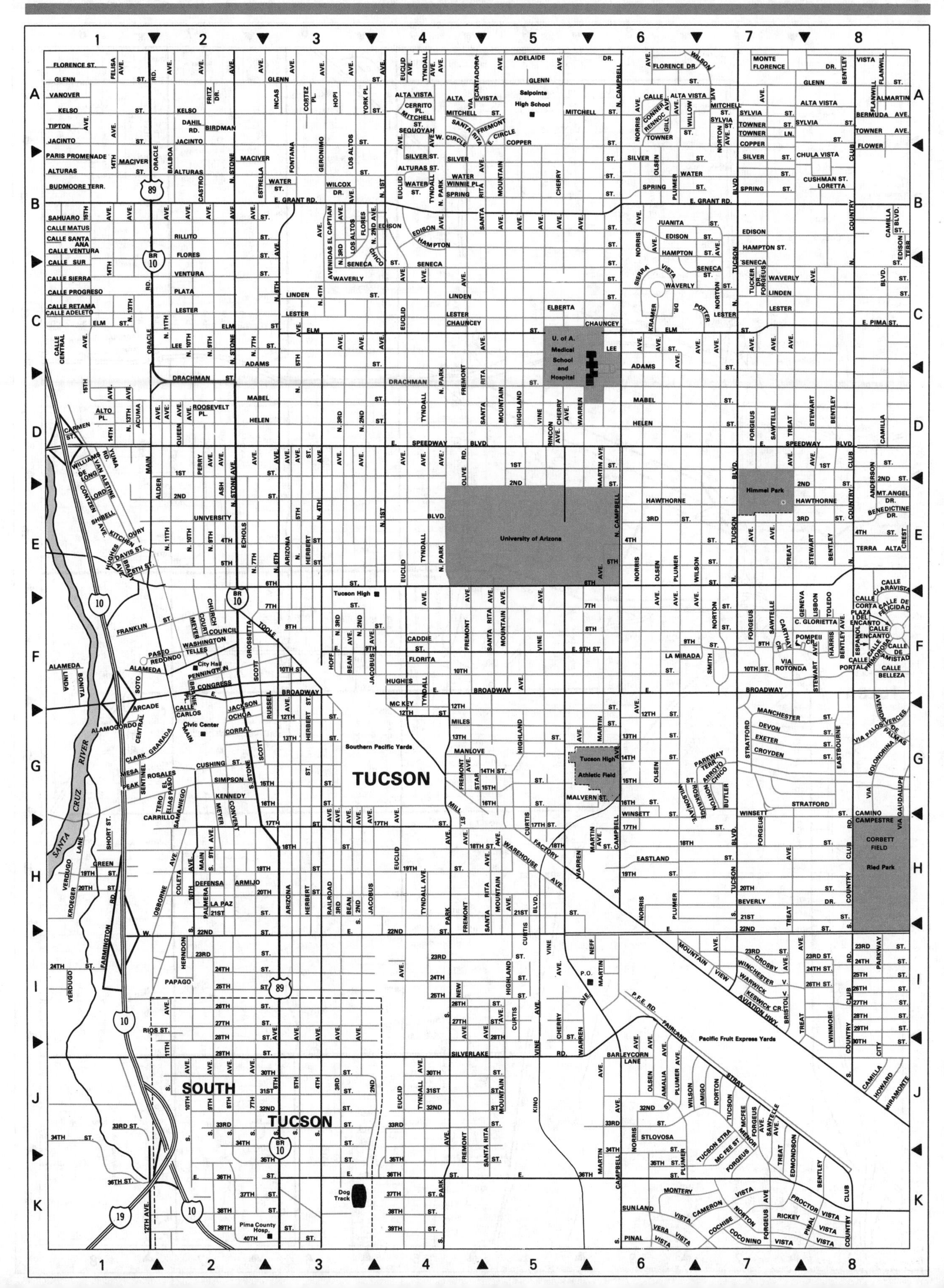

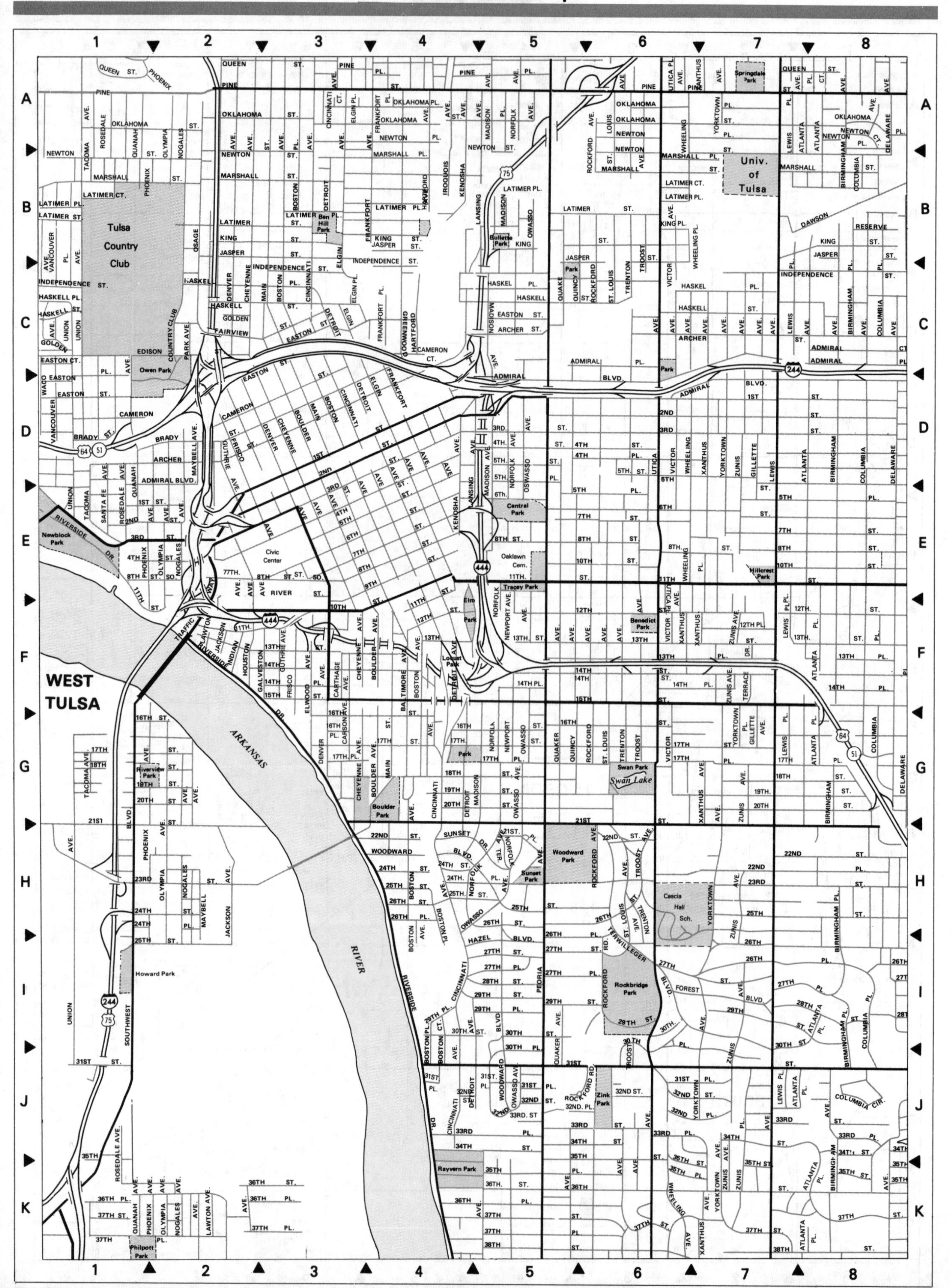
Scale of Miles
0
.2
.4
.6
N
1
2
3
4
5
6
7
8
A
B
C
D
E
F
G
H
I
J
K
Tulsa Country Club
Owen Park
Newblock Park
Ben Hill Park
Bullette Park
Springdale Park
Univ. of Tulsa
Civic Center
Central Park
Oaklawn Cem.
Tracey Park
Hillcrest Park
Elm Park
Benedict Park
Locust Park
Swan Park
Swan Lake
Riverview Park
Boulder Park
Woodward Park
Sunset Park
Cascia Hall Sch.
Rockbridge Park
Howard Park
Zink Park
Rayvern Park
Philpott Park
WEST TULSA
ARKANSAS
RIVER
RIVERSIDE DR.
ADMIRAL BLVD.
244
444
75
64
51

TULSA

- Admiral Blvd. C-4,C-6,D-1,D-2
- Admiral Ct. C-7
- Admiral Pl. C-7
- Archer St. C-4, C-6,D-1,D-2
- Atlanta Ave. D-7,K-7
- Atlanta Pl. B-7, F-7,J-7,K-7
- Baltimore Ave. F-4
- Birmingham Ave. B-7,D-7,J-7,K-7
- Birmingham Pl. C-7,H-7,I-7,K-7
- Boston Ave. C-3,F-4,G-4
- Boston Ct. I-4
- Boston Pl. B-3,I-4
- Boulder Ave. D-3,F-3,G-3
- Brady St. D-1,D-2
- Cameron D-1
- Cameron St. C-3,D-2
- Carson Ave. F-3
- Cheyenne Ave. C-2, D-3,F-3,G-3
- Cincinnati Ave. C-3,H-4,K-4
- College Ave. A-8,D-8,G-8
- Columbia Ave. D-7,K-7
- Columbia Cir. I-7
- Columbia Pl. C-8,F-8, I-7,K-7
- Country Club C-2
- Dawson Rd. B-7
- Delaware Ave. A-8, D-8,K-8
- Delaware Pl. C-8, G-8,J-8
- Denver Ave. C-2,D-2
- Detroit Ave. B-3, C-3,F-4,I-4,K-4
- Easton Ct. C-1
- Easton Pl. C-1
- Easton St. C-1-4
- Edison St. C-1,C-2
- Elgin Ave. B-3,C-3
- Elwood Ave. F-3,K-3
- Evanston Ave. A-8, D-8,I-8
- Evanston Pl. J-8
- Fairview St. C-2
- Florence Ave. D-8,J-8,K-8
- Florence Dr. H-8
- Florence Pl. C-8,G-8,K-8
- Frankfort Ave. B-3,C-3
- Frisco Ave. D-2
- Galveston Ave. F-2,K-2
- Gary Ave. B-8,D-8, G-8,J-8
- Gary Dr. I-8
- Gillette Ave. D-6
- Golden C-1
- Greenwood Ave. C-4
- Guthrie Ave. D-2
- Hartford Ave. C-4
- Haskell Pl. B-2,C-1
- Haskell St. C-1,C-2
- Hazel Blvd. H-4
- Houston Ave. F-2
- Independence St. B-1, B-2, B-7
- Indian Ave. F-2
- Iroquois Ave. B-4
- Jackson Ave. H-2,K-2
- Jasper St. B-2,B-7
- Kenosha Ave. B-4,D-4
- King St. B-2,B-3
- Lansing Ave. B-4,D-4
- Latimer Ct. B-1
- Latimer Pl. B-1,B-3
- Latimer St. B-1,B-2
- Lawton Ave. J-2
- Lewis Ave. D-7,K-7
- Lewis Pl. A-7,C-7,F-7
- Madison Ave. B-4,D-4,K-4
- Madison Pl. B-4,K-4
- Main St. C-2,D-3,F-3
- Marshall Pl. A-6
- Marshall St. B-1, B-2,B-5
- Maybell Ave. D-2, H-2,K-2
- Newport Ave. F-4
- Newton Ct. A-7
- Newton Pl. A-5,A-6
- Newton St. A-1,A-2,A-5
- Nogales Ave. E-2,G-2,J-2
- Norfolk Ave. B-5,F-4, G-4,K-4
- Oak Rd. K-6
- Oklahoma Pl. A-5, A-6,A-8
- Oklahoma St. A-2,A-5
- Olympia Ave. E-2, G-2,J-2
- Osage B-2
- Owasso Ave. B-5,H-4,K-5
- Park Ave. C-2
- Peoria Ave. H-5
- Phoenix Ave. E-1, G-1,J-1
- Pine Pl. A-3,A-4
- Pine St. A-2,A-6
- Quaker Ave. C-5,F-5
- Quanah Ave. D-1,J-1
- Queen St. A-2,A-7
- Quincy Ave. C-5, F-5,K-5
- Reserve B-7
- River St. E-2,E-3
- Riverside Dr. E-2,H-4
- Rockford Ave. C-5, F-5,G-5,K-5
- Rockford Rd. H-5,I-5
- Rosedale Ave. D-1,J-1
- St. Louis AVe. C-5, F-5,H-5,K-5
- Santa Fe Ave. D-1
- Southwest Blvd. I-1
- Sunset Dr. G-4
- Tacoma Ave. D-1,G-1
- Terwilleger Blvd. H-5,K-6
- Traffic Way D-2,E-2
- Trenton Ave. C-5, F-5,K-5
- Troost Ave. B-6, F-6,G-6,K-6
- Union Ave. D-1,H-1,K-1
- Utica Ave. D-6,K-6
- Utica Pl. A-6
- Vancouver Ave. D-1
- Victor Ave. D-6,F-6,K-6
- Waco Ave. C-1
- Wheeling Ave. D-6, J-6,K-6
- Woodward Blvd. G-3, G-4,I-4
- Xanthus Ave. D-6, G-6,J-6
- Yorktown Ave. D-6, H-6,I-6,J-6
- Zunis Ave. D-6,G-6, I-6,J-6,K-6
- Zinus Pl. J-6
- 1st St. C-6,C-7,D-1,D-3
- 2nd St. D-1,D-3,D-6
- 3rd St. D-3,D-4,D-6,E-1
- 4th Pl. D-5
- 4th St. D-3,D-4,D-5,E-1
- 5th Pl. D-4,D-5,D-7
- 5th St. D-3,D-4,D-6
- 6th St. D-4,D-6,E-3
- 7th St. D-5,E-3,E-7
- 8th St. E-3,E-4,E-5,E-7
- 8th St., S. E-1,E-2
- 9th St. E-3,E-4
- 10th St. E-3,E-5,E-7
- 11th St. E-1,E-4,E-6
- 12th St. E-4,E-5
- 13th Pl. F-4,F-6,F-7
- 13th St. E-2,E-4,E-6
- 14th Pl. F-2,F-7
- 14th St. F-2,F-5
- 15th St. F-2,F-5
- 16th St. F-1,F-3,F-5,F-8
- 17th Pl. F-4,F-6,F-8
- 17th St. F-1,F-3,F-6
- 18th St. F-1,F-2,F-4
- 19th St. G-1,G-4,G-8
- 20th St. G-1,G-4,G-6
- 21st Pl. G-8
- 21st St. G-1,G-5
- 22nd Pl. G-6,G-7
- 22nd St. G-3,G-7
- 23rd St. G-1,G-6,G-7,H-8
- 24th Pl. H-1,H-2
- 24th St. G-3,G-4,H-1,H-2
- 25th St. G-3,H-1,H-4, H-5,H-6
- 26th Pl. H-3,H-5,H-6,H-8
- 26th St. H-3,H-4,H-5,H-6
- 27th St. H-4-8
- 28th St. H-4,H-7,H-8
- 29th Pl. H-4
- 29th St. H-4-6,I-5
- 30th Pl. I-4,I-5
- 30th St. H-6,I-4,I-7
- 31st Pl. I-5,I-6
- 31st St. I-1,I-5
- 32nd Pl. I-6
- 32nd St. I-5,I-6,I-8
- 33rd Pl. I-4,I-6,I-7
- 33rd St. I-5,I-7,I-8
- 34th St. I-4-6,J-7-8
- 35th Pl. J-4,J-6,J-8
- 35th St. J-1,J-5-8
- 36th Pl. J-1,J-2,J-4
- 36th St. J-2,J-5
- 37th Pl. J-1,J-2,J-4
- 37th St. J-1,J-4,J-6,J-7
- 38th Pl. J-4
- 38th St. J-4,J-7
- 39th St. J-1,K-4,K-8
- 41st Pl. K-5
- 41st St. K-1,K-5
- 42nd Pl. K-5
- 42nd St. K-4,K-6,K-8
- 43rd St. K-1,K-4-6
- 44th St. K-7,K-8

POINTS OF INTEREST

- Archer Park C-8
- Ben Hill Park B-3
- Benedict Park E-6
- Boulder Park G-3
- Bullette Park B-4
- Central Park D-4,D-5
- Elm Park E-4
- Hillcrest Park E-7
- Howard Park H-1
- Locust Park F-4
- Newblock Park E-1
- Owen Park C-1
- Philpott Park J-8
- Rayvern Park J-4
- F. H. Reed Park K-1
- Riverview Park F-1
- Rockbridge Park H-5
- Springdale Park A-6
- Sunset Park G-5
- Swan Lake G-5
- Swan Park F-5,F-6
- Tracey Park E-4,E-5
- Woodward Park G-5
- Zink Park I-5

WASHINGTON, D.C.

- A St., N.E. F-8
- A St., S.E. F-8
- Adams Dr. F-4
- Army Navy Dr. J-1
- Bancroft Pl. A-1
- Bates St. B-6,B-7
- Belmont Rd. A-1
- C St., N.E. E-8
- C St., N.W. E-2,E-6
- C St., S.W. F-4,F-6
- California Pl. A-1
- California St. A-2,A-3
- Canal St. F-7,I-7
- Capitol St. H-7,E-7
- Caroline St. A-3
- Carroll St. F-8
- Carrollburg Pl. H-7
- Church St. B-3
- Columbia St. B-5
- Connecticut Ave. B-2
- Constitution Ave. E-2
- Corcoran St. B-3
- Cushing Pl. H-7
- D St., N.E. E-8
- D St., N.E. E-8
- D St., N.W. E-5
- D St., S.W. G-6
- Decatur Pl. B-1
- Defrees D-7
- Delaware AVe. H-6
- Duddington Pl. G-8
- DuPont Circle B-2
- E St., N.E. E-8
- E St., N.W. E-4
- E St., S.E. G-8
- Eads St. K-1
- East Capitol St. F-8
- Eaton Rd. K-8
- Eckington Pl. B-8
- Elm St. A-6
- F St., N.E. E-8
- F St., N.W. E-2,E-4
- Fenton Pl. C-7
- Fern St. K-1
- Firth Rd. K-8
- Flager St. A-6
- Florida Ave. A-2,A-6,B-6
- Franklin St. B-6
- French St. B-5
- George Washington Memorial Pkwy. H-1,J-3
- G Pl. D-7
- G St., N.E. D-8
- G St., N.W. D-2,D-4
- G St., S.E. G-8
- G St., S.W. G-5
- H St., N.E. D-8
- H St., N.W. D-2,D-4
- H St., S.W. G-6
- Half St. J-7
- Howard Rd. J-8
- Howison Pl. H-7
- I St., N.E. D-8
- I St., N.W. D-2,D-5
- I St., S.E. H-8
- I St., S.W. H-6
- Independence Ave. F-5
- Indian St. E-5
- Ivy St. G-7
- Jefferson Davis Hwy. K-1
- Jefferson Dr. F-4
- K St., N.E. D-8
- K St., N.W. D-2,D-6
- K St., S.E. H-7
- K St., S.W. H-6
- Kalorama Rd. A-1
- Kirby St. C-6
- L St., N.E. C-8
- L St., N.W. D-2,D-5
- L St., S.E. H-7
- L St., S.W. H-6
- Lafayette Square D-3
- Lincoln Rd. B-7
- Logan Circle B-4
- Louisiana Ave. E-7
- M St., N.E. C-8
- M St., N.W. C-2,C-5
- M St., S.E. H-8
- M St., S.W. H-6
- Madison Dr. F-4
- Maine Ave. H-5
- Marion St. B-5
- Maryland Ave. E-8,F-6
- Massachusetts Ave. C-3,D-6
- Morris St. C-8
- Mt. Vernon Pl. D-5
- Myrtle St. D-7
- N St., N.W. C-2,C-5
- N St., S.W. H-6
- Neal Pl. B-8,C-6
- New Hampshire Ave. B-3,C-2
- New Jersey Ave. C-6,G-7
- New York Ave. C-6
- North Carolina Ave. G-8
- O St., N.W. C-1,C-2,C-5
- O St., S.W. I-6
- Oakdale Pl. A-6
- Ohio Dr. G-2,I-4
- P St., N.W. B-2,B-5
- P St., S.W. I-6
- Patterson St. C-7
- Pennsylvania Ave. D-2,E-5
- Phelps Pl. A-2
- Pierce St. C-7
- Portner Pl. A-4
- Potomac Ave. I-7
- Q St., N.W. B-2,B-5
- Q St., S.W. I-6
- Quincy St. B-7
- R St., N.E. B-8
- R St., N.W. B-3,B-7
- R St., S.W. I-7
- Randolph Pl. B-7
- Rhode Island Ave. B-6,C-3
- Ridge St. C-6
- Riggs Pl. B-3,B-4
- Rock Creek and Potomac Pkwy. E-1
- S St., N.E. B-8
- S St., N.W. B-3,B-7
- S St., S.W. J-7
- Scott Circle C-3
- Seaton Pl. A-7
- Sheridan Circle B-1
- South Carolina Ave. G-8
- Stanton Square E-8
- Stevens Rd. K-8
- Swann St. A-3
- T St., N.E. A-8
- T St., N.W. A-3,A-7
- T St., S.W. J-7
- Tenton Pl. A-1
- Thomas Circle C-4
- Todd Pl. A-7
- Tracey Pl. A-1
- U St., N.E. A-8
- U St., N.W. A-3,A-6,A-7
- U St., N.E. A-8
- V St., N.E. A-8
- V St., N.W. A-3,A-6,A-7
- V St., S.W. J-7
- Van St. H-7
- Vermont Ave. C-4
- Vrginia Ave. E-2,G-6
- Wallach Pl. A-4
- Warner St. B-6
- Washington Circle D-1
- Washington Dr. F-4
- Water St. J-7
- Waterside Dr. B-1
- Westminster St. A-5
- Willard St. A-3
- Wyoming Pl. A-2
- 1st St., N.E. C-7
- 1st, N.W. E-7
- 1st St., S.E. I-7
- 1st, S.W. J-7
- 2nd St., N.E. F-8
- 2nd St., N.W. B-6
- 2nd St., S.E. H-8
- 2nd St., S.W. G-6,J-6
- 3rd St., N.E. A-8,E-8
- 3rd St., N.W. D-6,E-6
- 3rd St., S.E. H-8
- 4th St., N.E. A-8,E-8
- 4th St., N.W. A-6,E-6
- 4th St., S.E. H-8
- 4th St., S.W. F-6,G-6
- 5th St., N.E. A-8,E-8
- 5th St., N.W. A-6,E-6
- 5th St., S.E. H-8
- 6th St., N.W. E-6
- 6th St., S.W. H-6
- 6½ St., N.W. C-5
- 7th St., N.W. E-5
- 7th St., S.W. H-5
- 8th St., N.W. E-5
- 9th St., N.W. E-5
- 9th St., S.W. G-5
- 10th St., N.W. E-5
- 10th St., S.W. G-5
- 11th St., N.W. E-5
- 11th St., S.W. G-5
- 12th St., N.W. E-4
- 12th St., S.W. G-4
- 13th St., N.W. E-4
- 13th St., S.W. G-4
- 14th St., N.W. E-4
- 15th St., N.W. E-4
- 16th St., N.W. D-3
- 17th St., N.W. E-3
- 18th St., N.W. E-3
- 19th St., N.W. E-2
- 20th St., N.W. E-2
- 21st St., N.W. E-2
- 22nd St., N.W. D-2
- 24th St., N.W. D-1
- 25th St., N.W. D-1
- 26th St., N.W. D-1
- 27th St., N.W. C-1

POINTS OF INTEREST

- Arlington Memorial Bridge F-1
- Buzzard Point K-7
- Carnegie Institute B-3,B-4
- Center Hwy. Bridge H-3
- Douglass Bridge I-7,J-8
- DuPont Circle B-2
- East Potomac Park I-5
- Fort McNair J-6
- George Manson Mem. Bridge H-3
- George Washington University D-2
- Greenleaf Point K-6
- Jefferson Memorial G-3
- Kennedy Center E-1
- Lady Bird Johnson Park H-1
- Lafayette Square D-3
- Lincoln Memorial F-1
- Logan Circle B-4
- Mt. Vernon Pl. D-5
- Reflecting Pool F-2
- Rochambeau Bridge H-3
- Rock Creek Park C-1
- Scott Circle C-3
- Sheridan Circle B-1
- Stanton Square E-8
- The Mall F-5
- The White House D-3
- Thomas Circle C-4
- U. S. Capitol F-7
- Washington Circle D-1
- Washington Monument F-3
- Washington National Airport K-3
- Waterfowl Sanctuary J-2
- West Potomac Park F-2

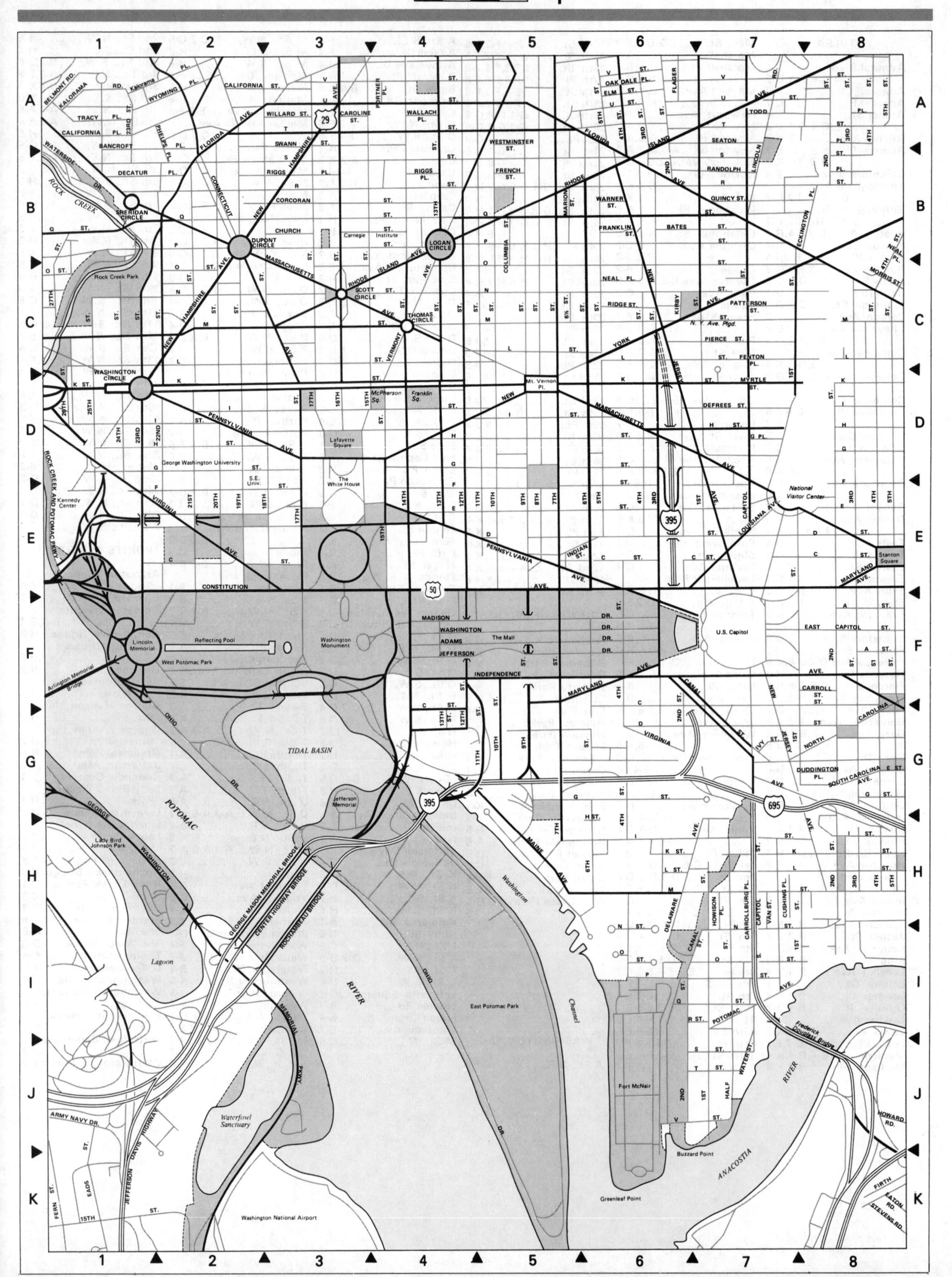
Scale of Miles
0 .1 .2 .3 .4
N
Rock Creek Park
Sheridan Circle
Dupont Circle
Scott Circle
Logan Circle
Thomas Circle
Washington Circle
Carnegie Institute
Mt. Vernon Pl.
McPherson Sq.
Franklin Sq.
Lafayette Square
The White House
George Washington University
S.E. Univ.
Kennedy Center
Rock Creek and Potomac Pkwy.
N. Y. Ave. Plgd.
National Visitor Center
Stanton Square
U.S. Capitol
The Mall
Washington Monument
Lincoln Memorial
Reflecting Pool
West Potomac Park
Arlington Memorial Bridge
Tidal Basin
Jefferson Memorial
East Potomac Park
Washington Channel
Potomac River
Lady Bird Johnson Park
Lagoon
Waterfowl Sanctuary
Washington National Airport
George Mason Memorial Bridge
Center Highway Bridge
Rochambeau Bridge
Memorial Pkwy.
Jefferson Davis Highway
Army Navy Dr.
Fort McNair
Buzzard Point
Greenleaf Point
Anacostia River
Frederick Douglass Bridge
Howard Rd.
Firth
Eaton Rd.
Stevens Rd.
Pennsylvania Ave.
Constitution Ave.
Independence Ave.
Massachusetts Ave.
New Hampshire Ave.
Connecticut Ave.
Rhode Island Ave.
New York Ave.
New Jersey Ave.
Virginia Ave.
Maryland Ave.
Louisiana Ave.
Indian Ave.
Maine Ave.
Delaware Ave.
Canal St.
South Carolina Ave.
North Carolina Ave.
East Capitol St.
Madison Dr.
Washington Dr.
Adams Dr.
Jefferson Dr.
Ohio Dr.
George Washington Pkwy.
29
50
395
695

INDEX

To The United States

Index to Canadian Cities and Towns on Pages 8-9.
Index to Mexican Cities and Towns on Page 11.

ALABAMA

Page 13

Population: 3,893,888
Capital: Montgomery
Land Area: 50,767 sq. mi.

ALASKA

Page 14

Population: 401,851
Capital: Juneau
Land Area: 570,883 sq. mi.

Donnelly D-7
Dot Lake D-8
Douglas G-9
Dutch Harbor F-2
Eagle D-8
Eagle River E-6
Inset Map B-2
Edna Bay H-9
Eek E-4
Egavic C-5
Elfin Cove G-9
Emmonak C-4
English Bay F-6
Eureka D-7
Fairbanks D-7
False Pass F-3
Farewell D-6
Flat D-5
Fort Yukon C-8
Gakona E-7
Galena C-6
Gambell C-3
Glen Alps E-6
Glennallen E-7
Golovin C-5
Goodnews Bay E-4
Gordon B-8
Grayling D-5
Gulkana E-7
Gustavus G-9
Haines F-9
Hawk Inlet G-9
Healy D-7
Homer E-6
Inset Map C-1
Hoonah G-9
Hooper Bay D-3
Hope E-6
Inset Map B-2
Houston A-2
Hughes C-6
Huslia C-6
Hydaburg H-9
Itulilick E-5
Juneau G-9
Kachemak E-6
Kake G-9
Kaktovik B-8
Kalla C-6
Kaltag C-5
Karluk F-5
Kashegelok E-5
Kasilof E-6
Inset Map B-1
Kaskanak E-5
Katalla F-7
Kenai B-1
Ketchikan H-10
Kiana B-5
Kinegnak E-4
King Cove F-3
King Salmon F-5
Kipnuk E-4
Kivalina B-5
Klawock H-9
Knik B-2
Kobuk C-6
Kodiak F-6
Kokrines C-6
Koliganek E-5
Kotlick C-4
Kotzebue B-5
Koyuk C-5
Koyukuk C-5
Kuliuk F-5
Kwethluk D-4
Kwigillingok E-4
Larsen Bay F-5
Latouche F-7
Levelock E-5
Lime Village E-5
Livengood C-7
Long D-6
Lower Tonsina E-7
Manley Hot Springs D-6
Manokotak E-4
McCarthy E-8
McGrath D-5
McKinley Park D-7
Medfra D-6
Mekoryuk D-3
Mentasta Lake E-7
Metlakata H-10
Minto D-7
Moose Pass B-2
Mount Edgecumber G-9
Mountain Village D-4
Naknek F-5
Napakiak D-4
Napaskiak E-4
Nenana D-7
New Stuyahok E-5
Newhalen E-5
Nikolai D-6
Nikolski F-1
Noatak B-5
Nogamut E-5
Nome C-4
Nondalton E-5
Noorvik B-5
North Pole D-7
Northeast Cape C-4
Northway E-8
Nulato C-5
Old Harbor F-5
Ooliktok A-7
Ouzinkie F-6
Palmer E-7
Inset Map A-2
Pedro Bay E-5
Pelican G-9
Perryville F-4
Petersburg E-6
Pilgrim Springs C-4
Pilot Point F-4
Inset Map D-4
Platinum E-4
Point Hope A-5
Point Lay A-5
Porcupine F-9
Port Alexander G-9
Port Lions F-6
Portlock F-6
Prudhoe Bay B-8
Quinhagak E-4
Rampart C-7
Rampart House C-8
Richardson D-7
Ruby C-6
Russian Mission D-4
Sagwon B-7
Salchaket D-7
Sand Pt F-4
Saroonga C-3
Selawik B-5
Seward E-6
Inset Map C-2
Shageluk D-5
Shaktoolik C-5
Shishmaref B-4
Shungnak B-6
Sinuk C-4
Sitka G-9
Skagway F-9
Sleetmute D-5
Soldotna E-6
Inset Map B-1
Squaw Harbor G-3
St. George E-2
St. Marys D-4
St. Michael C-5
St. Paul E-2
Stebbins C-4
Stevens Village C-7
Stony River D-5
Stuyahok D-5
Susitna B-2
Takotna D-5
Talkeetna E-7
Inset Map A-1
Tanalian Pt E-5
Tanana C-6
Tatitlek E-7
Teller C-4
Tetlin E-8
Tetlin Junction E-8
Togiak E-4
Tok D-8
Toksook Bay D-4
Tonsina E-7
Tuluksak D-4
Tununak D-4
Tyonek E-6
Inset Map B-1
Ugashik F-5
Umiat B-7
Unalakleet C-5
Unalaska F-2
Valdez E-7
Venetie C-8
Wainwright A-6
Wales B-4
Wasila A-2
Wasilla E-7
White Mountain C-5
Whittier B-2
Willow E-6
Inset Map A-1
Wiseman C-7
Wrangell G-10
Yakutat F-8

ARIZONA
Pages 16-17

Population: 2,718,215
Capital: Phoenix
Land Area: 113,508 sq. mi.

Adobe Pg. 171, B-4
Aguila J-3
Ajo L-4
Ak Chin K-5
Alpine J-10
Amado N-7
Anegam L-5
Angell F-7
Apache Jct. K-6
Arivaca Jct. N-7
Arizona City L-6
Arlington K-4
Artesa M-6
Ash Fork F-5
Ashurst K-9
Avondale K-5
Vicinity Pg. 171, E-2
Bagdad G-4
Bapchule K-6
Beaverhead J-10
Bellemont F-6
Benson M-8
Bisbee N-9
Black Canyon City H-5
Bonita L-9
Bouse J-2
Bowie M-10
Brenda J-2
Buckeye K-5
Bullhead City F-2
Bylas K-9
Cameron E-7
Camp Verde H-6
Carefree J-6
Carmen N-7
Carrizo J-8
Casa Grande K-6
Vicinity Pg. 171, E-2
Catalina L-7
Cave Creek J-6
Cedar Creek J-9
Cedar Ridge D-7
Cedar Springs F-8
Central L-9
Chambers F-9
Chandler K-6
Vicinity Pg. 171, F-7
Chandler Heights K-6
Chilchinbito D-9
Childs L-4
Chinle E-10
Chino Valley G-5
Chloride F-2
Christmas K-8
Christopher Creek H-7
Chuichu L-6
Circle City J-4
Clarkdale G-5
Clay Spring H-8
Claypool K-7
Clifton K-10
Cochise M-9
Coconino Pg. 126, E-3
Colorado City C-4
Concho H-9
Congress H-4
Continental M-7
Coolidge K-6
Cordes Jct. H-5
Cornville G-6
Cortaro L-7
Cottonwood G-6
Covered Wells M-5
Cow Springs D-8
Cross Canyon E-10
Currys Corner
Pg. 171, B-6
Curtis M-8
Dateland K-3
Desert View E-6
Dewey H-5
Dilkon F-8
Dinnehotso D-9
Dolan Springs F-2
Dome L-2
Dos Cabezas M-10
Douglas N-10
Dragoon M-9
Dudleyville L-8
Duncan L-10
Eager H-10
East Flagstaff
Pg. 126, B-3
Eden K-9
Ehrenberg J-2
Elfrida N-9
Eloy L-6
Emery Park Pg. 195, H-3
Fairbank N-8
Flagstaff F-6
Vicinity Pg. 126
Florence K-7
Florence Jct. K-7
Fort Apache J-9
Fort Defiance E-10
Fort Grant L-9
Fort McDowell J-6
Fort Thomas K-8
Fort Valley Pg. 126, A-1
Fountain Hills J-6
Francisco Grande K-5
Franklin L-10
Fredonia C-5
Gadsden L-1
Ganado F-9
Gila Bend K-4
Gilbert K-6
Glendale J-5
Vicinity Pg. 171, C-3
Globe K-8
Golden Shores G-2
Goldroad F-2
Goodyear K-5
Vicinity Pg. 171, E-1
Grand Canyon E-5
Gray Mountain E-7
Greasewood (Lower) F-9
Green Valley M-7
Greer J-10
Guadalupe K-5
Vicinity Pg. 171, F-5
Hackberry F-3
Hannigan Meadow K-10
Harcuvar J-3
Hawley Lake J-9
Hayden K-7
Heber H-8
Higley K-6
Hillside H-4
Holbrook G-8
Hope J-3
Hotevilla E-8
Houck F-10
Huachuca City N-8
Humboldt H-5
Indian Wells F-8
Jacob Lake D-5
Jaynes Pg. 195, D-2
Jerome G-5
Johnson M-8
Joseph City G-8
Katherine F-2
Kayenta D-8
Keams Canyon E-9
Kearney K-7
Kelvin K-7
Kingman F-2
Kirkland H-4
Kirkland Jct. H-5
Kohls Ranch H-7
Kyrene Pg. 171, F-6
La Palma L-6
Lake Havasu City G-2
Lake Montezuma G-6
Lakeside H-9
Laveen Pg. 171, E-3
Leupp F-7
Litchfield J-5
Littlefield C-3
Littletown Pg. 195, H-5
Lukeville M-4
Lupton F-10
Mammoth L-8
Many Farms D-10
Marana L-6
Marble Canyon D-6
Maricopa K-5
Mayer H-5
McGuireville G-6
McNeal N-9
Meadview E-2
Mesa J-6
Mexican Water C-9
Miami K-7
Moenkopi E-7
Moqui E-5
Morenci K-10
Morristown J-4
Mountainaire G-6
Vicinity Pg. 126, E-2
Mt. View M-7
Munds Park G-6
Naco N-9
Navajo G-10
Nelson F-3
New River J-5
Nicksville N-8
Nogales N-7
North Rim E-6
Nutrioso J-10
Oatman G-2
Ocotillo K-6
Olberg K-6
Old Oraibi E-8
Oracle L-7
Oracle Jct. L-7
Oraibi E-8
Oro Valley L-7
Overgaard H-8
Page C-7
Palm Springs K-6
Palo Verde K-4
Palominas N-8
Pantano M-8
Paradise City
Pg. 171, B-5
Paradise Valley J-5
Parker H-2
Parks F-6
Patagonia N-7
Paul Spur N-9
Paulden G-5
Payson H-6
Peach Springs F-4
Peeples Valley H-4
Peoria J-5
Vicinity Pg. 171, C-3
Peridot K-8
Phoenix J-6
Vicinity Pg. 171
Detailed City Pg. 272
Picacho L-6
Pima L-9
Pine H-6
Pine Springs F-5
Pinedale H-8
Pinetop H-9
Pirtleville N-9
Polacca E-8
Pomerene M-8
Prescott G-5
Prescott Valley G-5
Punkin Center J-7
Quartzite J-2
Queen Creek K-6
Quijotoa M-5
Randolph K-6
Red Lake D-7
Red Rock L-7
Rillito L-7
Riverside K-7
Riviera F-2
Rock Point D-10
Rock Springs J-5
Roosevelt J-7
Rough Rock D-9
Round Rock D-10
Rye H-7
Sacaton K-6
Safford L-9
Sahuarita M-7
Saint David M-9
Saint Johns H-10
Saint Michaels F-10
Salome J-3
San Carlos K-8
San Luis L-1
San Luis L-5
San Manuel L-8
San Simon M-10
Sanders F-10
Sasabe N-6
Scottsdale J-6
Vicinity Pg. 171, D-6
Seba Dalkai F-8
Second Mesa E-8
Sedona G-6
Seligman F-4
Sells M-5
Sentinel K-3
Sheep Hill Pg. 126, A-4
Shonto D-8
Show Low H-9
Shumway H-9
Sierra Vista N-8
Silver Bell N-6
Snowflake H-9
Solomon L-9
Somerton L-1
Sonoita N-7
South Tuscon M-7
Vicinity Pg. 195, G-2
Springerville H-10
Stanfield L-6
Star Valley H-7
Stargo K-10
Strawberry H-6
Sun City J-5
Vicinity Pg. 171, C-2
Sun City West J-5
Sun Lake K-6
Sun Valley G-9
Sunflower J-6
Sunizona M-9
Sunrise F-7
Sunsites M-9
Superior K-7
Suprise J-5
Tacna L-2
Taylor H-9
Teec Nos Pas D-10
Tempe K-6
Vicinity Pg. 171, E-6
Tes Nez Iha C-9
Thatcher L-9
The Gap D-7
Three Points (Robles)
M-6
Tolleson K-5
Vicinity Pg. 171, E-2
Tombstone N-9
Tonalea D-7
Tonopah J-4
Topock G-2
Tortilla Flat J-7
Tracy M-5
Truxton F-3
Tsaile D-10
Tsegi D-8
Tuba City E-7
Tubac N-7
Tucson M-7
Vicinity Pg. 195
Detailed City Pg. 297
Tumacacori N-7
Tusayan E-5
Vail M-8
Valentine F-3
Valle F-5
Vernon H-9
Vicksburg J-2
Walnut Canyon
Pg. 126, C-3
Wellton L-2
Wenden J-3
Whiteriver J-9
Why M-4
Wickenburg J-4
Wikieup G-3
Wilhoit H-5
Willcox M-9
Williams F-5
Window Rock F-10
Winkelman K-8
Winona F-7
Winslow H-7
Wintersburg K-4
Wittmann J-4
Yampai F-4
Yarnell H-4
Young H-7
Youngtown J-5
Vicinity Pg. 171, C-2
Yucca G-2
Yuma L-1

ARKANSAS
Page 15

Population: 2,286,435
Capital: Little Rock
Land Area: 52,078 sq. mi.

Alpena A-4
Altheimer E-6
Amity E-4
Arkadelphia E-4
Arkansas City F-7
Arkoma Pg. 127, D-2
Ash Flat A-7
Ashdown F-3
Athens E-3
Atkins C-4
Augusta C-7
Bald Knob C-7
Banks F-5
Barling C-2
Vicinity Pg. 127, D-4
Basne Pg. 127, D-2
Batesville B-6
Baucum Pg. 143, C-7
Bay B-8
Bearden F-5
Beebe D-6
Benton D-5
Bentonville A-3
Bergman A-4
Berryville A-3
Bismark E-4
Black Rock B-7
Blue Springs D-4
Bluff City F-4
Blytheville B-9
Booker Pg. 143, B-6
Booneville C-3
Boxley B-4
Bradley G-3
Brashears B-3
Brinkley D-7
Browns Gin Pg. 143, E-6
Bryant D-5
Buell Pg. 127, D-3
Bull Shoals A-5
Cabot D-6
Calico Rock B-6
Calion F-5
Camden F-4
Carlisle D-6
Carthage E-5
Cash B-8
Caulksville C-3
Cave City B-7
Cedars Pg. 127, E-1
Cedarville C-2
Centerville C-4
Charity Pg. 143, C-6
Charleston C-3
Cherry Valley C-8
Clarksville C-4
Clinton C-5
Conway C-5
Corning A-8
Cotton Plant D-7
Cove E-2
Crossett G-6
Crossroads E-7
Crows D-5
Cushman B-6
Dalark E-4
Damascus C-5
Danville D-4
Dardanelle C-4
DeQueen E-2
DeWitt E-7
Decatur A-2
Delight E-3
Deluce E-3
Dermott F-7
Des Arc D-7
Dierks E-3
Dodridge G-3
Dora Pg. 127, A-1
Douglas Corner
Pg. 143, D-2
Dover C-4
Dumas F-7
Earle C-8
El Dorado G-5
Elaine E-8
Emerson G-4
England D-6
Enterprise Pg. 127, E-2
Eudora G-8
Eureka Springs A-3
Falcon F-4
Fallsville B-4
Farmington B-2
Fayetteville B-3
Fisher C-7
Flippin A-5
Florence F-6
Fordyce F-5
Foreman F-2
Forrest City D-8
Fouke G-3
Fountain Hill F-6
Ft. Smith C-2
Vicinity Pg. 127
Furry Pg. 127, A-4
Galloway Pg. 143, B-7
Garland G-3
Geneva Pg. 143, D-5
Gentry A-2
Gillett E-7
Gilmore C-8
Glenwood E-3
Gosnell B-9
Gould E-7
Grady E-6
Grapevine E-5
Gravette A-2
Green Forest A-4
Greenbrier C-5
Greenwich Village
Pg. 191, K-7
Greenwood C-2
Greers Ferry C-6
Guiford Pg. 143, E-3
Gurdon E-4
Hackett C-2
Hagarville C-4
Halstead Pg. 143, D-2
Hamburg G-6
Hampton F-5
Hardy A-7
Harrisburg C-8
Harrison A-4
Hartford D-2
Haskell E-5
Hazen D-7
Heber Springs C-6
Helena D-8
Hermitage F-6
Higgins Pg. 143, E-5
Holly Grove D-7
Hope F-3
Horatio E-2
Horseshoe Bend A-6
Hot Springs D-4
Hoxie B-7
Hughes D-8
Humnoke E-7
Humphrey E-6
Huntsville B-3
Huttig G-6
Jacksonville D-6
Jasper B-4
Jonesboro B-8
Judsonia C-6
Junction City G-5
Keiser B-9
Kensett C-7
Kingsland F-5
Lake City B-8
Lake Village F-7
Lavaca C-3
LePanto C-9
Lead Hill E-5
Leslie B-5
Lewisville F-3
Lincoln B-2
Little Rock D-5
Vicinity Pg. 143
Detailed City Pg. 246
Lockesburg E-3
Lonoke D-6
Lowell A-3
Luxora B-9
Mabelvale Pg. 143, E-3
Madison D-8
Magnolia G-4
Malvern E-5
Mammoth Spring A-7
Mandeville Pg. 191, H-7
Manila B-9
Mansfield C-2
Marianna D-8
Marion C-9
Pg. 151, C-1
Marked Tree C-8
Marmaduke B-8
Marshall B-5
Martindale Pg. 143, C-1
Marvell D-8
Massard Pg. 127, D-3
Maumelle D-5
Vicinity Pg. 143, B-2
Maynard A-7
McAlmont Pg. 143, B-6
McCrory C-7
McGehee F-7
Melbourne B-6
Mena D-2
Mineral Springs F-3
Moffet Pg. 127, C-1
Monette B-8
Monticello F-6
Montrose G-8
Morrilton C-5
Mountain Home A-5
Mountain View B-6
Mt. Ida D-3
Mt. Pleasant G-3
Vicinity Pg. 191, K-9
Murfreesboro E-3
N. Little Rock D-5
Vicinity Pg. 143, B-5
Nashville E-3
Natural Steps
Pg. 143, A-1

CALIFORNIA
Pages 18-21

Population: 23,667,902
Capital: Sacramento
Land Area: 156,299 sq. mi.

COLORADO
Pages 22-23

Population: 2,889,964
Capital: Denver
Land Area: 103,595 sq. mi.

CONNECTICUT

Page 24

Population: 3,107,576
Capital: Hartford
Land Area: 4,872 sq. mi.

DELAWARE

Page 43

Population: 594,338
Capital: Dover
Land Area: 1,932 sq. mi.

DIST. OF COLUMBIA

Page 43

Population: 638,333
Capital: Washington
Land Area: 63 sq. mi.

FLORIDA

Pages 26-27

Population: 9,746,324
Capital: Tallahassee
Land Area: 58,560 sq. mi.

GEORGIA
Pages 28-29

Population: 5,463,105
Capital: Atlanta
Land Area: 54,153 sq. mi.

HAWAII
Page 30

Population: 964,691
Capital: Honolulu
Land Area: 6,425 sq. mi.

IDAHO
Page 31

Population: 943,935
Capital: Boise
Land Area: 82,413 sq. mi.

ILLINOIS
Page 32-33

Population: 11,426,518
Capital: Springfield
Land Area: 55,645 sq. mi.

ILLINOIS

INDIANA
Pages 34-35

Population: 5,490,224
Capital: Indianapolis
Land Area: 35,932 sq. mi.

IOWA
Page 36

Population: 2,913,808
Capital: Des Moines
Land Area: 55,965 sq. mi.

IOWA

KANSAS

KANSAS
Page 37

Population: 2,363,679
Capital: Topeka
Land Area: 81,781 sq. mi.

KANSAS

KENTUCKY

KENTUCKY
Pages 38-39

Population: 3,660,777
Capital: Frankfort
Land Area: 39,669 sq. mi.

KENTUCKY

LOUISIANA

LOUISIANA
Page 40

Population: 4,205,9000
Capital: Baton Rouge
Land Area: 44,521 sq. mi.

MAINE

MAINE
Page 41

Population: 1,124,660
Capital: Augusta
Land Area: 30,995 sq. mi.

MARYLAND

MARYLAND
Pages 42-43

Population: 4,216,975
Capital: Annapolis
Land Area: 9,837 sq. mi.

MARYLAND

MASSACHUSETTS

Pages 24-25

Population: 5,737,037
Capital: Boston
Land Area: 7,824 sq. mi.

MICHIGAN

Pages 44-45

Population: 9,262,078
Capital: Lansing
Land Area: 56,954 sq. mi.

MINNESOTA

Pages 46-47

Population: 4,075,970
Capital: St. Paul
Land Area: 79,548 sq. mi.

MINNESOTA

MISSISSIPPI

MISSISSIPPI
Page 50

Population: 2,520,638
Capital: Jackson
Land Area: 47,233 sq. mi.

MISSISSIPPI

MISSOURI

MISSOURI
Pages 48-49

Population: 4,916,686
Capital: Jefferson City
Land Area: 68,945 sq. mi.

MISSOURI

MONTANA

MONTANA
Page 51

Population: 786,690
Capital: Helena
Land Area: 145,398 sq.mi.

NEBRASKA

Pages 52-53

Population: 1,569,825
Capital: Lincoln
Land Area: 76,664 sq. mi.

NEVADA

Page 54

Population: 800,493
Capital: Carson City
Land Area: 109,893 sq. mi.

NEW HAMPSHIRE

Page 55

Population: 920,610
Capital: Concord
Land Area: 8,993 sq. mi.

NEW HAMPSHIRE

NEW JERSEY

NEW JERSEY
Pages 56-57

Population: 7,364,823
Capital: Trenton
Land Area: 7,468 sq. mi.

NEW JERSEY

NEW MEXICO
Page 62

Population: 1,302,894
Capital: Santa Fe
Land Area: 121,335 sq. mi.

NEW YORK

NEW YORK
Pages 58-61

Population: 17,558,072
Capital: Albany
Land Area: 47,377 sq. mi.

NEW YORK

NEW YORK

NORTH CAROLINA

Pages 64-65

Population: 5,881,766
Capital: Raleigh
Land Area: 48,843 sq. mi.

NORTH CAROLINA

NORTH DAKOTA

Page 63

Population: 652,717
Capital: Bismarck
Land Area: 69,300 sq. mi.

OHIO

OHIO

Pages 66-67

Population: 10,797,630
Capital: Columbus
Land Area: 41,004 sq. mi.

OHIO

OHIO

OHIO

OKLAHOMA

Pages 68-69

Population: 3,025,487
Capital: Oklahoma City
Land Area: 68,655 sq. mi.

OREGON

Pages 70-71

Population: 2,633,105
Capital: Salem
Land Area: 98,184 sq. mi.

PENNSYLVANIA

Pages 72-73

Population: 11,863,895
Capital: Harrisburg
Land Area: 44,888 sq. mi.

RHODE ISLAND

Page 25

Population: 947,154
Capital: Providence
Land Area: 1,055 sq. mi.

SOUTH CAROLINA

Pages 64-65

Population: 3,121,820
Capital: Columbia
Land Area: 30,203 sq. mi.

SOUTH DAKOTA

Page 74

Population: 690,768
Capital: Pierre
Land Area: 75,952 sq. mi.

Oldham D-9
Olivet F-9
Onaka C-7
Onida D-6
Orient C-7
Ortley C-9
Parker F-9
Parkston F-8
Parmelee E-5
Philip D-4
Pierpont C-9
Pierre D-6
Pine Ridge F-3
Plainview D-3
Plankinton E-8
Platte F-7
Pollock B-6
Prairie City B-3
Presho E-6
Pringle E-2
Provo E-1
Pukwana E-7
Quinn D-4
Ramona E-9
Rapid City D-3
Ravinia F-8
Raymond C-9
Redfield C-8
Redig B-2
Ree Heights D-7
Reliance E-7
Revillo C-10
Ridgeview C-5
Rockerville D-2
Rockham C-7
Roscoe C-7
Rosebud F-5
Roslyn C-9
Roswell E-9
Saint Charles F-7
Saint Francis F-5
Saint Lawrence D-7
Salem E-9
Scenic E-3
Scotland F-9
Selby B-6
Seneca C-7
Shadehill B-4
Sherman E-10
Sioux Falls E-9
Vicinity Pg. 191
Sisseton B-10
Sorum B-3
South Shore C-10
Spearfish D-2
Springfield F-8
Stephan D-7
Stickney E-8
Stockholm C-10
Stoneville C-3
Stratford C-8
Sturgis D-2
Summit C-10
Tabor F-9
Tea F-10
Timber Lake C-5
Tolstoy C-7
Toronto D-10
Trail City B-5
Trent E-10
Tripp F-8
Tulare D-8
Turton C-8
Tyndall F-9
Union Center D-3
Veblen B-9
Verdon C-8
Vermillion G-9
Viborg F-9
Vienna D-9
Vilas E-9
Vivian E-6
Wagner F-8
Wakonda F-9
Wall D-4
Wallace C-9
Ward E-10
Wasta D-3
Watertown D-9
Waubay C-9
Webster C-9
Wessington D-7
Wessington Springs E-8
Westport B-8
Wetonka B-8
Wewala F-6
White D-10
White Butte B-4
White Lake E-8
White River E-5
Whitewood D-2
Willow Lake D-9
Wilmot C-10
Winner F-6
Wood E-5
Woonsocket E-8
Worthing F-10
Yale D-9
Yankton F-9
Zeona C-3

TENNESSEE

Pages 38-39

Population: 4,591,120
Capital: Nashville
Land Area: 41,155 sq. mi.

Adams F-6
Adamsville J-4
Aetna H-5
Alamo H-3
Alcoa H-11
Vicinity Pg. 140, J-3
Alexandria G-8
Algood G-8
Allardt G-9
Allens H-3
Allisona H-7
Allred G-9
Alnwick Pg. 140, K-2
Alpine G-9
Altamont J-8
Alto J-7
Amerene Pg. 140, K-5
Amherst Pg. 140, F-2
Antioch Pg. 158, F-7
Arlington J-2
Vicinity Pg. 151, A-10
Armona Pg. 140, J-3
Arrington H-7
Arthur F-11
Asbury Pg. 140, F-6
Ashland City G-6
Ashport H-2
Athens H-10
Atoka H-2
Atwood G-4
Avondale Pg. 158, A-9
Bailey Pg. 151, F-9
Banner Hill G-13
Bartlebaugh Pg. 104, A-6
Bartlett J-2
Vicinity Pg. 151, C-7
Beacon Pg. 158, G-6
Bean Station G-12
Bearden Pg. 140, F-3
Beardstown H-5
Beech Grove
Pg. 140, A-4
Beersheba Springs H-8
Belfast J-7
Bell Bridge Pg. 140, D-2
Bell Buckle H-7
Belle Meade G-6
Vicinity Pg. 158, E-4
Bellevue Pg. 158, F-2
Bells H-3
Belvidere J-7
Benjestown Pg. 151, B-3
Benton J-10
Berry Hill Pg. 158, E-6
Bethel J-6
Vicinity Pg. 140, A-2
Bethel Springs J-4
Big Sandy G-4
Big Springs J-9
Vicinity Pg. 140, K-1
Birchwood J-9
Blaine G-11
Blue Grass Pg. 140, G-2
Bluff City F-13
Bogota G-3
Bolivar J-3
Bolton Pg. 151, A-9
Bon Air H-8
Boynton Pg. 104, E-5
Braden H-2
Bradford G-4
Brentwood G-6
Vicinity Pg. 158, G-5
Briceville G-10
Brighton H-2
Bristol F-13
Brooklin Pg. 158, G-8
Brownsville H-3
Bruceton G-4
Brunswick Pg. 151, B-8
Bulls Gap G-12
Bumpus Mills F-5
Burlington Pg. 140, E-6
Burlison H-2
Burns G-6
Byington Pg. 140, E-1
Byrdstown F-9
Calhoun J-9
Camden G-5
Camelot F-12
Capleville J-2
Vicinity Pg. 151, G-6
Carlisle G-5
Carlton Pg. 140, H-3
Carthage G-8
Caryville G-10
Cash Point J-6
Cedar Bluff Pg. 140, F-1
Cedar Hill F-6
Celina F-8
Center Point J-6
Centertown H-7
Centerville H-6
Central G-3
Charleston J-9
Charlotte G-5
Chattanooga J-8
Vicinity Pg. 104
Church Hill F-12
Clarksburg H-4
Clarksville F-5
Claxton Pg. 140, D-1
Cleveland J-9
Clifton J-5
Clinton G-10
Coalmont J-8
Collegedale J-9
Vicinity Pg. 104, B-7
Collierville J-2
Vicinity Pg. 151, F-10
Collinwood J-5
Columbia H-6
Concord Pg. 140, G-1
Condon Pg. 140, A-6
Cookeville G-8
Cooper Pg. 104, A-1
Copper Ridge
Pg. 140, B-4
Copperhill J-10
Cordova Pg. 151, D-8
Corinth Pg. 140, C-5
Cornersville J-6
Corryton Pg. 140, A-7
Cosby H-11
Covington H-2
Cowan Springs
Pg. 140, K-6
Crab Orchard H-9
Crenshaw Pg. 140, G-4
Cross Plains F-6
Crossville H-9
Culleoka H-6
Cumberland City G-5
Cumberland Furnace G-5
Cumberland Gap F-11
Dancyville H-3
Dandridge G-11
Darden H-4
Dayton H-9
Decatur H-9
Decaturville H-5
Decherd J-8
Dibrell H-8
Dickson G-5
Dixon Springs G-8
Donelson G-7
Vicinity Pg. 158, D-7
Dover F-5
Dowalltown G-8
Doyle H-8
Dresden G-4
Ducktown J-10
Dunlap J-9
Dyer G-3
Dyersburg G-3
Eads Pg. 151, C-10
Eagan F-10
Eagle Cliff Pg. 104, E-2
East Brainard
Pg. 104, C-6
East Ridge J-9
Vicinity Pg. 104, D-5
Eastview J-4
Ebenezer Pg. 140, G-2
Egypt Pg. 151, B-6
Elizabethton G-13
Elkton J-6
Ellendale Pg. 151, B-7
Ellejoy Pg. 140, J-7
Elmore Park Pg. 151, C-7
Elora J-7
Elza G-10
Englewood J-10
Erin G-5
Erwin G-13
Estill Springs J-7
Ethridge J-6
Etowah J-10
Evensville H-9
Fairview G-6
Vicinity Pg. 104, E-4
Farmington H-6
Fayetteville J-6
Finley G-3
Fisherville Pg. 151, D-10
Flat Woods H-5
Flatcreek J-7
Flintstone Pg. 104, E-3
Forbus F-9
Forest Hill Pg. 151, F-8
Forest Hills Pg. 158, F-4
Fort Oglethorpe
Pg. 104, E-5
Foster Corners
Pg. 158, F-8
Fountain City
Pg. 140, D-3
Frankewing J-7
Franklin H-6
Friendship G-3
Fruitvale H-3
Gadsden H-3
Gainesboro G-8
Gallatin G-7
Gates H-3
Gatlinburg H-11
Germanton Pg. 158, B-4
Germantown J-2
Vicinity Pg. 151, E-8
Gibson H-4
Gildfield Pg. 151, A-8
Gilt Edge H-2
Gleason G-4
Glendale Pg. 104, A-3
Goodlettsville G-7
Vicinity Pg. 158, A-6
Goodspring J-6
Gooseneck Pg. 140, K-1
Gordonsville G-7
Gower Pg. 158, E-3
Grand Junction J-3
Grandview H-9
Gravelly Hills
Pg. 140, H-1
Graveston Pg. 140, A-6
Gray G-13
Graysville H-9
Vicinity Pg. 104, D-6
Green Hill Pg. 158, C-9
Greenback H-10
Greenbrier G-6
Greeneville G-12
Greenfield G-4
Grimsley G-9
Habersham F-10
Haletown J-8
Halls G-3
Halls Crossroads G-11
Vicinity Pg. 140, C-4
Hampshire H-5
Hampton G-13
Hanging Limb G-9
Hanillville Pg. 104, A-4
Harbison Pg. 140, B-5
Harriman H-10
Harrison Pg. 104, A-6
Harrogate F-11
Hartsville G-7
Heiskel Pg. 140, C-2
Henderson H-4
Hendersonville G-7
Vicinity Pg. 158, A-8
Henning H-2
Henry G-4
Hermitage Hills
Pg. 158, C-8
Hickory Valley J-3
Hicks Crossing
Pg. 140, G-7
Highland J-5
Highland Park
Pg. 104, C-4
Hillvale Pg. 140, A-1
Hohenwald H-5
Hopewell Pg. 158, C-8
Hornsby J-3
Hubbard Pg. 140, K-5
Humboldt H-3
Hunter G-13
Huntingdon G-4
Huntland J-7
Huntsville G-10
India G-4
Inglewood G-7
Vicinity Pg. 158, C-6
Iron City J-5
Isham F-10
Ivydell G-10
Jacksboro G-10
Jackson H-3
Jamestown G-9
Jasper J-8
Jefferson City G-11
Jellico F-10
Jersey Pg. 104, B-5
Joelton G-6
Vicinity Pg. 158, A-4
John Sevier Pg. 140, D-6
Johnson City G-13
Jonesboro G-12
Karns Pg. 140, E-1
Keeling H-2
Kelso J-7
Kenton G-3
Kimberlin Heights
Pg. 140, F-7
Kimbro Pg. 158, G-8
Kingsport F-13
Kingston H-10
Kirkland H-7
Knobcreek Pg. 140, H-7
Knoxville H-10
Vicinity Pg. 140
Detailed City Pg. 243
Kodak H-11
Kyles Ford F-12
La Vergne G-7
Vicinity Pg. 158, G-9
LaFollette G-10
Laager J-8
Lafayette F-8
Lake City G-10
Lake Mont H-10
Lakesite J-9
Lakeview Pg. 104, D-4
Lancing G-9
Lascassas H-7
Latham G-4
Lawrenceburg J-5
Leach G-4
Lebanon G-7
Leipers Fork H-6
Lenoir City H-10
Lenow Pg. 151, C-9
Lewisburg H-6
Lexington H-4
Liberty G-8
Lickton Pg. 158, A-5
Linden H-5
Little Creek Pg. 158, B-5
Little River Pg. 140, H-4
Livingston G-8
Lobelville H-5
Locke J-2
Loretto J-5
Loudon H-10
Louisville Pg. 140, H-2
Lucy Pg. 151, A-6
Lupton City Pg. 104, A-4
Luttrell G-11
Lutts J-5
Lyles H-6
Lynchburg J-7
Lynnville J-6
Lytle Pg. 104, E-4
Madisonville H-10
Mahoney Mill
Pg. 140, J-1
Malesus H-3
Maloneyville Pg. 140, C-5
Manchester H-7
Marrowbone Pg. 158, C-1
Martin G-4
Maryville H-10
Vicinity Pg. 140, K-3
Mascot G-11
Vicinity Pg. 140, C-7
Mason H-2
Maury City H-3
Mayland G-8
Maynardville G-10
McBurg J-6
McEwen G-5
McKenzie G-4
McMinnville H-8
Meadowbrook
Pg. 140, E-2
Medina H-4
Medon H-3
Melbourne Pg. 140, A-3
Melrose Pg. 140, K-5
Memphis J-1
Vicinity Pg. 151
Detailed City Pg. 250
Mentor Pg. 140, J-3
Mercer H-3
Michie J-4
Middleton J-3
Midtown H-10
Milan H-4
Milledgeville J-4
Milligan College G-13
Millington J-2
Minor Hill J-6
Miser Station
Pg. 140, K-2
Miston G-2
Monteagle J-8
Monterey G-9
Moralfa Pg. 140, J-1
Morgantown H-9
Morris Chapel J-4
Morrison H-7
Morristown G-11
Moscow J-3
Mosheim G-12
Moss F-8
Mound City Pg. 151, C-2
Mount Carmel F-12
Mount Crest H-9
Mount Juliet G-7
Vicinity Pg. 158, C-10
Mount Olive H-11
Vicinity Pg. 140, G-4
Mount Pleasant H-6
Mount Tabor Pg. 140, K-3
Mount Vernon J-10
Vicinity Pg. 140, J-1
Mount Zion Pg. 158, A-2
Mountain City F-14
Mule Hollow Pg. 140, E-6
Munford H-2
Murfreesboro H-7
Nashville G-6
Vicinity Pg. 158
Detailed City Pg. 257
Neubert Pg. 140, G-6
New Hope K-8
New Johnsonville G-5
New Market G-11
New River G-10
New Tazewell G-11
Newbern G-3
Newport G-12
Niota H-10
Nixon J-4
Nobles G-4
Norris G-10
North Chattanooga
Pg. 104, B-4
Norwood Pg. 140, E-3
Nunnelly H-5
Oak Hill Pg. 158, F-6
Oak Ridge G-10
Oakdale G-9
Oakland J-2
Oakville Pg. 151, F-5
Oakwood F-5
Obion G-3
Ocoee J-9
Offutt Pg. 140, A-1
Oglesby Pg. 158, G-6
Old Glory Pg. 140, J-3
Old Hickory Pg. 158, B-8
Olivehill J-4
Oliver Springs G-10
Oneida F-10
Ooltewah Pg. 104, B-7
Palmer J-8
Palmyra F-5
Parago Mill Pg. 158, F-7
Paris G-4
Parrottsville G-12
Parsons H-4
Pasquo Pg. 158, F-3
Peak Pg. 140, C-1
Peakland H-9
Pedigo Pg. 140, B-3
Pegram G-6
Pelham J-8
Petersburg J-7
Philadelphia H-10
Picner Pg. 140, G-7
Pigeon Forge H-11
Pikeville H-9
Pinewood H-5
Pinson H-4
Pisgah Pg. 151, D-10
Pleasant Gap
Pg. 140, B-3
Pleasant Hill G-8
Pleasant View G-6
Portland F-7
Powell Pg. 140, D-2
Prospect Pg. 140, J-6
Providence G-7
Vicinity Pg. 158, F-7
Pulaski J-6
Pumpkin Center
Pg. 140, J-2
Puncheon Camp G-11
Puryear F-4
Quebeck H-8
Raleigh J-2
Vicinity Pg. 151, B-6
Ramer J-4
Ramsey Pg. 140, E-6
Ramsey Pg. 151, A-4
Raus J-7
Rayon City Pg. 158, B-7
Red Bank J-9
Red Boiling Springs F-8
Riceville J-10
Richland Pg. 158, D-3
Rickman G-8
Ridgely G-3
Ridgetop G-6
Ringold Pg. 104, E-7
Ripley H-2
Ritta Pg. 140, C-5
Riverdale Pg. 140, E-7
Rives G-3
Roan Mountain G-13
Robbins G-9
Rockford Pg. 140, H-4
Rockford Station
Pg. 140, H-4
Rockwood H-9
Rocky Branch
Pg. 140, K-6
Rocky Fork G-12
Rogersville G-12
Rossville J-2
Rural Hill Pg. 158, F-8
Russelville G-12
Rutherford G-3
Rutledge G-11
Ryall Springs
Pg. 104, C-6
Sadlersville F-6
Saint Bethlehem F-5
Saint Joseph J-5
Sale Creek J-9
Saltillo J-4
Sardis H-4
Saundersville
Pg. 158, A-9
Savannah J-4
Scotts Hill H-4
Selmer J-4
Sequoyah Hills
Pg. 140, F-3
Seven Points
Pg. 158, E-9
Sevier Home
Pg. 140, G-5
Sevierville H-11
Sewanee J-8
Shacklett G-6
Shady Valley F-13
Sharon G-3
Shelby Farms
Pg. 151, D-7
Shelbyville H-7
Shook Pg. 140, F-6
Shooks Gap H-11
Vicinity Pg. 140, G-6
Shop Springs G-7
Signal Mtn. J-8
Vicinity Pg. 104, A-3
Silver Point G-8
Skiem J-6
Smartt H-8
Smith Springs
Pg. 158, F-8
Smithville G-8
Smyrna G-7
Sneedville F-11
Soddy-Daisy J-9
Somerville J-2
South Carthage G-8
South Dyersburg G-3
South Fulton F-3
South Pittsburg J-8
Sparta H-8
Spencer H-8
Spencer Hill Pg. 104, E-3
Spring City H-9
Spring Creek H-4
Spring Hill H-6
Spring Lake Pg. 151, B-7
Springfield F-6
Stanton H-2
Stokes Mill Pg. 140, A-3
Sullivan Gdn. F-12
Summertown J-5
Summit Pg. 104, B-6
Sunbright G-10
Sunrise Pg. 140, C-6
Surgoinsville F-12
Sweetwater H-10
Sycamore Pg. 158, A-1
Taft J-6
Tarpley J-6
Tazewell F-11
Tekoa Pg. 140, E-2
Tellico Plains J-10
Tennessee Ridge G-5
Three Points
Pg. 140, C-7
Tiftona Pg. 104, C-2
Timberlake H-4
Tipton Pg. 140, G-4
Tiptonville F-3
Toone J-3
Tracy City J-8
Trenton G-3
Trezevant G-4
Triune H-7
Troy G-3
Tullahoma J-7
Tusculum G-13
Vicinity Pg. 158, F-6
Tyner Pg. 104, B-6
Una Pg. 158, F-7
Unicoi G-13
Union City G-3
Union Grove J-9
Union Hill Pg. 158, A-5
Vaughlans Gap
Pg. 158, F-4
Victoria J-8
Viola H-8
Violet J-10
Vonore H-10
Waco J-6
Walden J-9
Walkertown J-4
Walling H-8
Wartburg G-10
Wartrace H-7
Watertown G-8
Vicinity Pg. 140, K-7
Waverly G-5
Waynesboro J-5
Webber City J-5
West Emory Pg. 140, G-1
West Meade Pg. 158, E-4
West Shiloh J-4
Westmoreland F-7
White Bluff G-6
White House F-7
White Pine G-12
Whites Creek
Pg. 158, B-4
Whites Village
Pg. 140, F-5
Whitesburg G-12
Whiteville J-3
Whitwell J-8
Wilder G-9
Wildwood Pg. 140, J-5
Williston J-3
Winchester J-7
Winfield F-10
Woodale Pg. 140, D-7
Woodbury H-8
Woodland Mills F-3
Woodstock Pg. 151, B-5
Wrencoe Pg. 158, G-7
Wyanoke Pg. 151, E-1
Wynnburg G-3

TEXAS

Pages 75-79

Population: 14,229,191
Capital: Austin
Land Area: 262,017 sq. mi.

Abernathy E-7
Abilene G-10
Vicinity Pg. 91
Acala J-2
Ackerly G-7
Acme D-10
Acuff Pg. 149, C-7
Adamsville J-12
Addicks L-15
Vicinity Pg. 132, D-2

TEXAS

TEXAS

UTAH

Pages 80-81

Population: 1,461,037
Capital: Salt Lake City
Land Area: 82,073 sq. mi.

VERMONT

Page 55

Population: 511,456
Capital: Montpelier
Land Area: 9,273 sq. mi.

VIRGINIA

Pages 82-83

Population: 5,346,818
Capital: Richmond
Land Area: 39,703 sq. mi.

WASHINGTON

Pages 84-85

Population: 4,132,156
Capital: Olympia
Land Area: 66,511 sq. mi.

WEST VIRGINIA

Pages 82-83

Population: 1,949,644
Capital: Charleston
Land Area: 24,119 sq. mi.

WISCONSIN

Pages 86-87

Population: 4,705,767
Capital: Madison
Land Area: 54,426 sq. mi.

WISCONSIN

WYOMING

WYOMING
Pages 88-89

Population: 469,559
Capital: Cheyenne
Land Area: 96,989 sq. mi.